TM 9-2320-272-24P-2
5 Ton M939 Series Truck
Direct and General Support
Maintenance Manual
Repair Parts and Special Tools List
Vol 2 of 2
February 1999

This manual contains parts and special tool information for the 5 ton M939 US Military Trucks. This is volume 2 of 2 in the Direct / General Support Manual Parts and Special Tools Series. M939 series trucks are a 5 ton heavy duty 6x6 truck. Cargo versions were designed to transport 10,000 pounds of cargo in all terrain and all weather conditions. Originally designed in the 1970's to replace the M39 and M809 series of vehicles. 44,590 units were produced. This manual is printed to help private owners in the maintenance of their vehicles.

Should you have suggestions or feedback on ways to improve this book please send email to Books@OcotilloPress.com

Edited 2021 Ocotillo Press
ISBN 978-1-954285-70-5
No rights reserved. This content of this book is in the public domain as it is a work of the US Government. It is reproduced by the publisher as a convenience to enthusiasts and others who may wish to own a quality copy of it. It has been adjusted to accomodate the printing and binding process.

Ocotillo Press
Houston, TX 77017
Books@OcotilloPress.com

Disclaimer: The user of this book is responsible for following safe and lawful practices at all times. The publisher assumes no responsibility for the use of the content of this book. The publisher has made an effort to ensure that the text is complete and properly typeset, however omissions, errors, and other issues may exist that the publisher is unaware of.

TECHNICAL MANUAL
UNIT, DIRECT SUPPORT, AND GENERAL SUPPORT MAINTENANCE
REPAIR PARTS AND SPECIAL TOOLS LIST
FOR
TRUCK, 5-TON, 6X6, M939, M939A1, M939A2
SERIES TRUCKS (DIESEL)

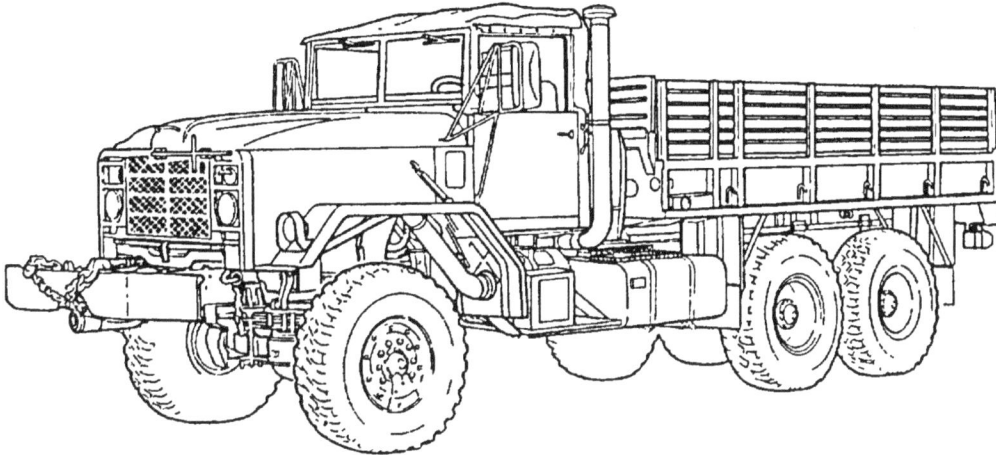

TRUCK, CARGO: 5-TON, 6X6, DROPSIDE,
M923 (2320-01-050-2084) (EIC: BRY); M923A1 (2320-01-206-4087) (EIC: BSS); M923A2 (2320-01-230-0307) (EIC: BS7);
M925 (2320-01-047-8769) (EIC: BRT); M925A1 (2320-01-206-4088) (EIC: BST); M925A2 (2320-01-230-0308) (EIC: BS8);

TRUCK, CARGO: 5-TON, 6X6 XLWB,
M927 (2320-01-047-8771) (EIC: BRV); M927A1 (2320-01-206-4089) (EIC: BSW); M927A2 (2320-01-230-0309) (EIC: BS9);
M928 (2320-01-047-8770) (EIC: BRU); M928A1 (2320-01-206-4090) (EIC: BSX); M928A2 (2320-01-230-0310) (EIC: BTM);

TRUCK, DUMP: 5-TON, 6X6,
M929 (2320-01-047-8756) (EIC: BTH); M929A1 (2320-01-206-4079) (EIC: BSY); M929A2 (2320-01-230-0305) (EIC: BTN);
M930 (2320-01-047-8755) (EIC: BTG); M930A1 (2320-01-206-4080) (EIC: BSZ); M930A2 (2320-01-230-0306) (EIC: BTO);

TRUCK, TRACTOR: 5-TON, 6X6,
M931 (2320-01-047-8753) (EIC: BTE); M931A1 (2320-01-206-4077) (EIC: BS2); M931A2 (2320-01-230-0302) (EIC: BTP);
M932 (2320-01-047-8752) (EIC: BTD); M932A1 (2320-01-205-2684) (EIC: BS55); M932A2 (2320-01-230-0303) (EIC: BTQ);

TRUCK, VAN, EXPANSIBLE: 5-TON, 6X6,
M934 (2320-01-047-8750) (EIC: BTB); M934A1 (2320-01-205-2682) (EIC: BS4); M934A2 (2320-01-230-0300) (EIC: BTR);

TRUCK, MEDIUM WRECKER: 5-TON, 6X6,
M936 (2320-01-047-8754) (EIC: BTF); M936A1 (2320-01-206-4078) (EIC: BS6); M936A2 (2320-01-230-0304) (EIC: BTT).

DEPARTMENTS OF THE ARMY AND THE AIR FORCE
FEBRUARY 1999

TECHNICAL MANUAL
NO. 9-2320-272-24P-2

TECHNICAL ORDER
NO. 36A12-IC-1155-4-2

HEADQUARTERS
DEPARTMENTS OF THE ARMY AND THE AIR FORCE
Washington D.C., 1 February 1999

UNIT, DIRECT SUPPORT, AND GENERAL SUPPORT MAINTENANCE
REPAIR PARTS AND SPECIAL TOOLS LIST
FOR
TRUCK, 5-TON, 6X6, M939, M939A1, AND M939A2 SERIES TRUCKS (DIESEL)

TRUCK	MODEL	EIC	NSN WITHOUT WINCH	NSN WITH WINCH
Cargo, Dropside	M923	BRY	2320-01-050-2084	
Cargo, Dropside	M923A1	BSS	2320-01-206-4087	
Cargo, Dropside	M923A2	BS7	2320-01-230-0307	
Cargo, Dropside	M925	BRT		2320-01-047-8769
Cargo, Dropside	M925A1	BST		2320-01-206-4088
Cargo, Dropside	M925A2	BS8		2320-01-230-0308
Cargo	M927	BRV	2320-01-047-8771	
Cargo	M927A1	BSW	2320-01-206-4089	
Cargo	M927A2	BS9	2320-01-230-0309	
Cargo	M928	BRU		2320-01-047-8770
Cargo	M928A1	BSX		2320-01-206-4090
Cargo	M928A2	BTM		2320-01-230-0310
Dump	M929	BTH	2320-01-047-8756	
Dump	M929A1	BSY	2320-01-206-4079	
Dump	M929A2	BTN	2320-01-230-0305	
Dump	M930	BTG		2320-01-047-8755
Dump	M930A1	BSZ		2320-01-206-4080
Dump	M930A2	BTO		2320-01-230-0306
Tractor	M931	BTE	2320-01-047-8753	
Tractor	M931A1	BS2	2320-01-206-4077	
Tractor	M931A2	BTP	2320-01-230-0302	
Tractor	M932	BTD		2320-01-047-8752
Tractor	M932A1	BS5		2320-01-205-2684
Tractor	M932A2	BTQ		2320-01-230-0303
Van, Expansible	M934	BTB	2320-01-047-8750	
Van, Expansible	M934A1	BS4	2320-01-205-2682	
Van, Expansible	M934A2	BTR	2320-01-230-0300	
Medium Wrecker	M936	BTF		2320-01-047-8754
Medium Wrecker	M936A1	BS6		2320-01-206-4078
Medium Wrecker	M936A2	BTT		2320-01-230-0304

REPORTING OF ERRORS AND RECOMMENDING IMPROVEMENTS
You can help improve this manual. If you find any mistakes or if you know of a way to improve the procedures, please let us know. Mail your letter or DA Form 2028 (Recommended Changes to Publications and Blank Forms), or DA Form 2028-2 located in back of this manual, directly to: Director, Armament and Chemical Acquisition and Logistics Activity, ATTN: AMSTA-AC-NML, Rock Island, IL 61299-7630. A reply will be furnished to you. You may also provide DA Form 2028-2 information via datafax or e-mail: ·
- E-mail: amsta-ac-nml.@ria-emh2.army.mil
- Fax: DSN 783-0726 or commercial (309) 782-0726

TABLE OF CONTENTS

TABLE OF CONTENTS (Cont'd)

TABLE OF CONTENTS (Cont'd)

TABLE OF CONTENTS (Cont'd)

TABLE OF CONTENTS (Cont'd)

TABLE OF CONTENTS (Cont'd)

TABLE OF CONTENTS (Cont'd)

TABLE OF CONTENTS (Cont'd)

TABLE OF CONTENTS (Cont'd)

TABLE OF CONTENTS (Cont'd)

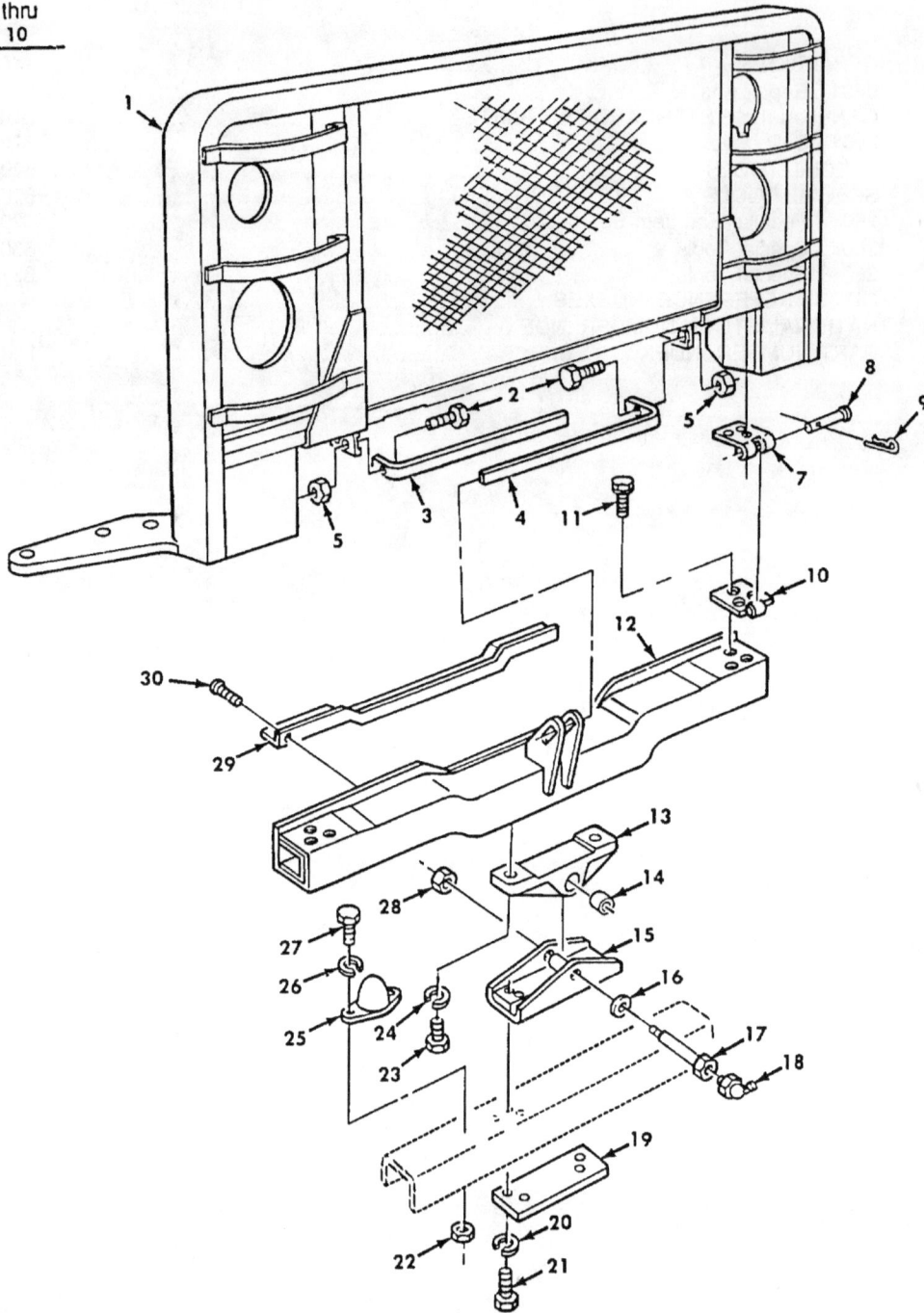

Figure 351. Front Guard Assembly and Hood Mounting Hardware.

(1) ITEM NO	(2) SMR CODE	(3) NSN	(4) CAGEC	(5) PART NUMBER	(6) DESCRIPTION AND USABLE ON CODES (UOC)	(7) QTY
					GROUP 18 BODY, CAB, HOOD, AND HULL 1801 BODY, CAB, HOOD, AND HULL ASSEMBLIES FIG. 351 FRONT GUARD ASSEMBLY AND HOOD MOUNTING HARDWARE	
1	PAOFF	2510010831149	19207	12255890	GRILLE, METAL	1
2	PAOZZ	5305002693239	80204	B1821BH038F138N	SCREW, CAP, HEXAGON H	4
3	PAOZZ	5340011044327	19207	12255901-1	BRACKET, ANGLE	1
4	PAOZZ	5340011044328	19207	12255901	BRACKET, ANGLE	1
5	PAOZZ	5310008140672	96906	MS51943-36	NUT, SELF-LOCKING, HE	4
6	PAOZZ	2510011038688	19207	12256579	HINGE, HOOD, VEHICULA...........................	2
7	PAOZZ	5340011147655	19207	12256589	.LEAF, BUTT HINGE	1
8	PAOZZ	5315000528492	96906	MS20392-4C97	.PIN, STRAIGHT, HEADED...........................	1
9	PAOZZ	5315010819991	88044	AN415-6	.PIN, LOCK	1
10	PAOZZ	5340011150615	19207	12256590	.LEAF, BUTT HINGE	1
11	PAOZZ	5305010907626	19207	7372083-1	SCREW, ASSEMBLED WAS	9
12	PFOZZ	5340010853594	19207	12256095	BRACKET, MOUNTING	1
13	PBOZZ	5340010853593	19207	12255820	BRACKET, MOUNTING	1
14	PAOZZ	5340011047843	19207	12256707	MOUNT, RESILIENT	1
15	PAOZZ	5340010831143	19207	12256705	BRACKET, MOUNTING	1
16	PAOZZ	5310011641642	19207	12302669	WASHER , FLAT	1
17	PAOZZ	4730011043805	19207	11669021-1	BOLT, FLUID PASSAGE...............................	1
18	PAOZZ	4730000504205	96906	MS15001-3	FITTING, LUBRICATION	1
19	PFOZZ	5340010911634	19207	12256577	PLATE, MENDING.......................................	1
20	PAOZZ	5310008347606	96906	MS35340-48	WASHER , LOCK	4
21	PAOZZ	5305007195221	80204	B1821BH050F150N	SCREW, CAP, HEXAGON H	4
22	PAOZZ	5310008807746	96906	MS51968-5	NUT, PLAIN, HEXAGON..............................	4
23	PAOZZ	5305007195219	96906	MS90727-111	SCREW, CAP, HEXAGON H	2
24	PAOZZ	5310000034094	01276	210104-8S	WASHER, LOCK	2
25	PAOZZ	5340002099377	19207	8331864	BUMPER, NONMETALLIC	2
26	PAOZZ	5310004079566	19207	7410218	WASHER, LOCK HOOD BUMPER.................	4
27	PAOZZ	5306000501238	96906	MS90727-32	BOLT, MACHINE	4
28	PAOZZ	5310004883888	96906	MS51943-40	NUT, SELF-LOCKING, HE	1
29	PBOZZ	2590010831115	19207	12256003	PAD, CUSHIONING	1
30	PAOZZ	5305004324251	96906	MS51861-65	SCREW, TAPPING.......................................	12

END OF FIGURE

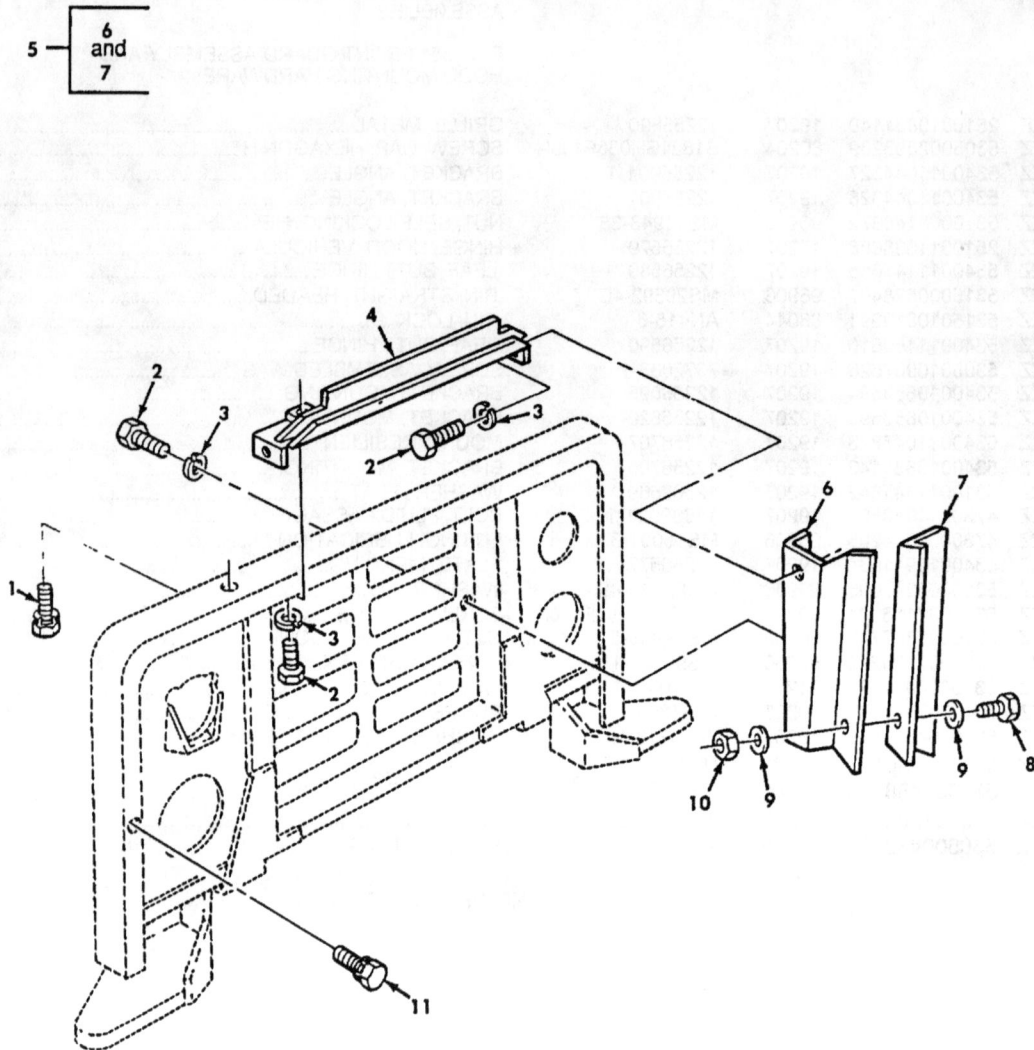

Figure 352. Radiator Baffle Assemblies and Mounting Hardware.

(1) ITEM NO	(2) SMR CODE	(3) NSN	(4) CAGEC	(5) PART NUMBER	(6) DESCRIPTION AND USABLE ON CODES (UOC)	(7) QTY
					GROUP 1801 BODY, CAB, HOOD, AND HULL ASSEMBLIES	
					FIG. 352 RADIATOR BAFFLE ASSEMBLIES AND MOUNTING HARDWARE	
1	PAFZZ	5305005698909	19207	7372083	SCREW,ASSEMBLED WA	10
2	PAFZZ	5305002678953	80204	B1821BH025F063N	SCREW,CAP,HEXAGON H	12
3	PAFZZ	5310005825965	96906	MS35338-44	WASHER,LOCK	12
4	PAFZZ	2930010831141	19207	12255797	SHROUD,FAN,RADIATOR	1
5	PAFZZ	2930010858137	19207	12256283-1	BAFFLE ASSEMBLY,RAD LEFT HAND	1
5	PAFZZ	2930010833016	19207	12256283-2	BAFFLE,RADIATOR RIGHT HAND	1
6	PAFZZ	2930010858138	19207	12256283-3	.BAFFLE,RADIATOR LEFT HAND	1
6	PFFZZ	2930010833017	19207	12256283-4	.BAFFLE,RADIATOR,RIG RIGHT HAND	1
7	PAFZZ	5330011044329	19207	12256283-5	.SEAL,NONMETALLIC ST LEFT AND	1
7	PAFZZ	5330011044330	19207	12256283-6	.SEAL,NON METALLIC ST RIGHT HAND	1
8	PAFZZ	5306000680513	60285	6893-2	BOLT,MACHINE	6
9	PAFZZ	5310008094058	96906	MS27183-1C	WASHER,FLAT	12
10	PAFZZ	5310008775796	96906	MS21044N4	NUT,SELF-LOCKING,HE	6
11	PAFZZ	5305010907626	19207	7372083-1	SCREW,ASSEMBLED WAS	12

END OF FIGURE

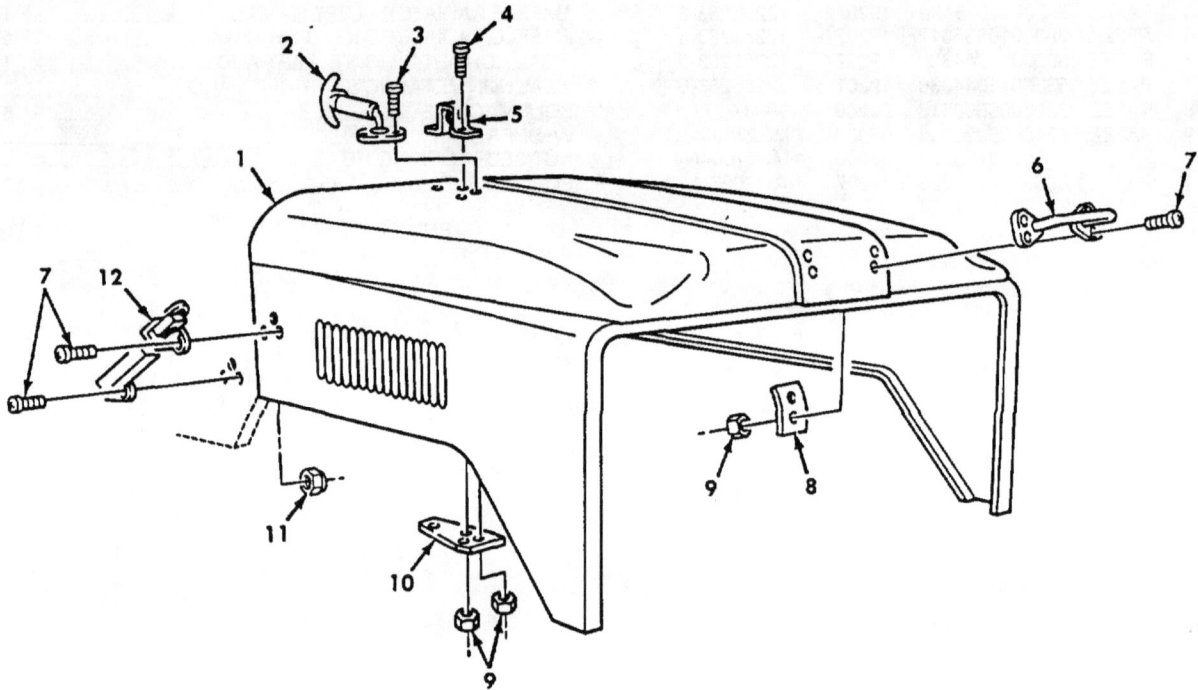

Figure 353. Hood and Related Parts.

(1) ITEM NO	(2) SMR CODE	(3) NSN	(4) CAGEC	(5) PART NUMBER	(6) DESCRIPTION AND USABLE ON CODES (UOC)	(7) QTY
					GROUP 1801 BODY, CAB, HOOD, AND HULL ASSEMBLIES	
					FIG. 353 HOOD AND RELATED PARTS	
1	PAOZZ	2510010823824	19207	12255776	HOOD,ENGINE COMPART	1
2	PAOZZ	5340007539214	19207	7539214	FASTENER,CYLINDER,S	2
3	PAOZZ	5305002678953	80204	B1821BH025F063N	SCREW,CAP,HEXAGON H	4
4	PAOZZ	5306000680513	60285	6893-2	BOLT,MACHINE	4
5	PAOZZ	2510010835442	19207	12255897	BRACKET,WINDSHIELD	2
6	PAOZZ	5340011087263	19207	12277126	HANDLE,BOW	1
7	PAOZZ	5305009571497	96906	MS35191-293	SCREW,MACHINE	16
8	PAOZZ	5340011089123	19207	12277127	PLATE,REINFORCEMENT	2
9	PAOZZ	5310009359022	96906	MS51943-32	NUT,SELF-LOCKING,HE	12
10	PFOZZ	5340010833015	19207	12255902	PLATE,MOUNTING	2
11	PAOZZ	5310008775796	96906	MS21044N4	NUT,SELF-LOCKING,HE	4
12	PAOZZ	5340011718267	19207	12302702	CATCH,CLAMPING	2

END OF FIGURE

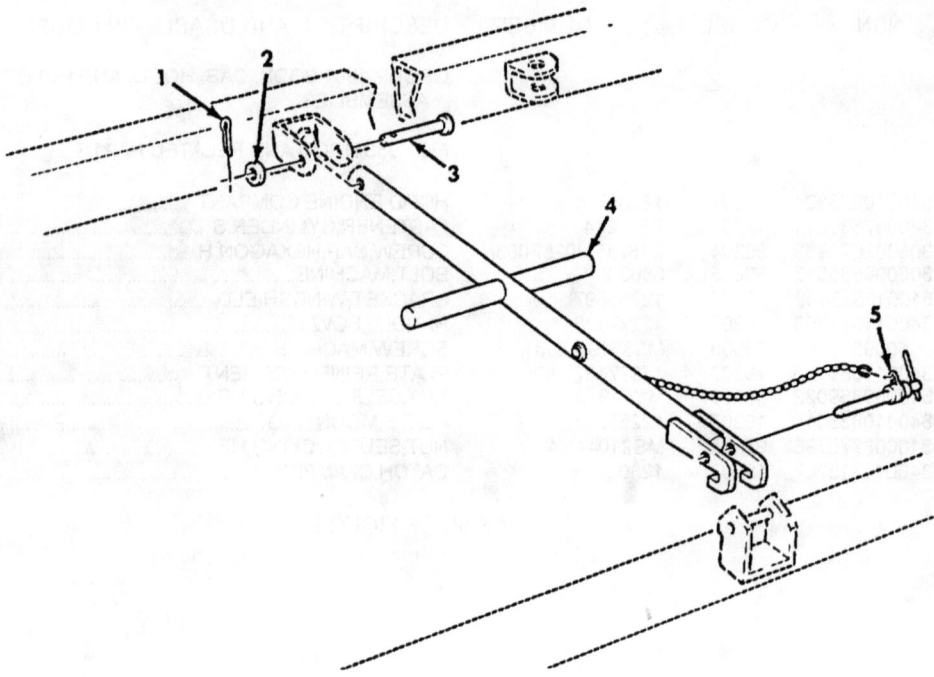

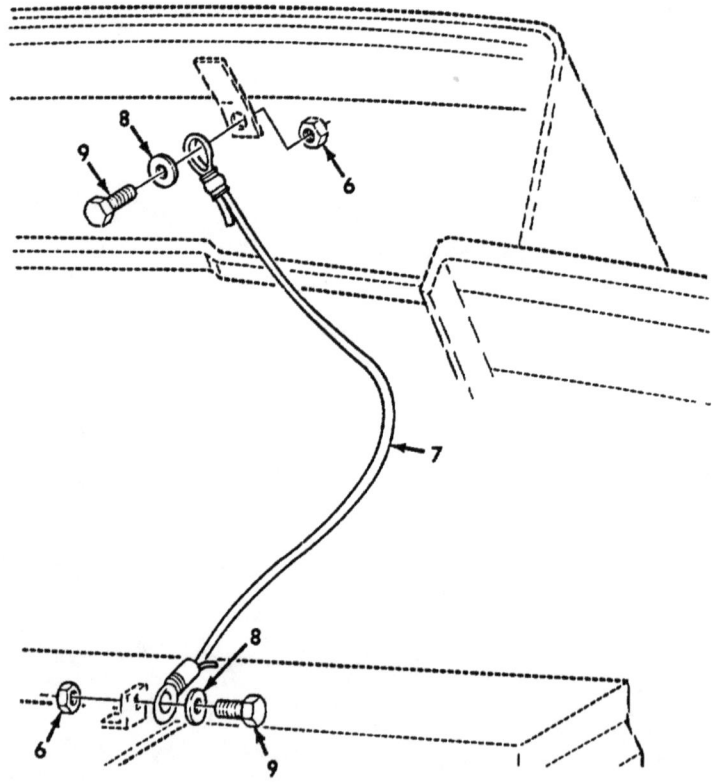

Figure 354. Hood Support and Stop Cables.

(1) ITEM NO	(2) SMR CODE	(3) NSN	(4) CAGEC	(5) PART NUMBER	(6) DESCRIPTION AND USABLE ON CODES (UOC)	(7) QTY
					GROUP 1801 BODY, CAB, HOOD, AND HULL ASSEMBLIES	
					FIG. 354 HOOD SUPPORT AND STOP CABLES	
1	PAOZZ	5315001879370	96906	MS24665-172	PIN,COTTER	1
2	PAOZZ	5310000806004	96906	MS27183-14	WASHER,FLAT	1
3	PAOZZ	5315009517542	96906	MS20392-5C67	PIN,STRAIGHT,HEADED	1
4	PAOZZ	2510010831126	19207	12256629-1	SUPPORT,HOOD UOC:DAA,DAB,DAC,DAE,DAG,DAJ,DAK,DAW, DAX,V12,V13,V14,V15,V17,V20,V22,V24, V25,ZAA,ZAB,ZAC,ZAE,ZAG,ZAJ,ZAK	1
4	PAOZZ	2510010831125	19207	12256629-2	SUPPORT,HOOD UOC:DAD,DAF,DAH,DAL,V16,V18,V19,V21, V39,ZAD,ZAF,ZAH,ZAL	1
5	PAOZZ	5315011259084	96906	MS17989-C624	PIN,QUICK RELEASE	1
6	PAOZZ	5310008140672	96906	MS51943-36	NUT,SELF-LOCKING,HE	4
7	PAOZZ	4010011147614	19207	12256425-3	WIRE ROPE ASSEMBLY	2
8	PAOZZ	5310008094061	96906	MS27183-15	WASHER,FLAT	4
9	PAOZZ	5305002692804	96906	MS90726-61	SCREW,CAP,HEXAGON H	4

END OF FIGURE

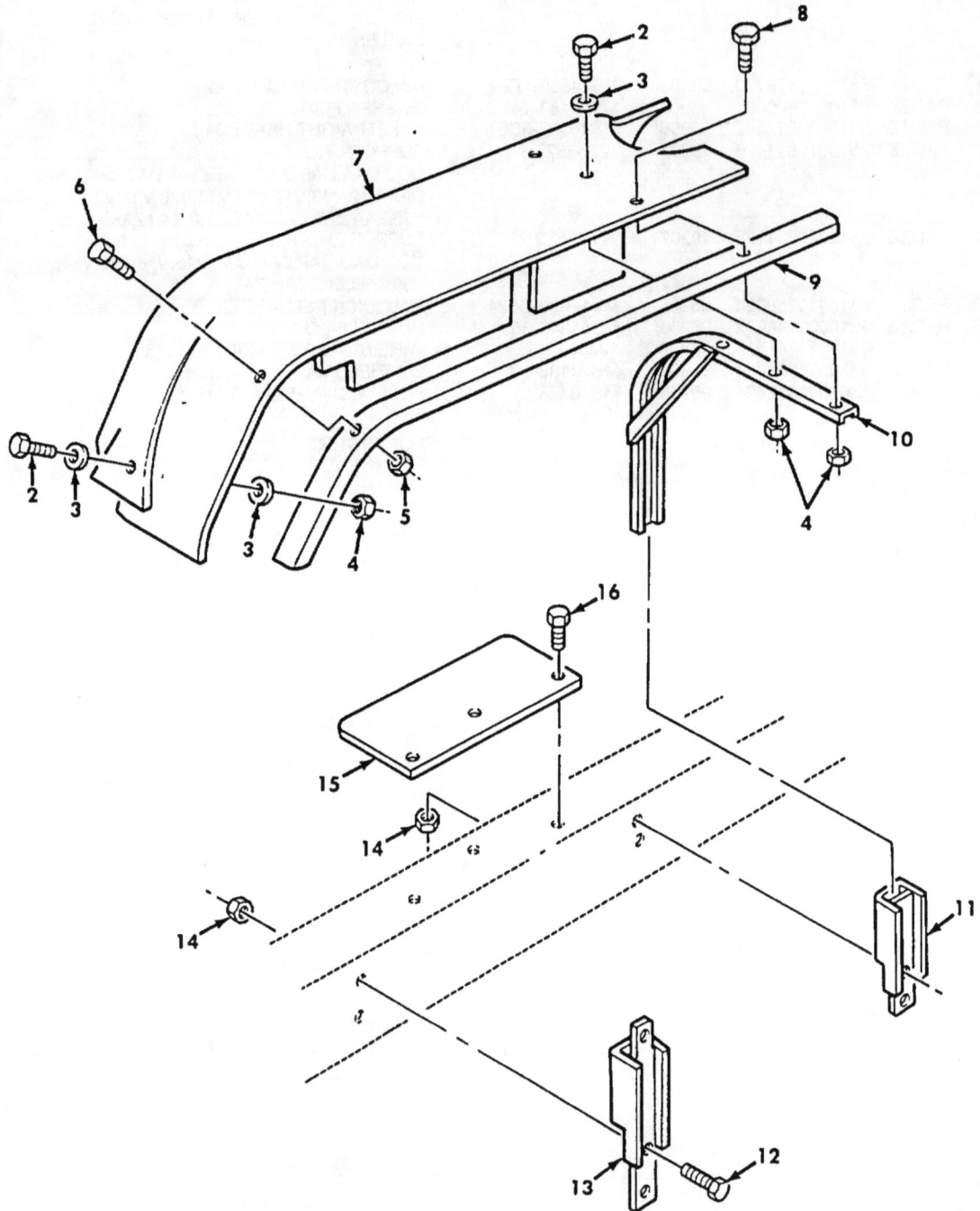

Figure 355. Splash Guard Assembly and Related Parts, L.H.

SECTION II   **TM 9-2320-272-24P-2**

(1) ITEM NO	(2) SMR CODE	(3) NSN	(4) CAGEC	(5) PART NUMBER	(6) DESCRIPTION AND USABLE ON CODES (UOC)	(7) QTY
					GROUP 1801 BODY, CAB, HOOD, AND HULL ASSEMBLIES	
					FIG. 355 SPLASH GUARD ASSEMBLY AND RELATED PARTS, L.H.	
1	PFOZZ	2540010934305	19207	12255849	GUARD,SPLASH,VEHICU	1
2	PAOZZ	5306000680514	80204	B1821BH025F088N	.BOLT,MACHINE	10
3	PAOZZ	5310008094058	96906	MS27183-10	.WASHER,FLAT	12
4	PAOZZ	5310000617325	96906	MS21045-4	.NUT,SELF-LOCKING,HE	12
5	PAOZZ	5310008775797	96906	MS21044N3	.NUT,SELF-LOCKING,HE	9
6	PAOZZ	5306000425570	24617	425570	.BOLT,ASSEMBLED WASH	9
7	XAOZZ		19207	12255895	.SHIELD,SPLASH,LH	1
8	PAOZZ	5305002693239	80204	B1821BH038F138N	.SCREW,CAP,HEXAGON H	2
9	XAOZZ		19207	12255847-1	.REINFORCEMENT LEFT HAND	1
10	XAOZZ		19207	12255850	.SUPPORT ASSY,SPLASH LEFT HAND	2
11	PFOZZ	2540010911613	19207	12256136	HOLDER,SPLASH GUARD	1
12	PAOZZ	5305002693239	80204	B1821BH038F138N	SCREW,CAP,HEXAGON H	4
13	PFOZZ	2540010911614	19207	12256137	HOLDER,SPLASH GUARD	1
14	PAOZZ	5310009591488	96906	MS51922-21	NUT,SELF-LOCKING	7
15	PFOZZ	2510010911687	19207	12276958	TREAD,METALLIC,NONS	1
16	PAOZZ	5305002692803	96906	MS90726-60	SCREW,CAP,HEXAGON H	3

END OF FIGURE

355-1

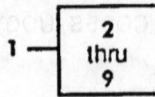

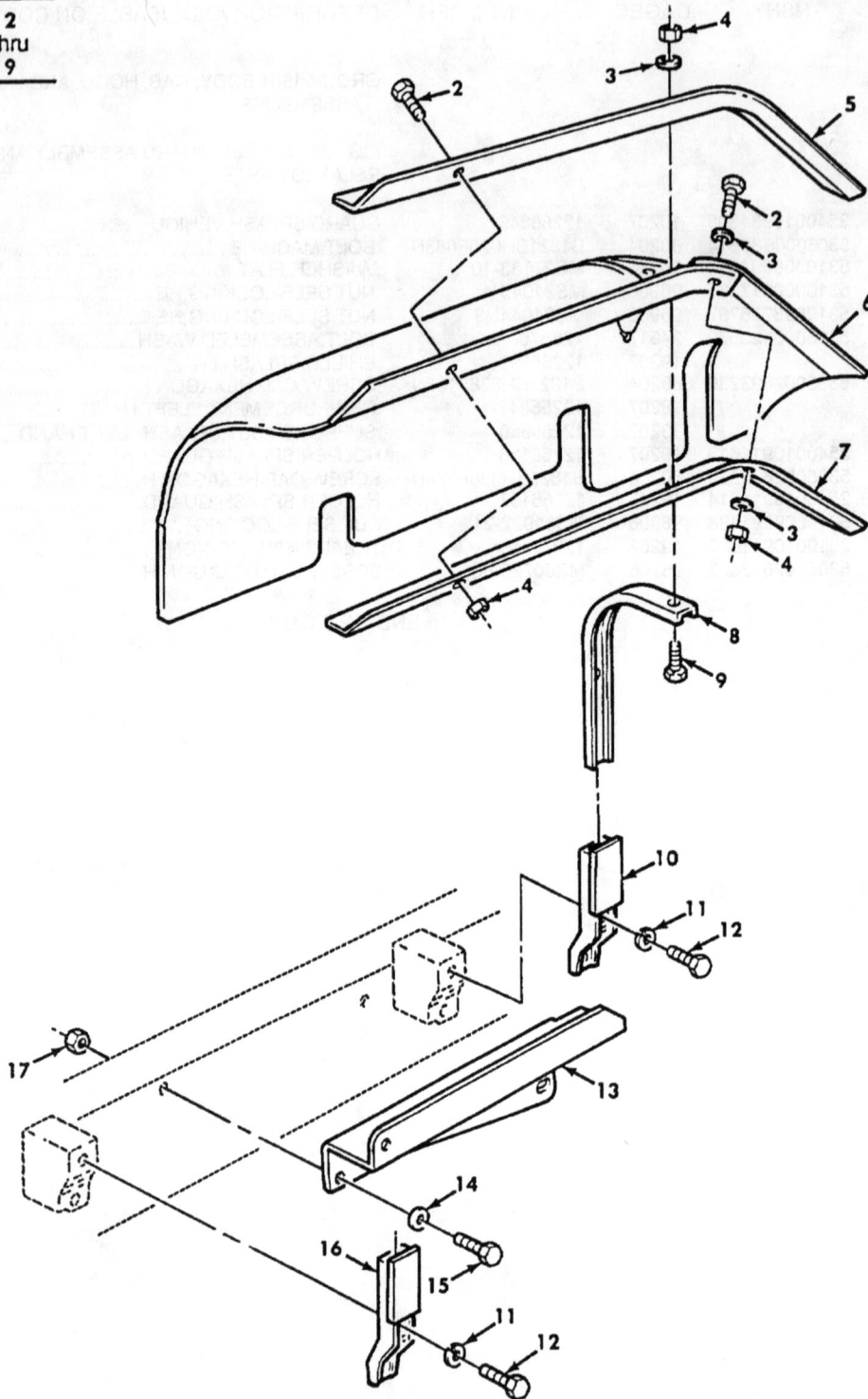

Figure 356. Splash Guard Assembly and Related Parts, R.H.

(1) ITEM NO	(2) SMR CODE	(3) NSN	(4) CAGEC	(5) PART NUMBER	(6) DESCRIPTION AND USABLE ON CODES (UOC)	(7) QTY
					GROUP 1801 BODY, CAB, HOOD, AND HULL ASSEMBLIES	
					FIG. 356 SPLASH GUARD ASSEMBLY AND RELATED PARTS, R.H.	
1	PAOZZ	2540011089124	19207	12277239	GUARD,SPLASH,VEHICU	1
2	PAOZZ	5306000680514	80204	B1821BH025FO88N	.BOLT,MACHINE	18
3	PAOZZ	5310008094058	96906	MS27183-10	.WASHER,FLAT	8
4	PAOZZ	5310000617325	96906	MS21045-4	.NUT,SELF-LOCKING,HE	18
5	XAOZZ		19207	12277235	.REINFORCEMENT	1
6	XAOZZ		19207	12277238	.SHIELD SPLASH R.H	1
7	XAOZZ		19207	12277236	.REINFORCEMENT	1
8	XAOZZ		19207	12277233	.SUPPORT BRACKET	2
9	PAOZZ	5305002693239	96906	MS90727-63	.SCREW,CAP,HEXAGON	2
10	PAOZZ	5340011307941	19207	12277237	BRACKET,DOUBLE ANGL	1
11	PAOZZ	5310006379541	96906	MS35338-46	WASHER,LOCK	4
12	PAOZZ	5305009125113	96906	MS51096-359	SCREW,CAP,HEXAGON H	4
13	PFOZZ	2540011307940	19207	12277234	GUARD,SPLASH,VEHICU	1
14	PAOZZ	5310000806004	96906	MS27183-14	WASHER ,FLAT	1
15	PAOZZ	5305002692803	96906	MS90726-60	SCREW,CAP,HEXAGON H	1
16	PAOZZ	5340011687904	19207	12302630	BRACKET,MOUNTING	1
17	PAOZZ	5310000617325	96906	MS21045-4	NUT,SELF-LOCKING,HE	1

END OF FIGURE

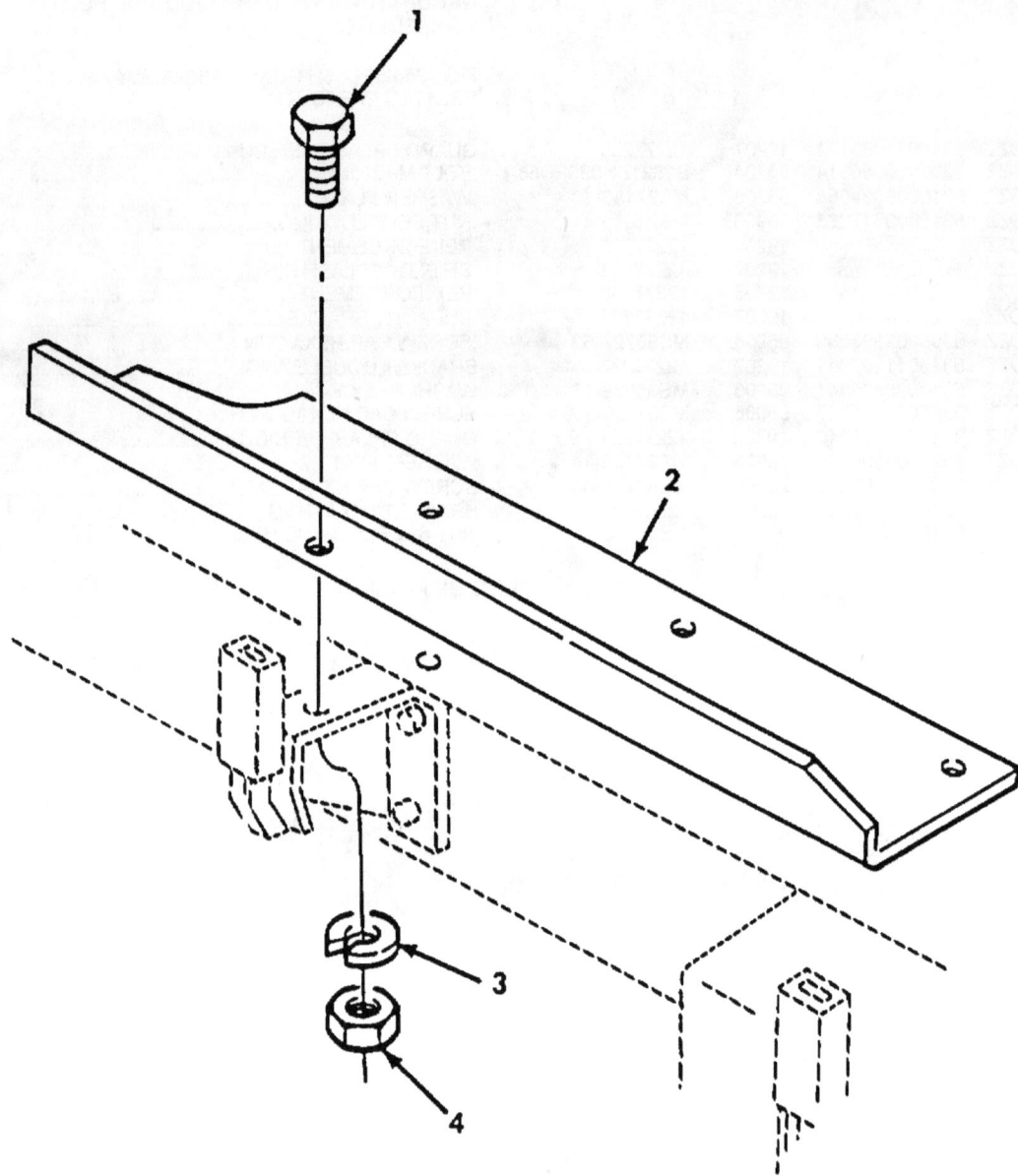

Figure 357. Splash Panel Extension (M939A2).

(1) ITEM NO	(2) SMR CODE	(3) NSN	(4) CAGEC	(5) PART NUMBER	(6) DESCRIPTION AND USABLE ON CODES (UOC)	(7) QTY
					GROUP 1801 BODY, CAB, HOOD, AND HULL ASSEMBLIES	
					FIG. 357 SPLASH PANEL EXTENSION (M939A2)	
1	PAOZZ	5305002693235	80204	B1821BH038F088N	SCREW,CAP,HEXAGON H .... UOC:ZAA,ZAB,ZAC,ZAD,ZAE,ZAF, ZAG,ZAH, ZAJ,ZAK,ZAL	4
2	PFOZZ	2540012794593	2W567	20510104	EXTENSION,SPLASH PA .... UOC:ZAA,ZAB,ZAC,ZAD,ZAE,ZAF,ZAG,ZAH, ZAJ, ZAK, ZAL	1
3	PAOZZ	5310006379541	96906	MS35338-46	WASHER,LOCK .... UOC:ZAA,ZAB,ZAC,ZAD,ZAE,ZAF,ZAG,ZAH, ZAJ, ZAK, ZAL	4
4	PAOZZ	5310008140672	96906	MS51943-36	NUT,SELF-LOCKING,HE .... UOC:ZAA,ZAB,ZAC,ZAD,ZAE,ZAF,ZAG,ZAH, ZAJ, ZAK, ZAL	4

END OF FIGURE

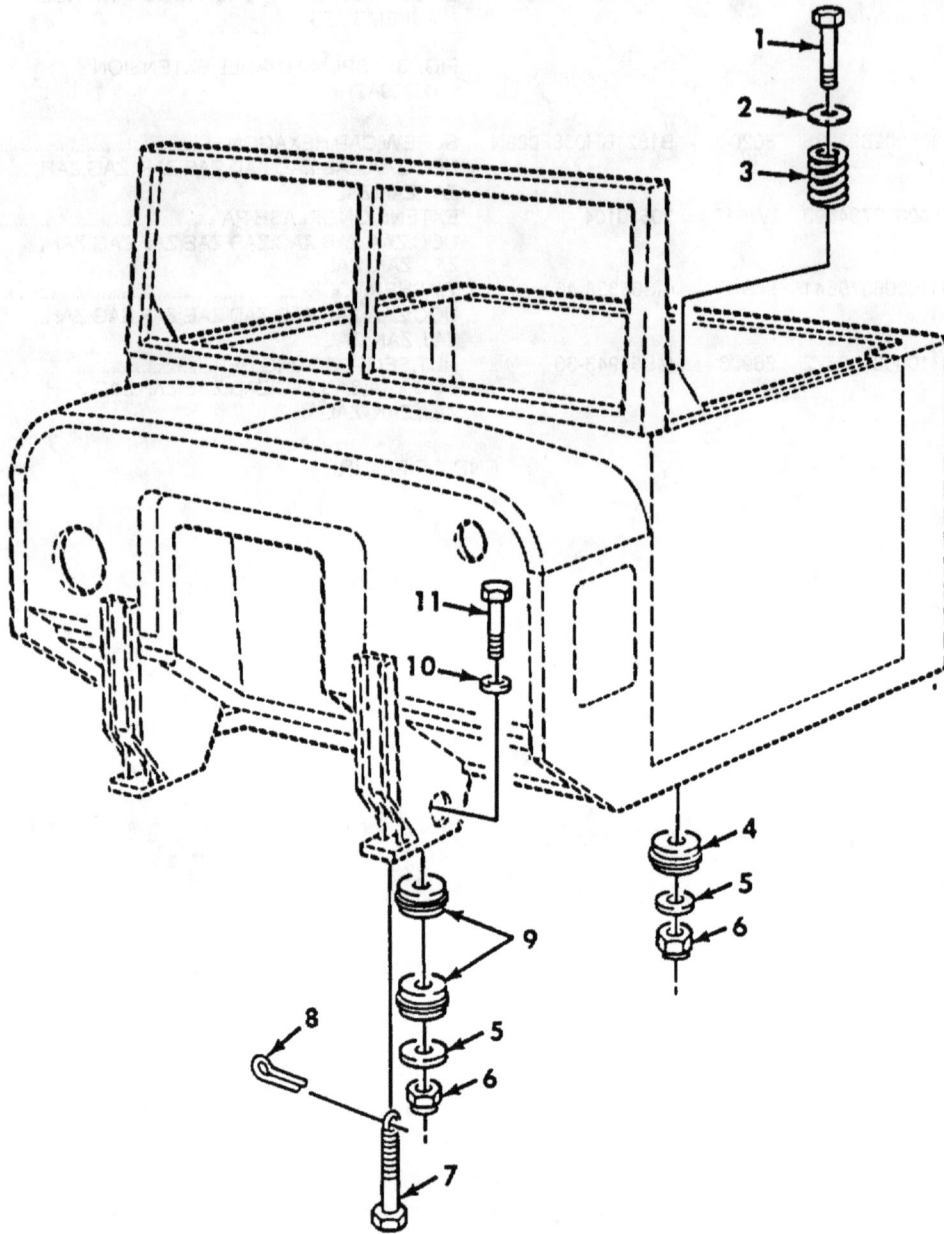

Figure 358. Cab Assembly Mounting Hardware.

(1) ITEM NO	(2) SMR CODE	(3) NSN	(4) CAGEC	(5) PART NUMBER	(6) DESCRIPTION AND USABLE ON CODES (UOC)	(7) QTY
					GROUP 1801 BODY, CAB, HOOD, AND HULL ASSEMBLIES	
					FIG. 358 CAB ASSEMBLY MOUNTING HARDWARE	
1	PAOZZ	5306011836970	19207	12302700	BOLT,SHOULDER	2
2	PAOZZ	5310008238803	96906	MS27183-21	WASHER,FLAT	2
3	PAOZZ	5360004769339	19207	7529293	SPRING,HELICAL,COMP	2
4	PAOZZ	5340000402073	19207	7521436	MOUNT,RESILIENT	2
5	PAOZZ	5310011096753	19207	12256962-1	WASHER,FLAT	4
6	PAOZZ	5310004883888	96906	MS51943-40	NUT,SELF-LOCKING,HE	4
7	PAOZZ	5306011346540	19207	12277353	BOLT,MACHINE	2
8	PAFZZ	5315002368359	96906	MS24665-370	PIN,COTTER	2
9	PAOZZ	5340013195383	57839	K50	MOUNT,RESILIENT	4
10	PAOZZ	5310011096754	19207	12256962	WASHER,FLAT	2
11	PAOZZ	5306011815018	19207	12256674-3	BOLT,SHOULDER	2

END OF FIGURE

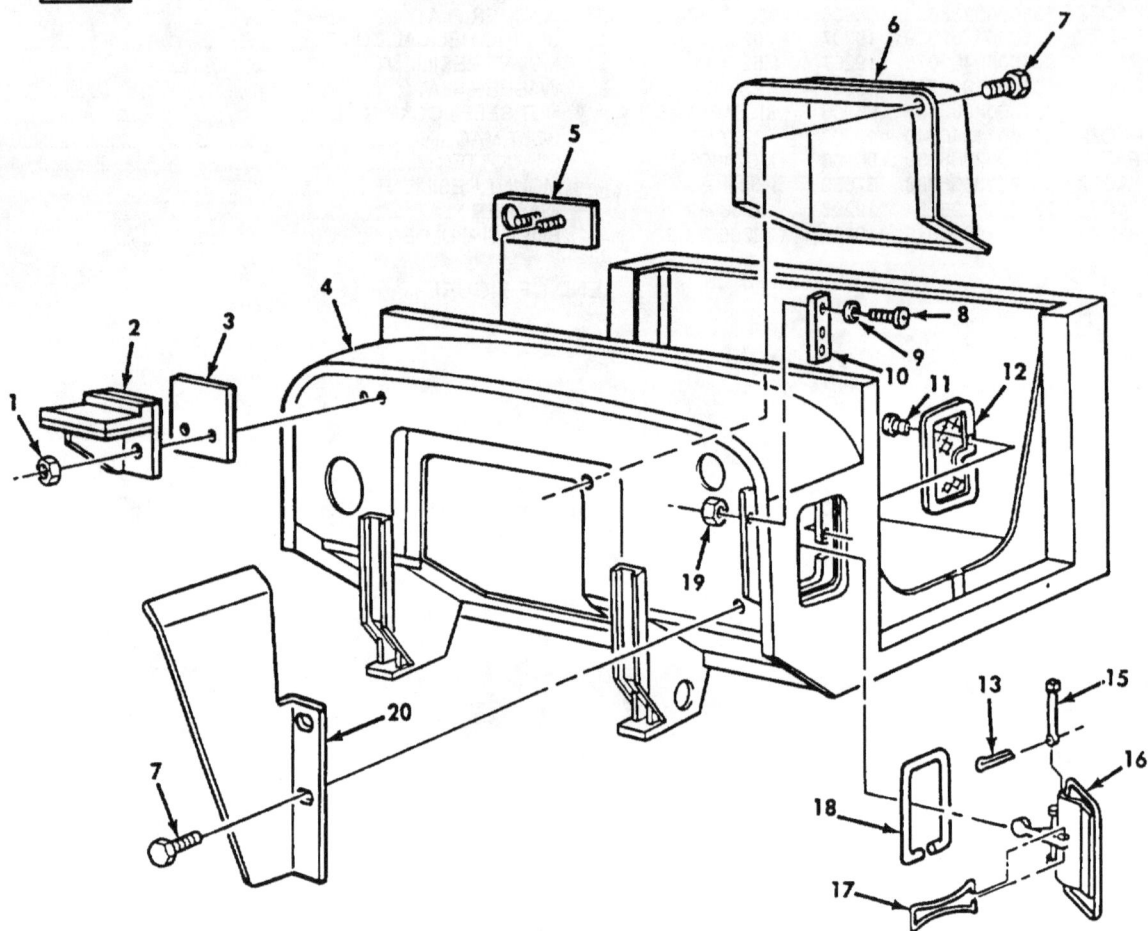

Figure 359. Cab Body Assembly and Related Parts.

(1) ITEM	(2) SMR	(3)	(4)	(5) PART	(6)	(7)

SECTION II

GROUP 1801 BODY, CAB, HOOD, AND HULL
ASSEMBLIES

FIG. 359 CAB BODY ASSEMBLY AND
RELATED PARTS

NO	CODE	NSN	CAGEC	NUMBER	DESCRIPTION AND USABLE ON CODES (UOC)	QTY
1	PAOZZ	5310008140672	96906	MS51943-36	NUT, SELF-LOCKING, HE	4
2	PAOZZ	5340011208444	19207	12256736	MOUNT, RESILIENT	2
3	PAOZZ	5365011108183	19207	12256738-1	SPACER, PLATE RIGHT	1
3	PAOZZ	5365011084814	19207	12256738-2	SPACER, PLATE LEFT	1
4	PAFZZ	2510012765729	19207	12255991-2	BODY, CAB ASSEMBLY	1
5	PAOZZ	2510011364442	19207	12256737	REINFORCEMENT, INNER	2
6	XDOZZ		19207	12255886	COVER, ACCESS	1
					UOC: DAA, DAB, DAC, DAD, DAE, DAF, DAG, DAH, DAJ, DAK, DAL, DAW, DAX, V12, V13, V14, V15, V16, V17, V18, V19, V20, V21, V22, V24, V25, V39	
7	PAOZZ	5305010907626	19207	7372083-1	SCREW, ASSEMBLED WAS	20
8	PAOZZ	5305009931851	96906	MS35207-267	SCREW, MACHINE	6
9	PAOZZ	5310000145850	96906	MS27183-42	WASHER, FLAT	6
10	PAOZZ	2590011089121	19207	12277032	PAD, CUSHIONING	2
11	PAOZZ	5305004324163	96906	MS51861-24	SCREW, TAPPING	18
12	PFOZZ	2510006930607	19207	7373316	VENTILATOR, AIR CIRC	2
13	PAOZZ	5315008395820	96906	MS24665-134	PIN, COTTER	2
14	PFOZZ	2510005464759	19207	7529312	VENTILATOR, COWL, CAB	2
15	PAOZZ	5315004539349	19207	7397723	.PIN, STRAIGHT, HEADED	1
16	XAOZZ		19207	7397725	.DOOR	1
17	PAOZZ	5340006227700	19207	7397726	CLIP, RETAINING	2
18	PAOZZ	9390007373317	19207	7373317	NONMETALLIC SPECIAL	2
19	PAOZZ	5310000617326	96906	MS21045-3	NUT, SELF-LOCKING, HE	6
20	PAOZZ	5340010831120	19207	12256012-2	BRACKET, MOUNTING RIGHT	1
20	PFOZZ	5340010831121	19207	12256012-1	BRACKET, MOUNTING LEFT	1

END OF FIGURE

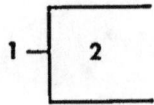

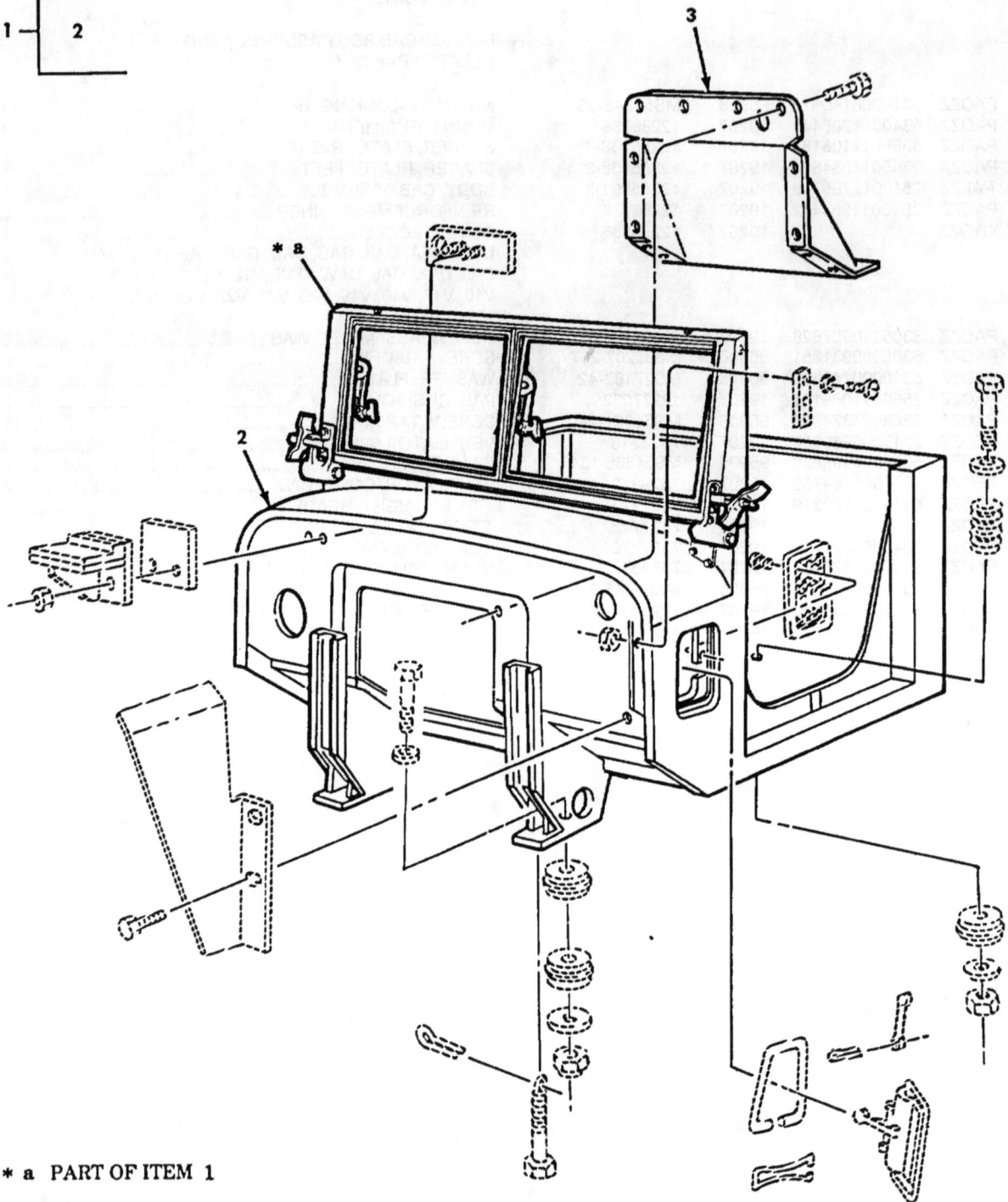

* a   PART OF ITEM 1

Figure 360. Basic Cab Assembly and Engine Cover (M939A2).

(1) ITEM NO	(2) SMR CODE	(3) NSN	(4) CAGEC	(5) PART NUMBER	(6) DESCRIPTION AND USABLE ON CODES (UOC)	(7) QTY
					GROUP 1801 BODY, CAB, HOOD, AND HULL ASSEMBLIES	
					FIG. 360 BASIC CAB ASSEMBLY AND ENGINE COVER (M939A2)	
1	PAFZZ	2510012765729	19207	12255991-2	CAB ASSEMBLY COMPLETE .... UOC: ZAA, ZAB, ZAC, ZAD, ZAE, ZAF, ZAG, ZAH, ZAJ, ZAK, ZAL	1
2	PBFFF		2P971	12363518	.CAB ASSEMBLY BASIC .... UOC:ZAA, ZAB, ZAC, ZAD, ZAE, ZAF, ZAG, ZAH, ZAJ, ZAK, ZAL	1
3	PFOZZ		2P971	M504130	COVER, ACCESS .... UOC: ZAA, ZAB, ZAC, ZAD, ZAE, ZAF, ZAG, ZAH, ZAJ, ZAK, ZAL	1

END OF FIGURE

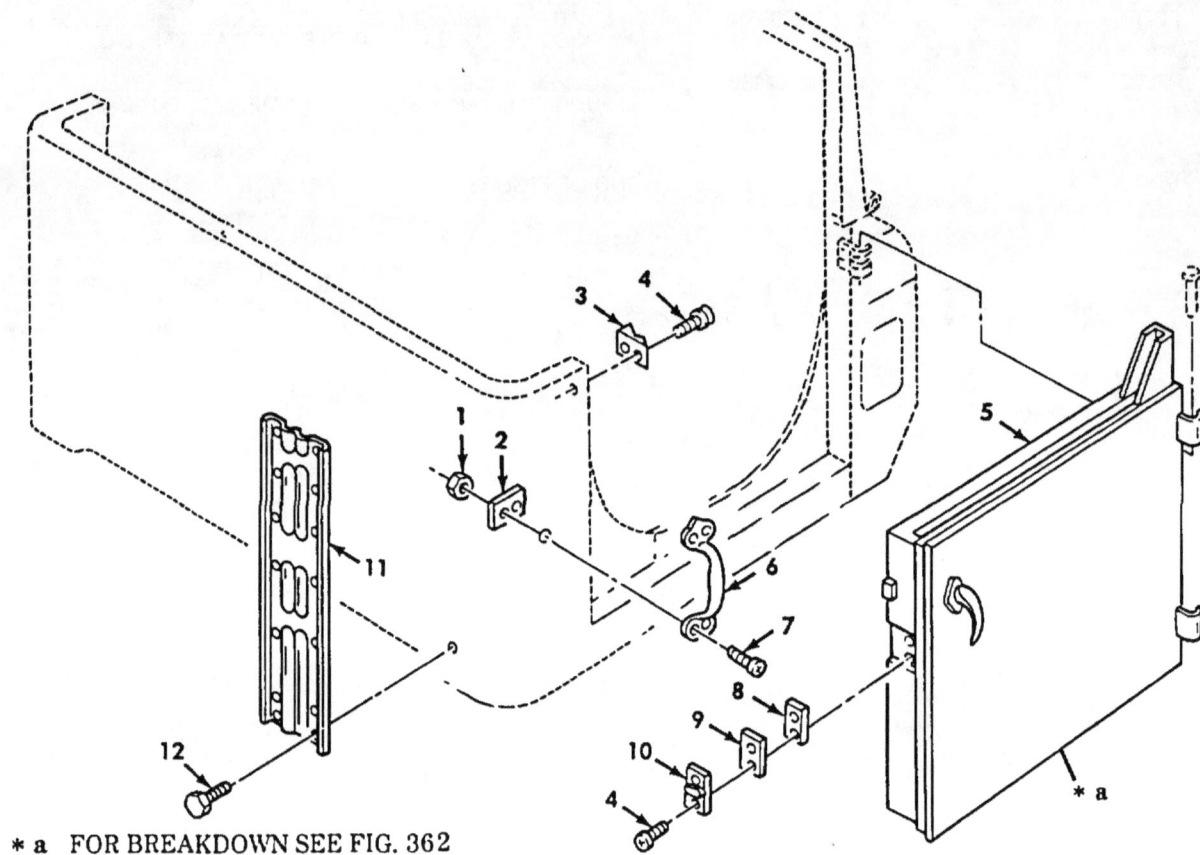

Figure 361. Door Assemblies, Machine Gun Mounting Bracket, and Related Parts.

* a   FOR BREAKDOWN SEE FIG. 362

(1) ITEM NO	(2) SMR CODE	(3) NSN	(4) CAGEC	(5) PART NUMBER	(6) DESCRIPTION AND USABLE ON CODES (UOC)	(7) QTY
					GROUP 1801 BODY, CAB, HOOD, AND HULL ASSEMBLIES	
					FIG. 361 DOOR ASSEMBLIES, MACHINE GUN MOUNTING BRACKET, AND RELATED PARTS	
1	PAOZZ	5310009359022	96906	MS51943-32	NUT, SELF-LOCKING, HE	8
2	PAOZZ	5340011089123	19207	12277127	PLATE, REINFORCEMENT	2
3	PAOZZ	5340007373302	19207	7373302	STRIKE, CATCH	2
4	PAOZZ	5305006965285	24617	187995	SCREW, ASSEMBLED WAS	8
5	PAOOO	2510007373293	19207	7373293	DOOR, VEHICULAR CAB, LEFT	1
5	PAOOO	2510007373294	5U403	7373294	DOOR, VEHICULAR CAB, RIGHT	1
6	PAOZZ	5340011087263	19207	12277126	HANDLE, BOW	2
7	PAOZZ	5305009571497	96906	MS35191-293	SCREW, MACHINE	8
8	PFOZZ	5365011290399	19207	7529300	SPACER, PLATE	2
9	PFOZZ	5340006212563	19207	7529310	COVER, ACCESS	2
10	PFOZZ	2540007540419	19207	7540419	WEDGE, DOOR DOVETAIL	2
11	PFOZZ	2510007539657	19207	7539657	BRACKET, REAR GUN MO	2
12	PAOZZ	5306011341966	21450	425647	BOLT, ASSEMBLED WASH	28

END OF FIGURE

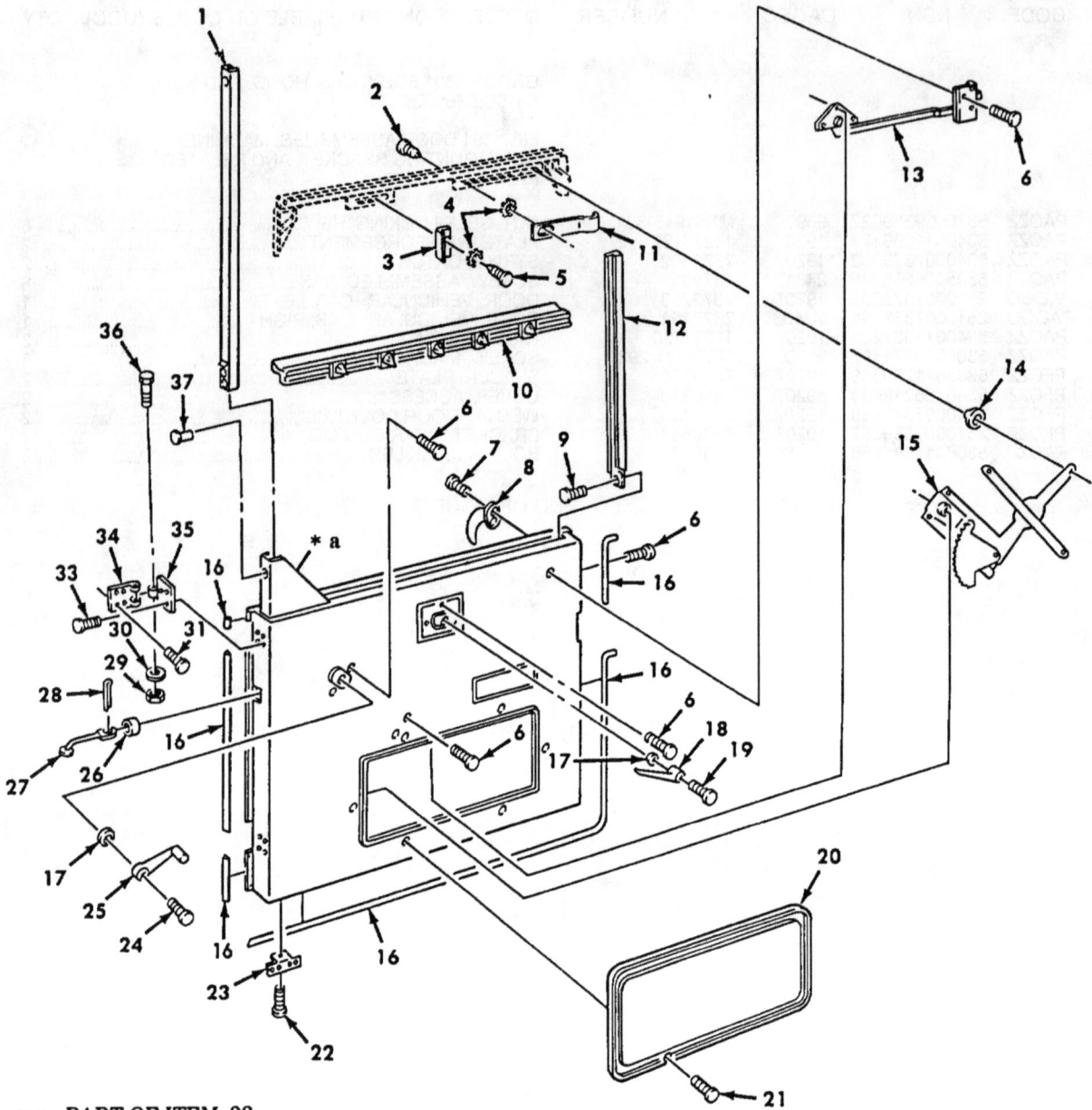

Figure 362. Door, Cab Assembly.

* a PART OF ITEM 33

(1) ITEM NO	(2) SMR CODE	(3) NSN	(4) CAGEC	(5) PART NUMBER	(6) DESCRIPTION AND USABLE ON CODES (UOC)	(7) QTY
					GROUP 1801 BODY, CAB, HOOD, AND HULL ASSEMBLIES	
					FIG. 362 DOOR, CAB ASSEMBLY	
1	PAOZZ	2540007373298	16662	7373298	CHANNEL, LIFT, VEHICL	1
2	PAOZZ	5305009953569	96906	MS24629-22	SCREW, TAPPING	1
3	PFOZZ	5340006259617	19207	7529306	BRACKET, ANGLE	2
4	PAOZZ	5310002090788	96906	MS35335-30	WASHER, LOCK	4
5	PAOZZ	5305008550973	96906	MS24629-24	SCREW, TAPPING	4
6	PAOZZ	5305012252106	21450	455176	SCREW, ASSEMBLED WAS	10
7	PAOZZ	5305004832339	19207	8743908	SCREW, SELF-LOCKING	2
8	PAOZZ	2540007410715	19207	7410715	HANDLE, DOOR, VEHICUL	2
9	PAOZZ	5305008550957	96906	MS24629-46	SCREW, TAPPING	1
10	PAOZZ	2510007373287	19207	7373287	WEATHERSTRIP	1
11	PAOZZ	2510010225783	19207	11682321	BRACKET, STOP, SASH...........................	1
12	PAOZZ	2540007373286	19207	7373286	CHANNEL, LIFT, VEHICL	1
13	PAOZZ	2540007373276	19207	7373276	LATCH, DOOR, VEHICULA	1
13	PAOZZ	2540007373277	19207	7373277	LATCH, DOOR, VEHICULA RIGHT SIDE.........	1
14	PAOZZ	2540005620422	19207	7529309	FASTENER, REGULATOR	2
15	PAOZZ	2540006930602	19207	7373289	REGULATOR, VEHICLE W LEFT DOOR GLASS.........	1
15	PAOZZ	2540006930603	19207	7373290	REGULATOR, VEHICLE W RIGHT DOOR GLASS..........	1
16	PAOZZ	9390012859623	19207	12368265	NONMETALLIC SPECIAL	1
17	PAOZZ	5330006212565	19207	7397852	GASKET	2
18	PAOZZ	2540007975609	19207	7975609	HANDLE, DOOR, VEHICUL	1
19	PAOZZ	5305000881302	19207	8743909	SCREW, SELF-LOCKING	1
20	PFOZZ	5340006930604	19207	7373299	COVER, ACCESS	1
21	PAOZZ	5305004324201	96906	MS51861-45	SCREW, TAPPING	6
22	PAOZZ	5305004324163	96906	MS51861-24	SCREW, TAPPING	5
23	PAOZZ	5340007373283	19207	7373283	BRACKET, ANGLE	5
24	PAOZZ	5305001450602	19207	8743910	SCREW, SELF-LOCKING	1
25	PAOZZ	2540007373304	19207	7975607	HANDLE, WINDOW REGUL	1
26	PAOZZ	2510000571630	19207	8757843	PAD, DOOR CHECK	2
27	PAOZZ	2510010849633	19207	8757842-1	ROD, DOOR CHECK, VEHI	2
28	PAOZZ	5315000590205	96906	MS24665-490	PIN, COTTER	2
29	PAOZZ	5310000617325	96906	MS21045-4	NUT, SELF-LOCKING, HE	4
30	PAOZZ	5310008094058	96906	MS27183-10	WASHER, FLAT	4
31	PAOZZ	5305001441514	19207	8743917	SCREW, MACHINE	12
32	XAOZZ		19207	7397673	DOOR, VEHICULAR LEFT HAND.................	1
32	XAOZZ		19207	7397674	DOOR, VEHICULAR RIGHT HAND...............	1
33	PAOZZ	5305000787021	19207	8743911	SCREW, SELF-LOCKING	8
34	PAOZZ	5340011147656	19207	7529301	LEAF, BUTT HINGE DOOR ASSEMBLY, RIGHT HAND	1
35	PAOZZ	5340011147657	19207	7529302	LEAF, BUTT HINGE DOOR ASSEMBLY, LEFT HAND	1
36	PAOZZ	5306004094066	19207	11608931	BOLT, SHOULDER....................................	4
37	PFOZZ	5320002626492	11083	7B5049	RIVET, TUBULAR	1

END OF FIGURE

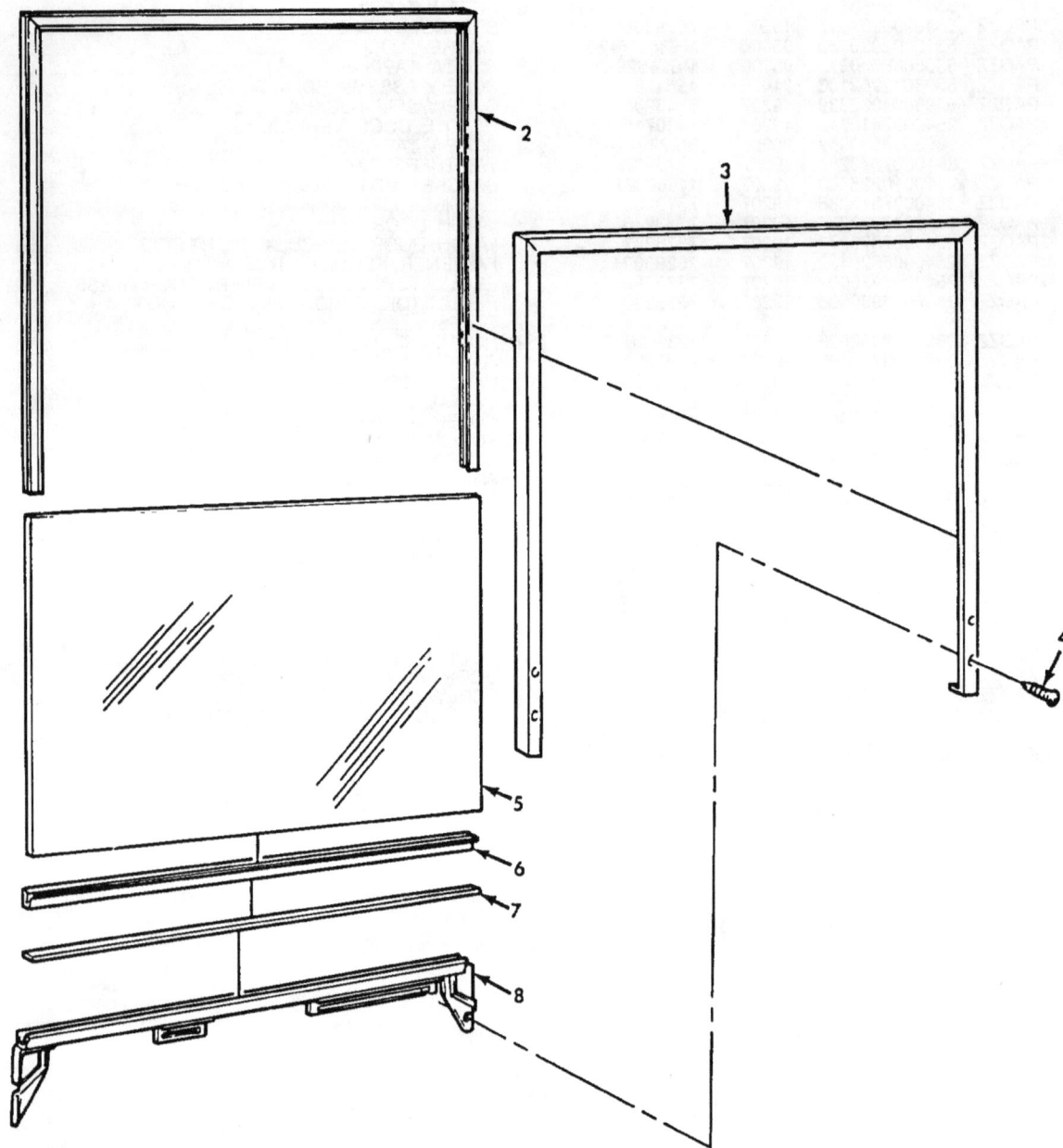

Figure 363. Window Frame Assembly, Cab Door.

(1) ITEM NO	(2) SMR CODE	(3) NSN	(4) CAGEC	(5) PART NUMBER	(6) DESCRIPTION AND USABLE ON CODES (UOC)	(7) QTY
					GROUP 1801 BODY, CAB, HOOD, AND HULL ASSEMBLIES	
					FIG. 363 WINDOW FRAME ASSEMBLY, CAB DOOR	
1	PAOZZ	2510006744487	72286	B410-1L	WINDOW, VEHICULAR WINDOW GLASS L.H	1
1	PAOOO	2510006501015	19207	7529305	FRAME, STRUCTURAL, VE WINDOW GLASS R.H.	1
2	PFOZZ	9390007373300	19207	7373300	.NONMETALLIC CHANNEL	1
3	PAOZZ	2510007373295	19207	7373295	.FRAME, WINDOW, VEHICU	1
4	PFOZZ	5305004024211	96906	MS51862-23	.SCREW, TAPPING	4
5	PFOZZ	9340010474100	19207	7529307	.GLASS, LAMINATED	1
6	PFOZZ	9390007373301	19207	7373301	.NONMETALLIC CHANNEL	1
7	PFOZZ	2510001795708	19207	10906350	.FILLER, CAB DOOR	1
8	PFOZZ	2540007373296	19207	7373296	.CHANNEL, LIFT, VEHICL DOOR WINDOW GLASS, LEFT HAND	1
8	PFOZZ	2540007373297	19207	7373297	.CHANNEL, LIFT, VEHICL DOOR WINDOW GLASS, RIGHT HAND	1

END OF FIGURE

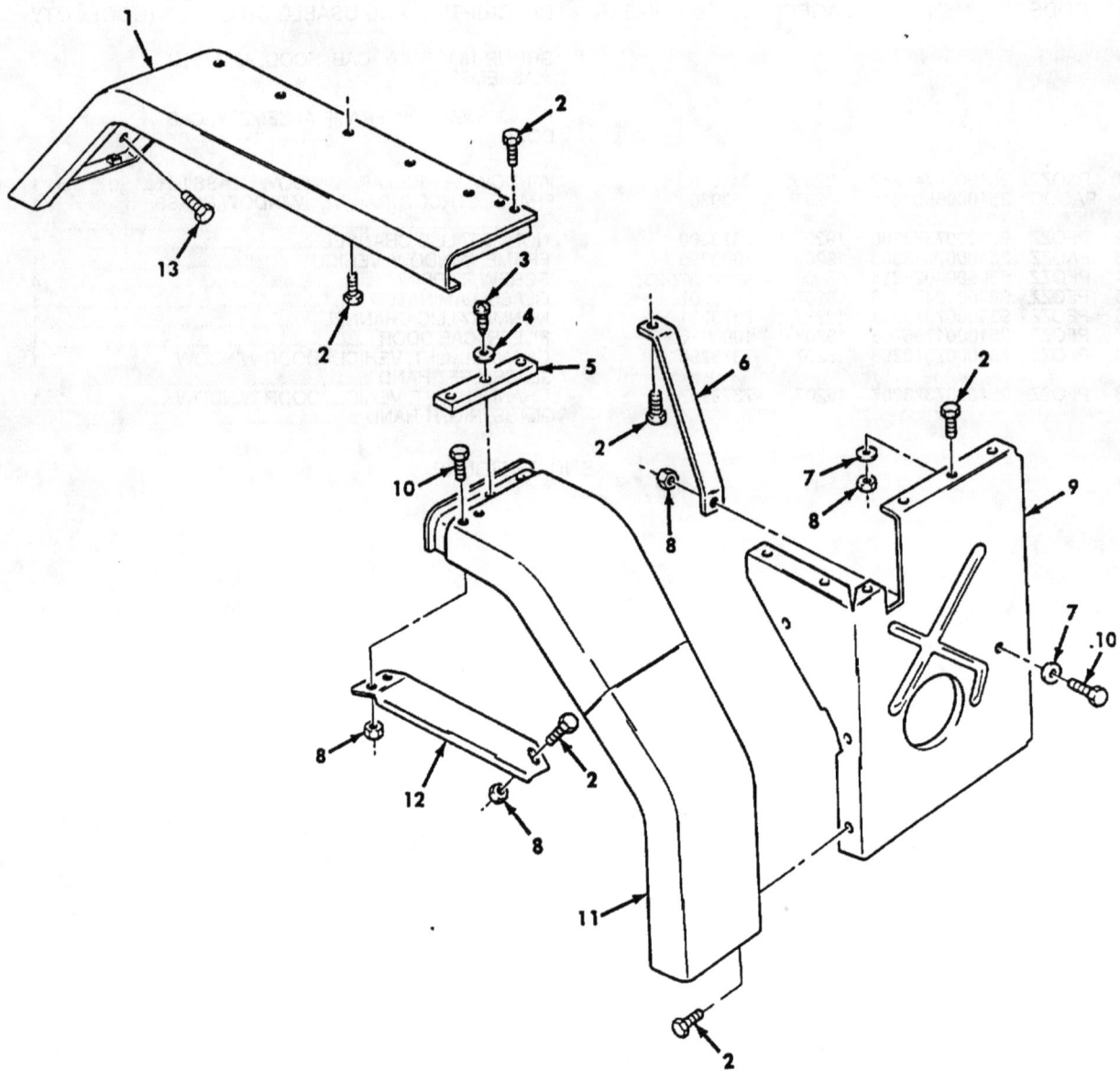

Figure 364. Fender, Fender Extension, and Splash Panel Assemblies.

SECTION II

(1) ITEM NO	(2) SMR CODE	(3) NSN	(4) CAGEC	(5) PART NUMBER	(6) DESCRIPTION AND USABLE ON CODES (UOC)	(7) QTY
					GROUP 1802 FENDERS, RUNNING BOARDS, AND WINDSHIELD	
					FIG. 364 FENDER, FENDER EXTENSION, AND SPLASH PANEL ASSEMBLIES	
1	PFFZZ	2510010823629	19207	12255918-1	FENDER, VEHICULAR LEFT HAND	1
1	PAFZZ	2510010823625	19207	12255918-2	FENDER, VEHICULAR RIGHT HAND	1
2	PAFZZ	5305010907626	19207	7372083-1	SCREW, ASSEMBLED WAS	32
3	PAFZZ	5305004324203	96906	MS51861-47	SCREW, TAPPING	6
4	PAOZZ	5310008098546	96906	MS27183-8	WASHER, FLAT	3
5	PAOZZ	5330011102462	19207	12255913	SEAL, NONMETALLIC ST	2
6	PFFZZ	5340010917627	19207	12256032	BRACKET, DOUBLE ANGL	2
7	PAFZZ	5310000806004	96906	MS27183-14	WASHER, FLAT	4
8	PAFZZ	5310008140672	96906	MS51943-36	NUT, SELF-LOCKING, HE	7
9	PAFZZ	2540010929323	19207	12255829-1	GUARD, SPLASH, VEHICU LEFT HAND	1
9	PAFZZ	2540010929324	19207	12255829-2	GUARD, SPLASH, VEHICU RIGHT HAND	1
10	PAFZZ	5305002692803	96906	MS90726-60	SCREW, CAP, HEXAGON H	2
11	PAFZZ	2510010823620	19207	12256316-1	FENDER, VEHICULAR LEFT HAND	1
11	PFFZZ	2510010823619	19207	12256316-2	FENDER, VEHICULAR RIGHT HAND	1
12	PFFZZ	2510010866802	19207	12256046-1	BRACE, FENDER	1
12	PFFZZ	2510010823623	19207	12256046-2	BRACE, FENDER	1
13	PAFZZ	5305011049018	19207	11663070	SCREW, ASSEMBLED WAS	2

END OF FIGURE

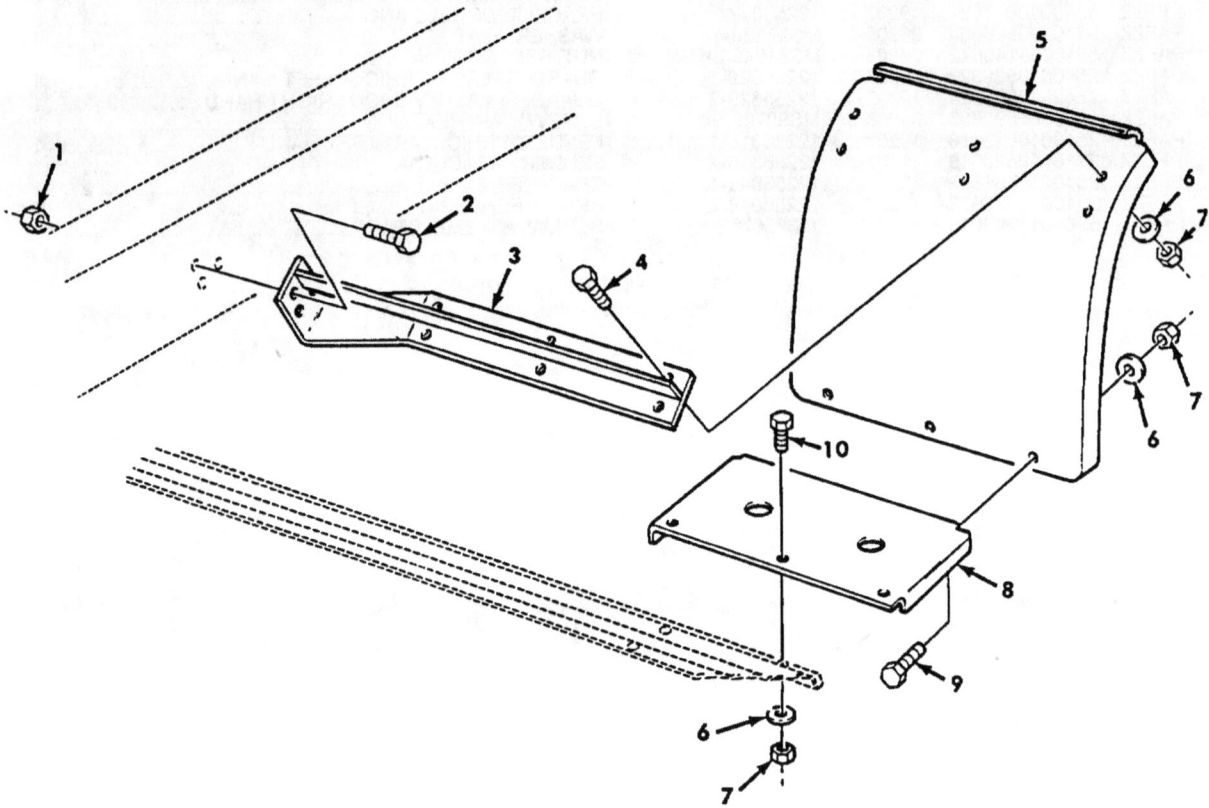

Figure 365. Splash Guards, Dump.

(1) ITEM NO	(2) SMR CODE	(3) NSN	(4) CAGEC	(5) PART NUMBER	(6) DESCRIPTION AND USABLE ON CODES (UOC)	(7) QTY
					GROUP 1802 FENDERS, RUNNING BOARDS, AND WINDSHIELD	
					FIG. 365 SPLASH GUARDS, DUMP	
1	PAOZZ	5310008775795	96906	MS21044-N8	NUT, SELF-LOCKING, HE .... UOC: V21, V22	6
1	PAOZZ	5310004883888	96906	MS51943-40	NUT, SELF-LOCKING, HE .... UOC: DAG, DAH, ZAG, ZAH	6
2	PAOZZ	5305007195219	96906	MS90727-111	SCREW, CAP, HEXAGON H .... UOC: DAG, DAH, V21, V22, ZAG, ZAH	6
3	PFOZZ	2590004084618	19207	10883101	BRACKET, VEHICULAR C .... UOC: DAG, DAH, V21, V22, ZAG, ZAH	1
3	PFOZZ	2510004084634	19207	10883102	BRACKET, SPLASH GUAR .... UOC: DAG, DAH, V21, V22, ZAG, ZAH	1
4	PAOZZ	5305009125113	96906	MS51096-359	SCREW, CAP, HEXAGON H .... UOC: V21, V22	12
4	PAOZZ	5305002693235	80204	B1821BH038F088N	SCREW, CAP, HEXAGON H .... UOC: DAG, DAH, ZAG, ZAH	12
5	PBOFF	2540011332150	19207	12277394	GUARD, SPLASH, VEHICU .... UOC: DAG, DAH, V21, V22, ZAG, ZAH	2
6	PAOZZ	5310000806004	96906	MS27183-14	WASHER, FLAT .... UOC: DAG, DAH, V21, V22, ZAG, ZAH	28
7	PAOZZ	5310008140672	96906	MS51943-36	NUT, SELF-LOCKING, HE .... UOC: DAG, DAH, ZAG, ZAH	24
7	PAOZZ	5310009591488	96906	MS51922-21	NUT, SELF-LOCKING, HE .... UOC: V21, V22	24
3	PFOZZ	5340011317452	19207	12277393	PLATE, MOUNTING .... UOC: V21, V22	2
8	PFOZZ	5340012104658	19207	12302956	PLATE, MENDING .... UOC: DAG, DAH, ZAG, ZAH	2
9	PAOZZ	5305002693234	96906	MS90727-58	SCREW, CAP, HEXAGON .... UOC: DAG, DAH, V21, V22	6
10	PAOZZ	5305002692803	96906	MS90726-60	SCREW, CAP, HEXAGON H .... UOC: DAG, DAH, ZAG, ZAH	6
10	PAOZZ	5305002693235	80204	B1821BH038F088N	SCREW, CAP, HEXAGON H .... UOC: V21, V22	6

END OF FIGURE

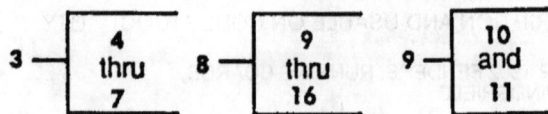

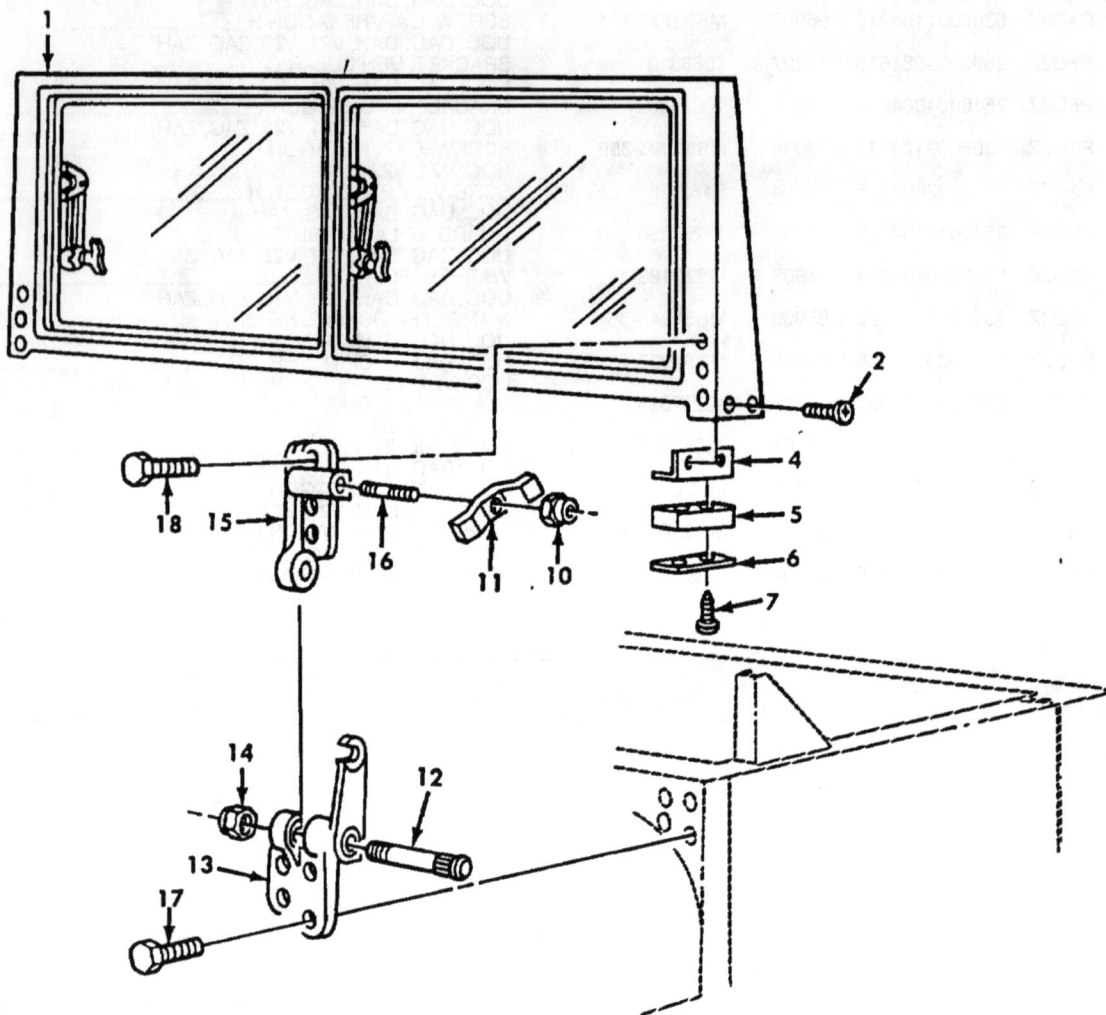

Figure 366. Windshield Assembly and Mounting Hardware.

(1) ITEM NO	(2) SMR CODE	(3) NSN	(4) CAGEC	(5) PART NUMBER	(6) DESCRIPTION AND USABLE ON CODES (UOC)	(7) QTY
					GROUP 1802 FENDERS, RUNNING BOARDS, AND WINDSHIELD	
					FIG. 366 WINDSHIELD ASSEMBLY AND MOUNTING HARDWARE	
1	PAOFF	2510011089122	19207	12277058	WINDSHIELD ASSEMBLY	1
2	PAOZZ	5305008550960	96906	MS24629-36	SCREW, TAPPING	4
3	PAOZZ	2510007368622	19207	7368622	BRACKET ASSEMBLY, WI	2
4	PFOZZ	5340011047400	19207	7368625	.BRACKET, ANGLE	1
5	PFOZZ	2510004093993	19207	7368623	.WEATHERSEAL, WINDSHI	1
6	PFOZZ	5340010823596	19207	7368624	.PLATE, MOUNTING	1
7	PAOZZ	5305008550960	96906	MS24629-36	.SCREW, TAPPING	2
8	PBOZZ	5340006960264	19207	7373321	HINGE, BUTT LEFT HAND....................................	1
8	PBOZZ	5340006960265	19207	7373322	HINGE, BUTT RIGHT HAND	1
9	PAOZZ	5310007373338	19207	7373338	.NUT, PLAIN, WING	1
10	XAOZZ		19207	7373335-XA	..NUT, PLAIN, SPLINE	1
11	XAOZZ		19207	7397729-XA	..KNOB	1
12	PAOZZ	5306011048389	19207	7373318	.BOLT, RIBBED SHOULDE	1
13	PAOZZ	5340011049075	19207	7529319	.LEAF, BUTT HINGE L.H. LOWER.........................	1
13	PAOZZ	5340011695686	19207	7529320	.HINGE, SPECIAL R.H. LOWER	1
14	PAOZZ	5310009390783	96906	MS21083N12	.NUT, SELF-LOCKING, HE	1
15	PAOZZ	5340011049076	19207	7529317	.HINGE, BUTT L.H. UPPER	1
15	PAOZZ	5340006223941	19207	7529318	.HINGE, HALF, WINDSHIE R.H. UPPER	1
16	PAOZZ	5307002068510	19207	7373319	.STUD, PLAIN	1
17	PAOZZ	5305011049018	19207	11663070	SCREW, ASSEMBLED WAS	8
18	PAOZZ	5306002861476	24617	9409115	BOLT, ASSEMBLED WASH	6

END OF FIGURE

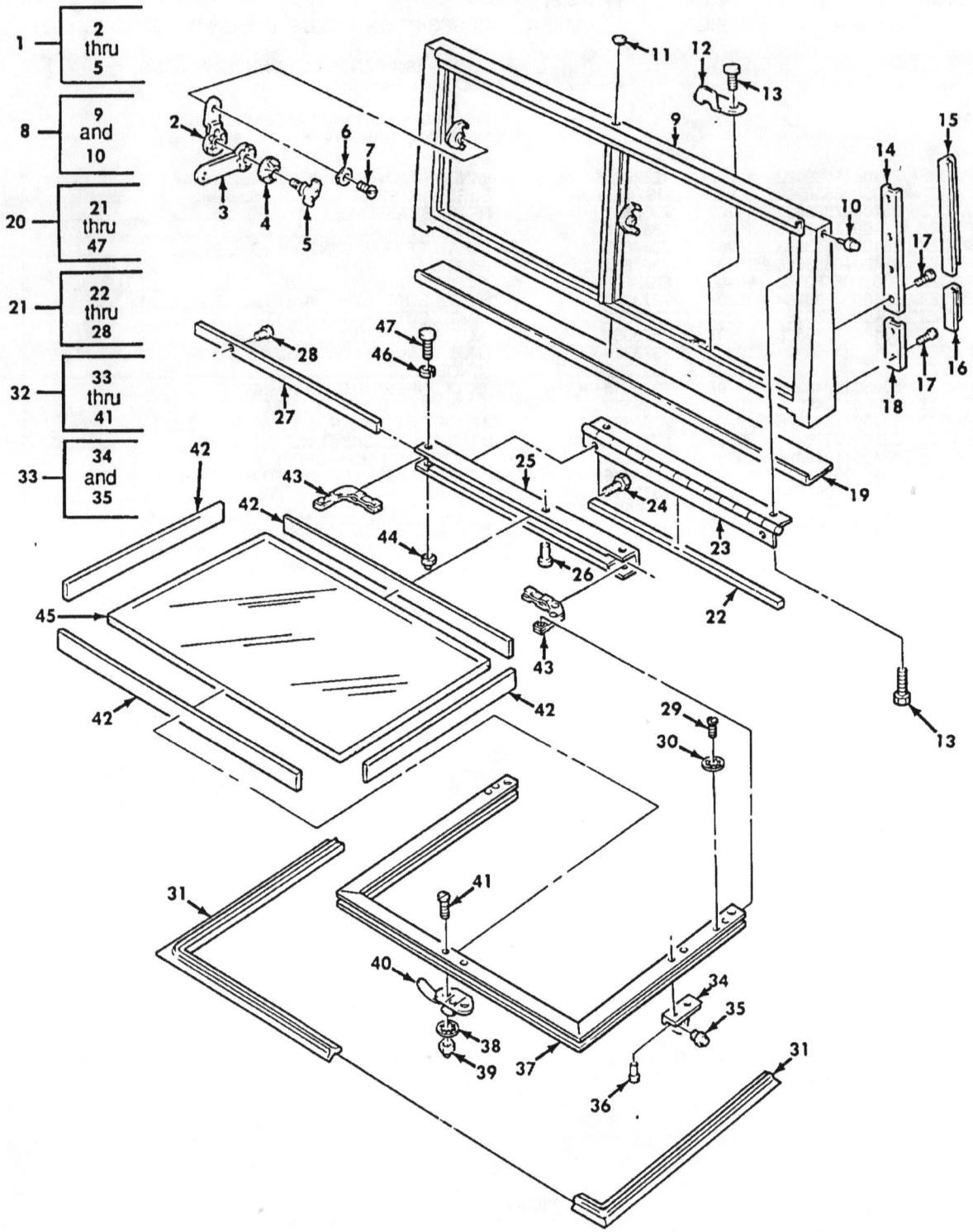

Figure 367. Windshield and Frame Assemblies.

(1) ITEM NO	(2) SMR CODE	(3) NSN	(4) CAGEC	(5) PART NUMBER	(6) DESCRIPTION AND USABLE ON CODES (UOC)	(7) QTY
					GROUP 1802 FENDERS, RUNNING BOARDS, AND WINDSHIELD	
					FIG. 367 WINDSHIELD AND FRAME ASSEMBLIES	
1	PAFZZ	2510011528812	19207	7373327-1	ARM, WINDSHIELD GLAS WINDSHIELD	4
2	PAOZZ	3040006223946	19207	7529327	.LEVER, REMOTE CONTRO	1
3	PAFZZ	3040011873620	19207	7529324-1	.LEVER, REMOTE CONTRO	1
4	PAFZZ	5310002812180	19207	7351289	.WASHER, SPRING TENSI	1
5	PAFZZ	5305006380957	19207	7529322	.THUMBSCREW	1
6	PAFZZ	5310002644083	19207	7371886	WASHER, SPRING TENSI	8
7	PAFZZ	5305003501210	19207	7374809	SCREW, MACHINE	8
8	PAFZZ	2510011307942	19207	12277061	FRAME, WINDOW, VEHICU	1
9	XAFZZ		19207	12277059	.FRAME ASSY, OUTER	1
10	PAFZZ	5325009051492	96906	MS27980-21B	.STUD, SNAP FASTENER	2
11	PAFZZ	5340002409228	19207	11621624-2	CAP, PROTECTIVE, DUST	5
12	PFFZZ	2510007373326	19207	7373326	PLATE, WINDSHIELD	2
13	PAFZZ	5305000425567	24617	425567	SCREW, ASSEMBLED WAS	20
14	PFFZZ	2510006930608	19207	7373324	RETAINER, WINDSHIELD	2
15	PAFZZ	9390005996405	19207	7373333	NONMETALLIC SPECIAL LEFT	1
15	PAFZZ	5330011219886	19207	7373332	RUBBER SEAL RIGHT	1
16	PAFZZ	9390011209864	19207	7373334	NONMETALLIC SPECIAL	2
17	PAFZZ	5305008550961	96906	MS24629-35	SCREW, TAPPING	12
18	PFFZZ	2510006930610	19207	7373336	RETAINER, WEATHERSTR	2
19	PAFZZ	5330011525943	19207	12302607	SEAL, NONMETALLIC ST	1
20	PAOFF	2510011307943	19207	12277069-1	WINDOW, VEHICULAR LEFT HAND	1
20	PAOFF	2510011307944	19207	12277069-2	WINDOW, VEHICULAR RIGHT HAND	1
21	PFFZZ	2510011143693	19207	12277075	.FRAME, WINDOW, VEHICU	1
22	PAFZZ	5680011225214	19207	12277124	..WEATHER STRIP	1
23	PFFZZ	5340011143694	19207	12277074	..HINGE, WINDSHIELD	1
24	PAFZZ	5305004846186	21450	423533	..SCREW, ASSEMBLED WAS	7
25	XAFZZ		19207	12277073 XA	..FRAME, WINDOW, VEHICU	1
26	PAFZZ	5365011495416	19207	7413375	..SPACER, SLEEVE	2
27	PAFZZ	9510011354762	19207	12277182	..BAR, METAL	1
28	PFFZZ	5320005823276	96906	MS20600AD6W4	..RIVET, BLIND	2
29	PAFZZ	5305009846192	96906	MS35206-244	.SCREW, MACHINE	4
30	PAFZZ	5310005590070	96906	MS35333-38	.WASHER, LOCK	4
31	PAFZZ	5330012570750	19207	7373325-1	.SEAL, NONMETALLIC SP	1
32	PFFZZ	2510011104060	19207	12277062-1	.FRAME, WINDOW, VEHICU	1
32	PFFZZ	2510011143689	19207	12277062-2	.FRAME, WINDOW, VEHICU	1
33	PFFZZ	2510011307945	19207	8720328	..BRACKET, ENGINE MOUN	2
34	XAFZZ		19207	8686994	...ADJUSTING ARM BRACK	1
35	PFFZZ	2510011474992	19207	7748089	...ADAPTER BUSHING, WIP	1
36	PFFZZ	5320011483706	21450	135290	..RIVET, SOLID	4
37	XAFZZ		19207	12277060-1	..FRAME ASSEMBLY	1
37	XAFZZ		19207	12277060-2	..FRAME ASSEMBLY	1
38	PAFZZ	5310005765752	96906	MS35333-39	..WASHER, LOCK	2
39	PAOZZ	5310000131498	96906	MS24679-63	.NUT, PLAIN, CAP	2
40	PAFZZ	5340011410832	19207	7373330-1	..LATCH, THUMB	1
40	PFFZZ	5340007373330	19207	7373330	..FASTENER, CASEMENT	1

(1) ITEM NO	(2) SMR CODE	(3) NSN	(4) CAGEC	(5) PART NUMBER	(6) DESCRIPTION AND USABLE ON CODES (UOC)	(7) QTY
41	PAFZZ	5305009953444	96906	MS35207-266	..SCREW, MACHINE	2
42	PAFZZ	9320011095696	19207	12277066	.RUBBER STOCK MOLDIN	1
43	PAFZZ	2510011421294	19207	7529332	.FRAME SECTION, STRUC	2
44	PAFZZ	5310000131498	96906	MS24679-63	.NUT, PLAIN, CAP	2
45	PAFZZ	9340011095934	19207	12277072	.GLASS, LAMINATED	1
46	PAFZZ	5310005765752	96906	MS35333-39	.WASHER, LOCK	2
47	PAFZZ	5305009953444	96906	MS35207-266	.SCREW, MACHINE	2

END OF FIGURE

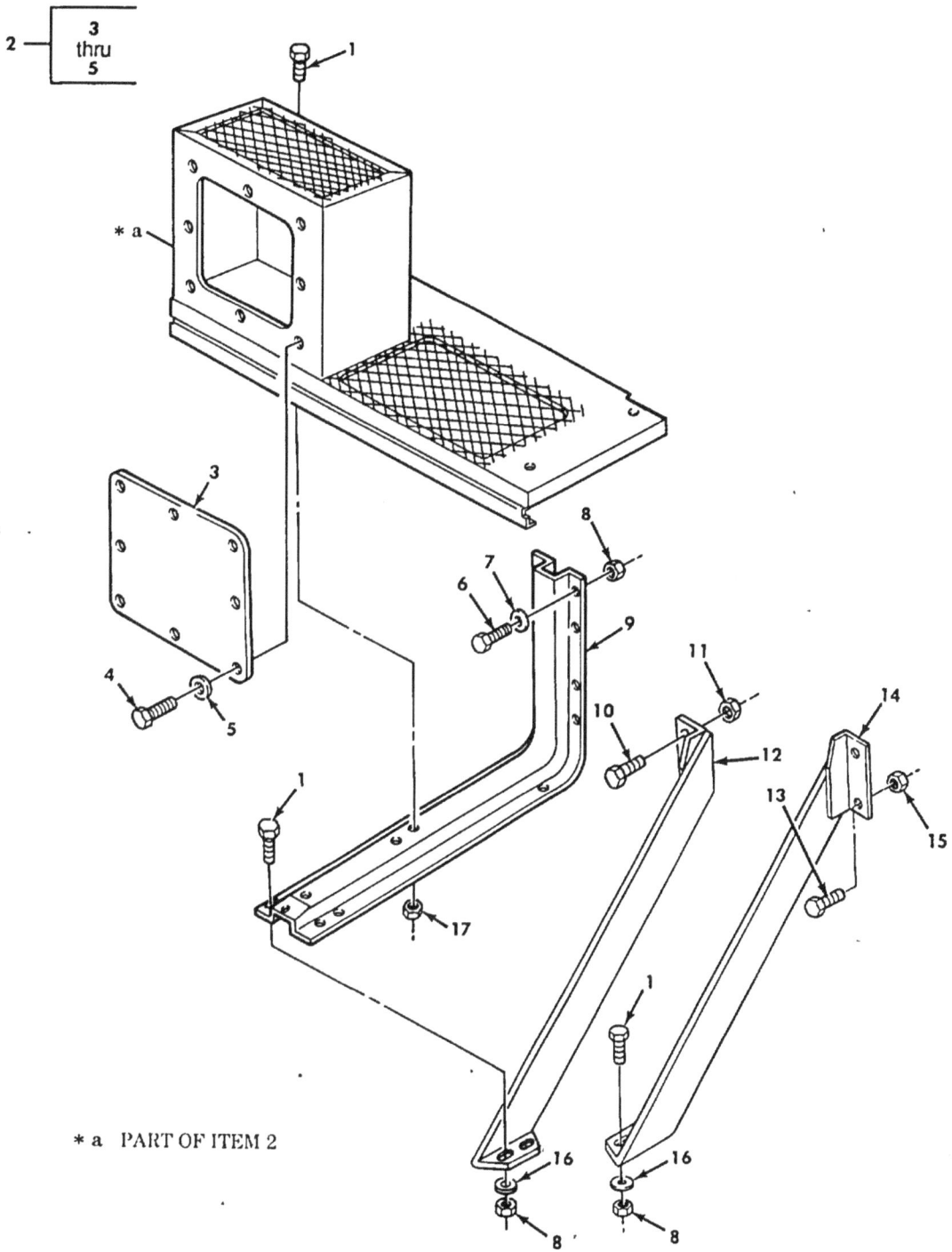

Figure 368. Running Board, Step, and Mounting Hardware.

* a   PART OF ITEM 2

(1) ITEM NO	(2) SMR CODE	(3) NSN	(4) CAGEC	(5) PART NUMBER	(6) DESCRIPTION AND USABLE ON CODES (UOC)	(7) QTY
					GROUP 1802 FENDERS, RUNNING BOARDS, AND WINDSHIELD	
					FIG. 368 RUNNING BOARD, STEP, AND MOUNTING HARDWARE	
1	PAOZZ	5305002692803	96906	MS90726-60	SCREW, CAP, HEXAGON H	8
2	PFOOO	2510010827460	19207	12255858	TREAD, METALLIC, NONS	1
3	PFOZZ	5340011036010	19207	12276933	.COVER, ACCESS	1
4	PAOZZ	5305002692803	96906	MS90726-60	.SCREW, CAP, HEXAGON H	8
5	PAOZZ	5310006379541	96906	MS35338-46	.WASHER, LOCK	8
6	PAOZZ	5305002692804	96906	MS90726-61	SCREW, CAP, HEXAGON H	8
					UOC: DAA, DAB, DAE, DAF, DAG, DAH, DAW, DAX, V12, V13, V14, V15, V19, V20, V21, V22, ZAB, ZAE, ZAF, ZAG, ZAH, ZAW, ZAX	
6	PAOZZ	5305002693240	80204	B1821BH038F150N	SCREW, CAP, HEXAGON H	8
					UOC: DAC, DAJ, DAX, DAL, V16, V17, V18, V24, V25, V39, ZAC, ZAJ, ZAK, ZAL	
7	PAOZZ	5310000806004	96906	MS27183-14	WASHER, FLAT	3
					UOC: DAC, DAD, DAE, DAF, DAG, DAH, DAJ, DAK, DAL, V16, V17, V18, V19, V20, V21, V22, V24, V25, V39, ZAC, ZAD, ZAE, ZAF, ZAG, ZAH, ZAJ, ZAK, ZAL	
8	PAOZZ	5310009591488	96906	MS51922-21	NUT, SELF-LOCKING...	8
9	PBOZZ	2510010907641	19207	12255911	SUPPORT, RUNNING BOA	2
10	PAOZZ	5305007195235	80204	B1821BH050F175N	SCREW, CAP, HEXAGON H	2
					UOC: V39	
10	PAOZZ	5305007195221	80204	B1821BH050F150N	SCREW, CAP, HEXAGON H	2
					UOC: DAA, DAB, DAC, DAD, DAE, DAF, DAG, DAH, DAJ, DAK, DAL, DAW, DAX, V12, V13, V14, V15, V16, V17, V24, V25, ZAA, ZAB, ZAC, ZAD, ZAE, ZAF, ZAG, ZAH, ZAJ, ZAK, ZAL	
11	PAOZZ	5310008775795	96906	MS21044-N8	NUT, SELF-LOCKING, HE	2
					UOC: DAB, DAC, DAD, DAF, DAG, DAJ, DAK, DAL, DAW, DAX, V12, V13, V14, V15, V16, V17, V18, V21, V22, V24, V25, V39, ZAB, ZAC, ZAD, ZAF, ZAG, ZAH, ZAJ, ZAK, ZAL	
11	PAOZZ	5310004883888	96906	MS51943-40	NUT, SELF-LOCKING, HE	2
					UOC: DAA, DAE, V19, V20, ZAA, ZAE	
12	PFOZZ	2540010911612	19207	12256044	TOOL BOX SUPPORT...	1
13	PAOZZ	5305007195221	80204	B1821BH050F150N	SCREW, CAP, HEXAGON H	2
					UOC: DAA, DAB, DAC, DAD, DAE, DAF, DAG, DAH, DAJ, DAK, DAL, DAW, DAX, V12, V13, V14, V15, V16, V17, V18, V19, V20, V21, V22, V24, ZAA, ZAB, ZAC, ZAD, ZAE, ZAF, ZAG, ZAH, ZAJ, ZAK, ZAL, ZAW, ZAX	
14	PAOZZ	5340010911609	19207	12256030-1	BRACKET, MOUNTING...	1
15	PAOZZ	5310004883888	96906	MS51943-40	NUT, SELF-LOCKING, HE	2
16	PAOZZ	5310000806004	96906	MS27183-14	WASHER, FLAT	2
					UOC: DAD, DAE, DAF, DAG, DAH, DAJ, DAK, DAL, V16, V17, V18, V19, V20, V21, V22, V24, V25, ZAD, ZAE, ZAF, ZAG, ZAH, ZAJ, ZAK, ZAL	

(1) ITEM NO	(2) SMR CODE	(3) NSN	(4) CAGEC	(5) PART NUMBER	(6) DESCRIPTION AND USABLE ON CODES (UOC)	(7) QTY
17	PAOZZ	5310009591488	96906	MS51922-21	NUT, SELF-LOCKING.	2

END OF FIGURE

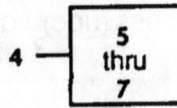

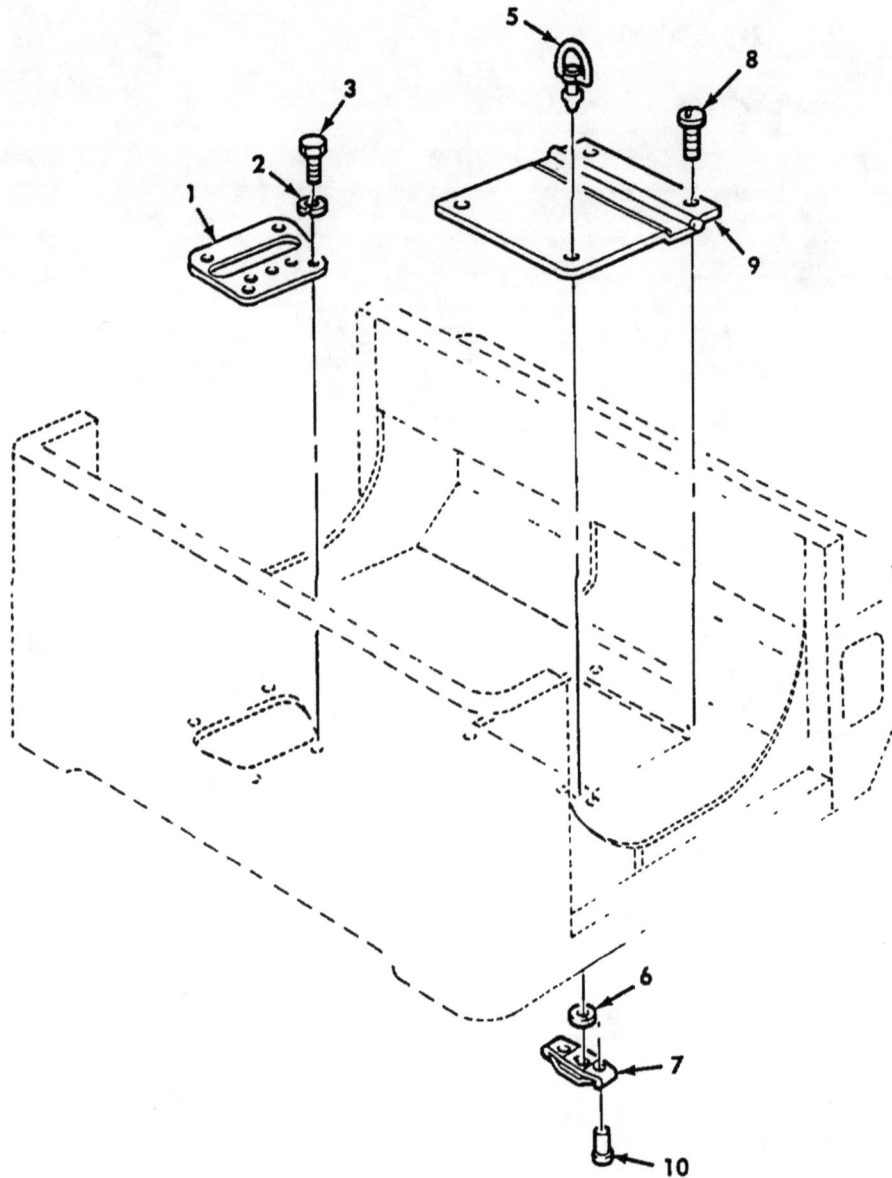

Figure 369. Transmission Access Door and Spring Brake Plate.

(1) ITEM NO	(2) SMR CODE	(3) NSN	(4) CAGEC	(5) PART NUMBER	(6) DESCRIPTION AND USABLE ON CODES (UOC)	(7) QTY
					GROUP 1805 FLOORS, SUBFLOORS, AND RELATED COMPONENTS	
					FIG. 369 TRANSMISSION ACCESS DOOR AND SPRING BRAKE PLATE	
1	PFOZZ	5340010831147	19207	12276930	PLATE, MENDING	1
2	PAOZZ	5310005825965	96906	MS35338-44	WASHER, LOCK	4
3	PAOZZ	5305002678953	80204	B1821BH025F063N	SCREW, CAP, HEXAGON H	4
4	PAOZZ	5325011087375	19207	12277043	FASTENER ASSEMBLY, T	2
5	PAOZZ	5325003435531	94222	85-15-140-16	.STUD, TURNLOCK FASTE	1
6	PAOZZ	5310008228525	94222	85-34-101-20	.WASHER, SPLIT	1
7	PAOZZ	5325002827471	94222	85-35-295-15	.RECEPTACLE, TURNLOCK..	1
8	PAOZZ	5305011049019	19207	7373279	SCREW, ASSEMBLED WAS	2
9	PFOZZ	5340011037839	19207	12256070	DOOR, ACCESS	1
10	PFOZZ	5320009958907	80205	NAS1399B4-4	RIVET, BLIND	4

END OF FIGURE

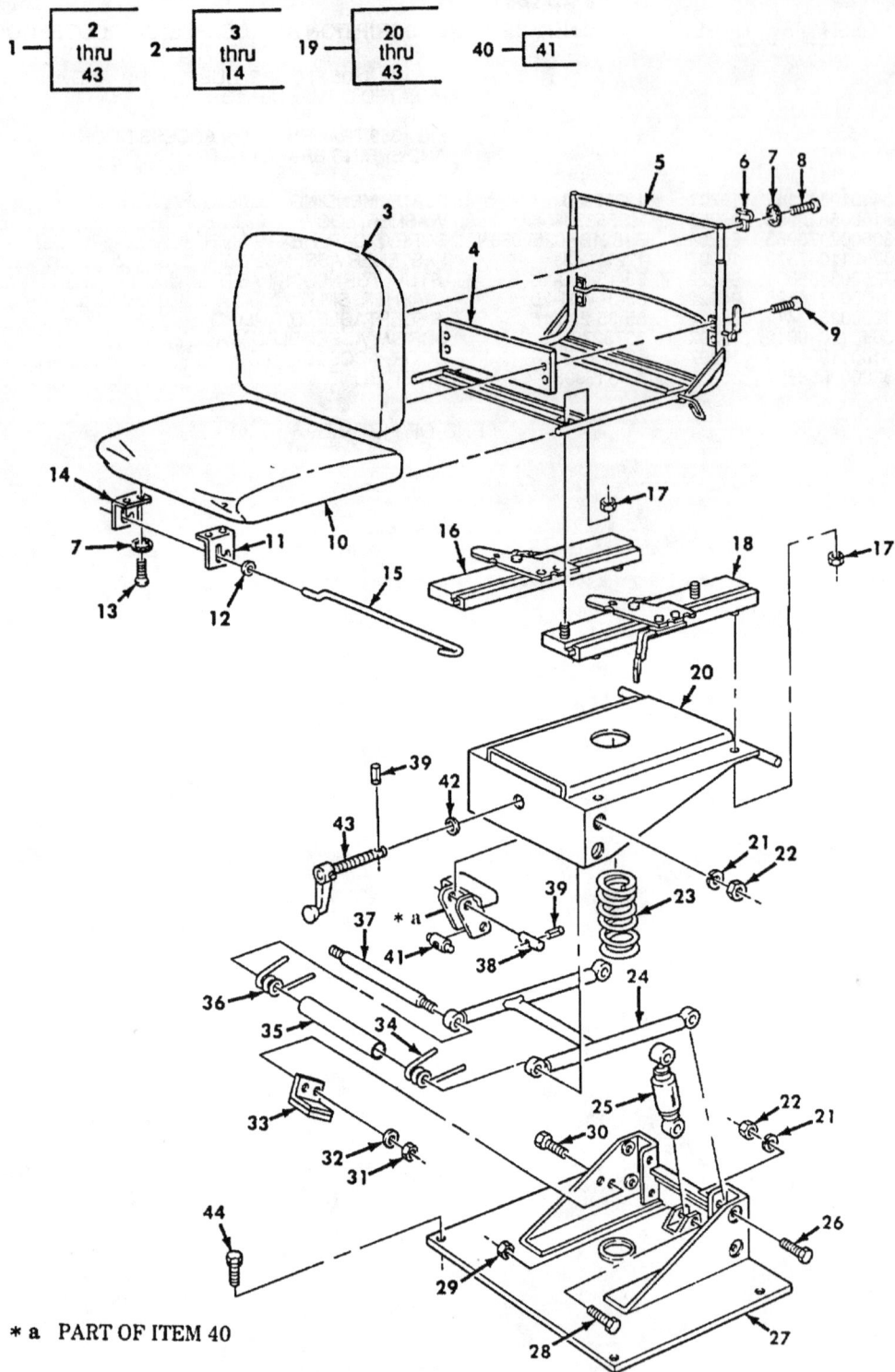

Figure 370. Operator's Seat Assembly.

* a PART OF ITEM 40

(1) ITEM NO	(2) SMR CODE	(3) NSN	(4) CAGEC	(5) PART NUMBER	(6) DESCRIPTION AND USABLE ON CODES (UOC)	(7) QTY
					GROUP 1806 UPHOLSTERY SEATS AND CARPETS	
					FIG. 370 OPERATOR'S SEAT ASSEMBLY	
1	AOOFF		19207	11640442-3	SEAT, VEHICULAR	1
2	PAOFF	2540011089114	19207	11663385-1	SEAT, VEHICULAR GREEN	1
2	XDOFF		19207	11663385-3	SEAT, VEHICULAR TAN	1
2	XDOFF		19207	11663385-5	SEAT, VEHICULAR WHITE	1
3	PAOZZ	2540004605826	19207	11640526	.CUSHION, SEAT BACK, V GREEN	1
3	XDOZZ		19207	11640526-1	.CUSHION, SEAT BACK, V TAN	1
3	XDOZZ		19207	11640526-2	.CUSHION, SEAT BACK, V WHITE	1
4	PFOZZ	2540004368289	19207	11640513	.SUPPORT, SEAT, VEHICU	1
5	PAOZZ	2540012255863	19207	11640521-1	.FRAME, SEAT, VEHICULA	1
6	PFOZZ	5340000046854	19207	11640507	.STRAP, RETAINING	2
7	PAOZZ	5310005501130	96906	MS35333-40	.WASHER, LOCK	6
8	PAOZZ	5305009881725	96906	MS35206-281	.SCREW, MACHINE	2
9	PAOZZ	5305009544295	96906	MS35190-287	.SCREW, MACHINE	4
10	PAOZZ	2540004605815	19207	11640522	.CUSHION, SEAT, VEHICU GREEN	1
10	XDOZZ		19207	11640522-1	.CUSHION, SEAT, VEHICU TAN	1
10	XDOZZ		19207	11640522-2	.CUSHION, SEAT, VEHICU WHITE	1
11	PFOZZ	5340007575877	19207	11640504-1	.BRACKET, ANGLE	1
12	PAOZZ	531G08093078	96906	MS27183-11	.WASHER, FLAT	2
13	PAOZZ	5305009881726	96906	MS35206-282	.SCREW, MACHINE	4
14	PFOZZ	5340007575901	19207	11640504-2	.BRACKET, ANGLE	1
15	PFOZZ	5340011173793	19207	11669053	CLIP, SPRING TENSION	1
16	PBOZZ	2540011174882	19207	11669167-2	ADJUSTER ASSEMBLY	1
17	PAOZZ	5310009843807	96906	MS51922-13	NUT, SELF-LOCKING, HE	8
18	PBOZZ	2540011173025	19207	11669167-1	ADJUSTER ASSEMBLY	1
19	PFOFF	2540012919037	19207	11640523-3	PEDESTAL, SEAT	1
20	PFFZZ	2540012867673	19207	11640524-1.	.FRAME, SEAT, VEHICULA	1
21	PAFZZ	5310000034094	01276	210104-8S	.WASHER, LOCK	8
22	PAFZZ	5310007680318	24617	9413509	.NUT, PLAIN, HEXAGON	8
23	PFFZZ	5360008320178	19207	11640508	.SPRING, HELICAL, COMP	1
24	PFFZZ	2540004701564	19207	11640520	.FRAME, SEAT, VEHICULA	2
25	PFFZZ	2510004897104	6B719	615116	.SHOCK ABSORBER, DIRE	1
26	PAFZZ	5306000200857	19207	11663341	.BOLT, SHOULDER	6
27	XDFZZ		19207	11640525-1	.FRAME ASSEMBLY, SEAT	1
28	PAFZZ	5306000201058	19207	11662487	.BOLT, SHOULDER	2
29	PAFZZ	5310008807744	96906	MS51967-5	.NUT, PLAIN, HEXAGON	2
30	PAFZZ	5305000680502	96906	MS90725-6	.SCREW, CAP, HEXAGON H	4
31	PAFZZ	5310007616882	96906	MS51967-2	.NUT, PLAIN, HEXAGON	4
32	PAFZZ	5310005501130	96906	MS35333-40	.WASHER, LOCK	4
33	PFFZZ	5340004701537	19207	11662489	.BRACKET, ANGLE	2
34	PFFZZ	5360004824813	19207	11640515-1	.SPRING, HELICAL, TORS RIGHT HAND TORQUE ROD	1
35	PFFZZ	3120012806050	19207	12302671	.BEARING, SLEEVE	1
36	PFFZZ	5360004824814	19207	11640515-2	.SPRING, HELICAL, TORS LEFT HAND TORQUE ROD	1
37	PFFZZ	5307001020962	19207	11640510	.STUD, SHOULDERED	1
38	PAFZZ	5315001712590	19207	11640506	.PIN, STRAIGHT, HEADLE	1

(1) ITEM NO	(2) SMR CODE	(3) NSN	(4) CAGEC	(5) PART NUMBER	(6) DESCRIPTION AND USABLE ON CODES (UOC)	(7) QTY
39	PAFZZ	5315008414443	96906	MS16562-225	.PIN, SPRING	2
40	PFFFF	2540011307946	19207	11640512	.SUPPORT, SEAT, VEHICU	1
41	PAFZZ	5315012592517	19207	11641043	.PIN, SHOULDER, HEADLE	1
42	PAFZZ	5310000014719	19207	11663036	.WASHER, SADDLE	1
43	PFFZZ	5340004701543	19207	11640516	.CONTROL ROD	1
44	PAOZZ	5305011049018	19207	11663070	SCREW, ASSEMBLED	4

END OF FIGURE

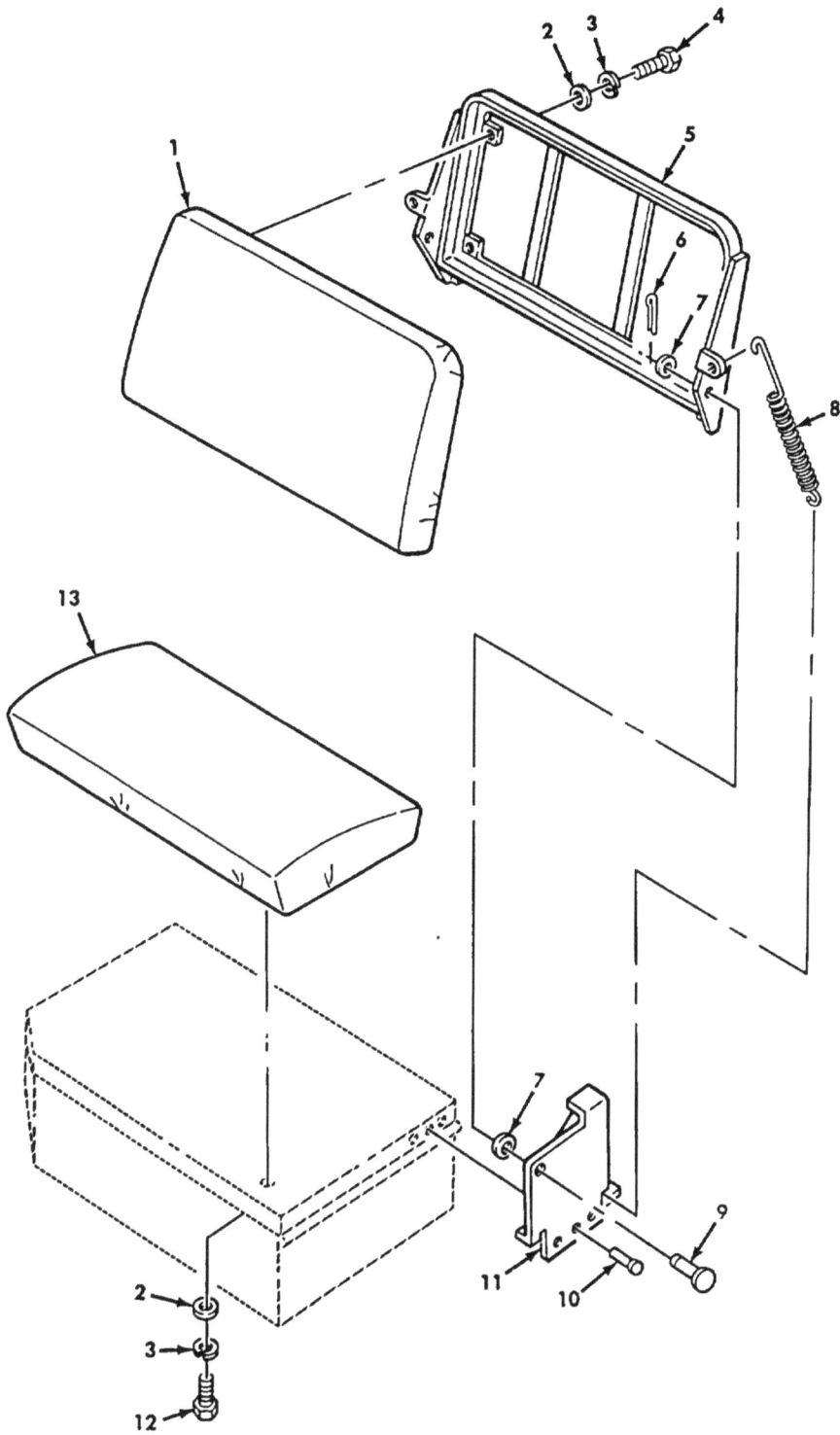

Figure 371. Companion Seat Assembly.

(1) ITEM NO	(2) SMR CODE	(3) NSN	(4) CAGEC	(5) PART NUMBER	(6) DESCRIPTION AND USABLE ON CODES (UOC)	(7) QTY
					GROUP 1806 UPHOLSTERY SEATS AND CARPETS	
					FIG. 371 COMPANION SEAT ASSEMBLY	
1	PAOZZ	2540010849644	19207	12255960	CUSHION, SEAT, VEHICU GREEN	1
1	XDOZZ		19207	12255960-4	CUSHION, SEAT, VEHICU TAN	1
1	XDOZZ		19207	12255960-5	CUSHION, SEAT, VEHICU WHITE	1
2	PAOZZ	5310008094058	96906	MS27183-10	WASHER, FLAT	14
3	PAOZZ	5310005825965	96906	MS35338-44	WASHER, LOCK	14
4	PAOZZ	5305002253843	80204	B1821BH025C100N	SCREW, CAP, HEXAGON H	8
5	PAOZZ	2540010823624	19207	12255922	FRAME, SEAT, VEHICULA	1
6	PAOZZ	5315008423044	96906	MS24665-283	PIN, COTTER	2
7	PAOZZ	5310000806004	96906	MS27183-14	WASHER, FLAT	4
8	PAOZZ	5360011080828	19207	12255925	SPRING, HELICAL, EXTE	2
9	PAOZZ	5315011091443	19207	12255915	PIN, STRAIGHT, HEADED	2
10	PFFZZ	5320011453191	96906	MS35743-55	RIVET, SOLID	12
11	PAFZZ	2540011307947	19207	12255807-1	SUPPORT, SEAT, VEHICU COMPANION, LEFT HAND	1
11	PAFZZ	2540011307948	19207	12255807-2	SUPPORT, SEAT, VEHICU COMPANION, RIGHT HAND	1
12	PAOZZ	5305000680502	96906	MS90725-6	SCREW, CAP, HEXAGON H	6
13	PAOZZ	2540010827510	19207	12255961	CUSHION, SEAT, VEHICU GREEN	1
13	XDOZZ		19207	12255961-4	CUSHION, SEAT, VEHICU TAN	1
13	XDOZZ		19207	12255961-5	CUSHION, SEAT, VEHICU WHITE	1

END OF FIGURE

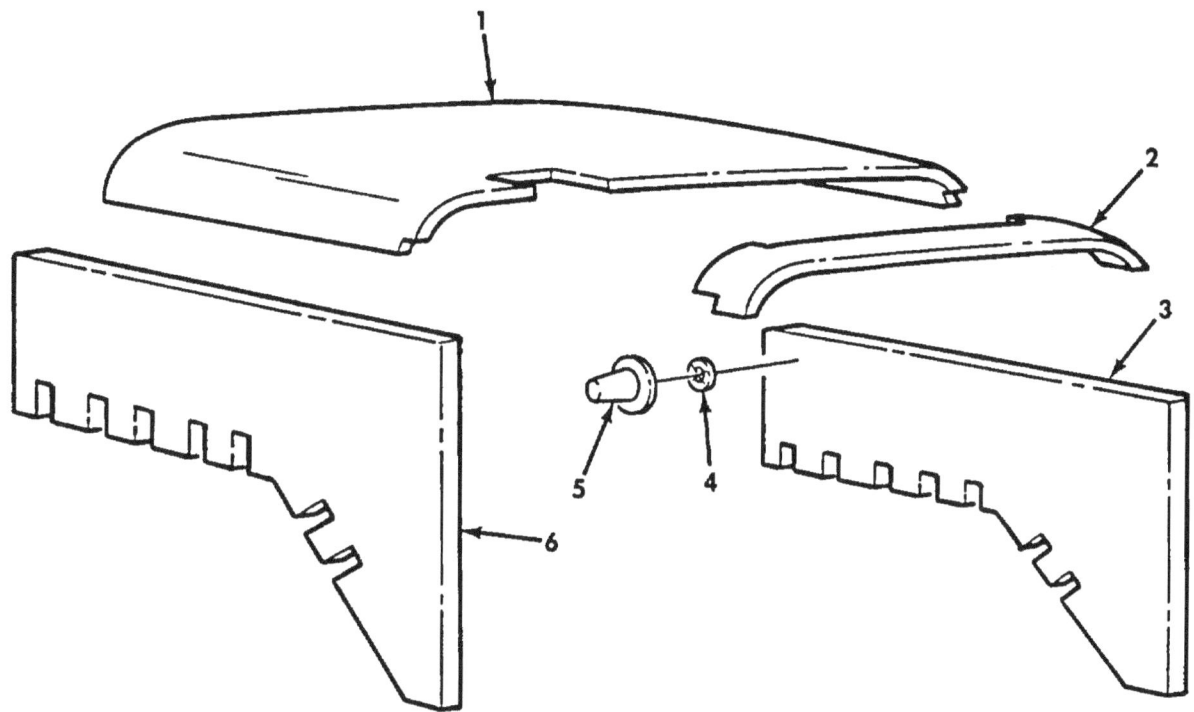

Figure 372. Hood Insulation.

(1) ITEM NO	(2) SMR CODE	(3) NSN	(4) CAGEC	(5) PART NUMBER	(6) DESCRIPTION AND USABLE ON CODES (UOC)	(7) QTY
					GROUP 1806 UPHOLSTERY SEATS AND CARPET	
					FIG. 372 HOOD INSULATION	
1	PAOZZ	5640013182812	19207	12277367	SOUND CONTROLLING B	1
2	PAOZZ	5640013192376	19207	12277369	INSULATION, TOP REAR	1
3	PAOZZ	5640013182815	19207	12277368-2	SOUND CONTROLLING B RIGHT HOOD	1
4	PAOZZ	5310012604937	85105	101301104	WASHER, RECESSED	34
5	PAOZZ	5340012607895	19207	12277375	CAP-PLUG, PROTECTIVE	34
6	PAOZZ	5640013182816	19207	12277368-1	SOUND CONTROLLING B LEFT HOOD	1

END OF FIGURE

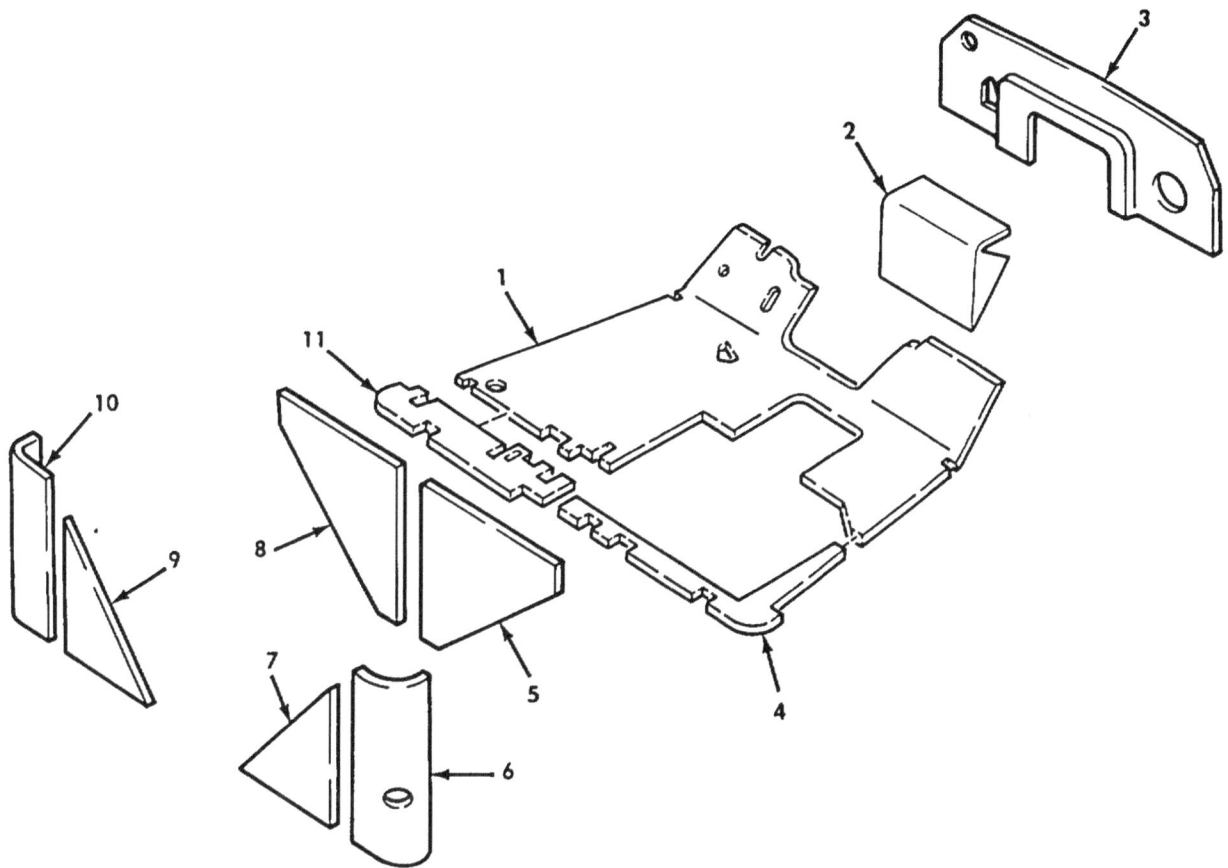

Figure 373. Cab Insulation.

(1) ITEM NO	(2) SMR CODE	(3) NSN	(4) CAGEC	(5) PART NUMBER	(6) DESCRIPTION AND USABLE ON CODES (UOC)	(7) QTY
					GROUP 1806 UPHOLSTERY SEATS AND CARPETS	
					FIG. 373 CAB INSULATION	
1	PAOZZ	2510013182814	19207	12356832-1	INSULATION, BLANKET .... UOC:DAA, DAB, DAC, DAD, DAE, DAF, DAG, DAH, DAJ, DAK, DAL, DAW, DAX, V12, V13, V14, V15, V16, V17, V18, V19, V20, V2 1, V22, V24, V25, V39	1
1	PAOZZ	2510013638981	19207	12356832-4	INSULATION, BLANKET .... UOC:ZAA, ZAB, ZAC, ZAD, ZAE, ZAF, ZAG, ZAH, ZAJ, ZAK, ZAL	1
2	PAOZZ	2510010831140	19207	12256296	INSULATION, ENGINE H .... UOC:DAA, DAB, DAC, DAD, DAE, DAF, DAG, DAH, DAJ, DAK, DAL, DAW, DAX, V12, V13, V14, V15, V16, V17, V18, V19, V20, V21, V22, V24, V25, V39	1
2	PAOZZ	2590012925707	24825	20510859	COVER, INSULATION ENGINE..................... UOC:ZAA, ZAB, ZAC, ZAD, ZAE, ZAF, ZAG, ZAH, ZAJ, ZAK, ZAL	1
3	PAOZZ	2510010831146	19207	12256303	INSULATION, VEHICULA.............................	1
4	PAOZZ	2510013182818	19207	12356832-3	INSULATION, BLANKET	1
5	PAOZZ	2510010823603	19207	12256295-2	INSULATION, VEHICULA	1
6	PAOZZ	2510010827455	19207	12256279	INSULATION, VEHICULA	1
7	PAOZZ	2510010823621	19207	12256294-2	INSULATION, VEHICULA	1
8	PAOZZ	2510010823604	19207	12256295-1	INSULATION, VEHICULA	1
9	PAOZZ	2510010823622	19207	12256294-1	INSULATION, VEHICULA	1
10	PAOZZ	2510010840446	19207	12256293	INSULATION, VEHICULA.............................	1
11	PAOZZ	2510013182817	19207	12356832-2	INSULATION, BLANKET	1

END OF FIGURE

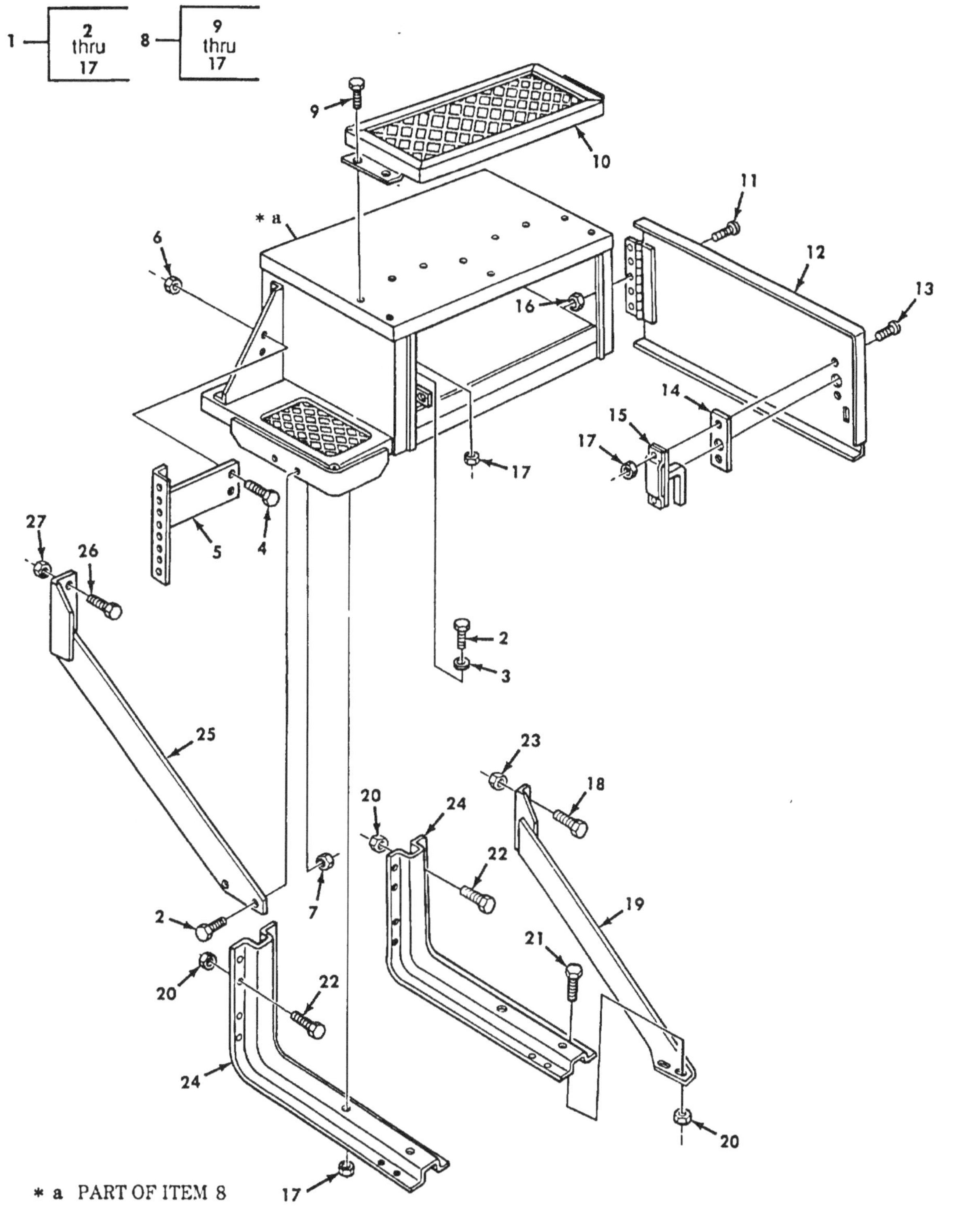

Figure 374. Toolbox, Step Assembly, and Mounting Hardware.

* a PART OF ITEM 8

(1) ITEM NO	(2) SMR CODE	(3) NSN	(4) CAGEC	(5) PART NUMBER	(6) DESCRIPTION AND USABLE ON CODES (UOC)	(7) QTY
					GROUP 1808 STOWAGE RACKS, BOXES, AND STRAPS	
					FIG. 374 TOOLBOX, STEP ASSEMBLY, AND MOUNTING HARDWARE	
1	AOOOO		19207	12255874	TOOL BOX AND STEP	1
2	PAOZZ	5305002692803	96906	MS90726-60	.SCREW, CAP, HEXAGON H	4
3	PAOZZ	5310000877493	96906	MS27183-13	.WASHER, FLAT	4
4	PAOZZ	5306000680513	60285	6893-2	.BOLT, MACHINE	2
5	PFOZZ	5340011307949	19207	12256370	.BRACKET, ANGLE	1
6	PAOZZ	5310009359022	96906	MS51943-32	.NUT, SELF-LOCKING, HE	2
7	PAOZZ	5310008140672	96906	MS51943-36	.NUT, SELF-LOCKING, HE	6
8	PAOZZ	2510010827458	19207	12256213	.STEP ASSEMBLY, TOOL	1
9	PAOZZ	5306000680513	60285	6893-2	..BOLT, MACHINE	10
10	PFOZZ	2590011361438	19207	12255857	..STEP ASSEMBLY, STOWA	1
11	PAOZZ	5305009897434	96906	MS35207-263	..SCREW, MACHINE	5
12	PFOZZ	5340011260176	19207	12255909	..DOOR, ACCESS	1
13	PAOZZ	5305009932738	96906	MS35207-280	..SCREW, MACHINE	2
14	PFOZZ	5365011079967	19207	7397843	..SPACER, PLATE	1
15	PFOZZ	5340010387759	19207	8376629-1	..LOCK, RIM	1
16	PAOZZ	5310008775797	96906	MS21044-N3	..NUT, SELF-LOCKING, HE	5
17	PAOZZ	5310009359022	96906	MS51943-32	..NUT, SELF-LOCKING, HE	10
18	PAOZZ	5305007195235	80204	B1821BH050F175N	SCREW, CAP, HEXAGON H	2
19	PAOZZ	5340010911610	19207	12256030-2	BRACKET, MOUNTING......................	1
20	PAOZZ	5310009591488	96906	MS51922-21	NUT, SELF-LOCKING, HE	10
21	PAOZZ	5305002692803	96906	MS90726-60	SCREW, CAP, HEXAGON H	4
22	PAOZZ	5305002693240	80204	B1821BH038F150N	SCREW, CAP, HEXAGON H .... UOC:DAJ, DAK, DAL, DAW, DAX, V16, V17, V18, V24, V25, V39, ZAJ, ZAK, ZAL	8
22	PAOZZ	5305002692804	96906	MS90726-61	SCREW, CAP, HEXAGON H .... UOC:DAA, DAB, DAE, DAF, DAG, DAH, DAW, DAX, V12, V13, V14, V15, V19, V20, V21, V22, ZAA, ZAB, ZAE, ZAF, ZAG, ZAH	8
23	PAOZZ	5310004883888	96906	MS51943-40	NUT, SELF-LOCKING, HE TOOL BOX STRAP.....	1
24	PFOZZ	2510010907641	19207	12255911	SUPPORT, RUNNING BOARD	2
25	PBOZZ	5340010907640	19207	12255871	BRACKET, ANGLE .... UOC:DAJ, DAX, DAL, DAW, DAX, V16, V17, V18, V24, V25, V39, ZAJ, ZAK, ZAL	1
26	PAOZZ	5305007195221	80204	B1821BH050F150N	SCREW, CAP, HEXAGON H .... UOC:DAJ, DAK, DAL, DAW, DAX, V16, V17, V18, V24, V25, V39, ZAJ, ZAK, ZAL	2
27	PAOZZ	5310008775795	96906	MS21044-N8	NUT, SELF-LOCKING, HE .... UOC:DAJ, DAK, DAL, DAW, DAX, V16, V17, V18, V19, V24, V25, V39, ZAJ, ZAK, ZAL	2

END OF FIGURE

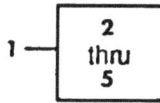

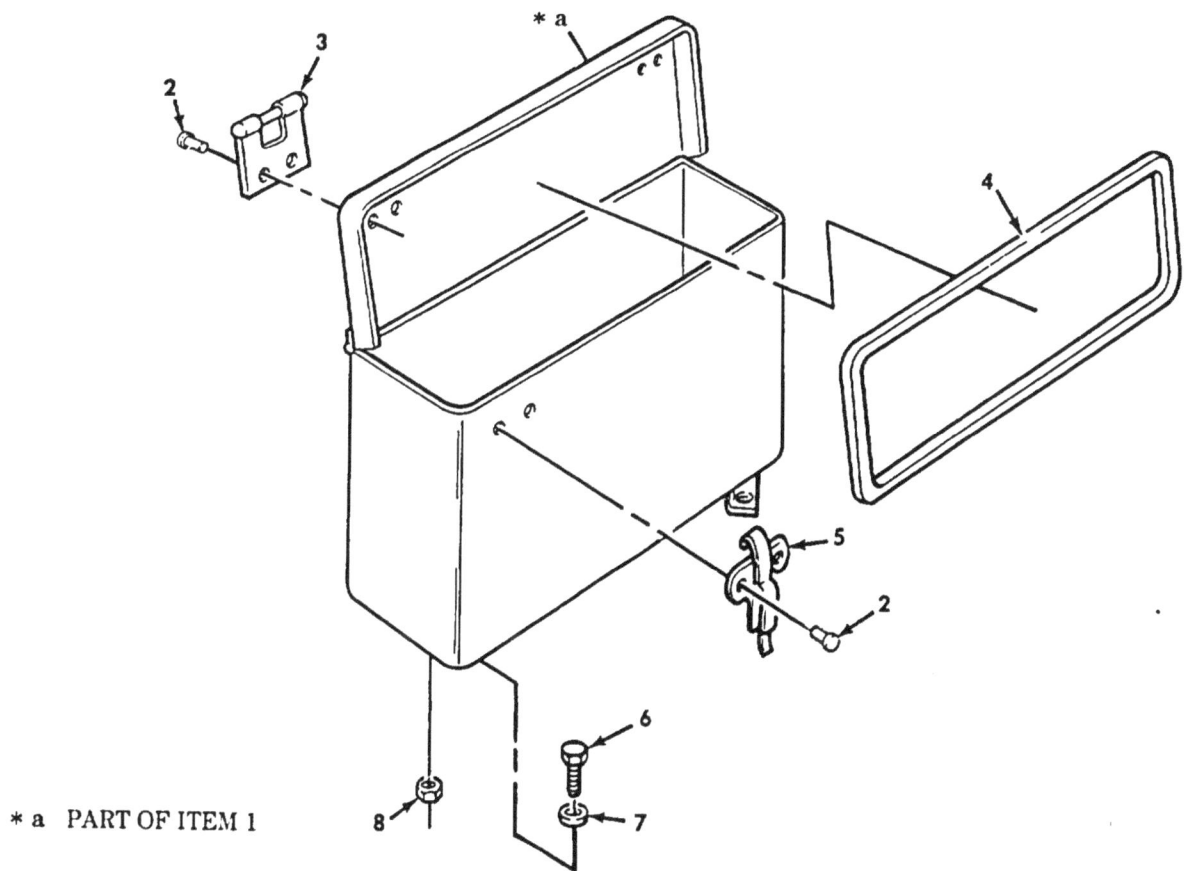

Figure 375. Map Compartment and Mounting Hardware.

* a   PART OF ITEM 1

(1) ITEM NO	(2) SMR CODE	(3) NSN	(4) CAGEC	(5) PART NUMBER	(6) DESCRIPTION AND USABLE ON CODES (UOC)	(7) QTY
					GROUP 1808 STOWAGE RACKS, BOXES, AND STRAPS	
					FIG. 375 MAP COMPARTMENT AND MOUNTING HARDWARE	
1	PAOOO	2540010849630	19207	12257065	COMPARTMENT, MAP	1
2	PFOZZ	5320012249157	24617	193780	.RIVET, SOLID	8
3	PBOZZ	5340011347635	19207	7047876	.LEAF, BUTT HINGE	2
4	PAOZZ	5330011066735	19207	12257062	.SEAL, NONMETALLIC ST	1
5	PAOZZ	5340011438312	19207	7413565-1	.CATCH, CLAMPING	2
6	PAOZZ	5305002692803	96906	MS90726-60	SCREW, CAP, HEXAGON H	4
7	PAOZZ	5310000877493	96906	MS27183-13	WASHER, FLAT	4
8	PAOZZ	5310008140672	96906	MS51943-36	NUT, SELF-LOCKING, HE	4

END OF FIGURE

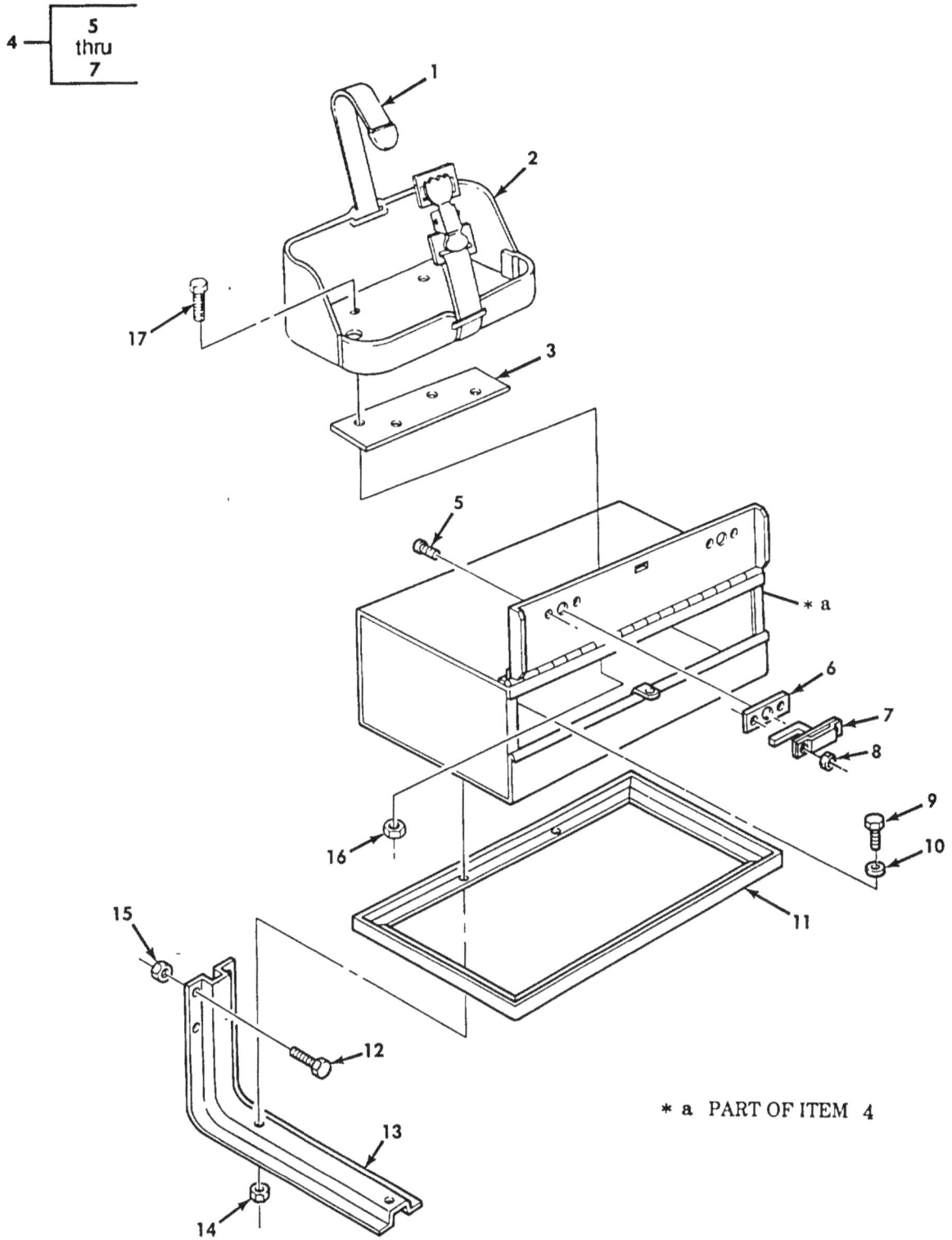

Figure 376. Toolbox Fuel Can Bracket and Mounting Hardware.

* a PART OF ITEM 4

(1) ITEM NO	(2) SMR CODE	(3) NSN	(4) CAGEC	(5) PART NUMBER	(6) DESCRIPTION AND USABLE ON CODES (UOC)	(7) QTY
					GROUP 1808 STOWAGE RACKS, BOXES, AND STRAPS	
					FIG. 376 TOOLBOX FUEL CAN BRACKET AND MOUNTING HARDWARE	
1	PAOZZ	5340009684060	19207	8690527	STRAP, WEBBING................................... UOC:DAA, DAB, DAC, DAD, DAW, DAX, V12, V13, V14, V15, V16, V17, ZAA, ZAB, ZAC, ZAD	1
2	PFOZZ	2590004736331	19207	6566675	BRACKET ASSEMBLY................................ UOC:DAA, DAB, DAC, DAD, DAW, DAX, V12, V13, V14, V15, V16, V17, ZAA, ZAB, ZAC, ZAD	1
3	PFOZZ	5340010893057	19207	12276967	PLATE, MENDING................................... UOC:DAA, DAB, DAC, DAD, DAW, DAX, V12, V13, V14, V15, V16, V17, ZAA, ZAB, ZAC, ZAD	1
4	PAOZZ	2540004721696	19207	10871462	BOX, ACCESSORIES STO .... UOC:DAA, DAB, DAC, DAD, DAW, DAX, V12, V13, V14, V15, V16, V17, ZAA, ZAB, ZAC, ZAD	1
5	PAOZZ	5305009932461	96906	MS35207-281	.SCREW, MACHINE................................. UOC:DAA, DAB, DAC, DAD, DAW, DAX, V12, V13, V14, V15, V16, V17, ZAA, ZAB, ZAC, ZAD	4
6	PFOZZ	5365011079967	19207	7397843	.SPACER, PLATE................................... UOC:DAA, DAB, DAC, DAD, DAW, DAX, V12, V13, V14, V15, V16, V17	2
7	PAOZZ	5340010387759	19207	8376629-1	.LOCK, RIM....................................... UOC:DAA, DAB, DAC, DAD, DAW, DAX, V12, V13, V14, V15, V16, V17	2
8	PAOZZ	5310009597600	96906	MS51922-5	NUT, SELF-LOCKING, HE .... UOC:DAA, DAB, DAC, DAD, DAW, DAX, V12, V13, V14, V15, V16, V17	4
9	PAOZZ	5305002692803	96906	MS90726-60	SCREW, CAP, HEXAGON H .... UOC:DAA, DAB, DAC, DAD, DAW, DAX, V12, V13, V14, V15, V16, V17	4
10	PAOZZ	5310008094061	96906	MS27183-15	WASHER, FLAT.................................... UOC:DAA, DAB, DAC, DAD, DAW, DAX, V12, V13, V14, V15, V16, V17, ZAA, ZAB, ZAC, ZAD	4
11	PFOZZ	2540004172583	19207	10871456	FRAME, REINFORCING T .... UOC:DAA, DAB, DAC, DAD, DAW, DAX, V12, V13, V14, V15, V16, V17, ZAA, ZAB, ZAC, ZAD	1
12	PAOZZ	5305007195238	80204	B1821BH050F200N	SCREW, CAP, HEXAGON H .... UOC:DAA, DAB, DAC, DAD, DAW, DAX, V12, V13, V14, V15, V16, V17, ZAA, ZAB, ZAC, ZAD	8
13	PFOZZ	9520011254082	19207	10871458-1	CHANNEL, STRUCTURAL .... UOC:DAA, DAB, DAC, DAD, DAW, DAX, V12, V13, V14, V15, V16, V17, ZAA, ZAB, ZAC, ZAD	2
14	PAOZZ	5310009824908	96906	MS21045-6	NUT, SELF-LOCKING, HE .... UOC:DAA, DAB, DAC, DAD, DAW, DAX, V12, V13, V14, V15, V16, V17, ZAA, ZAB, ZAC, ZAD	4
15	PAOZZ	5310000679507	96906	MS51922-37	NUT, SELF-LOCKING .... UOC:DAA, DAB, DAC, DAD, DAW, DAX, V12, V13, V14, V15, V16, V17, ZAA, ZAB, ZAC, ZAD	8
16	PAOZZ	5310008140672	96906	MS51943-36	NUT, SELF-LOCKING, HE .... UOC:DAA, DAB, DAC, DAD, DAW, DAX, V12, V13, V14, V15, V16, V17, ZAA, ZAB, ZAC, ZAD	4

(1) ITEM NO	(2) SMR CODE	(3) NSN	(4) CAGEC	(5) PART NUMBER	(6) DESCRIPTION AND USABLE ON CODES (UOC)	(7) QTY
17	PAOZZ	5305002693235	80204	B1821BH038F088N	SCREW, CAP, HEXAGON H .... UOC:DAA, DAB, DAC, DAD, DAW, DAX, V12, V13, V14, V15, V16, V17, ZAA, ZAB, ZAC, ZAD	4

END OF FIGURE

Figure 377. Toolbox and Mounting Hardware, Tractor.

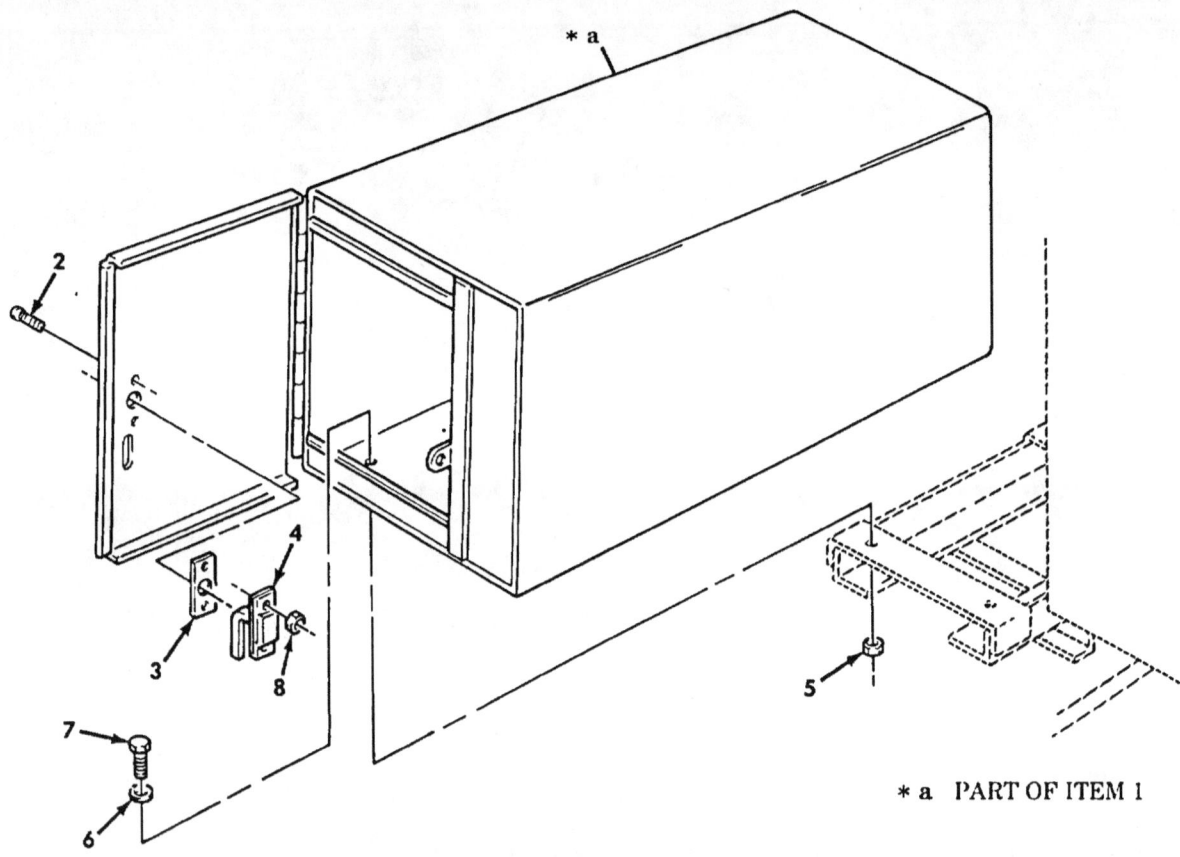

*a   PART OF ITEM 1

(1) ITEM NO	(2) SMR CODE	(3) NSN	(4) CAGEC	(5) PART NUMBER	(6) DESCRIPTION AND USABLE ON CODES (UOC)	(7) QTY
					GROUP 1808 STOWAGE RACKS, BOXES, AND STRAPS	
					FIG. 377 TOOLBOX AND MOUNTING HARDWARE, TRACTOR	
1	PAOZZ	2540011089129	19207	8758436-1	BOX, ACCESSORIES STO .... UOC:DAG, DAH, V21, V22, ZAG, ZAH	1
2	PAOZZ	5305009932461	96906	MS35207-281	.SCREW, MACHINE .... UOC:DAG, DAH, V21, V22, ZAG, ZAH	2
3	PFOZZ	5365011079967	19207	7397843	.SPACER, PLATE .... UOC:DAG, DAH, V21, V22, ZAG, ZAH	1
4	PBOZZ	5340010387759	19207	8376629-1	.LOCK, RIM... UOC:DAG, DAH, V21, V22, ZAG, ZAH	1
5	PAOZZ	5310008140672	96906	MS51943-36	NUT, SELF-LOCKING, HE .... UOC:DAG, DAH, V21, V22, ZAG, ZAH	6
6	PAOZZ	5310008094061	96906	MS27183-15	WASHER, FLAT .... UOC:DAG, DAH, V21, V22, ZAG, ZAH	6
7	PAOZZ	5305002692803	96906	MS90726-60	SCREW, CAP, HEXAGON H .... UOC:DAG, DAH, V21, V22, ZAG, ZAH	6
8	PAOZZ	5310009597600	96906	MS51922-5	NUT, SELF-LOCKING, HE .... UOC:DAG, DAH, V21, V22, ZAG, ZAH	2

END OF FIGURE

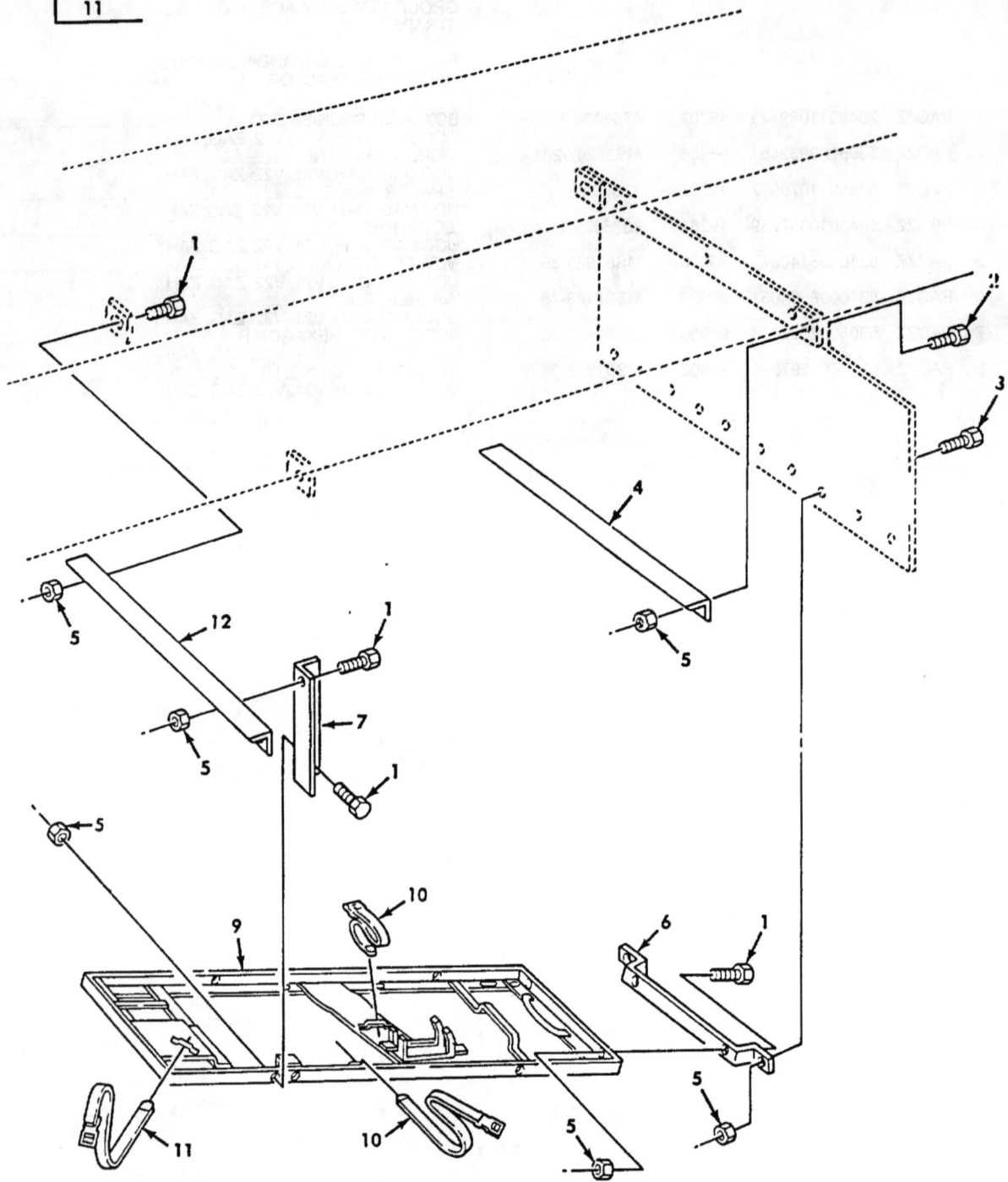

Figure 378. Tool Frame Assembly, Pioneer.

(1) ITEM NO	(2) SMR CODE	(3) NSN	(4) CAGEC	(5) PART NUMBER	(6) DESCRIPTION AND USABLE ON CODES (UOC)	(7) QTY
					GROUP 1808 STOWAGE RACKS, BOXES, AND STRAPS	
					FIG. 378 TOOL FRAME ASSEMBLY, PIONEER	
1	PAOZZ	5305002693234	96906	MS90727-58	SCREW, CAP, HEXAGON H	11
					UOC:DAJ, DAK, V24, V25, ZAJ, ZAK	
2	PAOZZ	5305002692803	96906	MS90726-60	SCREW, CAP, HEXAGON H	2
					UOC:DAJ, DAK, V24, V25, ZAJ, ZAK	
3	PAOZZ	5305002693238	80204	B1821BH038F125N	SCREW, CAP, HEXAGON H	2
					UOC:DAJ, DAK, V24, V25, ZAJ, ZAK	
4	PFOZZ	5340011794303	19207	11677693	BRACKET, ANGLE	1
					UOC:DAJ, DAK, V24, V25, ZAJ, ZAK	
5	PAOZZ	5310008140672	96906	MS51943-36	NUT, SELF-LOCKING, HE	15
					UOC:DAJ, DAK, V24, V25, ZAJ, ZAK	
6	PFOZZ	5340011587094	19207	8345033	BRACKET, MOUNTING	1
					UOC:DAJ, DAK, V24, V25, ZAJ, ZAK	
7	PFOZZ	5340011587095	19207	10896925-2	BRACKET, ANGLE RIGHT	1
					UOC:DAJ, DAK, V24, V25, ZAJ, ZAK	
7	PFOZZ	5340011587096	19207	10896925-3	BRACKET, ANGLE LEFT	1
					UOC:DAJ, DAK, V24, V25, ZAJ, ZAK	
8	PAOZZ	2540004098891	96906	MS53053-1	BRACKET ASSEMBLY, TO	1
					UOC:DAJ, DAK, V24, V25, ZAJ, ZAK	
9	PAOZZ	5340011573752	19204	10872111	.BRACKET, TOOL	1
					UOC:DAJ, DAK, V24, V25, ZAJ, ZAK	
10	PBOZZ	5340007533741	19200	7550233-1	.STRAP, WEBBING	2
					UOC:DAJ, DAK, V24, V25, ZAJ, ZAK	
11	XBOZZ	5340010328448	19200	7550233-2	.STRAP, WEBBING	1
					UOC:DAJ, DAK, V24, V25, ZAJ, ZAK	
12	PFOZZ	5340011810840	19207	11677701	BRACKET, ANGLE	1
					UOC:DAJ, DAK, V24, V25, ZAJ, ZAK	

END OF FIGURE

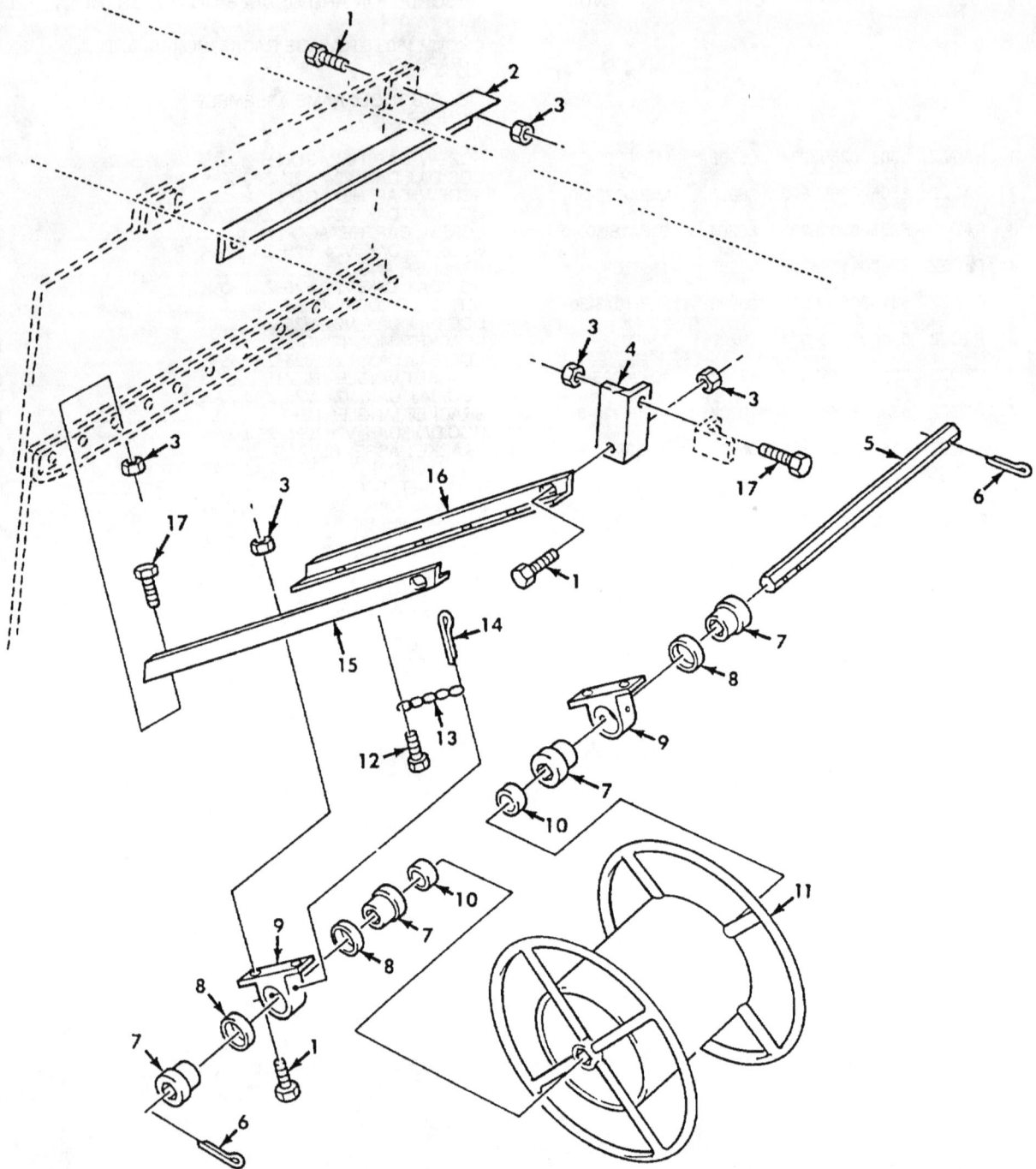

Figure 379. Van Cable Reel, Cable, and Mounting Hardware.

(1) ITEM NO	(2) SMR CODE	(3) NSN	(4) CAGEC	(5) PART NUMBER	(6) DESCRIPTION AND USABLE ON CODES (UOC)	(7) QTY
					GROUP 1808 STOWAGE RACKS, BOXES, AND STRAPS	
					FIG. 379 VAN CABLE REEL, CABLE, AND MOUNTING HARDWARE	
1	PAOZZ	5305002692803	96906	MS90726-60	SCREW, CAP, HEXAGON H .... UOC:DAJ, DAK, V24, V25, ZAJ, ZAK	8
2	PAOZZ	5340011791476	19207	11677693-1	BRACKET, ANGLE .... UOC:DAJ, DAK, V24, V25, ZAJ, ZAK	1
3	PAOZZ	5310008140672	96906	MS51943-36	NUT, SELF-LOCKING, HE .... UOC:DAJ, DAK, V24, V25, ZAJ, ZAK	12
4	PFOZZ	5340011843464	19207	11677677	BRACKET, ANGLE .... UOC:DAJ, DAK, V24, V25, ZAJ, ZAK	1
5	PAOZZ	3040004213966	19207	8735059	SHAFT, STRAIGHT .... UOC:DAJ, DAK, V24, V25, ZAJ, ZAK	1
6	PAOZZ	5315000137238	96906	MS24665-425	PIN, COTTER .... UOC:DAJ, DAK, V24, V25, ZAJ, ZAK	2
7	PAOZZ	3120007663327	19207	8735057	BUSHING, SLEEVE .... UOC:DAJ, DAK, V24, V25, ZAJ, ZAK	4
8	PAOZZ	5365004199465	19207	8735050	SPACER, RING .... UOC:DAJ, DAK, V24, V25, ZAJ, ZAK	3
9	PAOZZ	3040004434839	19207	8735060	COLLAR, SHAFT .... UOC:DAJ, DAK, V24, V25, ZAJ, ZAK	2
10	PAOZZ	5365004439901	19207	8735058	SPACER, SLEEVE .... UOC:DAJ, DAK, V24, V25, ZAJ, ZAK	2
11	PAOZZ	3830001796635	19207	8734961	REEL, CABLE .... UOC:DAJ, DAK, V24, V25, ZAJ, ZAK	1
12	PAOZZ	5305008550957	96906	MS24629-46	SCREW, TAPPING .... UOC:DAJ, DAK, V24, V25, ZAJ, ZAK	1
13	MOOZZ		80244	42-C-16570-6	CHAIN, WELDLESS MAKE FROM CHAIN P/N .... XB-196, 6 INCHES LONG UOC:DAJ, DAK, V24, V25, ZAJ, ZAK	1
14	PAOZZ	5315002341848	96906	MS24665-629	PIN, COTTER... UOC:DAJ, DAK, V24, V25, ZAJ, ZAK	1
15	PFOZZ	5340012276479	19207	11677720-2	BRACKET, ANGLE .... UOC:DAJ, DAK, V24, V25, ZAJ, ZAK	1
16	PFOZZ	5340011804819	19207	11677720-1	BRACKET, ANGLE .... UOC:DAJ, DAK, V24, V25, ZAJ, ZAK	1
17	PAOZZ	5305002693234	96906	MS90727-58	SCREW, CAP, HEXAGON .... UOC:DAJ, DAK, V24, V25, ZAJ, ZAK	4

END OF FIGURE

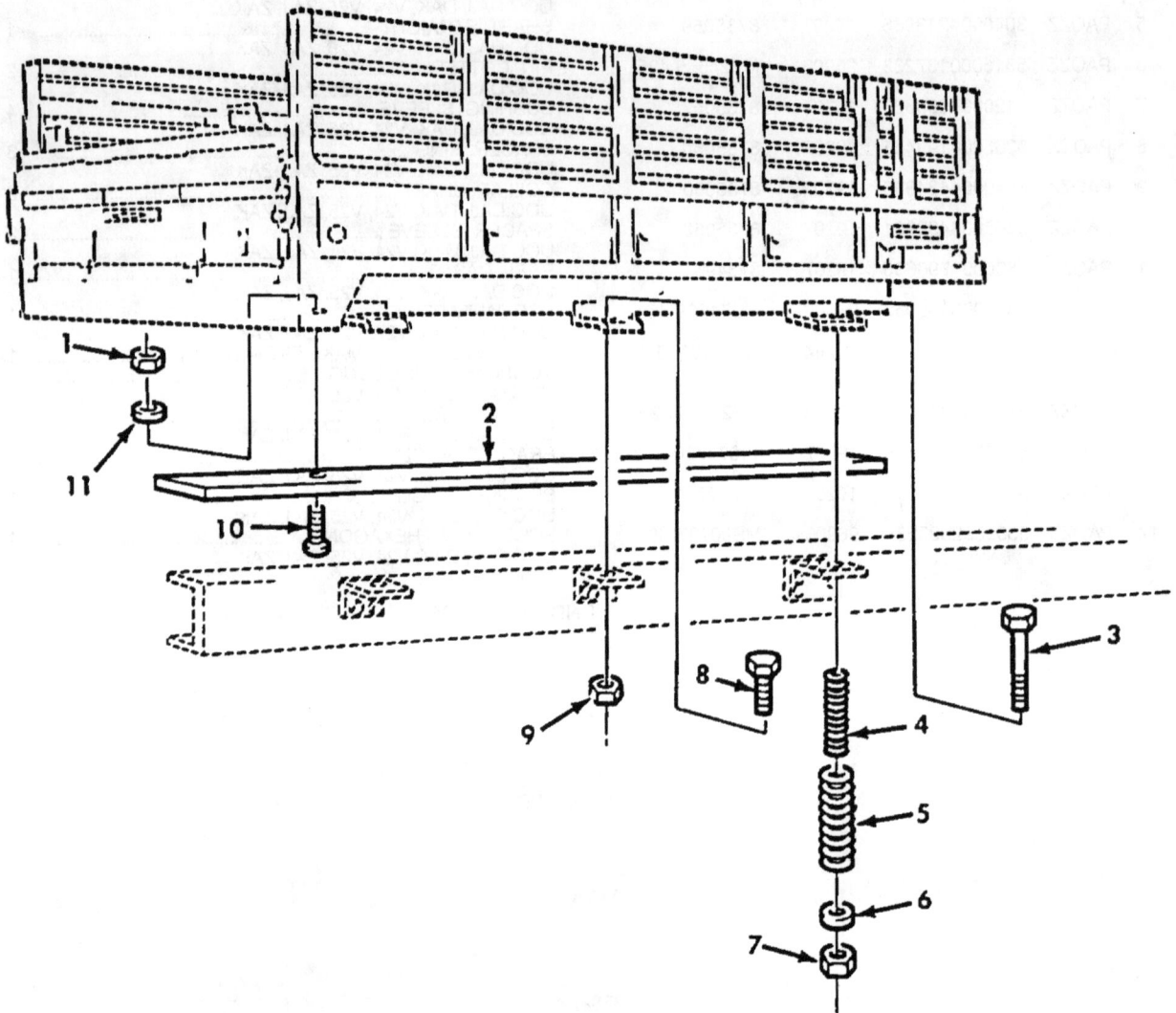

Figure 380. Standard Cargo and Dropside Cargo Mounting Hardware

(1) ITEM NO	(2) SMR CODE	(3) NSN	(4) CAGEC	(5) PART NUMBER	(6) DESCRIPTION AND USABLE ON CODES (UOC)	(7) QTY
					GROUP 1810 CARGO BODY	
					FIG. 380 STANDARD CARGO AND DROPSIDE CARGO MOUNTING HARDWARE	
1	PAFZZ	5310007320558	96906	MS51967-8	NUT, PLAIN, HEXAGON .... UOC:DAA, DAB, DAW, DAX, V12, V13, V14, V15, ZAA, ZAB	10
2	MFFZZ	2510011806164	19207	11665738	FRAME, WINDOW, VEHICU MAKE FROM WOOD, ... P/N 13219E0079 .... UOC:DAA, DAB, DAW, DAX, V12, V13, V14, V15, ZAA, ZAB	2
3	PAOZZ	5306005129218	19207	7971653	BOLT, MACHINE .... UOC:DAA, DAB, DAW, DAX, V12, V13, V14, V15, ZAA, ZAB	4
4	PAOZZ	5360007372792	19207	7372792	SPRING, HELICAL, COMP .... UOC:DAA, DAB, DAW, DAX, V12, V13, V14, V15, ZAA, ZAB	4
5	PAOZZ	5360007372793	19207	7372793	SPRING, HELICAL, COMP .... UOC:DAA, DAB, DAW, DAX, V12, V13, V14, V15, ZAA, ZAB	4
6	PAOZZ	5310009517209	96906	MS27183-22	WASHER, FLAT .... UOC:DAW, DAX, V14, V15, ZAA, ZAB	4
6	PAOZZ	5310009517209	96906	MS27183-22	WASHER, FLAT .... UOC:DAA, DAB, V12, V13	8
7	PAOZZ	5310009825009	96906	MS21045-10	NUT, SELF-LOCKING, HE .... UOC:DAA, DAB, V12, V13	4
7	PAOZZ	5310002416664	96906	MS51943-44	NUT, SELF-LOCKING, HE .... UOC:DAW, DAX, V14, V15, ZAA, ZAB	4
8	PAOZZ	5305009261826	80204	B1821BH075F275N	SCREW, CAP, HEXAGON H .... UOC:DAA, DAB, DAW, DAX, V12, V13, V14, V15, ZAA, ZAB	4
9	PAOZZ	5310009825012	96906	MS21045-12	NUT, SELF-LOCKING, HE .... UOC:DAA, DAB, DAW, DAX, V12, V13, V14, V15	4
9	PAOZZ	5310009353569	96906	MS51943-46	NUT, SELF-LOCKING, HE .... UOC:DAW, DAX, V14, V15, ZAA, ZAB	4
10	PAFZZ	5306008162441	96906	MS35751-71	BOLT, SQUARE NECK .... UOC:DAA, DAB, DAW, DAX, V12, V13, V14, V15, ZAA, ZAB	10
11	PAFZZ	5310006379541	96906	MS35338-46	WASHER, LOCK .... UOC:DAA, DAB, DAW, DAX, V12, V13, V14, V15, ZAA, ZAB	10

END OF FIGURE

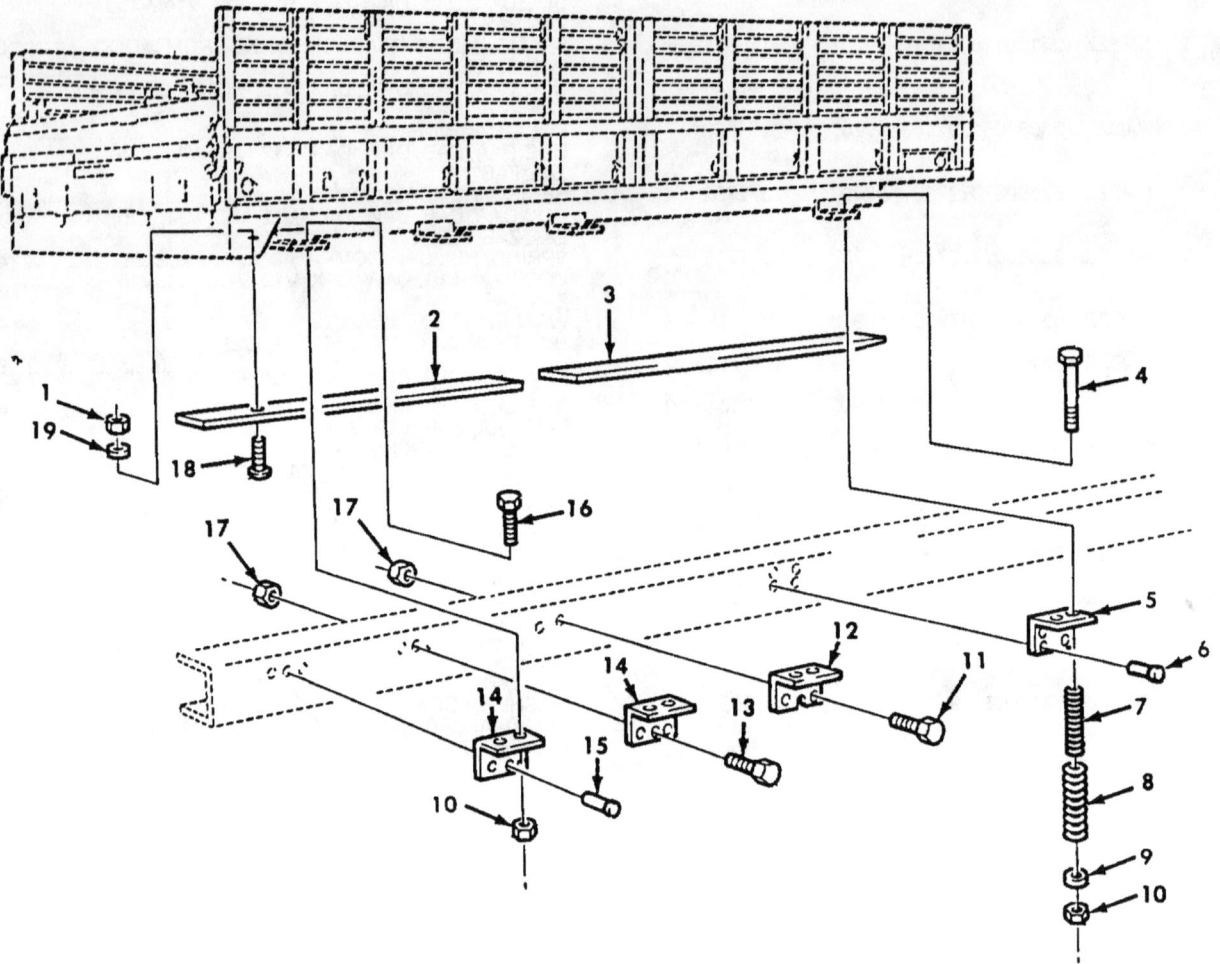

Figure 381. Cargo Body, Long Wheel Base Mounting Hardware.

(1) ITEM NO	(2) SMR CODE	(3) NSN	(4) CAGEC	(5) PART NUMBER	(6) DESCRIPTION AND USABLE ON CODES (UOC)	(7) QTY
					GROUP 1810 CARGO BODY	
					FIG. 381 CARGO BODY, LONG WHEEL BASE MOUNTING HARDWARE	
1	PAFZZ	5310007320558	96906	MS51967-8	NUT, PLAIN, HEXAGON .... UOC:DAC, DAD, V16, V17, ZAC, ZAD	20
2	MFFZZ		19207	11682324-1	SILL, WOOD MAKE FROM WOOD, P/N .... 13219E0079.......................... UOC:DAC, DAD, V16, V17, ZAC, ZAD	1
2	MFFZZ		19207	11682324-2	SILL, WOOD MAKE FROM WOOD, P/N .... 13219E0079.......................... UOC:DAC, DAD, V16, V17, ZAC, ZAD	1
3	MFFZZ		19207	11682323-2	SILL, WOOD MAKE FROM WOOD, P/N .... 13219E0079.......................... UOC:DAC, DAD, V16, V17, ZAC, ZAD	1
3	MFFZZ		19207	11682323-1	SILL, WOOD MAKE FROM WOOD, P/N .... 13219E0079.......................... UOC:DAC, DAD, V16, V17, ZAC, ZAD	1
4	PAOZZ	5306005129218	19207	7971653	BOLT, MACHINE .... UOC:DAC, DAD, V16, V17, ZAC, ZAD	4
5	PFFZZ	5340004094019	19207	10871208	BRACKET, ANGLE .... UOC:DAC, DAD, V16, V17, ZAC, ZAD	2
6	PFFZZ	5320000189512	19207	189512	RIVET, SOLID .... UOC:DAC, DAD, V16, V17, ZAC, ZAD	8
7	PAOZZ	5360007372792	19207	7372792	SPRING, HELICAL, COMP .... UOC:DAC, DAD, V16, V17, ZAC, ZAD	4
8	PAOZZ	5360007372793	19207	7372793	SPRING, HELICAL, COMP .... UOC:DAC, DAD, V16, V17, ZAC, ZAD	4
9	PAOZZ	5310009517209	96906	MS27183-22	WASHER, FLAT .... UOC:DAC, DAD, V16, ZAC, ZAD	8
10	PAOZZ	5310002416664	96906	MS51943-44	NUT, SELF-LOCKING, HE .... UOC:DAC, DAD, V16, ZAC, ZAD	16
11	PAFZZ	5305007195238	80204	B1821BH050F200N	SCREW, CAP, HEXAGON H .... UOC:DAC, DAD, V16, ZAC, ZAD	4
12	PFFZZ	5340002310278	19207	10871207	BRACKET, ANGLE .... UOC:DAC, DAD, V16, V17, ZAC, ZAD	2
13	PAFZZ	5305007195235	80204	B1821BH050F175N	SCREW, CAP, HEXAGON H .... UOC:DAC, DAD, V16, ZAC, ZAD	6
14	PFFZZ	5340004094018	19207	10871210	BRACKET, ANGLE .... UOC:DAC, DAD, V16, V17, ZAC, ZAD	4
15	PFFZZ	5320000104131	10001	319119PC12	RIVET, SOLID .... UOC:DAC, DAD, V16, V17, ZAC, ZAD	6
16	PAOZZ	5305007262552	80204	B1821BH063F225N	SCREW, CAP, HEXAGON H .... UOC:DAC, DAD, V16, V17, ZAC, ZAD	12
17	PAFZZ	5310004883888	96906	MS51943-40	NUT, SELF-LOCKING, HE .... UOC:DAC, DAD, V16, V17, ZAC, ZAD	10
18	PAFZZ	5306000109115	96906	MS35751-72	BOLT, SQUARE NECK .... UOC:DAC, DAD, V16, V17, ZAC, ZAD	20
19	PAFZZ	5310010992550	57328	65003-S	WASHER, BEVEL .... UOC:DAC, DAD, V16, V17, ZAC, ZAD	20

END OF FIGURE

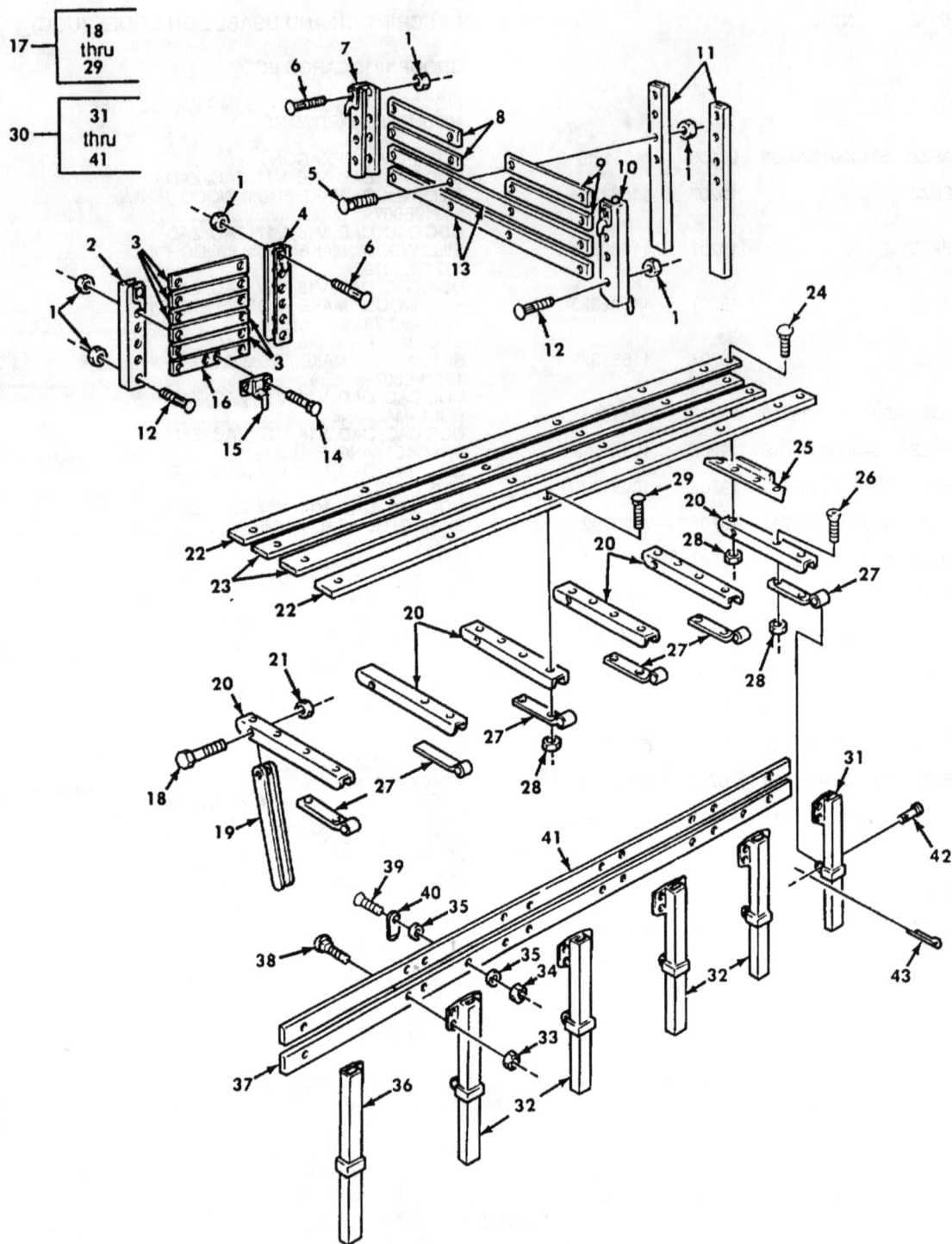

Figure 382. Cargo Body Side Rack, Gate Assemblies and Troop Seats - Fixed Side (M939, M939A1)

(1) ITEM NO	(2) SMR CODE	(3) NSN	(4) CAGEC	(5) PART NUMBER	(6) DESCRIPTION AND USABLE ON CODES (UOC)	(7) QTY
					GROUP 1810 CARGO BODY	
					FIG. 382 CARGO BODY SIDE RACK, GATE ASSEMBLIES, AND TROOP SEATS - FIXED SIDE (M939, M939A1)	
1	PAOZZ	5310008807744	96906	MS51967-5	NUT, PLAIN, HEXAGON .... UOC:DAA, DAB, V12, V13	30
2	PFOZZ	2510011307952	19207	8758065	STAKE, VEHICLE BODY.......... UOC:DAA, DAB, V12, V13	2
3	PAOZZ	2510014590259	19207	12450346-1	BOARD, SIDE RACK .... UOC:DAA, DAB, V12, V13	10
4	PAOZZ	2510011310122	19207	8758067	CHANNEL, RACK .... UOC:DAA, DAB, V12, V13	2
5	PAOZZ	5306000548024	96906	MS35751-47	BOLT, SQUARE NECK FRONT RACK .... UOC:DAA, DAB, V12, V13.	8
6	PAOZZ	5306011083185	19207	8758096	BOLT, RIBBED SHOULDE .... UOC:DAA, DAB, V12, V13	8
7	PFOZZ	2510011307951	19207	8758086	STAKE, VEHICLE BODY .... UOC:DAA, DAB, V12, V13	1
8	PAOZZ	2510014579827	19207	12450346-2	BOARD, FRONT RACK .... UOC:DAA, DAB, V12, V13	2
9	PAOZZ	2510014579788	19207	12450346-3	BOARD, FRONT RACK .... UOC:DAA, DAB, V12, V13	2
10	PBOZZ	2540011310111	19207	8758090	CHANNEL, FRONT SIDE .... UOC:DAA, DAB, V12, V13	1
11	PBOZZ	2510011307950	19207	8758083	STAKE, VEHICLE BODY .... UOC:DAA, DAB, V12, V13	2
12	PAOZZ	5306011083186	19207	8758095	BOLT, RIBBED SHOULDE .... UOC:DAA, DAB, V12, V13	16
13	PAOZZ	2510014579011	19207	12450346-7	BOARD, FRONT RACK .... UOC:DAA, DAB, V12, V13	2
14	PAOZZ	5306009115005	96906	MS35751-42	BOLT, SQUARE NECK .... UOC:DAA, DAB, V12, V13	2
15	PAOZZ	40300 10823597	19207	8758068	HOOK, CARGO .... UOC:DAA, DAB, V12, V13	1
16	PAOZZ	2510014584512	19207	12450353	BOARD, SIDE RACK .... UOC:DAA, DAB, V12, V13	1
17	PFOOO	2540001797094	19207	8758125	SEAT, TROOP SIDE RACK LEFT .... UOC:DAA, DAB, V12, V13	1
17	PFOOO	2540010853588	19207	8758139	SEAT, TROOP SIDE RACK RIGHT .... UOC:DAA, DAB, V12, V13	1
18	PAOZZ	5305002693244	80204	B1821BH1038F250N	.SCREW, CAP, HEXAGON H .... UOC:DAA, DAB, V12, V13	6
19	PAOZZ	2540007372790	19207	7372790	.LEG, SEAT .... UOC:DAA, DAB, V12, V13	6
20	PFOZZ	2540005921823	19207	7061093	.SUPPORT, SEAT, VEHICU .... UOC:DAA, DAB, V12, V13	6
21	PAOZZ	5310009591488	96906	MS51922-21	.NUT, SELF-LOCKING .... UOC:DAA, DAB, V12, V13	6

(1) ITEM NO	(2) SMR CODE	(3) NSN	(4) CAGEC	(5) PART NUMBER	(6) DESCRIPTION AND USABLE ON CODES (UOC)	(7) QTY
22	PAOZZ	2510001779137	19207	12450347-5	.BOARD, SIDE RACK, VEH.LEFT .... UOC:DAA, DAB, V12, V13	2
22	PAOZZ		19207	12450347-4	.BOARD, TROOP SEAT RIGHT .... UOC:DAA, DAB, V12, V13	2
23	PAOZZ	2510006544611	19207	12450345-5	.BOARD, SIDE RACK LEFT .... UOC:DAA, DAB, V12, V13	2
23	PAOZZ		19207	12450345-4	.BOARD, TROOP SEAT RIGHT .... UOC:DAA, DAB, V12, V13	2
24	PAOZZ	5306007536996	96906	MS35751-43	.BOLT, SQUARE NECK .... UOC:DAA, DAB, V12, V13	15
25	PFOZZ	2590006227757	19207	7370149	.ANGLE, SEAT SUPPORT .... UOC:DAA, DAB, V12, V13	1
26	PAOZZ	5305009845676	96906	MS35206-296	.SCREW, MACHINE .... UOC:DAA, DAB, V12, V13	1
27	PAOZZ	2540005216179	19207	7061094	.LEAF, TEE HINGE .... UOC:DAA, DAB, V12, V13	6
28	PAOZZ	5310008807744	96906	MS51967-5	.NUT, PLAIN, HEXAGON .... UOC:DAA, DAB, V12, V13	27
29	PAOZZ	5306000120231	96906	MS35751-44	.BOLT, SQUARE NECK .... UOC:DAA, DAB, V12, V13	11
30	PBOOO	2510010822642	19207	8758143	SIDE RACK, VEHICLE B RIGHT........ UOC:DAA, DAB, V12, V13	1
30	PBOOO	2510001777852	19207	8758128	SIDE RACK, VEHICLE B LEFT........ UOC:DAA, DAB, V12, V13	1
31	PFOZZ	2510001777903	19207	8758129	.STAKE, VEHICLE BODY SIDE RACK........ UOC:DAA, DAB, V12, V13	1
32	PFOZZ	2510001062200	19207	7370384	.POCKET, BOW, BODY RAC .... UOC:DAA, DAB, V12, V13	4
33	PAOZZ	5310008807744	96906	MS51967-5	.NUT, PLAIN, HEXAGON .... UOC:DAA, DAB, V12, V13	20
34	PAOZZ	5310009843806	96906	MS51922-9	.NUT, SELF-LOCKING, HE .... UOC:DAA, DAB, V12, V13	2
35	PAOZZ	5310000814219	96906	MS27183-12	.WASHER, FLAT .... UOC:DAA, DAB, V12, V13	4
36	PFOZZ	2510011104061	19207	8758132	.STAKE, VEHICLE BODY .... UOC:DAA, DAB, V12, V.13	1
37	PAOZZ		19207	12450347-16	.BOARD, SIDE RACK LEFT .... UOC:DAA, DAB, V12, V13	1
37	PAOZZ		19207	12450347-17	.BOARD, SIDE RACK RIGHT .... UOC:DAA, DAB, V12, V13	1
38	PAOZZ	5306009115005	96906	MS35751-42	.BOLT, SQUARE NECK .... UOC:DAA, DAB, V12, V13	20
39	PAOZZ	5306002264832	80204	B1821BH031C175N	.BOLT, MACHINE .... UOC:DAA, DAB, V12, V13	2
40	PAOZZ	5340007372788	19207	7372788	.BRACKET, ANGLE .... UOC:DAA, DAB, V12, V13	2
41	PAOZZ	2510001777893	19207	12450347-14	.BOARD, SIDE RACK LEFT .... UOC:DAA, DAB, V12, V13	1

(1) ITEM NO	(2) SMR CODE	(3) NSN	(4) CAGEC	(5) PART NUMBER	(6) DESCRIPTION AND USABLE ON CODES (UOC)	(7) QTY
41	PAOZZ		19207	12450347-10	.BOARD, SIDE RACK RIGHT .... UOC:DAA, DAB, V12, V13	1
42	PAOZZ	5315007370134	19207	7370134	PIN, STRAIGHT, HEADED.......... UOC:DAA, DAB, V12, V13	6
43	PAOZZ	5315008423044	96906	MS24665-283	PIN, COTTER............. UOC:DAA, DAB, V12, V13	6

END OF FIGURE

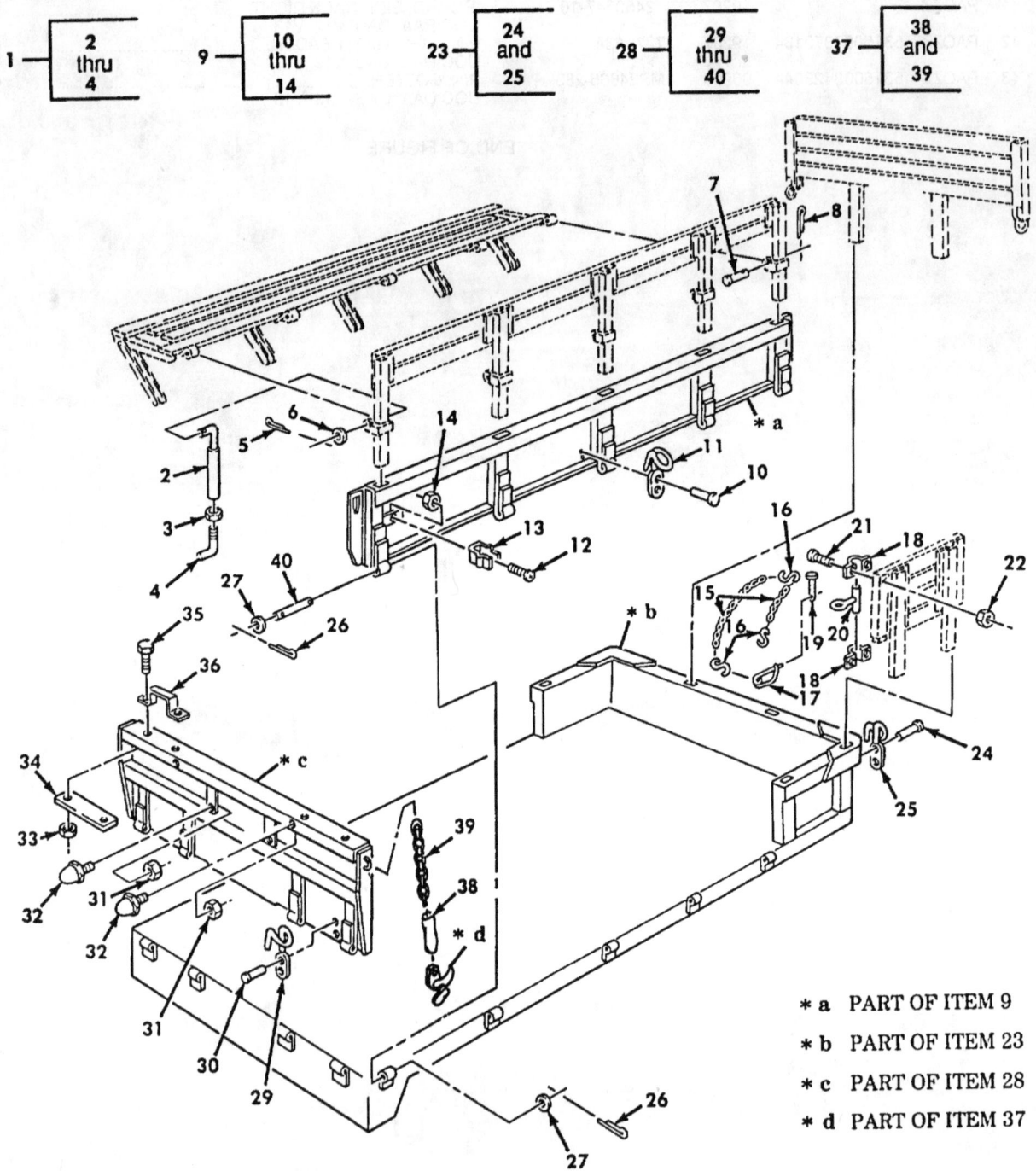

Figure 383. Cargo Body Side Gates, Tailgate, and Related Parts, Basic and Dropside.

* a   PART OF ITEM 9
* b   PART OF ITEM 23
* c   PART OF ITEM 28
* d   PART OF ITEM 37

(1) ITEM NO	(2) SMR CODE	(3) NSN	(4) CAGEC	(5) PART NUMBER	(6) DESCRIPTION AND USABLE ON CODES (UOC)	(7) QTY
					GROUP 1810 CARGO BODY	
					FIG. 383 CARGO BODY SIDE GATES, TAILGATE, AND RELATED PARTS, BASIC AND DROPSIDE	
1	PFOZZ	2510001230278	19207	10937885	ROD ASSEMBLY STABIL .... UOC:DAW, DAX, V14, V15, ZAA, ZAB	2
2	XAOZZ	3040011454284	19207	10937936	.CONNECTING LINK, RIG DROPSIDE .... STABILIZER .... UOC:DAW, DAX, V14, V15, ZAA, ZAB	1
3	PAOZZ	5310008348734	96906	MS35691-37	.NUT, PLAIN, HEXAGON .... UOC:DAW, DAX, V14, V15, ZAA, ZAB	1
4	PAOZZ	5306010908620	19207	10937933	.BOLT, HOOK .... UOC:DAW, DAX, V14, V15, ZAA, ZAB	1
5	PAOZZ	5315000803503	96906	MS24665-214	PIN, COTTER .... UOC:DAW, DAX, V14, V15, ZAA, ZAB	2
6	PAOZZ	5310008093079	96906	MS27183-19	WASHER, FLAT .... UOC:DAW, DAX, V14, V15, ZAA, ZAB	2
7	PAOZZ	5315007370134	19207	7370134	PIN, STRAIGHT, HEADED .... UOC:DAW, DAX, V14, V15, ZAA, ZAB	10
8	PAOZZ	5315008423044	96906	MS24665-283	PIN, COTTER .... UOC:DAW, DAX, V14, V15, ZAA, ZAB	10
9	PAOOO	2510009302714	19207	10937462	SIDE GATE, VEHICLE B .... UOC:DAW, DAX, V14, V15, ZAA, ZAB	2
10	PFOZZ	5320008892632	96906	MS35743-38	.RIVET, SOLID .... UOC:DAW, DAX, V14, V15, ZAA, ZAB	16
11	PAOZZ	5340004561011	61465	BP2415	.HOOK, SUPPORT .... UOC:DAW, DAX, V14, V15, ZAA, ZAB	8
12	PAOZZ	5305008893000	96906	MS35206-230	.SCREW, MACHINE .... UOC:DAW, DAX, V14, V15, ZAA, ZAB	2
13	PBOZZ	5340006998463	19207	7539185	.CLIP, SPRING TENSION .... UOC:DAW, DAX, V14, V15, ZAA, ZAB	2
14	PAOZZ	5310009824908	96906	MS21045-6	.NUT, SELF-LOCKING .... UOC:DAW, DAX, V14, V15, ZAA, ZAB	2
15	PAOZZ	4010010831159	19207	11593372-1	CHAIN, WELDLESS .... UOC:DAW, DAX, V14, V15, ZAA, ZAB	4
16	PAOZZ	4030009487315	96906	MS87006-33	HOOK, CRAIN, S .... UOC:DAW, DAX, V14, V15, ZAA, ZAB	6
17	PAOZZ	5315007418971	19207	7418971	PIN, RETAINING .... UOC:DAW, DAX, V14, V15, ZAA, ZAB	2
18	PFOZZ	5340004215083	19207	11611614	STRAP, RETAINING .... UOC:DAW, DAX, V14, V15, ZAA, ZAB	4
19	PAOZZ	5315010839387	19207	10937889	PIN, STRAIGHT, HEADED .... UOC:DAW, DAX, V14, V15, ZAA, ZAB	2
20	PAOZZ	2510004094005	19207	11611615	ROD, VEHICLE SIDE RA .... UOC:DAW, DAX, V14, V15, ZAA, ZAB	2
21	PAOZZ	5305009881170	96906	MS35206-284	SCREW, MACHINE .... UOC:DAW, DAX, V14, V15, ZAA, ZAB	8
22	PAOZZ	5310007616882	96906	MS51967-2	NUT, PLAIN, HEXAGON .... UOC:DAW, DAX, V14, V15, ZAA, ZAB	8
23	PFFFF	2510010831158	19207	11593374-1	BODY, CARGO TRUCK UOC:V14, V15	1

(1) ITEM NO	(2) SMR CODE	(3) NSN	(4) CAGEC	(5) PART NUMBER	(6) DESCRIPTION AND USABLE ON CODES (UOC)	(7) QTY
23	PBFFF	2510012106229	19207	11593374-2	BODY, CARGO TRUCK .... UOC:DAW, DAX, ZAA, ZAB	1
24	PFOZZ	5320008892632	96906	MS35743-38	.RIVET, SOLID UOC:DAW, DAX, V14, V15, ZAA, ZAB	4
25	PAOZZ	5340004561011	61465	BP2415	.HOOK, SUPPORT....... UOC:DAW, DAX, V14, V15, ZAA, ZAB	2
26	PAOZZ	5315008395822	96906	MS24665-353	PIN, COTTER .... UOC:DAW, DAX, V14, V15, ZAA, ZAB	28
27	PAOZZ	5310008238803	96906	MS27183-21	WASHER, FLAT .... UOC:DAW, DAX, V14, V15, ZAA, ZAB	28
28	PAOOO	2510008985415	19207	11611570	TAILGATE, VEHICLE BO UOC:V14, V15	1
28	PAOFF	2510012269408	19207	11611570-1	TAILGATE, VEHICLE BO .... UOC:DAW, DAX, ZAA, ZAB	1
29	PAOZZ	5340004561011	61465	BP2415	.HOOK, SUPPORT .... UOC:DAW, DAX, V14, V15, ZAA, ZAB	1
30	PFOZZ	5320010326530	96906	MS35743-36	.RIVET, SOLID .... UOC:DAW, DAX, V14, V15, ZAA, ZAB	4
31	PAOZZ	5310009591488	96906	MS51922-21	.NUT, SELF-LOCKING .... UOC:DAW, DAX, V14, V15, ZAA, ZAB	4
32	PAOZZ	5340006897213	19207	7529296	.BUMPER, NONMETALLIC .... UOC:DAW, DAX, V14, V15, ZAA, ZAB	4
33	PAOZZ	5310004883888	96906	MS51943-40	.NUT, SELF-LOCKING, HE .... UOC:DAW, DAX, V14, V15, ZAA, ZAB	4
34	PAOZZ	5340001193906	19207	8758106	.PLATE, MENDING .... UOC:DAW, DAX, V14, V15, ZAA, ZAB	2
35	PAOZZ	5305007195219	96906	MS90727-111	.SCREW, CAP, HEXAGON H .... UOC:DAW, DAX, V14, V15, ZAA, ZAB	4
36	PAOZZ	2510001193903	19207	7064165	.STEP, TAILGATE .... UOC:DAW, DAX, V14, V15, ZAA, ZAB	2
37	PAOZZ	5340001098212	19207	11609666	.HANDLE, MANUAL , CONTROL .... UOC:DAW, DAX, V14, V15, ZAA, ZAB	2
38	PAOZZ	4030005940475	19207	7061088	.PROTECTOR, GUY .... UOC:DAW, DAX, V14, V15, ZAA, ZAB	2
39	XAOZZ		80244	42-C-14490	.CHAIN, CLOSE STRAIGHT 5/16 NOMINAL .... 16 LINKS LONG........ UOC:DAW, DAX, V14, V15, ZAA, ZAB	2
40	PAOZZ	5315002376341	96906	MS51932-154	.PIN, STRAIGHT, HEADLE .... UOC:DAW, DAX, V14, V15, ZAA, ZAB	14

END OF FIGURE

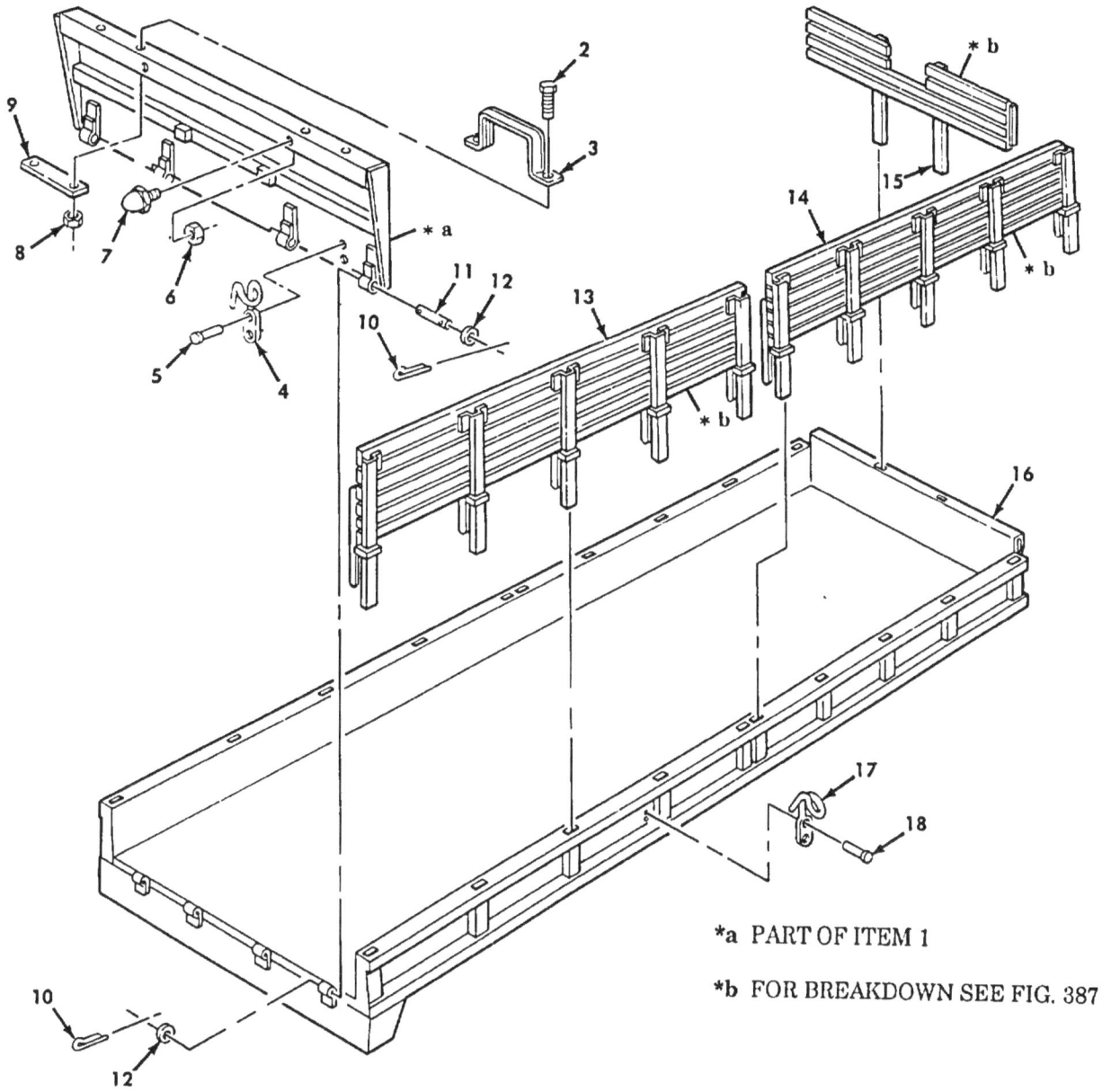

Figure 384. Cargo Body Tailgate and Side Rack Assemblies, Long Wheel Base.

*a PART OF ITEM 1

*b FOR BREAKDOWN SEE FIG. 387

(1) ITEM NO	(2) SMR CODE	(3) NSN	(4) CAGEC	(5) PART NUMBER	(6) DESCRIPTION AND USABLE ON CODES (UOC)	(7) QTY
					GROUP 1810 CARGO BODY	
					FIG. 384 CARGO BODY TAILGATE AND SIDE RACK ASSEMBLIES, LONG WHEEL BASE	
1	PAOOO	2510004982392	19207	8758038	TAILGATE, VEHICLE BO UOC:V16, V17,	1
1	PFOFF	2510012269407	19207	8758038-1	TAILGATE, VEHICLE BO CARGO .... UOC:DAC, DAD, ZAC, ZAD	1
2	PAOZZ	5305007195219	96906	MS90727-111	.SCREW, CAP, HEXAGON H .... UOC:DAC, DAD, V16, V17, ZAC, ZAD	4
3	PAOZZ	2510001193903	19207	7064165	.STEP, TAILGATE.............. UOC:DAC, DAD, V16, V17, ZAC, ZAD	2
4	PAOZZ	5340004561011	24617	2173982	.HOOK, SUPPORT .... UOC:DAC, DAD, V16, V17, ZAC, ZAD	2
5	PFOZZ	5320008892632	96906	MS35743-38	.RIVET, SOLID .... UOC:DAC, DAD, V16, V17, ZAC, ZAD	4
6	PAOZZ	5310009591488	96906	MS51922-21	.NUT, SELF-LOCKING .... UOC:DAC, DAD, V16, V17, ZAC, ZAD	2
7	PAOZZ	5340006897213	19207	7529296	.BUMPER, NONMETALLIC TAILGATE .... UOC:DAC, DAD, V16, V17, ZAC, ZAD	2
8	PAOZZ	5310004883888	96906	MS51943-40	.NUT, SELF-LOCKING, HE .... UOC:DAC, DAD, V16, V17, ZAC, ZAD	4
9	PAOZZ	5340001193906	19207	8758106	.PLATE, MENDING .... UOC:DAC, DAD, V16, V17, ZAC, ZAD	2
10	PAOZZ	5315008395822	96906	MS24665-353	PIN, COTTER .... UOC:DAC, DAD, V16, V17, ZAC, ZAD	8
11	PAOZZ	5315002376341	96906	MS51932-154	PIN, STRAIGHT, HEADLE .... UOC:DAC, DAD, V16, V17, ZAC, ZAD	4
12	PAOZZ	5310008238803	96906	MS27183-21	WASHER, FLAT .... UOC:DAC, DAD, V16, V17, ZAC, ZAD	8
13	PAOOO	2510004082452	19207	10871296	SIDE RACK, VEHICLE B RIGHT REAR SIDE...... UOC:DAC, DAD, V16, V17, ZAC, ZAD	1
13	PAOOO	2510004094020	19207	10871295	SIDE RACK, VEHICLE B LEFT REAR SIDE ...... UOC:DAC, DAD, V16, V17, ZAC, ZAD	1
14	PAOOO	2510004082453	19207	10871290	SIDE RACK, VEHICLE B RIGHT FRONT........ SIDE UOC:DAC, DAD, V16, V17, ZAC, ZAD	1
14	PAOOO	2510004082439	19207	10871289	SIDE RACK, VEHICLE B LEFT FRONT SIDE...... UOC:DAC, DAD, V16, V17, ZAC, ZAD	1
15	PAOOO	2510004082448	19207	10871297	SIDE RACK, VEHICLE B FRONT .... UOC:DAC, DAD, V16, V17, ZAC, ZAD	1
16	PFFFF	2510012106230	19207	10871356-2	BODY, CARGO TRUCK .... UOC:DAC, DAD, ZAC, ZAD	1
16	PFFFF	2510010827508	19207	10871356-1	BODY, CARGO TRUCK UOC:V16, V17,	1
17	PAOZZ	5340004561011	61465	BP2415	HOOK, SUPPORT................ UOC:DAC, DAD, V16, V17, ZAC, ZAD	18
18	PFOZZ	5320008892632	96906	MS35743-38	RIVET, SOLID .... UOC:DAC, DAD, V16, V17, ZAC, ZAD	36

END OF FIGURE

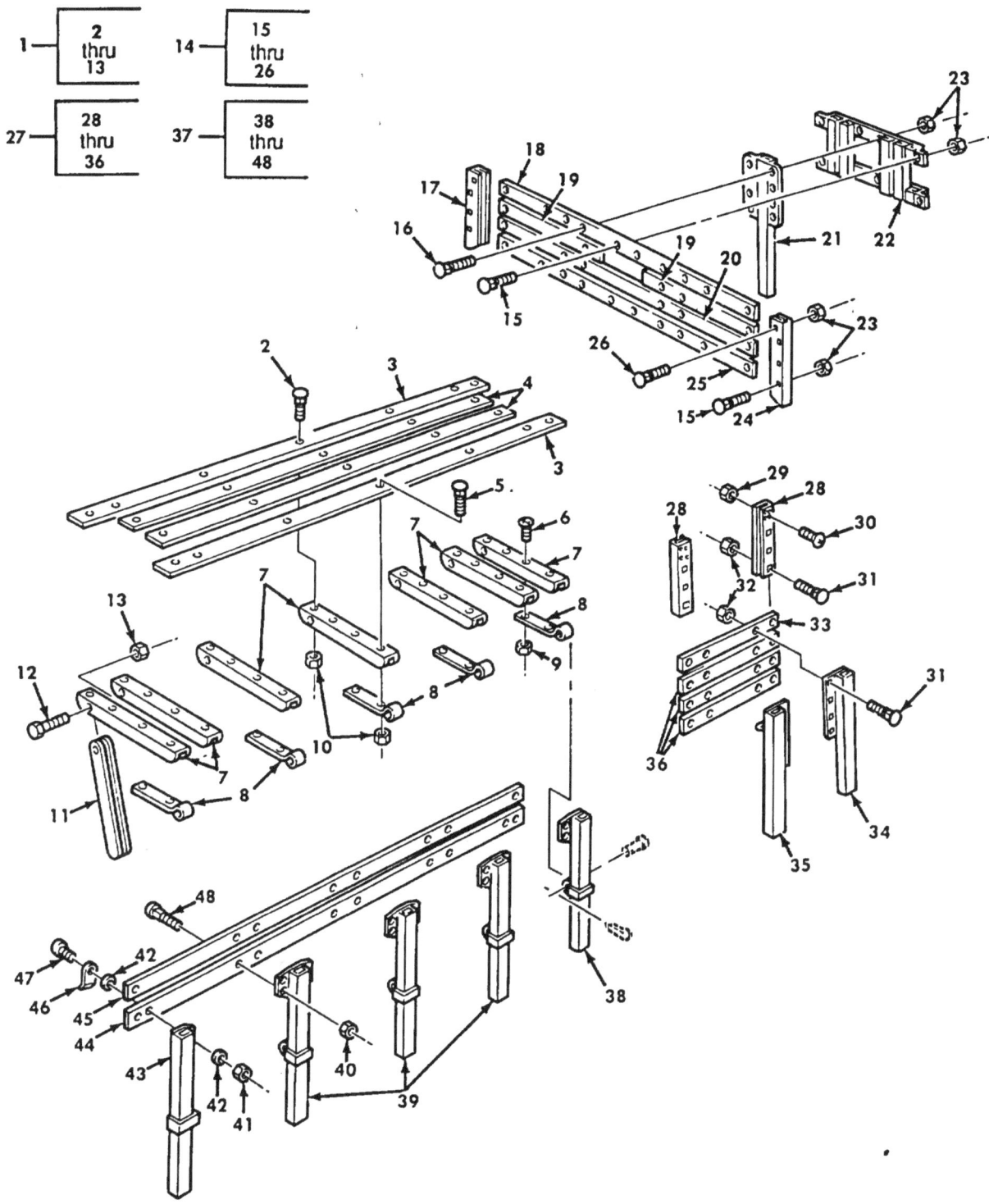

Figure 385. Cargo Body Personnel Seat Assembly, Side Rack and Gate Assemblies, Dropside.

(1) ITEM NO	(2) SMR CODE	(3) NSN	(4) CAGEC	(5) PART NUMBER	(6) DESCRIPTION AND USABLE ON CODES (UOC)	(7) QTY
					GROUP 1810 CARGO BODY	
					FIG. 385 CARGO BODY PERSONNEL SEAT ASSEMBLY, SIDE RACK AND GATE ASSEMBLIES, DROPSIDE	
1	PAOOO	2540005911108	19207	7370383	SEAT, VEHICULAR .... UOC:DAW, DAX, V14, V15, ZAA, ZAB	2
2	PAOZZ	5306009115005	96906	MS35751-42	.BOLT, SQUARE NECK UOC:DAW, DAX, V14, V15, ZAA, ZAB	16
3	PAOZZ	2540006227750	19207	12450345-6	.BOARD, SEAT, RACK .... UOC:DAW, DAX, V14, V15, ZAA, ZAB	2
4	PAOZZ		19207	12450345-1	.SLAT, TROOP SEAT .... UOC:DAW, DAX, V14, V15, ZAA, ZAB	2
5	PAOZZ	5306007536996	96906	MS35751-43	.BOLT, SQUARE NECK .... UOC:DAW, DAX, V14, V15, ZAA, ZAB	8
6	PAOZZ	5305009845677	96906	MS35206-297	.SCREW, MACHINE .... UOC:DAW, DAX, V14, V15	2
7	PFOZZ	2540005921823	19207	7061093	.SUPPORT, SEAT, VEHICU .... UOC:DAW, DAX, V14, V15, ZAA, ZAB	7
8	PAOZZ	2540005216179	19207	7061094	.LEAF, TEE HINGE .... UOC:DAW, DAX, V14, V15, ZAA, ZAB	5
9	PAOZZ	5310009843806	96906	MS51922-9	.NUT, SELF-LOCKING, HE 1/2" .... UOC:DAW, DAX, V14, V15, ZAA, ZAB	2
10	PAOZZ	5310008807744	96906	MS51967-5	.NUT, PLAIN, HEXAGON .... UOC:DAW, DAX, V14, V15, ZAA, ZAB	24
11	PAOZZ	2540007372790	19207	7372790	.LEG, SEAT .... UOC:DAW, DAX, V14, V15, ZAA, ZAB	5
12	PAOZZ	5305002693244	80204	B1821BH038F250N	.SCREW, CAP, HEXAGON .... UOC:DAW, DAX, V14, V15, ZAA, ZAB	5
13	PAOZZ	5310010583353	72582	192481	.NUT, SELF-LOCKING, HE .... UOC:DAW, DAX, V14, V15, ZAA, ZAB	5
14	PAOOO	2510004093991	19207	11611612	SIDE RACK, VEHICLE B .... UOC:DAW, DAX, V14, V15, ZAA, ZAB	1
15	PAOZZ	5306009115005	96906	MS35751-42	.BOLT, SQUARE NECK .... UOC:DAW, DAX, V14, V15, ZAA, ZAB	22
16	PAOZZ	5306000548024	96906	MS35751-47	.BOLT, SQUARE NECK .... UOC:DAW, DAX, V14, V15, ZAA, ZAB	8
17	PFOZZ	2510011310118	19207	11611602-2	.STAKE, VEHICLE BODY .... UOC:DAW, DAX, V14, V15, ZAA, ZAB	1
18	PAOZZ	2510014589670	19207	12450347-12	.BOARD, FRONT RACK .... UOC:DAW, DAX, V14, V15, ZAA, ZAB	1
19	PAOZZ		19207	12450346-5	.BOARD, FRONT RACK .... UOC:DAW, DAX, V14, V15, ZAA, ZAB	2
20	PAOZZ		19207	12450346-11	.BOARD, FRONT RACK .... UOC:DAW, DAX, V14, V15, ZAA, ZAB	1
21	PFOZZ	2510011310119	19207	11611618	.STAKE, VEHICLE BODY .... UOC:DAW, DAX, V14, V15, ZAA, ZAB	2
22	PFOZZ	2510001193904	19207	10937932	.RETAINER, STAKE BOW .... UOC:DAW, DAX, V14, V15, ZAA, ZAB	2

(1) ITEM NO	(2) SMR CODE	(3) NSN	(4) CAGEC	(5) PART NUMBER	(6) DESCRIPTION AND USABLE ON CODES (UOC)	(7) QTY
23	PAOZZ	5310008807744	96906	MS51967-5	.NUT, PLAIN, HEXAGON .... UOC:DAW, DAX, V14, V15, ZAA, ZAB	32
24	PBOZZ	9520011310117	19207	11611602-1	.CHANNEL, STRUCTURAL .... UOC:DAW, DAX, V14, V15, ZAA, ZAB	1
25	PAOZZ		19207	12450347-11	.BOARD, FRONT RACK .... UOC:DAW, DAX, V14, V15, ZAA, ZAB	1
26	PAOZZ	5306007536996	96906	MS35751-43	.BOLT, SQUARE NECK .... UOC:DAW, DAX, V14, V15, ZAA, ZAB	2
27	PAOZZ	2510009307778	19207	11611598	GATE ASSEMBLY, SIDE .... UOC:DAW, DAX, V14, V15, ZAA, ZAB-	2
28	XDOZZ		19207	11611597	.CHANNEL, STRUCTURAL .... UOC:DAW, DAX, V14, V15, ZAA, ZAB	2
29	PAOZZ	5310007616882	96906	MS51967-2	.NUT, PLAIN, HEXAGON .... UOC:DAW, DAX, V14, V15, ZAA, ZAB	2
30	PAOZZ	5305009881170	96906	MS35206-284	.SCREW, MACHINE .... UOC:DAW, DAX, V14, V15, ZAA, ZAB	2
31	PAOZZ	5306009115005	96906	MS35751-42	.BOLT, SQUARE NECK .... UOC:DAW, DAX, V14, V15, ZAA, ZAB	14
32	PAOZZ	5310008807744	96906	MS51967-5	.NUT, PLAIN, HEXAGON .... UOC:DAW, DAX, V14, V15, ZAA, ZAB	14
33	PAOZZ	2510014590564	19207	12450354	.BOARD, SIDE RACK .... UOC:DAW, DAX, V14, V15, ZAA, ZAB	1
34	PFOZZ	2510011310115	19207	11611609-2	.STAKE, VEHICLE BODY .... UOC:DAW, DAX, V14, V15, ZAA, ZAB	1
35	PFOZZ	2510011310114	19207	11611609-1	.STAKE, VEHICLE BODY .... UOC:DAW, DAX, V14, V15, ZAA, ZAB	1
36	PAOZZ		19207	12450346-6	.BOARD, SIDE RACK .... UOC:DAW, DAX, V14, V15, ZAA, ZAB	3
37	PAOZZ	2510001241297	19207	10937880	SIDE RACK, VEHICLE B .... UOC:DAW, DAX, V14, V15, ZAA, ZAB	2
38	PFOZZ	2510011310121	19207	10937496	.STAKE, VEHICLE BODY .... UOC:DAW, DAX, V14, V15, ZAA, ZAB	1
39	PFOZZ	2510001062200	19207	7370384	.POCKET, BOW, BODY RAC .... UOC:DAW, DAX, V14, V15, ZAA, ZAB	3
40	PAOZZ	5310008807744	96906	MS51967-51	.NUT, PLAIN, HEXAGON .... UOC:DAW, DAX, V14, V15, ZAA, ZAB	18
41	PAOZZ	5310009843806	96906	MS51922-9	.NUT, SELF-LOCKING, HE .... UOC:DAW, DAX, V14, V15, ZAA, ZAB	2
42	PAOZZ	5310000814219	96906	MS27183-12	.WASHER, FLAT .... UOC:DAW, DAX, V14, V15, ZAA, ZAB	4
43	PFOZZ	2510011310120	19207	10937495	.STAKE, VEHICLE BODY .... UOC:DAW, DAX, V14, V15, ZAA, ZAB	1
44	PAOZZ	2510006223931	19207	12450345-11	.BOARD, SIDE RACK .... UOC:DAW, DAX, V14, V15, ZAA, ZAB	1
45	PAOZZ	2510011774434	19207	12450345-8	.BOARD, SIDE RACK .... UOC:DAW, DAX, V14, V15, ZAA, ZAB	1

(1) ITEM NO	(2) SMR CODE	(3) NSN	(4) CAGEC	(5) PART NUMBER	(6) DESCRIPTION AND USABLE ON CODES (UOC)	(7) QTY
46	PAOZZ	5340007372788	19207	7372788	.BRACKET, ANGLE .... UOC:DAW, DAX, V14, V15, ZAA, ZAB	2
47	PAOZZ	5306002264832	80204	B1821BH031C175N	.BOLT, MACHINE .... UOC:DAW, DAX, V14, V15, ZAA, ZAB	2
48	PAOZZ	5306007215944	96906	MS35751-41	.BOLT, SQUARE NECK .... UOC:DAW, DAX, V14, V15, ZAA, ZAB	16

END OF FIGURE

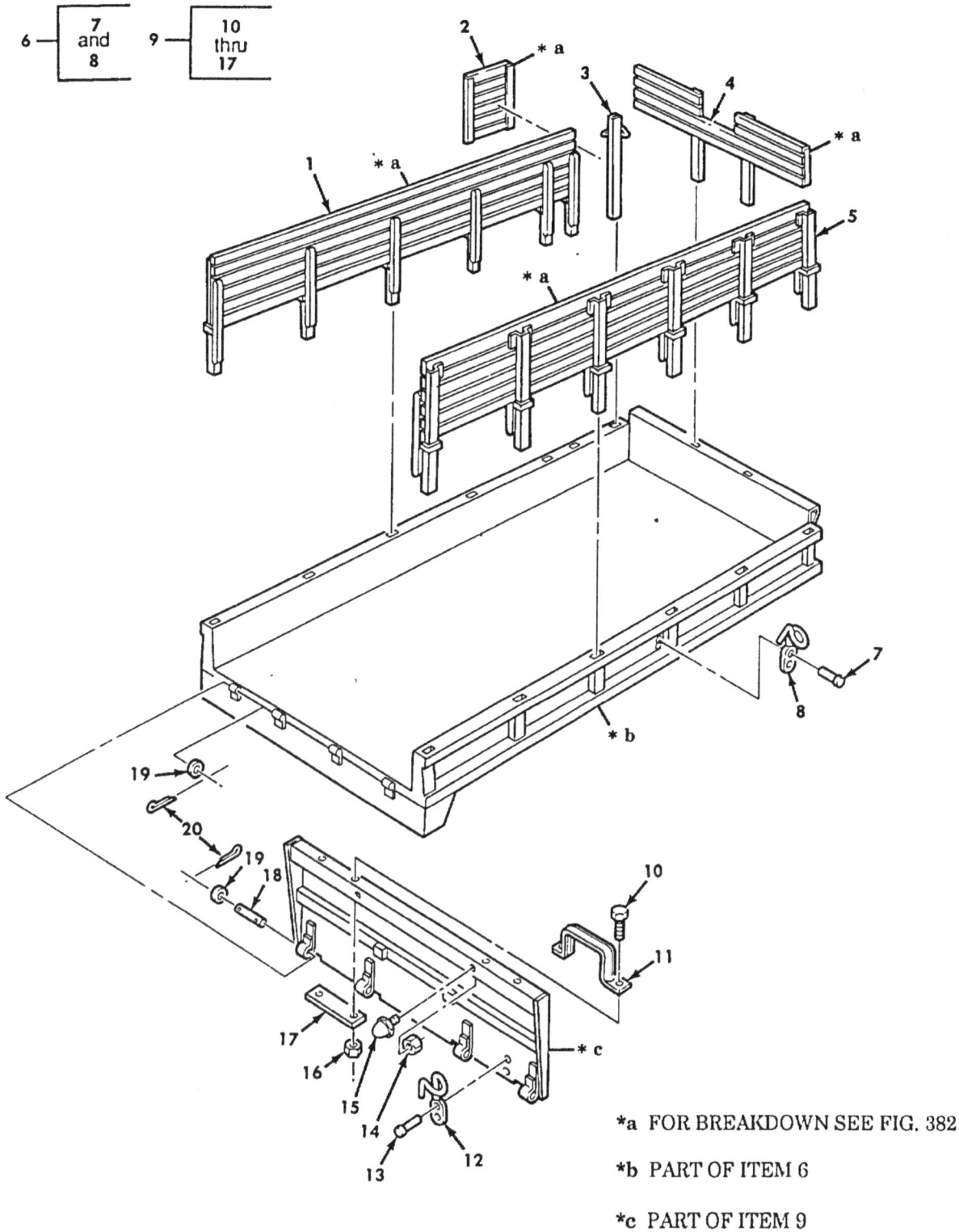

Figure 386. Cargo Body Assembly, Basic, Side Racks, Tailgate Assembly, and Related Parts.

*a FOR BREAKDOWN SEE FIG. 382

*b PART OF ITEM 6

*c PART OF ITEM 9

(1) ITEM NO	(2) SMR CODE	(3) NSN	(4) CAGEC	(5) PART NUMBER	(6) DESCRIPTION AND USABLE ON CODES (UOC)	(7) QTY
					GROUP 1810 CARGO BODY	
					FIG. 386 CARGO BODY ASSEMBLY, BASIC, SIDE RACKS, TAILGATE ASSEMBLY, AND RELATED PARTS	
1	PAOOO	2510001797093	19207	8758124	SIDE RACK, COMPLETE LEFT.SIDE .... UOC:DAA, DAB, V12, V13	1
2	PAOOO	2510006544606	19207	8758064	GATE, SIDE RACK LEFT SIDE .... UOC:DAA, DAB, V12, V13	1
3	PAOZZ	2510004097973	19207	8758177	STAKE, VEHICLE BODY LEFT SIDE .... UOC:DAA, DAB, V12, V13	1
4	PAOOO	2510010827507	19207	12256171	SIDE RACK, FRONT .... UOC:DAA, DAB, V12, V13	1
5	PAOOO	2510001795706	19207	8758138	SIDE RACK, RIGHT .... UOC:DAA, DAB, V12, V13	1
6	PFFFF	2510010822645	19207	12256166	BODY, CARGO TRUCK UOC:V12, V13	1
6	PFFFF	2510012106228	19207	12256166-1	BODY, CARGO TRUCK .... UOC:DAA, DAB	1
7	PFOZZ	5320008892632	96906	MS35743-38	.RIVET, SOLID .... UOC:DAA, DAB, V12, V13	24
8	PAOZZ	5340013302622	19207	7064246	.HOOK, SUPPORT .... UOC:DAA, DAB, V12, V13	12
9	PAOOO	2510004982392	19207	8758038	TAILGATE, VEHICLE BO UOC:V12, V13	1
9	PFOFF	2510012269407	19207	8758038-1	TAILGATE, VEHICLE BO .... UOC:DAA, DAB	1
10	PAOZZ	5305007195219	96906	MS90727-111	.SCREW, CAP, HEXAGON H .... UOC:DAA, DAB, V12, V13	4
11	PAOZZ	2510001193903	19207	7064165	.STEP, TAILGATE .... UOC:DAA, DAB, V12, V13	2
12	PAOZZ	5340004561011	61465	BP2415	.HOOK, SUPPORT .... UOC:DAA, DAB, V12, V13	2
13	PFOZZ	5320008892632	96906	MS35743-38	.RIVET, SOLID .... UOC:DAA, DAB, V12, V13	4
14	PAOZZ	5310009591488	96906	MS51922-21	.NUT, SELF-LOCKING .... UOC:DAA, DAB, V12, V13	2
15	PAOZZ	5340006897213	19207	7529296	.BUMPER, NONMETALLIC .... UOC:DAA, DAB, V12, V13	2
16	PAOZZ	5310004883888	96906	MS51943-40	.NUT, SELF-LOCKING, HE .... UOC:DAA, DAB, V12, V13	4
17	PAOZZ	5340001193906	19207	8758106	.PLATE, MENDING .... UOC:DAA, DAB, V12, V13	2
18	PAOZZ	5315002376341	96906	MS51932-154	PIN, STRAIGHT, HEADLE .... UOC:DAA, DAB, V12, V13	4
19	PAOZZ	5310008238803	96906	MS27183-21	WASHER, FLAT .... UOC:DAA, DAB, V12, V13	8
20	PAOZZ	5315008395822	96906	MS24665-353	PIN, COTTER .... UOC:DAA, DAB, V12, V13	8

END OF FIGURE

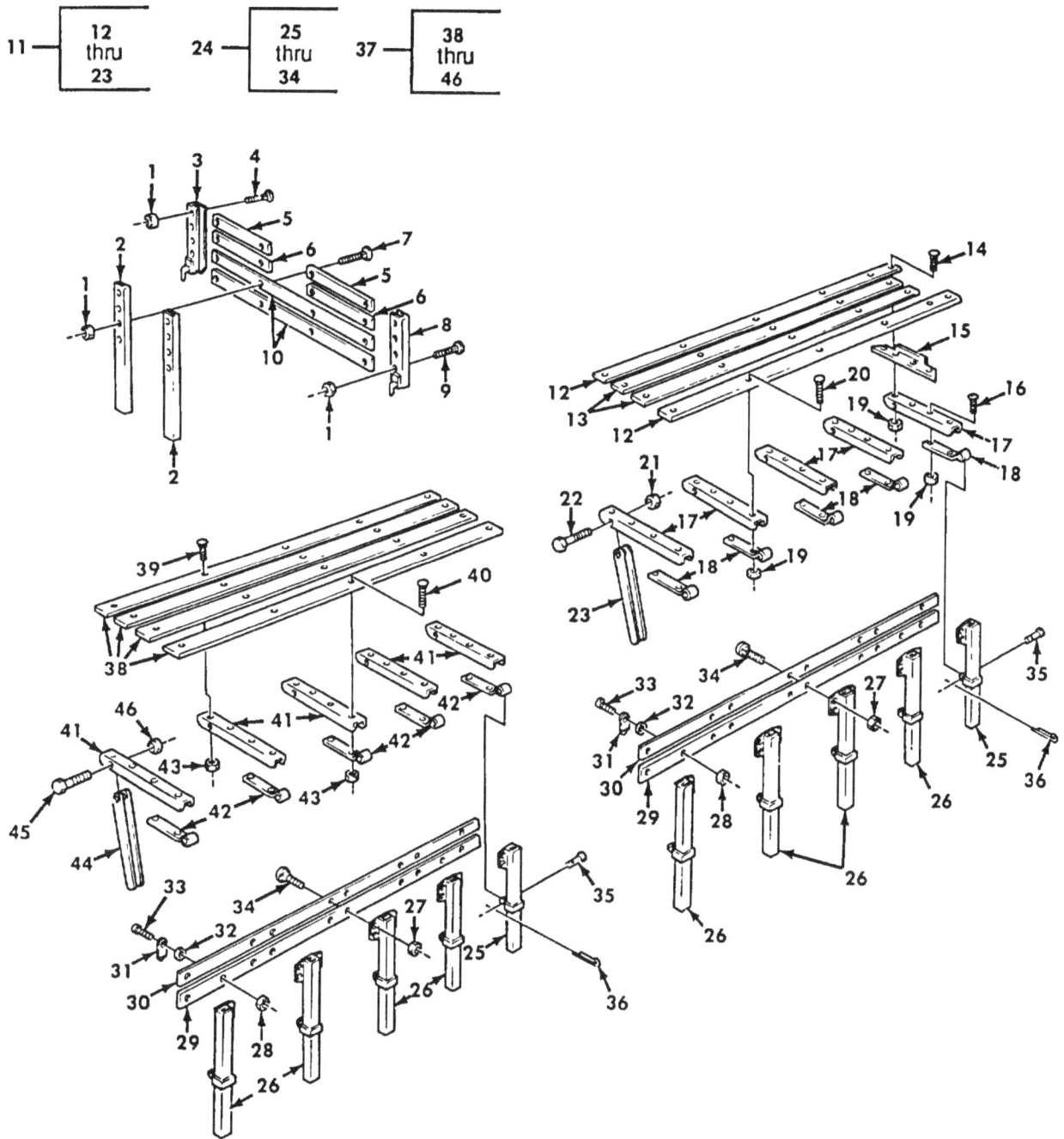

Figure 387. Cargo Body Rack Assemblies, and Troop Seats, Long Wheel Base.

(1) ITEM NO	(2) SMR CODE	(3) NSN	(4) CAGEC	(5) PART NUMBER	(6) DESCRIPTION AND USABLE ON CODES (UOC)	(7) QTY
					GROUP 1810 CARGO BODY	
					FIG. 387 CARGO BODY RACK ASSEMBLIES AND TROOP SEATS, LONG WHEEL BASE	
1	PAOZZ	5310008807744	96906	MS51967-5	NUT, PLAIN, HEXAGON UOC:DAC, DAD, V16, V17, ZAC, ZAD	16
2	PBOZZ	2510011307950	19207	8758083	STAKE, VEHICLE BODY UOC:DAC, DAD, V16, V17, ZAC, ZAD	2
3	PFOZZ	2540011310111	19207	8758090	CHANNEL, FRONT SIDE UOC:DAC, DAD, V16, V17, ZAC, ZAD	1
4	PAOZZ	5306011083185	19207	8758096	BOLT, RIBBED SHOULDE UOC:DAC, DAD, V16, V17, ZAC, ZAD	2
5	PAOZZ	2510014579827	19207	12450346-2	BOARD, SIDE RACK UOC:DAC, DAD, V16, V17, ZAC, ZAD	2
6	PAOZZ	2510014579788	19207	8758094	BOARD, FRONT RACK UOC:DAC, DAD, V16, V17, ZAC, ZAD	2
7	PAOZZ	5306000548024	96906	MS35751-47	BOLT, SQUARE NECK UOC:DAC, DAD, V16, V17, ZAC, ZAD	8
8	PFOZZ	2510011307951	19207	8758086	STAKE, VEHICLE BODY UOC:DAC, DAD, V16, V17, ZAC, ZAD	1
9	PAOZZ	5306011083186	19207	8758095	BOLT, RIBBED SHOULDE UOC:DAC, DAD, V16, V17, ZAC, ZAD	6
10	PAOZZ	2510014579011	19207	12450346-7	BOARD, SIDE RACK UOC:DAC, DAD, V16, V17, ZAC, ZAD	2
11	PAOOO	2510011296074	19207	10871294	SIDE RACK, VEHICLE B.RIGHT UOC:DAC, DAD, V16, V17, ZAC, ZAD	1
11	PAOOO	2540010823592	19207	10871293	SEAT, TROOP, SIDE RAC.LEFT UOC:DAC, DAD, V16, V17, ZAC, ZAD	1
12	PAOZZ		19207	12450346-14	.BOARD, TROOP SEAT UOC:DAC, DAD, V16, V17, ZAC, ZAD	2
13	PAOZZ	2540014590567	19207	12450345-2	.BOARD, TROOP SEAT UOC:DAC, DAD, V16, V17, ZAC, ZAD	2
14	PAOZZ	5306007536996	96906	MS35751-43	.BOLT, SQUARE NECK UOC:DAC, DAD, V16, V17, ZAC, ZAD	13
15	PFOZZ	2590006227757	19207	7370149	.ANGLE, SEAT SUPPORT UOC:DAC, DAD, V16, V17, ZAC, ZAD	1
16	PAOZZ	5305009845676	96906	MS35206-296	.SCREW, MACHINE UOC:DAC, DAD, V16, V17, ZAC, ZAD	1
17	PFOZZ	2540005921823	19207	7061093	.SUPPORT, SEAT, VEHICU UOC:DAC, DAD, V16, V17, ZAC, ZAD	5
18	PAOZZ	2540005216179	19207	7061094	.LEAF, TEE HINGE UOC:DAC, DAD, V16, V17, ZAC, ZAD	5
19	PAOZZ	5310008807744	96906	MS51967-5	.NUT, PLAIN, HEXAGON UOC:DAC, DAD, V16, V17, ZAC, ZAD	23
20	PAOZZ	5306000120231	96906	MS35751-44	.BOLT, SQUARE NECK UOC:DAC, DAD, V16, V17, ZAC, ZAD	9

(1) ITEM NO	(2) SMR CODE	(3) NSN	(4) CAGEC	(5) PART NUMBER	(6) DESCRIPTION AND USABLE ON CODES (UOC)	(7) QTY
21	PAOZZ	5310009591488	96906	MS51922-21	.NUT, SELF-LOCKING .... UOC:DAC, DAD, V16, V17	5
22	PAOZZ	5305002693244	80204	B1821BH038F250N	.SCREW, CAP, HEXAGON H .... UOC:DAC, DAD, V16, V17, ZAC, ZAD	5
23	PAOZZ	2540007372790	19207	7372790	.LEG, SEAT .... UOC:DAC, DAD, V16, V17, ZAC, ZAD	5
24	PAOOO	2510010831156	19207	10871284	SIDE RACK, VEHICLE B.LEFT .... UOC:DAC, DAD, V16, V17, ZAC, ZAD	2
24	PAOOO	2510010823788	19207	10871285	SIDE RACK, VEHICLE B.RIGHT .... UOC:DAC, DAD, V16, V17, ZAC, ZAD	2
25	PFOZZ	2510011310112	19207	10871282	.STAKE, VEHICLE BODY R.H .... UOC:DAC, DAD, V16, V17, ZAC, ZAD	1
25	PFOZZ	2510011310113	19207	10871283	.STAKE, VEHICLE BODY L.H .... UOC:DAC, DAD, V16, V17, ZAC, ZAD	1
26	PFOZZ	2510001062200	19207	7370384	.POCKET, BOW, BODY RAC .... UOC:DAC, DAD, V16, V17, ZAC, ZAD	4
27	PAOZZ	5310010643422	96906	MS35690-504	.NUT, PLAIN, HEXAGON .... UOC:DAC, DAD, V16, V17, ZAC, ZAD	18
28	PAOZZ	5310009843806	96906	MS51922-9	.NUT, SELF-LOCKING, HE .... UOC:DAC, DAD, V16, V17, ZAC, ZAD	2
29	PAOZZ	2510014579780	19207	12450347-15	.BOARD, SIDE RACK .... UOC:DAC, DAD, V16, V17, ZAC, ZAD	1
30	PAOZZ	2510014589668	19207	12450347-9	.BOARD SIDE RACK .... UOC:DAC, DAD, V16, V17, ZAC, ZAD	1
31	PAOZZ	5340007372788	19207	7372788	.BRACKET, ANGLE .... UOC:DAC, DAD, V16, V17, ZAC, ZAD	2
32	PAOZZ	5310000814219	96906	MS27183-12	.WASHER, FLAT .... UOC:DAC, DAD, V16, V17, ZAC, ZAD	4
33	PAOZZ	5306002264832	80204	B1821BH031C175N	.BOLT, MACHINE .... UOC:DAC, DAD, V16, V17, ZAC, ZAD	2
34	PAOZZ	5306009115005	96906	MS35751-42	.BOLT, SQUARE NECK .... UOC:DAC, DAD, V16, V17, ZAC, ZAD	18
35	PAOZZ	5315007370134	19207	7370134	PIN, STRAIGHT, HEADED .... UOC:DAC, DAD, V16, V17, ZAC, ZAD	20
36	PAOZZ	5315008423044	96906	MS24665-283	PIN, COTTER .... UOC:DAC, DAD, V16, V17, ZAC, ZAD	20
37	PAOOO	2540010827457	19207	10871287	SEAT, VEHICULA R .... UOC:DAC, DAD, V16, V17, ZAC, ZAD	2
38	PAOZZ		19207	12450346-10	.BOARD, TROOP SEAT .... UOC:DAC, DAD, V16, V17, ZAC, ZAD	4
39	PAOZZ	5306007536996	96906	MS35751-43	.BOLT, SQUARE NECK .... UOC:DAC, DAD, V16, V17	10
40	PAOZZ	5306000120231	96906	MS35751-44	.BOLT, SQUARE NECK .... UOC:DAC, DAD, V16, V17, ZAC, ZAD	10
41	PFOZZ	2540005921823	19207	7061093	.SUPPORT, SVEHICU .... UOC:DAC, DAD, V16, V17, ZAC, ZAD	5
42	PAOZZ	2540005216179	19207	7061094	.LEAF, TEE HINGE .... UOC:DAC, DAD, V16, V17, ZAC, ZAD	5
43	PAOZZ	5310008807744	96906	MS51967-5	.NUT, PLAIN, HEXAGON .... UOC:DAC, DAD, V16, V17, ZAC, ZAD	20
44	PAOZZ	2540007372790	19207	7372790	.LEG, SEAT .... UOC:DAC, DAD, V16, V17, ZAC, ZAD	5

(1) ITEM NO	(2) SMR CODE	(3) NSN	(4) CAGEC	(5) PART NUMBER	(6) DESCRIPTION AND USABLE ON CODES (UOC)	(7) QTY
45	PAOZZ	5305002693244	80204	B1821BH03BF250N	.SCREW, CAP, HEXAGON H .... UOC:DAC, DAD, V16, V17, ZAC, ZAD	5
46	PAOZZ	5310009591488	96906	MS51922-21	.NUT, SELF-LOCKING .... UOC:DAC, DAD, V16, V17	5

END OF FIGURE

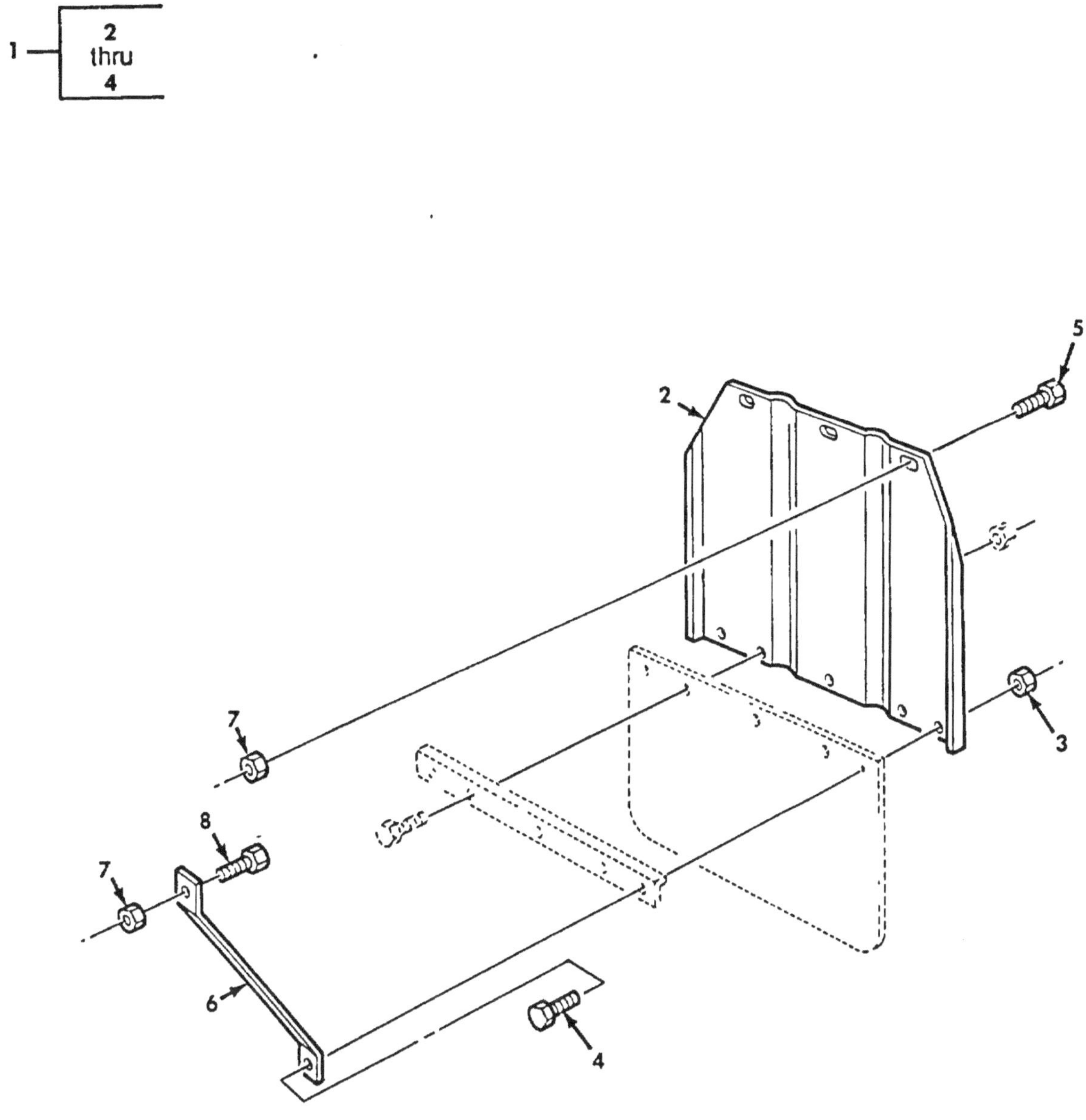

Figure 388. Cargo Body Splash Guard Assembly and Related Parts.

(1) ITEM NO	(2) SMR CODE	(3) NSN	(4) CAGEC	(5) PART NUMBER	(6) DESCRIPTION AND USABLE ON CODES (UOC)	(7) QTY
					GROUP 1810 CARGO BODY	
					FIG. 388 CARGO BODY SPLASH GUARD ASSEMBLY AND RELATED PARTS	
1	PAOZZ	2540011821309	19207	8758159	GUARD, SPLASH, VEHICU .... UOC:DAA, DAB, V12, V13	4
1	PAOZZ	2540001692855	19207	11608778	GUARD, SPLASH, VEHICU .... UOC:DAC, DAD, DAW, DAX, V14, V15, V16, V17, ZAA, ZAB, ZAC, ZAD	4
2	PFOZZ	2540008576332	19207	8758160	.GUARD, SPLASH, VEHICU UPPER .... UOC:DAA, DAB, V12, V13	1
2	PFOZZ	2540010886036	19207	11608763	.GUARD, SPLASH, VEHICU .... UOC:DAC, DAD, DAW, DAX, V14, V15, V16, V17, ZAA, ZAB, ZAC, ZAD	1
3	PAOZZ	5310008140672	96906	MS51943-36	.NUT, SELF-LOCKING, HE .... UOC:DAC, DAD, DAW, DAX, V14, V15, V16, V17, ZAA, ZAB, ZAC, ZAD	2
3	PAOZZ	5310009591488	96906	MS51922-21	.NUT, SELF-LOCKING, HE .... UOC:DAA, DAB, V12, V13	2
4	PAOZZ	5305002693240	80204	B1821BH038F150N	.SCREW, CAP, HEXAGON H .... UOC:DAA, DAB, DAC, DAD, DAW, DAX, V12, V13, V14, V15, V16, V17, ZAA, ZAB, ZAC, ZAD	2
5	PAOZZ	5305002692803	96906	MS90726-60	SCREW, CAP, HEXAGON H .... UOC:DAA, DAB, DAC, DAD, DAW, DAX, V12, V13, V14, V15, V16, V17, ZAA, ZAB, ZAC, ZAD	12
6	PFOZZ	5340010906407	19207	12256169	BRACKET, DOUBLE ANGL .... UOC:DAA, DAB, DAC, DAD, DAW, DAX, V12, V13, V14, V15, V16, V17	8
7	PAOZZ	5310008140672	96906	MS51943-36	NUT, SELF-LOCKING, HE .... UOC:DAA, DAB, DAC, DAD, DAW, DAX, V12, V13, V14, V15, V16, V17, ZAA, ZAB, ZAC, ZAD	20
8	PAOZZ	5305002693238	80204	B1821BH038F125N	SCREW, CAP, HEXAGON H .... UOC:DAA, DAB, DAC, DAD, DAW, DAX, V12, V13, V14, V15, V16, V17, ZAA, ZAB, ZAC, ZAD	8

END OF FIGURE

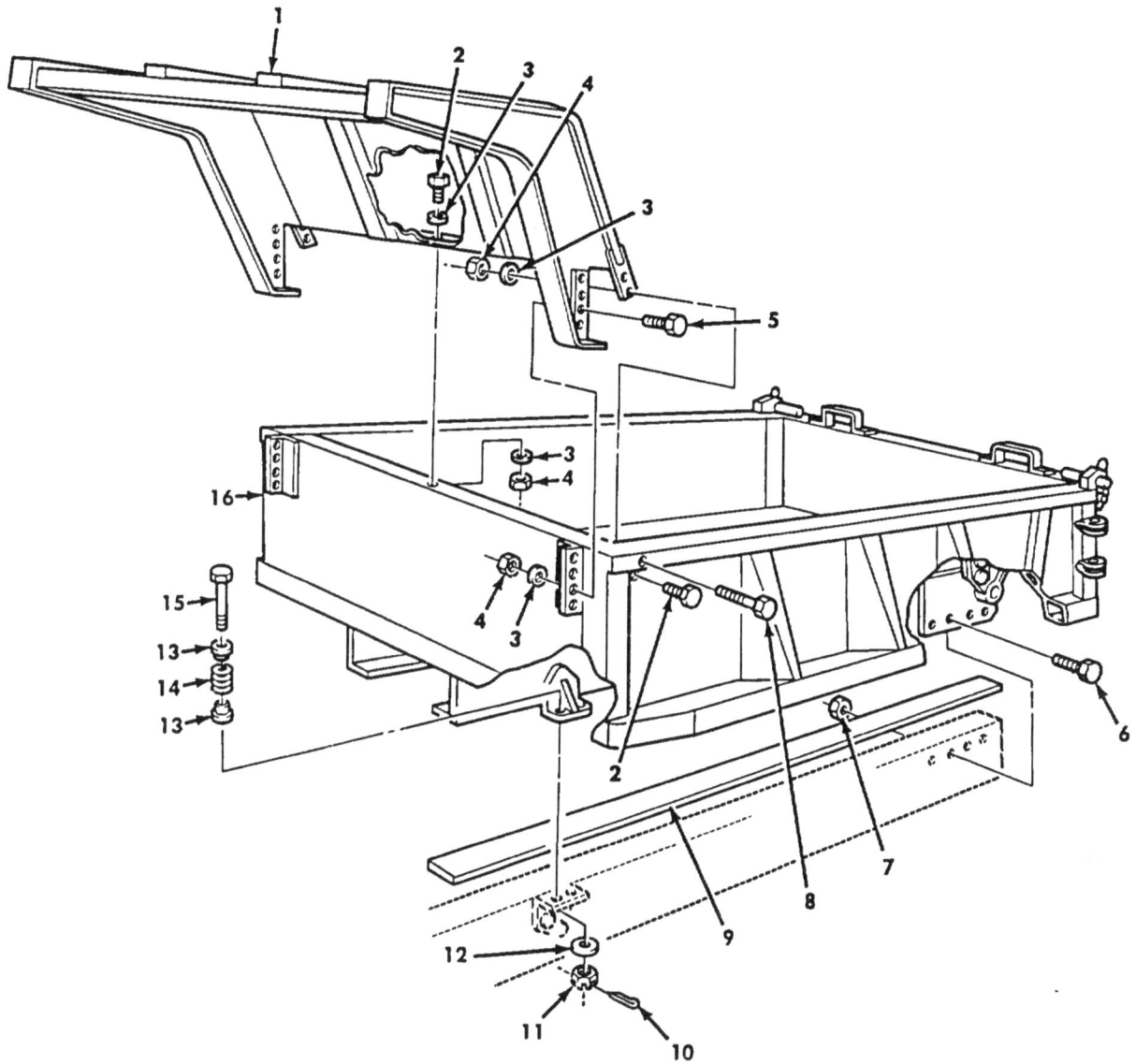

Figure 389. Dump Body and Hoist Assembly, Cab Shield, and Mounting Hardware.

(1) ITEM NO	(2) SMR CODE	(3) NSN	(4) CAGEC	(5) PART NUMBER	(6) DESCRIPTION AND USABLE ON CODES (UOC)	(7) QTY
					GROUP 1810 CARGO BODY	
					FIG. 389 DUMP BODY AND HOIST ASSEMBLY, CAB SHIELD, AND MOUNTING HARDWARE	
1	PBFFF	2510011431265	19207	12432457	BODY, CARGO DUMP TRU UOC:V19, V20	1
1	PBFFF	2510012106197	19207	12303034	PANEL, BODY, VEHICULA .... UOC:DAE, DAF, ZAE, ZAF	1
2	PAFZZ	5305007195219	96906	MS90727-111	SCREW, CAP, HEXAGON H .... UOC:DAE, DAF, V19, V20, ZAE, ZAF	9
3	PAFZZ	5310008095998	96906	MS27183-18	WASHER, FLAT .... UOC:DAE, DAF, V19, V20, ZAE, ZAF	26
4	PAFZZ	5310004883888	96906	MS51943-40	NUT, SELF-LOCKING, HE .... UOC:DAE, DAF, V19, V20, ZAE, ZAF	19
5	PAFZZ	5305007254183	96906	MS90726-113	SCREW, CAP, HEXAGON H .... UOC:DAE, DAF, V19, V20, ZAE, ZAF	8
6	PAFZZ	5305007262550	80204	B1821BH063F175N	SCREW, CAP, HEXAGON H .... UOC:DAE, DAF, V19, V20, ZAE, ZAF	8
7	PAFZZ	5310002416664	96906	MS51943-44	NUT, SELF-LOCKING, HE .... UOC:DAE, DAF, V19, V20, ZAE, ZAF	8
8	PAFZZ	5305007168174	96906	MS90726-124	SCREW, CAP, HEXAGON H .... UOC:DAE, DAF, V19, V20, ZAE, ZAF	2
9	MFFZZ		19207	11608809	SILL, WOOD..MAKE FROM WOOD, P/N .... 13219E0079... UOC:V19, V20	2
9	MFFZZ		19207	12302980	SILL, WOOD..MAKE FROM WOOD, P/N .... 13219E0079... UOC:DAE, DAF, ZAE, ZAF	2
10	PAFZZ	5315000137228	96906	MS24665-423	PIN, COTTER .... UOC:DAE, DAF, V19, V20, ZAE, ZAF	4
11	PAFZZ	5310009980608	96906	MS35692-61	NUT, PLAIN, SLOTTED, H .... UOC:DAE, DAF, V19, V20, ZAE, ZAF	4
12	PAFZZ	5310008098533	96906	MS27183-23	WASHER, FLAT .... UOC:DAE, DAF, V19, V20, ZAE, ZAF	4
13	PAFZZ	5340004890363	19207	7409030	SEAT, HELICAL COMPRE .... UOC:DAE, DAF, V19, V20, ZAE, ZAF	8
14	PAFZZ	5360004112511	19207	7409029	SPRING, HELICAL, COMP .... UOC:DAE, DAF, V19, V20, ZAE, ZAF	4
15	PAFZZ	5305004821035	19207	8758719	SCREW, CAP, HEXAGON H .... UOC:DAE, DAF, V19, V20, ZAE, ZAF	4
16	XBFZZ		19207	11608806-1	BODY AND HOIST ASSY.................................. UOC:V19, V20	1
16	XBFZZ		19207	11608806-2	BODY AND HOIST ASSY.................................. UOC:DAE, DAF, ZAE, ZAF	1

END OF FIGURE

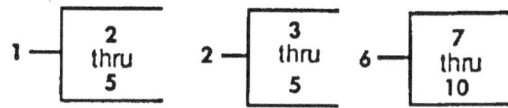

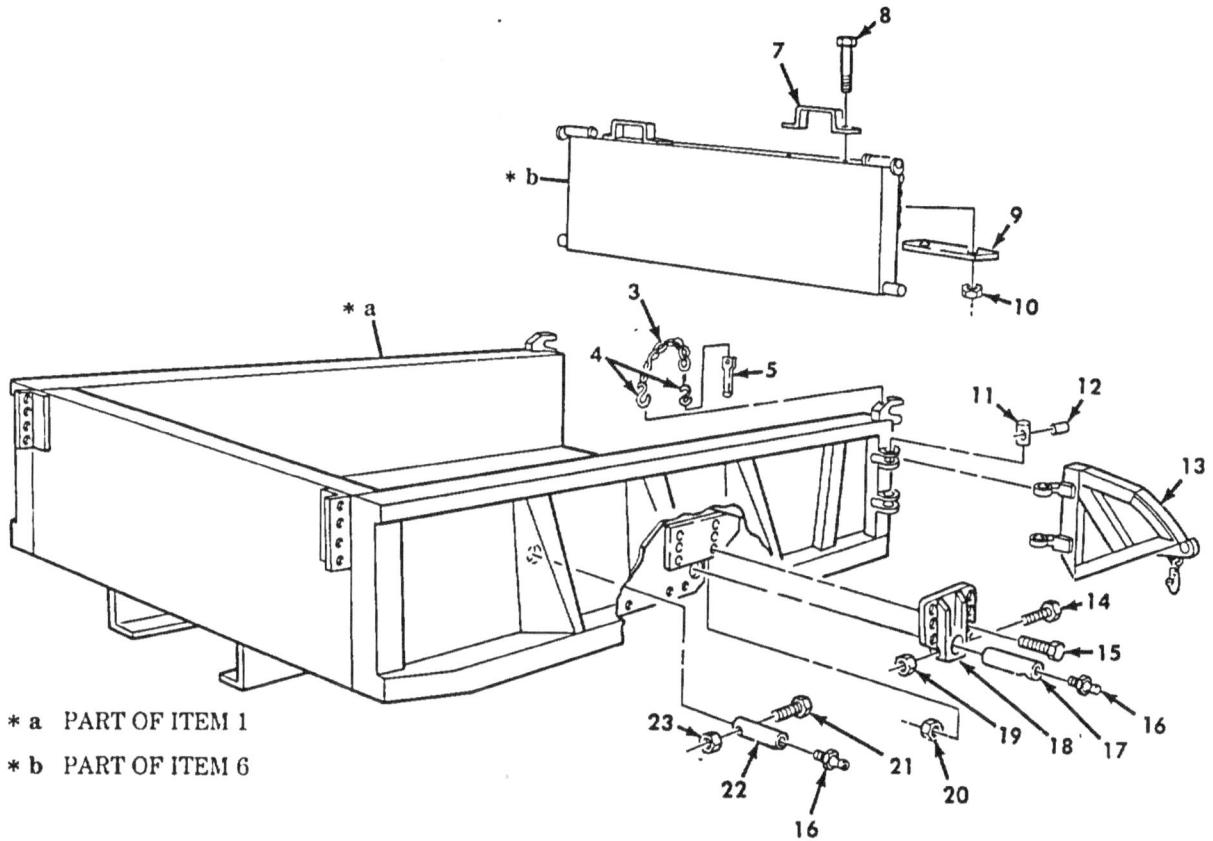

Figure 390. Dump Body, Basic, Tailgate, and Related Parts.

(1) ITEM NO	(2) SMR CODE	(3) NSN	(4) CAGEC	(5) PART NUMBER	(6) DESCRIPTION AND USABLE ON CODES (UOC)	(7) QTY
					GROUP 1810 CARGO BODY	
					FIG. 390 DUMP BODY, BASIC, TAILGATE, AND RELATED PARTS	
1	PFFFF	2510011431265	19207	12432457	BODY, CARGO DUMP TRU .... UOC:DAE, DAF, V19, V20, ZAE, ZAF	1
2	PAOZZ	4010008096294	19207	8720887	.CHAIN ASSEMBLY, SING .... UOC:DAE, DAF, V19, V20, ZAE, ZAF	2
3	PAOZZ	4010010909352	19207	8720887-1	..CHAIN, WELDED .... UOC:DAE, DAF, V19, V20, ZAE, ZAF	1
4	PAOZZ	4030009487315	96906	MS87006-33	..HOOK, CHAIN, S .... UOC:DAE, DAF, V19, V20, ZAE, ZAF	2
5	PAOZZ	5315007409017	19207	7409017	..PIN, STRAIGHT, HEADED .... UOC:DAE, DAF, V19, V20, ZAE, ZAF	1
6	PAOFF	2510000402264	19207	7409213	TAILGATE, VEHICLE BO .... UOC:DAE, DAF, V19, V20, ZAE, ZAF	1
7	PAOZZ	2510001193903	19207	7064165	.STEP, TAILGATE .... UOC:DAE, DAF, V19, V20, ZAE, ZAF	2
8	PAOZZ	5305007195274	96906	MS90727-125	.SCREW, CAP, HEXAGON H .... UOC:DAE, DAF, V19, V20, ZAE, ZAF	4
9	PAOZZ	5340001193906	19207	8758106	.PLATE, MENDING .... UOC:DAE, DAF, V19, V20, ZAE, ZAF	2
10	PAOZZ	5310008775795	96906	MS21044-N8	.NUT, SELF-LOCKING, HE .... UOC:DAE, D AF, V19, V20, ZAE, ZAF	4
11	PAOZZ	5315001504146	19207	8758706	PIN, STRAIGHT, HEADLE .... UOC:DAE, DAF, V19, V20, ZAE, ZAF	4
12	PAOZZ	5315011323569	96906	MS35672-46	PIN, GROOVED, HEADLES .... UOC:DAE, DAF, V19, V20, ZAE, ZAF	4
13	PBOZZ	2510014474754	19207	8758628-1	WING, TAILGATE, LEFT L.H .... UOC:DAE, DAF, V19, V20, ZAE, ZAF	1
13	PBOZZ	2510014397799	19207	8758629-1	WING, TAILGATE EXTEN R.H .... UOC:DAE, DAF, V19, V20, ZAE, ZAF	1
14	PAFZZ	5305002694528	96906	MS90727-75	SCREW, CAP, HEXAGON H .... UOC:DAE, DAF, V19, V20, ZAE, ZAF	2
15	PAFZZ	5305007262551	80204	B1821BH063F200N	SCREW, CAP, HEXAGON H .... UOC:DAE, DAF, V19, V20, ZAE, ZAF	12
16	PAOZZ	4730000504208	96906	MS15003-1	FITTING, LUBRICATION .... UOC:DAE, DAF, V19, V20, ZAE, ZAF	4
17	PAFZZ	5315007409045	19207	7409045	PIN, STRAIGHT, HEADLE .... UOC:DAE, DAF, V19, V20, ZAE, ZAF	2
18	PFFZZ	5340009996467	19207	8758705	BRACKET, MOUNTING .... UOC:DAE, DAF, V19, V20, ZAE, ZAF	2
19	PAFZZ	5310008140672	96906	MS51943-36	NUT, SELF-LOCKING, HE .... UOC:DAE, DAF, V19, V20, ZAE, ZAF	2
20	PAFZZ	5310002416664	96906	MS51943-44	NUT, SELF-LOCKING, HE .... UOC:DAE, DAF, V19, V20, ZAE, ZAF	12
21	PAFZZ	5306002259100	96906	MS90726-45	BOLT, MACHINE .... UOC:DAE, DAF, V19, V20, ZAE, ZAF	2
22	PAFZZ	5315007409043	19207	7409043	PIN, STRAIGHT, HEADLE .... UOC:DAE, DAF, V19, V20, ZAE, ZAF	2
23	PAFZZ	5310002416658	96906	MS51943-34	NUT, SELF-LOCKING, HE .... UOC:DAE, DAF, V19, V20, ZAE, ZAF	2

END OF FIGURE

390-1

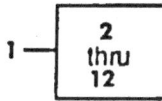

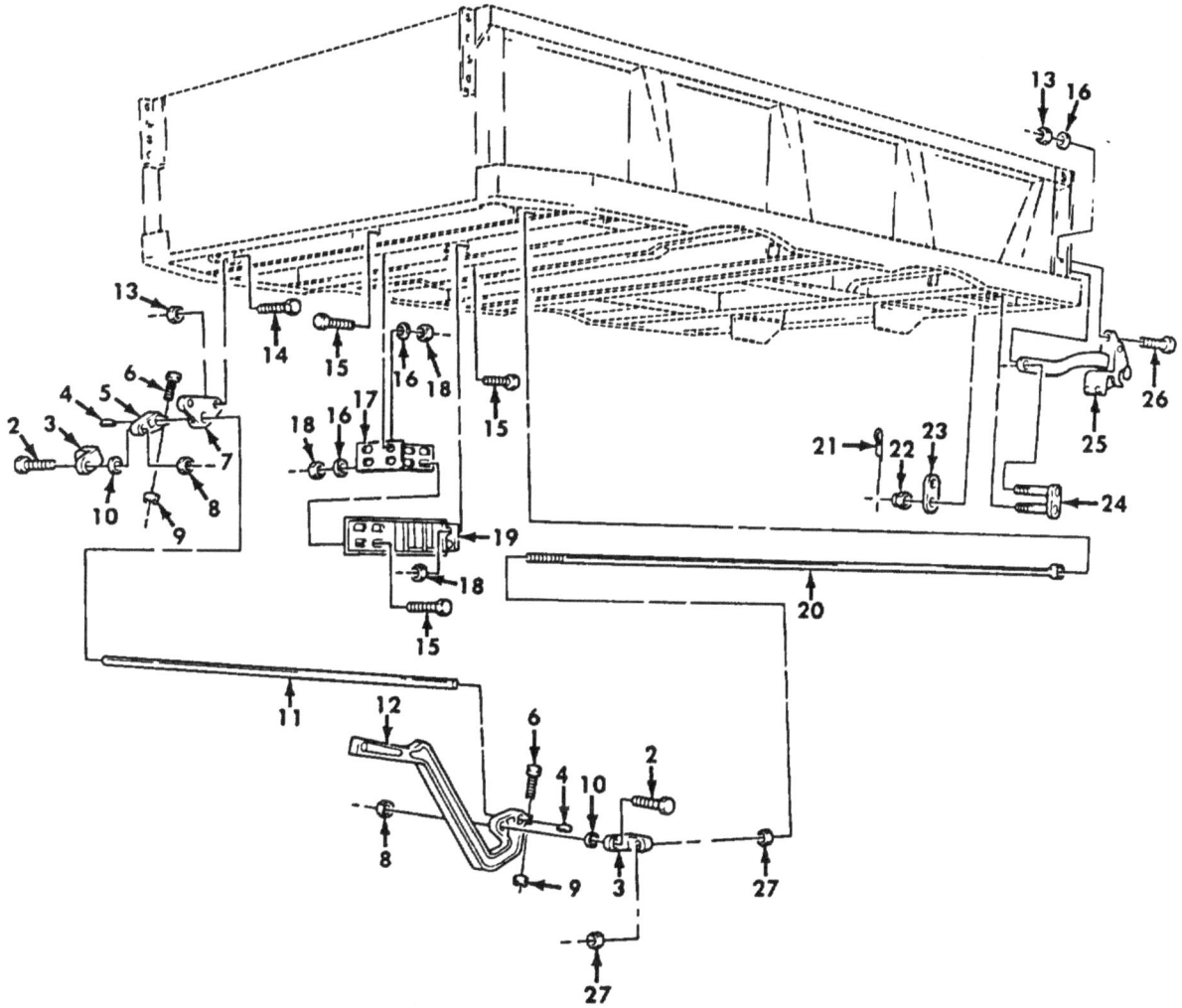

Figure 391. Dump Tailgate Controls, Safety Catch, and Related Parts.

(1) ITEM NO	(2) SMR CODE	(3) NSN	(4) CAGEC	(5) PART NUMBER	(6) DESCRIPTION AND USABLE ON CODES (UOC)	(7) QTY
					GROUP 1810 CARGO BODY	
					FIG. 391 DUMP TAILGATE CONTROLS, SAFETY CATCH, AND RELATED PARTS	
1	PFOOO	3040004097974	19207	8758598	LEVER, MANUAL CONTRO .... UOC:DAE, DAF, V19, V20, ZAE, ZAF	1
2	PAOZZ	5305004221161	19207	7971324	.SCREW, CAP, HEXAGON H .... UOC:DAE, DAF, V19, V20, ZAE, ZAF	2
3	PAOZZ	3040004094021	19207	7409012	.CONNECTING LINK, RIG .... UOC:DAE, DAF, V19, V20, ZAE, ZAF	2
4	PAOZZ	5315006165520	96906	MS35756-14	.KEY, WOODRUFF .... UOC:DAE, DAF, V19, V20, ZAE, ZAF	2
5	PFOZZ	3040004222008	19207	7409011	.LEVER, REMOTE CONTRO .... UOC:DAE, DAF, V19, V20, ZAE, ZAF	1
6	PAOZZ	5305007254183	96906	MS90726-113	.SCREW, CAP, HEXAGON H .... UOC:DAE, DAF, V19, V20, ZAE, ZAF	2
7	PAOZZ	2510007409013	19207	7409013	.BRACKET, EYE, ROTATIN .... UOC:DAE, DAF, V19, V20, ZAE, ZAF	1
8	PAOZZ	5310002256408	96906	MS51922-53	.NUT, SELF-LOCKING, HE .... UOC:DAE, DAF, V19, V20, ZAE, ZAF	2
9	PAOZZ	5310008775795	96906	MS21044-N8	.NUT, SELF-LOCKING, HE .... UOC:DAE, DAF, V19, V20, ZAE, ZAF	2
10	PAOZZ	5310008238803	96906	MS27183-21	.WASHER, FLAT .... UOC:DAE, DAF, V19, V20, ZAE, ZAF	2
11	PFOZZ	2510004098941	19207	8758599	.SHAFT, STRAIGHT .... UOC:DAE, DAF, V19, V20, ZAE, ZAF	1
12	PFOZZ	5340004222019	19207	7409010	.LEVER, MANUAL CONTRO .... UOC:DAE, DAF, V19, V20, ZAE, ZAF	1
13	PAOZZ	5310008775795	96906	MS21044-N8	NUT, SELF-LOCKING, HE .... UOC:DAE, DAF, V19, V20, ZAE, ZAF	12
14	PAOZZ	5305007254183	96906	MS90726-113	SCREW, CAP, HEXAGON H .... UOC:DAE, DAF, V19, V20, ZAE, ZAF	6
15	PAOZZ	5305007195221	80204	B1821BH050F150N	SCREW, CAP, HEXAGON H .... UOC:DAE, DAF, V19, V20, ZAE, ZAF	12
16	PAOZZ	5310008095998	96906	MS27183-18	WASHER, FLAT .... UOC:DAE, DAF, V19, V20, ZAE, ZAF	14
17	PBOZZ	5340011276920	19207	12300634	STRIKE, CATCH .... UOC:DAE, DAF, V19, V20, ZAE, ZAF	1
18	PAOZZ	5310004883888	96906	MS51943-40	NUT, SELF-LOCKING, HE .... UOC:DAE, DAF, V19, V20, ZAE, ZAF	12
19	PAOZZ	5340011256078	19207	12300641	BRACKET, MOUNTING .... UOC:DAE, DAF, V19, V20, ZAE, ZAF	1
20	PAOZZ	2510007409212	19207	7409212	ROD TAIL GATE .... UOC:DAE, DAF, V19, V20, ZAE, ZAF	2
21	PAOZZ	5315002981481	96906	MS24665-357	PIN, COTTER .... UOC:DAE, DAF, V19, V20, ZAE, ZAF	4
22	PAOZZ	5310008427783	96906	MS35692-53	NUT, PLAIN, SLOTTED .... UOC:DAE, DAF, V19, V20, ZAE, ZAF	4
23	PAOZZ	5340011530890	19207	7409009	CLAMP, LOOP .... UOC:DAE, DAF, V19, V20, ZAE, ZAF	2
24	PAOZZ	2510007409008	19207	7409008	LINK ASSEMBLY, DUMP .... UOC:DAE, DAF, V19, V20, ZAE, ZAF	2

(1) ITEM NO	(2) SMR CODE	(3) NSN	(4) CAGEC	(5) PART NUMBER	(6) DESCRIPTION AND USABLE ON CODES (UOC)	(7) QTY
25	PAOZZ	2510007409002	19207	7409002	LATCH TAIL GATE .... UOC:DAE, DAF, V19, V20, ZAE, ZAF	2
26	PAOZZ	5305007195235	80204	B1821BH050F175N	SCREW, CAP, HEXAGON H .... UOC:DAE, DAF, V19, V20, ZAE, ZAF	6
27	PAOZZ	5310007638905	96906	MS51968-20	NUT, PLAIN, HEXAGON TAILGATE CONTROL. .... UOC:DAE, DAF, V19, V20, ZAE, ZAF	4

END OF FIGURE

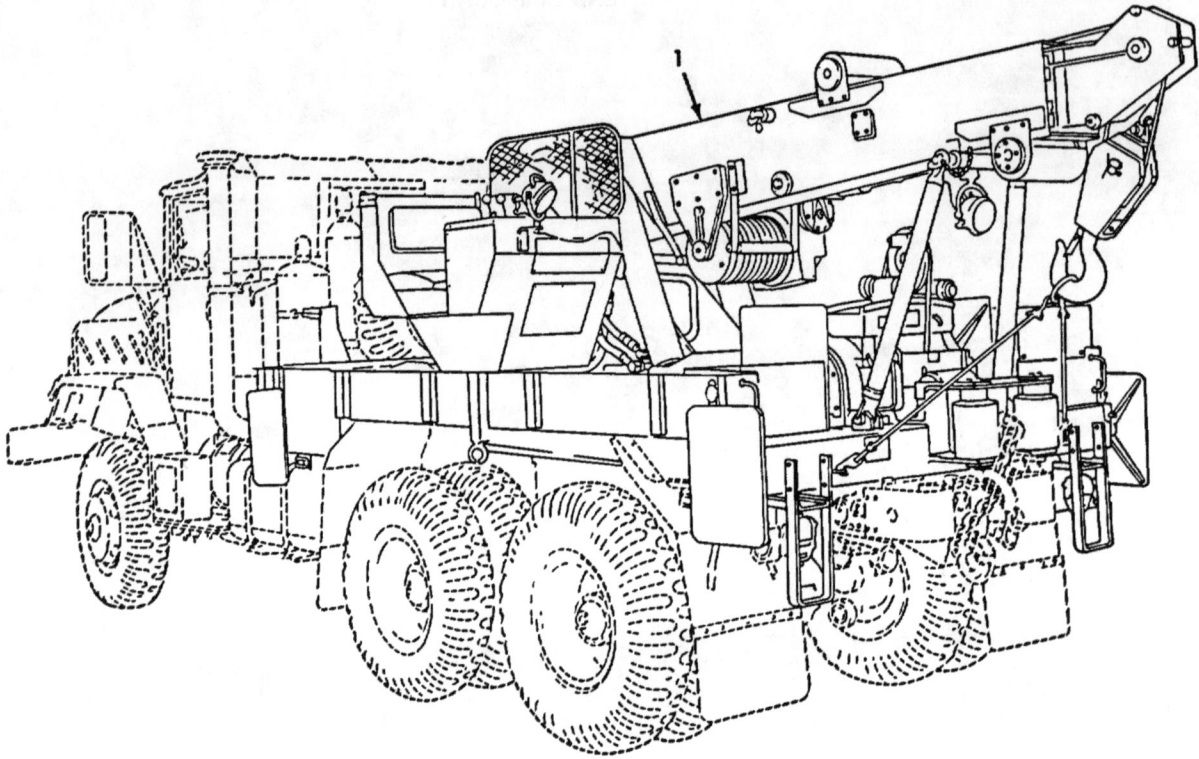

Figure 392. Wrecker Crane and Body Assembly.

(1) ITEM NO	(2) SMR CODE	(3) NSN	(4) CAGEC	(5) PART NUMBER	(6) DESCRIPTION AND USABLE ON CODES (UOC)	(7) QTY
					GROUP 1812 SPECIAL PURPOSE BODIES	
					FIG. 392 WRECKER CRANE AND BODY ASSEMBLY	
1	XAFZZ		19207	12256615	CRANE AND BODY ASSY UOC:V18	1
1	XAFZZ		19207	12256615-1	CRANE AND BODY ASSY .... UOC:DAL, ZAL	1

END OF FIGURE

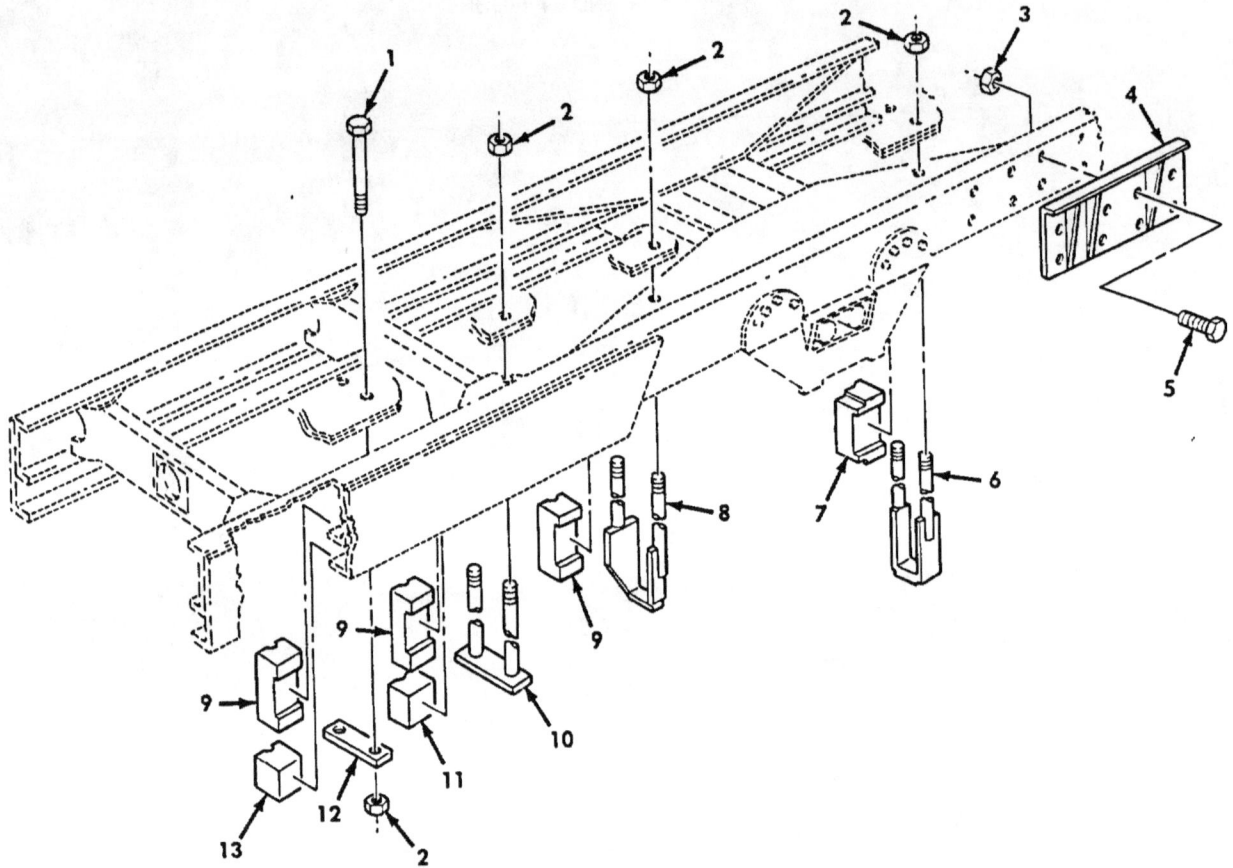

Figure 393. Wrecker Crane and Body Mounting Hardware.

(1) ITEM NO	(2) SMR CODE	(3) NSN	(4) CAGEC	(5) PART NUMBER	(6) DESCRIPTION AND USABLE ON CODES (UOC)	(7) QTY
					GROUP 1812 SPECIAL PURPOSE BODIES	
					FIG. 393 WRECKER CRANE AND BODY MOUNTING HARDWARE	
1	PAFZZ	5305011436534	19207	12255630	SCREW, CAP, HEXAGON H .... UOC:DAL, V18, ZAL	4
2	PAFZZ	5310003251900	80205	NAS1021-N17	NUT, SELF-LOCKING, HE .... UOC:DAL, V18, ZAL	16
3	PAFZZ	5310009825009	96906	MS21045-10	NUT, SELF-LOCKING, HE .... UOC:DAL, V18, ZAL	18
4	PAFZZ	5340004442108	19207	10876397	BRACKET, MOUNTING .... UOC:DAL, V18, ZAL	1
4	PAFZZ	5340004442107	19207	10876396	BRACKET, ANGLE .... UOC:DAL, V18, ZAL	1
5	PAFZZ	5305007262550	80204	B1821BH063F175N	SCREW, CAP, HEXAGON H .... UOC:DAL, V18, ZAL	18
6	PAFZZ	5306003517842	19207	10900134	BOLT, BODY, SECURING .... UOC:DAL, V18, ZAL	2
7	PAFZZ	2590011310123	19207	10876329	BLOCK, FILLER, BOGIE .... UOC:DAL, V18, ZAL	2
8	PAFZZ	5306001515726	19207	10876332	BOLT, U L.H .... UOC:DAL, V18, ZAL	1
8	PAFZZ	5306004195878	19207	10876333	BOLT, U R.H .... UOC:DAL, V18, ZAL	1
9	PAFZZ	5510011328746	19207	10876331	BLOCK, FILLER, WOOD .... UOC:DAL, V18, ZAL	6
10	PAFZZ	5306002310211	19207	10900130	BOLT, U .... UOC:DAL, V18, ZAL	2
11	PAFZZ	5510011327138	19207	10876325	BLOCK, FILLER, WOOD .... UOC:DAL, V18, ZAL	1
11	PAFZZ	5510011327139	19207	10876326	BLOCK, FILLER, WOOD .... UOC:DAL, V18, ZAL	1
12	PAFZZ	5340011310124	19207	10900137	PLATE, MOUNTING .... UOC:DAL, V18, ZAL	2
13	PAFZZ	5510011324879	19207	10876324	BLOCK, FLOOR, WOOD .... UOC:DAL, V18, ZAL	2

END OF FIGURE

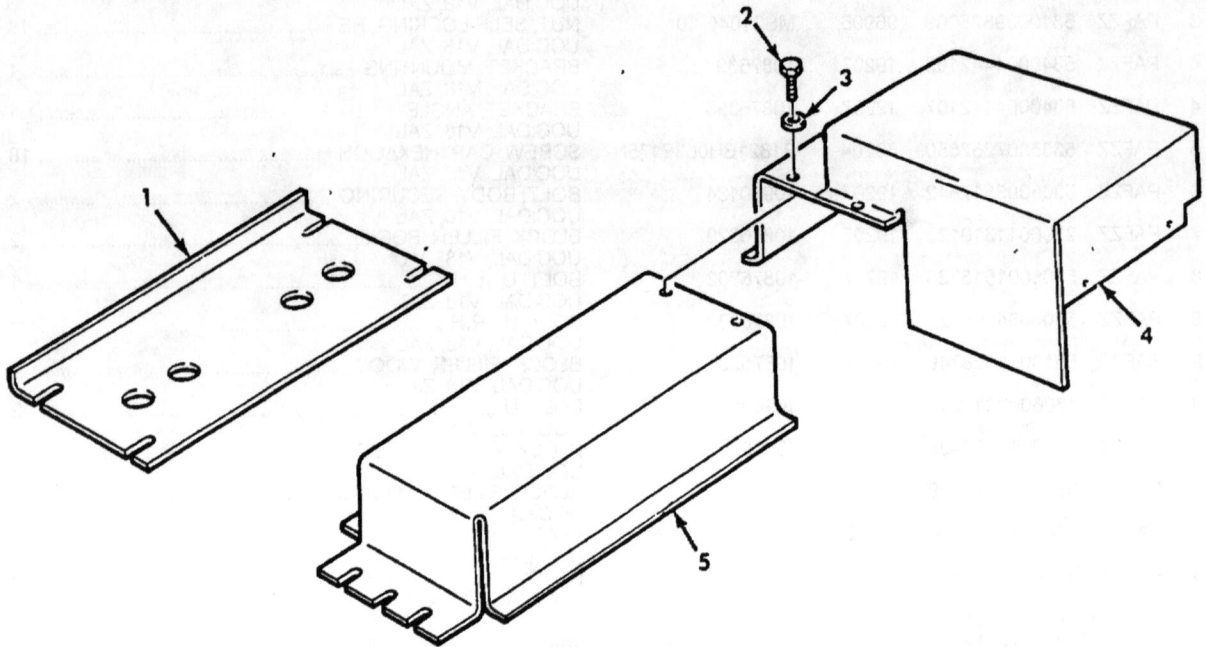

Figure 394. Wrecker Crane and Body Floor Plate, Cover, and Housing Assembly.

(1) ITEM NO	(2) SMR CODE	(3) NSN	(4) CAGEC	(5) PART NUMBER	(6) DESCRIPTION AND USABLE ON CODES (UOC)	(7) QTY
					GROUP 1812 SPECIAL PURPOSE BODIES	
					FIG. 394 WRECKER CRANE AND BODY FLOOR PLATE, COVER, AND HOUSING ASSEMBLY	
1	PBOZZ	9515011402379	19207	12256694	PLATE, FLOOR, METAL	1
					UOC:DAL, V18, ZAL	
2	PAOZZ	5305002693234	96906	MS90727-58	SCREW, CAP, HEXAGON	16
					UOC:DAL, V18, ZAL	
3	PAOZZ	5310006379541	96906	MS35338-46	WASHER, LOCK	16
					UOC:DAL, V18, ZAL	
4	PFOZZ	5340010823610	19207	12256720	COVER, ACCESS	1
					UOC:V18	
4	PAOZZ	5340012104659	19207	12256720-2	COVER, ACCESS	1
					UOC:DAL, ZAL	
5	PFOZZ	5340010823609	19207	12256812	COVER, ACCESS	1
					UOC:DAL, V18, ZAL	

END OF FIGURE

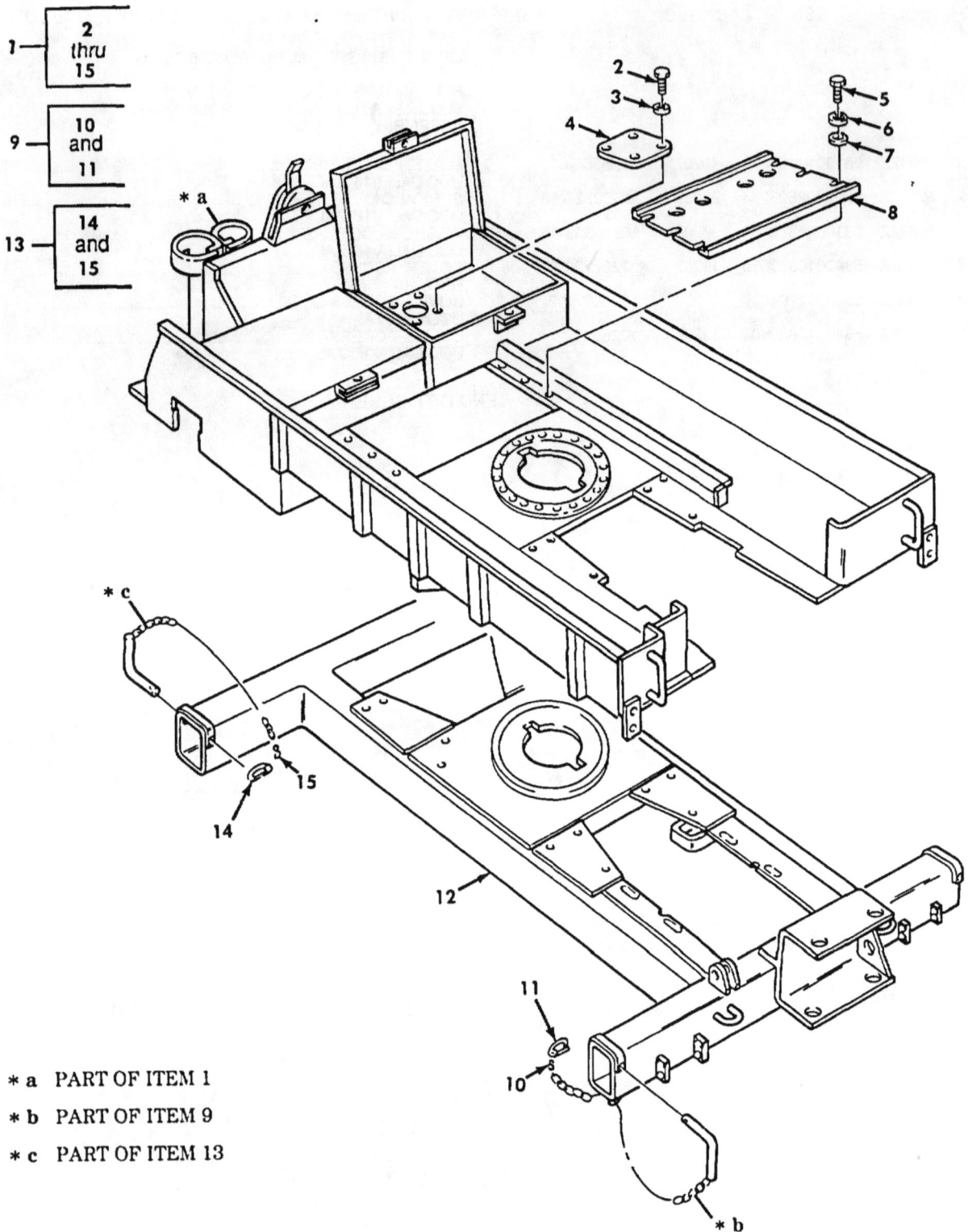

Figure 395. Wrecker Body Assembly.

* a   PART OF ITEM 1
* b   PART OF ITEM 9
* c   PART OF ITEM 13

(1) ITEM NO	(2) SMR CODE	(3) NSN	(4) CAGEC	(5) PART NUMBER	(6) DESCRIPTION AND USABLE ON CODES (UOC)	(7) QTY
					GROUP 1812 SPECIAL PURPOSE BODIES	
					FIG. 395 WRECKER BODY ASSEMBLY	
1	XAFZZ		19207	12256616	BODY ASSY, WRECKER UOC:V18	1
1	XAFZZ		19207	12303002	BODY ASSY, WRECKER .... UOC:DAL, ZAL	1
2	PAFZZ	5305000526922	96906	MS24629-58	.SCREW, TAPPING .... UOC:DAL, V18, ZAL	4
3	PAFZZ	5310005825965	96906	MS35338-44	.WASHER, LOCK .... UOC:DAL, V18, ZAL	4
4	PAFZZ	5340004217235	19207	10900086	.COVER, ACCESS .... UOC:DAL, V18, ZAL	1
5	PAFZZ	5305002693234	96906	MS90727-58	.SCREW, CAP, HEXAGON .... UOC:DAL, V18, ZAL	6
6	PAFZZ	5310006379541	96906	MS35338-46	.WASHER, LOCK .... UOC:DAL, V18, ZAL	6
7	PAFZZ	5310000806004	96906	MS27183-14	.WASHER, FLAT .... UOC:DAL, V18, ZAL	6
8	PAFZZ	2510004213956	19207	10876502	.FLOOR PLATE, VEHICUL .... UOC:DAL, V18, ZAL	1
9	PAFZZ	5315004956497	19207	10915207	.PIN, ASSEMBLY, OUTRIG .... UOC:DAL, V18, ZAL	2
10	PAFZZ	4030007809350	96906	MS87006-13	..HOOK, CHAIN, S .... UOC:DAL, V18, ZAL	1
11	PAFZZ	5315002901349	19207	7358098	..PIN, RETAINING .... UOC:DAL, V18, ZAL	1
12	XAFZZ		19207	12256680	.UNDERSTRUCTURE ASSY .... UOC:DAL, V18, ZAL	1
13	PAFZZ	5340004468732	19207	10876503	.HANDLE, MANUAL CONTR .... UOC:DAL, V18, ZAL	2
14	PAFZZ	5315007418971	19207	7418971	..PIN, RETAINING .... UOC:DAL, V18, ZAL	1
15	PAFZZ	4030007809350	96906	MS87006-13	..HOOK, CHAIN, S .... UOC:DAL, V1B, ZAL	1

END OF FIGURE

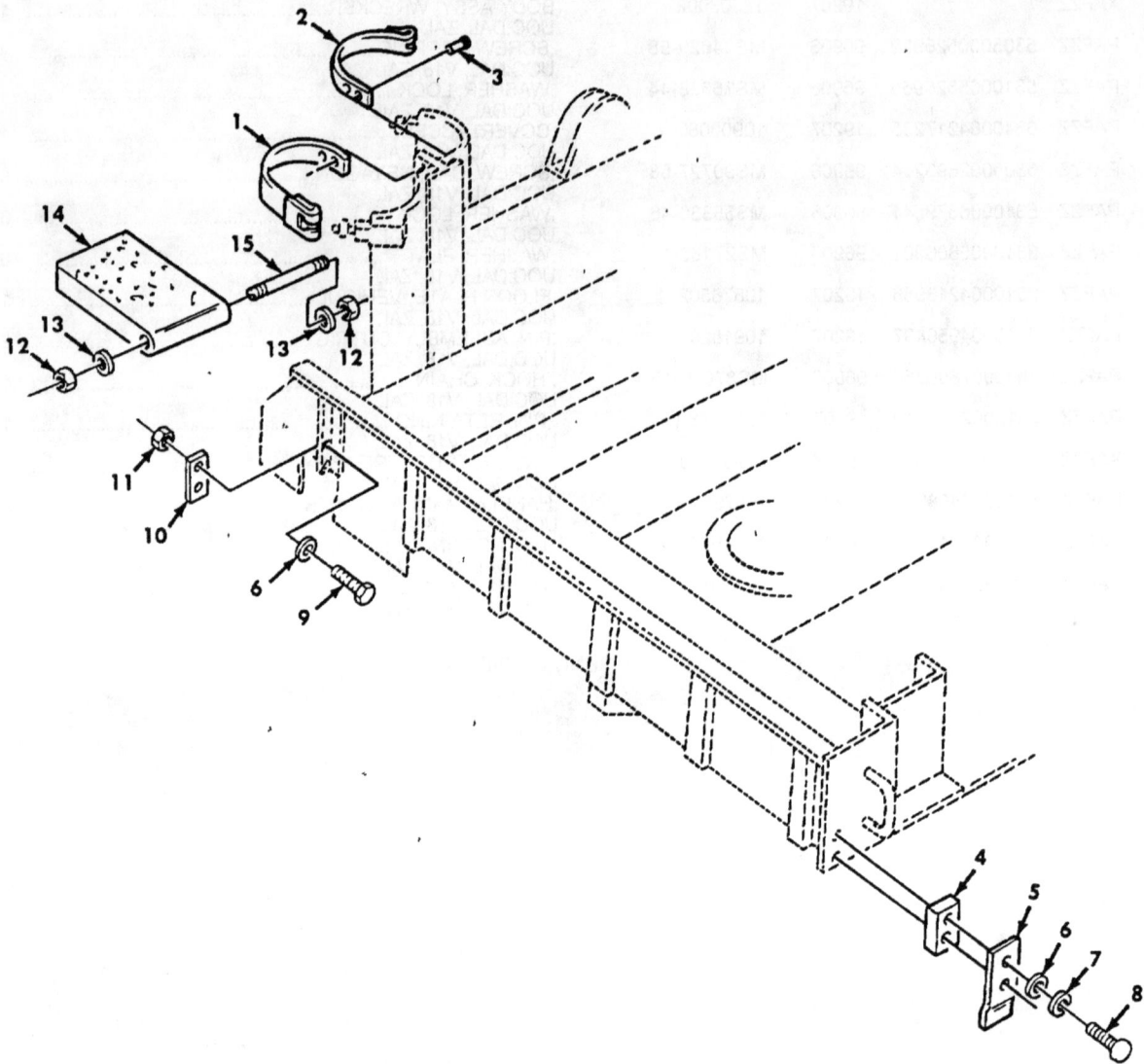

Figure 396. Wrecker Step and Welding Tank Straps.

(1) ITEM NO	(2) SMR CODE	(3) NSN	(4) CAGEC	(5) PART NUMBER	(6) DESCRIPTION AND USABLE ON CODES (UOC)	(7) QTY
					GROUP 1812 SPECIAL PURPOSE BODIES	
					FIG. 396 WRECKER STEP AND WELDING TANK STRAPS	
1	PAFZZ	5340010835402	19207	12256619	BAND, RETAINING .... UOC:DAL, V18, ZAL	1
2	PAFZZ	5340010835403	19207	12256620	BAND, RETAINING .... UOC:DAL, V18, ZAL	1
3	PAFZZ	5320004783313	96906	MS35743-39	RIVET, SOLID .... UOC:DAL, V18, ZAL	4
4	PAFZZ	5340010822516	19207	10900129	PLATE, MENDING .... UOC:DAL, V18, ZAL	2
5	PAFZZ	5340010831107	19207	10900175	PLATE, MOUNTING .... UOC:DAL, V18, ZAL	2
6	PAFZZ	5310008095998	96906	MS27183-18	WASHER, FLAT .... UOC:DAL, V18, ZAL	8
7	PAFZZ	5310000034094	01276	210104-8S	WASHER, LOCK .... UOC:DAL, V18, ZAL	4
8	PAFZZ	5305002267767	96906	MS90726-109	SCREW, CAP, HEXAGON H .... UOC:DAL, V1B, ZAL	4
9	PAFZZ	5305007254183	96906	MS90726-113	SCREW, CAP, HEXAGON H .... UOC:DAL, V18, ZAL	4
10	PAFZZ	5340011310125	19207	10900133	PLATE, MOUNTING .... UOC:DAL, V18, ZAL	2
11	PAFZZ	5310008775795	96906	MS21044-N8	NUT, SELF-LOCKING, HE .... UOC:DAL, V18, ZAL	4
12	PAFZZ	5310009591488	96906	MS51922-21	NUT, SELF-LOCKING .... UOC:DAL, V18, ZAL	4
13	PAFZZ	5310000806004	96906	MS27183-14	WASHER, FLAT .... UOC:DAL, V18, ZAL	4
14	PFFZZ	2510010819226	19207	10876579	STEP .... UOC:DAL, V18, ZAL	2
15	PAFZZ	5307011073675	19207	10900171-1	STUD, PLAIN .... UOC:DAL, V18, ZAL	2

END OF FIGURE

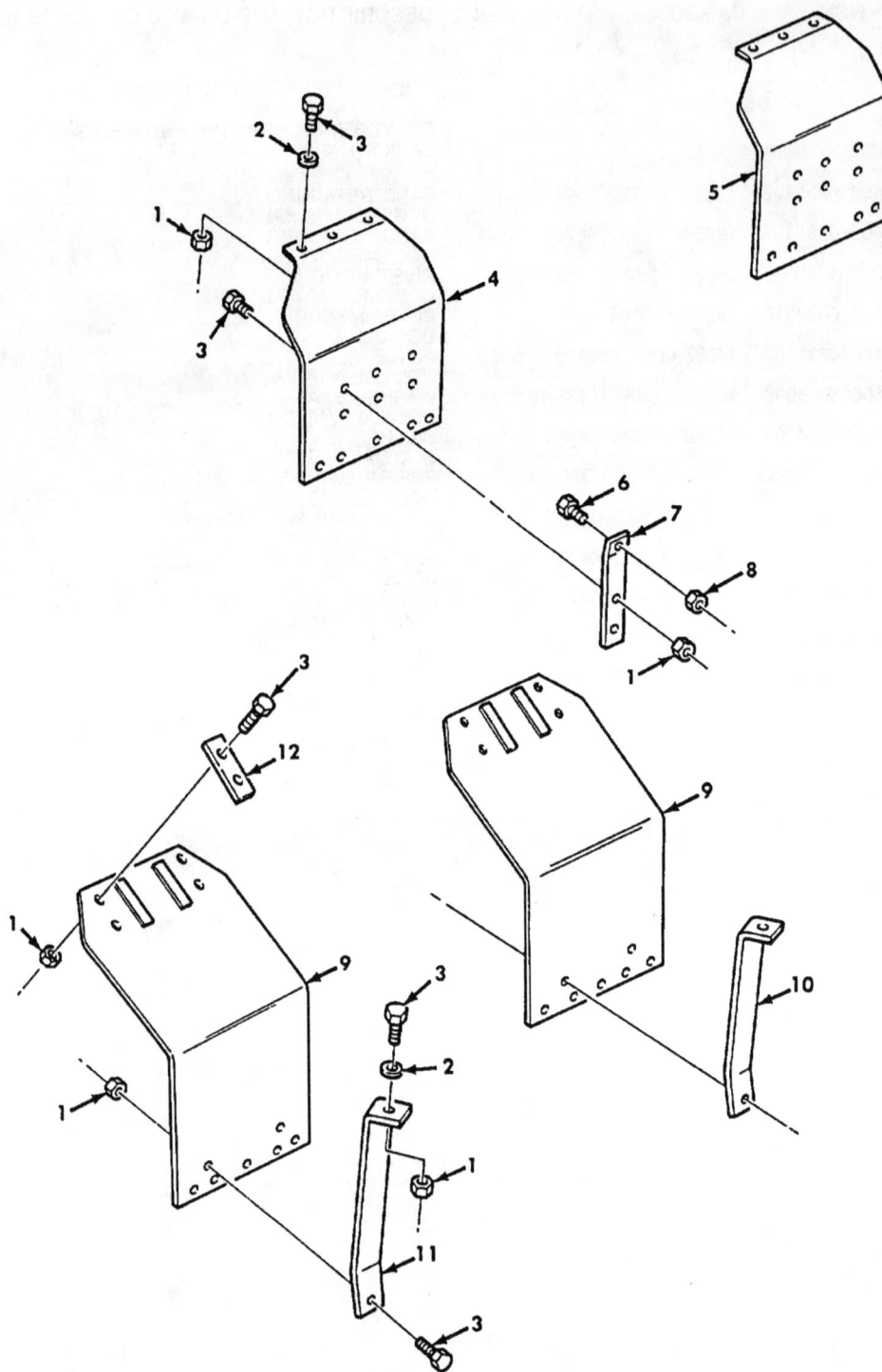

Figure 397. Wrecker Splash Guards.

(1) ITEM NO	(2) SMR CODE	(3) NSN	(4) CAGEC	(5) PART NUMBER	(6) DESCRIPTION AND USABLE ON CODES (UOC)	(7) QTY
					GROUP 1812 SPECIAL PURPOSE BODIES	
					FIG. 397 WRECKER SPLASH GUARDS	
1	PAOZZ	5310002416658	96906	MS51943-34	NUT, SELF-LOCKING, HE .... UOC:DAL, V18, ZAL	34
2	PAOZZ	5310000814219	96906	MS27183-12	WASHER, FLAT .... UOC:DAL, V18, ZAL	10
3	PAOZZ	5306002259088	96906	MS90726-33	BOLT, MACHINE .... UOC:DAL, V18, ZAL	34
4	PFOZZ	2540010915449	19207	11648465-2	GUARD, SPLASH, VEHICU L.H., FRONT .... UOC:DAL, V18, ZAL	1
5	PFOZZ	2540010895017	19207	11648465-1	GUARD, SPLASH, VEHICU R.H., FRONT .... UOC:DAL, V18, ZAL	1
6	PAOZZ	5305002692804	96906	MS90726-61	SCREW, CAP, HEXAGON H .... UOC:DAL, V18, ZAL	6
7	PFOZZ	5340011075220	19207	11648456	BRACKET, ANGLE FRONT .... UOC:DAL, V18, ZAL	6
8	PAOZZ	5310008140672	96906	MS51943-36	NUT, SELF-LOCKING, HE .... UOC:DAL, V18, ZAL	6
9	PFOZZ	2540010895018	19207	11648466	GUARD, SPLASH, VEHICU REAR .... UOC:DAL, V18, ZAL	2
10	PFOZZ	5340011075219	19207	11648458-1	BRACKET, ANGLE REAR .... UOC:DAL, V18, ZAL	2
11	PFOZZ	5340011097553	19207	11648458-2	BRACKET, ANGLE REAR .... UOC:DAL, V18, ZAL	2
12	PFOZZ	5365011092472	19207	11648462-1	SPACER, PLATE REAR .... UOC:DAL, V18, ZAL	2

END OF FIGURE

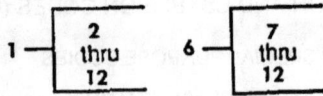

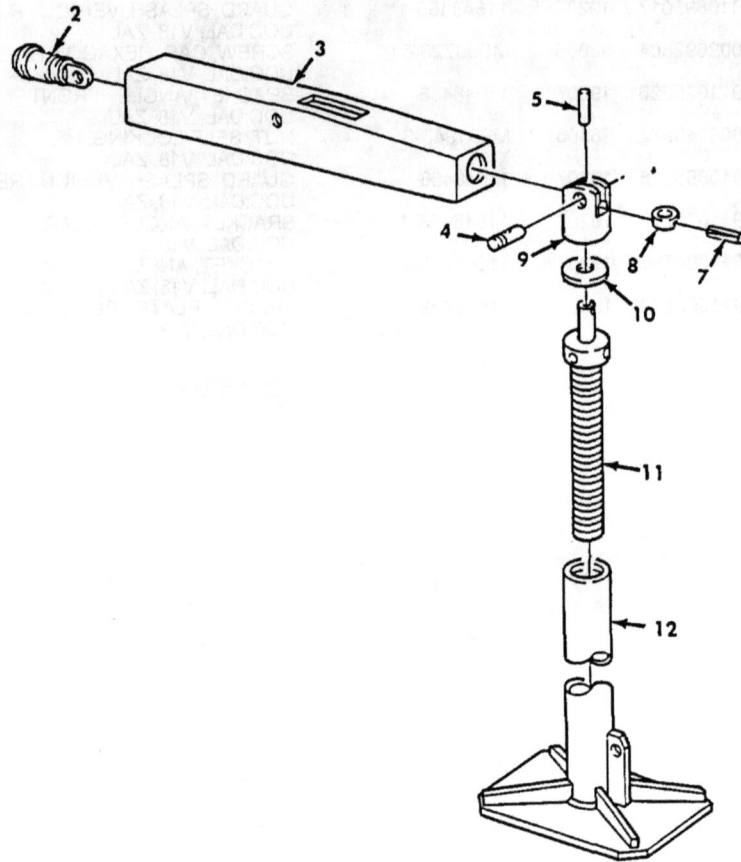

Figure 398. Wrecker Outrigger Jack Assembly and Related Parts.

(1) ITEM NO	(2) SMR CODE	(3) NSN	(4) CAGEC	(5) PART NUMBER	(6) DESCRIPTION AND USABLE ON CODES (UOC)	(7) QTY
					GROUP 1812 SPECIAL PURPOSE BODIES	
					FIG. 398 WRECKER OUTRIGGER JACK ASSEMBLY AND RELATED PARTS	
1	PFOFF	2590010853596	19207	10876254	JACK ASSEMBLY, OUTRI .... UOC:DAL, V18, ZAL	2
2	PAFZZ	2590003517835	19207	10876395	.SOCKET, BEAM, CRANE .... UOC:DAL, V18, ZAL	1
3	PAFZZ	3950003517834	19207	10876240	.OUTRIGGER, CRANE .... UOC:DAL, V18, ZAL	1
4	PAFZZ	5315007409835	19207	7409835	.PIN, STRAIGHT, HEADLE .... UOC:DAL, V18, ZAL	1
5	PAFZZ	5315009994238	96906	MS35672-47	.PIN, GROOVED, HEADLES .... UOC:DAL, V18, ZAL	1
6	PAOOZ	2590002317418	19207	10876244	.JACK, LEVELING SUPPO .... UOC:DAL, V18, ZAL	1
7	PAFZZ	5315009994238	96906	MS35672-47	..PIN, GROOVED, HEADLES .... UOC:DAL, V18, ZAL	1
8	PAFZZ	5365007409784	19207	7409784	..SPACER, SLEEVE .... UOC:DAL, V18, ZAL	1
9	PAFZZ	5340002990069	19207	10876418	..CLEVIS, ROD END .... UOC:DAL, V18, ZAL	1
10	PAFZZ	5310007409862	19207	7409862	..WASHER, FLAT .... UOC:DAL, V18, ZAL	1
11	PAFZZ	2590003517836	19207	10876416	..SCREW ASSEMBLY, JACK .... UOC:DAL, V18, ZAL	1
12	PFFZZ	2590011310126	19207	10876243	..SHOE, JACK SUPPORT .... UOC:DAL, V18, ZAL	1

END OF FIGURE

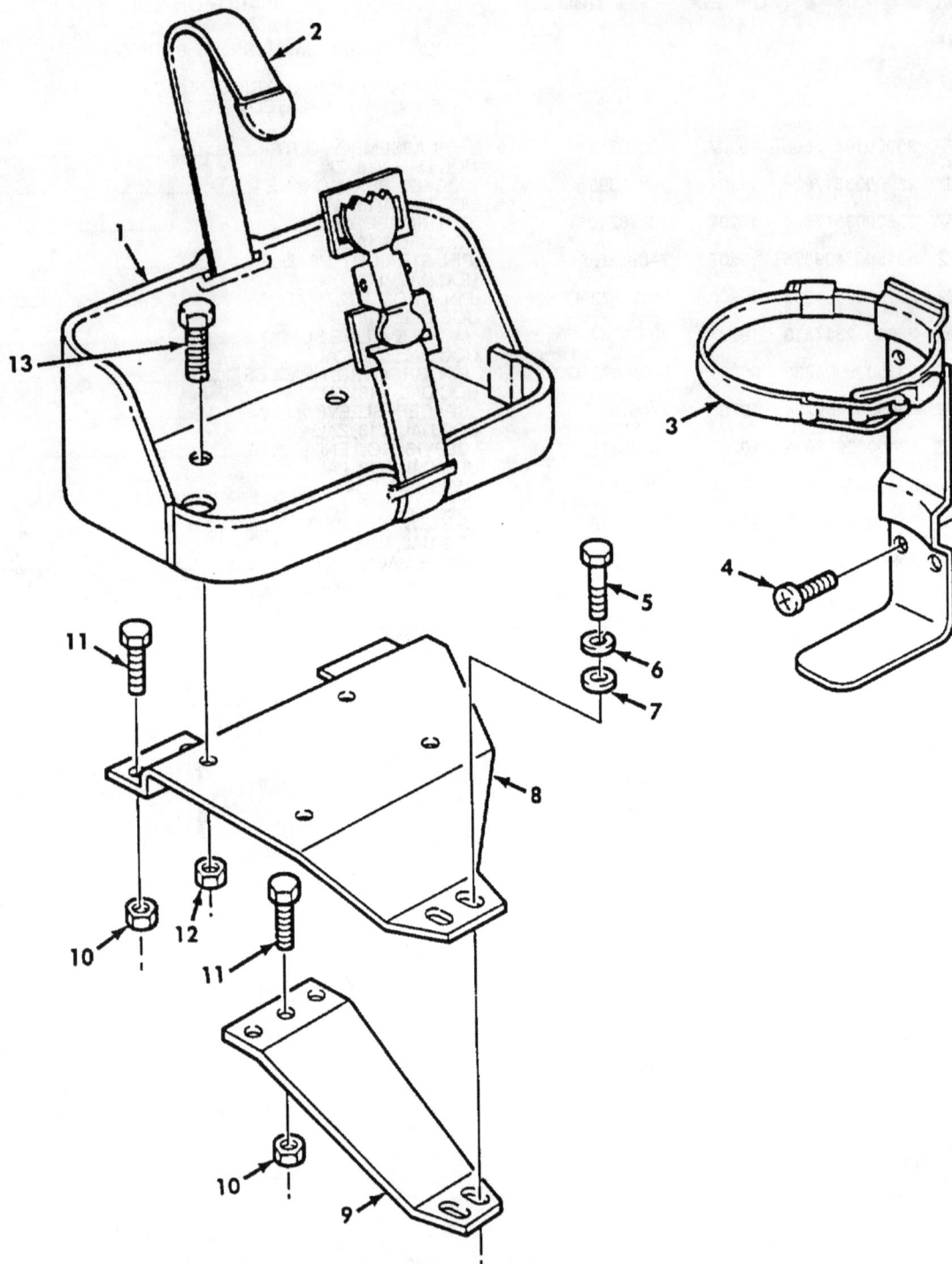

Figure 399. *Wrecker Fuel Can Bracket and Fire Extinguisher Bracket*

(1) ITEM NO	(2) SMR CODE	(3) NSN	(4) CAGEC	(5) PART NUMBER	(6) DESCRIPTION AND USABLE ON CODES (UOC)	(7) QTY
					GROUP 1812 SPECIAL PURUOSE BODIES	
					FIG. 399 WRECKER FUEL CAN BRACKET AND FIRE EXTINGUISHER BRACKET	
1	PFOZZ	2590004736331	19207	6566675	BRACKIET, VEHICULAR C .... UOC:DAL, V18, ZAL	1
2	PAOZZ	5340009684060	19207	8690527	STRAP, WEBBITNG .... UOC:DAL, V1B, ZAL	1
3	PAOZZ	4210011834822	19207	12255634	BRACKET, FIRE EXTING .... UOC:DAL, V18, ZAL	2
4	PAOZZ	5305000888946	96906	MS35207-278	SCREW, MACHINE .... UOC:DAL, V18, ZAL	8
5	PAOZZ	5305005434372	80204	B1821BH038C075N	SCREW, CAP, HEXAGON H .... UOC:DAL, V18, ZAL	2
6	PAOZZ	5310006379541	96906	MS35338-46	WASHER, LOCK .... UOC:DAL, V18, ZAL	2
7	PAOZZ	5310000806004	96906	MS27183-14	WASHER, FLAT .... UOC:DAL, V18, ZAL	2
8	PFOZZ	5340004832163	19207	10938449	BRACKET, MOUNTING .... UOC:DAL, V1B, ZAL	1
9	PFOZZ	5340001583774	19207	10899416	BRACKET, DOUBLE ANGL .... UOC:DAL, V18, ZAL	1
10	PAOZZ	5310002416685	96906	MS51943-34	NUT, SELF-LOCKING .... UOC:DAL, V18, ZAL	7
11	PAOZZ	5306000501238	96906	MS90727-32	BOLT, MACHINE .... UOC:DAL, V18, ZAL	4
12	PAOZZ	5310008140672	96906	MS51943-36	NUT, SELF-LOCKING, HE .... UOC:DAL, V1B, ZAL	4
13	PAOZZ	5305002693234	96906	MS90727-58	SCREW, CAP, HEXAGON .... UOC:DAL, V18, ZAL	4

END OF FIGURE

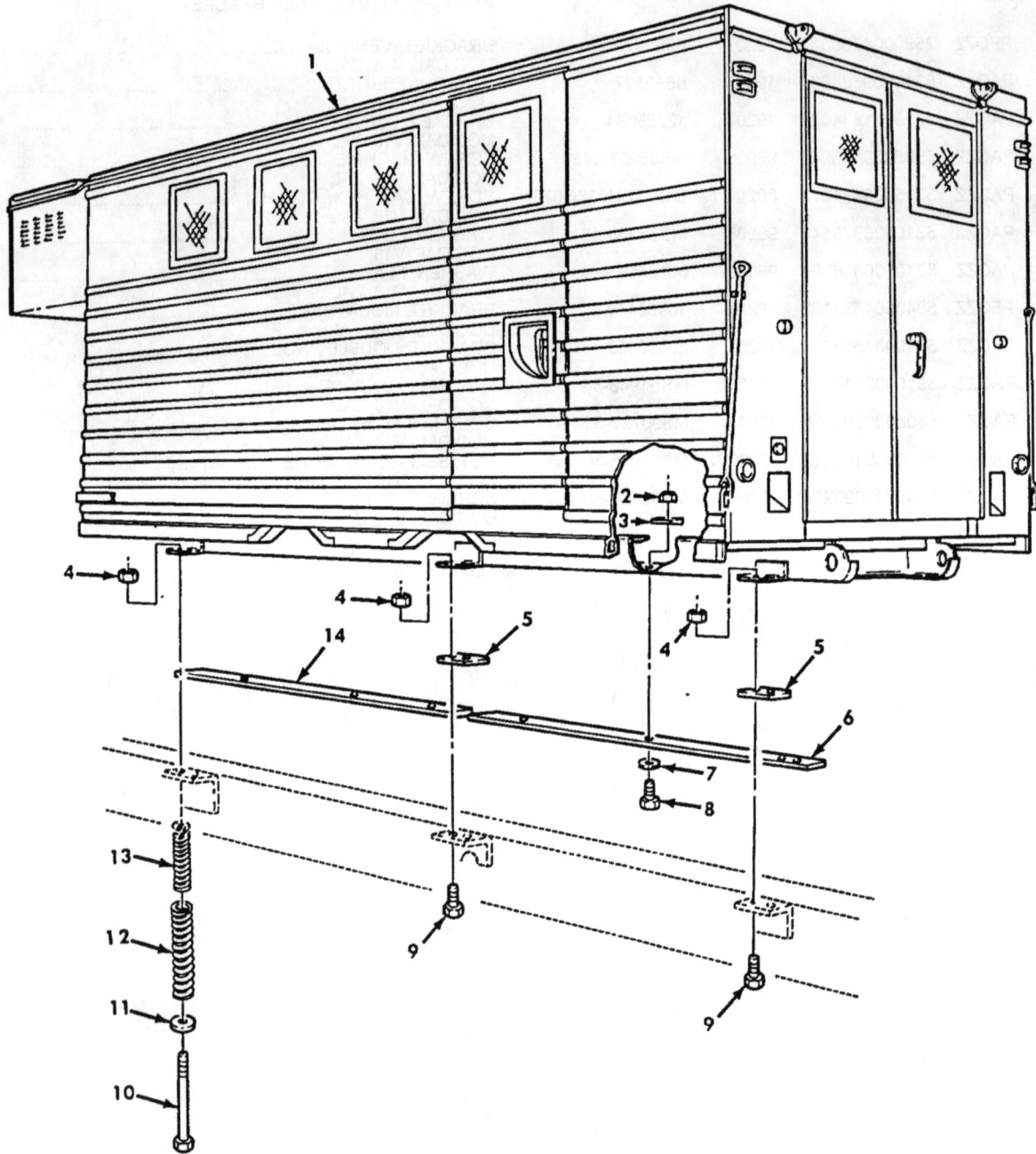

Figure 400. Van Body Assembly and Mounting Hardware.

(1) ITEM NO	(2) SMR CODE	(3) NSN	(4) CAGEC	(5) PART NUMBER	(6) DESCRIPTION AND USABLE ON CODES (UOC)	(7) QTY

GROUP 1812 SPECIAL PURPOSE BODIES

FIG. 400 VAN BODY ASSEMBLY AND MOUNTING HARDWARE

(1) ITEM NO	(2) SMR CODE	(3) NSN	(4) CAGEC	(5) PART NUMBER	(6) DESCRIPTION AND USABLE ON CODES (UOC)	(7) QTY
1	XAHZZ		19207	12256591-1	VAN BODY ASSY EXPANDIBLE UOC:V24	1
1	XAHHH		19207	12256591-3	BODY, VAN TRUCK .... UOC:DAJ, ZAJ	1
2	PAHZZ	5310008140672	96906	MS51943-36	NUT, SELF-LOCKING, HE .... UOC:DAJ, V24, ZAJ	10
3	PAHZZ	5310010992550	57328	65003-S	WASHER, BEVEL .... UOC:DAJ, V24, ZAJ	10
4	PAOZZ	5310002416664	96906	MS51943-44	NUT, SELF-LOCKING, HE .... UOC:DAJ, V24, ZAJ	12
5	PAOZZ	5340004718635	19207	11593258	BRACKET, MOUNTING .... UOC:DAJ, V24, ZAJ	4
6	MHHZZ		19207	11677828-1	SILL, WOOD RIGHT SIDE, MAKE FROM.................. WOOD, P/N 13219E0079 .... UOC:DAJ, V24, ZAJ	1
6	MHHZZ		19207	11677828-2	SILL, WOOD LEFT SIDE, MAKE FROM WOOD, ... P/N 13219E0079 .... UOC:DAJ, V24, ZAJ	1
7	PAHZZ	5310000806004	96906	MS27183-14	WASHER, FLAT .... UOC:DAJ, V24, ZAJ	10
8	PAHZZ	5305002693240	80204	B1821BH038F150N	SCREW, CAP, HEXAGON H .... UOC:DAJ, V24, ZAJ	10
9	PAOZZ	5305007262552	80204	B1821BH063F225N	SCREW, CAP, HEXAGON H UOC:V24	8
9	PAOZZ	5305007262553	96906	MS90727-166	SCREW, CAP, HEXAGON H .... UOC:DAJ, ZAJ	8
10	PAOZZ	5306005129218	19207	7971653	BOLT, MACHINE .... UOC:DAJ, V24, ZAJ	4
11	PAOZZ	5310009517209	96906	MS27183-22	WASHER, FLAT .... UOC:DAJ, V24, ZAJ	4
12	PAOZZ	5360007372793	19207	7372793	SPRING, HELICAL, COMP .... UOC:DAJ, V24, ZAJ	4
13	PAOZZ	5360007372792	19207	7372792	SPRING, HELICAL, COMP .... UOC:DAJ, V24, ZAJ	4
14	NHHZZ		19207	11677829-1	SILL, WOOD RIGHT SIDE, MAKE FROM .... WOOD, P/N 13219E0079.................................... UOC:DAJ, V24, ZAJ	1
14	MHHZZ		19207	11677829-2	SILL, WOOD LEFT SDIE, MAKE FROM WOOD, ... P/N 13219E0079 .... UOC:DAJ, V24, ZAJ	1

END OF FIGURE

1 —[2] 6 —[7]

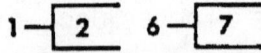

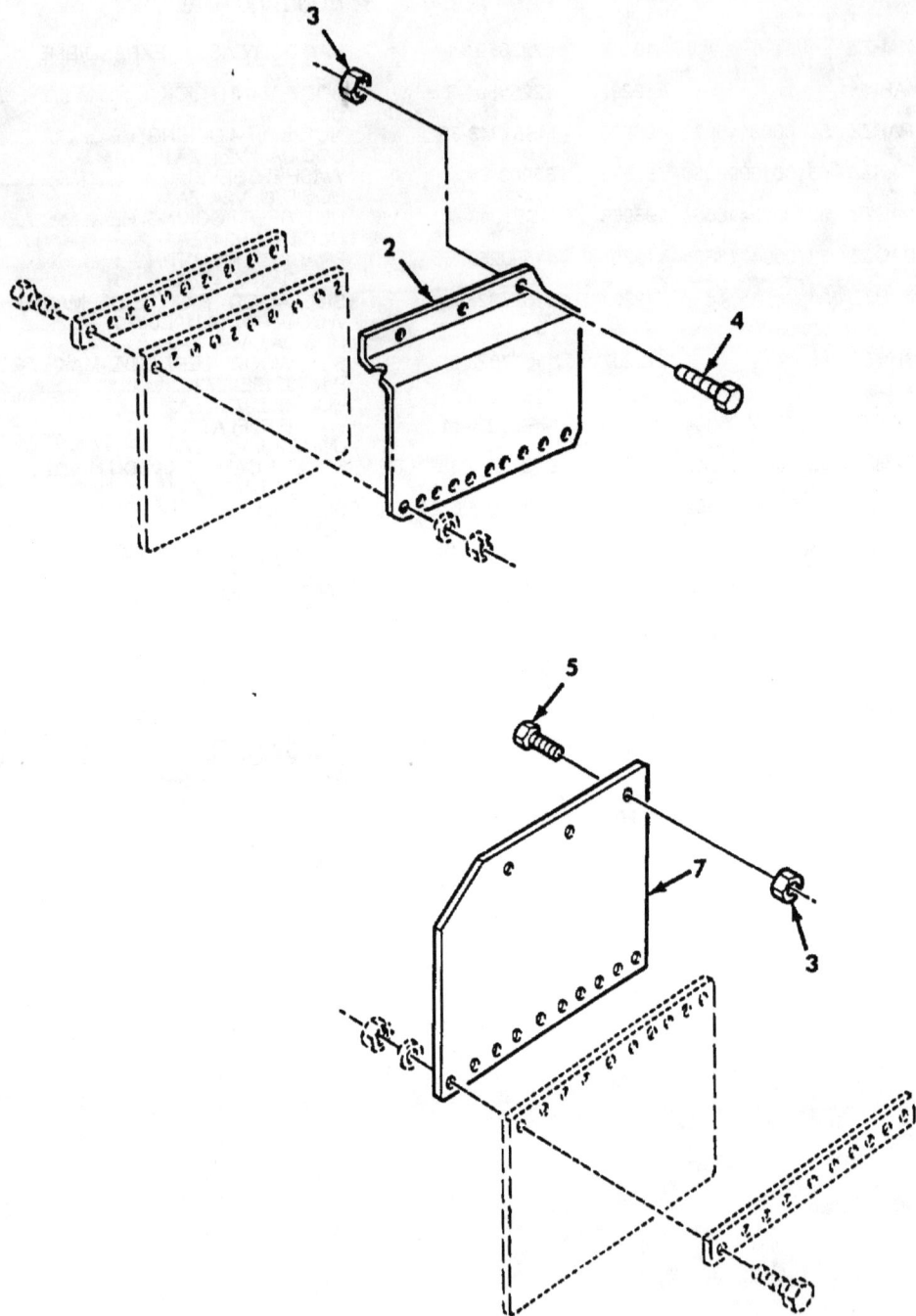

Figure 401. Van Body Splash Guard Assemblies.

(1) ITEM NO	(2) SMR CODE	(3) NSN	(4) CAGEC	(5) PART NUMBER	(6) DESCRIPTION AND USABLE ON CODES (UOC)	(7) QTY
					GROUP 1812 SPECIAL PURPOSE BODIES	
					FIG. 401 VAN BODY SPLASH GUARD ASSEMBLIES	
1	PFOZZ	2540011587169	19207	11677681	GUARD, SPLASH, VEHICU FORWARD REAR, ... LEFT SIDE.. UOC:DAJ, V24, ZAJ	1
1	PFOZZ	2540011587170	19207	11677718	GUARD, SPLASH, VEHICU FORWARD REAR, ... RIGHT HAND .... UOC:DAJ, V24, ZAJ	1
2	PFOZZ	2540011607915	19207	11677726	.GUARD, SPLASH, VEHICU LEFT SIDE .... UOC:DAJ, V24, ZAJ	1
2	PFOZZ	2540011587092	19207	11677724 '	.GUARD, SPLASH, VEHICU RIGHT SIDE .... UOC:DAJ, V24, ZAJ	1
3	PAOZZ	5310008140672	96906	MS51943-36	.NUT, SELF-LOCKING, HE .... UOC:DAJ, V24, ZAJ	12
5	PAOZZ	5305002693240	80204	B1821BH038F150N	SCREW, CAP, HEXAGON H REAR WHEELS .... UOC:DAJ, V24, ZAJ	6
6	AOOOZ	2540004050212	19207	11593275-2	GUARD, SPLASH, VEHICU REAR WHEELS, ... LEFT SIDE .... UOC:DAJ, V24, ZAJ	1
6	PFOZZ	2540011587171	19207	11677678	GUARD, SPLASH, VEHICU REAR WHEELS, ... RIGHT SIDE .... UOC:DAJ, V24, ZAJ	1
7	PFOZZ	2540011104057	19207	11593277	.GUARD, SPLASH, VEHICU LEFT SIDE .... UOC:DAJ, V24, ZAJ	1
7	PFOZZ	2540011587093	19207	11677679	.GUARD, SPLASH, VEHICU RIGHT SIDE .... UOC:DAJ, V24, ZAJ	1

END OF FIGURE

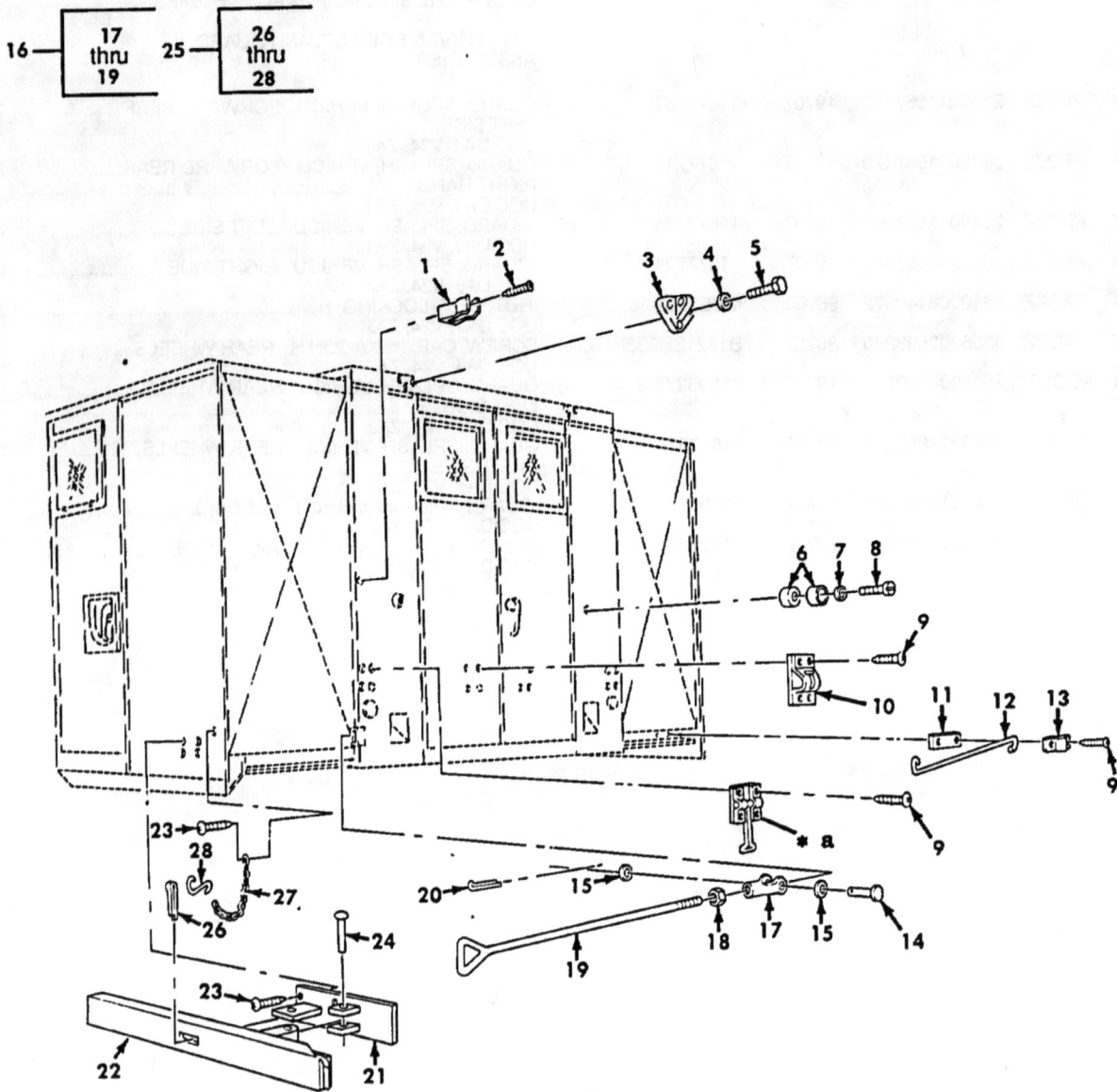

Figure 402. Van Side Lock Handle, Lifting Brackets, Rod Retainer, and Related Parts.

(1) ITEM NO	(2) SMR CODE	(3) NSN	(4) CAGEC	(5) PART NUMBER	(6) DESCRIPTION AND USABLE ON CODES (UOC)	(7) QTY

GROUP 1812 SPECIAL PURPOSE BODIES

FIG. 402 VAN SIDE LOCK HANDLE, LIFTING BRACKETS, ROD RETAINER, AND RELATED PARTS

(1) ITEM NO	(2) SMR CODE	(3) NSN	(4) CAGEC	(5) PART NUMBER	(6) DESCRIPTION AND USABLE ON CODES (UOC)	(7) QTY
1	PAOZZ	5340006998463	19207	7539185	CLIP, SPRING TENSION .... UOC:DAJ, V24, ZAJ	4
2	PAOZZ	5305004324163	96906	MS51861-24	SCREW, TAPPING .... UOC:DAJ, V24, ZAJ	4
3	PAOZZ	5340002310210	19207	8380499	PAD EYE .... UOC:DAJ, V24, ZAJ	2
4	PAOZZ	5310008206653	96906	MS35338-50	WASHER, LOCK .... UOC:DAJ, V24, ZAJ	6
5	PAOZZ	5305007245910	96906	MS90725-162	SCREW, CAP, HEXAGON H .... UOC:DAJ, V24, ZAJ	6
6	PAOZZ	5340007663330	19207	7535643	BUMPER, NONMETALLIC .... UOC:DAJ, V24, ZAJ	4
7	PAOZZ	5310000814219	96906	MS27183-12	WASHER, FLAT .... UOC:DAJ, V24, ZAJ	2
8	PAOZZ	5305011441625	96906	MS51851-112	SCREW, TAPPING .... UOC:DAJ, V24, ZAJ	2
9	PAOZZ	5305004324203	96906	MS51861-47	SCREW, TAPPING .... UOC:DAJ, V24, ZAJ	24
10	PAOZZ	5340003021840	19207	8376986	HOLDER, DOOR .... UOC:DAJ, V24, ZAJ	2
11	PAOZZ	5365004219697	19207	7084772	SPACER, PLATE .... UOC:DAJ, V24, ZAJ	4
12	PAOZZ	5340004213990	19207	8735435	HOLDER, DOOR .... UOC:DAJ, V24, ZAJ	4
13	PAOZZ	5340001583889	19207	8735425	STRAP, RETAINING .... UOC:DAJ, V24, ZAJ	4
14	PAOZZ	5315009572399	96906	MS20392-5C35	PIN, STRAIGHT, HEADED .... UOC:DAJ, V24, ZAJ	4
15	PAOZZ	5310000877493	96906	MS27183-13	WASHER, FLAT .... UOC:DAJ, V24, ZAJ	8
16	PBOOO	2540004172758	19207	10937607	HANDLE ASSEMBLY, LOC VAN BODY REAR DOOR .... UOC:DAJ, V24, ZAJ	4
17	PAOZZ	5340009850823	96906	MS35812-4	.CLEVIS, ROD END .... UOC:DAJ, V24, ZAJ	1
18	PAOZZ	5310007320559	96906	MS51968-8	.NUT, PLAIN, HEXAGON .... UOC:DAJ, V24, ZAJ	1
19	PAOZZ	5340011819445	19207	10937560	.CONTROL ROD .... UOC:DAJ, V24, ZAJ	1
20	PAOZZ	5315008423044	96906	MS24665-283	PIN, COTTER .... UOC:DAJ, V24, ZAJ	4
21	PFOZZ	2510004213944	19207	10937568	BASE, SIDE LOCK, VAN .... UOC:DAJ, V24, ZAJ	4
22	PAOZZ	2540004172722	19207	11607374	HANDLE ASSEMBLY .... UOC:DAJ, V24, ZAJ	4
23	PAOZZ	5305004324254	96906	MS51861-69	SCREW, TAPPING	16

(1) ITEM NO	(2) SMR CODE	(3) NSN	(4) CAGEC	(5) PART NUMBER	(6) DESCRIPTION AND USABLE ON CODES (UOC)	(7) QTY
					UOC:DAJ, V24, ZAJ	
24	PFOZZ	5320004180985	19207	10937579-1	RIVET, SOLID	4
					UOC:DAJ, V24, ZAJ	
25	PAOOO	4010004457212	19207	10937632	CHAIN ASSEMBLY, SING	4
					UOC:DAJ, V24, ZAJ	
26	PAOZZ	5315002341848	96906	MS24665-629	.PIN, COTTER	1
					UOC:DAJ, V24, ZAJ	
27	MOOZZ		16003	C43974 X-8	.CHAIN MAKE FROM CHAIN, P/N C43974, ... 8 INCHES LONG...	1
					UOC:DAJ, V24, ZAJ	
28	PAOZZ	4030005144420	21450	593416	.HOOK, CHAIN, S	1
					UOC:DAJ, V24, ZAJ	

END OF FIGURE

J

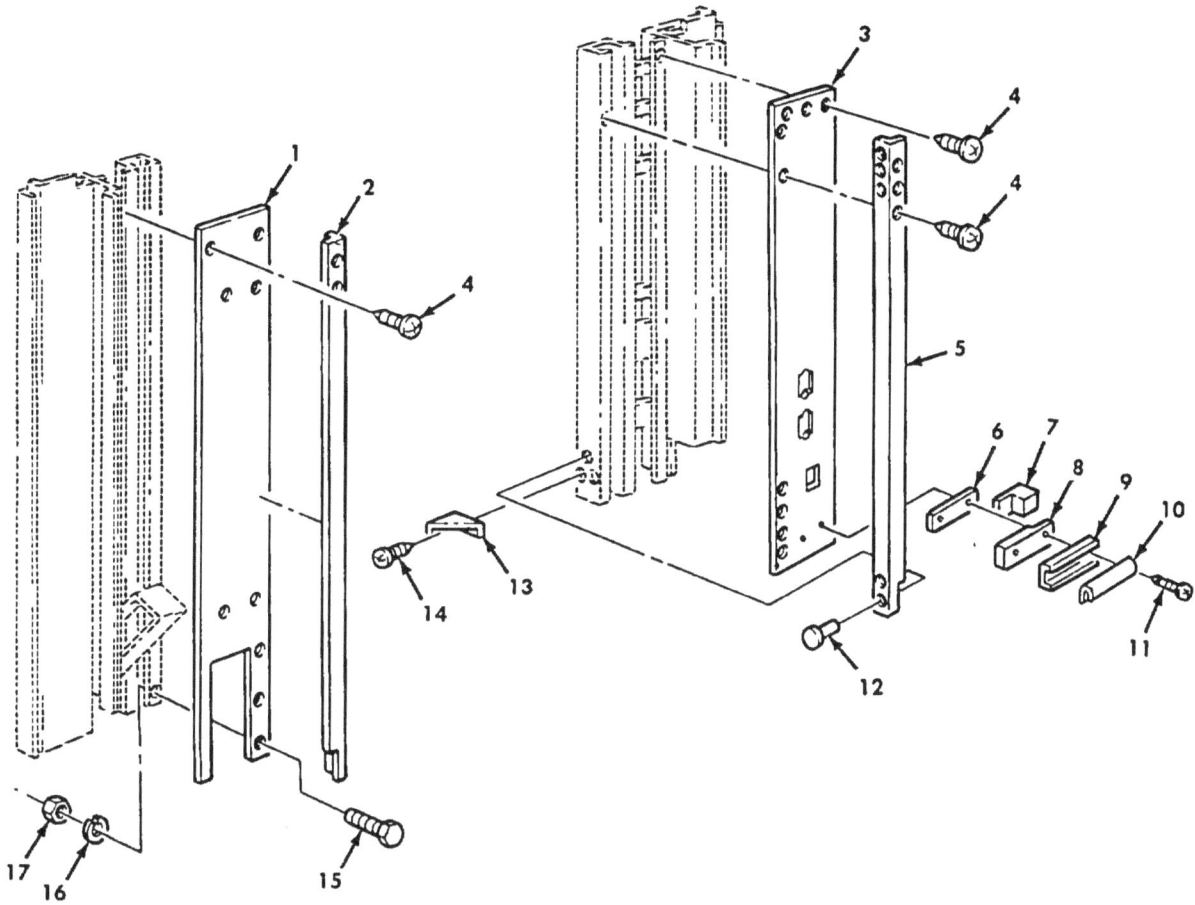

Figure 403. Van Rear Wall Panels and Mounting Hardware.

(1) ITEM NO	(2) SMR CODE	(3) NSN	(4) CAGEC	(5) PART NUMBER	(6) DESCRIPTION AND USABLE ON CODES (UOC)	(7) QTY
					GROUP 1812 SPECIAL PURPOSE BODIES	
					FIG. 403 VAN REAR WALL PANELS AND MOUNTING HARDWARE	
1	PFHZZ	2510011208448	19207	11677712	PANEL, VAN BODY .... UOC:DAJ, V24, ZAJ	1
2	PBHZZ	2590011863722	19207	8380482	BEZEL, AUTOMOTIVE TR LEFT REAR WALL .... UOC:DAJ, V24, ZAJ	1
3	PFHZZ	2510011208449	19207	11677680	PANEL, BODY, VEHICULA .... UOC:DAJ, V24, ZAJ	1
4	PAHZZ	5305004333685	96906	MS51861-37C	SCREW, TAPPING .... UOC:DAJ, V24, ZAJ	78
5	PBHZZ	2590011869585	19207	8380462	BEZEL, AUTOMOTIVE TR RIGHT REAR WALL ...... UOC:DAJ, V24, ZAJ	1
6	MOFZZ		19207	7535591	SPACER .... UOC:DAJ, V24, ZAJ	4
7	PFHZZ	5330011586289	19207	7535590	SEAL, NONMETALLIC SP .... UOC:DAJ, V24, ZAJ	4
8	PFHZZ	5340011946474	19207	7535593	PLATE, MENDING .... UOC:DAJ, V24, ZAJ	4
9	PFHZZ	5340011945298	19207	7535592	STRAP, RETAINING .... UOC:DAJ, V24, ZAJ	4
10	MFHZZ		91340	10608E44S-4	SEAL RUBBER STRIP MAKE FROM RUBBER, P/N 10608E44S, 4 INCHES LONG......... UOC:DAJ, V24, ZAJ	4
11	PFFZZ	5305004324172	96906	MS51861-37	SCREW, TAPPING .... UOC:DAJ, V24, ZAJ	8
12	PFFZZ	5320005841285	53551	RV200-6-2	RIVET, BLIND .... UOC:DAJ, V24, ZAJ	14
13	PAHZZ	2590011208447	19207	8380412	GUSSET, CORNER, WALL .... UOC:DAJ, V24, ZAJ	2
14	PAHZZ	5305004333711	96906	MS51861-35C	SCREW, TAPPING .... UOC:DAJ, V24, ZAJ	8
15	PAHZZ	5305007247219	80204	B1821BH063C125N	SCREW, CAP, HEXAGON H .... UOC:DAJ, V24, ZAJ	4
16	PAHZZ	5310008206653	96906	MS35338-50	WASHER, LOCK .... UOC:DAJ, V24, ZAJ	4
17	PAHZZ	5310007638920	96906	MS51967-20	NUT, PLAIN, HEXAGON .... UOC:DAJ, V24, ZAJ	4

END OF FIGURE

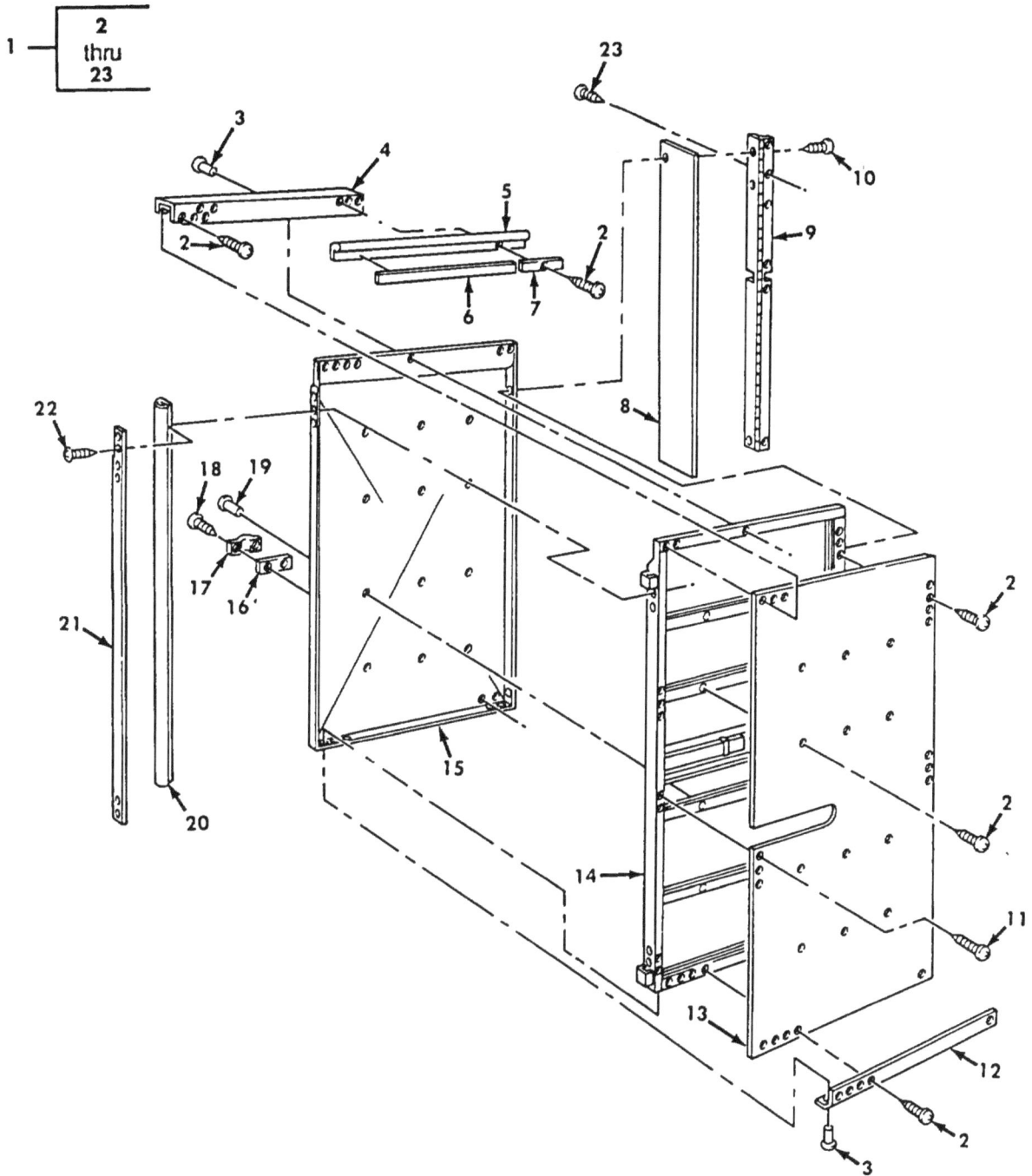

Figure 404. Van Hinged End Panel Assemblies.

(1) ITEM NO	(2) SMR CODE	(3) NSN	(4) CAGEC	(5) PART NUMBER	(6) DESCRIPTION AND USABLE ON CODES (UOC)	(7) QTY
					GROUP 1812 SPECIAL PURPOSE BODIES	
					FIG. 404 VAN HINGED END PANEL ASSEMBLIES	
1	PAFFF	2510011018358	19207	11672543-2	DOOR, VEHICULAR RIGHT SIDE .... UOC:DAJ, V24, ZAJ	2
1	PAFFF	2510011018359	19207	11672543-1	DOOR ASSEMBLY END P LEFT SIDE PANEL.................... UOC:DAJ, V24, ZAJ	2
2	PAFFF	5305004324172	96906	MS51861-37	.SCREW, TAPPING .... UOC:DAJ, V24, ZAJ	60
3	PFFZZ	5320005823268	53551	RV200-6-3	.RIVET, BLIND .... UOC:DAJ, V24, ZAJ	18
4	PFFZZ	3040011787087	19207	7084771	.BRACKET, EYE, ROTATIN RIGHT SIDE DOOR.. UOC:DAJ, V24, ZAJ	1
4	PFFZZ	2510011787086	19207	7084985	.CHANNEL, RETAINER LEFT DOOR .... UOC:DAJ, V24, ZAJ	1
5	MFFZZ		19207	10937683-2X35	.NONMETALLIC SPECIAL MAKE FROM .... WEATHERSEAL, P/N 10937683-2, 35 INCHES LONG.. UOC:UOC:DAJ, V24, ZAJ	1
6	PFFZZ	2510011844765	19207	10937693-2	.RETAINER DOOR SEAL, UPPER .... UOC:DAJ, V24, ZAJ	1
7	PAFZZ	2510011974200	19207	10937693-3	.RETAINER .... UOC:DAJ, V24, ZAJ	1
8	MFFZZ		19207	8380420-78	.RUBBER STRIP .... UOC:DAJ, V24, ZAJ	1
9	PAFZZ	5340000413126	19207	7084988	.HINGE, BUTT .... UOC:DAJ, V24, ZAJ	1
10	PAFZZ	5305004324253	96906	MS51861-67	.SCREW, TAPPING .... UOC:DAJ, V24, ZAJ	16
11	PAFZZ	5305004324201	96906	MS51861-45	.SCREW, TAPPING .... UOC:DAJ, V24, ZAJ	18
12	PFFZZ	5340011783734	19207	7084770	.BRACKET, ANGLE LOWER RIGHT PANEL .... UOC:DAJ, V24, ZAJ	1
12	PFFZZ	5340011863505	19207	7084984	.BRACKET, ANGLE LEFT LOWER HINGE .... PANEL.. UOC:DAJ, V24, ZAJ	1
13	PFFZZ	2510011212541	19207	7084768	.PANEL, BODY, VEHICULA INNER RIGHT.......... DOOR.. UOC:DAJ, V24, ZAJ	1
13	PFFZZ	2510012047704	19207	7084769	.DOOR, VEHICULAR LEFT INNER HINGED........ DOOR.. UOC:DAJ, V24, ZAJ	1
14	XAFZZ		19207	12300967-2	.FRAME ASSEMBLY RIGHT HINGE PANEL .... UOC:DAJ, V24, ZAJ	1
14	XAFZZ		19207	12300967-1	.FRAME ASSEMBLY LEFT HINGE PANEL .... UOC:DAJ, V24, ZAJ	1
15	PFFZZ	5670011774460	19207	12300936-2	.DOOR FRAME, METAL RIGHT SIDE .... UOC:DAJ, V24, ZAJ	1
15	XDFZZ		19207	12300936-1	.DOOR FRAME, METAL LEFT SIDE	1

(1) ITEM NO	(2) SMR CODE	(3) NSN	(4) CAGEC	(5) PART NUMBER	(6) DESCRIPTION AND USABLE ON CODES (UOC)	(7) QTY
					UOC:DAJ, V24, ZAJ	
16	PAFZZ	5365004219697	19207	7084772	.SPACER, PLATE	1
					UOC:DAJ, V24, ZAJ	
17	PAOZZ	5340001583889	19207	8735425	.STRAP, RETAINING	1
					UOC:DAJ, V24, ZAJ	
18	PAOZZ	5305004324203	96906	MS51861-47	.SCREW, TAPPING	2
					UOC:DAJ, V24, ZAJ	
19	PFFZZ	5320005823302	96906	MS20600AD5W2	.RIVET, BLIND	12
					UOC:DAJ, V24, ZAJ	
20	MFFZZ		91340	10608E44S BULK	.SEAL RUBBER STRIP MAKE FROM................ RUBBER, P/N 10608E44S, 78 INCHES LONG	1
					UOC:DAJ, V24, ZAJ	
21	PAFZZ	2590011801006	19207	8380421	.BRACKET, VEHICULAR C	1
					UOC:DAJ, V24, ZAJ	
22	PAFZZ	5305004324173	96906	MS51861-15	.SCREW, TAPPING	27
					UOC:DAJ, V24, ZAJ	
23	PAFZZ	5305004770144	96906	MS51861-68	.SCREW, TAPPING	16
					UOC:DAJ, V24, ZAJ	

END OF FIGURE

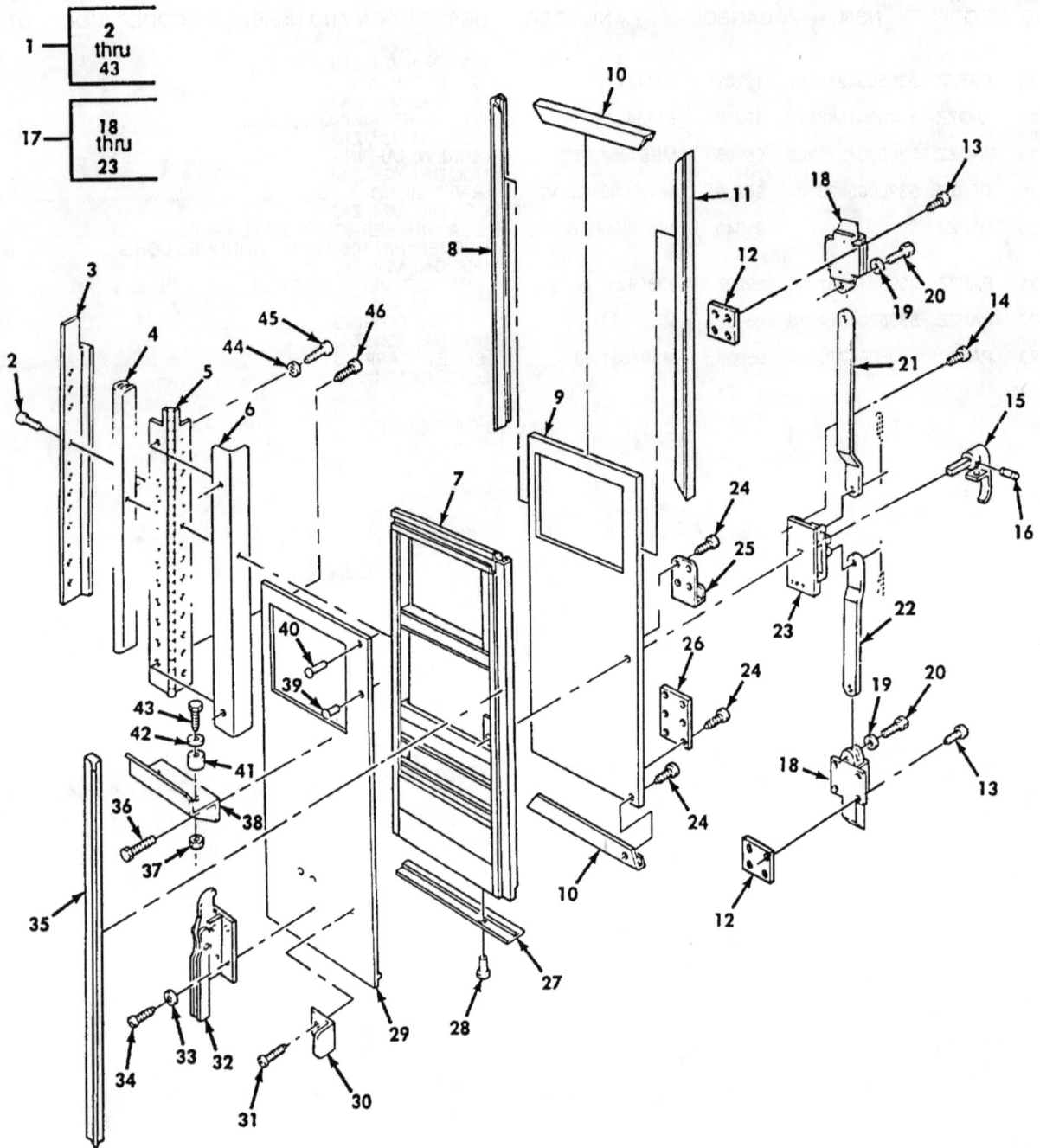

Figure 405. Van Body Door Assembly, Left Rear.

(1) ITEM NO	(2) SMR CODE	(3) NSN	(4) CAGEC	(5) PART NUMBER	(6) DESCRIPTION AND USABLE ON CODES (UOC)	(7) QTY
					GROUP 1812 SPECIAL PURPOSE BODIES	
					FIG. 405 VAN BODY DOOR ASSEMBLY, LEFT REAR	
1	PFFFF	2510011431267	19207	11607385-1	DOOR ASSEMBLY .... UOC:DAK, V25, ZAK	1
1	PFFFF	2510004051970	19207	11607385	DOOR, VEHICULAR .... UOC:DAJ, V24, ZAJ	1
2	PAFZZ	5305004770144	96906	MS51861-68	.SCREW, TAPPING .... UOC:DAJ, DAK, V24, V25, ZAJ, ZAK	16
3	XBFZZ		19207	7535627	.RETAINER, SEAL, VEHIC .... UOC:DAJ, DAK, V24, V25, ZAJ, ZAK	1
4	MFFZZ		19207	11607302-78	.SEAL, NONMETALLIC SP MAKE FROM SEAL, P/N 11607302, 78 INCHES LONG .... UOC:DAJ, DAK, V24, V25, ZAJ, ZAK	1
5	PAOZZ	5340001584077	19207	7084990	.HINGE, BUTT .... UOC:DAJ, DAK, V24, V25, ZAJ, ZAK	1
6	MOOZZ		19207	8380420-77	.RUBBER STRIP MAKE FROM RUBBER, P/N 8380420, 77 INCHES LONG .... UOC:DAJ, DAK, V24, V25, ZAJ, ZAK	1
7	XAOZZ		19207	11607388	.FRAME ASSEMBLY .... UOC:DAJ, DAK, V24, V25, ZAJ, ZAK	1
8	PBFZZ	2590011794911	19207	11607395	.BEZEL, AUTOMOTIVE TR LEFT REAR .... DOOR ASSY UOC:DAJ, DAK, V24, V25, ZAJ, ZAK	1
9	PBFZZ	2510011620564	19207	11607392	.PANEL, BODY, VEHICULA INNER .... UOC:DAJ, DAK, V24, V25, ZAJ, ZAK	1
10	PBFZZ	2590004172735	19207	8380444	.BEZEL, AUTOMOTIVE TR LEFT REAR........... UOC:DAJ, DAK, V24, V25, ZAJ, ZAK	2
11	PBFZZ	2540009241296	19207	8380443	.MOULDINGXREAR DOOR .... UOC:DAJ, DAK, V24, V25, ZAJ, ZAK	1
12	PFOZZ	5365012199172	19207	12300732	.SPACER, PLATE .... UOC:DAJ, DAK, V24, V25, ZAJ, ZAK	5
13	PAOZZ	5305004324205	96906	MS51861-49	.SCREW, TAPPING .... UOC:DAJ, DAK, V24, V25, ZAJ, ZAK	8
14	PAOZZ	5305004324203	96906	MS51861-47	.SCREW, TAPPING .... UOC:DAJ, DAK, V24, V25, ZAJ, ZAK	4
15	PAOZZ	2540008097793	19207	11607269	.HANDLE, DOOR, VEHICUL .... UOC:DAJ, DAK, V24, V25, ZAJ, ZAK	1
16	PAOZZ	5315008662673	96906	MS35677-48	.PIN, GROOVED, HEADLES .... UOC:DAJ, DAK, V24, V25, ZAJ, ZAK	1
17	PBFZZ	2540002310207	19207	11607265-4	.LOCK ASSEMBLY REAR DOOR ASSY .... UOC:DAJ, DAK, V24, V25, ZAJ, ZAK	1
18	PAOZZ	5340008390098	19207	7748911	..BOLT, FLUSH .... UOC:DAJ, DAK, V24, V25, ZAJ, ZAK	2
19	PAOZZ	5310006379541	96906	MS35338-46	..WASHER, LOCK .... UOC:DAJ, DAK, V24, V25, ZAJ, ZAK	2
20	PAFZZ	5305002693234	96906	MS90727-58	..SCREW, CAP, HEXAGON .... UOC:DAJ, DAK, V24, V25, ZAJ, ZAK	2
21	PBOZZ	3040011978555	19207	8722186-21	..CONNECTING LINK, RIG .... UOC:DAJ, DAK, V24, V25, ZAJ, ZAK	1
22	PBOZZ	3040011978556	19207	8722186-22	..CONNECTING LINK, RIG .... UOC:DAJ, DAK, V24, V25, ZAJ, ZAK	1

(1) ITEM NO	(2) SMR CODE	(3) NSN	(4) CAGEC	(5) PART NUMBER	(6) DESCRIPTION AND USABLE ON CODES (UOC)	(7) QTY
					UOC:DAJ, V24, ZAJ	
23	PBOZZ	2510011662016	19207	11592573-2	..CENTER CASE	1
					UOC:DAJ, V24, ZAJ	
24	PAOZZ	5305004324172	96906	MS51861-37	.SCREW, TAPPING	45
					UOC:DAJ, V24, ZAJ	
25	PAOZZ	5340002310216	19207	7084860	.CLIP, RATCHET, WRENCH	1
					UOC:DAJ, V24, ZAJ	
26	PAOZZ	2590004059771	19207	7084840	.BRACKET ASSEMBLY, WR	1
					UOC:DAJ, V24, ZAJ	
27	PBOZZ	5330011062067	19207	10937627-1	.SEAL, BUMPER	1
					UOC:DAJ, V24, ZAJ	
28	PAOZZ	5320005823276	96906	MS20600AD6W4	.RIVET, BLIND	7
					UOC:DAJ, V24, ZAJ	
29	XAFZZ	2510011787046	19207	10937619	.SKIN, DOOR, LH REAR	1
					UOC:DAJ, V24, ZAJ	
30	PAOZZ	5340004199464	19207	7084861	.BRACKET, ANGL E	2
					UOC:DAJ, V24, ZAJ,	
31	PAOZZ	5305004830554	96906	MS51862-26	.SCREW, TAPPING	4
					UOC:DAJ, V24, ZAJ	
32	PAOZZ	2540002310200	19207	7084828	.CLAMP, HOLD DOWN	1
					UOC:DAJ, V24, ZAJ	
33	PAOZZ	5310005825965	96906	MS35338-44	.WASHER, LOCK	4
					UOC:DAJ, V24, ZAJ	
34	PAOZZ	5305009881724	96906	MS35206-280	.SCREW, M ACHINE	4
					UOC:DAJ, V24, ZAJ	
35	PBFZZ	9390012859623	19207	12368265	.NONMETALLIC SPECIAL	1
					UOC:DAJ, V24, ZAJ	
36	PAOZZ	5305004291552	21450	171104	.SCREW, TAPPIN G	4
					UOC:DAJ, V24, ZAJ	
37	PAOZZ	5310000617325	96906	MS21045-4	.NUT, SEL-LOCKING, HE	2
					UOC:DAJ, V24, ZAJ	
38	PAOZZ	2510002317444	19207	8759465	.RACK, LADDER, VEHICLE	1
					UOC:DAJ, V24, ZAJ	
39	PAOZZ	5320005841285	53551	RV200-6-2	.RIVET, BLIND	39
					UOC:DAJ, V24, ZAJ	
40	PAOZZ	5320002421582	96906	MS20470-A6-4	.RIVET, SOLID	13
					UOC:DAJ, V24, ZAJ	
41	PAOZZ	5365002309682	19207	7535620-1	.BUSHING, NONMETALLIC	2
					UOC:DAJ, V24, ZAJ	
42	PAOZZ	5310008094058	96906	MS27183-10	.WASHER, FLAT	2
					UOC:DAJ, V24, ZAJ,	
43	PAOZZ	5305002678958	80204	B1821BH025F200N	.SCREW, CAP, HEXAGON H	2
					UOC:DAJ, V24, ZAJ	
44	PAOZZ	5310005825965	96906	MS35338-44	WASHER, LOCK	2
					UOC:DAJ, V24, ZAJ	
45	PAOZZ	5305004324252	96906	MS51861-66	SCREW, TAPPING	2
					UOC:DAJ, V24, ZAJ	
46	PAOZZ	5305004770144	96906	MS51861-68	SCREW, TAPPING	15
					UOC:DAJ, V24, ZAJ	

END OF FIGURE

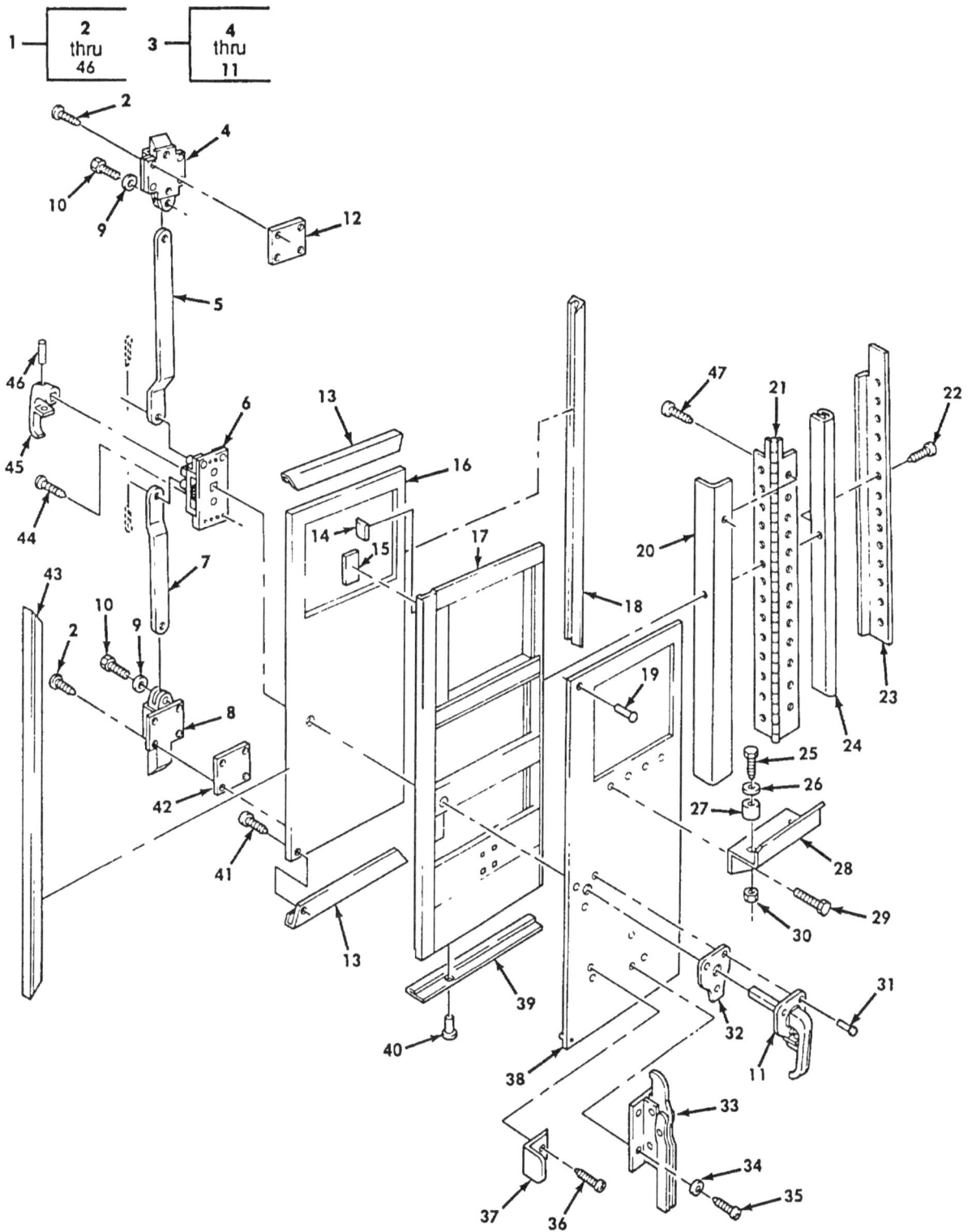

Figure 406. Van Body Door Assembly, Right Rear.

(1) ITEM NO	(2) SMR CODE	(3) NSN	(4) CAGEC	(5) PART NUMBER	(6) DESCRIPTION AND USABLE ON CODES (UOC)	(7) QTY
					GROUP 1812 SPECIAL PURPOSE BODIES	
					FIG. 406 VAN BODY DOOR ASSEMBLY, RIGHT REAR	
1	PAFFF	5670011873639	19207	12300923-1	DOOR, METAL, SWINGING .... UOC:DAJ, V24, ZAJ	1
2	PAOZZ	5305004324205	96906	MS51861-49	.SCREW, TAPPING .... UOC:DAJ, V24, ZAJ	6
3	PAOZZ	2540002310206	19207	11607265-3	.LOCK ASSEMBLY .... UOC:DAJ, V24, ZAJ	1
4	PAOZZ	5340001786080	19207	11607402	..BOLT, EXIT .... UOC:DAJ, V24, ZAJ	1
5	PAOZZ	3040011978555	19207	8722186-21	..CONNECTING LINK, RIG .... UOC:DAJ, V24, ZAJ	1
6	PAOZZ	2510011794084	19207	11592573-1	..CENTER CASE .... UOC:DAJ, V24, ZAJ	1
7	PFOZZ	3040011978556	19207	8722186-22	..CONNECTING LINK, RIG .... UOC:DAJ, V24, ZAJ	1
8	PAOZZ	5340008390098	19207	7748911	..BOLT, FLUSH .... UOC:DAJ, V24, ZAJ	1
9	PAOZZ	5310006379541	96906	MS35338-46	..WASHER, LOCK .... UOC:DAJ, V24, ZAJ	2
10	PAOZZ	5305002693234	80204	B1821BH038F075N	..SCREW, CAP, HEXAGON H .... UOC:DAJ, V24, ZAJ	2
11	PFOZZ	2540002872571	19207	7264749	..HANDLE, DOOR, VEHICUL .... UOC:DAJ, V24, ZAJ	1
12	PAOZZ	5365012399381	19207	12300731	.SPACER, PLATE .... UOC:DAJ, V24, ZAJ	3
13	PBFZZ	2590009241425	19207	8380445	.BEZEL, AUTOMOTIVE TR TOP AND BOTTOM DOOR .... UOC:DAJ, V24, ZAJ	2
14	PAOZZ	5330004151481	19207	11607263	.GASKET .... UOC:DAJ, V24, ZAJ	2
15	PAOZZ	5330002433571	19207	11607262	.GASKET .... UOC:DAJ, V24, ZAJ	2
16	XAFZZ		19207	11607393-1	.PANEL .... UOC:DAJ, V24, ZAJ	1
17	XAFZZ		19207	12300914	.FRAME ASSEMBLY .... UOC:DAJ, V24, ZAJ	1
18	PAFZZ	2540009241296	19207	8380443	.MOULDINGXREAR DOOR .... UOC:DAJ, V24, ZAJ	1
19	PAOZZ	5320002421582	96906	MS20470A6-4	.RIVET, SOLID .... UOC:DAJ, V24, ZAJ	52
20	MOOZZ		19207	8380420-77	.RUBBER STRIP MAKE FROM RUBBER STRIP, P/N 8380420, 77 INCHES LONG............................ UOC:DAJ, V24, ZAJ	1
21	PAOZZ	5340001584077	19207	7084990	.HINGE, BUTT .... UOC:DAJ, V24, ZAJ	1
22	PAOZZ	5305004770144	96906	MS51861-68	.SCREW, TAPPING .... UOC:DAJ, V24, ZAJ	16

(1) ITEM NO	(2) SMR CODE	(3) NSN	(4) CAGEC	(5) PART NUMBER	(6) DESCRIPTION AND USABLE ON CODES (UOC)	(7) QTY
23	PAOZZ	2510004051946	19207	7535564	.RETAINER, SEAL, DOOR .... UOC:DAJ, V24, ZAJ	1
24	MOOZZ		19207	11607302-78	.SEAL, NONMETALLIC SP MAKE FROM .... RUBBER, P/N 11607302, 78 INCHES LONG UOC:DAJ, V24, ZAJ	1
25	PAOZZ	5305002678958	80204	B1821BH025F200N	.SCREW, CAP, HEXAGON H .... UOC:DAJ, V24, ZAJ	2
26	PAOZZ	5310008094059	96906	MS27183-10	.WASHER, FLAT .... UOC:DAJ, V24, ZAJ	2
27	PAOZZ	5340007666336	19207	7535620	.BUMPER, NONMETALLIC .... UOC:DAJ, V24, ZAJ	2
28	PAOZZ	2510002317444	19207	8759465	.RACK, LADDER, VEHICLE .... UOC:DAJ, V24, ZAJ	1
29	PAOZZ	5305004291552	21450	171104	.SCREW, TAPPING .... UOC:DAJ, V24, ZAJ	4
30	PAOZZ	5310000617325	96906	MS21045-4	.NUT, SELF-LOCKING, HE .... UOC:DAJ, V24, ZAJ	2
31	PAOZZ	5320010682340	96906	MS20600-MP8W/4	.RIVET, BLIND .... UOC:DAJ, V24, ZAJ	3
32	PAOZZ	5330004146695	19207	11592566	.GASKET .... UOC:DAJ, V24, ZAJ	1
33	PAOZZ	2540002310200	19207	7084828	.CLAMP, HOLD DOWN .... UOC:DAJ, V24, ZAJ	1
34	PAOZZ	5310005825965	96906	MS35338-44	.WASHER, LOCK .... UOC:DAJ, V24, ZAJ	4
35	PAOZZ	5305009881724	96906	MS35206-280	.SCREW, MACHINE .... UOC:DAJ, V24, ZAJ	4
36	PAOZZ	5305004830554	96906	MS51862-26	.SCREW, TAPPING .... UOC:DAJ, V24, ZAJ	4
37	PAOZZ	5340004199464	19207	7084861	.BRACKET, ANGLE .... UOC:DAJ, V24, ZAJ	2
38	XAFZZ		19207	12300920-XA	.SKIN, DOOR, RIGHT REA .... UOC:DAJ, V24, ZAJ	1
39	PFOZZ	9515011245055	19207	10937627-2	.STRIP, METAL .... UOC:DAJ, V24, ZAJ	1
40	PAOZZ	5320005823276	96906	MS20600AD6W4	.RIVET, BLIND .... UOC:DAJ, V24, ZAJ	7
41	PAFZZ	5305004324172	96906	MS51861-37	.SCREW, TAPPING .... UOC:DAJ, V24, ZAJ	34
42	PBOZZ	5365012199172	19207	12300732	.SPACER, PLATE .... UOC:DAJ, V24, ZAJ	2
43	PBFZZ	2590011780759	19207	11607394	.BEZEL, AUTOMOTIVE TR .... UOC:DAJ, V24, ZAJ	1
44	PAOZZ	5305004324203	96906	MS51861-47	.SCREW, TAPPING .... UOC:DAJ, V24, ZAJ	4
45	PAOZZ	2540008097796	19207	10882484	.HANDL E, DOOR, VEHICUL .... UOC:DAJ, V24, ZAJ	1
46	PAOZZ	5315008662673	96906	MS35677-48	.PIN, GROOVED, HEADLES .... UOC:DAJ, V24, ZAJ	1
47	PAFZZ	5305004770144	96906	MS51861-68	SCREW, TAPPING .... UOC:DAJ, V24, ZAJ	16

END OF FIGURE

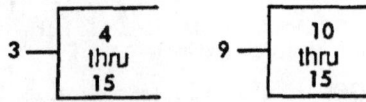

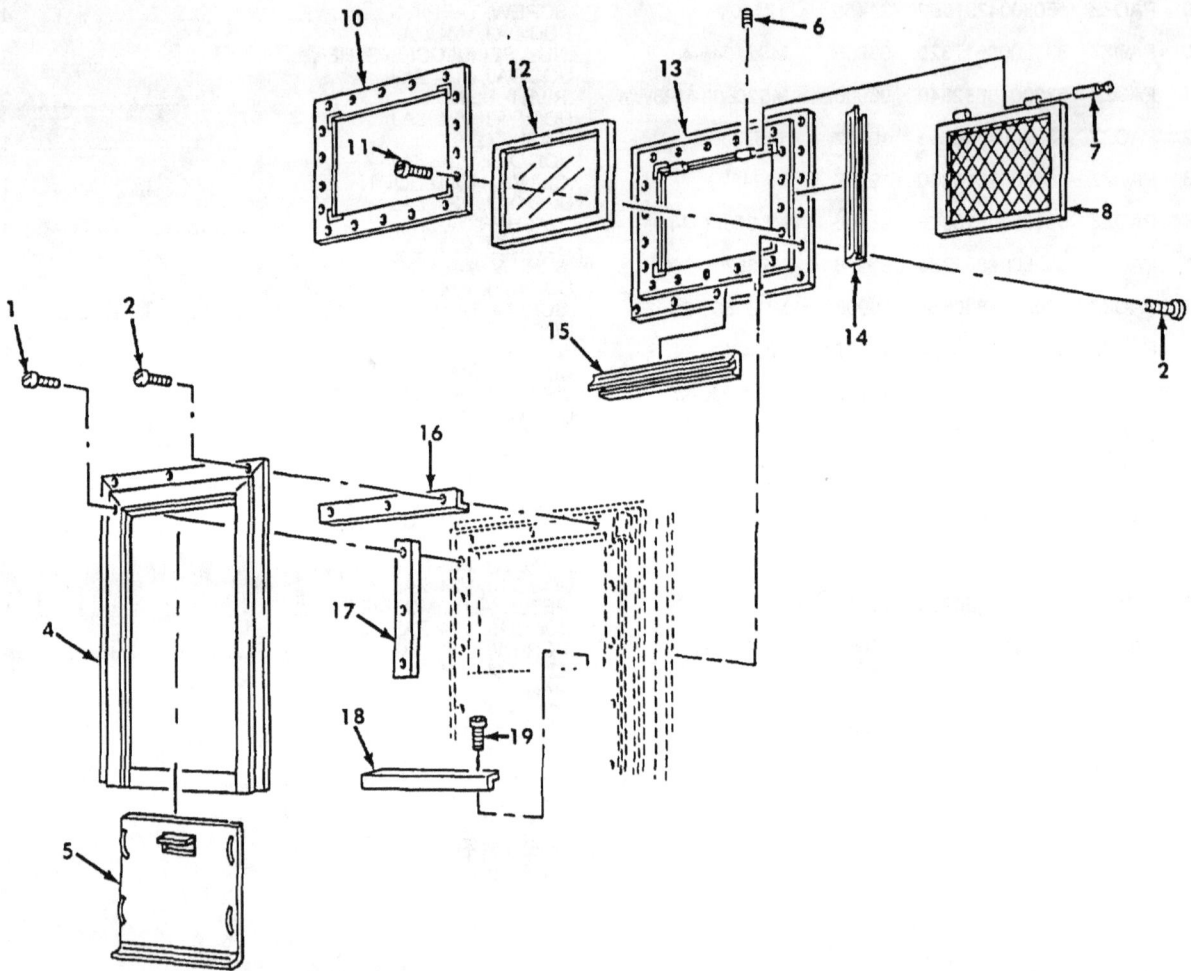

Figure 407. Van Sash Assembly, Rear Doors.

(1) ITEM NO	(2) SMR CODE	(3) NSN	(4) CAGEC	(5) PART NUMBER	(6) DESCRIPTION AND USABLE ON CODES (UOC)	(7) QTY
					GROUP 1812 SPECIAL PURPOSE BODIES	
					FIG. 407 VAN SASH ASSEMBLY, REAR DOORS	
1	PAOZZ	5305004324172	96906	MS51861-37	SCREW, TAPPING .... UOC: DAJ, V24, ZAJ	18
2	PAOZZ	5305004324203	96906	MS51861-47	SCREW, TAPPING .... UOC: DAJ, V24, ZAJ	18
3	PAOOO	2510002351888	19207	7047096	WINDOW SASH VEHICUL SIDE DOOR ASSY .... LEFT AND RIGHT .... UOC: DAJ, V24, ZAJ	1
4	PFOZZ	2590011787043	19554	13521G2	.FRAME, BLACKOUT .... UOC: DAJ, V24, , ZAJ	1
5	PAOZZ	2510011794083	19207	10896815	.BLACKOUT, PANEL .... UOC: DAJ, V24, ZAJ	1
6	PAOZZ	5305007195342	96906	MS51963-34	.SETSCREW .... UOC: DAJ, V24, ZAJ	2
7	PAOZZ	5315011053318	19207	10896789	.PIN, GROOVED, HEADLES .... UOC: DAJ, V24, ZAJ	2
8	PFOZZ	5410011285529	19207	10896813	.SCREEN, WINDOW, METAL .... UOC: DAJ, V24, ZAJ	1
9	XAOZZ		19207	10896816-1	.WINDOW, VEHICULAR .... UOC: DAJ, V24, ZAJ	1
10	XAOZZ		19207	10896883	..FRAME .... UOC: DAJ, V24, ZAJ	1
11	PAOZZ	5305009012135	96906	MS35493-52	..SCREW, WOOD .... UOC: DAJ, V24, ZAJ	18
12	PFOZZ	2510008098046	19207	10896799-1	..WINDOW, VEHICULAR .... UOC: DAJ, V24, ZAJ	1
13	XAOZZ	2510011617675	19207	10896881	..FRAME .... UOC: DAJ, V24, ZAJ	1
14	PFOZZ	5680011771525	19207	10896885-2	..WEATHER STRIP .... UOC: DAJ, V24, ZAJ	2
15	PFOZZ	5680011771526	19207	10896885-1	..WEATHER STRIP .... UOC: DAJ, V24, ZAJ	2
16	PFOZZ	2510010243618	19207	7084792	TRIM, PLYWOOD .... UOC: DAJ, V24, ZAJ	1
17	PFOZZ	2510010222580	19207	7084794	TRIM, PLYWOOD .... UOC: DAJ, V24, ZAJ	2
18	PAOZZ	2510010243619	19207	7084793	TRIM, PLYWOOD .... UOC: DAJ, V24, ZAJ	1
19	PAOZZ	5305004324172	96906	MS51861-37	SCREW, TAPPING SIDE DOOR SASH .... UOC: DAJ, V24, ZAJ	3

END OF FIGURE

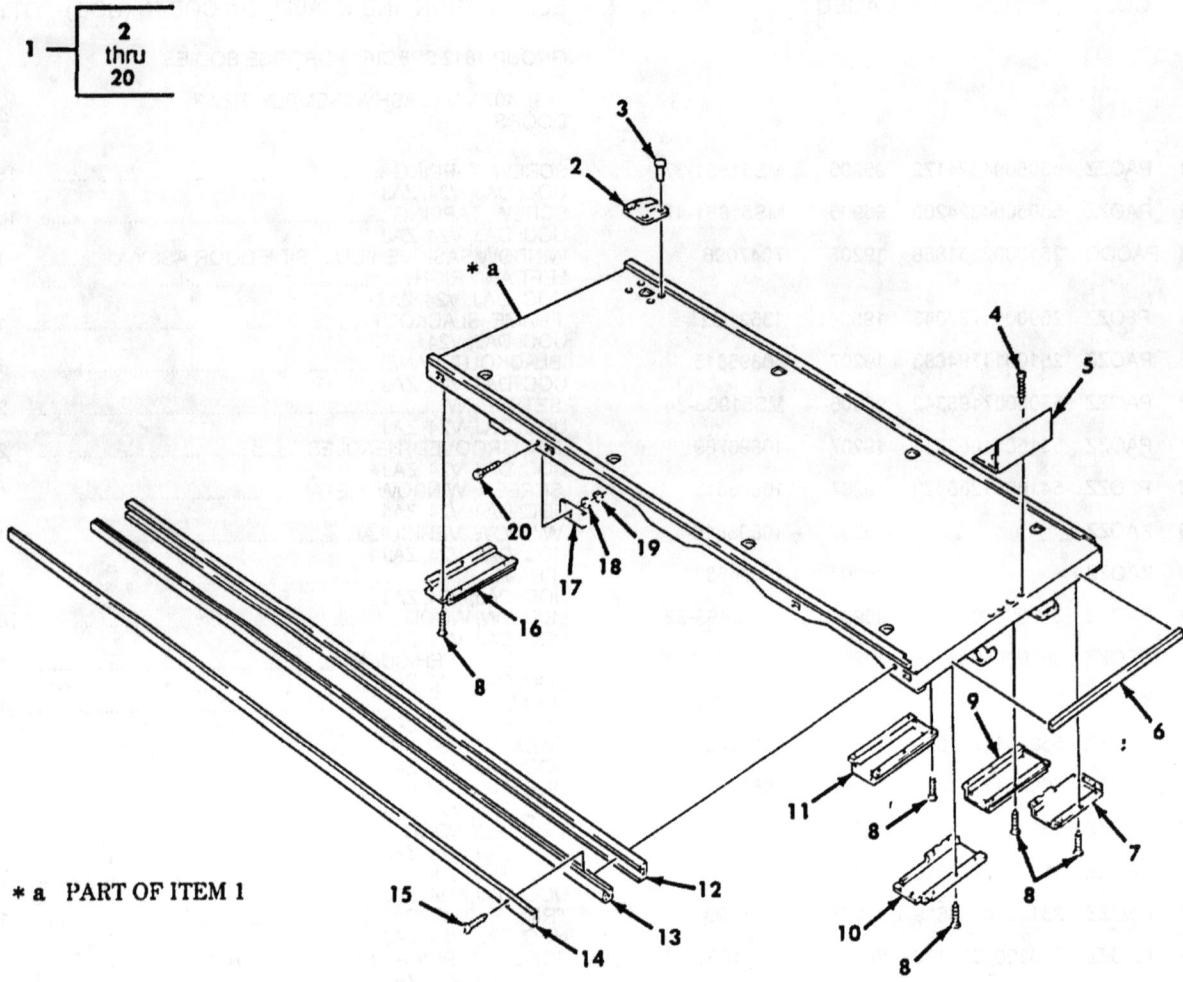

*a   PART OF ITEM 1

Figure 408. Van Underframe Assembly.

(1) ITEM NO	(2) SMR CODE	(3) NSN	(4) CAGEC	(5) PART NUMBER	(6) DESCRIPTION AND USABLE ON CODES (UOC)	(7) QTY
					GROUP 1812 SPECIAL PURPOSE BODIES	
					FIG. 408 VAN UNDERFRAME ASSEMBLY	
1	PFHHH	2510011048966	19207	11677754-9	FRAME, STRUCTURAL, VE .... UOC: DAJ, V24, ZAJ	1
1	PFHHH		19207	12300727-2	EDGING, LINOLEUM .... UOC: DAJ, V24, ZAJ	1
2	PBHZZ	3990011769359	19207	11607401	.TIE DOWN, CARGO, VEHI .... UOC: DAJ, V24, ZAJ	8
3	PAHZZ	5320002313663	96906	MS24661-226	.RIVET, BLIND .... UOC: DAJ, V24, ZAJ	32
4	PAHZZ	5305002690770	96906	MS51862-56C	.SCREW, TAPPING .... UOC: DAJ, V24, ZAJ	4
5	PFHZZ	5340004199484	19207	8735437	.BRACKET, ANGLE .... UOC: DAJ, V24, ZAJ	1
6	PAHZZ	5330001523217	19207	7373291-3	.SEAL, NONMETALLIC SP .... UOC: DAJ, V24, ZAJ	1
7	PFHZZ	5340012269161	19207	8735038-1	.COVER, ACCESS RIGHT SIDE UNDERFRAME ...... UOC: DAJ, V24, ZAJ	1
8	PAHZZ	5305004324201	96906	MS51861-45	.SCREW, TAPPING .... UOC: DAJ, V24, ZAJ	108
9	PFHZZ	2510004172789	19207	10944429-2	.COVER, BOX, MECHANISM .... UOC: DAJ, V24, ZAJ	1
10	PBHZZ	5340004195860	19207	8735038	.COVER, ACCESS LEFT SIDE .... UOC: DAJ, V24, ZAJ	1
11	PFHZZ	5340004172788	19207	10944429-1	.COVER, ACCESS .... UOC: DAJ, V24, ZAJ	4
12	PFHZZ	2590011471517	19207	8380424	.BEZEL, AUTOMOTIVE TR .... UOC: DAJ, V24, ZAJ	2
13	MHHZZ		91340	10608E44S BULK	.SEAL RUBBER STRIP MAKE FROM RUBBER, P/N 10608E44S, 206 INCHES LONG UOC: DAJ, V24, ZAJ	1
14	PBHZZ	2510011801004	19207	8380423	.RETAINER, FRAME, SEAL .... UOC: DAJ, V24, ZAJ	2
15	PAHZZ	5305004324170	96906	MS51861.-35	.SCREW, TAPPING .... UOC: DAJ, V24, ZAJ	138
16	PFHZZ	5340004195859	19207	875037	.COVER, ACCESS .... UOC: DAJ, V24, ZAJ	8
17	PAHZZ	2510002317465	19207	8734990	.PLATE, STOP .... UOC: DAJ, V24, ZAJ	4
18	PAHZZ	5310005825965	96906	MS35338-44	.WASHER, LOCK .... UOC: DAJ, V24, ZAJ	8
19	PAHZZ	5310007616882	96906	MS51967-2	.NUT, PLAIN, HEXAGON .... UOC: DAJ, V24, ZAJ	8
20	PAHZZ	5305000680502	96906	MS90725-6	.SCREW, CAP, HEXAGON H .... UOC: DAJ, V24, ZAJ	8

END OF FIGURE

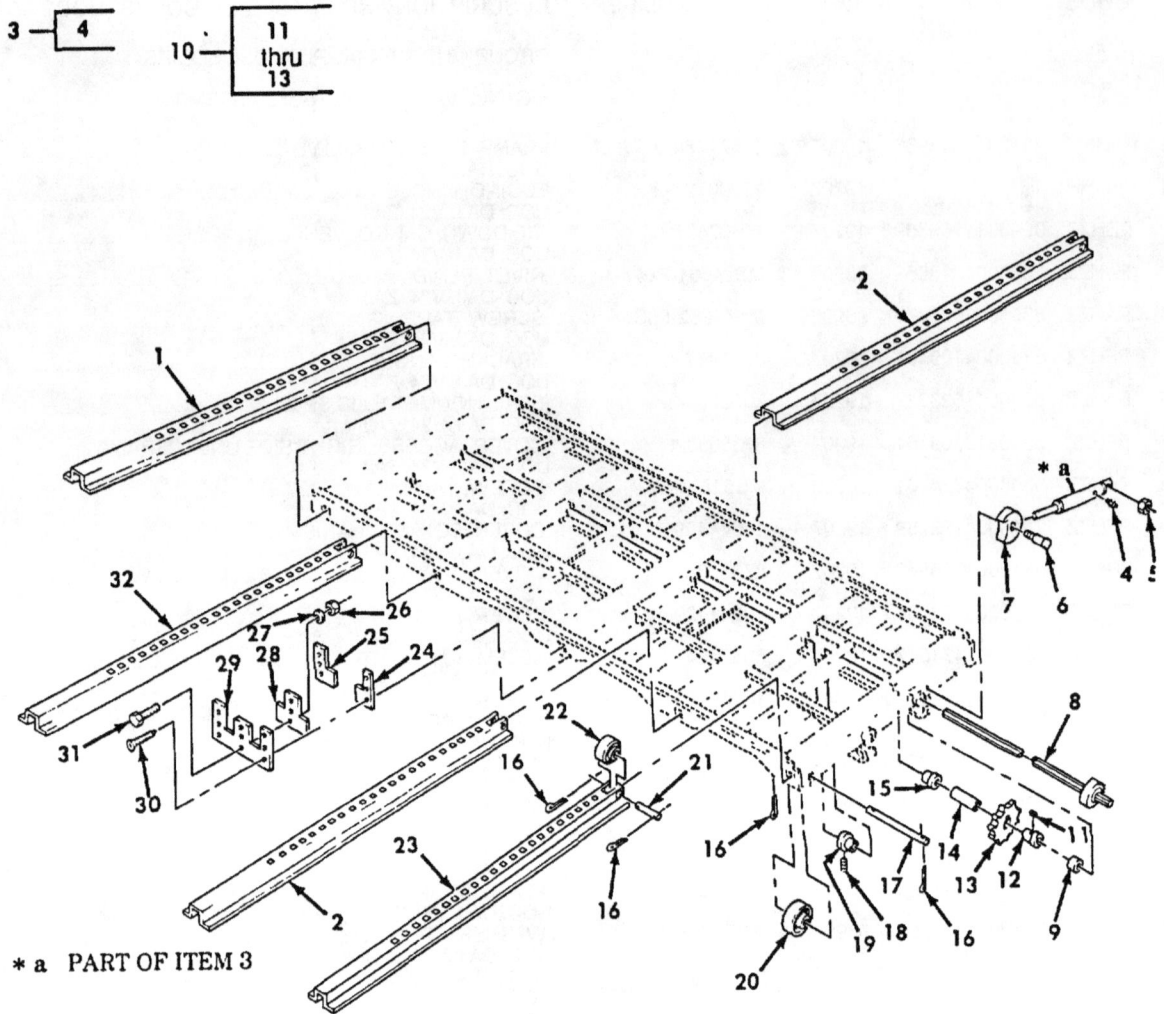

Figure 409. Van Retractable Beams, Ratchet Shaft, and Rollers.

(1) ITEM NO	(2) SMR CODE	(3) NSN	(4) CAGEC	(5) PART NUMBER	(6) DESCRIPTION AND USABLE ON CODES (UOC)	(7) QTY
					GROUP 1812 SPECIAL PURPOSE BODIES	
					FIG. 409 VAN RETRACTABLE BEAMS, RATCHET SHAFT, AND ROLLERS	
1	PFHZZ	2510012251001	19207	8735018-1	BEAM ASSEMBLY, BODY LEFT FRONT FRAME .... UOC: DAJ, V24, ZAJ	2
2	PFHZZ	2510011634903	19207	8735027	FRAME SECTION, STRUC RIGHT SIDE .... RETRACTABLE .... UOC: DAJ, V24, ZAJ	4
3	PFHHH	5340007992218	19207	7084725	PLUNGER, DETENT .... UOC :DAJ, V24, ZAJ	2
4	PAHZZ	4730000504203	96906	MS15001-1	.FITTING, LUBRICATION .... UOC: DAJ, V24, ZAJ	1
5	PAHZZ	5310000881251	96906	MS51922-1	NUT, SELF-LOCKING, HE .... UOC: DAJ, V24, ZAJ	2
6	PAHZZ	5305009144171	96906	MS51975-28	SCREW, SHOULDER .... UOC: DAJ, V24, ZAJ	2
7	PAHZZ	3040002310212	19207	8735056	PAWL .... UOC: DAJ, V24, ZAJ	2
8	PAHZZ	2520004151479	19207	8735022	SHAFT, AND CLUTCH HA .... UOC: DAJ, V24, ZAJ	2
9	PBHZZ	5365004551382	19207	8735024	SPACER, SLEEVE .... UOC: DAJ, V24, ZAJ	10
10	PAHHH	2590002317484	19207	8735074	SPROCKET ASSEMBLY .... UOC: DAJ, V24, ZAJ	10
11	PAHZZ	5315012172269	19207	7535631	.KEY, MACHINE .... UOC: DAJ, V24, ZAJ	1
12	PFHZZ	3120012166699	19207	8735026	.BUSHING, SLEEVE .... UOC: DAJ, V24, ZAJ	1
13	PFHZZ	3020011617710	19207	8735025	.GEAR, SPUR .... UOC: DAJ, V24, ZAJ	1
14	PAHZZ	5365002990067	19207	8735028	SPACER, SLEEVE .... UOC: DAJ, V24, ZAJ	10
15	PAHZZ	3120007663327	19207	8735057	BUSHING, SLEEVE .... UOC: DAJ, V24, ZAJ	18
16	PAHZZ	5315000590206	96906	MS24665-491	PIN, COTTER .... UOC: DAJ, V24, ZAJ	40
17	PAHZZ	3040002310221	19207	8735029	SHAFT, STRAIGHT .... UOC: DAJ, V24, ZAJ	10
18	PAHZZ	5305007263091	96906	MS51965-43	SETSCREW .... UOC: DAJ, V24, ZAJ	8
19	PAHZZ	3040002317470	19207	8735023	COLLAR, SHAFT .... UOC: DAJ, V24, ZAJ	8
20	PAHZZ	2510006791733	19207	7047098	ROLLER, SHAFT .... UOC: DAJ, V24, ZAJ	10
21	PAHZZ	3040004199431	19207	8735030	SHAFT, STRAIGHT .... UOC: DAJ, V24, ZAJ	10
22	PAHZZ	2510006791420	19207	7047097	ROLLER, BODY FRAME B .... UOC: DAJ, V24, ZAJ	10
23	PFHZZ	2590012264587	19207	8735423-1	BRACKET, VEHICULAR C LEFT REAR FRAME .... UOC: DAJ, V24, ZAJ	2

(1) ITEM NO	(2) SMR CODE	(3) NSN	(4) CAGEC	(5) PART NUMBER	(6) DESCRIPTION AND USABLE ON CODES (UOC)	(7) QTY
24	PAHZZ	5330004702115	19207	8735035	SEAL, NONMETALLIC ST .... UOC: DAJ, V24, ZAJ	2
25	PAHZZ	5330004199468	19207	8735034	GASKET .... UOC: DAJ, V24, ZAJ	10
26	PAHZZ	5310000617326	96906	MS21045-3	NUT, SELF-LOCKING, HE .... UOC: DAJ, V24, ZAJ	20
27	PAHZZ	5310001670765	88044	AN970-3	WASHER, FLAT .... UOC: DAJ, V24, ZAJ	20
28	PAHZZ	5330004199469	19207	8735036	GASKET .... UOC: DAJ, V24, ZAJ	10
29	PFHZZ	2510004172736	19207	8735033	RETAINER, SEAL, BEAM LOWER LEFT PANEL .... UOC: DAJ, V24, ZAJ	10
30	PAHZZ	5305004324201	96906	MS51861-45	SCREW, TAPPING .... UOC: DAJ, V24, ZAJ	70
31	PAHZZ	5305010062052	96906	MS51849-65	SCREW, MACHINE .... UOC: DAJ, V24, ZAJ	20
32	PBHZZ	9520011634902	19207	10872414	CHANNEL, STRUCTURAL LEFT SIDE FRAME .... UOC: DAJ, V24, ZAJ	2

END OF FIGURE

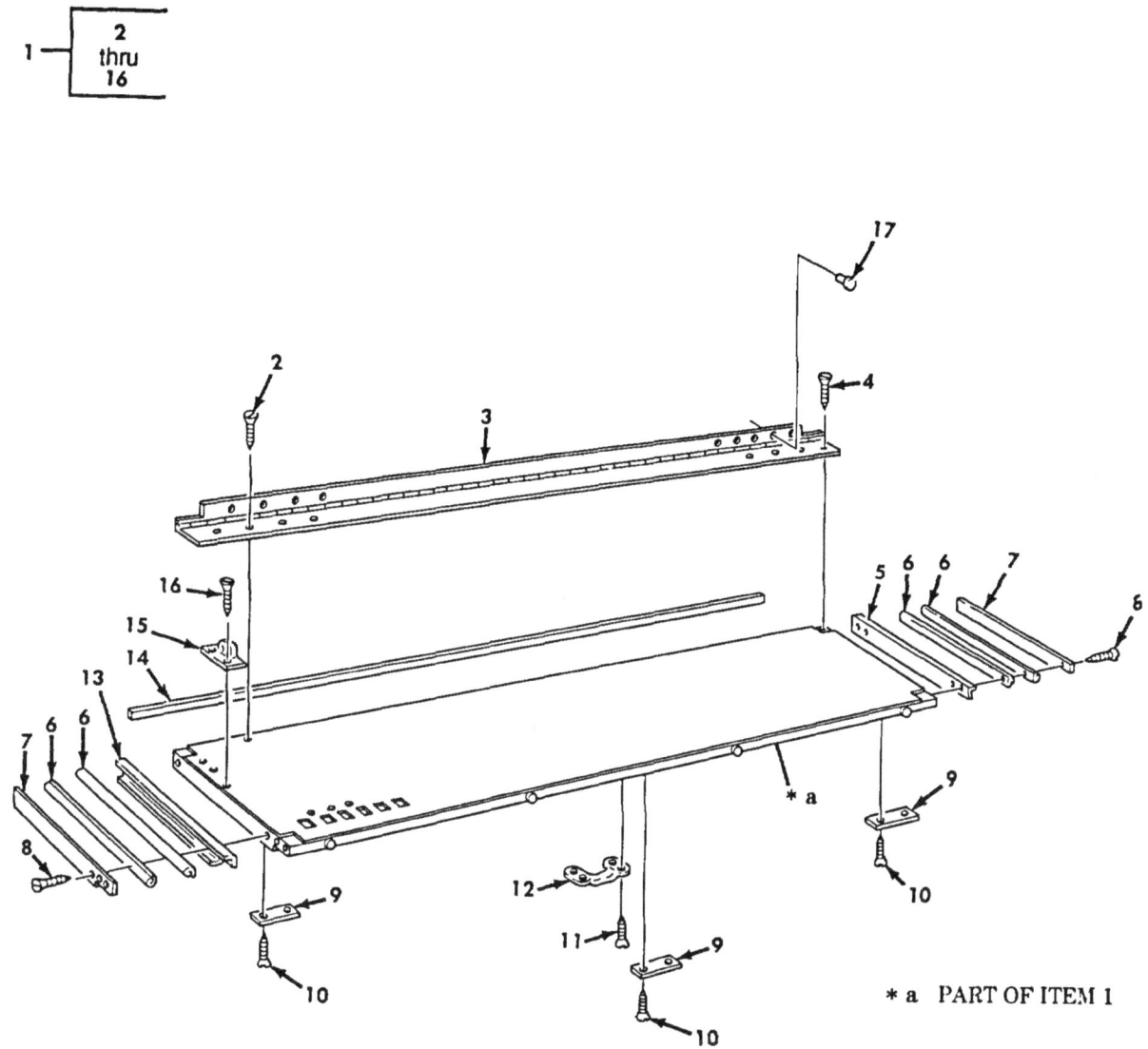

Figure 410. Van Hinged Floor Assembly.

* a   PART OF ITEM 1

(1) ITEM NO	(2) SMR CODE	(3) NSN	(4) CAGEC	(5) PART NUMBER	(6) DESCRIPTION AND USABLE ON CODES (UOC)	(7) QTY
					GROUP 1812 SPECIAL PURPOSE BODIES	
					FIG. 410 VAN HINGED FLOOR ASSEMBLY	
1	PFHHH	9515012859855	19207	11677745-5	PLATE, FLOOR, METAL LEFT	1
					UOC: DAJ, V24, ZAJ	
1	PFHHH	9515012859856	19207	11677745-6	PLATE, FLOOR, METAL RIGHT	1
					UOC: DAJ, V24, ZAJ	
2	PAHZZ	5305000527494	96906	MS24629-63	.SCREW, TAPPING	55
					UOC: DAJ, V24, ZAJ	
3	PAHZZ	5340004454561	19207	7045151	.HINGE, BUTT	1
					UOC: DAJ, V24, ZAJ	
4	PAHZZ	5305000526921	96906	MS24629-57	.SCREW, TAPPING	4
					UOC: DAJ, V24, ZAJ	
5	XDHZZ		19207	7084744	.CHANNEL, STRUCTURAL	1
					UOC: DAJ, V24, ZAJ	
6	MHHZZ		91340	10608E44S BULK	.SEAL RUBBER STRIP MAKE FROM...............	4
					RUBBER STRIP, P/N 10608E44S, 39	
					INCHES LONG..	
					UOC: DAJ, V24, ZAJ	
7	PAHZZ	2540002228883	19207	8380401	.RETAINER, SEAL	2
					UOC: DAJ, V24, ZAJ	
8	PAHZZ	5305004324171	96906	MS51861-36	.SCREW, TAPPING	30
					UOC: DAJ, V24, ZAJ	
9	PAHZZ	5365002317440	19207	7084876	.SPACER, PLATE	5
					UOC: DAJ, V24, ZAJ	
10	PAHZZ	5305010232428	96906	MS51862-26C	.SCREW, TAPPING	10
					UOC: DAJ, V24, ZAJ	
11	PAHZZ	5305002690770	96906	MS51862-56C	.SCREW, TAPPING	4
					UOC: DAJ, V24, ZAJ	
12	PAHZZ	5340006212591	19207	7397853	.HANDLE, BOW	1
					UOC: DAJ, V24, ZAJ	
13	XDHZZ		19207	7084744	.CHANNEL, STRUCTURAL LEFT HAND	1
					FRAME ASSEMBLY	
					UOC: DAJ, V24, ZAJ	
13	XDHZZ		19207	8380402	.MOLDING, METAL RIGHT HAND FRAME	1
					ASSEMBLY	
					UOC: DAJ, V24, ZAJ	
14	MHHZZ		19207	11607267-2-203	.RUBBER STRIP MAKE FROM RUBBER	1
					STRIP, P/N 11607267-2, 203 INCHES LONG	
					UOC: DAJ, V24, ZAJ	
15	PAHZZ	2520012218893	19207	10897028-1	.SUPPORT, COUNTER BAL	2
					UOC: DAJ, V24, ZAJ	
16	PAHZZ	5305009585258	96906	MS35190-317	.SCREW, MACHINE	4
					UOC: DAJ, V24, ZAJ	
17	PAHZZ	5320009307865	96906	MS24662-234	.RIVET, BLIND	112
					UOC: DAJ, V24, ZAJ	

END OF FIGURE

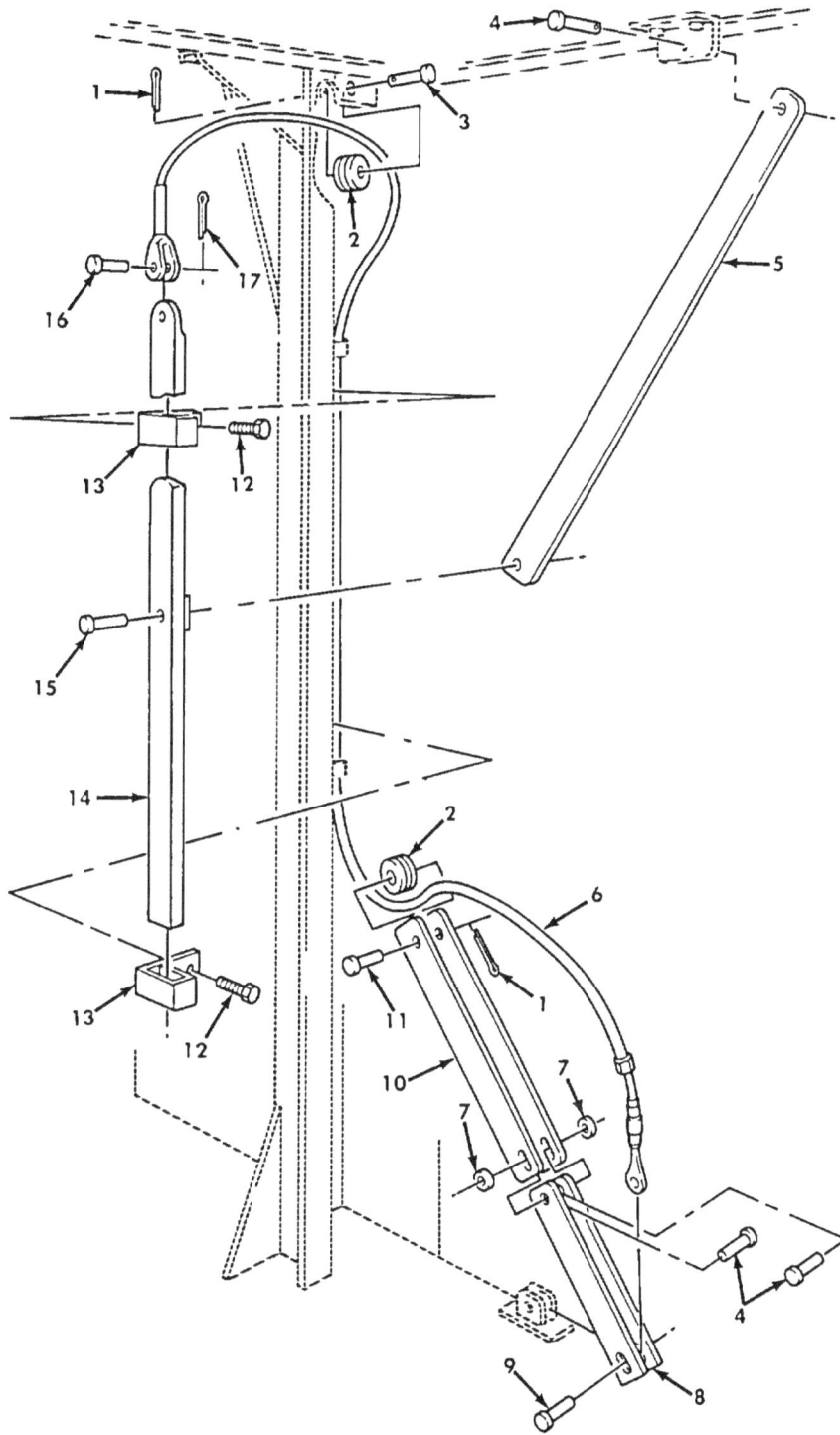

Figure 411. Van Body Counterbalance Assembly and Mounting Hardware.

(1) ITEM NO	(2) SMR CODE	(3) NSN	(4) CAGEC	(5) PART NUMBER	(6) DESCRIPTION AND USABLE ON CODES (UOC)	(7) QTY
					GROUP 1812 SPECIAL PURPOSE BODIES	
					FIG. 411 VAN BODY COUNTERBALANCE ASSEMBLY AND MOUNTING HARDWARE	
1	PAOZZ	5315008395821	96909	MS24665-351	PIN, COTTER .... UOC: DAJ, V24, ZAJ	2
2	PAOZZ	3020004217240	19207	7084969	PULLEY, GROOVE .... UOC: DAJ, V24, ZAJ	2
3	PAOZZ	5315004156294	19207	11637822-1	PIN, STRAIGHT, HEADED LOWER ARM .... UOC: DAJ, V24, ZAJ	1
4	PFOZZ	5320011945034	21450	106912	RIVET, SOLID .... UOC: DAJ, V24, ZAJ	3
5	PFOZZ	3040011774468	19207	12300938	CONNECTING LINK, RIG .... UOC: DAJ, V24, ZAJ	1
6	PAOZZ	4010012035687	19207	8342292-3	WIRE ROPE ASSEMBLY .... UOC: DAJ, V24, ZAJ	1
7	PAOZZ	5310008093078	96906	MS27183-11	WASHER, FLAT .... UOC: DAJ, V24, ZAJ	2
8	PAOZZ	2520011774462	19207	12300939	ARM, COUNTERBALANCE .... UOC: DAJ, V24, ZAJ	1
9	PFOZZ	5320011861277	96906	MS35743-60	RIVET, SOLID .... UOC: DAJ, V24, ZAJ	1
10	PAOZZ	3040011774463	19207	12300937	ARM, COUNTERBALANCE .... UOC: DAJ, V24, ZAJ	1
11	PAOZZ	5315004156294	19207	11637822-1	PIN, STRAIGHT, HEADED .... UOC: DAJ, V24, ZAJ	1
12	PAOZZ		96906	MS51862-56	SCREW, TAPPING .... UOC: DAJ, V24, ZAJ	2
13	PFOZZ	5340004151494	19207	8735084	CLIP ASSEMBLY, GUIDE .... UOC: DAJ, V24, ZAJ	2
14	PAOZZ	2510004250512	19207	7084961	BAR, COUNTER BALANCE .... UOC: DAJ, V24, ZAJ	1
15	PFOZZ	5320010696364	96906	MS35743-57	RIVET, SOLID .... UOC: DAJ, V24, ZAJ	1
16	PAOZZ	5315004175223	96906	MS35810-1	PIN, STRAIGHT, HEADED .... UOC: DAJ, V24, ZAJ	1
17	PAOZZ	5315008392325	96906	MS24665-132	PIN, COTTER .... UOC: DAJ, V24, ZAJ	1

END OF FIGURE

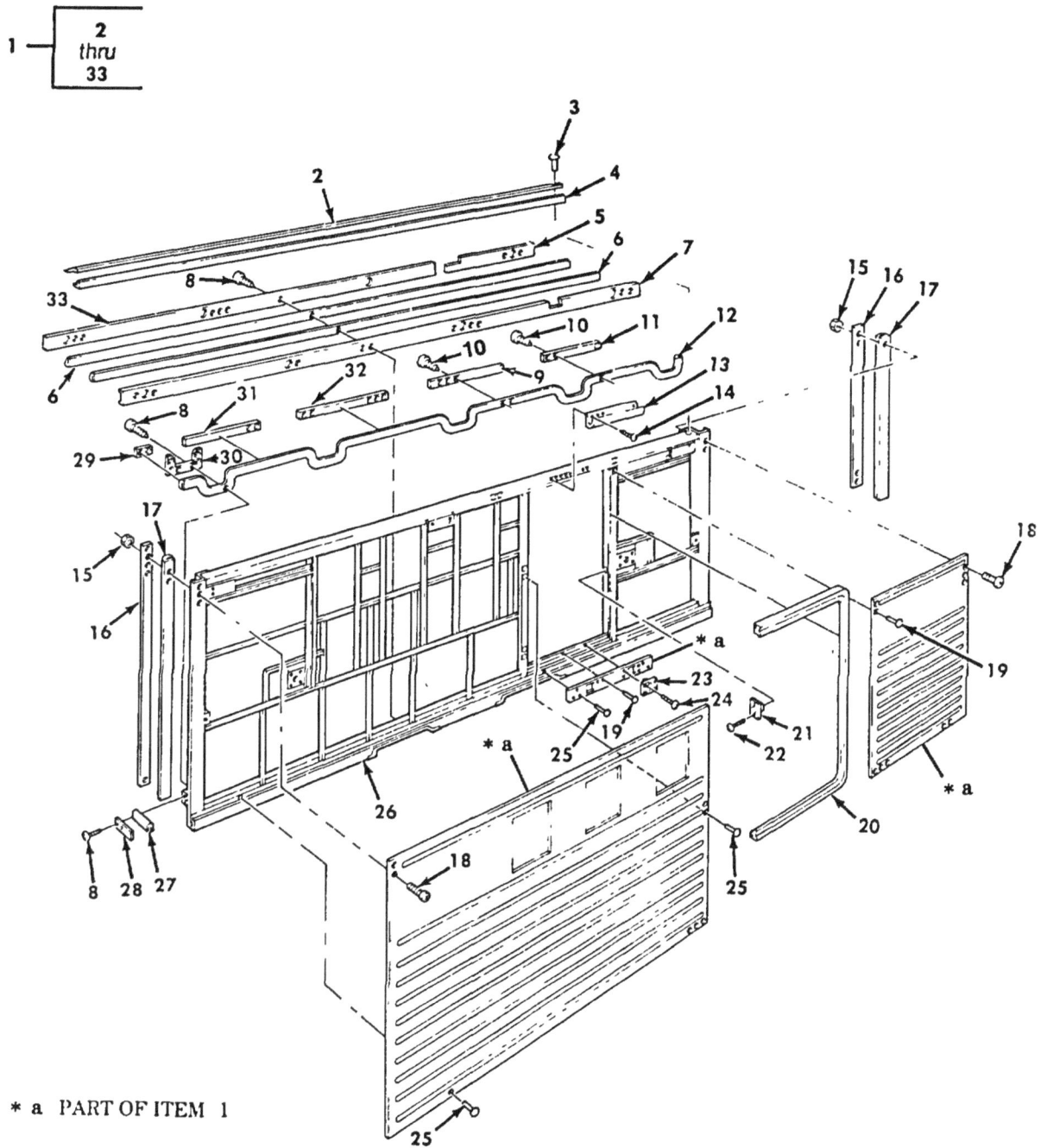

* a PART OF ITEM 1

Figure 412. Van Side Assemblies, Right and Left Side.

(1) ITEM NO	(2) SMR CODE	(3) NSN	(4) CAGEC	(5) PART NUMBER	(6) DESCRIPTION AND USABLE ON CODES (UOC)	(7) QTY
					GROUP 1812 SPECIAL PURPOSE BODIES	
					FIG. 412 VAN SIDE ASSEMBLIES, RIGHT AND LEFT SIDE	
1	MHHHH		19207	11593267	SIDE ASSEMBLY, BODY RIGHT SIDE .... UOC: DAJ, V24, ZAJ	1
1	MHHHH		19207	11611587	PANEL, BODY, VEHICULA VAN BODY .... UOC: DAJ, V24, ZAJ	1
2	PFHZZ	5340011689267	19207	11592574	.BRACKET, ANGLE .... UOC: DAJ, V24, ZAJ	1
3	PAHZZ	5320005823276	53551	RV200-6-4	.RIVET, BLIND .... UOC: DAJ, V24, ZAJ	189
4	MHHZZ		19207	10937640-204	.SEAL, NONMETALLIC AN MAKE FROM............... RUBBER, P/N 10937640, 204 INCHES LONG. UOC: DAJ, V24, ZAJ	1
5	PFHZZ	2510011787044	19207	11607334	.RETAINER, BODY SEAL .... UOC: DAJ, V24, ZAJ	1
6	MHHZZ		91340	10608E44S BULK	.SEAL RUBBER STRIP MAKE FROM............... RUBBER, P/N 10608E44S, 206 INCHES LONG UOC: DAJ, V24, ZAJ	2
7	PFHZZ	2510011801005	19207	8380425	.CHANNEL RETAINER, SE .... UOC: DAJ, V24, ZAJ	1
7	PFHZZ	2590011869584	19207	8380425-1	.BEZEL, AUTOMOTIVE TR .... UOC: DAJ, V24, ZAJ	
8	PAHZZ	5305004324170	96906	MS51861-35	.SCREW, TAPPING 	158
9	PFHZZ	5365011873590	19207	8380470	.SPACER, PLATE .... UOC: DAJ, V24, ZAJ	1
10	PAHZZ	5305004324170	96906	MS51861-35	.SCREW, TAPPING .... UOC: DAJ, V24, ZAJ	25
11	PBHZZ	5365011873591	19207	8380469	.SPACER, PLATE .... UOC: DAJ, V24, ZAJ	1
12	MHHZZ		91340	10608E44S BULK	.SEAL RUBBER STRIP MAKE FROM .... RUBBER, P/N 10608E44S, 242 INCHES LONG. UOC: DAJ, V24, ZAJ	1
13	PBOZZ	2590011810309	19207	7084960	.BEZEL, AUTOMOTIVE TR .... UOC: DAJ, V24, ZAJ	1
14	PAHZZ	5305001498610	96906	MS51862-36	.SCREW, TAPPING SIDE DOOR .... UOC: DAJ, V24, ZAJ	9
15	PAHZZ	5310009349757	96906	MS35649-282	.NUT, PLAIN, HEXAGON .... UOC: DAJ, V24, ZAJ	82
16	PBHZZ	5340011654506	19207	8380422	.STRAP, RETAINING .... UOC: DAJ, V24, ZAJ	2
17	MHHZZ		91340	10608E44S BULK	.SEAL RUBBER STRIP MAKE FROM .... RUBBER, P/N 10608E44S, 86 INCHES LONG. UOC: DAJ, V24, ZAJ	2
18	PAOZZ	5305009846191	96906	MS35206-243	.SCREW, MACHINE .... UOC: DAJ, V24, ZAJ	82

(1) ITEM NO	(2) SMR CODE	(3) NSN	(4) CAGEC	(5) PART NUMBER	(6) DESCRIPTION AND USABLE ON CODES (UOC)	(7) QTY
19	PFHZZ	5320005841285	53551	RV200-6-2	.RIVET, BLIND .... UOC: DAJ, V24, ZAJ	46
20	PAHZZ	5330011267512	19207	7373291-2	.SEAL, NONMETALLIC SP LEFT SIDE .... UOC: DAJ, V24, ZAJ	1
21	PAHZZ	2510003517851	19207	8735338	.RETAINER, LOCK .... UOC: DAJ, V24, ZAJ	1
22	PAHZZ	5305004324203	96906	MS51861-47	.SCREW, TAPPING .... UOC: DAJ, V24, ZAJ	2
23	PBOZZ	2510003517852	19207	8735349	.HANGER, LADDER, VAN B .... UOC: DAJ, V24, ZAJ	2
24	PAHZZ	5305000526874	96906	MS24627-50	.SCREW, TAPPING .... UOC: DAJ, V24, ZAJ	8
25	PFHZZ	5320005823268	53551	RV200-6-3	.RIVET, BLIND .... UOC: DAJ, V24, ZAJ	248
26	XAHZZ		19207	11611588	.FRAME ASSY, SOL SIDE LEFT SIDE .... UOC: DAJ, V24, ZAJ	1
26	XAHZZ		19207	11593266	.FRAME ASSEMBLY RIGHT SIDE .... UOC: DAJ, V24, ZAJ	1
27	MHHZZ		91340	10608E44S BULK	.SEAL RUBBER STRIP MAKE FROM .... RUBBER, P/N 10608E44S, 6 INCHES LONG...................... UOC: DAJ, V24, ZAJ	2
28	PBHZZ	5365011867247	19207	8380475	.SPACER, PLATE LEFT SIDE LOWER .... FRONT AND REAR ENDS .... UOC: DAJ, V24, ZAJ	2
29	PAHZZ	5365011313396	19207	8380474	.SPACER, PLATE LEFT SIDE .... UOC: DAJ, V24, ZAJ	2
30	PFHZZ	2510011860867	19207	8380473	.RETAINER, U-PLATE RIGHT SIDE .... UOC: DAJ, V24, ZAJ	5
31	PFOZZ	5365011863773	19207	8380472	.SPACER, PLATE RIGHT SIDE LOWER .... UOC: DAJ, V24, ZAJ	1
32	PBHZZ	5365011873589	19207	8380471	.SPACER, PLATE LEFT SIDE LOWER .... UOC: DAJ, V24, ZAJ	1
33	PFHZZ	5330011925788	19207	12301007	.GASKET RIGHT SIDE .... UOC: DAJ, V24, ZAJ	1

END OF FIGURE

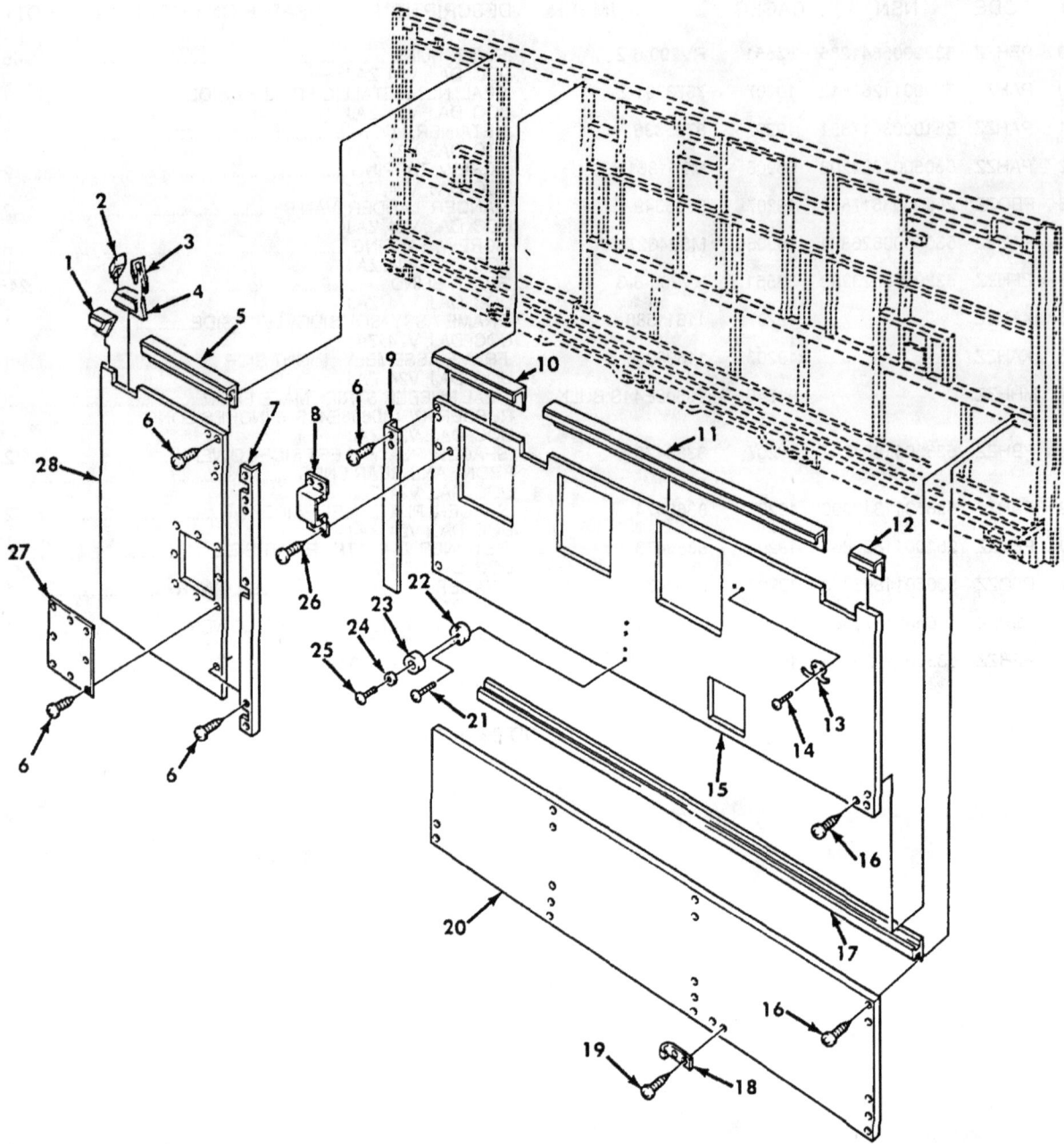

Figure 413. Van Interior Side Panels, Right and Left Side.

(1) ITEM NO	(2) SMR CODE	(3) NSN	(4) CAGEC	(5) PART NUMBER	(6) DESCRIPTION AND USABLE ON CODES (UOC)	(7) QTY
					GROUP 1812 SPECIAL PURPOSE BODIES	
					FIG. 413 VAN INTERIOR SIDE PANELS, RIGHT AND LEFT SIDE	
1	PBHZZ	2590011863723	19207	8380476	BEZEL, AUTOMOTIVE TR VAN BODY REAR PANEL UPPER RIGHT SIDE UOC: DAJ, V24, ZAJ	1
1	PBHZZ	2590011810311	19207	8380437	BEZEL, AUTOMOTIVE TR VAN BODY REAR PANEL UPPER LEFT SIDE UOC: DAJ, V24, ZAJ	1
2	PBHZZ	2590011821847	19207	8380438	BEZEL, AUTOMOTIVE TR RIGHT SIDE UOC: DAJ, V24, ZAJ	3
3	PBHZZ	2590011851278	19207	8380439	BEZEL, AUTOMOTIVE TR VAN BODY RIGHT HAND UPPER PANEL UOC: DAJ, V24, ZAJ	3
4	PBHZZ	2590011863720	19207	8380440	BEZEL, AUTOMOTIVE TR VAN BODY RIGHT HAND UPPER PANEL UOC: DAJ, V24, ZAJ	3
5	PBHZZ	2590011866096	19207	8380442	BEZEL, AUTOMOTIVE TR VAN BODY REAR PANEL UPPER LEFT SIDE UOC: DAJ, V24, ZAJ	1
6	PAHZZ	5305004324172	96906	MS51861-37	SCREW, TAPPING RIGHT SIDE UOC: DAJ, V24, ZAJ	76
7	PBHZZ	2590011942149	19207	8380479	MOLDING, METAL VAN BODY REAR PANEL RIGHT SIDE UOC: DAJ, V24, ZAJ	1
7	PBHZZ	2590011863721	19207	8380480	BEZEL, AUTOMOTIVE TR VAN BODY REAR PANEL LEFT SIDE UOC: DAJ, V24, ZAJ	1
8	PAOZZ	5975004146466	19207	7018109	JUNCTION BOX RIGHT SIDE UOC: DAJ, V24, ZAJ	1
9	PBHZZ	2590011880317	19207	8380481	MOLDING, METAL RIGHT SIDE UOC: DAJ, V24, ZAJ	1
9	PBHZZ	2590011863722	19207	8380482	BEZEL, AUTOMOTIVE TR VAN BODY LEFT SIDE DOOR HINGE UOC: DAJ, V24, ZAJ	1
10	PBHZZ	2590011863723	19207	8380476	BEZEL, AUTOMOTIVE TR VAN BODY RIGHT HAND UPPER PANEL UOC: DAJ, V24, ZAJ	1
10	PBHZZ	2590011926089	19207	8380477	BEZEL, AUTOMOTIVE TR SOLID SIDE LEFT HAND UPPER PANEL TOP UOC: DAJ, V24, ZAJ	1
11	PBHZZ	2590011880316	19207	8380446	MOLDING, METAL VAN BODY RIGHT SIDE UPPER PANEL TOP UOC: DAJ, V24, ZAJ	1
12	PBHZZ	2590011810311	19207	8380437	BEZEL, AUTOMOTIVE TR VAN BODY RIGHT HAND UPPER PANEL UOC: DAJ, V24, ZAJ	1
12	PBHZZ	2590011863723	19207	8380476	BEZEL, AUTOMOTIVE TR SOLID SIDE LEFT HAND UPPER PANEL END UOC: DAJ, V24, ZAJ	1

(1) ITEM NO	(2) SMR CODE	(3) NSN	(4) CAGEC	(5) PART NUMBER	(6) DESCRIPTION AND USABLE ON CODES (UOC)	(7) QTY
13	XDHZZ		19207	7017450	HOOK, VAN BODY COAT .... UOC: DAJ, V24, ZAJ	4
14	PAOZZ	5305004324390	96906	MS51862-16	SCREW, TAPPING .... UOC: DAJ, V24, ZAJ	8
15	MHHZZ		19207	7084887-140	PANEL, SOLID SIDE, TO RIGHT SIDE, ... MAKE FROM PLYWOOD, P/N NN-P-530TYIGROUPB, 140 INCHES LONG .... UOC: DAJ, V24, ZAJ	1
16	PAHZZ	5305008550965	96906	MS24629-38	SCREW, TAPPING RIGHT SIDE .... UOC: DAJ, V24, ZAJ	91
17	PBHZZ	2590012025769	19207	8380454	MOLDING, METAL VAN BODY RIGHT SIDE .... PANEL .... UOC: DAJ, V24, ZAJ	1
18	PAOZZ	5340004219696	19207	7084896	CLIP, SPRING TENSION RIGHT SIDE .... UOC: DAJ, V24, ZAJ	2
19	PAOZZ	5305004324170	96906	MS51861-35	SCREW, TAPPING .... UOC: DAJ, V24, ZAJ	4
20	MHHZZ		19207	7084891-139	PANEL, SOLID SIDE, LO RIGHT SIDE, ... MAKE FROM PLYWOOD, P/N NN-P-530TYIGROUPB, 139 INCHES LONG .... UOC: DAJ, V24, ZAJ	1
20	MHHZZ		19207	7084886-139	PANEL, SOLID SIDE, LO LEFT SIDE, MAKE .... FROM PLYWOOD, P/N NN-P-530TYIGROUPB, 139 INCHES LONG............ UOC: DAJ, V24, ZAJ	1
21	PAHZZ	5305000526879	96906	MS24627-55	SCREW, TAPPING RIGHT SIDE .... UOC: DAJ, V24, ZAJ	8
22	PAHZZ	5365004976718	19207	11607335	SPACER, WOOD RIGHT SIDE .... UOC: DAJ, V24, ZAJ	4
23	PAHZZ	5340007663330	19207	7535643	BUMPER, NONMETALLIC RIGHT SIDE .... UOC: DAJ, V24, ZAJ	4
24	PAHZZ	5310008238804	96906	MS27183-9	WASHER, FLAT RIGHT SIDE .... UOC: DAJ, V24, ZAJ	4
25	PAHZZ	5305011212696	96906	MS51861-72	SCREW, TAPPING RIGHT SIDE .... UOC: DAJ, V24, ZAJ	4
26	PAOZZ	5305004324203	96906	MS51861-47	SCREW, TAPPING .... UOC: DAJ, V24, ZAJ	4
27	PAHZZ	5340011910596	19207	7084892	COVER, ACCESS .... UOC: DAJ, V24, ZAJ	1
28	MHHZZ		19207	7084889	PANEL, SOL SIDE REAR RIGHT SIDE, ... MAKE FROM PLYWOOD, P/N NN-P-530TYGROUPB .... UOC: DAJ, V24, ZAJ	1
28	MHHZZ		19207	7084890	PANEL, SOL SIDE REAR LEFT SIDE, ... MAKE FROM PLYWOOD, P/N NN-P-530TYIGROUPB. UOC: DAJ, V24, ZAJ,	1

END OF FIGURE

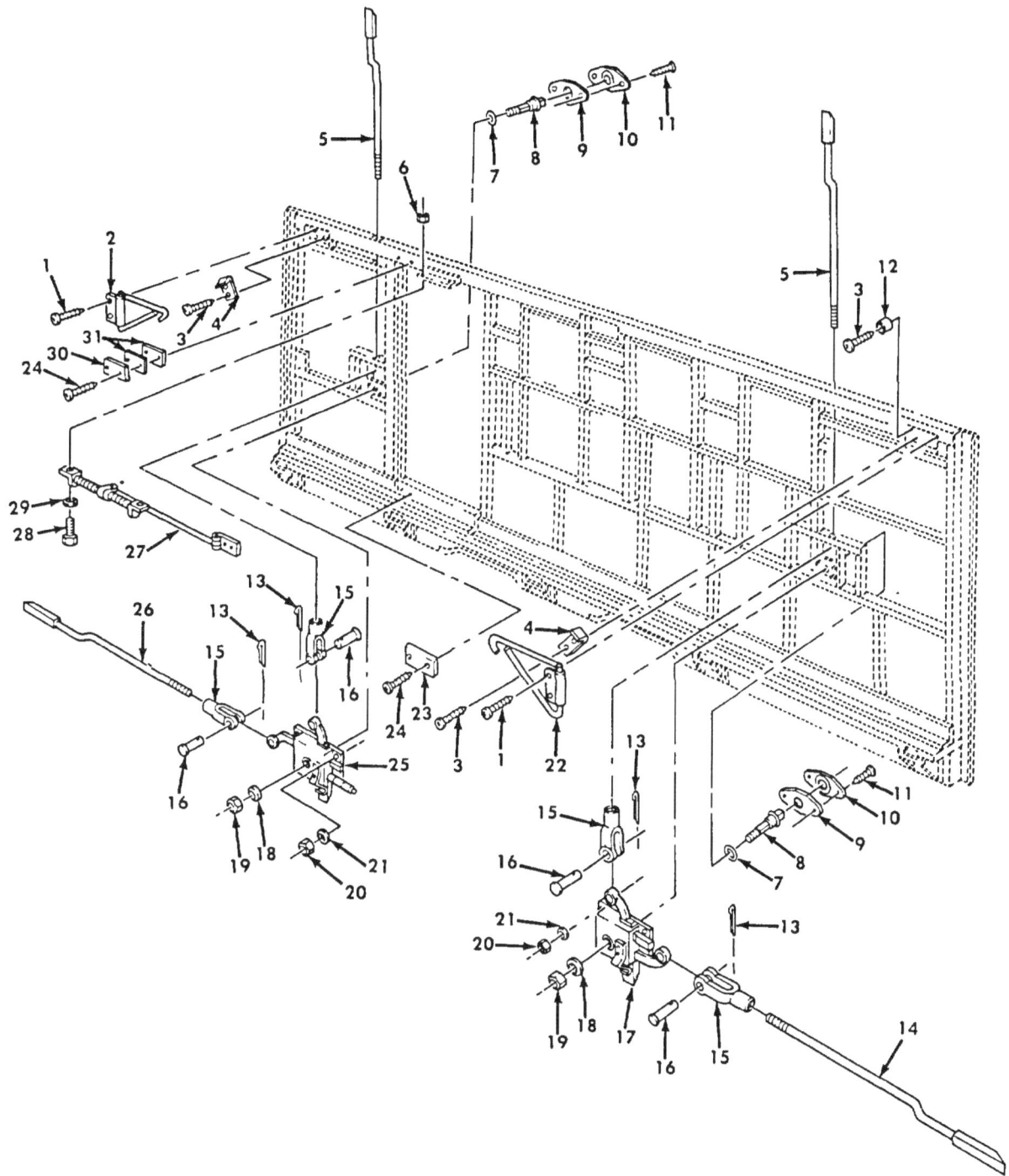

Figure 414. Van Side Panel Door Checks, Interior Latch Assemblies, and Related Parts, Right and Left Side.

(1) ITEM NO	(2) SMR CODE	(3) NSN	(4) CAGEC	(5) PART NUMBER	(6) DESCRIPTION AND USABLE ON CODES (UOC)	(7) QTY
					GROUP 1812 SPECIAL PURPOSE BODIES	
					FIG. 414 VAN SIDE PANEL DOOR CHECKS, INTERIOR LATCH ASSEMBLIES, AND RELATED PARTS, RIGHT AND LEFT SIDE	
1	PAOZZ	5306004219402	21450	425346	BOLT, ASSEMBLED WASH .... UOC: DAJ, V24, ZAJ	6
2	PAOZZ	2510002343259	19207	7084882	HOOK AND HANGER ASS RIGHT SIDE .... UOC: DAJ, V24, ZAJ	1
2	PAOZZ	2510002343258	19207	7084881	HOOK ASSEMBLY LEFT SIDE .... UOC: DAJ, V24, ZAJ	1
3	PAOZZ	5305004324170	96906	MS51861-35	SCREW, TAPPING .... UOC: DAJ, V24, ZAJ	9
4	PAOZZ	5340004219696	19207	7084896	CLIP, SPRING TENSION .... UOC: DAJ, V24, ZAJ	3
5	PBHZZ	5340010529024	19207	7084895	BAR, LOCKING SIDE PANEL TOP .... UOC: DAJ, V24, ZAJ	2
6	PAOZZ	5310007616882	96906	MS51967-2	NUT, PLAIN, HEXAGON .... UOC: V24, DAJ, ZAJ	2
7	PAOZZ	5330012170734	24510	582818	GASKET .... UOC: V24, DAJ, ZAJ	2
8	PAOZZ	5306012097114	19207	12301004	BOLT, SHOULDER .... UOC: DAJ, V24, ZAJ	2
9	PFOZZ	5330004151484	19207	7535583	GASKET .... UOC: DAJ, V24	2
10	PFOZZ	5340011650469	19207	7535572	PLATE, RETAINING, SEA .... UOC: DAJ, V24, ZAJ	2
11	PAOZZ	5305004324172	96906	MS51861-37	SCREW, TAPPING .... UOC: DAJ, V24, ZAJ	4
12	PAOZZ	5340002647182	75543	747R	BUMPER, NONMETALLIC .... UOC: DAJ, V24, ZAJ	3
13	PAOZZ	5315008161794	89749	F316	PIN, COTTER .... UOC: DAJ, V24, ZAJ	4
14	PAOZZ	5340010529023	19207	7084894	BOLT, FLUSH .... UOC: DAJ, V24, ZAJ	1
15	PAHZZ	5340005508070	96906	MS35812-6	CLEVIS, ROD END .... UOC: DAJ, V24, ZAJ	4
16	PAHZZ	5315000817042	96906	MS20392-7C37	PIN, STRAIGHT, HEADED .... UOC: DAJ, V24, ZAJ	4
17	PAHZZ	5340006791494	19207	7084997	LATCH, MORTISE RIGHT FRONT SIDE .... PANEL UOC: DAJ, V24, ZAJ	1
17	PAHZZ	5340006791495	19207	8380467	LATCH, MORTISE LEFT SIDE PANEL .... UOC: DAJ, V24, ZAJ	1
18	PAHZZ	5310012026775	24617	9417794	WASHER, FLAT .... UOC: DAJ, V24, ZAJ	2
19	PAHZZ	5310008348732	96906	MS35691-33	NUT, PLAIN, HEXAGON .... UOC: DAJ, V24, ZAJ	2
20	PAHZZ	5310008807744	96906	MS51967-5	NUT, PLAIN, HEXAGON .... UOC: DAJ, V24, ZAJ	8
21	PAHZZ	5310004079566	96906	MS35338-45	WASHER, LOCK .... UOC: DAJ, V24, ZAJ	8
22	PAOZZ	2510002343259	19207	7084882	HOOK AND HANGER ASS LEFT SIDE	1

(1) ITEM NO	(2) SMR CODE	(3) NSN	(4) CAGEC	(5) PART NUMBER	(6) DESCRIPTION AND USABLE ON CODES (UOC)	(7) QTY
22	PAOZZ	251000234325B	19207	7084881	UOC: DAJ, V24, ZAJ HOOK ASSEMBLY RIGHT SIDE	1
23	PAOZZ	5340002326056	19207	11607430	UOC: DAJ, V24, ZAJ STRIKE, CATCH	1
24	PAOZZ	5305000526881	96906	MS24627-63	UOC: DAJ, V24, ZAJ SCREW, TAPPING	2
25	PAHZZ	5340013110225	19207	7084996	UOC: DAJ, V24, ZAJ LOCK SET, MORTISE RIGHT SIDE	1
25	PAHZZ	5340006791492	19207	8380468	UOC: DAJ, V24, ZAJ LATCH, MORTISE LEFT SIDE	1
26	PAHZZ	5340010529022	19207	7084893	UOC: DAJ, V24, ZAJ BOLT, FLUSH	1
27	PAOZZ	5340006791446	19207	8698434	UOC: DAJ, V24, ZAJ CHECK ASSEMBLY, DOOR	1
28	PAOZZ	5305002253843	80204	B1821BH025C100N	UOC: DAJ, V24, ZAJ SCREW, CAP, HEXAGON H	2
29	PAOZZ	5310005825965	96906	MS35338-44	UOC: DAJ, V24, ZAJ WASHER, LOCK	2
30	PAOZZ	5340004950236	19207	10882221	UOC: DAJ, V24, ZAJ STRIKE, CATCH	1
31	PAOZZ	5365001594668	19207	10891476	UOC: DAJ, V24, ZAJ SPACER, PLATE .... UOC: DAJ, V24, ZAJ	2

END OF FIGURE

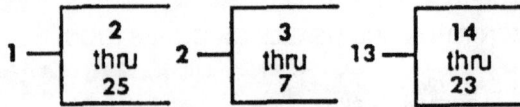

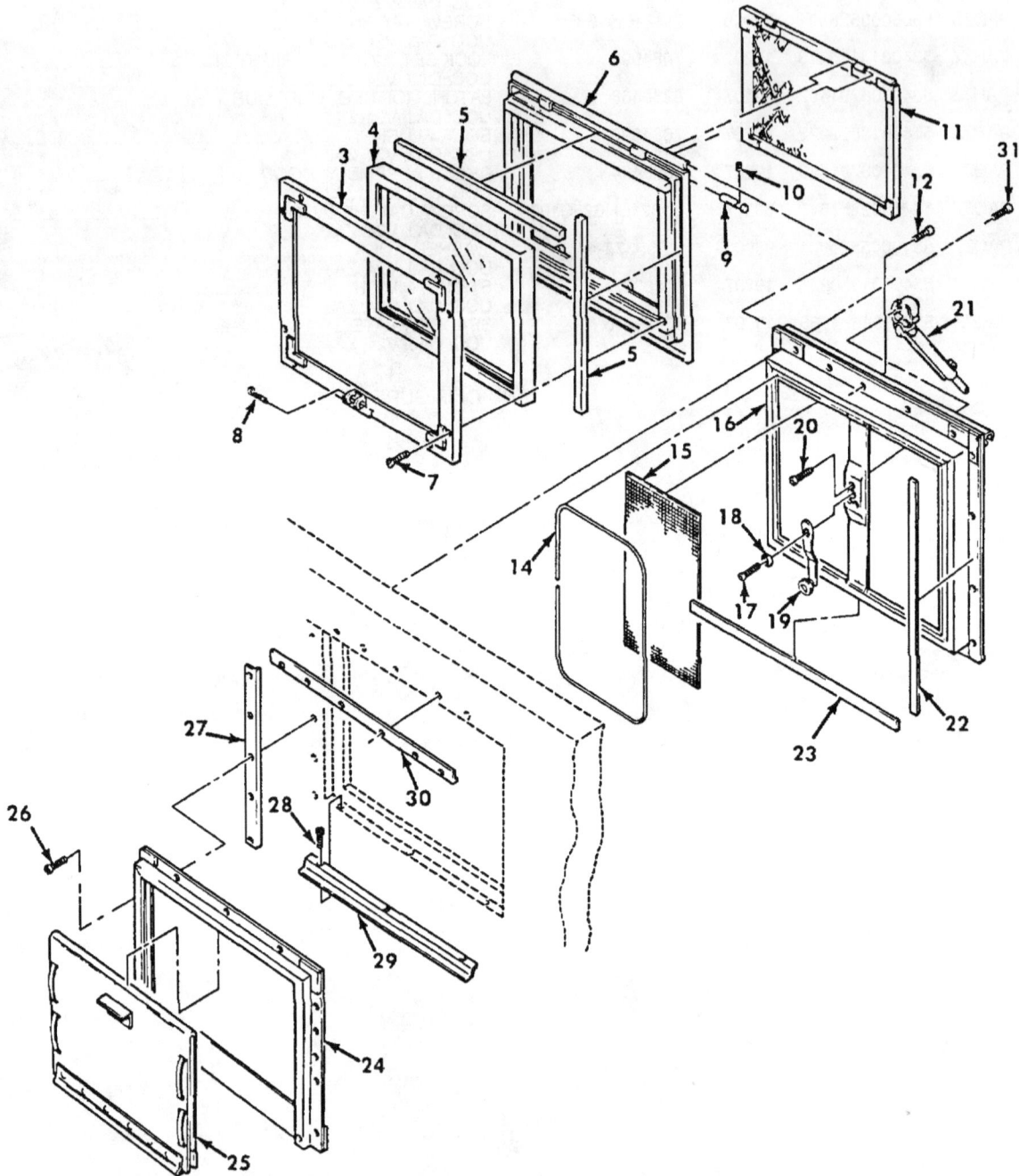

Figure 415. Van Retractable Window Sash Assembly and Related Parts, Right and Left Side.

(1) ITEM NO	(2) SMR CODE	(3) NSN	(4) CAGEC	(5) PART NUMBER	(6) DESCRIPTION AND USABLE ON CODES (UOC)	(7) QTY
					GROUP 1812 SPECIAL PURPOSE BODIES	
					FIG. 415 VAN RETRACTABLE WINDOW SASH ASSEMBLY, AND RELATED PARTS, RIGHT AND LEFT SIDE	
1	PAOOO	2510010429692	19207	8729078	SASH ASSEMBLY .... UOC: DAJ, V24, ZAJ	3
2	PFOOO	2510011617631	19207	10896779-1	.WINDOW, VEHICULAR .... UOC: DAJ, V24, ZAJ	1
3	XAOZZ		19207	10896780	..FRAME, INNER .... UOC: DAJ, V24, ZAJ	1
4	PFOZZ	2510008098046	19207	10896799-1	.WINDOW, VEHICULAR .... UOC: DAJ, V24, ZAJ	1
5	MOOZZ		19207	10937682-81	..WEATHERSEAL MAKE FROM RUBBER, P/N .... 10937682 UOC: DAJ, V24, ZAJ	1
6	XAOZZ		19207	10896781	..FRAME, OUTER .... UOC: DAJ, V24, ZAJ	1
7	PAOZZ	5305001801964	96906	MS35493-56	..SCREW, WOOD .... UOC: DAJ, V24, ZAJ	18
8	PAOZZ	5315008395821	96906	MS24665-351	.PIN, COTTER .... UOC: DAJ, V24, ZAJ	1
9	PBOZZ	5315011053318	19207	10896789	.PIN, GROOVED, HEADLES .... UOC: DAJ, V24, ZAJ	2
10	PAOZZ	5305007195342	96906	MS51963-34	.SETSCREW .... UOC: DAJ, V24, ZAJ	2
11	PBOZZ	5410011285529	19207	10896813	.SCREEN, WINDOW, METAL .... UOC: DAJ, V24, ZAJ	1
12	PAOZZ	5305008893116	96906	MS35206-213	.SCREW, MACHINE .... UOC: DAJ, V24, ZAJ	1
13	PAOOO	5410011617703	19207	10896812	.SCREEN, WINDOW, METAL .... UOC: DAJ, V24, ZAJ	1
14	MOOZZ		19207	10921722-2-52	..TUBING MAKE FROM TUBING, P/N .... 10921722-2, 52 INCHES LONG. UOC: DAJ, V24, ZAJ	2
15	MOOZZ		19207	10896867	..SCREEN MAKE FROM P/N RR-W- .... 365ATYPE2 20X20 (81348). UOC: DAJ, V24, ZAJ	1
16	XAOZZ		19207	10896856	..FRAME ASSEMBLY .... UOC: DAJ, V24, ZAJ	1
17	PAOZZ	5305000594568	96906	MS35190-253	.SCREW, MACHINE .... UOC: DAJ, V24, ZAJ	1
18	PAOZZ	5310011854675	19207	10896868	..WASHER, RECESSED .... UOC: DAJ, V24, ZAJ	1
19	PFOZZ	2540011018453	19207	10896841	.HANDLE, DOOR, VEHIC .... UOC: DAJ, V24, ZAJ	1
20	PAOZZ	5305004324170	96906	MS51861-35	..SCREW, TAPPING .... UOC: DAJ, V24, ZAJ	3
21	PAOZZ	2540010469402	19207	10896845	.REGULATOR, VEHICLE W .... UOC: DAJ, V24, ZAJ	1
22	PAOZZ	5680011854944	19207	10896870-2	..WEATHER STRIP FRAME UPPER	2

(1) ITEM NO	(2) SMR CODE	(3) NSN	(4) CAGEC	(5) PART NUMBER	(6) DESCRIPTION AND USABLE ON CODES (UOC)	(7) QTY
					UOC: DAJ, V24, ZAJ	
23	PAOZZ	5680011839754	19207	10896870-1	..WEATHER STRIP	2
					UOC: DAJ, V24, ZAJ	
24	PFOZZ	2590011787043	19554	13521G2	.FRAME, BLACKOUT	2
					UOC: DAJ, V24, ZAJ	
25	PAOZZ	2510011794083	19207	10896815	.BLACKOUT, PANEL	1
					UOC: DAJ, V24, ZAJ	
26	PAOZZ	5305004324170	96906	MS51861-35	SCREW, TAPPING	15
					UOC: DAJ, V24, ZAJ	
27	PAOZZ	2510010222580	19207	7084794	TRIM, PLYWOOD	2
					UOC: DAJ, V24, ZAJ	
28	PAOZZ	5305004324172	96906	MS51861-37	SCREW, TAP PING	3
					UOC: DAJ, V24, ZAJ	
29	PAOZZ	2510010243619	19207	7084793	TRIM, PLYWOOD	1
					UOC: DAJ, V2 4, ZAJ	
30	PAOZZ	2510010243618	19207	7084792	TRIM, PLYWOOD	1
					UOC: DAJ, V24, ZAJ	
31	PAOZZ	5305004324203	96906	MS51861-47	SCREW, TAPPING	18
					UOC: DAJ, V24, ZAJ	

END OF FIGURE

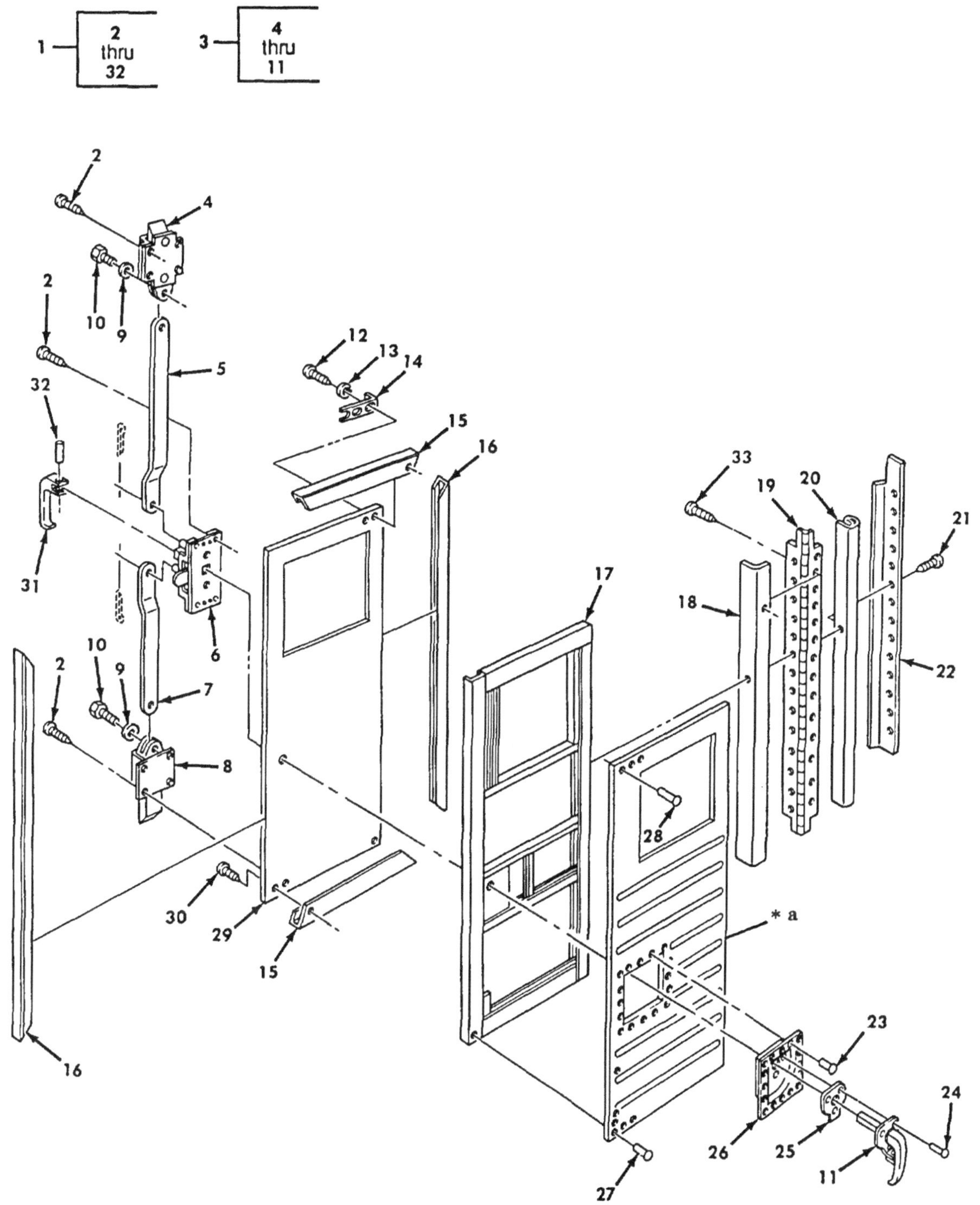

* a PART OF ITEM 1

Figure 416. Van Body Door Assemblies, Right and Left Side.

(1) ITEM NO	(2) SMR CODE	(3) NSN	(4) CAGEC	(5) PART NUMBER	(6) DESCRIPTION AND USABLE ON CODES (UOC)	(7) QTY
					GROUP 1812 SPECIAL PURPOSE BODIES	
					FIG. 416 VAN BODY DOOR ASSEMBLIES, RIGHT AND LEFT SIDE	
1	PAFFF	2510011730092	19207	11607383-3	DOOR, VAN BODY, VEHIC RIGHT .... UOC:DAJ, V24, ZAJ	1
1	PAFFF	2510011730091	19207	11607383-4	DOOR, VEHICULAR LEFT .... UOC:DAJ, V24, ZAJ	1
2	PAOZZ	5305004324203	96906	MS51861-47	.SCREW, TAPPING .... UOC:DAJ, V24, ZAJ	12
3	PAOOO	5340004785877	19207	11607265-1	.LOCK SET, RIM LEFT SIDE .... UOC:DAJ, V24, ZAJ	1
3	PAOOO	5340004785878	19207	11607265-2	.LOCK SET, RIM, RIGHT SIDE .... UOC:DAJ, V24, ZAJ	1
4	PAOZZ	5340008390098	19207	7748911	..BOLT, FLUSH........... UOC:DAJ, V24, ZAJ	1
5	PFOZZ	3040012337768	19207	8722186-23	..CONNECTING LINK, RIG UPPER .... UOC:DAJ, V24, ZAJ	1
6	PFOZZ	2590006301567	19207	10911036-1	..LOCK, VAN DOOR RIGHT SIDE .... UOC:DAJ, V24, ZAJ	1
6	PFOZZ	2540009184184	19207	10911036-2	..LATCH, DOOR, VEHICULA LEFT SIDE .... UOC:DAJ, V24, ZAJ	1
7	PFOZZ	3040012059264	19207	8722186-20	..CONNECTING LINK, RIG LOWER .... UOC:DAJ, V24, ZAJ	1
8	PAOZZ	5306008097824	19207	11607266	..BOLT, DOOR, VAN BODY LOWER .... UOC:DAJ, V24, ZAJ	1
9	PAOZZ	5310006379541	96906	MS35338-46	..WASHER, LOCK .... UOC:DAJ, V24, ZAJ	2
11	PFOZZ	2540002872571	19207	7264749	..HANDLE, DOOR, VEHICUL .... UOC:DAJ, V24, ZAJ	1
12	PAFZZ	5305000527492	96906	MS24629-61	.SCREW, TAPPING .... UOC:DAJ, V24, ZAJ	2
13	PAOZZ	5310005825965	96906	MS35338-44	.WASHER, LOCK . ... UOC:DAJ, V24, ZAJ	2
14	PAOZZ	5340002385606	19207	7084799	.BRACKET, ANGLE . ... UOC:DAJ, V24, ZAJ	1
15	PBFZZ	2590004172735	19207	8380444	.BEZEL, AUTOMOTIVE TR UPPER SIDE DOOR .... UOC:DAJ, V24, ZAJ	2
16	PBOZZ	2590011880318	19207	8380448	.MOLDING, METAL INNER PANEL HINGE LEFT AND RIGHT HAND .... UOC:DAJ, V24, ZAJ	2
17	XAOZZ		19207	12300968-2	.FRAME LEFT........... UOC:DAJ, V24, ZAJ	1
17	XAOZZ		19207	12300968-1	.FRAME RIGHT........... UOC:DAJ, V24, ZAJ	1
18	MOOZZ		19207	8380420-77	RUBBER STRIP MAKE FROM RUBBER .STRIP, P/N 8380420, 77 INCHES LONG........... UOC:DAJ, V24, ZAJ	1
19	PAOZZ	5340001077769	19207	7084989	HINGE, BUTT . ... UOC:DAJ, V24, ZAJ	1

(1) ITEM NO	(2) SMR CODE	(3) NSN	(4) CAGEC	(5) PART NUMBER	(6) DESCRIPTION AND USABLE ON CODES (UOC)	(7) QTY
20	MFFZZ		19207	11607302-78	.SEAL, NONMETALLIC SP MAKE FROM SEAL, P/N 11607302, 78 INCHES LONG UOC:DAJ, V24, ZAJ	1
21	PAOZZ	5305004770144	96906	MS51861-68	.SCREW, TAPPING UOC:DAJ, V24, ZAJ	14
22	PAOZZ	2590004172611	19207	8380484	BEZEL, AUTOMOTIVE TR UOC:DAJ, V24, ZAJ	1
23	PAOZZ	5320005823268	53551	RV200-6-3	RIVET, BLIND UOC:DAJ, V24, ZAJ	16
24	PAOZZ	5320007215384	96906	MS20600AD8W7	RIVET, BLIND UOC:DAJ, V24, ZAJ	3
25	PAOZZ	5330004146695	19207	11592566	GASKET UOC:DAJ, V24, ZAJ	1
26	PBFZZ	2510011787045	19207	11592567-1	POCKET, DOOR, LOCK SIDE DOOR ASSY LEFT SIDE UOC:DAJ, V24, ZAJ	1
26	PAOZZ	2510004213949	19207	11592567-2	POCKET, DOOR LOCK, VE SIDE DOOR ASSY RIGHT SIDE UOC:DAJ, V24, ZAJ	1
27	PAOZZ	5320002421582	96906	MS20470A6-4	RIVET, SOLID UOC:DAJ, V24, ZAJ	66
28	PFHZZ	5320005841285	53551	RV200-6-2	RIVET, BLIND UOC:DAJ, V24	104
29	PFOZZ	2510011774466	19207	12300934-1	PANEL, BODY, VEHICULA RIGHT PANEL VAN BODY UOC:DAJ, V24, ZAJ	1
29	PFOZZ	2510011774467	19207	12300934-2	PANEL, VAN BODY LEFT PANEL VAN BODY UOC:DAJ, V24, ZAJ	1
30	PAOZZ	5305004324172	96906	MS51861-37	SCREW, TAPPING. UOC:DAJ, V24, ZAJ	50
31	PAOZZ	2540008097792	19207	11592462	HANDLE, DOOR, VEHICUL UOC:DAJ, V24, ZAJ	1
32	PAOZZ	5315006822207	96906	MS35677-46	PIN, GROOVED, HEADLES UOC:DAJ, V24, ZAJ	1
33	PAOZZ	5305004770144	96906	MS51861-68	SCREW, TAPPING UOC:DAJ, V24, ZAJ	30

END OF FIGURE

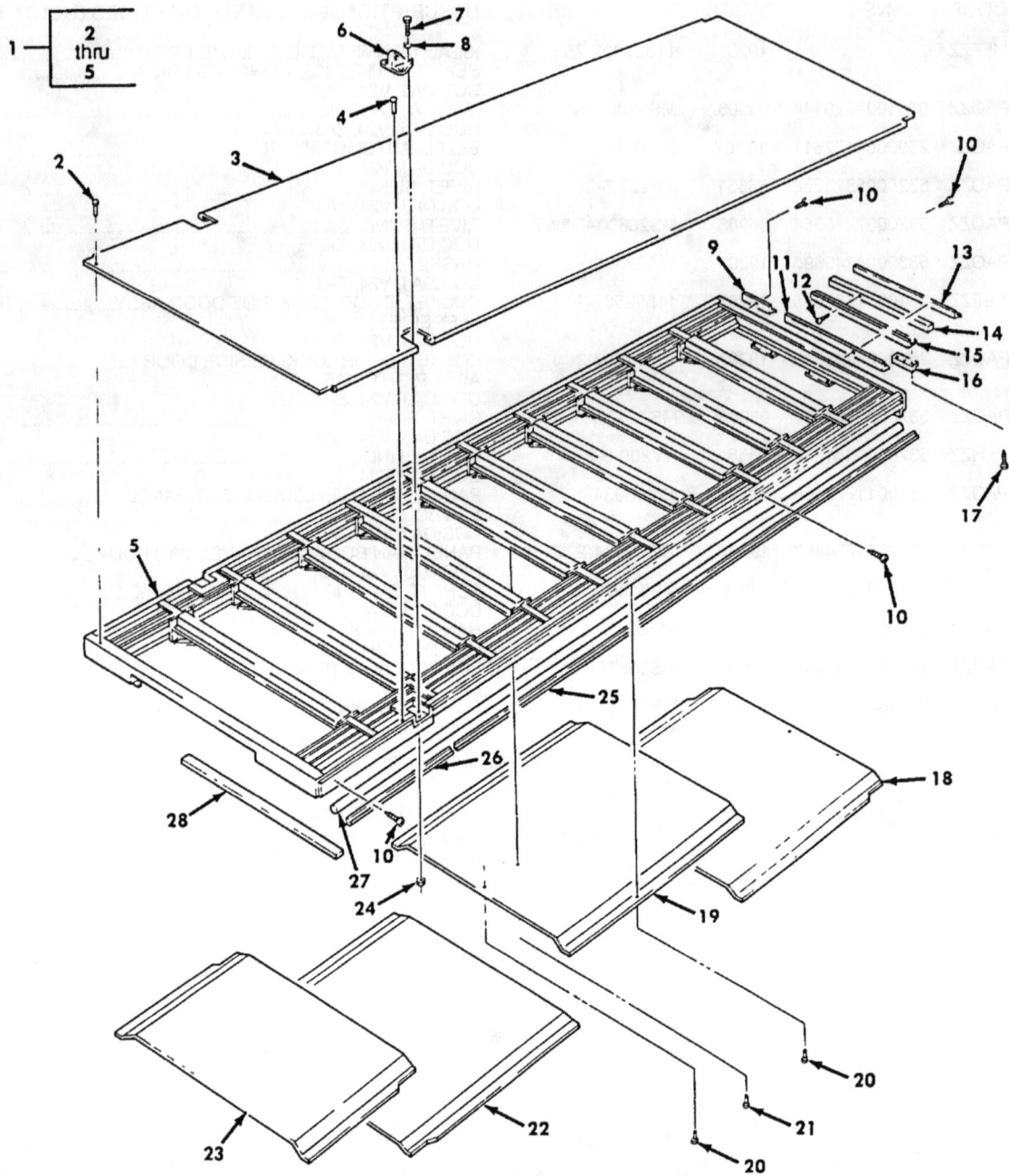

Figure 417. Van Interior and Exterior Ceiling, Frame, and Related Parts.

(1) ITEM NO	(2) SMR CODE	(3) NSN	(4) CAGEC	(5) PART NUMBER	(6) DESCRIPTION AND USABLE ON CODES (UOC)	(7) QTY
					GROUP 1812 SPECIAL PURPOSE BODIES	
					FIG. 417 VAN INTERIOR AND EXTERIOR CEILING, FRAME, AND RELATED PARTS	
1	XDHHH		19207	12300868	ROOF ASSEMBLY .... UOC:DAJ, V24, ZAJ	1
2	PAHZZ	5320002421580	96906	MS20470A6-6	.RIVET, SOLID.. UOC:DAJ, V24, ZAJ	116
3	PFHZZ	2510012251000	19207	12300881	.PANEL, ROOF, VEHICULA.. UOC:DAJ, V24, ZAJ	1
4	PAHZZ	5320009567355	90030	AD64H	.RIVET, BLIND... UOC:DAJ, V24, ZAJ	144
5	XAHZZ		19207	12300882	.FRAME .... UOC:DAJ, V24, ZAJ	1
6	PAHZZ	5340002310210	19207	8380499	PAD EYE .... UOC:DAJ, V24, ZAJ	2
7	PAHZZ	5305007254138	96906	MS90726-170	SCREW, CAP, HEXAGON H .... UOC:DAJ, V24, ZAJ	6
8	PAHZZ	5310008206653	96906	MS35338-50	WASHER, LOCK .... UOC:DAJ, V24, ZAJ	6
9	PBHZZ	2590011801092	19207	7084956	BEZEL, AUTOMOTIVE TR CEILING, RIGHT .... REARE... UOC:DAJ, V24, ZAJ	1
9	PBHZZ	2590011816059	19207	7084957	BEZEL, AUTOMOTIVE TR CEILING, LEFT......................... REAR .... UOC:DAJ, V24, ZAJ	1
10	PAHZZ	5305001498610	96906	MS51862-36	SCREW, TAPPING. .... UOC:DAJ, V24, ZAJ	141
11	PBRZZ	2590011795802	19207	7084958	BEZEL, AUTOMOTIVE TR CEILING, CENTER. .... UOC:DAJ, V24, ZAJ	1
12	PAHZZ	5305001380069	96906	MS51861-44	SCREW, TAPPING .... UOC:DAJ, V24, ZAJ	8
13	PAHZZ	5330001523217	19207	7373291-3	SEAL, NONMETALLIC SP .... UOC:DAJ, V24, ZAJ	1
14	XDHZZ	5510012040084	19207	12300897	BLOCK, FILLER, WOOD .... UOC:DAJ, V24, ZAJ	1
15	PFHZZ	2510011873614	19207	12300895	MOLDING, METAL .... UOC:DAJ, V24, ZAJ	1
16	PFHZZ	5340012170818	19207	11665767	BRACKET, ANGLE .... UOC:DAJ, V24, ZAJ	1
17	PAHZZ	5305001965570	96906	MS51862-61	SCREW, TAPPING .... UOC:DAJ, V24, ZAJ	21
18	PFHZZ	2510011620561	19207	11665761	CEILING, REAR .... UOC:DAJ, V24, ZAJ	1
19	PAHZZ	2510011620562	19207	11665760	CEILING, CENTER .... UOC:DAJ, V24, ZAJ	1
20	PAHZZ	5320005841285	53551	RV200-6-2	RIVET, BLIND .... UOC:DAJ, V24, ZAJ	85
21	PAHZZ	5320008134144	96906	MS20600-AD6W7	RIVET, BLIND.. UOC:DAJ, V24, ZAJ	280
22	PFHZZ	9535011062065	19207	11665759	SHEET, METAL	1

(1) ITEM NO	(2) SMR CODE	(3) NSN	(4) CAGEC	(5) PART NUMBER	(6) DESCRIPTION AND USABLE ON CODES (UOC)	(7) QTY
					UOC:DAJ, V24, ZAJ	
23	PFHZZ	2510012372945	19207	12300867	PANEL, BODY, VEHICULA..	1
					UOC:DAJ, V24, ZAJ	
24	PAHZZ	5310007638905	96906	MS51968-20	NUT, PLAIN, HEXAGON...	6
					UOC:DAJ, V24, ZAJ	
25	PBHZZ	2590011810310	19207	7084955	BEZEL, AUTOMOTIVE TR	2
					UOC:DAJ, V24, ZAJ	
26	PBHZZ	2590011896415	19207	12302855-1	BEZEL, AUTOMOTIVE TR CEILING, LEFT........................... SIDE	1
					UOC:DAJ, V24, ZAJ	
26	PBHZZ	2590011896416	19207	12302855-2	MOLDING, METAL CEILING, LEFT SIDE............................	1
					UOC:DAJ, V24, ZAJ	
27	PAHZZ	9390004050215	19207	10915159	NONMETALLIC SPECIAL	2
					UOC:DAJ, V24, ZAJ	
28	MHHZZ		19207	8735226	PLYWOOD, SOFTWOOD, CO...................................	10
					UOC:DAJ, V24, ZAJ	

END OF FIGURE

417-2

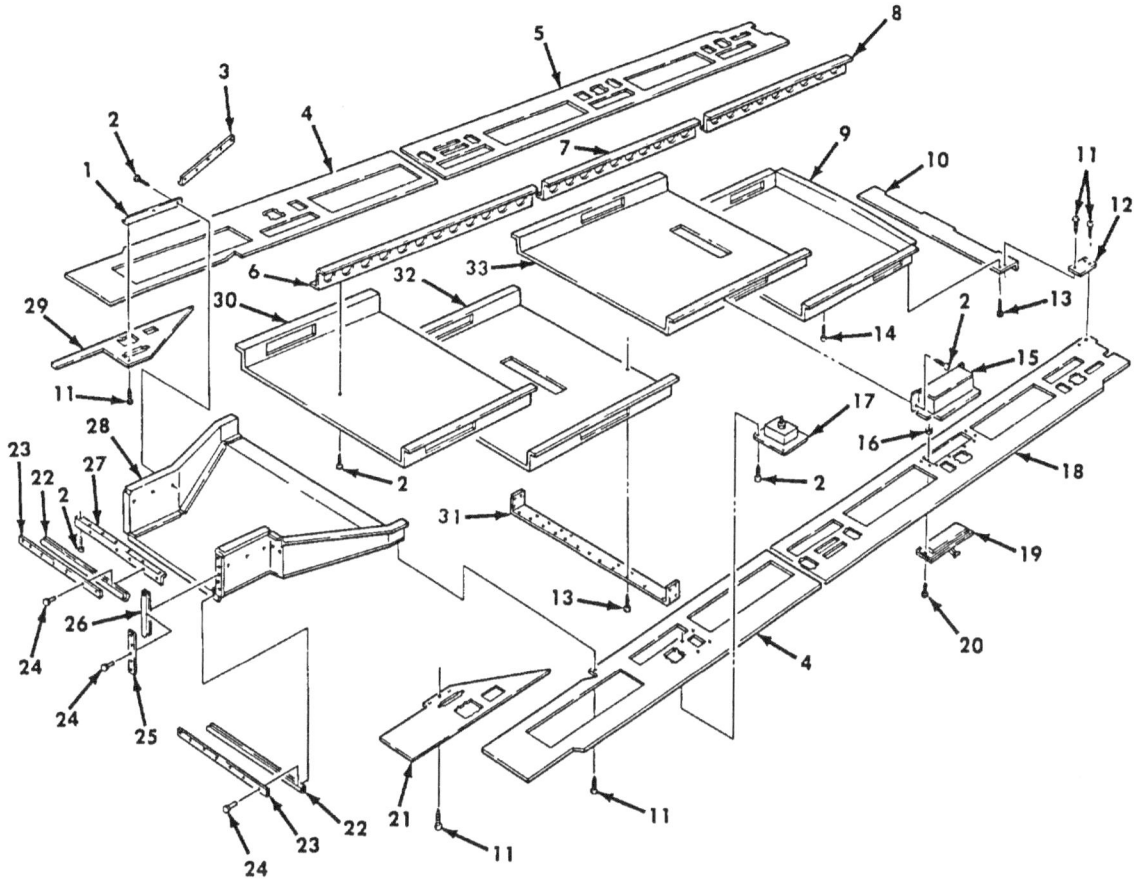

Figure 418. Van Air Conditioning Duct and Related Parts.

(1) ITEM NO	(2) SMR CODE	(3) NSN	(4) CAGEC	(5) PART NUMBER	(6) DESCRIPTION AND USABLE ON CODES (UOC)	(7) QTY

GROUP 1812 SPECIAL PURPOSE BODIES

FIG. 418 VAN AIR CONDITIONING DUCT AND RELATED PARTS

(1) ITEM NO	(2) SMR CODE	(3) NSN	(4) CAGEC	(5) PART NUMBER	(6) DESCRIPTION AND USABLE ON CODES (UOC)	(7) QTY
1	PFHZZ	2510011896417	19207	12302853	ANGLE, FILLER FRONT CEILING.................................... UOC:DAJ, V24, ZAJ	2
2	PAHZZ	5305004333685	96906	MS51861-37C	SCREW, TAPPING .... UOC:DAJ, V24, ZAJ	276
3	PFHZZ	5340012319290	19207	12302854	BRACKET, ANGLE .... UOC:DAJ, V24, ZAJ	2
4	XAHZZ		19207	12300870	PANEL .... UOC:DAJ, V24, ZAJ	2
5	XBHZZ		19207	12300869	PANEL LEFT SIDE .... UOC:DAJ, V24, ZAJ	1
6	PFOZZ	5340013107071	19207	12300873	BRACKET, DOUBLE ANGL FRONT.............................. UOC:DAJ, V24, ZAJ	1
7	PFOZZ	5340013107070	19207	1665775	BRACKET, DOUBLE ANGL CENTER............................ UOC:DAJ, V24, ZAJ	1
9	PFHZZ	4520011803577	19207	11677737	PIPE, AIR CONDITION REAR .... UOC:DAJ, V24, ZAJ	1
10	PBHZZ	2510011808503	19207	11677736	GRILLE, METAL CEILING, REAR AIR DUCT................. UOC:DAJ, V24, ZAJ	1
11	PAHZZ	5305004324170	96906	MS51861-35	SCREW, TAPPING. .... UOC:DAJ, V24, ZAJ	170
12	PBHZZ	5365011913575	19207	11677580	SPACER, PLATE CEILING.. UOC:DAJ, V24, ZAJ	2
13	PAHZZ	5305004324163	96906	MS51861-24	SCREW, TAPPING .... UOC:DAJ, V24, ZAJ	117
14	PAHZZ	5305004324171	96906	MS51861-36	SCREW, TAPPING .... UOC:DAJ, V24, ZAJ	26
15	PFHZZ	2530012131040	19207	11677714	DEFLECTOR, DIRT AND .... UOC:DAJ, V24, ZAJ	8
16	PAHZZ	5310001413062	80205	NAS1329A3-130	NUT, PLAIN, BLIND RIV .... UOC:DAJ, V24, ZAJ	16
17	XAHZZ		19207	8735151	BOX ASSEMBLY TELEPHONE JACK............................ UOC:DAJ, V24, ZAJ	8
18	XBHZZ		19207	12300872	PANEL, RIGHT .... UOC:DAJ, V24, ZAJ	1
19	PFHZZ	4520011786680	19207	11640377-9	REGISTER, METAL .... UOC:DAJ, V24, ZAJ	10
20	PAHZZ	5305009931848	96906	MS35207-265	SCREW, MACHINE .... UOC:DAJ, V24, ZAJ	20
21	XBHZZ		19207	12300866	PANEL, CEILING FILLE .... UOC:DAJ, V24, ZAJ	1
22	MHHZZ		91340	10608E44S BULK	SEAL RUBBER STRIP, MAKE FROM RUBBER, ... P/N 10608E44S, 28 INCHES LONG .... UOC:DAJ, V24, ZAJ	2
23	PFHZZ	5340012053548	19207	12300723	PLATE, MENDING .... UOC:DAJ, V24, ZAJ	2

(1) ITEM NO	(2) SMR CODE	(3) NSN	(4) CAGEC	(5) PART NUMBER	(6) DESCRIPTION AND USABLE ON CODES (UOC)	(7) QTY
24	PAHZZ	5320005823276	96906	MS20600AD6W4	RIVET, BLIND .... UOC:DAJ, V24, ZAJ	24
25	PFHZZ	5340012053549	19207	12300724	PLATE, MENDING .... UOC:DAJ, V24, ZAJ	2
26	MHHZZ		91340	10608E44S BULK	SEAL RUBBER STRIP MAKE FROM RUBBER, ... P/N 10608E44S, 15 INCHES LONG .... UOC:DAJ, V24, ZAJ	2
27	PFHZZ	5340012085371	19207	12300722	BRACKET, ANGLE . ... UOC:DAJ, V24, ZAJ	1
28	XBHZZ		19207	12300875	TRANSITION ASSY .... UOC:DAJ, V24, ZAJ	1
29	XBHZZ		19207	12300865	PANEL, CEILING FILLE .... UOC:DAJ, V24, ZAJ	1
30	PBHZZ	4520012578938	19207	12300871	PIPE, AIR CONDITION .... UOC:DAJ, V24, ZAJ	1
31	PAHZZ	2510011978562	19207	11677758	LAP STRIP, DUCT .... UOC:DAJ, V24, ZAJ	4
32	PFHZZ	2540011836796	19207	11677747	VENTILATOR, AIR CIRC .... UOC:DAJ, V24, ZAJ	1
33	PFHZZ	2540011803579	19207	11677741	VENTILATOR, AIR CIRC .... UOC:DAJ, V24, ZAJ	1

END OF FIGURE

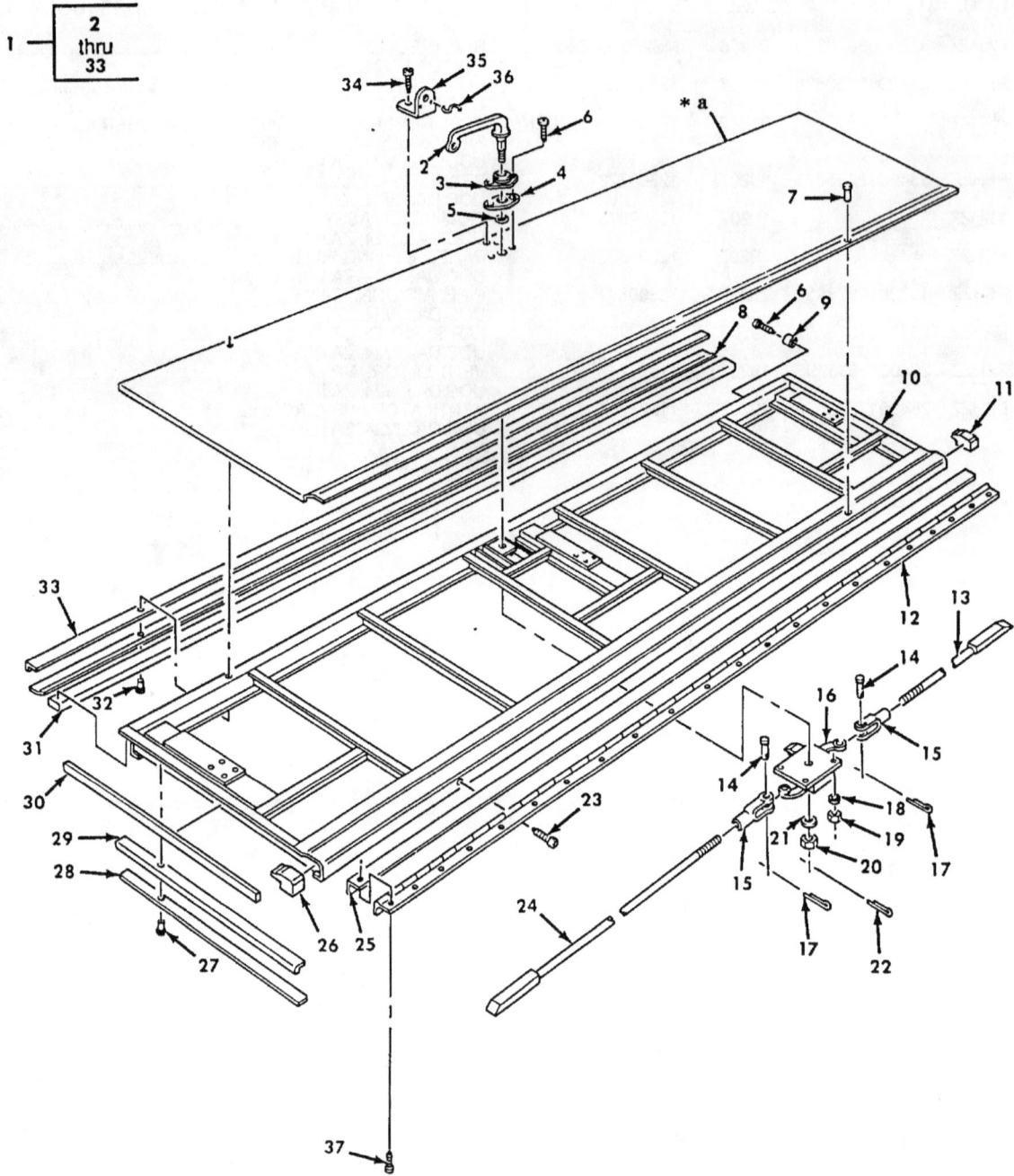

* a PART OF ITEM 1

Figure 419. Van Hinged Roof Assembly, Latch, and Related Parts, Right and Left Side.

(1) ITEM NO	(2) SMR CODE	(3) NSN	(4) CAGEC	(5) PART NUMBER	(6) DESCRIPTION AND USABLE ON CODES (UOC)	(7) QTY
					GROUP 1812 SPECIAL PURPOSE BODIES	
					FIG. 419 VAN HINGED ROOF ASSEMBLY, LATCH, AND RELATED PARTS, RIGHT AND LEFT SIDE	
1	PBHHH	2510011774458	19207	12300975-2	ROOF ASSEMBLY, HINGE RIGHT SIDE .... UOC:DAJ, V24, ZAJ	1
1	PBHHH	2510011794112	19207	12300975-1	PANEL, BODY, VEHICULA LEFT SIDE .... UOC:DAJ, V24, ZAJ	1
2	PFOZZ	2540001598823	19207	8380419	.HANDLE, DOOR, VEHICUL .... UOC:DAJ, V24, ZAJ	1
3	PAOZZ	5340002317428	19207	8698433	.ESCUTCHEON PLATE .... UOC:DAJ, V24, ZAJ	1
4	PAOZZ	5330004151488	19207	8380431	.GASKET............. UOC:DAJ, V24, ZAJ	1
5	PAOZZ	5330005228544	21450	582826	.GASKET............. UOC:DAJ, V24, ZAJ	1
6	PAOZZ	5305004324170	96906	MS51861-35	.SCREW, TAPPING .... UOC:DAJ, V24, ZAJ	6
7	PAHZZ	5320009567355	90030	AD64H	.RIVET, BLIND .... UOC:DAJ, V24, ZAJ	157
8	PAHZZ	5340010904479	19207	10937677	.PLATE, MOUNTING VAN BODY HINGE............... ROOF END SEAL RIGHT AND LEFT SIDE............... UOC:DAJ, V24, ZAJ	1
9	PAHZZ	5340002647182	19207	7035447	.BUMPER, NONMETALLIC .... UOC:DAJ, V24, ZAJ	4
10	XAHZZ		19207	12300971-1	.FRAME ASSEMBLY LEFT SIDE........ UOC:DAJ, V24, ZAJ	1
10	XAHZZ		19207	12300971-2	.FRAME ASSEMBLY RIGHT SIDE........ UOC:DAJ, V24, ZAJ	1
11	PAHZZ	2590004715344	19207	8380498	.PAD, CUSHIONING RIGHT SIDE........ UOC:DAJ, V24, ZAJ	1
11	PAHZZ	2590004715343	19207	7084738	.PAD, CUSHIONING LEFT SIDE........ UOC:DAJ, V24, ZAJ	1
12	PAHZZ	5340001583877	19207	7084987	.HINGE, BUTT........ UOC:DAJ, V24, ZAJ	1
13	PBOZZ	5306011835954	19207	7084729	.ROD, THREADED END .... UOC:DAJ, V24, ZAJ	1
14	PAOZZ	5315001401938	96906	MS35810-6	.PIN, STRAIGHT, HEADED .... UOC:DAJ, V24, ZAJ	2
15	PAOZZ	5340004664948	71843	2708-6A	.CLEVIS, ROD END .... UOC:DAJ, V24, ZAJ	2
16	PAOZZ	5340006791492	71843	2392A	.LATCH, MORTISE .... UOC:DAJ, V24, ZAJ	1
17	PAOZZ	5315008395822	96906	MS24665-353	.PIN, COTTER .... UOC:DAJ, V24, ZAJ	2
18	PAOZZ	5310004079566	96906	MS35338-45	.WASHER, LOCK. .... UOC:DAJ, V24, ZAJ	4
19	PAOZZ	5310008807744	96906	MS51967-5	.NUT, PLAIN, HEXAGON .... UOC:DAJ, V24, ZAJ	4
20	PAOZZ	5310002256993	96906	MS51922-33	.NUT, SELF-LOCKING, HE	1

(1) ITEM NO	(2) SMR CODE	(3) NSN	(4) CAGEC	(5) PART NUMBER	(6) DESCRIPTION AND USABLE ON CODES (UOC)	(7) QTY
					UOC:DAJ, V24, ZAJ	
21	PAOZZ	5310008093079	96906	MS27183-19	.WASHER, FLAT	1
					UOC:DAJ, V24, ZAJ	
22	PAOZZ	5315008395821	96906	MS24665-351	.PIN, COTTER	1
					UOC:DAJ, V24, ZAJ	
23	PAHZZ	5305004324253	96906	MS51861-67	.SCREW, TAPPING	69
					UOC:DAJ, V24, ZAJ	
24	PBOZZ	5306011835953	19207	7534663	.ROD, THREADED END	1
					UOC:DAJ, V24, ZAJ	
25	MHHZZ		19207	8380420-206	.RUBBER STRIP MAKE FROM RUBBER, P/N 8380420, 206 INCHES LONG	1
					UOC:DAJ, V24, ZAJ	
26	PAHZZ	2590004715343	19207	7084738	.PAD, CUSHIONING LEFT SIDE	1
					UOC:DAJ, V24, ZAJ	
26	PAHZZ	2590004715344	19207	8380498	.PAD, CUSHIONING RIGHT SIDE	1
					UOC:DAJ, V24, ZAJ	
27	PAHZZ	5320002348557	96906	MS20470AS-8	.RIVET, SOLID	24
					UOC:DAJ, V24, ZAJ	
28	PBHZZ	5365012083878	19207	10937693-4	.SPACER, PLATE	2
					UOC:DAJ, V24, ZAJ	
29	MHHZZ		19207	10937683-3-35	.NONMETALLC SPECIAL MAKE FROM RUBBER, P/N 10937683-3, 35 INCHES LONG	2
					UOC:DAJ, V24, ZAJ	
30	MHHZZ		19207	10937691-35	.RUBBER STRIP MAKE FROM RUBBER, P/N 10937691, 35 INCHES LONG	2
					UOC:DAJ, V24, ZAJ	
31	MHHZZ		19207	10937727-206	WEATERSEAL MAKE FROM RUBBER, P/N 10937727, 206 INCHES LONG	1
					UOC:DAJ, V24, ZAJ	
32	PAHZZ	5320002643266	96906	MS20470A6-9	.RIVET, SOLID	52
					UOC:DAJ, V24, ZAJ	
33	MHHZZ		19207	10937640-206	.SEAL MAKE FROM .RUBBER, P/N 10937640, 206 INCHES LONG	1
					UOC:DAJ, V24, ZAJ	
34	PAOZZ	5305010668646	96906	MS51862-33	SCREW, TAPPING	6
					UOC:DAJ, V24, ZAJ	
35	PAOZZ	5340004195881	19207	10896774	BRACKET, ANGLE	2
					UOC:DAJ, V24, ZAJ	
36	PAOZZ	4030007809350	96906	MS87006-13	HOOK, CHAIN, S.	2
					UOC:DAJ, V24, ZAJ	
37	PAHZZ	5305004324253	96906	MS51861-67	SCREW, TAPPING	136
					UOC:DAJ, V24, ZAJ	

END OF FIGURE

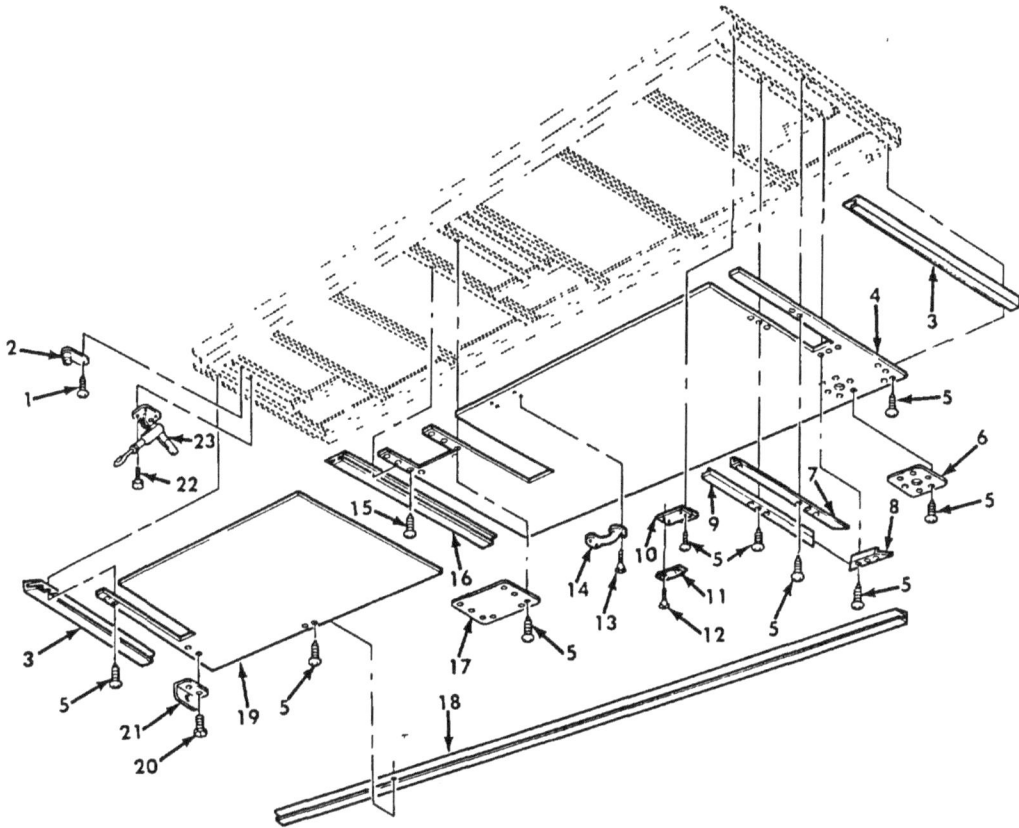

Figure 420. Van Hinged Roof Assembly, Holder Assemblies, Panels, and Related Parts.

(1) ITEM NO	(2) SMR CODE	(3) NSN	(4) CAGEC	(5) PART NUMBER	(6) DESCRIPTION AND USABLE ON CODES (UOC)	(7) QTY
					GROUP 1812 SPECIAL PURPOSE BODIES	
					FIG. 420 VAN HINGED ROOF ASSEMBLY, HOLDER ASSEMBLIES, PANELS, AND RELATED PARTS	
1	PAHZZ	5305004324253	96906	MS51861-67	SCREW, TAPPING .... UOC:DAJ, V24, ZAJ	6
2	PAOZZ	2510004702091	19207	7084726	HOLDER ASSEMBLY .... UOC:DAJ, V24, ZAJ	3
3	PBOZZ	2590011774455	19207	12300973-2	MOLDING, METAL RIGHT HINGED REAR .... PANEL .... UOC:DAJ, DAK, V24, ZAJ	1
3	PFOZZ	2590011774456	19207	12300973-1	MOLDING, METAL LEFT HINGED REAR .... PANEL .... UOC:DAJ, V24, ZAJ	1
4	MHHZZ		19207	12300961-2	PANEL, HINGED ROOF RIGHT, MAKE FROM...... PLYWOOD, P/N NN-P-530TY1GROUPB .... UOC:DAJ, V24, ZAJ	1
4	MHHZZ		19207	12300961-1	PANEL, HINGED ROOF LEFT, MAKE FROM....... PLYWOOD, P/N NN-P-530TY1GROUPB .... UOC:DAJ, V24, ZAJ	1
5	PAHZZ	5305004324170	96906	MS51861-35	SCREW, TAPPING....................................... UOC:DAJ, V24, ZAJ	154
6	XBHZZ		19207	7084732	PLATE, MOUNTING, HI .... UOC:DAJ, V24, ZAJ	1
8	PFHZZ	2590011977239	19207	8380465	BEZEL, AUTOMOTIVE TR .... UOC:DAJ, V24, ZAJ	3
9	PBOZZ	2590011946994	19207	8380463	BEZEL, AUTOMOTIVE TR .... UOC:DAJ, V24, ZAJ	3
10	PAHZZ	2510011774457	19207	12300960	FILLER .... UOC:DAJ, V24, ZAJ	3
11	PAHZZ	5340012088080	19207	7084724	PLATE, MENDING .... UOC-DAJ, V24, ZAJ	3
12	PAHZZ	5305011576794	96906	MS24627-62	SCREW, TAPPING .... UOC:DAJ, V24, ZAJ	6
13	XDOZZ		96906	MS51862-56	SCREW, TAPPING .... UOC:DAJ, V24, ZAJ	4
14	PAOZZ	5340006212591	19207	7397853	HANDLE, BOW .... UOC:DAJ, V24, ZAJ	1
15	PAHZZ	5305004324171	96906	MS51861-36	SCREW, TAPPING .... UOC:DAJ, V24, ZAJ	2
16	PFOZZ	2510011739200	19207	8380453-1	MOLDING, METAL .... UOC:DAJ, V24, ZAJ	1
17	PAHZZ	5340011900373	19207	12300947	COVER, ACCESS . ... UOC:DAJ, V24, ZAJ	1
18	PBOZZ	2590011863725	19207	8380447	BEZEL, AUTOMOTIVE TR .... UOC:DAJ, V24, ZAJ	1
19	MHHZZ		19207	12300963-2	PANEL, ROOF RIGHT SIDE, MAKE FROM......... PLYWOOD, P/N NN-P-530TY1GROUPB .... UOC:DAJ, V24, ZAJ	1
19	MHHZZ		19207	12300963-1	PANEL, BODY, VEHICULA LEFT SIDE, MAKE.....	1

(1) ITEM NO	(2) SMR CODE	(3) NSN	(4) CAGEC	(5) PART NUMBER	(6) DESCRIPTION AND USABLE ON CODES (UOC)	(7) QTY
					FROM PLYWOOD, P/N NN-P-550TY1GROUPB .... UOC:DAJ, V24, ZAJ	
20	PAHZZ	5305000039255	96906	MS51851-10B	SCREW, TAPPING .... UOC:DAJ, V24, ZAJ	4
21	PAHZZ	5340004199474	19207	7084730	BRACKET, ANGLE .... UOC:DAJ, V24, ZAJ	2
22	PAOZZ	5305010903012	96906	MS51851-106	SCREW, TAPPING .... UOC:DAJ, V24, ZAJ	12
23	PAOZZ	2510004702090	19207	7084716	CLAMP ASSEMBLY, ROOF .... UOC:DAJ, V24, ZAJ	3

END OF FIGURE

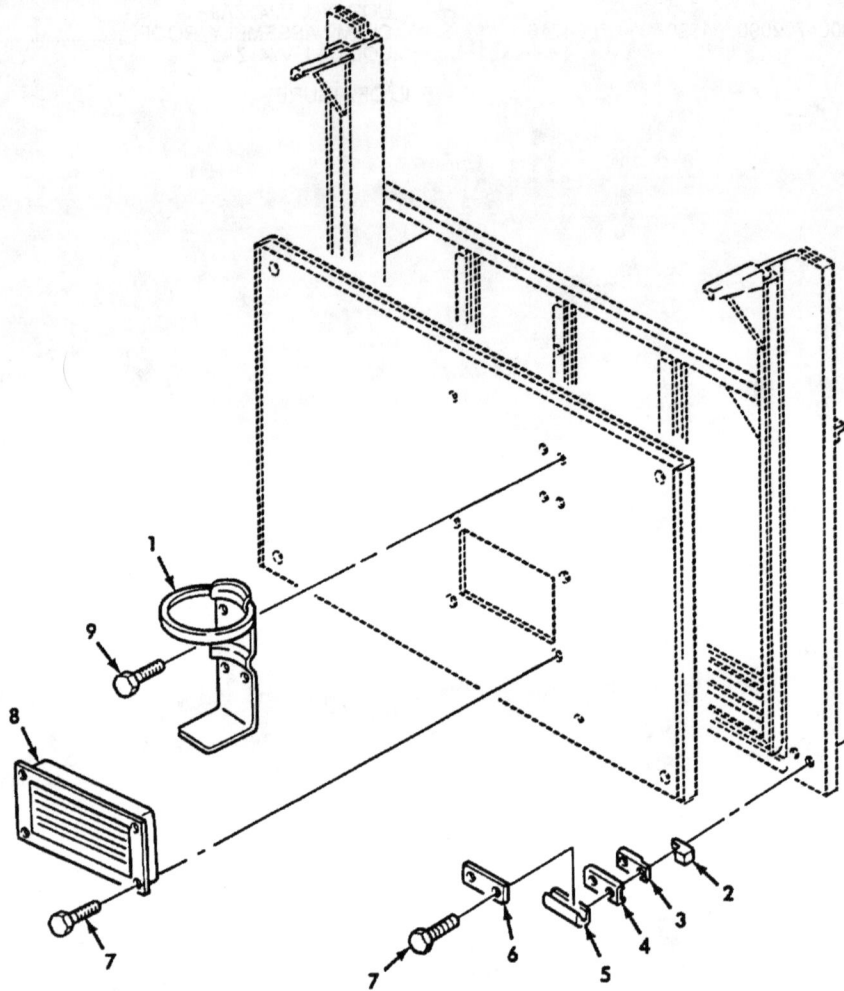

Figure 421. Van Front Wall and Fire Extinguisher Bracket.

(1) ITEM NO	(2) SMR CODE	(3) NSN	(4) CAGEC	(5) PART NUMBER	(6) DESCRIPTION AND USABLE ON CODES (UOC)	(7) QTY
					GROUP 1812 SPECIAL PURPOSE BODIES	
					FIG. 421 VAN FRONT WALL AND FIRE EXTINGUISHER BRACKET	
1	PFOZZ	4210011834822	19207	12255634	BRACKET, FIRE EXTING .... UOC:DAJ, V24, ZAJ	2
2	MOHZZ		19207	7535591	SPACER MAKE FROM NN-P-530.......................... UOC:DAJ, V24, ZAJ	2
3	PFHZZ	5330011586289	19207	7535590	SEAL, NONMETALLIC SP .... UOC:DAJ, V24, ZAJ	2
4	PFHZZ	5340011945298	19207	7535592	STRAP, RETAINING .... UOC:DAJ, V24, ZAJ	2
5	MHHZZ		19207	7534653-4	SEAL, REAR DOOR MAKE FROM RUBBER, P/ .... N 7534653, 4 INCHES LONG........... UOC:DAJ, V24, ZAJ	2
6	PFHZZ	5340011946474	19207	7535593	PLATE, MENDING...................... UOC:DAJ, V24, ZAJ	2
7	PAHZZ	5305004324172	96906	MS51861-37	SCREW, TAPPING .... UOC:DAJ, V24, ZAJ	53
8	PAHZZ	4520011865897	19207	7535647	REGISTER, METAL . ... UOC:DAJ, V24, ZAJ	4
9	PAOZZ	5305004324253	96906	MS51861-67	SCREW, TAPPING .... UOC:DAJ, V24, ZAJ	4

END OF FIGURE

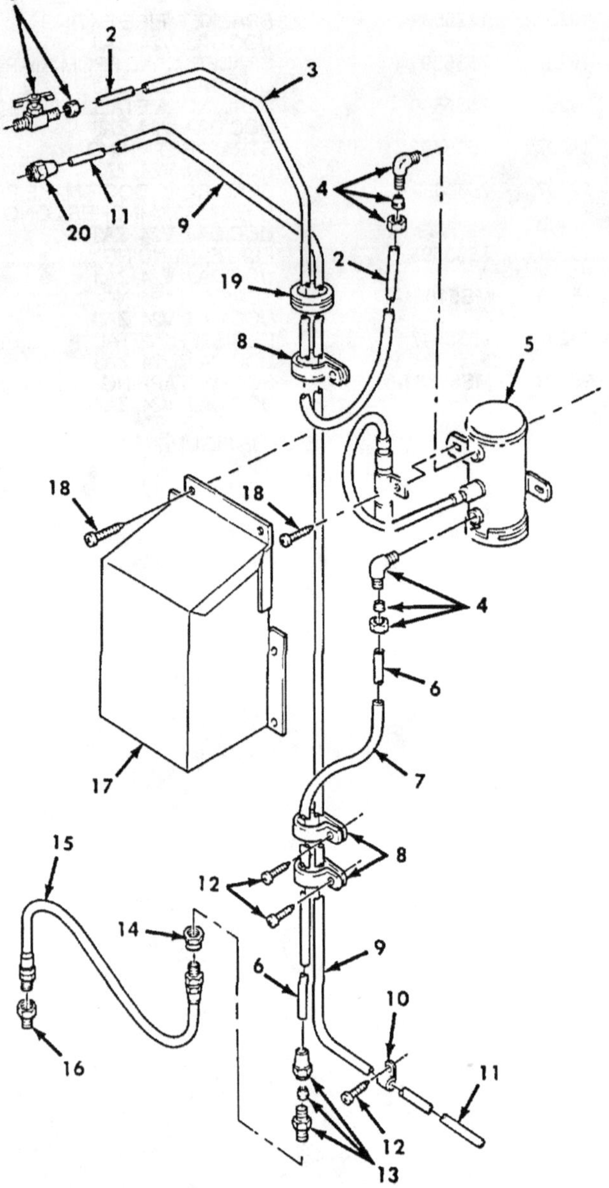

Figure 422. Van Heater Fuel Pump, Cover, Lines, and Fittings.

(1) ITEM NO	(2) SMR CODE	(3) NSN	(4) CAGEC	(5) PART NUMBER	(6) DESCRIPTION AND USABLE ON CODES (UOC)	(7) QTY
					GROUP 1812 SPECIAL PURPOS BODIES	
					FIG. 422 VAN HEATER FUEL PUMP, COVER, LINES, AND FITTINGS	
1	PAOZZ	4820002723351	21450	592786	COCK, SHUTOFF, SCREW .... UOC:DAJ, V24, ZAJ	1
2	MOOZZ		19207	8689206-40	TUBE MAKE FROM TUBING, P/N 8689206, 40 INCHES LONG .... UOC:DAJ, V24, ZAJ	1
3	MOOZZ		80244	17-C-18035-60-38	CONDUIT, NONMETALLIC MAKE FROM CONDUIT, P/N 17-C-18035-60, 38 INCHES LONG .... UOC:DAJ, V24, ZAJ	1
4	PAOZZ	4730000968756	96906	MS51815-3	ELBOW, PIPE TO TUBE .... UOC:DAJ, V24, ZAJ	2
5	PAOZZ	2910007106054	96906	MS51321-1-24N1	PUMP, FUEL, ELECTRICA .... UOC:DAJ, V24, ZAJ	1
6	MOOZZ		19207	8689206-54	TUBE MAKE FROM TUBING, P/N 8689206, 54 INCHES LONG .... UOC:DAJ, V24, ZAJ	1
7	MOOZZ		80244	17-C-18035-60-52	CONDUIT, NONMETALLIC MAKE FROM CONDUIT, P/N 17-C-18035-60, 52 INCHES LONG .... UOC:DAJ, V24, ZAJ	1
8	PAOZZ	5340007247038	96906	MS21333-76	CLAMP, LOOP .... UOC:DAJ, V24, ZAJ	3
9	MOOZZ		80244	17-C-18035-60-92	CONDUIT, NONMETALLIC MAKE FROM CONDUIT, P/N 17-C-18035-60, 92 INCHES LONG .... UOC:DAJ, V24, ZAJ	1
10	PAOZZ	5340008272453	96906	MS35150-5	STRAP, RETAINING .... UOC:DAJ, V24, ZAJ	1
11	MOOZZ		19207	8689206-92	TUBE MAKE FROM TUBING, P/N 8689206, 92 INCHES LONG .... UOC:DAJ, V24, ZAJ	1
12	PAOZZ	5305004324203	96906	MS51861-47	SCREW, TAPPING .... UOC:DAJ, V24, ZAJ	3
13	PAOZZ	4730002778750		81343 4-2	120102BA ADAPTER, STRAIGHT, PI .... UOC:DAJ, V24, ZAJ	1
14	PAOZZ	4730012406112	79470	252X5	ADAPTER, STRAIGHT, PI .... UOC:DAJ, V24, ZAJ	1
15	PAOZZ	4720004828956	19207	7539107-2	HOSE ASSEMBLY, NONME .... UOC:DAJ, V24, ZAJ	1
16	PAOZZ	4730002889953	79470	202X5X4	ADAPTER, STRAIGHT, PI .... UOC:DAJ, V24, ZAJ	1
17	PFOZZ	5340011819449	19207	11608871	COVER, ACCESS . ... UOC:DAJ, V24, ZAJ	1
18	PAOZZ	5305004324201	96906	MS51861-45	SCREW, TAPPING .... UOC:DAJ, V24, ZAJ	9
19	PAOZZ	5325012427083	96906	MS35489-17	GROMMET, NONMETALLIC .... UOC:DAJ, V24, ZAJ	1
20	PAOZZ	4730000116452	79470	111OX4	NUT, TUBE COUPLING .... UOC:DAJ, V24, ZAJ	1

END OF FIGURE

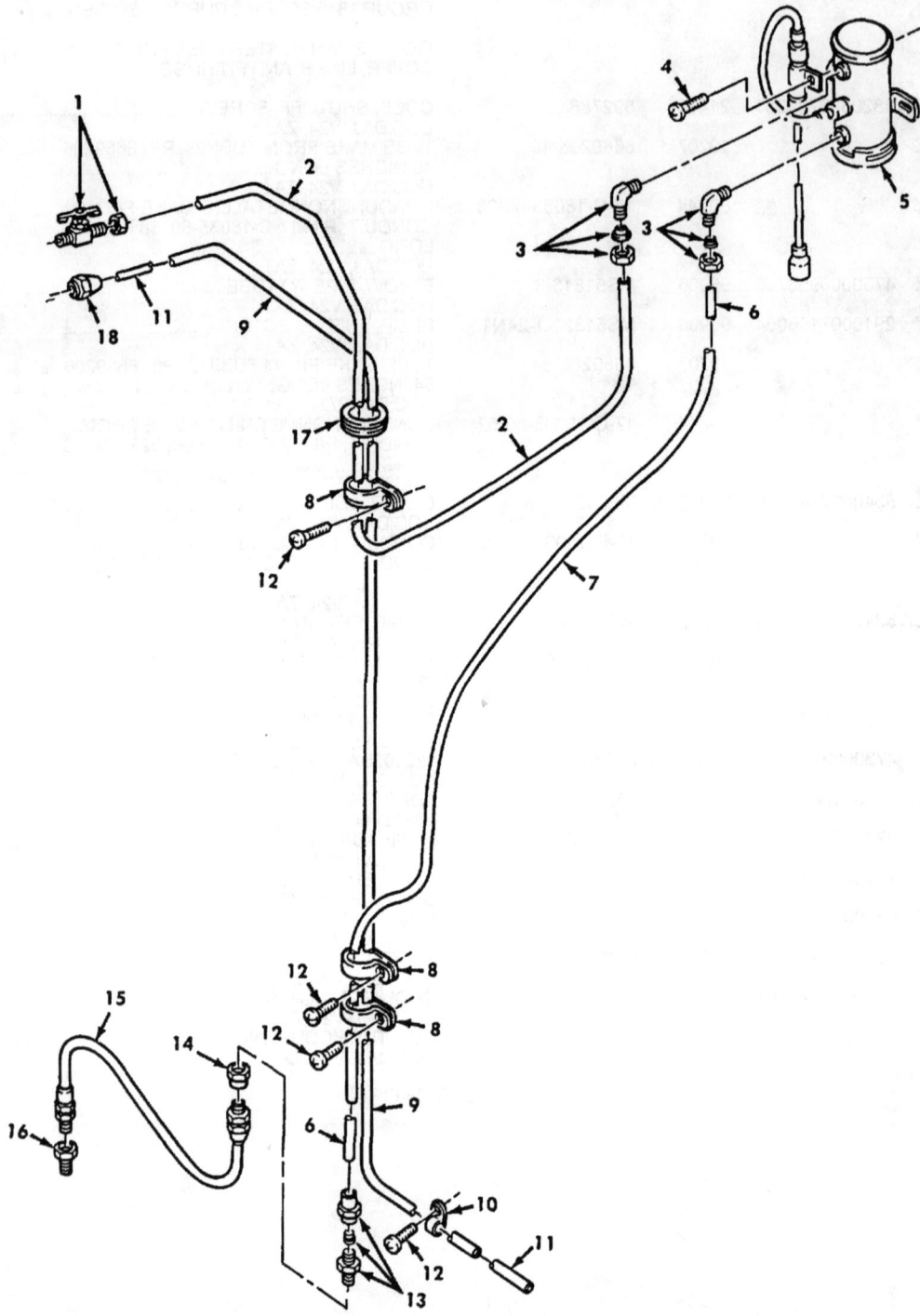

Figure 423. Van Heater Fuel Pump, Lines, and Fittings.

(1) ITEM NO	(2) SMR CODE	(3) NSN	(4) CAGEC	(5) PART NUMBER	(6) DESCRIPTION AND USABLE ON CODES (UOC)	(7) QTY
					GROUP 1812 SPECIAL PURPOSE BODIES	
					FIG. 423 VAN HEATER FUEL PUMP, LINES, AND FITTINGS	
1	XDOZZ		79470	A555	COCK, SHUTOFF, SCREW .... UOC: DAJ, ZAJ	1
2	PAOZZ	4710012106198	19207	12303046	TUBE, BENT, METALLIC .... UOC: DAJ, ZAJ	1
3	PAOZZ	4730000968756	96906	MS51815-3	ELBOW, PIPE TO TUBE .... UOC:DAJ, ZAJ	2
4	PAOZZ	5305004324201	96906	MS51861-45	SCREW, TAPPING .... UOC:DAJ, ZAJ	2
5	PAOZZ	2910007106054	96906	MS51321-1-24N1	PUMP, FUEL, ELECTRICA .... UOC:DAJ, ZAJ	1
6	MOOZZ		19207	8689206-70	TUBE, METALLIC MAKE FROM TUBE, P/N .... 8682906, 70 INCHES LONG........................ UOC:DAJ, ZAJ	1
7	MOOZZ		80244	17-C-18035-60-68	CONDUIT MAKE FROM CONDUIT, P/N 17-C-........................ 18035-60, 68 INCHES LONG........................ UOC:DAJ, ZAJ	1
8	PAOZZ	5340007247038	96906	MS21333-76	CLAMP, LOOP........................ UOC: DAJ, ZAJ	3
9	MOOZZ		80244	17-C-18035-60-90	CONDUIT MAKE FROM CONDUIT, P/N 17-C-........................ 18035-60, 90 INCHES LONG........................ UOC:DAJ, ZAJ	1
10	PAOZZ	5340008272453	96906	MS35150-5	STRAP, RETAINING .... UOC:DAJ, ZAJ	1
11	MOOZZ		19207	8689206-92	TUBE, METALLIC MAKE FROM TUBE, P/N .... 8689206, 92 INCHES LONG........................ UOC:DAJ, ZAJ	1
12	PAOZZ	5305004324203	96906	MS51861-47	SCREW, TAPPING. .... UOC:DAJ, ZAJ	4
13	PAOZZ	4730002778750	81343	4-2 120102BA	ADAPTER, STRAIGHT, PI........................ UOC:DAJ, ZAJ	1
14	PAOZZ	4730012406112	79470	252X5	ADAPTER, STRAIGHT, PI........................ UOC:DAJ, ZAJ	1
15	PAOZZ	4720004828956	19207	7539107-2	HOSE ASSEMBLY, NONME .... UOC:DAJ, ZAJ	1
16	PAOZZ	4730002889953	21450	187343	ADAPTER, STRAIGHT, PI .... UOC:DAJ, ZAJ	1
17	PAOZZ	5325012427083	96906	MS35489-17	GROMMET, NONMETALLIC .... UOC:DAJ, ZAJ	1
18	PAOZZ	4730000116452	79470	111OX4	NUT, TUBE COUPLING........................ UOC:DAJ, ZAJ	1

END OF FIGURE

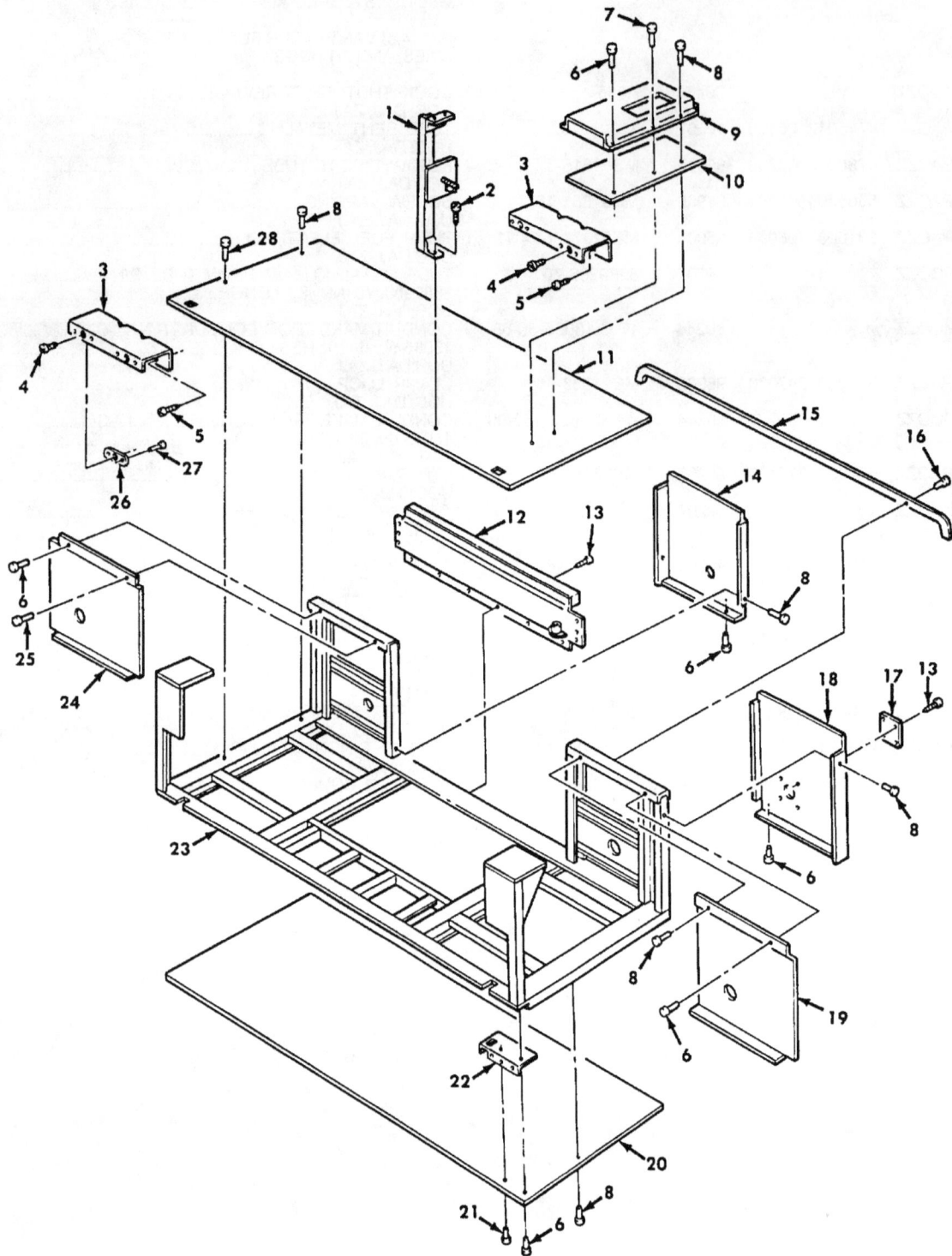

Figure 424. Van Bonnet Frame and Related Parts.

(1) ITEM NO	(2) SMR CODE	(3) NSN	(4) CAGEC	(5) PART NUMBER	(6) DESCRIPTION AND USABLE ON CODES (UOC)	(7) QTY
					GROUP 1812 SPECIAL PURPOSE BODIES	
					FIG. 424 VAN BONNET FRAME AND RELATED PARTS	
1	PAHZZ	2510011870379	19207	12300756	POST ASSEMBLY, CONTA..................... UOC:DAJ, V24, ZAJ	1
2	PAHZZ	5305004324253	96906	MS51861-67	SCREW, TAPPING UOC:DAJ, V24, ZAJ	3
3	PFHZZ	5340011870376	19207	12302732	BRACKET, DOUBLE ANGL .... UOC:DAJ, V24, ZAJ	2
4	PAHZZ	5305004213986	19207	7336058	SETSCREW .... UOC:DAJ, V24, ZAJ	4
5	PAHZZ	5305001913640	96906	MS51851-85	SCREW, TAPPING .... UOC:DAJ, V24, ZAJ	6
6	PFHZZ	5320005823276	96906	MS20600AD6W4	RIVET, BLIND............................ UOC:DAJ, V24, ZAJ	142
7	PFHZZ	5320008134144	96906	MS20600AD6-W7	RIVET, BLIND............................ UOC:DAJ, V24, ZAJ	4
8	PFHZZ	5320005823268	53551	RV200-6-3	RIVET, BLIND............................ UOC:DAJ, V24, ZAJ	195
9	XBHZZ		19207	12302722	DUCT .... UOC:DAJ, V24, ZAJ	2
10	XBHZZ		19207	12302774	INSULATION . ... UOC:DAJ, V24, ZAJ	2
11	PFHZZ	2510012255864	19207	12302731	PANEL, BONNET FLOOR .... UOC:DAJ, V24, ZAJ	1
12	PBHZZ	2510011870375	19207	12302748	PANEL, BODY, VEHICULA.............. UOC:DAJ, V24, ZAJ	1
13	PAHZZ	5305004324170	96906	MS51861-35	SCREW, TAPPING .... UOC:DAJ, V24, ZAJ	21
14	PBHZZ	2510011870372	19207	12302705-1	PANEL, BODY, VEHICULA .... UOC:DAJ, V24, ZAJ	1
15	PFFZZ	2590011808571	19207	7084952	BEZEL, AUTOMOTIVE TR BOX, BONNET............. ASSEMBLY .... UOC:DAJ, V24, ZAJ	1
16	PFHZZ	5305001498610	96906	MS51862-36	SCREW, TAPPING .... UOC:DAJ, V24, ZAJ	27
17	PFHZZ	2510001598822	19207	7084949	PLATE, BELLOW RETAIN .... UOC:DAJ, V24, ZAJ	1
18	PBHZZ	2510011870378	19207	12302705-2	PANEL, BODY, VEHICULA .... UOC:DAJ, V24, ZAJ	1
19	PBHZZ	2510011870374	19207	12302706-2	PANEL, BODY, VEHICULA .... UOC:DAJ, V24, ZAJ	1
20	PBHZZ	2510011870377	19207	12302730	PANEL, BODY, VEHICULA UOC:V24	1
20	PBHZZ	2510012116613	19207	12302730-1	PANEL, BODY, VEHICULA .... UOC:DAJ, ZAJ	1
22	XBHZZ		19207	8735125	FILLER RIGHT................................ UOC:DAJ, V24, ZAJ	1
22	XAHZZ		19207	8735124	FILLER LEFT .... UOC:DAJ, V24, ZAJ	1

(1) ITEM NO	(2) SMR CODE	(3) NSN	(4) CAGEC	(5) PART NUMBER	(6) DESCRIPTION AND USABLE ON CODES (UOC)	(7) QTY
23	XAHZZ		19207	12302720	FRAME ASSEMBLY UOC:V24, V25	1
23	XAHZZ		19207	12302720-1	FRAME .... UOC:DAJ, DAK, ZAJ, ZAK	1
24	PBHZZ	2510011870373	19207	12302706-1	PANEL, BODY, VEHICULA FRONT .... UOC:DAJ, V24, ZAJ	1
25	PFHZZ	5320005823301	53551	RV200-6-5	RIVET, BLIND .... UOC:DAJ, V24, ZAJ	21
26	PFHZZ	5310008857734	94222	17-10015-13	NUT, SHEET SPRING .... UOC:DAJ, V24, ZAJ	4
27	PFHZZ	5320006164350	53551	RV200-6-1	RIVET, BLIND .... UOC:DAJ, V24, ZAJ	8
28	PFHZZ	5320005823499	11815	CR9163-6-6	RIVET, BLIND .... UOC:DAJ, V24, ZAJ	35

END OF FIGURE

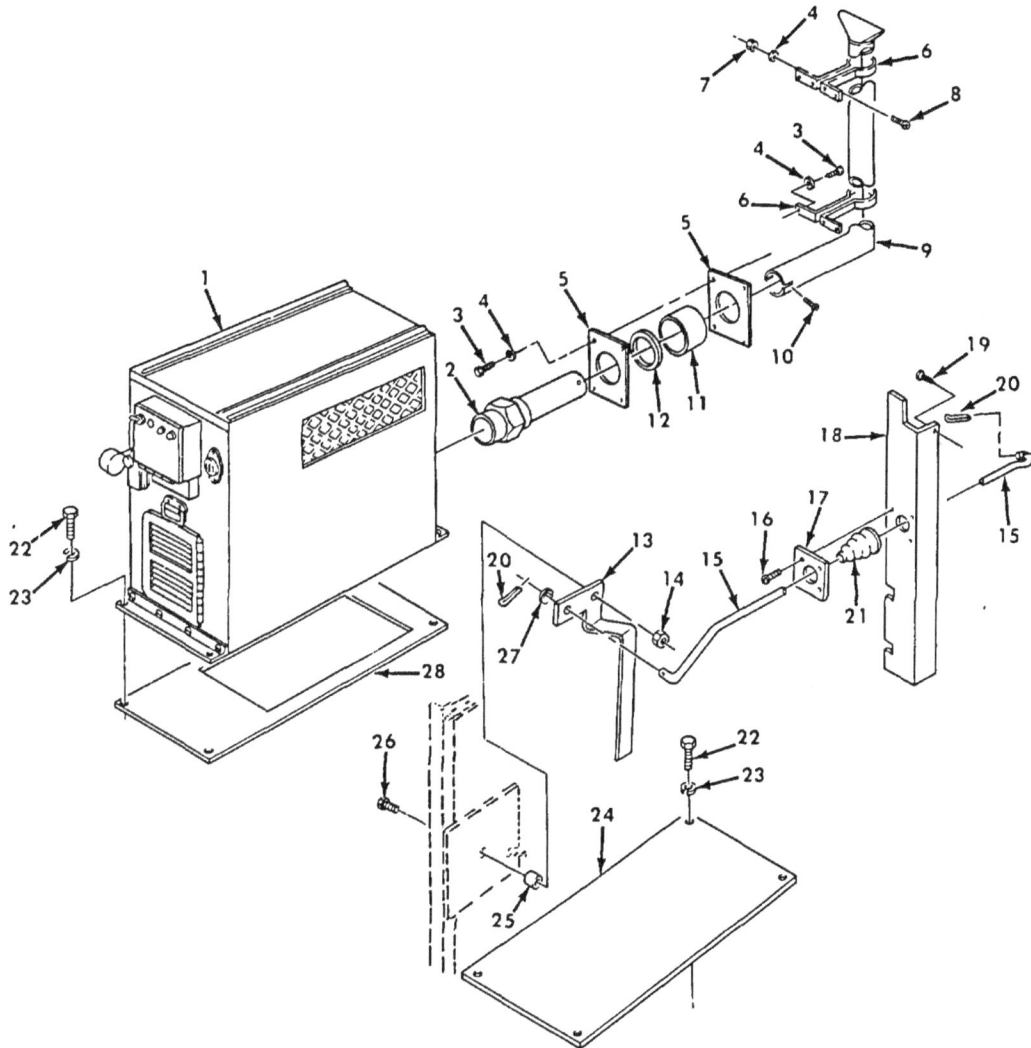

Figure 425. Van Heater, Exhaust, and Vent Door Controls.

(1) ITEM NO	(2) SMR CODE	(3) NSN	(4) CAGEC	(5) PART NUMBER	(6) DESCRIPTION AND USABLE ON CODES (UOC)	(7) QTY
					GROUP 1812 SPECIAL PURPOSE BODIES	
					FIG. 425 VAN HEATER, EXHAUST, AND VENT DOOR CONTROLS	
1	PFOHH	4520001141055	92878	UH68D	HEATER, SPACE .... UOC: DAJ, V24, ZAJ	1
2	PFOZZ	4730012024101	19207	11681649	UNION, PIPE TO TUBE .... UOC: DAJ, V24, ZAJ	2
3	PAOZZ	5305004324251	96906	MS51861-65	SCREW, TAPPING .... UOC: DAJ, V24, ZAJ	16
4	PAOZZ	5310005825965	96906	MS35338-44	WASHER, LOCK .... UOC: DAJ, V24, ZAJ	18
5	PFOZZ	4520011698680	19207	10919646	PLATE, FLUE PIPE, HEA .... UOC: DAJ, V24, ZAJ	2
6	PAOZZ	5340004234080	19207	10919652	CLAMP, LOOP .... UOC:DAJ, DAK, V24,ZAJ	2
7	PAOZZ	5310008775796	96906	MS21044N4	NUT, SELF-LOCKING, HE .... UOC: DAJ, V24, ZAJ	2
8	PAOZZ	5305009932461	96906	MS35207-281	SCREW, MACHINE .... UOC: DAJ, V24, ZAJ	2
9	PFOZZ	2540003789049	19207	10919648	EXHAUST STACK, HEATE .... UOC: DAJ, V24, ZAJ	1
10	PAOZZ	5305000581082	96906	MS51861-34	SCREW, TAPPING .... UOC: DAJ, V24, ZAJ	2
11	PFOZZ	5365003890317	19207	10919647	SPACER, SLEEVE . ... UOC: DAJ, V24, ZAJ	1
12	PAOZZ	5330006268281	19207	10923475	O-RING .... UOC: DAJ, V24, ZAJ	4
13	PAOZZ	3040011928356	19207	12300745	LEVER, MANUAL CONTRO .... UOC: DAJ, V24, ZAJ	1
14	PAOZZ	5310009843807	96906	MS51922-13	NUT, SELF-LOCKING, HE .... UOC: DAJ, V24, ZAJ	1
15	PFOZZ	5340012319313	19207	12302747	CONTROL ROD .... UOC: DAJ, V24, ZAJ	1
16	PAOZZ	5305004324170	96906	MS51861-35	SCREW, TAPPING .... UOC: DAJ, V24, ZAJ	4
17	PFOZZ	2510001598822	19207	7084949	PLATE, BELLOW RETAIN .... UOC: DAJ, V24, ZAJ	1
18	PBOZZ	2510011870382	19207	12302766	PANEL, BODY, VEHICULA RIGHT SIDE .... UOC: DAJ, V24, ZAJ	1
19	PAOZZ	5305004324172	96906	MS51861-37	SCREW, TAPPING .... UOC: DAJ, V24, ZAJ	6
20	PAOZZ	5315008395821	96906	MS24665-351	PIN, COTTER .... UOC: DAJ, V24, ZAJ	2
21	PAOZZ	5340006791447	19207	7535612	BELLOWS, PROTECTION .... UOC: DAJ, V24, ZAJ	1
22	PAOZZ	5306002258499	96906	MS90725-34	BOLT, MACHINE .... UOC: DAJ, V24, ZAJ	8
23	PAOZZ	5310004079566	96906	MS35338-45	WASHER, LOCK . ... UOC: DAJ, V24, ZAJ	8
24	PFOZZ	5340012097447	19207	12300747	COVER, ACCESS RIGHT SIDE	1

(1) ITEM NO	(2) SMR CODE	(3) NSN	(4) CAGEC	(5) PART NUMBER	(6) DESCRIPTION AND USABLE ON CODES (UOC)	(7) QTY
25	PAOZZ	5365012319280	19207	12300744	UOC: DAJ, V24, ZAJ SPACER, SLEEVE	1
26	PAOZZ	5306010282443	80205	NAS1297-5-10	UOC: DAJ, V24, ZAJ BOLT, SHOULDER	1
27	PAOZZ	5310000806004	96906	MS27183-14	UOC: DAJ, V24, ZAJ WASHER, FLAT	1
28	PFOZZ	5330002228992	19207	8735318	UOC: DAJ, V24, ZAJ GASKET UOC: DAJ, V24, ZAJ	1

END OF FIGURE

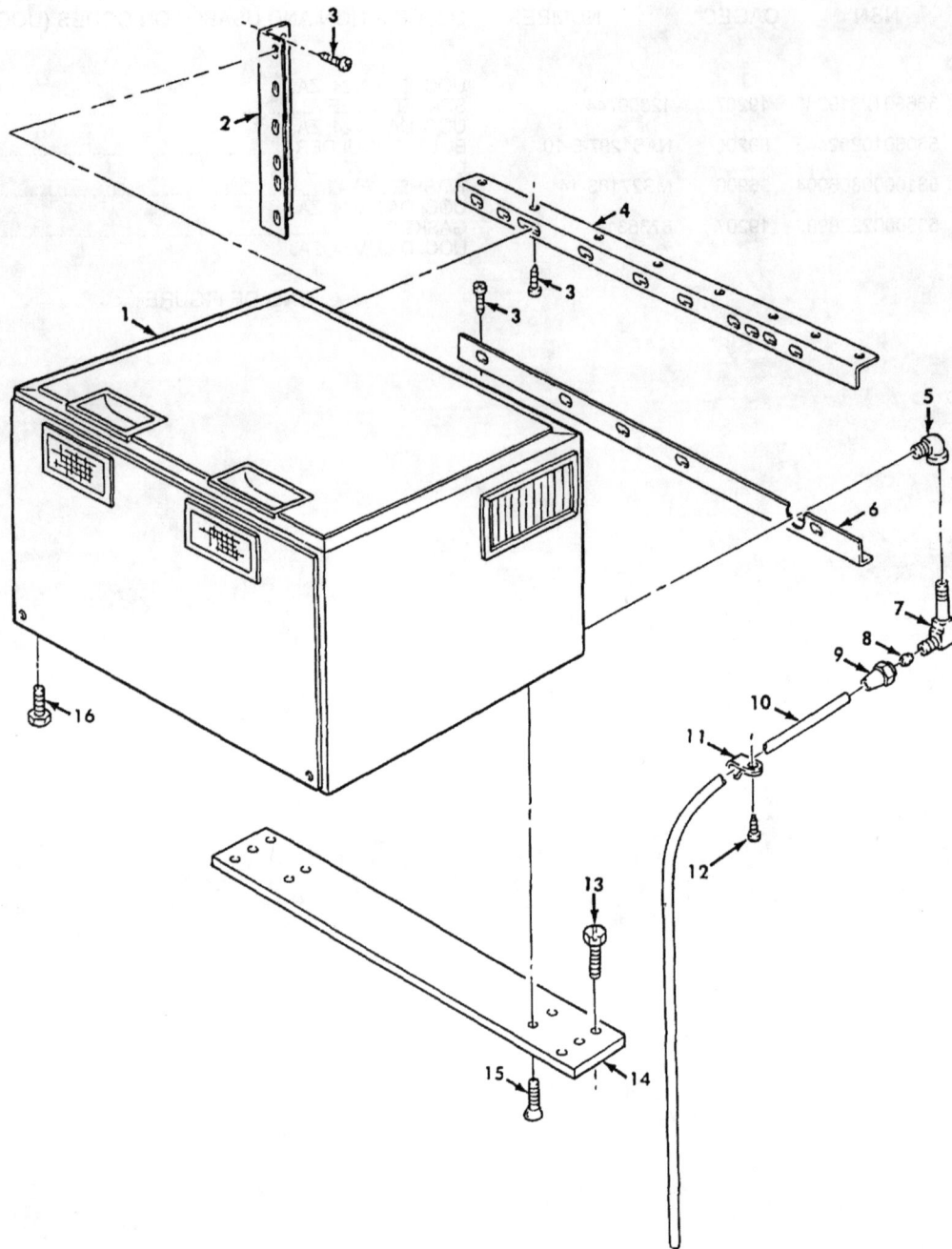

Figure 426. Van Air Conditioner, Drain Tube, and Mounting Hardware.

SECTION II

(1) ITEM NO	(2) SMR CODE	(3) NSN	(4) CAGEC	(5) PART NUMBER	(6) DESCRIPTION AND USABLE ON CODES (UOC)	(7) QTY
					GROUP 1812 SPECIAL PURPOSE BODIES	
					FIG. 426 VAN AIR CONDITIONER, DRAIN TUBE, AND MOUNTING HARDWARE	
1	PAHHH	4120013306543	63702	MH-40-MP	AIR CONDITIONER INSTALL WITH KIT NSN, 2540-01-325-1933 UOC: DAJ, V24, ZAJ	1
2	PAFZZ	2510011870381	19207	12302765	STRIP, PANEL LEFT FRONT UOC: DAJ, V24, ZAJ	1
3	PAHZZ	5305004324172	96906	MS51861-37	SCREW, TAPPING UOC: DAJ, V24, ZAJ	12
4	PFHZZ	5340011896402	19207	12302764	BRACKET, NGLE. UOC: DAJ, V24, ZAJ	1
5	PAOZZ	4730002775553	19207	444166	ELBOW, PIPE UOC: DAJ, V24, ZAJ	1
6	PAHZZ	2510011870380	19207	12302763	FRAME SECTION, STRUC UOC: DAJ, V24, ZAJ	1
7	PAOZZ	4730011654647	19207	11669790	ELBOW, PIPE UOC: DAJ, V24, ZAJ	1
8	PAOZZ	4730009696941	79470	C5165X4	SLEEVE, FLARED, TUBE UOC: DAJ, V24, ZAJ	1
9	PAOZZ	4730003148366	96906	MS51531-B4	NUT, TUBE COUPLING UOC: DAJ, V24, ZAJ	1
10	MOOZZ		19207	8684206-68	TUBE MAKE FROM TUBE, P/N 8689206,68 INCHES LONG UOC: DAJ, V24, ZAJ	1
11	PAOZZ	5340000797837	96906	MS21333-67	CLAMP, LOOP UOC: DAJ, V24, ZAJ	3
12	PAOZZ	5305004324170	96906	MS51861-35	SCREW, TAPPING UOC: DAJ, V24, ZAJ	3
13	PAHZZ	5305012097112	96906	MS51851-129	SCREW, TAPPING UOC: DAJ, V24, ZAJ	6
14	PAHZZ	5340011896403	19207	12300751	PLATE, MOUNTING UOC: DAJ, V24, ZAJ	1
15	PAHZZ	5305009592703	96906	MS35191-322	SCREW, MACHINE UOC: DAJ, V24, ZAJ	4
16	PAHZZ	5305005698909	19207	7372083	SCREW, ASSEMBLED WA UOC: DAJ, V24, ZAJ	2

END OF FIGURE

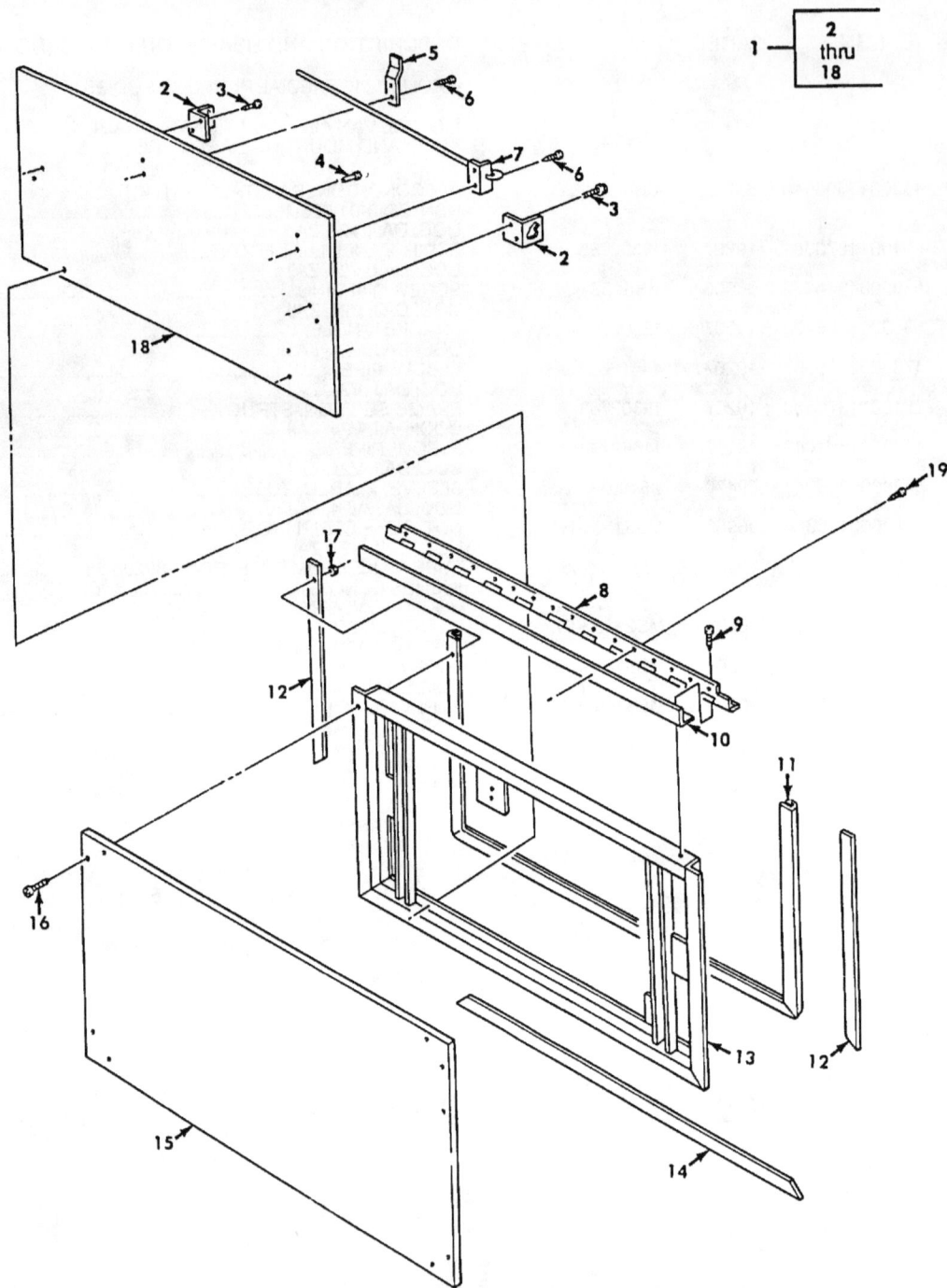

Figure 427. Van Bonnet Access Door Assembly.

(1) ITEM NO	(2) SMR CODE	(3) NSN	(4) CAGEC	(5) PART NUMBER	(6) DESCRIPTION AND USABLE ON CODES (UOC)	(7) QTY
					GROUP 1812 SPECIAL PURPOSE BODIES	
					FIG. 427 VAN BONNET ACCESS DOOR ASSEMBLY	
1	PAFFF	2510011870566	19207	12302740	DOOR, ACCESS .... UOC: DAJ, V24, ZAJ	1
2	PFFZZ	5340003285458	19207	10937687	.BRACKET, ANGLE............................. UOC: DAJ, V24, ZAJ	1
3	PAFZZ	5305002253843	80204	B1821BH025C100N	.SCREW, CAP, HEXAGON H .... UOC: DAJ, V24, ZAJ	2
4	PFFZZ	5320005841285	53551	RV200-6-2	.RIVET, BLIND .... UOC: DAJ, V24, ZAJ	66
5	PFFZZ	5340011639912	19207	7084942	.BRACKET, DOUBLE ANGL .... UOC: DAJ, V24, ZAJ	1
6	PAFZZ	5305004324201	96906	MS51861-45	.SCREW, TAPPING .... UOC: DAJ, V24, ZAJ	2
7	PFFZZ	510011639752	19207	7084939	.ROD ASSEMBLY, ACCES .... UOC: DAJ, V24, ZAJ	1
8	PAFZZ	5340010657287	19207	7534654	.HINGE, STRAP .... UOC: DAJ, V24, ZAJ	1
9	PAFZZ	5305004324252	96906	MS51861-66	.SCREW, TAPPING .... UOC: DAJ, V24, ZAJ	12
10	MFFZZ		19207	8380420-51	.RUBBER STRIP MAKE FROM RUBBER .... STRIP, P/N 8380420, 51 INCHES LONG............................ UOC: DAJ, V24, ZAJ	1
11	PAFZZ	5330012097354	19207	12302744	.SEAL, NONMETALLIC SP .... UOC: DAJ, V24, ZAJ	1
12	PFFZZ	2590011870363	19207	12302742	.MOLDING, METAL .... UOC: DAJ, V24, ZAJ	2
13	XAFZZ		19207	12302738	.FRAME .... UOC: DAJ, V24, ZAJ	1
14	PFFZZ	2590011639755	19207	8759398	.MOLDING, METAL .... UOC: DAJ, V24, ZAJ	1
15	XAFZZ		19207	12302739	.PANEL, BODY, VEHICULA .... UOC: DAJ, V24, ZAJ	1
16	PFFZZ	5305009846191	96906	MS35206-243	.SCREW, MACHINE .... UOC: DAJ, V24, ZAJ	33
17	PAFZZ	5310009349757	96906	MS35649-282	.NUT, PLAIN, HEXAGON .... UOC: DAJ, V24, ZAJ	33
18	PFFZZ	2510011870364	19207	12302750	.PANEL, BODY, VEHICULA .... UOC: DAJ, V24, ZAJ	1
19	PAFZZ	5305004324252	96906	MS51861-66	.SCREW, TAPPING .... UOC: DAJ, V24, ZAJ	12

END OF FIGURE

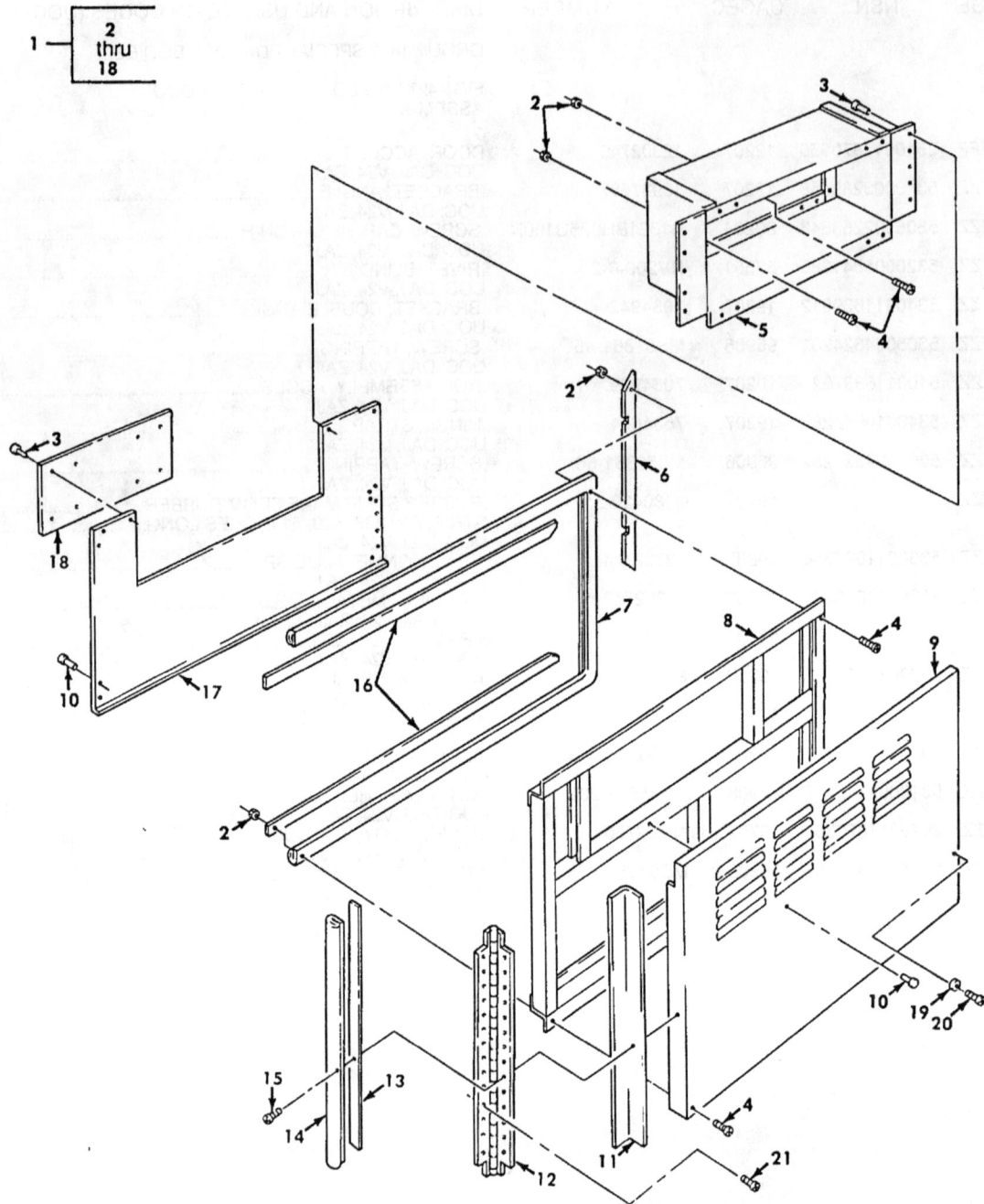

Figure 428. Van Bonnet Door Assembly, Right and Left Side.

(1) ITEM NO	(2) SMR CODE	(3) NSN	(4) CAGEC	(5) PART NUMBER	(6) DESCRIPTION AND USABLE ON CODES (UOC)	(7) QTY
					GROUP 1812 SPECIAL PURPOSE BODIES	
					FIG. 428 VAN BONNET DOOR ASSEMBLY, RIGHT AND LEFT SIDE	
1	PAFFF	2510011870567	19207	12302848-1	DOOR, ACCESS RIGHT SIDE .... UOC: DAJ, V24, ZAJ	1
1	PAFFF	2510011870568	19207	12302848-2	DOOR, ACCESS LEFT SIDE .... UOC: DAJ, V24, ZAJ	1
2	PAFZZ	5310009349757	96906	MS35649-282	.NUT, PLAIN, HEXAGON .... UOC: DAJ, V24, ZAJ	25
3	PAFZZ	5320005823268	53551	RV200-6-3	.RIVET, BLIND .... UOC: DAJ, V24, ZAJ	50
4	PAFZZ	5305009846191	96906	MS35206-243	.SCREW, MACHINE . ... UOC: DAJ, V24, ZAJ	25
5	PFFZZ	2510011870368	19207	12302805	.PANEL ASSEMBLY, AIR .... UOC: DAJ, V24, ZAJ	1
6	PFFZZ	2590012251071	19207	12302846	.BEZEL, AUTOMOTIVE TR .... UOC: DAJ, V24, ZAJ	1
7	MFFZZ		19207	11607302-78	.SEAL, NONMETALLIC SP MAKE FROM............................ RUBBER, P/N 11607302, 78 INCHES LONG........................ UOC: DAJ, V24, ZAJ	1
8	XAFZZ		19207	12302817-1	.FRAME ASSEMBLY, DOOR RIGHT SIDE............................ UOC: DAJ, V24, ZAJ	1
8	XAFZZ		19207	12302817-2	.FRAME ASSEMBLY, DOOR LEFT SIDE............................ UOC: DAJ, V24, ZAJ	1
9	PFFZZ	2510011870366	19207	12302811-1	.PANEL, BODY, VEHICULA RIGHT SIDE .... UOC: DAJ, V24, ZAJ	1
9	PBFZZ	2510011870367	19207	12302811-2	.PANEL, BODY, VEHICULA LEFT SIDE .... UOC: DAJ, V24, ZAJ	1
10	PAFZZ	5320005841285	53551	RV200-6-2	.RIVET, BLIND .... UOC: DAJ, V24, ZAJ	50
11	MFFZZ		19207	8380420-27	.RUBBER STRIP MAKE FROM RUBBER .... STRIP, P/N 8380420, 27 INCHES LONG........................ UOC: DAJ, V24, ZAJ	1
12	PAFZZ	5340011968113	19207	12302819	.HINGE, BUTT... UOC: DAJ, V24, ZAJ	1
13	MFFZZ		19207	11607302-24	.SEAL, HINGE MAKE FROM SEAL, P/N .... 11607302, 24 INCHES LONG........................ UOC: DAJ, V24, ZAJ	1
14	PFFZZ	5340011870370	19207	12302842	.PLATE, RETAINING, SEA .... UOC: DAJ, V24, ZAJ	1
15	PAFZZ	5305004770144	96906	MS51861-68	.SCREW, TAPPING .... UOC: DAJ, V24, ZAJ	6
16	PFFZZ	2510011870371	19207	12302844	.MOLDING, METAL .... UOC: DAJ, V24, ZAJ	2
17	PFFZZ	2510012020965	19207	12302810	.PANEL, BODY, VEHICULA .... UOC: DAJ, V24, ZAJ	1
18	PAFZZ	2510011870369	19207	12302809	.PANEL, AIR INTAKE BONNET ASSEMBLY........................ UOC: DAJ, V24, ZAJ	1
19	PAFZZ	5310000877493	96906	MS27183-13	.WASHER, FLAT .... UOC: DAJ, V24, ZAJ	4

(1) ITEM NO	(2) SMR CODE	(3) NSN	(4) CAGEC	(5) PART NUMBER	(6) DESCRIPTION AND USABLE ON CODES (UOC)	(7) QTY
20	PAFZZ	5305002264831	80204	B1821BH031C150N	SCREW, CAP, HEXAGON H .... UOC: DAJ, V24, ZAJ	4
21	PAOZZ	5305004770144	96906	MS51861-68	SCREW, TAPPING .... UOC: DAJ, V24, ZAJ	14

END OF FIGURE

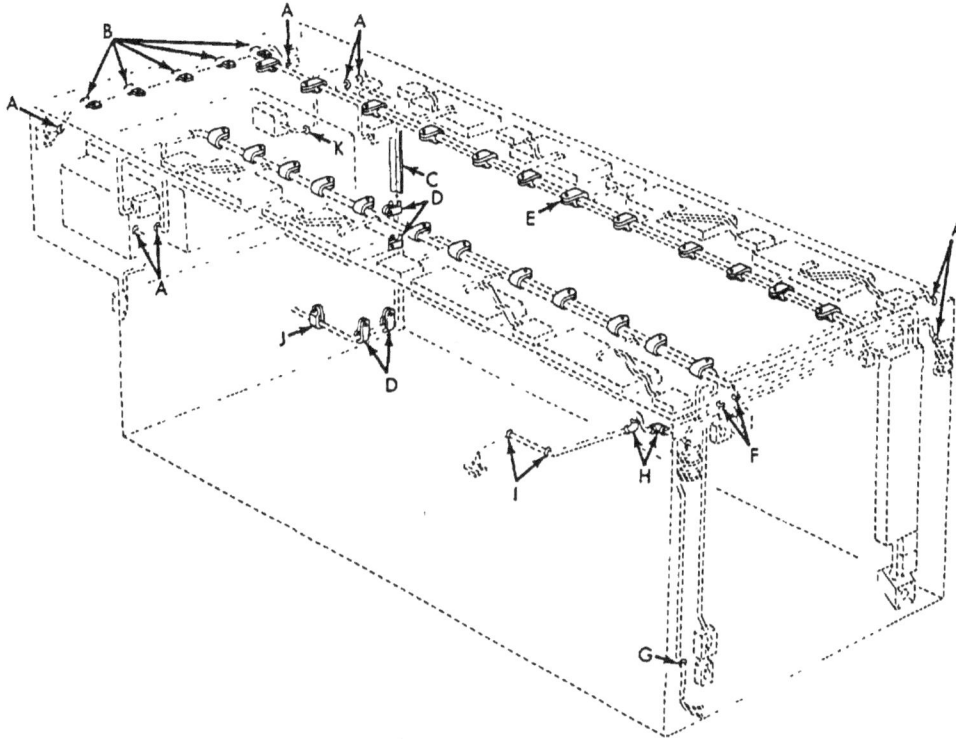

Figure 429. Van Main Wiring Harness Mounting Hardware.

SECTION II

(1) ITEM NO	(2) SMR CODE	(3) NSN	(4) CAGEC	(5) PART NUMBER	(6) DESCRIPTION AND USABLE ON CODES (UOC)	(7) QTY
					GROUP 1812 SPECIAL PURPOSE BODIES	
					FIG. 429 VAN MAIN WIRING HARNESS MOUNTING HARDWARE	
1	PAHZZ	5325012427083	96906	MS35489-17	GROMMET, NONMETALLIC .... UOC: DAJ, V24, ZAJ	9
2	PAHZZ	5305004324201	96906	MS51861-45	SCREW, TAPPING .... UOC: DAJ, V24, ZAJ	29
3	PAHZZ	5340001906783	96906	MS35140-4	STRAP, RETAINING .... UOC: DAJ, V24, ZAJ	5
4	PAHZZ	5305004324170	96906	MS51861-35	SCREW, TAPPING .... UOC: DAJ, V24, ZAJ	6
5	PAHZZ	5340004195866	19207	7059241	STRAP, RETAINING .... UOC: DAJ, V24, ZAJ	2
6	PAHZZ	5305004324172	96906	MS51861-37	SCREW, TAPPING .... UOC: DAJ, V24, ZAJ	4
7	PAHZZ	5340001501658	17773	11176106-5	CLAMP, LOOP .... UOC: DAJ, V24, ZAJ	4
8	PAHZZ	5340000502622	96906	MS21334-36	CLAMP, LOOP .... UOC: DAJ, V24, ZAJ	24
9	PAHZZ	5325001716387	96906	MS35489-51	GROMMET, NONMETALLIC .... UOC: DAJ, V24, ZAJ	2
10	PAHZZ	5325002766089	96906	MS35489-16	GROMMET, NONMETALLIC .... UOC: DAJ, V24, ZAJ	8
11	PFHZZ	5975001521127	83879	ACV-938	BOX CONNECTOR, ELECT .... UOC: DAJ, V24, ZAJ	4
12	PAHZZ	5325001849846	81349	C3030	GROMMET, NONMETALLIC .... UOC: DAJ, V24, ZAJ	4
13	PAHZZ	5340009645267	96906	MS21333-120	CLAMP, LOOP .... UOC: DAJ, V24, ZAJ	1
14	PAHZZ	5325002708890	96906	MS35489-22	GROMMET, NONMETALLIC .... UOC: DAJ, V24, ZAJ	1

END OF FIGURE

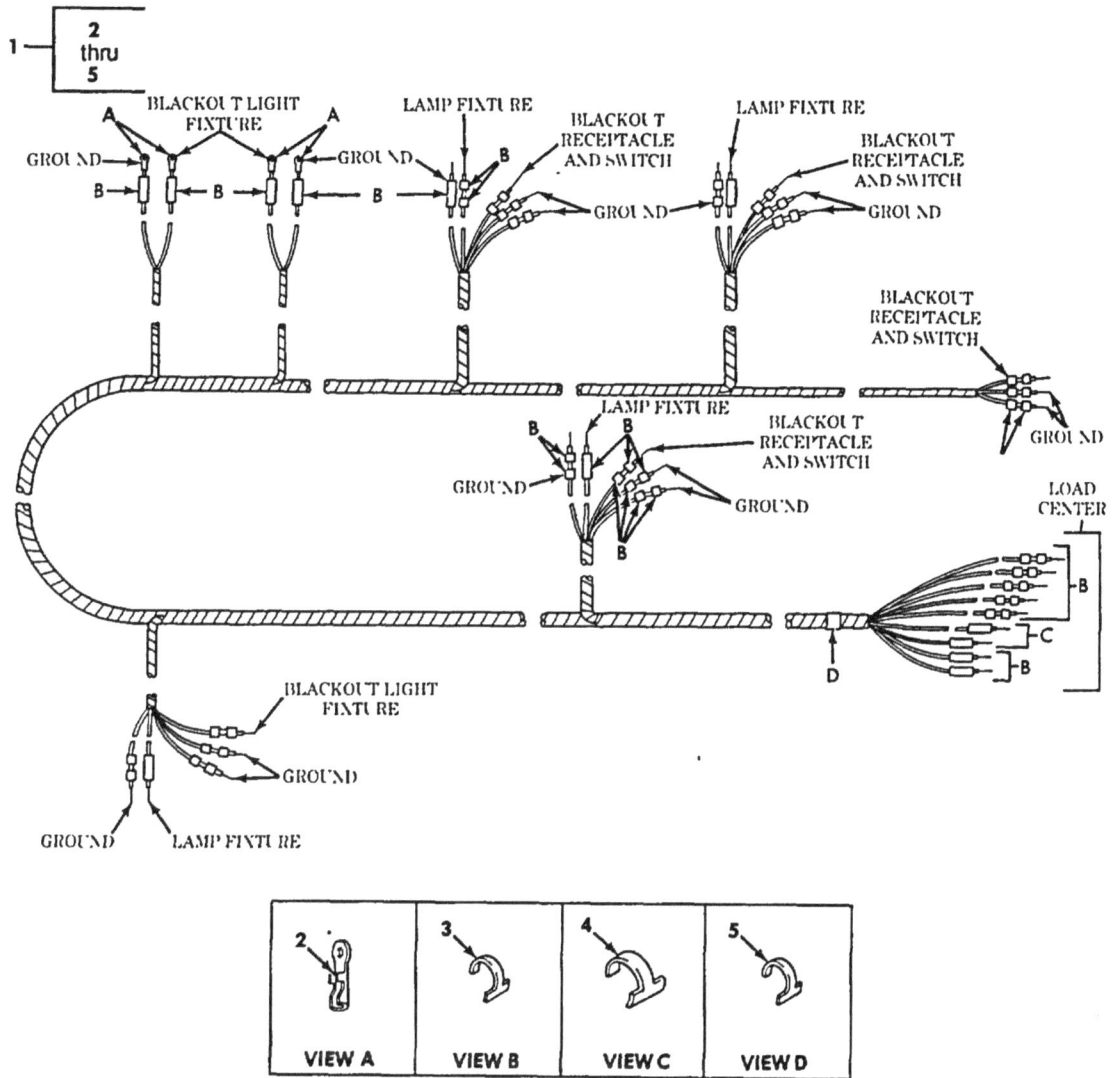

Figure 430. Van Main Wiring Harness, Right and Left Side.

(1) ITEM NO	(2) SMR CODE	(3) NSN	(4) CAGEC	(5) PART NUMBER	(6) DESCRIPTION AND USABLE ON CODES (UOC)	(7) QTY

SECTION II

GROUP 1812 SPECIAL PURPOSE BODIES

FIG. 430 VAN MAIN WIRING HARNESS, RIGHT AND LEFT SIDE

(1) ITEM NO	(2) SMR CODE	(3) NSN	(4) CAGEC	(5) PART NUMBER	(6) DESCRIPTION AND USABLE ON CODES (UOC)	(7) QTY
1	PFHHH	2920011832693	19207	1677738	WIRING HARNESS, BRAN MAIN, RIGHT SIDE................... UOC: DAJ, V24, ZAJ	1
1	PFHHH	6150011354478	19207	11677740	WIRING HARNESS, BRAN MAIN, LEFT SIDE. .... UOC: DAJ, V24, ZAJ	1
2	PAHZZ	5940001519361	96906	MS35436-44	.TERMINAL, LUG .... UOC: DAJ, V24, ZAJ	4
3	PAHZZ	9905010697222	81349	M43436/2-1	.BAND, MARKER .... UOC: DAJ, V24, ZAJ	60
4	PAHZZ	9905007524649	81349	M43436/1-1	.BAND, MARKER .... UOC: DAJ, V24, ZAJ	2
5	PAHZZ	9905009357777	81349	M43436/4-2	.BAND, MARK ER .... UOC: DAJ, V24, ZAJ	1

END OF FIGURE

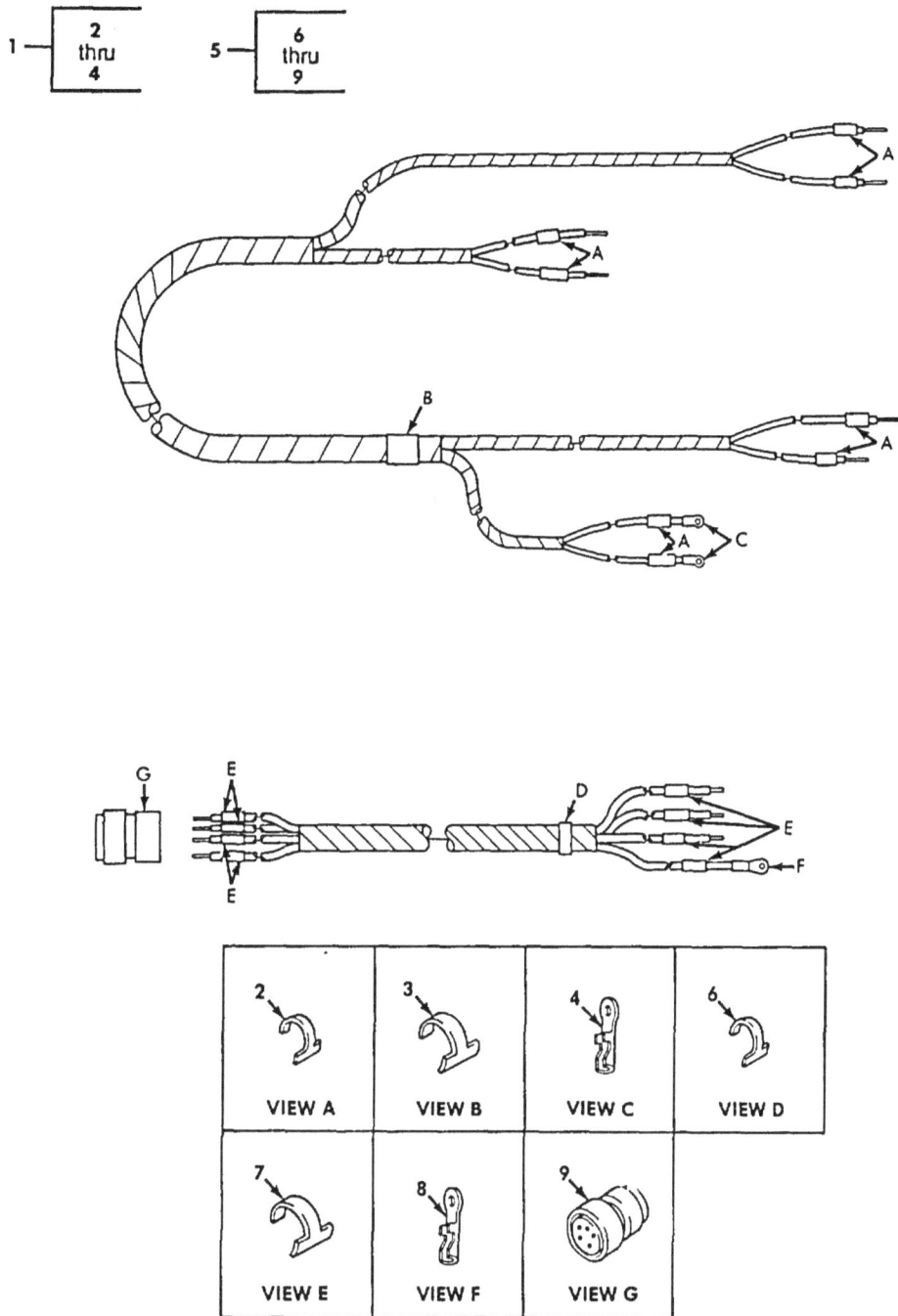

Figure 431. Van Air Conditioning and Heater Wiring Harnesses.

(1) ITEM NO	(2) SMR CODE	(3) NSN	(4) CAGEC	(5) PART NUMBER	(6) DESCRIPTION AND USABLE ON CODES (UOC)	(7) QTY
					GROUP 1812 SPECIAL PURPOSE BODIES	
					FIG. 431 VAN AIR CONDITIONING AND HEATER WIRING HARNESSES	
1	PFFFF	5995003517868	19207	10937519	WIRING HARNESS, BRAN UOC: DAJ, V24, ZAJ	1
2	PAFZZ	9905008933570	81349	M43436/1-3	.BAND, MARKER UOC: DAJ, V24, ZAJ	8
3	PAFZZ	9905007524649	81349	M43436/1-1	.BAND, MARKER UOC: DAJ, V24, ZAJ	1
4	PAFZZ	5940005340991	96906	MS35436-6	.TERMINAL, LUG UOC: DAJ, V24, ZAJ	2
5	PFFFF	6150011434527	19207	12300743	WIRING HARNESS, BRAN UOC: DAJ, V24, ZAJ	1
6	PAFZZ	9905008933570	81349	M43436/1-3	.BAND, MARKER UOC: DAJ, V24, ZAJ	1
7	PAFZZ	9905008414445	81349	M43436/1-2	.BAND, MARKER UOC: DAJ, V24, ZAJ	8
8	PAFZZ	5940005574344	96906	MS25036-120	.TERMINAL, LUG UOC: DAJ, V24, ZAJ	1
9	PAFZZ	5935010400463	96906	MS3456W24-22S	.CONNECTOR, PLUG, ELEC UOC: DAJ, V24, ZAJ	1

END OF FIGURE

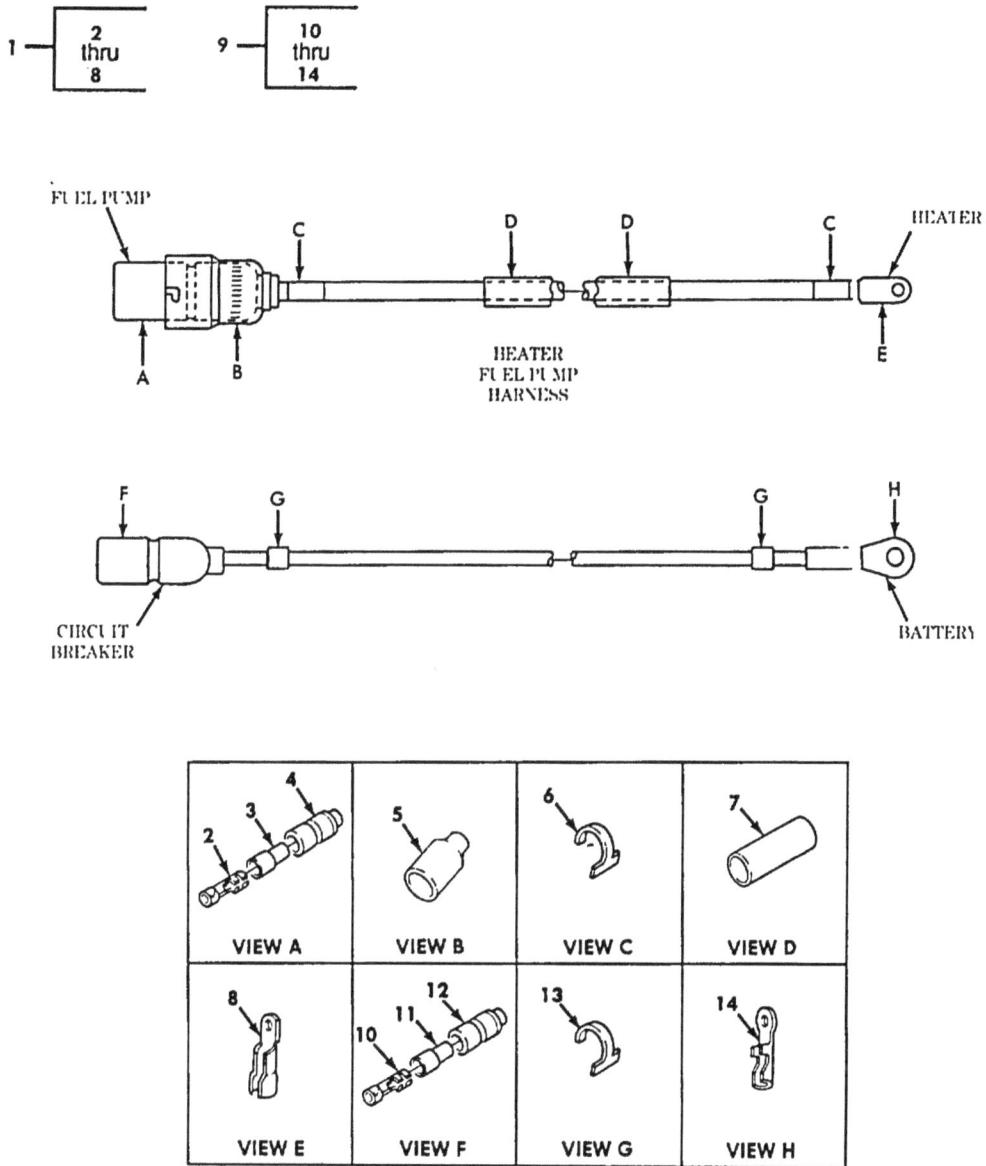

Figure 432. Van Heater Fuel Pump and Circuit Breaker Harnesses.

(1) ITEM NO	(2) SMR CODE	(3) NSN	(4) CAGEC	(5) PART NUMBER	(6) DESCRIPTION AND USABLE ON CODES (UOC)	(7) QTY
					GROUP 1812 SPECIAL PURPOSE BODIES	
					FIG. 432 VAN HEATER FUEL PUMP AND CIRCUIT BREAKER HARNESSES	
1	PFFFF	6150011354401	19207	11677729	WIRING HARNESS .... UOC: DAJ, V24, ZAJ	1
2	PAFZZ	1015007982997	19207	7982997	.TERMINAL, SOLDERED F .... UOC: DAJ, V24, ZAJ	1
3	PAFZZ	5970008338562	19207	8338562	.INSULATOR, BUSHING .... UOC: DAJ, V24, ZAJ	1
4	PAFZZ	5935008338561	19207	8338561	.SHELL, ELECTRICAL CO .... UOC: DAJ, V24, ZAJ	1
5	PAFZZ	5935008682606	18876	7982403	.SHELL, ELECTRICAL CO .... UOC: DAJ, V24, ZAJ	1
6	PAFZZ	9905010138723	81349	M43436/3-1	.BAND, MARKER .... UOC: DAJ, V24, ZAJ	2
7	MFFZZ		80244	17-C-18035-50-50	.CONDUIT, NONMETALLIC MAKE FROM.................... CONDUIT, P/N 17-C-18035-50, 50 INCHES LONG.................... UOC: DAJ, V24, ZAJ	1
8	PAFZZ	5940007056701	19207	7056701	.TERMINAL, LUG .... UOC: DAJ, V24, ZAJ	1
9	PFFFF	2590002228906	19207	11608875	.LEAD, ELECTRICAL .... UOC: DAJ, V24, ZAJ	1
10	PAFZZ	1015007982997	19207	7982997	.TERMINAL, SOLDERED F .... UOC: DAJ, V24, ZAJ	1
11	PAFZZ	5970008338562	19207	8338562	.INSULATOR, BUSHING .... UOC: DAJ, V24, ZAJ	1
12	PAFZZ	5935008338561	19207	8338561	.SHELL, ELECTRICAL CO .... UOC: DAJ, V24, ZAJ	1
13	PAFZZ	9905010138723	81349	M43436/3-1	.BAND, MARKER .... UOC: DAJ, V24, ZAJ	2
14	PAFZZ	5325002766228	96906	MS35489-9	.GROMMET, NONMETALLIC .... UOC: DAJ, V24, ZAJ	1

END OF FIGURE

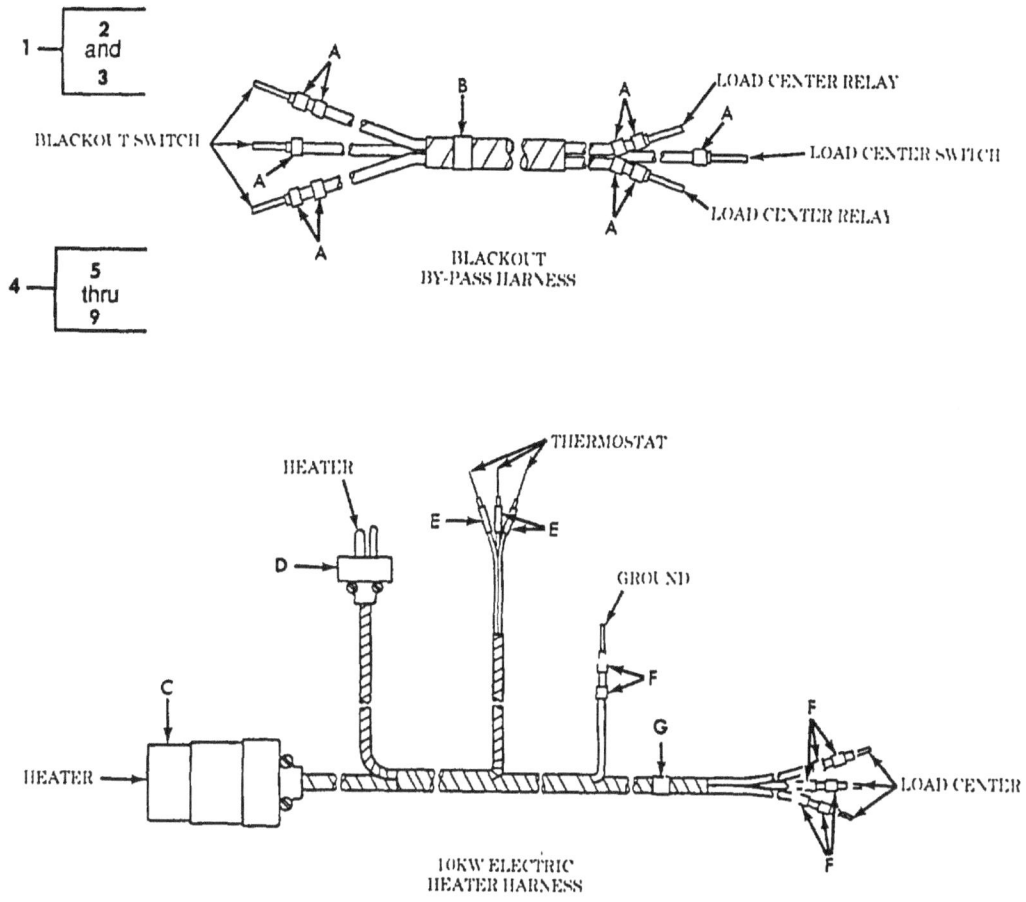

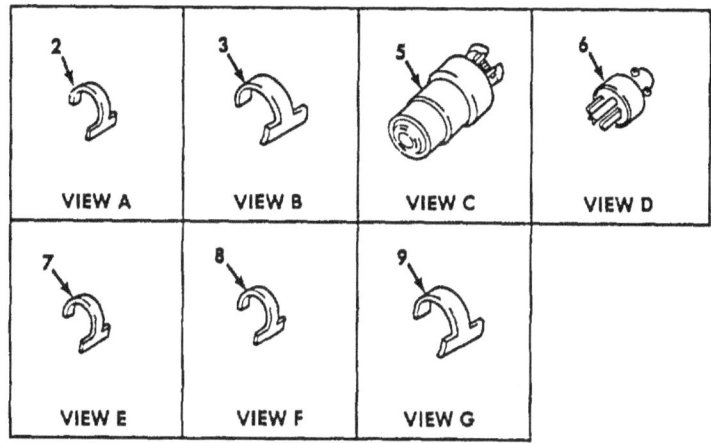

Figure 433. Van Blackout By-Pass and 10KW Electric Heater Wiring Harnesses.

(1) ITEM NO	(2) SMR CODE	(3) NSN	(4) CAGEC	(5) PART NUMBER	(6) DESCRIPTION AND USABLE ON CODES (UOC)	(7) QTY
					GROUP 1812 SPECIAL PURPOSE BODIES	
					FIG. 433 VAN BLACKOUT BY-PASS AND 10KW ELECTRIC HEATER WIRING HARNESSES	
1	PFFFF	6150001580066	19207	11677686	WIRING HARNESS BLACKOUT BY-PASS .... UOC: DAJ, V24, ZAJ	1
2	PAFZZ	9905010697222	81349	M43436/2-1	.BAND, MARKER .... UOC: DAJ, V24, ZAJ	10
3	PAFZZ	9905008933570	81349	M43436/1-3	.BAND, MARKER .... UOC: DAJ, V24, ZAJ	10
4	PFFFF	6150011354482	19207	11677708	WIRING HARNESS, BRAN 10KW ELECTRIC .... HEATER .... UOC: DAJ, V24, ZAJ	1
5	PAFZZ	5935011147615	19207	11663288	.CONNECTOR, PLUG, ELEC . ... UOC: DAJ, V24, ZAJ	1
6	PAFZZ	5935010053579	81348	WC596/13-3	.CONNECTOR, PLUG, ELEC .... UOC: DAJ, V24, ZAJ	1
7	PAFZZ	9905010138723	81349	M43436/3-1	.BAND, MARKER .... UOC: DAJ, V24, ZAJ	5
8	PAFZZ	9905007524649	81349	M43436/1-1	.BAND, MARKER .... UOC: DAJ, V24, ZAJ	8
9	PAFZZ	9905008933570	81349	M43436/1-3	.BAND, MARKER .... UOC: DAJ, V24, ZAJ	1

END OF FIGURE

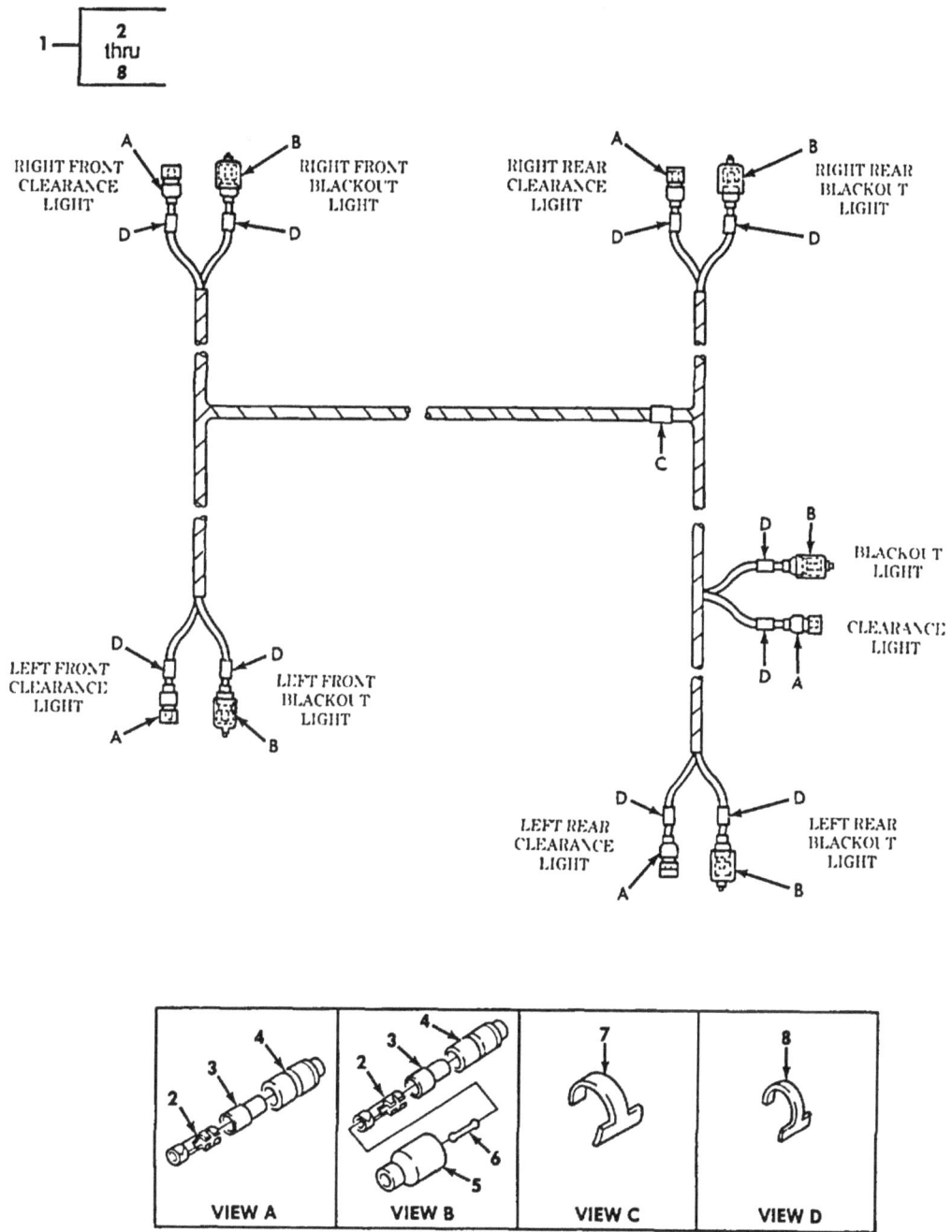

Figure 434. Van Blackout and Clearance Light Wiring Harness.

(1) ITEM NO	(2) SMR CODE	(3) NSN	(4) CAGEC	(5) PART NUMBER	(6) DESCRIPTION AND USABLE ON CODES (UOC)	(7) QTY
					GROUP 1812 SPECIAL PURPOSE BODIES	
					FIG. 434 VAN BLACKOUT AND CLEARANCE LIGHT WIRING HARNESS	
1	PFFFF	6150002228988	19207	7535589	WIRING HARNESS, BRAN BLACKOUT AND .... CLEARANCE .... UOC: DAJ, V24, ZAJ	1
2	PAFZZ	5940003996676	19207	8338564	.TERMINAL ASSEMBLY .... UOC: DAJ, V24, ZAJ	10
3	PAFZZ	5970008338562	19207	8338562	.INSULATOR, BUSHING .... UOC: DAJ, V24, ZAJ	10
4	PAFZZ	5935008338561	19207	8338561	.SHELL, ELECTRICAL CO.. UOC: DAJ, V24, ZAJ	10
5	PAFZZ	5935005729180	19207	8338566	.SHELL, ELECTRICAL CO.. UOC: DAJ, V24, ZAJ	5
6	PAFZZ	5935002140904	19207	7982907	.DUMMY CONNECTOR, PLU..................................... UOC: DAJ, V24, ZAJ	5
7	PAFZZ	9905008414445	81349	M43436/1-2	.BAND, MARKER .... UOC: DAJ, V24, ZAJ	1
8	PAFZZ	9905010138723	81349	M43436/3-1	.BAND, MARKER .... UOC: DAJ, V24, ZAJ	10

END OF FIGURE

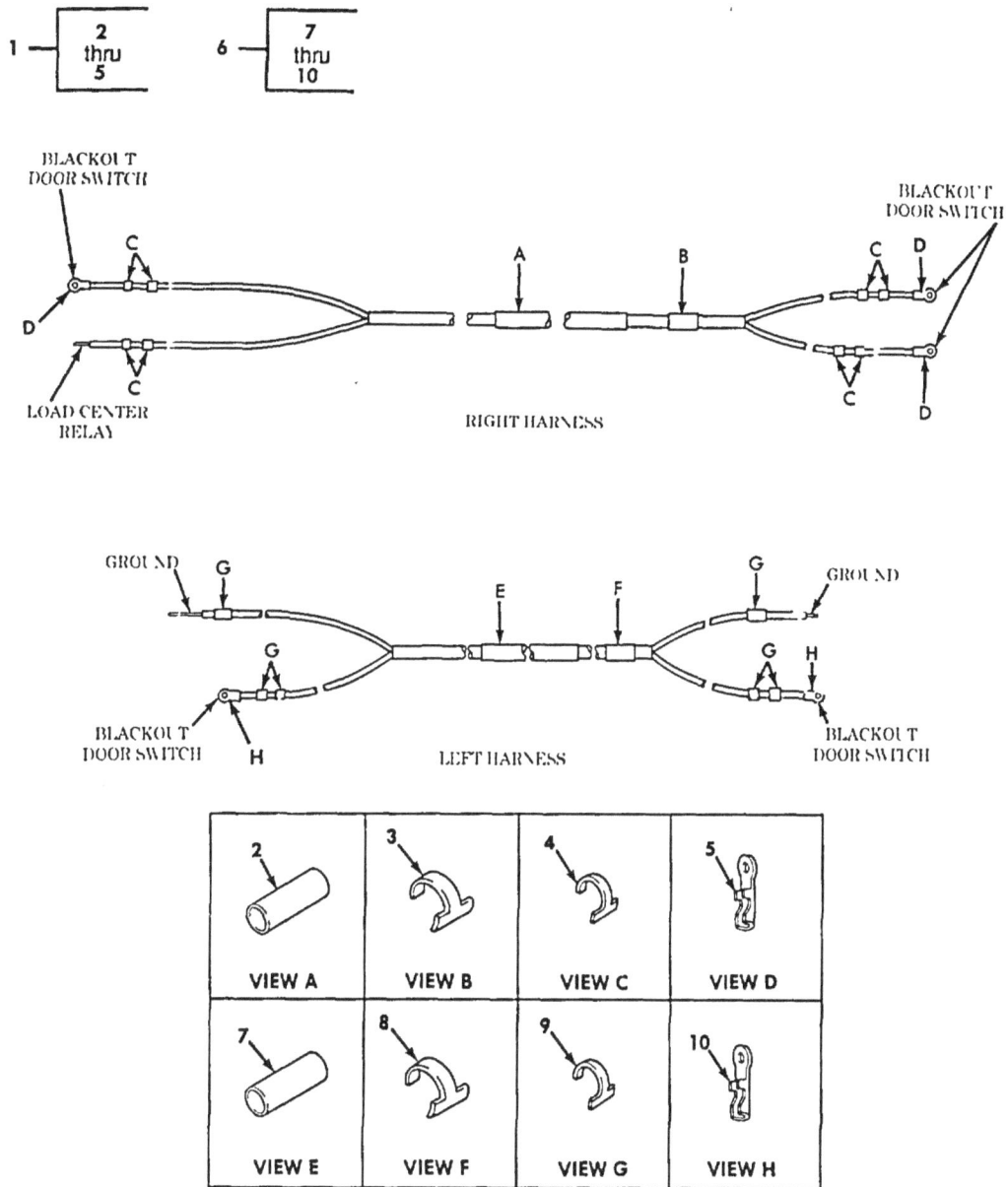

Figure 435. Van Blackout Wiring Harness, Right and Left Side.

(1) ITEM NO	(2) SMR CODE	(3) NSN	(4) CAGEC	(5) PART NUMBER	(6) DESCRIPTION AND USABLE ON CODES (UOC)	(7) QTY
					GROUP 1812 SPECIAL PURPOSE BODIES	
					FIG. 435 VAN BLACKOUT WIRING HARNESS, RIGHT AND LEFT SIDE	
1	PFFFF	5995002317454	19207	10937533	WIRING HARNESS, BRAN BLACKOUT, RIGHT. .... UOC: DAJ, V24, ZAJ	1
2	MFFZZ	5975001771930	80244	17-C-18035-60-15	.CONDUIT, NONMETALLIC MAKE FROM.......................... CONDUIT, P/N 17-C-18035-60, 15 INCHES LONG... UOC: DAJ, V24, ZAJ	1
3	PAFZZ	9905007524649	81349	M43436/1-1	.BAND, MARKER .... UOC: DAJ, V24, ZAJ	1
4	PAFZZ	9905010138723	81349	M43436/3-1	.BAND, MARKER. .... UOC: DAJ, V24, ZAJ	8
5	PAFZZ	5940005340986	96906	MS35436-4	.TERMINAL, LUG .... UOC: DAJ, V24, ZAJ	3
6	PFFFF	6150002228943	19207	10937532	WIRING HARNESS BLACKOUT, LEFT............................. UOC: DAJ, V24, ZAJ	1
7	MFFZZ	5975001771930	80244	17-C-18035-60-15	.CONDUIT, NONMETALLIC MAKE FROM.......................... CONDUIT, P/N 17-C-18035-60, 15 INCHES LONG... UOC: DAJ, V24, ZAJ	1
8	PAFZZ	9905008414445	81349	M43436/1-2	.BAND, MARKER .... UOC: DAJ, V24, ZAJ	1
9	PAFZZ	9905010138723	81349	M43436/3-1	.BAD, MARKER. .... UOC: DAJ, V24, ZAJ	6
10	PAFZZ	5940001519361	96906	MS35436-44	.TERMINAL, LUG... UOC: DAJ, V24, ZAJ	2

END OF FIGURE

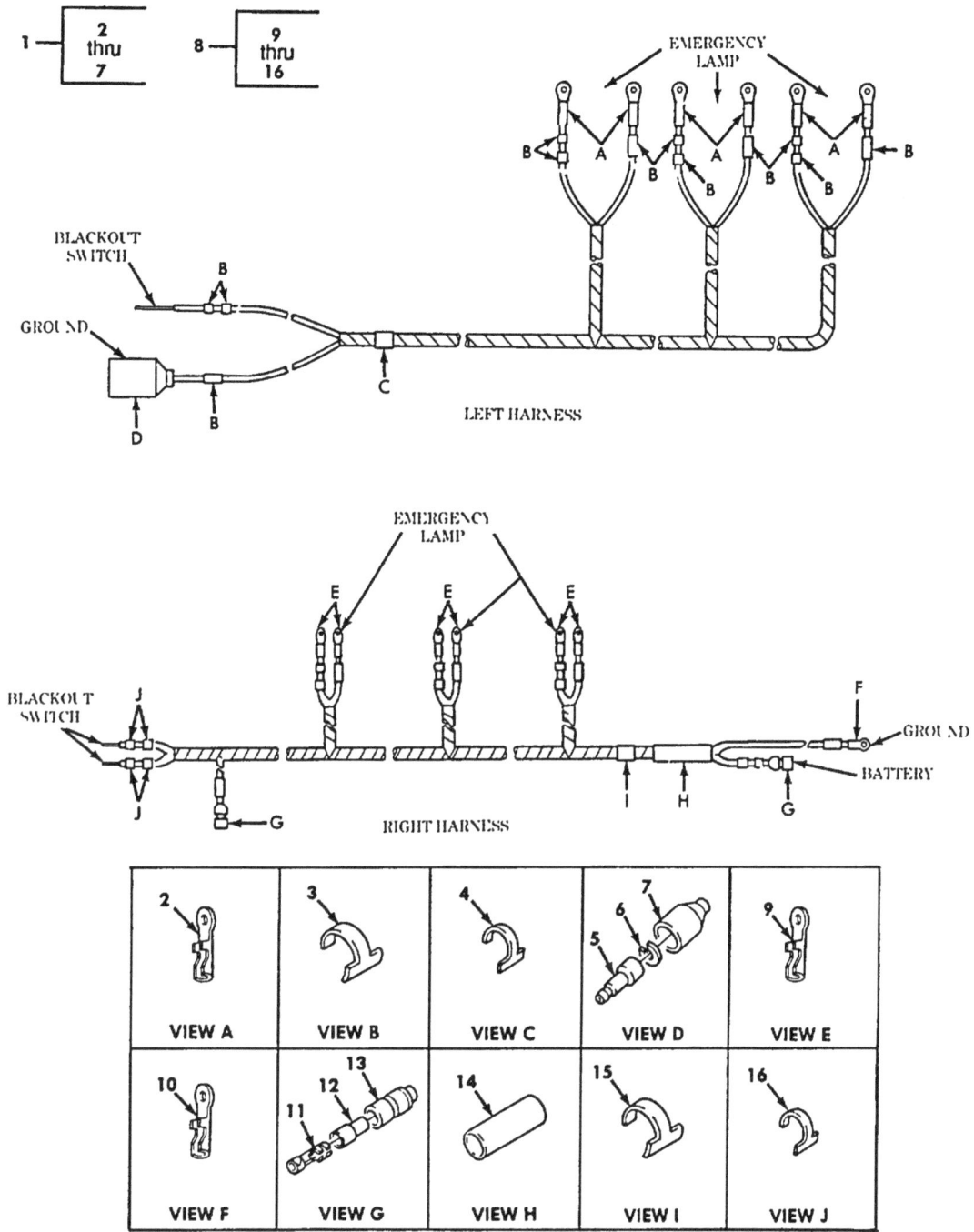

Figure 436. Van Emergency Lamp Wiring Harness, Right and Left Side.

(1) ITEM NO	(2) SMR CODE	(3) NSN	(4) CAGEC	(5) PART NUMBER	(6) DESCRIPTION AND USABLE ON CODES (UOC)	(7) QTY

GROUP 1812 SPECIAL PURPOSE BODIES

FIG. 436 VAN EMERGENCY LAMP WIRING HARNESS, RIGHT AND LEFT SIDE

(1) ITEM NO	(2) SMR CODE	(3) NSN	(4) CAGEC	(5) PART NUMBER	(6) DESCRIPTION AND USABLE ON CODES (UOC)	(7) QTY
1	PFFFF	6150011354479	19207	11677731	WIRING HARNESS, BRAN . ... UOC: DAJ, V24, ZAJ	1
2	PAFZZ	5940005340986	96906	MS35436-4	.TERMINAL, LUG LEFT SIDE .... UOC: DAJ, V24, ZAJ	6
3	PAFZZ	9905010138723	81349	M43436/3-1	.BAND, MARKER LEFT SIDE .... UOC: DAJ, V24, ZAJ	12
4	PAFZZ	9905007524649	81349	M43436/1-1	.BAND, MARKER LEFT SIDE .... UOC: DAJ, V24, ZAJ	1
5	PAFZA	5999000572929	19204	572929	.CONTACT, ELECTRICAL LEFT SIDE .... UOC: DAJ, V24, ZAJ	1
6	PAFZZ	5310008338567	19207	8338567	.WASHER, SLOTTED LEFT SIDE .... UOC: DAJ, V24, ZAJ	1
7	PAFZZ	5935005729180	19207	8338566	.SHELL, ELECTRICAL CO LEFT SIDE .... UOC: DAJ, V24, ZAJ	1
8	PFFFF	6150011354480	19207	11677723	WIRING HARNESS, BRAN RIGHT SIDE .... UOC: DAJ, V24, ZAJ.	1
9	PAFZZ	5940005340986	96906	MS35436-4	.TERMINAL, LUG RIGHT SIDE .... UOC: DAJ, V24, ZAJ	6
10	PAFZZ	5940008923151	96906	MS35436-9	.TERMINAL, LUG RIGHT SIDE .... UOC: DAJ, V24, ZAJ	1
11	PAFZZ	1015007982997	19207	7982997	.TERMINAL, SOLDERED F RIGHT SIDE ...... UOC: DAJ, V24, ZAJ	2
12	PAFZZ	5970008338562	9207	8338562	.INSULATOR, BUSHING RIGHT SIDE .... UOC: DAJ, V24, ZAJ	2
13	PAFZZ	5935008338561	19207	8338561	.SHELL, ELECTRICAL CO RIGHT SIDE .... UOC: DAJ, V24, ZAJ	2
14	MFFZZ		80244	17-C-18035-90-50	.CONDUIT, NONMETALLIC MAKE FROM.......... CONDUIT, P/N LOOM 3/8 ID, 50 INCHES LONG.. UOC: DAJ, V24, ZAJ	1
15	PAFZZ	9905008414445	81349	M43436/1-2	.BAND, MARKER RIGHT SIDE .... UOC: DAJ, V24, ZAJ	1
16	PAFZZ	9905010138723	81349	M43436/3-1	.BAND, MARKER RIGHT SIDE .... UOC: DAJ, V24, ZAJ	17

END OF FIGURE

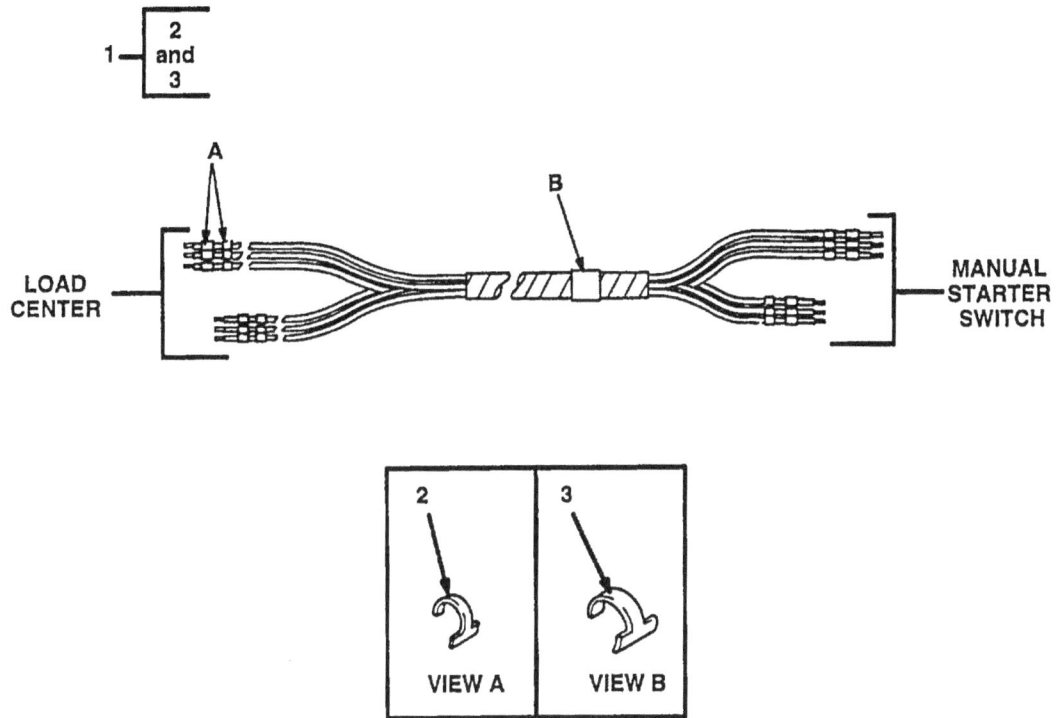

Figure 437. 400 HZ Converter Wiring Harness

(1) ITEM NO	(2) SMR CODE	(3) NSN	(4) CAGEC	(5) PART NUMBER	(6) DESCRIPTION AND USABLE ON CODES (UOC)	(7) QTY
					GROUP 1812 SPECIAL PURPOSE BODIES	
					FIG. 437 400HZ CONVERTER WIRING HARNESS	
1	PFFFF	2590011354405	19207	11677709	WIRING HARNESS MANUAL STARTER .... SWITCHES UOC: DAJ, V24, ZAJ	1
2	PAFZZ	9905010697222	81349	M43436/2-1	.BAND, MARKER .... UOC: DAJ, V24, ZAJ	24
3	PAFZZ	9905008933570	81349	M43436/1-3	.BAND, MARKER .... UOC: DAJ, V24, ZAJ	1

END OF FIGURE

400 HZ OUTPUT
HARNESS

400 HZ
FREQUENCY
CONVERTER

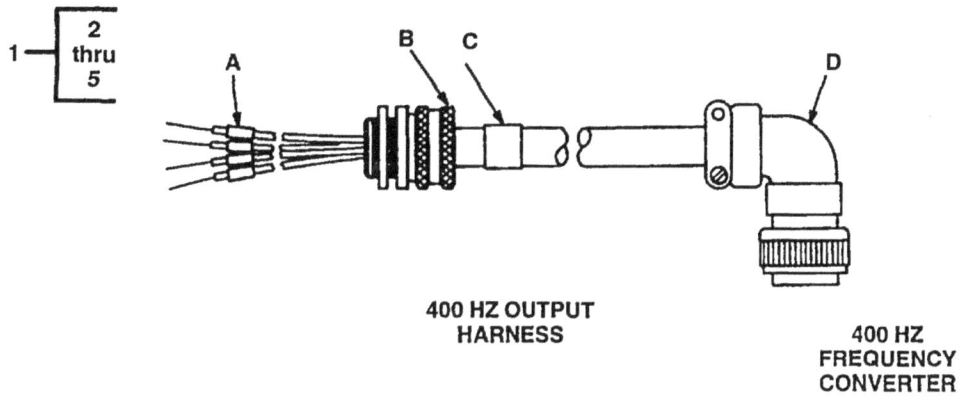

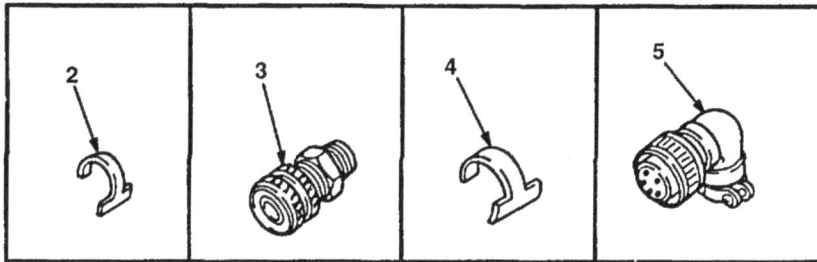

Figure 438. Van 400Hz Converter Wiring Harness

(1) ITEM NO	(2) SMR CODE	(3) NSN	(4) CAGEC	(5) PART NUMBER	(6) DESCRIPTION AND USABLE ON CODES (UOC)	(7) QTY
					GROUP 1812 SPECIAL PURPOSE BODIES	
					FIG. 438 VAN AUXILIARY HYDRAULIC PUMP WIRING HARNESS	
1	PFFFF	2590011343777	19207	11677696	WIRING HARNESS 1 UOC:DAJ,V24,ZAJ	1
2	PAFZZ	9905010697222	81349	M43436/2-1	.BAND,MARKER 4 UOC:DAJ,V24,ZAJ	4
3	PAFZZ	5975004560627	19207	11663369	.CONNECTOR 1 UOC:DAJ,V24,ZAJ	1
4	PAFZZ	9905008933570	81349	M43436/1-3	.BAND,MARKER 1 UOC:DAJ,V24,ZAJ	1
5	PAFZZ	5935008434561	96906	MS3108R20-4S	.CONNECTOR,PLUG 1 UOC:DAJ,V24,ZAJ	1

END OF FIGURE

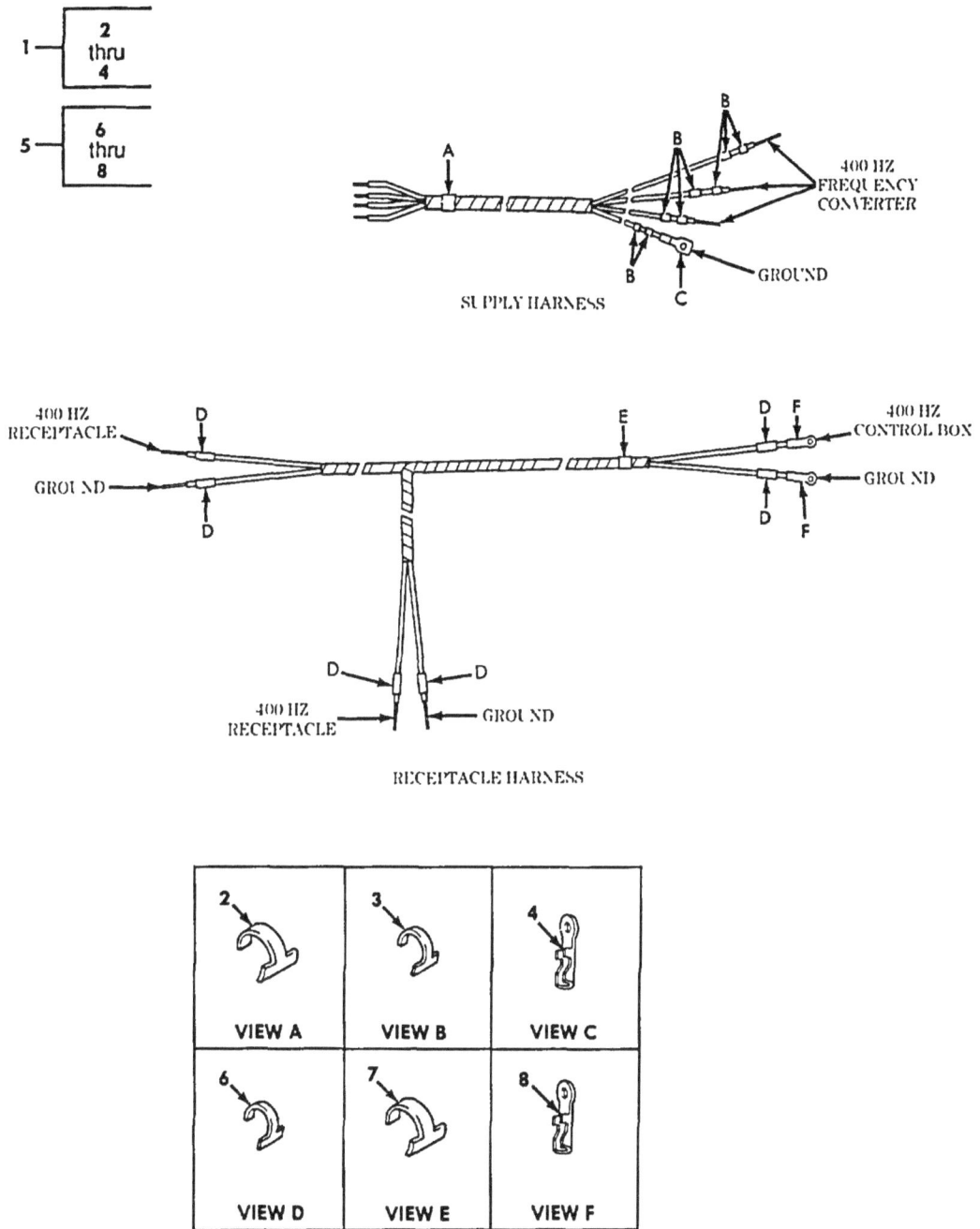

Figure 439. Van 400Hz Wiring Harness.

(1) ITEM NO	(2) SMR CODE	(3) NSN	(4) CAGEC	(5) PART NUMBER	(6) DESCRIPTION AND USABLE ON CODES (UOC)	(7) QTY
					GROUP 1812 SPECIAL PURPOSE BODIES	
					FIG. 439 VAN 400HZ WIRING HARNESS	
1	PFFFF	2590011439543	19207	11677730	HARNESS,WIRING UOC:DAJ,V24,ZAJ	1
2	PAFZZ	990500841445	81349	M43436/1-2	.BAND,MARKER RIGHT SIDE UOC:DAJ,V24,ZAJ	1
3	PAFZZ	9905010697222	81349	M43436/2-1	.BAND,MARKER UOC:DAJ,DAK,V24,V25,ZAJ,ZAK	8
4	PAFZZ	5940005341028	96906	MS35436-10	.TERMINAL,LUG UOC:DAJ,V24,ZAJ	1
5	PFFFF	2590011354481	19207	11677717	HARNESS,WIRING UOC:DAJ,V24,ZAJ	1
6	PAFZZ	9905010697222	81349	M43436/2-1	.BAND,MARKER UOC:DAJ,V24,ZAJ	6
7	PAFZZ	990500841445	81349	M43436/1-2	.BAND,MARXER RIGHT SIDE UOC:DAJ,V24,ZAJ	1
8	PAFZZ	5940005341028	96906	MS35436-10	.TERMINAL,LUG UOC:DAJ,V24,ZAJ	2

END OF FIGURE

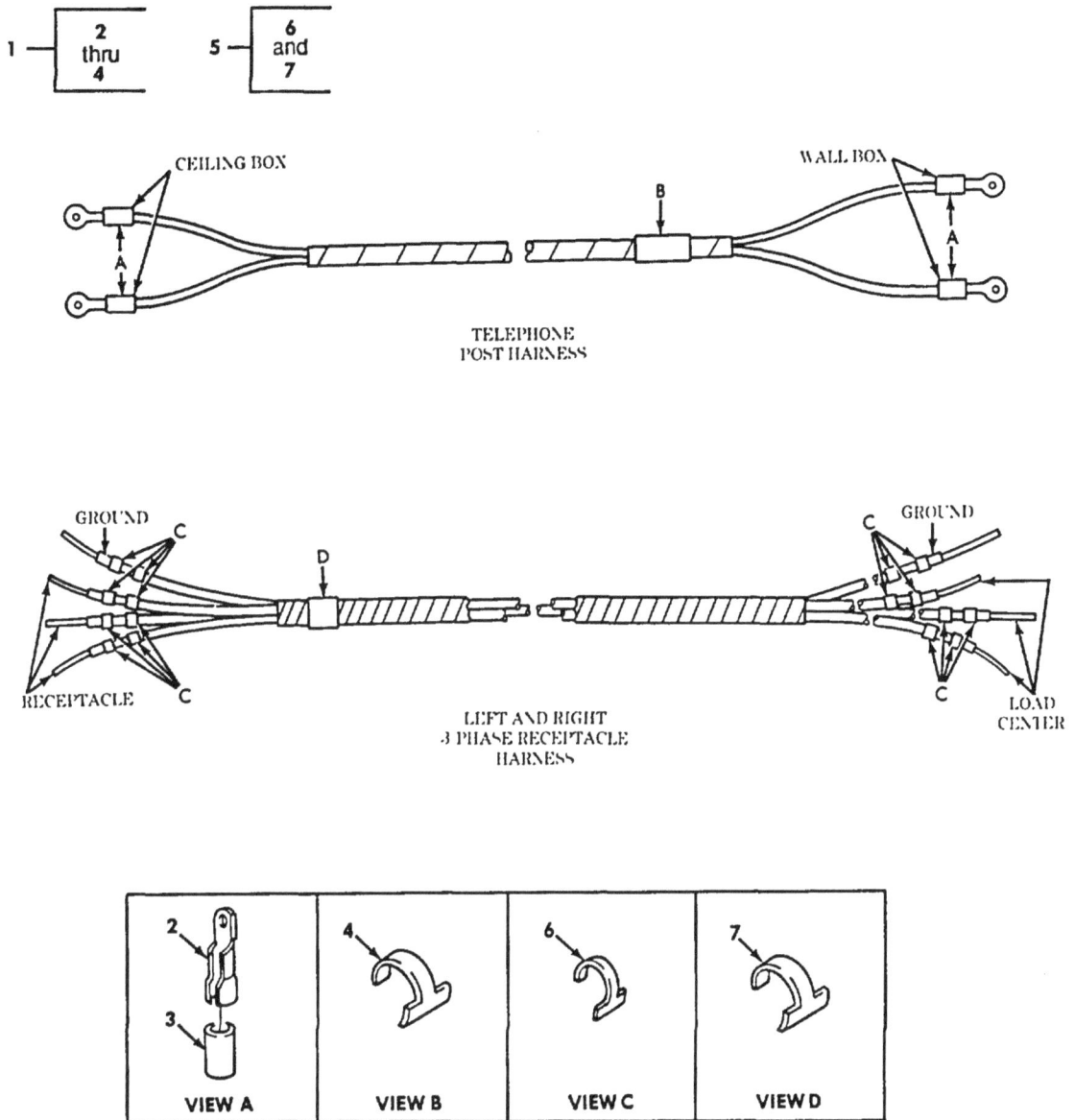

TELEPHONE
POST HARNESS

LEFT AND RIGHT
3 PHASE RECEPTACLE
HARNESS

VIEW A   VIEW B   VIEW C   VIEW D

Figure 440. Van Telephone Jack and 3-Phase Receptacle Harness.

(1) ITEM NO	(2) SMR CODE	(3) NSN	(4) CAGEC	(5) PART NUMBER	(6) DESCRIPTION AND USABLE ON CODES (UOC)	(7) QTY
					GROUP 1812 SPECIAL PURPOSE BODIES	
					FIG. 440 VAN TELEPHONE JACK AND 3-PHASE RECEPTACLE HARNESS	
1	PAFFF	2590002343248	19207	7535639	WIRING HARNESS UOC:DAJ,V24,ZAJ	1
2	PAFZZ	5940001071481	96906	MS20659-104	.TERMINAL,LUG UOC:DAJ,V24,ZAJ	4
3	PAFZZ	5970002966078	80244	17-1-1725-56	.INSULATION,SLEEVE UOC:DAJ,V24,ZAJ	4
4	PAFZZ	9905007524649	81349	M43436/1-1	.BAND,MARKER LEFT SIDE UOC:DAJ,V24,ZAJ	1
5	PFFFF	2590011354403	19207	11677719	WIRING,HARNESS RIGHT UOC:DAJ,V24,ZAJ	1
5	PFFFF	2590011354402	19207	11677728	WIRING,HARNESS LEFT UOC:DAJ,V24,ZAJ	1
6	PAFZZ	9905010697222	81349	M43436/2-1	.BAND,MARKER UOC:DAJ,V24,ZAJ	16
7	PAFZZ	9905009353863	81349	M43436/4-1	.BAND,MARKER UOC:DAJ,V24,ZAJ	1

END OF FIGURE

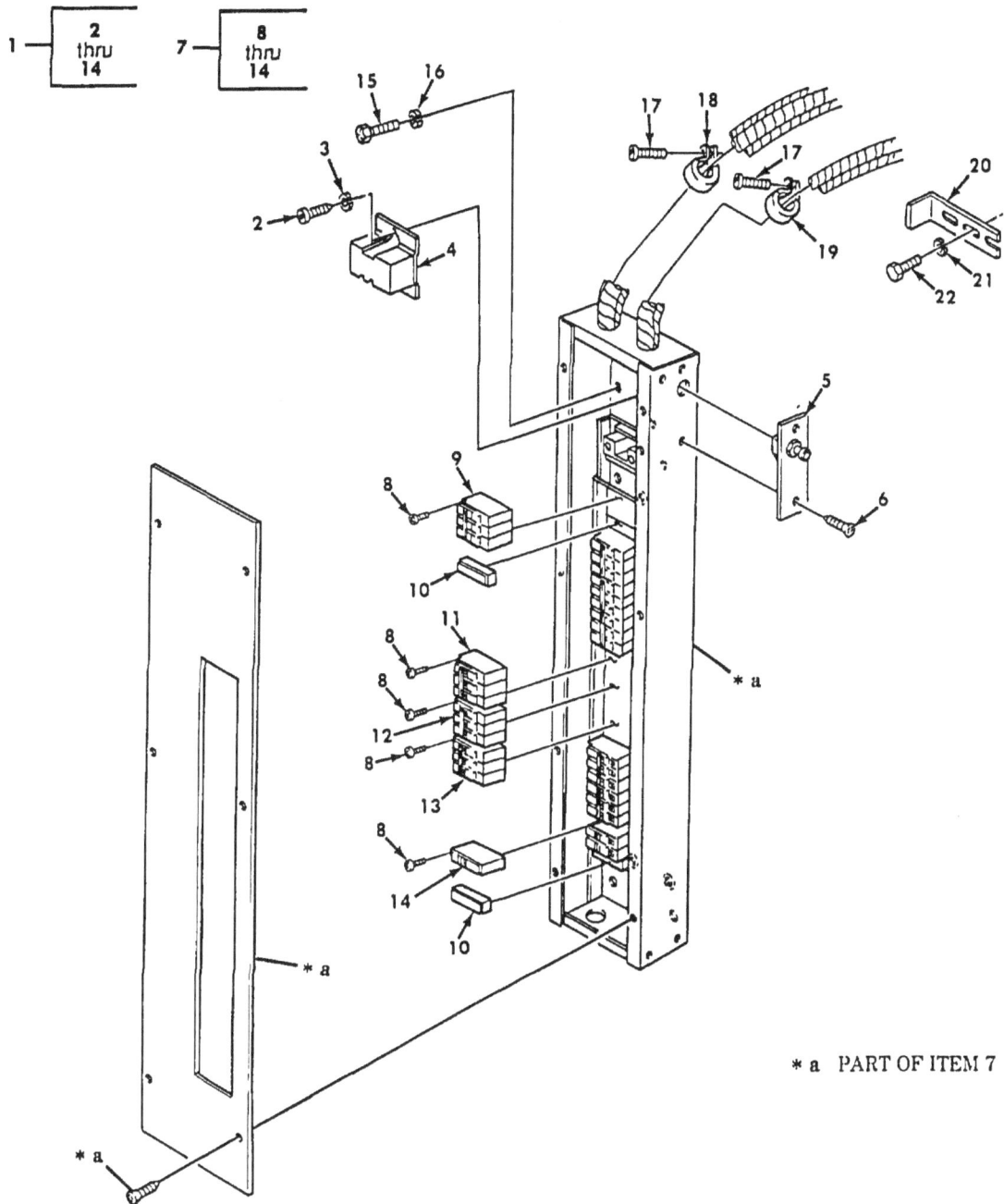

Figure 441. Van Electrical Load Center and Mounting Hardware.

* a   PART OF ITEM 7

(1) ITEM NO	(2) SMR CODE	(3) NSN	(4) CAGEC	(5) PART NUMBER	(6) DESCRIPTION AND USABLE ON CODES (UOC)	(7) QTY
					GROUP 1812 SPECIAL PURPOSE BODIES	
					FIG. 441 VAN ELECTRICAL LOAD CENTER AND MOUNTING HARDWARE	
1	AFFFF		19207	11677744	CIRCUIT BREAKER UOC:DAJ,V24,ZAJ	1
2	PAFZZ	5305008550961	96906	MS24629-35	.SCREW,WOOD UOC:DAJ,V24,ZAJ	3
3	PAFZZ	5310005590070	96906	MS35233-38	.WASHER,LOCK UOC:DAJ,V24,ZAJ	3
4	PAFZZ	5945004969708	19207	10937542	.RELAY,DISTRIBUTION UOC:DAJ,V24,ZAJ	1
5	PAFZZ	5930001160531	19207	10937541	.SWITCH,PUSH BLACKOUT UOC:DAJ,V24,ZAJ	1
6	PAFZZ	5305010284831	96906	MS51862-12	.SCREW,TAPPING UOC:DAJ,V24,ZAJ	2
7	PFFFF	5925011150557	19207	11677749	.CIRCUIT BREAKER BOX...................... UOC:DAJ,V24,ZAJ	1
8	PAFZZ	5305000889044	96906	MS35207-260	..SCREW,MACHINE UOC:DAJ,V24,ZAJ	29
9	PAFZZ	5925004979661	19207	10937537	..CIRCUIT BREAKER UOC:DAJ,V24,ZAJ	1
10	PAFZZ	5925011030996	19207	10937550	..PLATE UOC:DAJ,V24,ZAJ	6
11	PAFZZ	5925013847883	66842	BQ3B020H	..CIRCUIT BREAKER UOC:DAJ,V24,ZAJ	4
12	PAFZZ	5925004979659	19207	10937535	..CIRCUIT BREAKER UOC:DAJ,V24,ZAJ	1
13	PAFZZ	5925011147584	19207	11677698	..CIRCUIT BREAKER UOC:DAJ,V24,ZAJ	1
14	PAFZZ	5925004979658	19207	10937540	..CIRCUIT BREAKER UOC:DAJ,V24,ZAJ	9
15	PAFZZ	5305002692803	96906	MS90726-60	SCREW,CAP,HEXAGON UOC:DAJ,V24,ZAJ	4
16	PAFZZ	5310006379541	96906	MS35338-46	WASHER,LOCK UOC:DAJ,V24,ZAJ	4
17	PAFZZ	5305004324201	96906	MS51861-45	SCREW,TAPPING UOC:DAJ,V24,ZAJ	2
18	PAFZZ	5340001502772	96906	MS21333-84	CLAMP,LOOP UOC:DAJ,V24,ZAJ	1
19	PAFZZ	5340009226302	96906	MS21333-81	CLAMP,LOOP UOC:DAJ,V24,ZAJ	1
20	PAOZZ	5340002385606	19207	7084799	BRACKET,ANGLE BLACKOUT SWITCH........... UOC:DAJ,V24,ZAJ	1
21	PAOZZ	5310005825965	96906	MS35338-44	WASHER,LOCK. UOC:DAJ,V24,ZAJ	2
22	PAOZZ	5305011442190	96906	MS24628-67	SCREW,TAPPING UOC:DAJ,V24,ZAJ	2

END OF FIGURE

441-1

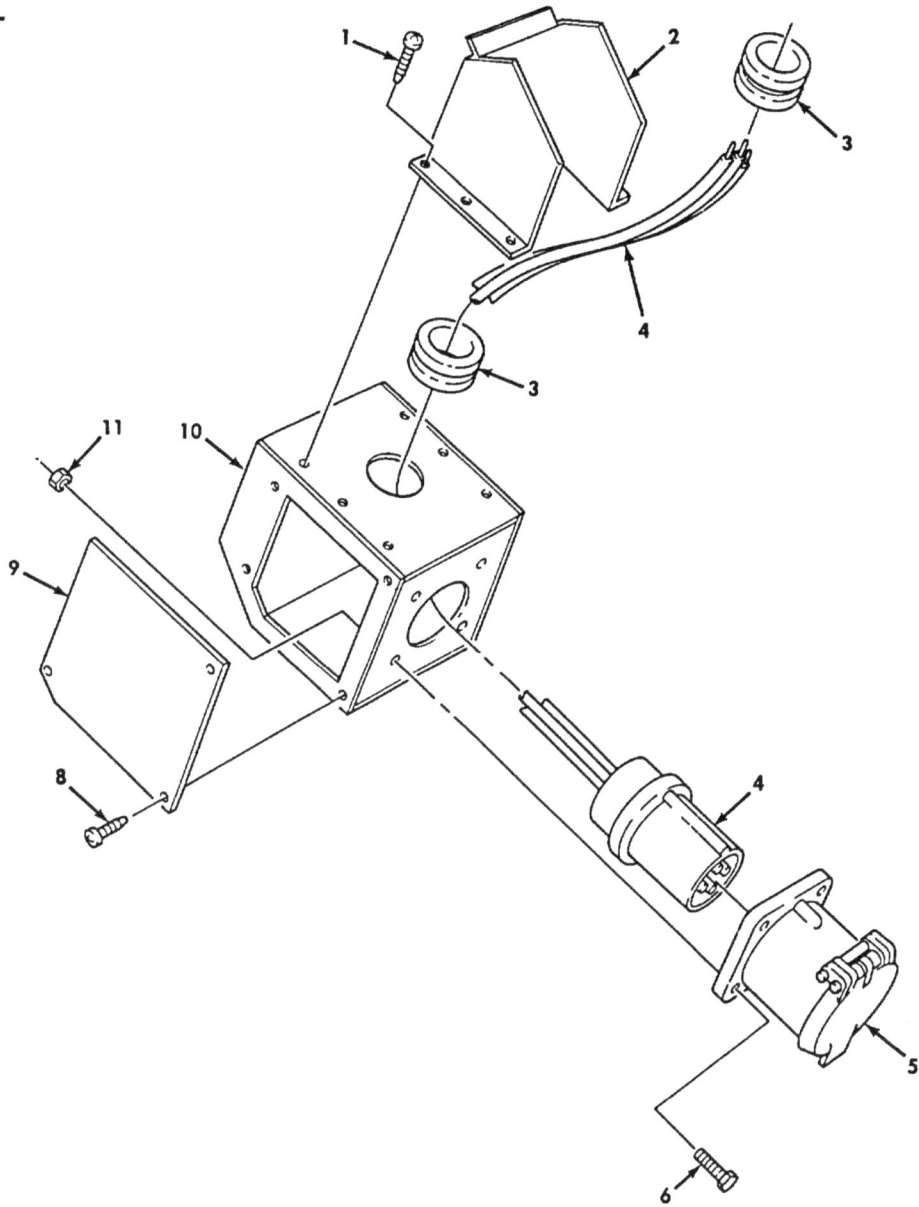

Figure 442. Van Electrical Junction Box and Related Parts.

(1) ITEM NO	(2) SMR CODE	(3) NSN	(4) CAGEC	(5) PART NUMBER	(6) DESCRIPTION AND USABLE ON CODES (UOC)	(7) QTY
					GROUP 1812 SPECIAL PURPOSE BODIES	
					FIG. 442 VAN ELECTRICAL JUNCTION BOX AND RELATED PARTS	
1	PAFZZ	5305000526908	96906	MS24629-4	SCREW,TAPPING UOC:DAJ,V24,ZAJ	6
2	PFFZZ	5420011174884	19207	11677707	GUSSET,BRIDGE UOC:DAJ,V24,ZAJ	2
3	PAFZZ	5325002901960	96906	MS35489-27	GROMMET,NONMETALLIC UOC:DAJ,V24,ZAJ	2
4	PAFZZ	5935004702118	19207	11601640	CONNECTOR,PLUG UOC:DAJ,V24,ZAJ	1
5	PAFZZ	5935002229013	19207	11601642	COVER AND SHELL UOC:DAJ,V24,ZAJ	1
6	PAFZZ	5306000501239	96906	MS90727-38	SCREW,CAP,HEXAGON UOC:DAJ,V24,ZAJ	4
7	PAFFF	5975010996418	19207	10937846-1	JUNCTION BOX UOC:DAJ,V24,ZAJ	1
8	PAFZZ	5305008550958	96906	MS24629-45	.SCREW,TAPPING UOC:DAJ,V24,ZAJ	4
9	PFFZZ	2590011704947	19207	10937841-1	.PLATE UOC:DAJ,V24,ZAJ	1
10	PFFZZ	2590011704948	19207	10937847-1	.HOUSING,RECEPTACLE UOC:DAJ,V24,ZAJ	1
11	PAFZZ	5310009824912	96906	MS21045-5	NUT,SELF-LOCKING UOC:DAJ,V24,ZAJ	4

END OF FIGURE

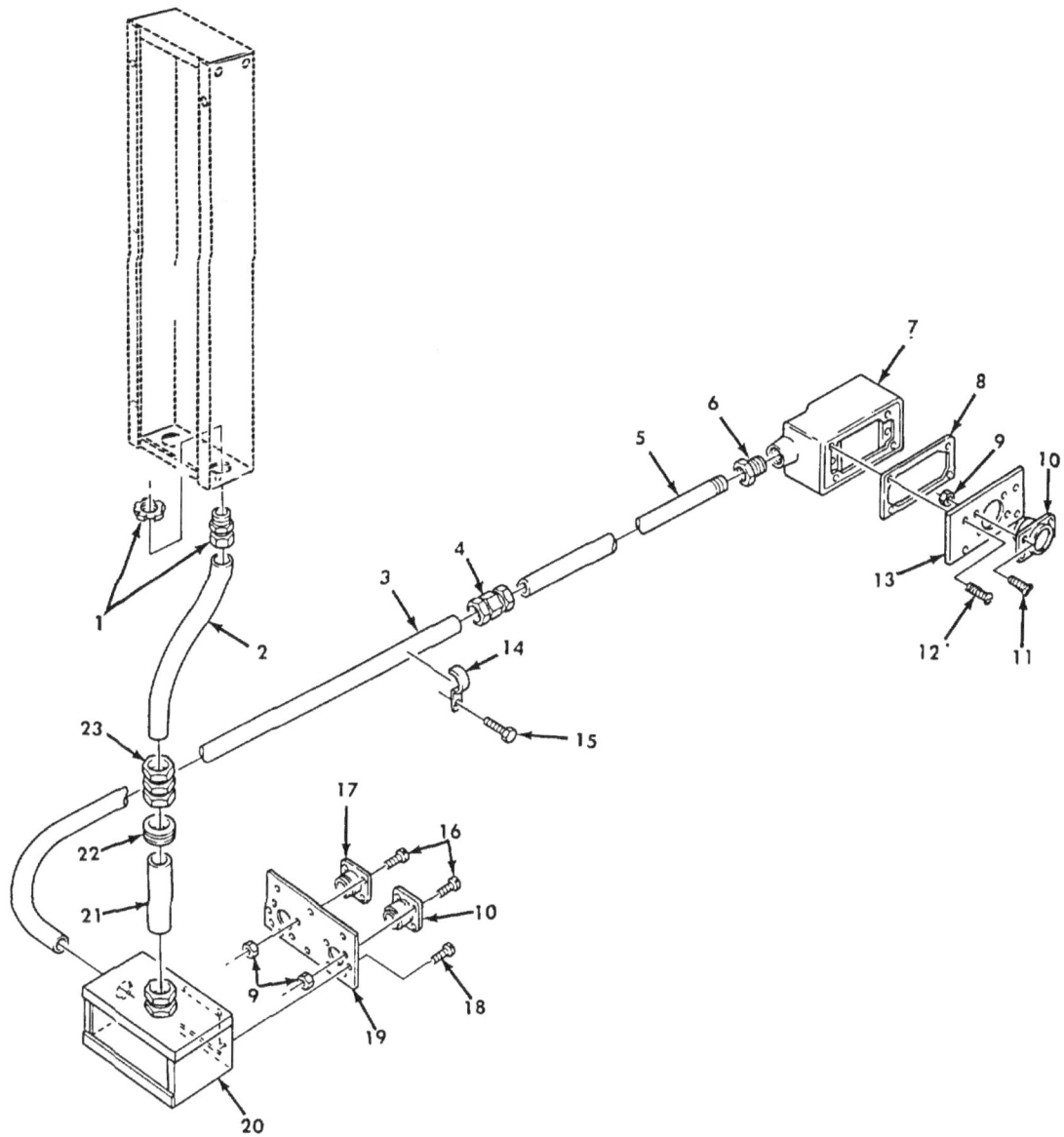

Figure 443. Van 400Hz Converter Junction Box, Auxiliary Hydraulic Pump, Electrical Connector Box, and Related Parts.

(1) ITEM NO	(2) SMR CODE	(3) NSN	(4) CAGEC	(5) PART NUMBER	(6) DESCRIPTION AND USABLE ON CODES (UOC)	(7) QTY
					GROUP 1812 SPECIAL PURPOSE BODIES	
					FIG. 443 VAN 400HZ CONVERTER JUNCTION BOX, AUXILIARY HYDRAULIC PUMP, ELECTRICAL CONNECTOR BOX, AND RELATED PARTS	
1	PAFZZ	5975011245053	19207	876985-5	CONNECTOR,CONDUIT UOC:DAJ,V24,ZAJ	1
2	PFFZZ	5975011157087	19207	11677704	CONDUIT,METAL,RIGID UOC:DAJ,V24,ZAJ	1
3	PFFZZ	5975011234562	19207	1167706	CONDUIT,METAL,RIGID UOC:DAJ,V24,ZAJ	1
4	PAFZZ	5975011147661	19207	11663099	COUPLING,ELECTRIC .... UOC:DAJ,V24, ZAJ	1
5	PFFZZ	5975011203728	19207	11677578	CONDUIT,METAL,RIGID UOC:DAJ,V24,ZAJ	1
6	PAFZZ	5975002846167	03743	TWAD75	ADAPTER UOC:DAJ,V24,ZAJ	1
7	PAFZZ	5975011295739	19207	11669043	BOX,JUNCTION UOC:DAJ,V24,ZAJ	1
8	PAFZZ		19207	12375494	GASKET UOC:DAJ,V24,ZAJ	1
9	PAFZZ	5310000617325	96906	MS21045-4	NUT,SELF-LOCKING .... UOC:DAJ,V24, ZAJ	4
10	PAFZZ	5935012288537	96906	MS3452W22-9S	CONNECTOR,RECEPTACLE UOC:DAJ,V24,ZAJ	2
11	PAFZZ	5305009580656	96906	MS35207-218	SCREW,MACHINE UOC:DAJ,V24,ZAJ	4
12	PAFZZ	5305008893001	96906	MS35206-231	SCREW,TAPPING UOC:DAJ,V24,ZAJ	4
13	XDFZZ		19207	12300742	COVER UOC:DAJ,V24,ZAJ	1
14	PAFZZ	5340005433915	96906	MS35140-2	STRAP,RETAINING UOC:DAJ,V24,ZAJ	6
15	PAFZZ	5305011193606	96906	MS24629-75	SCREW,TAPPING UOC:DAJ,V24,ZAJ	6
16	PAFZZ	5305009584347	9690	MS35207-216	SCREW,MACHINE UOC:DAJ,V24,ZAJ	8
17	PAFZZ	5935010853305	96906	MS3452W20-4P	CONNECTOR,RECEPTACLE UOC:DAJ,V24,ZAJ	1
18	PAFZZ	5305008550961	96906	MS24629-35	SCREW,TAPPING UOC:DAJ,V24,ZAJ	8
19	PFFZZ	5975011256777	19207	11677687	COVER,JUNCTION UOC:DAJ,V24,ZAJ	1
20	PFFZZ	5975011166269	19207	11677739	BOX,CONNECTOR,ELECTRICAL UOC:DAJ,V24,ZAJ	1
21	PFFZZ	5975012319293	19207	11677576	CONDUIT,METAL,RIGID UOC:DAJ,V24,ZAJ	1
22	PAFZZ	5315010546989	96906	MS35489-116	GROMMET,ELECTRICAL UOC:DAJ,V24,ZAJ	1
23	PAFZZ	5975006611003	03743	95T150	COUPLING,ELECTRICAL UOC:DAJ,V24,ZAJ	1

END OF FIGURE

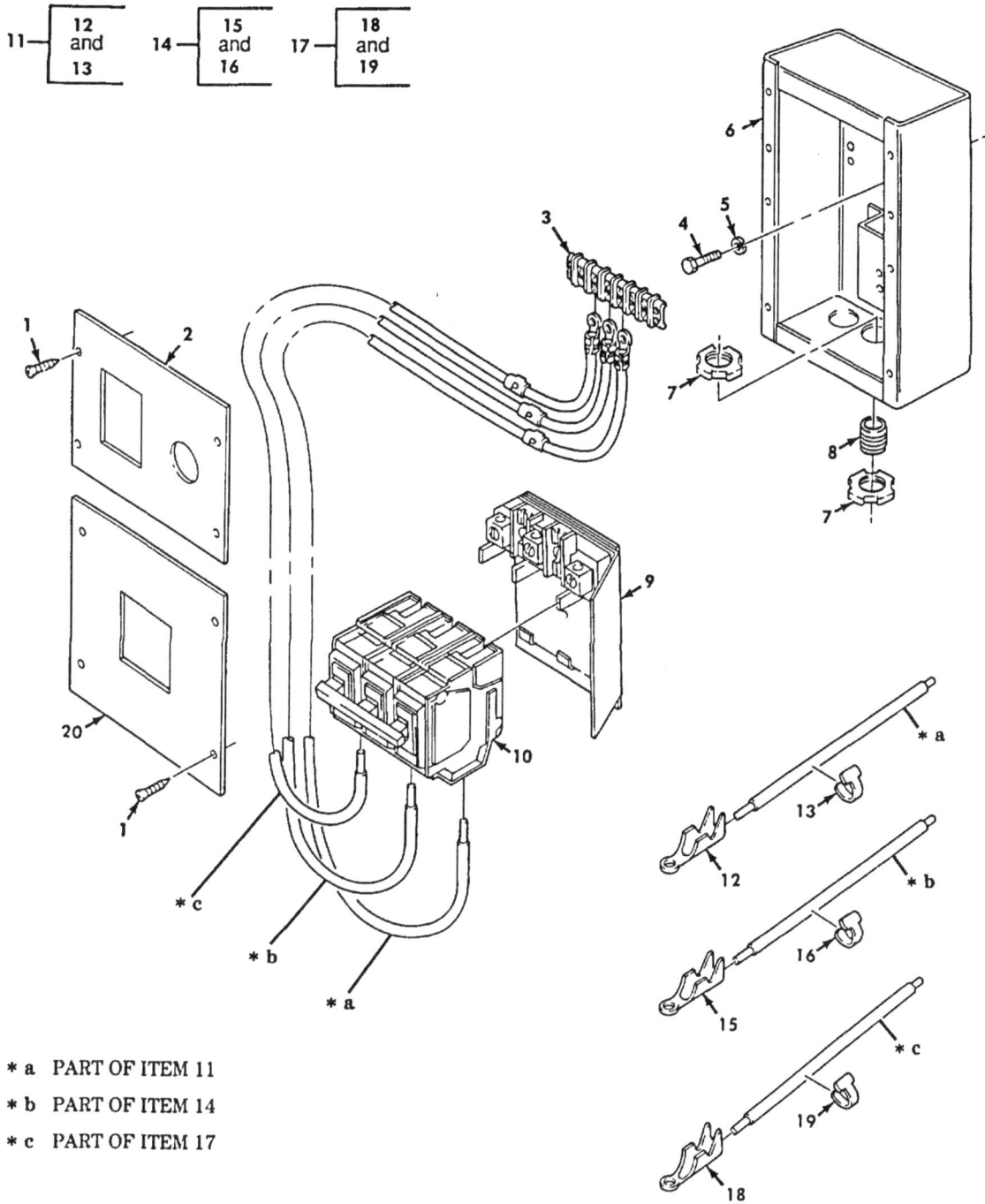

* a PART OF ITEM 11
* b PART OF ITEM 14
* c PART OF ITEM 17

Figure 444. Van Circuit Breakers, Control Box, and Related Parts.

(1) ITEM NO	(2) SMR CODE	(3) NSN	(4) CAGEC	(5) PART NUMBER	(6) DESCRIPTION AND USABLE ON CODES (UOC)	(7) QTY
					GROUP 1812 SPECIAL PURPOSE BODIES	
					FIG. 444 VAN CIRCUIT BREAKERS, CONTROL BOX, AND RELATED PARTS	
1	PAFZZ	5305008550961	96906	MS24629-35	SCREW,TAPPING UOC:DAJ,V24,ZAJ	8
2	PAFZZ	5340011147747	19207	11677700	COVER,ACCESS UOC:DAJ,V24,ZAJ	1
3	PFFZZ	5940011174895	19207	876875	TERMINAL BOARD UOC:DAJ,V24,ZAJ	1
4	PAFZZ	5305002693239	96906	MS90727-63	SCREW,CAP,HEXAGON UOC:DAJ,V24,ZAJ	4
5	PAFZZ	5310005825965	96906	MS35338-44	WASHER,LOCK UOC:DAJ,V24,ZAJ	4
6	PFFZZ	6110011147701	19207	11677727	DISTRIBUTION BOX UOC:DAJ,V24,ZAJ	1
7	PAFZZ	5975007935550	88044	AN3066-12	LOCKNUT,ELECTRICAL UOC:DAJ,V24,ZAJ	4
8	PAFZZ	4730001962058	96906	MS51953-97B	NIPPLE,PIPE UOC:DAJ,V24,ZAJ	2
9	PFFZZ	5925011289524	19207	1663055	BASE,MOUNTING CIRCUIT UOC:DAJ,V24,ZAJ	1
10	PAFZZ	5925011147582	19207	11663054	CIRCUIT BREAKER UOC:DAJ,V24,ZAJ	1
11	PFFFF	2590011354428	19207	11677641-3	LEAD,ELECTRICAL UOC:DAJ,V24,ZAJ	1
12	PAFZZ	5940005341028	96906	MS35436-10	.TERMINAL,LUG UOC:DAJ,V24,ZAJ	1
13	PAFZZ	9905007524649	81349	M43436/1-1	.BAND,MARKER UOC:DAJ,V24,ZAJ	1
14	PFFFF	2590011354427	19207	11677641-2	LEAD,ELECTRICAL UOC:DAJ,V24,ZAJ	1
15	PAFZZ	5940005341028	96906	MS35436-10	.TERMINAL,LUG UOC:DAJ,V24,ZAJ	1
16	PAFZZ	9905007524649	81349	M43436/1-1	.BAND,MARKER UOC:DAJ,V24,ZAJ	1
17	PFFFF	2590011354426	19207	11677641-1	LEAD,ELECTRICAL UOC:DAJ,V24,ZAJ	1
18	PAFZZ	5940005341028	96906	MS35436-10	.TERMINAL,LUG UOC:DAJ,V24,ZAJ	1
19	PAFZZ	9905007524649	81349	M43436/1-1	.BAND,MARKER UOC:DAJ,V24,ZAJ	1
20	PFFZZ	5340011147748	19207	11677716	COVER,ACCESS UOC:DAJ,V24,ZAJ	1

END OF FIGURE

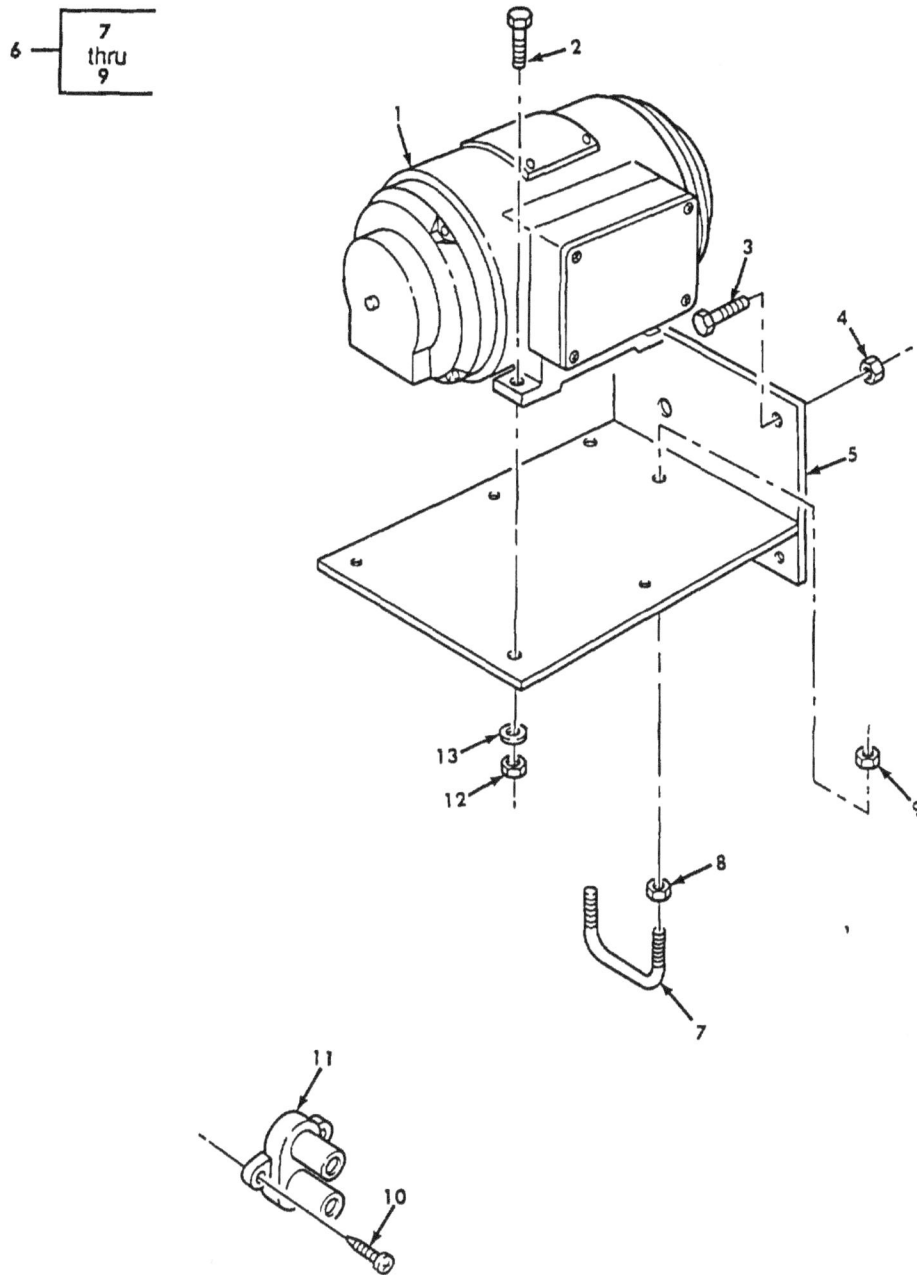

Figure 445. Van 400Hz Converter, Circuit Breaker, and Mounting Hardware.

(1) ITEM NO	(2) SMR CODE	(3) NSN	(4) CAGEC	(5) PART NUMBER	(6) DESCRIPTION AND USABLE ON CODES (UOC)	(7) QTY
					GROUP 1812 SPECIAL PURPOSE BODIES	
					FIG. 445 VAN 400HZ CONVERTER, CIRCUIT BREAKER, AND MOUNTING HARDWARE	
1	PFFZZ	6125010207268	91723	30-154	MOTOR-GENERATOR 1 UOC:DAJ,V24,ZAJ	
2	PAFZZ	5305002693241	96906	MS90727-65	SCREW,CAP,HEXAGON 4 UOC:DAJ,V24,ZAJ	
3	PAFZZ	5305007254183	96906	MS90726-113	SCREW,CAP,HEXAGON 4 UOC:DAJ,V24,ZAJ	
4	PAFZZ	5310000624954	96906	MS21045-8	NUT,SELF-LOCKING 4 UOC:DAJ,V24,ZAJ	
5	PFFZZ	3010011681490	19207	11677732	SUPPORT,CONVERTER 1 UOC:DAJ,V24,ZAJ	
6	PAFFF	5306004713273	19207	10883223	BOLT,ASSEMBLY U 1 UOC:DAJ,V24,ZAJ	
7	PAFZZ	5306004019561	19207	7373215	.U-BOLT 1 UOC:DAJ,V24,ZAJ	
8	PAFZZ	5310007320560	96906	MS51968-14	.NUT,PLAIN,HEXAGON 2 UOC:DAJ,V24,ZAJ	
9	PAFZZ	5310000679507	96906	MS51922-37	.NUT,SELF-LOCKING 2 UOC:DAJ,V24,ZAJ	
10	PAFZZ	5305004324172	96906	MS51861-37	SCREW,TAPPING 2 UOC:DAJ,V24,ZAJ	
11	PAFZZ	5925000264767	81349	M13516/1-1	CIRCUIT BREAKER 1 UOC:DAJ,V24,ZAJ	
12	PAFZZ	5310009824908	96906	MS21045-6	NUT,SELF-LOCKING 4 UOC:DAJ,V24,ZAJ	
13	PAFZZ	5310000806004	96906	MS27183-14	WASHER,FLAT 4 UOC:DAJ,V24,ZAJ	

END OF FIGURE

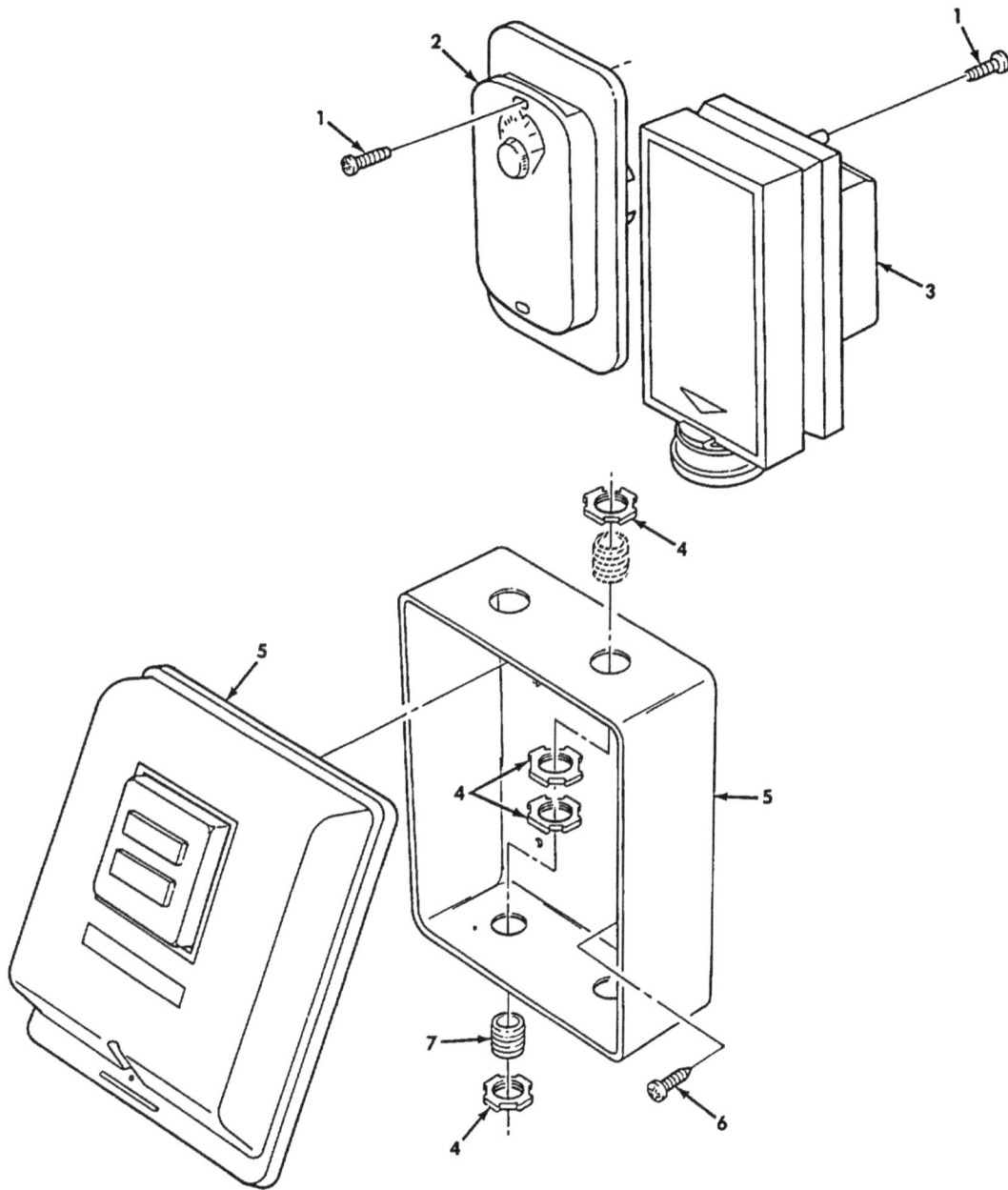

Figure 446. Van Heater Thermostatic Switch and Fluorescent Starter.

(1) ITEM NO	(2) SMR CODE	(3) NSN	(4) CAGEC	(5) PART NUMBER	(6) DESCRIPTION AND USABLE ON CODES (UOC)	(7) QTY
					GROUP 1812 SPECIAL PURPOSE BODIES	
					FIG. 446 VAN HEATER THERMOSTATIC SWITCH AND FLUORESCENT STARTER	
1	PAFZZ	5305008550961	96906	MS24629-35	SCREW,TAPPING UOC:DAJ,V24,ZAJ	4
2	PAFZZ	5930001746224	19207	7336056	SWITCH,THERMOSTATIC UOC:DAJ,V24,ZAJ	1
3	PAFZZ	5390011152322	19207	11663213	SWITCH,THERMOSTATIC UOC:DAJ,V24,ZAJ	1
4	PAFZZ	5975007935550	88044	AN3066-12	LOCKNUT,ELECTRICAL UOC:DAJ,V24,ZAJ	4
5	PAFZZ	5930041147581	19207	11663315	SWITCH,BOX UOC:DAJ,V24,ZAJ	1
6	PAFZZ	5305004324203	96906	MS51861-47	SCREW,TAPPING UOC:DAJ,V24,ZAJ	4
7	PAFZZ	4730001962058	96906	MS51953-978	NIPPLE,PIE UOC:DAJ,V24,ZAJ	2

END OF FIGURE

Figure 447. Van Telephone Jack and Related Parts, Interior and Exterior.

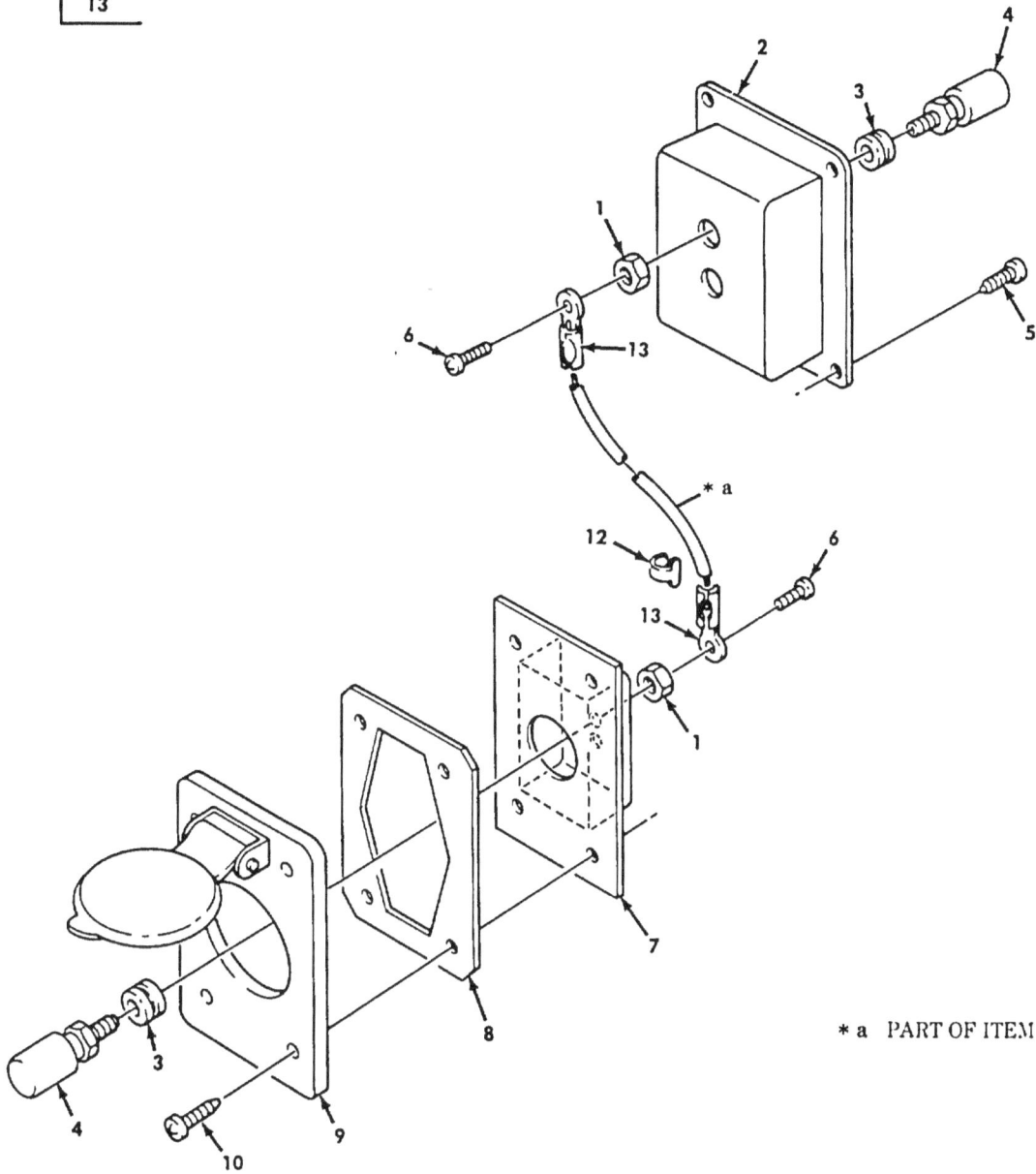

* a   PART OF ITEM 11

(1) ITEM NO	(2) SMR CODE	(3) NSN	(4) CAGEC	(5) PART NUMBER	(6) DESCRIPTION AND USABLE ON CODES (UOC)	(7) QTY
					GROUP 1812 SPECIAL PURPOSE BODIES	
					FIG. 447 VAN TELEPHONE JACK AND RELATED PARTS,INTERIOR AND EXTERIOR	
1	PAOZZ	5310007680319	96906	MS51968-2	NUT,PLAIN,HEXAGON UOC:DAJ,V24,ZAJ	6
2	PAOZZ	5975004180861	19207	8735431	JUNCTION BOX UOC:DAJ,V24,ZAJ	2
3	PAOZZ	5325002636632	96906	MS35489-6	GROMMET,NONMETALLIC UOC:DAJ,V24,ZAJ	6
4	PAOZZ	5940002721477	80063	SCC1360116P1	POST,BINDING,ELECTR UOC:DAJ,V24,ZAJ	6
5	PAOZZ	5305008550961	96906	MS24629-35	SCREW,TAPPING UOC:DAJ,V24,ZAJ	4
6	PAOZZ	5305009846189	96906	MS35206-241	SCREW,MACHINE UOC:DAJ,V24,ZAJ	4
7	PFOZZ	5340011470861	19207	10937077	PANEL ASSEMBLY UOC:DAJ,V24,ZAJ	1
8	PAOZZ	5330004146754	19207	7059240	GASKET UOC:DAJ,V24,ZAJ	1
9	PAOZZ	5935002423488	19207	7759184	COVER,ELECTRICAL UOC:DAJ,V24,ZAJ	1
10	PAOZZ	5305004324170	96906	MS51861-35	SCREW,TAPPING UOC:DAJ,V24,ZAJ	4
11	PAOOO	2590004260766	19207	7059245	LEAD,ELECTRICAL UOC:DAJ,V24,ZAJ	2
12	PAOZZ	9905007524649	81349	M43436/1-1	.BAND,MARKER UOC:DAJ,V24,ZAJ	1
13	PAOZZ	5940001133138	96906	MS20659-102	.TERMINAL,LUG UOC:DAJ,V24,ZAJ	2

END OF FIGURE

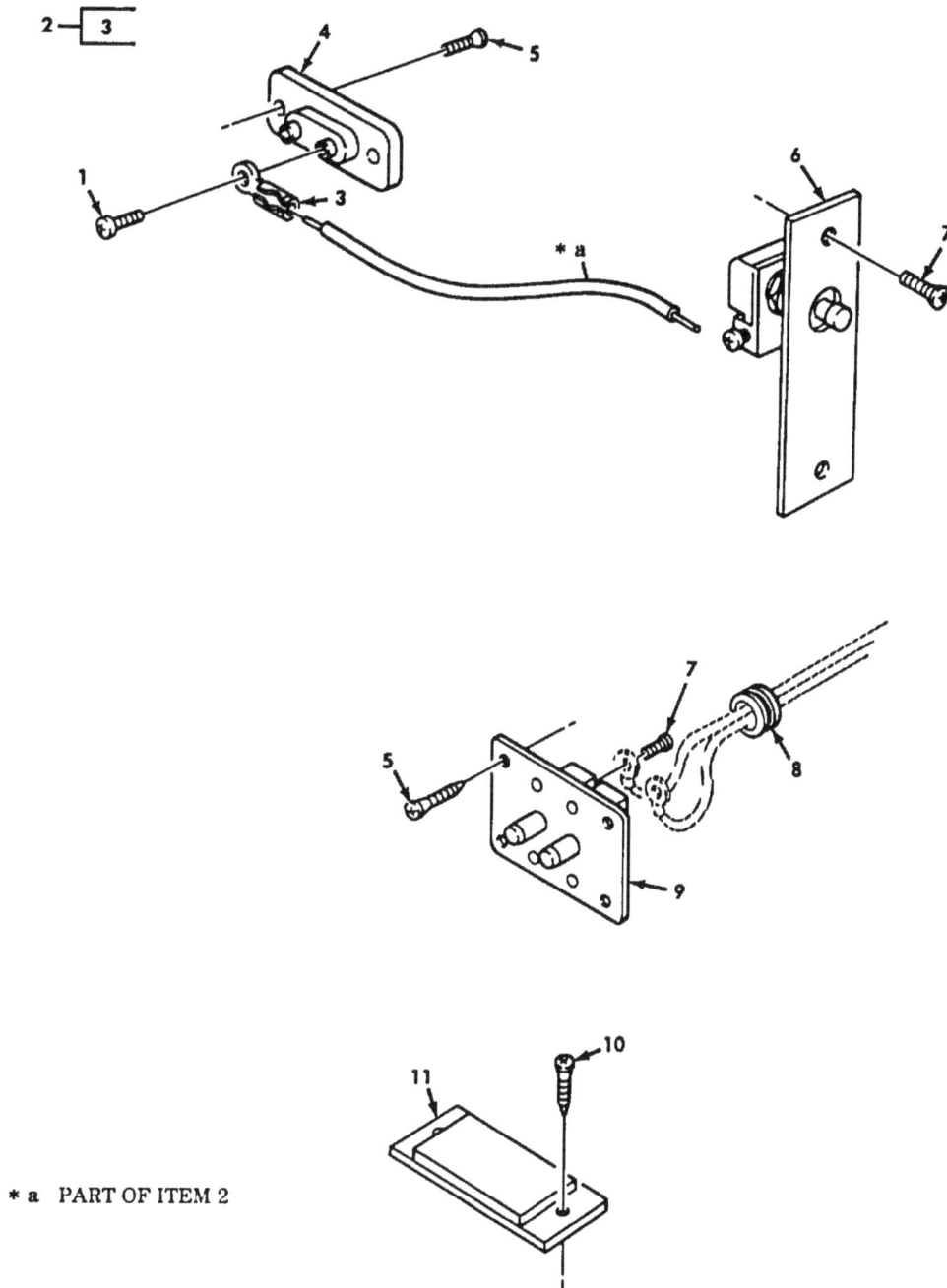

Figure 448. Van Hinged Roof Plunger and Blackout Light Switch.

(1) ITEM NO	(2) SMR CODE	(3) NSN	(4) CAGEC	(5) PART NUMBER	(6) DESCRIPTION AND USABLE ON CODES (UOC)	(7) QTY
					GROUP 1812 SPECIAL PURPOSE BODIES	
					FIG. 448 VAN HINGED ROOF PLUNGER AND BLACKOUT LIGHT SWITCH	
1	PAOZZ	5305009585453	96906	MS35190-236	SCREW,MACHINE .... UOC: DAJ,V24,ZAJ	4
2	PAOOO	2590002310201	19207	7535587	LEAD,ELECTRICAL .... UOC: DAJ,V24,ZAJ	4
3	PAOZZ	5940001519361	96906	MS35436-44	.TERMINAL,LUG UOC:DAJ,V24,ZAJ	1
4	PAOZZ	5940001504396	19207	7336049	TERMINAL BOARD UOC:DAJ,V24,ZAJ	2
5	PAOZZ	5305004324387	96906	MS51862-13	SCREW,TAPPING .... UOC:DAJ,V24, ZAJ	12
6	PAOZZ	5930001160531	19207	10937541	SWITCH,PUSH UOC:DAJ,V24,ZAJ	2
7	PAOZZ	5305009844983	9690	MS35206-226	SCREW,MACHINE UOC:DAJ,V24,ZAJ	8
8	PAOZZ	5325002766228	96906	MS35489-9	GROMMET,NONMETALLIC UOC:DAJ,V24,ZAJ	2
9	PAOZZ	5999004580730	19207	7336048	CONTACT,ELECTRICAL UOC:DAJ,V24,ZAJ	2
10	PAOZZ	5305004324394	96906	MS51862-24	SCREW,TAPPING .... UOC: DAJ,V24, ZAJ	4
11	PAOZZ	2510004051921	19207	7759189	PLATE,FLOOR HINGE UOC:DAJ,V24,ZAJ	2

END OF FIGURE

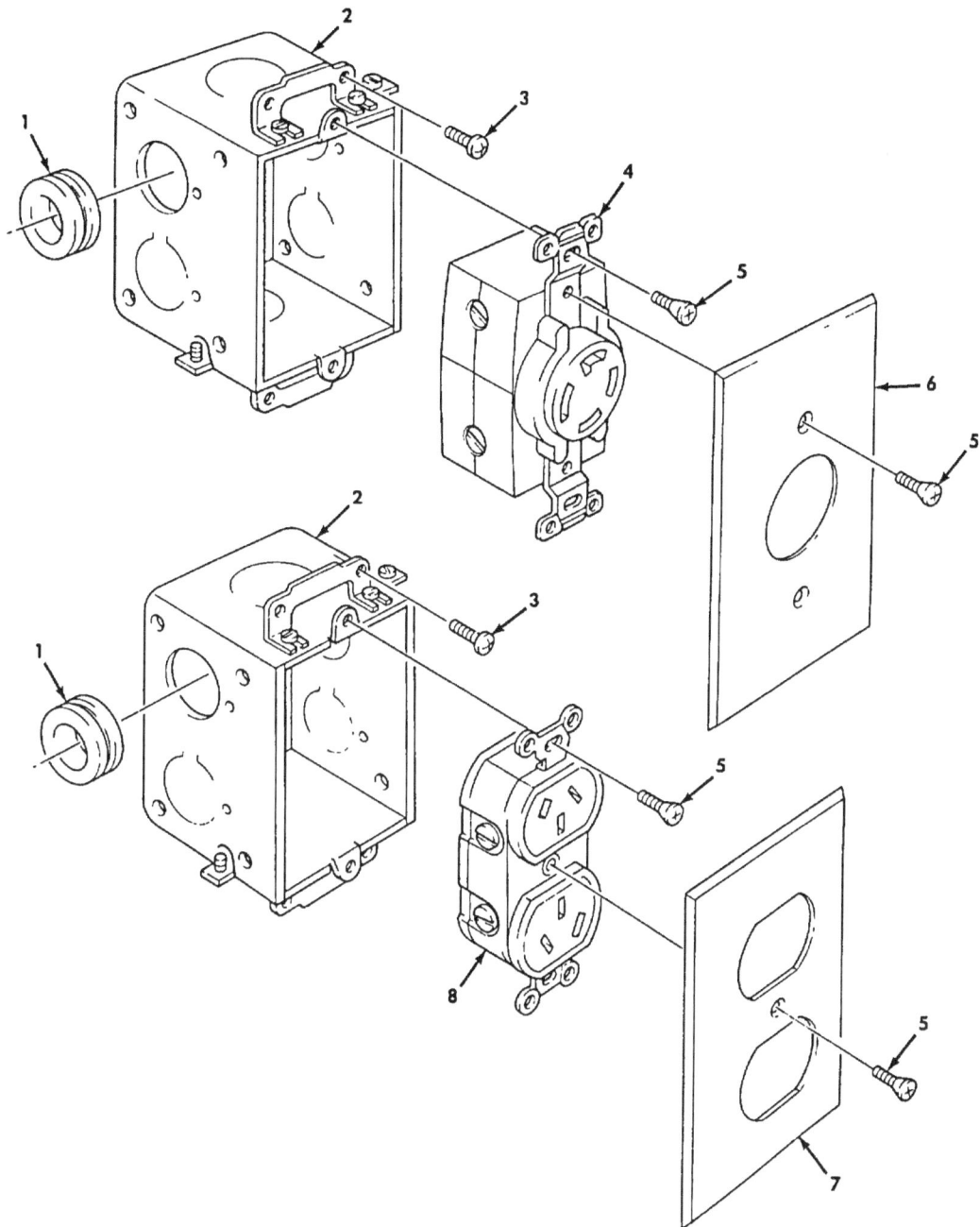

Figure 449. Van 3-Phase and 400Hz Receptacles.

(1) ITEM NO	(2) SMR CODE	(3) NSN	(4) CAGEC	(5) PART NUMBER	(6) DESCRIPTION AND USABLE ON CODES (UOC)	(7) QTY
					GROUP 1812 SPECIAL PURPOSE BODIES	
					FIG. 449 VAN 3-PHASE AND 400HZ RECEPTACLES	
1	PAOZZ	5325002811557	96906	MS35489-17	GROMMET,NONMETALLIC UOC:DAJ,V24,ZAJ	4
2	PFOZZ	5975002543141	19207	8743871	JUNCTION BOX UOC:DAJ,V24,ZAJ	4
3	PAOZZ	5305004324170	96906	MS51861-35	SCREW,TAPPING UOC:DAJ,V24,ZAJ	16
4	PFOZZ	5935011147678	19207	11663108	CONNECTOR,RECEPTACLE UOC:DAJ,V24,ZAJ	2
5	PAOZZ	5305009585453	96906	MS35190-236	SCREW,MACHINE UOC:DAJ,V24,ZAJ	14
6	PFOZZ	5975010445419	71183	7321	PLATE,WALL,ELECTRICAL UOC:DAJ,V24,ZAJ	2
7	PFOZZ	5975011150616	19207	8380747	PLATE,WALL,ELECTRICAL UOC:DAJ,V24,ZAJ	2
8	PFOZZ	5935011147677	19207	11663109	CONNECTOR,RECEPTACLE UOC:DAJ,V24,ZAJ	2

END OF FIGURE

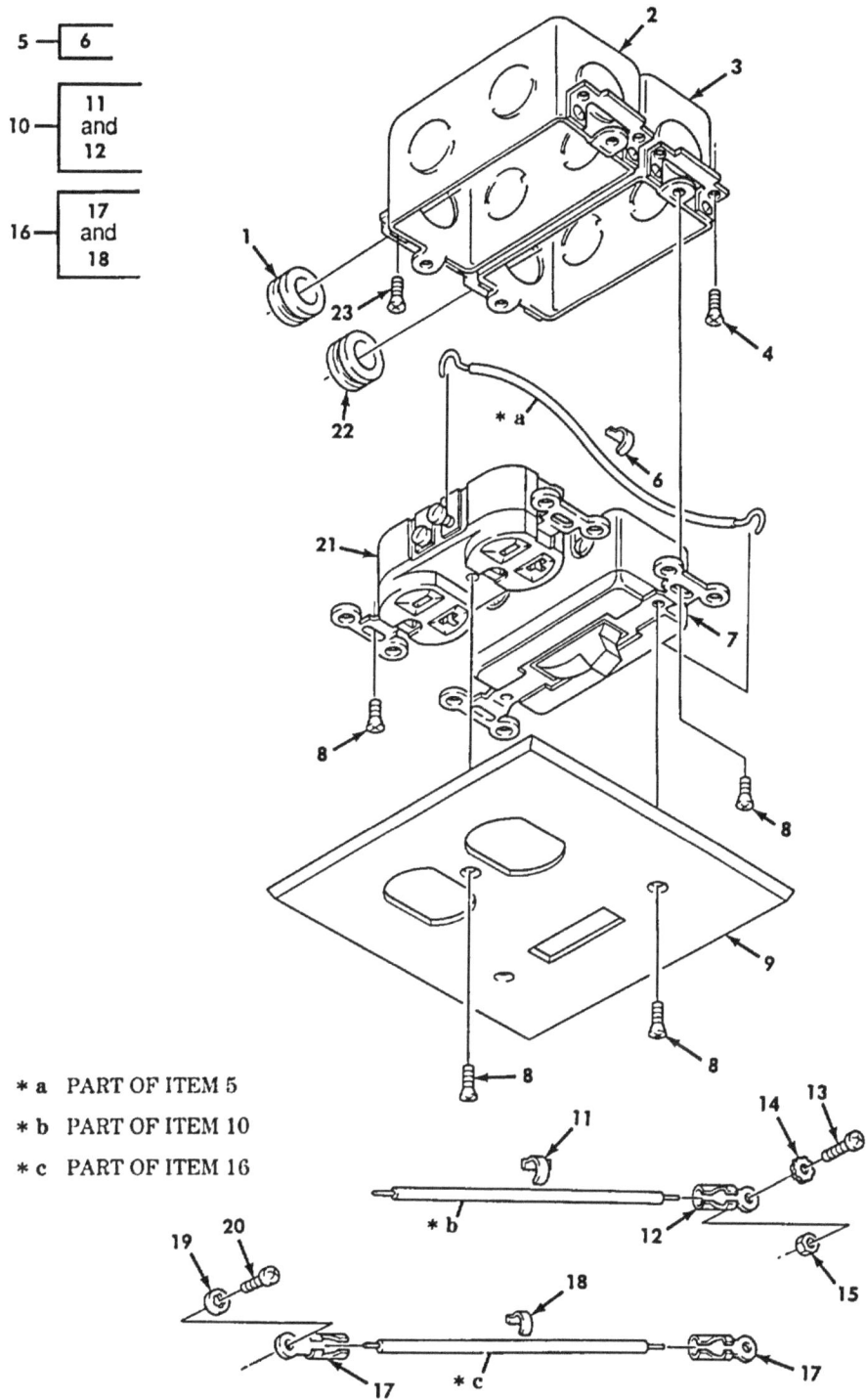

*a PART OF ITEM 5
*b PART OF ITEM 10
*c PART OF ITEM 16

Figure 450. Van Ceiling Switch, Front Wall Switch, Receptacle Assembly, and Related Parts.

(1) ITEM NO	(2) SMR CODE	(3) NSN	(4) CAGEC	(5) PART NUMBER	(6) DESCRIPTION AND USABLE ON CODES (UOC)	(7) QTY
					GROUP 1812 SPECIAL PURPOSE BODIES	
					FIG. 450 VAN CEILING SWITCH, FRONT WALL SWITCH, RECEPTACLE ASSEMBLY, AND RELATED PARTS	
1	PAOZZ	5325002766228	96906	MS35489-9	GROMMET,NONMETALLIC UOC:DAJ,V24,ZAJ	4
2	PAOZZ	5975001689172	19207	10937536	JUNCTION BOX UOC:DAJ,V24,ZAJ	4
3	PFOZZ	5975002543141	19207	8743871	JUNCTION BOX UOC:DAJ,V24,ZAJ	16
4	PAOZZ	5305008550961	96906	MS24629-35	SCREW,TAPPING UOC:DAJ,V24,ZAJ	32
5	PAOOO	2590004442094	19207	10937500	LEAD,ELECTRICAL UOC:DAJ,V24,ZAJ	10
6	PAOZZ	9905007524649	8134	M43436/1-1	.BAND,MARKER UOC:DAJ,V24,ZAJ	1
7	PAOZZ	5930007245417	81349	WS896/2-08A	SWITCH,TOGGLE UOC:DAJ,V24,ZAJ	10
8	PAOZZ	5305009585453	96906	MS35190-236	SCREW,MACHINE UOC:DAJ,V24,ZAJ	70
9	PAOZZ	5975004039490	19207	8342290	PLATE,WALL,ELECTRICAL UOC:DAJ,V24,ZAJ	10
10	PFOOO	2590011354423	19207	11677692	LEAD,ELECTRICAL UOC:DAJ,V24,ZAJ	8
11	PAOZZ	9905010138723	81349	M43436/3-1	.BAND,MARKER UOC:DAJ,V24,ZAJ	1
12	PAOZZ	5940006603632	96906	MS35346-7	.TERMINAL LUG UOC:DAJ,V24,ZAJ	1
13	PAOZZ	5305009844988	96906	MS35206-228	SCREW,MACHINE UOC:DAJ,V24,ZAJ	8
14	PAOZZ	5310002090788	96906	MS35335-30	WASHER,LOCK UOC:DAJ,V24,ZAJ	8
15	PAOZZ	5310009349747	80045	MS35649-262	NUT,PLAIN,HEXAGON UOC:DAJ,V24,ZAJ	8
16	PFOOO	2590011354425	19207	11677685	LEAD,ELECTRICAL UOC:DAJ,V24,ZAJ	2
17	PAOZZ	5940005572343	96906	MS35436-11	.TERMINAL,LUG UOC:DAJ,V24,ZAJ	2
18	PAOZZ	9905010697222	81349	M43436/2-1	.BAND,MARKER UOC:DAJ,V24,ZAJ	1
19	PAOZZ	5310005765752	96906	MS35333-39	WASHER,LOCK UOC:DAJ,V24,ZAJ	2
20	PAOZZ	5305008550957	96906	MS24629-46	SCREW,TAPPING UOC:DAJ,V24,ZAJ	2
21	PAOZZ	5935011079924	19207	10937538	CONNECTOR,RECEPTACLE UOC:DAJ,V24,ZAJ	10
22	PAOZZ	5325002811557	96906	MS35489-17	GROMMET,NONMETALLIC UOC:DAJ,V24,ZAJ	16
23	PAOZZ	5305004324170	96906	MS51861-35	SCREW,TAPPING UOC:DAJ,V24,ZAJ	8

END OF FIGURE

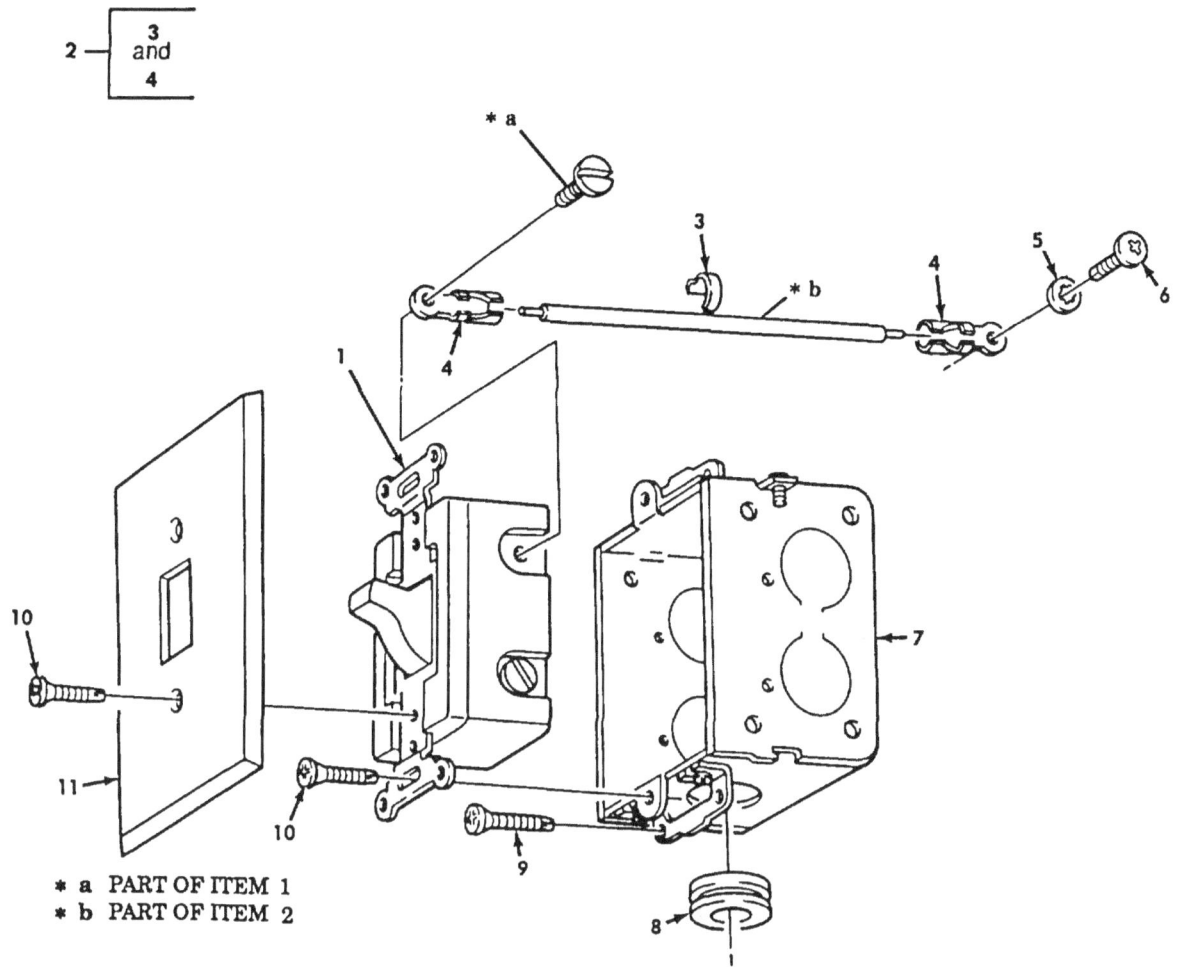

Figure 451. Van Blackout and Emergency Switches.

(1) ITEM NO	(2) SMR CODE	(3) NSN	(4) CAGEC	(5) PART NUMBER	(6) DESCRIPTION AND USABLE ON CODES (UOC)	(7) QTY

GROUP 1812 SPECIAL PURPOSE BODIES

FIG. 451 VAN BLACKOUT AND EMERGENCY SWITCHES

(1) ITEM NO	(2) SMR CODE	(3) NSN	(4) CAGEC	(5) PART NUMBER	(6) DESCRIPTION AND USABLE ON CODES (UOC)	(7) QTY
1	PAOZZ	5930007245417	81348	WS89612-08A	SWITCH,TOGGLE BLACKOUT UOC:DAJ,V24,ZAJ	1
1	PAOZZ	5930006605584	81349	WS896/2-02A	SWITCH,TOGGLE EMERGENCY UOC:DAJ,V24,ZAJ	1
2	PFOOO	2590011354425	19207	11677685	LEAD,ELECTRICAL UOC:DAJ,V24,ZAJ	2
3	PAOZZ	9905010697222	81349	M43436/2-1	.BAND,MARKER UOC:DAJ,V24,ZAJ	1
4	PAOZZ	5940005572343	19207	MS35436-11	.TERMINAL,LUG UOC:DAJ,V24,ZAJ	2
5	PAOZZ	5310005765752	96906	MS35333-39	WASHER,LOCK UOC:DAJ,V24,ZAJ	2
6	PAOZZ	5305008550957	96906	MS24629-46	SCREW,TAPPING UOC:DAJ,V24,ZAJ	2
7	PFOZZ	5975002543141	19207	8743871	JUNCTION BOX UOC:DAJ,V24,ZAJ	2
8	PAOZZ	5325002811557	96906	MS35489-17	GROMMET,NONMETALLIC .... UOC: DAJ,V24,ZAJ	2
9	PAOZZ	5305008550961	96906	MS24629-35	SCREW,TAPPING UOC:DAJ,V24,ZAJ	8
10	PAOZZ	5305009585453	96906	MS35190-236	SCREW,MACHINE UOC:DAJ,V24,ZAJ	8
11	PAOZZ	5975002431275	71183	CATALOG 91071	PLATE,WALL,ELECTRICAL UOC:DAJ,V24,ZAJ	2

END OF FIGURE

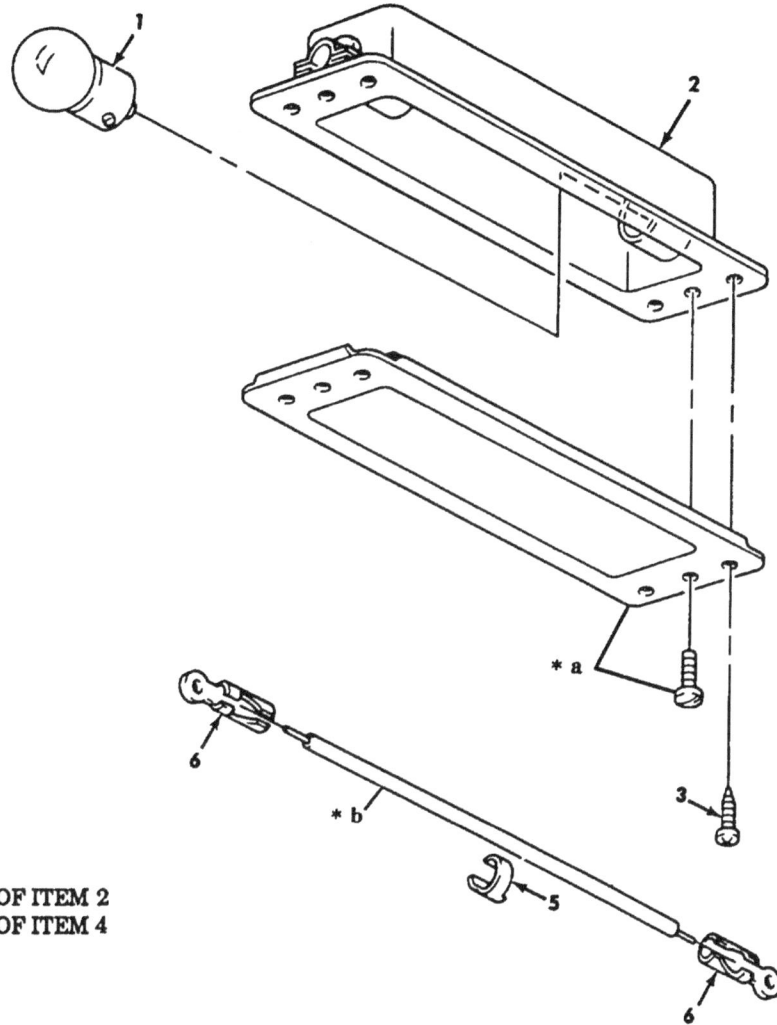

* a PART OF ITEM 2
* b PART OF ITEM 4

Figure 452. Van Blackout and Emergency Lamps.

(1) ITEM NO	(2) SMR CODE	(3) NSN	(4) CAGEC	(5) PART NUMBER	(6) DESCRIPTION AND USABLE ON CODES (UOC)	(7) QTY
					GROUP 1812 SPECIAL PURPOSE BODIES	
					FIG. 452 VAN BLACKOUT AND EMERGENCY LAMPS	
1	PAOZZ	6240001558725	96909	MS15584-6	LAMP,INCANDESCENT EMERGENCY UOC:DAJ,V24,ZAJ	12
1	PAOZZ	6240005426219	96906	MS16123-1	LAMP,INCANDESCENT BLACKOUT .... UOC:DAJ,V24, ZAJ	4
2	PFFZZ	6210009700330	17744	L-16166	FIXTURE,LIGHTING EMERGENCY UOC:DAJ,V24,ZAJ	6
2	PFFZZ	6210000629121	17744	L-16167	FIXTURE,LIGHTING BLACKOUT UOC:DAJ,V24,ZAJ	2
3	PAFZZ	5305008550961	96906	MS24629-35	SCREW,TAPPING UOC:DAJ,V24,ZAJ	32
4	PFFFF	2950011354424	19207	11677689	LEAD,ELECTRICAL UOC:DAJ,V24,ZAJ	12
5	PAFZZ	9905010138723	81349	M43436/3-1	.BAND,MARKER .... UOC: DAJ,V24,ZAJ	2
6	PAFZZ	5940005340986	96906	MS35436-4	.TERMINAL,LUG .... UOC:DAJ,V24 ,ZAJ	2

END OF FIGURE

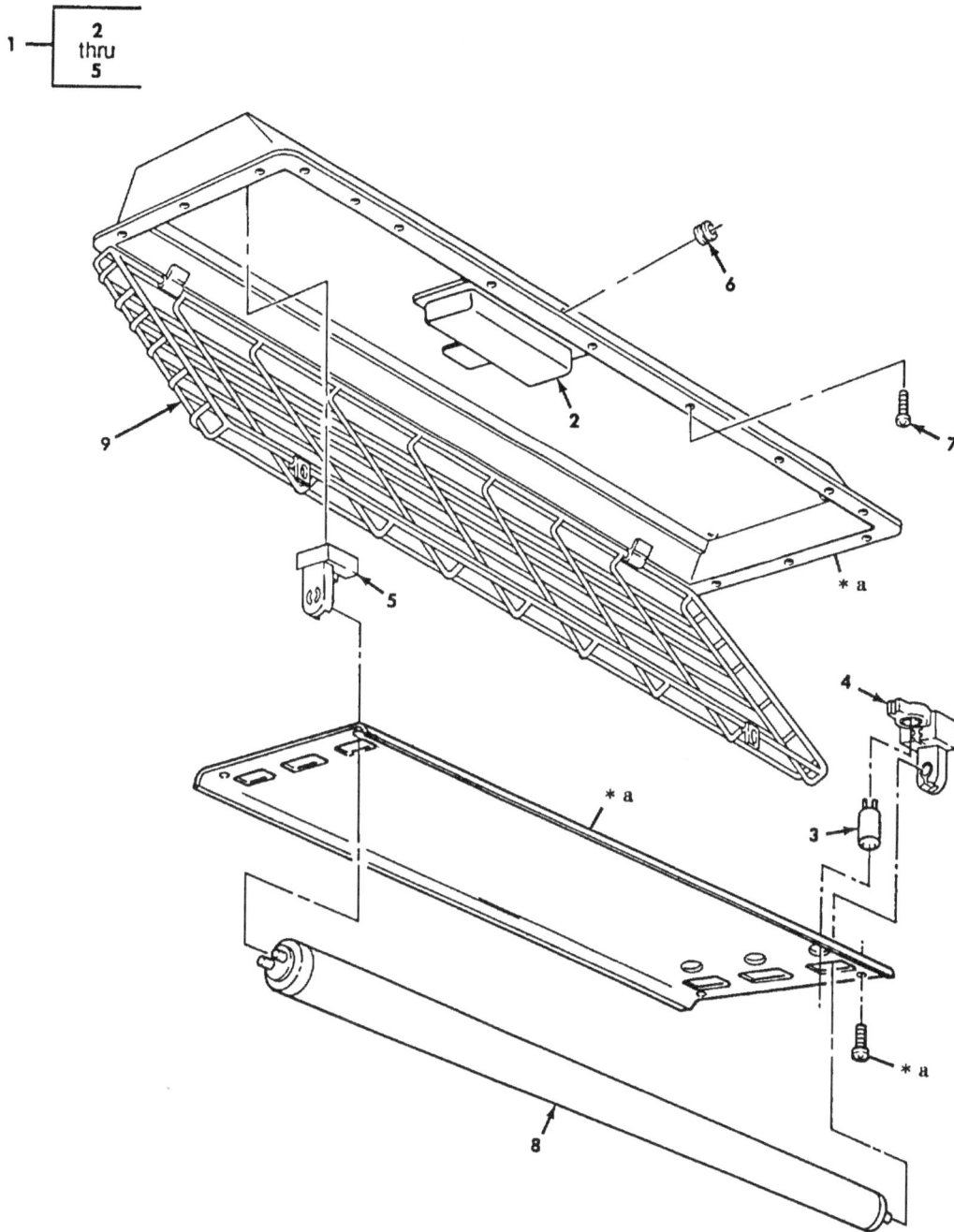

* a   PART OF ITEM 1

Figure 453. Van Ceiling Lamp Fixtures and Bulb.

(1) ITEM NO	(2) SMR CODE	(3) NSN	(4) CAGEC	(5) PART NUMBER	(6) DESCRIPTION AND USABLE ON CODES (UOC)	(7) QTY
					GROUP 1812 SPECIAL PURPOSE BODIES	
					FIG. 453 VAN CEILING LAMP FIXTURES AND BULB	
1	PAOOO	6210011475823	19207	11667981	FIXTURE,FLUORESCENT 27 INCHES LONG UOC:DAJ,V24,ZAJ	8
2	XDOZZ		08595	6G1042	BALLAST,LAMP UOC:DAJ,V24,ZAJ	1
3	PFOZZ	6250003444274	08805	FS25	.STARTER,FLUORESCENT UOC:DAJ,V24,ZAJ	3
4	PFOZZ	6250011147548	04074	226	LAMPHOLDER UOC:DAJ,V24,ZAJ	3
5	PFOZZ	6250011147549	19207	11667980-2	.LAMPHOLDER UOC:DAJ,V24,ZAJ	3
6	PAOZZ	5325002919366	96906	MS35489-11	GROMMET,NONMETALLIC UOC:DAJ,V24,ZAJ	8
7	PAOZZ	5305008550961	96906	MS24629-35	SCREW,TAPPING UOC:DAJ,V24,ZAJ	160
8	PAOZZ	6240000610315	08108	F25T12CW28	LAMP,FLUORESCENT 27 INCHES LONG UOC:DAJ,V24,ZAJ	24
9	XAOZZ		19207	11667981	COVER UOC:DAJ,V24,ZAJ	8

END OF FIGURE

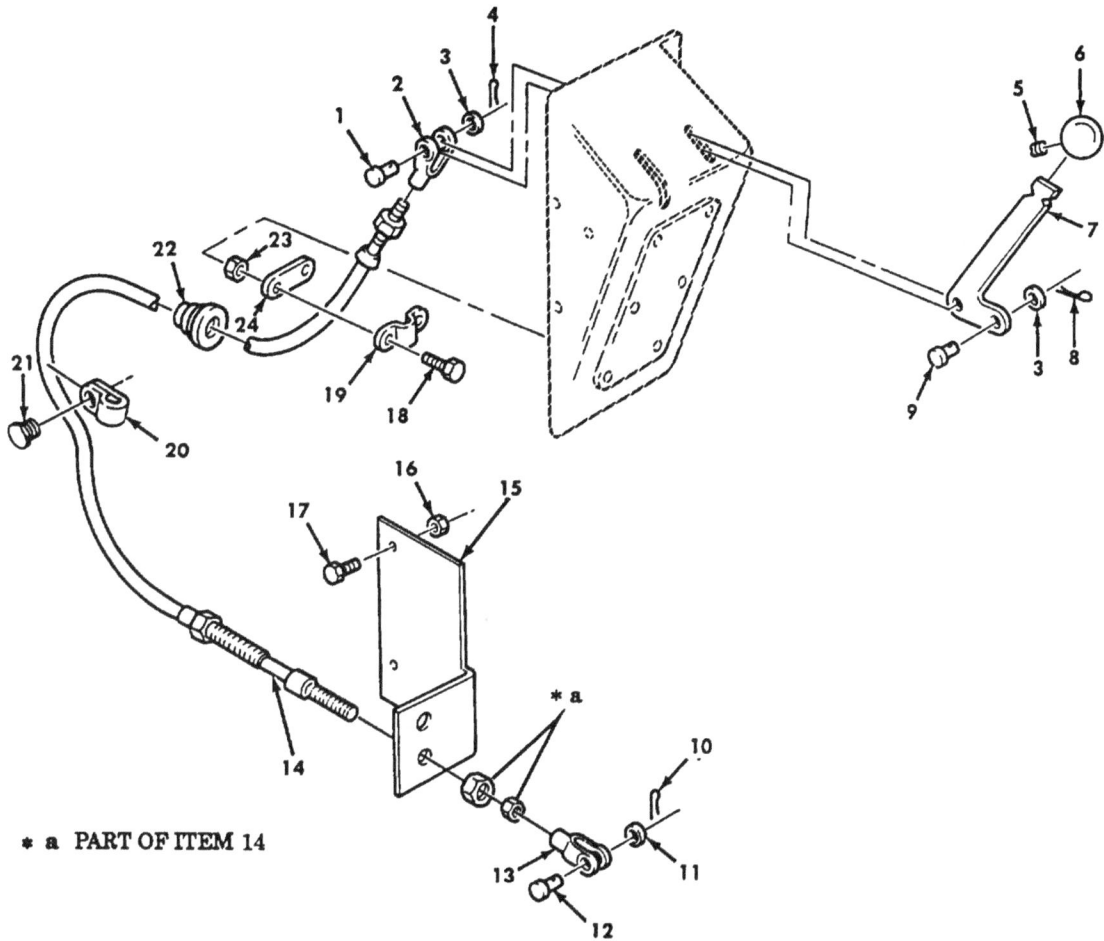

Figure 454. Front Winch Controls and Mounting Hardware.

* a PART OF ITEM 14

(1) ITEM NO	(2) SMR CODE	(3) NSN	(4) CAGEC	(5) PART NUMBER	(6) DESCRIPTION AND USABLE ON CODES (UOC)	(7) QTY
					GROUP 20 HOIST, WINCH, CAPSTAN, WINDLASS, POWER CONTROL UNIT, AND POWER TAKE OFF 2001 HOIST, CAPSTAN, WINDLASS, CRANE OR WINCH ASSEMBLY	
					FIG. 454 FRONT WINCH CONTROLS AND MOUNTING HARDWARE	
1	PAFZZ	5315000817874	96906	MS20392-3C23	PIN,STRAIGHT,HEADED .... UOC:DAB,DAD,DAF,DAH,DAL,DAX,V12,V14, V16,V18,V19,V21,V39,ZAB,ZAD,ZAF,ZAH, ZAL	1
2	PAFZZ	5340008659496	96906	MS35812-2	CLEVIS,ROD END .... UOC:DAB,DAD,DAF,DAH,DAL,DAX,V12,V14, V16,V18,V19,V21,V39,ZAB,ZAD,ZAF,ZAH, ZAL	1
3	PAOZZ	5310008094058	96906	MS27183-10	WASHER,FLAT .... UOC:DAB,DAD,DAF,DAH,DAL,DAX,V12,V14, V16,V18,V19,V21,V39,ZAB,ZAD,ZAF,ZAH, ZAL	2
4	PAFZZ	5315002341854	96906	MS24665-153	PIN,COTTER .... UOC:DAB,DAD,DAF,DAH,DAL,DAX,V12,V14, V16,V18,V19,V21,V39,ZAB,ZAD,ZAF,ZAR, ZAL	1
5	PAFZZ	5305007239383	96906	MS51963-67	SETSCREW. .... UOC:DAB,DAD,DAF,DAH,DAL,DAX,V12,V14, V16,V18,V19,V21,V39,ZAB,ZAD,ZAF,ZAH, ZAL	2
6	PAFZZ	5355011074178	19207	12276921-1	KNOB .... UOC:DAB,DAD,DAF,DAH,DAL,DAX,V12,V14, V16,V18,V19,V21,V39,ZAB,ZAD,ZAF,ZAH, ZAL	1
7	PFFZZ	3040011049152	19207	12276914	LEVER,MANUAL CONTRO.. UOC:DAB,DAD,DAF,DAH,DAL,DAX,V12,V14, V16,V18,V19,V21,V39,ZAB,ZAD,ZAF,ZAH, ZAL	1
8	PAFZZ	5315008151405	96906	MS24665-151	PIN,COTTER .... UOC:DAB,DAD,DAF,DAH,DAL,DAX,V12,V14, V16,V18,V19,V21,V39,ZAB,ZAD,ZAF,ZAH, ZAL	1
9	PAFZZ	5315000637366	96906	MS20392-3C17	PIN,STRAIGHT,HEADED .... UOC:DAB,DAD,DAF,DAH,DAL,DAX,V12,V14, V16,V18,V19,V21,V39,ZAB,ZAD,ZAF,ZAH, ZAL	1
10	PAFZZ	5315002341863	96906	MS24665-300	PIN,COTTER.. UOC:DAB,DAD,DAF,DAH,DAL,DAX,V12,V14, V16,V18,V19,V21,V39,ZAB,ZAD,ZAF,ZAH, ZAL	1
11	PAFZZ	5310000806004	96906	MS27183-14	WASHER,FLAT.. UOC:DAB,DAD,DAF,DAH,DAL,DAX,V12,V14, V16,V18,V19,V21,V39,ZAB,ZAD,ZAF,ZAH, ZAL	1

(1) ITEM NO	(2) SMR CODE	(3) NSN	(4) CAGEC	(5) PART NUMBER	(6) DESCRIPTION AND USABLE ON CODES (UOC)	(7) QTY
12	PAFZZ	5315009902889	96906	MS20392-5C31	PIN,STRAIGHT,HEADED .... UOC:DAB,DAD,DAF,DAH,DAL,DAX,V12,V14, V16,V18,V19,V21,V39,ZAB,ZAD,ZAF,ZAH, ZAL	1
13	PAFZZ	5340011049005	19207	12276916	CLEVIS,ROD END .... UOC:DAB,DAD,DAF,DAH,DAL,DAX,V12,V14, V16,V18,V19,V21,V39,ZAB,ZAD,ZAF,ZAH, ZAL	1
14	PAFZZ	4010011126562	19207	11669464	WIRE ROPE ASSEMBLY, ... UOC:DAB,DAD,DAF,DAH,DAL,DAX,V12,V14, V16,V18,V19,V21,V39,ZAB,ZAD,ZAF,ZAH, ZAL	1
15	PFFZZ	5340012107510	19207	12302753	BRACKET,ANGLE UOC:DAL,V18,V39,ZAL	1
15	PFFZZ	5340011075221	19207	12276906	BRACKET,ANGLE FRONT WINCH CONTROL .... ASSEMBLY.. UOC:DAB,DAD,DAF,DAH,DAX,V12,V14,V16, V19,V21,ZAB,ZAD,ZAF,ZAH	1
16	PAFZZ	5310008140672	96906	MS51943-36	NUT,SELF-LOCKING,HE .... UOC:DAB,DAD,DAF,DAH,DAL,DAX,V12,V14, V16,V18,V19,V21,V39,ZAB,ZAD,ZAF,ZAH, ZAL	2
17	PAFZZ	5305002692803	96906	MS90726-60	SCREW,CAP,HEXAGON H.. UOC:DAB,DAD,DAF,DAH,DAL,DAX,V12,V14, V16,V18,V19,V21,V39,ZAB,ZAD,ZAF,ZAH, ZAL	2
18	PAFZZ	5306000680513	60285	6893-2	BOLT,MACHINE. .... UOC:DAB,DAD,DAF,DAH,DAL,DAX,V12,V14, V16,V18,V19,V21,V39,ZAB,ZAD,ZAF,ZAH, ZAL	2
19	PAFZZ	5340011285302	19207	12256285	STRAP,RETAINING... UOC:DAB,DAD,DAF,DAH,DAL,DAX,V12,V14, V16,V18,V19,V21,V39,ZAB,ZAD,ZAF,ZAH, ZAL	1
20	PAFZZ	5340011179876	19207	7397785	CLAMP,LOOP UOC:DAB,DAD,DAF,DAH,DAL,DAX,V12,V14, V16,V18,V19,V21,V39	1
21	PAFZZ	4730002873281	72582	127950	PLUG,PIPE .... UOC:DAB,DAD,DAF,DAH,DAL,DAX,V12,V14, V16,V18,V19,V21,V39,ZAB,ZAD,ZAF,ZAH, ZAL	1
22	PAFZZ	5325011147763	19207	12256578-4	GROMMET,NONMETALLIC .... UOC:DAB,DAD,DAF,DAH,DAL,DAX,V12,V14, V16,V18,V19,V21,V39,ZAB,ZAD,ZAF,ZAH, ZAL	1
23	PAFZZ	5310009359022	96906	MS51943-32	NUT,SELF-LOCKING,HE .... UOC:DAB,DAD,DAF,DAH,DAL,DAX,V12,V14, V16,V18,V19,V21,V39,ZAB,ZAD,ZAF,ZAH, ZAL	2
24	PAFZZ	5365011093353	19207	12256286	SPACER,PLATE.. UOC:DAB,DAD,DAF,DAH,DAL,DAX,V12,V14, V16,V18,V19,V21,V39,ZAB,ZAD,ZAF,ZAH, ZAL	1

END OF FIGURE

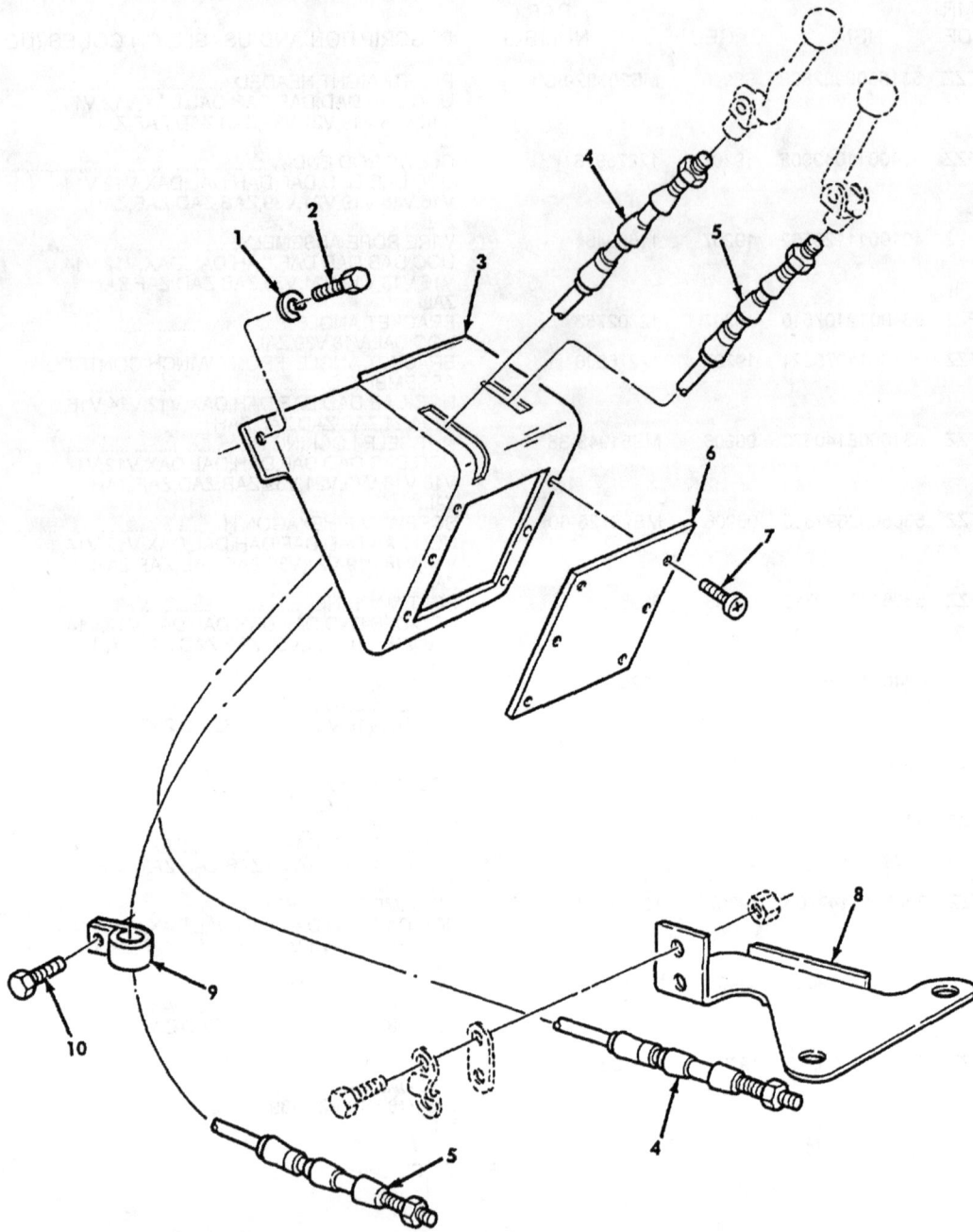

Figure 455. P.T.O. and Winch Controls (M939A2).

(1) ITEM NO	(2) SMR CODE	(3) NSN	(4) CAGEC	(5) PART NUMBER	(6) DESCRIPTION AND USABLE ON CODES (UOC)	(7) QTY
					GROUP 2001 HOIST, CAPSTAN, WINDLASS, CRANE OR WINCH ASSEMBLY	
					FIG. 455 P.T.O. AND WINCH CONTROLS	
1	PAFZZ	5310002090786	96906	MS35335-33	WASHER,LOCK .... UOC:ZAA,ZAB,ZAD,ZAE,ZAF,ZAH,ZAK,ZAL	3
2	PAFZZ	5305000680515	80204	B1821BH025F100N	SCREW,CAP,HEXAGON H .... UOC:ZAA,ZAB,ZAD,ZAE,ZAF,ZAH,ZAK,ZAL	3
3	PAFZZ	2590010849632	19207	12276907	PANEL ASSEMBLY,CONT .... UOC:ZAA,ZAB,ZAD,ZAE,ZAF,ZAH,ZAK,ZAL	1
4	PAFZZ	3040012824336	60602	35841-42	CONTROL ASSEMBLY,PU .... UOC:ZAA,ZAB,ZAD,ZAE,ZAF,ZAH,ZAK,ZAL	1
5	PAFZZ	2590012854600	60602	45114-60	CONTROL ASSEMBLY,PU .... UOC:ZAA,ZAB,ZAD,ZAF,ZAH,ZAL	1
5	PAFZZ	2520012848240	60602	45114-1	CONTROL ASSEMBLY,GA .... UOC:ZAE,ZAK	1
6	PAFZZ	5340011038772	19207	12276908	COVER,ACCESS .... UOC:ZAA,ZAB,ZAD,ZAE,ZAF,ZAH,ZAK,ZAL	1
7	PAFZZ	5305004324201	96906	MS51861-45	SCREW,TAPPING .... UOC:ZAA,ZAB,ZAD,ZAE,ZAF,ZAH,ZAK,ZAL	6
8	PAFZZ	5340012822229	47457	12277391A	BRACKET,MOUNTING CONTROL .... UOC:ZAA,ZAB,ZAD,ZAE,ZAF,ZAH,ZAK,ZAL	1
9	PAFZZ	5340009018132	96906	MS21334-26	CLAMP,LOOP .... UOC:ZAA,ZAB,ZAD,ZAE,ZAF,ZAH,ZAK,ZAL	1
10	PAFZZ	5305012139852	72582	440502	SCREW,MACHINE .... UOC:ZAA,ZAB,ZAD,ZAE,ZAF,ZAH,ZAK,ZAL	1

END OF FIGURE

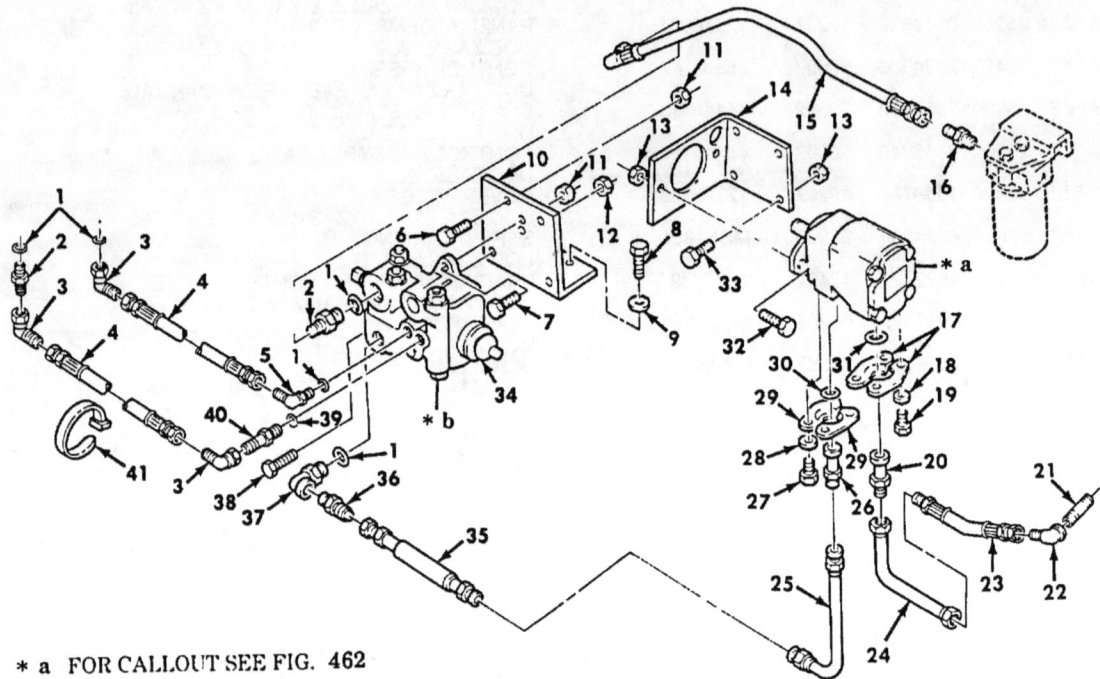

* a FOR CALLOUT SEE FIG. 462
* b FOR BREAKDOWN SEE FIG. 463

Figure 456. Front Winch Pump, Linear Valve, and Related Hoses.

(1) ITEM NO	(2) SMR CODE	(3) NSN	(4) CAGEC	(5) PART NUMBER	(6) DESCRIPTION AND USABLE ON CODES (UOC)	(7) QTY
					GROUP 2001 HOIST, CAPSTAN, WINDLASS, CRANE OR WINCH ASSEMBLY	
					FIG. 456 FRONT WINCH PUMP, LINEAR VALVE, AND RELATED HOSES	
1	PAFZZ	5331002518839	96906	MS28778-12	O-RING................................ UOC:DAB,DAD,DAH,DAX,V12,V14,V16,V21, V39,ZAB,ZAD,ZAH	4
2	PAFZZ	4730011165969	96906	MS51525A10-12S	ADAPTER,STRAIGHT,TU................ UOC:DAB,DAD,DAH,DAX,V12,V14,V16,V21, V39,ZAB,ZAD,ZAH	3
3	PAFZZ	4730011391585	96906	MS51521A10	ELBOW,TUBE.......................... UOC:DAB,DAD,DAH,DAX,V12,V14,V16,V21, V39,ZAB,ZAD,ZAH	3
4	PAFZZ	4720010890766	19207	12257110-1	HOSE ASSEMBLY,NONME............... UOC:DAB,DAD,DAH,DAX,V12,V14,V16,V21, V39,ZAB,ZAD,ZAH	2
5	PAFZZ	4730009747313	96906	MS51527A10	ELBOW,TUBE TO BOSS................. UOC:DAB,DAD,DAH,DAX,V12,V14,V16,V21, V39,ZAB,ZAD,ZAH	1
6	PAFZZ	5305002693240	80204	B1821BH038F150N	SCREW,CAP,HEXAGON H............... UOC:DAB,DAD,DAH,DAX,V12,V14,V16,V21, V39,ZAB,ZAD,ZAH	1
7	PAFZZ	5305002693245	80204	B1821BH038F275N	SCREW,CAP,HEXAGON H............... UOC:DAB,DAD,DAX,V12,V14,V16,V39,ZAB, ZAD	2
7	PAFZZ	5305002693240	80204	B1821BH038F150N	SCREW,CAP,HEXAGON H VALVE BRACKET......... UOC:DAH,V21,ZAH	4
8	PAFZZ	5305002692803	96906	MS90726-60	SCREW,CAP,HEXAGON H............... UOC:DAB,DAD,DAX,V12,V14,V16,V39,ZAB, ZAD	2
9	PAFZZ	5310000806004	96906	MS27183-14	WASHER,FLAT........................ UOC:DAB,DAD,DAX,V12,V14,V16,V39,ZAB, ZAD	2
10	PFFZZ	5340011043828	19207	12257128	BRACKET,ANGLE UOC:DAH,V21,ZAH	1
10	PFFZZ	5340011978215	19207	12302754	BRACKET,ANGLE...................... UOC:DAB,DAD,DAX,V12,V14,V16,V39,ZAB, ZAD	1
11	PAFZZ	5310008140672	96906	MS51943-36	NUT,SELF-LOCKING,HE................ UOC:DAB,DAD,DAH,DAX,V12,V14,V16,V21, V39,ZAB,ZAD,ZAH	5
12	PAFZZ	5310009941006	96906	MS51943-38	NUT,SELF-LOCKING,HE................ UOC:DAB,DAD,DAH,DAX,V12,V14,V16,V21, V39,ZAB,ZAD,ZAH	2
13	PAFZZ	5310004883888	96906	MS51943-40	NUT,SELF-LOCKING,HE................ UOC:DAB,DAD,DAH,DAX,V12,V14,V16,V21, V39,ZAB,ZAD,ZAH	6
14	PFFZZ	5340011044326	19207	12257127	BRACKET,ANGLE...................... UOC:DAB,DAD,DAH,DAX,V12,V14,V16,V21, V39,ZAB,ZAD,ZAH	1

(1) ITEM NO	(2) SMR CODE	(3) NSN	(4) CAGEC	(5) PART NUMBER	(6) DESCRIPTION AND USABLE ON CODES (UOC)	(7) QTY
15	PAFZZ	4720011054067	19207	12257109-2	HOSE ASSEMBLY,NONME DIRECT LINEAR .... VALVE UOC:DAB,DAD,DAH,DAX,V12,V14,V16,V21, V39,ZAB,ZAD,ZAH	1
16	PAFZZ	4730010904919	19207	12255651-2	ADAPTER,STRAIGHT ,PI... UOC:DAB,DAD,DAH,DAX,V12,V14,V16,V21, V39,ZAB,ZAD,ZAH	1
17	PAFZZ	4730000893406	81349	M52525/16-24	FLANGE,PIPE,SWIVEL.. UOC:DAB,DAD,DAH,DAX,V12,V14,V16,V21, V39,ZAB,ZAD,ZAH	2
18	PAFZZ	5310005845272	96906	MS35338-48	WASHER,LOCK UOC:DAB,DAD,DAH,DAX,V12,V14,V16,V21, V39,ZAB,ZAD,ZAH	4
19	PAFZZ	5305000712069	80204	B1821BH050C150N	SCREW,CAP,HEXAGON H.. UOC:DAB,DAD,DAH,DAX,V12,V14,V16,V21, V39,ZAB,ZAD,ZAH	4
20	PAFZZ	4730010893829	19207	12257119	ADAPTER,STRAIGHT,FL... UOC:DAB,DAD,DAH,DAX,V12,V14,V16,V21, V39,ZAB,ZAD,ZAH	1
21	PAFZZ	4730002873790	21450	219831	NIPPLE,PIPE... UOC:DAB,DAD,DAH,DAX,V12,V14,V16,V21, V39,ZAB,ZAD,ZAH	1
22	PAFZZ	4730000810311	96906	MS51506A24	ELBOW,PIPE TO TUBE... UOC:DAB,DAD,DAH,DAX,V12,V14,V16,V21, V39,ZAB,ZAD,ZAH	1
23	PAFZZ	4720011054069	19207	12276970-3	HOSE ASSEMBLY,NONME UOC:V39	1
23	PAFZZ	4720011054068	19207	12276970-2	HOSE ASSEMBLY,NONME UOC:DAB,DAD,DAX,V12,V14,V16,ZAB,ZAD	1
23	PAFZZ	4720010889651	19207	12276970-1	HOSE ASSEMBLY,NONME UOC:DAH,V21,ZAH	1
24	PAFZZ	4710010899375	19207	12276920	TUBE ASSEMBLY,METAL.. UOC:DAH,V21,V39,ZAH	1
24	PAFZZ	4710011060914	19207	12257112	TUBE ASSEMBLY,METAL.. UOC:DAB,DAD,DAX,V12,V14,V16,ZAB,ZAD	1
25	PAFZZ	4710010892060	19207	12257108	TUBE ASSEMBLY,METAL.. UOC:DAB,DAD,DAH,DAX,V12,V14,V16,V21, V39,ZAB,ZAD,ZAH	1
26	PAFZZ	4730010904924	19207	12257120	ADAPTER,STRAIGHT,FL... UOC:DAB,DAD,DAH,DAX,V12,V14,V16,V21, V39,ZAB,ZAD,ZAH	1
27	PAFZZ	5305000680511	80204	B1821BH038C125N	SCREW,CAP,HEXAGON H.. UOC:DAB,DAD,DAH,DAX,V12,V14,V16,V21, V39,ZAB,ZAD,ZAH	4
28	PAFZZ	5310006379541	96906	MS35338-46	WASHER,LOCK UOC:DAB,DAD,DAH,DAX,V12,V14,V16,V21, V39,ZAB,ZAD,ZAH	4
29	PAFZZ	4730007557609	81349	M52525/16-16	FLANGE,PIPE.. UOC:DAB,DAD,DAH,DAX,V12,V14,V16,V21, V39,ZAB,ZAD,ZAH	2
30	PAFZZ	5330005797925	96906	MS28775-219	O-RING... UOC:DAB,DAD,DAH,DAX,V12,V14,V16,V21, V39,ZAB,ZAD,ZAH	1

(1) ITEM NO	(2) SMR CODE	(3) NSN	(4) CAGEC	(5) PART NUMBER	(6) DESCRIPTION AND USABLE ON CODES (UOC)	(7) QTY
31	PAFZZ	5331005797927	96906	MS28775-225	O-RING... UOC:DAB,DAD,DAH,DAX,V12,V14,V16,V21, V39,ZAB,ZAD,ZAH	1
32	PAFZZ	5305007195235	80204	B1821BH050F175N	SCREW,CAP,HEXAGON H................................. UOC:DAB,DAD,DAH,DAX,V12,V14,V16,V21, V39,ZAB,ZAD,ZAH	2
33	PAFZZ	5305007195238	80204	B1821BH050F200N	SCREW,CAP,HEXAGON H................................. UOC:DAB,DAD,DAH,DAX,V12,V14,V16,V21, V39,ZAB,ZAD,ZAH	4
34	PAFZZ	4820011049159	19207	11669457	VALVE,LINEAR,DIRECT..................................... UOC:DAB,DAD,DAH,DAX,V12,V14,V16,V21, V39	1
35	PAFZZ	4720010892016	19207	12276973	HOSE ASSEMBLY,NONME............................... UOC:DAB,DAD,DAH,DAX,V12,V14,V16,V21, V39,ZAB,ZAD,ZAH	1
36	PAFZZ	4730000211802	96906	MS51500A16-12	ADAPTER,STRAIGHT,PI..................................... UOC:DAB,DAD,DAH,DAX,V12,V14,V16,V21, V39,ZAB,ZAD,ZAH	1
37	PAFZZ	4730010906474	19207	12257113	ELBOW,PIPE TO BOSS..................................... UOC:DAB,DAD,DAH,DAX,V12,V14,V16,V21, V39,ZAB,ZAD,ZAH	1
38	PAFZZ	5305009146133	96906	MS18153-88	SCREW,CAP,HEXAGON H................................. UOC:DAB,DAD,DAH,DAX,V12,V14,V16,V21, V39,ZAB,ZAD,ZAH	2
39	PAFZZ	5331002859842	96906	MS28778-10	O-RING... UOC:DAB,DAD,DAH,DAX,V12,V14,V16,V21, V39,ZAB,ZAD	2
40	PAFZZ	4730011929590	96906	MS51526A10	ADAPTER,STRAIGHT,TU..................................... UOC:DAB,DAD,DAH,DAX,V12,V14,V16,V21, V39,ZAB,ZAD,ZAH	1
41	PAFZZ	5975008994606	96906	MS3367-2-0	STRAP,TIEDOWN,ELECT..................................... UOC:DAB,DAD,DAH,DAX,V12,V14,V16,V21, V39,ZAB,ZAD,ZAH	1

END OF FIGURE

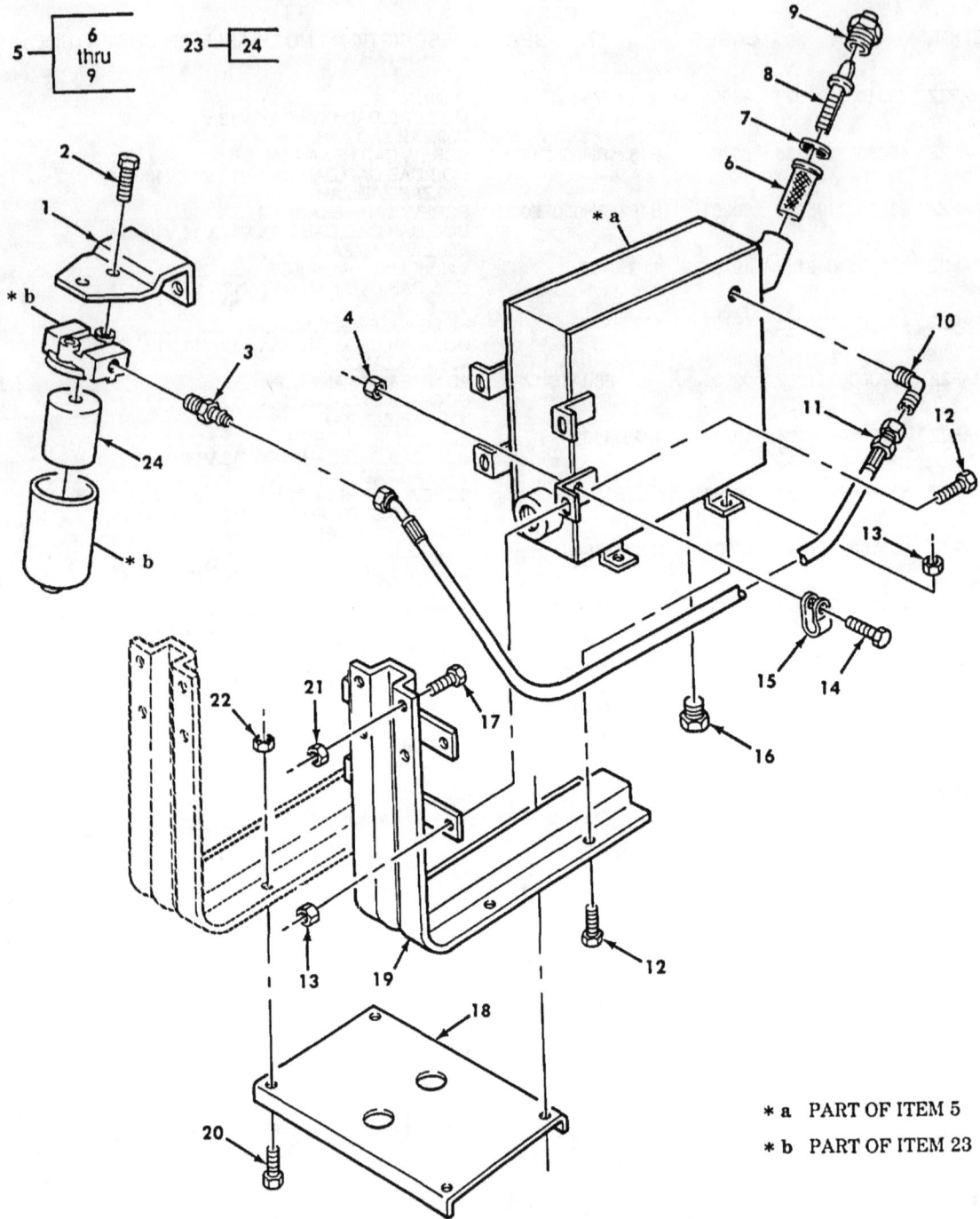

Figure 457. Front Winch Filter and Reservoir Assembly.

* a   PART OF ITEM 5
* b   PART OF ITEM 23

(1) ITEM NO	(2) SMR CODE	(3) NSN	(4) CAGEC	(5) PART NUMBER	(6) DESCRIPTION AND USABLE ON CODES (UOC)	(7) QTY

GROUP 2001 HOIST, CAPSTAN, WINDLASS, CRANE OR WINCH ASSEMBLY

FIG. 457 FRONT WINCH FILTER AND RESERVOIR ASSEMBLY

1	PFOZZ	5340010849631	19207	12257125	BRACKET,MOUNTING................................... UOC:DAB,DAD,DAF,DAH,DAL,DAX,V12,V14, V16,V18,V19,V21,V39,ZAB,ZAD,ZAF,ZAH, ZAL	1
2	PAOZZ	5305005434372	80204	B1821BH038C075N	SCREW,CAP,HEXAGON H............................... UOC:DAB,DAD,DAF,DAH,DAL,DAX,V12,V14, V16,V18,V19,V21,V39,ZAB,ZAD,ZAF,ZAH, ZAL	2
3	PAOZZ	4730010904919	19207	12255651-2	ADAPTER,STRAIGHT,PI................................ UOC:DAB,DAD,DAF,DAH,DAL,DAX,V12,V14, V16,V18,V19,V21,V39,ZAB,ZAD,ZAF,ZAH, ZAL	1
4	PAOZZ	5310000617326	96906	MS21045-3	NUT,SELF-LOCKING,HE................................. UOC:DAB,DAD,DAF,DAH,DAL,DAX,V12,V14, V16,V18,V19,V21,V39,ZAB,ZAD,ZAF,ZAH, ZAL	1
5	PAOZZ	2590010855352	19207	12257126	TANK,OIL,HYDRAULIC.................................. UOC:DAB,DAD,DAF,DAH,DAL,DAX,V12,V14, V16,V18,V19,V21,V39,ZAB,ZAD,ZAF,ZAH, ZAL	1
6	PAOZZ	4330011686891	19207	7409025	FILTER ELEMENT,FLUI................................ UOC:DAB,DAD,DAF,DAH,DAL,DAX,V12,V14, V16,V18,VI9,V21,V39,ZAB,ZAD,ZAF,ZAH, ZAL	1
7	PAOZZ	5365001829635	19207	7409080	SPACER,RING... UOC:DAB,DAD,DAF,DAH,DAL,DAX,V12,V14, V16,V18,V19,V21,V39,ZAB,ZAD,ZAF,ZAH, ZAL	1
8	PAOZZ	6680007409026	19207	7409026	GAGE ROD,LIQUID LEV.............................. UOC:DAB,DAD,DAF,DAH,DAL,DAX,V12,V14, V16,V18,V19,V21,V39,ZAB,ZAD,ZAF,ZAH, ZAL	1
9	PAOZZ	4730011234546	19207	7409027	PLUG,PIPE... UOC:DAB,DAD,DAF,DAH,DAL,DAX,V12,V14, V16,V18,V19,V21,V39,ZAB,ZAD,ZAF,ZAH, ZAL	1
10	PAOZZ	4730003228339	96906	MS51504A10S	ELBOW,PIPE TO TUBE................................. UOC:DAB,DAD,DAF,DAH,DAL,DAX,V12,V14, V16,V18,V19,V21,V39,ZAB,ZAD,ZAF,ZAH, ZAL	1
11	PAOZZ	4720011068289	19207	12276972-1	HOSE ASSEMBLY,NONME........................... UOC:DAB,DAD,DAF,DAH,DAL,DAX,V12,V14, V15,V16,V18,V21,V39,ZAB,ZAD,ZAF,ZAH, ZAL	1
12	PAOZZ	5305002693238	80204	B1821BH038F125N	SCREW,CAP,HEXAGON H............................... UOC:DAB,DAD,DAF,DAH,DAL,DAX,V12,V14, V16,V18,V19,V21,V39,ZAB,ZAD,ZAF,ZAH, ZAL	8

(1) ITEM NO	(2) SMR CODE	(3) NSN	(4) CAGEC	(5) PART NUMBER	(6) DESCRIPTION AND USABLE ON CODES (UOC)	(7) QTY
13	PAOZZ	5310008140672	96906	MS51943-36	NUT,SELF-LOCKING,HE .... UOC:DAB,DAD,DAF,DAH,DAL,DAX,V12,V14, V16,V18,V19,V21,V39,ZAB,ZAD,ZAF,ZAH, ZAL	8
14	PAOZZ	5306000425570	21450	425570	BOLT,ASSEMBLED WASH.. UOC:DAB,DAD,DAF,DAH,DAL,DAX,V12,V14, V16	1
15	PAOZZ	5340006850567	96906	MS9024-18	CLAMP,LOOP .... UOC:DAB,DAD,DAF,DAH,DAL,DAX,V12,V14, V16,V18,V19,V21,V39,ZAB,ZAD,ZAF,ZAH, ZAL	1
16	PAOZZ	4730012066162	81349	WW-P-471ACABCD	PLUG,PIPE.. UOC:DAB,DAD,DAH,DAX,V12,V14,V16,V21, V39,ZAB,ZAD,ZAH	1
17	PAOZZ	5305007195238	80204	B1821BH050F200N	SCREW,CAP,HEXAGON H....................................... UOC:DAB,DAD,DAF,DAH,DAL,DAX,V12,V14, V16,V18,V19,V21,V39,ZAB,ZAD,ZAF,ZAH, ZAL	4
18	PFOZZ	5340010822519	19207	12276902-2	PLATE,MENDING... UOC:DAB,DAD,DAF,DAH,DAX,V12,V14,V16, V19,V21,ZAB,ZAD,ZAF,ZAH	1
18	XBOZZ	5340010822518	19207	12276902-1	PLATE,MOUNTING... UOC:DAV,V39	1
19	PAOZZ	5340010822510	19207	12255648	BRACKET,DOUBLE ANGL .... UOC:DAB,DAD,DAF,DAH,DAL,DAX,V12,V14, V16,V18,V19,V21,V39,ZAB,ZAD,ZAF,ZAH, ZAL	1
20	PAOZZ	5305002692803	96906	MS90726-60	SCREW,CAP,HEXAGON H....................................... UOC:DAB,DAD,DAF,DAH,DAL,DAX,V12,V14, V16,V18,V19,V21,V39,ZAB,ZAD,ZAF,ZAH, ZAL	1
21	PAOZZ	5310004883888	96906	MS51943-40	NUT,SELF-LOCKING,HE .... UOC:DAB,DAD,DAF,DAH,DAL,DAX,V12,V14, V16,V18,V19,V21,V39,ZAB,ZAC,ZAF,ZAH, ZAL	4
22	PAOZZ	5310008140672	96906	MS51943-36	NUT,SELF-LOCKING,HE .... UOC:DAB,DAD,DAF,DAH,DAL,DAX,V12,V14, V16,V18,V19,V21,V39,ZAB,ZAD,ZAF,ZAH, ZAL	2
23	PAOOO	4330004888613	62983	OFM101-25	FILTER,FLUID .... UOC:DAB,DAD,DAF,DAH,DAL,DAX,V12,V14, V16,V18,V19,V21,V39,ZAB,ZAD,ZAF,ZAH, ZAL	1
24	PAOZZ	2940009508410	70040	PF297	FILTER ELEMENT,FLUI... UOC:DAB,DAD,DAF,DAH,DAL,DAX,V12,V14, V16,V18,V19,V21,V39,ZAB,ZAD,ZAF,ZAH, ZAL	1

END OF FIGURE

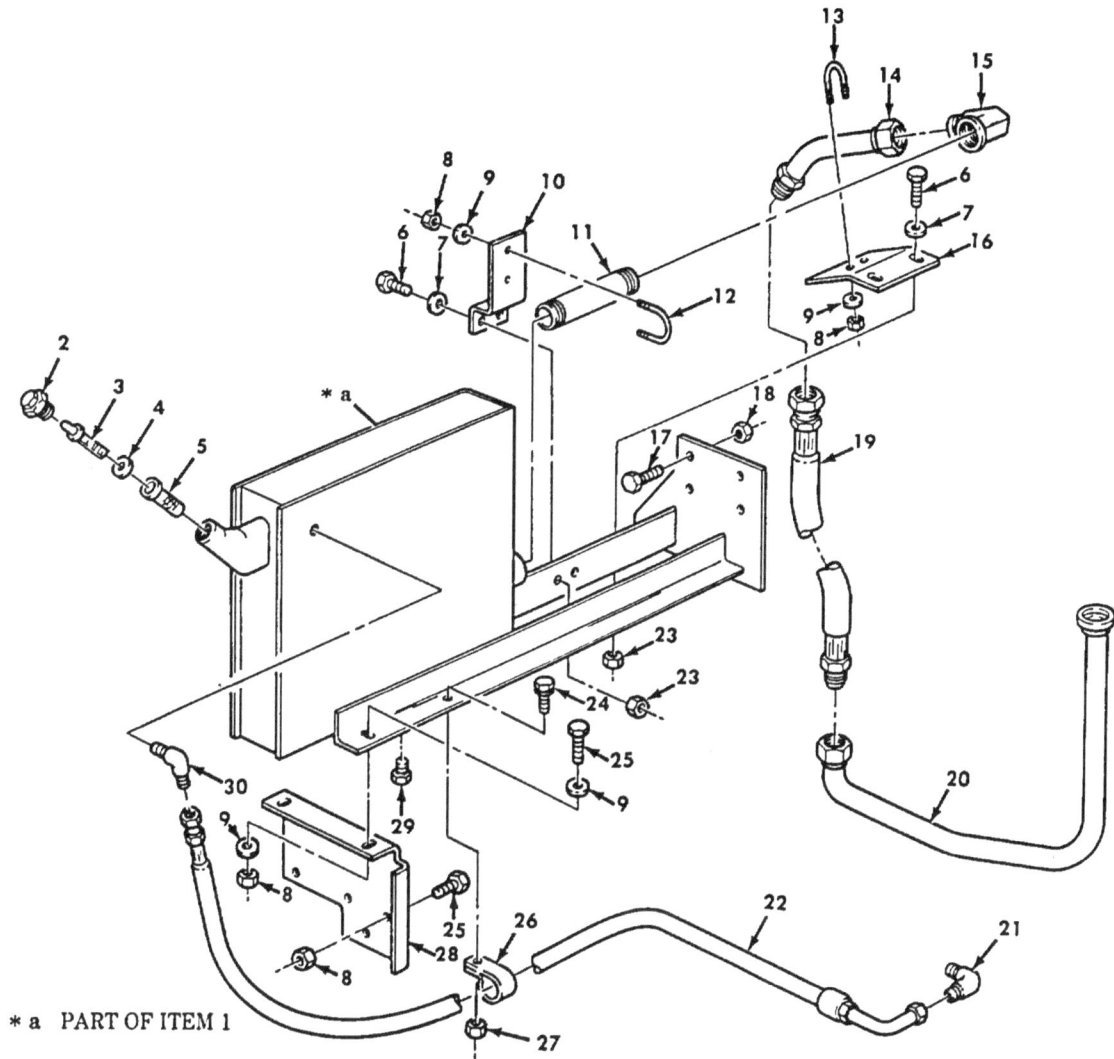

Figure 458. Front Winch Reservoir Assembly and Mounting Hardware.

*a PART OF ITEM 1

(1) ITEM NO	(2) SMR CODE	(3) NSN	(4) CAGEC	(5) PART NUMBER	(6) DESCRIPTION AND USABLE ON CODES (UOC)	(7) QTY
					GROUP 2001 HOIST, CAPSTAN, WINDLASS, CRANE OR WINCH ASSEMBLY	
					FIG. 458 FRONT WINCH RESERVOIR ASSEMBLY AND MOUNTING HARDWARE	
1	PAOOO	2590012106199	19207	12302992	TANK,OIL,HYDRAULIC .... UOC:DAH,ZAH	1
2	PAOZZ	4730011234546	19207	7409027	.PLUG,PIPE... UOC:DAH,ZAH	1
3	PBOZZ	6680007409026	19207	7409026	.GAGE ROD,LIQUID LEV .... UOC:DAH,ZAH	1
4	PBOZZ	5365001829635	19207	7409080	.SPACER,RING .... UOC:DAH,ZAH	1
5	PBOZZ	4330011686891	19207	7409025	.FILTER ELEMENT,FLUI .... UOC:DAH,ZAH	1
6	PAOZZ	5305007195219	96906	MS90727-111	SCREW,CAP,HEXAGON H.................................... UOC:DAH,ZAH	4
8	PAOZZ	5310004883888	96906	MS51943-40	NUT,SELF-LOCKING,HE .... UOC:DAH,ZAH	10
9	PAOZZ	5310008095998	96906	MS27183-18	WASHER,FLAT .... UOC:DAH,ZAH	8
10	PAOZZ	5340012106238	19207	12302994	BRACKET,DOUBLE ANGL .... UOC:DAH,ZAH	1
11	PAOZZ	4730000542027	96906	MS51953-181	NIPPLE,PIPE .... UOC:DAH,ZAH	1
12	PAOZZ	5306000097723	80205	NAS3104-16-14	BOLT,U .... UOC:DAH	1
13	PAOZZ	5306010460553	80205	NAS3104-12-12	BOLT,U .... UOC:DAH,ZAH	1
14	PAOZZ	4710012116614	19207	12302993	TUBE ASSEMBLY,METAL .... UOC:DAH,ZAH	1
15	PAOZZ	4730000810311	96906	MS51506A24	ELBOW,PIPE TO TUBE....................................... UOC:DAH,ZAH	1
16	PAOZZ	5340013097782	19207	12302988	BRACKET,ANGLE .... UOC:DAH,ZAH	1
17	PAOZZ	5305000712069	80204	B1821BH050C150N	SCREW,CAP,HEXAGON H.................................... UOC:DAH,ZAH	4
18	PAOZZ	5310004883889	96906	MS51943-39	NUT,SELF-LOCKING,HE .... UOC:DAH,ZAH	2
19	PAOZZ	4720012765923	19207	12276970-5	HOSE ASSEMBLY,NONME .... UOC:DAH,ZAH	1
20	PAOZZ	4720012106201	19207	12303015	HOSE ASSEMBLY,NONME .... UOC:DAH,ZAH	1
21	PAOZZ	4730011391585	96906	MS51521-B10	ELBOW,TUBE... UOC:DAH,ZAH	1
22	PAOZZ	4720012106200	19207	12303014	HOSE ASSEMBLY,NONME .... UOC:DAH,ZAH	1
23	PAOZZ	5310001436102	96906	MS51922-6	NUT,SELF-LOCKING,HE .... UOC:DAH,ZAH	4
24	PAOZZ	5306000425570	21450	425570	BOLT,ASSEMBLED WASH.................................... UOC:DAH,ZAH	1

(1) ITEM NO	(2) SMR CODE	(3) NSN	(4) CAGEC	(5) PART NUMBER	(6) DESCRIPTION AND USABLE ON CODES (UOC)	(7) QTY
25	PAOZZ	5305002267767	96906	MS90726-109	SCREW,CAP,HEXAGON H... UOC:DAH,ZAH	6
26	PAOZZ	5340006850567	96906	MS9024-18	CLAMP,LOOP .... UOC:DAH,ZAH	1
27	PAOZZ	5310000617326	96906	MS21045-3	NUT,SELF-LOCKING,HE .... UOC:DAH,ZAH	1
28	PAOZZ	5340012179141	19207	12302957	BRACKET,ANGLE .... UOC:DAH,ZAH	1
29	PAOZZ	4730012066162	81349	WW-P-471ACABCD	PLUG,PIPE.. UOC:DAH,ZAH	1
30	PAOZZ	4730008338230	96906	MS51504-A10	ELBOW,PIPE TO TUBE.. UOC:DAH,ZAH	1

END OF FIGURE

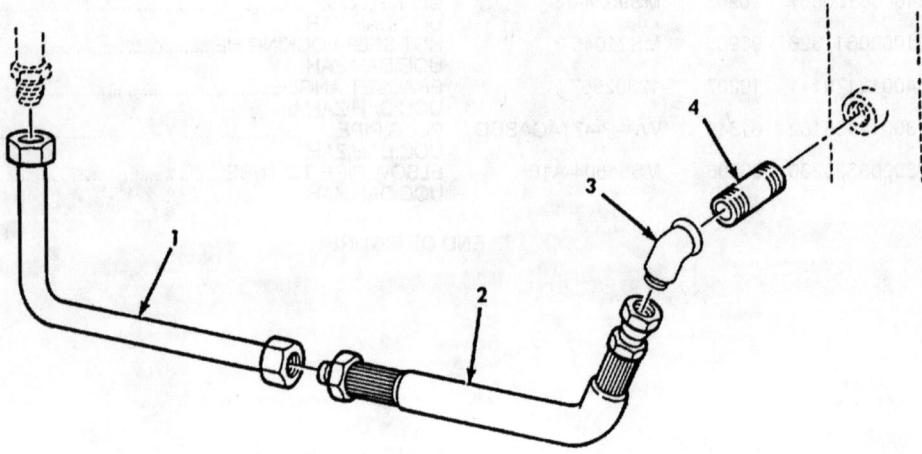

Figure 459. Winch Pump Drive, Hydraulic Tubing (M1939A2).

(1) ITEM NO	(2) SMR CODE	(3) NSN	(4) CAGEC	(5) PART NUMBER	(6) DESCRIPTION AND USABLE ON CODES (UOC)	(7) QTY
					GROUP 2001 HOIST, CAPSTAN, WINDLASS, CRANE OR WINCH ASSEMBLY	
					FIG. 459 WINCH PUMP DRIVE, HYDRAULIC TUBING (M939A2)	
1	PAFZZ	4710012853007	47457	20510863	TUBE ASSEMBLY,METAL .... UOC:ZAB,ZAD	1
2	PAFZZ	4720012810994	98441	A6036-2	HOSE ASSEMBLY,NONME .... UOC:ZAB,ZAD	1
3	PAFZZ	4730000810311	96906	MS51506-B24	ELBOW,PIPE TO TUBE.. UOC:ZAB,ZAD	1
4	PAFZZ	4730001961534	96906	MS51953-175	NIPPLE,PIPE .... UOC:ZAB,ZAD	1

END OF FIGURE

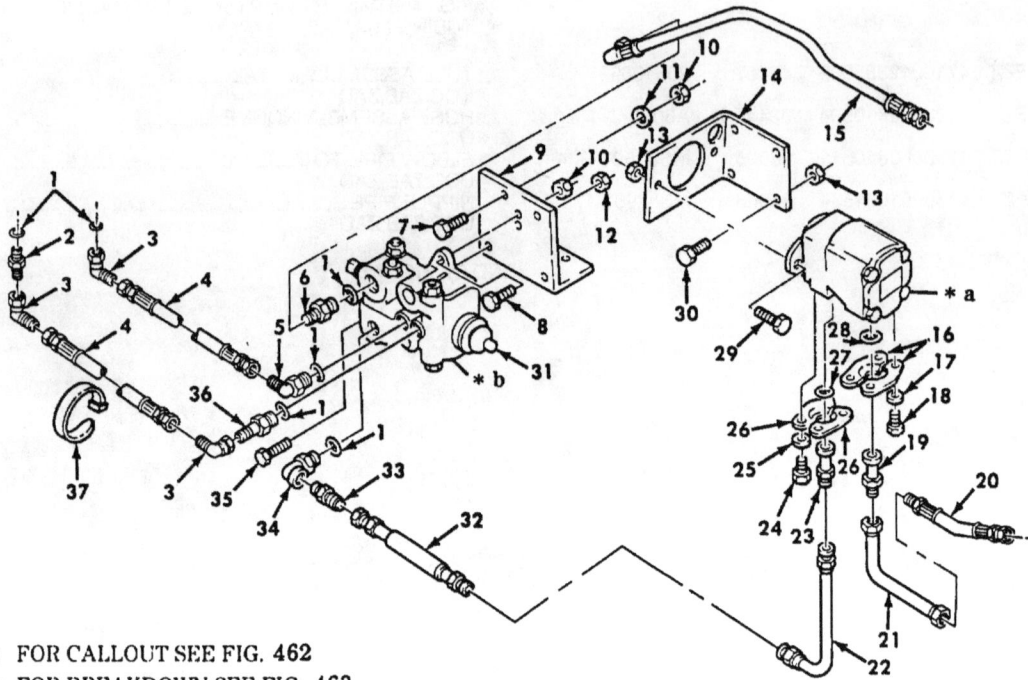

* a FOR CALLOUT SEE FIG. 462
* b FOR BREAKDOWN SEE FIG. 463

Figure 460. Front Winch Pump, Linear Valve, Related Hoses and Fittings, Wrecker.

(1) ITEM NO	(2) SMR CODE	(3) NSN	(4) CAGEC	(5) PART NUMBER	(6) DESCRIPTION AND USABLE ON CODES (UOC)	(7) QTY
					GROUP 2001 HOIST, CAPSTAN, WINDLASS, CRANE OR WINCH ASSEMBLY	
					FIG. 460 FRONT WINCH PUMP, LINEAR VALVE, RELATED HOSES AND FITTINGS, WRECKER	
1	PAFZZ	5331002518839	96906	MS28778-12	O-RING UOC:DAL,V18,ZAL	6
2	PAFZZ	4730011165969	96906	MS51525A10-12S	ADAPTER,STRAIGHT,TU UOC:DAL,V18,ZAL	3
3	PAFZZ	4730011391585	96906	MS51521A10	ELBOW,TUBE UOC:DAL,V18,ZAL	3
4	PAFZZ	4720010890766	19207	12257110-1	HOSE ASSEMBLY,NONME UOC:DAL,V18,ZAL	2
5	PAFZZ	4730009747313	96906	MS51527-A10	ELBOW UOC:DAL,V18,ZAL	1
6	PAFZZ	4730007105571	96906	MS51525-A12	ADAPTER,STRAIGHT,TU UOC:DAL,V18,ZAL	1
7	PAFZZ	5305002693240	80204	B1821BH038F150N	SCREW,CAP,HEXAGON H UOC:DAL,V18,ZAL	3
8	PAFZZ	5305002693244	80204	B1821BH038F250N	SCREW,CAP,HEXAGON H UOC:DAL,V18,ZAL	2
9	PFFZZ	5340011978215	19207	12302754	BRACKET,ANGLE UOC:DAL,V18,ZAL	1
10	PAFZZ	5310008140672	96906	MS51943-36	NUT,SELF-LOCKING,HE UOC:DAL,V18,ZAL	5
11	PAFZZ	5310000806004	96906	MS27183-14	WASHER,FLAT UOC:DAL,V18,ZAL	2
12	PAFZZ	5310002416665	96906	MS51943-48	NUT,SELF-LOCKING,HE UOC:DAL,V18,ZAL	2
13	PAFZZ	5310004883888	96906	MS51943-40	NUT,SELF-LOCKING,HE UOC:DAL,V18,ZAL	6
14	PFFZZ	5340011044326	19207	12257127	BRACKET,ANGLE UOC:DAL,V18,ZAL	1
15	PAFZZ	4720010899887	19207	12257109-3	HOSE ASSEMBLY,NONME UOC:DAL,V18,ZAL	1
16	PAFZZ	4730000893406	81349	M52525/16-24	FLANGE,PIPE,SWIVEL UOC:DAF,V18,ZAL	2
17	PAFZZ	5310001670680	96906	MS35338-49	WASHER,LOCK UOC:DAL,V18,ZAL	4
18	PAFZZ	5305000712069	80204	B1821BH050C150N	SCREW,CAP,HEXAGON H UOC:DAL,V18,ZAL	4
19	PAFZZ	4730010893829	19207	12257119	ADAPTER,STRAIGHT,FL UOC:DAL,V18,ZAL	1
20	PAFZZ	4720011324859	19207	12276970-4	HOSE ASSEMBLY,NONME UOC:DAL,V18,ZAL	1
21	PAFZZ	4710011060914	19207	12257112	TUBE ASSEMBLY,METAL UOC:DAL,V18,ZAL	1
22	PAFZZ	4710010892060	19207	12257108	TUBE ASSEMBLY,METAL UOC:DAL,V18,ZAL	1
23	PAFZZ	4730010904924	19207	12257120	ADAPTER,STRAIGHT,FL UOC:DAL,V18,ZAL	1

(1) ITEM NO	(2) SMR CODE	(3) NSN	(4) CAGEC	(5) PART NUMBER	(6) DESCRIPTION AND USABLE ON CODES (UOC)	(7) QTY
					UOC:DAL,V18,ZAL	
24	PAFZZ	5305000680511	80204	B1821BH038C125N	SCREW,CAP,HEXAGON H..	4
					UOC:DAL,V18,ZAL	
25	PAFZZ	5310006379541	96906	MS35338-46	WASHER,LOCK..	4
					UOC:DAL,V18,ZAL	
26	PAFZZ	4730007557609	81349	M52525/16-16	FLANGE,PIPE..	2
					UOC:DAL,V18,ZAL	
27	PAFZZ	5330005797925	96906	MS28775-219	O-RING..	1
					UOC:DAL,V18,ZAL	
28	PAFZZ	5331005797927	96906	MS28775-225	O-RING..	1
					UOC:DAL,V18,ZAL	
29	PAFZZ	5305007195235	80204	B1821BH050F175N	SCREW,CAP,HEXAGON H..	2
					UOC:DAL,V18,ZAL	
30	PAFZZ	5305009640589	96906	MS51095-416	SCREW,CAP,HEXAGON H..	4
					UOC:DAL,V18,ZAL	
31	PAFZZ	4820011049159	62983	CM11-ND2-R22-BL- 21-066	VALVE,LINEAR,DIRECT.. UOC:DAL,V18,ZAL	1
32	PAFZZ	4720010892016	19207	12276973	HOSE ASSEMBLY,NONME UOC:DAL,V18,ZAL	1
33	PAFZZ	4730000211802	96906	MS51500A16-12	ADAPTER,STRAIGHT,PI UOC:DAL,V18,ZAL	1
34	PAFZZ	4730010906474	19207	12257113	ELBOW,PIPE TO BOSS UOC:DAL,V18,ZAL	1
35	PAFZZ	5305009146133	96906	MS18153-88	SCREW,CAP,HEXAGON H.. UOC:DAL,V18,ZAL	2
36	PAFZZ	4730011929590	96906	MS51526A10	ADAPTER,STRAIGHT,TU UOC:DAL,V18,ZAL	1
37	PAFZZ	5975008994606	96906	MS3367-2-0	STRAP,TIEDOWN,ELECT.. UOC:DAL,V18,ZAL	1

END OF FIGURE

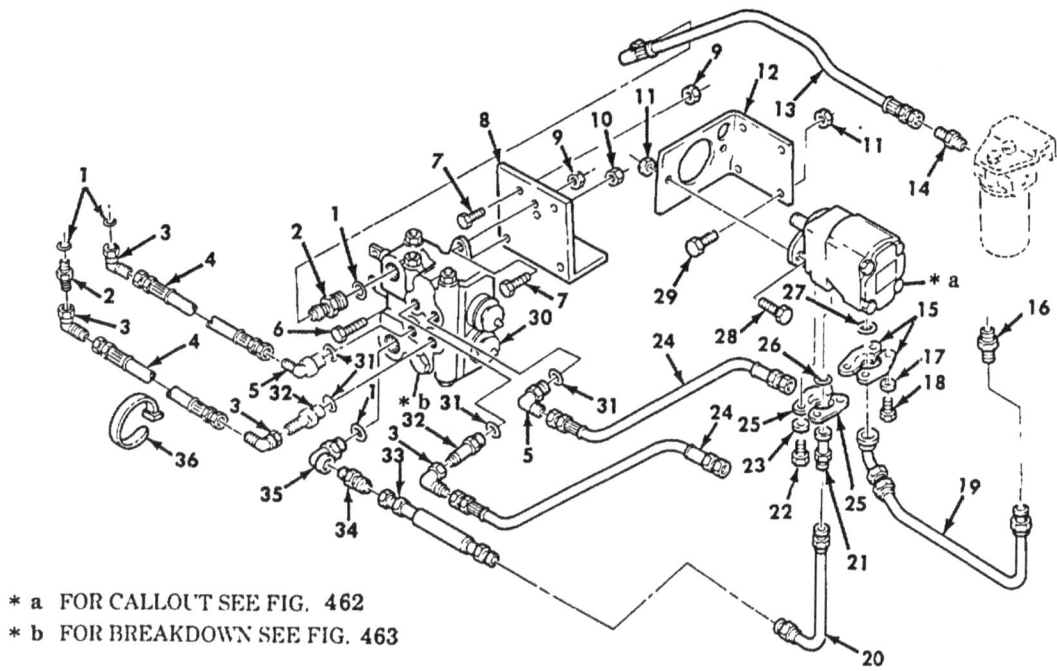

* a FOR CALLOUT SEE FIG. 462
* b FOR BREAKDOWN SEE FIG. 463

Figure 461. Front Winch Pump, Linear Valve, Related Hoses and Fittings, Dump.

(1) ITEM NO	(2) SMR CODE	(3) NSN	(4) CAGEC	(5) PART NUMBER	(6) DESCRIPTION AND USABLE ON CODES (UOC)	(7) QTY
					GROUP 2001 HOIST, CAPSTAN, WINDLASS, CRANE OR WINCH ASSEMBLY	
					FIG. 461 FRONT WINCH PUMP, LINEAR VALVE, RELATED HOSES AND FITTINGS, DUMP	
1	PAFZZ	5330002518839	96906	MS28778-12	O-RING UOC:DAF,V19,ZAF	4
2	PAFZZ	4730011165969	96906	MS51525A10-12S	ADAPTER,STRAIGHT,TU UOC:DAF,V19,ZAF	2
3	PAFZZ	4730011391585	96906	MS51521A10	ELBOW,TUBE UOC:DAF,V19,ZAF	4
4	PAFZZ	4720010890766	19207	12257110-1	HOSE ASSEMBLY,NONME UOC:DAF,V19,ZAF	2
5	PAFZZ	4730009747313	96906	MS51527-A10	ELBOW,TUBE TO BOSS UOC:DAF,V19,ZAF	2
6	PAFZZ	5305009146133	96906	MS18153-88	SCREW,CAP,HEXAGON H UOC:DAF,V19,ZAF	2
7	PAFZZ	5305002693240	80204	B1821BH038F150N	SCREW,CAP,HEXAGON H UOC:DAF,V19,ZAF	5
8	PFFZZ	5340011043828	19207	12257128	BRACKET,ANGLE UOC:DAF,V19,ZAF	1
9	PAFZZ	5310008140672	96906	MS51943-36	NUT,SELF-LOCKING,HE UOC:DAF,V19,ZAF	5
10	PAFZZ	5310009941006	96906	MS51943-38	NUT,SELF-LOCKING,HE UOC:DAF,V19,ZAF	2
11	PAFZZ	5310004883888	96906	MS51943-40	NUT,SELF-LOCKING,HE UOC:DAF,V19,ZAF	6
12	PFFZZ	5340011044326	19207	12257127	BRACKET,ANGLE UOC:DAF,V19,ZAF	1
13	PAFZZ	4720011054067	19207	12257109-2	HOSE ASSEMBLY,NONME DIRECT LINEAR .... VALVE UOC:DAF,V19	1
14	PAFZZ	4730010904919	19207	12255651-2	ADAPTER,STRAIGHT,PI UOC:DAF,V19,ZAF	1
15	PAFZZ	4730000893406	81349	M52525/16-24	FLANGE,PIPE,SWIVEL UOC:DAF,V19,ZAF	2
16	PAFZZ	4730006767566	96906	MS51500A24	ADAPTER,STRAIGHT,PI UOC:DAF,V19,ZAF	1
17	PAFZZ	5310000034094	01276	210104-8S	WASHER,LOCK UOC:DAF,V19,ZAF	4
18	PAFZZ	5305000712069	80204	B1821BH050C150N	SCREW,CAP,HEXAGON H UOC:DAF,V19,ZAF	4
19	PAFZZ	4720011923504	19207	12302871	HOSE ASSEMBLY,NONME UOC:DAF,V19,ZAF	1
20	PAFZZ	4710010892060	19207	12257108	TUBE ASSEMBLY,METAL UOC:DAF,V19,ZAF	1
21	PAFZZ	4730010904924	19207	12257120	ADAPTER,STRAIGHT,FL UOC:DAF,V19,ZAF	1
22	PAFZZ	5305000680511	80204	B1821BH038C125N	SCREW,CAP,HEXAGON H UOC:DAF,V19,ZAF	4

(1) ITEM NO	(2) SMR CODE	(3) NSN	(4) CAGEC	(5) PART NUMBER	(6) DESCRIPTION AND USABLE ON CODES (UOC)	(7) QTY
23	PAFZZ	5310006379541	96906	MS35338-46	WASHER,LOCK UOC:DAF,V19,ZAF	4
24	PAFZZ	4720011063982	19207	12257110-2	HOSE ASSEMBLY,NONMETALLIC UOC:DAF,V19,ZAF	2
25	PAFZZ	4730007557609	81349	M52525/16-16	FLANGE,PIPE... UOC:DAF,V19,ZAF	2
26	PAFZZ	5330005797925	96906	MS28775-219	O-RING.. UOC:DAF,V19,ZAF	1
27	PAFZZ	5330005797927	96906	MS28775-225	O-RING.. UOC:DAF,V19,ZAF	1
28	PAFZZ	5305007195235	80204	B1821BH050F175N	SCREW,CAP,HEXAGON H UOC:DAF,V19,ZAF	2
29	PAFZZ	5305007195238	80204	B1821BH050F200N	SCREW,CAP,HEXAGON H UOC:DAF,V19,ZAF	4
30	PAFFF	4820012859851	62983	CM11-N02-R12-DL5	VALVE,REGULATING,FL .... -21-233 UOC:DAF,V19,ZAF	1
31	PAFZZ	5330002859842	96906	MS28778-10	O-RING.. UOC:DAF,V19,ZAF	4
32	PAFZZ	4730011929590	96906	MS51526A10	ADAPTER,STRAIGHT,TU UOC:DAF,V19,ZAF	2
33	PAFZZ	4720010892016	19207	12276973	HOSE ASSEMBLY,NONME UOC:DAF,V19,ZAF	1
34	PAFZZ	4730000211802	96906	MS51500A16-12	ADAPTER,STRAIGHT,PI UOC:DAF,V19,ZAF	1
35	PAFZZ	4730010906474	19207	12257113	ELBOW,PIPE TO BOSS UOC:DAF,V19,ZAF	1
36	PAFZZ	5975008994606	96906	MS3367-2-0	STRAP,TIEDOWN,ELECT................................ UOC:DAF,V19,ZAF	1

END OF FIGURE

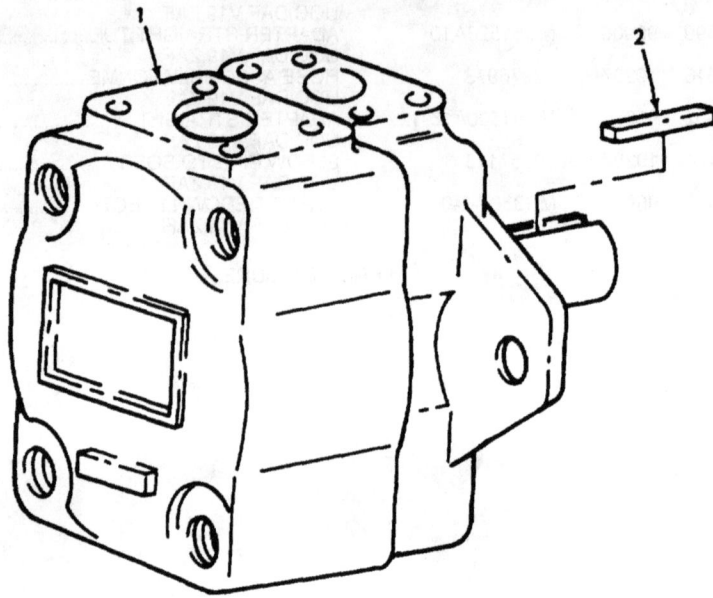

Figure 462. Front Winch Pump Assembly.

(1) ITEM NO	(2) SMR CODE	(3) NSN	(4) CAGEC	(5) PART NUMBER	(6) DESCRIPTION AND USABLE ON CODES (UOC)	(7) QTY
					GROUP 2001 HOIST, CAPSTAN, WINDLASS, CRANE OR WINCH ASSEMBLY	
					FIG. 462 FRONT WINCH PUMP ASSEMBLY	
1	PAFZZ	4320011062061	62983	26VQ14A-1C-20	PUMP, ROTARY .... UOC:DAB, DAD, DAF, DAH, DAL, DAX, V12, V14, V16, V18, V19, V21, V39, ZAB, ZAD, ZAF, ZAH, ZAL	1
2	PAFZZ	5315002643099	96906	MS20066-187	KEY, MACHINE .... UOC:DAB, DAD, DAF, DAH, DAL, DAX, V12, V14, V16, V18, V19, V21, V39, ZAB, ZAD, ZAF, ZAH, ZAL	1

END OF FIGURE

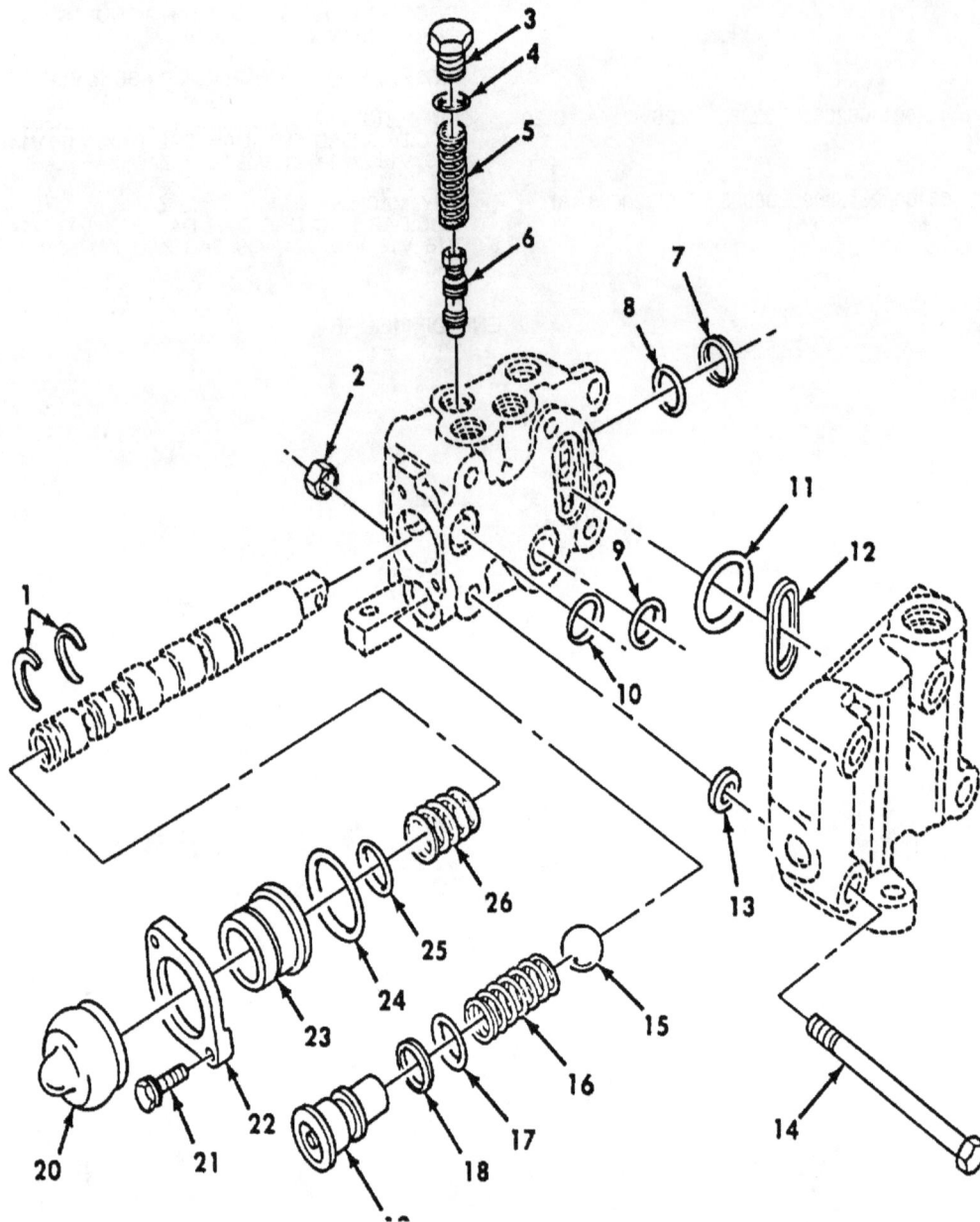

Figure 463. Front Winch Linear Value Assembly.

(1) ITEM NO	(2) SMR CODE	(3) NSN	(4) CAGEC	(5) PART NUMBER	(6) DESCRIPTION AND USABLE ON CODES (UOC)	(7) QTY
					GROUP 2001 HOIST, CAPSTAN, WINDLASS, CRANE OR WINCH ASSEMBLY	
					FIG. 463 FRONT WINCH LINEAR VALVE ASSEMBLY	
1	PAFZZ	5310008671465	62983	284156	WASHER, SLOTTED................................ UOC:DAB, DAD, DAF, DAH, DAL, DAX, V12, V14, V16, V18, V19, V21, V39, ZAB, ZAD, ZAF, ZAH, ZAL	2
2	PAFZZ	5310007320558	62983	1454	NUT, PLAIN, HEXAGON............................. UOC:DAB, DAD, DAF, DAH, DAL, DAX, V12, V14, V16, V18, V19, V21, V39, ZAB, ZAD, ZAF, ZAH, ZAL	4
3	PAFZZ	5365006746831	62983	186580	PLUG, MACHINE THREAD........................ UOC:DAB, DAD, DAF, DAH, DAL, DAX, V12, V14, V16, V18, V19, V21, V39, ZAB, ZAD, ZAF, ZAH, ZAL	1
4	PAFZZ	5331009486482	62983	154129	O-RING.. UOC:DAB, DAD, DAF, DAH, DAL, DAX, V12, V14, V16, V18, V19, V21, V39, ZAB, ZAD, ZAF, ZAH, ZAL	1
5	PAFZZ	5360009181920	62983	259871	SPRING, HELICAL, COMP........................ UOC:DAB, DAD, DAF, DAH, DAL, DAX, V12, V14, V16, V18, V19, V21, V39, ZAB, ZAD, ZAF, ZAH, ZAL	1
6	PAFZZ	4820011320582	62983	233019	VALVE, VENT FIRST 2000 VEHICLES........ UOC:DAB, DAD, DAF, DAH, DAL, DAX, V12, V14, V16, V18, V19, V21, V39, ZAB, ZAD, ZAF, ZAH, ZAL	1
6	PAFZZ	4820000341690	62983	232798	VALVE, VENT AFTER 2000 VEHICLES........ UOC:DAB, DAD, DAF, DAH, DAL, DAX, V12, V14, V16, V18, V19, V21, V39, ZAB, ZAD, ZAF, ZAH, ZAL	1
7	PAFZZ	5330002460158	62983	286669	PACKING, PREFORMED PART OF KIT P/N........ 920278........ UOC:DAB, DAD, DAF, DAH, DAL, DAX, V12, V14, V16, V18, V19, V21, V39, ZAB, ZAD, ZAF, ZAH, ZAL	1
8	PAFZZ	5365002363825	62983	680701	RING, QUADRANT PART OF KIT P/N 920278........ UOC:DAB, DAD, DAF, DAH, DAL, DAX, V12, V14, V16, V18, V19, V21, V39, ZAB, ZAD, ZAF, ZAH, ZAL	1
9	PAFZZ	5330009535206	62983	223489	SEAL, CONTROL VALVE PART OF KIT P/N........ 920278........ UOC:DAB, DAD, DAF, DAH, DAL, DAX, V12, V14, V16, V18, V19, V21, V39, ZAB, ZAD, ZAF, ZAH, ZAL	1
10	PAFZZ	5330009535207	62983	223493	GASKET PART OF KIT P/N 920278................ UOC:DAB, DAD, DAF, DAH, DAL, DAX, V12, V14, V16, V18, V19, V21, V39, ZAB, ZAD, ZAF, ZAH, ZAL	1
11	PAFZZ	5330009535208	62983	226161	PACKING, CONTROL VAL PART OF KIT P/N........ 920278........ UOC:DAB, DAD, DAF, DAH, DAL, DAX, V12, V14, V16, V18, V19, V21, V39, ZAB, ZAD, ZAF, ZAH,	1

(1) ITEM NO	(2) SMR CODE	(3) NSN	(4) CAGEC	(5) PART NUMBER	(6) DESCRIPTION AND USABLE ON CODES (UOC)	(7) QTY
					ZAL	
12	PAFZZ	2590000527504	62983	237736	RETAINER, OIL SEAL PART OF KIT P/N 920278 .... UOC:DAB, DAD, DAF, DAH, DAL, DAX, V12, V14, V16, V18, V19, V21, V39, ZAB, ZAD, ZAF, ZAH, ZAL	1
13	PAFZZ	5365001826713	62983	307198	SHIM.................. UOC:DAB, DAD, DAF, DAH, DAL, DAX, V12, V14, V16, V18, V19, V21, V39, ZAB, ZAD, ZAF, ZAH, ZAL	4
14	PAFZZ	5305010200709	62983	146835	SCREW.................. UOC:DAB, DAD, DAF, DAH, DAL, DAX, V12, V14, V16, V18, V19, V21, V39, ZAB, ZAD, ZAF, ZAH, ZAL	4
15	PAFZZ	3110008994353	62983	1656	BALL, BEARING.................. UOC:DAB, DAD, DAF, DAH, DAL, DAX, V12, V14, V16, V18, V19, V21, V39, ZAB, ZAD, ZAF, ZAH, ZAL	1
16	PAFZZ	5360009535205	62983	223388	SPRING, HELICAL, COMP.................. UOC:DAB, DAD, DAF, DAH, DAL, DAX, V12, V14, V16, V18, V19, V21, V39, ZAB, ZAD, ZAF, ZAH, ZAL	1
17	PAFZZ	5331005798108	62983	154008	O-RING PART OF KIT P/N 920278.................. UOC:DAB, DAD, DAF, DAH, DAL, DAX, V12, V14, V16, V18, V19, V21, V39, ZAB, ZAD, ZAF, ZAH, ZAL	1
18	PAFZZ	5330010487346	62983	271722	RETAINER, PACKING PART OF KIT P/N 920278.................. UOC:DAB, DAD, DAF, DAH, DAL, DAX, V12, V14, V16, V18, V19, V21, V39, ZAB, ZAD, ZAF, ZAH, ZAL	1
19	PAFZZ	2530000870327	62983	222640	POWER STEERING PLUG.................. UOC:DAB, DAD, DAF, DAH, DAL, DAX, V12, V14, V16, V18, V19, V21, V39, ZAB, ZAD, ZAF, ZAH, ZAL	1
20	KFFZZ		62983	307951	DUST COVER PART OF KIT P/N 923080.................. UOC:DAB, DAD, DAF, DAH, DAL, DAX, V12, V14, V16, V18, V19, V21, V39, ZAB, ZAD, ZAF, ZAH, ZAL	1
21	PAFZZ	5305004713909	62983	282027	SCREW, MACHINE .... UOC:DAB, DAD, DAF, DAH, DAL, DAX, V12, V14, V16, V18, V19, V21, V39, ZAB, ZAD, ZAF, ZAH, ZAL	2
22	PAFZZ	5342008895209	62983	284154	RETAINER.................. UOC:DAB, DAD, DAF, DAH, DAL, DAX, V12, V14, V16, V18, V19, V21, V39, ZAB, ZAD, ZAF, ZAH, ZAL	1
23	PAFZZ	2530000870369	62983	284155	SLEEVE PART OF KIT P/N 923080.................. UOC:DAB, DAD, DAF, DAH, DAL, DAX, V12, V14, V16, V18, V19, V21, V39, ZAB, ZAD, ZAF, ZAH, ZAL	1
24	PAFZZ	5330002761933	62983	187000	PACKING, PREFORMED PART OF KIT P/N..................	1

(1) ITEM NO	(2) SMR CODE	(3) NSN	(4) CAGEC	(5) PART NUMBER	(6) DESCRIPTION AND USABLE ON CODES (UOC)	(7) QTY
					923080.. UOC:DAB, DAD, DAF, DAH, DAL, DAX, V12, V14, V16, V18, V19, V21, V39, ZAB, ZAD, ZAF, ZAH, ZAL	
25	PAFZZ	5365002416916	62983	680702	RING, QUADRANT PART OF KIT P/N 923080..................... UOC:DAB, DAD, DAF, DAH, DAL, DAX, V12, V14, V16, V18, V19, V21, V39, ZAB, ZAD, ZAF, ZAH, ZAL	1
26	PAFZZ	5360009609326	62983	246632	SPRING, HELICAL, COMP.. UOC:DAB, DAD, DAF, DAH, DAL, DAX, V12, V14, V16, V18, V19, V21, V39, ZAB, ZAD, ZAF, ZAH, ZAL	1

END OF FIGURE

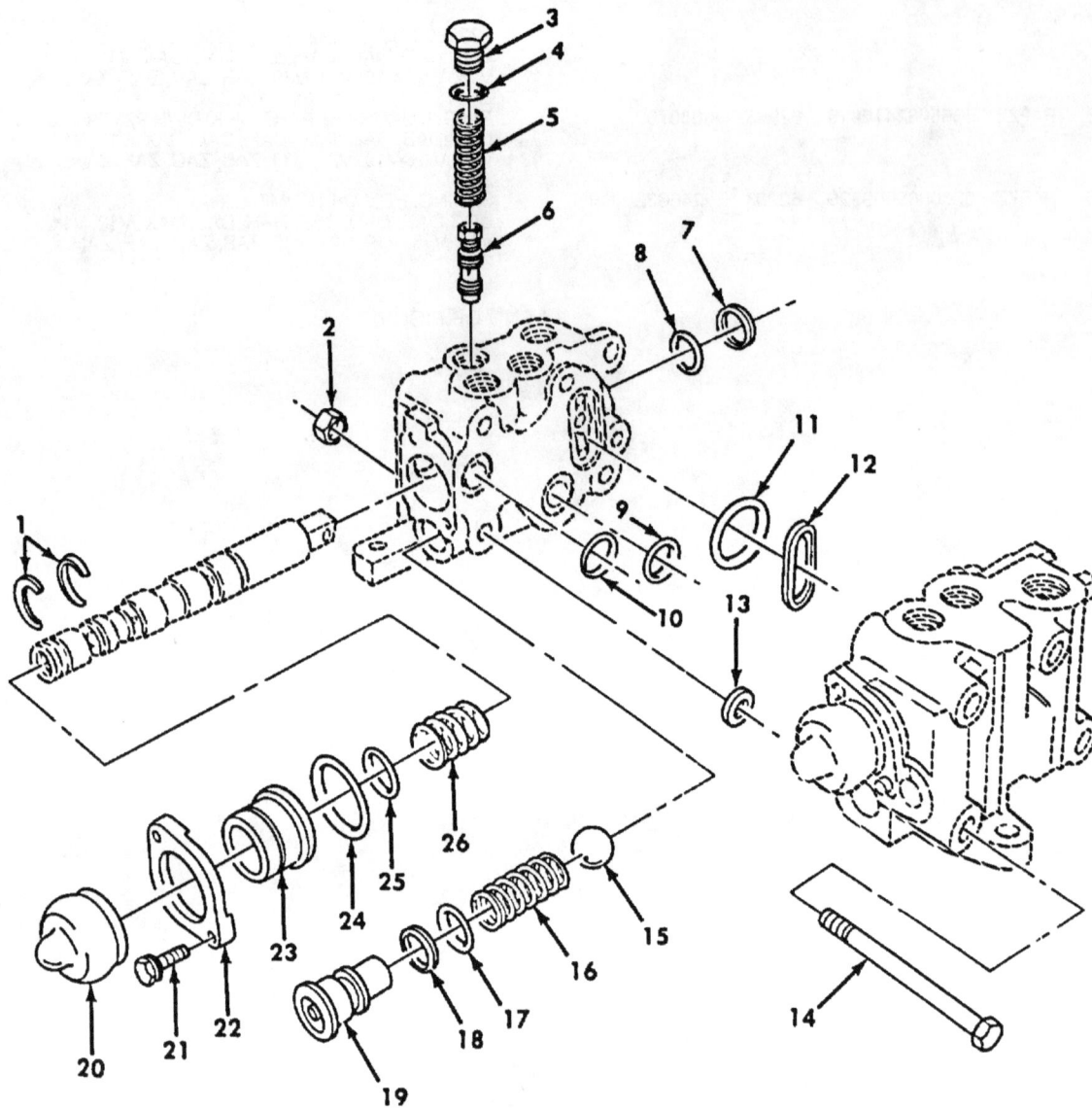

Figure 464. Front Winch Linear Valve Assembly, Dump.

(1) ITEM NO	(2) SMR CODE	(3) NSN	(4) CAGEC	(5) PART NUMBER	(6) DESCRIPTION AND USABLE ON CODES (UOC)	(7) QTY
					GROUP 2001 HOIST, CAPSTAN, WINDLASS, CRANE OR WINCH ASSEMBLY	
					FIG. 464 FRONT WINCH LINEAR VALVE ASSEMBLY, DUMP	
1	PAFZZ	5310008671465	62983	284156	WASHER, SLOTTED.................................... UOC:DAF, V19, ZAF	4
2	PAFZZ	5310007320558	62983	1454	NUT, PLAIN, HEXAGON .... UOC:DAF, V19, ZAF	4
3	PAFZZ	5365006746831	62983	186580	PLUG, MACHINE THREAD.......................... UOC:DAF, V19, ZAF	1
4	PAFZZ	5331009486482	62983	154129	O-RING... UOC:DAF, V19, ZAF	1
5	PAFZZ	5360009181920	62983	259871	SPRING, HELICAL, COMP.......................... UOC:DAF, V19, ZAF	1
6	PAFZZ	4820009609329	62983	232797	PISTON, VALVE... UOC:DAF, V19, ZAF	1
7	PAFZZ	5330002460158	62983	286669	PACKING, PREFORMED PART OF KIT P/N.......................... 920278... UOC:DAF, V19, ZAF	2
8	PAFZZ	5365002363825	62983	680701	RING, QUADRANT PART OF KIT P/N 920278............. UOC:DAF, V19, ZAF	1
9	PAFZZ	5330009535206	62983	223489	SEAL, CONTROL VALVE PART OF KIT P/N .... 920278... UOC:DAF, V19, ZAF	1
10	PAFZZ	5330009535207	62983	223493	GASKET PART OF KIT P/N 920278................. UOC:DAF, V19, ZAF	1
11	PAFZZ	5330009535208	62983	226161	PACKING, CONTROL VAL PART OF KIT P/N .... 920278... UOC:DAF, V19, ZAF	1
12	PAFZZ	2590000527504	62983	237736	RETAINER, OIL SEAL PART OF KIT P/N .... 920278... UOC:DAF, V19, ZAF	1
13	PAFZZ	5365001826713	62983	307198	SHIM... UOC:DAF, V19, ZAF	4
14	PAFZZ	5305010200709	62983	146835	SCREW... UOC:DAF, V19, ZAF	4
15	PAFZZ	3110008994353	62983	1656	BALL, BEARING... UOC:DAF, V19, ZAF	2
16	PAFZZ	5360009535205	62983	223388	SPRING, HELICAL, COMP.......................... UOC:DAF, V19, ZAF	2
17	PAFZZ	5331005798108	62983	154008	O-RING PART OF KIT P/N 920278................. UOC:DAF, V19, ZAF	2
18	PAFZZ	5330010487346	62983	271722	RETAINER, PACKING PART OF KIT P/N .... 920278... UOC:DAF, V19, ZAF	2
19	PAFZZ	2530000870327	62983	222640	POWER STEERING PLUG .... UOC:DAF, V19, ZAF	2
20	KFFZZ		62983	307951	DUST COVER PART OF KIT P/N 923080........................... PART OF KIT P/N 680702 .... UOC:DAF, V19, ZAF	2

(1) ITEM NO	(2) SMR CODE	(3) NSN	(4) CAGEC	(5) PART NUMBER	(6) DESCRIPTION AND USABLE ON CODES (UOC)	(7) QTY
21	PAFZZ	5305004713909	62983	282027	SCREW, MACHINE .... UOC:DAF, V19, ZAF	4
22	PAFZZ	5342008895209	62983	284154	RETAINER .... UOC:DAF, V19, ZAF	2
23	PAFZZ	2530000870369	62983	284155	SLEEVE PART OF KIT P/N 923080 PART .... OF KIT P/N 680702 .... UOC:DAF, V19, ZAF	2
24	PAFZZ	5330002761933	62983	187000	PACKING, PREFORMED PART OF KIT P/N .... 923080 PART OF KIT P/N 920278 PART OF KIT P/N 680702 .... UOC:DAF, V19, ZAF	2
25	PAFZZ	5365002416916	62983	680702	RING, QUADRANT PART OF KIT P/N 923080 .... PART OF KIT P/N 920278 PART OF KIT P/N 680702 .... UOC:DAF, V19	2
26	PAFZZ	5360009609326	62983	246632	SPRING, HELICAL, COMP .... UOC:DAF, V19, ZAF	2

END OF FIGURE

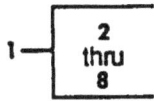

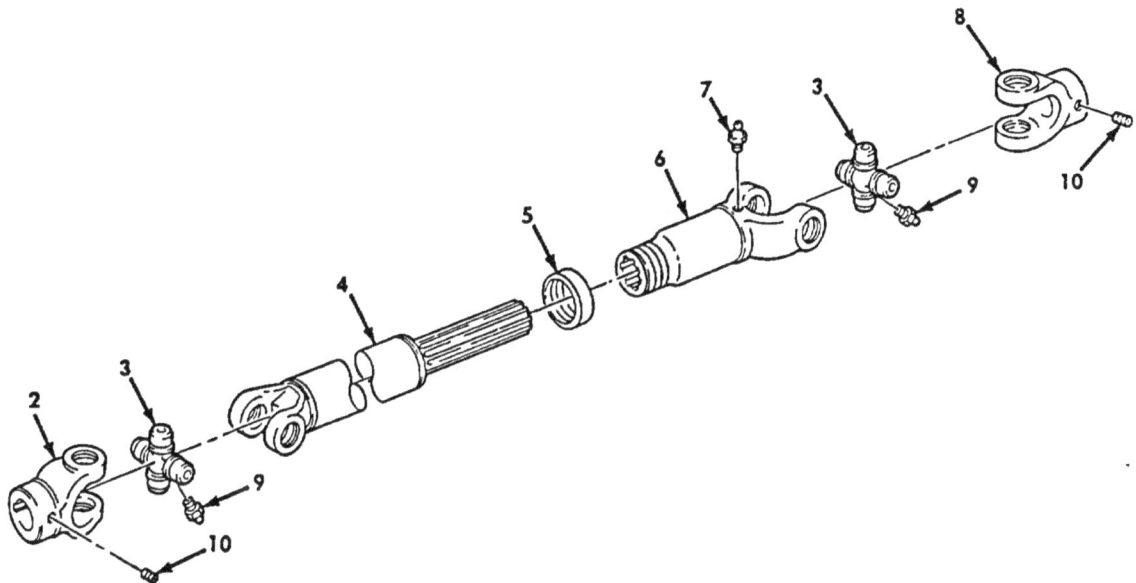

Figure 465. Front Winch Propeller Driveshaft Assembly (MI939, A1939AI).

(1) ITEM NO	(2) SMR CODE	(3) NSN	(4) CAGEC	(5) PART NUMBER	(6) DESCRIPTION AND USABLE ON CODES (UOC)	(7) QTY
					GROUP 2001 HOIST, CAPSTAN, WINDLASS, CRANE OR WINCH ASSEMBLY	
					FIG. 465 FRONT WINCH PROPELLER DRIVESHAFT ASSEMBLY (M939, M939A1)	
1	PAOOO	2520011112280	19207	11669461	PROPELLER SHAFT WIT.................... UOC:DAB, DAD, DAF, DAH, DAL, DAX, V12, V14, V16, V18, V19, V21, V39	1
2	PAOZZ	2520011341089	78500	L6NYR20-4	.YOKE, UNIVERSAL JOIN.................... UOC:DAB, DAD, DAF, DAH, DAL, DAX, V12, V14, V16, V1B, V19, V21, V39	1
3	PAOZZ	2520003522168	78500	CPL6N8	.SPIDER, UNIVERSAL JO.................... UOC:DAB, DAD, DAF, DAH, DAL, DAX, V12, V14, V16, V18, V19, V21, V39, ZAB, ZAD, ZAF, ZAH, ZAL	2
4	XAOZZ		70960	9035544862	.SHAFT ASSEMBLY.................... UOC:DAB, DAD, DAF, DAH, DAL, DAX, V12, V14, V16, V18, V19, V21, V39	1
5	PAOZZ	2520011343706	78500	DCL6N-3	.CAP, DUST, PROPELLER.................... UOC:DAB, DAD, DAF, DAH, DAL, DAX, V12, V14, V16, V18, V19, V21, V39	1
6	XAOZZ		70960	L6NLS20-22	.SLIP YOKE.................... UOC:DAB, DAD, DAF, DAH, DAL, DAX, V12, V14, V16, V18, V19, V21, V39, ZAB, ZAD, ZAF, ZAH, ZAL	1
7	PAOZZ	4730000504203	96906	MS15001-1	.FITTING, LUBRICATION.................... UOC:DAB, DAD, DAF, DAH, DAL, DAX, V12, V14, V16, V18, V19, V21, V39	1
8	PAOZZ	2520011343471	78500	L6NYR14-19	YOYE, UNIVERSAL JOIN.................... UOC:DAB, DAD, DAF, DAH, DAL, DAX, V12, V14, V16, V18, V19, V21, V39, ZAB, ZAD, ZAF, ZAH, ZAL	1
9	PAOZZ	4730000504203	96906	MS15001-1	.FITTING, LUBRICATION.................... UOC:DAB, DAD, DAF, DAH, DAL, DAX, V12, V14, V16, V18, V19, V21, V39	2
10	PAOZZ	5305000589389	96906	MS51977-74	.SETSCREW.................... UOC:DAB, DAD, DAF, DAH, DAL, DAX, V12, V14, V16, V18, V19, V21, V39	2

END OF FIGURE

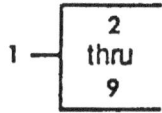

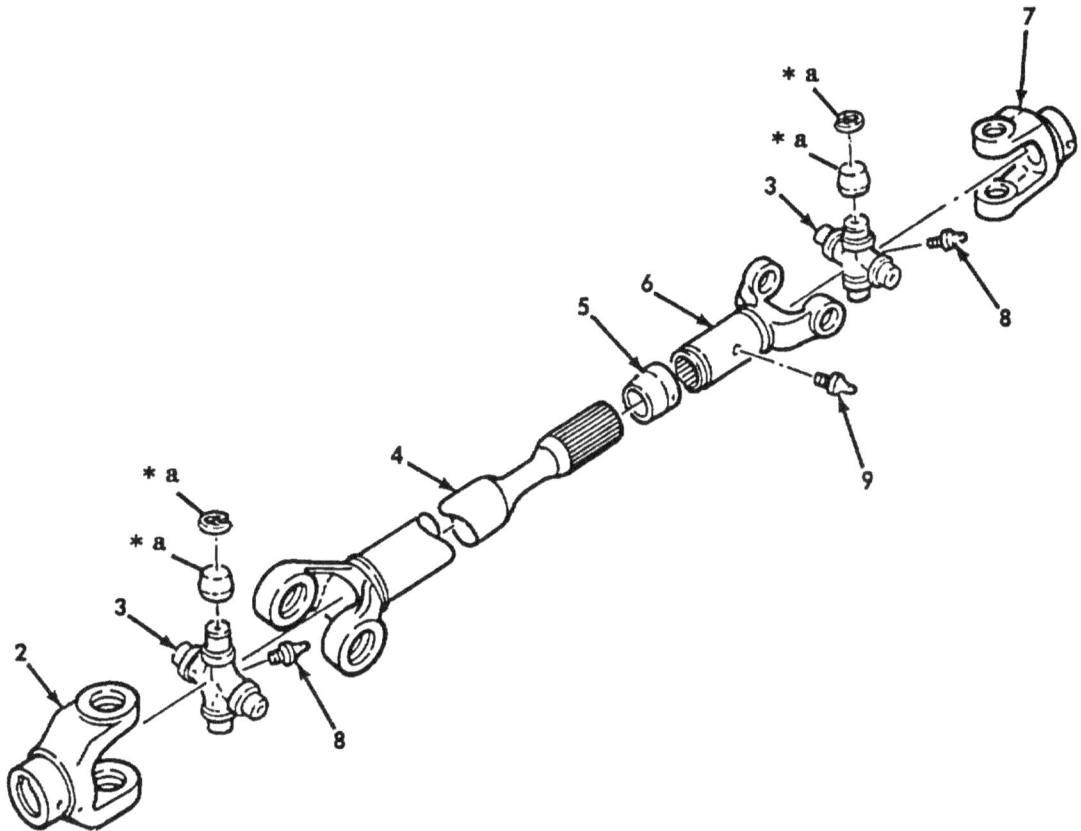

* a PART OF ITEM 3

Figure 466. Propeller Drive Shaft, Front Winch (7M939A2).

(1) ITEM NO	(2) SMR CODE	(3) NSN	(4) CAGEC	(5) PART NUMBER	(6) DESCRIPTION AND USABLE ON CODES (UOC)	(7) QTY
					GROUP 2001 HOIST, CAPSTAN, WINDLASS, CRANE OR WINCH ASSEMBLY	
					FIG. 466 PROPELLER DRIVE SHAFT, FRONT WINCH (M939A2)	
1	PAOOO	2520012727767	78500	903-04-48-785	PROPELLER SHAFT WIT ASSEMBLY UOC:ZAB, ZAD, ZAE, ZAF, ZAH, ZAL	1
2	PAOZZ	2520012856282	70960	L6NYR20-87	.YOKE, UNIVERSAL JOIN UOC:ZAB, ZAD, ZAE, ZAF, ZAH, ZAL	1
3	PAOZZ	2520003522168	95019	5-92X	.SPIDER, UNIVERSAL JO UOC:ZAB, ZAD, ZAE, ZAF, ZAH, ZAL	2
4	XAOZZ		70960	903-55	.SHAFT ASSEMBLY UOC:ZAB, ZAD, ZAE, ZAF, ZAH, ZAL	1
5	PAOZZ	2530012860108	70960	SERUR14-16	.BOOT, VEHICULAR COM UOC:ZAB, ZAD, ZAE, ZAF, ZAH, ZAL	1
6	XAOZZ		70960	L6NLS20-22	.YOKE, SLIP UOC:ZAB, ZAD, ZAE, ZAF, ZAH, ZAL	1
7	PAOZZ	2520011343471	78500	L6NYR14-19	YOKE, UNIVERSAL JOIN UOC:ZAB, ZAD, ZAE, ZAF, ZAH, ZAL	1
8	PAOZZ	4730001720028	96906	MS15003-4	.FITTING, LUBRICATION UOC:ZAB, ZAD, ZAE, ZAF, ZAH, ZAL	2
9	PAOZZ	4730001720010	96906	MS15002-1	.FITTING, LUBRICATION UOC:ZAB, ZAD, ZAE, ZAF, ZAH, ZAL	1

END OF FIGURE

3 — 4

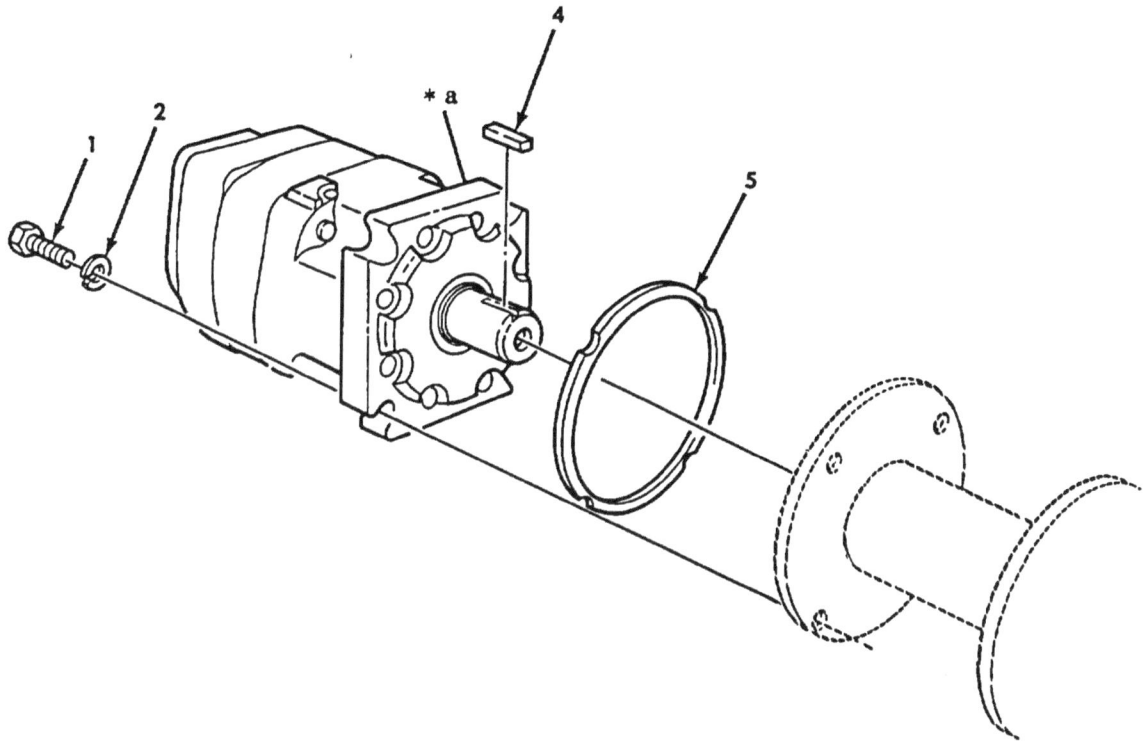

* a PART OF ITEM 3

Figure 467. Front Winch Motor and Mounting Hardware.

SECTION II

(1) ITEM NO	(2) SMR CODE	(3) NSN	(4) CAGEC	(5) PART NUMBER	(6) DESCRIPTION AND USABLE ON CODES (UOC)	(7) QTY
					GROUP 2001 HOIST, CAPSTAN, WINDLASS, CRANE OR WINCH ASSEMBLY	
					FIG. 467 FRONT WINCH MOTOR AND MOUNTING HARDWARE	
1	PAOZZ	5306012265917	81349	M24240-4-20916	BOLT, MACHINE.. UOC:DAB, DAD, DAF, DAH, DAL, DAX, V12, V14, V16, V18, V19, V21, V39, ZAB, ZAD, ZAF, ZAH, ZAL	4
2	PAOZZ	5310000034094	01276	210104-8S	WASHER, LOCK... UOC:DAB, DAD, DAF, DAH, DAL, DAX, V12, V14, V16, V18, V19, V21, V39, ZAB, ZAD, ZAF, ZAH, ZAL	4
3	PAOZZ	4320011376293	96151	109-1094-106	MOTOR, HYDRAULIC... UOC:DAB, DAD, DAF, DAH, DAL, DAX, V12 , V14, V16, V18, V19, V21, V39, ZAB, ZAB, ZAD, ZAF, ZAH, ZAL	1
4	PAOZZ	5315002420818	96906	MS20067-221	KEY, MACHINE... UOC:DAB, DAD, DAF, DAH, DAL, DAX, V12, V14, V16, V18, V19, V21, V39, ZAB, ZAD, ZAF, ZAH, ZAL	1
5	PAOZZ	5330011150604	19207	12277128	GASKET.. UOC:DAB, DAD, DAF, DAH, DAL, DAX, V12, V14, V16, V18, V19, V21, V39, ZAB, ZAD, ZAF, ZAH, ZAL	1

END OF FIGURE

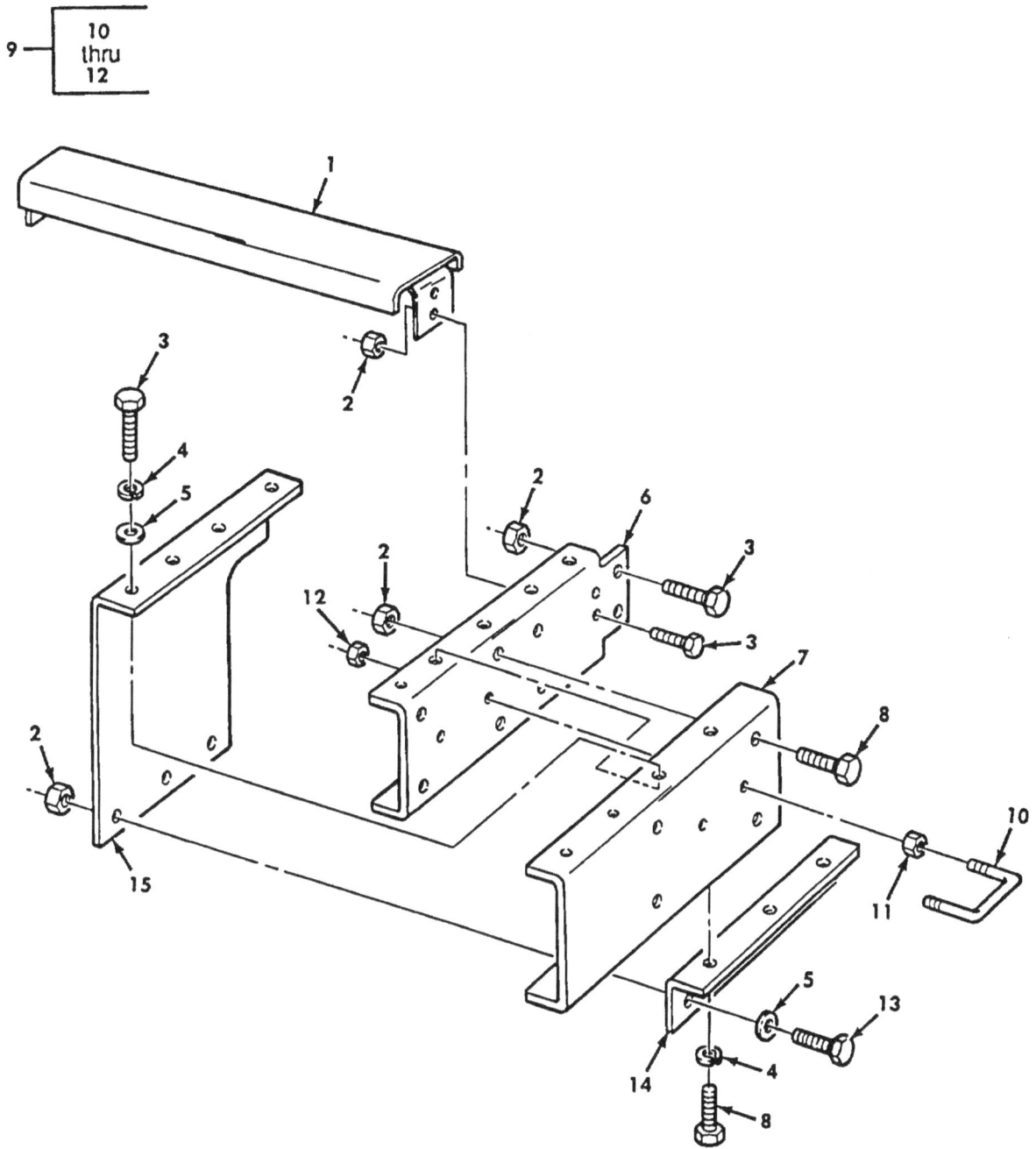

Figure 468. Front Winch Brackets and Mounting Hardware.

(1) ITEM NO	(2) SMR CODE	(3) NSN	(4) CAGEC	(5) PART NUMBER	(6) DESCRIPTION AND USABLE ON CODES (UOC)	(7) QTY
					GROUP 2001 HOIST, CAPSTAN, WINDLASS, CRANE OR WINCH ASSEMBLY	
					FIG. 468 FRONT WINCH BRACKETS AND MOUNTING HARDWARE	
1	PFOZZ	2510012127620	19207	12302674-2	FRAME SECTION, STRUC UOC:V12, V14, V16, V18, V19, V21, V39	1
2	PAOZZ	5310004883888	96906	MS51943-40	NUT, SELF-LOCKING, HE .... UOC:DAB, DAD, DAF, DAH, DAL, DAX, V12, V14, V16, V18, V19, V21, V39, ZAB, ZAD, ZAF, ZAH, ZAL	30
3	PAOZZ	5305007195235	80204	B1821BH050F175N	SCREW, CAP, HEXAGON H .... UOC:DAB, DAD, DAF, DAH, DAX, V12, V14, V16, V19, V21, V39, ZAB, ZAD, ZAF, ZAH	22
4	PAOZZ	5310000034094	01276	210104-8S	WASHER, LOCK .... UOC:DAL, V18, ZAL	16
4	PAOZZ	5310000034094	01276	210104-8S	WASHER, LOCK .... UOC:DAB, DAD, DAF, DAH, DAX, V12, V14, V16, V19, V21, V39, ZAB, ZAD, ZAF, ZAH	22
5	PAOZZ	5310008095998	96906	MS27183-18	WASHER, FLAT .... UOC:DAL, V18, ZAL	18
5	PAOZZ	5310008095998	96906	MS27183-18	WASHER, FLAT .... UOC:DAB, DAD, DAF, DAH, DAX, V12, V14, V16, V19, V21, V39, ZAB, ZAD, ZAF, ZAH	20
6	PFOZZ	2510010855353	19207	12255919-1	CHANNEL, FRAME .... UOC:DAB, DAD, DAF, DAH, DAL, DAX, V12, V14, V16, V18, V19, V21, V39, ZAB, ZAD, ZAF, ZAH, ZAL	1
6	PFOZZ	2510010835404	19207	12255919-2	CHANNEL, FRAME .... UOC:DAB, DAD, DAF, DAH, DAL, DAX, V12, V14, V16, V18, V19, V21, V39, ZAB, ZAD, ZAF, ZAH, ZAL	1
7	PAOZZ	2510010822515	19207	12255926	FRAME SECTION, STRUC .... UOC:DAB, DAD, DAF, DAH, DAL, DAX, V12, V14, V16, V18, V19, V21, V39, ZAB, ZAD, ZAF, ZAH, ZAL	2
8	PAOZZ	5305007250154	96906	MS90727-112	SCREW, CAP, HEXAGON H .... UOC:DAB, DAD, DAF, DAH, DAL, DAX, V12, V14, V16, V18, V19, V21, V39, ZAB, ZAD, ZAF, ZAH, ZAL	14
9	PAOZZ	2590012811271	19207	12356842	KIT, LOOP TIE DOWN .... UOC:DAB, DAD, DAF, DAH, DAL, DAX, V12, V14, V16, V18, V19, V21, V39, ZAB, ZAD, ZAF, ZAH, ZAL	2
10	PAOZZ	2540012811272	19207	12356845	.KIT-VEHICLE TIEDOWN .... UOC:DAB, DAD, DAF, DAH, DAX, V12, V14, V16, V18, V19, V21, V39, ZAB, ZAD, ZAF, ZAH, ZAL	1
11	PAOZZ	5310007320560	96906	MS51968-14	.NUT, PLAIN, HEXAGON .... UOC:DAB, DAD, DAF, DAH, DAL, DAX, V12, V14, V16, V18, V19, V21, V39, ZAB, ZAD, ZAF, ZAH,	2

(1) ITEM NO	(2) SMR CODE	(3) NSN	(4) CAGEC	(5) PART NUMBER	(6) DESCRIPTION AND USABLE ON CODES (UOC)	(7) QTY
12	PAOZZ	5310008775795	96906	MS21044-N8	ZAL .NUT, SELF-LOCKING, HE UOC:DAB, DAD, DAF, DAH, DAL, DAX, V12, V14, V16, V18, V19, V21, V39, ZAB, ZAD, ZAF, ZAH, ZAL	2
13	PAOZZ	5305007254183	96906	MS90726-113	SCREW, CAP, HEXAGON H UOC:DAB, DAD, DAF, DAH, DAL, DAX, V12, V14, V16, V18, V19, V21, V39, ZAB, ZAD, ZAF, ZAH, ZAL	12
14	PFOZZ	5340011109205	19207	12255916	BRACKET, ANGLE UOC:DAB, DAD, DAF, DAH, DAL, DAX, V12, V14, V16, V18, V19, V21, V39, ZAB, ZAD, ZAF, ZAH, ZAL	2
15	PAOZZ	2590010822520	19207	12255917-1	PLATE, MOUNTING FRONT WINCH ASSEMBLY, LEFT HAND UOC:DAB, DAD, DAF, DAH, DAL, DAX, V12, V14, V16, V18, V19, V21, V39, ZAB, ZAD, ZAF, ZAH, ZAL	1
15	PAOZZ	2590010822521	19207	12255917-2	PLATE, MOUNTING FRONT WINCH ASSEMBLY, RIGHT HAND UOC:DAB, DAD, D AF, DAH, DAL, DAX, V12, V14, V16, V18, V19, V21, V39, ZAB, ZAD, ZAF, ZAH, ZAL	1

END OF FIGURE

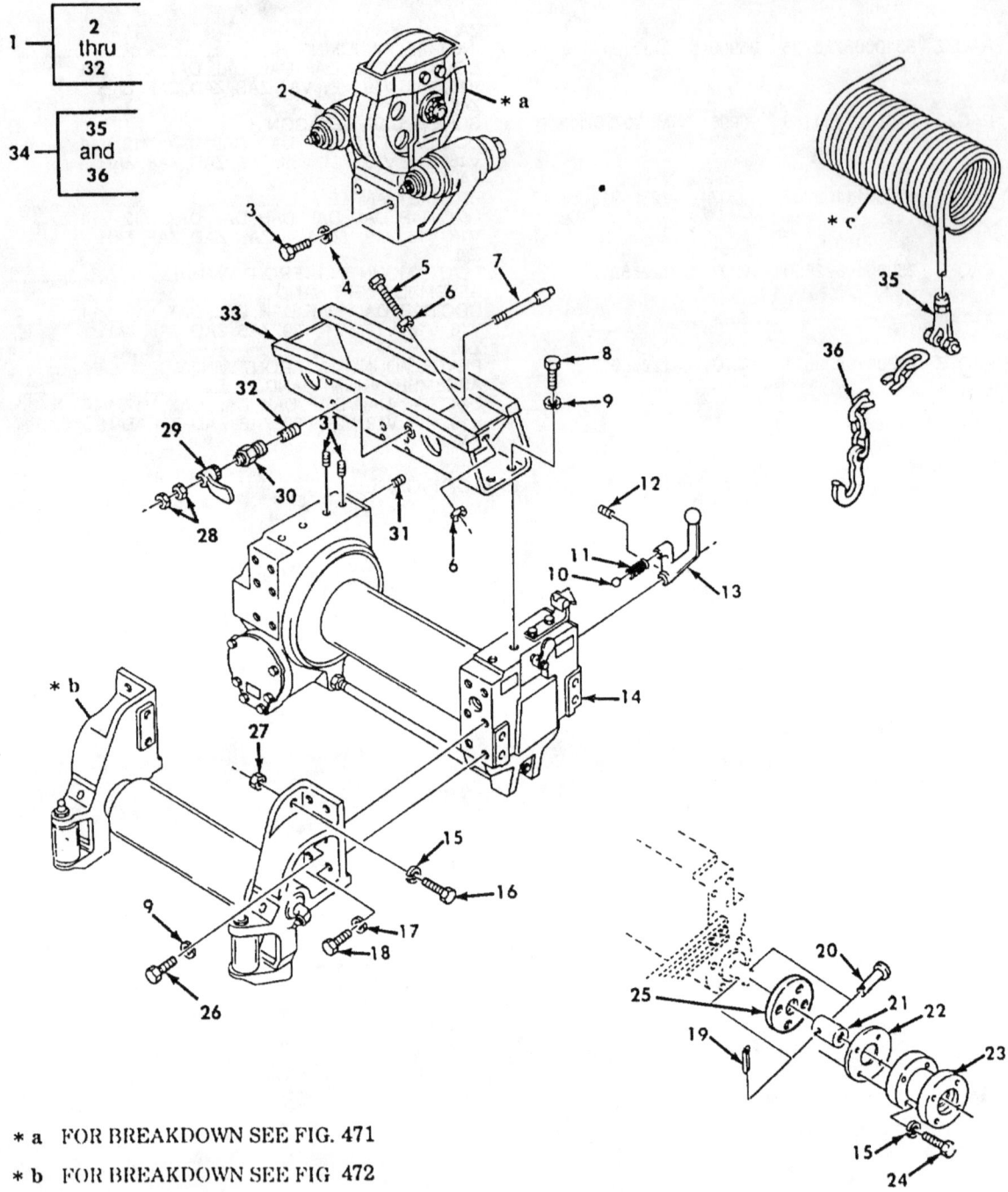

*a   FOR BREAKDOWN SEE FIG. 471

*b   FOR BREAKDOWN SEE FIG 472

*c   PART OF ITEM 34

Figure 469. Front Winch Assembly Complete and Wire Rope Assembly, Wrecker.

(1) ITEM NO	(2) SMR CODE	(3) NSN	(4) CAGEC	(5) PART NUMBER	(6) DESCRIPTION AND USABLE ON CODES (UOC)	(7) QTY
					GROUP 2001 HOIST, CAPSTAN, WINDLASS, CRANE OR WINCH ASSEMBLY	
					FIG. 469 FRONT WINCH ASSEMBLY COMPLETE AND WIRE ROPE ASSEMBLY, WRECKER	
1	PAOFH	2590007411122	19207	7411122	WINCH ASY FRONT .... UOC:DAL, V18, ZAL	1
2	PAFZZ	2590001210707	19207	7954483	.SHEAVE ASSEMBLY, SWI .... UOC:DAL, V18, ZAL	1
3	PAFZZ	5305000522218	96906	MS35265-107	.SCREW, MACHINE .... UOC:DAL, V18, ZAL	4
4	PAFZZ	5310006379541	96906	MS35338-46	.WASHER, LOCK .... UOC:DAL, V18, ZAL	4
5	PAFZZ	5305000549285	96906	MS51955-71	.SCREW .... UOC:DAL, V18, ZAL	2
6	PAFZZ	5310008348732	96906	MS35691-33	.NUT, PLAIN, HEXAGON .... UOC:DAL, V18, ZAL	4
7	PAFZZ	4820006525548	19207	8741768	.POPPET, LOCK, TROLLEY .... UOC:DAL, V18, ZAL	1
8	PAFZZ	5305007247219	80204	B1821BH063C125N	.SCREW, CAP, HEXAGON H .... UOC:DAL, V18, ZAL	4
9	PAFZZ	5310008206653	80045	23MS35338-50	.WASHER, LOCK .... UOC:DAL, V18, ZAL	12
10	PAFZZ	3110009489796	96906	MS19061-20013	.BALL, BEARING .... UOC:DAL, V18, ZAL	1
11	PAFZZ	5360007409727	61465	304677	.SPRING, HELICAL, COMP .... UOC:DAL, V18, ZAL	1
12	PAFZZ	5305007246810	96906	MS51963-101	.SETSCREW .... UOC:DAL, V18, ZAL	1
13	PAFZZ	5340003517831	19207	7954471	.LEVER, MANUAL CONTRO .... UOC:DAL, V18, ZAL	1
14	PAFFH	2590003517865	19207	7954486	.WINCH, DRUM, VEHICLE .... UOC:DAL, V18, ZAL	1
15	PAFZZ	5310000034094	01276	210104-8S	.WASHER, LOCK .... UOC:DAL, V18, ZAL	8
16	PAFZZ	5305007195235	80204	B1821BH050F175N	.SCREW, CAP, HEXAGON H .... UOC:DAL, V18, ZAL	4
17	PAFZZ	5310005847888	96906	MS35338-51	.WASHER, LOCK .... UOC:DAL, V18, ZAL	4
18	PAFZZ	5305009399204	96906	MS90725-187	.SCREW, CAP, HEXAGON H .... UOC:DAL, V18, ZAL	4
19	PAFZZ	5315008423044	96906	MS24665-283	.PIN, COTTER .... UOC:DAL, V18, ZAL	1
20	PAFZZ	5315011096846	19207	12276978	.PIN, STRAIGHT, HEADED .... UOC:DAL, V1B, ZAL	1
21	PAFZZ	3010011016712	19207	12276915	.COUPLING, SHAFT, RIGI .... UOC:DAL, V18, ZAL	1
22	PAFZZ	5330000573823	19207	7409822	.GASKET .... UOC:DAL, V18, ZAL	1
23	PAFZZ	5340010841232	19207	12276904	.BRACKET, MOUNTING	2

(1) ITEM NO	(2) SMR CODE	(3) NSN	(4) CAGEC	(5) PART NUMBER	(6) DESCRIPTION AND USABLE ON CODES (UOC)	(7) QTY
					UOC:DAL, V18, ZAL	
24	PAFZZ	5305000712073	80204	B1821BH050C250N	.SCREW, CAP, HEXAGON H	4
					UOC:DAL, V18, ZAL	
25	PAFZZ	5330011150651	19207	12277028	.RETAINER, OIL SEAL	4
					UOC:DAL, V18, ZAL	
26	PAFZZ	5305007247222	80204	B1821BH063C200N	.SCREW, CAP, HEXAGON H	8
					UOC:DAL, V18, ZAL	
27	PAFZZ	5310007320560	96906	MS51968-14	.NUT, PLAIN, HEXAGON	4
					UOC:DAL, V18, ZAL	
28	PAFZZ	5310008807744	96906	MS51967-5	.NUT, PLAIN, HEXAGON	2
					UOC:DAL, V18, ZAL	
29	PAFZZ	2540007409150	19207	7409150	.LATCH, DRUM LOCK	1
					UOC:DAL, V18, ZAL	
30	PAFZZ	2590002262347	19207	8344240	.NUT, WINCH POPPET	1
					UOC:DAL, V18, ZAL	
31	PAFZZ	5305007247206	96906	MS51963-155	.SETSCREW	10
					UOC:DAL, V18, ZAL	
32	PAFZZ	5360007538706	19207	7538706	.SPRING, HELICAL, COMP	1
					UOC:DAL, V18, ZAL	
33	PAFZZ	2590010822526	19207	8344234	.TRACK ASSEMBLY, TROL	1
					UOC:DAL, V18, ZAL	
34	MOOZZ		19207	12253105-15	WIRE ROPE ASSY, SING MAKE FROM ROPE, ... P/N 7699769.	1
					UOC:DAL, V18, ZAL	
35	PAOZZ	4030001582409	19207	7074517	.SOCKET, WIRE ROPE	1
					UOC:DAL, V18, ZAL	
36	PAOZZ	4010010270356	19207	12253104-4	.CHAIN ASSEMBLY, SING	1
					UOC:DAL, V18, ZAL	

END OF FIGURE

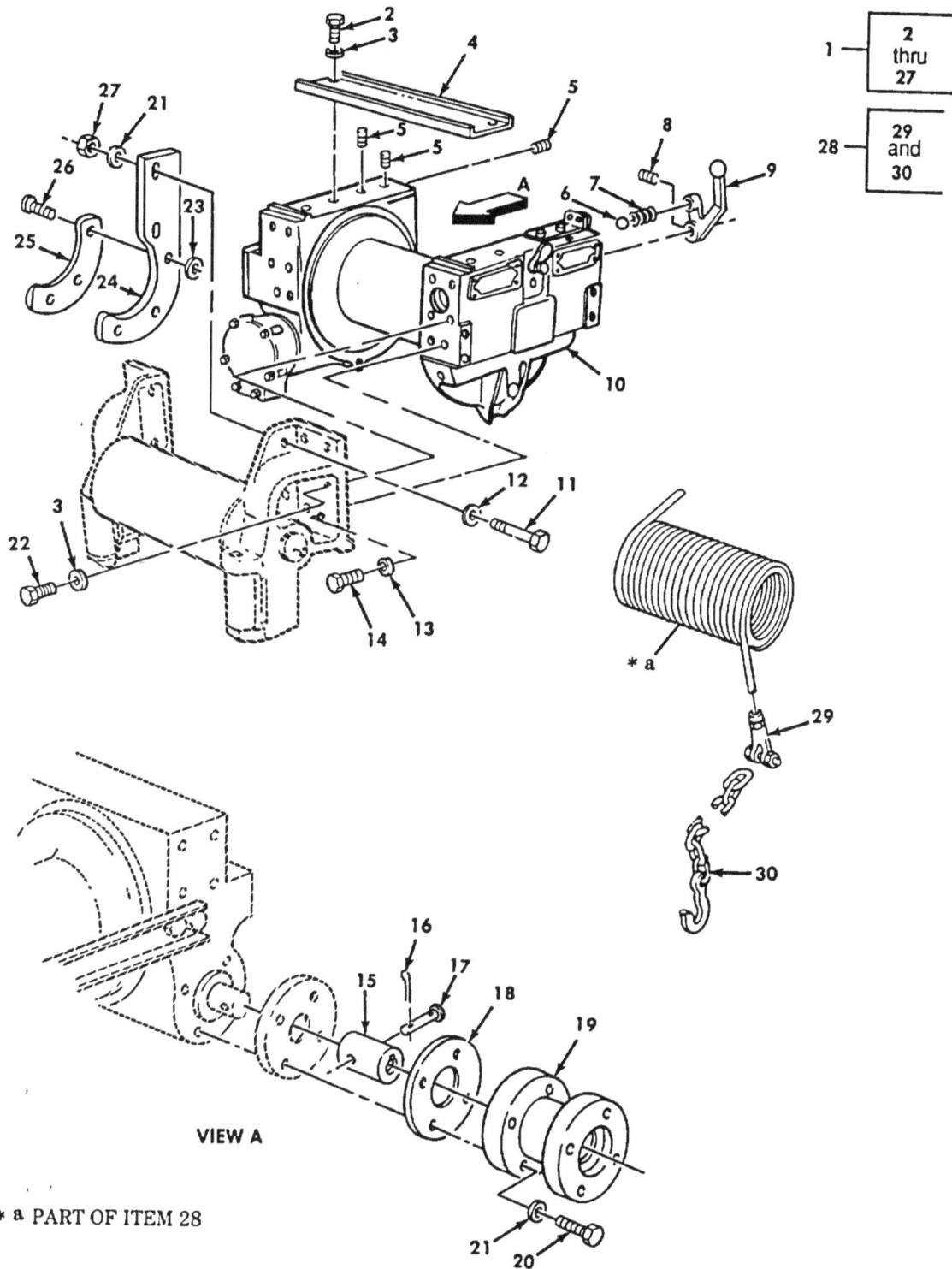

Figure 470. Front Winch Assembly Complete and Wire Rope Assembly.

* a PART OF ITEM 28

VIEW A

(1) ITEM NO	(2) SMR CODE	(3) NSN	(4) CAGEC	(5) PART NUMBER	(6) DESCRIPTION AND USABLE ON CODES (UOC)	(7) QTY
					GROUP 2001 HOIST, CAPSTAN, WINDLASS, CRANE OR WINCH ASSEMBLY	
					FIG. 470 FRONT WINCH ASSEMBLY COMPLETE AND WIRE ROPE ASSEMBLY	
1	PAOFH	2590010822644	19207	7412382-1	WINCH, DRUM, VEHICLE .... UOC:DAB, DAD, DAF, DAH, DAL, DAX, V12, V14, V16, V18, V19, V21, V39, ZAB, ZAD, ZAF, ZAH, ZAL	1
2	PAFZZ	5305007247219	80204	B1821BH063C125N	.SCREW, CAP, HEXAGON H .... UOC:DAB, DAD, DAF, DAH, DAL, DAX, V12, V14, V16, V18, V19, V21, V39, ZAB, ZAD, ZAF, ZAH, ZAL	4
3	PAFZZ	5310008206653	80045	23MS35338-50	.WASHER, LOCK .... UOC:DAB, DAD, DAF, DAH, DAL, DAX, V12, V14, V16, V18, V19, V21, V39, ZAB, ZAD, ZAF, ZAH, ZAL	12
4	PAFZZ	2520007409813	19207	7409813	.TOP CHANNEL .... UOC:DAB, DAD, DAF, DAH, DAL, DAX, V12, V14, V16, V18, V19, V21, V39, ZAB, ZAD, ZAF, ZAH, ZAL	1
5	PAFZZ	5305007247206	96906	MS51963-155	.SETSCREW .... UOC:DAB, DAD, DAF, DAH, DAL, DAX, V12, V14, V16, V18, V19, V21, V39, ZAB, ZAD, ZAF, ZAH, ZAL	10
6	PAFZZ	3110009489796	96906	MS19061-20013	.BALL, BEARING .... UOC:DAB, DAD, DAF, DAH, DAL, DAX, V12, V14, V16, V18, V19, V21, V39, ZAB, ZAD, ZAF, ZAH, ZAL	1
7	PAFZZ	5360007409727	61465	304677	.SPRING, HELICAL, COMP .... UOC:DAB, DAD, DAF, DAH, DAL, DAX, V12, V14, V16, V18, V19, V21, V39, ZAB, ZAD, ZAF, ZAH, ZAL	1
8	PAFZZ	5305007246810	96906	MS51963-101	.SETSCREW .... UOC:DAB, DAD, DAF, DAH, DAL, DAX, V12, V14, V16, V18, V19, V21, V39, ZAB, ZAD, ZAF, ZAH, ZAL	1
9	PAFZZ	5340003517831	19207	7954471	.LEVER, MANUAL CONTRO .... UOC:DAB, DAD, DAF, DAH, DAL, DAX, V12, V14, V16, V18, V19, V21, V39, ZAB, ZAD, ZAF, ZAH, ZAL	1
10	PAFFH	2590003517865	19207	7954486	.WINCH, DRUM, VEHICLE .... UOC:DAB, DAD, DAF, DAH, DAL, DAX, V12, V14, V16, V18, V19, V21, V39, ZAB, ZAD, ZAF, ZAH, ZAL	1
11	PAFZZ	5305000712071	80204	B1821BH050C200N	.SCREW, CAP, HEXAGON H ROLLER AND .... TENSIONER ASSEMBLY TO PLATE ASSEMBLY UOC:DAB, DAD, DAF, DAH, DAL, DAX, V12, V14, V16, V18, V19, V21, V39, ZAB, ZAD, ZAF, ZAH, ZAL	4
12	PAFZZ	5310001848992	96906	MS9320-14	.WASHER, FLAT	4

(1) ITEM NO	(2) SMR CODE	(3) NSN	(4) CAGEC	(5) PART NUMBER	(6) DESCRIPTION AND USABLE ON CODES (UOC)	(7) QTY
					UOC:DAB, DAD, DAF, DAH, DAL, DAX, V12, V14, V16, V18, V19, V21, V39, ZAB, ZAD, ZAF, ZAH, ZAL	
13	PAFZZ	5310005847888	96906	MS35338-51	.WASHER, LOCK	4
					UOC:DAB, DAD, DAF, DAH, DAL, DAX, V12, V14, V16, V18, V19, V21, V39, ZAB, ZAD, ZAF, ZAH, ZAL	
14	PAFZZ	5305009399204	96906	MS90725-187	.SCREW, CAP, HEXAGON H	4
					UOC:DAB, DAD, DAF, DAH, DAL, DAX, V12, V14, V16, V18, V19, V21, V39, ZAB, ZAD, ZAF, ZAH, ZAL	
15	PAFZZ	3010011016712	19207	12276915	.COUPLING, SHAFT, RIGI	1
					UOC:DAB, DAD, DAF, DAH, DAL, DAX, V12, V14, V16, V18, V19, V21, V39, ZAB, ZAD, ZAF, ZAH, ZAL	
16	PAFZZ	5315008423044	96906	MS24665-283	.PIN, COTTER	1
					UOC:DAB, DAD, DAF, DAH, DAL, DAX, V12, V14, V16, V18, V19, V21, V39, ZAB, ZAD, ZAF, ZAH, ZAL	
17	PAFZZ	5315011096846	19207	12276978	.PIN, STRAIGHT, HEADED	1
					UOC:DAB, DAD, DAF, DAH, DAL, DAX, V12, V14, V16, V18, V19, V21, V39, ZAB, ZAD, ZAF, ZAH, ZAL	
18	PAFZZ	5330000573823	19207	7409822	.GASKET	1
					UOC:DAB, DAD, DAF, DAH, DAL, DAX, V12, V14, V16, V18, V19, V21, V39, ZAB, ZAD, ZAF, ZAH, ZAL	
19	PAFZZ	5340010841232	19207	12276904	.BRACKET, MOUNTING	1
					UOC:DAB, DAD, DAF, DAH, DAL, DAX, V12, V14, V16, V18, V19, V21, V39, ZAB, ZAD, ZAF, ZAH, ZAL	
20	PAFZZ	5305000712073	80204	B1821BH050C250N	.SCREW, CAP, HEXAGON H	4
					UOC:DAB, DAD, DAF, DAH, DAL, DAX, V12, V14, V16, V18, V19, V21, V39, ZAB, ZAD, ZAF, ZAH, ZAL	
21	PAFZZ	5310000034094	01276	210104-8S	.WASHER, LOCK	8
					UOC:DAB, DAD, DAF, DAH, DAL, DAX, V12, V14, V16, V18, V19, V21, V39, ZAB, ZAD, ZAF, ZAH, ZAL	
22	PAFZZ	5305007247222	80204	B1821BH063C200N	.SCREW, CAP, HEXAGON H	8
					UOC:DAB, DAD, DAF, DAH, DAL, DAX, V12, V14, V16, V18, V19, V21, V39, ZAB, ZAD, ZAF, ZAH, ZAL	
23	PAFZZ	5310001949213	96906	MS35336-39	.WASHER, LOCK	6
					UOC:DAB, DAD, DAF, DAH, DAL, DAX, V12, V14, V16, V18, V19, V21, V39, ZAB, ZAD, ZAF, ZAH, ZAL	
24	PAFZZ	2590011310128	19207	12300665-1	.PLATE, MOUNTING, WINC RIGHT	1
					UOC:DAB, DAD, DAF, DAH, DAL, DAX, V12, V14, V16, V18, V19, V21, V39, ZAB, ZAD, ZAF, ZAH, ZAL	
24	PAFZZ	2590011310129	19207	12300665-2	.PLATE, MOUNTING, WINC LEFT	1

(1) ITEM NO	(2) SMR CODE	(3) NSN	(4) CAGEC	(5) PART NUMBER	(6) DESCRIPTION AND USABLE ON CODES (UOC)	(7) QTY
					UOC:DAB, DAD, DAF, DAH, DAL, DAX, V12, V14, V16, V18, V19, V21, V39, ZAB, ZAD, ZAF, ZAH, ZAL	
25	PAFZZ	5340011310130	19207	12300666-1	.PLATE, MOUNTING RIGHT	1
					UOC:DAB, DAD, DAF, DAH, DAL, DAX, V12, V14, V16, V18, V19, V21, V39, ZAB, ZAD, ZAF, ZAH, ZAL	
25	PAFZZ	5340011310131	19207	12300666-2	.PLATE, MOUNTING LEFT	1
					UOC:DAB, DAD, DAF, DAH, DAL, DAX, V12, V14, V16, V18, V19, V21, V39, ZAB, ZAD, ZAF, ZAH, ZAL	
26	PAFZZ	5305009592723	96906	MS35190-319	.SCREW, MACHINE	6
					UOC:DAB, DAD, DAF, DAH, DAL, DAX, V12, V14, V16, V18, V19, V21, V39, ZAB, ZAD, ZAF, ZAH, ZAL	
27	PAFZZ	5310007680318	24617	9413509	.NUT, PLAIN, HEXAGON	4
					UOC:DAB, D AD, DAF, DAH, DAL, DAX, V12, V14, V16, V18, V19, V21, V39, ZAB, ZAD, ZAF, ZAH, ZAL	
28	MOOZZ		19207	12253105-13	WIRE ROPE ASSY, SING MAKE FROM WIRE .... ROPE, P/N 7699769	1
					UOC:DAB, DAD, DAF, DAH, DAL, DAX, V12, V14, V16, V18, V19, V21, V39, ZAB, ZAD, ZAF, ZAH, ZAL	
29	PAOZZ	4030001582409	19207	7074517	.SOCKET, WIRE ROPE	1
					UOC:DAB, DAD, DAF, DAH, DAL, DAX, V12, V14, V16, V18, V19, V21, V39, ZAB, ZAD, ZAF, ZAH, ZAL	
30	PAOZZ	4010010270356	19207	12253104-4	.CHAIN ASSEMBLY, SING	1
					UOC:DAB, DAD, DAF, DAH, DAL, DAX, V12, V14, V16, V18, V19, V21, V39, ZAB, ZAD, ZAF, ZAH, ZAL	

END OF FIGURE

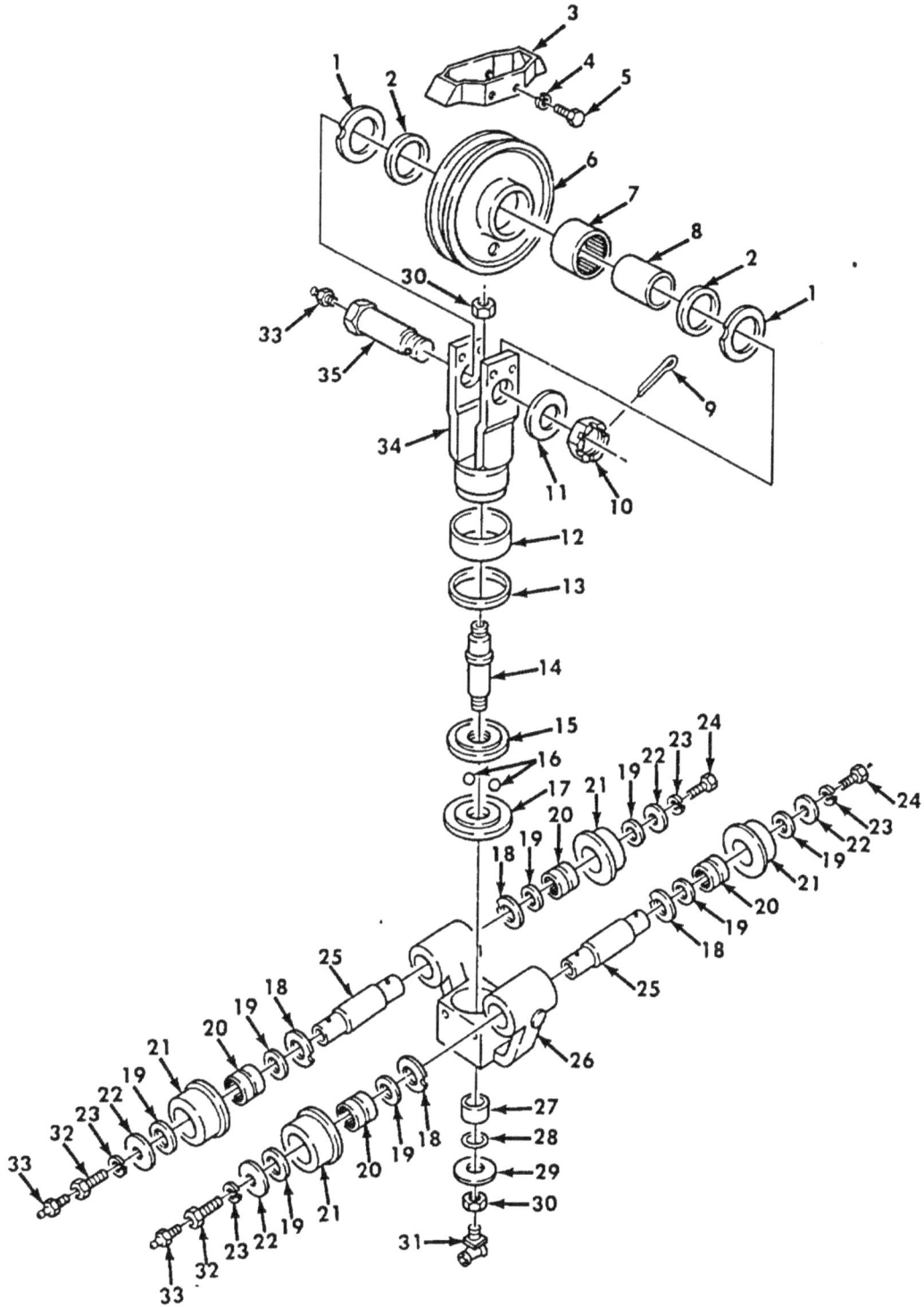

Figure 471. Front Winch Sheave Assembly, Wrecker.

(1) ITEM NO	(2) SMR CODE	(3) NSN	(4) CAGEC	(5) PART NUMBER	(6) DESCRIPTION AND USABLE ON CODES (UOC)	(7) QTY
					GROUP 2001 HOIST, CAPSTAN, WINDLASS, CRANE OR WINCH ASSEMBLY	
					FIG. 471 FRONT WINCH SHEAVE ASSEMBLY, WRECKER	
1	PAFZZ	3120010853338	19207	8344242	BEARING, WASHER, THRU .... UOC:DAL, V18, ZAL	2
2	PAFZZ	5330007411154	19207	7411154	FELT, MECHANICAL, PRE .... UOC:DAL, V18, ZAL	2
3	PAFZZ	3020010823598	19207	8344505	GUARD, MECHANICAL DR .... UOC:DAL, V18, ZAL	1
4	PAFZZ	5310006379541	96906	MS35338-46	WASHER, LOCK .... UOC:DAL, V18 , ZAL	4
5	PAFZZ	5305005434372	80204	B1821BH038C075N	SCREW, CAP, HEXAGON H .... UOC:DAL, V18, ZAL	4
6	PAFZZ	3020007411121	19207	7411121	PULLEY, GROOVE .... UOC:DAL, V18, ZAL	1
7	PAFZZ	3110001981080	96906	MS17131-54	BEARING, ROLLER, NEED .... UOC:DAL, V1B, ZAL	1
8	PAFZZ	3i20007411125	19207	7411125	BUSHING, SLEEVE .... UOC:DAL, V18, ZAL	1
9	PAFZZ	5315002981498	96906	MS24665-362	PIN, COTTER .... UOC:DAL, V18, ZAL	1
10	PAFZZ	5310004199476	19207	8344241	NUT, PLAIN, SLOTTED, H .... UOC:DAL, V18, ZAL	1
11	PAFZZ	5310004213994	19207	8690892	WASHER, FLAT .... UOC:DAL, V18, ZAL	1
12	PAFZZ	3120007411156	19207	7411156	BEARING, SLEEVE .... UOC:DAL, V18, ZAL	1
13	PAFZZ	5330007411160	19207	7411160	FELT, MECHANICAL, PRE .... UOC:DAL, V18, ZAL	1
14	PAFZZ	3040011949884	19207	7994976-1	SHAFT, SHOULDERED .... UOC:DAL, V18, ZAL	1
15	PAFZZ	3110007411123	19207	7411123	RETAINER, BALL, BEARI .... UOC:DAL, V18, ZAL	1
16	PAFZZ	3110001006158	96906	MS19059-2422	BALL, BEARING .... UOC:DAL, V18, ZAL	26
17	PAFZZ	3110007411124	19207	7411124	SEAT, BEARING .... UOC:DAL, V18, ZAL	1
18	PAFZZ	3120001596992	19207	7411152	BEARING, WASHER, THRU .... UOC:DAL, V18, ZAL	4
19	PAFZZ	5330007409929	19207	7409929	PACKING WITH RETAIN .... UOC:DAL, V18, ZAL	8
20	PAFZZ	3110002273245	21450	707728	BEARING, ROLLER, NEED .... UOC:DAL, V18, ZAL	4
21	PAFZZ	3120010829008	19207	8344243	ROLLER, LINEAR-ROTAR .... UOC:DAL, V18, ZAL	4
22	PAFZZ	5310007411153	19207	7411153	WASHER , FLAT .... UOC:DAL, V18, ZAL	4
23	PAFZZ	5310008206653	80045	23MS35338-50	WASHER, LOCK .... UOC:DAL, V18, ZAL	4

(1) ITEM NO	(2) SMR CODE	(3) NSN	(4) CAGEC	(5) PART NUMBER	(6) DESCRIPTION AND USABLE ON CODES (UOC)	(7) QTY
24	PAFZZ	5305007262525	96906	MS90727-158	SCREW, CAP, HEXAGON H .... UOC:DAL, V18, ZAL	2
25	PAFZZ	2530010853592	19207	7994954	AXLE, VEHICULAR, NOND .... UOC:DAL, V18, ZAL	2
26	PAFZZ	3040010823599	19207	8344245	BRACKET, EYE, NONROTA .... UOC:DAL, V18, ZAL	1
27	PAFZZ	3110001570531	96906	MS17131-29	BEARING, ROLLER, NEED .... UOC:DAL, V18, ZAL	1
28	PAFZZ	5330007411159	19207	7411159	FELT, MECHANICAL, PRE .... UOC:DAL, V18, ZAL	1
29	PAFZZ	5310008098536	96906	MS27183-24	WASHER, FLAT .... UOC:DAL, V18, ZAL	1
30	PAFZZ	5310009234219	96906	MS21083C12	NUT, SELF-LOCKING, HE .... UOC:DAL, VI1, ZAL	2
31	PAOZZ	4730001720034	96906	MS15003-6	FITTING, LUBRICATION .... UOC:DAL, V18, ZAL	1
32	PAFZZ	5306007411155	19207	7411155	BOLT, INTERNALLY REL .... UOC:DAL, V18, ZAL	2
33	PAOZZ	4730000504208	96906	MS15003-1	FITTING, LUBRICATION .... UOC:DAL, V18, ZAL	3
34	PAFZZ	5340011016713	19207	8344246	CLEVIS, ROD END .... UOC:DAL, V18, ZAL	1
35	PAFZZ	4730001864967	19207	8344244	BOLT, FLUID PASSAGE .... UOC:DAL, V18, ZAL	1

END OF FIGURE

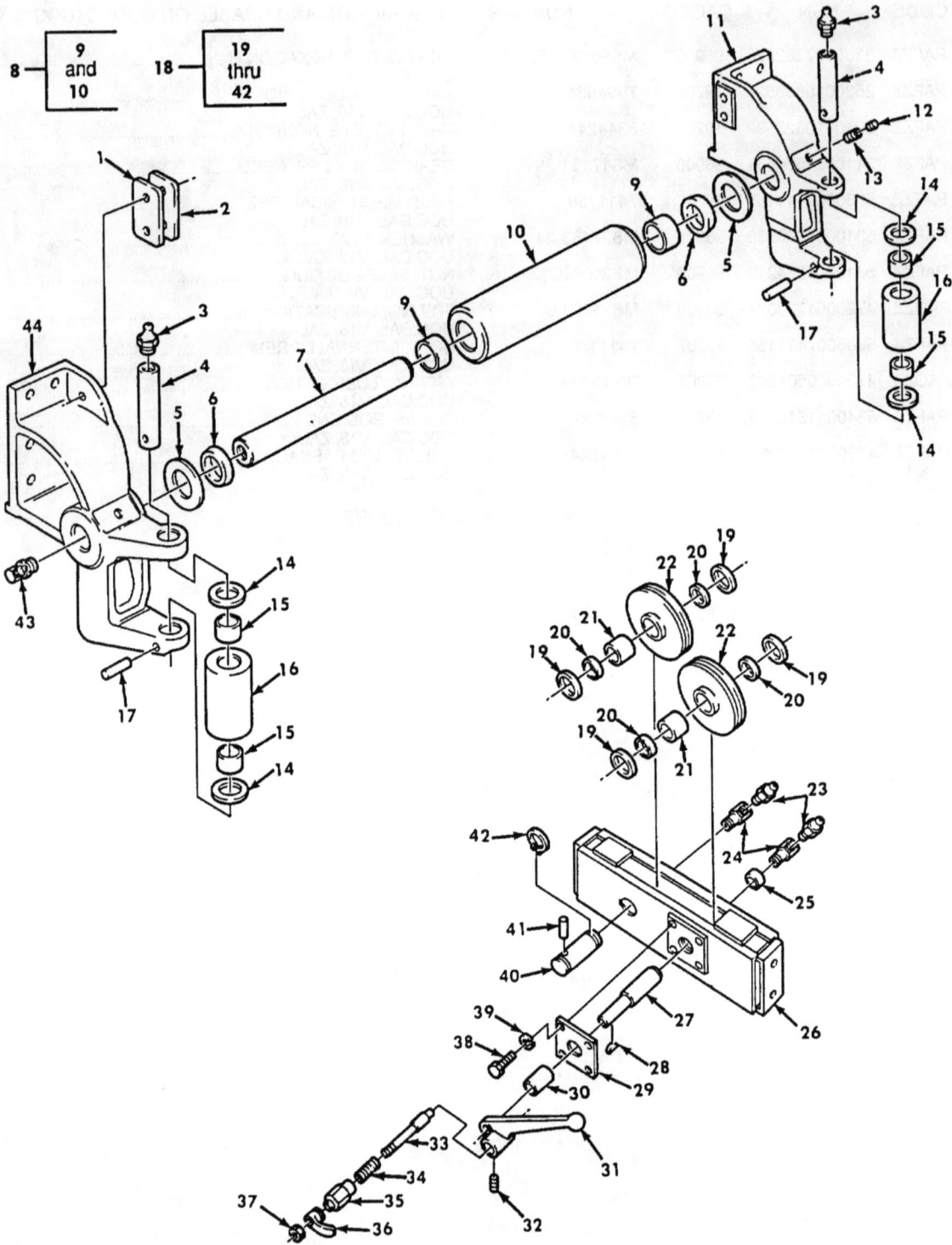

Figure 472. Front Winch Tensioner Assembly, Roller Assembly, and Related Parts.

(1) ITEM NO	(2) SMR CODE	(3) NSN	(4) CAGEC	(5) PART NUMBER	(6) DESCRIPTION AND USABLE ON CODES (UOC)	(7) QTY
					GROUP 2001 HOIST, CAPSTAN, WINDLASS, CRANE OR WINCH ASSEMBLY	
					FIG. 472 FRONT WINCH TENSIONER ASSEMBLY, ROLLER ASSEMBLY, AND RELATED PARTS	
1	PAOZZ	5365011064286	19207	7411162	SPACER, PLATE .... UOC:DAL, V18, ZAL	1
2	PAOZZ	5365011064282	19207	8344247	SHIM .... UOC:DAL, V18, ZAL	1
3	PAOZZ	4730000504208	96906	MS15003-1	FITTING, LUBRICATION .... UOC:DAB, DAD, DAL, DAX, V12, V14, V16, V18, V39, ZAB, ZAD, ZAL	2
4	PAFZZ	5315007409609	19207	7409609	PIN, ROLLER WINCH .... UOC:DAB, DAD, DAL, DAX, V12, V14, V16, V18, V39, ZAB, ZAD, ZAL	2
5	PAFZZ	3120007409729	19207	7409729	BEARING, WASHER, THRU .... UOC:DAB, DAD, DAL, DAX, V12, V14, V16, V18, V39, ZAB, ZAD, ZAL	2
6	PAFZZ	5330007417094	19207	7417094	PACKING ASSEMBLY .... UOC:DAB, DAD, DAL, DAX, V12, V14, V16, V18, V39, ZAB, ZAD, ZAL	2
7	PAFZZ	3040007409725	19207	7409725	SHAFT, SHOULDERED .... UOC:DAB, DAD, DAL, DAX, V12, V14, V16, V18, V39, ZAB, ZAD, ZAL	1
8	PAFFF	2520007409694	19207	7409694	ROLLER ASSEMBLY, WIN .... UOC:DAB, DAD, DAL, DAX, V12, V14, V16, V18, V39, ZAB, ZAD, ZAL	1
9	PAFZZ	3110002273249	96906	MS51961-22	.BEARING, ROLLER, NEED .... UOC:DAB, DAD, DAL, DAX, V12, V14, V16, V18, V39, ZAB, ZAD, ZAL	2
10	PAFZZ	3990010835407	19207	7954470	.ROLLER, MATERIAL HAN .... UOC:DAB, DAD, DAL, DAX, V12, V14, V16, V18, V39, ZAB, ZAD, ZAL	1
11	PFFZZ	5340012782190	19207	7409804-1	BRACKET, MOUNTING .... UOC:DAB, DAD, DAL, DAX, V12, V14, V16, V18, V39, ZAB, ZAD, ZAL	1
12	PAFZZ	5305007250168	96906	MS51963-137	SETSCREW .... UOC:DAB, DAD, DAL, DAX, V12, V14, V16, V18, V39, ZAB, ZAD, ZAL	2
13	PAFZZ	5305009390576	96906	MS51977-84	SETSCREW .... UOC:DAB, DAD, DAL, DAX, V12, V14, V16, V18, V39, ZAB, ZAD, ZAL	2
14	PAFZZ	5310007409728	19207	7409728	WASHER, FLAT .... UOC:DAB, DAD, DAL, DAX, V12, V14, V16, V18, V39, ZAB, ZAD, ZAL	4
15	PAFZZ	3120007409807	19207	7409807	BEARING, SLEEVE .... UOC:DAB, DAD, DAL, DAX, V12, V14, V16, V18, V39, ZAB, ZAD, ZAL	4
16	PFFZZ	3120007409695	19207	7409695	BEARING, SLEEVE .... UOC:DAB, DAD, DAL, DAX, V12, V14, V16, V18,	2

(1) ITEM NO	(2) SMR CODE	(3) NSN	(4) CAGEC	(5) PART NUMBER	(6) DESCRIPTION AND USABLE ON CODES (UOC)	(7) QTY
					V39, ZAB, ZAD, ZAL	
17	PAFZZ	5315000142972	96906	MS35671-42	PIN, GROOVED, HEADLES	2
					UOC:DAB, DAD, DAL, DAX, V12, V14, V16, V18, V39, ZAB, ZAD, ZAL	
18	PAFFF	2530009984711	19207	8690884	TENSION DEVICE, TRAC	1
					UOC:DAL, V18, ZAL	
19	PAFZZ	3120007411170	19207	7411170	.BEARING, WASHER, THRU	4
					UOC:DAL, V18, ZAL	
20	PAFZZ	5330007417093	19207	7417093	.PACKING ASSEMBLY	4
					UOC:DAL, V18, ZAL	
21	PAFZZ	3110002273241	96906	MS51961-9	.BEARING, ROLLER, NEED	2
					UOC:DAL, V18, ZAL	
22	PAFZZ	3020004845831	19207	8344250	.PULLEY, GROOVE	2
					UOC:DAL, V18, ZAL	
23	PAOZZ	4730001720022	96906	MS15003-2	.FITTING, LUBRICATION	2
					UOC:DAL, V18, ZAL	
24	PAFZZ	4730002784814	11083	2D1683	.ELBOW, PIPE	2
					UOC:DAL, V18, ZAL	
25	PAFZZ	3120007411164	19207	7411164	.BEARING, SLEEVE	1
					UOC:DAL, V18, ZAL	
26	PAFZZ	3040001793540	19207	8344248	.HOUSING, MECHANICAL	1
					UOC:DAL, V18, ZAL	
27	PAFZZ	3040004172756	19207	7994965	.CAM, CONTROL	1
					UOC:DAL, V18, ZAL	
28	PAFZZ	5315006165519	96906	MS35756-1	.KEY, WOODRUFF	1
					UOC:DAL, V18, ZAL	
29	PAFZZ	3130010853595	19207	8344506	.HOUSING, BEARING UNI	1
					UOC:DAL, V18, ZAL	
30	PAFZZ	3120006619026	19207	7411163	.BEARING , SLEEVE	1
					UOC:DAL, V18, ZAL	
31	PAFZZ	2590002343262	19207	8344251	.LEVER, MANUAL CONTRO	1
					UOC:DAL, V18, ZAL	
32	PAFZZ	5305007245898	96906	MS51963-83	.SETSCREW	1
					UOC:DAL, V18, ZAL	
33	PAFZZ	2520007411166	19207	7411166	.POPPET WINCH TROLLE	1
					UOC:DAL, V18, ZAL	
34	PAFZZ	5360007538706	19207	7538706	.SPRING, HELICAL, COMP	1
					UOC:DAL, V18, ZAL	
35	PAFZZ	2590002262347	19207	8344240	.NUT, WINCH POPPET	1
					UOC:DAL, V18, ZAL	
36	PAFZZ	2540007409150	19207	7409150	.LATCH, DRUM LOCK	1
					UOC:DAL, V18, ZAL	
37	PAFZZ	5310008911709	96906	MS35691-9	.NUT, PLAIN, HEXAGON	1
					UOC:DAL, V18, ZAL	
38	PAFZZ	5305002693234	80204	B1821BH038F075N	.SCREW, CAP, HEXAGON H	4
					UOC:DAA, DAL, V18, ZAL	
39	PAFZZ	5310006379541	96906	MS35338-46	.WASHER, LOCK	4
					UOC:DAL, V18, ZAL	
40	PAFZZ	3040007411161	19207	7411161	.SHAFT, STRAIGHT	1
					UOC:DAL, V18, ZAL	
41	PAFZZ	5315000544028	96906	MS51838-147	.PIN, STRAIGHT, HEADLE	1
					UOC:DAL, V18, ZAL	

(1) ITEM NO	(2) SMR CODE	(3) NSN	(4) CAGEC	(5) PART NUMBER	(6) DESCRIPTION AND USABLE ON CODES (UOC)	(7) QTY
42	PAFZZ	5325005307968	96906	MS16624-1100	.RING, RETAINING .... UOC:DAL, V18, ZAL	1
43	PAOZZ	4730001720034	96906	MS15003-6	FITTING, LUBRICATION .... UOC:DAB, DAD, DAL, DAX, V12, V14, V16, V18, V39, ZAB, ZAD, ZAL	2
44	PFFZZ	2520007409805	61465	305478	BRACKET ROLLER .... UOC:DAB, DAD, DAL, DAX, V12, V14, V16, V18, V39, ZAB, ZAD, ZAL	1

END OF FIGURE

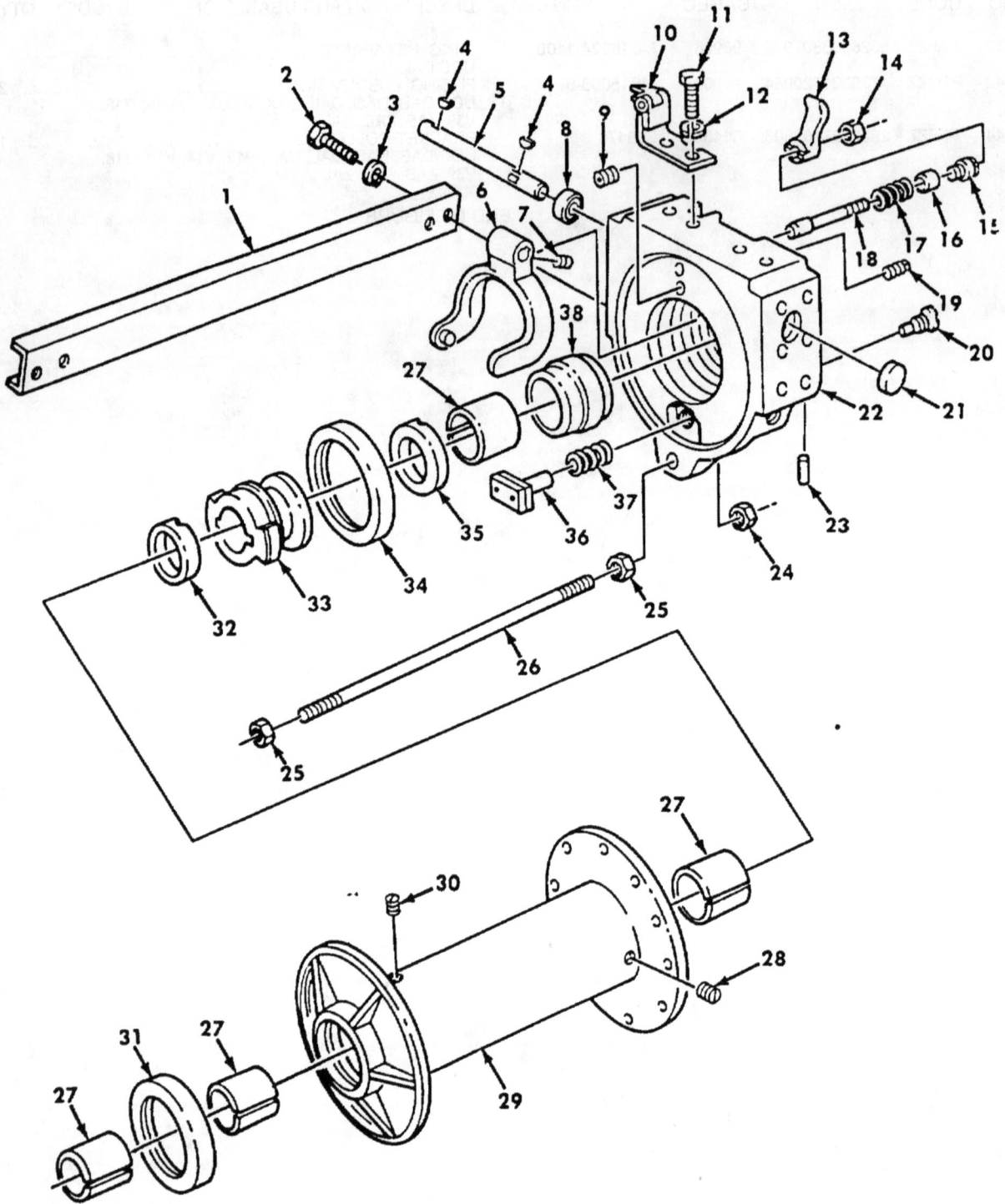

Figure 473. Front Winch Basic, Drum and Related Parts.

(1) ITEM NO	(2) SMR CODE	(3) NSN	(4) CAGEC	(5) PART NUMBER	(6) DESCRIPTION AND USABLE ON CODES (UOC)	(7) QTY
					GROUP 2001 HOIST, CAPSTAN, WINDLASS, CRANE OR WINCH ASSEMBLY	
					FIG. 473 FRONT WINCH BASIC, DRUM AND PELATED PARTS	
1	PFFZZ	9520007409812	19207	7409812	CHANNEL, STRUCTURAL .... UOC:DAB, DAD, DAF, DAH, DAL, DAX, V12, V14, V16, V18, V19, V21, V39, ZAB, ZAD, ZAF, ZAH, ZAL	1
2	PAFZZ	5305009001118	80204	B1821BH075C150 N	SCREW, CAP, HEXAGON H .... UOC:DAB, DAD, DAF, DAH, DAL, DAX, V12, V14, V16, V18, V19, V21, V39, ZAB, ZAD, ZAF, ZAH, ZAL	4
3	PAFZZ	5310005847888	96906	MS35338-51	WASHER, LOCK .... UOC:DAB, DAD, DAF, DAH, DAL, DAX, V12, V14, V16, V18, V19, V21, V39, ZAB, ZAD, ZAF, ZAH, ZAL	4
4	PAFZZ	5315006165521	96906	MS35756-13	KEY, WOODRUFF .... UOC:DAB, DAD, DAF, DAH, DAL, DAX, V12, V14, V16, V18, VI9, V21, V39, ZAB, ZAD, ZAF, ZAH, ZAL	2
5	PAFZZ	3040007457685	19207	7017190	SHAFT, STRAIGHT .... UOC:DAB, DAD, DAF, DAH, DAL, DAX, V12, V14, V16, V18, V19, V21, V39, ZAB, ZAD, ZAF, ZAH, ZAL	1
6	PAFZZ	2520008607340	19207	7017195	SHIFTER FORK, VEHICU .... UOC:DAB, DAD, DAF, DAH, DAL, DAX, V12, V14, V16, V18, V19, V21, V39, ZAB, ZAD, ZAF, ZAH, ZAL	1
7	PAFZZ	5305000589378	96906	MS51977-50	SETSCREW .... UOC:DAB, DAD, DAF, DAH, DAL, DAX, V12, V14, V16, V18, V19, V21, V39, ZAB, ZAD, ZAF, ZAH, ZAL	1
8	PAFZZ	5330008666236	19207	7418774	SEAL, PLAIN ENCASED .... UOC:DAB, DAD, DAF, DAH, DAL, DAX, V12, V14, V16, V18, V19, V21, V39, ZAB, ZAD, ZAF, ZAH, ZAL	1
9	PAFZZ	4730009541281	81348	WW-P-471ACABCB	PLUG, PIPE .... UOC:DAB, DAD, DAF, DAH, DAL, DAX, V12, V14, V16, V18 , V19, V21, V39, ZAB, ZAD, ZAF, ZAH, ZAL	1
10	PAFZZ	5340000402331	19207	8331202	HINGE, BUTT .... UOC:DAB, DAD, DAF, DAH, DAL, DAX, V12, V14, V16, V18, V19, V21, V39, ZAB, ZAD, ZAF, ZAH, ZAL	1
11	PAFZZ	5305007245910	96906	MS90725-162	SCREW, CAP, HEXAGON H .... UOC:DAB, DAD, DAF, DAH, DAL, DAX, V12, V14, V16, V1 B, V19, V21, V39, ZAB, ZAD, ZAF, ZAH, ZAL	2
12	PAFZZ	5310008206653	80045	23MS35338-50	WASHER, LOCK .... UOC:DAB, DAD, DAF, DAH, DAL, DAX, V12, V14,	2

(1) ITEM NO	(2) SMR CODE	(3) NSN	(4) CAGEC	(5) PART NUMBER	(6) DESCRIPTION AND USABLE ON CODES (UOC)	(7) QTY
					V16, V18, V19, V21, V39, ZAB, ZAD, ZAF, ZAH, ZAL	
13	PAFZZ	2540007409150	19207	7409150	LATCH, DRUM LOCK .... UOC:DAB, DAD, DAF, DAH, DAL, DAX, V12, V14, V16, V18, VI9, V21, V39, ZAB, ZAD, ZAF, ZAH, ... ZAL	1
14	PAFZZ	5310008807744	96906	MS51967-5	NUT, PLAIN, HEXAGON .... UOC:DAB, DAD, DAF, DAH, DAL, DAX, V12, V14, V16, V18, V19, V21, V39, ZAB, ZAD, ZAF, ZAH, ZAL	1
15	PAFZZ	5306007538725	19207	7538725	BOLT, INTERNALLY REL .... UOC:DAB, DAD, DAF, DAH, DAL, DAX, V12, V14, V16, V18, V19, V21, V39, ZAB, ZAD, ZAF, ZAH, ZAL	1
16	PAFZZ	5365004098987	19207	8343681	SPACER, SLEEVE .... UOC:DAB, DAD, DAF, DAH, DAL, DAX, V12, V14, V16, V18, V19, V21, V39, ZAB, ZAD, ZAF, ZAH, ZAL	1
17	PAFZZ	5360007538706	19207	7538706	SPRING, HELICAL, COMP .... UOC:DAB, DAD, DAF, DAH, DAL, DAX, V12, V16, V17, V18, V19, V21, V39, ZAB, ZAD, ZAF, ZAH, ZAL	1
18	PAFZZ	5315006960789	19207	8723828	PIN, SHOULDER, HEADLE .... UOC:DAB, DAD, DAF, DAH, DAL, DAX, V12, V16, V17, V18, V19, V21, V39, ZAB, ZAD, ZAF, ZAH, ZAL	1
19	PAFZZ	4730005558291	10001	12Z329PC93	PLUG, PIPE .... UOC:DAB, DAD, DAF, DAH, DAL, DAX, V12, V16, ... V17, V18, V19, V21, V39, ZAB, ZAD, ZAF, ZAH, ZAL	6
20	PAOZZ	5305008736946	19207	7538733	SETSCREW .... UOC:DAB, DAD, DAF, DAH, DAL, DAX, V12, V16, V17, V18, V19, V21, V39, ZAB, ZAD, ZAF, ZAH, ZAL	1
21	PAFZZ	5340007409684	19207	7409684	PLUG, EXPANSION .... UOC:DAB, DAD, DAF, DAH, DAL, DAX, V12, V16, V17, V18, V19, V21, V39, ZAB, ZAD, ZAF, ZAH, ZAL	1
22	PFFZZ	3040002228991	19207	7954482	HOUSING, MECHANICAL .... UOC:DAB, DAD, DAF, DAH, DAL, DAX, V12, V16, V17, V18, V19, V21, V39, ZAB, ZAD, ZAF, ZAH, ZAL	1
23	PAFZZ	5315000590029	21450	590029	PIN, STRAIGHT, HEADLE .... UOC:DAB, DAD, DAF, DAH, DAL, DAX, V12, V16, V17, V18, V19, V21, V39, ZAB, ZAD, ZAF, ZAH, ZAL	1
24	PAFZZ	5310000451081	21450	451081	NUT, SELF-LOCKING, HE .... UOC:DAB, DAD, DAF, DAH, DAL, DAX, V12, V16, V17, V18, V19, V21, V39, ZAB, ZAD, ZAF, ZAH, ZAL	1
25	PAFZZ	5310008381702	96906	MS35691-57	NUT, PLAIN, HEXAGON .... UOC:DAB, DAD, DAF, DAH, DAL, DAX, V12, V14,	2

(1) ITEM NO	(2) SMR CODE	(3) NSN	(4) CAGEC	(5) PART NUMBER	(6) DESCRIPTION AND USABLE ON CODES (UOC)	(7) QTY
					V16, V18, V19, V21, V39, ZAB, ZAD, ZAF, ZAH, ZAL	
26	PAFZZ	5306007409677	19207	7409677	ROD, THREADED END .... UOC:DAB, DAD, DAF, DAH, DAL, DAX, V12, V14, V16, V18, V19, V21, V39, ZAB, ZAD, ZAF, ZAH, ZAL	1
27	PAFZZ	3120007409800	19207	7409800	BEARING, SLEEVE .... UOC:DAB, DAD, DAF, DAH, DAL, DAX, V12, V14, V16, V18, V19, V21, V39, ZAB, ZAD, ZAF, ZAH, ZAL	4
28	PAFZZ	4730000575555	29930	444697	PLUG, PIPE .... UOC:DAB, DAD, DAF, DAH, DAL, DAX, V12, V14, V16, V18, V19, V21, V39, ZAB, ZAD, ZAF, ZAH, ZAL	1
29	PFFZZ	2590001185551	19207	7954477	REEL, CABLE .... UOC:DAB, DAD, DAF, DAH, DAL, DAX, V12, V14, V16, V18, V19, V21, V39, ZAB, ZAD, ZAF, ZAH, ZAL	1
30	PAFZZ	5305007250197	96906	MS51963-172	SETSCREW .... UOC:DAB, DAD, DAF, DAH, DAL, D AX, V12, V14, V16, V18, V19, V21, V39, ZAB, ZAD, ZAF, ZAH, ZAL	1
31	PAFZZ	5330011263469	19207	11640313	SEAL, PLAIN ENCASED .... UOC:DAB, DAD, DAF, DAH, DAL, DAX, V12, V14, V16, V18, V19, V21, V39, ZAB, ZAD, ZAF, ZAH, ZAL	1
32	PAFZZ	3120007409685	19207	7409685	BEARING, WASHER, THRU .... UOC:DAB, DAD, DAF, DAH, DAL, DAX, V12, V14, V16, V18, V19, V21, V39, ZAB, ZAD, ZAF, ZAH, ZAL	1
33	PAFZZ	2520007409814	19207	7409814	CLUTCH HALF, POSITIV .... UOC:DAB, D AD, DAF, DAH, DAL, DAX, V12, V14, V16, V18, V19, V21, V39, ZAB, ZAD, ZAF, ZAH, ZAL	1
34	PAFZZ	5330011195801	19207	10875107-7	SEAL, PLAIN ENCASED .... UOC:DAB, DAD, DAF, DAH, DAL, DAX, V12, V14, V16, V18, V19, V21, V39, ZAB, ZAD, ZAF, ZAH, ZAL	1
35	PAFZZ	3120007409591	19207	7409591	BEARING, WASHER, THRU .... UOC:DAB, DAD, DAF, DAH, DAL, DAX, V12, V14, V16, V18, V19, V21, V39, ZAB, ZAD, ZAF, ZAH, ZAL	1
36	PAFZZ	2530007538726	19207	7538726	BRAKE SHOE .... UOC:DAB, DAD, DAF, DAH, DAL, DAX, V12, V14, V16, V18, V19, V21, V39, ZAB, ZAD, ZAF, ZAH, ZAL	1
37	PAFZZ	5360010429532	19207	7538712	SPRING, HELICAL, COMP .... UOC:DAB, DAD, DAF, DAH, DAL, DAX, V12, V14, V16, V18, V19, V21, V39, ZAB, ZAD, ZAF, ZAH, ZAL	1
38	PAFZZ	3130009192915	19207	7954454	CAP, PILLOW BLOCK .... UOC:DAB, DAD, DAF, DAH, DAL, DAX, V12, V14,	1

(1) ITEM NO	(2) SMR CODE	(3) NSN	(4) CAGEC	(5) PART NUMBER	(6) DESCRIPTION AND USABLE ON CODES (UOC)	(7) QTY
					V16,V18,V19,V21,V39,ZAB,ZAD,ZAF,ZAH, ZAL	

END OF FIGURE

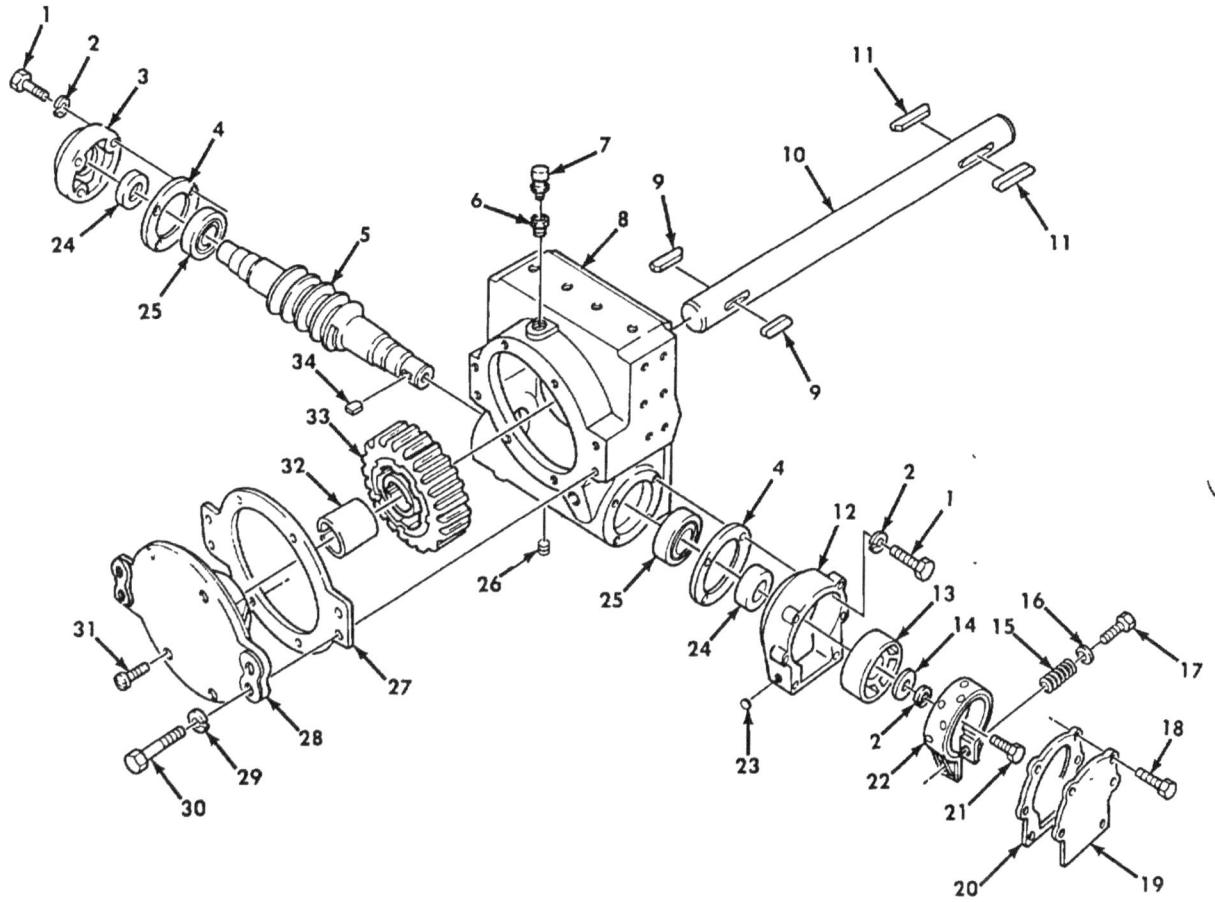

Figure 474. Front Winch Basic, Gear Case and Related Parts.

(1) ITEM NO	(2) SMR CODE	(3) NSN	(4) CAGEC	(5) PART NUMBER	(6) DESCRIPTION AND USABLE ON CODES (UOC)	(7) QTY
					GROUP 2001 HOIST, CAPSTAN, WINDLASS, CRANE OR WINCH ASSEMBLY	
					FIG. 474 FRONT WINCH BASIC, GEAR CASE AND RELATED PARTS	
1	PAFZZ	5305000712070	80204	B1821BH050C175N	SCREW, CAP, HEXAGON H .... UOC:DAB, DAD, DAF, DAH, DAL, DAX, V12, V14, V16, V1B, V19, V21, V39, ZAB, ZAD, ZAF, ZAH, ZAL	8
2	PAFZZ	5310000034094	01276	210104-8S	WASHER, LOCK .... UOC:DAB, DAD, DAF, DAH, DAL, DAX, V12, V14, V16, V18, V19, V21, V39, ZAB, ZAD, ZAF, ZAH, ZAL	9
3	PBFZZ	3110007409809	19207	7409809	PLATE, RETAINING, BEA .... UOC:DAB, DAD, DAF, DAH, DAL, DAX, V12, V14, V16, V18, V19, V21, V39, ZAB, ZAD, ZAF, ZAH, ZAL	1
4	PAFZZ	5330000573823	19207	7409822	GASKET .... UOC:DAB, DAD, DAF, DAH, DAL, DAX, V12, V14, V16, V18, V19, V21, V39, ZAB, ZAD, ZAF, ZAH, ZAL	2
5	PAFZZ	3040002310219	19207	7954478	WORM SHAFT .... UOC:DAB, DAD, DAF, DAH, DAL, DAX, V12, V14, V16, V18, V19, V21, V39, ZAB, ZAD, ZAF, ZAH, ZAL	1
6	PAFZZ	4730002026670	44185	F100753	BUSHING, PIPE .... UOC:DAB, DAD, DAF, DAH, DAL, DAX, V12, V14, V16, V18, V19, V21, V39, ZAB, ZAD, ZAF, ZAH, ZAL	1
7	PAFZZ	4820007264719	57733	5196397	VALVE, VENT .... UOC:DAB, DAD, DAF, DAH, DAL, DAX, V12, V14, V16, V18, V19, V21, V39, ZAB, ZAD, ZAF, ZAH, ZAL	1
8	PFFZZ	3040004051931	19207	7954484	HOUSING, MECHANICAL .... UOC:DAB, DAD, DAF, DAH, DAL, DAX, V12, V14, V16, V18, V19, V21, V39, ZAB, ZAD, ZAF, ZAH, ZAL	1
9	PAFZZ	5315002817652	19207	8327444	KEY, MACHINE .... UOC:DAB, DAD, DAF, DAH, DAL, DAX, V12, V14, V16, V18, V19, V21, V39, ZAB, ZAD, ZAF, ZAH, ZAL	2
10	PAFZZ	3040009210478	19207	7954473	SHAFT, STRAIGHT .... UOC:DAB, DAD, DAF, DAH, DAL, DAX, V12, V14, V16, V18, V19, V21, V39, ZAB, ZAD, ZAF, ZAH, ZAL	1
11	PAFZZ	5315002817650	19207	8327443	KEY, MACHINE .... UOC:DAB, DAD, DAF, DAH, DAL, DAX, V12, V14, V16, V18, V19, V21, V39, ZAB, ZAD, ZAF, ZAH, ZAL	2
12	PAFZZ	3040001185554	19207	7954475	HOUSING, MECHANICAL .... UOC:DAB, DAD, DAF, DAH, DAL, DAX, V12, V14,	1

(1) ITEM NO	(2) SMR CODE	(3) NSN	(4) CAGEC	(5) PART NUMBER	(6) DESCRIPTION AND USABLE ON CODES (UOC)	(7) QTY
					V16, V18, V19, V21, V39, ZAB, ZAD, ZAF, ZAH, ZAL	
13	PAFZZ	2530007409817	19207	7409817	BRAKE DRUM .... UOC:DAB, DAD, DAF, DAH, DAL, DAX, V12, V14, V16, V18, VI9, V21, V39, ZAB, ZAD, ZAF, ZAH, ZAL	1
14	PAFZZ	5310005955839	19207	8330414	WASHER, FLAT .... UOC:DAB, DAD, DAF, DAH, DAL, DAX, V12, V14, V16, V18, V19, V21, V39, ZAB, ZAD, ZAF, ZAH, ZAL	1
15	PAFZZ	5360006644374	19207	5277992	SPRING, HELICAL, COMP .... UOC:DAB, DAD, DAF, DAH, DAL, DAX, V12, V14, V16, V18, V19, V21, V39, ZAB, ZAD, ZAF, ZAH, ZAL	1
16	PAFZZ	5310000806004	96906	MS27183-14	WASHER, FLAT .... UOC:DAB, DAD, DAF, DAH, DAL, DAX, V12, V14, V16, V18, V19, V21, V39, ZAB, ZAD, ZAF, ZAH, ZAL	1
17	PAFZZ	5305002693250	96906	MS90727-74	SCREW, CAP, HEXAGON H .... UOC:DAB, DAD, DAF, DAH, DAL, DAX, V12, V14, V16, V18, V19, V21, V39, ZAB, ZAD, ZAF, ZAH, ZAL	1
18	PAFZZ	5306000187527	21450	187527	BOLT, ASSEMBLED WASH .... UOC:DAB, DAD, D AF, DAH, DAL, DAX, V12, V14, V16, V18, V19, V21, V39, ZAB, ZAD, ZAF, ZAH, ZAL	6
19	PAFZZ	2520007409816	19207	7409816	COVER, BRAKE CASE .... UOC:DAB, DAD, DAF, DAH, DAL, DAX, V12, V14, V16, V18, V19, V21, V39, ZAB, ZAD, ZAF, ZAH, ZAL	1
20	PAFZZ	5330008953424	19207	7973339	GASKET .... UOC:DAB, DAD, DAF, DAH, DAL, DAX, V12, V14, V16, V18, V19, V21, V39, ZAB, ZAD, ZAF, ZAH, ZAL	1
21	PAFZZ	5305002267767	96906	MS90726-109	SCREW, CAP, HEXAGON H .... UOC:DAB, DAD, DAF, DAH, DAL, DAX, V12, V14, V16, V18, V19, V21, V39, ZAB, ZAD, ZAF, ZAH, ZAL	1
22	PFFZZ	3040007409565	19207	7409565	BRAKE BAND AND LINI .... UOC:DAB, DAD, DAF, DAH, DAL, DAX, V12, V14, V16, V18, V19, V21, V39, ZAB, ZAD, ZAF, ZAH, ZAL	1
23	PAFZZ	53400005015B9	96906	MS35648-3	PLUG, EXPANSION .... UOC:DAB, DAD, DAF, DAH, DAL, DAX, V12, V14, V16, V18, V19, V21, V39, ZAB, ZAD, ZAF, ZAH, ZAL	1
24	PAFZZ	5330011509691	80201	17657/55-542465	SEAL, PLAIN ENCASED .... UOC:DAB, DAD, DAF, DAH, DAL, DAX, V12, V14, V16, V18, V19, V21, V39, ZAB, ZAD, ZAF, ZAH, ZAL	2
25	PAFZZ	3110005543929	21450	700287	BEARING, BALL, ANNULA .... UOC:DAB, DAD, DAF, DAH, DAL, DAX, V12, V14,	2

(1) ITEM NO	(2) SMR CODE	(3) NSN	(4) CAGEC	(5) PART NUMBER	(6) DESCRIPTION AND USABLE ON CODES (UOC)	(7) QTY
					V16, V18, V19, V21, V39, ZAB, ZAD, ZAF, ZAH, ZAL	
26	PAFZZ	4730010304950	24617	272977	PLUG, PIPE .... UOC:DAB, DAD, DAF, DAH, DAL, DAX, V12, V14, V16, V18, V19, V21, V39, ZAB, ZAD, ZAF, ZAH, ZAL	1
27	PAFZZ	5330007409821	19207	7409821	GASKET .... UOC:DAB, DAD, DAF, DAH, DAL, DAX, V12, V14, V16, V18, V19, V21, V39, ZAB, ZAD, ZAF, ZAH, ZAL	1
28	PBFZZ	5340009234233	19207	7954476	COVER, ACCESS .... UOC:DAB, DAD, DAF, DAH, DAL, DAX, V12, V14, V16, V18, V19, V21, V39, ZAB, ZAD, ZAF, ZAH, ZAL	1
29	PAFZZ	5310008206653	80045	23MS35338-50	WASHER, LOCK .... UOC:DAB, DAD, DAF, DAH, DAL, DAX, V12, V14, V16, V18, V19, V21, V39, ZAB, ZAD, ZAF, ZAH, ZAL	8
30	PAFZZ	5306007409803	19207	7409803	BOLT, MACHINE .... UOC:DAB, DAD, DAF, DAH, DAL, DAX, V12, V14, V16, V18, V19, V21, V39, ZAB, ZAD, ZAF, ZAH, ZAL	8
31	PAFZZ	5305009585267	96906	MS35190-343	SCREW, MACHINE .... UOC:DAB, DAD, DAF, DAH, DAL, DAX, V12, V14, V16, V18, V19, V21, V39, ZAB, ZAD, ZAF, ZAH, ZAL	1
32	PFFZZ	3120007409800	19207	7409800	BEARING, SLEEVE .... UOC:DAB, DAD, DAF, DAH, DAL, DAX, V12, V14, V16, V18, V19, V21, V39, ZAB, ZAD, ZAF, ZAH, ZAL	1
33	PAFZZ	3020007409823	19207	7409823	GEAR, WORM WHEEL .... UOC:DAB, DAD, DAF, DAH, DAL, DAX, V12, V14, V16, V18, V19, V21, V39, ZAB, ZAD, ZAF, ZAH, ZAL	1
34	PAFZZ	5315000423293	96906	MS20067-270	KEY, MACHINE .... UOC:DAB, DAD, DAF, DAH, DAL, DAX, V12, V14, V16, V18, V19, V21, V39, ZAB, ZAD, ZAF, ZAH, ZAL	1

END OF FIGURE

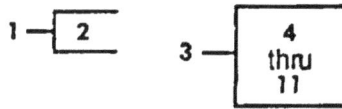

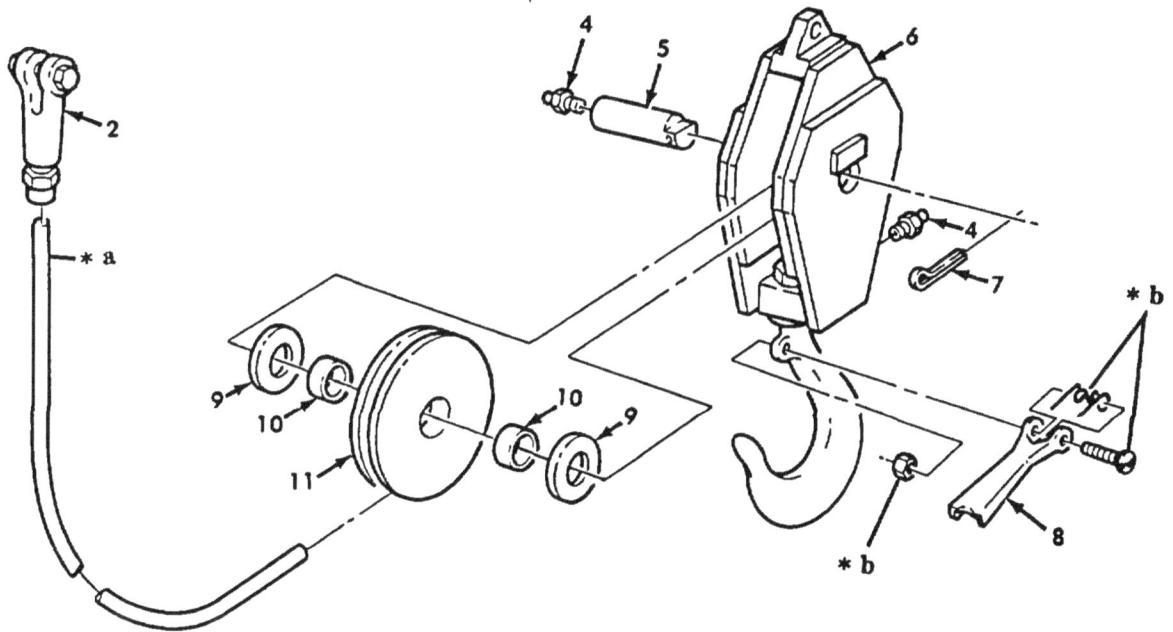

*a PART OF ITEM 1
*b PART OF ITEM 8

Figure 475. Wrecker Tackle Block and Wire Rope Assembly.

(1) ITEM NO	(2) SMR CODE	(3) NSN	(4) CAGEC	(5) PART NUMBER	(6) DESCRIPTION AND USABLE ON CODES (UOC)	(7) QTY
					GROUP 2001 HOIST, CAPSTAN, WINDLASS, CRANE OR WINCH ASSEMBLY	
					FIG. 475 WRECKER TACKLE BLOCK AND WIRE ROPE ASSEMBLY	
1	MOOZZ		19207	10900237	WIRE ROPE ASSEMBLY MAKE FROM WIRE ROPE, P/N 7699767 UOC:DAL, V1B, ZAL	1
2	PAOZZ	4030002623152	19207	7071882	.SOCKET, WIRE ROPE UOC:DAL, V18, ZAL	1
3	PAOOO	3940004146533	19207	10876250	BLOCK, TACKLE UOC:DAL, V18, ZAL	1
4	PAOZZ	4730000504208	96906	MS15003-1	.FITTING, LUBRICATION UOC:DAL, V18, ZAL	2
5	PAOZZ	3040004146535	19207	10900083	.SHAFT, STRAIGHT UOC:DAL, V18, ZAL	1
6	PAOZZ	2540004180603	19207	10876249	.PINTLE ASSEMBLY, TOW UOC:DAL, V18, ZAL	1
7	PAOZZ	5315008460126	96906	MS24665-628	.PIN, COTTER UOC:DAL, V18, ZAL	1
8	PAOZZ	2540012010968	19691	S-4055C-15T	.LATCH, PINTLE HOOK UOC:DAL, V18, ZAL	1
9	PAOZZ	3120007409729	19207	7409729	.BEARING, WASHER, THRU UOC:DAL, V18, ZAL	2
10	PAOZZ	3110009023757	27737	S2414	.BEARING, ROLLER, NEED UOC:DAL, V18, ZAL	2
11	PAOZZ	3020004195854	19207	10876423	.PULLEY, GROOVE UOC:DAL, V18, ZAL	1

END OF FIGURE

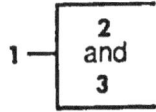

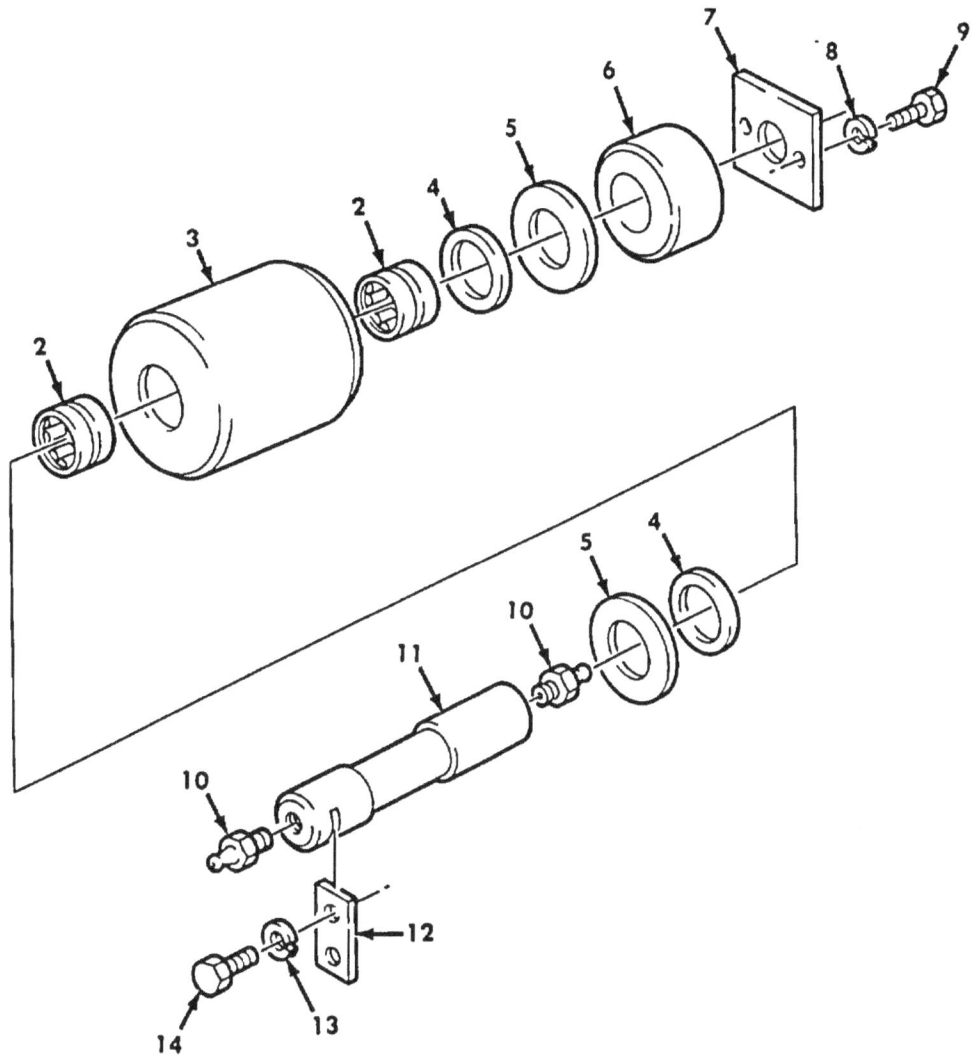

Figure 476. Wrecker Rear Winch Vertical Roller and Mounting Hardware.

(1) ITEM NO	(2) SMR CODE	(3) NSN	(4) CAGEC	(5) PART NUMBER	(6) DESCRIPTION AND USABLE ON CODES (UOC)	(7) QTY
					GROUP 2001 HOIST, CAPSTAN, WINDLASS, CRANE OR WINCH ASSEMBLY	
					FIG. 476 WRECKER REAR WINCH VERTICAL ROLLER AND MOUNTING HARDWARE	
1	PAOOO	2520007409939	61465	M305849	ROLLER ASSEMBLY .... UOC:DAL, V18, ZAL	2
2	PAFZZ	3110002273255	96906	MS51961-32	.BEARING, ROLLER, NEED .... UOC:DAL, V18, ZAL	2
3	PFFZZ	2590004172725	19207	10876554	.ROLLER, WINCH .... UOC:DAL, V18, ZAL	1
4	PAFZZ	5310007409928	19207	7409928	WASHER, FLAT .... UOC:DAL, V18, ZAL	4
5	PAFZZ	3120007409978	19207	7409978	BEARING, WASHER, THRU .... UOC:DAL, V18, ZAL	4
6	PAFZZ	5365011036006	19207	10900055	SPACER, SLEEVE .... UOC:DAL, V18	2
7	PFFZZ	5340011346535	19207	10900282	COVER, ACCESS .... UOC:DAL, V18	2
8	PAFZZ	5310008206653	80045	23MS35338-50	WASHER, LOCK .... UOC:DAL, V18, ZAL	5
9	PAFZZ	5305007262544	80204	B1821BH063F138N	SCREW, CAP, HEXAGON H .... UOC:DAL, V18	4
10	PAFZZ	4730000504208	96906	MS15003-1	FITTING, LUBRICATION .... UOC:DAL, V18, ZAL	4
11	PAFZZ	3040007409947	19207	7409947	SHAFT, SHOULDERED .... UOC:DAL, V18, ZAL	2
12	PFFZZ	2590004172787	19207	10900278	PLATE, RETAINER, ROLL .... UOC:DAL, V18, ZAL	2
13	PAFZZ	5310006379541	96906	MS35338-46	WASHER, LOCK .... UOC:DAL, V18, ZAL	4
14	PAFZZ	5305002693234	96906	MS90727-58	SCREW, CAP, HEXAGON H .... UOC:DAL, V18, ZAL	4

END OF FIGURE

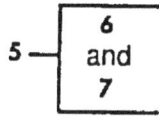

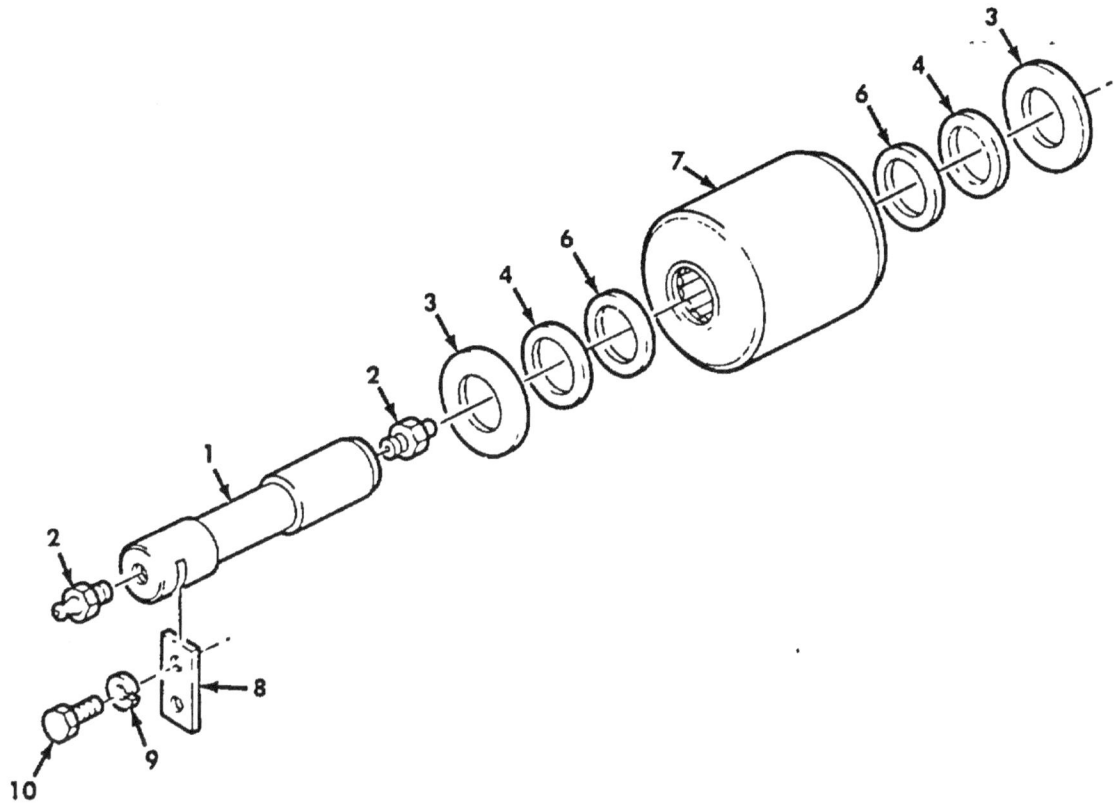

Figure 477. Wrecker Rear Winch Horizontal Roller and Mounting Hardware.

(1) ITEM NO	(2) SMR CODE	(3) NSN	(4) CAGEC	(5) PART NUMBER	(6) DESCRIPTION AND USABLE ON CODES (UOC)	(7) QTY
					GROUP 2001 HOIST, CAPSTAN, WINDLASS, CRANE OR WINCH ASSEMBLY	
					FIG. 477 WRECKER REAR WINCH HORIZONTAL ROLLER AND MOUNTING HARDWARE	
1	PAFZZ	3040007409947	19207	7409947	SHAFT, SHOULDERED .... UOC:DAL, V18, ZAL	2
2	PAFZZ	4730000504208	96906	MS15003-1	FITTING, LUBRICATION .... UOC:DAL, V18, ZAL	4
3	PAFZZ	3120007409978	19207	7409978	BEARING, WASHER, THRU .... UOC:DAL, V18, ZAL	4
4	PAFZZ	5310007409928	19207	7409928	WASHER, FLAT .... UOC:DAL, V18, ZAL	4
5	PAOOO	2520007409939	61465	M305849	ROLLER ASSEMBLY .... UOC:DAL, V18, ZAL	2
6	PAFZZ	3110002273255	96906	MS51961-32	.BEARING, ROLLER, NEED .... UOC:DAL, V18, ZAL	2
7	PFFZZ	2590004172725	19207	10876554	.ROLLER, WINCH .... UOC:DAL, V18, ZAL	1
8	PFFZZ	2590004172787	19207	10900278	PLATE, RETAINER, ROLL .... UOC:DAL, V18, ZAL	2
9	PAFZZ	5310006379541	96906	MS35338-46	WASHER, LOCK .... UOC:DAL, V18, ZAL	4
10	PAFZZ	5305002693234	96906	MS90727-58	SCREW, CAP, HEXAGON H .... UOC:DAL, V18, ZAL	4

END OF FIGURE

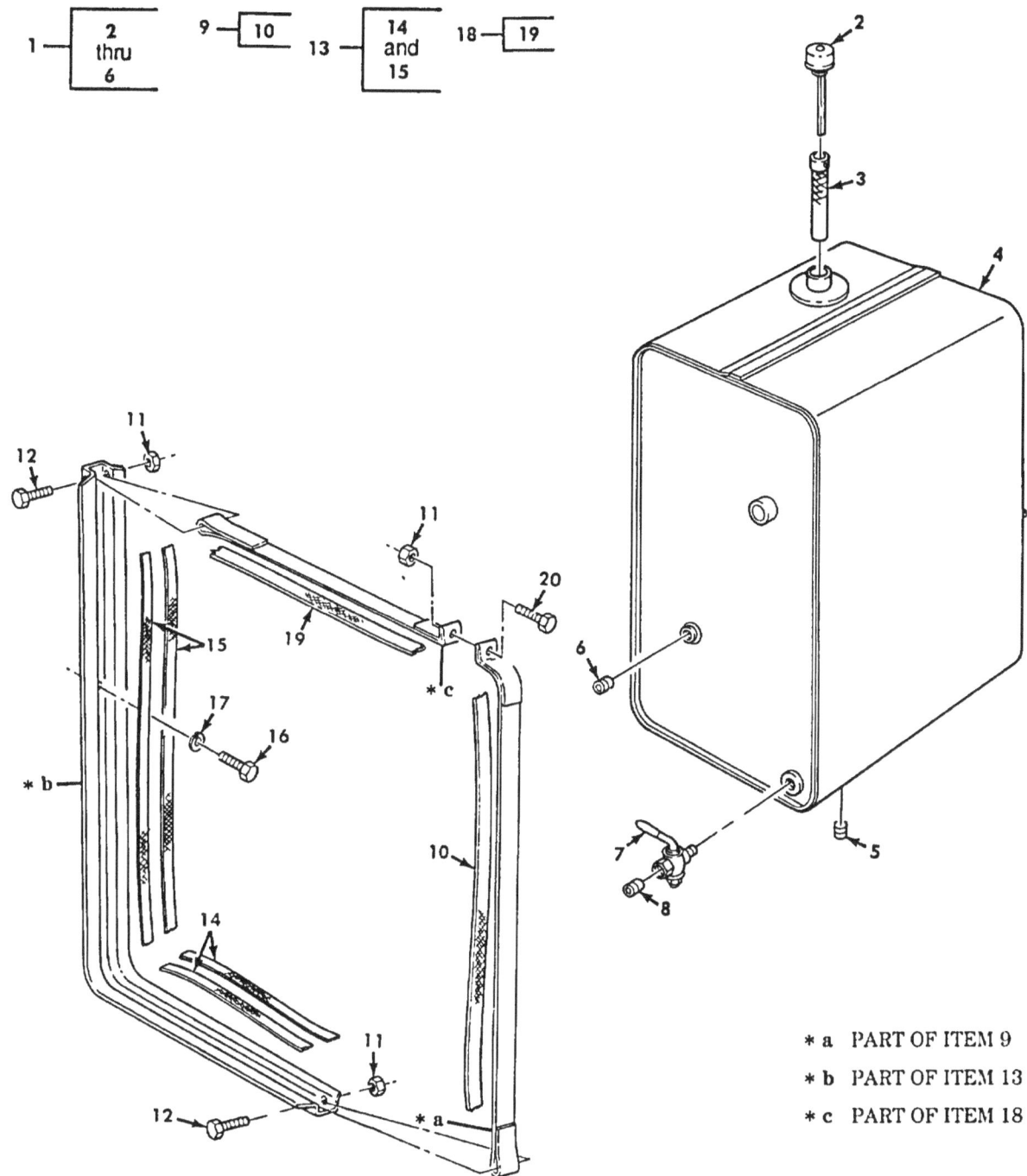

Figure 478. Wrecker Hydraulic Oil Tank Assembly and Mounting Hardware.

* a   PART OF ITEM 9
* b   PART OF ITEM 13
* c   PART OF ITEM 18

(1) ITEM NO	(2) SMR CODE	(3) NSN	(4) CAGEC	(5) PART NUMBER	(6) DESCRIPTION AND USABLE ON CODES (UOC)	(7) QTY
					GROUP 2001 HOIST, CAPSTAN, WINDLASS, CRANE OR WINCH ASSEMBLY	
					FIG. 478 WRECKER HYDRAULIC OIL TANK ASSEMBLY AND MOUNTING HARDWARE	
1	PAFFF	2590011297523	19207	12256603	TANK, OIL, HYDRAULIC .... UOC:DAL, V18, ZAL	1
2	PAFZZ	6680011472421	64829	9303557	.GAGE ROD-BREATHER, L .... UOC:DAL, V18, ZAL	1
3	PAFZZ	4730002317450	19207	10900106	.STRAINER ELEMENT, SE .... UOC:DAL, V18, ZAL	1
4	XAFZZ		19207	12256604	.BODY, TANK .... UOC:DAL, V18, ZAL	1
5	PAFZZ	4730003593872	81348	WW-P-471ACABCE	.PLUG, PIPE .... UOC:DAL, V18, ZAL	1
6	PAFZZ	4730000575555	29930	444697	.PLUG, PIPE .... UOC:DAL, V18, ZAL	1
7	PAFZZ	4820004614216	19207	10900185	COCK, SHUTOFF, SCREW .... UOC:DAL, V18, ZAL	1
8	PAFZZ	4730011072027	21450	444673	PLUG, PIPE .... UOC:DAL, V18, ZAL	1
9	PAFFF	2590011147551	19207	12256563	STRAP ASSEMBLY, OIL UOC:V18	2
9	PAFFF	5340012100201	19207	12256563-2	STRAP, RETAINING .... UOC:DAL, ZAL	2
10	MFFZZ		81348	27-W-921-40	.STRAP, WEBBING MAKE FROM WEBBING, P/N 27-W-921, 40 INCHES LONG. UOC:DAL, V18, ZAL	1
11	PAFZZ	5310008140672	96906	MS51943-36	NUT, SELF-LOCKING, HE .... UOC:DAL, V18, ZAL	6
12	PAFZZ	5305002693246	96906	MS90727-70	SCREW, CAP, HEXAGON H .... UOC:DAL, V18, ZAL	4
13	PAFFF	5340011376302	19207	12256592	BRACKET, MOUNTING UOC:V18	2
13	PAFFF	5340012100202	19207	12256592-2	BRACKET, MOUNTING .... UOC:DAL, ZAL	2
14	MFFZZ		81348	27-W-918-20	.STRAP, WEBBING MAKE FROM WEBBING, P/N 27-W-918, 20 INCHES LONG. UOC:DAL, V18, ZAL	2
15	MFFZZ		81348	27-W-918-31	.STRAP, WEBBING MAKE FROM WEBBING, P/N 27-W-918, 31 INCHES LONG. UOC:DAL, V18, ZAL	2
16	PAFZZ	5305009900695	80205	B1821BH050FO88N	SCREW, CAP, HEXAGON H .... UOC:DAL, V18, ZAL	1
17	PAFZZ	5310000034094	01276	210104-8S	WASHER, LOCK .... UOC:DAL, V18, ZAL	8
18	PAFFF	5340011147550	19207	12256562	BAND, RETAINING .... UOC:DAL, V18, ZAL	2
19	MFFZZ		81348	27-W-921-19	.STRAP, WEBBING MAKE FROM WEBBING, P/N 27-W-921, 19 INCHES LONG. UOC:DAL, V18, ZAL	1
20	PAFZZ	5305002693242	80204	B1821BH038F200N	SCREW, CAP, HEXAGON H .... UOC:DAL, V18, ZAL	2

END OF FIGURE

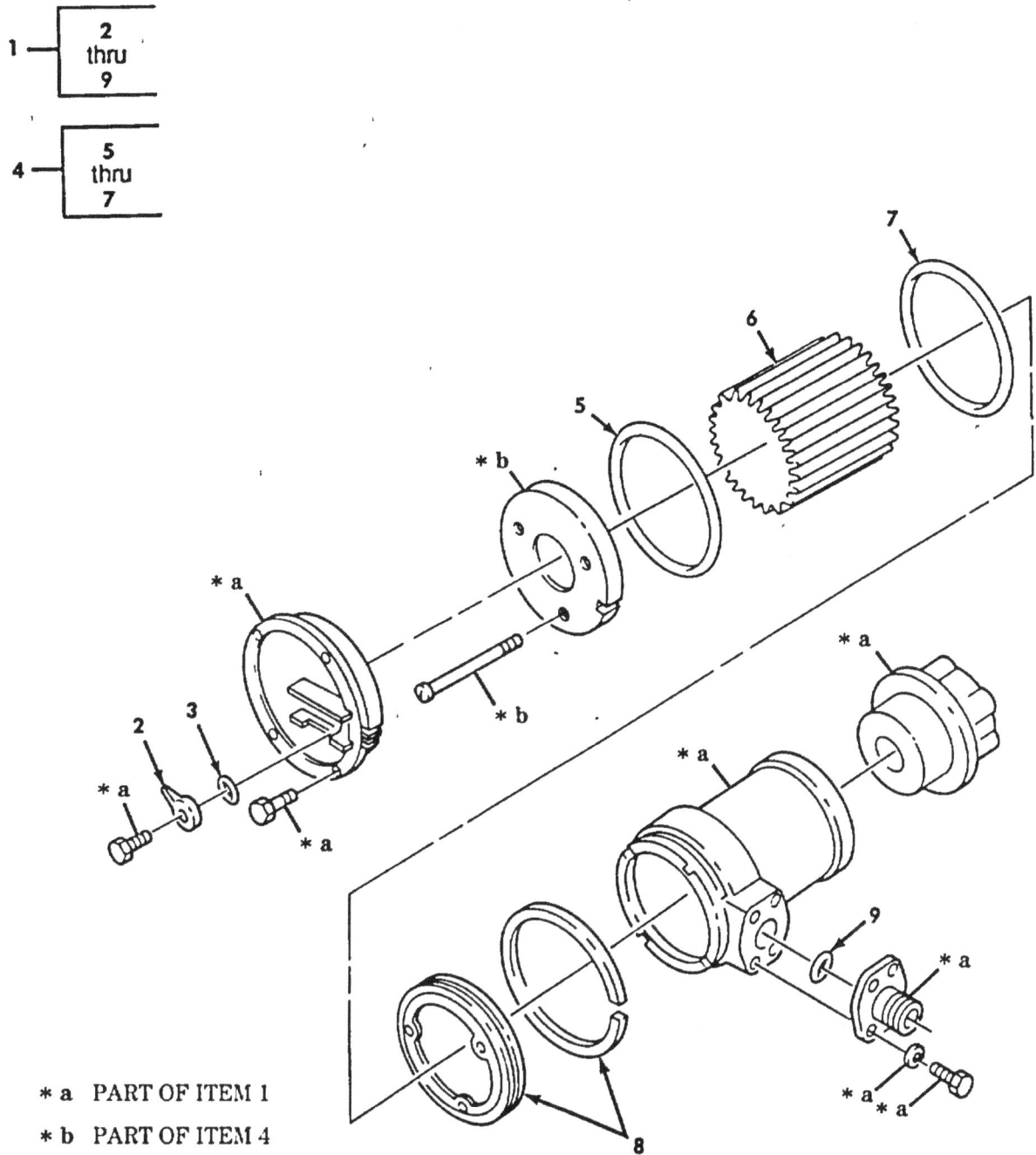

*a   PART OF ITEM 1
*b   PART OF ITEM 4

Figure 479. Wrecker Hydraulic Fluid Filter Assembly.

(1) ITEM NO	(2) SMR CODE	(3) NSN	(4) CAGEC	(5) PART NUMBER	(6) DESCRIPTION AND USABLE ON CODES (UOC)	(7) QTY
					GROUP 2001 HOIST, CAPSTAN, WINDLASS, CRANE OR WINCH ASSEMBLY	
					FIG. 479 WRECKER HYDRAULIC FLUID FILTER ASSEMBLY	
1	PAOOO	2940001356537	19207	10921633	FILTER, FLUID .... UOC:DAL, V18, ZAL	1
2	PAOZZ	2590011843938	05779	900719	.LEVER, OPERATING .... UOC:DAL, V18, ZAL	1
3	PAOZZ	5331008737214	03038	N72006	.O-RING .... UOC:DAL, V18, ZAL	1
4	PAOOO	4330011632733	05779	901422	.FILTER ELEMENT, FLUI .... UOC:DAL, V18, ZAL	1
5	PAOZZ	5330011663662	05779	N72260	..PACKING, PREFORMED .... UOC:DAL, V18, ZAL	1
6	PAOZZ	4330011851226	05779	927110	..FILTER ELEMENT, FLUI .... UOC:DAL, V18, ZAL	1
7	PAOZZ	5331001521759	05779	N72259	..O-RING .... UOC:DAL, V18, ZAL	
8	PAOZZ	2590011194103	05779	900877	.PISTON RING, FILTER .... UOC:DAL, V18, ZAL	1
9	PAOZZ	5331009824259	97907	981072	.O-RING .... UOC:DAL, V18, ZAL	1

END OF FIGURE

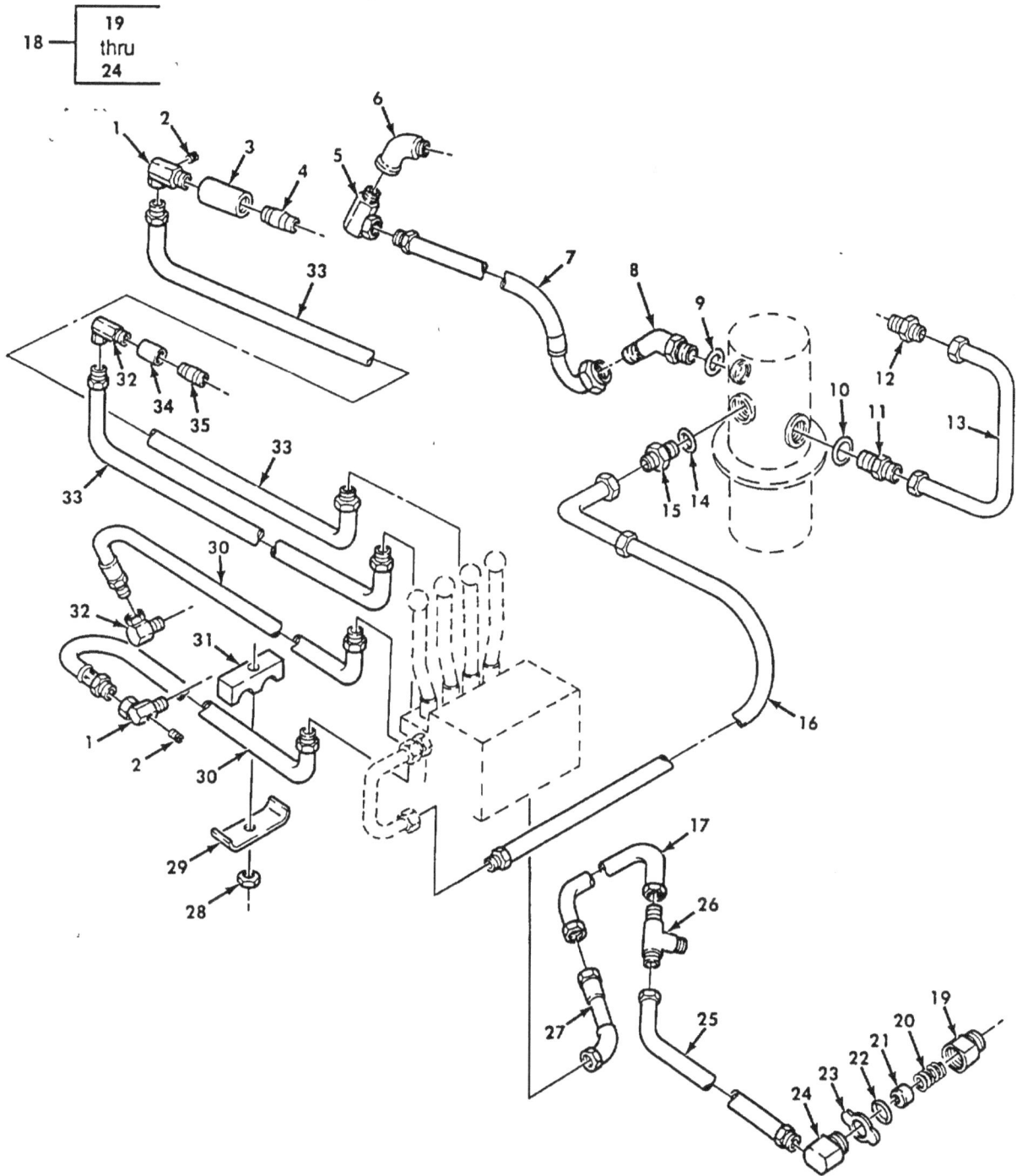

Figure 480. Wrecker Hydraulic Tubing, Swivel to Extension Cylinder and Control Valve.

(1) ITEM NO	(2) SMR CODE	(3) NSN	(4) CAGEC	(5) PART NUMBER	(6) DESCRIPTION AND USABLE ON CODES (UOC)	(7) QTY

GROUP 2001 HOIST, CAPSTAN, WINDLASS, CRANE OR WINCH ASSEMBLY

FIG. 480 WRECKER HYDRAULIC TUBING, SWIVEL TO EXTENSION CYLINDER AND CONTROL VALVE

(1) ITEM NO	(2) SMR CODE	(3) NSN	(4) CAGEC	(5) PART NUMBER	(6) DESCRIPTION AND USABLE ON CODES (UOC)	(7) QTY
1	PAFZZ	4730004396028	19207	10900108	ELBOW, PIPE .... UOC:DAL, V18, ZAL	2
2	PAFZZ	4730000189566	81348	WW-P-471ACABCA	PLUG, PIPE .... UOC:DAL, V18, ZAL	2
3	PAFZZ	2590004434847	19207	10900112	REGULATOR, HYDRAULIC .... UOC:DAL, V18, ZAL	1
4	PAFZZ	4730001961506	96906	MS51953-124	NIPPLE, PIPE .... UOC:DAL, V18, ZAL	1
5	PAFZZ	4730002891245	19207	7954727	ADAPTER, STRAIGHT, PI INLET HOSE .... UOC:DAL, V18, ZAL	1
6	PAFZZ	4730000179447	96906	MS39231-9	ELBOW, PIPE .... UOC:DAL, V18, ZAL	1
7	PAFZZ	4720011208516	19207	12256602-1	HOSE ASSEMBLY, NONME .... UOC:DAL, V18, ZAL	1
8	PAFZZ	4730011957339	96906	MS51528A32	ELBOW, TUBE TO BOSS .... UOC:DAL, V18, ZAL	1
9	PAFZZ	5331002859847	96906	MS28778-32	O-RING .... UOC:DAL, V18, ZAL	1
10	PAFZZ	5331008163546	96906	MS28778-20	O-RING .... UOC:DAL, V18, ZAL	1
11	PAFZZ	4730010661282	96906	MS51525A20	ADAPTER, STRAIGHT, TU .... UOC:DAL, V18, ZAL	2
12	PAFZZ	4730002028470	96906	MS51500A20	ADAPTER, STRAIGHT, PI .... UOC:DAL, V18, ZAL	1
13	PAFZZ	4710010892059	19207	12256568	TUBE ASSEMBLY, METAL .... UOC:DAL, V18, ZAL	1
14	PAFZZ	5331008045694	96906	MS28778-16	O-RING .... UOC:DAL, V18, ZAL	1
15	PAFZZ	4730009305392	96906	MS51525A16	ADAPTER, STRAIGHT, TU .... UOC:DAL, V18, ZAL	1
16	PFFZZ	4720011208518	19207	12256559	HOSE ASSEMBLY, NONME SWIVEL TEE TO CONTROLS .... UOC:DAL, V18, ZAL	1
17	PAFZZ	4710010589494	19207	11677309	TUBE ASSEMBLY, METAL .... UOC:DAL, V18, ZAL	1
18	PAFFF	4820001341122	19207	10900080	VALVE, CHECK HYDRAULIC TANK ... UOC:DAL, V18, ZAL	1
19	XDFZZ		19207	10900082	.CAP, VALVE, SNUBBER .... UOC:DAL, V18, ZAL	1
20	PFFZZ	5360011312063	19207	10900107	.SPRING, HELICAL, COMP .... UOC:DAL, V18, ZAL	1
21	PAFZZ	2590011311940	19207	10900091	.PLUNGER, VALVE .... UOC:DAL, V18, ZAL	1
22	PAFZZ	5331008087612	96906	MS28775-131	.O-RING	1

(1) ITEM NO	(2) SMR CODE	(3) NSN	(4) CAGEC	(5) PART NUMBER	(6) DESCRIPTION AND USABLE ON CODES (UOC)	(7) QTY
22	PAFZZ	5330008087612	96906	MS28775-131	.PACKING, PREFORMED .... UOC:DAL, V1B, ZAL	1
23	XAFZZ		19207	10900113	.WASHER .... UOC:DAL, V1B, ZAL	1
24	XAFZZ		19207	10900081	.SEAT .... UOC:DAL, V18, ZAL	
25	PAFZZ	4720008265610	19207	10876144	HOSE ASSEMBLY, NONME .... UOC:DAL, V18, ZAL	1
26	PAFZZ	4730002788935	96906	MS20825-16	TEE, PIPE TO TUBE .... UOC:DAL, V18, ZAL	1
27	PAFZZ	4720010589490	19207	11669207	HOSE ASSEMBLY, NONME .... UOC:DAL, V18, ZAL	1
28	PAFZZ	5310008140672	96906	MS51943-36	NUT, SELF-LOCKING, HE .... UOC:DAL, V18, ZAL	4
29	PAFZZ	5340004193084	19207	10910298	CLAMP, SYNCHRO HOSE SUPPORT STRAP .... UOC:DAL, V1B, ZAL	4
30	PAFZZ	4720008265607	19207	10876139	HOSE ASSEMBLY, NONME RIGHT SWING PORT CONNECTOR .... UOC:DAL, V18, ZAL	2
31	PAFZZ	5340004213992	19207	10910299	STRAP, RETAINING RIGHT SWING PORT CONNECTOR. .... UOC:DAL, V18, ZAL	4
32	PAFZZ	4730003594708	79470	9405-16-16	ELBOW, PIPE .... UOC:DAL, V18, ZAL	2
33	PAFZZ	4720008265606	19207	10876142	HOSE ASSEMBLY, NONME RETRACT AND EXTEND PORT. .... UOC:DAL, V1B, ZAL	2

END OF FIGURE

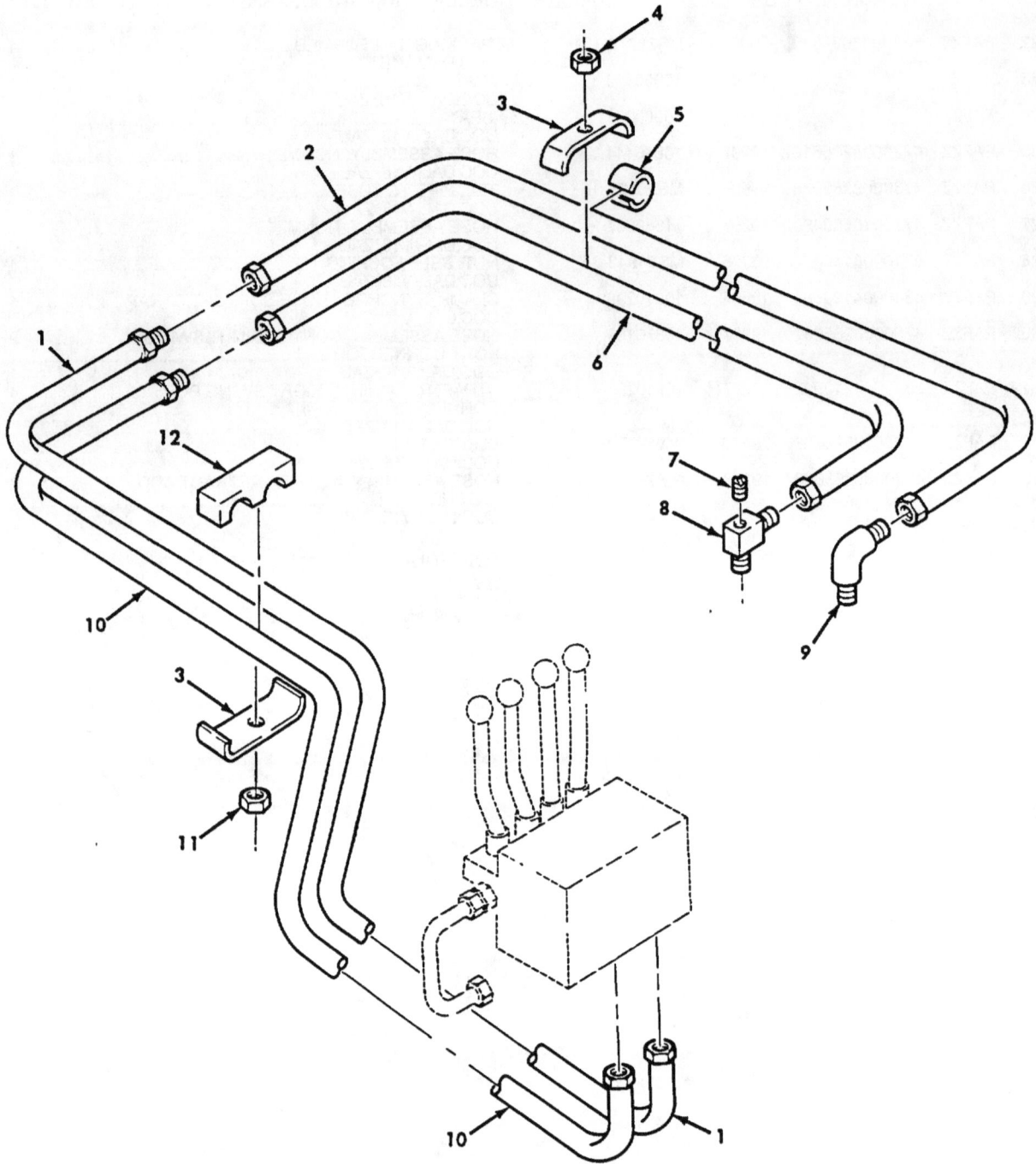

Figure 481. Wrecker Hydraulic Tubing, Control Valve to Hoist Motor.

(1) ITEM NO	(2) SMR CODE	(3) NSN	(4) CAGEC	(5) PART NUMBER	(6) DESCRIPTION AND USABLE ON CODES (UOC)	(7) QTY
					GROUP 2001 HOIST, CAPSTAN, WINDLASS, CRANE OR WINCH ASSEMBLY	
					FIG. 481 WRECKER HYDRAULIC TUBING, CONTROL VALVE TO HOIST MOTOR	
1	PAFZZ	4720012211448	19207	10876146-1	HOSE ASSEMBLY, NONME .... UOC:DAL, V18, ZAL	1
2	PAFZZ	4710000896193	19207	10876430	TUBE ASSEMBLY, METAL CONTROL VALVE .... ASSEMBLY.. UOC:DAL, V18, ZAL	1
3	PAFZZ	5340004193084	19207	10910298	CLAMP, SYNCHRO CONTROL VALVE .... ASSEMBLY HOSE.. UOC:DAL, V18, ZAL	6
4	PAFZZ	5310000874652	96906	MS51922-17	NUT, SELF-LOCKING, HE .... UOC:DAL, V18, ZAL	4
5	PAFZZ	2590004213942	19207	10906322	PAD, TUBE CONTROL VALVE ASSEMBLY........................ HOSE AT STRAP .... UOC:DAL, V18, ZAL	8
6	PAFZZ	4710009339587	19207	10876428	TUBE ASSEMBLY, METAL CONTROL VALVE .... ASSEMBLY.. UOC:DAL, V18, ZAL	1
7	PAHZZ	4730000428988	24491	10015	PLUG, PIPE.. UOC:DAL, V18, ZAL	1
8	PAFZZ	4730004569831	19207	10900266	ELBOW, PIPE TO TUBE... UOC:DAL, V18, ZAL	1
9	PAFZZ	4730002827609	96906	MS20822-16K	ELBOW, PIPE TO TUBE HOIST DOWN......................... OPERATION... UOC:DAL, V18, ZAL	1
10	PAFZZ	4720012359627	19207	10876137-1	HOSE ASSEMBLY, NONME .... UOC:DAL, V18, ZAL	1
11	PAFZZ	5310008140672	96906	MS51943-36	NUT, SELF-LOCKING, HE .... UOC:DAL, V18, ZAL	2
12	PAFZZ	5340004213992	19207	10910299	STRAP, RETAINING CONTROL VALVE................................ ASSEMBLY HOSE.. UOC:DAL, V18, ZAL	2

END OF FIGURE

7 —⌐ 8 thru 11

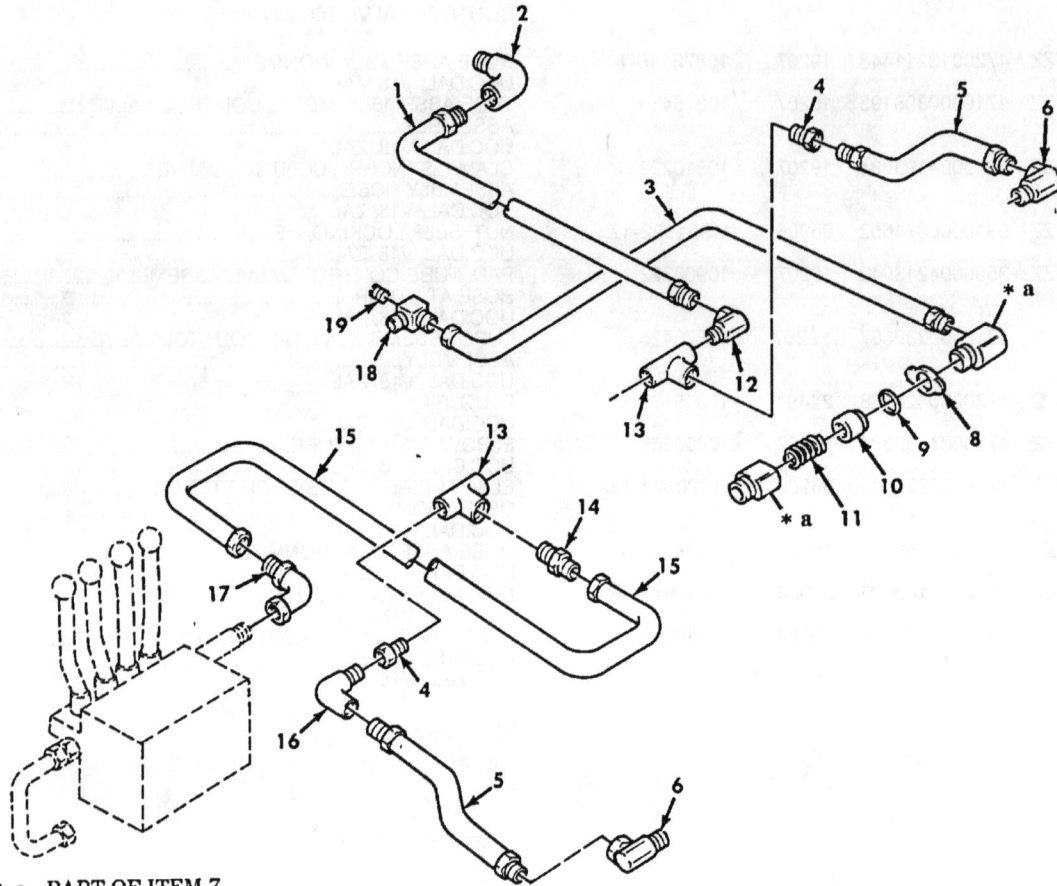

* a   PART OF ITEM 7

Figure 482. Wrecker Hydraulic Tubing, Control Valve to Elevating Cylinder and Filter to Elevating Cylinder.

(1) ITEM NO	(2) SMR CODE	(3) NSN	(4) CAGEC	(5) PART NUMBER	(6) DESCRIPTION AND USABLE ON CODES (UOC)	(7) QTY
					GROUP 2001 HOIST, CAPSTAN, WINDLASS, CRANE OR WINCH ASSEMBLY	
					FIG. 482 WRECKER HYDRAULIC TUBING, CONTROL VALVE TO ELEVATING CYLINDER AND FILTER TO ELEVATING CYLINDER	
1	PAFZZ	4720011208517	19207	12256561	HOSE ASSEMBLY, NONME OIL FILTER .... UOC:DAL, V18, ZAL	1
2	PAFZZ	4730000127823	96906	MS39230-7	ELBOW, PIPE OIL TANK RETURN LINE .... UOC:DAL, V18, ZAL	1
3	PAFZZ	4720008265610	19207	10876144	HOSE ASSEMBLY, NONME .... UOC:DAL, V18, ZAL	1
4	PAFZZ	4730001930883		21450 125915	BUSHING, PIPE ELEVATING CYLINDER .... HOSE .... UOC:DAL, V18, ZAL	2
5	PAFZZ	4720005221449	19207	10876145	HOSE ASSEMBLY, NONME TURNTABLE PUMP .... UOC:DAL, V18, ZAL	2
6	PAFZZ	4730003594708	19207	9405-16-16	ELBOW, PIPE OIL FILTER HOSE .... UOC:DAL, V18, ZAL	2
7	PAFFF	4820001341122	19207	10900080	VALVE, CHECK .... UOC:DAL, V18, ZAL	1
8	XAFZZ	5310011312085	19207	10900113	.WASHER, FLAT .... UOC:DAL, V18, ZAL	1
9	PAFZZ	5331008087612	96906	MS28775-131	.O-RING .... UOC:DAL, V18, ZAL	1
10	PAFZZ	2590011311940	19207	10900091	.PLUNGER, VALVE .... UOC:DAL, V18, ZAL	1
11	PFFZZ	5360011312063	19207	10900107	.SPRING, HELICAL, COMP .... UOC:DAL, V18, ZAL	1
12	PAFZZ	4730002891266	27618	11345P11	ELBOW, PIPE ELEVATING CYLINDER........................... RETURN HOSE .... UOC:DAL, V18, ZAL	1
13	PAFZZ	4730001388133	21450	218709	TEE, PIPE ELEVATING CYLINDER................................ CROSSOVER PIPE .... UOC:DAL, V18, ZAL	2
14	PAFZZ	4730002028470	79470	C5205X20	ADAPTER, STRAIGHT, PI .... UOC:DAL, V18, ZAL	1
15	PAFZZ	4720010589489	19207	11669206	HOSE ASSEMBLY, NONME .... UOC:DAL, V18, ZAL	1
16	XDFZZ		19207	10913152	ELBOW, PIPE .... UOC:DAL, V18, ZAL	1
17	PAFZZ	4730011637823	96906	MS51506-A20	ELBOW, PIPE TO TUBE .... UOC:DAL, V18, ZAL	1
18	PAFZZ	4730004569831	19207	10900266	ELBOW, PIPE TO TUBE ELEVATING.............................. CYLINDER .... UOC:DAL, V18, ZAL	1
19	PAFZZ	4730000189566	81348	WW-P-471ACABCA	PLUG, PIPE .... UOC:DAL, V18, ZAL	1

END OF FIGURE

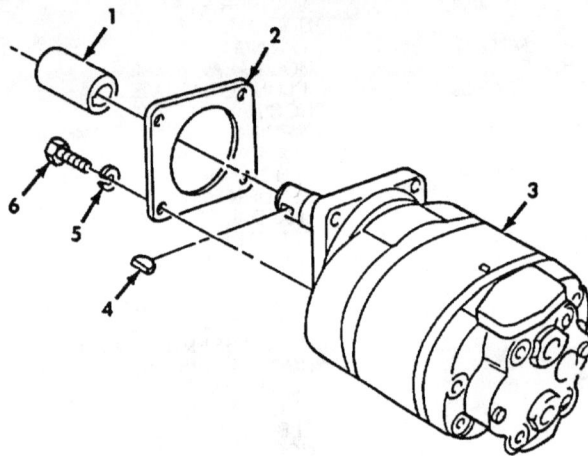

Figure 483. Wrecker Hoist Hydraulic Winch Motor and Crane Motor Assembly.

(1) ITEM NO	(2) SMR CODE	(3) NSN	(4) CAGEC	(5) PART NUMBER	(6) DESCRIPTION AND USABLE ON CODES (UOC)	(7) QTY
					GROUP 2001 HOIST, CAPSTAN, WINDLASS, CRANE OR WINCH ASSEMBLY	
					FIG. 483 WRECKER HOIST HYDRAULIC WINCH MOTOR AND CRANE MOTOR ASSEMBLY	
1	PAFZZ	3040001849720	19207	10900089	COLLAR, SHAFT HOIST WINCH MOTOR .... 1 UOC:DAL, V18, ZAL	1
2	PAFZZ	5330001823489	19207	10899995	GASKET HOIST WINCH MOTOR .... 2 UOC:DAL, V18, ZAL	2
3	PAFZZ	4320013204744	19207	12375388	MOTOR-PUMP, HYDRAULI CRANE .... 1 UOC:DAL, V18, ZAL	1
4	PAFZZ	5315006165500	96906	MS35756-21	KEY, WOODRUFF GEARSHAFT .... 2 UOC:DAL, V18, ZAL	2
5	PAFZZ	5310008206653	80045	23MS35338-50	WASHER, LOCK .... 8 UOC:DAL, V18, ZAL	8
6	PAFZZ	5305007245910	96906	MS90725-162	SCREW, CAP, HEXAGON H .... 8 UOC:DAL, V18, ZAL	8

END OF FIGURE

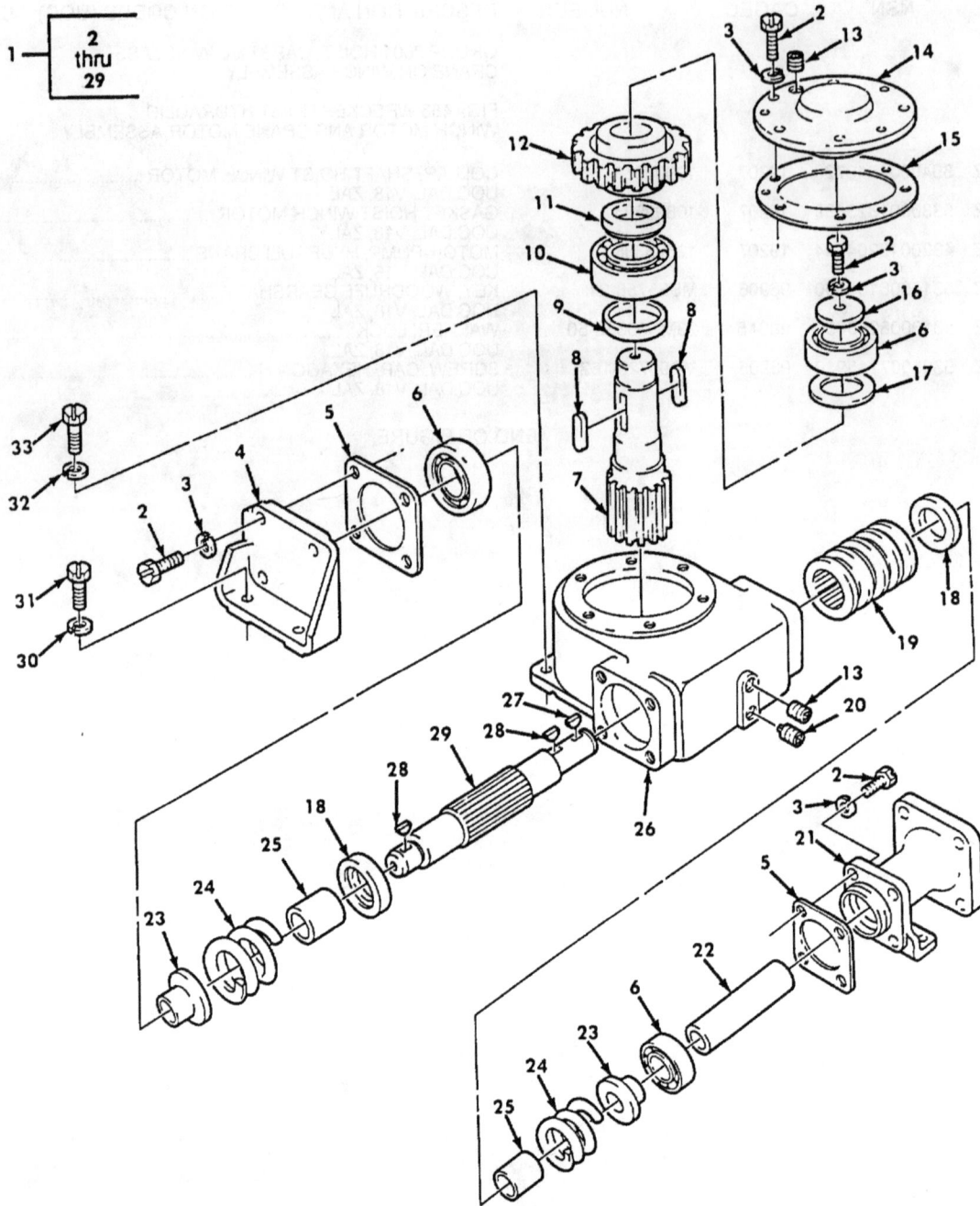

Figure 484. Wrecker Crane Gearcase Assembly and Mounting Hardware.

(1) ITEM NO	(2) SMR CODE	(3) NSN	(4) CAGEC	(5) PART NUMBER	(6) DESCRIPTION AND USABLE ON CODES (UOC)	(7) QTY
					GROUP 2001 HOIST, CAPSTAN, WINDLASS, CRANE OR WINCH ASSEMBLY	
					FIG. 484 WRECKER CRANE GEARCASE ASSEMBLY AND MOUNTING HARDWARE	
1	PAFHH	3020009339585	19207	10876198	GEAR SET, WORM AND W .... UOC:DAL, V18, ZAL	1
2	PAHZZ	5305007195219	96906	MS90727-111	.SCREW, CAP, HEXAGON H .... UOC:DAL, V18, ZAL	15
3	PAHZZ	5310000034094	01276	210104-8S	.WASHER, LOCK .... UOC:DAL, V18, ZAL	15
4	PFHZZ	2520011376294	19207	10876382	.CAP, DUST, PROPELLER .... UOC:DAL, V18, ZAL	1
5	PAHZZ	5330008265202	19207	10876133	.GASKET .... UOC:DAL, V18, ZAL	2
6	PAHZZ	3110005543468	64731	G40-5	.BEARING, BALL, ANNULA .... UOC:DAL, V18, ZAL	3
7	PBHZZ	3040005049038	19207	10876163	.SHAFT, SHOULDERED CRANE SWINGER GEAR CASE .... UOC:DAL, V18, ZAL	1
8	PAHZZ	5315002817651	19207	8328341	.KEY, MACHINE .... UOC:DAL, V18, ZAL	2
9	PAHZZ	5330001782191	21450	500163	.SEAL, PLAIN ENCASED .... UOC:DAL, V18, ZAL	1
10	PAHZZ	3110005543411	21450	714042	.BEARING, BALL, ANNULA .... UOC:DAL, V18, ZAL	1
11	PAHZZ	5310011010077	19207	10900232	.BEARING, WASHER, THRU .... UOC:DAL, V18, ZAL	1
12	PBHZZ	3020002627572	19207	7538688	.GEAR, WORM WHEEL SWINGER GEARCASE .... UOC:DAL, V18, ZAL	1
13	PAHZZ	4730011072027	21450	444673	.PLUG, PIPE .... UOC:DAL, V18, ZAL	2
14	PBHZZ	5340004146561	19207	10876226	.COVER, ACCESS SWINGER GEARCASE .... UOC:DAL, V18, ZAL	1
15	PAHZZ	5330008265203	19207	10876132	.GASKET .... UOC:DAL, V18, ZAL	1
16	PAHZZ	5310007538215	19207	10876166	.WASHER, FLAT .... UOC:DAL, V18 , ZAL	1
17	PAHZZ	3120011010076	19207	10900202	.BEARING, WASHER, THRU .... UOC:DAL, V18, ZAL	1
18	PAHZZ	5340011010090	19207	10900212	.SEAT, HELICAL COMPRE .... UOC:DAL, V18, ZAL	2
19	PBHZZ	3020009722641	19207	10876360	.GEAR, WORM CRANE SWINGER GEARCASE .... UOC:DAL, V18, ZAL	1
20	PAHZZ	4730009686129	96906	MS49006-10	.PLUG, PIPE, MAGNETIC .... UOC:DAL, V18, ZAL	1
21	PAHZZ	3040001779266	19207	10876232	.HOUSING, MECHANICAL .... UOC:DAL, V18, ZAL	1
22	PAHZZ	3010000693047	19207	10899996	.COUPLING, SHAFT, RIGI .... UOC:DAL, V18, ZAL	1
23	PAHZZ	3120009371164	19207	10899994	.BUSHING, SLEEVE	2

(1) ITEM NO	(2) SMR CODE	(3) NSN	(4) CAGEC	(5) PART NUMBER	(6) DESCRIPTION AND USABLE ON CODES (UOC)	(7) QTY
					UOC:DAL, V18, ZAL	
24	PAHZZ	5360001364759	19207	10900189	.SPRING, HELICAL, COMP	2
					UOC:DAL, V18, ZAL	
25	PAHZZ	5365014354806	19207	10900204	.SPACER, SLEEVE	2
					UOC:DAL, V18, ZAL	
26	XAHZZ		19207	10876186	.CASE, GEAR SWINGER	1
					UOC:DAL, V18, ZAL	
27	PAHZZ	5315006165500	96906	MS35756-21	.KEY, WOODRUFF GEARCASE WORM SHAFT	1
					UOC:DAL, V18, ZAL	
28	PAHZZ	5315006165501	96906	MS35756-20	.KEY, WOODRUFF	2
					UOC:DAL, V18, ZAL	
29	PAHZZ	3040009722640	19207	10876361	.SHAFT, SHOULDERED	1
					UOC:DAL, V18, ZAL	
30	PAFZZ	5310000034094	01276	210104-8S	WASHER, LOCK	4
					UOC:DAL, V18, ZAL	
31	PAFZZ	5305007254183	96906	MS90726-113	SCREW, CAP, HEXAGON H	4
					UOC:DAL, V18, ZAL	
32	PAFZZ	5310008206653	80045	23MS35338-50	WASHER, LOCK	2
					UOC:DAL, V18, ZAL	
33	PAFZZ	5305007262550	80204	B1821BH063F175N	SCREW, CAP, HEXAGON H	2
					UOC:DAL, V18, ZAL	

END OF FIGURE

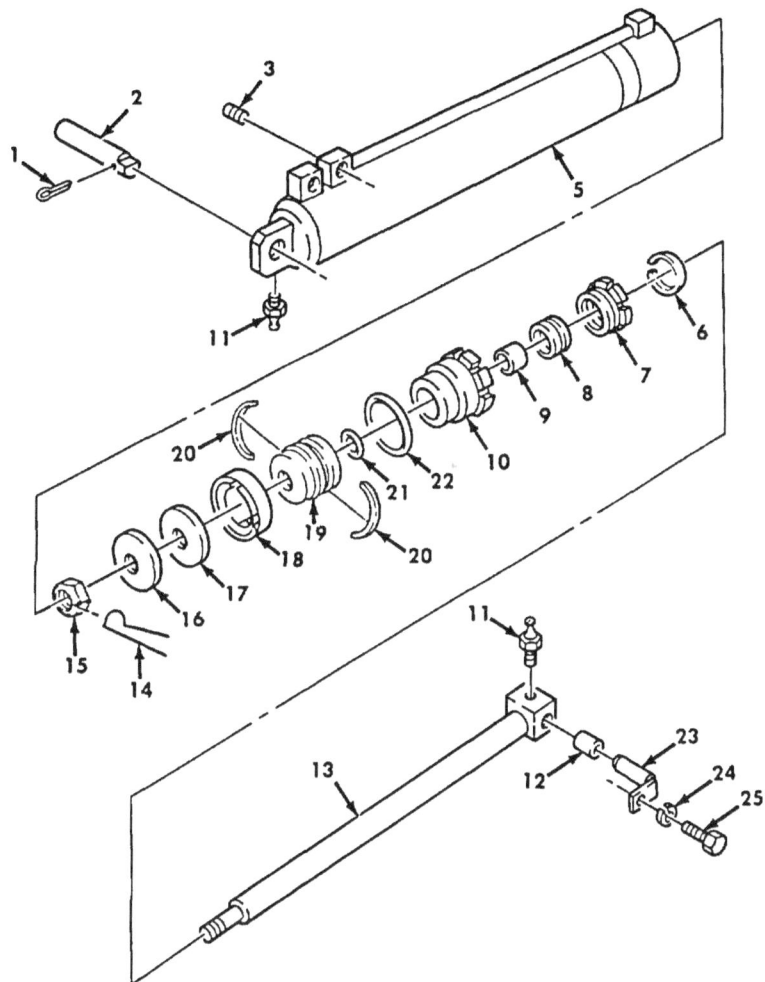

Figure 485. Wrecker Crane Boom Elevating Cylinder Assembly.

(1) ITEM NO	(2) SMR CODE	(3) NSN	(4) CAGEC	(5) PART NUMBER	(6) DESCRIPTION AND USABLE ON CODES (UOC)	(7) QTY
					GROUP 2001 HOIST, CAPSTAN, WINDLASS, CRANE OR WINCH ASSEMBLY	
					FIG. 485 WRECKER CRANE BOOM ELEVATING CYLINDER ASSEMBLY	
1	PAFZZ	5315008499854	96906	MS24665-498	PIN, COTTER	2
					UOC:DAL, V18, ZAL	
2	PAFZZ	5315011064036	19207	10900013	PIN, STRAIGHT, HEADLE	2
					UOC:DAL, V18, ZAL	
3	PAFZZ	4730002212141	96906	MS20913-8S	PLUG, PIPE	4
					UOC:DAL, V18, ZAL	
4	PAFHH	3040009722638	19207	10876213	CYLINDER ASSEMBLY, A	2
					UOC:DAL, V18, ZAL	
5	XAHZZ		19207	10876311	.CYLINDER HYDRAULIC CRANE BOOM	1
					UOC:DAL, V18, ZAL	
6	PAHZZ	5330009722635	61465	2012993-4	.SEAL, NONMETALLIC SP	1
					UOC:DAL, V18, ZAL	
7	PAHZZ	4730002385594	19207	10876461	.PACKING NUT	1
					UOC:DAL, V18, ZAL	
8	PAHZZ	5330005234235	19207	10900300	.PACKING ASSEMBLY	1
					UOC:DAL, V18, ZAL	
9	PAHZZ	3120008265630	19207	10876159	.BEARING, SLEEVE PISTON ROD	1
					UOC:DAL, V18, ZAL	
10	PAHZZ	2590001779196	19207	10876310	.HEAD, BOOM CYLINDER	1
					UOC:DAL, V18, ZAL	
11	PAOZZ	4730000504208	96906	MS15003-1	.FITTING, LUBRICATION	2
					UOC:DAL, V1B, ZAL	
12	PAHZZ	3120009511850	19207	10911102	.BEARING, SLEEVE	2
					UOC:DAL, V18, ZAL	
13	PBHZZ	3040011376264	19207	10876312	.ROD, PISTON, LINEAR A	1
					UOC:DAL, V18, ZAL	
14	MHHZZ		96906	MS20995F91-12	.WIRE, NONELECTRICAL MAKE FROM WIRE, P/N QQ-W-461	1
					UOC:DAL, V18, ZAL	
15	PAHZZ	5310004702107	19207	10900151	.NUT, PLAIN, HEXAGON PISTON ROD	1
					UOC:DAL, V18, ZAL	
16	PAHZZ	5310004193082	19207	10876506	.WASHER, FLAT HYDRAULIC CYLINDER PISTON	1
					UOC:DAL, V18, ZAL	
17	PAHZZ	5310004457238	19207	10876454	.WASHER, RECESSED HYDRAULIC CYLINDER PISTON SEAL	1
					UOC:DAL, V18, ZAL	
18	PAHZZ	5340005234305	19207	10900304	.CUP, COMPRESSION HYDRAULIC CYLINDER PISTON	1
					UOC:DAL, V18, ZAL	
19	PAHZZ	2590009722634	19207	10876570	.PISTON, HYDRAULIC CY	1
					UOC:DAL, V18, ZAL	
20	PAHZZ	4310005048923	19207	10876153	.RING, PISTON ELEVATING CYLINDER ASSEMBLY	2
					UOC:DAL, V18, ZAL	
21	PAHZZ	5331002979990	96906	MS28775-222	.O-RING	1

(1) ITEM NO	(2) SMR CODE	(3) NSN	(4) CAGEC	(5) PART NUMBER	(6) DESCRIPTION AND USABLE ON CODES (UOC)	(7) QTY
					UOC:DAL, V18, ZAL	
22	PAHZZ	5331010192448	96906	MS28775-249	.PACKING, PREFORMED	1
					UOC:DAL, V18, ZAL	
23	PAFZZ	5315004213931	19207	10899991	PIN, KEEPER	1
					UOC:DAL, V18, ZAL	
24	PAFZZ	5310000034094	01276	210104-8S	WASHER, LOCK	1
					UOC:DAL, V18, ZAL	
25	PAFZZ	5305002267767	96906	MS90726-109	SCREW, CAP, HEXAGON H	2
					UOC:DAL, V18, ZAL	

END OF FIGURE

485-2

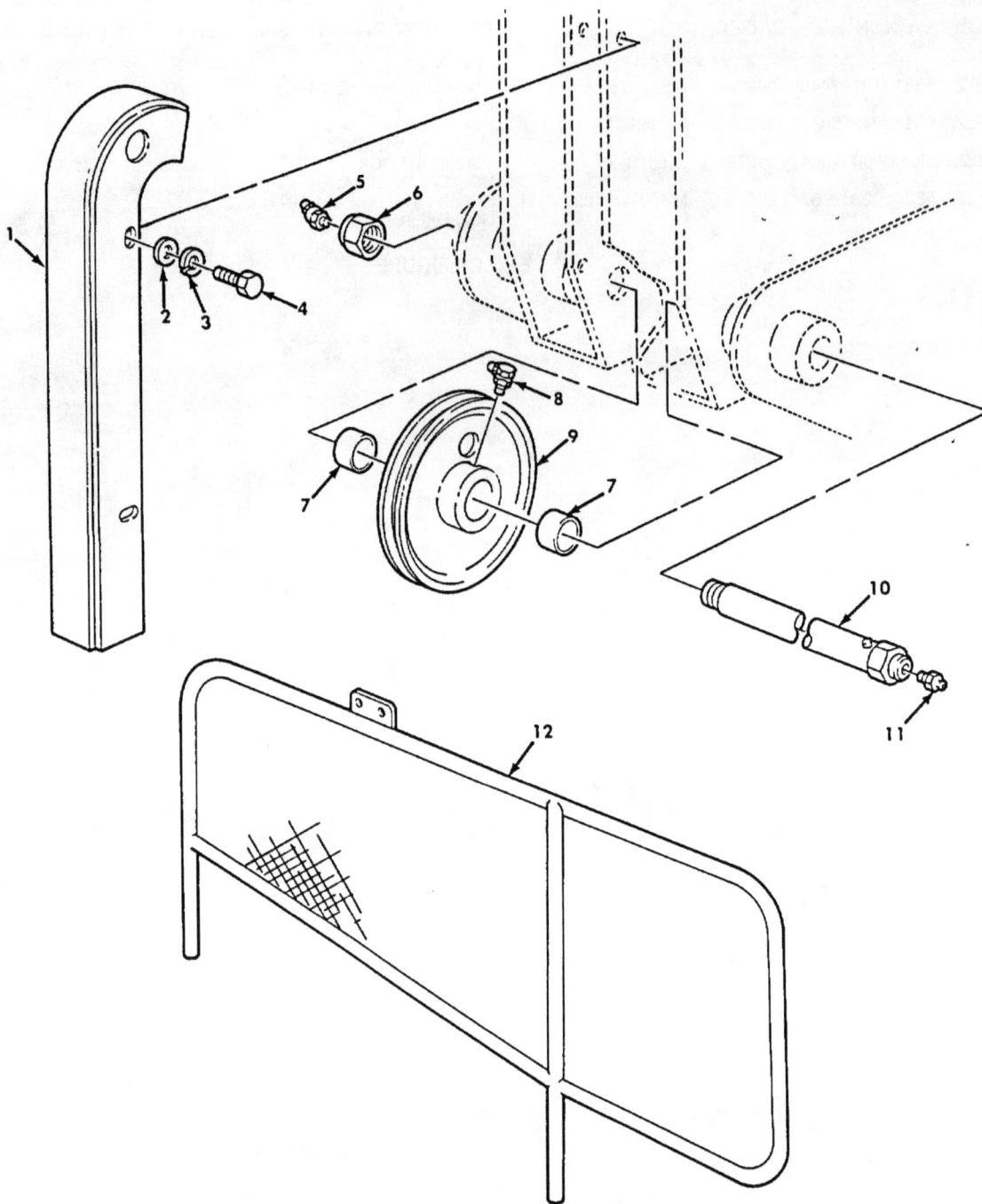

Figure 486. Wrecker Sheave and Operator Guard.

(1) ITEM NO	(2) SMR CODE	(3) NSN	(4) CAGEC	(5) PART NUMBER	(6) DESCRIPTION AND USABLE ON CODES (UOC)	(7) QTY
					GROUP 2001 HOIST, CAPSTAN, WINDLASS, CRANE OR WINCH ASSEMBLY	
					FIG. 486 WRECKER SHEAVE AND OPERATOR GUARD	
1	PAOZZ	3020013105493	19207	10876279	GUARD, MECHANICAL DR .... UOC:DAL, V18, ZAL	1
2	PAOZZ	5310000806004	96906	MS27183-14	WASHER, FLAT .... UOC:DAL, V18, ZAL	4
3	PAOZZ	5310006379541	96906	MS35338-46	WASHER, LOCK .... UOC:DAL, V18, ZAL	4
5	PAOZZ	4730001720031	96906	MS15003-5	FITTING, LUBRICATION .... UOC:DAL, V18, ZAL	1
6	PAFZZ	5310000624954	96906	MS21045-8	NUT, SELF-LOCKING, HE .... UOC:DAL, V1B, ZAL	1
7	PAFZZ	3120005048930	19207	10876158	BEARING, SLEEVE .... UOC:DAL, V18, ZAL	2
8	PAFZZ	4730000504208	96906	MS15003-1	FITTING, LUBRICATION .... UOC:DAL, V18, ZAL	1
9	PFFZZ	3020011150619	19207	10900056	PULLEY, FLAT .... UOC:DAL, V18, ZAL	1
10	PAFZZ	3040009339579	19207	10876292	SHAFT, SHOULDERED CRANE BOOM SHEAVE .... ASSEMBLY... UOC:DAL, V18, ZAL	1
11	PAFZZ	4730000504208	96906	MS15003-1	FITTING, LUBRICATION .... UOC:DAL, V18, ZAL	1
12	PAFZZ	3020010897333	19207	10883445	GUARD, MECHANICAL DR GONDOLA.............................. ASSEMBLY... UOC:DAL, V18, ZAL	1

END OF FIGURE

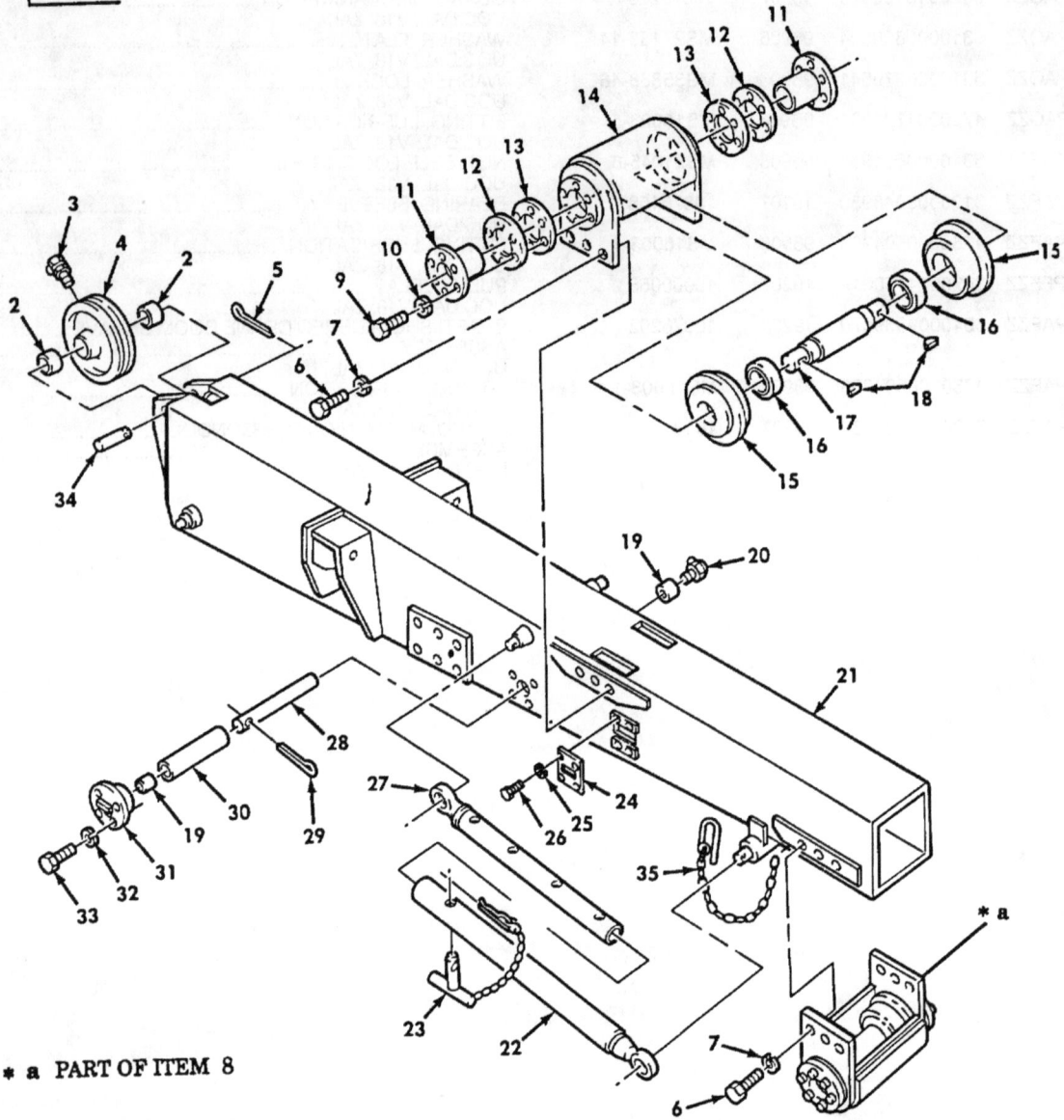

* a PART OF ITEM 8

Figure 487. Wrecker Boom Assembly.

(1) ITEM NO	(2) SMR CODE	(3) NSN	(4) CAGEC	(5) PART NUMBER	(6) DESCRIPTION AND USABLE ON CODES (UOC)	(7) QTY
					GROUP 2001 HOIST, CAPSTAN, WINDLASS, CRANE OR WINCH ASSEMBLY	
					FIG. 487 WRECKER BOOM ASSEMBLY	
1	PBFFF	2590011208506	19207	10876173	BOOM, CRANE .... UOC:DAL, V18, ZAL	1
2	PAFZZ	3120005048930	19207	10876158	.BEARING, SLEEVE .... UOC:DAL, V18, ZAL	2
3	PAOZZ	4730000504208	96906	MS15003-1	.FITTING, LUBRICATION .... UOC:DAL, V18, ZAL	1
4	PFFZZ	3020011150619	19207	10900056	.PULLEY, FLAT .... UOC:DAL, V18, ZAL	1
5	PAFZZ	5315000137308	96906	MS24665-627	.PIN, COTTER .... UOC:DAL, V18, ZAL	1
6	PAFZZ	5305007272283	80204	B1821BH063F150N	.SCREW, CAP, HEXAGON H .... UOC:DAL, V18, ZAL	12
7	PAFZZ	5310008206653	80045	23MS35338-50	.WASHER, LOCK .... UOC:DAL, V18, ZAL	12
8	PAFFF	2590001795581	19207	10876227	.ROLLER ASSEMBLY, BO .... UOC:DAL, V18, ZAL	2
9	PAFZZ	5305007250154	96906	MS90727-112	..SCREW, CAP, HEXAGON H .... UOC:DAL, V18, ZAL	8
10	PAFZZ	5310000034094	01276	210104-8S	..WASHER, LOCK .... UOC:DAL, V18, ZAL	8
11	PAFZZ	2590011147741	19207	10876412	..SUPPORT, SHAFT, ROLLE .... UOC:DAL, V18, ZAL	2
12	PAFZZ	5365004054378	19207	10900004	..SHIM .... UOC:DAL, V18, ZAL	2
13	PAFZZ	5365004221160	19207	10900005	..SHIM .... UOC:DAL, V18, ZAL	2
14	PFFZZ	2590002214824	19207	10876231	..HOUSING ROLLER, BOO .... UOC:DAL, V18, ZAL	1
15	PFFZZ	2590011465257	19207	10876359	..WHEEL, BOOM, WRECKER .... UOC:DAL, V18, ZAL	2
16	PFFZZ	3110002273381	21450	713806	..BEARING, BALL, ANNULA .... UOC:DAL, V18, ZAL	2
17	PAFZZ	3040001188694	19207	10876370	..SHAFT, SHOULDERED .... UOC:DAL, V18, ZAL	1
18	PAFZZ	5315000124553	96906	MS35756-17	..KEY, WOODRUFF .... UOC:DAL, V18, ZAL	2
19	PAFZZ	3120007409807	19207	7409807	.BEARING, SLEEVE .... UOC:DAL, V18, ZAL	2
20	PAFZZ	4730001720034	96906	MS15003-6	.FITTING, LUBRICATION .... UOC:DAL, V18, ZAL	1
21	PAFZZ	2590000867459	19207	10876190	.BOOM SECTION, OUTER, ... UOC:DAL, V18, ZAL	1
22	PAFZZ	4710008572782	19207	10876437	.TUBE, BOOM SUPPORT .... UOC:DAL, V18, ZAL	2
23	PAFZZ	5315000520110	19207	10899366	.PIN, STRAIGHT, HEADED BOOM SUPPORT..................... TUBE .... UOC:DAL, V18, ZAL	2

(1) ITEM NO	(2) SMR CODE	(3) NSN	(4) CAGEC	(5) PART NUMBER	(6) DESCRIPTION AND USABLE ON CODES (UOC)	(7) QTY
24	PFFZZ	2590002317452	19207	10899983	.PLATE, STOP, BOOM WIN .... UOC:DAL, V18, ZAL	2
25	PAFZZ	5310000034094	01276	210104-8S	.WASHER, LOCK .... UOC:DAL, V18, ZAL	8
26	PAFZZ	5305002267767	96906	MS90726-109	.SCREW, CAP, HEXAGON H .... UOC:DAL, V18, ZAL	8
27	PAFZZ	2590001015594	19207	10910347	.LEG, INNER, SHOE, JACK .... UOC:DAL, V18, ZAL	2
28	PAFZZ	3040002343250	19207	10899979	.SHAFT, STRAIGHT .... UOC:DAL, V18, ZAL	1
29	PAFZZ	5315000137228	96906	MS24665-423	.PIN, COTTER .... UOC:DAL, V18, ZAL	1
30	PAFZZ	2590002317451	19207	10899978	.ROLLER, BOOM WINCH .... UOC:DAL, V18, ZAL	1
31	PFFZZ	3040001507145	19207	10899980	.PLATE, RETAINING, SHA .... UOC:DAL, V18, ZAL	1
32	PAFZZ	5310000115093	96906	MS35338-65	.WASHER, LOCK .... UOC:DAL, V18, ZAL	4
33	PAFZZ	5305002693238	80204	B1821BH038F125N	.SCREW, CAP, HEXAGON H .... UOC:DAL, V18, ZAL	4
34	PAFZZ	3040001436390	19207	10900203	.SHAFT, STRAIGHT .... UOC:DAL, V18, ZAL	1

END OF FIGURE

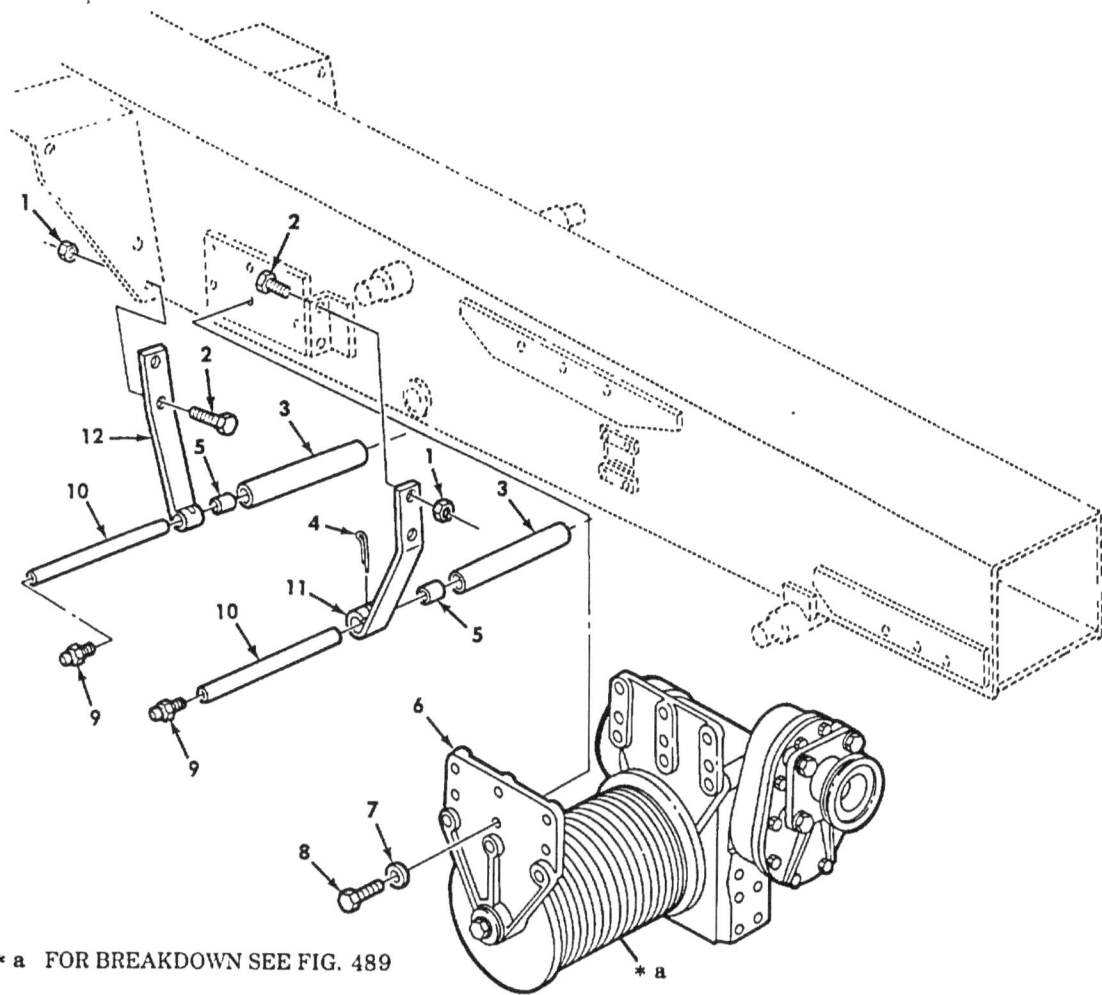

* a   FOR BREAKDOWN SEE FIG. 489

Figure 488. Crane Hoist Winch and Mounting Hardware.

(1) ITEM NO	(2) SMR CODE	(3) NSN	(4) CAGEC	(5) PART NUMBER	(6) DESCRIPTION AND USABLE ON CODES (UOC)	(7) QTY
					GROUP 2001 HOIST, CAPSTAN, WINDLASS, CRANE OR WINCH ASSEMBLY	
					FIG. 488 CRANE HOIST WINCH AND MOUNTING HARDWARE	
1	PAFZZ	5310000624954	96906	MS21045-8	NUT, SELF-LOCKING, HE .... UOC:DAL, V18, ZAL	8
2	PAFZZ	5305007195219	96906	MS90727-111	SCREW, CAP, HEXAGON H .... UOC:DAL, V18, ZAL	8
3	PAFZZ	2590002205104	19207	10900017	ROLLER, BOOM, WINCH .... UOC:DAL, V18, ZAL	2
4	PAFZZ	5315000137214	96906	MS24665-359	PIN, COTTER .... UOC:DAL, V18, ZAL	2
5	PAFZZ	3120007409807	19207	7409807	BEARING, SLEEVE .... UOC:DAL, V18, ZAL	4
6	PGFFH	2590009749670	19207	10876197	WINCH, DRUM, VEHICLE .... UOC:DAL, V18, ZAL	1
7	PAFZZ	5310005847888	96906	MS35338-51	WASHER, LOCK .... UOC:DAL, V18, ZAL	12
8	PAFZZ	5305007262551	80204	B1821BH063F200N	SCREW, CAP, HEXAGON H .... UOC:DAL, V18, ZAL	12
9	PAFZZ	4730000504208	96906	MS15003-1	FITTING, LUBRICATION .... UOC:DAL, V18, ZAL	2
10	PAFZZ	3040002555700	19207	10900018	SHAFT, STRAIGHT .... UOC:DAL, V18, ZAL	2
11	PAFZZ	2590000454206	19207	10876563	HANGER, ROLLER GUIDE .... UOC:DAL, V18, ZAL	2
12	PAFZZ	2590000454205	19207	10900021	HANGER, ROLLER GUIDE .... UOC:DAL, V18, ZAL	2

END OF FIGURE

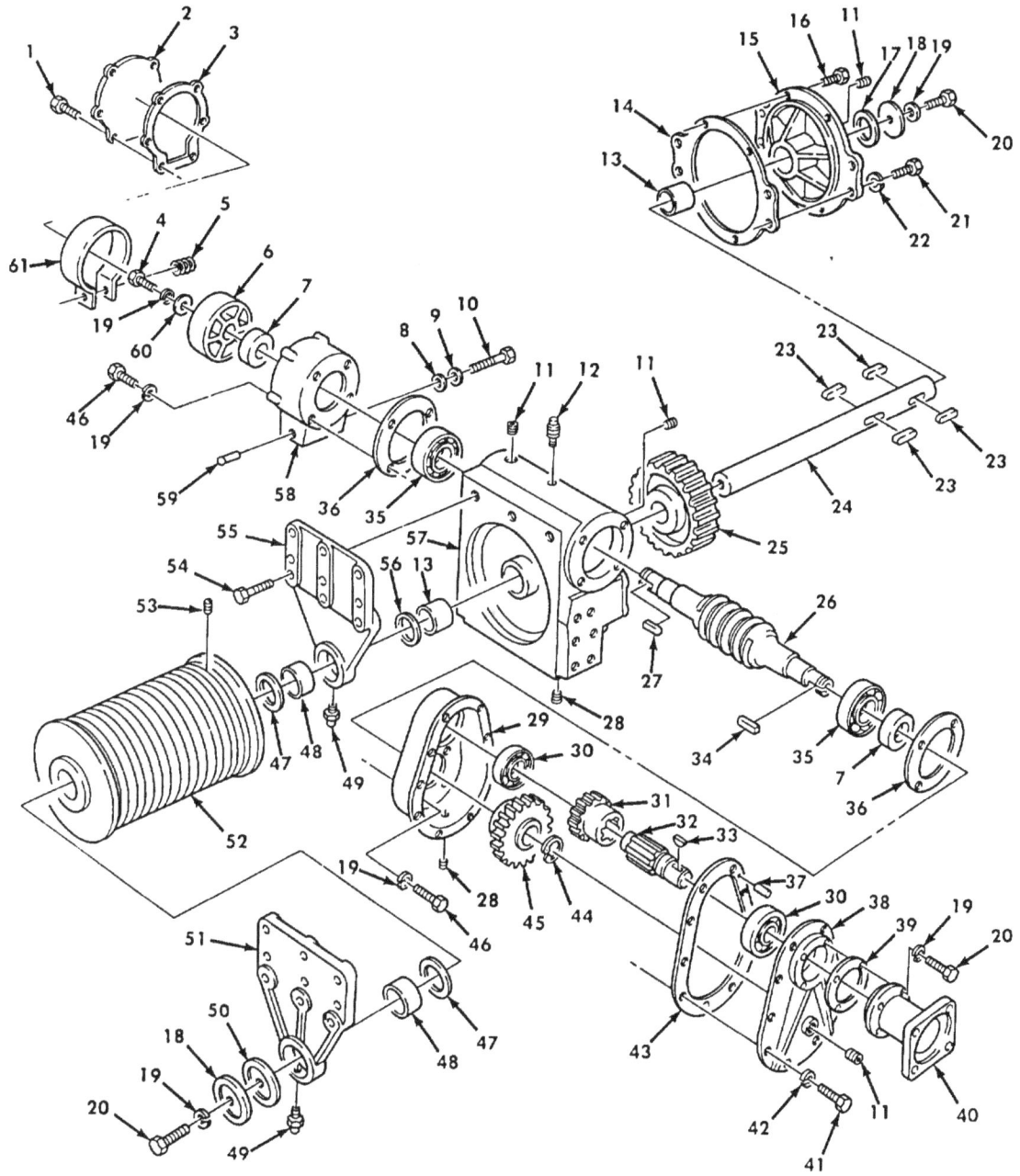

Figure 489. Wrecker Boom Hoist Winch Housing, Worm Gear, and Brake.

(1) ITEM NO	(2) SMR CODE	(3) NSN	(4) CAGEC	(5) PART NUMBER	(6) DESCRIPTION AND USABLE ON CODES (UOC)	(7) QTY
					GROUP 2001 HOIST, CAPSTAN, WINDLASS, CRANE OR WINCH ASSEMBLY	
					FIG. 489 WRECKER BOOM HOIST WINCH HOUSING, WORM GEAR, AND BRAKE	
1	PAFZZ	5306000187527	21450	187527	BOLT, ASSEMBLED WASH .... UOC:DAL, V18, ZAL	6
2	PFFZZ	2520007409816	19207	7409816	COVER, BRAKE CASE .... UOC:DAL, V18, ZAL	1
3	PAFZZ	5330008953424	19207	7973339	GASKET .... UOC:DAL, V18, ZAL	1
4	PAFZZ	5305002267767	96906	MS90726-109	SCREW, CAP, HEXAGON H .... UOC:DAL, V18, ZAL	1
5	PAFZZ	5360006644374	19207	5277992	SPRING, HELICAL, COMP .... UOC:DAL, V18, ZAL	1
6	PAFZZ	2530007409817	19207	7409817	BRAKE DRUM .... UOC:DAL, V18, ZAL	1
7	PAFZZ	5330011509691	80201	17657/55-542465	SEAL, PLAIN ENCASED .... UOC:DAL, V18, ZAL	2
8	PAFZZ	5330010463300	81349	M83461/1-012	O-RING .... UOC:DAL, V18, ZAL	1
9	PAFZZ	5310000806004	96906	MS27183-14	WASHER, FLAT .... UOC:DAL, V18, ZAL	1
10	PAFZZ	5305002693250	96906	MS90727-74	SCREW, CAP, HEXAGON H .... UOC:DAL, V18, ZAL	1
11	PAFZZ	4730002888555	96906	MS49005-9	PLUG, PIPE .... UOC:DAL, V18, ZAL	4
12	PAFZZ	4820007264719	57733	5196397	VALVE, VENT .... UOC:DAL, V18, ZAL	1
13	PAFZZ	3120007409800	19207	7409800	BEARING, SLEEVE .... UOC:DAL, V18, ZAL	2
14	PAFZZ	5330007409821	19207	7409821	GASKET .... UOC:DAL, V18, ZAL	1
15	PAFZZ	5340004718631	19207	10876253	COVER, ACCESS .... UOC:DAL, V18, ZAL	1
16	PAFZZ	5305009585267	96906	MS35190-343	SCREW, MACHINE .... UOC:DAL, V18, ZAL	4
17	PAFZZ	5330012546377	96906	MS51000-131-2	SEAL, PLAIN ENCASED .... UOC:DAL, V18, ZAL	1
18	PAFZZ	5310005125213	19207	10900102	WASHER, FLAT .... UOC:DAL, V18, ZAL	2
19	PAFZZ	5310000034094	01276	210104-8S	WASHER, LOCK .... UOC:DAL, V18, ZAL	15
20	PAFZZ	5305007195219	96906	MS90727-111	SCREW, CAP, HEXAGON H .... UOC:DAL, V18, ZAL	6
21	PAFZZ	5305007247221	80204	B1821BH063C175N	SCREW, CAP, HEXAGON H .... UOC:DAL, V18, ZAL	4
22	PAFZZ	5310008206653	96906	MS35338-50	WASHER, LOCK .... UOC:DAL, V18, ZAL	4
23	PAFZZ	5315002817652	19207	8327444	KEY, MACHINE .... UOC:DAL, V18, ZAL	4

(1) ITEM NO	(2) SMR CODE	(3) NSN	(4) CAGEC	(5) PART NUMBER	(6) DESCRIPTION AND USABLE ON CODES (UOC)	(7) QTY
24	PFFZZ	3040011010086	19207	10876355	SHAFT, STRAIGHT .... UOC:DAL, V18, ZAL	1
25	PAFZZ	3020007409823	19207	7409823	GEAR, WORM WHEEL .... UOC:DAL, V18, ZAL	1
26	PAFZZ	3040005049037	19207	10876164	WORM SHAFT .... UOC:DAL, V18, ZAL	1
27	PAFZZ	5315000423293	96906	MS20067-270	KEY, MACHINE .... UOC:DAL, V1B, ZAL	1
28	PAFZZ	4730009686129	96906	MS49006-10	PLUG, PIPE, MAGNETIC .... UOC:DAL, V18, ZAL	2
29	PFFZZ	3040004213961	19207	10876246	HOUSING, MECHANICAL .... UOC:DAL, V18, ZAL	1
30	PAFZZ	3110005543184	21450	700536	BEARING, BALL, ANNULA .... UOC:DAL, V18, ZAL	2
31	PAFZZ	3020005221175	19207	10876134	GEAR, HELICAL .... UOC:DAL, V18, ZAL	1
32	PAFZZ	3040005048913	19207	10876364	SHAFT, SHOULDERED .... UOC:DAL, V18, ZAL	1
33	PAFZZ	5315006165500	96906	MS35756-21	KEY, WOODRUFF REDUCTION GEAR SHAFT .... UOC:DAL, V18, ZAL	1
34	PAFZZ	5315002420818	96906	MS20067-221	KEY, MACHINE .... UOC:DAL, V18, ZAL	2
35	PAFZZ	3110005543929	21450	700287	BEARING, BALL, ANNULA .... UOC:DAL, V1B, ZAL	2
36	PAFZZ	5330000573823	19207	7409822	GASKET .... UOC:DAL, V18, ZAL	2
37	PAFZZ	5315000589931	21450	589931	PIN, STRAIGHT, HEADLE .... UOC:DAL, V18, ZAL	2
38	PFFZZ	5340011010007	19207	10876247	COVER, ACCESS .... UOC:DAL, V18, ZAL	1
39	PAFZZ	5330000643691	19207	10900090	GASKET .... UOC:DAL, V18, ZAL	1
40	PFFZZ	4730004213962	19207	10876252	ADAPTER, STRAIGHT, FL .... UOC:DAL, V18, ZAL	1
41	PAFZZ	5306002259088	96906	MS90726-33	BOLT, MACHINE .... UOC:DAL, V18, ZAL	10
42	PAFZZ	5310004079566	19207	7410218	WASHER, LOCK CRANKCASE COVER .... UOC:DAL, V18, ZAL	10
43	PAFZZ	5330005221174	19207	10876131	GASKET .... UOC:DAL, V18, ZAL	1
44	PAFZZ	5325008037299	96906	MS16624-1150	RING, RETAINING .... UOC:DAL, V18, ZAL	1
45	PAFZZ	3020005221177	19207	10876135	GEAR, HELICAL .... UOC:DAL, V18, ZAL	1
46	PAFZZ	5305000712070	80204	B1821BH050C175N	SCREW, CAP, HEXAGON H .... UOC:DAL, V18, ZAL	8
47	PAFZZ	3120010853338	19207	8344242	BEARING, WASHER, THRU .... UOC:DAL, V18, ZAL	2
48	PAFZZ	3110005048929	19207	10876157	BEARING, SLEEVE, WINC .... UOC:DAL, V18, ZAL	2
49	PAFZZ	4730000504208	96906	MS15003-1	FITTING, LUBRICATION .... UOC:DAL, V18, ZAL	2

(1) ITEM NO	(2) SMR CODE	(3) NSN	(4) CAGEC	(5) PART NUMBER	(6) DESCRIPTION AND USABLE ON CODES (UOC)	(7) QTY
50	PAFZZ	5365001957835	19207	10900101	SHIM .... UOC:DAL, V18, ZAL	1
51	PFFZZ	3040001777830	19207	10876356	BRACKET, EYE, ROTATIN .... UOC:DAL, V18, ZAL	1
52	PAFZZ	2590002354400	19207	10876174	REEL, CABLE .... UOC:DAL, V18, ZAL	1
53	PAFZZ	5305007245830	96906	MS51965-90	SETSCREW .... UOC:DAL, V18, ZAL	1
54	PAFZZ	5305000186475	96906	MS35237-176	SCREW, MACHINE .... UOC:DAL, V18, ZAL	1
55	PFFZZ	3040004213960	19207	10876357	PLATE, RETAINING, SHA .... UOC:DAL, V18, ZAL	1
56	PAFZZ	5330004195875	19207	10938454	GASKET .... UOC:DAL, V18, ZAL	1
57	PFFZZ	3040004051931	19207	7954484	HOUSING, MECHANICAL .... UOC:DAL, V18, ZAL	1
58	PFFZZ	3040001185554	19207	7954475	HOUSING, MECHANICAL .... UOC:DAL, V18 , ZAL	1
59	PAFZZ	5340000501589	96906	MS35648-3	PLUG, EXPANSION .... UOC:DAL, V18, ZAL	1
60	PAFZZ	5310005955839	19207	8330414	WASHER, FLAT .... UOC:DAL, V18, ZAL	1
61	PAFZZ	3040007409565	19207	7409565	BRAKE BAND AND LINI .... UOC:DAL, V18, ZAL	1

END OF FIGURE

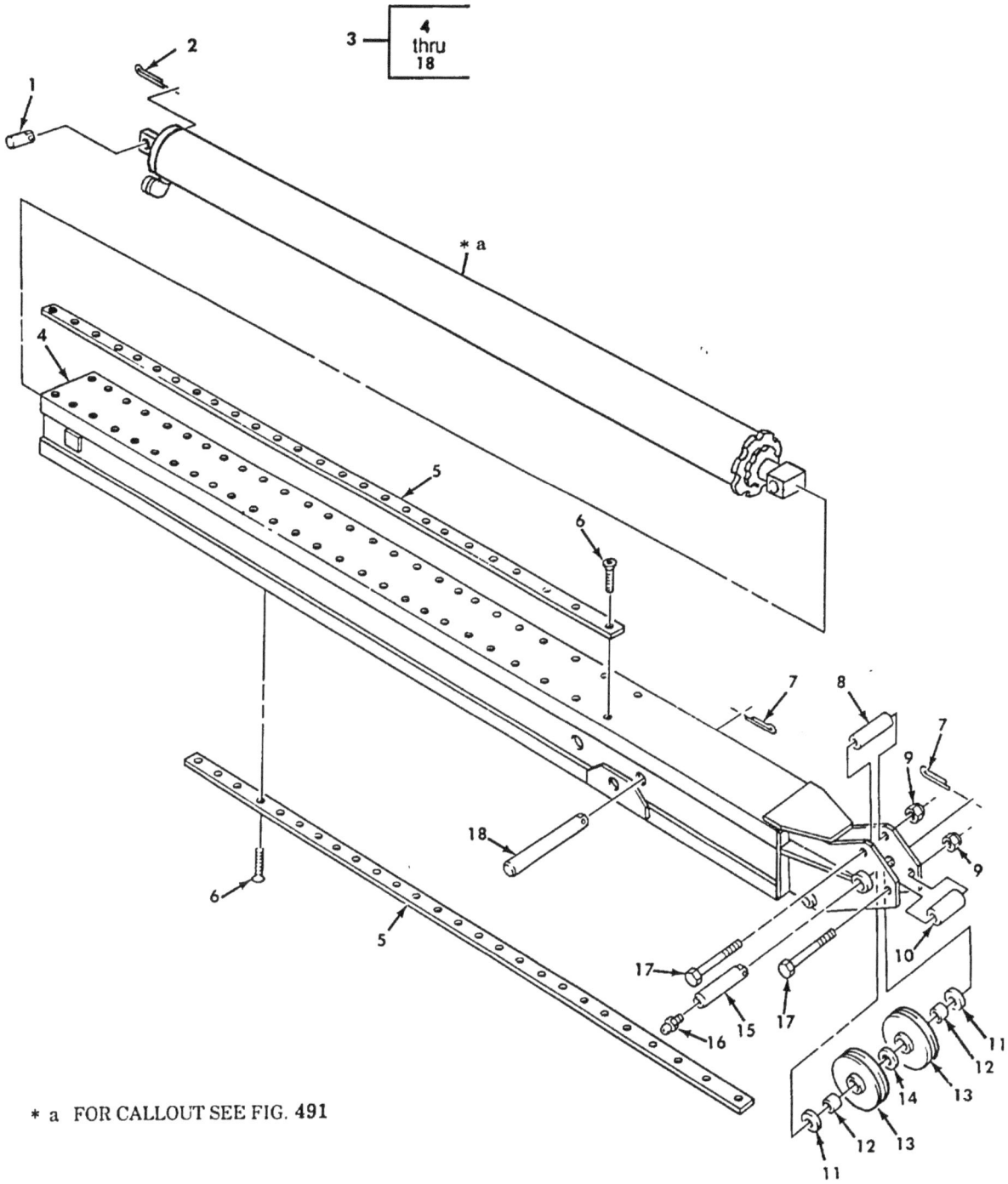

* a FOR CALLOUT SEE FIG. 491

Figure 490. Wrecker Inner Boom Extension Assembly.

(1) ITEM NO	(2) SMR CODE	(3) NSN	(4) CAGEC	(5) PART NUMBER	(6) DESCRIPTION AND USABLE ON CODES (UOC)	(7) QTY
					GROUP 2001 HOIST, CAPSTAN, WINDLASS, CRANE OR WINCH ASSEMBLY	
					FIG. 490 WRECKER INNER BOOM EXTENSION ASSEMBLY	
1	PAFZZ	3040001436390	19207	10900203	SHAFT, STRAIGHT UOC:DAL, V18, ZAL	1
2	PAFZZ	5315000137308	96906	MS24665-627	PIN, COTTER UOC:DAL, V18, ZAL	1
3	PAFHH	2590001797053	19207	10876205	BOOM EXTENSION, MIDD UOC:DAL, V18, ZAL	1
4	PAFZZ	2590000867460	19207	10876182	.BOOM SECTION, INNER UOC:DAL, V18, ZAL	1
5	PFFZZ	2590001341121	19207	10899986	.TRACK, ROLLER BOOM UOC:DAL, V18, ZAL	4
6	PAFZZ	5305000509221	96906	MS24667-41	.SCREW, CAP, SOCKET HE UOC:DAL, V18, ZAL	10
7	PAFZZ	5315008460126	96906	MS24665-628	.PIN, COTTER UOC:DAL, V18, ZAL	2
8	PAFZZ	5365001779194	19207	10900247	.SPACER, SLEEVE UOC:DAL, V18, ZAL	1
9	PAFZZ	5310009825012	96906	MS21045-12	.NUT, SELF-LOCKING, HE UOC:DAL, V18, ZAL	2
10	PAFZZ	5365004134371	19207	11611656	.SPACER, SLEEVE UOC:DAL, V18, ZAL	1
11	PAFZZ	3120007409729	19207	7409729	.BEARING, WASHER, THRU UOC:DAL, V18, ZAL	2
12	PAFZZ	3120005098270	19207	7064597	.BUSHING, SLEEVE UOC:DAL, V18, ZAL	2
13	PAFZZ	3020004468616	19207	10876383	.PULLEY, GROOVE UOC:DAL, V18, ZAL	2
14	PAFZZ	5310009848818	96906	MS27183-32	.WASHER, FLAT UOC:DAL, V18, ZAL	1
15	PAFZZ	5315002441340	19207	10899989	.PIN, STRAIGHT, HEADLE UOC:DAL, V18, ZAL	1
16	PAOZZ	4730001720034	96906	MS15003-6	.FITTING, LUBRICATION UOC:DAL, V18, ZAL	2
17	PAFZZ	5305007285475	96906	MS90727-200	.SCREW, CAP, HEXAGON H UOC:DAL, V18, ZAL	2
18	PAFZZ	5315004468740	19207	10899976	.PIN, STRAIGHT, HEADLE UOC:DAL, V18, ZAL	1

END OF FIGURE

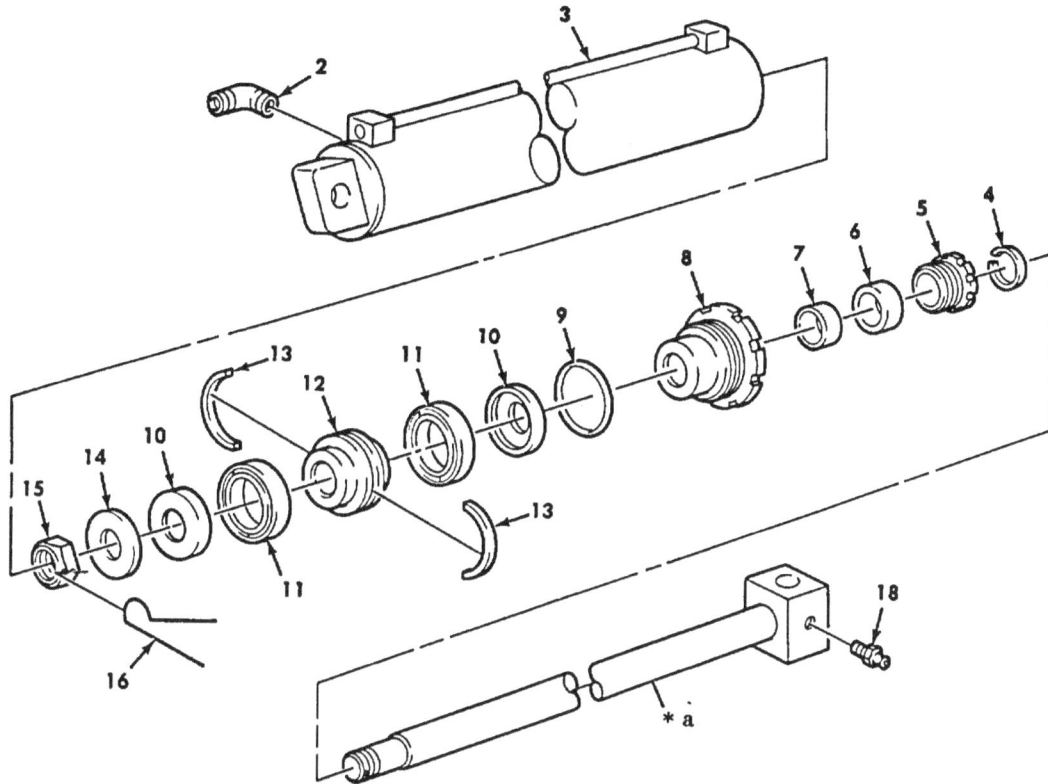

*a PART OF ITEM 17

Figure 491. Wrecker Inner Boom Extension Cylinder Assembly.

(1) ITEM NO	(2) SMR CODE	(3) NSN	(4) CAGEC	(5) PART NUMBER	(6) DESCRIPTION AND USABLE ON CODES (UOC)	(7) QTY
					GROUP 2001 HOIST, CAPSTAN, WINDLASS, CRANE OR WINCH ASSEMBLY	
					FIG. 491 WRECKER INNER BOOM EXTENSION CYLINDER ASSEMBLY	
1	PAFHH	3040009722639	61465	2046068	CYLINDER ASSEMBLY, A .... UOC:DAL, V18, ZAL	1
2	PAFZZ	4730002315596	88044	AN914-8	.ELBOW, PIPE.. UOC:DAL, V18, ZAL	1
3	PBHZZ	3040011010085	19207	10876307	.CYLINDER, ACTUATING, CRANE BOOM........................ EXTENSION.. UOC:DAL, V18, ZAL	1
4	PAHZZ	5330009722635	61465	2012993-4	.SEAL, NONMETALLIC SP .... UOC:DAL, V18, ZAL	1
5	PAHZZ	4730002385594	19207	10876461	.PACKING NUT .... UOC:DAL, V18, ZAL	1
6	PAHZZ	5330005234235	19207	10900300	.PACKING ASSEMBLY PISTON ROD .... UOC:DAL, V18, ZAL	1
7	PAHZZ	3120008265630	19207	10876159	.BEARING, SLEEVE PISTON ROD CYLINDER .... UOC:DAL, V18, ZAL	1
8	PAHZZ	2590001779196	19207	10876310	.HEAD, BOOM CYLINDER .... UOC:DAL, V18, ZAL	1
9	PAHZZ	5331005797544	96906	MS28775-243	.O-RING .... UOC:DAL, V18, ZAL	1
10	PAHZZ	5310004457238	19207	10876454	.WASHER, RECESSED CYLINDER, PISTON........................ SEAL... UOC:DAL, V18, ZAL	2
11	PAHZZ	5340005234305	19207	10900304	.CUP, COMPRESSION EXTENSION .... CYLINDER PISTON.. UOC:DAL, V18, ZAL	2
12	PAHZZ	2590009722632	19207	10876453	.PISTON, HYDRAULIC CY BOOM.................................. ACTUATING CYLINDER .... UOC:DAL, V18, ZAL	1
13	PAHZZ	4310005048923	19207	10876153	.RING, PISTON .... UOC:DAL, V18, ZAL	1
14	PAHZZ	5310004193082	19207	10876506	.WASHER, FLAT PISTON ROD .... UOC:DAL, V18, ZAL	1
15	PAHZZ	5310004702107	19207	10900151	.NUT, PLAIN, HEXAGON PISTON ROD .... UOC:DAL, V18, ZAL	1
16	MHHZZ		96906	MS20995F91-12	.WIRE, NONELECTRICAL .... UOC:DAL, V18, ZAL	1
17	PBHZZ	3040011010096	19207	10876309	.ROD, PISTON, LINEAR A ACTUATING.............................. CYLINDER.. UOC:DAL, V18, ZAL	1
18	PAHZZ	4730000504208	96906	MS15003-1	..FITTING, LUBRICATION .... UOC:DAL, V18, ZAL	2

END OF FIGURE

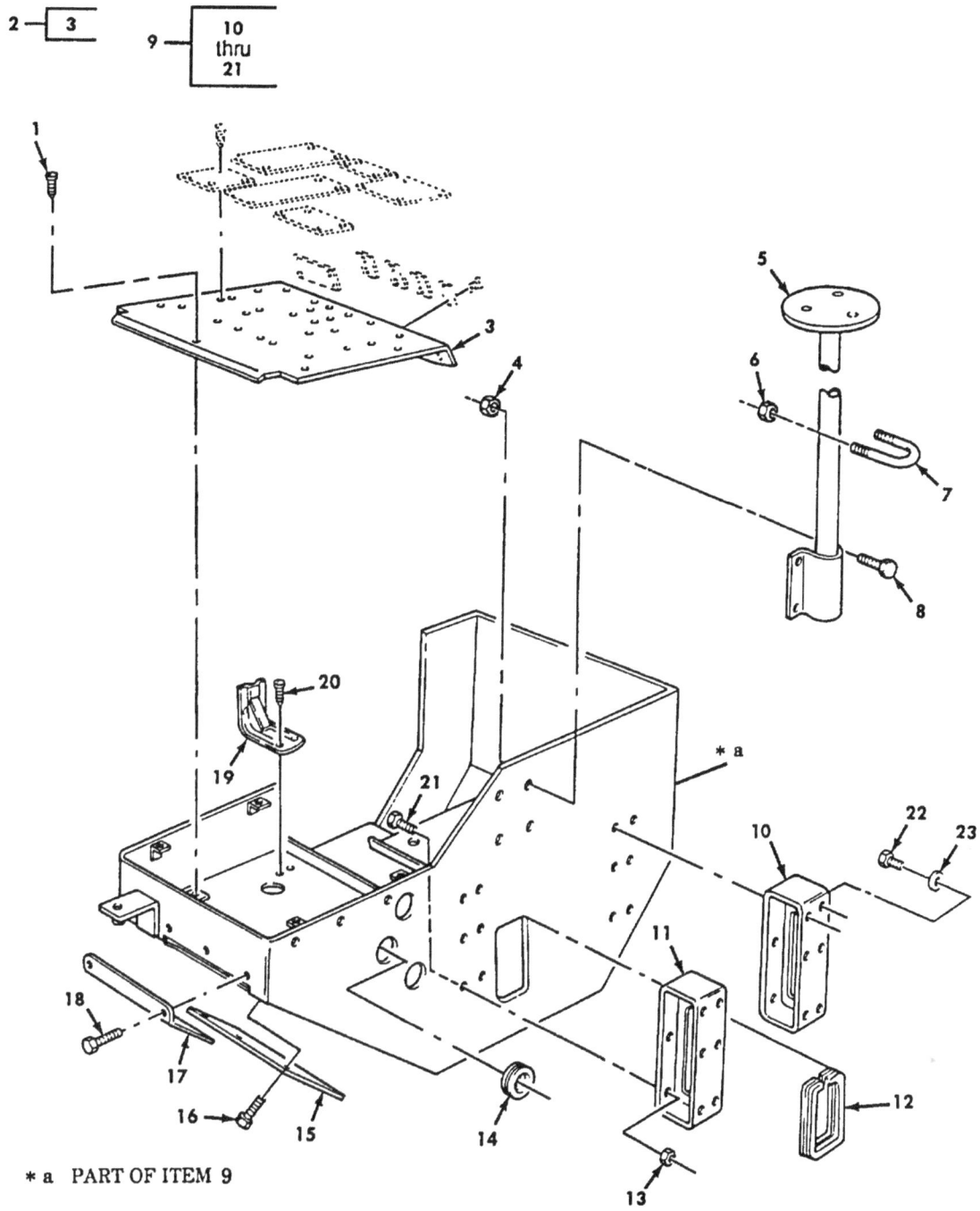

*a PART OF ITEM 9

Figure 492. Crane Gondola Assembly, Warning Light and Instruction Plate.

(1) ITEM NO	(2) SMR CODE	(3) NSN	(4) CAGEC	(5) PART NUMBER	(6) DESCRIPTION AND USABLE ON CODES (UOC)	(7) QTY
					GROUP 2001 HOIST, CAPSTAN, WINDLASS, CRANE OR WINCH ASSEMBLY	
					FIG. 492 CRANE GONDOLA ASSEMBLY, WARNING LIGHT AND INSTRUCTION PLATE	
1	PAOZZ	5305000526920	96906	MS24629-56	SCREW,TAPPING UOC:DAL,V18,ZAL	5
2	PBOZZ	9905011855762	19207	10900051-2	PANEL ASSY UOC:DAL,V18,ZAL	1
3	PBOZZ	5340012052503	19207	10876261-1	.PLATE,MOUNTING UOC:DAL,V18,ZAL	1
4	PAOZZ	5310008140672	96906	MS51943-36	NUT,SELF-LOCKING,HE UOC:DAL,V18,ZAL	4
5	PFOZZ	2590011465258	19207	12277333	MOUNT,LIGHT,WARNING UOC:DAL,V18,ZAL	1
6	PAOZZ	5310004883888	96906	MS51943-40	NUT,SELF-LOCKING,HE UOC:DAL,V18,ZAL	2
7	PFOZZ	5306011653272	19207	12277330	BOLT,U UOC:DAL,V18,ZAL	1
8	PAOZZ	5305002692804	96906	MS90726-61	SCREW,CAP,HEXAGON H UOC:DAL,V18,ZAL	4
9	PFFHH	2540004172732	19207	10876202	COVER,FITTED,VEHICU UOC:DAL,V18,ZAL	1
10	PFFZZ	5342010823612	19207	10876391-1	.BRACKET,ENGINE ACCE UOC:DAL,V18,ZAL	1
11	PFFZZ	5340010823611	19207	10876266	.BRACKET,DOUBLE ANGL UOC:DAL,V18,ZAL	1
12	PAFZZ	9390001660254	19207	10900472	.NONMETALLIC CHANNEL ...... UOC:DAL,V18 ,ZAL	1
13	PAFZZ	5310000624954	96906	MS21045-8	.NUT,SELF-LOCKING,HE UOC:DAL,V18,ZAL	12
14	PAFZZ	5330001911161	19207	10910268	.SEAL,NONMETALLIC CH CRANE CAB UOC:DAL,V18,ZAL	3
15	PFFZZ	5340004831107	19207	11593204	.COVER,ACCESS CRANE CAB UOC:DAL,V18,ZAL	1
16	PAFZZ	5306000187527	21450	187527	.BOLT,ASSEMBLED WASH UOC:DAL,V18,ZAL	6
17	PFFZZ	5340002373706	19207	10876552	.COVER,ACCESS CRANE CAB UOC:DAL,V18,ZAL	1
18	PAFZZ	5305004324201	96906	MS51861-45	.SCREW,TAPPING UOC:DAL,V18,ZAL	6
19	PFFZZ	5342012306702	19207	10910956-1	.BRACKET,OIL CAN UOC:DAL,V18,ZAL	1
20	PAFZZ	5305001380069	96906	MS51861-44	.SCREW,TAPPING UOC:DAL,V18,ZAL	2
21	PAFZZ	5305007195219	96906	MS90727-111	.SCREW,CAP,HEXAGON H UOC:DAL,V18,ZAL	12
22	PAHZZ	5305007250154	96906	MS90727-112	SCREW,CAP,HEXAGON H UOC:DAL,V18,ZAL	12
23	PAFZZ	5310000034094	01276	210104-8S	WASHER,LOCK UOC:DAL,V18,ZAL	12

END OF FIGURE

492-1

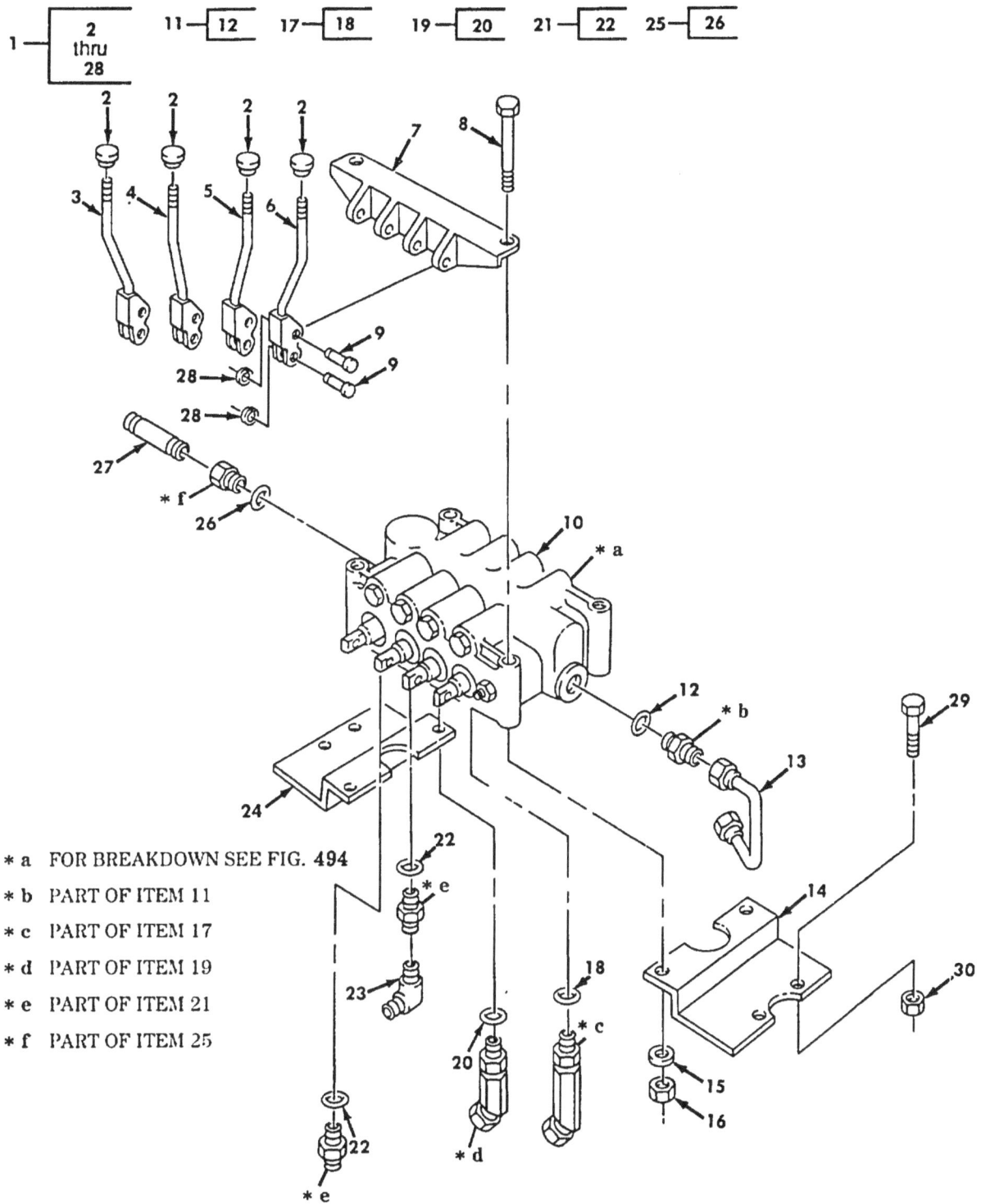

* a   FOR BREAKDOWN SEE FIG. 494

* b   PART OF ITEM 11

* c   PART OF ITEM 17

* d   PART OF ITEM 19

* e   PART OF ITEM 21

* f   PART OF ITEM 25

Figure 493. Gondola Hoist Control Valve Assembly.

(1) ITEM NO	(2) SMR CODE	(3) NSN	(4) CAGEC	(5) PART NUMBER	(6) DESCRIPTION AND USABLE ON CODES (UOC)	(7) QTY
					GROUP 2001 HOIST, CAPSTAN, WINDLASS, CRANE OR WINCH ASSEMBLY	
					FIG. 493 GONDOLA HOIST CONTROL VALVE ASSEMBLY	
1	PAFFF	4820004495059	19207	11621117	VALVE, LINEAR, DIRECT .... UOC:DAL, V18, ZAL	1
2	PAFZZ	5355001911029	19207	10900172	.KNOB MANUAL HOIST LEVER .... UOC:DAL, V18, ZAL	4
3	PBFZZ	3040004982386	19207	11621121-1	.LEVER, MANUAL CONTRO .... UOC:DAL, V18, ZAL	1
4	PBFZZ	2590004982388	19207	11640398-1	.LEVER, MANUAL CONTRO .... UOC:DAL, V18, ZAL	1
5	PBFZZ	2590004982387	19207	11640398-2	.LEVER, MANUAL CONTRO .... UOC:DAL, V18, ZAL	1
6	PBFZZ	3040004982389	19207	11621121-2	.LEVER, MANUAL CONTRO .... UOC:DAL, V18, ZAL	1
7	PFFZZ	2590004364601	19207	11621122-2	.BASE, MOUNTING .... UOC:DAL, V18, ZAL	1
8	PAFZZ	5305000712083	80204	B1821BH050C500N	.SCREW, CAP, HEXAGON H .... UOC:DAL, V18, ZAL	4
9	PAFZZ	5315004014383	19207	11621118	.PIN, GROOVED, HEADED .... UOC:DAL, V18, ZAL	8
10	PAFFF	4810011307930	19207	11621116	.VALVE, LINEAR, DIRECT .... UOC:DAL, V18, ZAL	1
11	PAFZZ	4730011208547	19207	11621892	.ADAPTER, STRAIGHT, TU .... UOC:DAL, V18, ZAL	1
12	PAFZZ	5331008163546	96906	MS28778-20	..O-RING .... UOC:DAL, V18, ZAL	1
13	PAFZZ	4710002354819	19207	11621120	.TUBE ASSEMBLY, METAL .... UOC:DAL, V18, ZAL	1
14	PFFZZ	5340004910331	19207	11621124	.BRACKET, DOUBLE ANGL .... UOC:DAL, V18, ZAL	1
15	PAFZZ	5310000034094	01276	210104-8S	.WASHER, LOCK .... UOC:DAL, V18, ZAL	4
16	PAFZZ	5310007680318	24617	9413509	.NUT, PLAIN, HEXAGON .... UOC:DAL, V18, ZAL	4
17	PAFZZ	4730004497356	19207	11621894-2	.ELBOW, PIPE TO BOSS .... UOC:DAL, V18, ZAL	2
18	PAFZZ	5331008045694	96906	MS28778-16	..O-RING .... UOC:DAL, V18, ZAL	1
19	PAFZZ	4730004509671	19207	11621894-1	.ELBOW, PIPE TO BOSS .... UOC:DAL, V18, ZAL	2
20	PAFZZ	5331008045694	96906	MS28778-16	..O-RING .... UOC:DAL, V18, ZAL	1
21	PAFZZ	4730011150646	19207	11621890-1	.ADAPTER, STRAIGHT, TU .... UOC:DAL, V18, ZAL	3
22	PAFZZ	5331008045694	96906	MS28778-16	..O-RING .... UOC:DAL, V18, ZAL	1
23	PAFZZ	4730001630236	24617	9410285	.ELBOW, TUBE .... UOC:DAL, V18, ZAL	1

(1) ITEM NO	(2) SMR CODE	(3) NSN	(4) CAGEC	(5) PART NUMBER	(6) DESCRIPTION AND USABLE ON CODES (UOC)	(7) QTY
24	PFFZZ	5340004910329	19207	11621123	.BRACKET,DOUBLE ANGL UOC:DAL,V18,ZAL	1
25	PAFZZ	4730002494416	19207	11621891	.ADAPTER,STRAIGHT,PI UOC:DAL,V18,ZAL	1
26	PAFZZ	5331007025643	96906	MS28775-128	..O-RING UOC:DAL,V18,ZAL	1
27	PAFZZ	4730002887495	96906	MS51953-154	.NIPPLE,PIPE UOC:DAL,V18,ZAL	1
28	PAFZZ	5325004193322	96906	MS90707-1050	.RING,RETAINING UOC:DAL,V18,ZAL	8

END OF FIGURE

Figure 494. Crane Gondola Control Valve Assembly, Basic.

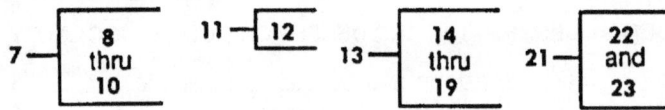

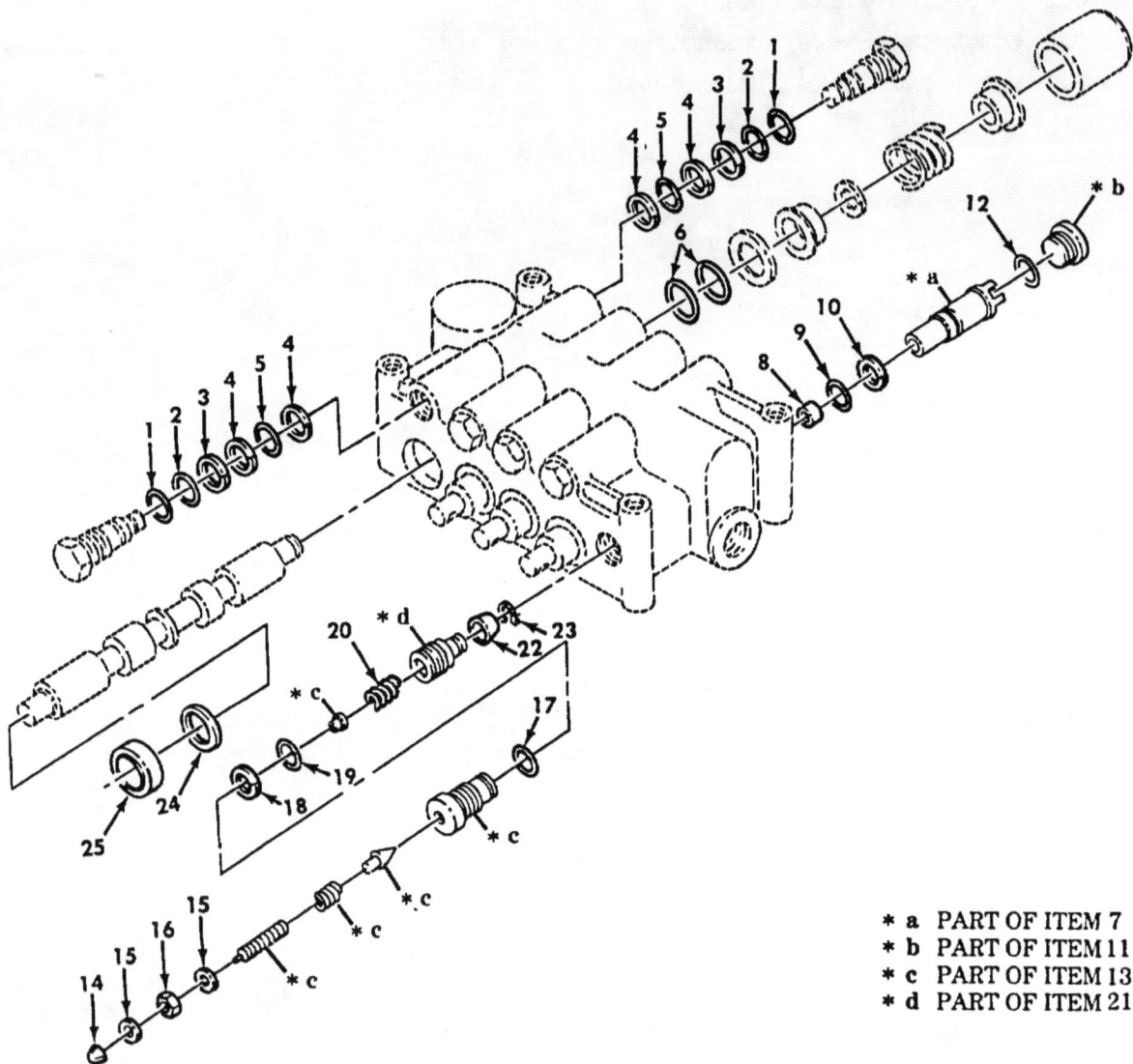

* a   PART OF ITEM 7
* b   PART OF ITEM 11
* c   PART OF ITEM 13
* d   PART OF ITEM 21

(1) ITEM NO	(2) SMR CODE	(3) NSN	(4) CAGEC	(5) PART NUMBER	(6) DESCRIPTION AND USABLE ON CODES (UOC)	(7) QTY
					GROUP 2001 HOIST, CAPSTAN, WINDLASS, CRANE OR WINCH ASSEMBLY	
					FIG. 494 CRANE GONDOLA CONTROL VALVE ASSEMBLY, BASIC	
1	PAHZZ	5331007025220	96906	MS28775-117	O-RING PART OF KIT P/N 5704273 .... UOC:DAL, V18, ZAL	8
2	PAHZZ	5331005797916	96906	MS28775-115	O-RING PART OF KIT P/N 5704273 .... UOC:DAL, V18, ZAL	8
3	PAHZZ	5330008013440	96906	MS28774-115	RETAINER, PACKING PART OF KIT P/N .... 5704273 .... UOC:DAL, V18, ZAL	8
4	PAHZZ	5330005763206	96906	MS28774-114	PACKING, PREFORMED PART OF KIT P/N .... 5704273 .... UOC:DAL, V18, ZAL	16
5	PAHZZ	5331006180801	96906	MS28775-114	O-RING PART OF KIT P/N 5704273 .... UOC:DAL, V18, ZAL	8
6	PAHZZ	5330005421329	96906	MS28775-120	O-RING .... UOC:DAL, V18, ZAL	8
7	PAHZZ	2590011416305	19207	11662536	SEAT, RETAINER ASSEM PART OF KIT P/N .... 5704274 .... UOC:DAL, V18, ZAL	1
8	PAHZZ	5342010624715	10988	L32618	.SEAT .... UOC:DAL, V18, ZAL	1
9	PAHZZ	5331007025220	96906	MS28775-117	.O-RING PART OF KIT P/N 5704273 .... UOC:DAL, V18, ZAL	1
10	PAHZZ	5330008357712	96906	MS28774-117	.RETAINER, PACKING .... UOC:DAL, V18, ZAL	1
11	PAHZZ	4730010762735	19207	11662758	PLUG, PIPE PART OF KIT P/N 5704274 .... UOC:DAL, V18, ZAL	1
12	PAHZZ	5330004722783	96906	MS28778-14	.PACKING, PREFORMED .... UOC:DAL, V18, ZAL	1
13	PAHZZ	4820011459140	19207	11662537	VALVE, REGULATING, FL PART OF KIT P/N .... 5704274 .... UOC:DAL, V18, ZAL	1
14	PAHZZ	5310010575518	96906	MS24679-66	.NUT, PLAIN, CAP .... UOC:DAL, V18, ZAL	1
15	PAHZZ	5310011399856	19207	11640433	.WASHER, FLAT .... UOC:DAL, V18, ZAL	2
16	PAHZZ	5310008539335	96906	MS35691-13	.NUT, PLAIN, HEXAGON .... UOC:DAL, V18, ZAL	1
17	PAHZZ	5330004722783	96906	MS28778-14	.PACKING, PREFORMED .... UOC:DAL, V18, ZAL	1
18	PAHZZ	5330008357712	96906	MS28774-117	.RETAINER, PACKING .... UOC:DAL, V18, ZAL	1
19	PAHZZ	5331007025220	96906	MS28775-117	.O-RING PART OF KIT P/N 5704273 .... UOC:DAL, V18, ZAL	1
20	PAHZZ	5360009340089	09990	612668	SPRING, SPECIAL PART OF KIT P/N .... 5704274 .... UOC:DAL, V18, ZAL	1
21	PAHZZ	4820010854762	19207	11662760	PLUNGER AND SCREEN PART OF KIT P/N	1

(1) ITEM NO	(2) SMR CODE	(3) NSN	(4) CAGEC	(5) PART NUMBER	(6) DESCRIPTION AND USABLE ON CODES (UOC)	(7) QTY
					5704274... UOC:DAL, V18, ZAL	
22	PAHZZ	4330011445557	09990	635439	.STRAINER ELEMENT, SE .... UOC:DAL, V18, ZAL	1
23	PAHZZ	5325011378828	08752	725197	.RING, RETAINING .... UOC:DAL, V18, ZAL	1
24	PAHZZ	5330011377089	19207	11640447	RETAINER, PACKING .... UOC:DAL, V18, ZAL	1
25	PAHZZ	2590011317453	19207	11640446	WIPER ASSEMBLY, SPOO .... UOC:DAL, V18, ZAL	1

END OF FIGURE

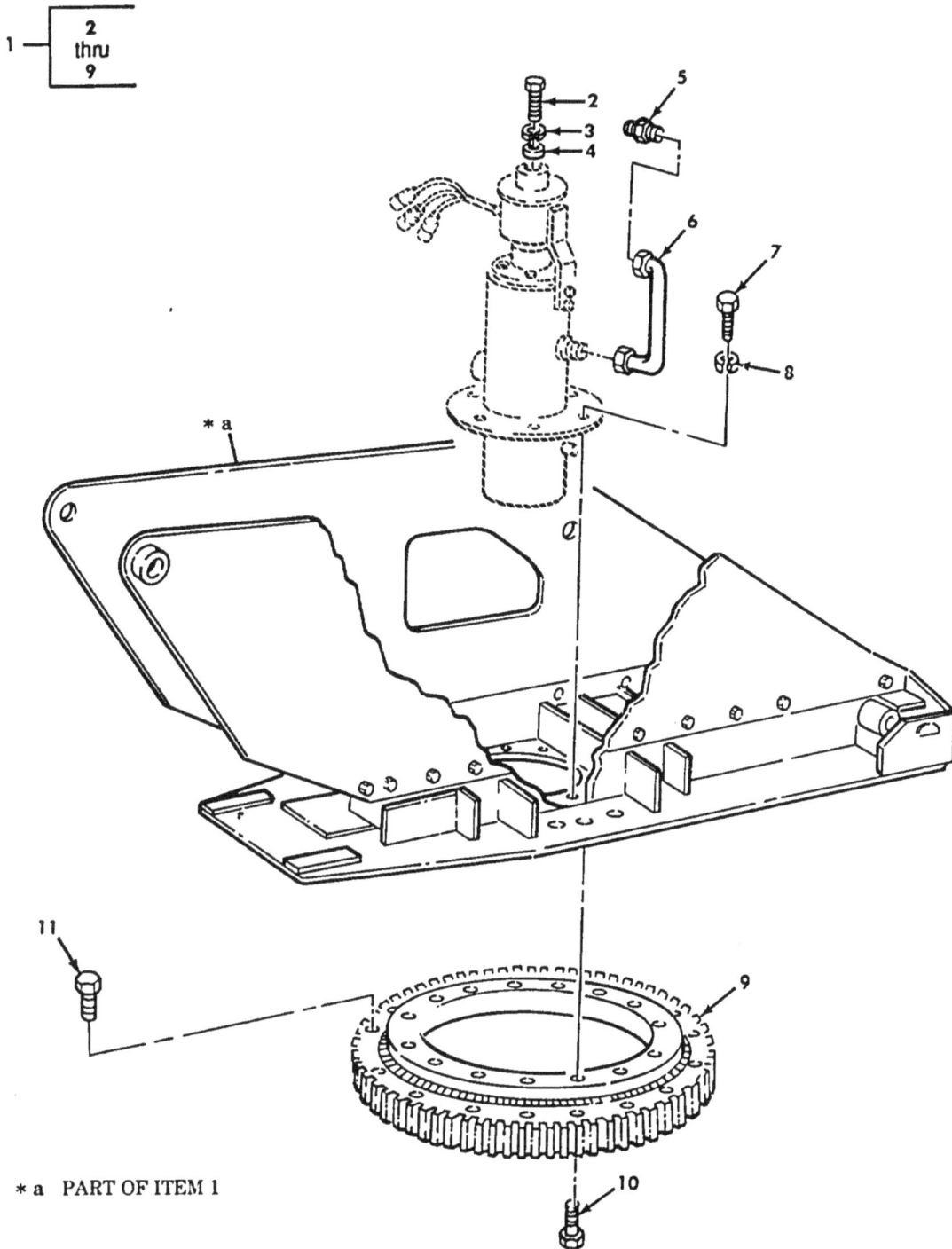

Figure 495. Wrecker Turntable Assembly.

*a PART OF ITEM 1

(1) ITEM NO	(2) SMR CODE	(3) NSN	(4) CAGEC	(5) PART NUMBER	(6) DESCRIPTION AND USABLE ON CODES (UOC)	(7) QTY
					GROUP 2001 HOIST, CAPSTAN, WINDLASS, CRANE OR WINCH ASSEMBLY	
					FIG. 495 WRECKER TURNTABLE ASSEMBLY	
1	PFFFF	2590010823607	19207	12256623	TURNTABLE ASSEMBLY .... UOC:DAL, V18, ZAL	1
2	PAFZZ	5305007195275	96906	MS90727-128	.SCREW, CAP, HEXAGON H .... UOC:DAL, V18, ZAL	1
3	PAFZZ	5310000034094	01276	210104-8S	.WASHER, LOCK .... UOC:DAL, V18, ZAL	1
4	PAFZZ	5310008095998	96906	MS27183-18	.WASHER, FLAT .... UOC:DAL, V18, ZAL	1
5	PAFZZ	4730002784311	96906	MS51500A10	.ADAPTER, STRAIGHT, PI .... UOC:DAL, V18, ZAL	1
6	PAFZZ	4710010892059	19207	12256568	.TUBE ASSEMBLY, METAL .... UOC:DAL, V18, ZAL	1
7	PAFZZ	5305002692803	96906	MS90726-60	.SCREW, CAP, HEXAGON .... UOC:DAL, V18, ZAL	6
8	PAFZZ	5310006379541	96906	MS35338-46	.WASHER, LOCK .... UOC:DAL, V18, ZAL	6
9	PBFZZ	2590005333398	19207	10900308	.GEAR AND BEARING .... UOC:DAL, V18, ZAL	1
10	PAFZZ	5305001829561	96906	MS35458-76	SCREW, CAP, SOCKET HE .... UOC:DAL, V18, ZAL	18
11	PAFZZ	5305009838082	96906	MS16998-112	SCREW, CAP, SOCKET .... UOC:DAL, V18, ZAL	18

END OF FIGURE

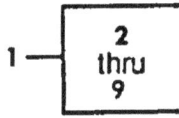

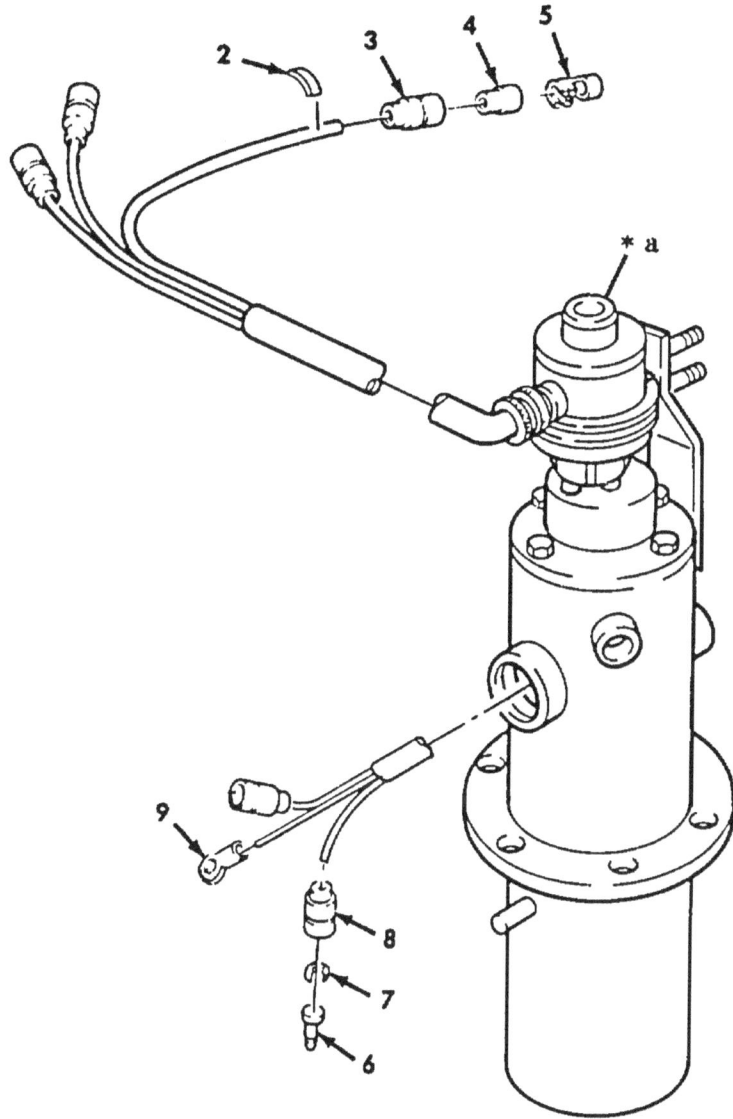

* a PART OF ITEM 1

Figure 496. Crane Hydraulic Swivel Assembly.

(1) ITEM NO	(2) SMR CODE	(3) NSN	(4) CAGEC	(5) PART NUMBER	(6) DESCRIPTION AND USABLE ON CODES (UOC)	(7) QTY
					GROUP 2001 HOIST, CAPSTAN, WINDLASS, CRANE OR WINCH ASSEMBLY	
					FIG. 496 CRANE HYDRAULIC SWIVEL ASSEMBLY	
1	PAFFF	4810011062062	19207	11669352	VALVE, LINEAR, DIRECT .... UOC:DAL, V18, ZAL	1
2	PAFZZ	9905007524649	81349	M43436/1-1	.BAND, MARKER .... UOC:DAL, V18, ZAL	6
3	PAFZZ	5935008338561	19207	8338561	.SHELL, ELECTRICAL CO .... UOC:DAL, V18, ZAL	1
4	PAFZZ	5970008338562	19207	8338562	.INSULATOR, BUSHING .... UOC:DAL, V18, ZAL	1
5	PAFZZ	5940003996676	19207	8338564	.TERMINAL ASSEMBLY .... UOC:DAL, V18, ZAL	1
6	PAFZA	5999000572929	19204	572929	.CONTACT, ELECTRICAL .... UOC:DAL, V18, ZAL	4
7	PAFZZ	5310008338567	19207	8338567	.WASHER, SLOTTED .... UOC:DAL, V18, ZAL	4
8	PAFZZ	5935005729180	19207	8338566	.SHELL, ELECTRICAL CO .... UOC:DAL, V18, ZAL	4
9	PAFZZ	5940007056709	19207	7056709	.TERMINAL, LUG .... UOC:DAL, V18, ZAL	1

END OF FIGURE

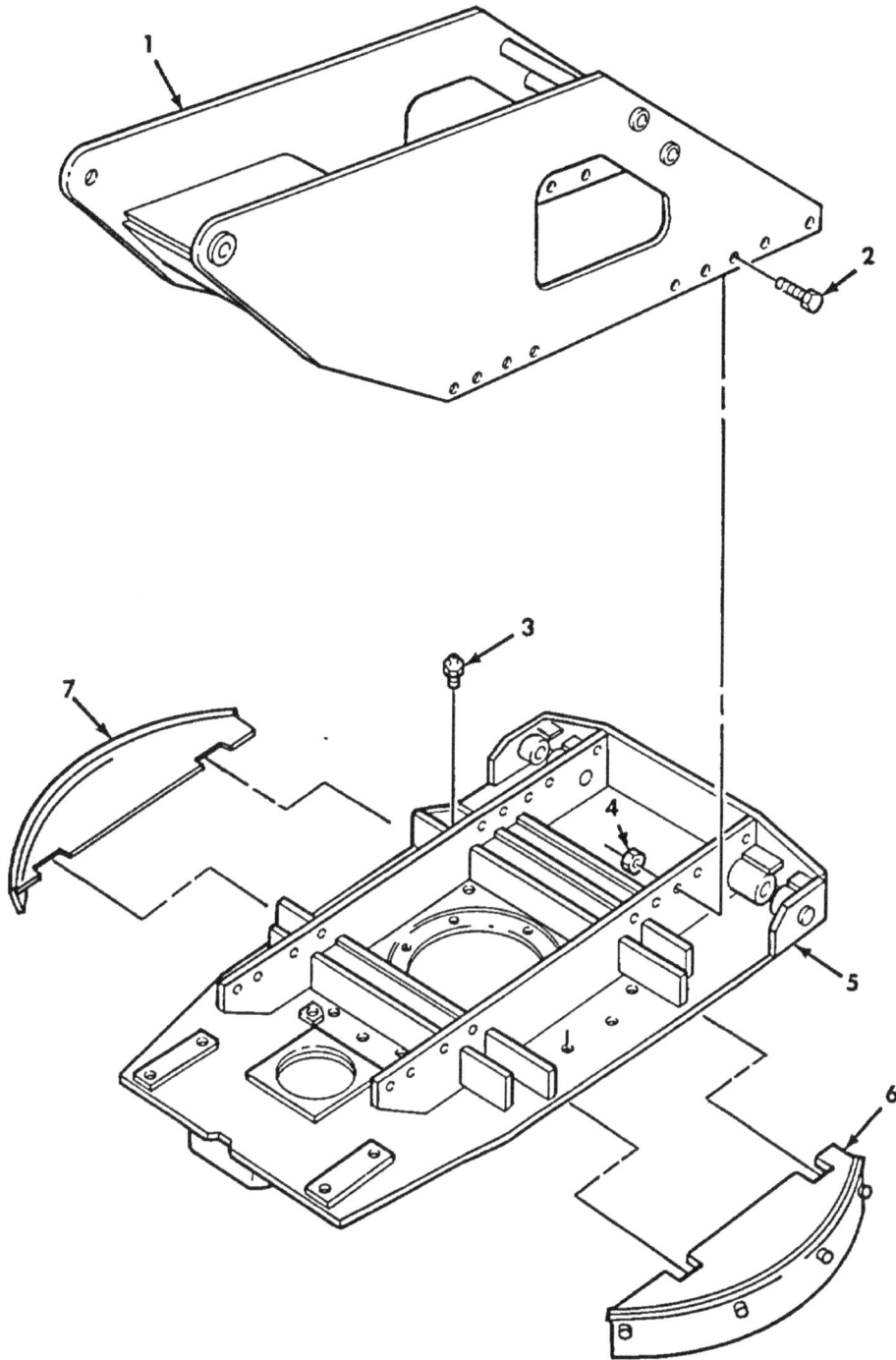

Figure 497. Crane Support, Turntable, and Bearing Guard.

(1) ITEM NO	(2) SMR CODE	(3) NSN	(4) CAGEC	(5) PART NUMBER	(6) DESCRIPTION AND USABLE ON CODES (UOC)	(7) QTY
					GROUP 2001 HOIST, CAPSTAN, WINDLASS, CRANE OR WINCH ASSEMBLY	
					FIG. 497 CRANE SUPPORT, TURNTABLE, AND BEARING GUARD	
1	PFFZZ	3950010823608	19207	12256622	SUPPORT, WINCH .... UOC:DAL, V18, ZAL	1
2	PAFZZ	5305000223843	80205	B1821BH088F250N	SCREW, CAP, HEXAGON H .... UOC:DAL, V18, ZAL	18
3	PAOZZ	4730000504208	96906	MS15003-1	FITTING, LUBRICATION .... UOC:DAL, V18, ZAL	1
4	PAFZZ	5310009825014	96906	MS21045-14	NUT, SELF-LOCKING, HE .... UOC:DAL, V18, ZAL	18
5	PFFZZ	2590010823606	19207	12256696	TURNTABLE, UNDERSTRU .... UOC:DAL, V18, ZAL	1
6	PAFZZ	2590004180673	19207	10910304	GUARD, MECHANICAL DR RIGHT HAND .... UOC:DAL, V18, ZAL	1
7	PAFZZ	2590004180668	19207	10910303	GUARD, MECHANICAL DR LEFT HAND .... UOC:DAL, V18, ZAL	1

END OF FIGURE

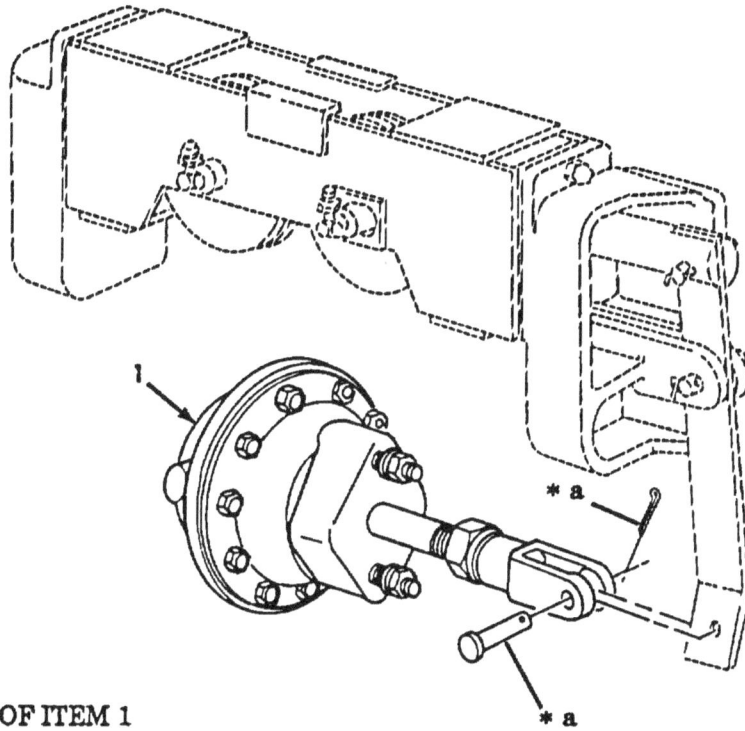

* a PART OF ITEM 1

Figure 498. Wrecker Rear Winch Tensioner Chamber Assembly.

(1) ITEM NO	(2) SMR CODE	(3) NSN	(4) CAGEC	(5) PART NUMBER	(6) DESCRIPTION AND USABLE ON CODES (UOC)	(7) QTY
					GROUP 2001 HOIST, CAPSTAN, WINDLASS, CRANE OR WINCH ASSEMBLY	
					FIG. 498 WRECKER REAR WINCH TENSIONER CHAMBER ASSEMBLY	
1	PAOZZ	2530010853787	19207	7974745-1	CHAMBER, AIR BRAKE .... UOC:DAL, V18, ZAL	1

END OF FIGURE

Figure 499. Wrecker Rear Winch Assembly Complete and Wire Rope Assembly.

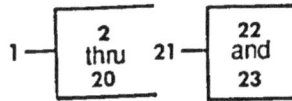

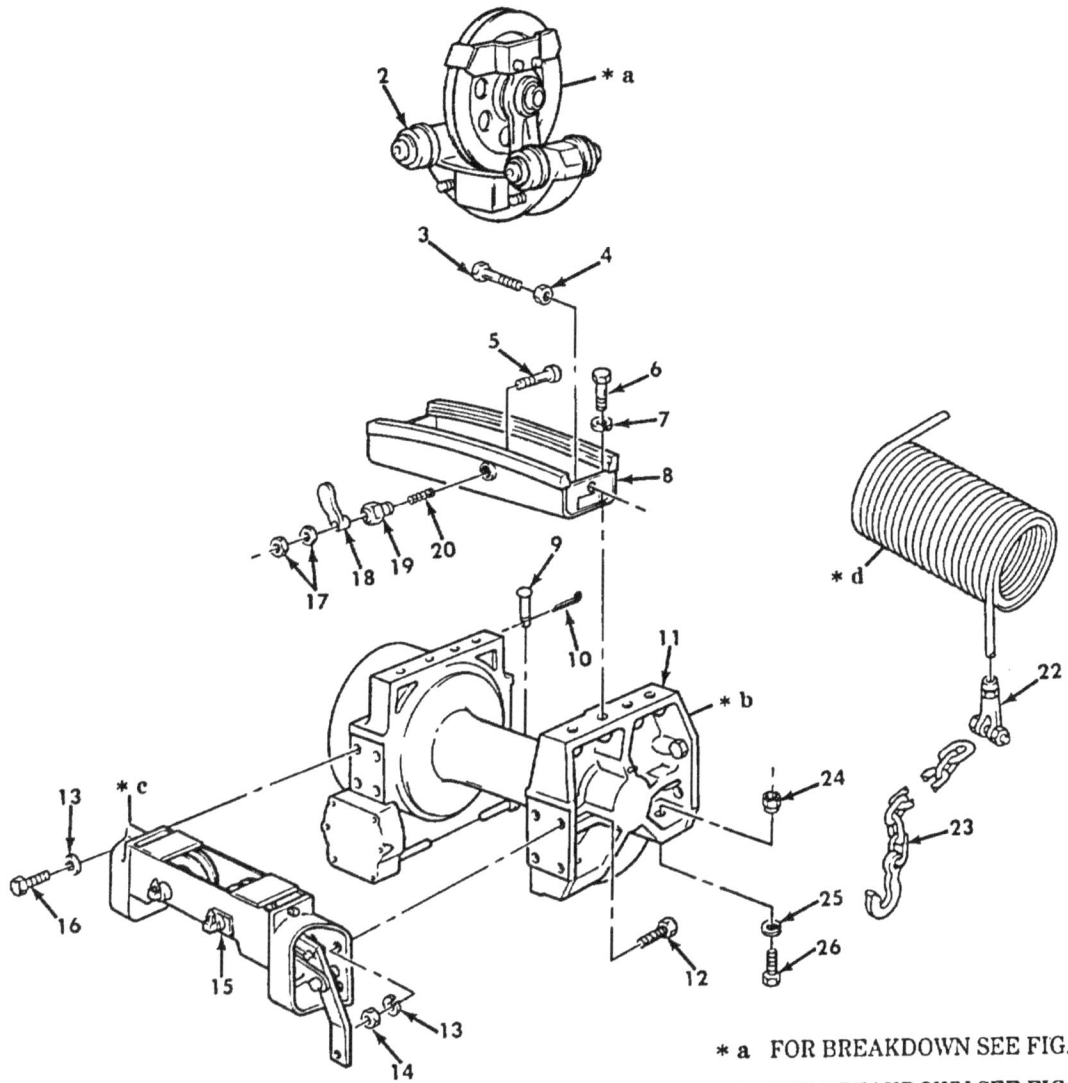

* a   FOR BREAKDOWN SEE FIG. 283
* b   FOR BREAKDOWN SEE FIG. 285
* c   FOR BREAKDOWN SEE FIG. 284
* d   PART OF ITEM 21

(1) ITEM NO	(2) SMR CODE	(3) NSN	(4) CAGEC	(5) PART NUMBER	(6) DESCRIPTION AND USABLE ON CODES (UOC)	(7) QTY
					GROUP 2001 HOIST, CAPSTAN, WINDLASS, CRANE, OR WINCH ASSEMBLY	
					FIG. 499 WRECKER REAR WINCH ASSEMBLY COMPLETE AND WIRE ROPE ASSEMBLY	
1	PAFFH	2540007409980	19207	7409980	WINCH ASY REAR .... UOC:DAL, V18, ZAL	1
2	PAFFZ	2540007409954	192077	409954	.TROLLEY ASSEMBLY WI .... UOC:DAL, V18, ZAL	1
3	PAFZZ	5306007417084	19207	7417084	.BOLT, MACHINE .... UOC:DAL, V18, ZAL	2
4	PAFZZ	5310008512677	96906	MS35691-49	.NUT, PLAIN, HEXAGON .... UOC:DAL, V18, ZAL	2
5	PAFZZ	4820006525548	19207	8741768	.POPPET, LOCK, TROLLEY .... UOC:DAL, V18, ZAL	1
6	PAFZZ	5305007247221	80204	B1821BH063C175N	.SCREW, CAP, HEXAGON H .... UOC:DAL, V18, ZAL	4
7	PAFZZ	5310008206653	80045	23MS35338-50	.WASHER, LOCK .... UOC:DAL, V18, ZAL	4
8	PFFZZ	2590004180887	19207	7954574	.TRACK ASSEMBLY, WINC .... UOC:DAL, V18, ZAL	1
9	PAOZZ	5315002822583	19207	8330478	.PIN, GROOVED, HEADED .... UOC:DAL, V18, ZAL	1
10	PAOZZ	5315002368345	96906	MS24665-5	.PIN, COTTER .... UOC:DAL, V18, ZAL	1
11	PFFFF	2590011208512	19207	7954590	.WINCH, DRUM, VEHICLE .... UOC:DAL, V18, ZAL	1
12	PAFZZ	5305009227994	80204	B1821BH075C250N	.SCREW, CAP, HEXAGON H .... UOC:DAL, V18, ZAL	2
13	PAFZZ	5310005847888	96906	MS35338-51	.WASHER, LOCK .... UOC:DAL, V18, ZAL	8
14	PAFZZ	5310007638921	96906	MS51967-23	.NUT, PLAIN, HEXAGON .... UOC:DAL, V18, ZAL	2
15	PFFFF	2590004040752	6146	M305820K	.TENSIONER, WINCH .... UOC:DAL, V18, ZAL	1
16	PAFZZ	5305009399204	96906	MS90725-187	.SCREW, CAP, HEXAGON H .... UOC:DAL, V18, ZAL	6
17	PAFZZ	5310008807744	96906	MS51967-5	.NUT, PLAIN, HEXAGON .... UOC:DAL, V18, ZAL	2
18	PAFZZ	2540007409150	19207	7409150	.LATCH, DRUM LOCK .... UOC:DAL, V18, ZAL	1
19	PAFZZ	2590002262347	19207	8344240	.NUT, WINCH POPPET .... UOC:DAL, V18, ZAL	1
20	PAFZZ	5360007538706	19207	7538706	.SPRING, HELICAL, COMP .... UOC:DAL, V18, ZAL	1
21	MOOZZ		19207	12253105-19	WIRE ROPE ASSEMBLY .... UOC:DAL, V18, ZAL	1
22	PAOZZ	4030007065553	19207	7065553	.SOCKET, WIRE ROPE .... UOC:DAL, V18, ZAL	1
23	PAOZZ	4010010350159	19207	12253104-5	.CHAIN ASSEMBLY, SING .... UOC:DAL, V18, ZAL	1

(1) ITEM NO	(2) SMR CODE	(3) NSN	(4) CAGEC	(5) PART NUMBER	(6) DESCRIPTION AND USABLE ON CODES (UOC)	(7) QTY
24	PAFZZ	5310003404953	96906	MS51943-50	NUT, SELF-LOCKING, HE .... UOC:DAL, V18, ZAL	4
25	PAFZZ	5310008098541	96906	MS27183-27	WASHER, FLAT .... UOC:DAL, V18, ZAL	4
26	PAFZZ	5305011284095	96906	MS90726-237	SCREW, CAP, HEXAGON H .... UOC:DAL, V18, ZAL	4

END OF FIGURE

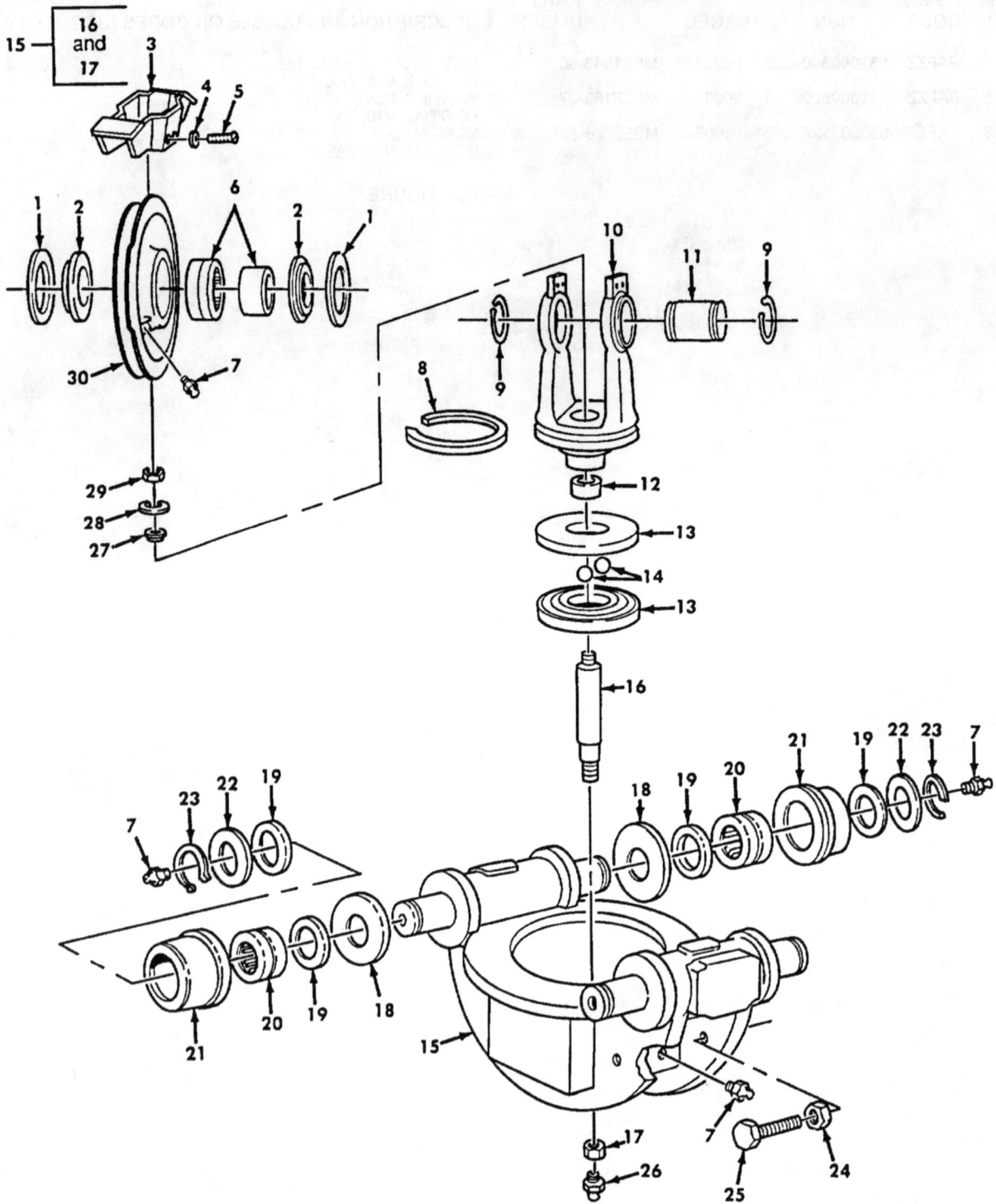

Figure 500. Wrecker Rear Winch Sheave Swivel and Trolley.

(1) ITEM NO	(2) SMR CODE	(3) NSN	(4) CAGEC	(5) PART NUMBER	(6) DESCRIPTION AND USABLE ON CODES (UOC)	(7) QTY
					GROUP 2001 HOIST, CAPSTAN, WINDLASS, CRANE OR WINCH ASSEMBLY	
					FIG. 500 WRECKER REAR WINCH SHEAVE SWIVEL AND TROLLEY	
1	PAFZZ	3120007409953	19207	7409953	BEARING, WASHER, THRU .... UOC:DAL, V18, ZAL	2
2	PAFZZ	5310011385516	19207	7954539-1	WASHER, SHOULDERED .... UOC:DAL, V18, ZAL	2
3	PAFZZ	3020007409893	19207	7409893	GUARD, MECHANICAL DR .... UOC:DAL, V18, ZAL	1
4	PAFZZ	5310006379541	96906	MS35338-46	WASHER, LOCK .... UOC:DAL, V18, ZAL	4
5	PAFZZ	5305002693234	96906	MS90727-58	SCREW, CAP, HEXAGON .... UOC:DAL, V18, ZAL	4
6	PAFZZ	3110007527760	96906	MS51961-40	BEARING, ROLLER, NEEDLE .... UOC:DAL, V18, ZAL	1
7	PAFZZ	4730000504208	96906	MS15003-1	FITTING, LUBRICATION .... UOC:DAL, V18, ZAL	7
8	PAFZZ	5330007409889	19207	7409889	FELT, MECHANICAL, PRE .... UOC:DAL, V18, ZAL	1
9	PAFZZ	5325002007234	96906	MS16624-1275	RING, RETAINING .... UOC:DAL, V18, ZAL	2
10	PFFZZ	5340007409892	19207	7409892	CLEVIS, ROD END .... UOC:DAL, V18, ZAL	1
11	PAFZZ	3040004702097	19207	8690893	SHAFT, STRAIGHT .... UOC:DAL, V18, ZAL	1
12	PAFZZ	3110002273239	96906	MS51961-6	BEARING, ROLLER, NEED .... UOC:DAL, V18, ZAL	1
13	PAFZZ	3110007409896	19207	7409896	RING, BEARING, INNER .... UOC:DAL, V18, ZAL	2
14	PAFZZ	3110001006159	96906	MS19059-2424	BALL, BEARING .... UOC:DAL, V18, ZAL	45
15	XAFZZ	2520011147402	19207	8690894	FRAME ASSY, TROLLY .... UOC:DAL, V18, ZAL	1
16	PBFZZ	3040007409950	19207	7409950	.SHAFT, SHOULDERED .... UOC:DAL, V18, ZAL	1
17	PAFZZ	5310000451031	21450	451031	.NUT, SELF-LOCKING, HE .... UOC:DAL, V18, ZAL	1
18	PAFZZ	3120007409958	19207	7409958	BEARING, WASHER, THRU .... UOC:DAL, V18, ZAL	4
19	PAFZZ	5330007409959	19207	7409959	FELT, MECHANICAL, PRE .... UOC:DAL, V18, ZAL	8
20	PAFZZ	3110002273253	96906	MS51961-30	BEARING, ROLLER, NEED .... UOC:DAL, V18, ZAL	4
21	PAFZZ	3120007409960	19207	7409960	BEARING, SLEEVE .... UOC:DAL, V18, ZAL	4
22	PAFZZ	5310007409962	19207	7409962	WASHER, FLAT .... UOC:DAL, V18, ZAL	4
23	PAFZZ	5365002524758	96906	MS16624-1200	RING, RETAINING .... UOC:DAL, V18, ZAL	4
24	PAFZZ	5310008348732	96906	MS35691-33	NUT, PLAIN, HEXAGON .... UOC:DAL, V18, ZAL	4

(1) ITEM NO	(2) SMR CODE	(3) NSN	(4) CAGEC	(5) PART NUMBER	(6) DESCRIPTION AND USABLE ON CODES (UOC)	(7) QTY
25	PAFZZ	5305011198889	21450	113038	SCREW, CAP, SOCKET HE	4
					UOC:DAL, V18, ZAL	
26	PAFZZ	4730001720034	96906	MS15003-6	FITTING, LUBRICATION	1
					UOC:DAL, V18, ZAL	
27	PAFZZ	5310005967753	72962	22NA797-82	NUT, SELF-LOCKING, PL	1
					UOC:DAL, V18, ZAL	
28	PAFZZ	5310007409955	19207	7409955	WASHER , FLAT	1
					UOC:DAL, V18, ZAL	
29	PAFZZ	5310004492378	96906	MS21245-L8	NUT, SELF-LOCKING, HE	1
					UOC:DAL, V18, ZAL	
30	PAFZZ	3020007409952	19207	7409952	PULLEY, GROOVE	1
					UOC:DAL, V18, ZAL	

END OF FIGURE

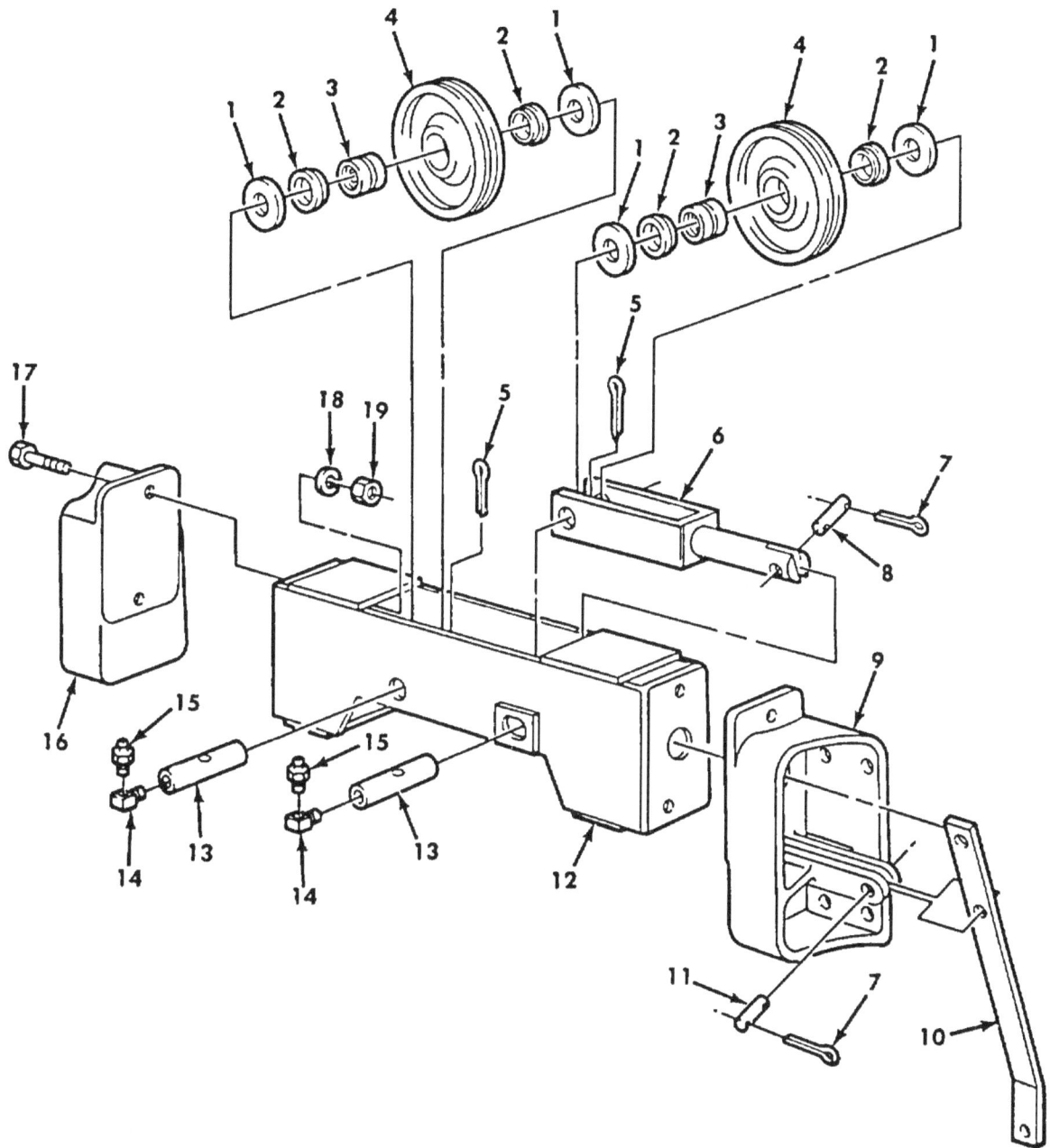

Figure 501. Wrecker Rear Winch Cable Tensioner.

(1) ITEM NO	(2) SMR CODE	(3) NSN	(4) CAGEC	(5) PART NUMBER	(6) DESCRIPTION AND USABLE ON CODES (UOC)	(7) QTY
					GROUP 2001 HOIST, CAPSTAN, WINDLASS, CRANE OR WINCH ASSEMBLY	
					FIG. 501 WRECKER REAR WINCH CABLE TENSIONER	
1	PAFZZ	5310007409975	19207	7409975	WASHER, FLAT .... UOC:DAL, V18, ZAL	4
2	PAFZZ	5330007409929	19207	7409929	PACKING WITH RETAIN .... UOC:DAL, V18, ZAL	4
3	PAFZZ	3110002273245	96906	MS51961-15	BEARING, ROLLER, NEED .... UOC:DAL, V18, ZAL	2
4	PAFZZ	3020007409948	19207	7409948	PULLEY, GROOVE .... UOC:DAL, V18, ZAL	2
5	PAFZZ	5315008495582	96906	MS24665-502	PIN, COTTER .... UOC:DAL, V18, ZAL	2
6	PFFZZ	2520007409930	19207	7409930	FRAME, ADJUSTABLE SH .... UOC:DAL, V18, ZAL	1
7	PAFZZ	5315008395821	96906	MS24665-351	PIN, COTTER .... UOC:DAL, V18, ZAL	4
8	PAFZZ	5315002366625	96906	MS51932-103	PIN, STRAIGHT, HEADLE .... UOC:DAL, V18, ZAL	1
9	PFFZZ	5340007409665	19207	7409665	BRACKET, MOUNTING .... UOC:DAL, V18, ZAL	1
10	PBFZZ	2520007409935	19207	7409935	LEVER, REMOTE CONTRO .... UOC:DAL, V18, ZAL	1
11	PAFZZ	5315011299190	96906	MS51932-135	PIN, STRAIGHT, HEADLE .... UOC:DAL, V18, ZAL	1
12	XAFZZ		19207	7954584	FRAME ASSY TENSIONE .... UOC:DAL, V18, ZAL	1
13	PAFZZ	3040007409945	19207	7409945	SHAFT, STRAIGHT .... UOC:DAL, V18, ZAL	2
14	PAFZZ	4730002784814	11083	2D1683	ELBOW, PIPE .... UOC:DAL, V18, ZAL	2
15	PAFZZ	4730001720022	96906	MS15003-2	FITTING, LUBRICATION .... UOC:DAL, V18, ZAL	2
16	PFFZZ	5340007409668	19207	7409668	BRACKET, MOUNTING .... UOC:DAL, V18, ZAL	1
17	PAFZZ	5305007254183	96906	MS90726-113	SCREW, CAP, HEXAGON H .... UOC:DAL, V18, ZAL	4
18	PAFZZ	5310000034094	01276	210104-8S	WASHER, LOCK .... UOC:DAL, V18, ZAL	4
19	PAFZZ	5310007320560	96906	MS51968-14	NUT, PLAIN, HEXAGON .... UOC:DAL, V18, ZAL	4

END OF FIGURE

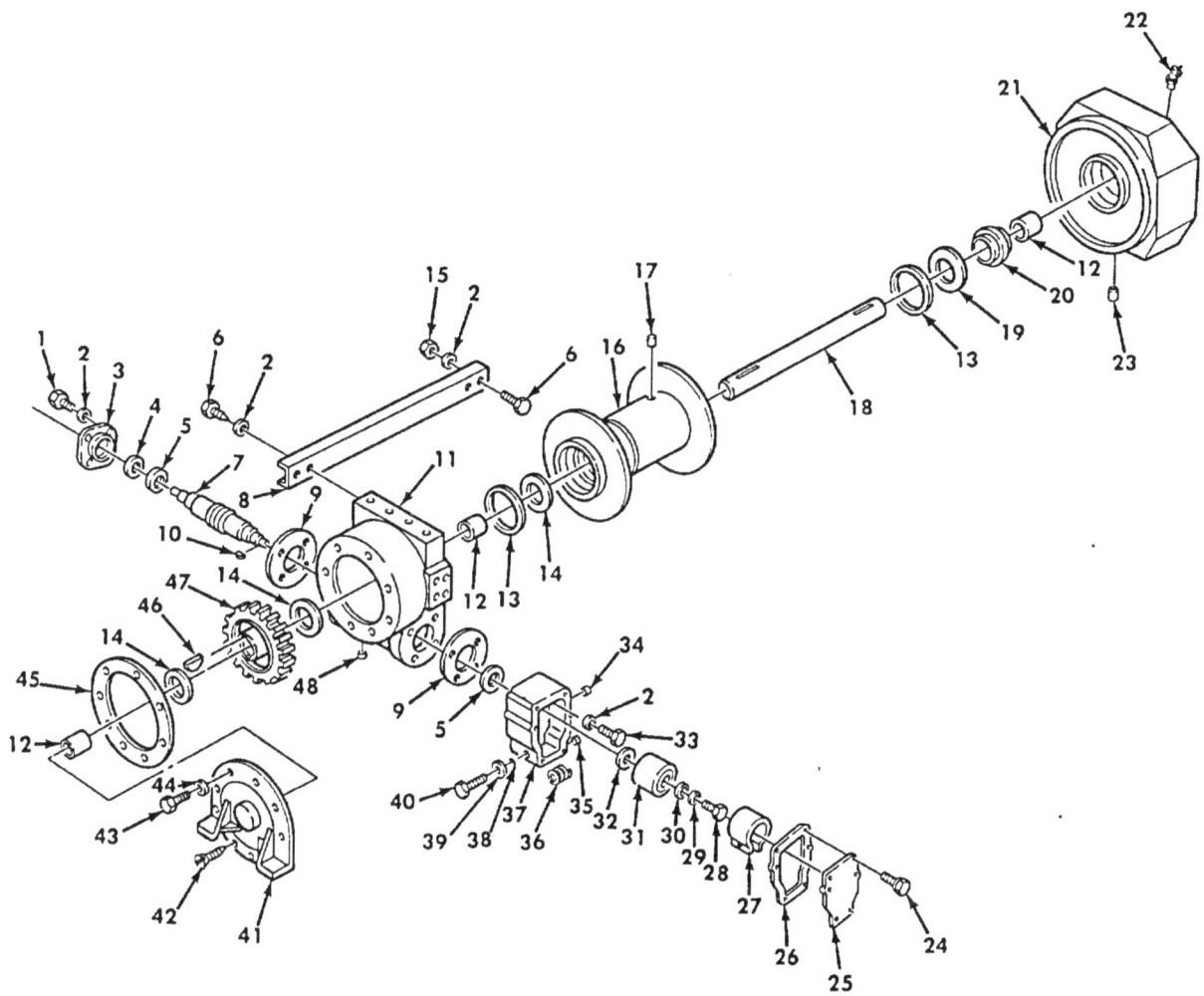

Figure 502. Wrecker Rear Winch Assembly, Basic.

(1) ITEM NO	(2) SMR CODE	(3) NSN	(4) CAGEC	(5) PART NUMBER	(6) DESCRIPTION AND USABLE ON CODES (UOC)	(7) QTY

GROUP 2001 HOIST, CAPSTAN, WINDLASS, CRANE OR WINCH ASSEMBLY

FIG. 502 WRECKER REAR WINCH ASSEMBLY, BASIC

(1)	(2)	(3)	(4)	(5)	(6)	(7)
1	PAFZZ	5305009227994	80204	B1821BH075C250N	SCREW, CAP, HEXAGON H .... UOC:DAL, V18, ZAL	4
2	PAFZZ	5310005847888	96906	MS35338-51	WASHER, LOCK .... UOC:DAL, V18, ZAL	12
3	PBFZZ	2520007409667	61465	M305815	CAP WORM GEAR .... UOC:DAL, V18, ZAL	1
4	PAFZZ	5330007409940	61465	M305832	SEAL, PLAIN ENCASED .... UOC:DAL, V18, ZAL	1
5	PAFZZ	3110002770476	43334	QH20312N01	BEARING, BALL, ANNULA .... UOC:DAL, V18, ZAL	2
6	PAFZZ	5305009399204	96906	MS90725-187	SCREW, CAP, HEXAGON H .... UOC:DAL, V18, ZAL	4
7	PAFZZ	3040003288862	19207	7954562	WORM SHAFT .... UOC:DAL, V18, ZAL	1
8	PFFZZ	9520011208551	19207	7409671	CHANNEL, STRUCTURAL .... UOC:DAL, V18, ZAL	1
9	PAFZZ	5330001664333	19207	7409931	GASKET .... UOC:DAL, V18, ZAL	2
10	PAF ZZ	5315000424950	96906	MS20067-305	KEY, MACHINE .... UOC:DAL, V18, ZAL	1
11	PAFZZ	3040007409669	19207	7409669	HOUSING, MECHANICAL .... UOC:DAL, V18, ZAL	1
12	PAFZZ	3120007409666	19207	7409666	BEARING, SLEEVE .... UOC:DAL, V18, ZAL	3
13	PAFZZ	5330005853210	21450	500207	SEAL, PLAIN ENCASED .... UOC:DAL, V18, ZAL	2
14	PAFZZ	5310007409977	19207	7409977	WASHER, KEY .... UOC:DAL, V18, ZAL	3
15	PAFZZ	5310007638922	96906	MS51967-24	NUT, PLAIN, HEXAGON .... UOC:DAL, V18, ZAL	1
16	PAFZZ	2590011208451	19207	7954586	REEL, CABLE .... UOC:DAL, V18, ZAL	1
17	PAFZZ	5305007250194	96906	MS51963-170	SETSCREW .... UOC:DAL, V18, ZAL	1
18	PAFZZ	3040005068451	19207	7954563	SHAFT, STRAIGHT .... UOC:DAL, V18, ZAL	1
19	PAFZZ	5310007409979	19207	7409979	WASHER, KEY .... UOC:DAL, V18, ZAL	1
20	PAFZZ	3120011208450	19207	7954534	BEARING, SLEEVE .... UOC:DAL, V18, ZAL	1
21	XAFZZ		19207	7954585	FRAME, END .... UOC:DAL, V18, ZAL	1
22	PAFZZ	4730000504208	96906	MS15003-1	FITTING, LUBRICATION .... UOC:DAL, V18, ZAL	1
23	PAFZZ	5315011396568	21450	589965	PIN, STRAIGHT, HEADLE .... UOC:DAL, V18, ZAL	1

(1) ITEM NO	(2) SMR CODE	(3) NSN	(4) CAGEC	(5) PART NUMBER	(6) DESCRIPTION AND USABLE ON CODES (UOC)	(7) QTY
24	PAFZZ	5306000187527	21450	187527	BOLT, ASSEMBLED WASH .... UOC:DAL, V18, ZAL	6
25	PBFZZ	2520007409673	19207	7409673	COVER, BRAKE CASE UOC:DAL, V18, ZAL	1
26	PAFZZ	5330007409932	19207	7409932	GASKET BRAKE COVER .... UOC:DAL, V18, ZAL	1
27	PAFZZ	2530007409663	19207	7409663	BRAKE BAND AND LINI .... UOC:DAL, V18, ZAL	1
28	PAFZZ	5305007195219	96906	MS90727-111	SCREW, CAP, HEXAGON H .... UOC:DAL, V18, ZAL	1
29	PAFZZ	5310000034094	01276	210104-8S	WASHER, LOCK UOC:DAL, V18, ZAL	1
30	PAFZZ	5310011036438	56529	78011-170-1	WASHER, FLAT UOC:DAL, V18, ZAL	1
31	PAFZZ	2530007409925	19207	7409925	BRAKE DRUM .... UOC:DAL, V18, ZAL	1
32	PAFZZ	5330011315416	01212	E-450121VG	SEAL, PLAIN ENCASED .... UOC:DAL, V18, ZAL	1
33	PAFZZ	5305009000576	80204	B1821BH075C225N	SCREW, CAP, HEXAGON H .... UOC:DAL, V18, ZAL	4
34	PAFZZ	4730002895176	02951	MS49005-8	PLUG, PIPE BRAKE CASE UOC:DAL, V18, ZAL	1
35	XDFZZ		96906	MS9380-09	PLUG, EXPANSION BRAKE CASE .... UOC:DAL, V18, ZAL	1
36	PAFZZ	5360005416501	19207	5416501	SPRING, HELICAL, COMP UOC:DAL, V18, ZAL	1
37	PBFZZ	3040011198711	9207	7954588	HOUSING, MECHANICAL .... UOC:DAL, V18, ZAL	1
38	PAFZZ	5330005822855	96906	MS28775-113	O-RING .... UOC:DAL, V18, ZAL	1
39	PAFZZ	5310008095998	96906	MS27183-18	WASHER, FLAT .... UOC:DAL, V18, ZAL	1
40	PAFZZ	5305007195275	96906	MS90727-128	SCREW, CAP, HEXAGON H .... UOC:DAL, V18, ZAL	1
41	PBFZZ	5340011443233	19207	7954587	COVER, ACCESS UOC:DAL, V18, ZAL	1
42	PAFZZ	5305009907168	96906	MS24667-85	SCREW, CAP, SOCKET HE .... UOC:DAL, V18, ZAL	2
43	PAFZZ	5305007247222	80204	B1821BH063C200N	SCREW, CAP, HEXAGON H .... UOC:DAL, V18, ZAL	6
44	PAFZZ	5310008206653	96906	MS35338-50	WASHER, LOCK .... UOC:DAL, V18, ZAL	6
45	PAFZZ	5330007409933	19207	7409933	GASKET UOC:DAL, V18, ZAL	1
46	PAFZZ	5315011195239	96906	MS20067-493	KEY, MACHINE .... UOC:DAL, V18, ZAL	4
47	PAFZZ	3020007409934	19207	7409934	GEAR, WORM WHEEL .... UOC:DAL, V18, ZAL	1
48	PAFZZ	4730005558291	10001	12Z329PC93	PLUG, PIPE .... UOC:DAL, V18, ZAL	2

END OF FIGURE

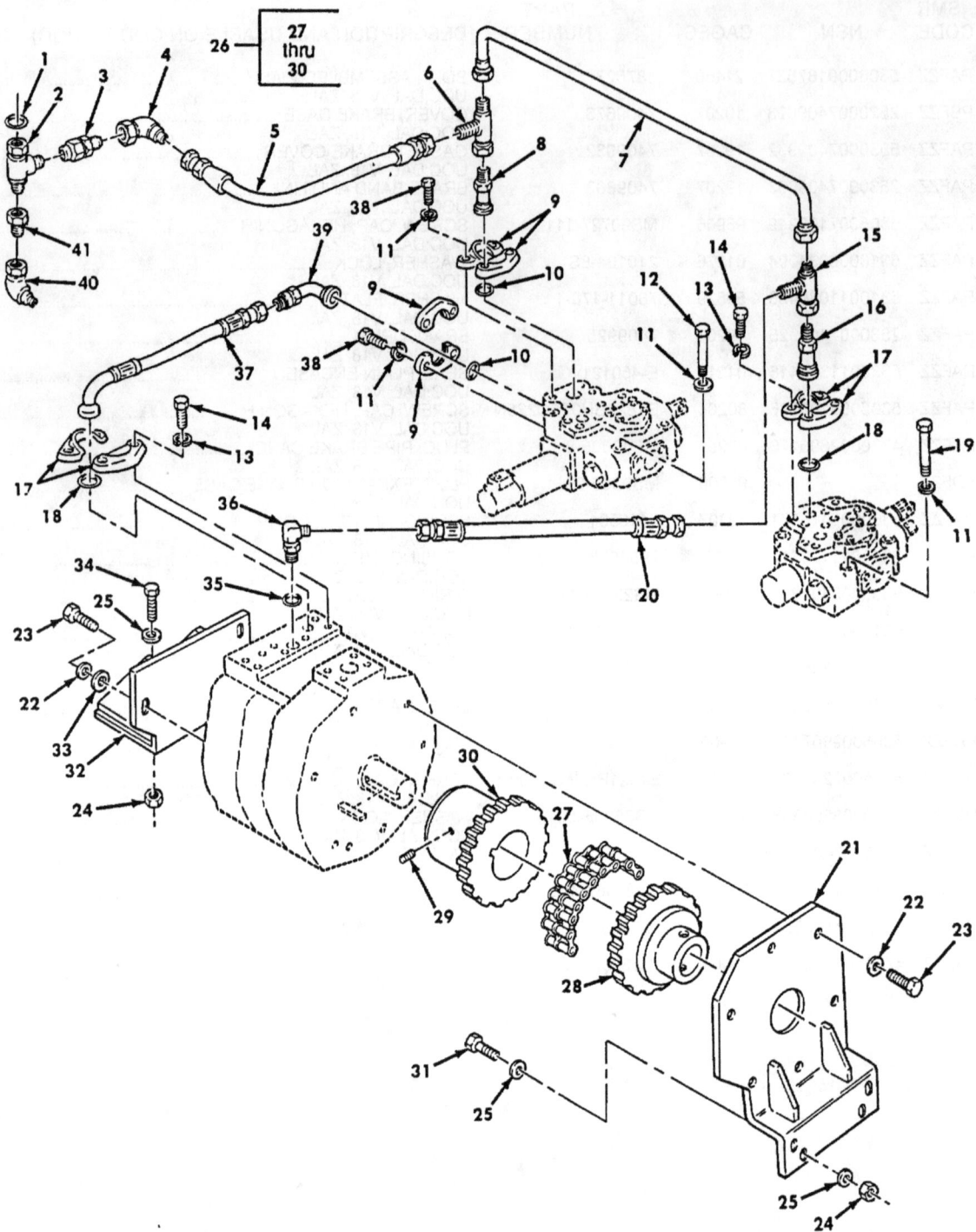

Figure 503. Wrecker Rear Winch Hydraulic Tubing and Control Valves.

(1) ITEM NO	(2) SMR CODE	(3) NSN	(4) CAGEC	(5) PART NUMBER	(6) DESCRIPTION AND USABLE ON CODES (UOC)	(7) QTY
					GROUP 2001 HOIST, CAPSTAN, WINDLASS, CRANE OR WINCH ASSEMBLY	
					FIG. 503 WRECKER REAR WINCH HYDRAULIC TUBING AND CONTROL VALVES	
1	PAFZZ	5330008163546	96906	MS28778-20	O-RING .... UOC:DAL, V18, ZAL	1
2	PAFZZ	4730010673932	96906	MS51523A20	TEE, TUBE .... UOC:DAL, V18, ZAL	1
3	PAFZZ	4730010661282	96906	MS51525A20	ADAPTER, STRAIGHT, TU .... UOC:DAL, V18, ZAL	1
4	PAFZZ	4730012386443	96906	MS51521A20	ELBOW, TUBE .... UOC:DAL, V18, ZAL	1
5	PAFZZ	4720011208522	19207	12256660	HOSE ASSEMBLY, NONME .... UOC:DAL, V18, ZAL	1
6	PAFZZ	4730010673932	96906	MS51523-B20	TEE, TUBE .... UOC:DAL, V18, ZAL	1
7	PFFZZ	4710011208540	19207	12256647	TUBE ASSEMBLY, METAL .... UOC:DAL, V18, ZAL	1
8	PAFZZ	2590011217606	19207	12256808-1	ADAPTER, CONTROL, HYD.................... UOC:DAL, V18, ZAL	1
9	PAFZZ	4730000893406	81349	M52525/16-24	FLANGE, PIPE, SWIVEL .... UOC:DAL, V18, ZAL	4
10	PAFZZ	5330005797927	96906	MS28775-225	O-RING .... UOC:DAL, V18, ZAL	2
11	PAFZZ	5310000034094	01276	210104-8S	WASHER , LOCK................................... UOC:DAL, V18, ZAL	16
12	PAFZZ	5305004106957	80204	B1821BH050F475N	SCREW, CAP, HEXAGON H .... UOC:DAL, V18, ZAL	2
13	PAFZZ	5310002090965	96906	MS35338-47	WASHER, LOCK .... UOC:DAL, V18, ZAL	8
14	PAFZZ	5305000711788	80204	B1821BH044C125N	SCREW, CAP, HEXAGON H .... UOC:DAL, V18, ZAL	8
15	PAFZZ	4730011231516	19207	12277045	TEE, TUBE CONTROL VALVE .... UOC:DAL, V18, ZAL	1
16	PAFZZ	2590011217605	19207	12256808-2	ADAPTER, CONTROL, HYD.................... UOC:DAL, V18, ZAL	1
17	PAFZZ	4730002550560	81349	M52525/16-20	FLANGE, PIPE .... UOC:DAL, V18, ZAL	4
18	PAFZZ	5330002979990	96906	MS28775-222	O-RING .... UOC:DAL, V18, ZAL	2
19	PAFZZ	5305007195274	96906	MS90727-125	SCREW, CAP, HEXAGON H .... UOC:DAL, V18, ZAL	2
20	PAFZZ	4720011296082	19207	12277044	HOSE ASSEMBLY, NONME .... UOC:DAL, V18, ZAL	1
21	PFFZZ	3040012158853	19207	12302840	BRACKET, EYE, ROTATIN .... UOC:DAL, V18, ZAL	1
22	PAFZZ	5310008206653	96906	MS35338-50	WASHER , LOCK.................................... UOC:DAL, V18, ZAL	9

(1) ITEM NO	(2) SMR CODE	(3) NSN	(4) CAGEC	(5) PART NUMBER	(6) DESCRIPTION AND USABLE ON CODES (UOC)	(7) QTY
23	PAFZZ	5305007247222	80204	B1821BH063C200N	SCREW, CAP, HEXAGON H .... UOC:DAL, V18, ZAL	9
24	PAFZZ	5310004883888	96906	MS51943-40	NUT, SELF-LOCKING, HE .... UOC:DAL, V18, ZAL	4
25	PAFZZ	5310008095998	96906	MS27183-18	WASHER, FLAT .... UOC:DAL, V18, ZAL	8
26	PAFFF	3010012156598	19207	11669859	COUPLING, SHAFT, FLEX .... UOC:DAL, V18, ZAL	1
27	PFFZZ	3020012162332	19207	11669856	CHAIN, ROLLER .... UOC:DAL, V18, ZAL	1
28	PFFZZ	3020012156599	19207	11669857	SPROCKET WHEEL .... UOC:DAL, V18, ZAL	1
29	PAFZZ	5305011867140	96906	MS51031-130	SETSCREW .... UOC:DAL, V18, ZAL	1
30	PFFZZ	3020012158827	19207	11669858	GEAR, SPUR .... UOC:DAL, V18, ZAL	1
31	PAFZZ	5305007254183	96906	MS90726-113	SCREW, CAP, HEXAGON H .... UOC:DAL, V18, ZAL	2
32	XDFZZ		19207	12256796	BRACKET, MOUNTING................................ UOC:DAL, V18, ZAL	1
33	PAFZZ	5310008238803	96906	MS27183-21	WASHER, FLAT .... UOC:DAL, V18, ZAL	3
34	PAFZZ	5305007195235	80204	B1821BH050F175N	SCREW, CAP, HEXAGON H .... UOC:DAL, V18, ZAL	2
35	PAFZZ	5330008080794	96906	MS28778-8	O-RING .... UOC:DAL, V18, ZAL	1
36	PAFZZ	4730008225609	96906	MS51527A8	ELBOW, TUBE TO BOSS .... UOC:DAL, V18, ZAL	1
37	PAFZZ	4720011147703	19207	12256811	HOSE ASSEMBLY, NONME .... CONTROL VALVE UOC:DAL, V18, ZAL	1
38	PAFZZ	5305000712069	80204	B1821BH050C150N	SCREW, CAP, HEXAGON H .... UOC:DAL, V18, ZAL	8
39	PFFZZ	4730011147743	19207	12256661-3	ELBOW, FLANGE TO PIP .... UOC:DAL, V18, ZAL	1
40	PAFZZ	4730011391585	96906	MS51521A10	ELBOW, TUBE .... UOC:DAL, V18, ZAL	1
41	PAFZZ	4730012359617	19207	12301152	ADAPTER, STRAIGHT, TU. .... UOC:DAL, V18, ZAL	1

END OF FIGURE

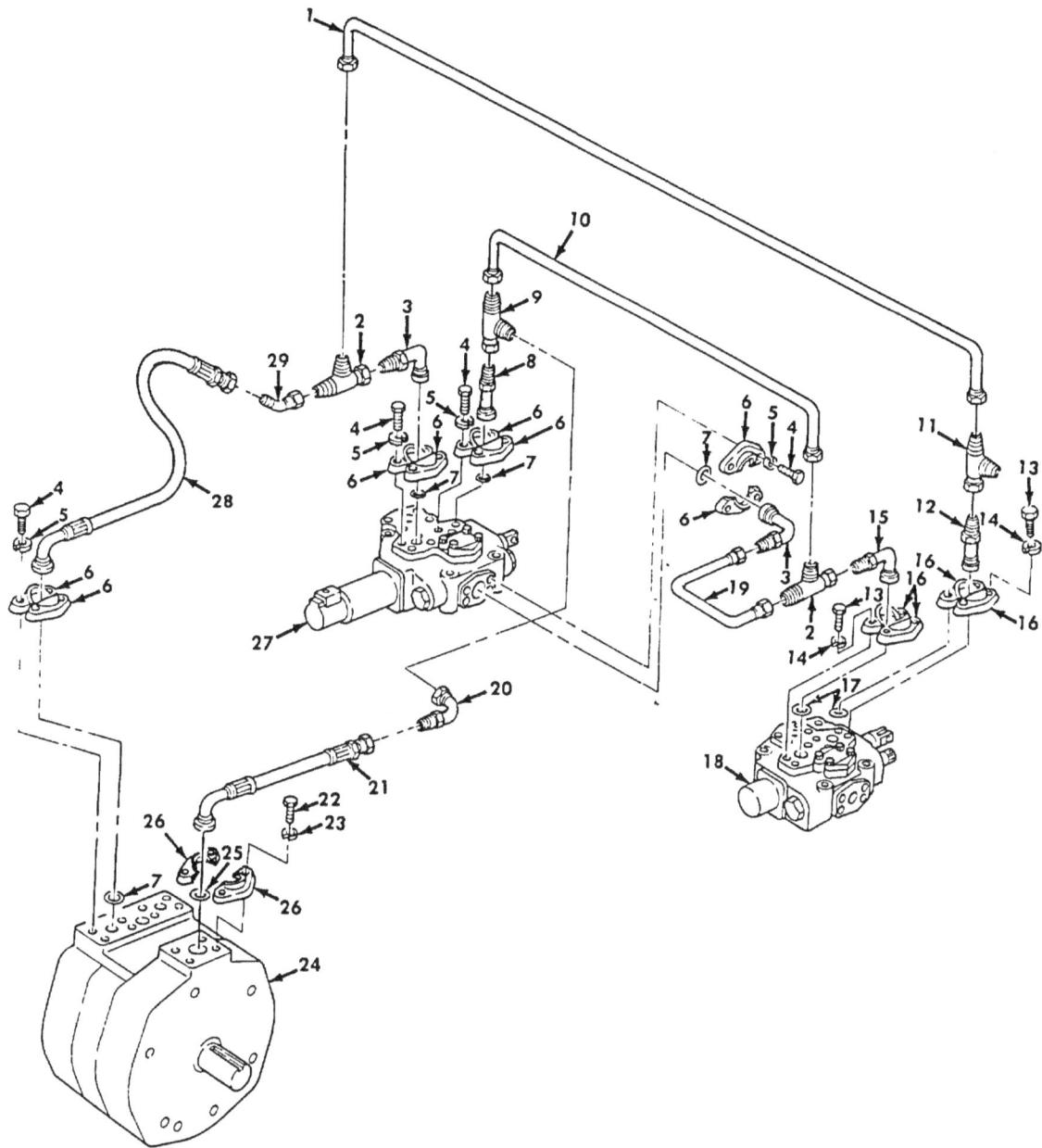

Figure 504. Wrecker Rear Winch Hydraulic Tubing, Control Valves, and Motor.

(1) ITEM NO	(2) SMR CODE	(3) NSN	(4) CAGEC	(5) PART NUMBER	(6) DESCRIPTION AND USABLE ON CODES (UOC)	(7) QTY
					GROUP 2001 HOIST, CAPSTAN, WINDLASS, CRANE OR WINCH ASSEMBLY	
					FIG. 504 WRECKER REAR WINCH HYDRAULIC TUBING, CONTROL VALVES, AND MOTOR	
1	PAFZZ	4710011208541	19207	12256642	TUBE ASSEMBLY, METAL UOC:DAL, V18, ZAL	1
2	PAFZZ	4730008275852	96906	MS51523-B16	TEE, TUBE UOC:DAL, V18, ZAL	2
3	PAFZZ	4730011147742	19207	12256661-2	ELBOW, FLANGE TO PIP UOC:DAL, V18, ZAL	8
4	PAFZZ	5305000711788	80204	B1821BH044C125N	SCREW, CAP, HEXAGON H UOC:DAL, V18, ZAL	16
5	PAFZZ	5310002090965	96906	MS35338-47	WASHER, LOCK UOC:DAL, V18, ZAL	16
6	PAFZZ	4730002550560	81349	M52525/16-20	FLANGE, PIPE SWIVEL UOC:DAL, V18, ZAL	8
7	PAFZZ	5330002979990	96906	MS28775-222	O-RING UOC:DAL, V18, ZAL	4
8	PAFZZ	4730011324858	19207	12256665	ADAPTER, STRAIGHT, FL UOC:DAL, V18, ZAL	1
9	PAFZZ	4730011281554	19207	12256640	TEE, TUBE CONTROL VALVE UOC:DAL, V18, ZAL	1
10	PFFZZ	4710011208539	19207	12256639	TUBE ASSEMBLY, METAL TO WINCH VALVE UOC:DAL, V18, ZAL	1
11	PAFZZ	4730008275852	96906	MS51523A16	TEE, TUBE UOC:DAL, V18, ZAL	1
12	PAFZZ	4730011217604	19207	12256665-1	ADAPTER, STRAIGHT, FL UOC:DAL, V18, ZAL	1
13	PAFZZ	5305000680511	80204	B1821BH038C125N	SCREW, CAP, HEXAGON H UOC:DAL, V18, ZAL	8
14	PAFZZ	5310006379541	96906	MS35338-46	WASHER, LOCK UOC:DAL, V18, ZAL	8
15	PAFZZ	4730011219889	19207	12256661-1	ELBOW, FLANGE TO PIP UOC:DAL, V18, ZAL	1
16	PAFZZ	4730007557609	81349	M52525/16-16	FLANGE, PIPE UOC:DAL, V18, ZAL	4
17	PAFZZ	5330005797925	96906	MS28775-219	O-RING UOC:DAL, V18, ZAL	2
18	PAFZZ	4820011276922	19207	11669330	VALVE, LINEAR, DIRECT UOC:DAL, V18, ZAL	1
19	PFFZZ	4710011208542	19207	12256644	TUBE ASSEMBLY, METAL UOC:DAL, V18, ZAL	1
20	PAFZZ	4730004098797	96906	MS51521-B24	ELBOW, TUBE UOC:DAL, V18, ZAL	1
21	PFFZZ	4720011208519	19207	12256810	HOSE ASSEMBLY, NONME UOC:DAL, V18, ZAL	1

(1) ITEM NO	(2) SMR CODE	(3) NSN	(4) CAGEC	(5) PART NUMBER	(6) DESCRIPTION AND USABLE ON CODES (UOC)	(7) QTY
22	PAFZZ	5305000712069	80204	B1821BH050C150N	SCREW, CAP, HEXAGON H .... UOC:DAL, V18, ZAL	4
23	PAFZZ	5310000034094	01276	210104-83S	WASHER, LOCK .... UOC:DAL, V18, ZAL	4
24	PBFZZ	4320011128365	19207	11669350	MOTOR, HYDRAULIC .... UOC:DAL, V18, ZAL	1
25	PAFZZ	5330005797927	96906	MS28775-225	O-RING .... UOC:DAL, V18, ZAL	1
26	PAFZZ	4730000893406	81349	M52525/16-24	FLANGE, PIPE, SWIVEL .... UOC:DAL, V18, ZAL	2
27	PAFZZ	4810011056966	19207	11669329-1	VALVE, REGULATING, FL .... UOC:DAL, V18, ZAL	1
28	PAFZZ	4720011147703	19207	12256811	HOSE ASSEMBLY, NONME .... UOC:DAL, V18, ZAL	1
29	PAFZZ	4730011929578	96906	MS51522-A16	ELBOW, TUBE .... UOC:DAL, V18, ZAL	1

END OF FIGURE

17 —[18]

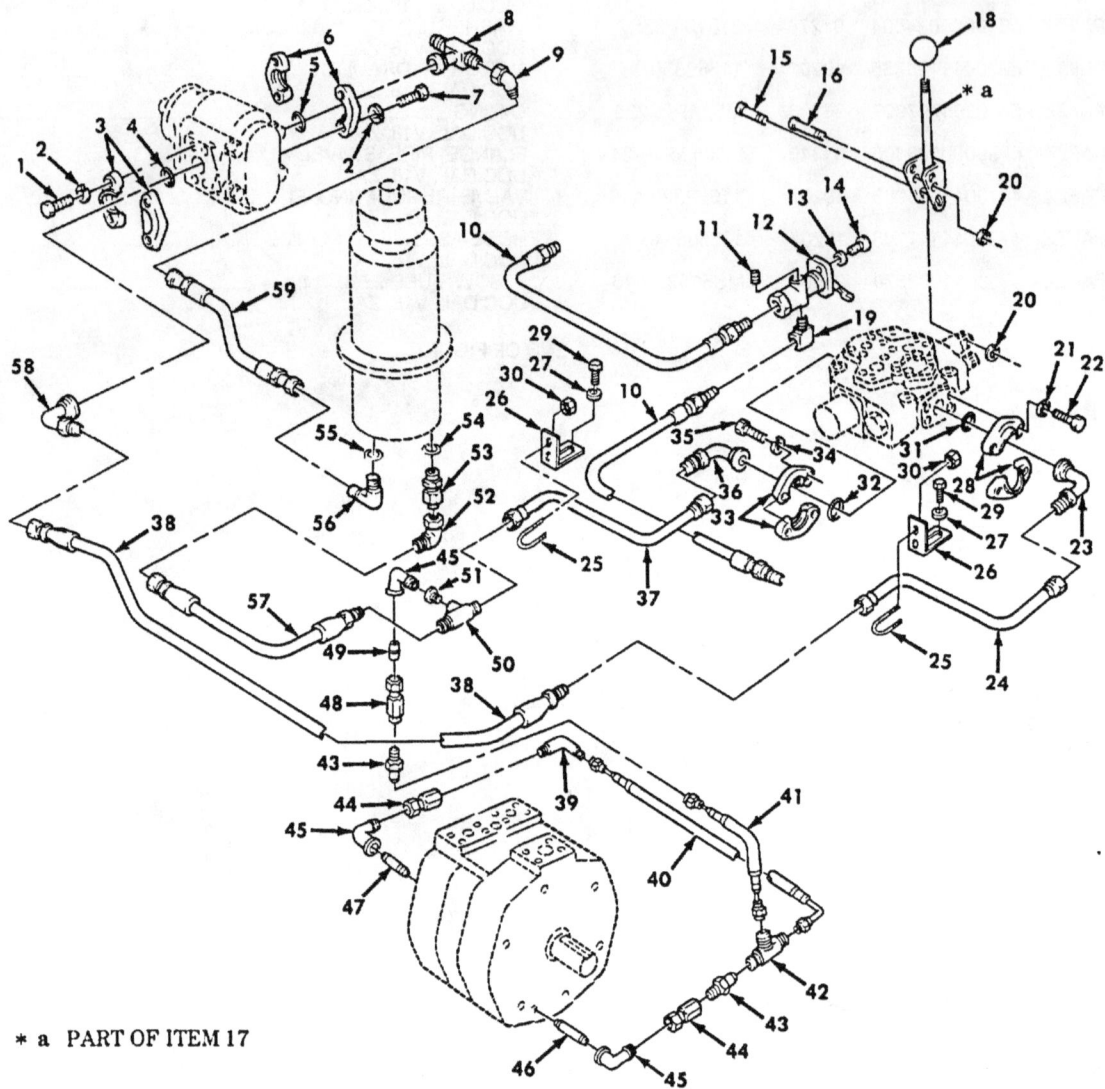

* a PART OF ITEM 17

Figure 505. Wrecker Rear Winch Hydraulic Tubing and Controls.

(1) ITEM NO	(2) SMR CODE	(3) NSN	(4) CAGEC	(5) PART NUMBER	(6) DESCRIPTION AND USABLE ON CODES (UOC)	(7) QTY
					GROUP 2001 HOIST, CAPSTAN, WINDLASS, CRANE OR WINCH ASSEMBLY	
					FIG. 505 WRECKER REAR WINCH HYDRAULIC TUBING AND CONTROLS	
1	PAFZZ	5305000712069	80204	B1821BH050C150N	SCREW, CAP, HEXAGON H .... UOC:DAL, V18, ZAL	4
2	PAFZZ	5310000034094	01276	210104-8S	WASHER, LOCK .... UOC:DAL, V18, ZAL	8
3	PAFZZ	4730000893406	81349	M52525/16-24	FLANGE, PIPE, SWIVEL .... UOC:DAL, V18, ZAL	2
4	PAFZZ	5330005797927	96906	MS28775-225	O-RING .... UOC:DAL, V18, ZAL	1
5	PAFZZ	5330005858247	96906	MS28775-232	O-RING .... UOC:DAL, V18, ZAL	1
6	PAFZZ	4730004006544	81349	M52525/16-40	FLANGE, PIPE, SWIVEL .... UOC:DAL, V18, ZAL	2
7	PAFZZ	5305000712070	80204	B1821BH050C175N	SCREW, CAP, HEXAGON H .... UOC:DAL, V18, ZAL	4
8	PAFZZ	4730011208495	19207	12257114	TEE, TUBE .... UOC:DAL, V18, ZAL	1
9	PAFZZ	4730012327159	96906	MS51522A32	ELBOW, TUBE .... UOC:DAL, V18, ZAL	1
10	PAFZZ	4720001085989	19207	8327011-4	HOSE ASSEMBLY, NONME VACUUM VALVE UOC:DAL, V18, ZAL	2
11	PAFZZ	4730000444655	19207	444655	PLUG, PIPE .... UOC:DAL, V18, ZAL	1
12	PAFZZ	4820005093036	19207	7003615	VALVE, VACUUM REGULA .... UOC:DAL, V18, ZAL	1
13	PAFZZ	5310000453296	96906	MS35338-43	WASHER, LOCK .... UOC:DAL, V18, ZAL	2
14	PAFZZ	5305009846210	96906	MS35206-263	SCREW, MACHINE .... UOC:DAL, V18, ZAL	2
15	PAFZZ	5315011217689	19207	11621118-1	PIN, GROOVED, HEADED....................................... UOC:DAL, V18, ZAL	2
16	PAFZZ	5315011217688	19207	11621118-2	PIN, GROOVED, HEADED....................................... UOC:DAL, V18, ZAL	2
17	PAFFF	5340011208452	19207	12256740	LEVER, MANUAL CONTRO .... UOC:DAL, V18, ZAL	2
18	PAFZZ	5355001911029	19207	10900172	KNOB CONTROL .... UOC:DAL, V18, ZAL	1
19	PAFZZ	4730002775553	24617	444040	ELBOW, PIPE .... UOC:DAL, V18, ZAL	1
20	PAFZZ	5325004193322	96906	MS90707-1050	RING, RETAINING .... UOC:DAL, V18, ZAL	4
21	PAFZZ	5310006379541	96906	MS35338-46	WASHER, LOCK .... UOC:DAL, V18, ZAL	4
22	PAFZZ	5305000680511	80204	B1821BH038C125N	SCREW, CAP, HEXAGON H .... UOC:DAL, V18, ZAL	4

(1) ITEM NO	(2) SMR CODE	(3) NSN	(4) CAGEC	(5) PART NUMBER	(6) DESCRIPTION AND USABLE ON CODES (UOC)	(7) QTY
23	PAFZZ	4730011219889	19207	12256661-1	ELBOW, FLANGE TO PIP .... UOC:DAL, V18, ZAL	1
24	PFFZZ	4710011208543	19207	12256637	TUBE ASSEMBLY, METAL .... UOC:DAL, V18, ZAL	1
25	PAFZZ	5306011733524	80205	NAS3104-9-10	BOLT, U .... UOC:DAL, V18, ZAL	2
26	PAFZZ	5340011765923	19207	12302665	BRACKET, ANGLE .... UOC:DAL, V18, ZAL	2
27	PAFZZ	5310000806004	96906	MS27183-14	WASHER , FLAT .... UOC:DAL, V18, ZAL	2
28	PAFZZ	4730007557609	81349	M52525/16-16	FLANGE, PIPE .... UOC:DAL, V18, ZAL	2
29	PAFZZ	5305002692804	96906	MS90726-61	SCREW, CAP, HEXAGON H .... UOC:DAL, V18, ZAL	2
30	PAFZZ	5310008775796	96906	MS21044N4	NUT, SELF-LOCKING, HE .... UOC:DAL, V18, ZAL	4
31	PAFZZ	5330005797925	96906	MS28775-219	O-RING .... UOC:DAL, V18, ZAL	1
32	PAFZZ	5330002979990	96906	MS28775-222	O-RING .... UOC:DAL, V18, ZAL	1
33	PAFZZ	4730002550560	81349	M52525/16-20	FLANGE, PIPE, SWIVEL .... UOC:DAL, V18, ZAL	2
34	PAFZZ	5310002090965	96906	MS35338-47	WASHER, LOCK .... UOC:DAL, V18, ZAL	4
35	PAFZZ	5305000711788	80204	B1821BH044C125N	SCREW, CAP, HEXAGON H .... UOC:DAL, V18, ZAL	4
36	PAFZZ	4730011147742	19207	12256661-2	ELBOW, FLANGE TO PIP .... UOC:DAL, V18, ZAL	1
37	PFFZZ	4710011208544	19207	12256638	TUBE ASSEMBLY, METAL .... SWIVEL MOTOR. UOC:DAL, V18, ZAL	1
38	PAFZZ	4720011208521	19207	12256658-2	HOSE ASSEMBLY, NONME .... UOC:DAL, V18, ZAL	1
39	PAFZZ	4730002736686	81343	8-8070202CA(CAD)	ELBOW, PIPE TO TUBE .... UOC:DAL, V18, ZAL	1
40	PAFZZ	4720011659531	19207	12302663	HOSE ASSEMBLY, NONME .... UOC:DAL, V18, ZAL	1
41	PAFZZ	4720011961166	19207	12302664	HOSE ASSEMBLY, NONME .... UOC:DAL, V18, ZAL	1
42	PAFZZ	4730000740713	96906	MS51523A8	TEE, TUBE .... UOC:DAL, V18, ZAL	1
43	PAFZZ	4730008099427	96906	MS51500A8-8	ADAPTER, STRAIGHT, PI .... UOC:DAL, V18, ZAL	2
44	PAFZZ	4820011574138	19207	12302667	VALVE, CHECK .... UOC:DAL, V18, ZAL	2
45	PAFZZ	4730002534414	96906	MS39230-4	ELBOW, PIPE .... UOC:DAL, V18, ZAL	3
46	PAFZZ	4730001961496	96906	MS51953-81	NIPPLE, PIPE .... UOC:DAL, V18, ZAL	2
47	PAFZZ	4730001961495	96906	MS51953-80	NIPPLE, PIPE .... UOC:DAL, V18, ZAL	2
48	PAFZZ	4820011600759	19207	12302666	VALVE, CALIBRATED FL	1

(1) ITEM NO	(2) SMR CODE	(3) NSN	(4) CAGEC	(5) PART NUMBER	(6) DESCRIPTION AND USABLE ON CODES (UOC)	(7) QTY
49	PAFZZ	4730001961467	72582	121208	UOC:DAL, V18, ZAL NIPPLE, PIPE	1
50	PAFZZ	4730002516827	18876	10164117	UOC:DAL, V18, ZAL TEE, PIPE TO TUBE	1
51	PAFZZ	4730002032836	81349	WW-P-471BDQBUFD	UOC:DAL, V18, ZAL BUSHING, PIPE	1
52	PAFZZ	4730001630236	96906	MS51521A16	UOC:DAL, V18, ZAL ELBOW, TUBE	1
53	PAFZZ	4730009305392	96906	MS51525A16	UOC:DAL, V18, ZAL ADAPTER, STRAIGHT, TU	1
54	PAFZZ	5330008045694	96906	MS28778-16	UOC:DAL, V18, ZAL O-RING	1
55	PAFZZ	5330002859847	96906	MS28778-32	UOC:DAL, V18, ZAL O-RING	1
56	PAFZZ	4730011477954	96906	MS51527A32	UOC:DAL, V18, ZAL ELBOW, TUBE TO BOSS	1
57	PAFZZ	4720011734609	19207	12256658-3	UOC:DAL, V18, ZAL HOSE ASSEMBLY, NONME	1
58	PFFZZ	4730011147743	19207	12256661-3	UOC:DAL, V18, ZAL ELBOW, FLANGE TO PIP	1
59	PAFZZ	4720011226166	19207	12256659	UOC:DAL, V18, ZAL HOSE ASSEMBLY, NONME	1

UOC:DAL, V18, ZAL

END OF FIGURE

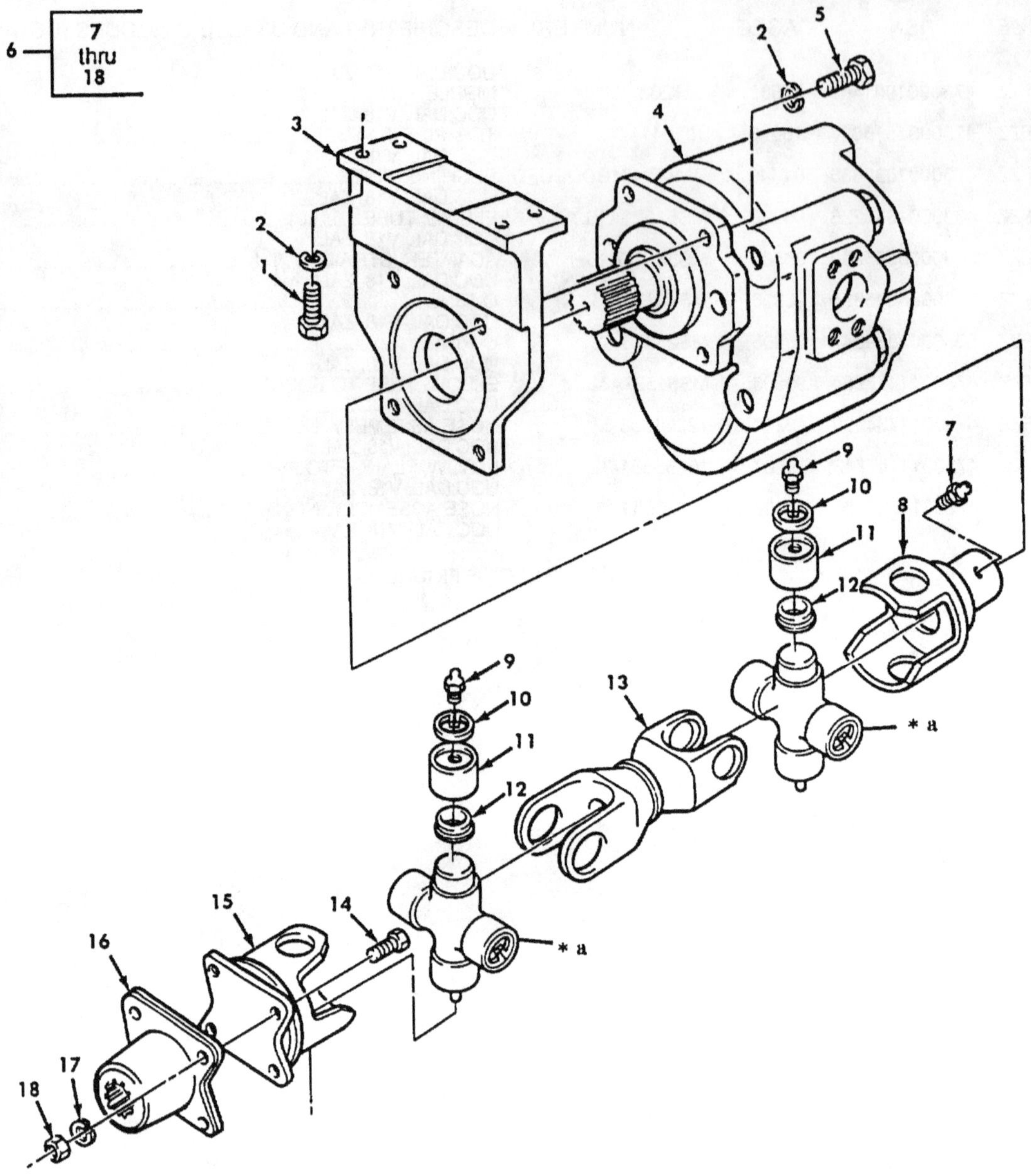

*a PART OF ITEM 6

Figure 506. Wrecker Hydraulic Pump and Drive Shaft.

SECTION II

(1) ITEM NO	(2) SMR CODE	(3) NSN	(4) CAGEC	(5) PART NUMBER	(6) DESCRIPTION AND USABLE ON CODES (UOC)	(7) QTY
					GROUP 2001 HOIST, CAPSTAN, WINDLASS, CRANE OR WINCH ASSEMBLY	
					FIG. 506 WRECKER HYDRAULIC PUMP AND DRIVE SHAFT	
1	PAOZZ	5305007195235	80204	B1821BH050F175N	SCREW, CAP, HEXAGON H .... UOC:DAL, V18, ZAL	4
2	PAOZZ	5310000034094	01276	210104-8S	WASHER, LOCK .... UOC:DAL, V18, ZAL	8
3	PFOZZ	5340010907631	19207	11669334	BRACKET, MOUNTING .... UOC:DAL, V18, ZAL	1
4	PAOZZ	4320010907632	19207	11669335	PUMP, HYDRAULIC, SING .... UOC:DAL, V18, ZAL	4
5	PAOZZ	5305000712070	80204	B1821BH050C175N	SCREW, CAP, HEXAGON H .... UOC:DAL, V18, ZAL	4
6	PAOOO	3010010907747	19207	11669333	UNIVERSAL JOINT .... UOC:DAL, V18, ZAL	1
7	PAOZZ	4730001720010	96906	MS15002-1	.FITTING, LUBRICATION .... UOC:DAL, V18, ZAL	1
8	XAOZZ		70960	L16SYS20-42	.FLANGE, CROSS MOUNT .... UOC:DAL, V18, ZAL	1
9	KFOZZ		70960	1641	.LUBRICATION FITTING PART OF KIT P/N .... CP35R-17 UOC:DAL, V18, ZAL	2
10	KFOZZ		70960	RR-1R22-2	.RETAINING RING PART OF KIT P/N .... CP35R-17 UOC:DAL, V18, ZAL	8
11	KFOZZ		70960	35R4-2A	.NEEDLE CUP ASSY PART OF KIT P/N .... CP35R-17 UOC:DAL, V18, ZAL	8
12	KFOZZ		70960	SE-RUR14-2	.OIL SEAL PART OF KIT P/N CP35R-17 .... UOC:DAL, V18, ZAL	8
13	XAOZZ		70960	3NDCA-4	.CENTER, SUPPORT, CROS .... UOC:DAL, V18, ZAL	2
14	PAOZZ	5305007098523	80204	B1821BH044F125N	.SCREW, CAP, HEXAGON H .... UOC:DAL, V18, ZAL	4
15	XAOZZ		70960	3NF20	.YOKE .... UOC:DAL, V18, ZAL	1
16	XAOZZ		70960	2WCS24-75	.FLANGE .... UOC:DAL, V18, ZAL	1
17	PAOZZ	5310004865355	70960	WA-LM7-2	.WASHER, LOCK .... UOC:DAL, V18, ZAL	4
18	PAOZZ	5310011354798	78500	NU-HX7-20-1	.NUT, PLAIN, HEXAGON .... UOC:DAL, V18, ZAL	4

END OF FIGURE

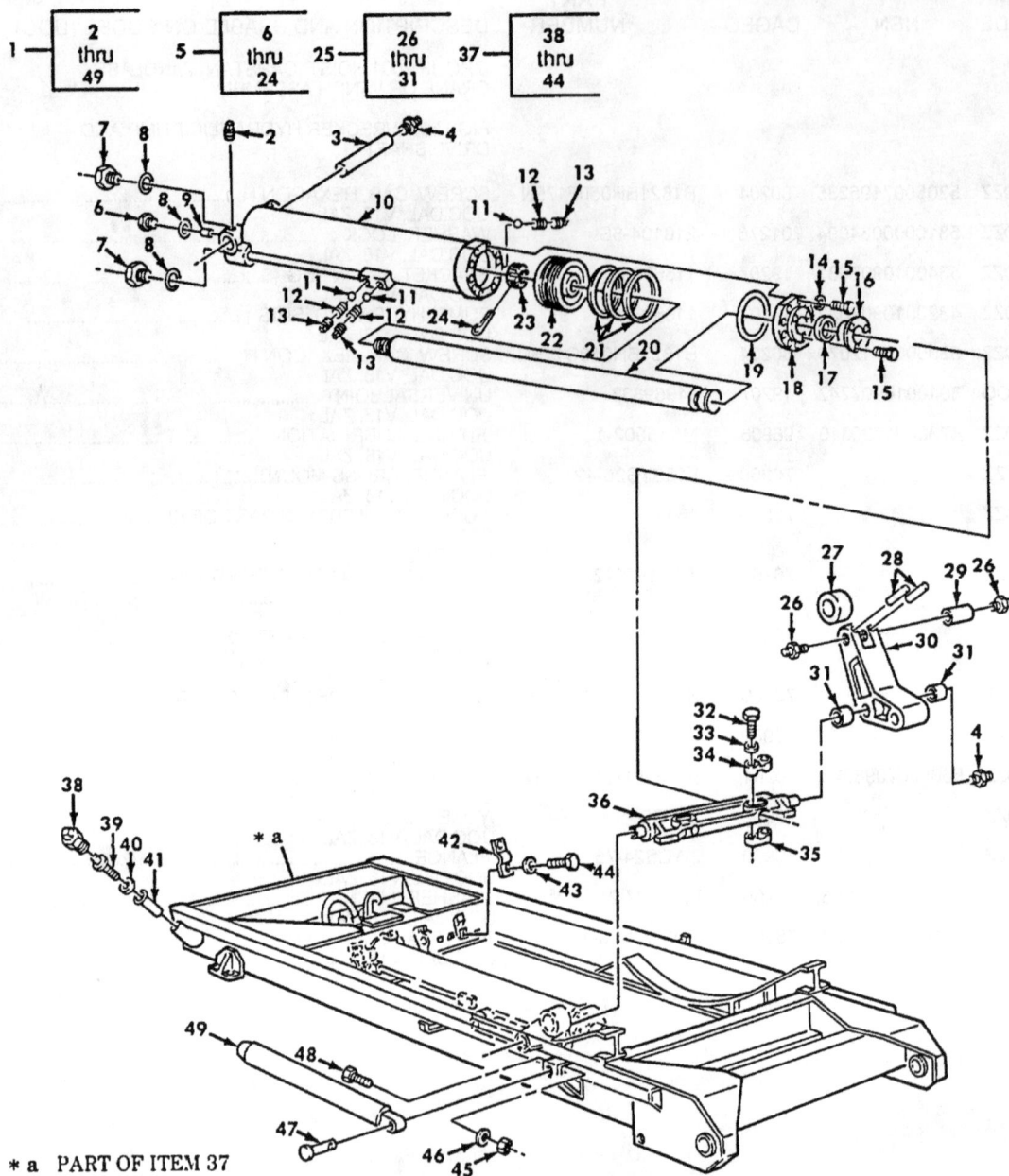

Figure 507. Dump Hydraulic Hoist Assembly.

*a   PART OF ITEM 37

(1) ITEM NO	(2) SMR CODE	(3) NSN	(4) CAGEC	(5) PART NUMBER	(6) DESCRIPTION AND USABLE ON CODES (UOC)	(7) QTY
					GROUP 2001 HOIST, CAPSTAN, WINDLASS, CRANE OR WINCH ASSEMBLY	
					FIG. 507 DUMP HYDRAULIC HOIST ASSEMBLY	
1	PAHHH	2590010853806	19207	12256407	HOIST UNIT, HYDRAULI UOC:V19, V20	1
1	PFHHH	2590012102174	19207	12256407-1	HOIST UNIT, HYDRAULI .... UOC:DAE, DAF, ZAE, ZAF	1
2	PAFZZ	5305000549271	96906	MS51955-36	.SETSCREW.. UOC:DAE, DAF, V19, V20, ZAE, ZAF	4
3	PAFZZ	3040005129223	27996	48B2071	.SHAFT, STRAIGHT .... UOC:DAE, DAF, V19, V20, ZAE, ZAF	2
4	PAFZZ	4730000504208	96906	MS15003-1	.FITTING, LUBRICATION .... UOC:DAE, DAF, V19, V20, ZAE, ZAF	4
5	PAFHH	3040005135786	19207	8327979	.CYLINDER ASSEMBLY, A .... UOC:DAE, DAF, V19, V20, ZAE, ZAF	2
6	PAHZZ	5365007409079	19207	7409079	..PLUG, MACHINE THREAD .... UOC:DAE, DAF, V19, V20, ZAE, ZAF	1
7	PAHZZ	4730007409058	19207	7409058	..ADAPTER BUSHING .... UOC:DAE, DAF, V19, V20, ZAE, ZAF	2
8	PAHZZ	5365001829635	19207	7409080	..SPACER, RING .... UOC:DAE, DAF, V19, V20, ZAE, ZAF	3
9	PAHZZ	4730003351812	19207	8327982	..RESTRICTOR, FLUID FL .... UOC:DAE, DAF, V19, V20, ZAE, ZAF	1
10	XBHZZ		19207	8720944	..CYLINDER, ACTUATING...................................... UOC:DAE, DAF, V19, V20, ZAE, ZAF	1
11	PAHZZ	3110001006159	96906	MS19059-2424	..BALL, BEARING .... UOC:DAE, DAF, V19, V20, ZAE, ZAF	3
12	PAHZZ	5360005974075	19207	7409054	..SPRING, HELICAL, COMP .... UOC:DAE, DAF, V19, V20, ZAE, ZAF	3
13	PAHZZ	4730008346187	81348	WW-P-471AASBUD	..PLUG, PIPE .... UOC:DAE, DAF, V19, V20, ZAE, ZAF	3
14	PAHZZ	5310006379541	96906	MS35338-46	..WASHER, LOCK .... UOC:DAE, DAF, V19, V20, ZAE, ZAF	10
15	PAHZZ	5305002693238	80204	B1821BH038F125N	..SCREW, CAP, HEXAGON H .... UOC:DAE, DAF, V19, V20, ZAE, ZAF	13
16	PAHZZ	5330007409050	19207	7409050	..RETAINER, PACKING .... UOC:DAE, DAF, V19, V20, ZAE, ZAF	1
17	PAHZZ	5330002694953	19207	11609215	..PACKING ASSEMBLY .... UOC:DAE, DAF, V19, V20, ZAE, ZAF	1
18	PAHZZ	3040005135787	19207	8327980	..CAP, LINEAR ACTUATIN .... UOC:DAE, DAF, V19, V20, ZAE, ZAF	1
19	PAHZZ	5330011830985	81349	M83461/1-427	..O-RING .... UOC:DAE, DAF, V19, V20, ZAE, ZAF	1
20	PAHZZ	2590003180912	19207	8327986	..ROD, PISTON, LINEAR .... UOC:DAE, DAF, V19, V20, ZAE, ZAF	1
21	PAHZZ	4310002878126	19207	8327985	..RING, PISTON .... UOC:DAE, DAF, V19, V20, ZAE, ZAF	3
22	PAHZZ	3040005135788	19207	8327983	..PISTON, LINEAR ACTUA .... UOC:DAE, DAF, V19, V20, ZAE, ZAF	1
23	PAHZZ	5310008496874	96906	MS35692-93	..NUT, PLAIN, SLOTTED, H .... UOC:DAE, DAF, V19, V20, ZAE, ZAF	1

(1) ITEM NO	(2) SMR CODE	(3) NSN	(4) CAGEC	(5) PART NUMBER	(6) DESCRIPTION AND USABLE ON CODES (UOC)	(7) QTY
24	PAHZZ	5315011364542	96906	MS24665-49	..PIN, COTTER .... UOC:DAE, DAF, V19, V20, ZAE, ZAF	1
25	PAFHH	2520007409040	19207	7409040	.ARM ROLLER CROSSHEA .... UOC:DAE, DAF, V19, V20, ZAE, ZAF	2
26	PAOZZ	4730000504208	96906	MS15003-1	..FITTING, LUBRICATION .... UOC:DAE, DAF, V19, V20, ZAE, ZAF	2
27	PAFZZ	3120007409042	19207	7409042	..ROLLER, LINEAR-ROTAR .... UOC:DAE, DAF, V19, V20, ZAE, ZAF	1
28	PAFZZ	5315000142976	21450	142976	..PIN, GROOVED, HEADLES .... UOC:DAE, DAF, V19, V20, ZAE, ZAF	2
29	PAFZZ	5315007409020	19207	7409020	..PIN, GROOVED, HEADLES .... UOC:DAE, DAF, V19, V20, ZAE, ZAF	1
30	PAFZZ	2520007409041	19207	7409041	..BELL CRANK .... UOC:DAE, DAF, V19, V20, ZAE, ZAF	1
31	PAFZZ	3120007409299	19207	7409299	..BEARING, SLEEVE .... UOC:DAE, DAF, V19, V20, ZAE, ZAF	2
32	PAFZZ	5305002693244	80204	B1821BH038F250N	.SCREW, CAP, HEXAGON H .... UOC:DAE, DAF, V19, V20, ZAE, ZAF	4
33	PAFZZ	5310006379541	96906	MS35338-46	.WASHER, LOCK .... UOC:DAE, DAF, V19, V20, ZAE, ZAF	4
34	PAFZZ	2520007409037	61465	M2020817	.RETAINER LOCK PISTO .... UOC:DAE, DAF, V19, V20, ZAE, ZAF	2
35	PAFZZ	5340012055957	19207	7409038	.FAIRLEAD HALF, TUBUL .... UOC:DAE, DAF, V19, V20, ZAE, ZAF	2
36	PAHZZ	2590003320095	19207	7971949	.CROSSHEAD, HYDRAULIC .... UOC:DAE, DAF, V19, V20, ZAE, ZAF	1
37	PAFHH	2590010853584	19207	12256409	.SUB-FRAME, HOIST UOC:V19, V20	1
37	PAFHH	2590012100203	19207	12256409-1	.HOIST UNIT, HYDRAULI .... UOC:DAE, DAF, ZAE, ZAF	1
38	PAFZZ	4730011234546	19207	7409027	.PLUG, PIPE .... UOC:DAE, DAF, V19, V20, ZAE, ZAF	1
39	PAOZZ	6680007409026	19207	7409026	.GAGE ROD, LIQUID LEV .... UOC:DAE, DAF, V19, V20, ZAE, ZAF	1
40	PAFZZ	5365001829635	19207	7409080	.SPACER, RING .... UOC:DAE, DAF, V19, V20, ZAE, ZAF	1
41	PAFZZ	4330011686891	19207	7409025	.FILTER ELEMENT, FLUI .... UOC:DAE, DAF, V19, V20, ZAE, ZAF	1
42	PAFZZ	3110009996469	19207	8758374	.BEARING, CAP AND BAS .... UOC:DAE, DAF, V19, V20, ZAE, ZAF	4
43	PAFZZ	5310005847888	96906	MS35338-51	.WASHER, LOCK .... UOC:DAE, DAF, V19, V20, ZAE, ZAF	8
44	PAFZZ	5305009162345	80204	B1821BH075F200N	.SCREW, CAP, HEXAGON H .... UOC:DAE, DAF, V19, V20, ZAE, ZAF	8
45	PAFZZ	5310007680319	96906	MS51968-2	.NUT, PLAIN, HEXAGON .... UOC:DAE, DAF, V19, V20, ZAE, ZAF	2
46	PAFZZ	5310005825965	96906	MS35338-44	.WASHER, LOCK .... UOC:DAE, DAF, V19, V20, ZAE, ZAF	2
47	PAFZZ	5315014161809	19207	12375688	.PIN, STRAIGHT, HEADED .... UOC:DAE, DAF, V19, V20, ZAE, ZAF	2
48	PAFZZ	5305002678957	80204	B1821BH025F175N	.SCREW, CAP, HEXAGON H .... UOC:DAE, DAF, V19, V20, ZAE, ZAF	2
49	PFFZZ	3040014361832	19207	7409023-1	.BRACKET, EYE, NONROTA .... UOC:DAE, DAF, V19, V20, ZAE, ZAF	2

END OF FIGURE

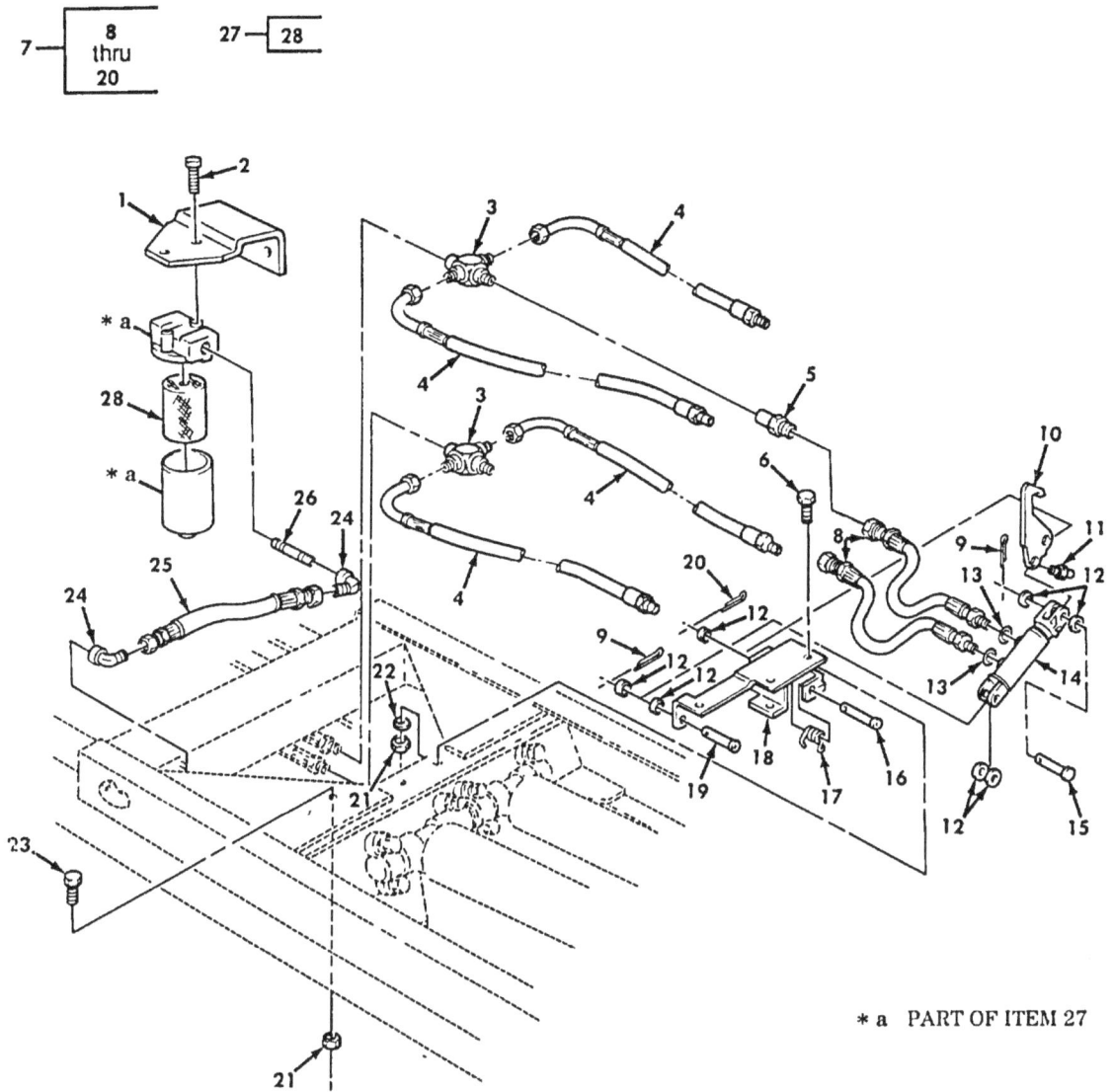

Figure 508. Dump Hoist Safety Latch Assembly, Hoses, and Filter.

*a PART OF ITEM 27

(1) ITEM NO	(2) SMR CODE	(3) NSN	(4) CAGEC	(5) PART NUMBER	(6) DESCRIPTION AND USABLE ON CODES (UOC)	(7) QTY
					GROUP 2001 HOIST, CAPSTAN, WINDLASS, CRANE OR WINCH ASSEMBLY	
					FIG. 508 DUMP HOIST SAFETY LATCH ASSEMBLY, HOSES, AND FILTER	
1	PFFZZ	5340010849631	19207	12257125	BRACKET, MOUNTING UOC:DAE, DAF, V19, V20, ZAE, ZAF	1
2	PAOZZ	5305005434372	80204	B1821BH038C075N	SCREW, CAP, HEXAGON H UOC:DAE, DAF, V19, V20, ZAE, ZAF	2
3	PAFZZ	4730012153218	19207	12302899	CROSS, PIPE TO TUBE UOC:DAE, DAF, V19, V20, ZAE, ZAF	2
4	PAFZZ	4720012154295	19207	12302900	HOSE ASSEMBLY, NONME UOC:DAE, DAF, V19, V20, ZAE, ZAF	4
5	PAFZZ	4730008099427	96906	MS51500A8-8	ADAPTER, STRAIGHT, PI UOC:DAE, DAF, V19, V20, ZAE, ZAF	2
6	PAFZZ	5305007195221	80204	B1821BH050F150N	SCREW, CAP, HEXAGON H UOC:DAE, DAF, V19, V20, ZAE, ZAF	4
7	PAFFF	5340011276921	19207	12300639	LATCH ASSEMBLY, SAFE UOC:DAE, DAF, V19, V20, ZAE, ZAF	1
8	PAFZZ	4720011280179	19207	12300644	.HOSE ASSEMBLY, NONME UOC:DAE, DAF, V19, V20, ZAE, ZAF	2
9	PAFZZ	5315000590491	96906	MS24665-372	.PIN, COTTER UOC:DAE, DAF, V19, V20, ZAE, ZAF	2
10	PAFZZ	2510011290278	19207	12300643	.HOOK, SAFETY LATCH UOC:DAE, DAF, V19, V20, ZAE, ZAF	1
11	PAFZZ	4730000504208	96906	MS15003-1	.FITTING, LUBRICATION UOC:DAE, DAF, V19, V20, ZAE, ZAF	2
12	PAFZZ	5310008098533	96906	MS27183-23	.WASHER, FLAT UOC:DAE, DAF, V19, V20, ZAE, ZAF	7
13	PAFZZ	5330008045695	25472	AN6920-6	.O-RING UOC:DAE, DAF, V19, V20, ZAE, ZAF	2
14	PAFZZ	2590011245052	19207	12300679	.CYLINDER, LATCH UOC:DAE, DAF, V19, V20, ZAE, ZAF	1
15	PAFZZ	5315009043408	96906	MS20392-10C67	.PIN, STRAIGHT, HEADED UOC:DAE, DAF, V19, V20, ZAE, ZAF	1
16	PAFZZ	5315011236812	19207	12300638	.PIN, STRAIGHT, HEADED UOC:DAE, DAF, V19, V20, ZAE, ZAF	1
17	PAFZZ	5360011256118	19207	12300640	.SPRING, HELICAL, EXTE UOC:DAE, DAF, V19, V20, ZAE, ZAF	1
18	PAFZZ	5340011245054	19207	12300635	.BRACKET, MOUNTING UOC:DAE, DAF, V19, V20, ZAE, ZAF	1
19	PAFZZ	5315009043412	96906	MS20392-10C57	.PIN, STRAIGHT, HEADED UOC:DAE, DAF, V19, V20, ZAE, ZAF	1
20	PAFZZ	5315002398032	96906	MS24665-513	.PIN, COTTER UOC:DAE, DAF, V19, V20, ZAE, ZAF	1
21	PAFZZ	5310004883888	96906	MS51943-40	NUT, SELF-LOCKING, HE UOC:DAE, DAF, V19, V20, ZAE, ZAF	5
22	PAFZZ	5310008095998	96906	MS27183-18	WASHER, FLAT UOC:DAE, DAF, V19, V20, ZAE, ZAF	5
23	PAFZZ	5305007250154	96906	MS90727-112	SCREW, CAP, HEXAGON H UOC:DAE, DAF, V19, V20, ZAE, ZAF	1

(1) ITEM NO	(2) SMR CODE	(3) NSN	(4) CAGEC	(5) PART NUMBER	(6) DESCRIPTION AND USABLE ON CODES (UOC)	(7) QTY
24	PAFZZ	4730008339315	96906	MS51504A16	ELBOW, PIPE TO TUBE .... UOC:DAE, DAF, V19, V20, ZAE, ZAF	2
25	PAFZZ	4720011053564	19207	12276917	HOSE ASSEMBLY, NONME .... UOC:DAE, DAF, V19, V20, ZAE, ZAF	1
26	PAFZZ	4730001388121	24617	219682	NIPPLE, PIPE .... UOC:DAE, DAF, V19, V20, ZAE, ZAF	1
27	PAOZZ	2940011126438	19207	11669456	FILTER, FLUID .... UOC:DAE, DAF, V19, V20, ZAE, ZAF	1
28	PAOZZ	2940009508410	70040	PF297	.FILTER ELEMENT, FLUI .... UOC:DAE, DAF, V19, V20, ZAE, ZAF	1

END OF FIGURE

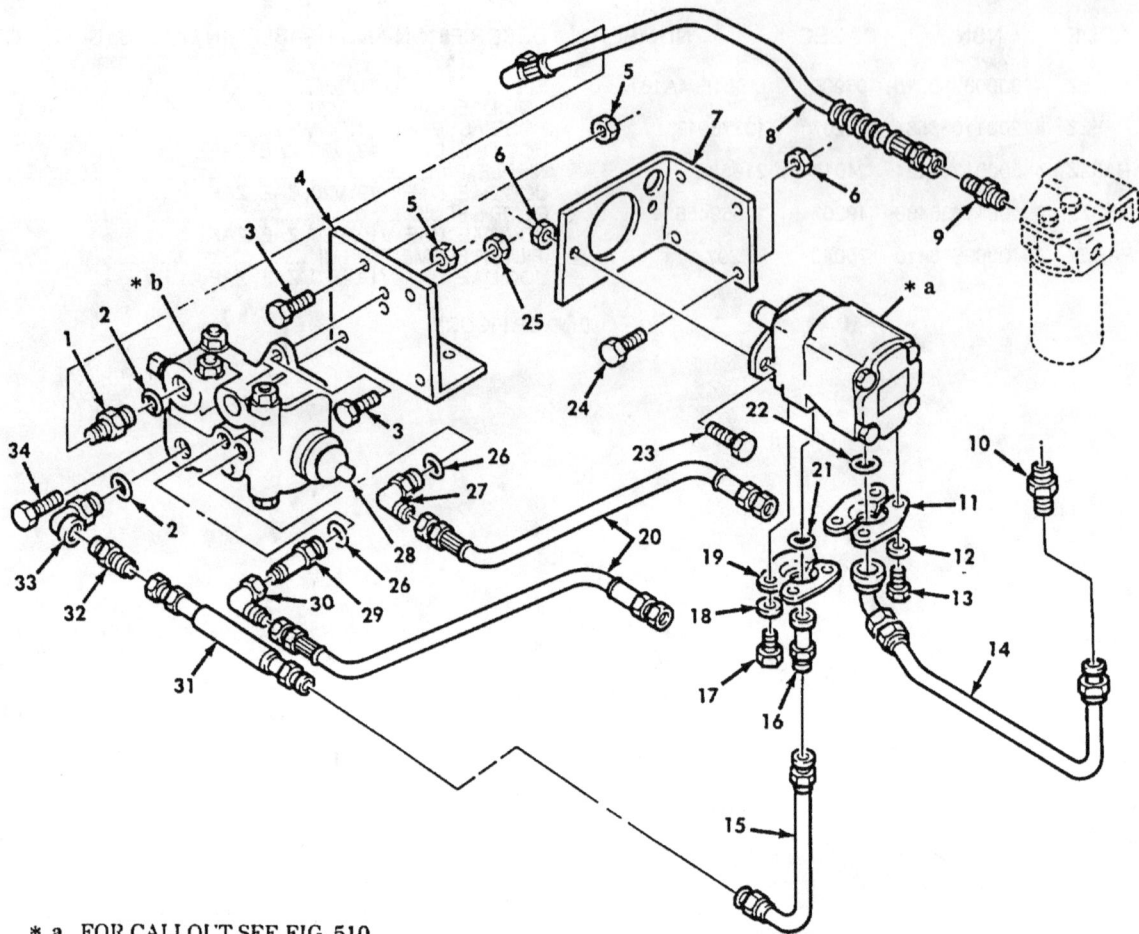

* a FOR CALLOUT SEE FIG 510
* b FOR BREAKDOWN SEE FIG. 511

Figure 509. Dump Control Pump, Valve, and Related Parts.

(1) ITEM NO	(2) SMR CODE	(3) NSN	(4) CAGEC	(5) PART NUMBER	(6) DESCRIPTION AND USABLE ON CODES (UOC)	(7) QTY
					GROUP 2001 HOIST, CAPSTAN, WINDLASS, CRANE OR WINCH ASSEMBLY	
					FIG. 509 DUMP CONTROL PUMP, VALVE, AND RELATED PARTS	
1	PAFZZ	4730011165969	96906	MS51525A10-12S	ADAPTER, STRAIGHT, TU UOC:DAE, V20, ZAE	1
2	PAFZZ	5330002518839	96906	MS28778-12	O-RING UOC:DAE, V20, ZAE	2
3	PAFZZ	5305002693240	80204	B1821BH038F150N	SCREW, CAP, HEXAGON H UOC:DAE, V20, ZAE	5
4	PFFZZ	5340011043828	19207	12257128	BRACKET, ANGLE UOC:DAE, V20, ZAE	1
5	PAFZZ	5310008140672	96906	MS51943-36	NUT, SELF-LOCKING, HE UOC:DAE, V20, ZAE	5
6	PAFZZ	5310004883888	96906	MS51943-40	NUT, SELF-LOCKING, HE UOC:DAE, V20, ZAE	6
7	PFFZZ	5340011044326	19207	12257127	BRACKET, ANGLE UOC:DAE, V20, ZAE	1
8	PAFZZ	4720011054067	19207	12257109-2	HOSE ASSEMBLY, NONME UOC:DAE, V20, ZAE	1
9	PAFZZ	4730010904919	19207	12255651-2	ADAPTER, STRAIGHT, PI UOC:DAE, V20, ZAE	1
10	PAFZZ	4730006767566	96906	MS51500A24	ADAPTER, STRAIGHT, PI UOC:DAE, V20, ZAE	1
11	PAFZZ	4730000893406	81349	M52525/16-24	FLANGE, PIPE, SWIVEL UOC:DAE, V20, ZAE	2
12	PAFZZ	5310000034094	01276	210104-8S	WASHER, LOCK UOC:DAE, V20, ZAE	4
13	PAFZZ	5305000712069	80204	B1821BH050C150N	SCREW, CAP, HEXAGON H UOC:DAE, V20, ZAE	4
14	PAFZZ	4720011923504	19207	12302871	HOSE ASSEMBLY, NONME UOC:DAE, V20, ZAE	1
15	PAFZZ	4710010892060	19207	12257108	TUBE ASSEMBLY, METAL UOC:DAE, V20, ZAE	1
16	PAFZZ	4730010904924	19207	12257120	ADAPTER, STRAIGHT , FL UOC:DAE, V20, ZAE	1
17	PAFZZ	5305000680511	80204	B1821BH038C125N	SCREW, CAP, HEXAGON H UOC:DAE, V20, ZAE	4
18	PAFZZ	5310006379541	96906	MS35338-46	WASHER, LOCK UOC:DAE, V20, ZAE	4
19	PAFZZ	4730007557609	81349	M52525/16-16	FLANGE, PIPE UOC:DAE, V20, ZAE	2
20	PAFZZ	4720011063982	19207	12257110-2	HOSE ASSEMBLY, NONMETALLIC UOC:DAE, V20, ZAE	2
21	PAFZZ	5330005797925	96906	MS28775-219	O-RING UOC:DAE, V20, ZAE	1
22	PAFZZ	5330005797927	96906	MS28775-225	O-RING UOC:DAE, V20, ZAE	1
23	PAFZZ	5305007195235	80204	B1821BH050F175N	SCREW, CAP, HEXAGON H UOC:DAE, V20, ZAE	2
24	PAFZZ	5305007195238	80204	B1821BH050F200N	SCREW, CAP, HEXAGON H UOC:DAE, V20, ZAE	4

(1) ITEM NO	(2) SMR CODE	(3) NSN	(4) CAGEC	(5) PART NUMBER	(6) DESCRIPTION AND USABLE ON CODES (UOC)	(7) QTY
25	PAFZZ	5310009941006	96906	MS51943-38	NUT, SELF-LOCKING, HE .... UOC:DAE, V20, ZAE	2
26	PAFZZ	5330002859842	96906	MS28778-10	O-RING .... UOC:DAE, V20, ZAE	1
27	PAFZZ	4730009747313	96906	MS51527A10	ELBOW, TUBE TO BOSS................................. UOC:DAE, V20, ZAE	1
28	PAFFF	4810011122162	19207	11669457-1	VALVE ASSEMBLY .... UOC:DAE, V20, ZAE	1
29	PAFZZ	4730011929590	96906	MS51526A10	ADAPTER, STRAIGHT, TU............................... UOC:DAE, V20, ZAE	1
30	PAFZZ	4730011391585	96906	MS51521A10	ELBOW, TUBE .... UOC:DAE, V20, ZAE	1
31	PAFZZ	4720010892016	19207	12276973	HOSE ASSEMBLY, NONME .... UOC:DAE, V20, ZAE	1
32	PAFZZ	4730000211802	96906	MS51500A16-12	ADAPTER, STRAIGHT, PI................................ UOC:DAE, V20, ZAE	1
33	PAFZZ	4730010906474	19207	12257113	ELBOW, PIPE TO BOSS .... UOC:DAE, V20, ZAE	1
34	PAFZZ	5305007098542	96906	MS90727-91	SCREW, CAP, HEXAGON .... UOC:DAE, V20, ZAE	2

END OF FIGURE

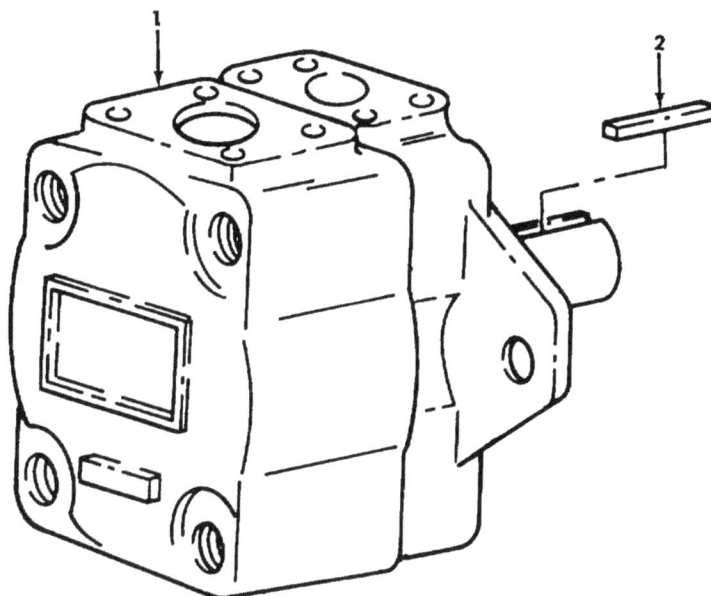

Figure 510. Dump Control Pump Assembly.

(1) ITEM NO	(2) SMR CODE	(3) NSN	(4) CAGEC	(5) PART NUMBER	(6) DESCRIPTION AND USABLE ON CODES (UOC)	(7) QTY
					GROUP 2001 HOIST, CAPSTAN, WINDLASS, CRANE OR WINCH ASSEMBLY	
					FIG. 510 DUMP CONTROL PUMP ASSEMBLY	
1	PAFZZ	4320011062061	62983	26VQ14A-1C-20	PUMP, ROTARY .... UOC:DAE, V20, ZAE	1
2	PAFZZ	5315002643099	96906	MS20066-187	KEY, MACHINE .... UOC:DAE, V20, ZAE	1

END OF FIGURE

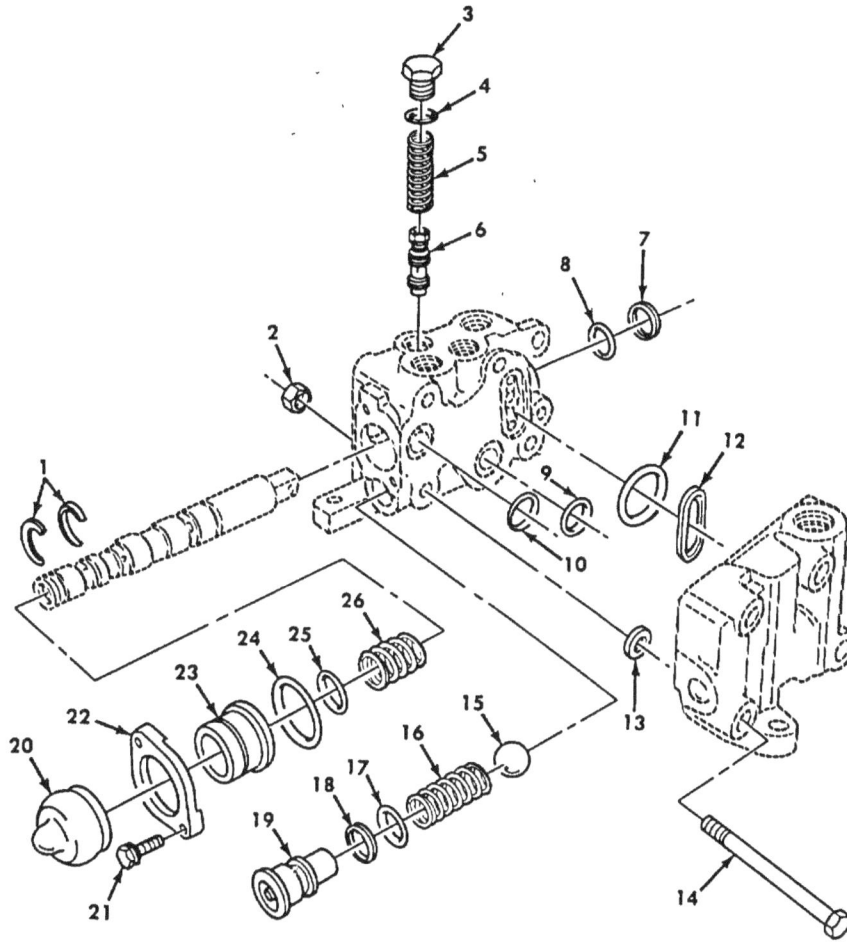

Figure 511. Dump Control Valve Assembly.

(1) ITEM NO	(2) SMR CODE	(3) NSN	(4) CAGEC	(5) PART NUMBER	(6) DESCRIPTION AND USABLE ON CODES (UOC)	(7) QTY
					GROUP 2001 HOIST, CAPSTAN, WINDLASS, CRANE OR WINCH ASSEMBLY	
					FIG. 511 DUMP CONTROL VALVE ASSEMBLY	
1	PAFZZ	5310008671465	62983	284156	WASHER, SLOTTED .... UOC:DAE, V20, ZAE	2
2	PAFZZ	5310007320558	62983	1454	NUT, PLAIN, HEXAGON.................................. UOC:DAE, V20, ZAE	4
3	PAFZZ	5365006746831	62983	186580	PLUG, MACHINE THREAD................................ UOC:DAE, V20, ZAE	1
4	PAFZZ	5330009486482	62983	154129	O-RING .... UOC:DAE, V20, ZAE	1
5	PAFZZ	5360009181920	62983	259871	SPRING, HELICAL, COMP .... UOC:DAE, V20, ZAE	1
6	PAFZZ	4820009609329	62983	232797	PISTON, VALVE .... UOC:DAE, V20, ZAE	1
7	KFFZZ	5330002460158	62983	286669	PACKING, PREFORMED PART OF KIT P/N.................... 920278 UOC:DAE, V20, ZAE	1
8	KFFZZ		62983	283856	RING, QUADRANT PART OF KIT P/N................................. 920278 UOC:DAE, V20, ZAE	1
9	KFFZZ	5330009535206	62983	223489	SEAL, CONTROL VALVE PART OF KIT P/N .... 920278 UOC:DAE, V20, ZAE	1
10	KFFZZ	5330009535207	62983	223493	GASKET PART OF KIT P/N 920278 .... UOC:DAE, V20, ZAE	1
11	KFFZZ	5330009535208	62983	226161	PACKING, CONTROL VAL PART OF KIT P/N .... 920278 UOC:DAE, V20, ZAE	1
12	KFFZZ	2590000527504	62983	237736	RETAINER, OIL SEAL PART OF KIT P/N .... 920278 UOC:DAE, V20, ZAE	1
13	PAFZZ	5365001826713	62983	307198	SHIM.. UOC:DAE, V20, ZAE	4
14	PAFZZ	5305010200709	62983	146835	SCREW... UOC:DAE, V20, ZAE	4
15	PAFZZ	3110008994353	62983	1656	BALL, BEARING... UOC:DAE, V20, ZAE	1
16	PAFZZ	5360009535205	62983 2	23388	SPRING, HELICAL, COMP .... UOC:DAE, V20, ZAE	1
17	KFFZZ	5330005798108	62983	154008	O-RING PART OF KIT P/N 920278..................... UOC:DAE, V20, ZAE	1
18	KFFZZ	5330010487346	62983	271722	RETAINER, PACKING PART OF KIT P/N .... 920278 UOC:DAE, V20, ZAE	1
19	PAFZZ	2530000870327	62983	222640	POWER STEERING PLUG .... UOC:DAE, V20, ZAE	1
20	KFFZZ		62983	307951	DUST COVER PART OF KIT P/N 923080................ PART OF KIT P/N 680702 .... UOC:DAE, V20, ZAE	1
21	PAFZZ	5305004713909	62983	282027	SCREW, MACHINE	2

(1) ITEM NO	(2) SMR CODE	(3) NSN	(4) CAGEC	(5) PART NUMBER	(6) DESCRIPTION AND USABLE ON CODES (UOC)	(7) QTY
					UOC:DAE, V20, ZAE	
22	PAFZZ	5340008895209	62983	284154	RETAINER	1
					UOC:DAE, V20, ZAE	
23	KFFZZ	2530000870369	62983	284155	SLEEVE PART OF KIT P/N 923080 PART	1
					OF KIT P/N 680702	
					UOC:DAE, V20, ZAE	
24	KFFZZ	5330002761933	62983	187000	PACKING, PREFORMED PART OF KIT P/N	1
					923080 PART OF KIT P/N 680702	
					UOC:DAE, V20, ZAE	
25	KFFZZ	5365002416916	62983	680702	RING, QUADRANT PART OF KIT P/N 923080	1
					PART OF KIT P/N 12255848	
					UOC:DAE, V20, ZAE	
26	PAFZZ	5360009609326	62983	246632	SPRING, HELICAL, COMP	1
					UOC:DAE, V20, ZAE	

END OF FIGURE

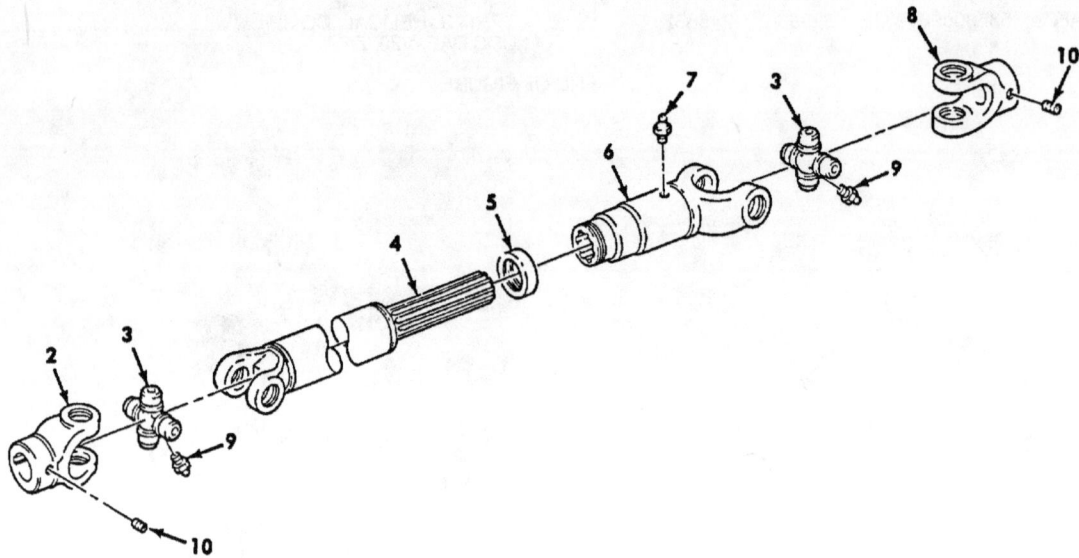

Figure 512. Dump Control Propeller Drive Shaft Assembly.

(1) ITEM NO	(2) SMR CODE	(3) NSN	(4) CAGEC	(5) PART NUMBER	(6) DESCRIPTION AND USABLE ON CODES (UOC)	(7) QTY
					GROUP 2001 HOIST, CAPSTAN, WINDLASS, CRANE OR WINCH ASSEMBLY	
					FIG. 512 DUMP CONTROL PROPELLER DRIVE SHAFT ASSEMBLY	
1	PAOOO	2520011112280	19207	11669461	PROPELLER SHAFT WIT .... UOC:DAE, V20, ZAE	1
2	PAOZZ	2520011341089	78500	L6NYR20-4	.YOKE, UNIVERSAL JOIN .... UOC:DAE, V20, ZAE	1
3	PAOZZ	2520011343455	78500	CPL6240	.UNIVERSAL JOINT .... UOC:DAE, V20, ZAE	2
4	XAOZZ		70960	9035544862	.SHAFT ASSEMBLY .... UOC:DAE, V20, ZAE	1
5	PAOZZ	2520011343706	78500	DCL6N-3	.CAP, DUST, PROPELLER .... UOC:DAE, V20, ZAE	1
6	XAOZZ		70960	L6NLS20-24	.SLIP YOKE .... UOC:DAE, V20, ZAE	1
7	PAOZZ	4730000504203	96906	MS15001-1	.FITTING, LUBRICATION .... UOC:DAE, V20, ZAE	1
8	PAOZZ	2520011343471	78500	L6NYR14-19	.YOKE, UNIVERSAL JOIN .... UOC:DAE, V20, ZAE	1
9	PAOZZ	4730000504203	96906	MS15001-1	.FITTING, LUBRICATION .... UOC:DAE, V20, ZAE	2
10	PAOZZ	5305000589389	96906	MS51977-74	.SETSCREW............... UOC:DAE, V20, ZAE	2

END OF FIGURE

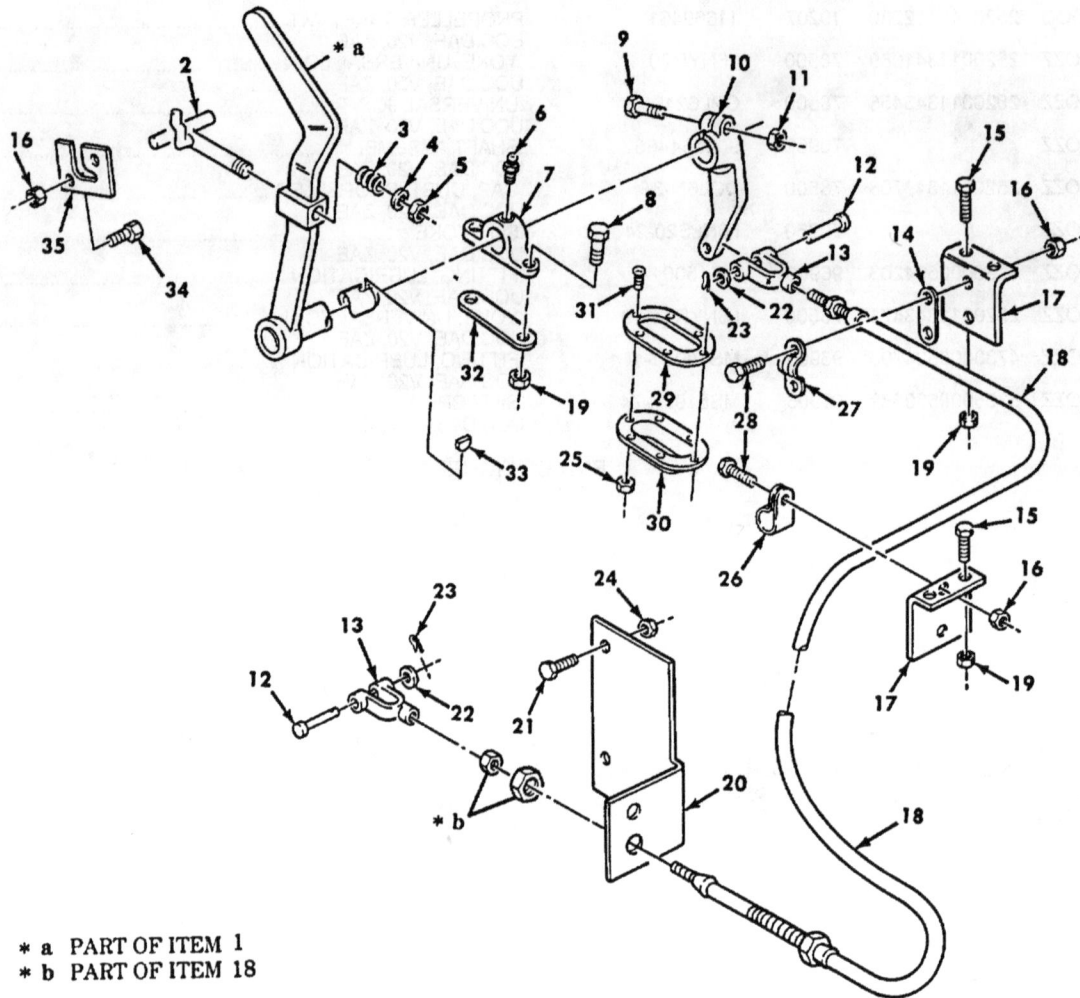

* a   PART OF ITEM 1
* b   PART OF ITEM 18

Figure 513. Dump Controls, Lever Assembly, Cable, and Mounting Hardware.

(1) ITEM NO	(2) SMR CODE	(3) NSN	(4) CAGEC	(5) PART NUMBER	(6) DESCRIPTION AND USABLE ON CODES (UOC)	(7) QTY

SECTION II

GROUP 2001 HOIST, CAPSTAN, WINDLASS, CRANE OR WINCH ASSEMBLY

FIG. 513 DUMP CONTROLS, LEVER ASSEMBLY, CABLE, AND MOUNTING HARDWARE

(1) ITEM NO	(2) SMR CODE	(3) NSN	(4) CAGEC	(5) PART NUMBER	(6) DESCRIPTION AND USABLE ON CODES (UOC)	(7) QTY
1	PAFZZ	3040011049154	19207	12256356-1	CONTROL LEVER ASSEM UOC:V19, V20	1
2	PAFZZ	5306010822524	19207	12256411	.BOLT, HOOK............. UOC:V19, V20	1
3	PAFZZ	5360007521975	19207	7521975	.SPRING, HELICAL, COMP UOC:V19, V20	1
4	PAFZZ	5310000806004	96906	MS27183-14	.WASHER, FLAT UOC:V19, V20	1
5	PAFZZ	5310008140672	96906	MS51943-36	.NUT, SELF-LOCKING, HE .... UOC:DAE, DAF, V19, V20, ZAE, ZAF	1
6	PAOZZ	4730000504208	96906	MS15003-1	FITTING, LUBRICATION .... UOC:DAE, DAF, V19, V20, ZAE, ZAF	2
7	PFFZZ	2520007411410	19207	7411410	BRACKET CONTROL LEV UOC:V19, V20	2
8	PAFZZ	5305002693238	80204	B1821BH038F125N	SCREW, CAP, HEXAGON H .... UOC:DAE, DAF, V19, V20, ZAE, ZAF	4
9	PAFZZ	5305007195239	96906	MS90727-116	SCREW, CAP, HEXAGON H .... UOC:DAE, DAF, V19, V20, ZAE, ZAF	1
10	PFFZZ	3040011043851	19207	12276954	LEVER, REMOTE CONTRO........ UOC:V19, V20	1
11	PAFZZ	5310004883888	96906	MS51943-40	NUT, SELF-LOCKING, HE .... UOC:DAE, DAF, V19, V20, ZAE, ZAF	1
12	PAFZZ	5315009902889	96906	MS20392-5C31	PIN, STRAIGHT, HEADED .... UOC:DAE, DAF, V19, V20, ZAE, ZAF	2
13	PAFZZ	5340011049005	19207	12276916	CLEVIS, ROD END........ UOC:DAE, DAF, V19, V20, ZAE, ZAF	2
14	PAFZZ	5365011093353	19207	12256286	SPACER, PLATE .... UOC:DAE, DAF, V19, V20, ZAE, ZAF	1
15	PAFZZ	5305002692804	96906	MS90726-61	SCREW, CAP, HEXAGON H .... UOC:DAE, DAF, V19, V20, ZAE, ZAF	4
16	PAFZZ	5310009359022	96906	MS51943-32	NUT, SELF-LOCKING, HE .... UOC:DAE, DAF, V19, V20, ZAE, ZAF	5
17	PFFZZ	5340011111351	19207	12276955	BRACKET, ANGLE .... UOC:DAE, DAF, V19, V20, ZAE, ZAF	2
18	PAFZZ	2590011421310	19207	11669464-1	CONTROL ASSEMBLY, PU........ UOC:DAE, DAF, V19, V20, ZAE, ZAF	1
19	PAFZZ	5310008140672	96906	MS51943-36	NUT, SELF-LOCKING, HE .... UOC:DAE, DAF, V19, V20, ZAE, ZAF	8
20	PFFZZ	5340011075221	19207	12276906	BRACKET, ANGLE .... UOC:DAE, DAF, V19, V20, ZAE, ZAF	1
21	PAFZZ	5305002692803	96906	MS90726-60	SCREW, CAP, HEXAGON H .... UOC:DAE, DAF, V19, V20, ZAE, ZAF	2

(1) ITEM NO	(2) SMR CODE	(3) NSN	(4) CAGEC	(5) PART NUMBER	(6) DESCRIPTION AND USABLE ON CODES (UOC)	(7) QTY
22	PAFZZ	5310000806004	96906	MS27183-14	WASHER, FLAT .... UOC:DAE, DAF, V19, V20, ZAE, ZAF	2
23	PAFZZ	5315002341863	96906	MS24665-300	PIN, COTTER............................. UOC:DAE, DAF, V19, V20, ZAE, ZAF	2
24	PAFZZ	5310008140672	96906	MS51943-36	NUT, SELF-LOCKING, HE .... UOC:DAE, DAF, V19, V20, ZAE, ZAF	2
25	PAFZZ	5310002081918	88044	AN365-1024A	NUT, SELF-LOCKING, HE .... UOC:DAE, DAF, V19, V20, ZAE, ZAF	6
26	PAFZZ	5340009848540	96906	MS21333-102	CLAMP, LOOP .... UOC:DAE, DAF, V19, V20, ZAE, ZAF	1
27	PAFZZ	5340011285302	19207	12256285	STRAP, RETAINING .... UOC:DAE, DAF, V19, V20, ZAE, ZAF	1
28	PAFZZ	5306000680513	60285	6893-2	BOLT, MACHINE . ... UOC:DAE, DAF, V19, V20, ZAE, ZAF	3
29	PAFZZ	5365010902074	19207	12256377	SPACER, PLATE .... UOC:DAE, DAF, V19, V20, ZAE, ZAF	1
30	PAFZZ	2530011048943	19207	12256378	BOOT, VEHICULAR COM .... UOC:DAE, DAF, V19, V20, ZAE, ZAF	1
31	PAFZZ	5306000440502	21450	440502	BOLT, MACHINE . ... UOC:DAE, DAF, V19, V20, ZAE, ZAF	6
32	PAFZZ	5365011099490	19207	12256382	SPACER, PLATE .... UOC:DAE, DAF, V19, V20, ZAAE, ZAF	2
33	PAFZZ	5315006165520	96906	MS35756-14	KEY, WOODRUFF .... UOC:DAE, DAF, V19, V20, ZAE, ZAF	1
34	PAFZZ	5305002693239	80204	B1821BH038F138N	SCREW, CAP, HEXAGON H .... UOC:DAE, DAF, V19, V20, ZAE, ZAF	2
35	PAFZZ	5340011119878	19207	12256376	STRIKE, CATCH . ... UOC:DAE, DAF, V19, V20, ZAE, ZAF	1

END OF FIGURE

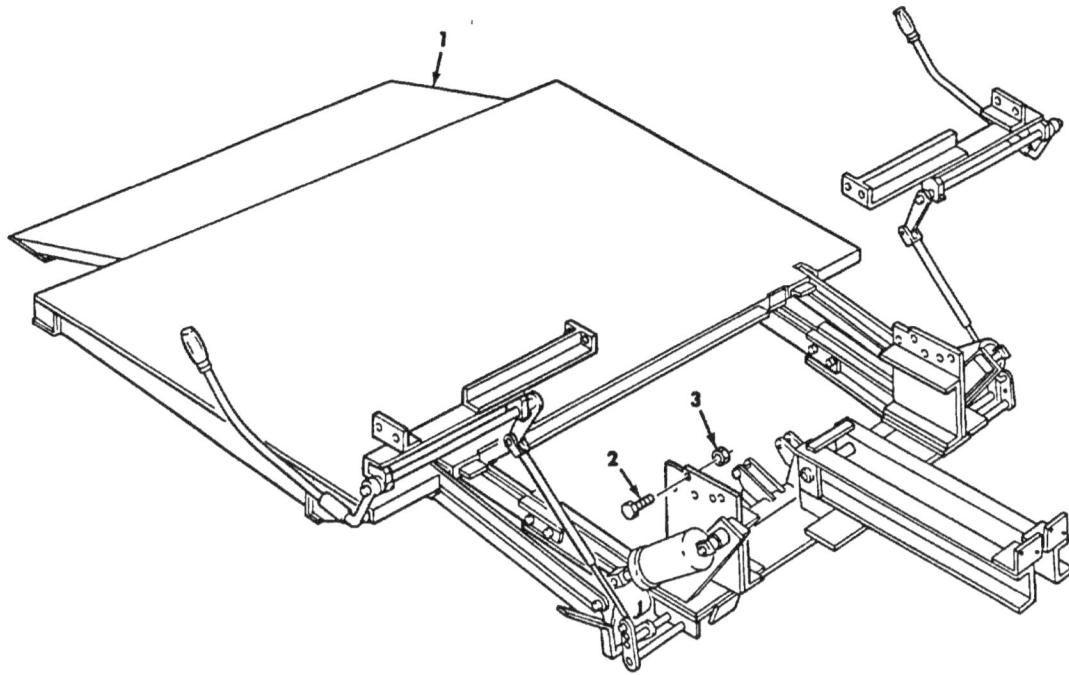

Figure 514. Van Hydraulic Liftgate Assembly and Mounting Hardware.

(1) ITEM NO	(2) SMR CODE	(3) NSN	(4) CAGEC	(5) PART NUMBER	(6) DESCRIPTION AND USABLE ON CODES (UOC)	(7) QTY
					GROUP 2001 HOIST, CAPSTAN, WINDLASS, CRANE OR WINCH ASSEMBLY	
					FIG. 514 VAN HYDRAULIC LIFTGATE ASSEMBLY AND MOUNTING HARDWARE	
1	PAFHH	3830011341946	64203	G-40-2	TAILGATE, HYDRAULIC .... UOC:DAK, V25, ZAK	1
2	PAFZZ	5305011354754	64203	G-18711	SCREW .... UOC:DAK, V25, ZAK	12
3	PAFZZ	5310011224595	64203	G-9333	NUT, PLAIN, HEXAGON .... UOC:DAK, V25, ZAK	12

END OF FIGURE

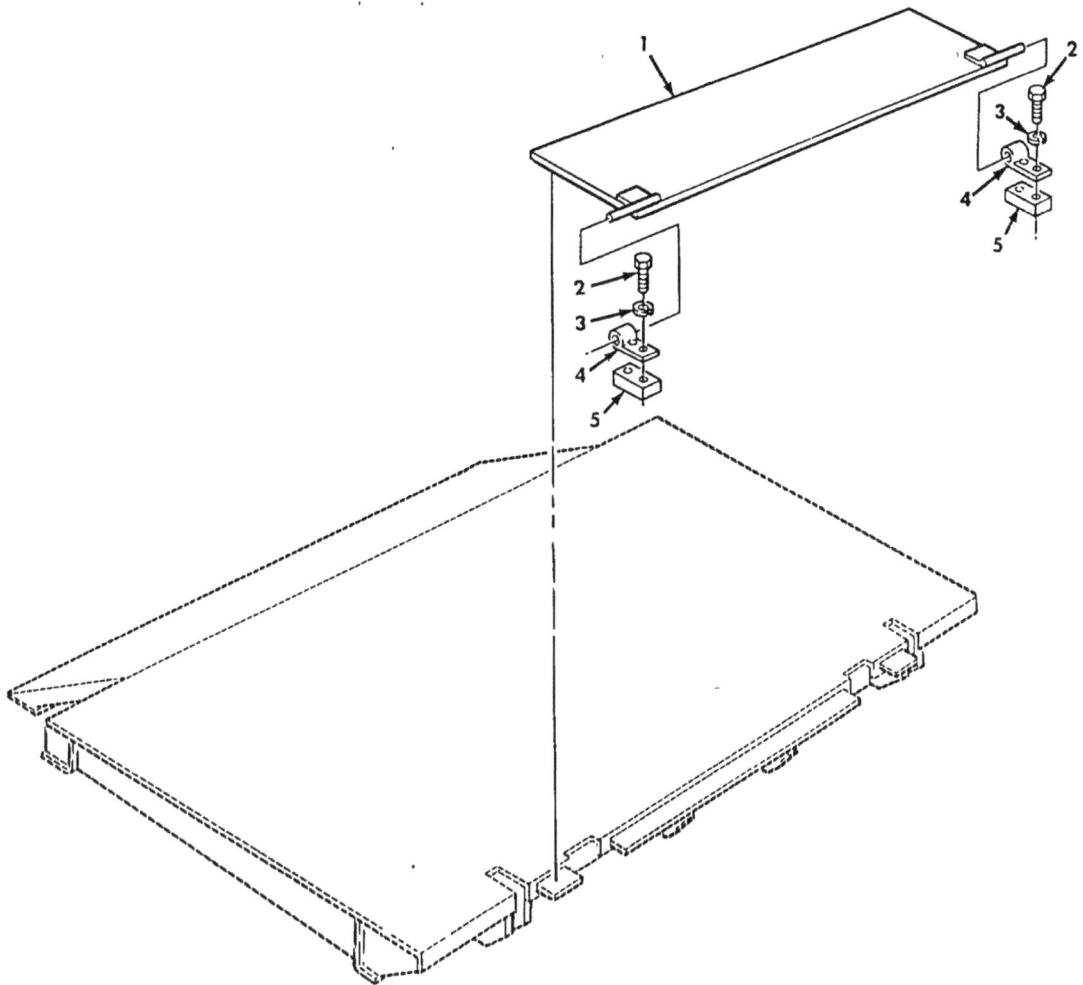

Figure 515. Liftgate Platform Assembly and Hinge Plate.

(1) ITEM NO	(2) SMR CODE	(3) NSN	(4) CAGEC	(5) PART NUMBER	(6) DESCRIPTION AND USABLE ON CODES (UOC)	(7) QTY
					GROUP 2001 HOIST, CAPSTAN, WINDLASS, CRANE OR WINCH ASSEMBLY	
					FIG. 515 LIFTGATE PLATFORM ASSEMBLY AND HINGE PLATE	
1	PAFZZ	5340012048720	64203	20-33433B	PLATE, HINGE .... UOC:DAK, V25, ZAK	1
2	PFFZZ	5305011354839	64203	10-42253	SCREW, CAP, HEXAGON H .... UOC:DAK, V25, ZAK	4
3	PFFZZ	5310011226148	64203	G-39250	WASHER, LOCK.. UOC:DAK, V25, ZAK	4
4	PAFZZ	5340012059479	64203	20-33607-A	LEAF, BUTT HINGE................................... UOC:DAK, V25, ZAK	2
5	PFFZZ	5340011224489	64203	10-40133-A	BAR... UOC:DAK, V25, ZAK	2

END OF FIGURE

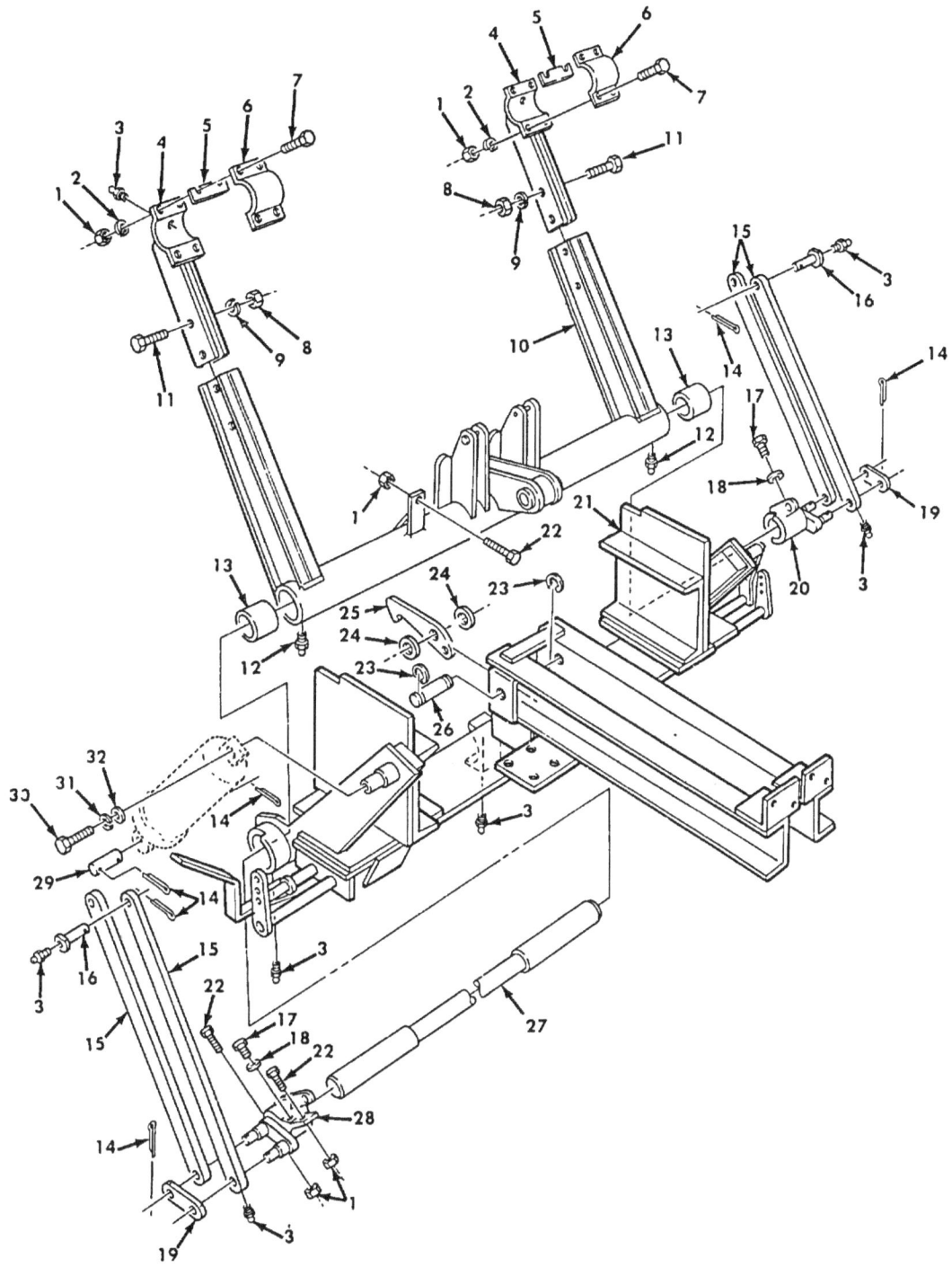

Figure 516. Liftgate Frame and Linkage.

(1) ITEM NO	(2) SMR CODE	(3) NSN	(4) CAGEC	(5) PART NUMBER	(6) DESCRIPTION AND USABLE ON CODES (UOC)	(7) QTY
					GROUP 2001 HOIST, CAPSTAN, WINDLASS, CRANE OR WINCH ASSEMBLY	
					FIG. 516 LIFTGATE FRAME AND LINKAGE	
1	PFFZZ	5310011226109	64203	G-9125	NUT, PLAIN, HEXAGON .... UOC:DAK, V25, ZAK	11
2	PFFZZ	5310011226148	64203	G-39250	WASHER, LOCK .... UOC:DAK, V25, ZAK	8
3	PAOZZ	4730011260218	64203	G-7030	FITTING, LUBRICATION .... UOC:DAK, V25, ZAK	10
4	PFFZZ	2510011222822	64203	20-33762-B	BEARING ARM, LIFT GA .... UOC:DAK, V25, ZAK	2
5	PFFZZ	5365011237082	64203	10-42651A	SHIM .... UOC:DAK, V25, ZAK	10
6	PFFZZ	3130011453943	64203	10-40397A	CAP, PILLOW BLOCK .... UOC:DAK, V25, ZAK	2
7	PFFZZ	5305011354839	64203	10-42253	SCREW, CAP, HEXAGON H .... UOC:DAK, V25, ZAK	8
8	PFFZZ	5340011224596	64203	G-9377	RETAINER, NUT AND BO .... UOC:DAK, V25, ZAK	4
9	PFFZZ	5310011226152	64203	G-39500	WASHER, LOCK .... UOC:DAK, V25, ZAK	4
10	XDFZZ		64203	20-33281-D	FRAME, LIFT ASSEMBLY .... UOC:DAK, V25, ZAK	1
11	PFFZZ	5305011458385	64203	10-42259-A	SCREW, CAP, HEXAGON H .... UOC:DAK, V25, ZAK	4
12	PAOZZ	4730000504208	96906	MS15003-1	FITTING, LUBRICATION .... UOC:DAK, V25, ZAK	2
13	PFFZZ	3120011285222	64203	10-42258-A	BUSHING, SLEEVE .... UOC:DAK, V25, ZAK	2
14	PAFZZ	5315011376858	64203	G-6161	PIN, COTTER .... UOC:DAK, V25, ZAK	12
15	PFFZZ	2590011431455	64203	10-42427-D	ROD, RADIUS.. UOC:DAK, V25, ZAK	4
16	PFFZZ	5315011222014	64203	10-40214A	PIN .... UOC:DAK, V25, ZAK	4
17	PFFZZ	5305011225474	64203	10-42469-A	SCREW, CAP, HEXAGON H .... UOC:DAK, V25, ZAK	2
18	PFFZZ	5310011227677	64203	G-74004	WASHER, LOCK .... UOC:DAK, V25, ZAK	2
19	PAFZZ	3040011982652	64203	10-40436-A	CONNECTING LINK, RIG .... UOC:DAK, V25, ZAK	2
20	PFFZZ	2520011260272	64203	20-33479-B	CLUTCH ASSEMBLY .... UOC:DAK, V25, ZAK	1
21	XAFZZ		64203	20-23214-D	MAIN FRAME ASSY .... UOC:DAK, V25, ZAK	1
22	PFFZZ	5305011224702	64203	10-42723-A	SCREW, CAP, HEXAGON H .... UOC:DAK, V25, ZAK	3
23	PAFZZ	5365008042025	64203	G-43028	RING, RETAINING . ... UOC:DAK, V25, ZAK	4
24	PFFZZ	3040011255529	64203	10-40446-A	COLLAR, SHAFT	4

(1) ITEM NO	(2) SMR CODE	(3) NSN	(4) CAGEC	(5) PART NUMBER	(6) DESCRIPTION AND USABLE ON CODES (UOC)	(7) QTY
					UOC:DAK, V25, ZAK	
25	PFFZZ	5340011224491	64203	10-42090-B	LATCH	2
					UOC:DAK, V25, ZAK	
26	PFFZZ	3040011260230	64203	10-42163A	SHAFT, STRAIGHT	10
					UOC:DAK, V25, EAK	
27	PFFZZ	3040011261749	64203	20-33506-B	SHAFT, SHOULDERED	1
					UOC:DAK, V25, ZAK	
28	PFFZZ	2520011263847	64203	20-33064-C	CLUTCH ASSEMBLY, FRI	1
					UOC:DAK, V25, ZAK	
29	PFFZZ	5315011269412	64203	10-40450-A	PIVOT	2
					UOC:DAK, V25, ZAK	
30	PFFZZ	5305011387624	64203	10-42276-A	SCREW	2
					UOC:DAK, V25, ZAK	
31	PFFZZ	5310011226150	64203	G-39300	WASHER, LOCK .	2
					UOC:DAK, V25, ZAK	
32	PFFZZ	5310011227676	64203	G-39475	WASHER, FLAT	2
					UOC:DAK, V25, ZAK	

END OF FIGURE

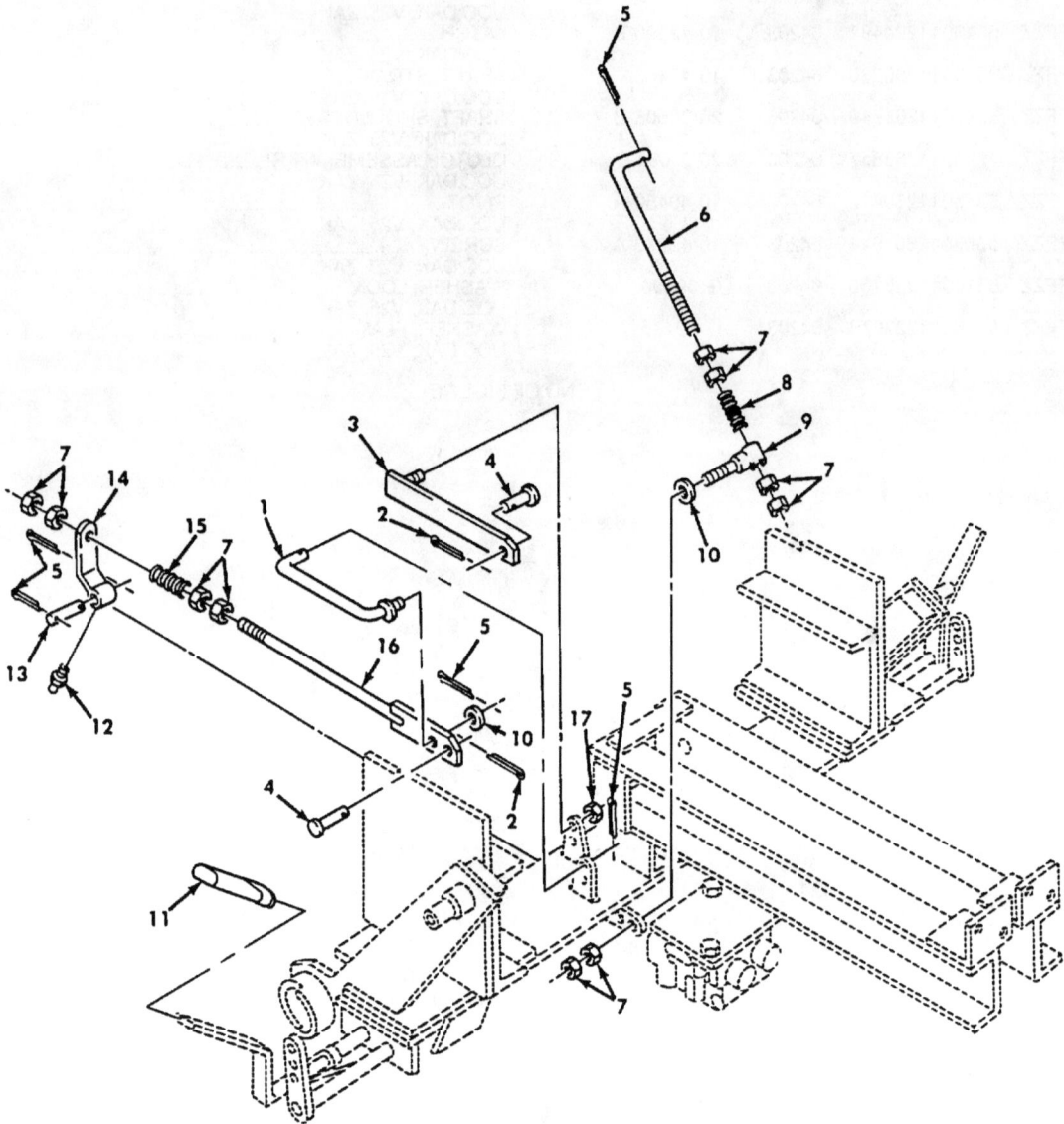

Figure 517. Liftgate Controls and Safety Latch.

(1) ITEM NO	(2) SMR CODE	(3) NSN	(4) CAGEC	(5) PART NUMBER	(6) DESCRIPTION AND USABLE ON CODES (UOC)	(7) QTY
					GROUP 2001 HOIST, CAPSTAN, WINDLASS, CRANE OR WINCH ASSEMBLY	
					FIG. 517 LIFTGATE CONTROLS AND SAFETY LATCH	
1	PAOZZ	3040011982653	64203	20-33739-A	CONNECTING LINK, RIG .... UOC:DAK, V25, ZAK	1
2	PAOZZ	5315011224603	64203	G-6005	PIN, COTTER .... UOC:DAK, V25, ZAK	2
3	PFOZZ	5340011356743	64203	20-33736-A	STANDOFF, THREADED, S .... UOC:DAK, V25, ZAK	1
4	PAOZZ	5315011224490	64203	10-42473-A	PIN .... UOC:DAK, V25, ZAK	2
5	PAOZZ	5315011224602	64203	G-6107	PIN, COTTER . ... UOC:DAK, V25, ZAK	6
6	PAOZZ	3040011465952	64203	10-40724-A	CONNECTING LINK, RIG .... UOC:DAK, V25, ZAK	2
7	PAOZZ	5310011225992	64203	G-9178	NUT .... UOC:DAK, V25, ZAK	16
8	PAOZZ	5360011270858	64203	10-42264-A	SPRING, HELICAL, COMP .... UOC:DAK, V25, ZAK	2
9	XDOZZ		64203	10-42296-A	BOLT, EYE .... UOC:DAK, V25, ZAK	2
10	PAOZZ	5310011227676	64203	G-39475	WASHER , FLAT . ... UOC:DAK, V25, ZAK	2
11	PAOZZ	5340012059057	64203	10-42460-A	GRIP, HANDLE .... UOC:DAK, V25, ZAK	2
12	PAOZZ	4730011260218	64203	G-7030	FITTING, LUBRICATION .... UOC:DAK, V25, ZAK	1
13	PAOZZ	5315011228482	64203	10-40447-A	PIN, STRAIGHT, HEADLE .... UOC:DAK, V25, ZAK	1
14	PFOZZ	5340011401156	64203	20-33660-A	CONNECTOR, ROD END .... UOC:DAK, V25, ZAK	1
15	PAOZZ	5360011220634	64203	10-42297-A	SPRING, HELICAL, COMP .... UOC:DAK, V25, ZAK	1
16	PAOZZ	5340012111622	64203	20-33738-A	CONNECTOR, ROD END .... UOC:DAK, V25, ZAK	1
17	PAOZZ	5310011226130	64203	G-9185	NUT, SELF-LOCKING, HE .... UOC:DAK, V25, ZAK	1

END OF FIGURE

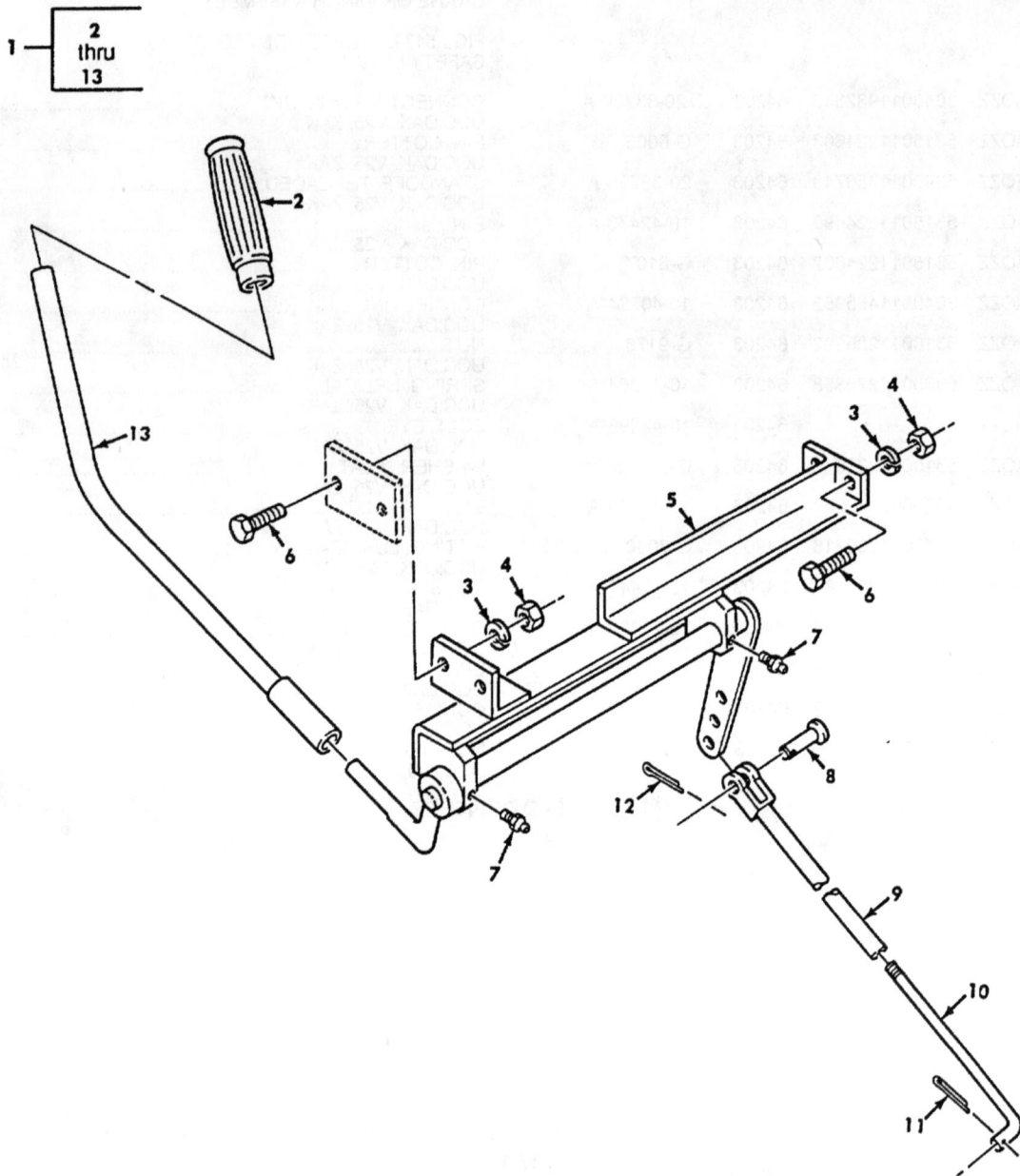

Figure 518. Liftgate Control Assembly, L.H. and R.H.

(1) ITEM NO	(2) SMR CODE	(3) NSN	(4) CAGEC	(5) PART NUMBER	(6) DESCRIPTION AND USABLE ON CODES (UOC)	(7) QTY
					GROUP 2001 HOIST, CAPSTAN, WINDLASS, CRANE OR WINCH ASSEMBLY	
					FIG. 518 LIFTGATE CONTROL ASSEMBLY, L.H. AND R.H.	
1	PFFZZ	2590012779100	64203	20-34197-B	CONTROL ASSEMBLY, LI LEFT HAND .... UOC:DAK, V25, ZAK	1
1	PFFZZ	3040012380863	64203	20-34198-B	LEVER, MANUAL CONTRO RIGHT HAND .... UOC:DAK, V25, ZAK	1
2	PAOZZ	5340011224632	64203	10-42266-A	.GRIP, HANDLE .... UOC:DAK, V25, ZAK	1
3	PFFZZ	5310011226148	64203	G-39250	.WASHER, LOCK .... UOC:DAK, V25, ZAK	4
4	PFFZZ	5310011226109	64203	G-9125	.NUT, PLAIN, HEXAGON .... UOC:DAK, V25, ZAK	4
5	XAFZZ		64203	20-33823-B	.CONTROL RELAY ASSY .... UOC:DAK, V25, ZAK	1
6	PFFZZ	5305011224703	64203	10-42451-A	.SCREW, CAP, HEXAGON H .... UOC:DAK, V25, ZAK	4
7	PAOZZ	4730011260218	64203	G-7030	.FITTING, LUBRICATION .... UOC:DAK, V25, ZAK	2
8	PAOZZ	5315011222033	64203	10-42265-A	.PIN, STRAIGHT, HEADED... UOC:DAK, V25, ZAK	1
9	PFOZZ	3040011350346	64203	20-33623-A	.CONNECTING LINK, RIG .... UOC:DAK, V25, ZAK	1
10	PAOZZ	3040011415032	64203	10-40730-A	.CONNECTING LINK, RIG .... UOC:DAK, V25, ZAK	1
11	PAOZZ	5315011224602	64203	G-6107	.PIN, COTTER .... UOC:DAK, V25, ZAK	1
12	PAOZZ	5315012065207	64203	G-6055	.PIN, COTTER .... UOC:DAK, V25, ZAK	1
13	PAOZZ	3040011259678	64203	20-33726A	.HANDLE, TAIL GATE BR .... UOC:DAK, V25, ZAK	1

END OF FIGURE

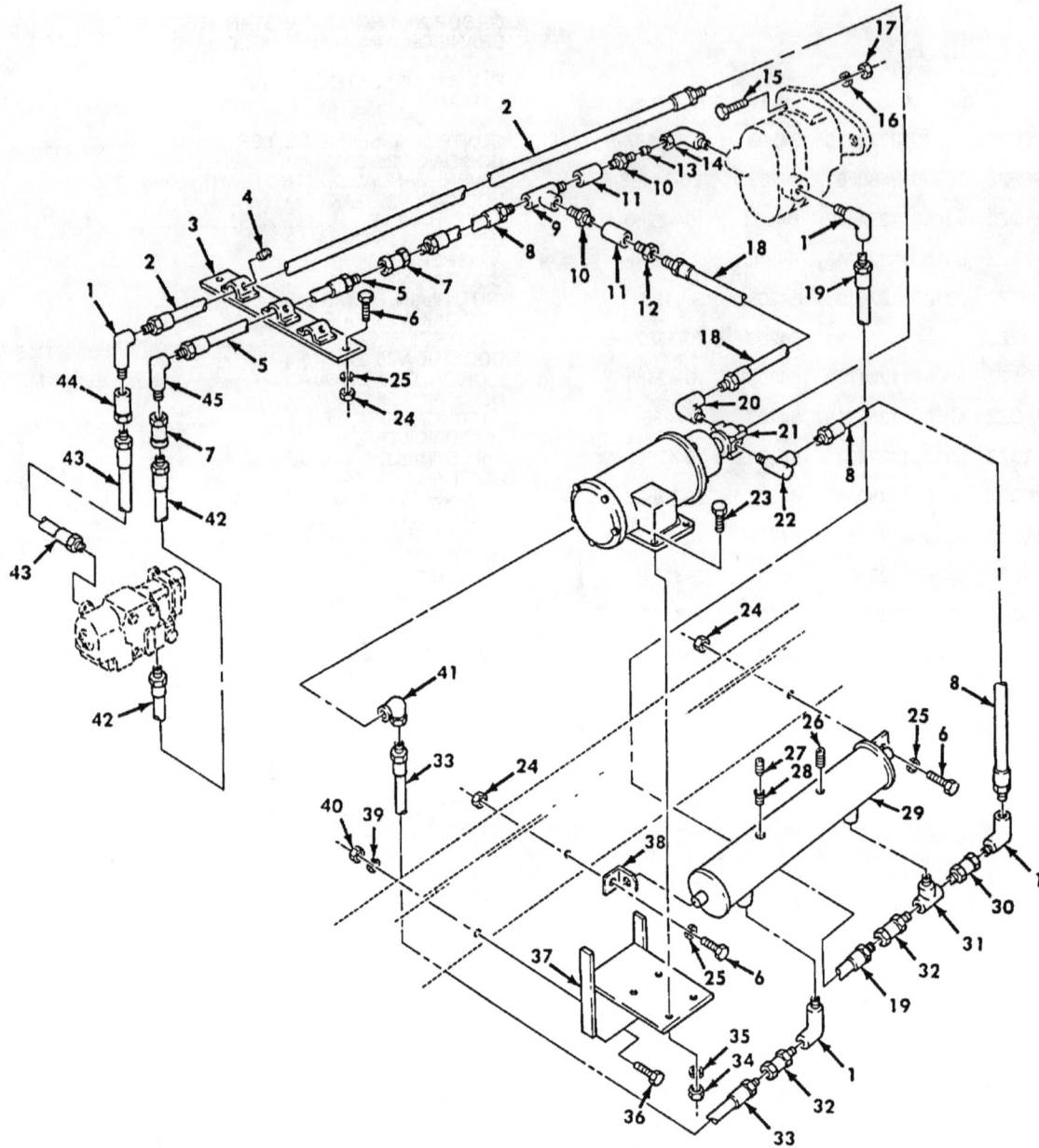

Figure 519. Liftgate Hydraulic Tubing, Oil Tank, and Pump Motor.

(1) ITEM NO	(2) SMR CODE	(3) NSN	(4) CAGEC	(5) PART NUMBER	(6) DESCRIPTION AND USABLE ON CODES (UOC)	(7) QTY
					GROUP 2001 HOIST, CAPSTAN, WINDLASS, CRANE OR WINCH ASSEMBLY	
					FIG. 519 LIFTGATE HYDRAULIC TUBING, OIL TANK, AND PUMP MOTOR	
1	PFFZZ	4730011259965	64203	G-8204	ELBOW, PIPE .... UOC:DAK, V25, ZAK	4
2	XBFZZ		64203	10-40771A	PIPE, METALLIC .... UOC:DAK, V25, ZAK	1
3	PFFZZ	5340011222015	64203	20-33820B	BRACKET, PUMP OUTLET TO CONTROL .... VALVE UOC:DAK, V25, ZAK	4
4	PFFZZ	5305011224693	64203	G-31051	SETSCREW .... UOC:DAK, V25, ZAK	8
5	XDFZZ		64203	10-40772A	PIPE, METALLIC .... UOC:DAK, V25, ZAK	1
6	PFFZZ	5305011354839	64203	10-42253	SCREW, CAP, HEXAGON H .... UOC:DAX, V25, ZAK	10
7	XDFZZ		64203	10-17118B-18	ADAPTER UNION .... UOC:DAK, V25, ZAK	2
8	PFFZZ	4720011259768	64203	20-24765B-15	HOSE ASSEMBLY, NONME .... UOC:DAK, V25, ZAK	2
9	PFFZZ	4730011324852	64203	10-42507A	TEE, TUBE .... UOC:DAK, V25, ZAK	1
10	PFFZZ	4730000528502	64203	10-41058A	NIPPLE, PIPE .... UOC:DAK, V25, ZAK	2
11	PFFZZ	4820011343457	64203	10-42508A	VALVE, CHECK .... UOC:DAK, V25, ZAK	2
12	XDFZZ		64203	G-46609	BUSHING, PIPE . ... UOC:DAK, V25, ZAK	1
13	PFFZZ	4730002783167	64203	G-46659	BUSHING, PIPE .... UOC:DAK, V25, ZAK	2
14	PFFZZ	4730011315944	64203	10-42505A	ELBOW, PIPE .... UOC:DAK, V25, ZAK	1
15	PFFZZ	5305011396484	64203	10-42284A	SCREW, CAP, HEXAGON H .... UOC:DAK, V25, ZAK	2
16	PFFZZ	5310011226150	64203	G-39300	WASHER, LOCK .... UOC:DAK, V25, ZAK	2
17	PFFZZ	5310011226111	64203	G-9177	NUT, PLAIN, HEXAGON .... UOC:DAK, V25, ZAK	2
18	PFFZZ	4720011312857	64203	20-24762-14	HOSE ASSEMBLY, NONME .... UOC:DAK, V25, ZAK	1
19	PFFZZ	4720011255865	64203	20-33139-11	HOSE ASSEMBLY, NONME .... UOC:DAK, V25, ZAK	1
20	PFFZZ	4730011328700	64203	10-42599A	ELBOW, TUBE. .... UOC:DAK, V25, ZAK	1
21	PFFZZ	6105011177944	64203	20-33191C	MOTOR AND PUMP .... UOC:DAK, V25, ZAK	1
22	PFFZZ	4730011263845	64203	10-42289A	ELBOW, TUBE. .... UOC:DAK, V25, ZAK	1
23	PFFZZ	5305011225639	64203	10-42417A	SCREW	1

(1) ITEM NO	(2) SMR CODE	(3) NSN	(4) CAGEC	(5) PART NUMBER	(6) DESCRIPTION AND USABLE ON CODES (UOC)	(7) QTY
24	PFFZZ	5310011226109	64203	G-9125	NUT, PLAIN, HEXAGON .... UOC:DAK, V25, ZAK	10
25	PFFZZ	5310011226148	64203	G-39250	WASHER, LOCK . ... UOC:DAK, V25, ZAK	10
26	PFFZZ	4730001874210	96906	MS51884-9	PLUG, PIPE .... UOC:DAK, V25, ZAK	1
27	PFFZZ	4730002026670	64203	G-46662	BUSHING, PIPE .... UOC:DAK, V25, ZAK	1
28	XDFZZ		64203	10-42466-A	BREATHER CAP .... UOC:DAK, V25, ZAK	1
29	PFFZZ	5430011260268	64203	20-33069C	TANK ASSEMBLY, OIL .... UOC:DAK, V25, ZAK	1
30	PAFZZ	4730002221840	96906	MS51846-77	NIPPLE, PIPE .... UOC:DAK, V25, ZAK	1
31	PFFZZ	4730011259969	64203	G-8211	TEE, PIPE .... UOC:DAK, V25, ZAK	1
32	PFFZZ	4730011260265	64203	10-17117-19	UNION, PIPE TO TUBE .... UOC:DAK, V25, ZAK	2
33	PFFZZ	4720011255521	64203	20-24768B-25	HOSE, NONMETALLIC .... UOC:DAK, V25, ZAK	1
34	PFFZZ	5310011226112	64203	G-9083	NUT, PLAIN, HEXAGON .... UOC:DAK, V25, ZAK	4
35	PFFZZ	5310011226151	64203	G-39203	WASHER, LOCK . ... UOC:DAK, V25, ZAK	4
36	PFFZZ	5305011225638	64203	10-42509A	SCREW .... UOC:DAK, V25, ZAK	4
37	PFFZZ	5340011559469	64203	20-33821B	BRACKET, ANGLE .... UOC:DAK, V25, ZAK	1
38	PFFZZ	5340011559468	64203	10-40723-A	BRACKET .... UOC:DAK, V25, ZAK	1
39	PFFZZ	5310011226153	64203	G-39350	WASHER, LOCK .... UOC:DAK, V25, ZAK	6
40	PFFZZ	5310011224492	64204	G-9225	NUT, HEXAGON .... UOC:DAK, V25, ZAK	6
41	PFFZZ	4730011315945	64203	G-8200	ELBOW, PIPE .... UOC:DAK, V25, ZAK	1
42	PFFZZ	4720011259767	64203	20-24765B-20	HOSE ASSEMBLY, NONME .... UOC:DAK, V25, ZAK	1
43	XBFZZ		64203 20-	33139B-10	HOSE, NONMETALLIC .... UOC:DAK, V25, ZAK	1
44	PFFZZ	4730011316099	64203	10-17118B-20	ADAPTER, UNION .... UOC:DAK, V25, ZAK	1
45	PFFZZ	4730011261723	64203	G-8154	ELBOW, PIPE .... UOC:DAK, V25, ZAK	1

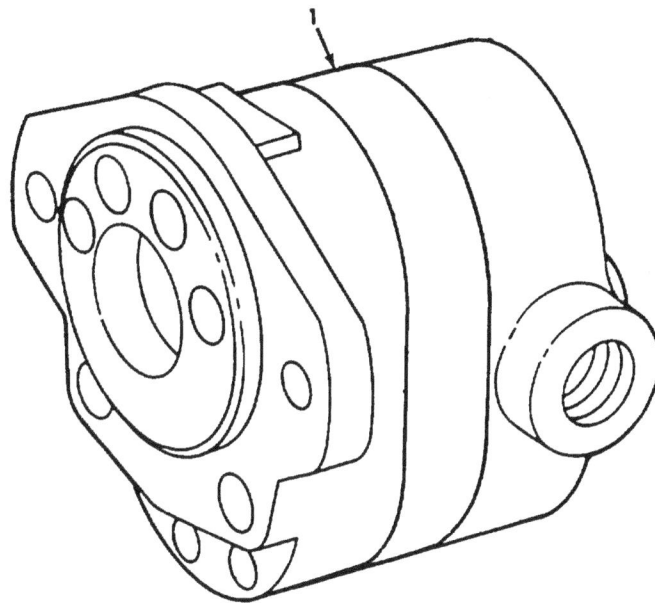

Figure 520. Liftgate Pump Assembly.

(1) ITEM NO	(2) SMR CODE	(3) NSN	(4) CAGEC	(5) PART NUMBER	(6) DESCRIPTION AND USABLE ON CODES (UOC)	(7) QTY
					GROUP 2001 HOIST, CAPSTAN, WINDLASS, CRANE OR WINCH ASSEMBLY	
					FIG. 520 LIFTGATE PUMP ASSEMBLY	
1	PFFZZ	4720011343422	64203	20-33140-A	PUMP, HYDRAULIC .... UOC:DAK, V25, ZAK	1

END OF FIGURE

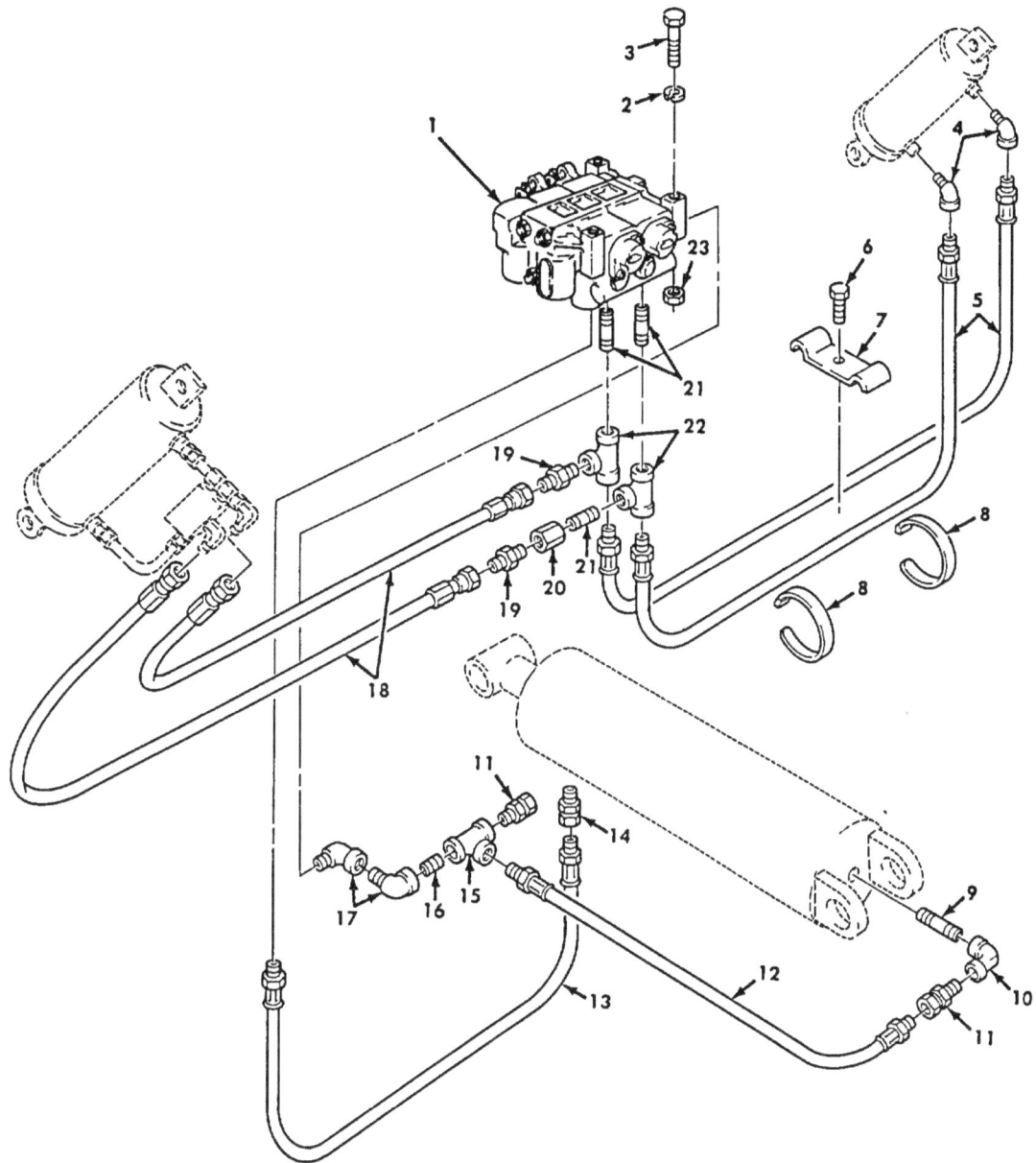

Figure 521. Liftgate Hydraulic Tubing and Flow Control Valve.

(1) ITEM NO	(2) SMR CODE	(3) NSN	(4) CAGEC	(5) PART NUMBER	(6) DESCRIPTION AND USABLE ON CODES (UOC)	(7) QTY
					GROUP 2001 HOIST, CAPSTAN, WINDLASS, CRANE OR WINCH ASSEMBLY	
					FIG. 521 LIFTGATE HYDRAULIC TUBING AND FLOW CONTROL VALVE	
1	PFFZZ	4820011355372	64203	20-33169	VALVE, LINEAR, DIRECT............ UOC:DAK, V25, ZAK	1
2	PFFZZ	5310011226149	64203	G-39200	WASHER, LOCK............ UOC:DAK, V25, ZAK	3
3	PFFZZ	5305011224634	64203	10-42256-A	SCREW, CAP, HEXAGON H............ UOC:DAK, V25, ZAK	3
4	PFFZZ	4730011260266	64203	10-42426	ELBOW, PIPE TO TUBE............ UOC:DAK, V25, ZAK	2
5	PFFZZ	4720011254467	64203	20-24761-21	HOSE ASSEMBLY, NONME............ UOC:DAK, V25, ZAK	2
6	PFFZZ	5305011224635	64203	10-42481	SCREW, CAP, HEXAGON H............ UOC:DAK, V25, ZAK	1
7	PFFZZ	5340011238045	64203	10-42512	CLIP............ UOC:DAK, V25, ZAK	1
8	PFFZZ	5340011238324	64203	10-42351	STRAP, RETAINING............ UOC:DAK, V25, ZAK	2
9	PFFZZ	4730001881896	32537	902466-113	NIPPLE, PIPE............ UOC:DAK, V25, ZAK	1
10	PFFZZ	4730011315945	64203	G-8200	ELBOW, PIPE............ UOC:DAK, V25, ZAK	1
11	PFFZZ	4730011260265	64203	10-17117-19	UNION, PIPE TO TUBE............ UOC:DAK, V25, ZAK	2
12	PFFZZ	4720011255865	64203	20-33139-11	HOSE ASSEMBLY, NONME............ UOC:DAK, V25, ZAK	1
13	PFFZZ	4720011259759	64203	20-24762-13	HOSE ASSEMBLY, NONME............ UOC:DAK, V25, ZAK	1
14	PFFZZ	4730011263541	64203	20-33662	RESTRICTOR, FLUID FL............ UOC:DAK, V25, ZAK	1
15	PFFZZ	4730011259969	64203	G-8211	TEE, PIPE............ UOC:DAK, V25, ZAK	1
16	PFFZZ	4730002221840	96906	MS51846-77	NIPPLE, PIPE............ UOC:DAK, V25, ZAK	1
17	PFFZZ	4730011259965	64203	G-8204	ELBOW, PIPE............ UOC:DAK, V25, ZAK	2
18	PFFZZ	4720011251365	64203	20-34206-11	HOSE ASSEMBLY, NONME............ UOC:DAK, V25, ZAK	2
19	PFFZZ	4730011259681	64203	10-42513	ADAPTER, STRAIGHT, PI............ UOC:DAK, V25, ZAK	2
20	PFFZZ	4820011434172	64203	20-33138	VALVE, FLOW CONTROL............ UOC:DAK, V25, ZAK	1
21	PFFZZ	4730000144593	44674	218406	NIPPLE, PIPE............ UOC:DAK, V25, ZAK	3
22	XBFZZ		64203	G-8061	TEE, PIPE............ UOC:DAK, V25, ZAK	2
23	PFFZZ	5310011226110	64203	G-9075	NUT, PLAIN, HEXAGON............ UOC:DAK, V25, ZAK	3

END OF FIGURE

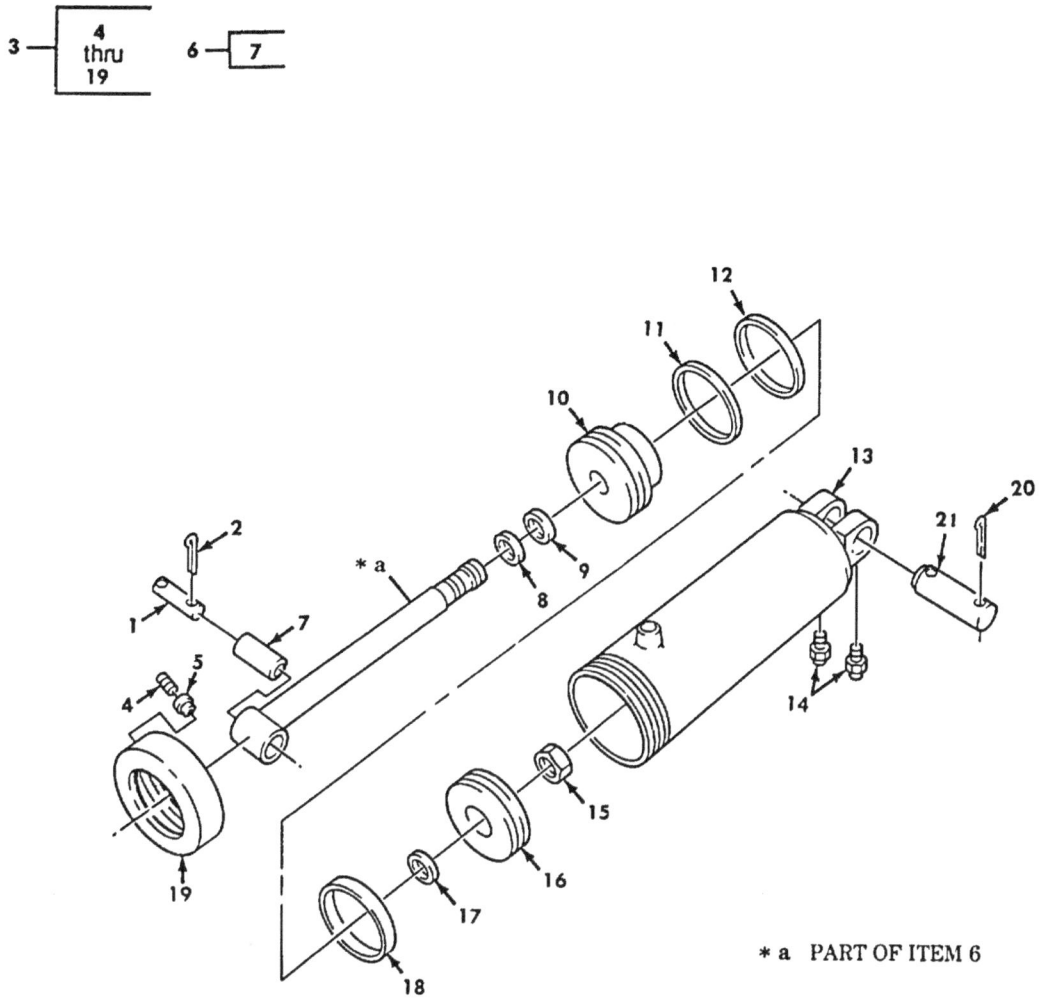

Figure 522. Liftgate Cylinder Assembly and Mounting Hardware.

(1) ITEM NO	(2) SMR CODE	(3) NSN	(4) CAGEC	(5) PART NUMBER	(6) DESCRIPTION AND USABLE ON CODES (UOC)	(7) QTY
					GROUP 2001 HOIST, CAPSTAN, WINDLASS, CRANE OR WINCH ASSEMBLY	
					FIG. 522 LIFTGATE CYLINDER ASSEMBLY AND MOUNTING HARDWARE	
1	PFFZZ	5315011269485	64203	10-40726-A	PIN, STRAIGHT, HEADLE .... UOC:DAK, V25, ZAK	1
2	PFFZZ	5315011269484	64203	10-41247	PIN, SPRING .... UOC:DAK, V25, ZAK	1
3	PAFFH	2590011176597	64203	20-34410D	CYLINDER ASSY .... UOC:DAK, V25, ZAK	1
4	PFFZZ	5305011361688	64203	10-27728	.SETSCREW .... UOC:DAK, V25, ZAK	2
5	PFFZZ	5325011224661	64203	10-11641A	.GROMMET, NONMETALLIC .... UOC:DAK, V25, ZAK	2
6	PFFZZ	3040011256491	64203	20-34409B	.ROD, PISTON, LINEAR A .... UOC:DAK, V25, ZAK	1
7	XDFZZ		64203	10-40897-A	..BEARING, SLEEVE .... UOC:DAK, V25, ZAK	1
8	PFFZZ	2590011259770	64203	10-41321A	.RING, WIPER. .... UOC:DAK, V25, ZAK	1
9	PFFZZ	5330011225636	64203	10-41320A	.SEAL........................ UOC:DAK, V25, ZAK	1
10	PFFZZ	2590011260373	64203	10-41361C	.HEAD, LINEAR ACTUATI .... UOC:DAK, V25, ZAK	1
11	PFFZZ	5365011226136	64203	10-9115A	.RING, RETAINING........................ UOC:DAK, V25, ZAK	1
12	PFFZZ	5330011436322	64203	G-63533	.O-RING .... UOC:DAK, V25, ZAK	1
13	XDFZZ		64203	20-34412B	.BODY, CYLINDER........................ UOC:DAK, V25, ZAK	1
14	PAOZZ	4730011259975	64203	G-7009	.FITTING, LUBRICATION .... UOC:DAK, V25, ZAK	2
15	PFFZZ	5310011225993	64203	10-40953-A	.NUT........................ UOC:DAK, V25, ZAK	1
16	PFFZZ	3040011259828	64203	10-41360C	.PISTON, LINEAR ACTU .... UOC:DAK, V25, ZAK	1
17	PFFZZ	5330011224612	64203	G-61513	.O-RING .... UOC:DAK, V25, ZAK	1
18	PFFZZ	5330011225392	64203	10-30326A	.SEAL, PLAIN ENCASED .... UOC:DAK, V25, ZAK	1
19	PFFZZ	4730011217718	64203	10-41540-C	.PACKING NUT........................ UOC:DAK, V25, ZAK	1
20	PFFZZ	5315011359402	64203	G-6365	PIN, COTTER........................ UOC:DAK, V25, ZAK	2
21	PFFZZ	5315011367307	64203	10-42685-A	PIN, STRAIGHT, HEADLE .... UOC:DAK, V25, ZAK	1

END OF FIGURE

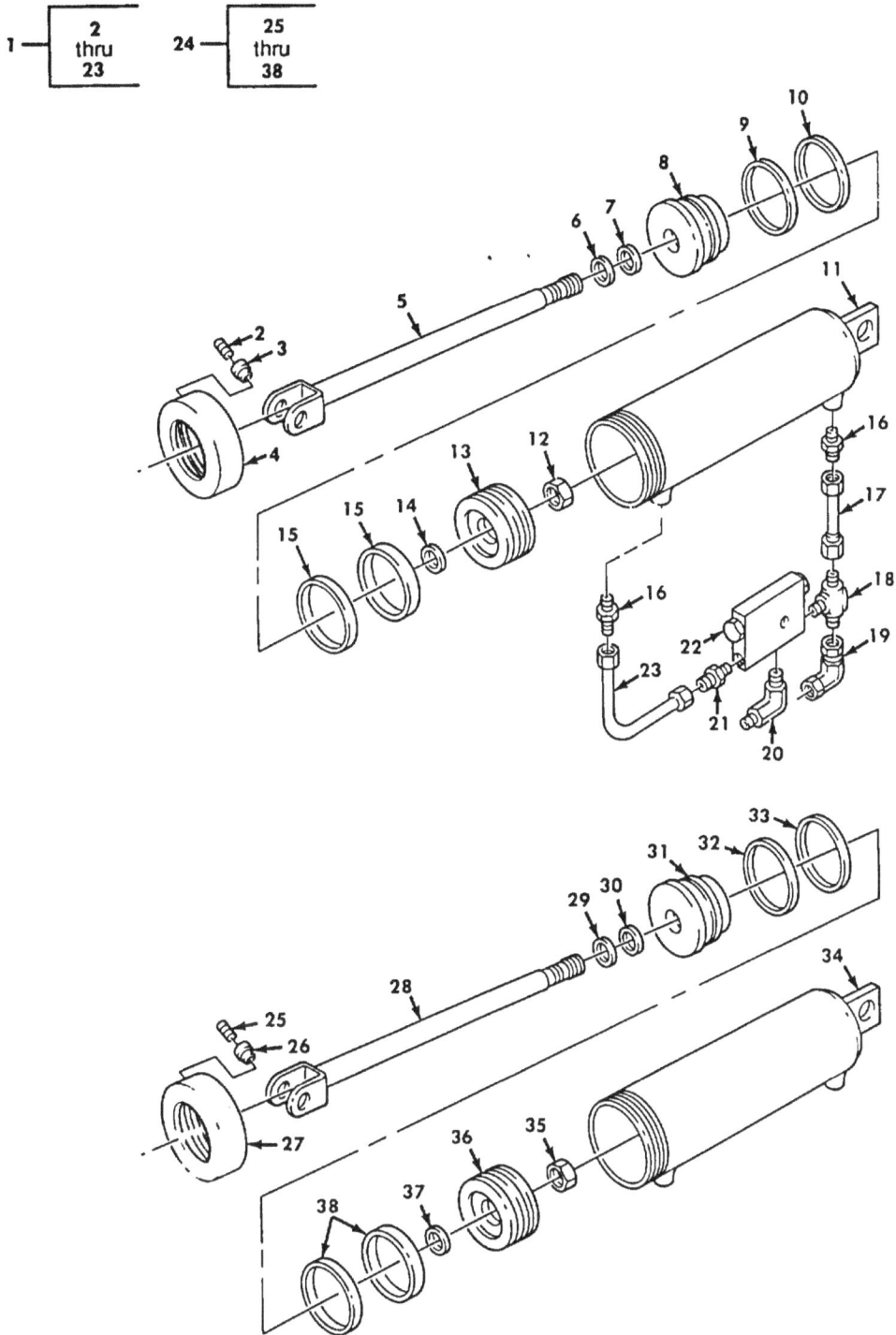

Figure 523. Liftgate Closing Cylinder Assemblies.

(1) ITEM NO	(2) SMR CODE	(3) NSN	(4) CAGEC	(5) PART NUMBER	(6) DESCRIPTION AND USABLE ON CODES (UOC)	(7) QTY
					GROUP 2001 HOIST, CAPSTAN, WINDLASS, CRANE OR WINCH ASSEMBLY	
					FIG. 523 LIFTGATE CLOSING CYLINDER ASSEMBLIES	
1	PFFHH	2590011176596	64203	20-33822	CLOSING CYLINDER .... UOC:DAK, V25, ZAK	1
2	PAHZZ	5305011361688	64203	10-27728	.SETSCREW .... UOC:DAK, V25, ZAK	2
3	PAHZZ	5325011224661	64203	10-11641A	.GROMMET, NONMETALLIC .... UOC:DAK, V25, ZAK	2
4	PAHZZ	5330011220864	64203	10-31091B	.RETAINER, PACKING .... UOC:DAK, V25, ZAK	1
5	PAHZZ	3040011334821	64203	20-34431B	.ROD, PISTON, LINEAR A .... UOC:DAK, V25, ZAK	1
6	PFHZZ	2590011260093	64203	10-41045A	.RING, WIPER .... UOC:DAK, V25, ZAK	1
7	PAHZZ	5330011224609	64203	10-41311A	.SEAL, PLAIN ENCASED .... UOC:DAK, V25, ZAK	1
8	PAHZZ	3040011423991	64203	10-41317B	.HEAD, LINEAR ACTUATI .... UOC:DAK, V25, ZAK	1
9	PAHZZ	5325011237063	64203	10-12295A	.RING, RETAINING .... UOC:DAK, V25, ZAK	1
10	PAHZZ	5330011220492	64203	G-63517	.O-RING .... UOC:DAK, V25, ZAK	1
11	XAHZZ		64203	20-34366B	.CYLINDER TUBE .... UOC:DAK, V25, ZAK	1
12	PAHZZ	5310011230905	64203	10-42322A	.NUT, PLAIN, HEXAGON .... UOC:DAK, V25, ZAK	1
13	PAHZZ	4310011259714	64203	10-41318B	.PISTON, COMPRESSOR .... UOC:DAK, V25, ZAK	1
14	PAHZZ	5330011224613	64203	G-63707	.O-RING .... UOC:DAK, V25, ZAK	1
15	PAHZZ	5330011225637	64203	10-31103A	.SEAL .... UOC:DAK, V25, ZAK	2
16	PFHZZ	4730011259681	64203	10-42513-A	.ADAPTER, STRAIGHT, PI .... UOC:DAK, V25, ZAK	2
17	XDHZZ		64203	20-34220-A	.TUBE ASSEMBLY, METAL .... UOC:DAK, V25, ZAK	1
18	PAHZZ	4730011259679	64203	10-42518A	.TEE, TUBE .... UOC:DAK, V25, ZAK	1
19	PFHZZ	4730011259680	64203	10-42517-A	.ELBOW, TUBE .... UOC:DAK, V25, ZAK	1
20	PFHZZ	4730011353009	64203	10-42516-A	.ELBOW, MALE .... UOC:DAK, V25, ZAK	1
21	PFHZZ	4730002890382	96906	MS51500A6-8	.ADAPTER, STRAIGHT, PI .... UOC:DAK, V25, ZAK	1
22	PFHZZ	4820011259698	64203	20-33170-A	.VALVE, CHECK .... UOC:DAK, V25, ZAK	1
23	XDHZZ		64203	20-34219-A	.TUBE ASSEMBLY, METAL .... UOC:DAK, V25, ZAK	1

(1) ITEM NO	(2) SMR CODE	(3) NSN	(4) CAGEC	(5) PART NUMBER	(6) DESCRIPTION AND USABLE ON CODES (UOC)	(7) QTY
24	PFFFF	3830011263886	64203	20-34363D	CYLINDER ASSEMBLY, A .... UOC:DAK, V25, ZAK	1
25	PFHZZ	5305011361688	64203	10-27728	.SETSCREW. UOC:DAK, V25, ZAK	2
26	PFHZZ	5325011224661	64203	10-11641A	.GROMMET, NONMETALLIC .... UOC:DAK, V25, ZAK	2
27	PFHZZ	5330011220864	64203	10-31091B	.RETAINER, PACKING .... UOC:DAK, V25, ZAK	1
28	PFHZZ	3040011334821	64203	20-34431B	.ROD, PISTON, LINEAR A UOC:DAK, V25, ZAK	1
29	PFHZZ	2590011260093	64203	10-41045A	.RING, WIPER . ... UOC:DAK, V25, ZAK	1
30	PFHZZ	5330011224609	64203	10-41311A	.SEAL, PLAIN ENCASED .... UOC:DAK, V25, ZAK	1
31	PFHZZ	3040011423991	64203	10-41317B	.HEAD, LINEAR ACTUATI .... UOC:DAK, V25, ZAK	1
32	PFHZZ	5325011237063	64203	10-12295A	.RING, RETAINING .... UOC:DAK, V25, ZAK	1
33	PFHZZ	5330011220492	64203	G-63517	.O-RING .... UOC:DAK, V25, ZAK	1
34	XAHZZ		64203	20-34366B	.CYLINDER TUBE .... UOC:DAK, V25, ZAK	1
35	PFHZZ	5310011230905	64203	10-42322A	.NUT, PLAIN, HEXAGON .... UOC:DAK, V25, ZAK	1
36	PFHZZ	4310011259714	64203	10-41318B	.PISTON, COMPRESSOR . ... UOC:DAK, V25, ZAK	1
37	PFHZZ	5330011224613	64203	G-63707	.O-RING .... UOC:DAK, V25, ZAK	1
38	PFHZZ	5330011225637	64203	10-31103A	.SEAL .... UOC:DAK, V25, ZAK	2

END OF FIGURE

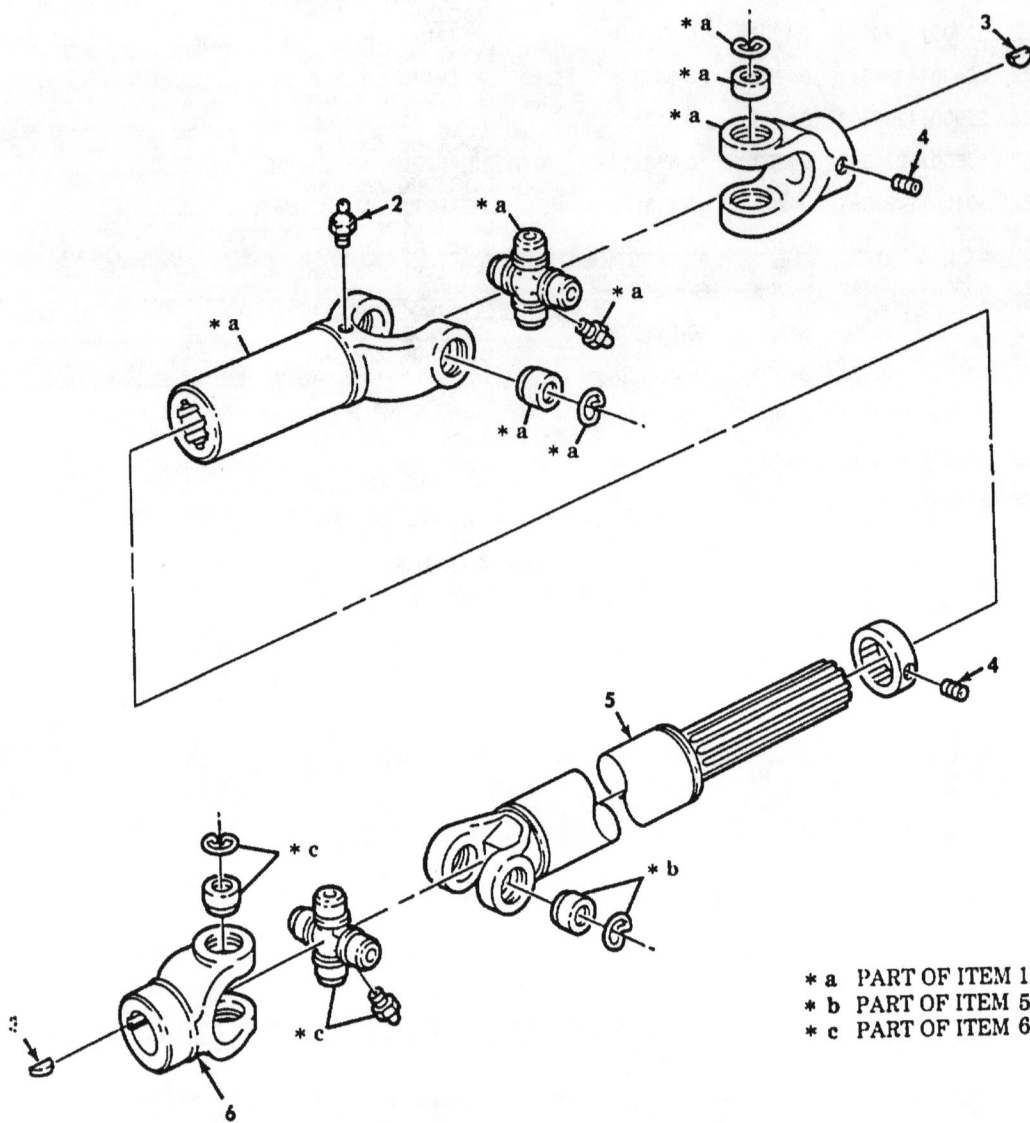

Figure 524. Liftgate Pump Drive Shafts.

* a   PART OF ITEM 1
* b   PART OF ITEM 5
* c   PART OF ITEM 6

(1) ITEM NO	(2) SMR CODE	(3) NSN	(4) CAGEC	(5) PART NUMBER	(6) DESCRIPTION AND USABLE ON CODES (UOC)	(7) QTY
					GROUP 2001 HOIST, CAPSTAN, WINDLASS, CRANE OR WINCH ASSEMBLY	
					FIG. 524 LIFTGATE PUMP DRIVE SHAFT	
1	PAOZZ	2520011259837	64203	10-42286A	UNIVERSAL JOINT, VEH .... UOC:DAK, V25, ZAK	1
2	PAOZZ	4730011259975	64203	G-7009	.FITTING, LUBRICATION .... UOC:DAK, V25, ZAK	1
3	PAOZZ	5315011222031	64203	G-40254	KEY, WOODRUFF................................. UOC:DAK, V25, ZAK	2
4	PAOZZ	5305011224692	64203	G-25102SET	SCREW .... UOC:DAK, V25, ZAK	2
5	PAOZZ	3040011385391	64203	10-40178A-28	SHAFT, DRIVE, FLEXIBL .... UOC:DAK, V25, ZAK	1
6	PAOZZ	2520011260443	64203	10-42288A	UNIVERSAL JOINT, VEH .... UOC:DAK, V25, ZAK	1

END OF FIGURE

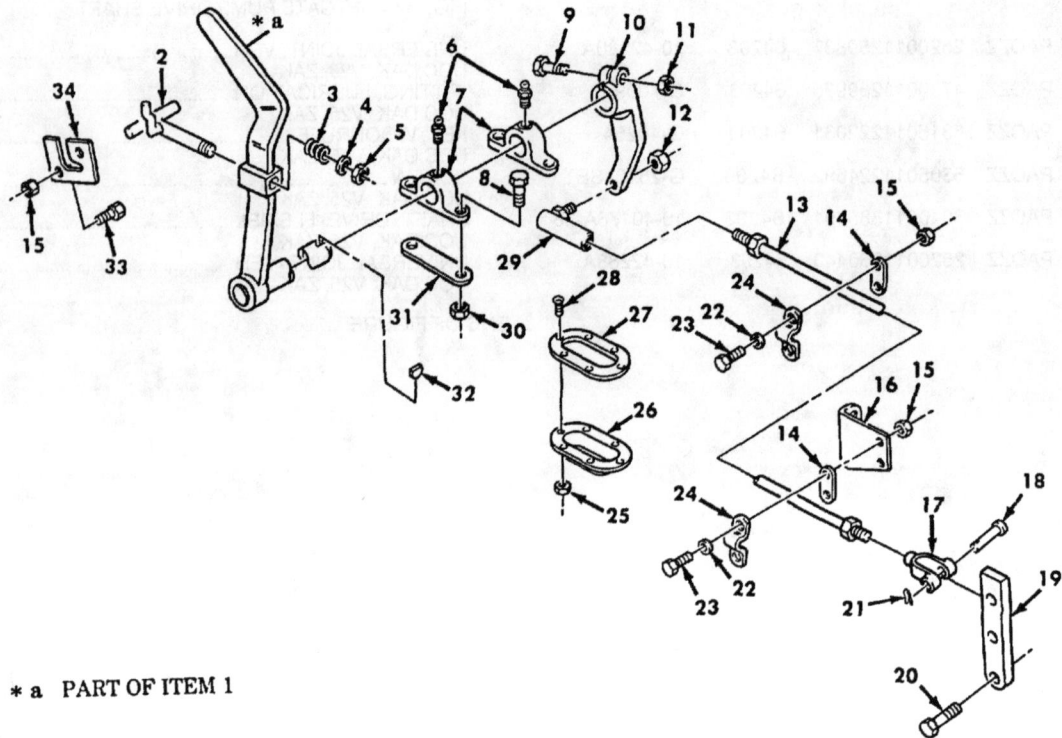

*a PART OF ITEM 1

Figure 525. Power Takeoff Controls, Lever Assembly, and Cable Mounting Hardware.

(1) ITEM NO	(2) SMR CODE	(3) NSN	(4) CAGEC	(5) PART NUMBER	(6) DESCRIPTION AND USABLE ON CODES (UOC)	(7) QTY
					GROUP 2004 POWER TAKEOFF ASSEMBLY	
					FIG. 525 POWER TAKEOFF CONTROLS, LEVER ASSEMBLY, AND CABLE MOUNTING HARDWARE	
1	PAFFF	3040011049155	19207	12256356-2	CONTROL LEVER ASSEM .... UOC:DAL, V18, ZAL	1
2	PAFZZ	5315011048942	19207	12256657	.PIN, THREADED END .... UOC:DAL, V18, ZAL	1
3	PAFZZ	5360007521975	19207	7521975	.SPRING, HELICAL, COMP .... UOC:DAL, V18	1
4	PAFZZ	5310000806004	96906	MS27183-14	.WASHER, FLAT .... UOC:DAL, V18, ZAL	1
5	PAFZZ	5310008140672	96906	MS51943-36	.NUT, SELF-LOCKING, HE .... UOC:DAL, V18, ZAL	1
6	PAFZZ	4730000504208	96906	MS15003-1	FITTING, LUBRICATION .... UOC:DAL, V18, ZAL	2
7	PFFZZ	2520007411410	19207	7411410	BRACKET CONTROL LEV .... UOC:DAL, V18	2
8	PAFZZ	5305002693238	80204	B1821BH038F125N	SCREW, CAP, HEXAGON H .... UOC:DAL, V18, ZAL	4
9	PAFZZ	5305007195239	96906	MS90727-116	SCREW, CAP, HEXAGON H .... UOC:DAL, V18, ZAL	1
10	PFFZZ	3040011036009	19207	12256689	LEVER, REMOTE CONTRO .... UOC:DAL, V18, ZAL	1
11	PAFZZ	5310004883888	96906	MS51943-40	NUT, SELF-LOCKING, HE .... UOC:DAL, V18, ZAL	1
12	PAFZZ	5310002416658	96906	MS51943-34	NUT, SELF-LOCKING, HE .... UOC:DAL, V18, ZAL	1
13	PAFZZ	2590011062060	19207	11669463	CONTROL ASSEMBLY, PU .... UOC:DAL, V18, ZAL	1
14	PAFZZ	5365011111520	19207	12256350	SPACER, PLATE .... UOC:DAL, V18, ZAL	2
15	PAFZZ	5310009359022	96906	MS51943-32	NUT, SELF-LOCKING, HE .... UOC:DAL, V18, ZAL	6
16	PFFZZ	5340012239799	19207	12302838	BRACKET, ANGLE .... UOC:DAL, V18, ZAL	1
17	PAFZZ	5340000573537	96906	MS35812-3	CLEVIS, ROD END .... UOC:DAL, V18, ZAL	1
18	PAFZZ	5315008123427	96906	MS20392-4C25	PIN, STRAIGHT, HEADED .... UOC:DAL, V18, ZAL	1
19	PAFZZ	5340011947036	19207	12302862	BRACKET, DOUBLE ANGL .... UOC:DAL, V18, ZAL	1
20	PAFZZ	5306011099384	19207	12256683	BOLT, SHOULDER .... UOC:DAL, V18, ZAL	1
21	PAFZZ	5315008392326	96906	MS24665-281	PIN, COTTER .... UOC:DAL, V18, ZAL	1
22	PAFZZ	5310005825965	96906	MS35338-44	WASHER, LOCK .... UOC:DAL, V18, ZAL	4
23	PAFZZ	5306000680513	60285	6893-2	BOLT, MACHINE . ... UOC:DAL, V18, ZAL	4
24	PAFZZ	5340011049012	19207	12256372	STRAP, RETAINING .... UOC:DAL, V18, ZAL	2

(1) ITEM NO	(2) SMR CODE	(3) NSN	(4) CAGEC	(5) PART NUMBER	(6) DESCRIPTION AND USABLE ON CODES (UOC)	(7) QTY
25	PAFZZ	5310002081918	88044	AN365-1024A	NUT, SELF-LOCKING, HE .... UOC:DAL, V18, ZAL	6
26	PAFZZ	2530011048943	19207	12256378	BOOT, VEHICULAR CON .... UOC:DAL, V18, ZAL	1
27	PAFZZ	5365010902074	19207	12256377	SPACER, PLATE .... UOC:DAL, V18, ZAL	1
28	PAFZZ	5306000440502	21450	440502	BOLT, MACHINE . ... UOC:DAL, V18, ZAL	6
29	PAFZZ	3040011079928	19207	11664437	BALL JOINT .... UOC:DAL, V18, ZAL	1
30	PAFZZ	5310008140672	96906	MS51943-36	NUT, SELF-LOCKING, HE .... UOC:DAL, V18, ZAL	4
31	PAFZZ	5365011099490	19207	12256382	SPACER, PLATE .... UOC:DAL, V18, ZAL	1
32	PAFZZ	5315006165520	96906	MS35756-14	KEY, WOODRUFF .... UOC:DAL, V18, ZAL	1
33	PAFZZ	5305002693239	96906	MS90727-63	SCREW, CAP, HEXAGON . ... UOC:DAL, V18, ZAL	2
34	PAFZZ	2590010909336	19207	12256656	PLATE, LOCKPIN .... UOC:DAL, V18, ZAL	1

END OF FIGURE

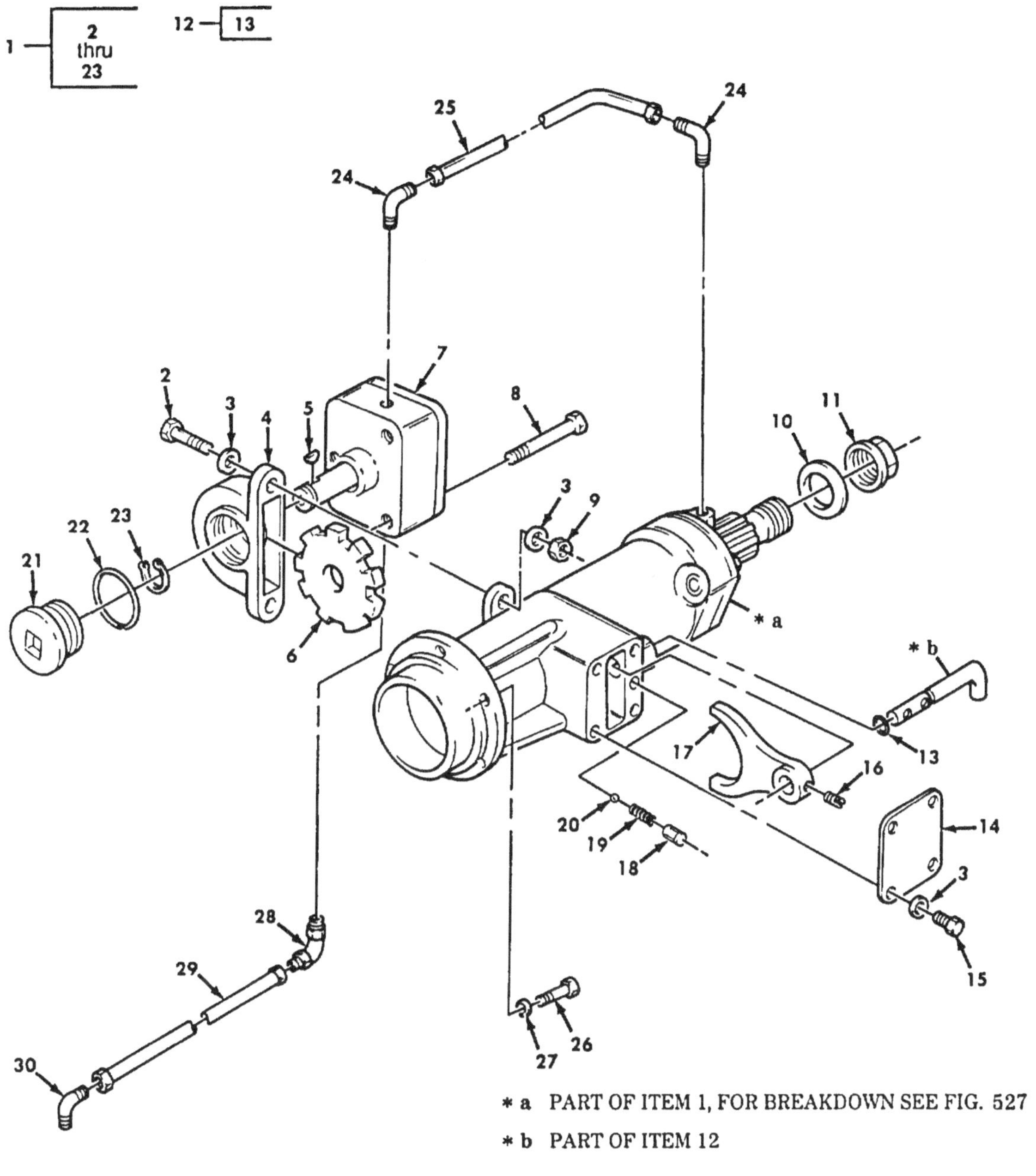
1
2
thru
23
12
13
24
25
24
7
2
3
4
5
8
11
10
3
9
22
23
21
6
* a
* b
13
17
16
20
19
18
14
3
15
28
29
26
27
30
* a PART OF ITEM 1, FOR BREAKDOWN SEE FIG. 527
* b PART OF ITEM 12

Figure 526. Transfer Power Takeoff Assembly and Related Parts.

(1) ITEM NO	(2) SMR CODE	(3) NSN	(4) CAGEC	(5) PART NUMBER	(6) DESCRIPTION AND USABLE ON CODES (UOC)	(7) QTY
					GROUP 2004 POWER TAKEOFF ASSEMBLY	
					FIG. 526 TRANSFER POWER TAKEOFF ASSEMBLY AND RELATED PARTS	
1	PAFFF	2520011185971	78500	MPS 2813	POWER TAKEOFF, TRANS .... UOC:DAL, V18, ZAL	1
2	PAFZZ	5306011320834	78500	15-X-1557	.BOLT .... UOC:DAL, V18, ZAL	2
3	PAFZZ	5310000791974	19207	11601651	.WASHER, FLAT .... UOC:DAL, V18, ZAL	8
4	PFFZZ	2520011312831	78500	3268-A-1067	.POWER TAKEOFF, TRANS .... UOC:DAL, V18, ZAL	1
5	PAFZZ	5315006165519	96906	MS35756-1	.KEY, WOODRUFF .... UOC:DAL, V18, ZAL	1
6	PAFZZ	3020011339037	78500	3892-J-4430	.GEAR, SPUR .... UOC:DAL, V18, ZAL	1
7	PAFZZ	2520011272626	78500	A-3303-E-5	.PUMP ASSEMBLY, POWER .... UOC:DAL, V18, ZAL	1
8	PAFZZ	5305000712514	80204	B1821BH025C275N	.SCREW, CAP, HEXAGON H .... UOC:DAL, V18, ZAL	3
9	PAFZZ	5310011260566	78500	NL-25-1-C	.NUT .... UOC:DAL, V18, ZAL	2
10	PAFZZ	5310002863727	78500	1229E1331	.WASHER, FLAT .... UOC:DAL, V18, ZAL	1
11	PAFZZ	5310011096056	78500	1227-C-939	.NUT, PLAIN, EXTENDED .... UOC:DAL, V18, ZAL	1
12	PAFZZ	3040011348909	78500	A2244-U-21	.CONNECTING LINK, RIG .... UOC:DAL, V18, ZAL	1
13	PAFZZ	5330010463300	81349	M83461/1-012	.O-RING .... UOC:DAL, V18, ZAL	1
14	PFFZZ	2520011277790	78500	2205-Q-43	.PLATE, COVER, POWER T .... UOC:DAL, V18, ZAL	1
15	PAFZZ	5305005432419	80204	B1821BH038C113N	.SCREW, CAP, HEXAGON H .... UOC:DAL, V18, ZAL	4
16	PAFZZ	5305002707328	61465	2069245	.SETSCREW .... UOC:DAL, V18, ZAL	1
17	PAFZZ	2520007349606	78500	2849-N-92	.SHIFTER FORK .... UOC:DAL, V18, ZAL	1
18	PAFZZ	2520011203673	78500	1846-X-258	.PLUG, LOCK, TRANSFER .... UOC:DAL, V18, ZAL	1
19	PAFZZ	5360003215710	78500	2858-T-20	.SPRING, HELICAL, COMP .... UOC:DAL, V18, ZAL	1
20	PAFZZ	2520011261493	78500	1898-H-34	.BALL, SHIFT SHAFT .... UOC:DAL, V18, ZAL	1
21	PAFZZ	5365005954948	78500	1250C3	.PLUG, MACHINE THREAD.............. UOC:DAL, V18, ZAL	1
22	PAFZZ	5330005497694	78500	2208-U-697	.GASKET................. UOC:DAL, V18, ZAL	1
23	PAFZZ	5325011267264	78500	1229-H-2816	.RING, RETAINING.............. UOC:DAL, V18, ZAL	1
24	PAFZZ	4730011305158	78500	A-1898-T-1164	ELBOW, PIPE	2

(1) ITEM NO	(2) SMR CODE	(3) NSN	(4) CAGEC	(5) PART NUMBER	(6) DESCRIPTION AND USABLE ON CODES (UOC)	(7) QTY
25	PAFZZ	4710011269565	78500	3196-H-8	UOC:DAL, V18, ZAL TUBE ASSEMBLY, METAL	1
26	PAFZZ	5305003385162	78500	S-2710-1-C	UOC:DAL, V18, ZAL SCREW, SPECIAL	6
27	PAFZZ	5310008933381	19207	7748744	UOC:DAL, V18, ZAL WASHER, FLAT	6
28	PAFZZ	4730011277346	78500	2206-J-88	UOC:DAL, V18, ZAL ELBOW, PIPE TO HOSE	1
29	PAFZZ	4720011343475	78500	A-2296-E-83	UOC:DAL, V18, ZAL HOSE ASSEMBLY, NONME	1
30	PAFZZ	4730011346988	78500	2206-F-58	UOC:DAL, V18, ZAL ELBOW, PIPE TO TUBE UOC:DAL, V18, ZAL	1

END OF FIGURE

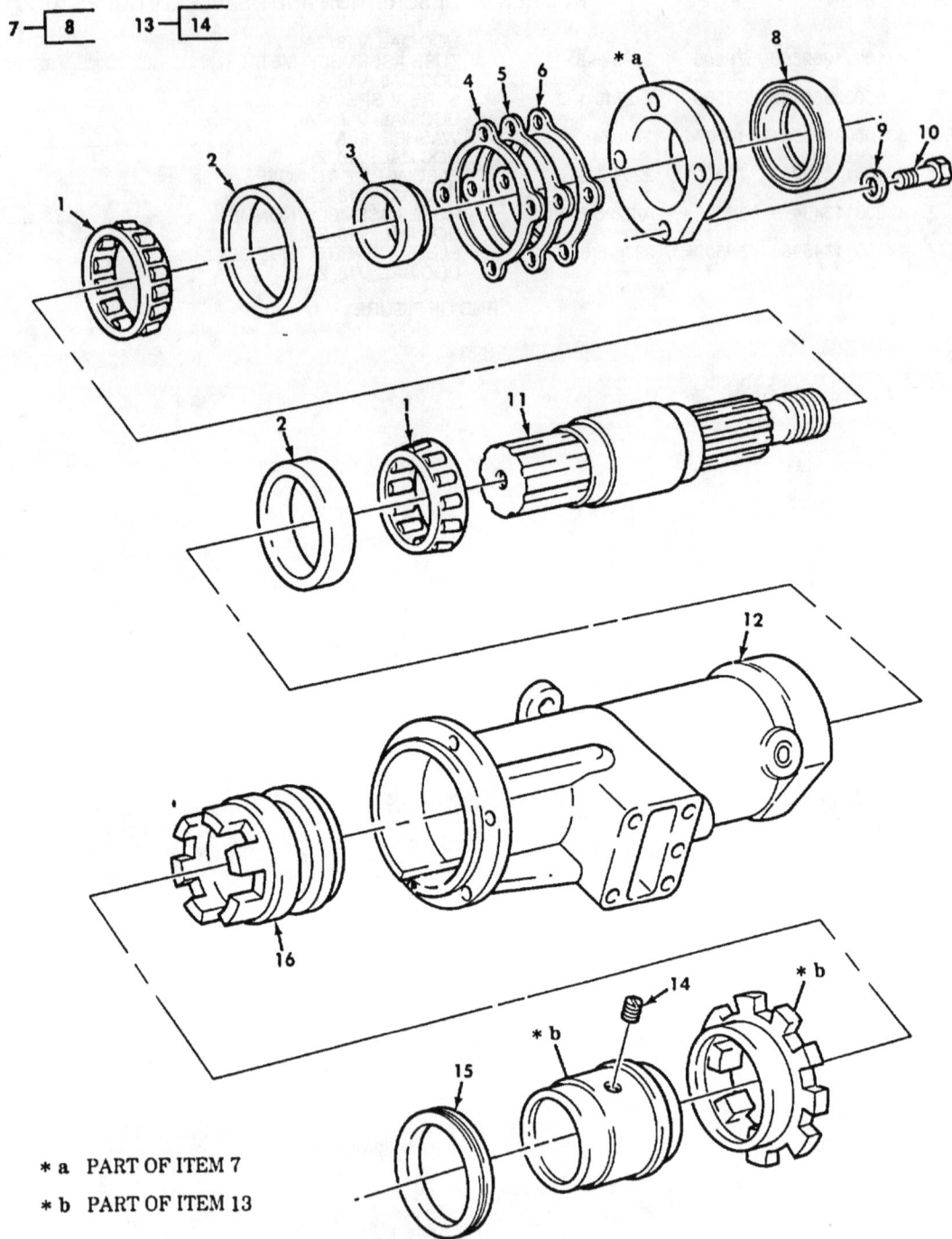

Figure 527. Transfer Power Takeoff Main Shaft and Drive Couplings.

* a   PART OF ITEM 7
* b   PART OF ITEM 13

(1) ITEM NO	(2) SMR CODE	(3) NSN	(4) CAGEC	(5) PART NUMBER	(6) DESCRIPTION AND USABLE ON CODES (UOC)	(7) QTY
					GROUP 2004 POWER TAKEOFF ASSEMBLY	
					FIG. 527 TRANSFER POWER TAKEOFF MAIN SHAFT AND DRIVE COUPLINGS	
1	PAFZZ	3110001875730	78500	LM102949	CONE AND ROLLERS, TA .... UOC:DAL, V1B, ZAL	2
2	PAFZZ	3110001712489	78500	LM102910	CUP, TAPERED ROLLER .... UOC:DAL, V18, ZAL	2
3	PAFZZ	2520011326847	78500	1844-J-634	SLEEVE, OIL SEAL .... UOC:DAL, V18, ZAL	1
4	PAFZZ	5365006143903	78500	2803-L-220	SHIM .... UOC:DAL, V18, ZAL	1
5	PAFZZ	5365005453723	78500	2803-M-221	SHIM, REAR .... UOC:DAL, V18, ZAL	1
6	PAFZZ	5365005186592	19207	5186592	SHIM .... UOC:DAL, V18, ZAL	1
7	PAFZZ	3110011429490	78500	A3266-Q-745	PLATE, RETAINING, BEA .... UOC:DAL, V18, ZAL	1
8	PAFZZ	5330011260565	78500	A-1205-U-1633	.SEAL .... UOC:DAL, V18, ZAL	1
9	PAFZZ	5310012009879	78500	1229-E-1669-C	WASHER, FLAT .... UOC:DAL, V18, ZAL	4
10	PAFZZ	5306011261618	78500	S-256-1-C	BOLT, MACHINE .... UOC:DAL, V18, ZAL	4
11	XAFZZ		78500	3297-C-55	SHAFT, POWER TAKEOFF .... UOC:DAL, V18, ZAL	1
12	XAFZZ		78500	3282-K-63	HOUSING, PTO .... UOC:DAL, V18, ZAL	1
13	PAFZZ	2520011341839	78500	A3107-B-28	COLLAR AND GEAR ASS .... UOC:DAL, V18, ZAL	1
14	PAFZZ	5305011422792	78500	S-853	.SETSCREW .... UOC:DAL, V18, ZAL	1
15	PAFZZ	5330011374799	78500	1205-Y-1663	SEAL, PACKING .... UOC:DAL, V18, ZAL	1
16	PAFZZ	2520011235556	78500	3107-V-22	COLLAR, SLIDING CLUT .... UOC:DAL, V18, ZAL	1

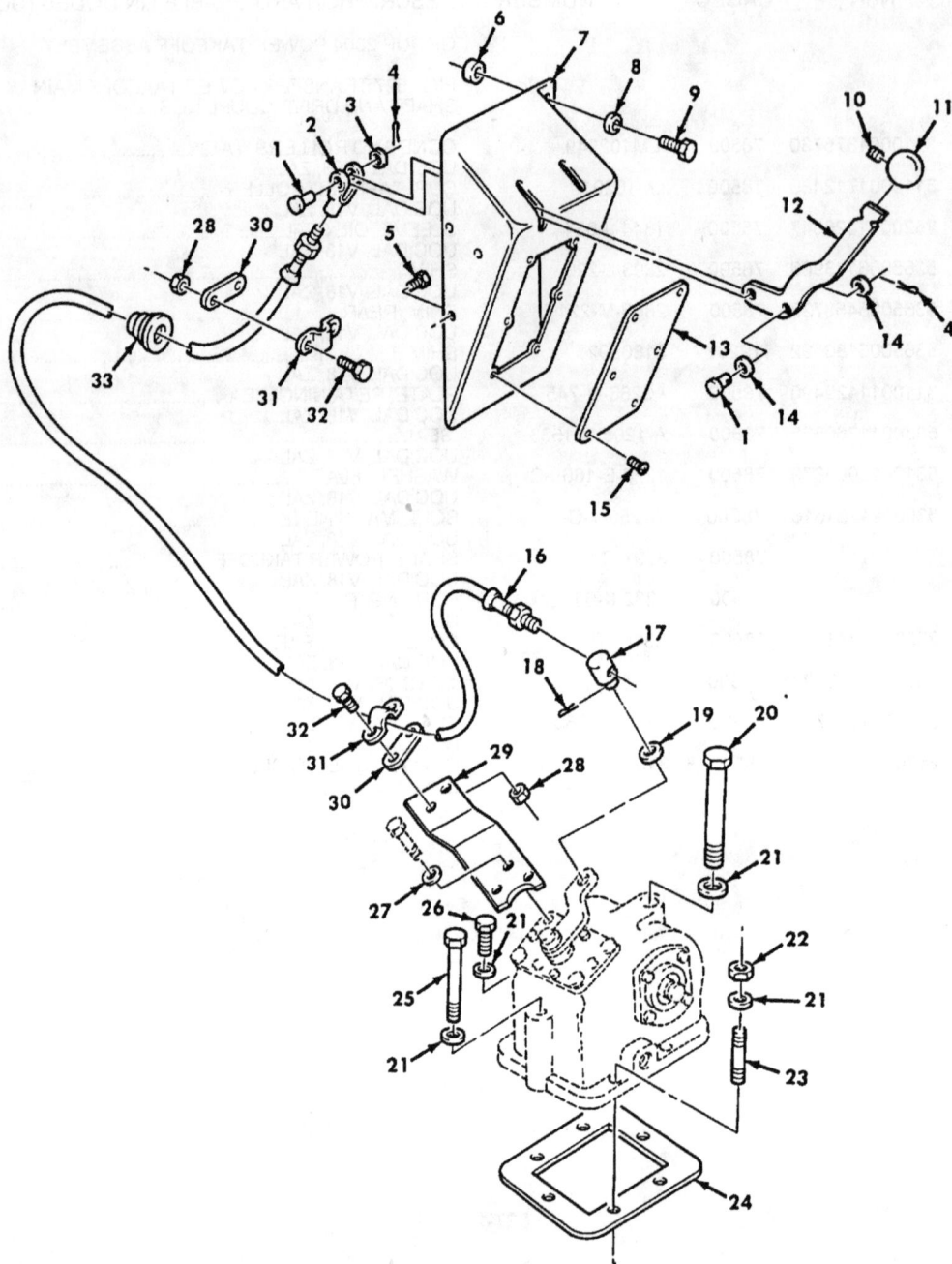

Figure 528. Transmission Power takeoff Shift Controls.

(1) ITEM NO	(2) SMR CODE	(3) NSN	(4) CAGEC	(5) PART NUMBER	(6) DESCRIPTION AND USABLE ON CODES (UOC)	(7) QTY
					GROUP 2004 POWER TAKEOFF ASSEMBLY	
					FIG. 528 TRANSMISSION POWER TAKEOFF SHIFT CONTROLS	
1	PAFZZ	5315008111241	96906	MS20392-2C17	PIN, STRAIGHT, HEADED .... UOC:DAB, DAD, DAE, DAF, DAH, DAK, DAL, DAX, V12, V14, V16, V18, V19, V20, V21, V25, V39, ZAB, ZAD, ZAE, ZAF, ZAH, ZAK, ZAL	2
2	PAFZZ	5340009696407	96906	MS35812-1	CLEVIS, ROD END .... UOC:DAB, DAD, DAE, DAF, DAH, DAX, DAL, DAX, V12, V14, V16, V18, V19, V20, V21, V25, V39, ZAB, ZAD, ZAE, ZAF, ZAH, ZAK, ZAL	1
3	PAFZZ	5310000145850	96906	MS27183-42	WASHER, FLAT .... UOC:DAB, DAD, DAE, DAF, DAH, DAK, DAL, DAX, V12, V14, V16, V18, V19, V20, V21, V25, V39, ZAB, ZAD, ZAE, ZAF, ZAH, ZAK, ZAL	1
4	PAFZZ	5315008151405	96906	MS24665-151	PIN, COTTER .... UOC:DAB, DAD, DABE, DAF, DAH, DAK, DAL, DAX, V12, V14, V16, V18, V19, V20, V21, V25, V39, ZAB, ZAD, ZAE, ZAF, ZAH, ZAK, ZAL	2
5	XDFZZ		96906	MS51861-65	SCREW, TAPPING .... UOC:DAB, DAD, D ABE, DAF, DAH, DAK, DAL, DAX, V12, V14, V16, V18, V19, V20, V21, V25, V39, ZAB, ZAD, ZAE, ZAF, ZAH, ZAK, ZAL	3
6	PAFZZ	5365011108163	19207	12256277	SPACER, SLEEVE .... UOC:DAB, DAD, DAE, DAF, DAH, DAK, DAL, DAX, V12, V14, V16, V18, V19, V20, V21, V25, V39, ZAB, ZAD, ZAE, ZAF, ZAH, ZAX, ZAL	2
7	PFFZZ	2590010849632	19207	12276907	PANEL ASSEMBLY, CONT .... UOC:DAB, DAD, DAE, DAF, DAH, DAK, DAL, DAX, V12, V14, V16, V18, V19, V20, V21, V25, V39 ZAB, ZAD, ZAE, ZAF, ZAH, ZAK, ZAL	1
8	PAFZZ	5310000806004	96906	MS27183-14	WASHER, FLAT .... UOC:DAB, DAD, DAE, DAF, DAH, DAK, DAL, DAX, V12, V14, V16, V18, V19, V20, V21, V25, V39, ZAB, ZAD, ZAE, ZAF, ZAH, ZAK, ZAL	2
9	PAFZZ	5305002693239	80204	B1821BH038F138N	SCREW, CAP, HEXAGON H .... UOC:DAB, DAD, DAE, DAF, DAH, DAK, DAL, DAX, V12, V14, V16, V18, V19, V20, V21, V25, V39, ZAB, ZAD, ZAE, ZAF, ZAH, ZAK, ZAL	2
10	PAFZZ	5305007239383	96906	MS51963-67	SETSCREW .... UOC:DAB, DAD, DAE, DAF, DAH, DAK, DAL, DAX, V12, V14, V16, V18, V19, V20, V21, V25, V39, ZAB, ZAD, ZAE, ZAF, ZAH, ZAK, ZAL	2
11	PAFZZ	5355011074178	19207	12276921-1	KNOB .... UOC:DAB, DAD, DAE, DAF, DAH, DAK, DAL, DAX, V12, V14, V16, V18, V19, V20, V21, V25, V39, ZAB, ZAD, ZAE, ZAF, ZAH, ZAK, ZAL	1
12	PFFZZ	3040010899326	19207	12276913	LEVER, REMOTE CONTRO .... UOC:DAB, DAD, DAE, DAF, DAH, DAK, DAL, DAX, V12, V14, V16, V18, V19, V20, V21, V25, V39, ZAB, ZAD, ZAE, ZAF, ZAH, ZAK, ZAL	1

(1) ITEM NO	(2) SMR CODE	(3) NSN	(4) CAGEC	(5) PART NUMBER	(6) DESCRIPTION AND USABLE ON CODES (UOC)	(7) QTY
13	PAFZZ	5340011038772	19207	12276908	COVER, ACCESS . UOC:DAB, DAD, DAE, DAF, DAH, DAK, DAL, DAX, V12, V14, V16, V18, V19, V20, V21, V25, V39 ZAB, ZAD, ZAE, ZAF, ZAH, ZAK, ZAL	1
14	PAFZZ	5310008094058	96906	MS27183-10	WASHER, FLAT UOC:DAB, DAD, DAE, DAF, DAH, DAK, DAL, DAX, V12, V14, V16, V18, V19, V20, V21, V25, V39 ZAB, ZAD, ZAE, ZAF, ZAH, ZAK, ZAL	2
15	PAFZZ	5305004324201	96906	MS51861-45	SCREW, TAPPING . UOC:DAB, DAD, DAE, DAF, DAH, DAK, DAL, DAX, V12, V14, V16, V18, V19, V20, V21, V25, V39, ZAB, ZAD, ZAE, ZAF, ZAH, ZAK, ZAL	6
16	PAFZZ	2590011062060	19207	11669463	CONTROL ASSEMBLY, PU UOC:DAB, DAD, DAE, DAF, DAH, DAK, DAL, DAX, V12, V14, V16, V18, V19, V20, V21, V25, V39 ZAB, ZAD, ZAE, ZAF, ZAH, ZAK, ZAL	1
17	PAFZZ	5315013852731	19207	12277392-1	PIN, SHOULDER, HEADLE UOC:DAB, DAD, DAE, DAF, DAH, DAK, DAL, DAX, V12, V14, V16, V18, V19, V20, V21, V25, V39, ZAB, ZAD, ZAE, ZAF, ZAH, ZAK, ZAL	1
18	PAFZZ	5315008392325	96906	MS24665-132	PIN, COTTER UOC:DAB, DAD, DAE, DAF, DAH, DAK, DAL, DAX, V12, V14, V16, V18, V19, V20, V21, V25, V39, ZAB, ZAD, ZAE, ZAF, ZAH, ZAK, ZAL	1
19	PAFZZ	5210000814219	96906	MS27183-12	WASHER, FLAT UOC:DAB, DAD, DAE, DAF, DAH, DAK, DAL, DAX, V12, V14, V16, V18, V19, V20, V21, V25, V39 ZAB, ZAD, ZAE, ZAF, ZAH, ZAK, ZAL	1
20	PAFZZ	5306011357202	95019	378041-4	BOLT, MACHINE UOC:DAB, DAD, DAE, DAF, DAH, DAK, DAL, DAX, V12, V14, V16, V18, V19, V20, V21, V25, V39, ZAB, ZAD, ZAE, ZAF, ZAH, ZAK, ZAL	1
21	PAFZZ	5310001939753	10001	265850PC88	WASHER, FLAT UOC:DAB, DAD, DAE, DAF, DAH, DAK, DAL, DAX, V12, V14, V16, V18, V19, V20, V21, V25, V39, ZAB, ZAD, ZAE, ZAF, ZAH, ZAK, ZAL	6
22	PAFZZ	5310007320559	96906	MS51968-8	NUT, PLAIN, HEXAGON UOC:DAB, DAD, DAE, DAF, DAH, DAK, DAL, DAX, V12, V14, V16, V18, V19, V20, V21, V25, V39, ZAB, ZAD, ZAE, ZAF, ZAH, ZAK, ZAL	2
23	PAFZZ	5307010558843	95019	379423-15	STUD, PLAIN UOC:DAB, DAD, DAE, DAF, DAH, DAK, DAL, DAX, V12, V14, V16, V18, V19, V20, V21, V25, V39, ZAB, ZAD, ZAE, ZAF, ZAH, ZAK, ZAL	2
24	PAFZZ	5330013932500	95019	35-P-41 NON-ASBE STOS	GASKET UOC:DAB, DAD, DAE, DAF, DAH, DAK, DAL, DAX, V12, V14, V16, V18, V19, V20, V21, V25, V39, ZAB, ZAD, ZAE, ZAF, ZAH, ZAK, ZAL	1
25	XBFZZ		95019	500398-30	SCREW, CAP, HEXAGON H UOC:DAB, DAD, DAE, DAF, DAH, DAK, DAL, DAX, V12, V14, V16, V18, V19, V20, V21, V25, V39, ZAB, ZAD, ZAE, ZAF, ZAH, ZAK, ZAL	1

(1) ITEM NO	(2) SMR CODE	(3) NSN	(4) CAGEC	(5) PART NUMBER	(6) DESCRIPTION AND USABLE ON CODES (UOC)	(7) QTY
26	PAFZZ	5305011328390	95019	500398-12	SCREW, CAP, HEXAGON H .... UOC:DAB, DAD, D AE , DAF, DAH, DAK, DAL, DAX, V12, V14, V16, V18, V19, V20, V21, V25, V39, ZAB, ZAD, ZAE, ZAF, ZAH, ZAK, ZAL	2
27	PAFZZ	5310011328275	24617	9418924	WASHER, FLAT .... UOC:DAB, DAD, DAE, DAF, DAH, DAK, DAL, DAX, V12, V14, V16, V18, V19, V20, V21, V25, V39, ZAB, ZAD, ZAE, ZAF, ZAH, ZAK, ZAL	1
28	PAFZZ	5310002081918	88044	AN365-1024A	NUT, SELF-LOCKING, HE .... UOC:DAB, DAD, DAE, DAF, DAH, DAK, DAL, DAX, V12, V14, V16, V18, V19, V20, V21, V25, V39, ZAB, ZAD, ZAE, ZAF, ZAH, ZAK, ZAL	4
29	PFFZZ	5340011683102	19207	12277391	BRACKET, DOUBLE ANGL .... UOC:DAB, DAD, DAE, DAF, DAH, DA , DAL, DAX, V12, V14, V16, V18, V19, V20, V21, V25, V39 ZAB, ZAD, ZAE, ZAF, ZAH, ZAK, ZAL	1
30	PAFZZ	5365011084815	19207	12256581	SPACER, PLATE .... UOC:DAB, DAD, DAE, DAF, DAH, DAK, DAL, DAX, V12, V14, V16, V18, V19, V20, V21, V25, V39, ZAB, ZAD, ZAE, ZAF, ZAH, ZAK, ZAL	2
31	PAFZZ	5340011049013	19207	12256582	STRAP, RETAINING . ... UOC:DAB, DAD, D AE , DAF, DAH, DAK, DAL, DAX, V12, V14, V16, V18, V19, V20, V21, V25, V39, ZAB, ZAD, ZAE, ZAF, ZAH, ZAK, ZAL	2
32	PAFZZ	5306000440502	21450	440502	BOLT, MACHINE . ... UOC:DAB, DAD, DAE, DAF, DAH, DAK, DAL, DAX, V12, V14, V16, V18, V19, V20, V21, V25, V39, ZAB, ZAD, ZAE, ZAF, ZAH, ZAK, ZAL	4
33	PAFZZ	5325011064125	19207	12256578-5	GROMMET, NONMETALLIC CONTROL CABLE .... UOC:DAB, DAD, DAE, DAF, DAM, DAK, DAL, DAX, V12, V14, V16, V18, V19, V20, V21, V25, V39 ZAB, ZAD, ZAE, ZAF, ZAH, ZAK, ZAL	1

END OF FIGURE

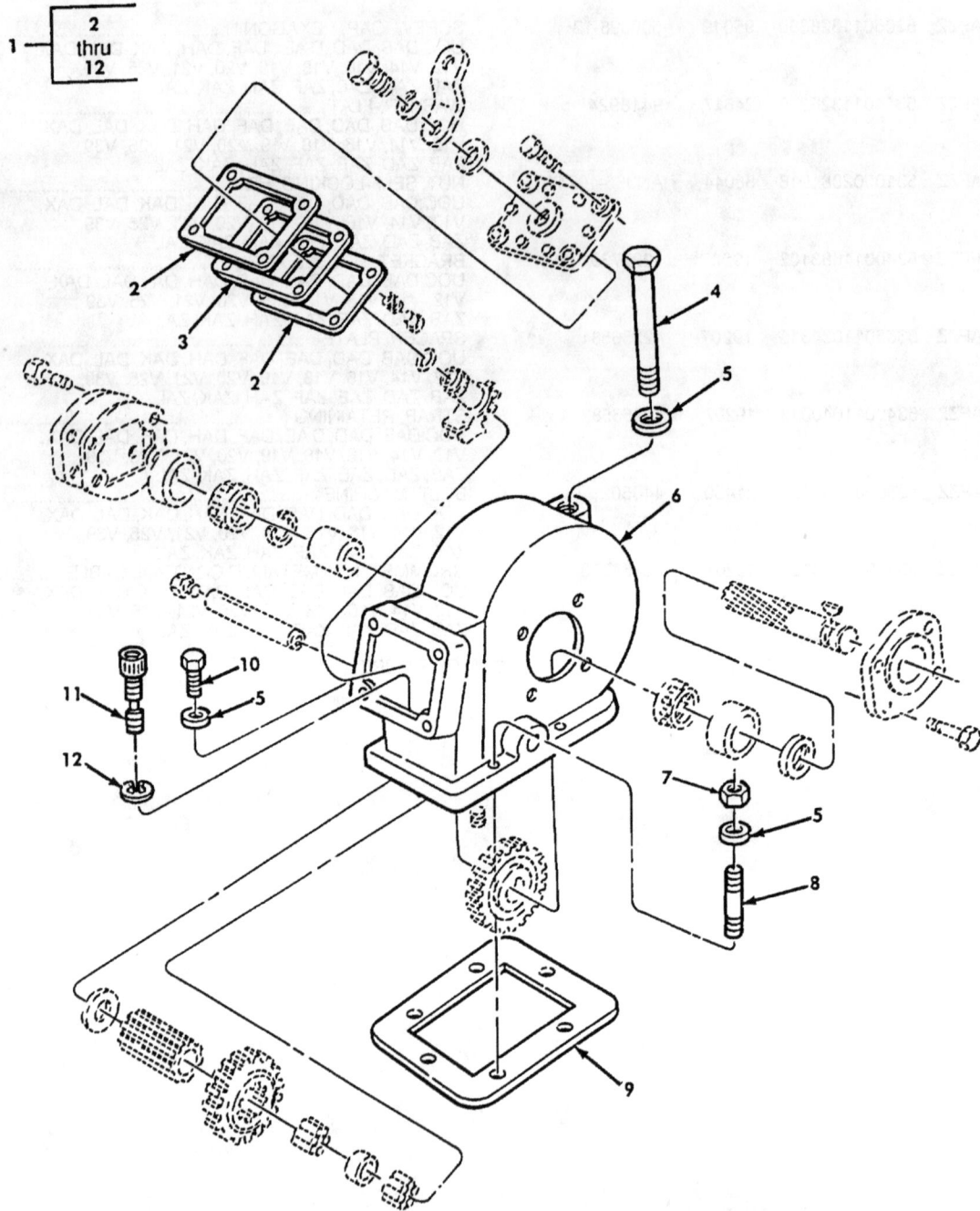

Figure 529. transmission P.T.O. Assembly (M939A2).

(1) ITEM NO	(2) SMR CODE	(3) NSN	(4) CAGEC	(5) PART NUMBER	(6) DESCRIPTION AND USABLE ON CODES (UOC)	(7) QTY
					GROUP 2004 POWER TAKEOFF ASSEMBLY	
					FIG. 529 TRANSMISSION P.T.O. ASSEMBLY (M939A2)	
1	PAFFF	2520012856295	95019	308594-1	POWER TAKEOFF, TRANS .... UOC:ZAB, ZAD, ZAE, ZAF, ZAH, ZAK, ZAL	1
2	PAFZZ	5330004850895	11757	35-P-8	.GASKET .... UOC:ZAB, ZAD, ZAE, ZAF, ZAH, ZAK, ZAL	2
3	PAFZZ	5330013029948	11757	5-A-062	.GASKET .... UOC:ZAB, ZAD, ZAE, ZAF, ZAH, ZAK, ZAL	1
4	PAFZZ	5306011357202	95019	378041-4	.BOLT, MACHINE.. UOC:ZAB, ZAD, ZAE, ZAF, ZAH, ZAX, ZAL	1
5	PAFZZ	5310001939753	10001	265850PC88	.WASHER, FLAT .... UOC:ZAB, ZAD, ZAE, ZAF, ZAH, ZAK, ZAL	5
6	XAFZZ		95019	1-P-455	.HOUSING.. UOC:ZAB, ZAD, ZAE, ZAF, ZAH, ZAK, ZAL	1
7	PAFZZ	5310011334481	11757	501146-3	.NUT, PLAIN, HEXAGON .... UOC:ZAB, ZAD, ZAE, ZAF, ZAH, ZAK, ZAL	2
8	PAFZZ	5307013085081	8N900	379423-18	.STUD, PLAIN .... UOC:ZAB, ZAD, ZAE, ZAF, ZAH, ZAK, ZAL	2
9	PAFZZ	5330013932500	95019	35-P-41 NON-ASBE STOS	.GASKET.. UOC:ZAB, ZAD, ZAE, ZAF, ZAH, ZAK, ZAL	1
10	PAFZZ	5305011328390	95019	500398-12	.SCREW, CAP, HEXAGON H .... UOC:ZAB, ZAD, ZAE, ZAF, ZAH, ZAK, ZAL	1
11	PAFZZ	5305011657541	95019	378766	.SCREW, CAP, HEXAGON H .... UOC:ZAB, ZAD, ZAE, ZAF, ZAH, ZAK, ZAL	2
12	PAFZZ	5325011652352	8N900	378767	.RING, RETAINING .... UOC:ZAB, ZAD, ZAE, ZAF, ZAH, ZAK, ZAL	1

END OF FIGURE

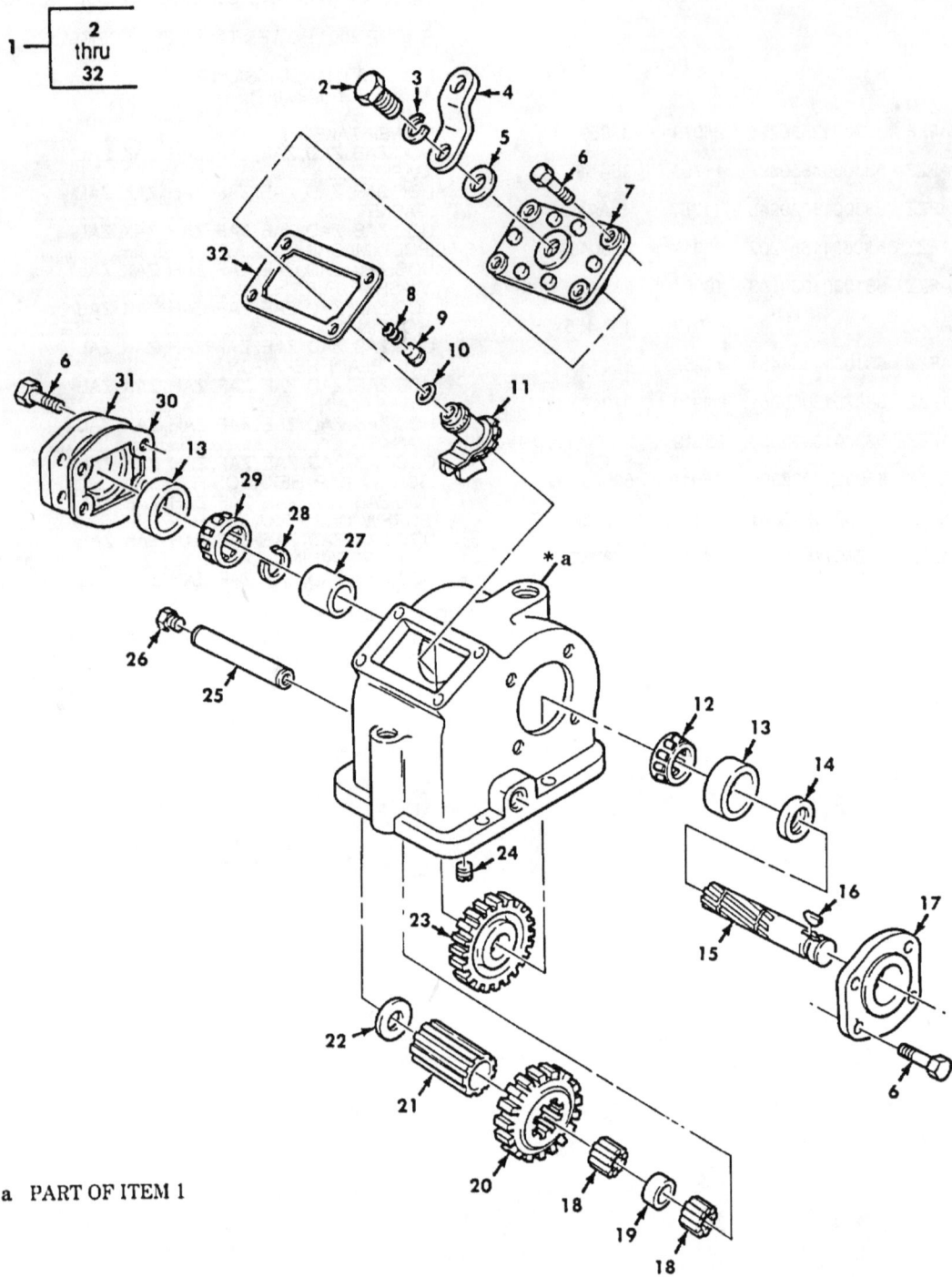

Figure 530. Transmission Power takeoff Assembly.

*a PART OF ITEM 1

(1) ITEM NO	(2) SMR CODE	(3) NSN	(4) CAGEC	(5) PART NUMBER	(6) DESCRIPTION AND USABLE ON CODES (UOC)	(7) QTY
					GROUP 2004 POWER TAKEOFF ASSEMBLY	
					FIG. 530 TRANSMISSION POWER TAKEOFF ASSEMBLY	
1	PAFFF	2520011056465	19207	11669313	POWER TAKEOFF, TRANS .... UOC:DAB, DAD, DAE, DAF, DAH, DAK, DAL, DAX, V12, V14, V16, V18, V19, V20, V21, V25, V39	1
2	PAFZZ	5305008857252	1175	500409-6	.SCREW............ UOC:DAB, DAD, DAE, DAF, DAH, DAK, DAL, DAX, V12, V14, V16, V18, V19, V20, V21, V25, V39, ZAB, ZAD, ZAE, ZAF, ZAH, ZAK, ZAL	1
3	PAFZZ	5310008381490	11757	378003	.WASHER, LOCK........... UOC:DAB, DAD, DAE, DAF, DAH, DAK, DAL, DAX, V12, V14, V16, V18, V19, V20, V21, V25, V39, ZAB, ZAD, ZAE, ZAF, ZAH, ZAK, ZAL	1
4	PAFZZ	3040008473169	11757	51-P-22	.CONNECTING LINK, RIG .... UOC:DAB, DAD, DAE, DAF, DAH, DAK, DAL, DAX, V12, V14, V16, V18, V19, V20, V21, V25, V39, ZAB, ZAD, ZAE, ZAF, ZAH, ZAK, ZAL	1
5	PAFZZ	5310004694073	11757	378004	.WASHER, FLAT .... UOC:DAB, DAD, DAE, DAF, DAH, DAK, DAL, DAX, V12, V14, V16, V18, V19, V20, V21, V25, V39, ZAB, ZAD, ZAE, ZAF, ZAH, ZAK, ZAL	1
6	PAFZZ	5306011041048	11757	378430-10	.BOLT, MACHINE .... UOC:DAB, DAD, DAE, DAF, DAH, DAK, DAL, DAX, V12, V14, V16, V18, V19, V20, V21, V25, V39, ZAB, ZAD, ZAE, ZAF, ZAH, ZAK, ZAL	12
7	PAFZZ	1450001759752	11757	34P17	.COVER, POWER TAKEOFF........... UOC:DAB, DAD, DAE , DAF, DAH, DAK, DAL, DAX, V12, V14, V16, V18B, V19, V20, V21, V25, V39, ZAB, ZAD, ZAE, ZAF, ZAH, ZAK, ZAL	1
8	PAFZZ	5360004726822	11757	37-P-20	.SPRING, HELICAL, COMP .... UOC:DAB, DAD, DAE, DAF, DAH, DAK, DAL, DAX, V12, V14, V16, V18, V19, V20, V21, V25, V39, ZAB, ZAD, ZAE, ZAF, ZAH, ZAK, ZAL	1
9	PAFZZ	2520011456820	11757	63-P-16	.POPPET SHIFTER .... UOC:DAB, DAD, DAE, DAF, DAH, DAK, DAL, DAX, V12, V14, V16, V18, V19, V20, V21, V25, V39, ZAB, ZAD, ZAE, ZAF, ZAH, ZAK, ZAL	1
10	PAFZZ	5330005822855	96906	MS28775-113	.O-RING .... UOC:DAB, DAD, DAE, DAF, DAH, DAK, DAL, DAX, V12, V14, V16, V18, V19, V20, V21, V25, V39	1
11	PAFZZ	2520002321938	18876	10161958	.PLATE, POWER TAKE-OF .... UOC:DAB, DAD, DAE, DAF, DAH, DAK, DAL, DAX, V12, V14, V16, V18, V19, V20, V21, V25, V39, ZAB, ZAD, ZAE, ZAF, ZAH, ZAK, ZAL	1
12	PAFZZ	3110001518636	11757	550397	.CONE AND ROLLERS, TA .... UOC:DAB, DAD, DAE, DAF, DAH, DAK, DAL, DAX, V12, V14, V16, V18, V19, V20, V21, V25, V39, ZAB, ZAD, ZAE, ZAF, ZAH, ZAK, ZAL	1
13	PAFZZ	3110001982170	11757	550221	.CUP, TAPERED ROLLER	2

(1) ITEM NO	(2) SMR CODE	(3) NSN	(4) CAGEC	(5) PART NUMBER	(6) DESCRIPTION AND USABLE ON CODES (UOC)	(7) QTY
					UOC:DAB, DAD, DAE, DAF, DAH, DAK, DAL, DAX, V12, V14, V16, V18, V19, V20, V21, V25, V39, ZAB, ZAD, ZAE, ZAF, ZAH, ZAK, ZAL	
14	PAFZZ-	5330002377828	11757	28-P-52	.SEAL, OIL	1
					UOC:DAB, DAD, DAE, DAF, DAH, DAK, DAL, DAX, V12, V14, V16, V18, V19, V20, V21, V25, V39, ZAB, ZAD, ZAE, ZAF, ZAH, ZAK, ZAL	
15	PAFZZ	3040001252961	11757	3-P-202	.SHAFT, SHOULDERED	1
					UOC:DAB, DAD, DAE, DAF, DAH, DAK, DAL, DAX, V12, V14, V16, V18, V19, V20, V21, V25, V39, ZAB, ZAD, ZAE, ZAF, ZAH, ZAK, ZAL	
16	PAFZZ	5315000431789	96906	MS35756-38	.KEY, WOODRUFF	1
					UOC:DAB, DAD, DAE, DAF, DAH, DAK, DAL, DAX, V12, V14, V16, V18, V19, V20, V21, V25, V39, ZAB, ZAD, ZAE, ZAF, ZAH, ZAK, ZAL	
17	PAFZZ	2520002321944	11757	328273X	.CAP, POWER TAKE-OFF	1
					UOC:DAB, DAD, DAE, DAF, DAH, DAK, DAL, DAX, V12, V14, V16, V18, V19, V20, V21, V25, V39, ZAB, ZAD, ZAE, ZAF, ZAH, ZAK, ZAL	
18	PAFZZ	3110011408880	11757	328024X	.ROLLER SET, BEARING	38
					UOC:DAB, DAD, DAE, DAF, DAH, DAK, DAL, DAX, V12, V14, V16, V18, V19, V20, V21, V25, V39, ZAB, ZAD, ZAE, ZAF, ZAH, ZAK, ZAL	
19	PAFZZ	5365001212776	11757	14-P-36	.SPACER, SLEEVE	1
					UOC:DAB, DAD, DAE, DAF, DAH, DAK, DAL, DAX, V12, V14, V16, V18, V19, V20, V21, V25, V39, ZAB, ZAD, ZAE, ZAF, ZAH, ZAK, ZAL	
20	PAFZZ	3020011328860	11757	5-P-569	.GEAR, SPUR	1
					UOC:DAB, DAD, DAE, DAF, DAH, DAK, DAL, DAX, V12, V14, V16, V18, V19, V20, V21, V25, V39, ZAB, ZAD, ZAE, ZAF, ZAH, ZAK, ZAL	
21	PAFZZ	3020000357894	11757	5-P-320	.GEAR CLUSTER	1
					UOC:DAB, DAD, DAE, DAF, DAH, DAK, DAL, DAX, V12, V14, V16, V18, V19, V20, V21, V25, V39, ZAB, ZAD, ZAE, ZAF, ZAH, ZAK, ZAL	
22	PAFZZ	5310004694039	11757	31-P-27	.WASHER, KEY	1
					UOC:DAB, DAD, DAE, DAF, DAH, DAK, DAL, DAX, V12, V14, V16, V18, V19, V20, V21, V25, V39, ZAB, ZAD, ZAE, ZAF, ZAH, ZAK, ZAL	
23	PAFFZ	3020004644438	11757	2-P-283	.GEAR, SPUR	1
					UOC:DAB, DAD, DAE, DAF, DAH, DAK, DAL, DAX, V12, V14, V16, V18, V19, V20, V21, V25, V39, .ZAB, ZAD, ZAE, ZAF, ZAH, ZAK, ZAL	
24	PAFZZ	5305011330163	11757	378452-3	SETSCREW...	1
					UOC:DAB, DAD, DAE, DAF, DAH, DAK, DAL, DAX, V12, V14, V16, V18, V19, V20, V21, V25, V39, ZAB, ZAD, ZAE, ZAF, ZAH, ZAK, ZAL	
25	PAFZZ	3040001252959	11757	9P35	.SHAFT, STRAIGHT	1
					UOC:DAB, DAD, DAE, DAF, DAH, DAK, DAL, DAX, V12, V14, V16, V18, V19, V20, V21, V25, V39, ZAB, ZAD, ZAE, ZAF, ZAH, ZAK, ZAL	
26	PAFZZ	4730009247886	11083	5M6214	.PLUG, PIPE	1
					UOC:DAB, DAD, DAE, DAF, DAH, DAK, DAL, DAX, V12, V14, V16, V18, V19, V20, V21, V25, V39,	

(1) ITEM NO	(2) SMR CODE	(3) NSN	(4) CAGEC	(5) PART NUMBER	(6) DESCRIPTION AND USABLE ON CODES (UOC)	(7) QTY
27	PAFZZ	5365001212780	11757	4-P-45	ZAB, ZAD, ZAE, ZAF, ZAH, ZAK, ZAL .SPACER, SLEEVE... UOC:DAB, DAD, DAE, DAF, DAH, DAK, DAL, DAX, V12, V14, V16, V18, V19, V20, V21, V25, V39,	1
28	PAFZZ	5325004770304	11757	378391	ZAB, ZAD, ZAE, ZAF, ZAH, ZAK, ZAL .RING, RETAINING.. UOC:DAB, DAD, DAE, DAF, DAH, DAK, DAL, DAX, V12, V14, V16, V18, V19, V20, V21, V25, V39,	1
29	PAFZZ	3110001009862	95019	550532	ZAB, ZAD, ZAE, ZAF, ZAH, ZAK, ZAL .CONE AND ROLLERS, TA .... UOC:DAB, DAD, DAE, DAF, DAH, DAK, DAL, DAX, V12, V14, V16, V18, V19, V20, V21, V25, V39,	1
30	PAFZZ	5330004850863	11757	22-P-24-1	ZAB, ZAD, ZAE, ZAF, ZAH, ZAK, ZAL .GASKET, STD... UOC:DAB, DAD, DAE, DAF, DAH, DAK, DAL, DAX, V12, V14, V16, V18, V19, V20, V21, V25, V39,	1
30	PAFZZ	5330004850865	11757	22-P-24-2	ZAB, ZAD, ZAE, ZAF, ZAH, ZAK, ZAL .GASKET.. UOC:DAB, DAD, DAE, DAF, DAH, DAK, DAL, DAX, V12, V14, V16, V18, V19, V20, V21, V25, V39,	1
31	PFFZA	3110002409897	11757	21-P-131	ZAB, ZAD, ZAE, ZAF, ZAH, ZAK, ZAL .PLATE, RETAINING, BEA .... UOC:DAB, DAD, DAE, DAF, DAH, DAK, DAL, DAX, V12, V14, V16, V18, V19, V20, V21, V25, V39,	1
32	PAFZZ	5330004850895	11757	35-P-8	ZAB, ZAD, ZAE, ZAF, ZAH, ZAK, ZAL .GASKET.. UOC:DAB, DAD, DAE, DAF, DAH, DAK, DAL, DAX, V12, V14, V16, V18, V19, V20, V21, V25, V39, ZAB, ZAD, ZAE, ZAF, ZAH, ZAK, ZAL	1

END OF FIGURE

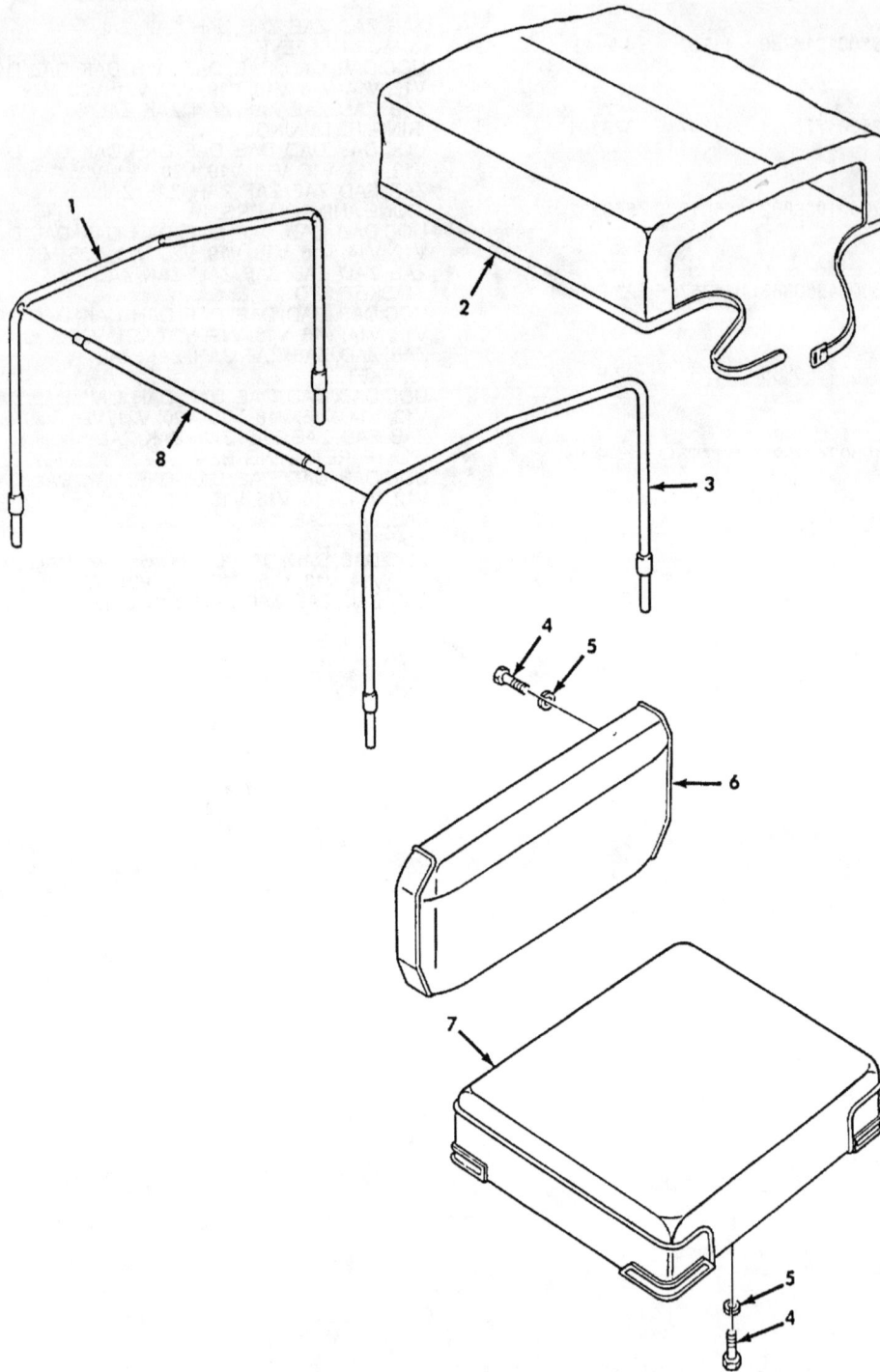

Figure 531. Wrecker Gondola Canopy and Seat.

(1) ITEM NO	(2) SMR CODE	(3) NSN	(4) CAGEC	(5) PART NUMBER	(6) DESCRIPTION AND USABLE ON CODES (UOC)	(7) QTY
					GROUP 22 BODY AND CHASSIS ACCESSORY ITEMS	
					2201 CANVAS OR PLASTIC ITEMS	
					FIG. 531 WRECKER GONDOLA CANOPY AND SEAT	
1	PAOZZ	2540008602357	19207	10876565	BOW, GONDOLA CANOPY FRONT .... UOC:DAL, V1B, ZAL	1
2	PAOZZ	2540008602355	19207	10876433	COVER, FITTED, VEHICU... UOC:DAL, V18, ZAL	1
3	PAOZZ	2540008602356	19207	10876566	BOW, GONDOLA CANOPY, REAR...................... UOC:DAL, V18, ZAL	1
4	PAOZZ	5305005434372	80204	B1821BH038C075N	SCREW, CAP, HEXAGON H .... UOC:DAL, V18, ZAL	4
5	PAOZZ	5310006379541	96906	MS35338-46	WASHER, LOCK .... UOC:DAL, V18, ZAL	4
6	PAOZZ	2540000634730	19207	10876402	CUSHION, SEAT BACK, V........................... UOC:DAL, V18, ZAL	1
7	PAOZZ	2540009722642	19207	10876401	SEAT, VEHICULAR .... UOC:DAL, V18, ZAL	1
8	PAOZZ	2540008602358	19207	10900249	CROSSMEMBER, GONDOLA............................... UOC:DAL, V18, ZAL	3

END OF FIGURE

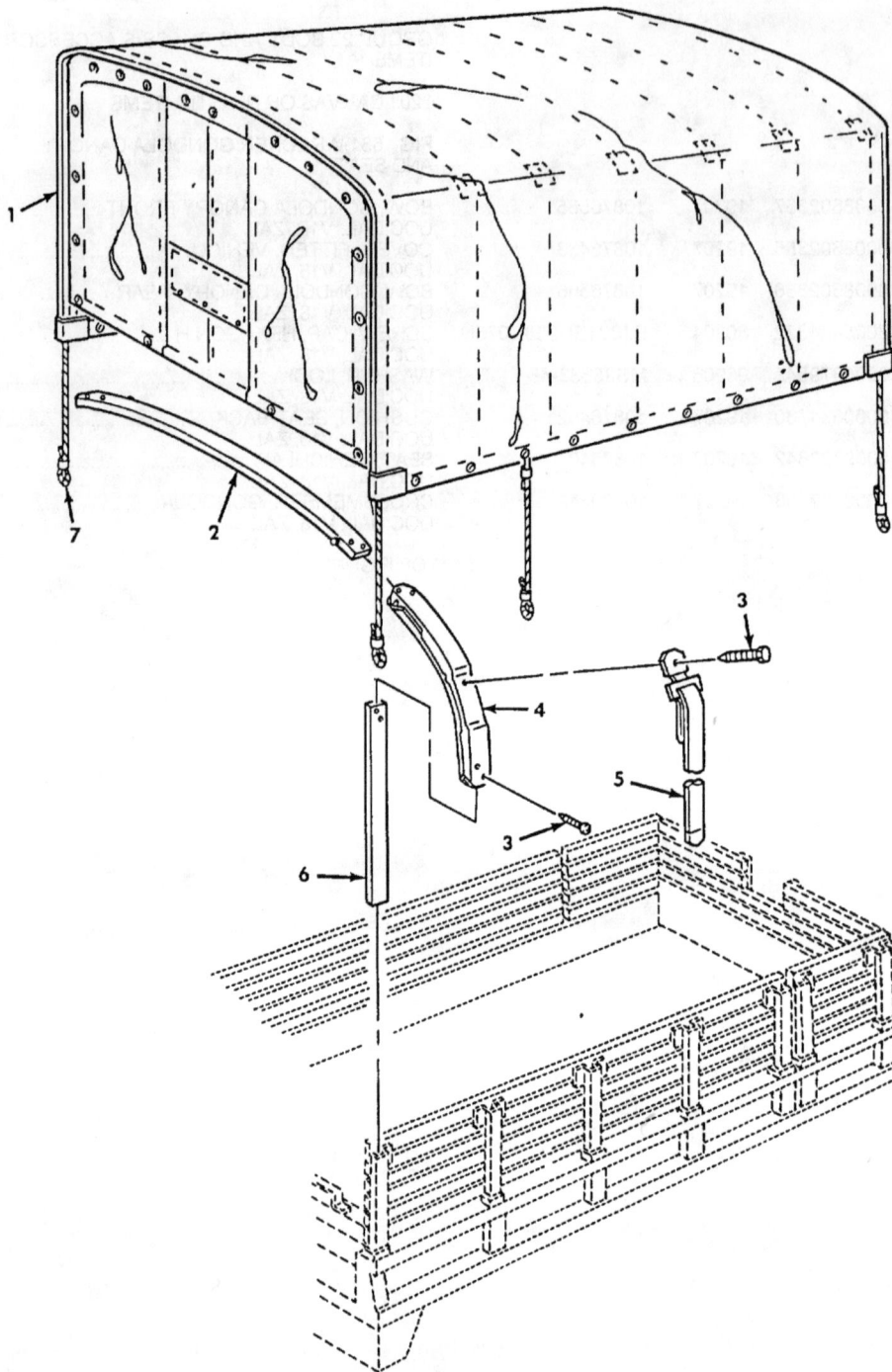

Figure 532. Cargo Bow and Stake Assembly and Related Parts.

(1) ITEM NO	(2) SMR CODE	(3) NSN	(4) CAGEC	(5) PART NUMBER	(6) DESCRIPTION AND USABLE ON CODES (UOC)	(7) QTY
					GROUP 2201 CANVAS OR PLASTIC ITEMS	
					FIG. 532 CARGO BOW AND STAKE ASSEMBLY AND RELATED PARTS	
1	PAOFF	2540014348725	19207	12450242-1	COVER, FITTED, VEHICU GREEN CAMO, XLWB PART OF KIT P/N 11672522 UOC:DAC, DAD, V16, V17, ZAC, ZAD	1
1	PAOFF	2540014354928	19207	12450242-2	COVER, FITTED, VEHICU TAN, XLWB PART OF KIT P/N 57K0166 UOC:DAC, DAD, V16, V17, ZAC, ZAD	1
1	PFOFF	2540014350568	19207	12450242-3	COVER, FITTED, VEHICU WHITE, XLWB PART OF KIT P/N 57K0167 UOC:DAC, DAD, V16, V17, ZAC, ZAD	1
1	PAOFF	2540009338645	19207	12450243-1	COVER, FITTED, VEHICU GREEN CAMO, DROPSIDE & FIXED SIDE PART OF-KIT P/N 11672523, 11672521 UOC:DAA, DAB, DAW, DAX, V12, V13, V14, V15, ZAA, ZAB	1
1	PAOFF	2540014354936	19207	12450243-2	COVER, FITTED, VEHICU TAN, DROPSIDE & FIXED SIDE PART OF KIT P/N 57K0164, 57K0168 UOC:DAA, DAB, DAW, DAX, V12, V13, V14, V15, ZAA, ZAB	1
1	PAOFF	2540014249440	19207	12450243-3	COVER, FITTED, VEHICU WHITE, DROP&FIXED SIDE PART OF KIT P/N 57K0165, 57K0169 UOC:DAA, DAB, DAW, DAX, V12, V13, V14, V15, ZAA, ZAB	1
2	PAOZZ	2540014358208	19207	12460216-1	BOW, VEHICULAR TOP PART OF KIT P/N 11672522, 57K0166, 57K0167 UOC:DAC, DAD, V16, V17, ZAC, ZAD	10
2	PAOZZ	2540014358208	19207	12460216-1	BOW, VEHICULAR TOP PART OF KIT P/N 11672523, 57K0168, 57K0169 UOC:DAW, DAX, V14, V15, ZAA, ZAB	6
2	PAOZZ	2540014358208	19207	12460216-1	BOW, VEHICULAR TOP PART OF KIT P/N 11672521, 57K0164, 57K0165 UOC:DAA, DAB, V12, V13, ZAA, ZAB	6
3	PAOZZ	5305009845680	96906	MS35206-300	SCREW, MACHINE PART OF KIT P/N 11672522, 57K0166, 57K0167 UOC:DAC, DAD, V16, V17, ZAC, ZAD	100
3	PAOZZ	5305009845680	96906	MS35206-300	SCREW, MACHINE PART OF KIT P/N 11672523, 57K0168, 57K0169 UOC:DAW, DAX, V14, V15, ZAA, ZAB	36
3	PAOZZ	5305009845680	96906	MS35206-300	SCREW, MACHINE PART OF KIT P/N 11672521, 57K0164, 57K0165 UOC:DAA, DAB, V12, V13, ZAA, ZAB	60
4	PAOZZ	2540003510145	19207	7064151	BOW ASSEMBLY PART OF KIT P/N 11672522, 57K0166, 57K0167 UOC:DAC, DAD, V16, V17, ZAC, ZAD	20
4	PAOZZ	2540004205036	19207	10937879	CORNER, BOW, VEHICULA PART OF KIT P/N 11672523, 57K0168, 57K0169 UOC:DAW, DAX, V14, V15, ZAA, ZAB	12
4	PAOZZ	2540003510145	19207	7064151	BOW ASSEMBLY PART OF KIT P/N	12

(1) ITEM NO	(2) SMR CODE	(3) NSN	(4) CAGEC	(5) PART NUMBER	(6) DESCRIPTION AND USABLE ON CODES (UOC)	(7) QTY
					11672521, 57K0 164, 57K0165 .... UOC:DAA, DAB, V12, V13, ZAA, ZAB	
5	PAOZZ		19207	12340208-6	STRAP, WEBBING GREEN PART OF KIT P/N 11672522, 11672521, 11672523 .... UOC:DAC, DAD, V16, V17, ZAC, ZAD	20
5	PAOZZ		19207	12340208-7	STRAP, WEBBING TAN PART OF KIT P/N 57K0164, 57K0166, 57K 0168 .... UOC:DAW, DAX, V14, V15, ZAA, ZAB	12
5	PFOZZ		19207	12340208-8	STRAP, WEBBING WHITE PART OF KIT P/N 57K0165, 57K0167, 57K0169 .... UOC:DAA, DAB, V12, V13, ZAA, ZAB	12
6	PAOZZ	2510011381157	19207	10871302-1	STAKE, VEHICLE BODY PART OF KIT P/N 11672522, 57K0166, 57K0167 .... UOC:DAC, DAD, V16, V17, ZAC, ZAD	20
6	PAOZZ	2510007372781	19207	7372781	STAKE, VEHICLE BODY PART OF KIT P/N 11672523, 57K0168, 57K0169 .... UOC:DAW, DAX, V14, V15, ZAA, ZAB	12
6	PAOZZ	2510007372781	19207	7372781	STAKE, VEHICLE BODY PART OF KIT P/N 11672521, 57K0164, 57K0165 .... UOC:DAA, DAB, V12, V13, ZAA, ZAB	12
7	PAOZZ		19207	12450217	CORD, ELASTIC .... UOC:DAA, DAB, DAC, DAD, DAW, DAX, V12, V13, V14, V15, V16, V17, ZAA, ZAB, ZAC, ZAD	V
	PAOFF	2540001219077	19207	11672521	ACCESSORY KIT, VEHIC GREEN CAMO, SHORT BODY SIDE .... UOC:DAA, DAB, V12, V13	1
	PAOFF	2540014231968	34623	57K0164	ACCESSORY KIT, VEHIC TAN, SHORT BODY SIDE .... UOC:DAA, DAB, V12, V13	1
	PFOFF	2540014231964	34623	57K0165	ACCESSORY KIT, VEHIC WHITE, SHORT BODY SIDE .... UOC:DAA, DAB, V12, V13	1
	PAOFF	2540001219081	19207	11672522	ACCESSORY KIT, VEHIC GREEN CAMO, LONG BODY SIDE .... UOC:DAC, DAD, V16, V17, ZAC, ZAD	1
	PAOFF	2540013652936	9C234	57K0166	ACCESSORY KIT, VEHIC TAN, LONG BODY SIDE .... UOC:DAC, DAD, V16, V17, ZAC, ZAD	1
	PFOFF	2540013652937	9C234	57K0167	ACCESSORY KIT, VEHIC WHITE, LONG, BODY SIDE .... UOC:DAC, DAD, V16, V17, ZAC, ZAD	1
	PAOFF	2540001219082	19207	11672523	ACCESSORY KIT, VEHIC GREEN CAMO, DROP BODY SIDE .... UOC:DAW, DAX, V14, V15, ZAA, ZAB	1
	PAOFF	2540013689848	34623	57K0168	ACCESSORY KIT, VEHIC TAN, DROP BODY, SIDE .... UOC:DAW, DAX, V14, V15, ZAA, ZAB	1
	PFOFF	2540013691392	34623	57K0169	ACCESSORY KIT, VEHIC WHITE, DROP BODY SIDE .... UOC:DAW, DAX, V14, V15, ZAA, ZAB	1

END OF FIGURE

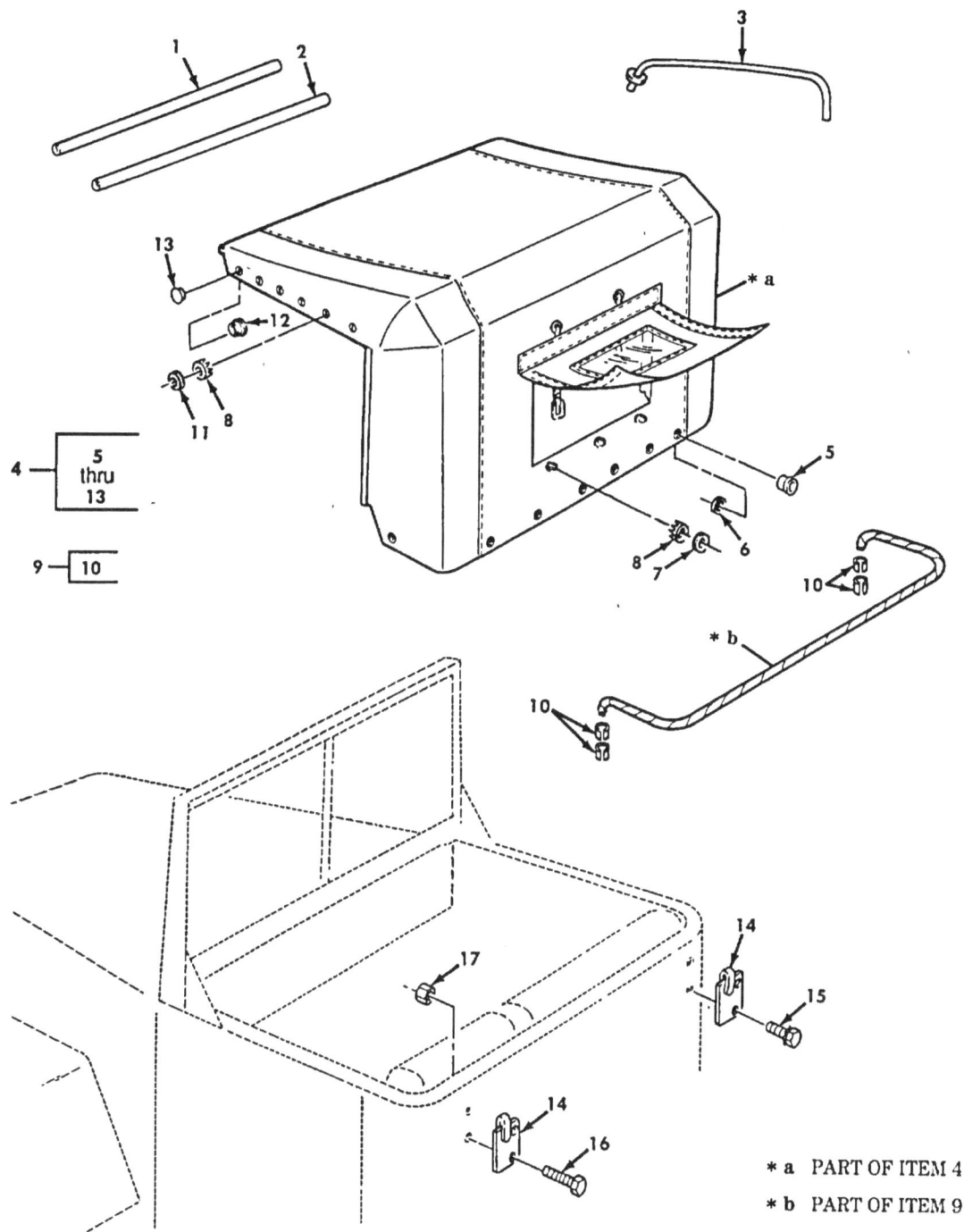

Figure 533. Soft Top Assembly and Related Parts.

* a   PART OF ITEM 4
* b   PART OF ITEM 9

(1) ITEM NO	(2) SMR CODE	(3) NSN	(4) CAGEC	(5) PART NUMBER	(6) DESCRIPTION AND USABLE ON CODES (UOC)	(7) QTY

GROUP 2201 CANVAS OR PLASTIC ITEMS

FIG. 533 SOFT TOP ASSEMBLY AND RELATED PARTS

(1) ITEM NO	(2) SMR CODE	(3) NSN	(4) CAGEC	(5) PART NUMBER	(6) DESCRIPTION AND USABLE ON CODES (UOC)	(7) QTY
1	PAOZZ	2510010819227	19207	12255938	FRAME SECTION, STRUC...............................	1
2	PAOZZ	9510011195679	19207	12255939-2	BAR, METAL	1
3	PAOZZ	2540010921264	19207	12255937	BOW, VEHICULAR	3
4	PAOFF	2540014176379	19207	12450238-1	COVER, FITTED, VEHICU GREEN...................	1
4	PAOFF	2540014354924	19207	12450238-2	COVER, FITTED, VEHICU TAN........................	1
4	PFOFF	2540014354931	19207	12450238-3	COVER, FITTED, VEHICU WHITE	1
5	PAFZZ	5325005262663	96906	MS20230-GB4	.EYELET, METALLIC STRAP ASSEMBLY.........	8
6	PAFZZ	5325005088901	96906	MS20230-WB4	.EYELET, METALLIC SOFT TOP.....................	8
7	PAFZZ	5325003034932	21450	549222	.CLINCH PLATE, TURNBU SOCKET...............	5
8	PAFZZ	5325002818643	21450	426687	.SOCKET, TURNBUTTON F SOFT TOP...........	17
9	MOOZZ		19207	12255947	.ROPE, ASSY MAKE FROM ROPE, P/N 21-R-..................... 358 OR 21-R-788S LONG OR NSN 4020012047039	1
10	PAOZZ	5340011043832	19207	7359274	..CLAMP, LOOP..	4
11	PAFZZ	5340001113605	21450	549182	.BUTTON, HOLD DOWN SNAP	12
12	PAFZZ	5325003718108	13940	BS 78505	.CLINCH PLATE, TURNBU	2
13	PAFZZ	5325002856250	88044	AN227-7B	.SOCKET, SNAP	2
14	PAOZZ	5340004561011	24617	2173982	HOOK, SUPPORT...	7
15	PAOZZ	5305000192417	21450	192417	SCREW, ASSEMBLED WAS	2
16	PAOZZ	5306000501238	96906	MS90727-32	BOLT, MACHINE..	12
17	PAOZZ	5310009843807	96906	MS51922-13	NUT, SELF-LOCKING, HE	12

END OF FIGURE

* a PART OF ITEM 2

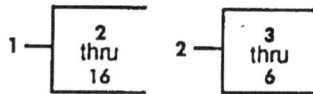

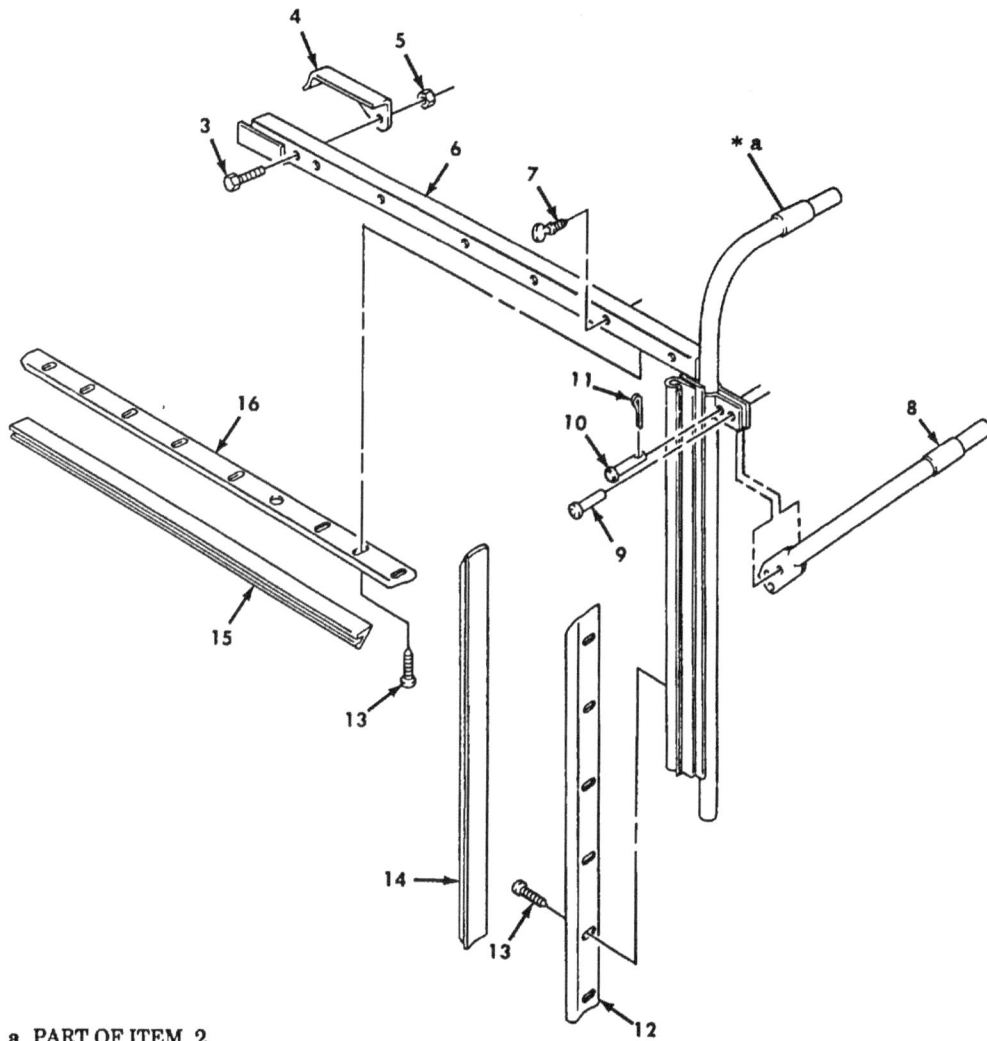

Figure 534.   So p Post Assembly.

(1) ITEM NO	(2) SMR CODE	(3) NSN	(4) CAGEC	(5) PART NUMBER	(6) DESCRIPTION AND USABLE ON CODES (UOC)	(7) QTY
					GROUP 2201 CANVAS OR PLASTIC ITEMS	
					FIG. 534 SOFT TOP POST ASSEMBLY	
1	PAOOO	2510007409596	19207	7409596	POST PILLAR CAB TOP L.H	1
1	PAOOO	2540007409597	19207	7409597	POST ASSEMBLY, TOP, T R.H	1
2	PAOOO	2510007409686	19207	7409686	.RAIL ASSEMBLY, SIDE L.H	1
2	PAOOO	2510007409687	19207	7409687	.FRAME SECTION, STRUC R.H	1
3	PAOZZ	5305002678955	80204	B1821BH025F138N	..SCREW, CAP, HEXAGON H	1
4	PAOZZ	2540008917830	19207	7005798	..CATCH, CAB TOP L.H	1
4	PAOZZ	2510000360298	19207	7005638	..CATCH, CAB TOP R.H	1
5	PAOZZ	5310008775796	96906	MS21044N4	..NUT, SELF-LOCKING, HE	1
6	PAOFF	2510011699850	19207	7005602	..SUPPORT, SOFT TOP, CA L.H. DOES NOT ..... INCLUDE VERTICAL SIDE SUPPORT AS DRAWN	1
6	PAOFF	2510011896412	19207	7005603	..SUPPORT, SOFT TOP, CA R.H. DOES NOT ..... INCLUDE VERTICAL SIDE SUPPORT AS DRAWN	1
7	PAOZZ	5325014456866	19207	12375525	..STUD, TURNBUTTON FAS	6
8	PAOZZ	2510007372712	19207	7372712	.BOW, SIDE RAIL L.H	1
8	PAOZZ	2510007372711	19207	7372711	.BOW, SIDE RAIL R.H	1
9	PFOZZ	5320008892632	96906	MS35743-38	.RIVET, SOLID	1
10	PAOZZ	5315010702168	96906	MS35810-11	.PIN, STRAIGHT, HEADED	1
11	PAOZZ	5315008392325	10001	12Z48PC611	.PIN, COTTER	1
12	PAOZZ	2510006930591	19207	7372718	.RETAINER, CAB TOP L.H	1
12	PAOZZ	2510006930592	19207	7372719	.RETAINER, CAB TOP R.H	1
13	PAOZZ	5305008797941	96906	MS24617-31	.SCREW, TAPPING	15
14	PAOZZ	5330007372720	19207	7372720	.SEAL, NONMETALLIC SP L.H	1
14	PAOZZ	5330010600992	19207	7372721	.SEAL, RUBBER SPECIAL R.H	1
15	PAOZZ	5330007372722	19207	7372722	.SEAL, NONMETALLIC SP	1
16	PAOZZ	2590006930589	19207	7372716	.BEZEL, AUTOMOTIVE TR	1

END OF FIGURE

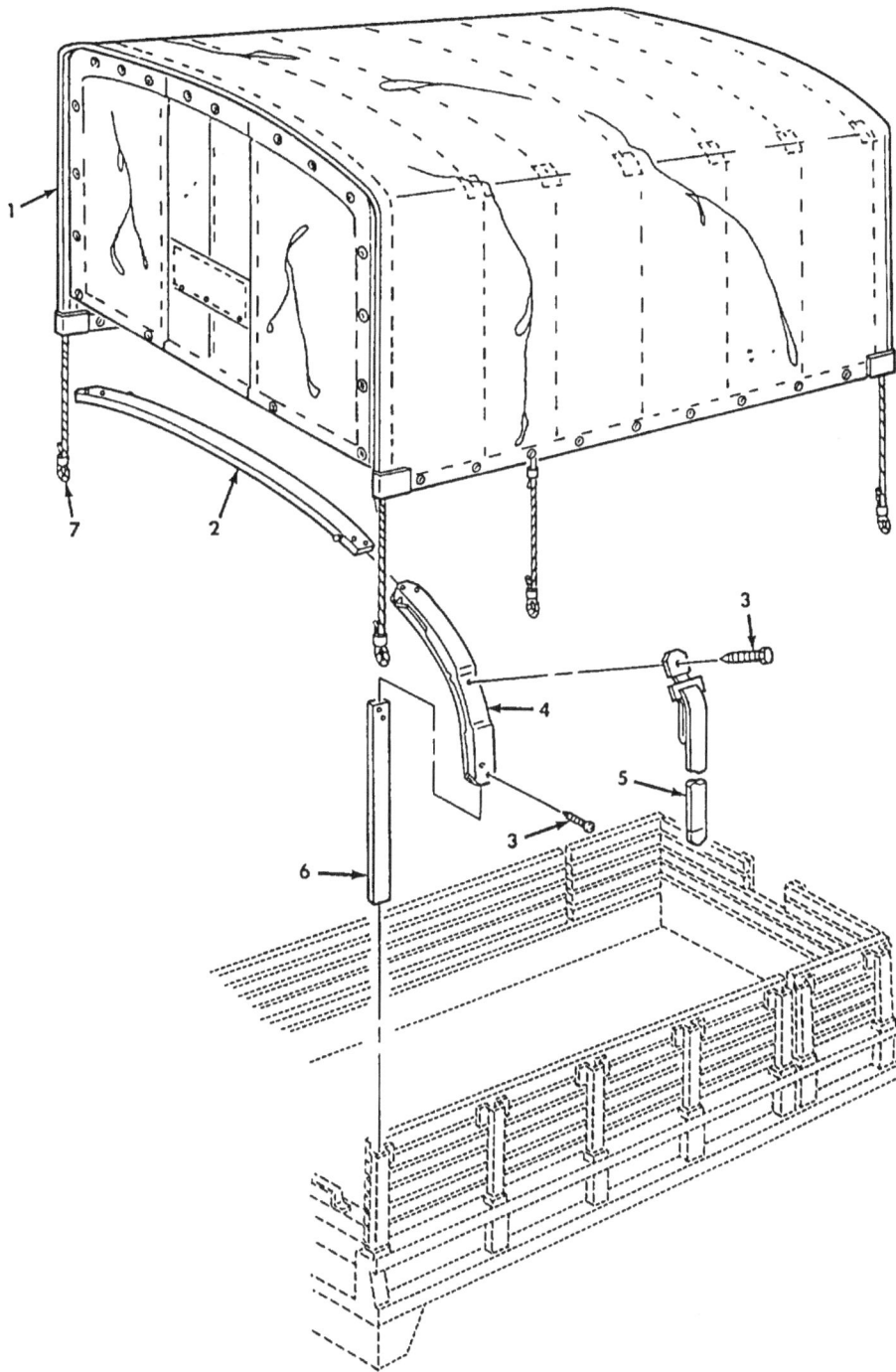

Figure 535. Tarpaulin Extension Kit, LWB Cargo Extra Height.

(1) ITEM NO	(2) SMR CODE	(3) NSN	(4) CAGEC	(5) PART NUMBER	(6) DESCRIPTION AND USABLE ON CODES (UOC)	(7) QTY
					GROUP 2201 CANVAS OR PLASTIC ITEMS	
					FIG. 535 TARPAULIN EXTENSION KIT, LWB CARGO EXTRA HEIGHT	
1	PAOFF		19207	12450242-6	COVER, FITTED, VEHICU PART OF KIT P/N 11672522-1, GREEN. UOC:DAC, DAD.V16, V17, ZAC, ZAD	1
1	PAOFF		19207	12450242-8	COVER, FITTED, VEHICU PART OF KIT P/N 57K3608, WHITE. UOC:DAC, DAD.V16, V17, ZAC, ZAD	1
1	PAOFF		19207	12450242-7	COVER, FITTED, VEHICU PART OF KIT P/N 57K3609, TAN UOC:DAC, DAD.V16, V17, ZAC, ZAD	1
2	PAOZZ	2540014358208	19207	12460216-1	BOW, VEHICULAR TOP PART OF KIT P/N 57K3608, 57K3609, 11672522-1 UOC:DAC, DAD.V16, V17, ZAC, ZAD	10
3	PAOZZ	5305009845680	96906	MS35206-300	SCREW, MACHINE PART OF KIT P/N 57K3608, 57K3609, 11682522-1 UOC:DAC, DAD.V16, V17, ZAC, ZAD	100
4	PAOZZ	2540003510145	24617	7064151	BOW ASSEMBLY PART OF KIT P/N 57K3608, 57K3609, 11682522-1 UOC:DAC, DAD, V16, V17, ZAC, ZAD	20
5	PAOZZ		19207	12340208-6	STRAP, WEBBING PART OF KIT P/N 11672522-1, GREEN. UOC:DAC, DAD, V16, V17, ZAC, ZAD	20
5	PAOZZ		19207	12340208-7	STRAP, WEBBING PART OF KIT P/N 57K3608, WHITE. UOC:DAC, DAD, V16, V17, ZAC, ZAD	20
5	PFOZZ		19207	12340208-8	STRAP, WEBBING PART OF KIT P/N 57K3609, TAN. UOC:DAC, DAD, V16, V17, ZAC, ZAD	20
6	PAOZZ	2510011381157	19207	10871302-1	STAKE, VEHICLE BODY PART OF KIT P/N 57K3608, 57K3609, 11672522-1 UOC:DAC, DAD, V16, V17, ZAC, ZAD	20
7	PAOZZ		19207	12450217	CORD, ELASTIC UOC:DAA, DAB, DAC, DAD, DAW, DAX, V12, V13, V14, V15, V16, V17, ZAA, ZAB, ZAC, ZAD	V
	PAOFF		19207	11672522-1	KIT, BOW AND COVER, GREEN UOC:DAC, DAD, V16, V17, ZAC, ZAD	1
	PFOFF		19207	57K3608	KIT, BOW AND COVER, WHITE. UOC:DAC, DAD, V16, V17, ZAC, ZAD	1
	PAOFF		19207	57K3609	KIT, BOW AND COVER, TAN. UOC:DAC, DAD, V16, V17, ZAC, ZAD	1

END OF FIGURE

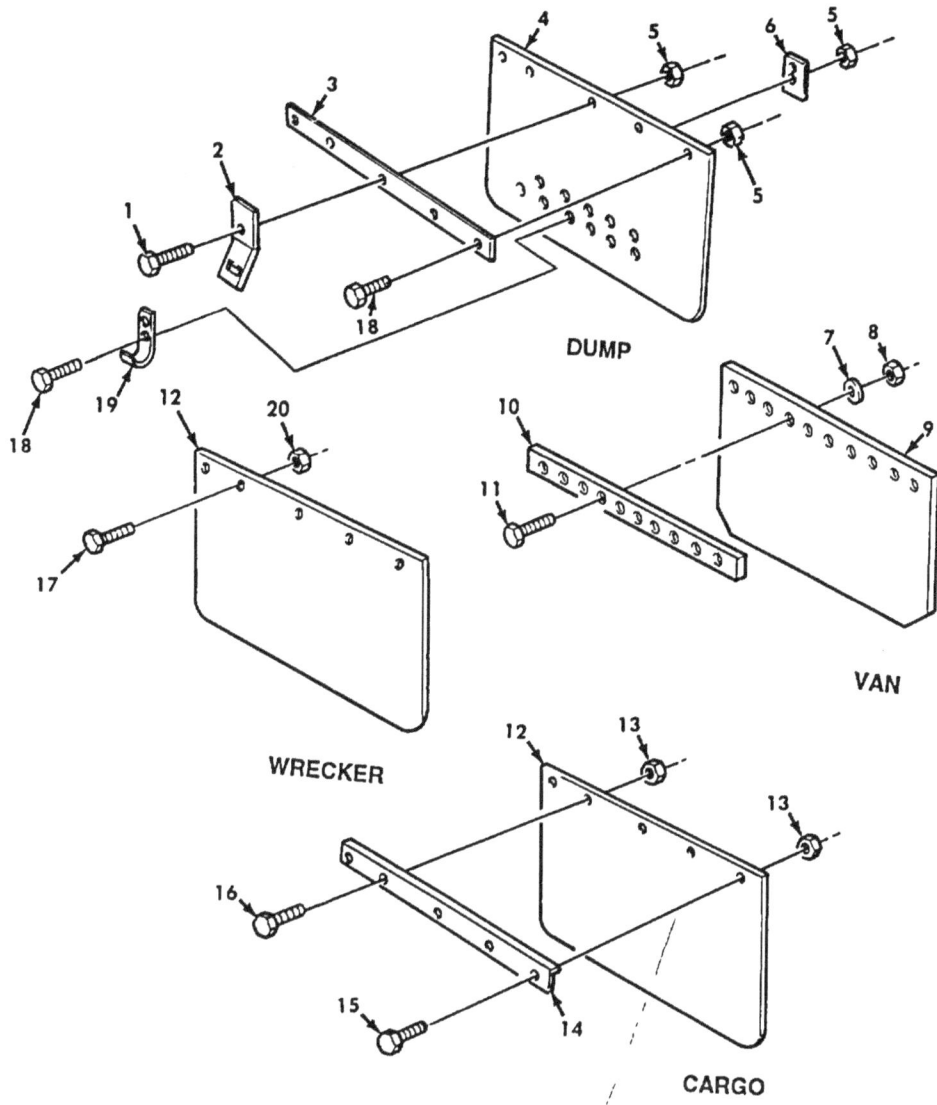

Figure 536. Splash Guard Assemblies.

(1) ITEM NO	(2) SMR CODE	(3) NSN	(4) CAGEC	(5) PART NUMBER	(6) DESCRIPTION AND USABLE ON CODES (UOC)	(7) QTY
					GROUP 2201 CANVAS OR PLASTIC ITEMS	
					FIG. 536 SPLASH GUARD ASSEMBLIES	
1	PAOZZ	5305002259092	96906	MS90726-37	SCREW, CAP, HEXAGON H .... UOC:DAE, DAF, V19, V20, ZAE, ZAF	2
2	PAOZZ	5340010978094	19207	11648463	BRACKET, ANGLE .... UOC:DAE, DAF, V19, V20, ZAE, ZAF	2
3	PFOZZ	5340010823595	19207	11648457	RETAINER STRAP, VEHI .... UOC:DAE, DAF, V19, V20, ZAE, ZAF	2
4	PAOZZ	2540010915450	19207	11648460	GUARD, SPLASH, VEHICU .... UOC:DAE, DAF, V19, V20, ZAE, ZAF	2
5	PAOZZ	5310002416658	96906	MS51943-34	NUT, SELF-LOCKING, HE .... UOC:DAE, DAF, V19, V20, ZAE, ZAF	3
6	PFOZZ	2510010831106	19207	11648464	RETAINER PLATE, SPLA .... UOC:DAE, DAF, V19, V20, ZAE, ZAF	2
7	PAOZZ	5310005825965	96906	MS35338-44	WASHER, LOCK LEFT REAR REAR .... 10 UOC:DAJ, DAK, V24, V25, ZAJ, ZAK	10
8	PAOZZ	5310007616882	96906	MS51967-2	NUT, PLAIN, HEXAGON LEFT REAR REAR .... UOC:DAJ, DAK, V24, V25, ZAJ, ZAK	10
8	PAOZZ	5310000617325	96906	MS21045-4	NUT, SELF-LOCKING, HE RIGHT AND LEFT .... 30 FORWARD REAR AND RIGHT REAR REAR .... UOC:DAJ, DAK, V24, V25, ZAJ, ZAK	30
9	PAOZZ	2540001692856	19207	11611632	GUARD, SPLASH, VEHICU RIGHT AND LEFT .... REAR REAR AND RIGHT FORWARD REAR .... UOC:DAJ, DAK, V24, V25, ZAJ, ZAK	3
9	PAOZZ	2540011635170	19207	11677690	GUARD, SPLASH, VEHICU LEFT FORWARD .... REAR........................ UOC:DAJ, DAK, V24, V25, ZAJ, ZAK	1
10	PFOZZ	2510004093992	19207	11611633	RETAINER, SPLASH GUA RIGHT AND LEFT .... REAR REAR .... UOC:DAJ, DAK, V24, V25, ZAJ, ZAK	2
10	PFOZZ	5340011634777	19207	11677688	RETAINER, SPLASH, LEFT FORWARD REAR .... UOC:DAJ, DAK, V24, V25, ZAJ, ZAK	1
10	PBOZZ	2590011757230	19207	11677713	RETAINER, SPLASH, RIGHT FORWARD REAR .... UOC:DAJ, DAK, V24, V25, ZAJ, ZAK	1
11	PAOZZ	5305002253843	80204	B1821BH025C100N	SCREW, CAP, HEXAGON H LEFT REAR REAR .... UOC:DAJ, DAK, V24, V25, ZAJ, ZAK	10
11	PAOZZ	5305002693239	80204	B1821BH038F138N	SCREW, CAP, HEXAGON H RIGHT AND LEFT .... FORWARD REAR AND RIGHT REAR REAR UOC:DAJ, DAK, V24, V25, ZAJ, ZAK	30
12	PAOZZ	2540007157407	19207	8758157	GUARD, SPLASH, VEHICU .... UOC:DAA, DAB, DAC, DAD, DAL, DAW, DAX, V12, V13, V14, V15, V16, V17, V18, ZAA, ZAB, ZAC, ZAD, ZAL	4
13	PAOZZ	5310008140672	96906	MS51943-36	NUT, SELF-LOCKING, HE .... UOC:DAA, DAB, DAC, DAD, DAL, DAW, DAX, V12, V13, V14, V15, V16, V17, V18, ZAA, ZAB, ZAC, ZAD, ZAL	20
14	PAOZZ	9520008576344	19207	8758158	ANGLE, STRUCTURAL .... UOC:DAA, DAB, DAC, DAD, DAW, DAX, V12, V13, V14, V15, V16, V17, ZAA, ZAB, ZAC, ZAD	4

(1) ITEM NO	(2) SMR CODE	(3) NSN	(4) CAGEC	(5) PART NUMBER	(6) DESCRIPTION AND USABLE ON CODES (UOC)	(7) QTY
15	PAOZZ	5305002693240	80204	B1821BH03BF150N	SCREW, CAP, HEXAGON H .... UOC:DAA, DAB, DAC, DAD, DAL, DAW, DAX, V12, V13, V14, V15, V16, V17, V18, ZAA, ZAB, ZAC, ZAD, ZAL	8
16	PAOZZ	5305002692803	96906	MS90726-60	SCREW, CAP, HEXAGON H .... UOC:DAA, DAB, DAC, DAD, DAL, DAW, DAX, V12, V13, V14, V15, V16, V17, V18, ZAA, ZAB, ZAC, ZAD, ZAL	12
17	PAOZZ	5305002692803	96906	MS90726-60	SCREW, CAP, HEXAGON H .... UOC:DAL, V18, ZAL	20
18	PAOZZ	5306000514077	80204	B1821BH031F113N	BOLT, MACHINE.. UOC:DAE, DAF, V19, V20, ZAE, ZAF	12
19	PBOZZ	2510010831105	19207	11648461	HOOK, SPLASH GUARD .... UOC:DAE, DAF, V19, V20, ZAE, ZAF	2
20	PAOZZ	5310008140672	96906	MS51943-36	NUT, SELF-LOCKING, HE .... UOC:DAL, V18, ZAL	20

END OF FIGURE

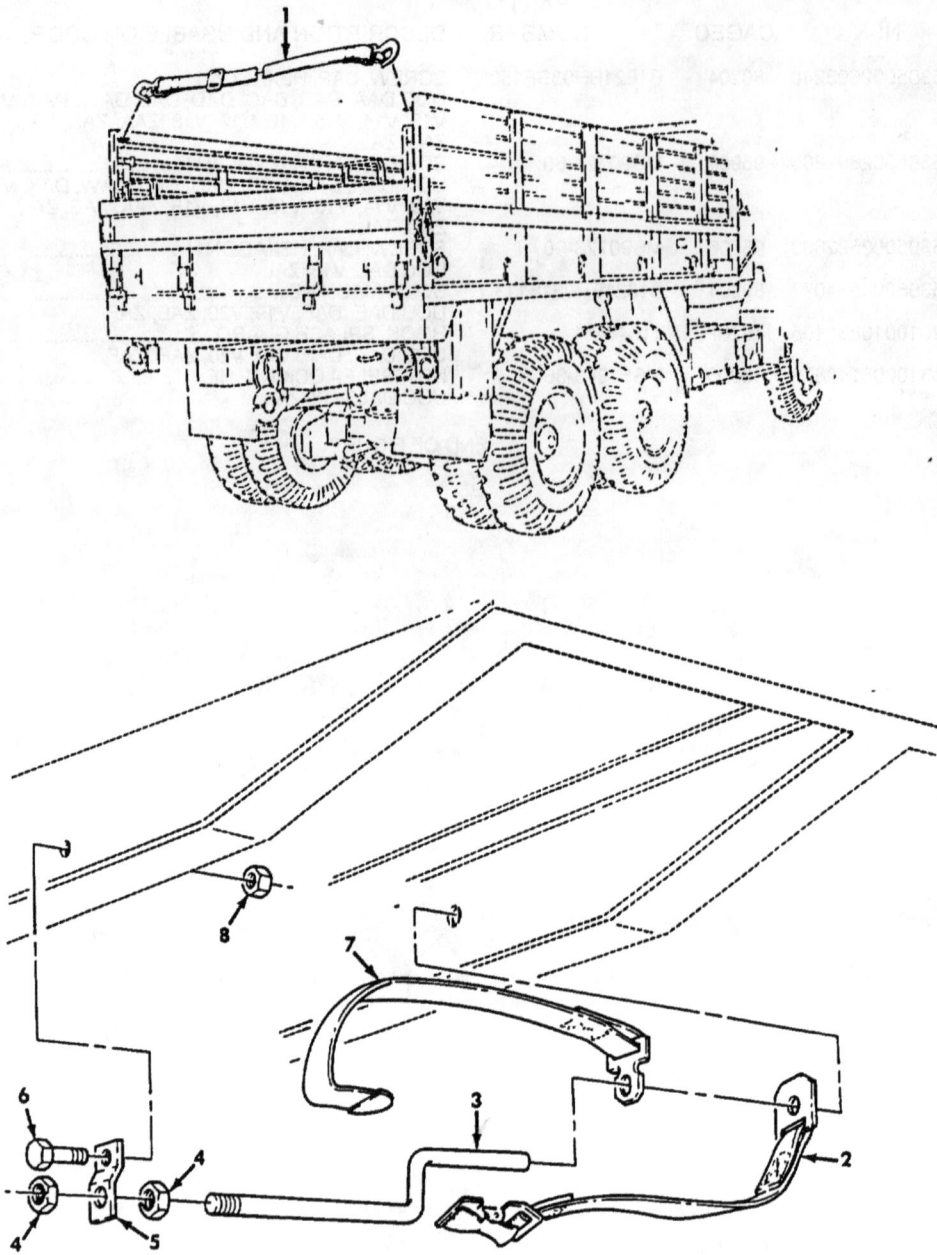

Figure 537. Adjustable Safety Strap and Bow Stowage Strap.

(1) ITEM NO	(2) SMR CODE	(3) NSN	(4) CAGEC	(5) PART NUMBER	(6) DESCRIPTION AND USABLE ON CODES (UOC)	(7) QTY
					SECTION II	
					GROUP 2201 CANVAS OR PLASTIC ITEMS	
					FIG. 537 ADJUSTABLE SAFETY STRAP AND BOW STOWAGE STRAP	
1	PAOZZ	5340011147712	19207	11682088-1	STRAP, WEBBING................................. UOC:DAA, DAB, DAC, DAD, DAW, DAX, V12, V13, V14, V15, V16, V17, ZAA, ZAB, ZAC, ZAD	1
2	PAOZZ	5340009302716	19207	10937881	STRAP, WEBBING . ... UOC:DAA, DAB, DAC, DAD, DAW, DAX, V12, V13, V14, V15, V16, V17, ZAA, ZAB, ZAC, ZAD	2
3	PAOZZ	5306004213950	19207	10937897	ROD, THREADED END............................. UOC:DAA, DAB, DAC, DAD, DAW, DAX, V12, V13, V14, V15, V16, V17, ZAA, ZAB, ZAC, ZAD	2
4	PAOZZ	5310008348734	96906	MS35691-37	NUT, PLAIN, HEXAGON............................. UOC:DAA, DAB, DAC, DAD, DAW, DAX, V12, V13, V14, V15, V16, V17, ZAA, ZAB, ZAC, ZAD	4
5	PBOZZ	5340004824339	19207	11611613	BRACKET, DOUBLE ANGL .... UOC:DAA, DAB, DAC, DAD, DAW, DAX, ZAA, ZAB, ZAC, ZAD	2
6	PAOZZ	5305009125113	96906	MS51096-359	SCREW, CAP, HEXAGON H .... UOC:DAA, DAB, DAC, DAD, DAW, DAX, V12, V13, V14, V15, V16, V17, ZAA, ZAB, ZAC, ZAD	2
7	PAOZZ	5340009302717	19207	10937882	STRAP, WEBBING................................... UOC:DAA, DAB, DAC, DAD, DAW, DAX, V12, V13, V14, V15, V16, V17, ZAA, ZAB, ZAC, ZAD	2
8	PAOZZ	5310008140672	96906	MS51943-36	NUT, SELF-LOCKING, HE .... UOC:DAA, DAB, DAC, DAD, DAW, DAX, V12, V13, V14, V15, V16, V17, ZAA, ZAB, ZAC, ZAD	2

END OF FIGURE

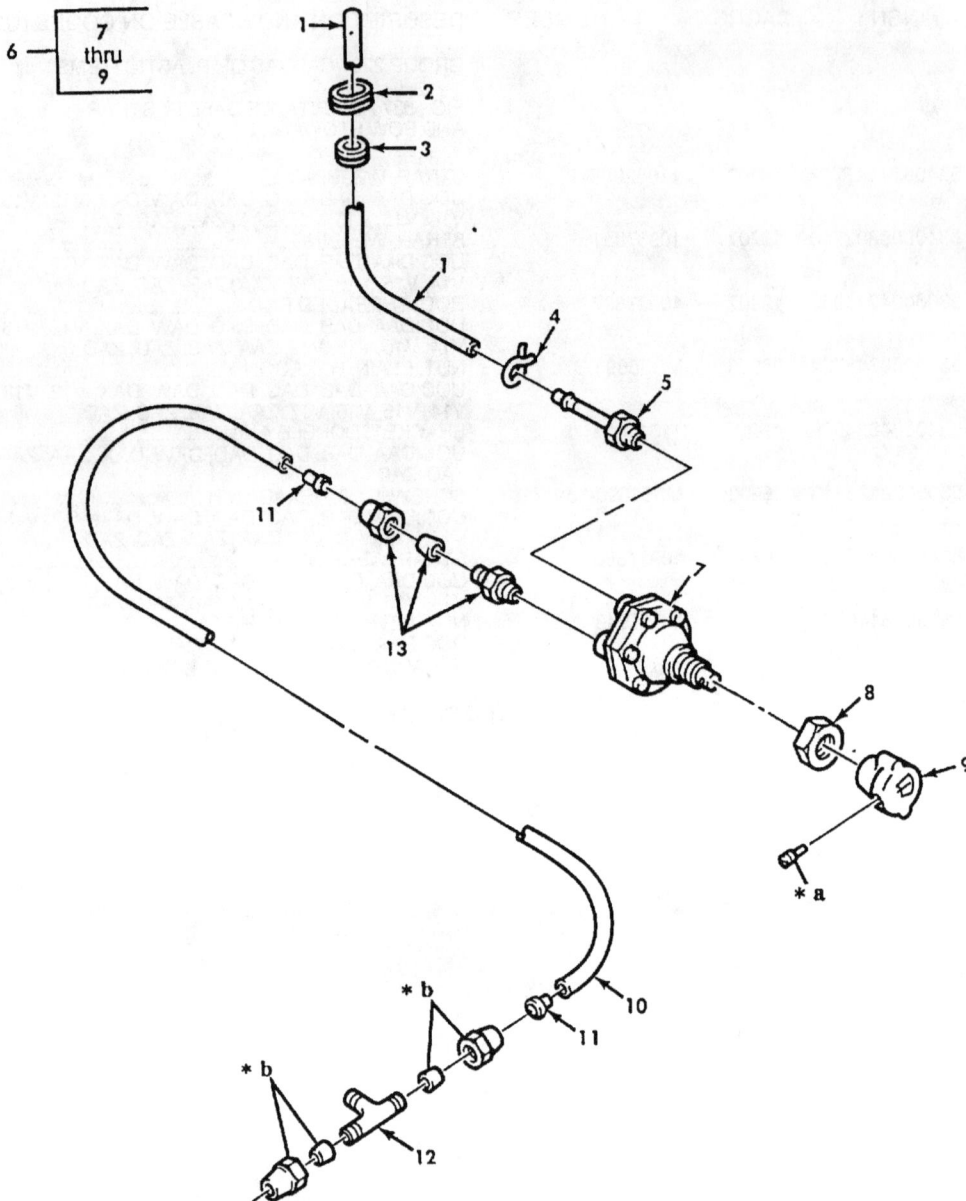

Figure 538. Windshield Wiper Air Lines and Control Assembly.

* a PART OF ITEM 9
* b PART OF ITEM 12

(1) ITEM NO	(2) SMR CODE	(3) NSN	(4) CAGEC	(5) PART NUMBER	(6) DESCRIPTION AND USABLE ON CODES (UOC)	(7) QTY
					GROUP 2202 ACCESSORY ITEMS	
					FIG. 538 WINDSHIELD WIPER AIR LINES AND CONTROL ASSEMBLY	
1	MOOZZ		19207	12277085-4	HOSE, WINDSHIELD WIP MAKE FROM HOSE, P/N A12876.R.H., 32.25" LONG.............	2
1	MOOZZ		19207	12277085-3	HOSE, NONMETALLIC MAKE FROM HOSE, P/N A12876.L.H., 27.25" LONG.............	2
2	PAOZZ	5325011569497	19207	12277374	GROMMET, NONMETALLIC, .L.H. CONTROL ASSEMBLY HOSE.............	1
3	PAOZZ	5325001850004	96906	MS35489-40	GROMMET, NONMETALLIC, R.H. CONTROL ASSEMBLY HOSE.............	2
4	PAOZZ	4730011098001	19207	11662913	CLAMP, HOSE.............	4
5	PAOZZ	4730007827102	96906	MS24522-2	ADAPTER, STRAIGHT, PI.............	4
6	PAOZZ	5930013214866	19207	12356924 PRIME	WIPER CONTROL PART OF KIT P/N 5705690.............	2
7	XAOZZ		82484	GK-9	.CONTROL ASSEMBLY, WI PART OF KIT P/N 5705690.............	2
8	PAOZZ	5310011354797	82484	S-710-3	.NUT, PLAIN, HEXAGON.............	2
9	PAOZZ	5355013868877	82484	C-2029	.KNOB.............	2
10	MOOZZ		19207	CPR104420-2-22	TUBE, R.H. MAKE FROM HOSE, P/N CPR104420-2, 22 INCHES LONG.............	1
10	MOOZZ		19207	CPR104420-2-32	TUBE, L.H. MAKE FROM HOSE, P/N CPR104420-2, 32 INCHES LONG.............	1
11	PAOZZ	4730010798821	19207	CPR102321-1	INSERT, TUBE FITTING.............	4
12	PAOZZ	4730004946580	81343	6-6-4 120425BA	TEE, PIPE TO TUBE.............	1
13	PAOZZ	4730001423075	81343	6-2 1 20102RA	ADAPTER, STRAIGHT, PI.............	2

END OF FIGURE

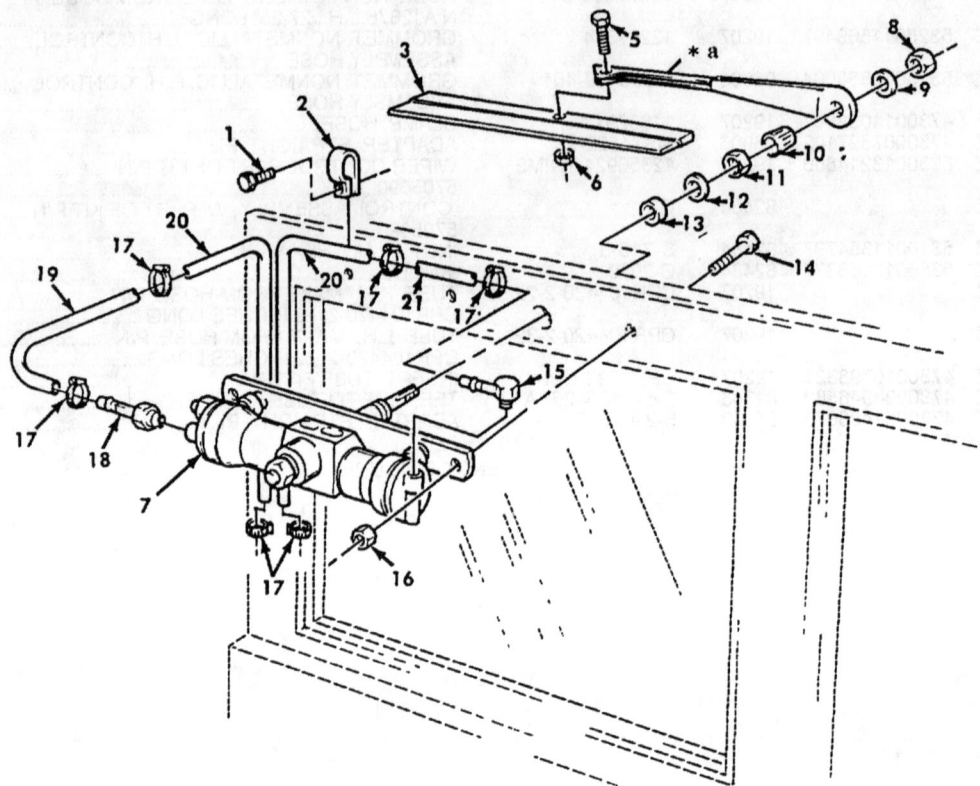

* a PART OF ITEM 4

Figure 539. Windshield Wiper Arm, Blade, and Motor Assembly (Old Style).

(1) ITEM NO	(2) SMR CODE	(3) NSN	(4) CAGEC	(5) PART NUMBER	(6) DESCRIPTION AND USABLE ON CODES (UOC)	(7) QTY
					GROUP 2202 ACCESSORY ITEMS	
					FIG. 539 WINDSHIELD WIPER ARM, BLADE, AND MOTOR ASSEMBLY(OLD STYLE)	
1	PAOZZ	5305008550957	96906	MS24629-46	SCREW, TAPPING	6
2	PAOZZ	5340004942234	96906	MS21334-3	CLAMP, LOOP	6
3	PAOZZ	2540004813637	60703	M874-44EC	BLADE, WINDSHIELD WI	2
4	PAOZZ	2540011236823	19207	11669624	ARM, WINDSHIELD WIPE	2
5	PAOZZ	5305011642310	60703	1951-7-BD	.SCREW, MACHINE	2
6	PAOZZ	5310011632472	82484	82845-JD	.NUT, SELF-LOCKING, HE	2
7	XBOZZ		19207	11669618-1	MOTOR, WINDSHIELD WI FOR MOTOR REPLACEMENT ORDER KIT P/N 5705690	2
8	PAOZZ	5310012042002	82484	91522-J	.NUT, PLAIN, HEXAGON	2
9	PAOZZ	5310001670721	96906	MS35333-41	.WASHER, LOCK	2
10	PAOZZ	5365009462231	60703	86324-1J	.BUSHING, TAPERED	2
11	PAOZZ	5310008041209	60703	76115J	.NUT, PLAIN, HEXAGON	2
12	PAOZZ	5310010284848	60703	77121-3JD	.WASHER, FLAT	2
13	PAOZZ	5310007607493	82484	77121-1	.WASHER, FLAT	2
14	PAOZZ	5305009953442	96906	MS35207-268	SCREW, MACHINE	4
15	PAOZZ	4730011916433	96906	MS24519-2	ELBOW, PIPE TO HOSE	2
16	PAOZZ	5310002632862	96906	MS21045-C3	NUT, SELF-LOCKING, HE	4
17	PAOZZ	4730011098001	19207	11662913	CLAMP, HOSE	12
18	PAOZZ	4730007827102	96906	MS24522-2	ADAPTER, STRAIGHT, PI	2
19	MOOZZ		19207	12277085-2	HOSE, NONMETALLIC MAKE FROM HOSE, P/N A12876	2
20	MOOZZ		19207	8689206-26	TUBE, METALLIC MAKE FROM TUBING, P/N 8689206, 26 INCHES LONG	4
21	MOOZZ		19207	12277085-1 XX	HOSE, NONMETALLIC MAKE FROM HOSE, P/N A12876	2

END OF FIGURE

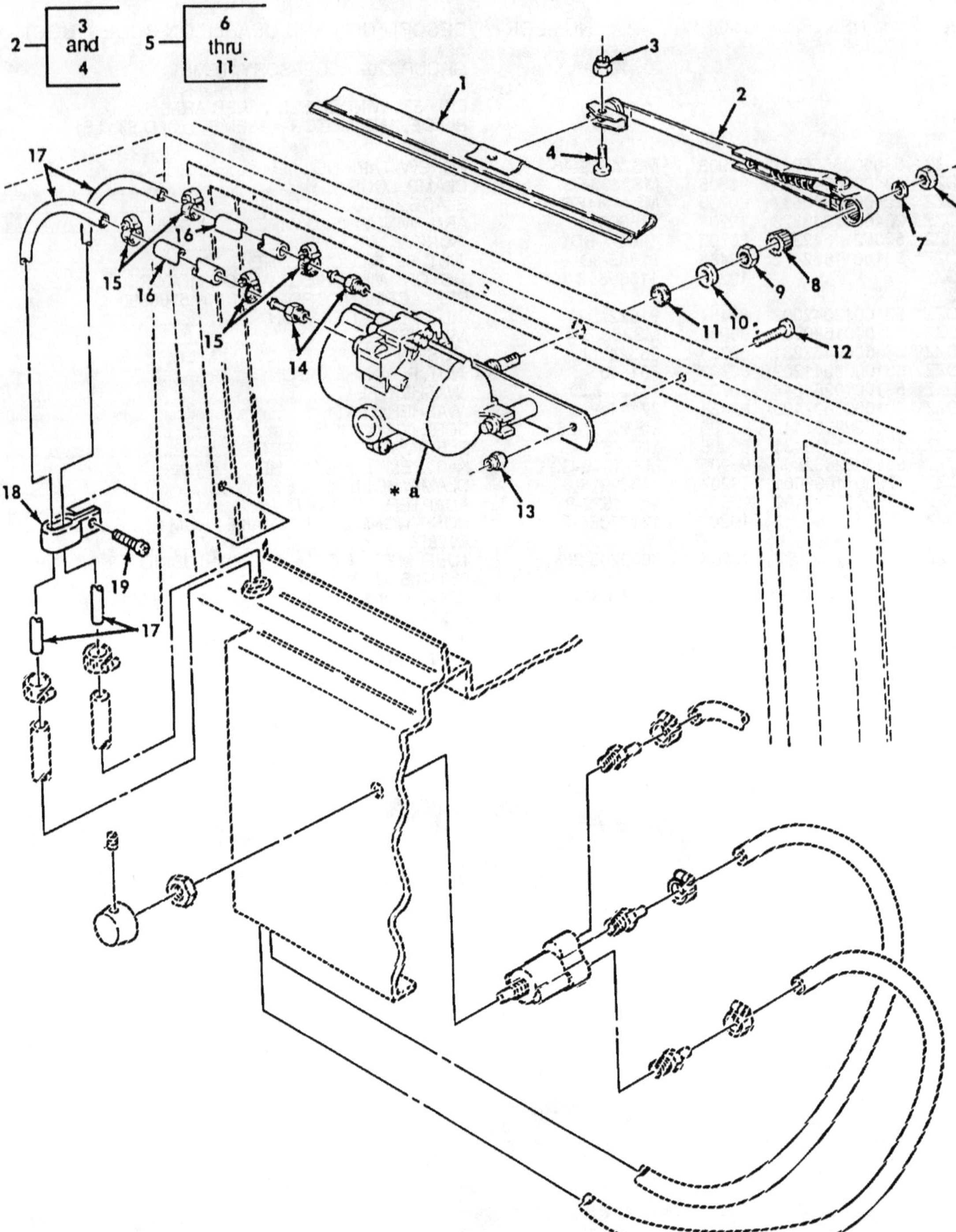

* a PART OF ITEM 5

Figure 540. Windshield Wiper Arm, Blade, and Motor Assembly (New Style).

(1) ITEM NO	(2) SMR CODE	(3) NSN	(4) CAGEC	(5) PART NUMBER	(6) DESCRIPTION AND USABLE ON CODES (UOC)	(7) QTY
					GROUP 2202 ACCESSORY ITEMS	
					FIG. 540 WINDSHIELD WIPER ARM, BLADE AND MOTOR ASSEMBLY (LATER VEHICLES)	
1	PAOZZ	2540004813637	60703	M874-44EC	BLADE, WINDSHIELD WI	2
2	PAOZZ	5305000546655	96906	MS51957-31	.SCREW, MACHINE 6-32 X 9/16	2
3	PAOZZ	5310001766341	96906	MS17830-06C	.NUT, SELF-LOCKING, HE 6-32	2
4	PAOZZ	2540011236823	19207	11669624	ARM, WINDSHIELD WIPE	2
5	PAOZZ	2540013104854	82484	12356925	MOTOR, WINDSHIELD PART OF KIT P/N 5705690	2
6	PAOZZ	5310008807746	96906	MS51968-5	.NUT, PLAIN, HEXAGON 5/16-24	2
7	PAOZZ	5310001670721	96906	MS35333-41	.WASHER, LOCK 5/16 IN SIZE	2
8	PAOZZ	5365009462231	82484	S-2280	.BUSHING, TAPERED	2
9	XAOZZ		82484	S-3725	.NUT	2
10	PFOZZ	5310008094085	96906	MS27183-16	.WASHER, FLAT	2
11	XAOZZ		82484	S-3723	.GASKET, LEATHER	2
12	PAOZZ	5305009953442	96906	MS35207-268	SCREW, MACHINE WIPER MOTOR ASSEMBLY.	4
13	PAOZZ	5310002632862	96906	MS21045-C3	NUT, SELF-LOCKING, HE	4
14	PAOZZ	4730007827102	96906	MS24522-2	ADAPTER, STRAIGHT, PI. PART OF KIT P/N 5705690	4
15	PAOZZ	4730011098001	19207	11662913	CLAMP, HOSE	12
16	MOOZZ		19207	12277085-1	HOSE, NONMETALLIC, R.H, MAKE FROM HOSE, P/N A12876 15 INCHES LONG PART OF KIT P/N 5705690	2
16	MOOZZ		19207	12277085-5	HOSE, NONMETALLIC, L.H., MAKE FROM. HOSE, P/N A12876 14 INCHES LONG PART OF KIT P/N 5705690	2
17	MOOZZ		19207	8689206-26	TUBE, METALLIC MAKE FROM HOSE, P/N 8689206	4
18	PAOZZ	5340004942234	96906	MS21334-3	CLAMP, LOOP	6
19	PAOZZ	5305008550957	96906	MS24629-46	SCREW, TAPPING	6

END OF FIGURE

540-1

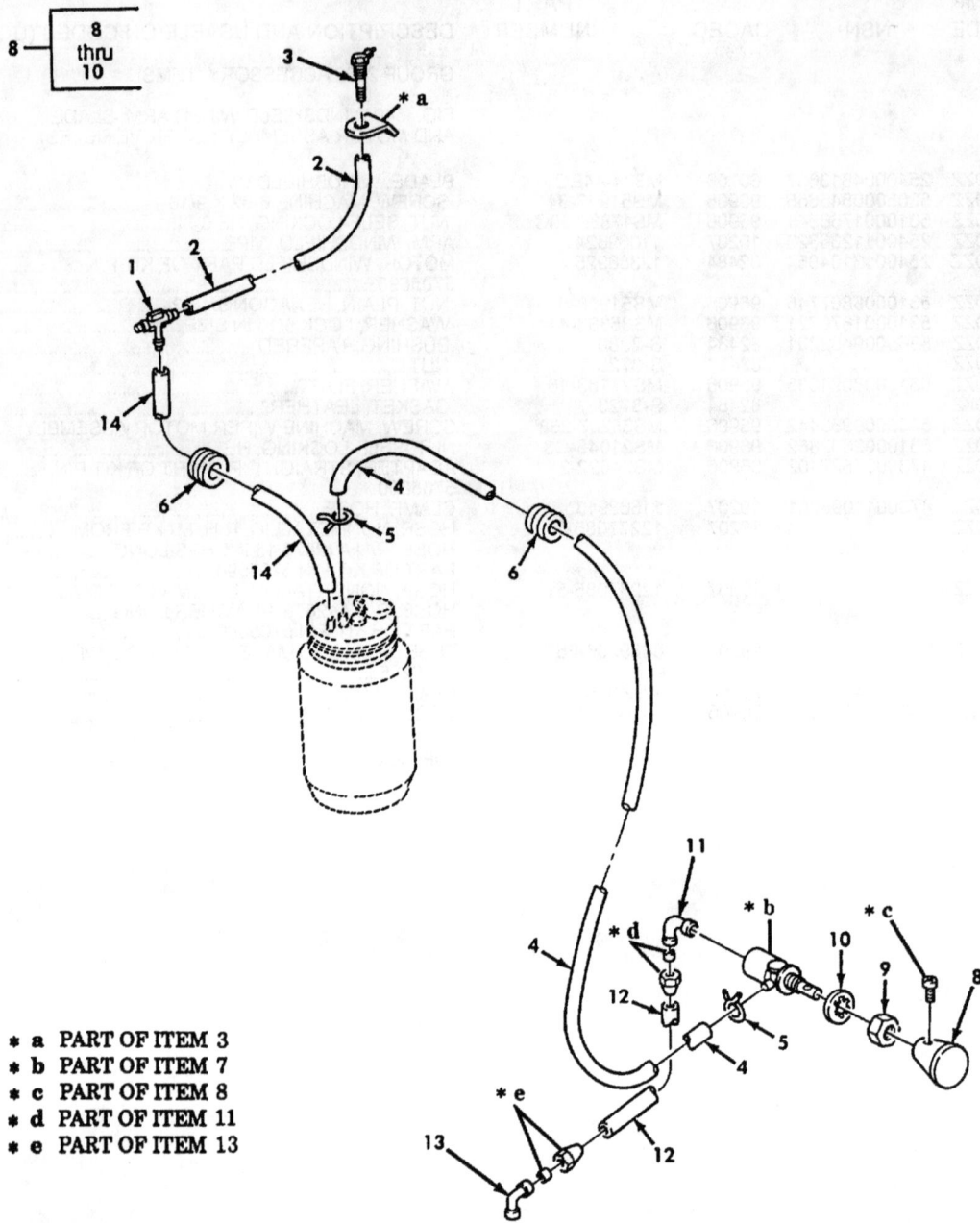

* a PART OF ITEM 3
* b PART OF ITEM 7
* c PART OF ITEM 8
* d PART OF ITEM 11
* e PART OF ITEM 13

Figure 541. Washer Bottle, Jet, and Control Assembly.

(1) ITEM NO	(2) SMR CODE	(3) NSN	(4) CAGEC	(5) PART NUMBER	(6) DESCRIPTION AND USABLE ON CODES (UOC)	(7) QTY
					GROUP 2202 ACCESSORY ITEMS	
					FIG. 541 WASHER BOTTLE, JET, AND CONTROL ASSEMBLY	
1	PAOZZ	4730010894370	19207	8738000	TEE, HOSE	1
2	MOOZZ		19207	12277246-2X 20	TUBING, NONMETALLIC MAKE FROM TUBING, P/N 211-0114-300, 20 INCHES LONG	2
3	PAOZZ	4730010831110	19207	8737999	NOZZLE, SPRAY, FLUID-	2
4	MOOZZ		19207	12277246-3X 33	HOSE, NONMETALLIC MAKE FROM TUBING, P/N 211-0114-300, 33 INCHES LONG	1
5	PAOZZ	4730010900258	19207	12256080	CLAMP, HOSE	2
6	PAOZZ	5325002496345	70485	2758	GROMMET, NONMETALLIC	2
7	PAOZZ	2540011067121	60703	89515-16	CONTROL, WASHER, WIND	1
8	XAOZZ		60703	88944-17	.KNOB	1
9	PAOZZ	5310011353464	60703	F1343-1ZT	.NUT, PLAIN, HEXAGON	1
10	PAOZZ	5310011354835	82484	87549-1	.WASHER, LOCK	1
11	PAOZZ	4730002871604	81343	6-2 120202BA	ELBOW, PIPE TO TUBE	1
12	MOOZZ	4720010144915	19207	CPR104420-2X42IN	HOSE, NONMETALLIC MAKE FROM HOSE, P/N CPR104420-2, 42 INCHES LONG	1
13	PAOZZ	4730000691187	81343	6-4 100202BA	ELBOW, PIPE TO TUBE	1
14	MOOZZ		19207	12277246-1X 37	TUBING, NONMETALLIC MAKE FROM TUBING, P/N 211-0114-300, 37 INCHES LONG	1

END OF FIGURE

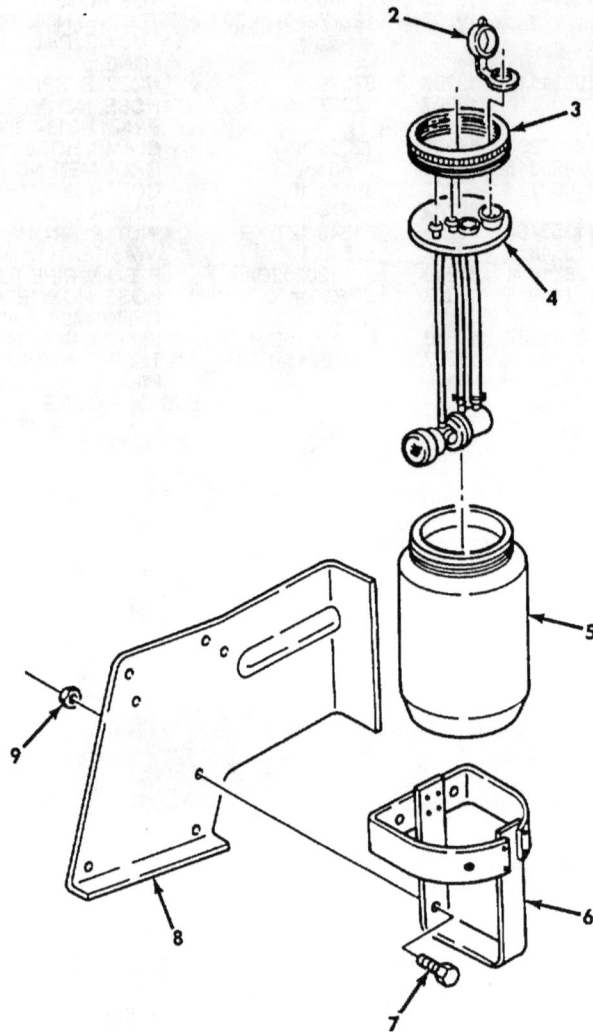

Figure 542. Washer Bottle and Mounting Brackets.

(1) ITEM NO	(2) SMR CODE	(3) NSN	(4) CAGEC	(5) PART NUMBER	(6) DESCRIPTION AND USABLE ON CODES (UOC)	(7) QTY
					GROUP 2202 ACCESSORY ITEMS	
					FIG. 542 WASHER BOTTLE AND MOUNTING BRACKETS	
1	PAOOO	2540011010010	60703	87900-112	WINDSHIELD WASHER A..	1
2	XAOZZ		19207	12375629	.CAP-PLUG, PROTECTIVE	1
3	PAOZZ	8125013413838	19207	12375627	.CAP, SCREW, JAR...	1
4	PAOZZ	2540013398594	19207	12375628	.WINDSHIELD WASHER A..	1
5	XAOZZ		19207	12375625	.RESERVOIR, WINDSHIEL..	1
6	XAOZZ		19207	12375626	.BRACKET, VEHICULAR C..	1
7	PAOZZ	5305009897435	96906	MS35207-264	SCREW, MACHINE	3
8	PFOZZ	5340011448676	19207	12277371	BRACKET, MOUNTING...	1
9	PAOZZ	5310008775797	96906	MS21044-N3	NUT, SELF-LOCKING, HE...	3

END OF FIGURE

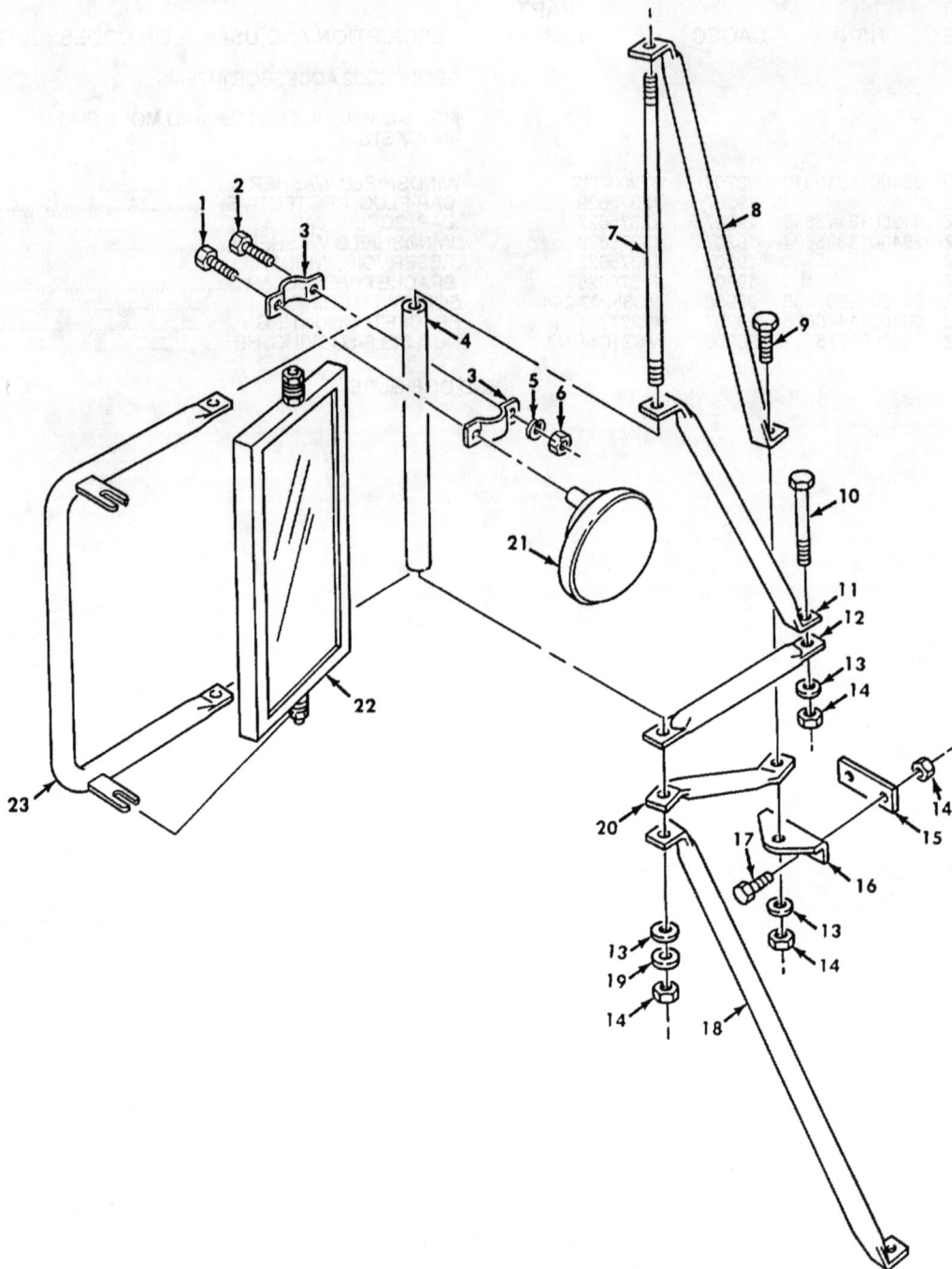

Figure 543. Rearview Mirrors.

(1) ITEM NO	(2) SMR CODE	(3) NSN	(4) CAGEC	(5) PART NUMBER	(6) DESCRIPTION AND USABLE ON CODES (UOC)	(7) QTY
					GROUP 2202 ACCESSORY ITEMS	
					FIG. 543 REARVIEW MIRRORS	
1	PAOZZ	5306000425841	24617	425B41	BOLT, ASSEMBLED WASH...	2
2	PAOZZ	5306000680513	60285	6893-2	BOLT, MACHINE..	2
3	PAOZZ	5340011654546	19207	12300828	STRAP, RETAINING...	4
4	PAOZZ	2540009336267	19207	11608925	ARM, REARVIEW MIRROR	2
5	PAOZZ	5310005825965	96906	MS35338-44	WASHER, LOCK..	2
6	PAOZZ	5310009359022	96906	MS51943-32	NUT, SELF-LOCKING, HE...	2
7	PAOZZ	5306001891775	19207	11608936	ROD, THREADED END...	2
8	PAOZZ	5340010827448	19207	12255995	BRACKET, MOUNTING..	2
9	PAOZZ	5305002693239	80204	B1821BH038F138N	SCREW, CAP, HEXAGON H VEHICULAR BRACE	2
10	PAOZZ	5306004094066	19207	11608931	BOLT, SHOULDER VEHICULAR BRACE...........................	4
11	PAOZZ	5340010822523	19207	12255997	BRACKET, MOUNTING..	2
12	PAOZZ	5340010822522	19207	12255999	BRACKET, MOUNTING..	2
13	PAOZZ	5310008094058	96906	MS27183-10	WASHER, FLAT...	10
14	PAOZZ	5310009359022	96906	MS51943-32	NUT, SELF-LOCKING, HE...	10
15	PAOZZ	2540009336262	19207	11608938	PLATE, COWL..	2
16	PAOZZ	254b009336263	19207	11608933	BRACKET, CLIP...	2
17	PAOZZ	5305000680516	80204	B1821BH025F113N	SCREW, CAP, HEXAGON H	4
18	PAOZZ	2510010823630	19207	12255994	FRAME SECTION, STRUC..	2
19	PAOZZ	5330004832408	19207	7372705	GASKET VEHICULAR BRACE...	4
20	PAOZZ	5340010858136	19207	12255996	BRACKET, MOUNTING..	2
21	PAOZZ	2540011654677	19207	12300829	MIRROR HEAD, VEHICUL...	2
22	PAOZZ	2540007885637	96906	MS53015-2	MIRROR ASSEMBLY, REA..	2
23	PAOZZ	5340010835406	19207	12255998	BRACKET, MOUNTING..	2

END OF FIGURE

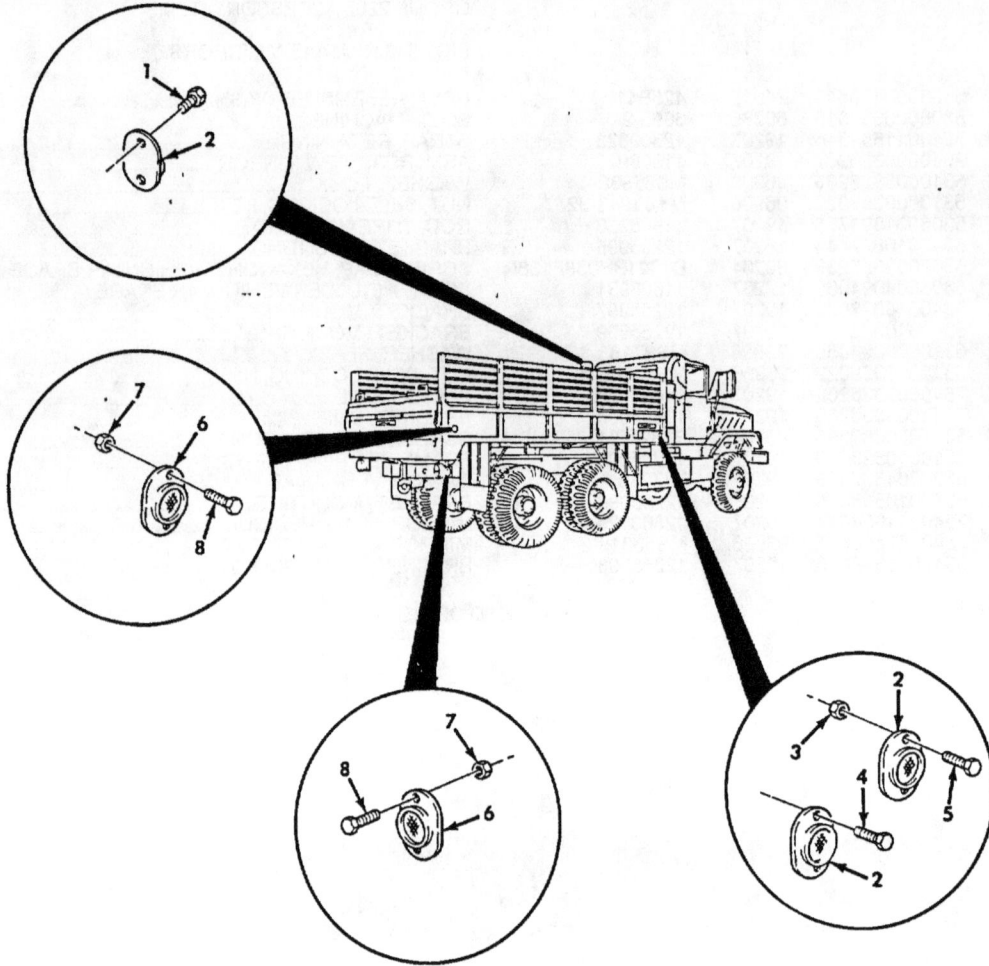

Figure 544. Cargo Body Reflectors and Mounting Hardware.

(1) ITEM NO	(2) SMR CODE	(3) NSN	(4) CAGEC	(5) PART NUMBER	(6) DESCRIPTION AND USABLE ON CODES (UOC)	(7) QTY
					GROUP 2202 ACCESSORY ITEMS	
					FIG. 544 CARGO BODY REFLECTORS AND MOUNTING HARDWARE	
1	PAOZZ	5305000526920	96906	MS24629-56	SCREW, TAPPING .... UOC:DAW, DAX, V14, V15, ZAA, ZAB	2
2	PAOZZ	9905002023639	96906	MS35387-2	REFLECTOR, INDICATIN AMBER .... UOC:DAA, DAB, DAC, DAD, DAW, DAX, V12, V13, V14, V15, V16, V17, ZAA, ZAB, ZAC, ZAD	4
3	PAOZZ	5310000881251	96906	MS51922-1	NUT, SELF-LOCKING, HE .... UOC:DAW, DAX, V14, V15, ZAA, ZAB	4
4	PAOZZ	5305000526920	96906	MS24629-56	SCREW, TAPPING .... UOC:DAW, DAX, V14, V15, ZAA, ZAB	8
5	PAOZZ	5305009881724	96906	MS35206-280	SCREW, MACHINE .... UOC:DAW, DAX, V14, V15, ZAA, ZAB	4
6	PAOZZ	9905002052795	96906	MS35387-1	REFLECTOR, INDICATIN RED .... UOC:DAA, DAB, DAC, DAD, DAW, DAX, V12, V13, V14, V15, V16, V17, ZAA, ZAB, ZAC, ZAD	4
7	PAOZZ	5310000881251	96906	MS51922-1	NUT, SELF-LOCKING, HE .... UOC:DAA, DAB, DAC, DAD, DAW, DAX, V12, V13, V14, V15, V16, V17, ZAA, ZAB, ZAC, ZAD	8
8	PAOZZ	5305009932738	96906	MS35207-280	SCREW, MACHINE .... UOC:DAA, DAB, DAC, DAD, DAW, DAX, V12, V13, V14, V15, V16, V17, ZAA, ZAB, ZAC, ZAD	8

END OF FIGURE

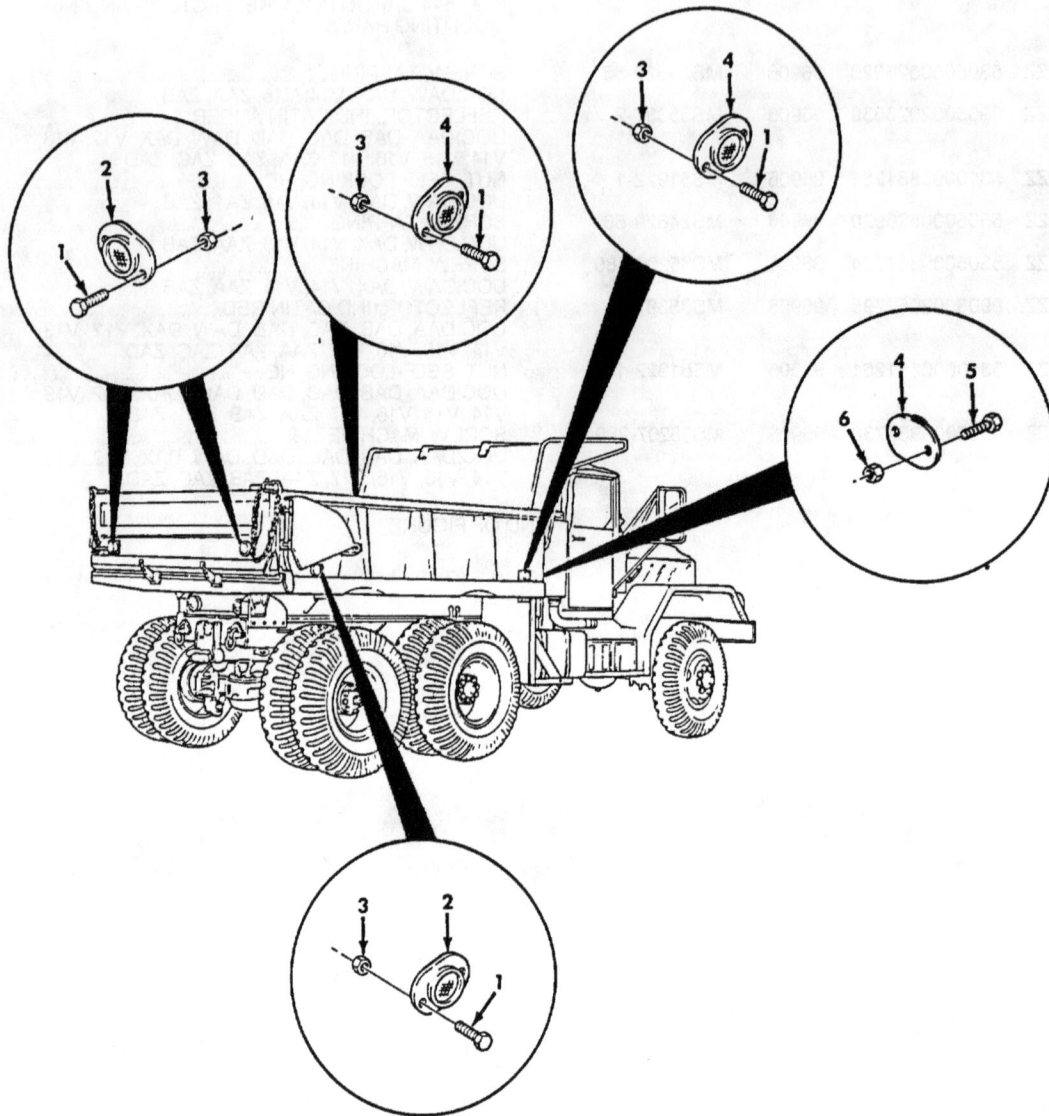

Figure 545. Dump Body Reflectors a Mounting Hardware.

(1) ITEM NO	(2) SMR CODE	(3) NSN	(4) CAGEC	(5) PART NUMBER	(6) DESCRIPTION AND USABLE ON CODES (UOC)	(7) QTY
					GROUP 2202 ACCESSORY ITEMS	
					FIG. 545 DUMP BODY REFLECTORS AND MOUNTING HARDWARE	
1	PAOZZ	5305009881724	96906	MS35206-280	SCREW, MACHINE .... UOC: DAE, DAF, V19, V20, ZAE, ZAF	16
2	PAOZZ	9905002052795	96906	MS35387-1	REFLECTOR, INDICATIN RED .... UOC: DAE, DAF, V19, V20, ZAE, ZAF	4
3	PAOZZ	5310000881251	96906	MS51922-1	NUT, SELF-LOCKING, HE .... UOC: DAE, DAF, V19, V20, ZAE, ZAF	16
4	PAOZZ	9905002023639	96906	MS35387-2	REFLECTOR, INDICATIN AMBER .... UOC: DAE, DAF, V19, V20, ZAE, ZAF	6
5	PAOZZ	5305009932738	96906	MS35207-280	SCREW, MACHINE .... UOC: DAE, DAF, V19, V20, ZAE, ZAF	4
6	PAOZZ	5310009359022	96906	MS51943-32	NUT, SELF-LOCKING, HE .... UOC: DAE, DAF, V19, V20, ZAE, ZAF	4

END OF FIGURE

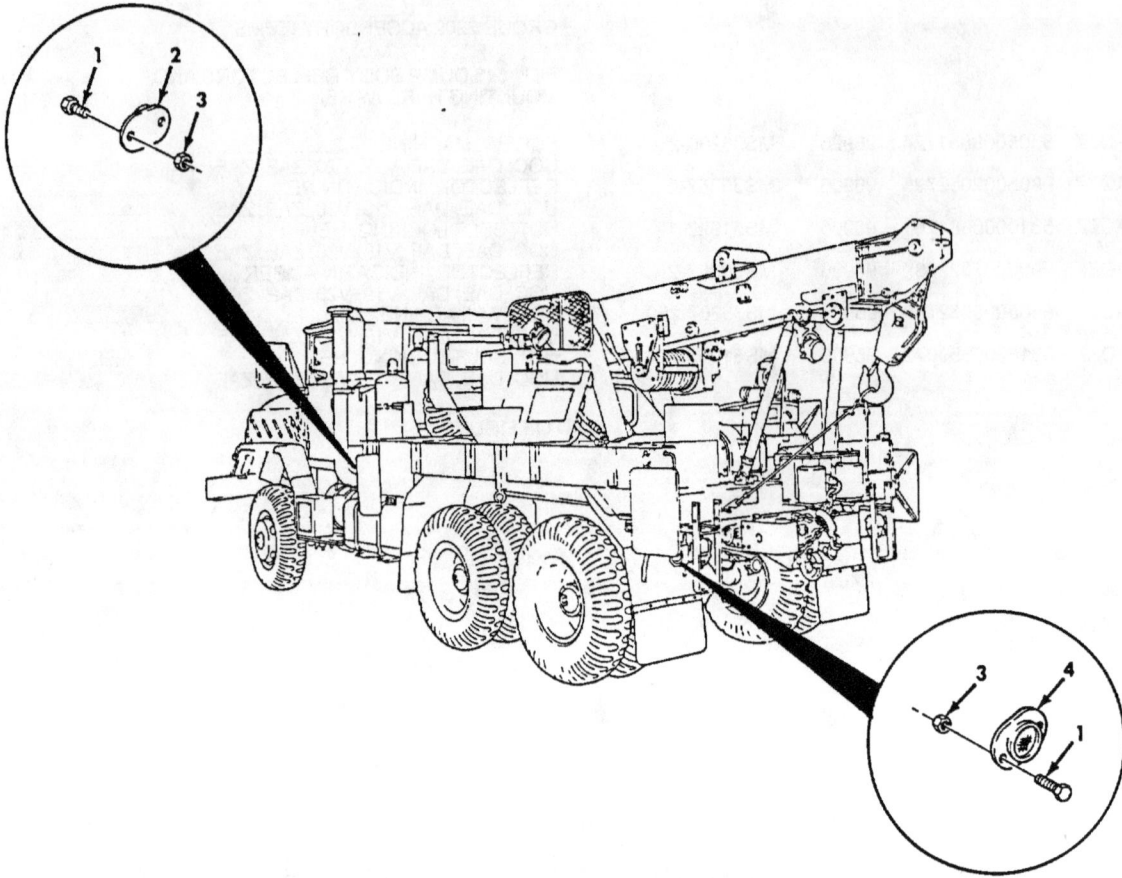

Figure 546. Wrecker Body Reflectors and Mounting Hardware.

(1) ITEM NO	(2) SMR CODE	(3) NSN	(4) CAGEC	(5) PART NUMBER	(6) DESCRIPTION AND USABLE ON CODES (UOC)	(7) QTY
					GROUP 2202 ACCESSORY ITEMS	
					FIG. 546 WRECKER BODY REFLECTORS AND MOUNTING HARDWARE	
1	PAOZZ	5305009881725	96906	MS35206-281	SCREW, MACHINE .... UOC: DAL, V18, ZAL	8
2	PAOZZ	9905002023639	96906	MS35387-2	REFLECTOR, INDICATIN AMBER .... UOC: DAL, V18, ZAL	2
3	PAOZZ	5310000881251	96906	MS51922-1	NUT, SELF-LOCKING, HE .... UOC: DAL, V18, ZAL	8
4	PAOZZ	9905002052795	96906	MS35387-1	REFLECTOR, INDICATIN RED .... UOC: DAL, V18, ZAL	2

END OF FIGURE

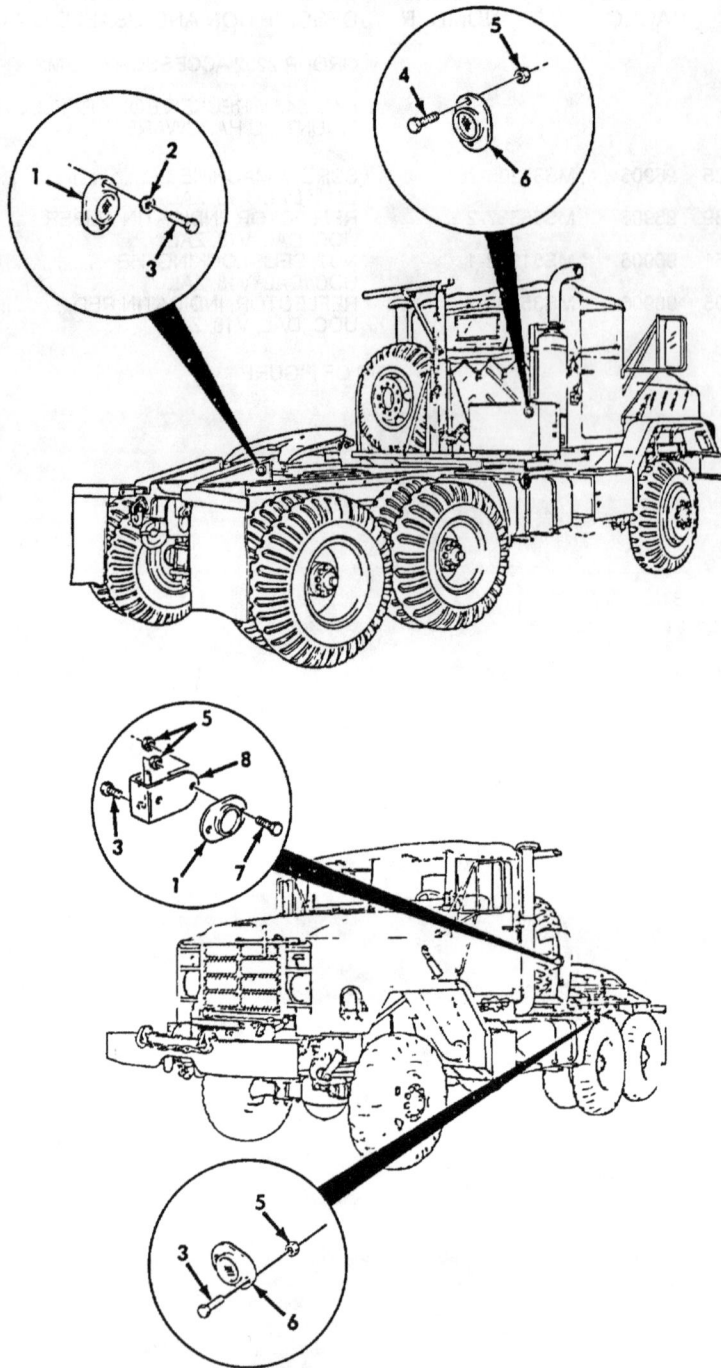

Figure 547. Tractor Reflectors and Mounting Hardware.

(1) ITEM NO	(2) SMR CODE	(3) NSN	(4) CAGEC	(5) PART NUMBER	(6) DESCRIPTION AND USABLE ON CODES (UOC)	(7) QTY
					GROUP 2202 ACCESSORY ITEMS	
					FIG. 547 TRACTOR REFLECTORS AND MOUNTING HARDWARE	
1	PAOZZ	9905002052795	96906	MS35387-1	REFLECTOR, INDICATIN .... UOC: DAG, DAH, V21, V22, ZAG, ZAH	2
2	PAOZZ	5310005825965	96906	MS35338-44	WASHER, LOCK .... UOC: DAG, DAH, V21, V22, ZAG, ZAH	4
3	PAOZZ	5305009932461	96906	MS35207-281	SCREW, MACHINE .... UOC: DAG, DAH, V21, V22, ZAG, ZAH	4
4	PAOZZ	5305009932738	96906	MS35207-280	SCREW, MACHINE .... UOC: DAG, DAH, V21, V22, ZAG, ZAH	8
5	PAOZZ	5310009359022	96906	MS51943-32	NUT, SELF-LOCKING, HE .... UOC: DAG, DAH, V21, V22, ZAG, ZAH	10
6	PAOZZ	9905002023639	96906	MS35387-2	REFLECTOR, INDICATIN ...... UOC: DAG, DAH, V21, V22, ZAG, ZAH	4
7	PAOZZ	5306000680513	60285	6893-2	BOLT, MACHINE .... UOC: DAG, DAH, V21, V22, ZAG, ZAH	2
8	PFOZZ	6210010893037	19207	8758198	HOLDER, REFLECTOR .... UOC: DAG, DAH, V21, V22, ZAG, ZAH	1

END OF FIGURE

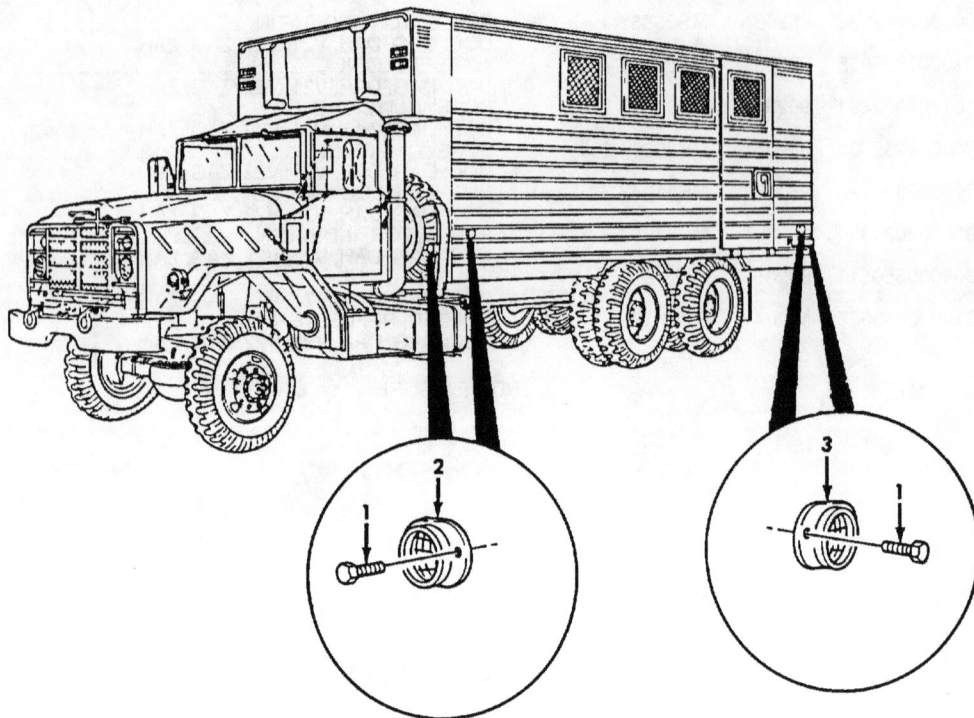

Figure 548. Van Body Reflectors and Mounting Hardware.

(1) ITEM NO	(2) SMR CODE	(3) NSN	(4) CAGEC	(5) PART NUMBER	(6) DESCRIPTION AND USABLE ON CODES (UOC)	(7) QTY
					GROUP 2202 ACCESSORY ITEMS	
					FIG. 548 VAN BODY REFLECTORS AND MOUNTING HARDWARE	
1	PAOZZ	5305004324252	96906	MS51861-66	SCREW, TAPPING .... UOC: DAJ, DAK, V24, V25, ZAJ, ZAK	16
2	PAOZZ	9905002023639	96906	MS35387-2	REFLECTOR, INDICATIN AMBER .... UOC: DAJ, DAK, V24, V25, ZAJ, ZAK	4
3	PAOZZ	9905002052795	96906	MS35387-1	REFLECTOR, INDICATIN RED .... UOC: DAJ, DAK, V24, V25, ZAJ, ZAK	4

END OF FIGURE

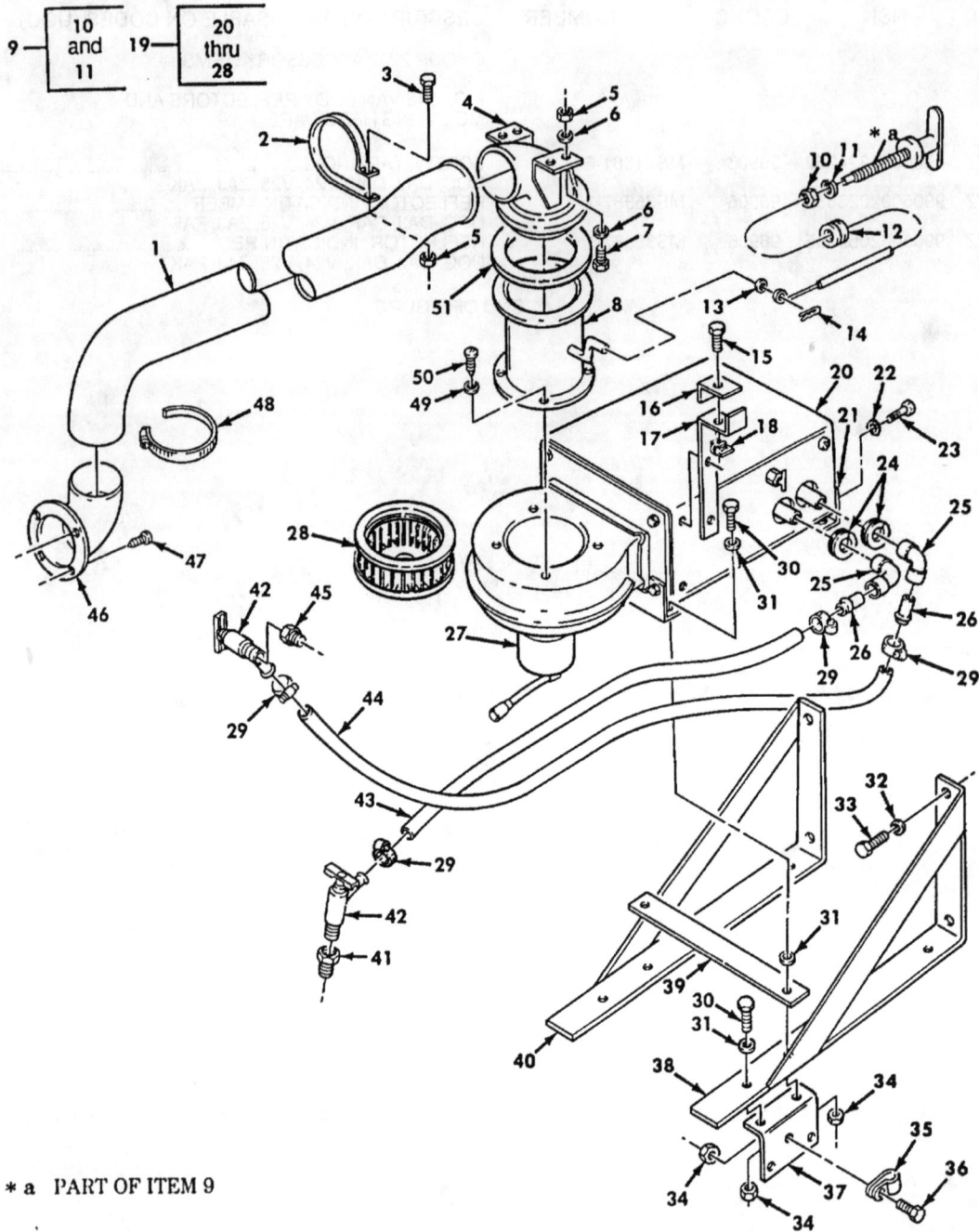

* a PART OF ITEM 9

Figure 549. Hot Water Heater, Hoses, Control Assembly, and Mounting Hardware.

SECTION II

(1) ITEM NO	(2) SMR CODE	(3) NSN	(4) CAGEC	(5) PART NUMBER	(6) DESCRIPTION AND USABLE ON CODES (UOC)	(7) QTY

GROUP 2207 PERSONNEL HEATER

FIG. 549 HOT WATER HEATER, HOSES, CONTROL ASSEMBLY, AND MOUNTING HARDWARE

1	PAOZZ	4720008092430	34623	976683	HOSE, AIR DUCT	1
2	PAOZZ	5340011278703	19207	7700263-2	CLAMP, LOOP	2
3	PAOZZ	5306000680513	60285	6893-2	BOLT, MACHINE	2
4	PAOZZ	2540010831109	19207	12255941	VENTILATOR, AIR CIRC	1
5	PAOZZ	5310009359022	96906	MS51943-32	NUT, SELF-LOCKING, HE	8
6	PAOZZ	5310008094058	96906	MS27183-10	WASHER, FLAT	8
7	PAOZZ	5305009932461	96906	MS35207-281	SCREW, MACHINE	4
8	PAOZZ	2540011380925	19207	12255935	CANISTER ASSEMBLY, F	1
9	PAOZZ	2590011097992	19207	12277189-3	CONTROL ASSEMBLY, PU	1
10	PAOZZ	5310000581626	96906	MS35650-3382	.NUT, PLAIN, HEXAGON	1
11	PAOZZ	5310006379541	96906	MS35338-46	.WASHER, LOCK	1
12	PAOZZ	5325007541072	96906	MS35489-138	GROMMET, NONMETALLIC	1
13	PAOZZ	5310011222060	79136	5305-18	PUSH ON NUT	1
14	PAOZZ	5315010578371	81352	AN415-4	PIN, LOCK	1
15	PAOZZ	5305009846212	96906	MS35206-265	SCREW, MACHINE	1
16	PAOZZ	5340011066751	19207	7700246	BRACKET, ANGLE	1
17	PFOZZ	5340010893076	19207	12256009	BRACKET, DOUBLE ANGL	1
18	PAOZZ	5310011277599	96906	MS90723-38	NUT, SHEET SPRING	1
19	PAOZZ	2540011908484	19207	12255989-1	HEATER, VEHICULAR, CO	1
20	PAOZZ	2540000208591	96906	MS51326-1	.HEATER	1
21	PAOZZ	2540010864674	19207	12255924	.SUPPORT, HOT WATER H	2
22	PAOZZ	5310008892527	96906	MS45904-72	.WASHER, LOCK	4
23	PAOZZ	5306002264822	80204	B1821BH031C050N	.BOLT, MACHINE	4
24	PAOZZ	5325012427083	96906	MS35489-17	.GROMMET	2
25	PAOZZ	4730002546450	96906	MS35917-5	.ELBOW	2
26	PAOZZ	4730001413164	19207	11648601	.ADAPTER, STRAIGHT, TU	2
27	PAOZZ	6105005129225	92878	30040-01	.MOTOR, DIRECT CURREN	1
28	PAOZZ	4140003180161	60399	424-228	.IMPELLER, FAN, CENTRI	1
29	PAOZZ	4730012448434	76599	4820SS	CLAMP, HOSE	4
30	PAOZZ	5306000514077	80204	B1821BH031F113N	BOLT, MACHINE	4
31	PAOZZ	5310000814219	96906	MS27183-12	WASHER, FLAT	6
32	PAOZZ	5310004079566	96906	MS35338-45	WASHER, LOCK	4
33	PAOZZ	5306000514081	80204	B1821BH031F175N	BOLT, MACHINE	4
34	PAOZZ	5310002416658	96906	MS51943-34	NUT, SELF-LOCKING, HE	5
35	PAOZZ	5340007255267	96906	MS21333-115	CLAMP, LOOP	1
36	PAOZZ	5305002259092	96906	MS90726-37	SCREW, CAP, HEXAGON H	5
37	PFOZZ	5340011479822	19207	12277397	BRACKET, ANGLE	1
38	PFOZZ	5340010823594	19207	12277395-1	BRACKET, MOUNTING	1
39	PFOZZ	5340011606919	19207	12277396	BAND, RETAINING	1
40	PFOZZ	5340010823593	19207	12277395-2	BRACKET, MOUNTING	1
41	PAOZZ	4730001960936	21450	115224	BUSHING, PIPE	1
42	PAOZZ	4820011010080	19207	10923515	COCK, DRAIN	2
43	MOOZZ		19207	8710557-32	HOSE, NONMETALLIC MAKE FROM HOSE, P/N 8710557, 32 INCHES LONG	1
44	MOOZZ		19207	8710557-28	HOSE, NONMETALLIC MAKE FROM HOSE, P/N 8710557, 22 INCHES LONG	1
45	PAOZZ	4730008176578	96906	MS14315-4	BUSHING, PIPE	1

(1) ITEM NO	(2) SMR CODE	(3) NSN	(4) CAGEC	(5) PART NUMBER	(6) DESCRIPTION AND USABLE ON CODES (UOC)	(7) QTY
46	PAOZZ	2540001778108	19207	7951057	ADAPTER, VEHICLE STO	1
47	XDOZZ		96906	MS51861-65	SCREW, TAPPING	4
48	PAOZZ	4730010674711	19207	11608950-16	CLAMP, HOSE	2
49	PAOZZ	5310005825965	96906	MS35338-44	WASHER, LOCK	4
50	PAOZZ	5305000680500	96906	MS90725-3	SCREW, CAP, HEXAGON H	5
51	PAOZZ	5330011089119	19207	12256004	SEAL, NONMETALLIC RO	1

END OF FIGURE

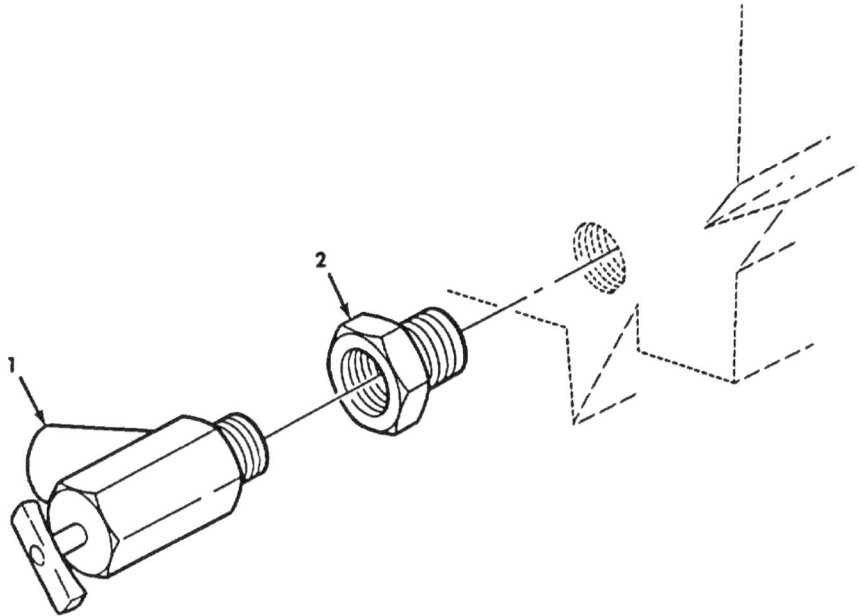

Figure 550. Hot Water Heater Value (M939A2).

(1) ITEM NO	(2) SMR CODE	(3) NSN	(4) CAGEC	(5) PART NUMBER	(6) DESCRIPTION AND USABLE ON CODES (UOC)	(7) QTY
					GROUP 2207 PERSONNEL HEATER	
					FIG. 550 HOT WATER HEATER VALVE (M939A2)	
1	PAOZZ	4820000268473	79470	131-60250	COCK, DRAIN .... UOC: ZAA, ZAB, ZAC, ZAD, ZAE, ZAF, ZAG, ZAH, ZAJ, ZAK, ZAL	2
2	PAOZZ	4730008176578	21450	120322	BUSHING, PIPE .... UOC: ZAA, ZAB, ZAC, ZAD, ZAE, ZAF, ZAG, ZAH, ZAJ, ZAK, ZAL	2

END OF FIGURE

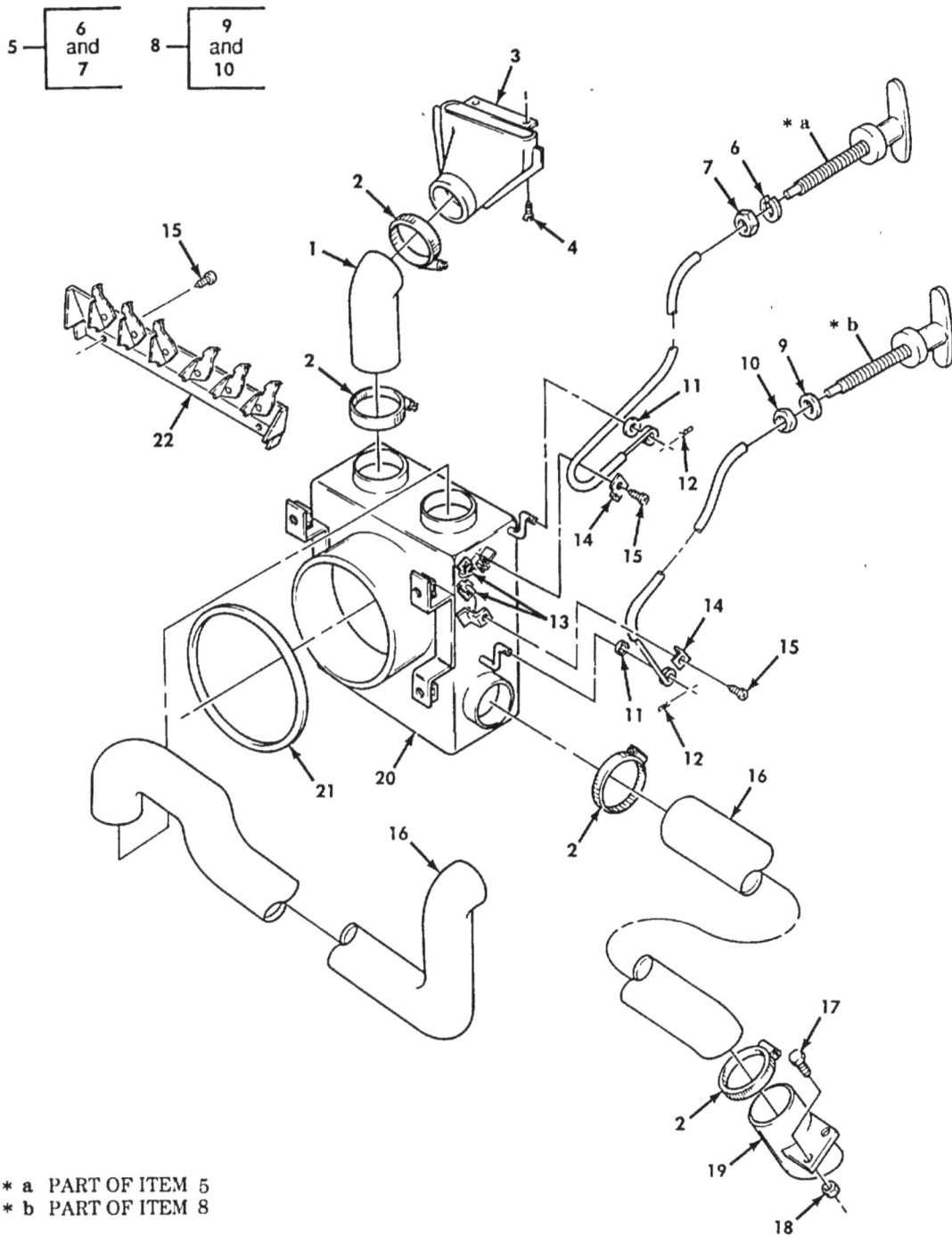

Figure 551. Hot Water Heater, Diverter, Air Ducts, and Controls.

* a PART OF ITEM 5
* b PART OF ITEM 8

(1) ITEM NO	(2) SMR CODE	(3) NSN	(4) CAGEC	(5) PART NUMBER	(6) DESCRIPTION AND USABLE ON CODES (UOC)	(7) QTY
					GROUP 2207 PERSONNEL HEATER	
					FIG. 551 HOT WATER HEATER, DIVERTER, AIR DUCTS, AND CONTROLS	
1	PAOZZ	4720010889680	19207	11648560-1	HOSE, AIR DUCT	1
2	XDOZZ		66295	C44PS	CLAMP, HOSE AIR DUCT	6
3	PFOZZ	2540010831116	19207	12255940	NOZZLE, DEFROSTER, VE	2
4	PAOZZ	5305009932738	96906	MS35207-280	SCREW, MACHINE	4
5	PAOZZ	2590011097990	19207	12277189-1	CONTROL ASSEMBLY, PU	1
6	PAOZZ	5310006379541	96906	MS35338-46	.WASHER, LOCK	1
7	PAOZZ	5310000581626	96906	MS35650-3382	.NUT, PLAIN, HEXAGON	1
8	PAOZZ	2590011097991	19207	12277189-2	CONTROL ASSEMBLY, PU	1
9	PAOZZ	5310006379541	96906	MS35338-46	.WASHER, LOCK	1
10	PAOZZ	5310000581626	96906	MS35650-3382	.NUT, PLAIN, HEXAGON	1
11	PAOZZ	5310011222060	19207	7951891	PUSH ON NUT	2
12	PAOZZ	5315010578371	81352	AN415-4	PIN, LOCK	2
13	PAOZZ	5310011277599	96906	MS90723-38	NUT, SHEET SPRING	2
14	PAOZZ	5340011066751	19207	7700246	BRACKET, ANGLE	2
15	PAOZZ	5305009846212	96906	MS35206-265	SCREW, MACHINE	2
16	PAOZZ	4720010889681	19207	11648560-2	HOSE, AIR DUCT	2
17	PAOZZ	5305000680504	96906	MS90726-3	SCREW, CAP, HEXAGON	2
18	PAOZZ	5305000680504	96906	MS51943-32	NUT, SELF-LOCKING	2
19	PAOZZ	2540010874741	19207	12256084	DEFLECTOR, VEHICULAR	1
20	PAOZZ	2540010831113	19207	12255949	DIVERTER, HEATER	1
21	PAOZZ	5330011087567	19207	12256005	GASKET	1
22	PAOZZ		19207	12277077	DEFROSTER ASSEMBLY	2

END OF FIGURE

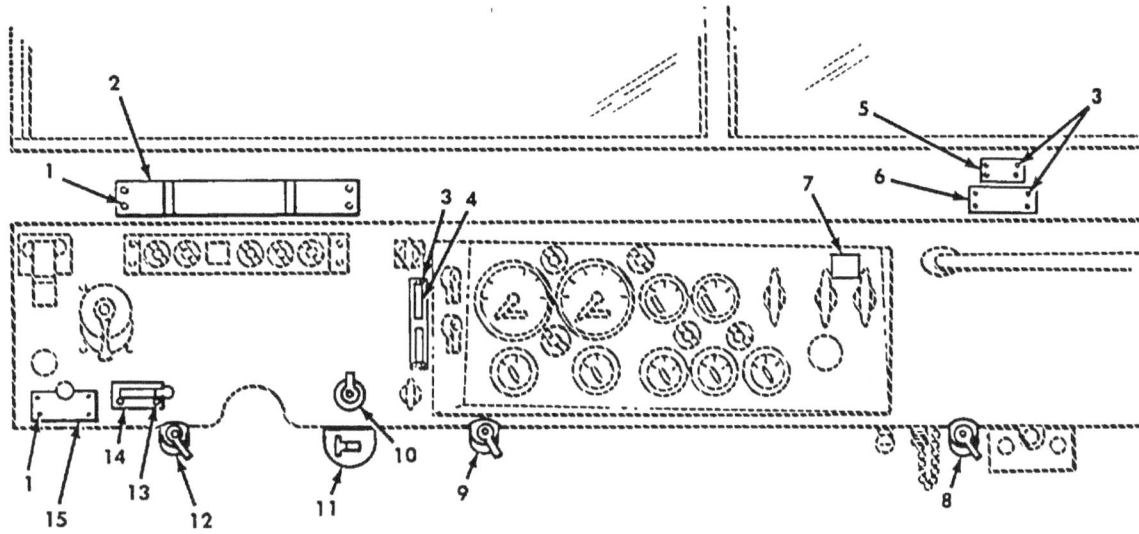

Figure 552. Data Plates, Instrument Panel.

(1) ITEM NO	(2) SMR CODE	(3) NSN	(4) CAGEC	(5) PART NUMBER	(6) DESCRIPTION AND USABLE ON CODES (UOC)	(7) QTY
					GROUP 2210 DATA PLATES	
					FIG. 552 DATA PLATES, INSTRUMENT PANEL	
1	PAOZZ	5305000526915	96906	MS24629-44	SCREW, TAPPING...........................	4
2	PAOZZ	9905011855787	19207	12277088	PLATE, INSTRUCTION V12, V13, V14, V15, V16, V17, V18, V19, V20, V21, V22, V24, V25, V39	1
2	PAOZZ	9905012269437	19207	12303061	PLATE, INSTRUCTION .... UOC: DAA, DAB, DAC, DAD, DAE, DAF, DAG, DAH, DAJ, DAK, DAL, DAW, DAX, ZAA, ZAB, ZAC, ZAD, ZAE, ZAF, ZAG, ZAH, ZAJ, ZAK, ZAL	1
3	PAOZZ	5305008550961	96906	MS24629-35	SCREW, TAPPING	10
4	PAOZZ	9905011195788	19207	12277147	PLATE, INSTRUCTION	1
5	PAOZZ	9905011150569	19207	12257056	PLATE, IDENTIFICATIO .... UOC: V39	1
5	PAOZZ	9905014189773	19207	12375718	PLATE, DESIGNATION .... UOC: V15	1
5	PAOZZ	9905014189774	19207	12375719	PLATE, DESIGNATION .... UOC: DAW	1
5	PAOZZ	9905014206755	19207	12375720	PLATE, DESIGNATION .... UOC: V13	1
5	PAOZZ	9905014189777	19207	12375722	PLATE, DESIGNATION .... UOC: V14	1
5	PAOZZ	9905014186628	19207	12375724	PLATE, DESIGNATION .... UOC: V12	1
5	PAOZZ	9905014200672	19207	12375725	PLATE, DESIGNATION .... UOC: DAB	1
5	PAOZZ	9905014189776	19207	12375727	PLATE, DESIGNATION .... UOC: DAC	1
5	PAOZZ	9905014188327	19207	12375728	PLATE, DESIGNATION .... UOC: V16	1
5	PAOZZ	9905014201451	19207	12375729	PLATE, DESIGNATION .... UOC: DAD	1
5	PAOZZ	9905014201764	19207	12375730	PLATE, DESIGNATION .... UOC: V20	1
5	PAOZZ	9905014189772	19207	12375731	PLATE, DESIGNATION .... UOC: DAE	1
5	PAOZZ	9905014188324	19207	12375732	PLATE, DESIGNATION .... UOC: V19	1
5	PAOZZ	9905014178434	19207	12375733	PLATE, DESIGNATION .... UOC: DAF	1
5	PAOZZ	9905014387063	19207	12375734	PLATE, DESIGNATION .... UOC: V22	1
5	PAOZZ		19207	12375736	PLATE, DESIGNATION .... UOC: V21	1
5	PAOZZ		19207	12375737	PLATE, DESIGNATION .... UOC: DAH	1
5	PAOZZ	9905014186622	19207	12375738	PLATE, DESIGNATION .... UOC: V24	1
5	PAOZZ	9905014188326	19207	12375739	PLATE, DESIGNATION .... UOC: DAJ	1

(1) ITEM NO	(2) SMR CODE	(3) NSN	(4) CAGEC	(5) PART NUMBER	(6) DESCRIPTION AND USABLE ON CODES (UOC)	(7) QTY
5	PAOZZ	9905014186633	19207	12375740	PLATE, DESIGNATION .... UOC: V25	1
5	PAOZZ	9905014200674	19207	12375741	PLATE, DESIGNATION .... UOC: DAK	1
5	PAOZZ	9905014186621	19207	12375742	PLATE, DESIGNATION .... UOC: V18	1
5	PAOZZ	9905014187651	19207	12375743	PLATE, DESIGNATION .... UOC: DAL	1
5	PAOZZ	9905012100229	19207	12302935	PLATE, IDENTIFICATIO .... UOC: DAX	1
6	PAOZZ	9905012104733	19207	12277090	PLATE, INSTRUCTION .... UOC: V19, V20, V21, V22	1
6	PAOZZ	9905012104734	19207	12277091	PLATE, IDENTIFICATIO .... UOC: V18	1
6	PAOZZ	9905011147598	19207	12257060	PLATE, IDENTIFICATIO .... UOC: V39	1
6	PAOZZ	9905011150570	19207	12256088	PLATE, IDENTIFICATIO .... UOC: V12, V13, V14, V15, V16, V17, V24, V25	1
6	PAOZZ	9905012100230	19207	12302944	PLATE, IDENTIFICATIO .... UOC: DAA, DAB, DAC, DAD, DAJ, DAK, DAW, DAX, ZAA, ZAC, ZAD, ZAJ, ZAK	1
6	PAOZZ	9905012100217	19207	12302946	PLATE, IDENTIFICATIO .... UOC: DAE, DAF, DAG, DAH, ZAE, ZAF, ZAG, ZAH	1
6	PAOZZ	9905012100231	19207	12302947	PLATE, IDENTIFICATIO .... UOC: DAL, ZAL	1
7	PAOZZ	7690013059103	19207	12356716	MARKER, IDENTIFICATIO	1
8	PAOZZ	9905011085164	19207	7409990-1	PLATE, INSTRUCTION .... UOC: DAL, V18, V39, ZAL	1
9	PAOZZ	9905002525587	19207	7409993	PLATE, IDENTIFICATIO .... UOC: DAL, V18, ZAL	1
10	PAOZZ	9905011081034	19207	12255840	PLATE, INSTRUCTION	1
11	PAOZZ	5340011098013	19207	12276976	BRACKET, ANGLE	1
12	PAOZZ	9905002525586	19207	7059462	PLATE, IDENTIFICATIO .... UOC: DAE, DAF, DAG, DAH, V19, V20, V21, V22, ZAE, ZAF, ZAG, ZAH	1
13	PAOZZ	5305009846208	96906	MS35206-261	SCREW, MACHINE	2
14	PAOZZ	9905001116662	19207	10937760	PLATE, INSTRUCTION	1
15	PAOZZ	9905011722393	19207	12277376	PLATE, INSTRUCTION	1

END OF FIGURE

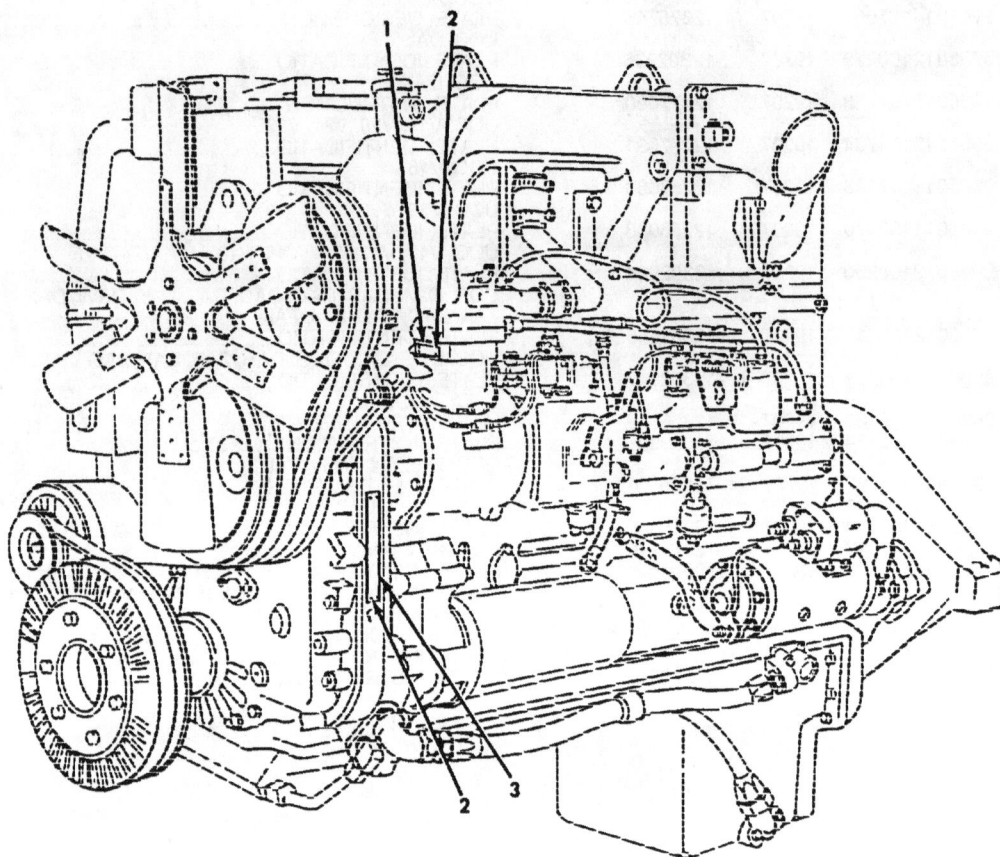

Figure 553. Data Plates, Engine (M939, M939A1).

(1) ITEM NO	(2) SMR CODE	(3) NSN	(4) CAGEC	(5) PART NUMBER	(6) DESCRIPTION AND USABLE ON CODES (UOC)	(7) QTY
					GROUP 2210 DATA PLATES	
					FIG. 553 DATA PLATES, ENGINE(M939, M939A1)	
1	PAOZZ	9905004737260	15434	136403	PLATE, MARKING, BLANK .... UOC: DAA, DAB, DAC, DAD, DAE, DAF, DAG, DAH, DAJ, DAK, DAL, DAW, DAX, V12, V13, V14, V15, V16, V17, V18, V19, V20, V21, V22, V24, V25, V39	1
2	PAOZZ	5305008046318	15434	S-2286	SCREW.. UOC: DAA, DAB, DAC, DAD, DAE, DAF, DAG, DAH, DAJ, DAK, DAL, DAW, DAX, V12, V13, V14, V15, V16, V17, V18, V19, V20, V21, V22, V24, V25, V39	8
3	PAOZZ	9905012293443	15434	3045552	PLATE, IDENTIFICATIO .... UOC: DAA, DAB, DAC, DAD, DAE, DAF, DAG, DAH, DAJ, DAK, DAL, DAW, DAX, V12, V13, V14, V15, V16, V17, V18, , V19, 20, V21, V22, V24, V25, V39	1

END OF FIGURE

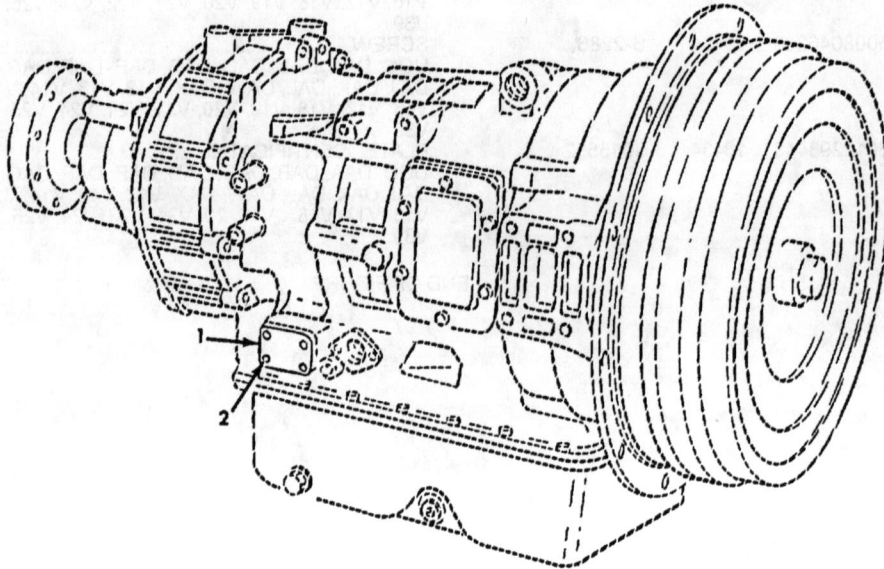

Figure 554. Data Plates, Transmission.

(1) ITEM NO	(2) SMR CODE	(3) NSN	(4) CAGEC	(5) PART NUMBER	(6) DESCRIPTION AND USABLE ON CODES (UOC)	(7) QTY
					GROUP 2210 DATA PLATES	
					FIG. 554 DATA PLATES, TRANSMISSION	
1	PAOZZ	9905013172715	73342	6838494	PLATE, IDENTIFICATIO	1
2	PAOZZ	5305005576612	11862	8622361	SCREW, DRIVE	1

END OF FIGURE

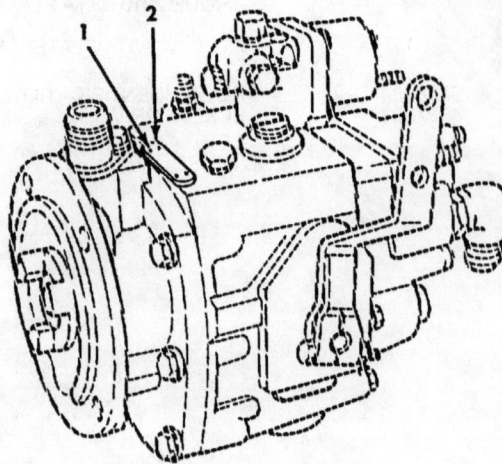

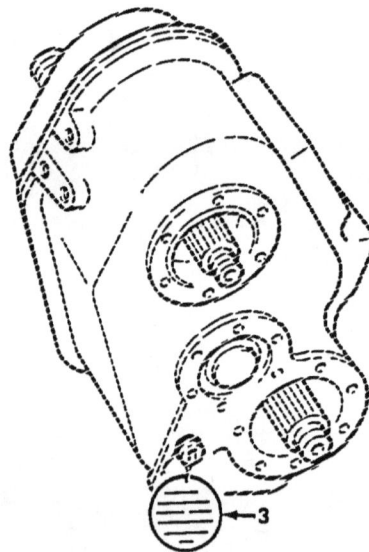

Figure 555. Data Plates, Fuel Pump and Transfer.

(1) ITEM NO	(2) SMR CODE	(3) NSN	(4) CAGEC	(5) PART NUMBER	(6) DESCRIPTION AND USABLE ON CODES (UOC)	(7) QTY
					GROUP 2210 DATA PLATES	
					FIG. 555 DATA PLATES, FUEL PUMP AND TRANSFER	
1	PAOZZ	5305008046318	15434	S-2286	SCREW .... UOC: DAA, DAB, DAC, DAD, DAE, DAF, DAG, DAH, DAJ, DAK, DAL, DAW, DAX, V12, V13, V14, V15, V16, V17, V18, V19, V20, V21, V22, V24, V25, V39	4
2	PAOZZ	9905007337622	15434	105375	PLATE, IDENTIFICATIO FUEL PUMP .... UOC: DAA, DAB, DAC, DAD, DAE, DAF, DAG, DAH, DAJ, DAK, DAL, DAW, DAX, V12, V13, V14, V15, V16, V17, V18, V19, V20, V21, V22, V24, V25, V39	1
3	PFOZZ	9905013538846	78500	1199-N-3082	TAG, INSTRUCTION	1

END OF FIGURE

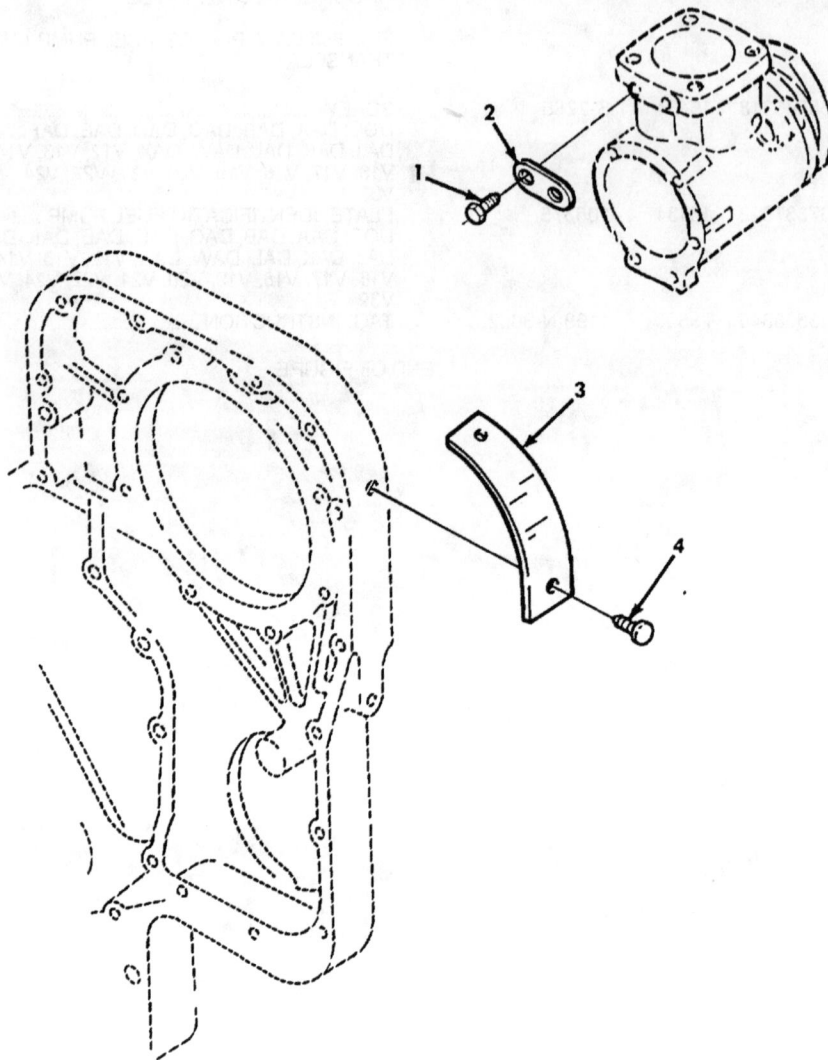

Figure 556. Engine Data Plates (M939A2).

(1) ITEM NO	(2) SMR CODE	(3) NSN	(4) CAGEC	(5) PART NUMBER	(6) DESCRIPTION AND USABLE ON CODES (UOC)	(7) QTY
					GROUP 2210 DATA PLATES	
					FIG. 556 ENGINE DATA PLATES (M939A2)	
1	PAFZZ	5305008046318	15434	S-2286	SCREW.. UOC: ZAA, ZAB, ZAC, ZAD, ZAE, ZAF, ZAG, ZAH, ZAJ, ZAK, ZAL	2
2	PAFZZ	9905004737260	15434	136403	PLATE, MARKING, BLANK.. UOC: ZAA, ZAB, ZAC, ZAD, ZAE, ZAF, ZAG, ZAH, ZAJ, ZAK, ZAL	1
3	PAOZZ	9905012794691	15434	3920703	PLATE, MARKING, BLANK .... UOC: ZAA, ZAB, ZAC, ZAD, ZAE, ZAF, ZAG, ZAH, ZAJ, ZAK, ZAL	1
4	PAOZZ	5305012760859	15434	3908612	SCREW, DRIVE... UOC: ZAA, ZAB, ZAC, ZAD, ZAE, ZAF, ZAG, ZAH, ZAJ, ZAK, ZAL	2

END OF FIGURE

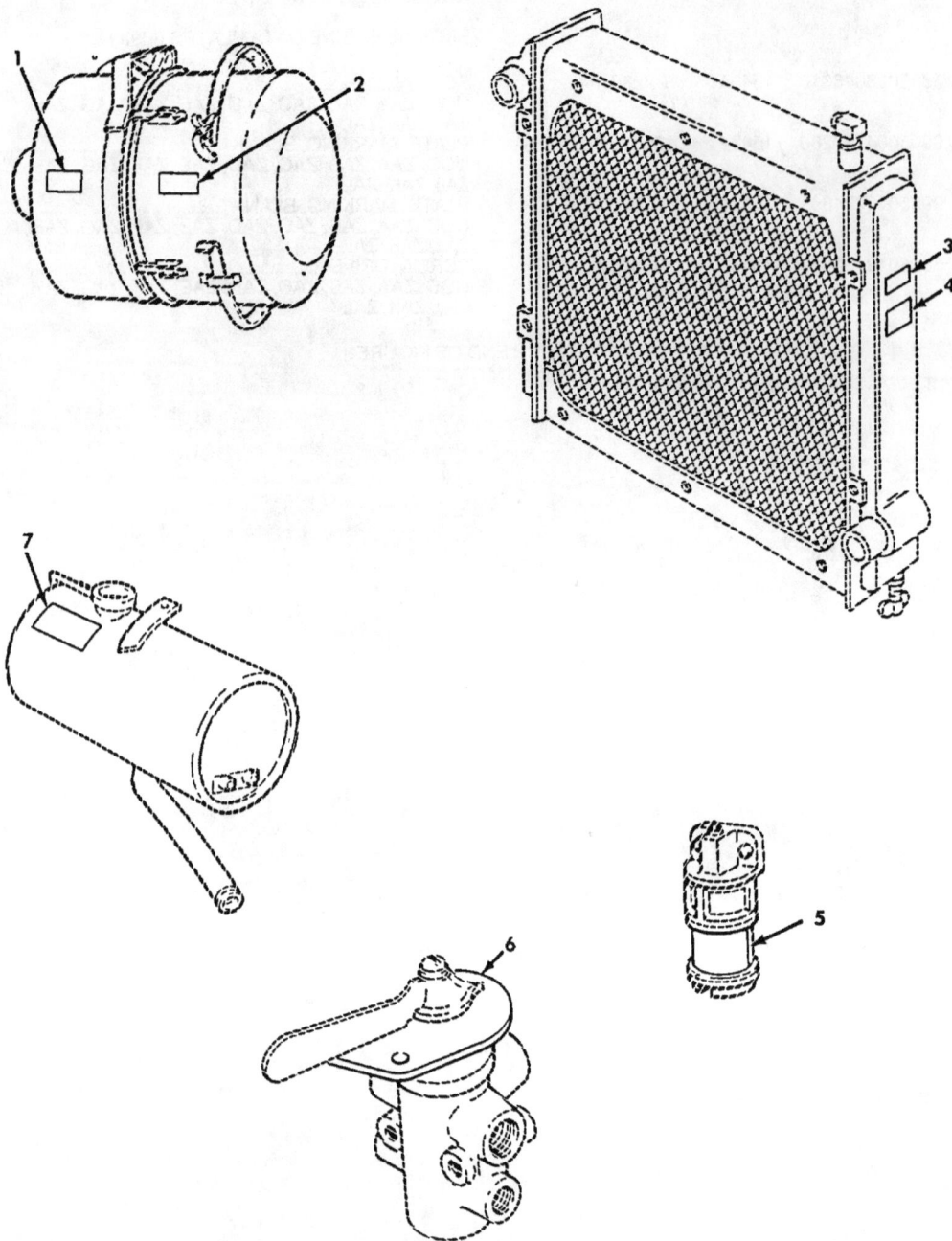

Figure 557. Data Plates, Air Cleaner, Radiator, Air Cleaner Indicator, Surge Tank, and Fuel Tank Selector.

(1) ITEM NO	(2) SMR CODE	(3) NSN	(4) CAGEC	(5) PART NUMBER	(6) DESCRIPTION AND USABLE ON CODES (UOC)	(7) QTY
					GROUP 2210 DATA PLATES	
					FIG. 557 DATA PLATES, AIR CLEANER, RADIATOR, AIR CLEANER INDICATOR, SURGE TANK, AND FUEL TANK SELECTOR	
1	PAOZZ	9905001975962	19207	11604607	PLATE, IDENTIFICATIO AIR CLEANER	1
2	PAOZZ	9905001975957	19207	11604608	PLATE, INSTRUCTION..	1
3	PFOZZ	7690011431270	19207	12302641	DECAL	2
4	PAOZZ	7690011317499	19207	12302640	DECAL	2
5	PAOZZ	7690004098937	19207	11648494	DECAL INDICATOR HOUSING	1
6	PAOZZ	9905011147589	19207	12256338	PLATE, IDENTIFICATIO FUEL TANK	1
					UOC: DAE, DAF, DAG, DAH, DAL, V18, V19, V20, V21, V22, ZAE, ZAF, ZAG, ZAH, ZAL	
7	PAOZZ	7690011966355	19207	12302869	MARKER, IDENTIFICATI SURGE TANK	1
					UOC: DAA, DAB, DAC, DAD, DAE, DAF, DAG, DAH, DAJ, DAK, DAL, DAW, DAX, V12, V13, V14, V15, V16, V17, V18, V19, V20, V21, V22, V24, V25, V39	

END OF FIGURE

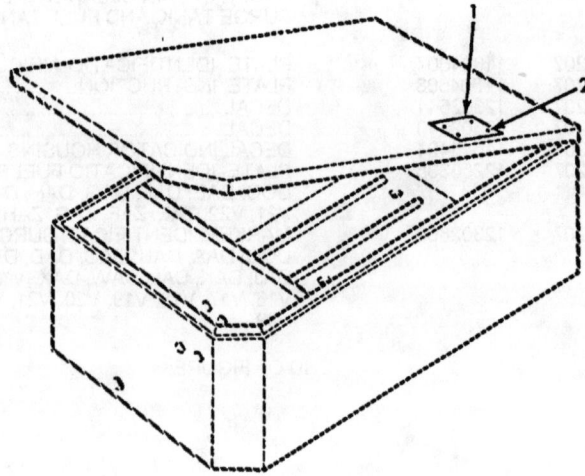

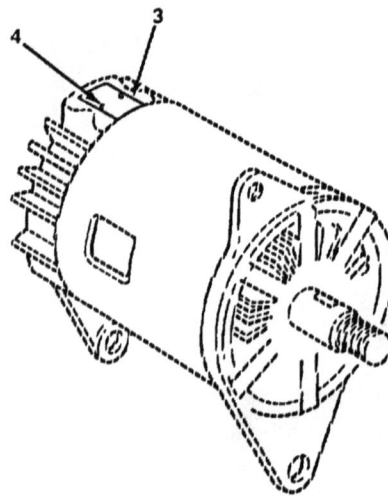

Figure 558. Data Plates, Battery Box, and Generator.

(1) ITEM NO	(2) SMR CODE	(3) NSN	(4) CAGEC	(5) PART NUMBER	(6) DESCRIPTION AND USABLE ON CODES (UOC)	(7) QTY
					GROUP 2210 DATA PLATES	
					FIG. 558 DATA PLATES, BATTERY BOX, AND GENERATOR	
1	PAOZZ	9905011147606	19207	12277125	PLATE, INSTRUCTION CONNECTING............................... BATTERY CABLES	1
2	PAOZZ	5320005107823	81349	M24243/1-B402	RIVET, BLIND..	4
3	PAOZZ	9905010435322	19207	11630585-1	PLATE, INSTRUCTION GENERATOR	1
4	PAOZZ	5320005823273	53551	RV250-4-1	RIVET, BLIND.	2

END OF FIGURE

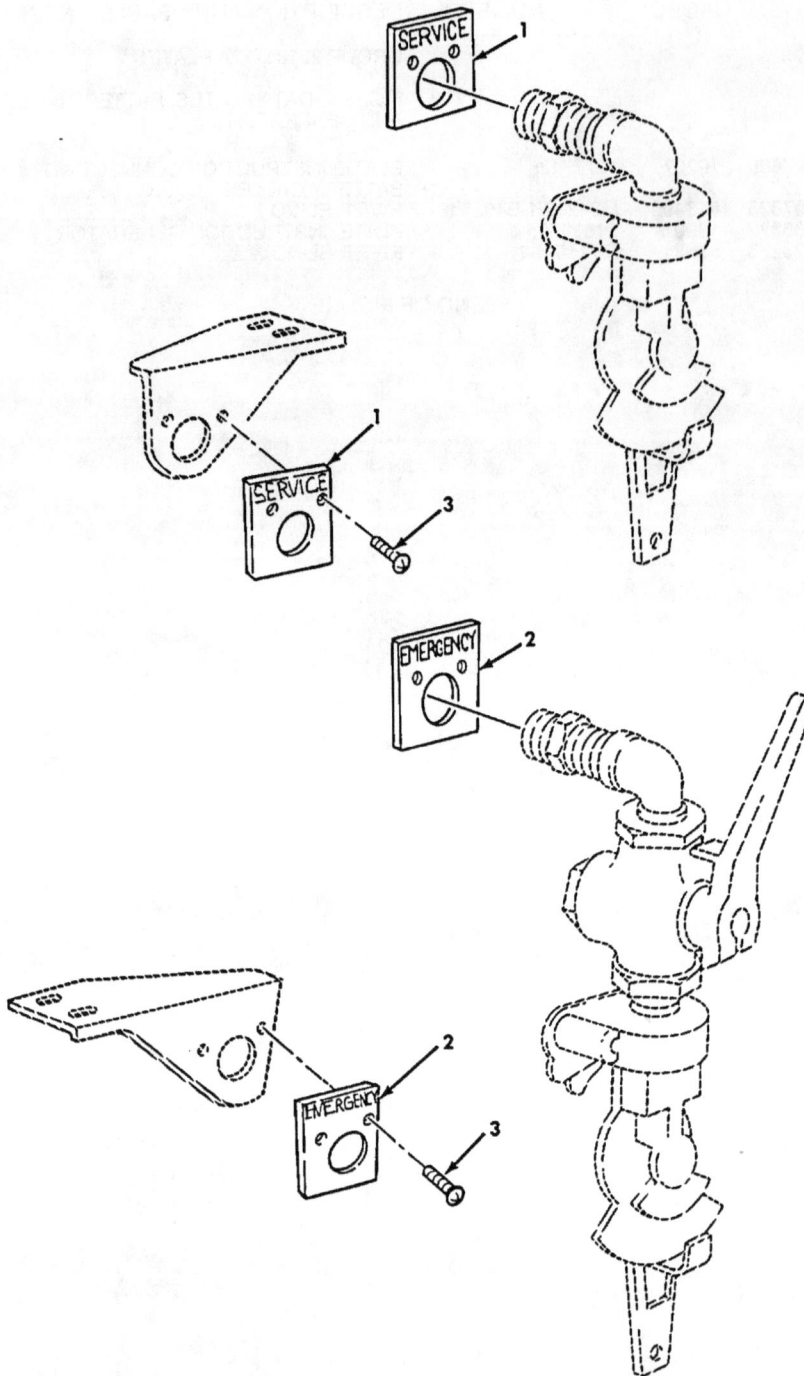

Figure 559. Data Plates, Service and Emergency Couplings.

(1) ITEM NO	(2) SMR CODE	(3) NSN	(4) CAGEC	(5) PART NUMBER	(6) DESCRIPTION AND USABLE ON CODES (UOC)	(7) QTY
					GROUP 2210 DATA PLATES	
					FIG. 559 DATA PLATES, SERVICE AND EMERGENCY COUPLINGS	
1	PAOZZ	9905009997370	96906	MS53007-1	PLATE, IDENTIFICATIO REAR SERVICE .... UOC:DAA,DAB,DAC,DAD,DAE,DAF,DAJ,DAK, DAL,DAX,DAW,V12,V13,V14,V15,V16,V17, V18,V19,V20,V24,V25,V39,ZAA,ZAB,ZAC, ZAD,ZAE,ZAF,ZAJ,ZAK,ZAL	2
1	PAOZZ	9905009997370	96906	MS53007-1	PLATE, IDENTIFICATIO UOC:DAG,DAH,V21,V22,ZAG,ZAH	2
2	PAOZZ	9905009997369	96906	MS53007-2	PLATE, IDENTIFICATIO FRONT EMERGENCY.................... UOC:DAG,DAH,V21,V22,ZAG,ZAH	2
2	PAOZZ	9905009997369	96906	MS53007-2	PLATE, IDENTIFICATIO FRONT EMERGENCY.................... UOC:DAA,DAB,DAC,DAD,DAE,DAF,DAJ,DAK, DAL,DAW,DAX,V12,V13,V14,V15,V16,V17, V18,V19,V20,V24,V25,V39,ZAA,ZAB,ZAC, ZAD,ZAE,ZAF,ZAJ,ZAK,ZAL	2
3	PAOZZ	5305008550971	96906	MS24629-21	SCREW, TAPPING... UOC:DAC,DAD,DAE,DAF,DAG,DAH,DAJ,DAK, DAL,V16,V17,V18,V19,V20,V21,V22,V24, V25,V39,ZAC,ZAD,ZAE,ZAF,ZAG,ZAH,ZAJ, ZAK,ZAL	4

END OF FIGURE

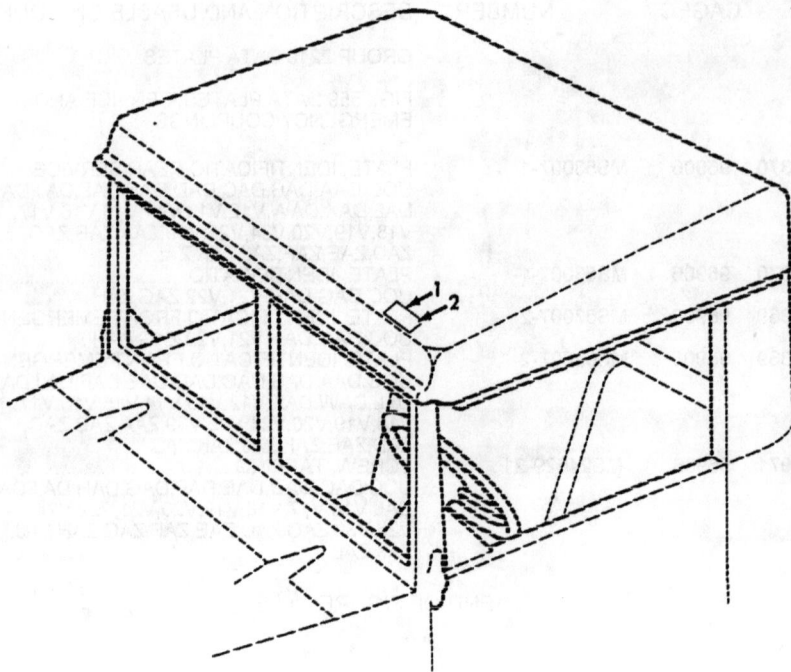

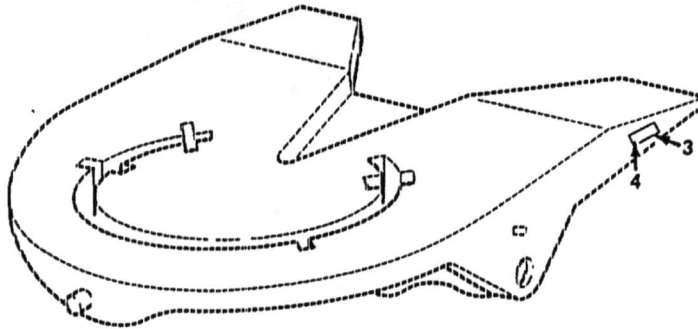

Figure 560. Data Plates, Hardtop and Fifth Wheel.

(1) ITEM NO	(2) SMR CODE	(3) NSN	(4) CAGEC	(5) PART NUMBER	(6) DESCRIPTION AND USABLE ON CODES (UOC)	(7) QTY
					GROUP 2210 DATA PLATES	
					FIG. 560 DATA PLATES, HARDTOP AND FIFTH WHEEL	
1	PAOZZ	9905011081032	19207	12256148	PLATE, IDENTIFICATIO	1
2	PAOZZ	5305002535618	61038	23046	SCREW, DRIVE..	4
3	PAOZZ	9905011226098	14731	59-002366	PLATE, INSTRUCTION.. UOC:DAG,DAH,V21,V22,ZAG,ZAH	1
4	PAOZZ	5305011277087	14371	08-203713	SCREW, DRIVE.. UOC:DAG,DAH,V21,V22,ZAG,ZAH	2

END OF FIGURE

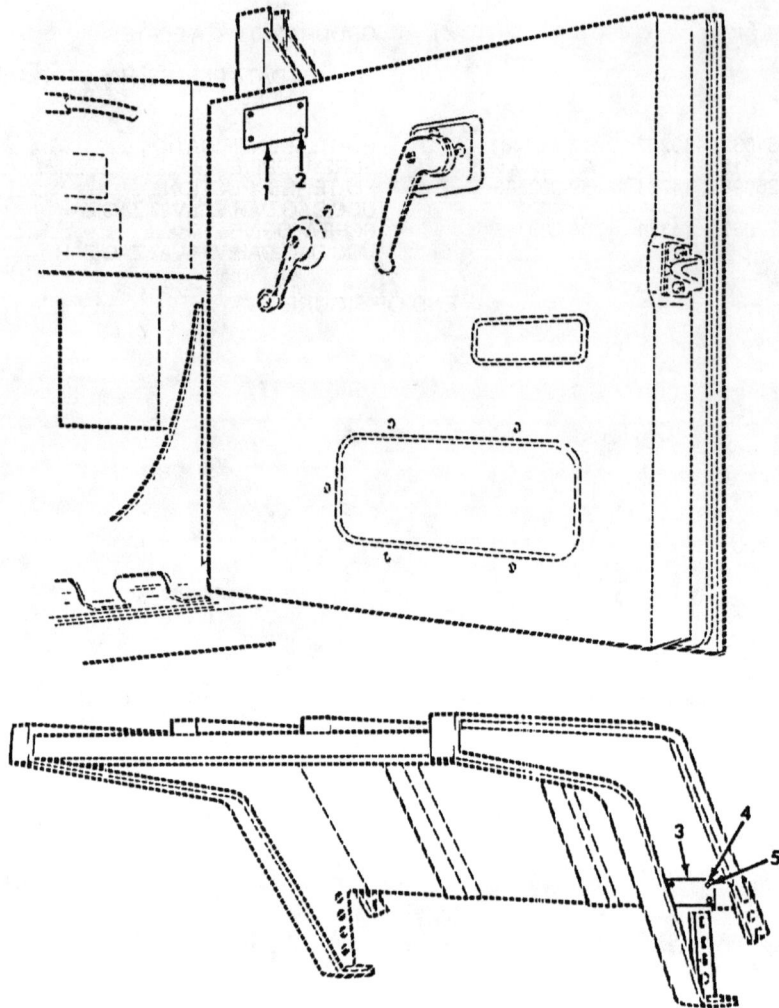

Figure 561. Data Plates, Cab, Dump Cab Protector, and Cab Door.

GROUP 2210 DATA PLATES

FIG. 561 DATA PLATES, CAB, DUMP CAB
PROTECTOR, AND CAB DOOR

ITEM NO	SMR CODE	NSN	CAGEC	PART NUMBER	DESCRIPTION AND USABLE ON CODES (UOC)	QTY
1	PAOZZ	9905011855781	19207	12302850	PLATE, INSTRUCTION UOC:DAL,V18,ZAL	1
2	PAOZZ	5305004328027	96906	MS51861-22	SCREW, TAPPING UOC:DAL,V18,ZAL	4
3	PAOZZ	9905011408219	19207	12255711	PLATE, INSTRUCTION UOC:DAE,DAF,V19,V20,ZAE,ZAF	1
4	PAOZZ	5305009932738	6906	MS35207-280	SCREW, MACHINE UOC:DAE,DAF,V19,V20,ZAE,ZAF	4
5	PAOZZ	5310009359022	96906	MS51943-32	NUT, SELF-LOCKING, HE UOC:DAE,DAF,V19,V20,ZAE,ZAF	4

END OF FIGURE

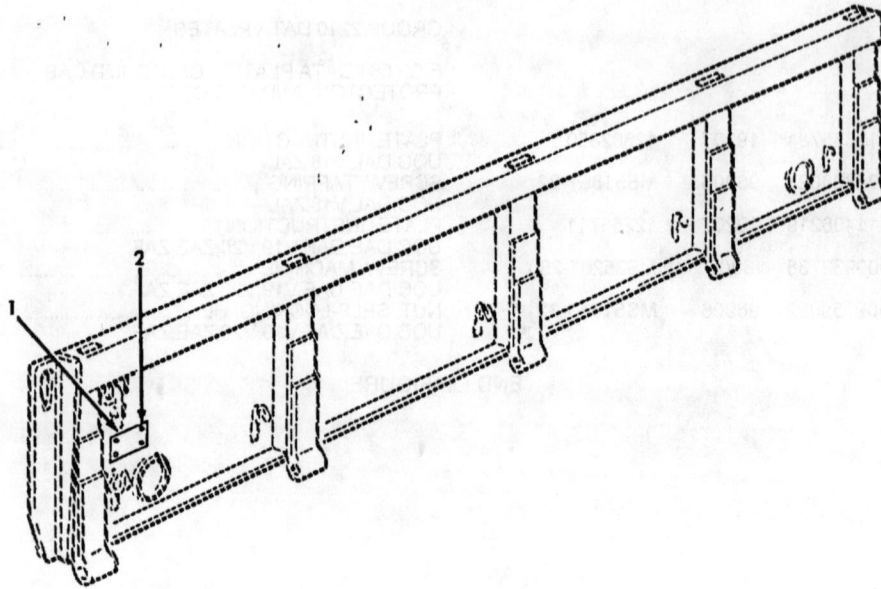

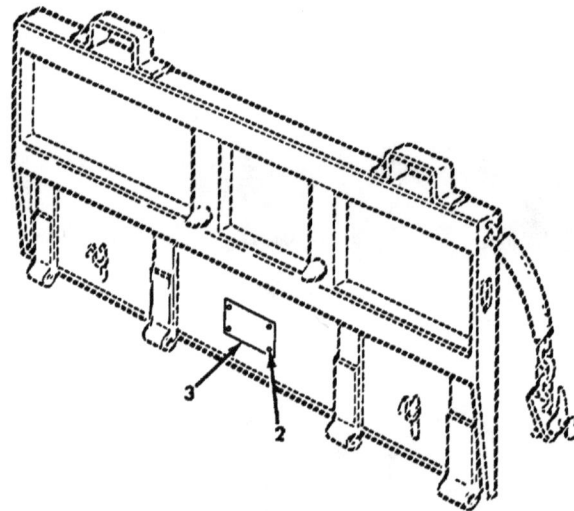

Figure 562. Data Plates, Cargo Body Side Gate and Tailgate.

(1) ITEM NO	(2) SMR CODE	(3) NSN	(4) CAGEC	(5) PART NUMBER	(6) DESCRIPTION AND USABLE ON CODES (UOC)	(7) QTY

GROUP 2210 DATA PLATES

FIG. 562 DATA PLATES, CARGO BODY
SIDE GATE AND TAILGATE

1	PAOZZ	9905001165294	19207	11608789	PLATE, INSTRUCTION	1
					UOC:DAW,DAX,V14,V15,ZAA,ZAB	
2	PFOZZ	5320011348671	19207	11621056-3	RIVET, BLIND	8
					UOC:DAW,DAX,V14,V15,ZAA,ZAB	
3	PAOZZ	9905001165295	19207	11608788	PLATE, INSTRUCTION CAUTION DROPSIDE.....................	1
					TAILGATE	
					UOC:DAW,DAX,V14,V15,ZAA,ZAB	

END OF FIGURE

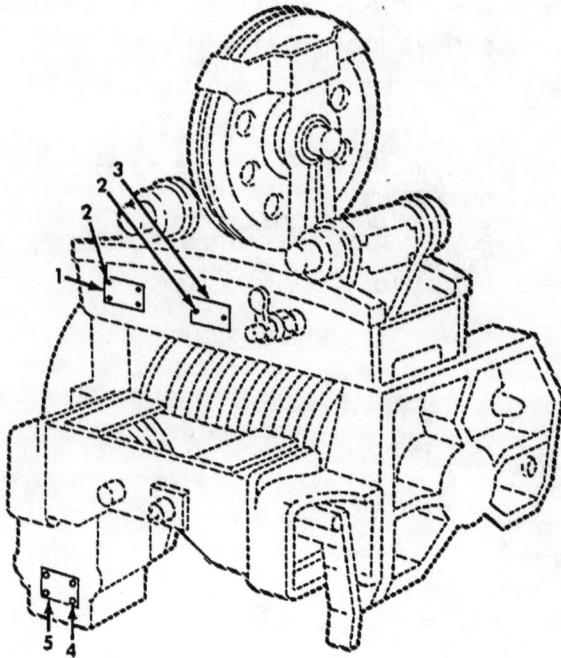

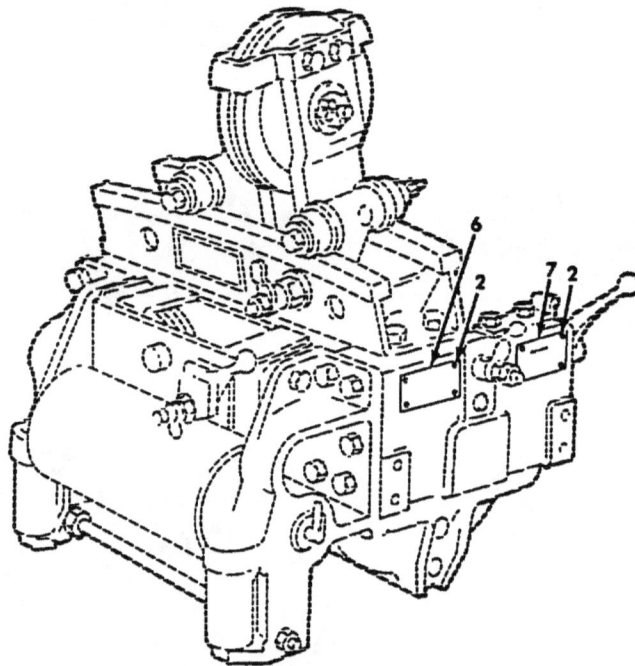

Figure 563. Data Plates, Front and Rear Winches.

(1) ITEM NO	(2) SMR CODE	(3) NSN	(4) CAGEC	(5) PART NUMBER	(6) DESCRIPTION AND USABLE ON CODES (UOC)	(7) QTY
					GROUP 2210 DATA PLATES	
					FIG. 563 DATA PLATES, FRONT AND REAR WINCHES	
1	PAOZZ	9905002629929	19207	8344235	PLATE, INSTRUCTION .... UOC:DAB,DAD,DAF,DAH,DAL,DAX,V12,V14, V16,V18,V19,V21,V39,ZAB,ZAD,ZAF,ZAR, ZAL	1
2	PAOZZ	5305002535618	96906	MS21318-27	SCREW, DRIVE . ... UOC:DAB,DAD,DAF,DAH,DAL,DAX,V12,V14, V16,V18,V19,V21,V39,ZAB,ZAD,ZAF,ZAH, ZAL	14
3	PAOZZ	9905011357476	19207	12300691-3	PLATE, IDENTIFICATIO REAR WINCH................ ASSEMBLY.. UOC:DAL,V18,ZAL	1
3	PAOZZ	9905011423114	19207	12300691-2	PLATE, IDENTIFICATIO .... UOC:DAB,DAD,DAF,DAH,DAL,DAX,V12,V14, V16,V18,V19,V21,V39,ZAB,ZAD,ZAF,ZAH, ZAL	1
4	PAOZZ	9905001411619	19207	8690900	PLATE, INSTRUCTION........................... UOC:DAB,DAD,DAF,DAH,DAL,DAX,V12,V14, V16,V18,V19,V21,V39,ZAB,ZAD,ZAF,ZAH, ZAL	1
5	PAOZZ	5305012032609	21450	142341	SCREW, CAP, HEXAGON H .... UOC:DAB,DAD,DAF,DAH,DAL,DAX,V12,V14, V16,V18,V19,V21,V39,ZAB,ZAD,ZAF,ZAH, ZAL	4
6	PAOZZ	9905006345269	19207	7994966	PLATE, INSTRUCTION .... UOC:DAB,DAD,DAF,DAH,DAL,DAX,V12,V14, V16,V18,V19,V21,V39,ZAB,ZAD,ZAF,ZAH, ZAL	1
7	PAOZZ	9905009136879	19207	8690899	PLATE, INSTRUCTION WARNING, WINCH .... UOC:DAB,DAD,DAF,DAH,DAL,DAX,V12,V14, V16,V18,V19,V21,V39,ZAB,ZAD,ZAF,ZAH, ZAL	1
7	PAOZZ	9905011147607	19207	12277131	PLATE, INSTRUCTION WINCH UOC:DAL,V18,ZAL	1

END OF FIGURE

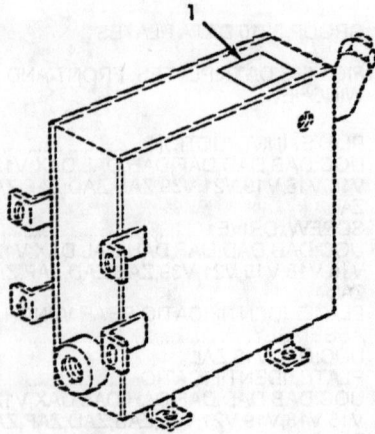

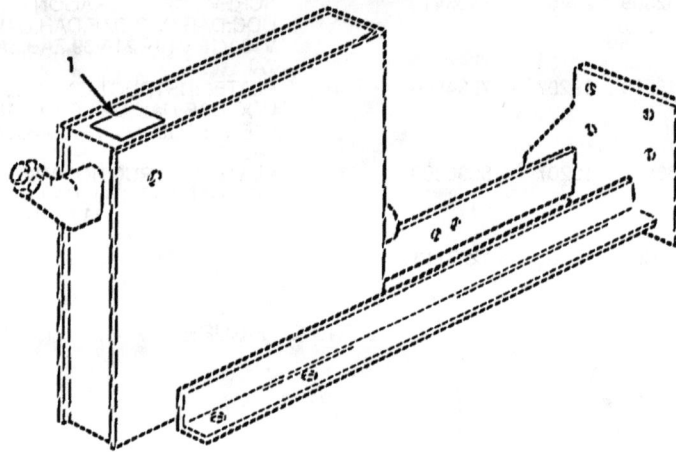

Figure 564. Data Plates, Hydraulic Reservoir Tank.

(1) ITEM NO	(2) SMR CODE	(3) NSN	(4) CAGEC	(5) PART NUMBER	(6) DESCRIPTION AND USABLE ON CODES (UOC)	(7) QTY
					GROUP 2210 DATA PLATES	
					FIG. 564 DATA PLATES, HYDRAULIC RESERVOIR TANK	
1	PAOZZ	7690011147620	19207	12276971	DECAL RESERVOIR WARNING, HYDRAULIC .... UOC:DAB,DAD,DAF,DAH,DAL,DAX,V12,V14, V16,V18,V19,V21,V39,ZAB,ZAD,ZAF,ZAH, ZAL	1

END OF FIGURE

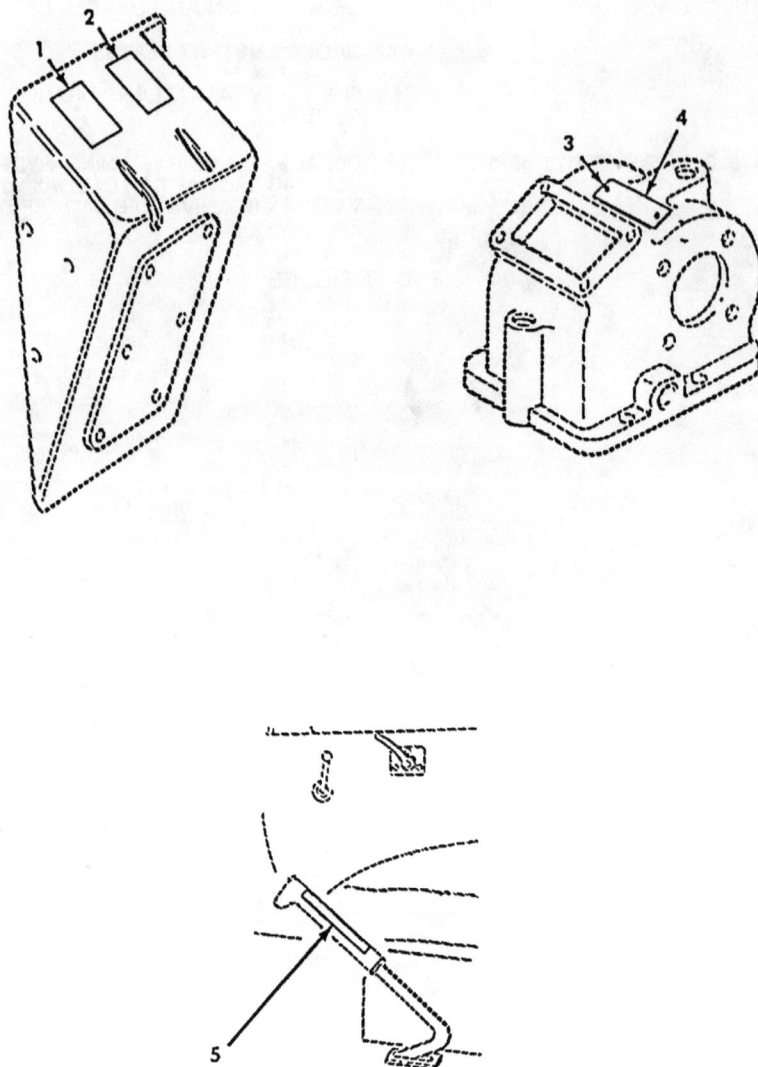

Figure 565. Data Plates, Power Takeoff and Transfer Control.

(1) ITEM NO	(2) SMR CODE	(3) NSN	(4) CAGEC	(5) PART NUMBER	(6) DESCRIPTION AND USABLE ON CODES (UOC)	(7) QTY
					GROUP 2210 DATA PLATES	
					FIG. 565 DATA PLATES, POWER TAKEOFF AND TRANSFER CONTROL	
1	PAOZZ	9905011089187	19207	12276924	PLATE, INSTRUCTION POWER TAKEOFF ENGAGEMENT. UOC:DAB,DAD,DAF,DAH,DAL,DAX,V12,V14, V16,V18,V19,V21,V39,ZAB,ZAD,ZAF,ZAH, ZAL	1
2	PAOZZ	9905011147603	19207	12276922	PLATE, INSTRUCTION WINCH UOC:DAB,DAD,DAF,DAH,DAL,DAX,V12,V14, V16,V18,V19,V21,V39,ZAB,ZAD,ZAF,ZAH, ZAL	1
3	PAOZZ	5305002535609	96906	MS21318-13	SCREW, DRIVE UOC:DAB,DAD,DAF,DAH,DAL,DAX,V12,V14, V16,V18,V19,V21,V39,ZAB,ZAD,ZAF,ZAH, ZAL	2
4	PAOZZ	9905001343558	11757	68 P 2	PLATE, IDENTIFICATIO DANA POWER TAKE -OFF. UOC:DAB,DAD,DAF,DAH,DAL,DAX,V12,V14, V16,V18,V19,V21,V39,ZAB,ZAD,ZAF,ZAH, ZAL	1
5	PFOZZ	7690012918971	19207	12301444	DECAL TRANSFER CONTROL	1

END OF FIGURE

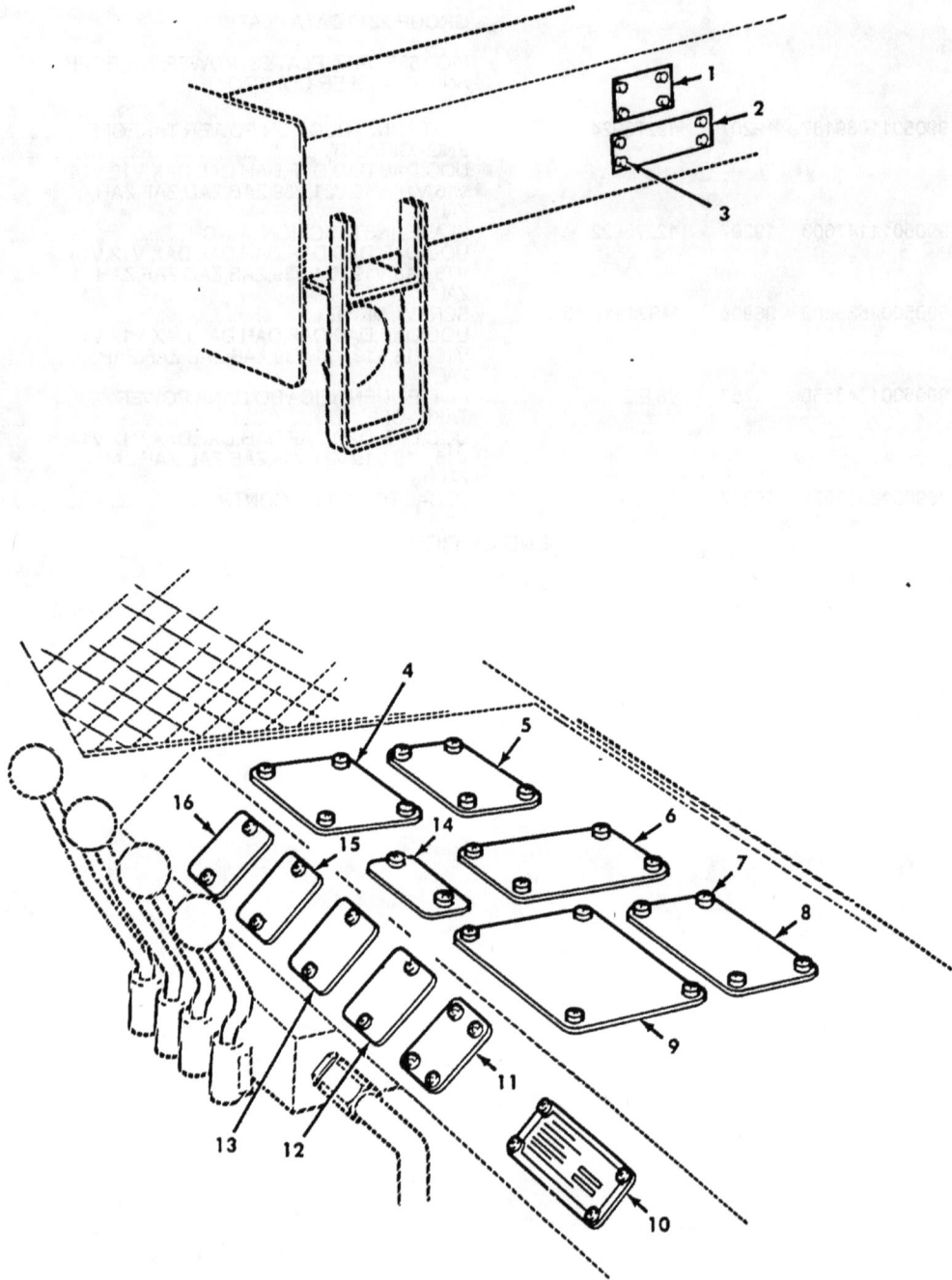

Figure 566. Data Plates, Wrecker Body and Gondola.

(1) ITEM NO	(2) SMR CODE	(3) NSN	(4) CAGEC	(5) PART NUMBER	(6) DESCRIPTION AND USABLE ON CODES (UOC)	(7) QTY
					GROUP 2210 DATA PLATES	
					FIG. 566 DATA PLATES, WRECKER BODY AND GONDOLA	
1	PFOZZ	9905011357474	19207	12300659	PLATE, INSTRUCTION REAR WRECKER BODY.................. UOC:DAL,V18,ZAL	1
2	PAOZZ	9905011491343	19207	12302634	PLATE, IDENTIFICATIO REAR WINCH................................ UOC:DAL,V18,ZAL	1
3	PAOZZ	5305002535618	96906	MS21318-27	SCREW, DRIVE. UOC:DAL,V18,ZAL	8
4	PAOZZ	9905001971742	19207	10900027	PLATE, INSTRUCTION SUB-ZERO....................................... OPERATING, CRANE. UOC:DAL,V18,ZAL	1
5	PAOZZ	9905011855783	19207	12302851	PLATE, INSTRUCTION OIL HYDRAULIC.............................. SYSTEM UOC:DAL,V18,ZAL	1
6	PAOZZ	9905001971746	19207	10900003	PLATE, INSTRUCTION CRANE CAPACITY........................... UOC:DAL,V18,ZAL	1
7	PAOZZ	5305004328027	96906	MS51861-22	SCREW, TAPPING. UOC:DAL,V18,ZAL	42
8	PAOZZ	9905001971744	19207	10900023	PLATE, INSTRUCTION CAUTION, WRECKER..................... UOC:DAL,V18,ZAL	1
9	PAOZZ	9905011855784	19207	12302852	PLATE, INSTRUCTION OPERATING DATA, ... WRECKER UOC:DAL,V18	1
10	PAOZZ	9905009334632	19207	10883351	PLATE, IDENTIFICATIO WRECKER BODY UOC:DAL,V18,ZAL	1
11	PAOZZ	9905011996809	19207	12302881	PLATE, INSTRUCTION BOOM SWING................................ UOC:DAL,V18,ZAL	1
12	PAOZZ	9905001971743	19207	10900182	PLATE, INSTRUCTION SWING.. UOC:DAL,V18,ZAL	1
13	PAOZZ	9905001971748	19207	10900186	PLATE, INSTRUCTION CROWD UOC:DAL,V18,ZAL	1
14	PAOZZ	9905011722394	19207	10900184	PLATE, INSTRUCTION WARNING, WRECKER UOC:DAL,V18,ZAL	1
15	PAOZZ	9905009334635	19207	10900181	PLATE, INSTRUCTION HOIST OPERATION........................ UOC:DAL,V18,ZAL	1
16	PAOZZ	9905001971741	19207	10900183	PLATE, INSTRUCTION BOOM OPERATIONS..................... UOC:DAL,V18,ZAL	1

END OF FIGURE

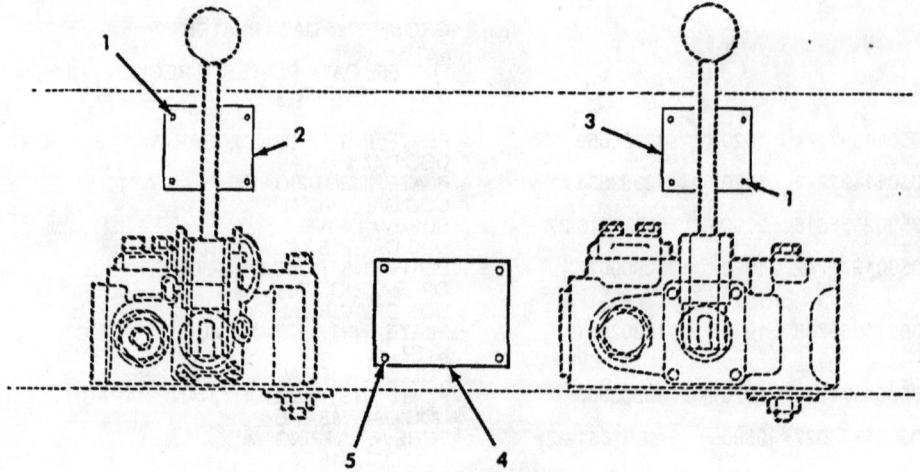

Figure 567. Data Plates, Hydraulic Control Values and Towbar.

(1) ITEM NO	(2) SMR CODE	(3) NSN	(4) CAGEC	(5) PART NUMBER	(6) DESCRIPTION AND USABLE ON CODES (UOC)	(7) QTY
					GROUP 2210 DATA PLATES	
					FIG. 567 DATA PLATES, HYDRAULIC CONTROL VALVES AND TOWBAR	
1	PAOZZ	5305000526913	96906	MS24629-33	SCREW, TAPPING UOC:DAL,V18,ZAL	8
2	PAOZZ	9905011855785	19207	12302860	PLATE, INSTRUCTION UOC:DAL,V18,ZAL	1
3	PAOZZ	9905011855786	19207	12302861	PLATE, INSTRUCTION WRECKER WARNING...................... UOC:DAL,V1B,ZAL	1
4	PAOZZ	9905011855782	19207	12302849	PLATE, INSTRUCTION UOC:DAL,V18,ZAL	1
5	PAOZZ	5305002535618	96906	MS21318-27	SCREW, DRIVE UOC:DAL,V18,ZAL	4
6	PAOZZ	9905011186092	19204	7551080	PLATE, INSTRUCTION UOC:DAL,V18,ZAL	1
7	PAOZZ	7690011147621	19207	7551081	DECAL TOWBAR UOC:DAL,V18,ZAL	1

END OF FIGURE

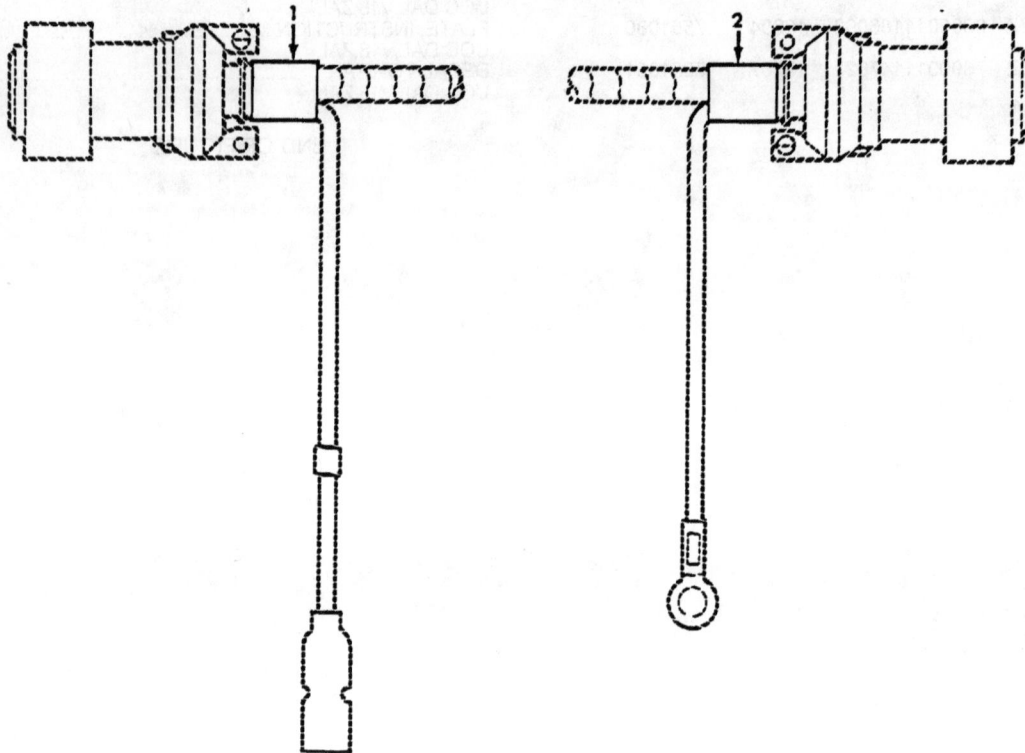

Figure 568. Data Plates, Control Box Harness.

(1) ITEM NO	(2) SMR CODE	(3) NSN	(4) CAGEC	(5) PART NUMBER	(6) DESCRIPTION AND USABLE ON CODES (UOC)	(7) QTY
					GROUP 2210 DATA PLATES	
					FIG. 568 DATA PLATES, CONTROL BOX HARNESS	
1	PAOZZ	7510010785855	19207	7951714	TAPE, IDENTIFICATION	1
2	PAOZZ	7510010764238	19207	7951713	TAPE, IDENTIFICATION	1

END OF FIGURE

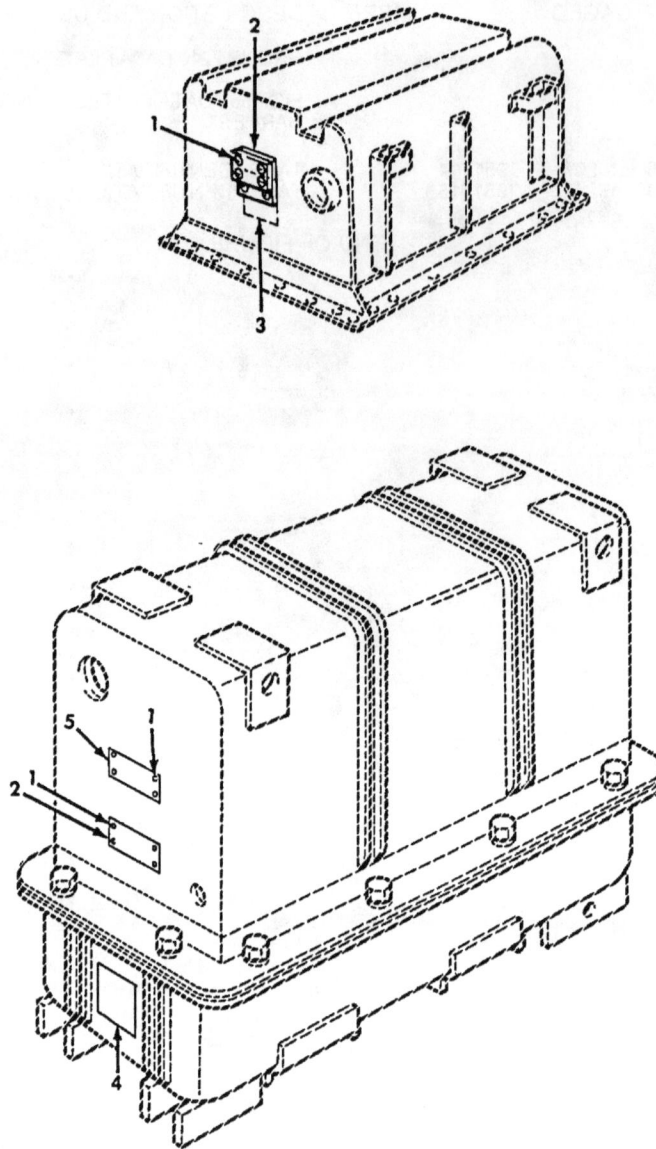

Figure 569. Data Plates, Transfer, Transmission and Engine Container.

(1) ITEM NO	(2) SMR CODE	(3) NSN	(4) CAGEC	(5) PART NUMBER	(6) DESCRIPTION AND USABLE ON CODES (UOC)	(7) QTY
					GROUP 2210 DATA PLATES	
					FIG. 569 DATA PLATES, TRANSFER, TRANSMISSION AND ENGINE CONTAINER	
1	PAOZZ	5305002535627	96906	MS21318-48	SCREW, DRIVE	8
2	PAOZZ	9905004098948	19207	7973325	PLATE, IDENTIFICATIO	2
3	PAOZZ	7690001712761	19207	10932091	DECAL	1
4	PFOZZ	7690011312061	19207	12302611	DECAL	1
5	PAFZZ	9905009012942	19207	7973326	PLATE, IDENTIFICATIO ENGINE .... UOC:DAA,DAB,DAC,DAD,DAE,DAF,DAG,DAH, DAJ,DAK,DAL,DAW,DAX,V12,V13,V14,V15, V16,V17,V18,V19,V20,V21,V22,V24,V25, V39	1

END OF FIGURE

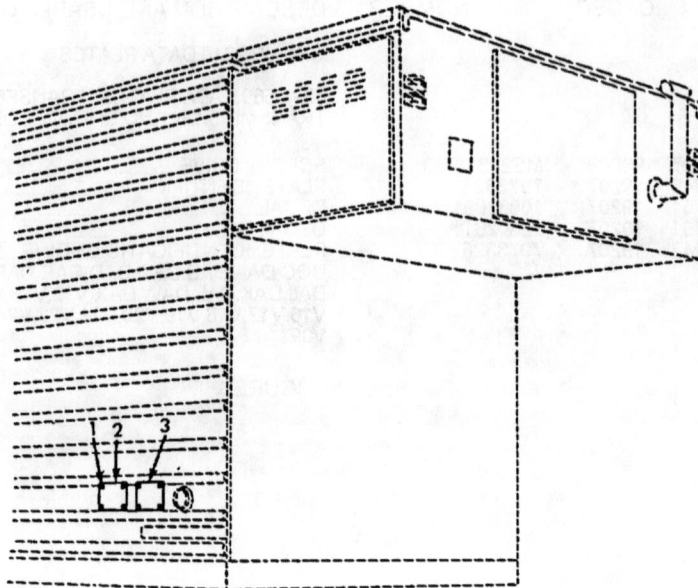

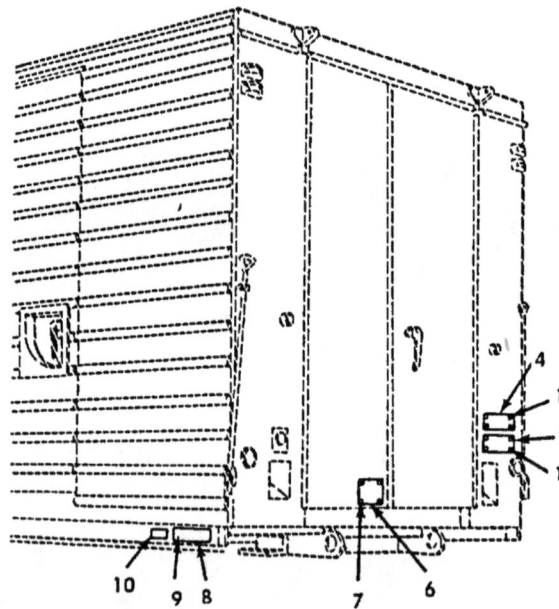

Figure 570. Data Plates, Van Body.

(1) ITEM NO	(2) SMR CODE	(3) NSN	(4) CAGEC	(5) PART NUMBER	(6) DESCRIPTION AND USABLE ON CODES (UOC)	(7) QTY
					GROUP 2210 DATA PLATES	
					FIG. 570 DATA PLATES, VAN BODY	
1	PAOZZ	5320005841285	53551	RV200-6-2	RIVET, BLIND DATA PLATES UOC:DAJ,DAK,V24,V25,ZAJ,ZAK	24
2	PAOZZ	9905011855778	19207	12302856	PLATE, IDENTIFICATIO UOC:DAJ,DAK,V24,V25,ZAJ,ZAK	1
3	PAOZZ	9905011855779	19207	12302857	PLATE, IDENTIFICATIO VAN DATA..................................... UOC:V24,V25	1
3	PAOZZ	9905012100219	19207	12302962	PLATE, IDENTIFICATIO .... UOC:DAK,ZAK	1
3	PAOZZ	9905012100220	19207	12302963	PLATE, IDENTIFICATIO .... UOC:DAJ,ZAJ	1
4	PAOZZ	9905006597755	19207	7535615	PLATE, INSTRUCTION ELECTRICAL, 3-............................. PHASE UOC:DAJ,DAK,V24,V25,ZAJ,ZAK	1
5	PAOZZ	9905004077000	19207	10937530	PLATE, INSTRUCTION WARNING, ... ELECTRICAL GROUND UOC:DAJ,DAK,V24,V25,ZAJ,ZAK	1
6	PAOZZ	9905006597757	192077	535600	PLATE, INSTRUCTION WARNING EXPANSION. UOC:DAJ,DAK,V24,V25,ZAJ,ZAK	1
7	PAOZZ	5320005823268	53551	RV200-6-3	RIVET, BLIND DATA PLATE UOC:DAJ,DAK,V24,V25,ZAJ,ZAK	4
8	PAOZZ	9905011867948	64203	10-41638	PLATE, INSTRUCTION VAN, LEFT REAR .... SIDE UOC:DAJ,DAK,V24,V25,ZAJ,ZAK	1
8	PAOZZ	9905011732183	64203	10-41535	PLATE, INSTRUCTION VAN, RIGHT REAR .... SIDE UOC:DAJ,DAK,V24,V25,ZAJ,ZAK	1
9	PAOZZ	5305011228516	64203	G-20151	SCREW, DRIVE DECAL UOC:DAJ,DAK,V24,V25,ZAJ,ZAK	8
10	PAOZZ	7690011739197	64203	10-42521	DECAL VAN, SIDE UOC:DAJ,DAK,V24,V25,ZAJ,ZAK	2

END OF FIGURE

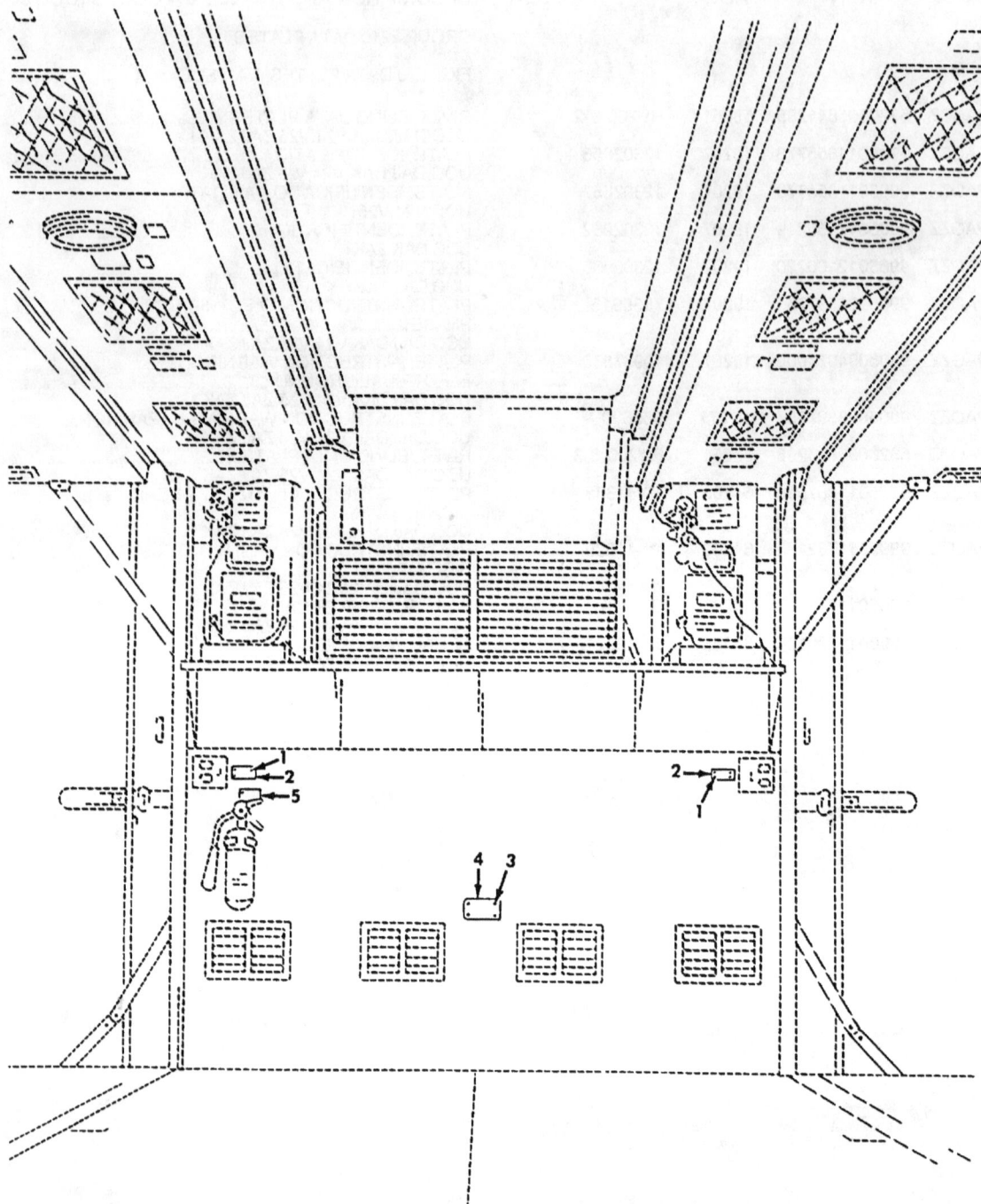

Figure 571. Data Plates, Van Body Interior, Front Wall.

(1) ITEM NO	(2) SMR CODE	(3) NSN	(4) CAGEC	(5) PART NUMBER	(6) DESCRIPTION AND USABLE ON CODES (UOC)	(7) QTY
					GROUP 2210 DATA PLATES	
					FIG. 571 DATA PLATES, VAN BODY INTERIOR, FRONT WALL	
1	PAOZZ	5305001408001	96906	MS51861-12	SCREW, TAPPING INSTRUCTION PLATE UOC:DAJ,DAX,V24,V25,ZAJ,ZAK	8
2	PAOZZ	9905001065746	19207	7535601	PLATE, INSTRUCTION BLACKOUT SWITCH UOC:DAJ,DAK,V24,V25,ZAJ,ZAK	2
3	PAOZZ	5305009012099	96906	MS35493-51	SCREW, WOOD DATA PLATE UOC:DAJ,DAK,V24,V25,ZAJ,ZAK	4
4	PAOZZ	9905010134599	19207	7535642	PLATE, INSTRUCTION REGISTER CONTROL. UOC:DAJ,DAX,V24,V25,ZAJ,ZAK	1
5	PAOZZ	7690004898322	19207	7053776	DECAL FIRE EXTINGUISHER UOC:DAJ,DAX,V24,V25,ZAJ,ZAK	1

END OF FIGURE

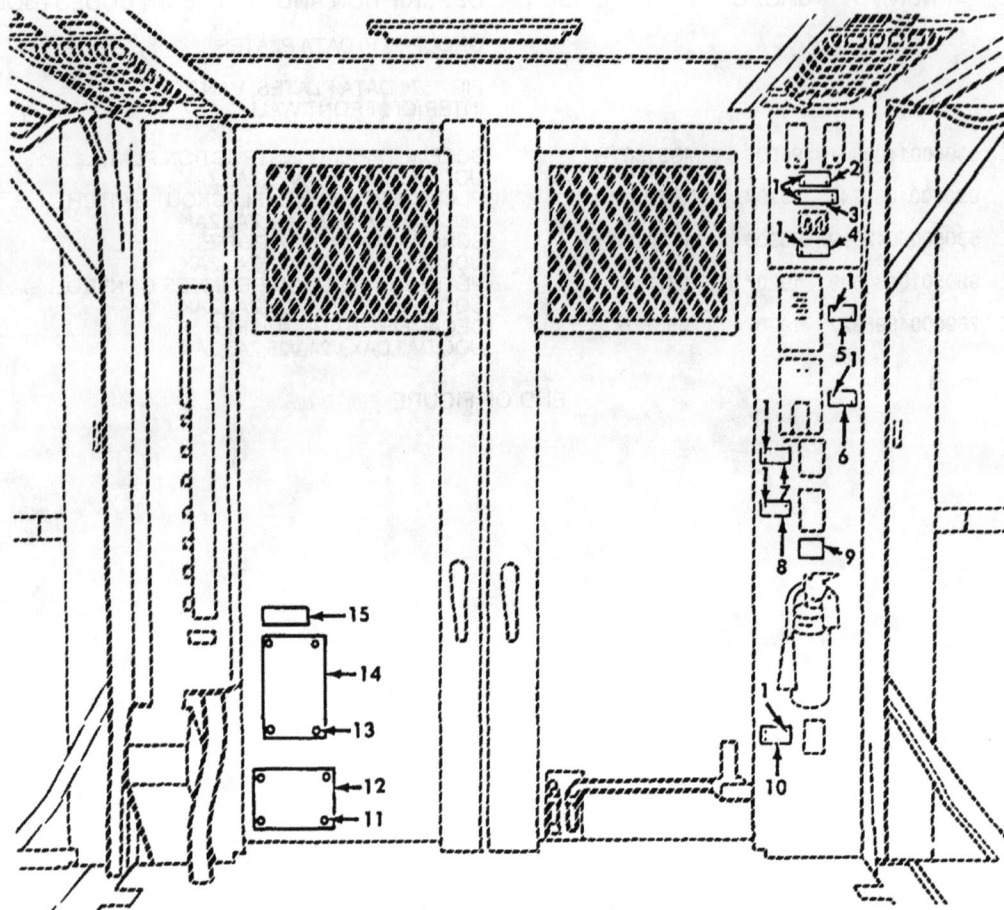

Figure 572. Data Plates, Van Body Interior, Rear Wall.

SECTION II

(1) ITEM NO	(2) SMR CODE	(3) NSN	(4) CAGEC	(5) PART NUMBER	(6) DESCRIPTION AND USABLE ON CODES (UOC)	(7) QTY
					GROUP 2210 DATA PLATES	
					FIG. 572 DATA PLATES, VAN BODY INTERIOR, REAR WALL	
1	PAOZZ	5305001408001	96906	MS51861-12	SCREW, TAPPING UOC:DAJ,DAK,V24,V25,ZAJ,ZAK	32
2	PAOZZ	9905001065744	19207	7535603	PLATE, IDENTIFICATIO HEATER THERMOSTATS.. UOC:DAJ,DAK,V24,V25,ZAJ,ZAK	1
3	PAOZZ	9905001065745	19207	7535641	PLATE, INSTRUCTION THERMOSTATS UOC:DAJ,DAK,V24,V25,ZAJ,ZAK	1
4	PAOZZ	9905010544002	19207	11677663	PLATE, IDENTIFICATIO 400 CYCLE CONVERTER.. UOC:DAJ,DAK,V24,V25,ZAJ,ZAK	1
5	PAOZZ	9905010543827	19207	11677584	PLATE, IDENTIFICATON 400 CYCLE CONVERTER.. UOC:DAJ,DAK,V24,V25,ZAJ,ZAK	1
6	PAOZZ	9905010464677	19207	11677675	PLATE, IDENTIFICATIO AUXILIARY PUMP UOC:DAJ,DAK,V24,V25,ZAJ,ZAK	1
7	PAOZZ	9905001065746	19207	7535601	PLATE, INSTRUCTION BLACKOUT CIRCUIT. UOC:DAJ,DAK,V24,V25,ZAJ,ZAK	1
8	PAOZZ	9905001065750	19207	7535597	PLATE, IDENTIFICATIO EMERGENCY LIGHT UOC:DAJ,DAK,V24,V25,ZAJ,ZAK	1
9	PAOZZ	7690004898322	19207	7053776	DECAL FIRE EXIT.. UOC:DAJ,DAK,V24,V25,ZAJ,ZAK	1
10	PAOZZ	9905004034814	19207	7535608	PLATE, IDENTIFICATIO TELEPHONE JACK. UOC:DAJ,DAK,V24,V25,ZAJ,ZAK	1
11	PAOZZ	5305001408001	96906	MS51861-12	SCREW, TAPPING DATA PLATE UOC:DAJ,DAK,V24,V25,ZAJ,ZAK	4
12	PAOZZ	9905011434526	19207	12300740	PLATE, INSTRUCTION WIRING DIAGRAM.......... UOC:DAJ,DAK,V24,V25,ZAJ,ZAK	1
13	PAOZZ	5305009003243	96906	MS35493-1	SCREW, WOOD INSTRUCTION PLATE UOC:DAJ,DAK,V24,V25,ZAJ,ZAK	4
14	PAOZZ	9905006597754	19207	7535644	PLATE, INSTRUCTION EXPANSION UOC:DAJ,DAK,V24,V25,ZAJ,ZAK	1
15	PAOZZ	7690005556073	19207	11681630	DECAL UOC:DAJ,DAK,V24,V25,ZAJ,ZAK	1

END OF FIGURE

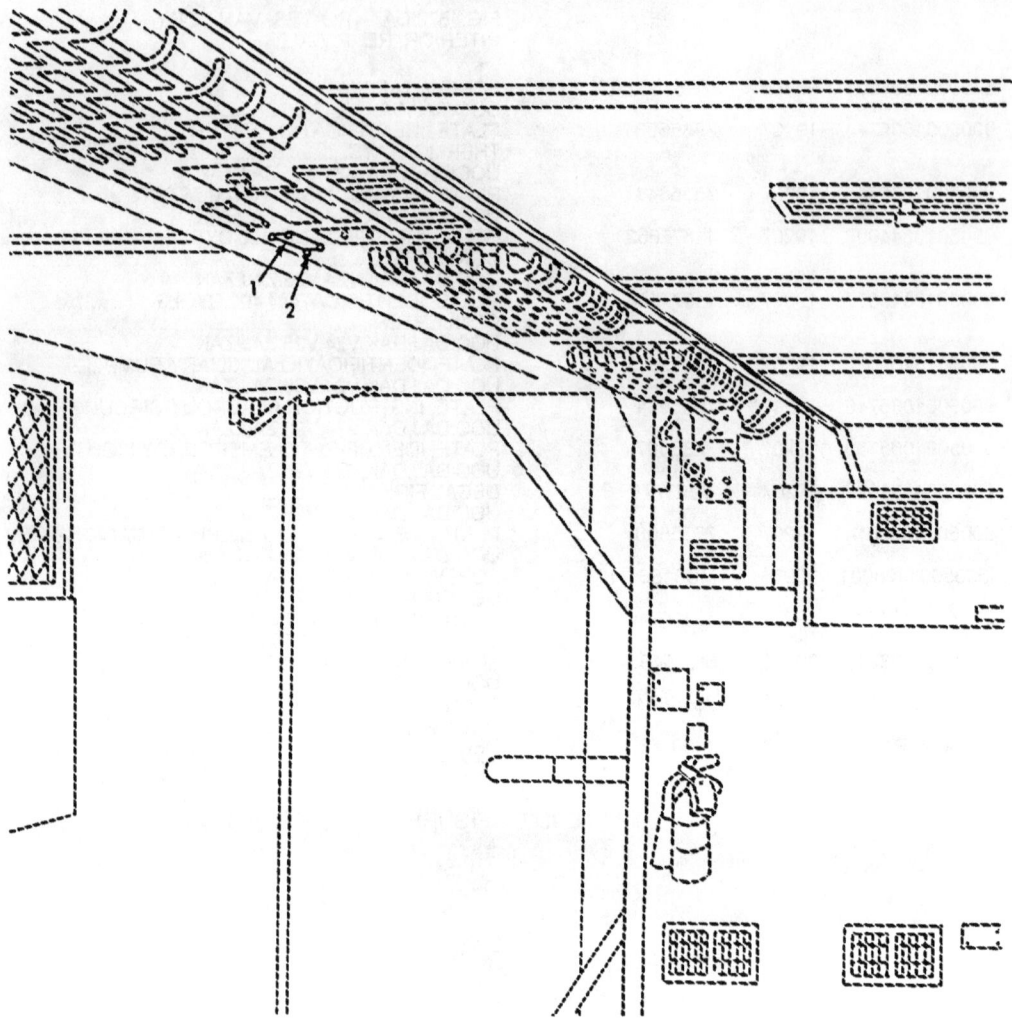

Figure 573. Data Plates, Van Body Ceiling, Left Side.

(1) ITEM NO	(2) SMR CODE	(3) NSN	(4) CAGEC	(5) PART NUMBER	(6) DESCRIPTION AND USABLE ON CODES (UOC)	(7) QTY
					GROUP 2210 DATA PLATES	
					FIG. 573 DATA PLATES, VAN BODY CEILING, LEFT SIDE	
1	PAOZZ	9905004034814	19207	7535608	PLATE, IDENTIFICATIO, TELEPHONE JACK. .... UOC:DAJ, DAK, V24, V25, ZAJ, ZAK	1
2	PAOZZ	5305008550968	96906	MS24629-10	SCREW, TAPPING, IDENTIFICATION PLATE. 4 UOC:DAJ, DAK, V24, V25, ZAJ, ZAK	
					END OF FIGURE	

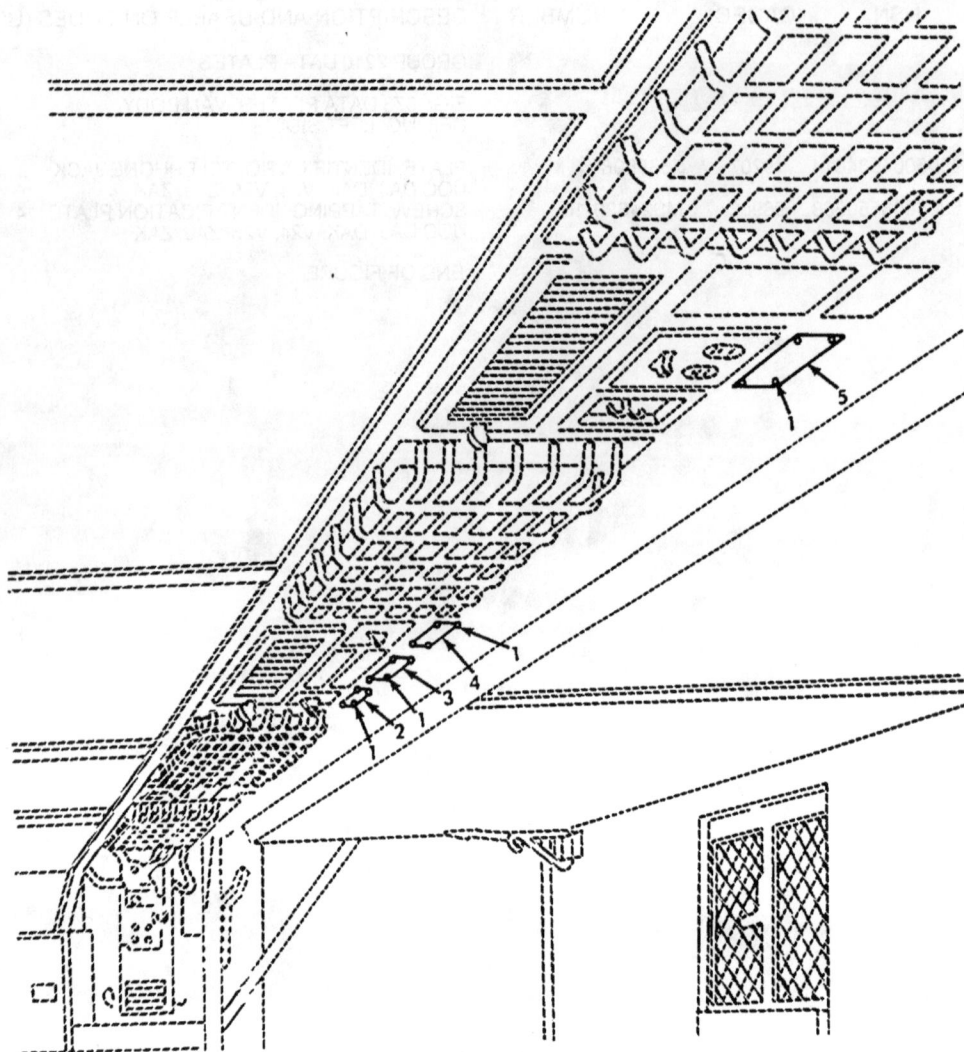

Figure 574. Data Plates, Van Body Ceiling, Right Side.

SECTION II

(1) ITEM NO	(2) SMR CODE	(3) NSN	(4) CAGEC	(5) PART NUMBER	(6) DESCRIPTION AND USABLE ON CODES (UOC)	(7) QTY
					GROUP 2210 DATA PLATES	
					FIG. 574 DATA PLATES, VAN BODY CEILING, RIGHT SIDE	
1	PAOZZ	5305008550968	96906	MS24629-10	SCREW, TAPPING, DATA PLATES . UOC:DAJ, DAK, V24, V25, ZAJ, ZAK	56
2	PAOZZ	9905010544002	19207	11677663	PLATE, IDENTIFICATIO 400 CYCLE UOC:DAJ, DAK, V24, V25, ZAJ, ZAK	2
3	PAOZZ	9905010464676	19207	11677668	PLATE, IDENTIFICATIO BLACKOUT AND EMERGENCY LIGHTS UOC:DAJ, DAK, V24, V25, ZAJ, ZAK	2
4	PAOZZ	9905010543828	19207	11677676	PLATE, IDENTIFICATIO 3-PHASE 60 CYCLE UOC:DAJ, DAK, V24, V25, ZAJ, ZAK	2
5	PAOZZ	9905001065746	19207	7535601	PLATE, INSTRUCTION, BLACKOUT CIRCUIT UOC:DAJ, DAK, V24, V25, ZAJ, ZAK	8

END OF FIGURE

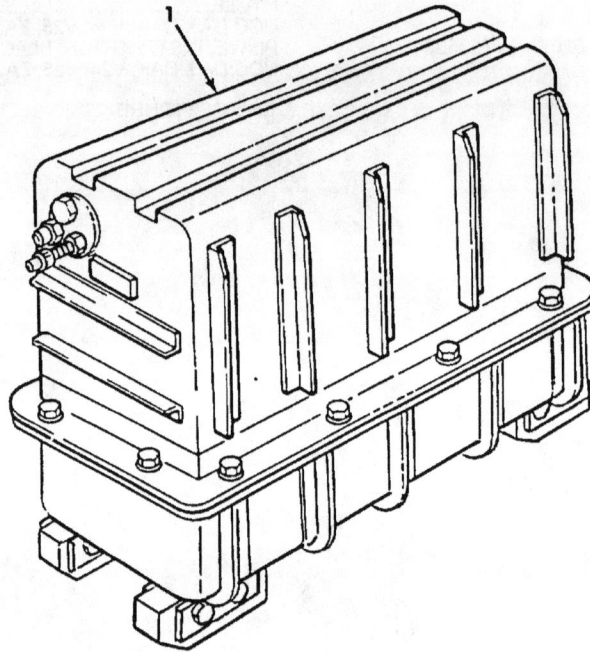

Figure 575. Engine Container Assemnbly (31939, A1939A 1).

(1) ITEM NO	(2) SMR CODE	(3) NSN	(4) CAGEC	(5) PART NUMBER	(6) DESCRIPTION AND USABLE ON CODES (UOC)	(7) QTY
					GROUP 33 SPECIAL PURPOSE KITS 3301 REUSEABLE SHIPPING CONTAINERS(M939, M939A1)	
					FIG. 575 ENGINE CONTAINER ASSEMBLY	
1	XBFFH		19207	11664574	SHIPPING AND STORAG .... UOC:DAA, DAB, DAC, DAD, DAE, DAF, DAG, DAH, DAJ, DAK, DAL, DAW, DAX, V12, V13, V14, V15, V16, V17, V18, V19, V20, V21, V22, V24, V25, V39	1
					END OF FIGURE	

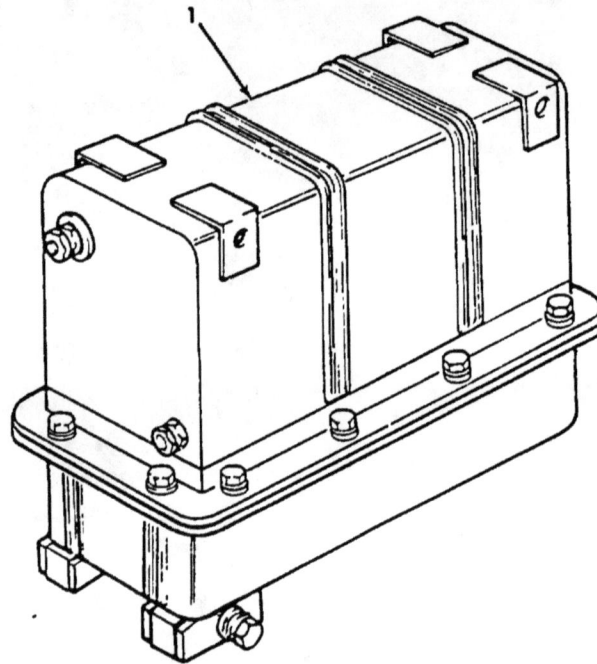

Figure 576. Transmission Container Assembly.

(1) ITEM NO	(2) SMR CODE	(3) NSN	(4) CAGEC	(5) PART NUMBER	(6) DESCRIPTION AND USABLE ON CODES (UOC)	(7) QTY
					GROUP 3301 REUSABLE SHIPPING CONTAINERS	
					FIG. 576 TRANSMISSION CONTAINER ASSEMBLY	
1	PFFFF	8145011174978	19207	12277382	SHIPPING AND STORAG	1
					END OF FIGURE	

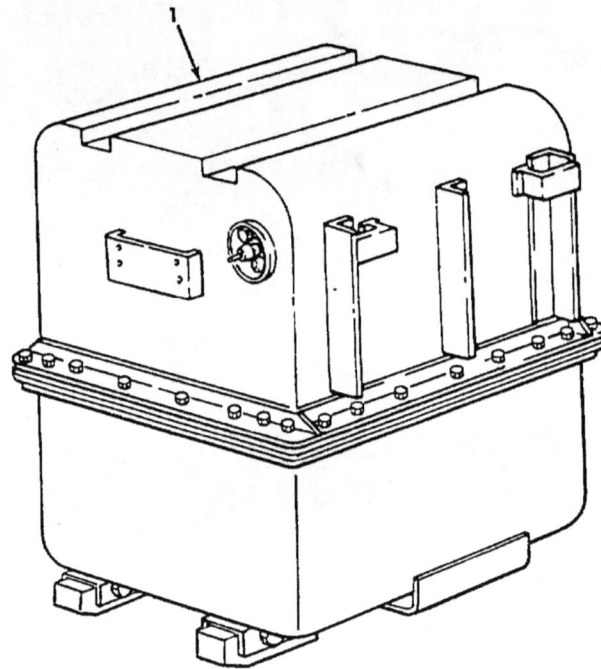

Figure 577. Transfer Case Container Assembly.

(1) ITEM NO	(2) SMR CODE	(3) NSN	(4) CAGEC	(5) PART NUMBER	(6) DESCRIPTION AND USABLE ON CODES (UOC)	(7) QTY
					GROUP 3301 REUSABLE SHIPPING CONTAINERS	
					FIG. 577 TRANSFER CASE CONTAINER ASSEMBLY	
1	PAFFF	8145011346538	19207	10932088-1	SHIPPING AND STORAG	1
					END OF FIGURE	

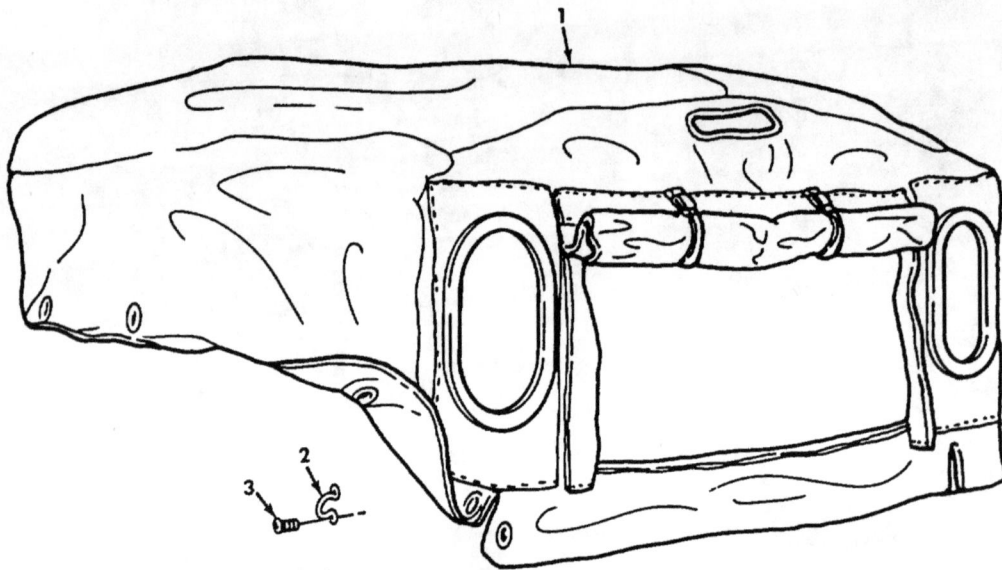

Figure 578. Radiator and Hood Cover Kit.

(1) ITEM NO	(2) SMR CODE	(3) NSN	(4) CAGEC	(5) PART NUMBER	(6) DESCRIPTION AND USABLE ON CODES (UOC)	(7) QTY
					GROUP 3303 WINTERIZATION KITS	
					FIG. 578 RADIATOR AND HOOD COVER KIT	
1	KFFZZ		19207	12256518	RADIATOR AND HOOD C PART OF KIT P/N 12302886	1
2	PAOZZ	5340007716428	19207	7716428	LOOP, STRAP FASTENER PART OF KIT P/N 12302886	16
3	PAOZZ	5305008550960	96906	MS24629-36	SCREW, TAPPING PART OF KIT P/N 12302886	32
					END OF FIGURE	

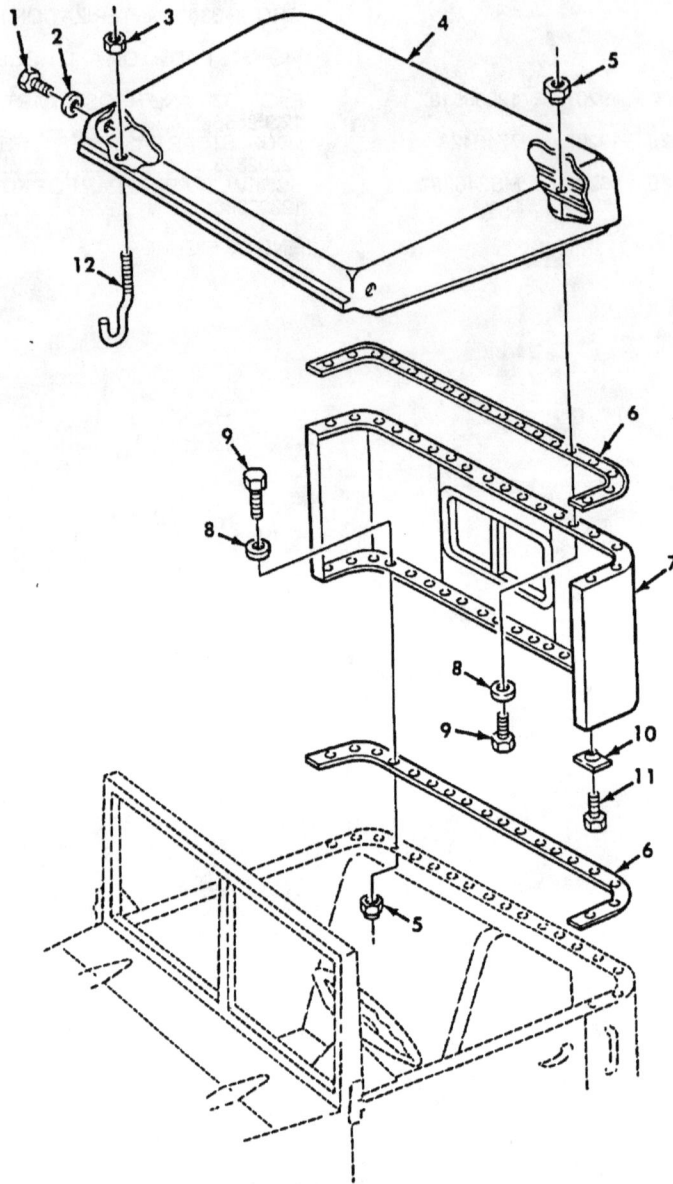

Figure 579. Hardtop Kit.

(1) ITEM NO	(2) SMR CODE	(3) NSN	(4) CAGEC	(5) PART NUMBER	(6) DESCRIPTION AND USABLE ON CODES (UOC)	(7) QTY
					GROUP 3303 WINTERIZATION KITS	
					FIG. 579 HARDTOP KIT	
1	PAOZZ	5305004324172	96906	MS51861-37	SCREW, TAPPING..	2
2	PAOZZ	5310005967693	96906	MS35335-31	WASHER, LOCK..	2
3	PAOZZ	5310011377062	96906	MS51943-12	NUT, SELF-LOCKING, HE PART OF KIT P/N...................... 12256227...	4
4	KFOZZ		19207	12256140	ROOF ASSY(COMPLETE) PART OF KIT P/N .... 12256227...	1
5	PAOZZ	5310008775796	96906	MS21044N4	NUT, SELF-LOCKING, HE PART OF KIT P/N1 .... 32 12256227...	
6	PAOZZ	5330011208454	19207	12256106	SEAL, NONMETALLIC ST PART OF KIT P/N .... 12256227...	2
7	KFOZZ		19207	12256145	BACK AND QUARTER AS PART OF KIT P/N....................... 12256227...	1
8	PAOZZ	5310008093078	96906	MS27183-11	WASHER, FLAT PART OF KIT P/N 12256227	48
9	PAOZZ	5306000680513	80204	B1821BH025F075N	BOLT, MACHINE..	32
10	PAOZZ	5310004933986	19207	7085367	NUT, SHEET SPRING PART OF KIT P/N .... 12256227...	2
11	PAOZZ	5305002150290	21450	162684	SCREW, TAPPING PART OF KIT P/N .... 12256227...	2
12	PAOZZ	5306011195834	19207	12256214	BOLT, HOOK PART OF KIT P/N 12256227	4
	PAOZZ	2540010965023	19207	12256227	PARTS KIT, VEHICULAR	1

```
                    BACK AND QUARTER AS      (    1)    579-7
                    BOLT, HOOK               (    4)    579-12
                    NUT, SELF-LOCKING, HE    (    4)    579-3
                    NUT, SELF-LOCKING, HE    (   32)    579-5
                    NUT, SHEET SPRING        (    2)    579-10
                    ROOF ASSY(COMPLETE)      (    1)    579-4
                    SCREW, TAPPING           (    2)    579-11
                    SEAL, NONMETALLIC ST     (    2)    579-6
                    WASHER, FLAT             (   48)    579-8
```

END OF FIGURE

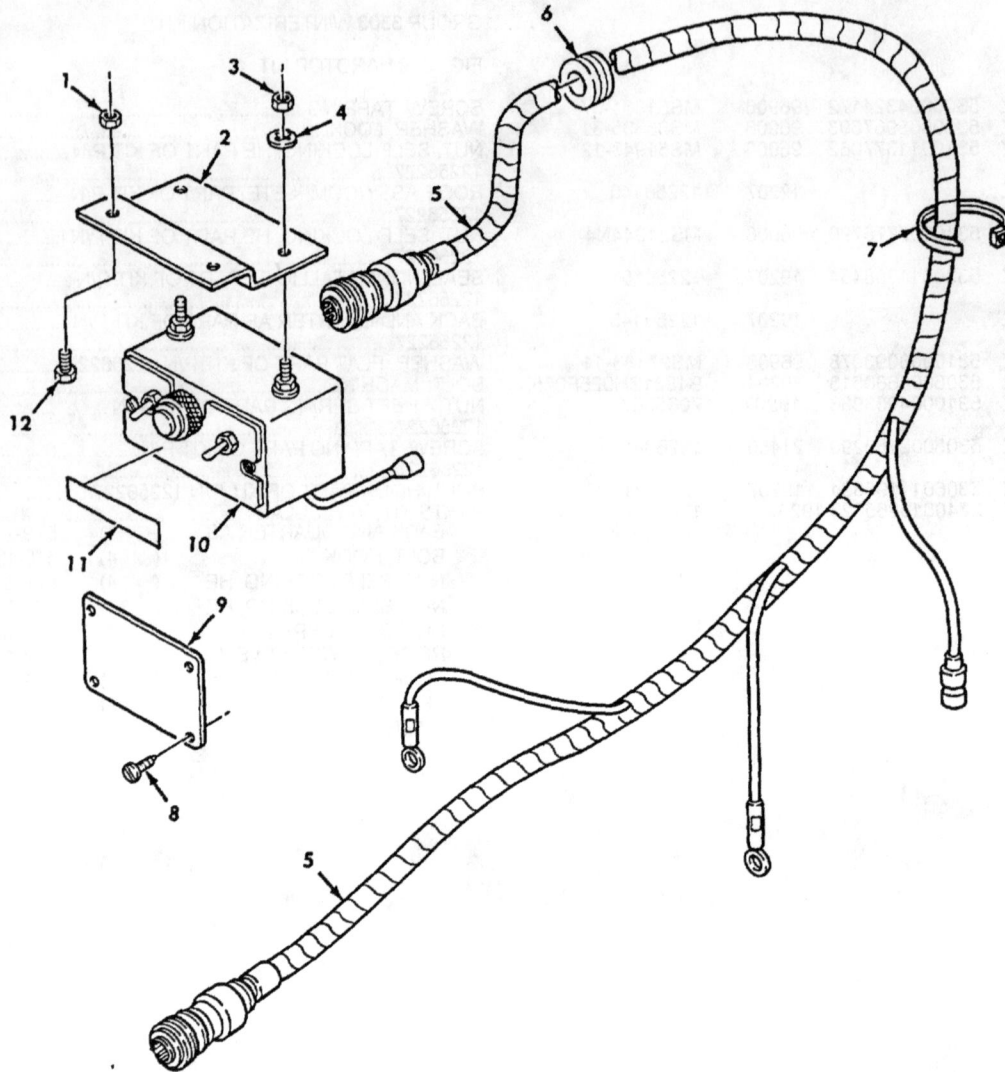

Figure 580. Engine Coolant Heater Kit, Heater Control Box, and Wiring Harness.

(1) ITEM NO	(2) SMR CODE	(3) NSN	(4) CAGEC	(5) PART NUMBER	(6) DESCRIPTION AND USABLE ON CODES (UOC)	(7) QTY
					GROUP 3303 WINTERIZATION KITS	
					FIG. 580 ENGINE COOLANT HEATER KIT, HEATER CONTROL BOX, AND WIRING HARNESS	
1	PAOZZ	5310009359022	96906	MS51943-32	NUT, SELF-LOCKING, HE PART OF KIT P/N 12256466..................... UOC:DAA, DAB, DAC, DAD, DAE, DAF, DAG, DAH, DAJ, DAK, DAL, DAW, DAX, V12, V13, V14, V15, V16, V17, V18, V19, V20, V21, V22, V24, V25, V39	2
2	PAOZZ	5340011097950	19207	11648534	BRACKET, MOUNTING PART OF KIT P/N 12256466..................... UOC:DAA, DAB, DAC, DAD, DAE, DAF, DAG, DAH, DAJ, DAK, DAL, DAW, DAX, V12, V13, V14, V15, V16, V17, V18, V19, V20, V21, V22, V24, V25, V39	1
3	PAOZZ	5310007616882	96906	MS51967-2	NUT, PLAIN, HEXAGON PART OF KIT P/N 12256466..................... UOC:DAA, DAB, DAC, DAD, DAE, DAF, DAG, DAH, DAJ, DAK, DAL, DAW, DAX, V12, V13, V14, V15, V16, V17, V18, V19, V20, V21, V22, V24, V25, V39	2
4	PAOZZ	5310005825965	96906	MS35338-44	WASHER, LOCK PART OF KIT P/N 12256466.............. UOC:DAA, DAB, DAC, DAD, DAE, DAF, DAG, DAH, DAJ, DAK, DAL, DAW, DAX, V12, V13, V14, V15, V16, V17, V18, V19, V20, V21, V22, V24, V25, V39	2
5	PAOZZ	6150010965053	19207	12256438	CABLE ASSEMBLY, SPEC PART OF KIT P/N 12256466..................... UOC:DAA, DAB, DAC, DAD, DAE, DAF, DAG, DAH, DAJ, DAK, DAL, DAW, DAX, V12, V13, V14, V15, V16, V17, V18, V19, V20, V21, V22, V24, V25, V39	1
6	PAOZZ	5325011247760	96906	MS35489-79	GROMMET, NONMETALLIC PART OF KIT P/N 12256466..................... UOC:DAA, DAB, DAC, DAD, DAE, DAF, DAG, DAH, DAJ, DAK, DAL, DAW, DAX, V12, V13, V14, V15, V16, V17, V18, V19, V20, V21, V22, V24, V25, V39	1
7	PAOZZ	5975008994606	96906	MS3367-2-0	STRAP, TIEDOWN, ELECT PART OF KIT P/N 12256466..................... UOC:DAA, DAB, DAC, DAD, DAE, DAF, DAG, DAH, DAJ, DAK, DAL, DAW, DAX, V12, V13, V14, V15, V16, V17, V18, V19, V20, V21, V22, V24, V25, V39	1
8	PAOZZ	5305007544355	21450	127851	SCREW, TAPPING PART OF KIT P/N 12256466..................... UOC:DAA, DAB, DAC, DAD, DAE, DAF, DAG, DAH, DAJ, DAK, DAL, DAW, DAX, V12, V13, V14, V15, V16, V17, V18, V19, V20, V21, V22, V24, V25,	4

(1) ITEM NO	(2) SMR CODE	(3) NSN	(4) CAGEC	(5) PART NUMBER	(6) DESCRIPTION AND USABLE ON CODES (UOC)	(7) QTY
					V39	
9	PAOZZ	9905010327002	19207	10896651	PLATE, INSTRUCTION PART OF KIT P/N .... 12256466 .... UOC:DAA, DAB, DAC, DAD, DAE, DAF, DAG, DAH, DAJ, DAK, DAL, DAW, DAX, V12, V13, V14, V15, V16, V17, V18, V19, V20, V21, V22, V24, V25, V39	1
10	PAOZZ	2590011256154	19207	11669705	CONTROL BOX, ELECTRI PART OF KIT P/N .... 12256466 .... UOC:DAA, DAB, DAC, DAD, DAE, DAF, DAG, DAH, DAJ, DAK, DAL, DAW, DAX, V12, V13, V14, V15, V16, V17, V18, V19, V20, V21, V22, V24, V25, V39	1
11	PAOZZ	7690000306615	19207	10896514	DECAL PART OF KIT P/N 12256466 .... UOC:DAA, DAB, DAC, DAD, DAE, DAF, DAG, DAH, DAJ, DAK, DAL, DAW, DAX, V12, V13, V14, V15, V16, V17, V18, V19, V20, V21, V22, V24, V25, V39	1
12	PAOZZ	5305002678953	80204	B1821BH025F063N	SCREW, CAP, HEXAGON H PART OF KIT P/N ...... 12256466 .... UOC:DAA, DAB, DAC, DAD, DAE, DAF, DAG, DAH, DAJ, DAK, DAL, DAW, DAX, V12, V13, V14, V15, V16, V17, V18, V19, V20, V21, V22, V24, V25, V39	2

END OF FIGURE

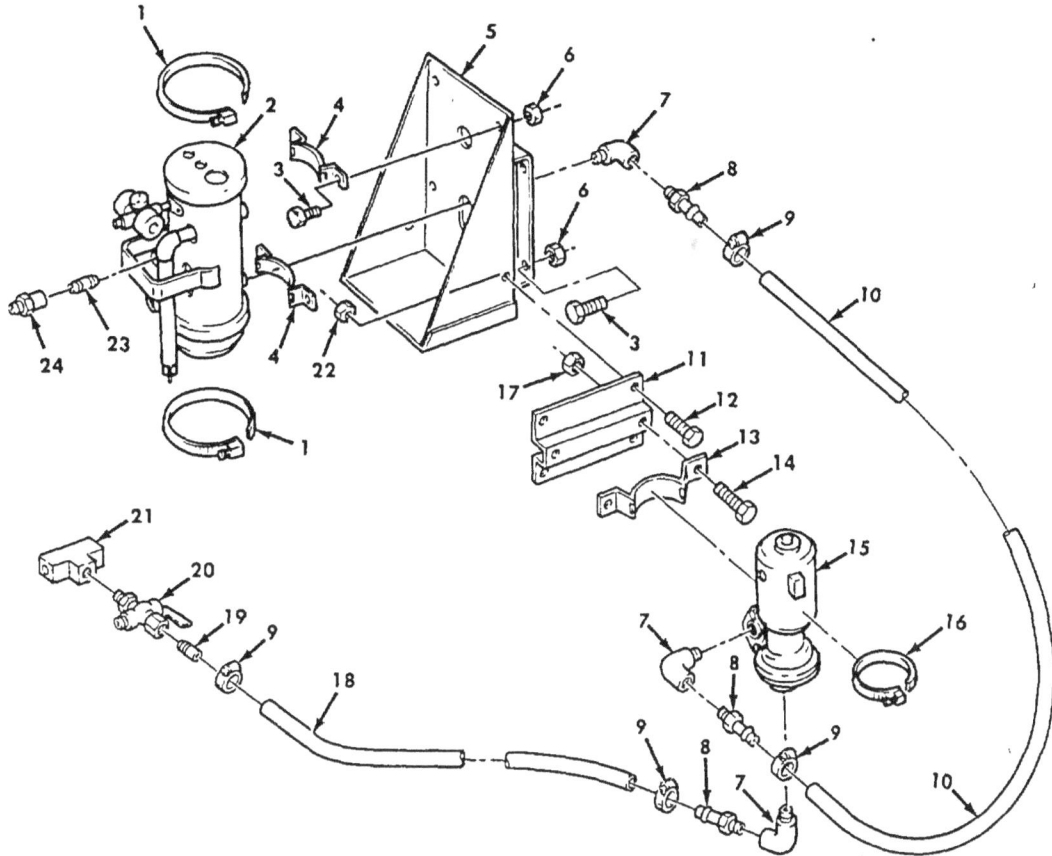

Figure 581. Engine Coolant Heater Kit, Heater Assembly, Pump, and Related Parts.

(1) ITEM NO	(2) SMR CODE	(3) NSN	(4) CAGEC	(5) PART NUMBER	(6) DESCRIPTION AND USABLE ON CODES (UOC)	(7) QTY
					GROUP 3303 WINTERIZATION KITS	
					FIG. 581 ENGINE COOLANT HEATER KIT, HEATER ASSEMBLY, PUMP, AND RELATED PARTS	
1	PAOZZ	4730009086294	75160	C16054	CLAMP, HOSE PART OF KIT P/N 12256466..................... UOC:DAA, DAB, DAC, DAD, DAE, DAF, DAG, DAH, DAJ, DAK, DAL, DAW, DAX, V12, V13, V14, V15, V16, V17, V18, V19, V20, V21, V22, V24, V25, V39	2
2	PAOFZ	2990009971532	19207	11601698	HEATER, COOLANT, ENGI PART OF KIT P/N..................... 12256466... UOC:DAA, DAB, DAC, DAD, DAE, DAF, DAG, DAH, DAJ, DAK, DAL, DAW, DAX, V12, V13, V14, V15, V16, V17, V18, V19, V20, V21, V22, V24, V25, V39	1
3	PAOZZ	5305002693234	96906	MS90727-58	SCREW, CAP, HEXT PART OF KIT P/N .... 12256466... UOC:DAA, DAB, DAC, DAD, DAE, DAF, DAG, DAH, DAJ, DAK, DAL, DAW, DAX, V12, V13, V14, V15, V16, V17, V18, V19, V20, V21, V22, V24, V25, V39	8
4	KFOZZ	5340012068586	19207	10896477	STRAP, RETAINING PART OF KIT P/N .... 12256466... UOC:DAA, DAB, DAC, DAD, DAE, DAF, DAG, DAH, DAJ, DAK, DAL, DAW, DAX, V12, V13, V14, V15, V16, V17, V18, V19, V20, V21, V22, V24, V25, V39	2
5	KFOZZ		19207	12256437	BRACKET ASSEMBLY PART OF KIT P/N............................ 12256466... UOC:DAA, DAB, DAC, DAD, DAE, DAF, DAG, DAH, DAJ, DAK, DAL, DAW, DAX, V12, V13, V14, V15, V16, V17, V18, V19, V20, V21, V22, V24, V25, V39	1
6	PAOZZ	5310008140672	96906	MS51943-36	NUT, SELF-LOCKING, HE PART OF KIT P/N...................... 12256466... UOC:DAA, DAB, DAC, DAD, DAE, DAF, DAG, DAH, DAJ, DAK, DAL, DAW, DAX, V12, V13, V14, V15, V16, V17, V18, V19, V20, V21, V22, V24, V25, V39	8
7	PAOZZ	4730012213565	81343	8-8 130239C	ELBOW, PIPE PART OF KIT P/N 12256466...................... UOC:DAA, DAB, DAC, DAD, DAE, DAF, DAG, DAH, DAJ, DAK, DAL, DAW, DAX, V12, V13, V14, V15, V16, V17, V18, V19, V20, V21, V22, V24, V25, V39	4
8	PAOZZ	4730009445888	96906	MS24522-7	ADAPTER, STRAIGHT, PI PART OF KIT P/N .... 12256466... UOC:DAA, DAB, DAC, DAD, DAE, DAF, DAG, DAH, DAJ, DAK, DAL, DAW, DAX, V12, V13, V14, V15, V16, V17, V18, V19, V20, V21, V22, V24, V25, V39	4

(1) ITEM NO	(2) SMR CODE	(3) NSN	(4) CAGEC	(5) PART NUMBER	(6) DESCRIPTION AND USABLE ON CODES (UOC)	(7) QTY
9	PAOZZ	4730009083194	96906	MS35842-11	CLAMP, HOSE PART OF KIT P/N 12256466. UOC:DAA, DAB, DAC, DAD, DAE, DAF, DAG, DAH, DAJ, DAK, DAL, DAW, DAX, V12, V13, V14, V15, V16, V17, V18, V19, V20, V21, V22, V24, V25, V39	2
10	MOOZZ		19207	8710557-23	HOSE, MAKE FROM HOSE, P/N 8710557 PART OF KIT P/N 12256466 UOC:DAA, DAB, DAC, DAD, DAE, DAF, DAG, DAH, DAJ, DAK, DAL, DAW, DAX, V12, V13, V14, V15, V16, V17, V18, V19, V20, V21, V22, V24, V25, V39	1
11	KFOZZ	5340012885156	19207	12302637	STRAP, RETAINING PART OF KIT P/N 12256466 UOC:DAA, DAB, DAC, DAD, DAE, DAF, DAG, DAH, DAJ, DAK, DAL, DAW, DAX, V12, V13, V14, V15, V16, V17, V18, V19, V20, V21, V22, V24, V25, V39	1
12	PAOZZ	5305000680512	80204	B1821BH025F056N	SCREW, CAP, HEXAGON H PART OF KIT P/N 12256466 UOC:DAA, DAB, DAC, DAD, DAE, DAF, DAG, DAH, DAJ, DAK, DAL, DAW, DAX, V12, V13, V14, V15, V16, V17, V18, V19, V20, V21, V22, V24, V25, V39	4
13	PFOZZ	5340011467667	19207	11592371	STRAP, RETAINING PART OF KIT P/N 12256466 UOC:DAA, DAB, DAC, DAD, DAE, DAF, DAG, DAH, DAJ, DAK, DAL, DAW, DAX, V12, V13, V14, V15, V16, V17, V18, V19, V20, V21, V22, V24, V25, V39	1
14	PAOZZ	5306002259086	96906	MS90726-31	BOLT, MACHINE PART OF KIT P/N 12256466 UOC:DAA, DAB, DAC, DAD, DAE, DAF, DAG, DAH, DAJ, DAK, DAL, DAW, DAX, V12, V13, V14, V15, V16, V17, V18, V19, V20, V21, V22, V24, V25, V39	2
15	PAOZZ	4320009302045	19207	10946835	PUMP UNIT, CENTRIFUG PART OF KIT P/N 12256466 UOC:DAA, DAB, DAC, DAD, DAE, DAF, DAG, DAH, DAJ, DAK, DAL, DAW, DAX, V12, V13, V14, V15, V16, V17, V18, V19, V20, V21, V22, V24, V25, V39	1
16	PAOZZ	4730009086292	96906	MS35842-14	CLAMP, HOSE PART OF KIT P/N 12256466. UOC:DAA, DAB, DAC, DAD, DAE, DAF, DAG, DAH, DAJ, DAK, DAL, DAW, DAX, V12, V13, V14, V15, V16, V17, V18, V19, V20, V21, V22, V24, V25, V39	1
17	PAOZZ	5310002416658	96906	MS51943-34	NUT, SELF-LOCKING, HE PART OF KIT P/N 12256466 UOC:DAA, DAB, DAC, DAD, DAE, DAF, DAG, DAH, DAJ, DAK, DAL, DAW, DAX, V12, V13, V14, V15, V16, V17, V18, V19, V20, V21, V22, V24, V25, V39	4

(1) ITEM NO	(2) SMR CODE	(3) NSN	(4) CAGEC	(5) PART NUMBER	(6) DESCRIPTION AND USABLE ON CODES (UOC)	(7) QTY
18	MOOZZ		19207	8710557-92	HOSE MAKE FROM HOSE, P/N .... MS521301B203R UOC:DAA, DAB, DAC, DAD, DAE, DAF, DAG, DAH, DAJ, DAK, DAL, DAW, DAX, V12, V1, V14, V15, V16, V17, V18, V19, V20, V2 1, V22, V24, V25, V39	1
19	PAOZZ	4730008176578	96906	MS1415-4	BUSHING, PIPE PART OF KIT P/N .... 12256466 UOC:DAA, DAB, DAC, DAD, DAE, DAF, DAG, DAH, DAJ, DAK, DAL, DAW, DAX, V12, V13, V14, V15, V16, V17, V18, V19, V20, V21, V22, V24, V25, V39	1
20	PAOZZ	4820002743646	21450	904030	COCK, PLUG PART OF KIT P/N 12256466................ UOC:DAA, DAB, DAC, DAD, DAE, DAF, DAG, DAH, DAJ, DAK, DAL, DAW, DAX, V12, V13, V14, V15, V16, V17, V18, V19, V20, V21, V22, V24, V25, V39	1
21	PAOZZ		26417	454147	TEE, PIPE PART OF KIT P/N 12256466................. UOC:DAA, DAB, DAC, DAD, DAE, DAF, DAG, DAH, DAJ, DAK, DAL, DAW, DAX, V12, V13, V14, V15, V16, V17, V18, V19, V20, V21, V22, V24, V25, V39	1
22	PAOZZ	5310009359022	96906	MS51943-32	NUT, SELF-LOCKING PART OF KIT P/N.............. 12256466 UOC:DAA, DAB, DAC, DAD, DAE, DAF, DAG, DAH, DAJ, DAK, DAL, DAW, DAX, V12, V13, V14, V15, V16, V17, V18, V19, V20, V21, V22, V24, V25, V39	4
23	PAOZZ	4730001961467	72582	121208	NIPPLE, PIPE PART OF KIT P/N .... 12256466 UOC:DAA, DAB, DAC, DAD, DAE, DAF, DAG, DAH, DAJ, DAK, DAL, DAW, DAX, V12, V13, V14, V15, V16, V17, V18, V19, V20, V21, V22, V24, V25, V39	1
24	PAOZZ	4730005425598	79470	C3309X8	COUPLING, PIPE PART OF KIT P/N .... 12256466 UOC:DAA, DAB, DAC, DAD, DAE, DAF, DAG, DAH, DAJ, DAK, DAL, DAW, DAX, V12, V13, V14, V15, V16, V17, V18, V19, V20, V21, V22, V24, V25, V39	1

END OF FIGURE

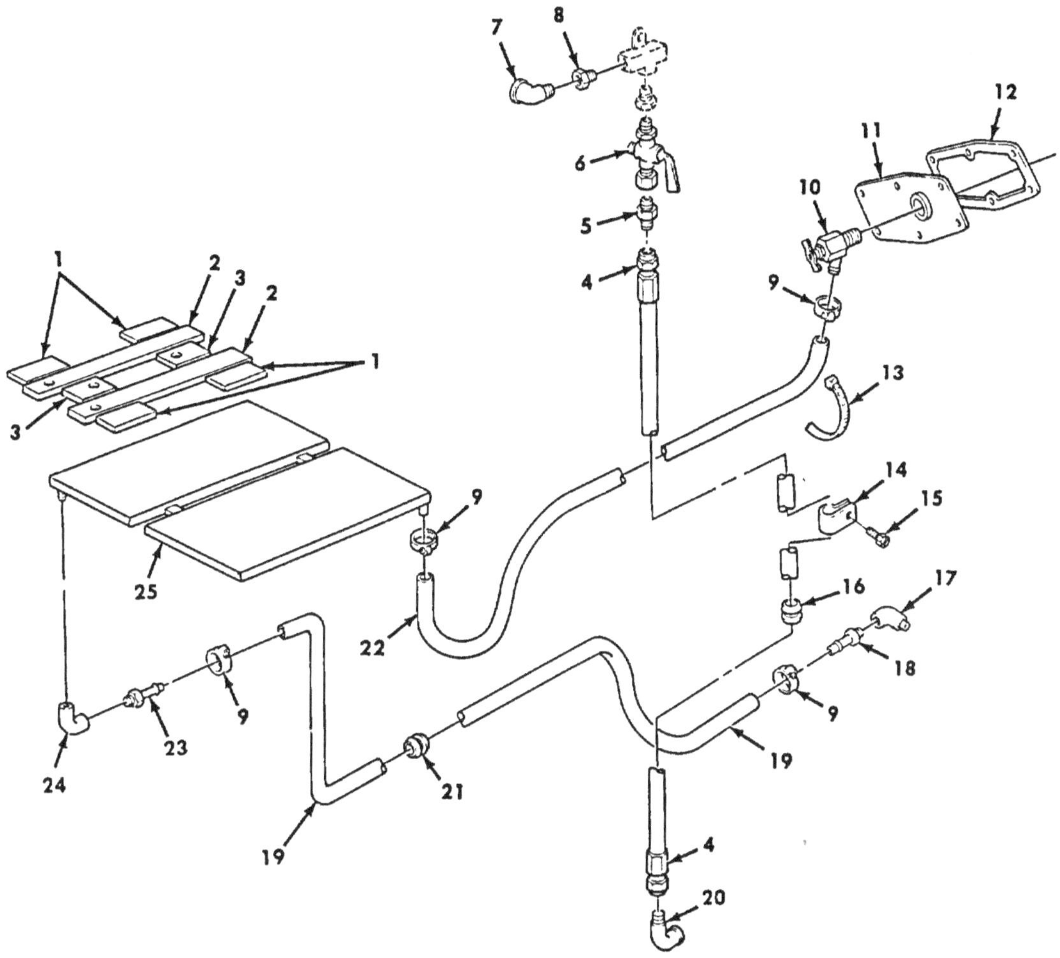

Figure 582. Engine Coolant Heater Kit, Hoses, Fittings, and Pad Assembly.

(1) ITEM NO	(2) SMR CODE	(3) NSN	(4) CAGEC	(5) PART NUMBER	(6) DESCRIPTION AND USABLE ON CODES (UOC)	(7) QTY
					GROUP 3303 WINTERIZATION KITS	
					FIG. 582 ENGINE COOLANT HEATER KIT, HOSES, FITTINGS, AND PAD ASSEMBLY	
1	PAOZZ	6140011256075	19207	12277132	BLOCK, SUPPORT, BATTE PART OF KIT P/N .... 12256466 .... UOC:DAA, DAB, DAC, DAD, DAE, DAF, DAG, DAH, DAJ, DAK, DAL, DAW, DAX, V12, V13, V14, V15, V16, V17, V18, V19, V20, V21, V22, V24, V25, V39	4
2	PAOZZ	6140011256073	19207	12277134	BLOCK, SUPPORT, BATTE PART OF KIT P/N .... 12256466 .... UOC:DAA, DAB, DAC, DAD, DAE, DAF, DAG, DAH, DAJ, DAK, DAL, DAW, DAX, V12, V13, V14, V15, V16, V17, V18, V19, V20, V21, V22, V24, V25, V39	2
3	PAOZZ	6140011256074	19207	12277133	BLOCK, SUPPORT, BATTE PART OF KIT P/N .... 12256466 .... UOC:DAA, DAB, DAC, DAD, DAE, DAF, DAG, DAH, DAJ, DAK, DAL, DAW, DAX, V12, V13, V14, V15, V16, V17, V18, V19, V20, V21, V22, V24, V25, V39	2
4	PAOZZ	4720011063986	19207	12256449	HOSE ASSEMBLY, NONME PART OF KIT P/N .... 12256466 .... UOC:DAA, DAB, DAC, DAD, DAE, DAF, DAG, DAH, DAJ, DAK, DAL, DAW, DAX, V12, V13, V14, V15, V16, V17, V18, V19, V20, V21, V22, V24, V25, V39	1
5	PAOZZ	4730009003296	07295	100740	ADAPTER, STRAIGHT, PI PART OF KIT P/N .... 12256466 .... UOC:DAA, DAB, DAC, DAD, DAE, DAF, DAG, DAH, DAJ, DAK, DAL, DAW, DAX, V12, V13, V14, V15, V16, V17, V18, V19, V20, V21, V22, V24, V25, V39	1
6	PAOZZ	4820002743646	21450	543852	COCK, PLUG PART OF KIT P/N 12256466 .... UOC:DAA, DAB, DAC, DAD, DAE, DAF, DAG, DAH, DAJ, DAK, DAL, DAW, DAX, V12, V13, V14, V15, V16, V17, V18, V19, V20, V21, V22, V24, V25, V39	1
7	PAOZZ	4730002775553	96906	MS51952-2	ELBOW, PIPE PART OF KIT P/N 12256466.......................... UOC:DAA, DAB, DAC, DAD, DAE, DAF, DAG, DAH, DAJ, DAK, DAL, DAW, DAX, V12, V13, V14, V15, V16, V17, V18, V19, V20, V21, V22, V24, V25, V39	1
8	PAOZZ	4730001960930	96906	MS14315-1XA	BUSHING, PIPE PART OF KIT P/N .... 12256466 .... UOC:DAA, DAB, DAC, DAD, DAE, DAF, DAG, DAH, DAJ, DAK, DAL, DAW, DAX, V12, V13, V14, V15, V16, V17, V18, V19, V20, V21, V22, V24, V25, V39	1
9	PAOZZ	4730009083194	96906	MS35842-11	CLAMP, HOSE PART OF KIT P/N 12256466.	6

(1) ITEM NO	(2) SMR CODE	(3) NSN	(4) CAGEC	(5) PART NUMBER	(6) DESCRIPTION AND USABLE ON CODES (UOC)	(7) QTY
					UOC:DAA, DAB, DAC, DAD, DAE, DAF, DAG, DAH, DAJ, DAK, DAL, DAW, DAX, V12, V13, V14, V15, V16, V17, V18, V19, V20, V21, V22, V24, V25, V39	
10	PAOZZ	4820011010080	19207	10923515	COCK, DRAIN PART OF KIT P/N 12256466	2
					UOC:DAA, DAB, DAC, DAD, DAE, DAF, DAG, DAH, DAJ, DAK, DAL, DAW, DAX, V12, V13, V14, V15, V16, V17, V18, V19, V20, V21, V22, V24, V25, V39	
11	PAOZZ	2990011929724	19207	12302639	COVER, WATER HEADER PART OF KIT P/N 12256466	1
					UOC:DAA, DAB, DAC, DAD, DAE, DAF, DAG, DAH, DAJ, DAK, DAL, DAW, DAX, V12, V13, V14, V15, V16, V17, V18, V19, V20, V21, V22, V24, V25, V39	
12	PAOZZ	5330011973228	19207	12302635	GASKET PART OF KIT P/N 12256466	1
					UOC:DAA, DAB, DAC, DAD, DAE, DAF, DAG, DAH, DAJ, DAK, DAL, DAW, DAX, V12, V13, V14, V15, V16, V17, V18, V19, V20, V21, V22, V24, V25, V39	
13	PAOZZ	5975008994606	96906	MS3367-2-0	STRAP, TIEDOWN, ELECT PART OF KIT P/N 12256466	14
					UOC:DAA, DAB, DAC, DAD, DAE, DAF, DAG, DAH, DAJ, DAK, DAL, DAW, DAX, V12, V13, V14, V15, V16, V17, V18, V19, V20, V21, V22, V24, V25, V39	
14	PAOZZ	5340007022848	96906	MS21333-128	CLAMP, LOOP PART OF KIT P/N 12256466	5
					UOC:DAA, DAB, DAC, DAD, DAE, DAF, DAG, DAH, DAJ, DAK, DAL, DAW, DAX, V12, V13, V14, V15, V16, V17, V18, V19, V20, V21, V22, V24, V25, V39	
15	PAOZZ	5305011013312	96906	MS51851-126	SCREW, TAPPING PART OF KIT P/N 12256466	5
					UOC:DAA, DAB, DAC, DAD, DAE, DAF, DAG, DAH, DAJ, DAK, DAL, DAW, DAX, V12, V13, V14, V15, V16, V17, V18, V19, V20, V21, V22, V24, V25, V39	
16	PAOZZ	5325002901960	96906	MS35489-27	GROMMET, NONMETALLIC. PART OF KIT P/N 12256466	1
					UOC:DAA, DAB, DAC, DAD, DAE, DAF, DAG, DAH, DAJ, DAK, DAL, DAW, DAX, V12, V13, V14, V15, V16, V17, V18, V19, V20, V21, V22, V24, V25, V39	
17	PAOZZ	4730012213565	81343	8-8 130239C	ELBOW, PIPE PART OF KIT P/N 12256466	1
					UOC:DAA, DAB, DAC, DAD, DAE, DAF, DAG, DAH, DAJ, DAK, DAL, DAW, DAX, V12, V13, V14, V15, V16, V17, V18, V19, V20, V21, V22, V24, V25, V39	
18	PAOZZ	4730009445888	96906	MS24522-7	ADAPTER, STRAIGHT, PI PART OF KIT P/N 12256466	1
					UOC:DAA, DAB, DAC, DAD, DAE, DAF, DAG, DAH, DAJ, DAK, DAI, DAW, DAX, V12, V13, V14, V15, V16, V17, V18, V19, V20, V2, V22 , V24, V25, V39	

(1) ITEM NO	(2) SMR CODE	(3) NSN	(4) CAGEC	(5) PART NUMBER	(6) DESCRIPTION AND USABLE ON CODES (UOC)	(7) QTY
19	MOOZZ		19207	8710557-46	HOSE, MAKE FROM HOSE, P/N 8710557, 46 INCHES LONG PART OF KIT P/N 12256466 .... UOC:DAA, DAB, DAC, DAD, DAE, DAF, DAG, DAH, DAJ, DAK, DAL, DAW, DAX, V12 , V13 , V14 , V15, V16, V17, V18, V19, V20, V21, V22, V24, V25, V39	2
20	PAOZZ	4730002546211	99199	A335	ELBOW, PIPE TO TUBE PART OF KIT P/N 12256466 .... UOC:DAA, DAB, DAC, DAD, DAE, DAF, DAG, DAH, DAJ, DAK, DAL, DAW, DAX, V12, V13, V14, V15, V16, V17, V18, V19, V20, V21, V22, V24, V25, V39	2
21	PAOZZ	5325002766343	10001	33G1724	GROMMET, NONMETALLIC PART OF KIT P/N 12256466 .... UOC:DAA, DAB, DAC, DAD, DAE, DAF, DAG, DAH, DAJ, DAK, DAL, DAW, DAX, V12, V13, V14, V15, V16, V17, V18, V19, V20, V21, V22, V24, V25, V39	1
22	MOOZZ		19207	8710557-84	HOSE, NONMETALLIC MAKE FROM HOSE P/ N 8710557 PART OF KIT P/N 12256466 .... UOC:DAA, DAB, DAC, DAD, DAE, DAF, DAG, DAH, DAJ, DAK, DAL, DAW, DAX, V12, V13, V14, V15, V16, V17, V18 , V19, V20, V21, V22, V24, V25, V39	1
23	PAOZZ	4730012864611	96906	MS24522-23	ADAPTER, STRAIGHT, PI PART OF KIT P/N 12256466 .... UOC:DAA, DAB, DAC, DAD, DAE, DAF, DAG, DAH, DAJ, DAK, DAL, DAW, DAX, V12, V13, V14, V15, V16, V17, V18, V19, V20, V21, V22 , V24, V25, V39	1
24	PAOZZ	4730010663071	24617	454086	ELBOW, PIPE PART OF KIT P/N 12256466 .... UOC:DAA, DAB, DAC, DAD, DAE, DAF, DAG, DAH, DAJ, DAK, DAL, DAW, DAX, V12, V13, V14, V15, V16, V17, V18, V19, V20, V21, V22, V24, V25, V39	2
25	KFOZZ	2540012361175	19207	12256434	PAD ASSEMBLY PART OF KIT P/N 12256466 .... UOC:DAA, DAB, DAC, DAD, DAE, DAF, DAG, DAH, DAJ, DAK, DAL, DAW, DAX, V12, V13, V14, V15, V16, V17, V18, V19, V20, V21, V22, V24, V25, V39	1

END OF FIGURE

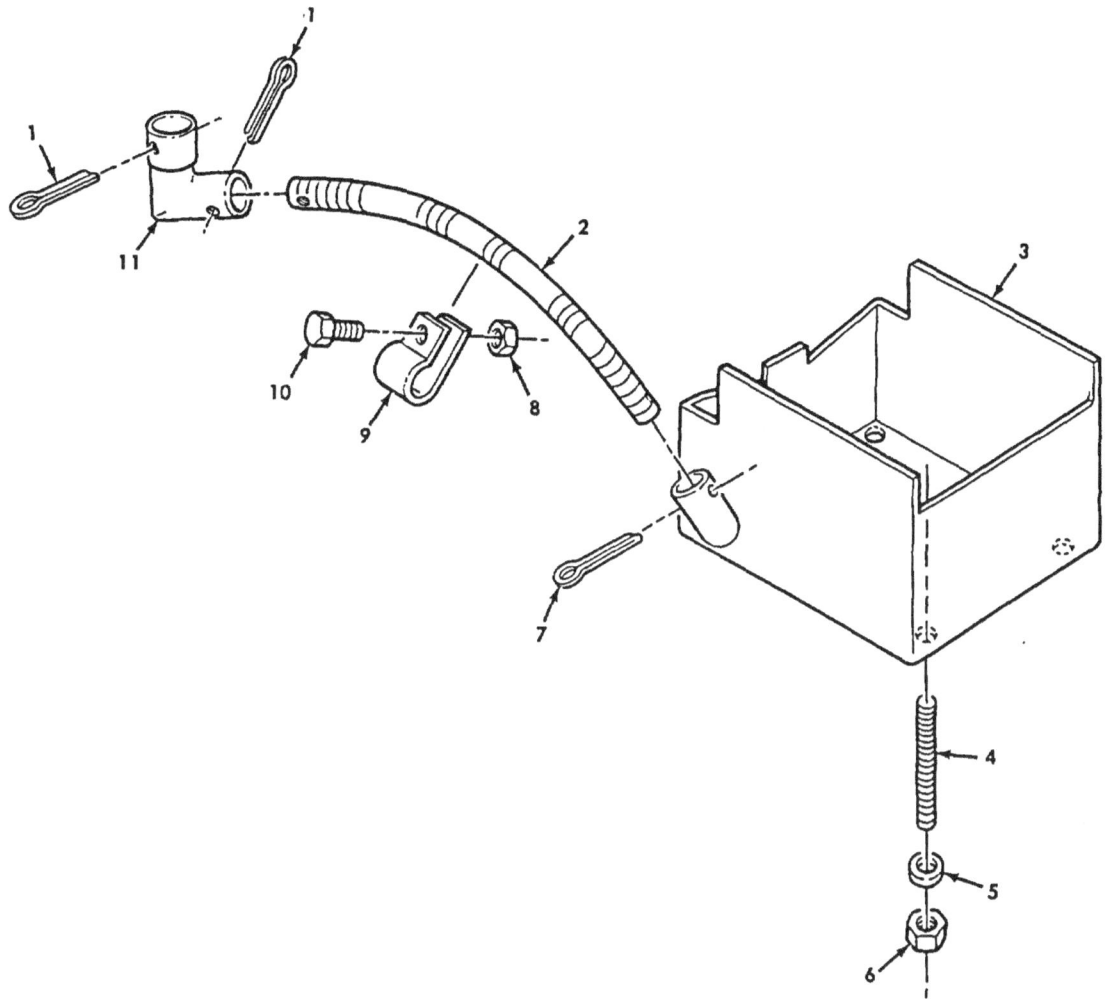

Figure 583. Engine Coolant Heater Kit, Oil Pan Shroud, and Related Parts.

(1) ITEM NO	(2) SMR CODE	(3) NSN	(4) CAGEC	(5) PART NUMBER	(6) DESCRIPTION AND USABLE ON CODES (UOC)	(7) QTY
					GROUP 3303 WINTERIZATION KITS	
					FIG. 583 ENGINE COOLANT HEATER KIT, OIL PAN SHROUD, AND RELATED PARTS	
1	PAOZZ	5315008499854	96906	MS24665-498	PIN, COTTER PART OF KIT P/N 12256466. UOC:DAA, DAB, DAC, DAD, DAE, DAF, DAG, DAH, DAJ, DAK, DAL, DAW, DAX, V12, V13, V14, V15, V16, V17, V18, V19, V20, V21, V22, V24, V25, V39	2
2	PAOZZ	4720011810098	19207	7986268	HOSE, METALLIC PART OF KIT P/N 12256466. UOC:DAA, DAB, DAC, DAD, DAE, DAF, DAG, DAH, DAJ, DAK, DAL, DAW, DAX, V12, V13, V14, V15, V16, V17, V18, V19, V20, V21, V22, V24, V25, V39	1
3	KFOZZ	2990011764801	19207	11648559	SHROUD, OIL PAN PART OF KIT P/N 12256466. UOC:DAA, DAB, DAC, DAD, DAE, DAF, DAG, DAH, DAJ, DAK, DAL, DAW, DAX, V12, V13, V14, V15, V16, V17, V18, V19, V20, V21, V22, V24, V25, V39	1
4	PAOZZ	5307001744863	19207	11648628	STUD, CONTINUOUS THR UOC:DAA, DAB, DAC, DAD, DAE, DAF, DAG, DAH, DAJ, DAK, DAL, DAW, DAX, V12, V13, V14, V15, V16, V17, V18, V19, V20, V21, V22, V24, V25, V39	4
5	PAOZZ	5310000877493	96906	MS27183-13	WASHER, FLAT PART OF KIT P/N 12256466 UOC:DAA, DAB, DAC, DAD, DAE, DAF, DAG, DAH, DAJ, DAK, DAL, DAW, DAX, V12, V1, VV14, V15, V16, V17, V18, V19, V20, V21, V22, V24, V25, V39	4
6	PAOZZ	5310008140673	96906	MS51943-33	NUT, SELF-LOCKING, HE UOC:DAA, DAB, DAC, DAD, DAE, DAF, DAG, DAH, DAJ, DAK, DAL, DAW, DAX, V12, V13, V14, V15, V16, V17, V18, V19, V20, V21, V22, V24, V25, V39	4
7	PAOZZ	5315000590187	96906	MS24665-363	PIN, COTTER PART OF KIT P/N 12256466. UOC:DAA, DAB, DAC, DAD, DAE, DAF, DAG, DAH, DAJ, DAK, DAL, DAW, DAX, V12, V13, V14, V15, V16, V17, V18, V19, V20, V21, V22, V24, V25, V39	1
8	PAOZZ	5310008140672	96906	MS51943-36	NUT, SELF-LOCKING, HE PART OF KIT P/N 12256466. UOC:DAA, DAB, DAC, DAD, DAE, DAF, DAG, DAH, DAJ, DAK, DAL, DAW, DAX, V12, V13, V14, V15, V16, V17, V18, V19, V20, V21, V22, V24, V25, V39	1

(1) ITEM NO	(2) SMR CODE	(3) NSN	(4) CAGEC	(5) PART NUMBER	(6) DESCRIPTION AND USABLE ON CODES (UOC)	(7) QTY
9	PAOZZ	5340011730241	19207	8707524	CLAMP, LOOP, PART OF KIT P/N .... 12256466 .... UOC:DAA, DAB, DAC, DAD, DAE, DAF, DAG, DAH, DAJ, DAK, DAL, DAW, DAX, V12, V13, V14, V15, V16, V17, V18, V19, V20, V21, V22, V24, V25, V39	1
10	PAOZZ	5305002692803	96906	MS90726-60	SCREW, CAP, HEXAGON H PART OF KIT P/N .... 12256466 .... UOC:DAA, DAB, DAC, DAD, DAE, DAF, DAG, DAH, DAJ, DAK, DAL, DAW, DAX, V12, V13, V14, V15, V16, V17, V18, V19, V20, V21, V22, V24, V25, V39	1
11	KFOZZ		19207	7951084-3	ELBOW, TUBE .... UOC:DAA, DAB, DAC, DAD, DAE, DAF, DAG, DAH, DAJ, DAK, DAL, DAW, DAX, V12, V13, V14, V15, V16, V17, V18, V19, V20, V21, V22, V24, V25, V39	1
	PAOZZ	2990010958287	19207	12256466	HEATER, COOLANT, ENGI .... UOC:DAA, DAB, DAC, DAD, DAE, DAF, DAG, DAH, DAJ, DAK, DAL, DAW, DAX, V12, V13, V14, V15, V16, V17, V18, V19, V20, V21, V22, V24, V25, V39	1

ADAPTER, STRAIGHT, PI	(3)	581-8
ADAPTER, STRAIGHT, PI	(1)	582-23
ADAPTER, STRAIGHT, PI	(1)	582-5
ADAPTER, STRAIGHT, PI	(1)	582-18
BLOCK, SUPPORT, BATTE	(4)	582-1
BLOCK, SUPPORT, BATTE	(2)	582-2
BLOCK, SUPPORT, BATTE	(2)	582-3
BOLT, MACHINE	(2)	581-14
BRACKRT	(1)	581-11
BRACKET ASSEMBLY	(1)	581-6
BRACKET, MOUNTING	(1)	580-2
BUSHING, PIPE	(1)	582-8
BUSHING, PIPE	(6)	581-19
CABLE ASSEMBLY, SPEC	(1)	580-5
CLAMP, HOSE	(2)	581-1
CLAMP, HOSE	(6)	582-9
CLAMP, HOSE	(1)	581-16
CLAMP, HOSE	(2)	581-9
CLAMP, LOOP	(1)	583-9
CLAMP, LOOP	(5)	582-14
COCK, DRAIN	(2)	582-10
COCK, PLUG	(1)	582-6
COCK, PLUG	(1)	581-20
CONTROL BOX, ELECTRI	(1)	580-10
COUPLING, PIPE	(1)	581-24
COVER, WATER HEADER	(1)	582-11

(1) ITEM NO	(2) SMR CODE	(3) NSN	(4) CAGEC	(5) PART NUMBER	(6) DESCRIPTION AND USABLE ON CODES (UOC)	(7) QTY
					DECAL	(1) 580-11
					ELBOW	(1) 583-11
					ELBOW	(2) 582-20
					ELBOW, PIPE	(1) 582-7
					ELBOW, PIPE	(4) 581-7
					ELBOW, PIPE	(2) 582-24
					ELBOW, PIPE	(1) 582-17
					GASKET	(1) 582-12
					GROMMET, NONMETALLIC	(1) 580-6
					GROMMET, NONMETALLIC	(1) 582-21
					HEATER, COOLANT, ENGI	(1) 581-2
					HOSE	(1) 582-19
					HOSE	(1) 581-10
					HOSE ASSEMBLY, NONME	(1) 582-4
					HOSE, METALLIC	(1) 583-2
					HOSE, NONMETALLIC	(1) 582-22
					NIPPLE, PIPE	(1) 581-23
					NUT, PLAIN, HEXAGON	(2) 580-3
					NUT, SELF-LOCKING, HE	(2) 580-1
					NUT, SELF-LOCKING, HE	(4) 581-17
					NUT, SELF-LOCKING, HE	(1) 583-8
					NUT, SELF-LOCKING, HE	(8) 581-6
					NUT, SELF-LOCKING, HE	(1) 581-22
					PAD ASSEMBLY	(1) 582-25
					PIN, COTTER	(1) 583-7
					PIN, COTTER	(2) 583-1
					PLATE, INSTRUCTION	(1) 580-9
					PUMP UNIT, CENTRIFUG	(1) 581-15
					SCREW, CAP, HEXAGON H	(4) 581-12
					SCREW, CAP, HEXAGON H	(2) 580-12
					SCREW, CAP, HEXAGON H	(1) 583-10
					SCREW, TAPPING	(4) 580-8
					SHROUD, OIL PAN	(1) 583-3
					STRAP, RETAINING	(2) 581-4
					STRAP, RETAINING	(1) 581-13
					STRAP, RETAINING	(1) 581-11
					STRAP, TIEDOWN, ELECT	(14) 582-13
					STRAP, TIEDOWN, ELECT	(1) 580-7
					TEE, PIPE	(1) 581-21
					WASHER, FLAT	(4) 583-5
					WASHER, LOCK	(2) 580-4

END OF FIGURE

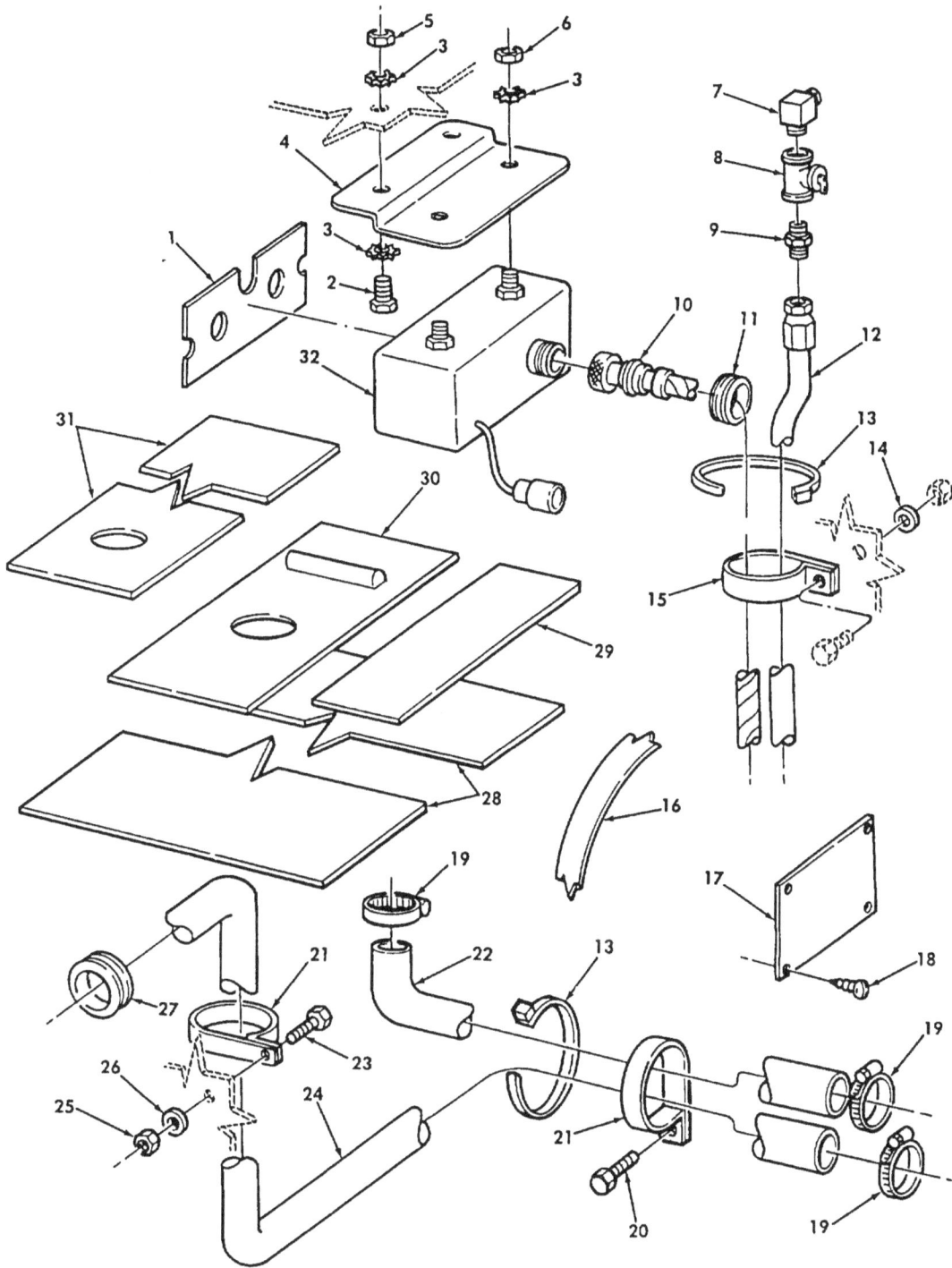

Figure 584. Engine Coolant Heater Kit, Control Box, and Related Parts (M939A2).

(1) ITEM NO	(2) SMR CODE	(3) NSN	(4) CAGEC	(5) PART NUMBER	(6) DESCRIPTION AND USABLE ON CODES (UOC)	(7) QTY
					GROUP 3303 WINTERIZATION KIT	
					FIG. 584 ENGINE COOLANT HEATER KIT, CONTROL BOX, AND RELATED PARTS(M939A2)	
1	PAOZZ	7690000306615	19207	10896514	DECAL PART OF KIT P/N 20511136	1
2	PAOZZ	5305002678953	96906	MS90727-5	SCREW, CAP, HEXAGON H PART OF KIT P/N 20511136	5
3	PAOZZ	5310002090786	96906	MS35335-33	WASHER, LOCK PART OF KIT P/N 20511136	6
4	PFOZZ	5340011097950	19207	11648534	BRACKET, MOUNTING PART OF KIT P/N 20511136	1
5	PAOZZ	5310009359022	96906	MS51943-32	NUT, SELF-LOCKING, HE PART OF KIT P/N 20511136	5
6	PAOZZ	5310007616882	96906	MS51967-2	NUT, PLAIN, HEXAGON PART OF KIT P/N 20511136	2
7	PAOZZ	4730011815777	98441	1/4CR-B	ELBOW, PIPE PART OF KIT P/N 20511136	1
8	PAOZZ	4820002633019	96906	MS35931-2	COCK, PLUG PART OF KIT P/N 20511136	1
9	PAOZZ	4730008036266	72582	8924145	ADAPTER, STRAIGHT, PI PART OF KIT P/N 20511136	1
10	PAOZZ	6150010965053	19207	12256438	CABLE ASSEMBLY, SPEC PART OF KIT P/N 20511136	1
11	PAOZZ	5325011247760	96906	MS35489-79	GROMMET, NONMETALLIC PART OF KIT P/N 20511136	1
12	PAOZZ	4720011063986	19207	12256449	HOSE ASSEMBLY, NONME PART OF KIT P/N 20511136	1
13	PAOZZ	5975001563253	96906	MS3367-2-9	STRAP, TIEDOWN, ELECT PART OF KIT P/N 20511136	14
14	PAOZZ	5310010925496	24617	9422845	WASHER, FLAT PART OF KIT P/N 20511136	1
15	PAOZZ	5340008091500	96906	MS21333-107	CLAMP, LOOP PART OF KIT P/N 20511136	1
16	PAOZZ	3030012873155	47457	8PK1730	BELT, V PART OF KIT P/N 20511136	1
17	PFOZZ	9905010327002	19207	10896651	PLATE, INSTRUCTION PART OF KIT P/N 20511136	1
18	PAOZZ	5305012962849	96906	MS24625-47	SCREW, TAPPING PART OF KIT P/N 20511136	4
19	PAOZZ	4730009083194	96906	MS35842-11	CLAMP, HOSE PART OF KIT P/N 20511136	3
20	PAOZZ	5306012899197	24617	11500713	BOLT, MACHINE PART OF KIT P/N 20511136	1
21	PAOZZ	5340007022848	96906	MS21333-128	CLAMP, LOOP PART OF KIT P/N 20511136	3
22	MOOZZ		19207	8710557 X 72 IN	HOSE, MAKE FROM HOSE, P/N 8710557, 72 INCHES LONG PART OF KIT P/N 20511136	1
23	PAOZZ	5305007195219	96906	MS90727-111	SCREW, CAP, HEXAGON H PART OF, KIT P/N 20511136	2
24	MOOZZ		19207	8710557 X 107 IN	HOSE, MAKE FROM HOSE, P/N 8710557, 107 INCHES LONG PART OF KIT P/N 20511136	1
25	PAOZZ	5310008140672	96906	MS51943-36	NUT, SELF-LOCKING, HE PART OF KIT P/N 20511136	2
26	PAOZZ	5310002748041	90407	12084P11	WASHER, FLAT PART OF KIT P/N 20511136	2
27	PAOZZ	5325002766343	96906	MS35489-23	GROMMET, NONMETALLIC PART OF KIT P/N 20511136	2

(1) ITEM NO	(2) SMR CODE	(3) NSN	(4) CAGEC	(5) PART NUMBER	(6) DESCRIPTION AND USABLE ON CODES (UOC)	(7) QTY
28	PFOZZ	2540012361175	19207	12256434	PAD ASSEMBLY PART OF KIT P/N .... 20511136..	1
29	PAOZZ	6140011256075	19207	12277132	BLOCK, SUPPORT, BATTE PART OF KIT P/N .... 20511136..	4
30	PAOZZ	6140011256074	19207	12277133	BLOCK, SUPPORT, BATTE PART OF KIT P/N .... 20511136..	2
31	PAOZZ	6140011256073	19207	12277134	BLOCK, SUPPORT, BATTE PART OF KIT P/N .... 20511136..	2
32	PAOFF	2590011256154	19207	11669705	CONTROL BOX, ELECTRI PART OF KIT P/N .... 20511136..	1

END OF FIGURE

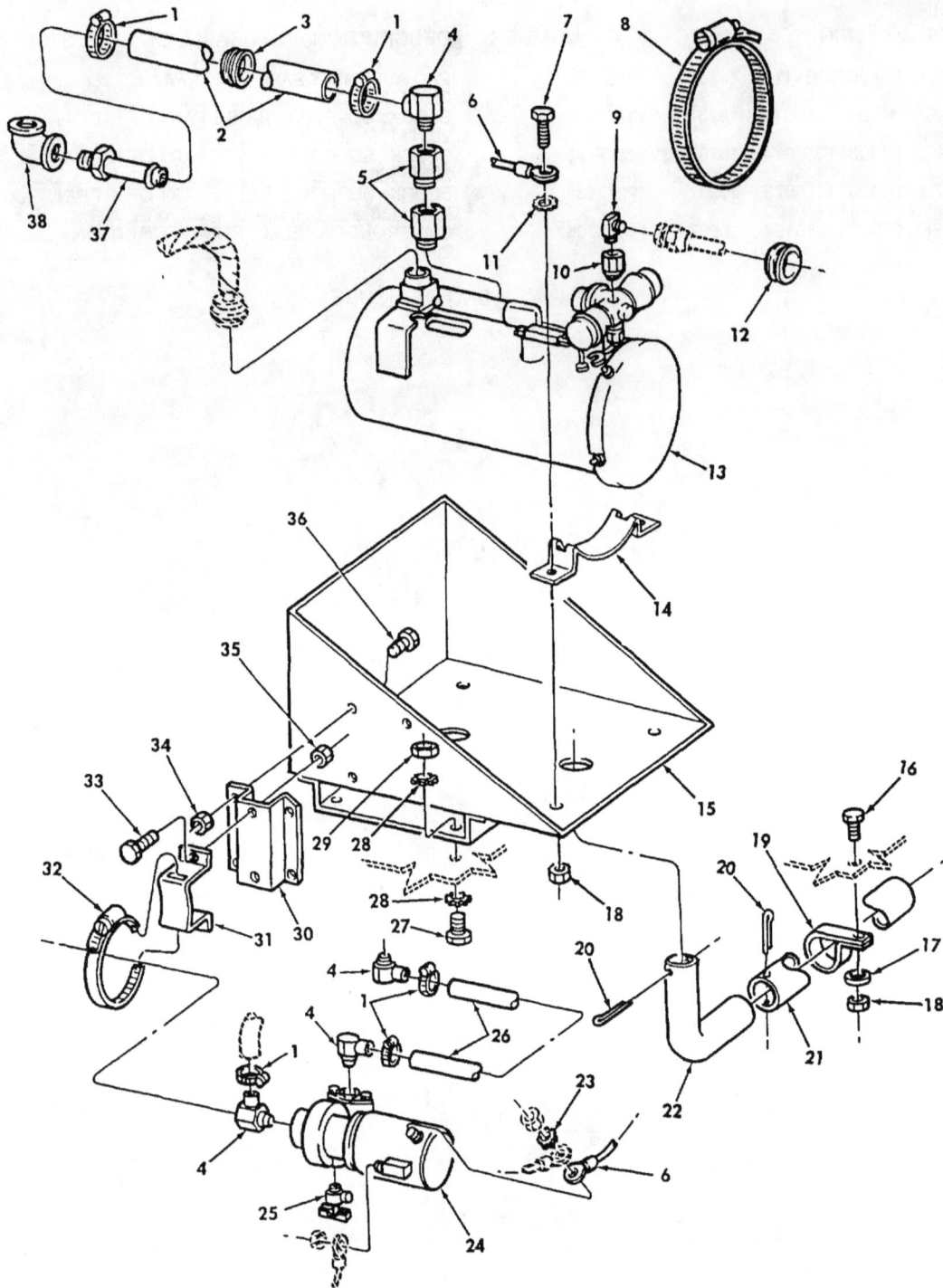

Figure 585. Engine Coolant Heater Kit, Heater Assembly, Pump Assembly,
and Related Parts (M939A2).

(1) ITEM NO	(2) SMR CODE	(3) NSN	(4) CAGEC	(5) PART NUMBER	(6) DESCRIPTION AND USABLE ON CODES (UOC)	(7) QTY
					GROUP 3303 WINTERIZATION KIT	
					FIG. 585 ENGINE COOLANT HEATER KIT, HEATER ASSEMBLY, PUMP ASSEMBLY, AND RELATED PARTS (M939A2)	
1	PAOZZ	4730009083194	96906	MS35842-11	CLAMP, HOSE PART OF KIT P/N 20511136.	5
2	MOOZZ		19207	8710557 X 44 IN	HOSE, MAKE FROM HOSE, P/N 8710557, 44 INCHES LONG PART OF KIT P/N 20511136	1
3	PAOZZ	5325002766343	96906	MS35489-23	GROMMET, NONMETALLIC PART OF KIT P/N 20511136	1
4	PAOZZ	4730009400947	96906	MS24519-7	ELBOW, PIPE TO HOSE PART OF KIT P/N 20511136	4
5	PAOZZ	4730004755168	24617	444004	COUPLING, PIPE PART OF KIT P/N 20511136	2
6	MOOZZ		47457	20511284 X BULK	CABLE ASSEMBLY, PART OF KIT P/N 20511136	2
7	PAOZZ	5305002693234	96906	MS90727-58	SCREW, CAP, HEXAGON H PART OF KIT P/N 20511136	4
8	PAOZZ	4730009086294	96906	MS35842-16	CLAMP, HOSE PART OF KIT P/N 20511136.	2
9	PAOZZ	4730002546211	21450	118753	ELBOW, PIPE TO TUBE PART OF KIT P/N 20511136	1
10	PAOZZ	4730004153172	46717	LA-519-9	COUPLING, PIPE PART OF KIT P/N 20511136	1
11	PAOZZ	5310006276128	96906	MS35335-35	WASHER, LOCK PART OF KIT P/N 20511136	4
12	PAOZZ	5325002901960	96906	MS35489-27	GROMMET, NONMETALLIC PART OF KIT P/N 20511136	1
13	PAOFF	2990009971532	19207	11601698	HEATER, COOLANT, ENGI PART OF KIT P/N 20511136	1
14	PFOZZ	5340012068586	19207	10896477	STRAP, RETAINING PART OF KIT P/N 20511136	2
15	KFOZZ		19207	12256437	BRACKET ASSEMBLY, PART OF KIT P/N 20511136	1
16	PAOZZ	5305007195219	96906	MS90727-111	SCREW, CAP, HEXAGON PART OF KIT P/N 20511136	2
17	PAOZZ	5310002748041	90407	12084P11	WASHER, FLAT PART OF KIT P/N 20511136	17
18	PAOZZ	5310008140672	96906	MS51943-39	NUT, SELF-LOCKING, HE PART OF KIT P/N 20511136	5
19	PAOZZ	5340005980225	88044	AN742-26	CLAMP, LOOP PART OF KIT P/N 20511136	1
20	PAOZZ	5315008792910	96906	MS24665-427	PIN, COTTER PART OF KIT P/N 20511136.	2
21	PAOZZ	2990012873189	47457	20511145-1	PIPE, EXHAUST PART OF KIT P/N 20511136	1
22	PAOZZ	2990012885844	47457	20511139-1	PIPE, EXHAUST PART OF KIT P/N 20511136	1
23	PAOZZ	5310002090788	96906	MS35335-30	WASHER, LOCK PART OF KIT P/N 20511136	1
24	PAOZZ	4320009302045	19207	10946835	PUMP UNIT, CENTRIFUG PART OF KIT P/N 20511136	1
25	PAOZZ	4820002752224	96906	MS35783-1	COCK, DRAIN PART OF KIT P/N 20511136.	1
26	MOOZZ		19207	8710557 X 22 IN	HOSE, MAKE FROM HOSE, P/N 8710557, 22 INCHES LONG PART OF KIT P/N 20511136	1

(1) ITEM NO	(2) SMR CODE	(3) NSN	(4) CAGEC	(5) PART NUMBER	(6) DESCRIPTION AND USABLE ON CODES (UOC)		(7) QTY
27	PAOZZ	5305012881417	72582	11505185	SCREW, CAP, HEXAGON H PART OF KIT P/N 20511136		4
28	PAOZZ	5310005146674	96906	MS35335-34	WASHER, LOCK PART OF KIT P/N 20511136		8
29	PAOZZ	5310012907456	72582	11500661	NUT, CLIP-ON PART OF KIT P/N 20511136		4
30	PAOZZ	5340012885156	19207	12302637	STRAP, RETAINING PART OF KIT P/N 20511136		1
31	PFOZZ	5340011467667	19207	11592371	STRAP, RETAINING PART OF KIT P/N 20511136		1
32	PAOZZ	4730009086292	96906	MS35842-14	CLAMP, HOSE PART OF KIT P/N 20511136		1
33	PAOZZ	5306002259086	96906	MS90726-31	BOLT, MACHINE PART OF KIT P/N 20511136		2
34	PAOZZ	5310009359022	96906	MS51943-32	NUT, SELF-LOCKING, HE PART OF KIT P/N 20511136		1
35	PAOZZ	5310002416658	96906	MS51943-34	NUT, SELF-LOCKING, HE PART OF KIT P/N 2051113 6		2
36	PAOZZ	5305002678953	96906	MS90727-5	SCREW, CAP, HEXAGON PART OF KIT P/N 20511136		1
37	PAOZZ	4730012864611	96906	MS24522-23	ADAPTER, STRAIGHT, PI PART OF KIT P/N 20511136		1
38	PAOZZ	4730012870953	7D408	44498	ELBOW, PIPE PART OF KIT P/N 20511136		1
	PDFZZ	2990012843218	47457	20511136	HEATER, COOLANT, ENGI ENGINE COOLANT HEATER		1
					ADAPTER, STRAIGHT, PI	(1)	584-9
					ADAPTER, STRAIGHT, PI	(1)	585-37
					BELT, V	(1)	584-16
					BLOCK, SUPPORT, BATTE	(4)	584-29
					BLOCK, SUPPORT, BATTE	(2)	584-30
					BLOCK, SUPPORT, BATTE	(2)	584-31
					BOLT, MACHINE	(1)	584-20
					BOLT, MACHINE	(2)	585-33
					BRACKET ASSEMBLY,	(1)	585-15
					BRACKET, MOUNTING	(1)	584-4
					CABLE ASSEMBLY, SPEC	(1)	584-10
					CABLE ASSEMBLY,	(1)	585-6
					CLAMP, HOSE	(3)	584-19
					CLAMP, HOSE	(1)	585-32
					CLAMP, HOSE	(2)	585-8
					CLAMP, HOSE	(5)	585-1
					CLAMP, LOOP	(1)	585-19
					CLAMP, LOOP	(1)	584-15
					CLAMP, LOOP	(3)	584-21
					COCK, DRAIN	(1)	585-25
					COCK, PLUG	(1)	584-8
					CONTROL BOX, ELECTRI	(1)	584-32
					COUPLING, PIPE	(1)	585-10
					COUPLING, PIPE	(2)	585-5
					DECAL	(1)	584-1
					ELBOW, PIPE TO TUBE	(1)	585-9
					ELBOW, PIPE TO HOSE	(4)	585-4
					ELBOW, PIPE	(1)	585-38
					ELBOW, PIPE	(1)	584-7
					GROMMET, NONMETALLIC	(2)	584-27

(1) ITEM NO	(2) SMR CODE	(3) NSN	(4) CAGEC	(5) PART NUMBER	(6) DESCRIPTION AND USABLE ON CODES (UOC)	(7) QTY
					GROMMET, NONMETALLIC	(1) 585-12
					GROMMET, NONMETALLIC	(1) 585-3
					GROMMET, NONMETALLIC	(1) 584-11
					HEATER, COOLANT, ENGI	(1) 585-13
					HOSE ASSEMBLY, NONME	(1) 584-12
					HOSE	(1) 585-26
					HOSE	(1) 585-2
					HOSE	(1) 584-22
					HOSE	(1) 584-24
					NUT, CLIP-ON	(4) 585-29
					NUT, PLAIN, HEXAGON	(2) 584-6
					NUT, SELF-LOCKING, HE	(5) 584-5
					NUT, SELF-LOCKING	(5) 585-18
					NUT, SELF-LOCKING, HE	(2) 585-35
					NUT, SELF-LOCKING	(1) 585-34
					NUT, SELF-LOCKING, HE	(2) 584-25
					PAD ASSEMBLY	(1) 584-28
					PIN, COTTER	(2) 585-20
					PIPE, EXHAUST	(1) 585-22
					PIPE, EXHAUST	(1) 585-21
					PLATE, INSTRUCTION	(1) 584-17
					PUMP UNIT, CENTRIFUG	(1) 585-24
					SCREW, CAP, HEXAGON H	(4) 585-27
					SCREW, CAP, HEXAGON H	(5) 584-2
					SCREW, CAP, HEXAGON	(2) 585-16
					SCREW, CAP, HEXAGON H	(4) 585-7
					SCREW, CAP, HEXAGON	(1) 585-36
					SCREW, CAP, HEXAGON H	(2) 584-23
					SCREW, TAPPING	(4) 584-18
					STRAP, RETAINING	(2) 585-14
					STRAP, RETAINING	(1) 585-31
					STRAP, RETAINING	(1) 585-30
					STRAP, TIEDOWN, ELECT	(14) 584-13
					WASHER, FLAT	(1) 584-14
					WASHER, FLAT	(2) 584-26
					WASHER, FLAT	(17) 585-17
					WASHER, LOCK	(8) 585-28
					WASHER, LOCK	(4) 585-11
					WASHER, LOCK	(6) 584-3
					WASHER, LOCK	(1) 585-23

END OF FIGURE

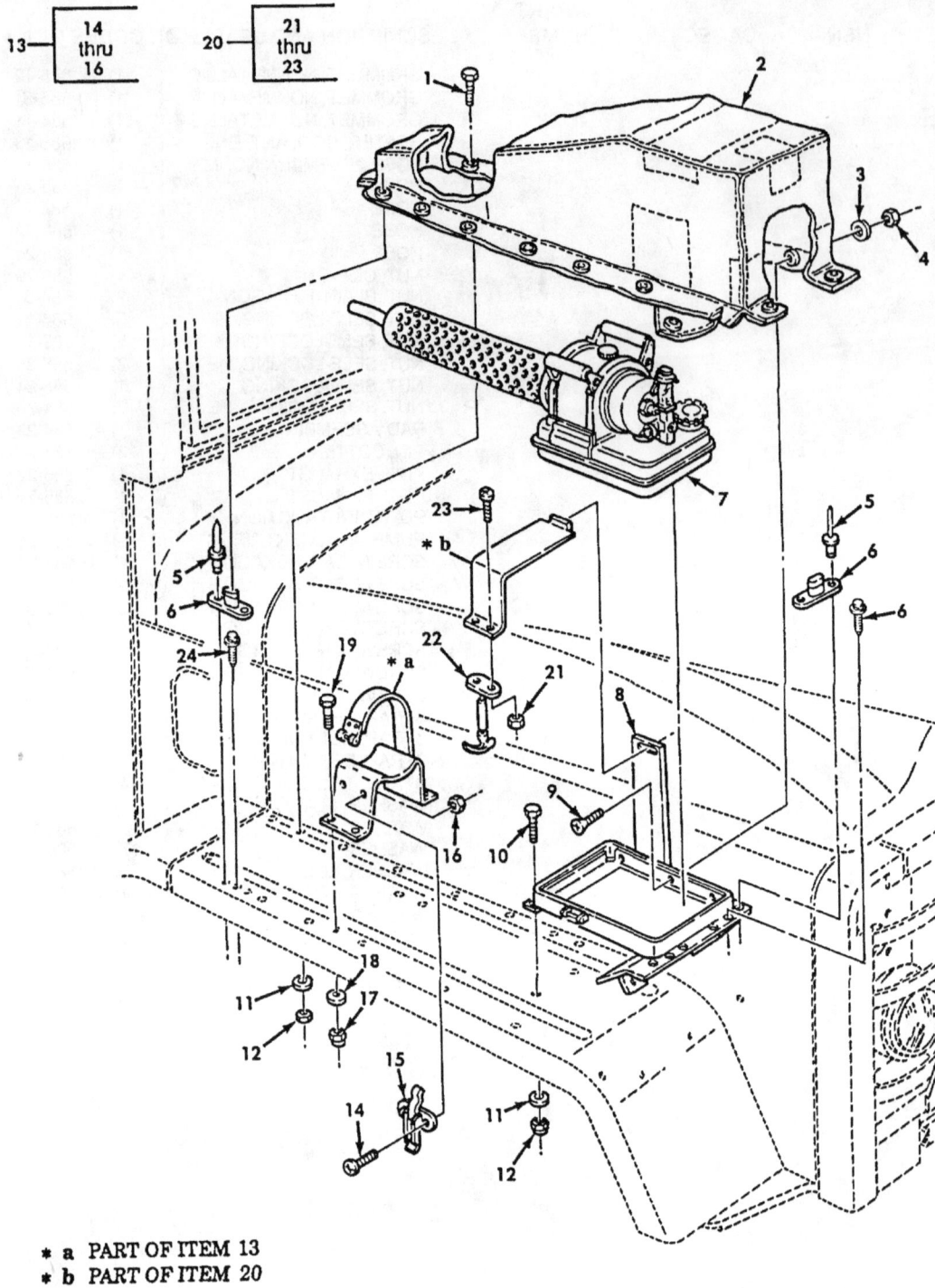

Figure 586. Swingfire Heater Kit, Heater, Cover, and Bracket.

* a PART OF ITEM 13
* b PART OF ITEM 20

(1) ITEM NO	(2) SMR CODE	(3) NSN	(4) CAGEC	(5) PART NUMBER	(6) DESCRIPTION AND USABLE ON CODES (UOC)	(7) QTY
					GROUP 3303 WINTERIZATION KITS	
					FIG. 586 SWINGFIRE HEATER KIT, HEATER, COVER, AND BRACKET	
1	PAFZZ	5305008893000	96906	MS35206-230	SCREW, MACHINE PART OF KIT P/N	6
2	PAFZZ	2590013846244	19207	12302812	COVER, VEHICULAR COM SWINGFIRE	1
3	PAFZZ	5310000454007	96906	MS35338-41	WASHER, LOCK	14
4	PAFZZ	5310009349747	96906	MS35649-262	NUT, PLAIN, HEXAGON.	14
5	PAFZZ	5320013542548	96906	MS20604R5W2	RIVET, BLIND UOC: ZAA, ZAB, ZAC, ZAD, ZAE, ZAF, ZAG, ZAH, ZAJ, ZAK, ZAL	22
6	PAFZZ	5325011985532	21450	549176	STUD, TURNBUTTON FAS UOC: DAA, DAB, DAC, DAD, DAE, DAF, DAG, DAH, DAJ, DAK, DAL, DAW, DAX, V12, V13, V14, V15, V16, V17, V18, V19, V20, V21, V22, V24, V25, V39	8
6	PAFZZ	5325008235999	90763	XB78323-05001	STUD, TURNBUTTON FAS UOC: ZAA, ZAB, ZAC, ZAD, ZAE, ZAF, ZAG, ZAH, ZAJ, ZAK, ZAL	2
7	PFFZZ	2990012024128	19207	11668950	HEATER, AIR DUCT, ENG	1
8	PFFZZ	5340011920626	19207	12302822	BRACKET, MOUNTING	1
9	PAFZZ	5305000594553	96906	MS35190-238	SCREW, MACHINE	3
10	PAFZZ	5305002692803	96906	MS90726-60	SCREW, CAP, HEXAGON H	5
11	PAFZZ	5310011022715	24617	9416095	WASHER, FLAT	5
12	PAFZZ	5310008140672	96906	MS51943-36	NUT, SELF-LOCKING, HE	5
13	KFFZZ		19207	12302825	BRACKET ASSEMBLY	1
14	KFFZZ	5305009897435	96906	MS35207-264	.SCREW, MACHINE	2
15	KFFZZ	5340011280191	19207	7413565	.CATCH, CLAMPING	1
16	KFFZZ	5310008175797	96906	MS21044N3	.NUT, SELF-LOCKING, HE	2
17	PAFZZ	5310009359022	96906	MS51943-32	NUT, SELF-LOCKING, HE	4
18	PAFZZ	5310011992293	11862	9421394	WASHER, FLAT	4
19	PAFZZ	5306000680513	60285	6893-2	BOLT, MACHINE	4
20	PAFZZ	5340011769443	19207	12250498	CLAMP, SYNCHRO	1
21	KFFZZ	5310008775797	96906	MS21044N3	.NUT, SELF-LOCKING, HE	4
22	KFFZZ		19207	7971783	.HOOK	1
23	KFFZZ	5305009897439	96906	MS35207-264	.SCREW, MACHINE	2
24	PAFZZ	5325012004035	21450	587646	STUD, TURNBUTTON FAS	3

END OF FIGURE

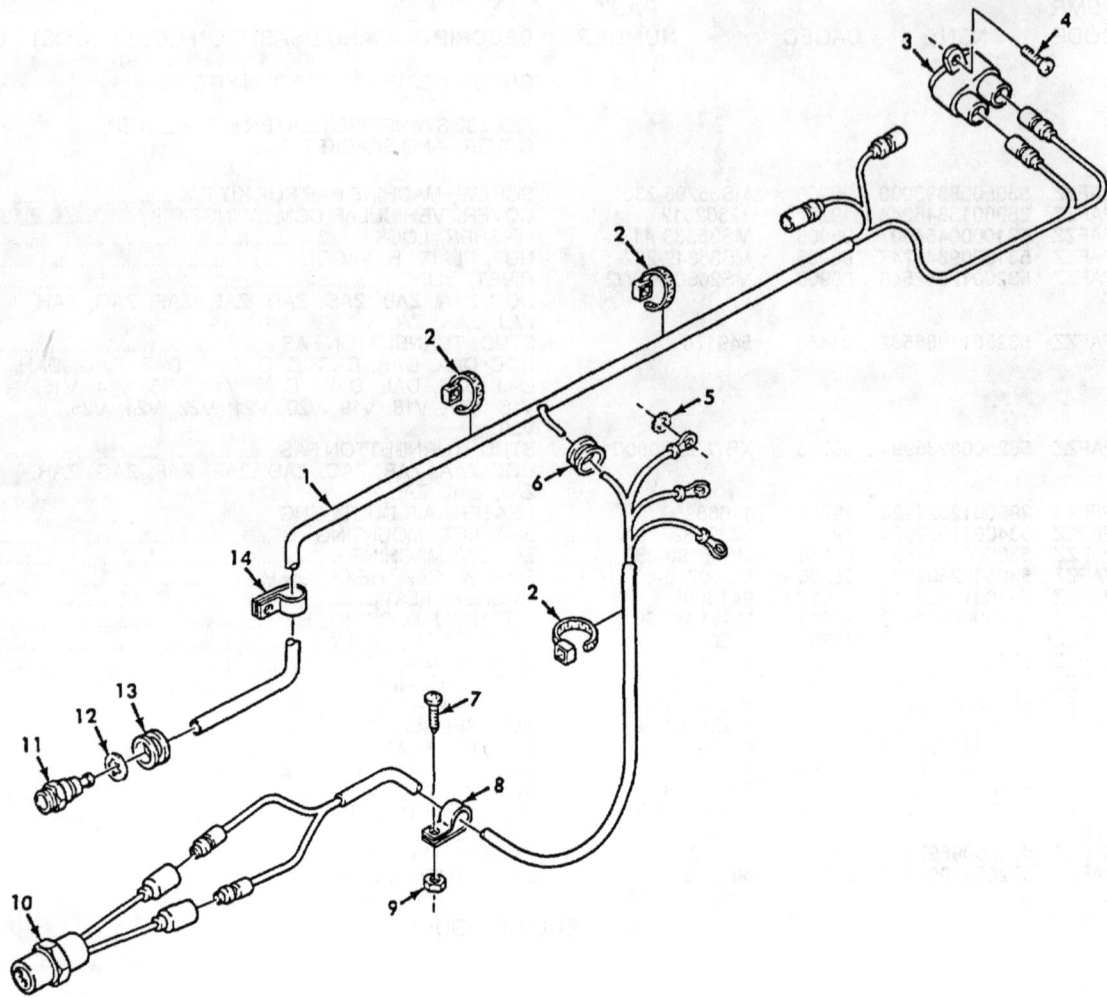

Figure 587. Swingfire Heater Kit, Wiring Harness (M939, M939A1).

(1) ITEM NO	(2) SMR CODE	(3) NSN	(4) CAGEC	(5) PART NUMBER	(6) DESCRIPTION AND USABLE ON CODES (UOC)	(7) QTY
					GROUP 3303 WINTERIZATION KITS	
					FIG. 587 SWINGFIRE HEATER KIT, WIRING HARNESS (M939, M939A1)	
1	PAFZZ	5995013036428	19207	12356785	WIRING HARNESS, BRAN PART OF KIT P/N .... 12302775 .... UOC: DAA, DAB, DAC, DAD, DAE, DAF, DAG, DAH, DAJ, DAK, DAL, DAW, DAX, V12, V13, V14, V15, V16, V17, V18, V19, V20, V21, V22, V24, V25, V39	11
2	PAFZZ	5975009846582	96906	MS3367-1-0	STRAP, TIEDOWN, ELECT PART OF KIT P/N .... 12302775 .... UOC: DAA, DAB, DAC, DAD, DAE, DAF, DAG, DAH, DAJ, DAK, DAL, DAW, DAX, V12, V13, V14, V15, V16, V17, V18, V19, V20, V21, V22, V24, V25, V39	11
3	PFFZZ	5925000264767	81349	M13516/1-1	CIRCUIT BREAKER PART OF KIT P/N .... 12302775 .... UOC: DAA, DAB, DAC, DAD, DAE, DAF, DAG, DAH, DAJ, DAK, DAL, DAW, DAX, V12, V13, V14, V15, V16, V17, V18, V19, V20, V21, V22, V24, V25, V39	1
4	PAFZZ	5305008550965	96906	MS24629-38	SCREW, TAPPING PART OF KIT P/N .... 12302775 .... UOC: DAA, DAB, DAC, DAD, DAE, DAF, DAG, DAH, DAJ, DAK, DAL, DAW, DAX, V12, V13, V14, V15, V16, V17, V18, V19, V20, V21, V22, V24, V25, V39	2
5	PAFZZ	5310008892528	96906	MS45904-68	WASHER, LOCK PART OF KIT P/N 12302775 .... UOC: DAA, DAB, DAC, DAD, DAE, DAF, DAG, DAH, DAJ, DAK, DAL, DAW, DAX, V12, V13, V14, V15, V16, V17, V18, V19, V20, V21, V22, V24, V25, V39	1
5	PAFZZ	5310005590070	96906	MS35333-38	WASHER, LOCK PART OF KIT P/N 12302775 .... UOC: DAA, DAB, DAC, DAD, DAE, DAF, DAG, DAH, DAJ, DAK, DAL, DAW, DAX, V12, V13, V14, V15, V16, V17, V18, V19, V20, V21, V22, V24, V25, V39	3
6	PAFZZ	5325011247760	96906	MS35489-79	GROMMET, NONMETALLIC PART OF KIT P/N .... 12302775 .... UOC: DAA, DAB, DAC, DAD, DAE, DAF, DAG, DAH, DAJ, DAK, DAL, DAW, DAX, V12, V13, V14, V15, V16, V17, V18, V19, V20, V21, V22, V24, V25, V39	1
7	PAFZZ	5305002678953	80204	B1821BH025F063N	SCREW, CAP, HEXAGON H PART OF KIT P/N .... 12302775 .... UOC: DAA, DAB, DAC, DAD, DAE, DAF, DAG, DAH, DAJ, DAK, DAL, DAW, DAX, V12, V13, V14, V15, V16, V17, V18, V19, V20, V2 1, V22, V24, V25, V39	1
8	PAFZZ	5340009936207	96906	MS21333-99	CLAMP, LOOP PART OF KIT P/N 12302775	1

(1) ITEM NO	(2) SMR CODE	(3) NSN	(4) CAGEC	(5) PART NUMBER	(6) DESCRIPTION AND USABLE ON CODES (UOC)	(7) QTY
					UOC: DAA, DAB, DAC, DAD, DAE, DAF, DAG, DAH, DAJ, DAK, DAL, DAW, DAX, V12, V13, V14, V15, V16, V17, V18, V19, V20, V21, V22, V24, V25, V39	
9	PAFZZ	5310009359022	96906	MS51943-32	NUT, SELF-LOCKING, HE PART OF KIT P/N 12302775	1
					UOC: DAA, DAB, DAC, DAD, DAE, DAF, DAG, DAH, DAJ, DAK, DAL, DAW, DAX, V12, V13, V14, V15, V16, V17, V18, V19, V20, V21, V22, V24, V25, V39	
10	PAFZZ	5930013182809	19207	12356766	SWITCH, THERMOSTATIC PART OF KIT P/N 12302775	1
					UOC: DAA, DAB, DAC, DAD, DAE, DAF, DAG, DAH, DAJ, DAK, DAL, DAW, DAX, V12, V13, V14, V15, V16, V17, V18, V19, V20, V21, V22, V24, V25, V39	
11	PAFZZ	5935011920627	19207	11669531-1	CONNECTOR, RECEPTACL PART OF KIT P/N 12302775	1
					UOC: DAA, DAB, DAC, DAD, DAE, DAF, DAG, DAH, DAJ, DAK, DAL, DAW, DAX, V12, V13, V14, V15, V16, V17, V18, V19, V20, V21, V22, V24, V25, V39	
12	PAFZZ	5310005503714	96906	MS35333-47	WASHER, LOCK PART OF KIT P/N 12302775	1
					UOC: DAA, DAB, DAC, DAD, DAE, DAF, DAG, DAH, DAJ, DAK, DAL, DAW, DAX, V12, V13, V14, V15, V16, V17, V18, V19, V20, V21, V22, V24, V25, V39	
13	PAFZZ	5325001745317	96906	MS35489-4	GROMMET, NONMETALLIC PART OF KIT P/N 12302775	1
					UOC: DAA, DAB, DAC, DAD, DAE, DAF, DAG, DAH, DAJ, DAK, DAL, DAW, DAX, V12, V13, V14, V15, V16, V17, V18, V19, V20, V21, V22, V24, V25, V39	
14	PAFZZ	5340009907610	96906	MS21333-66	CLAMP, LOOP PART OF KIT P/N 12302775	2
					UOC: DAA, DAB, DAC, DAD, DAE, DAF, DAG, DAH, DAJ, DAK, DAL, DAW, DAX, V12, V13, V14, V15, V16, V17, V18, V19, V20, V21, V22, V24, V25, V39	

END OF FIGURE

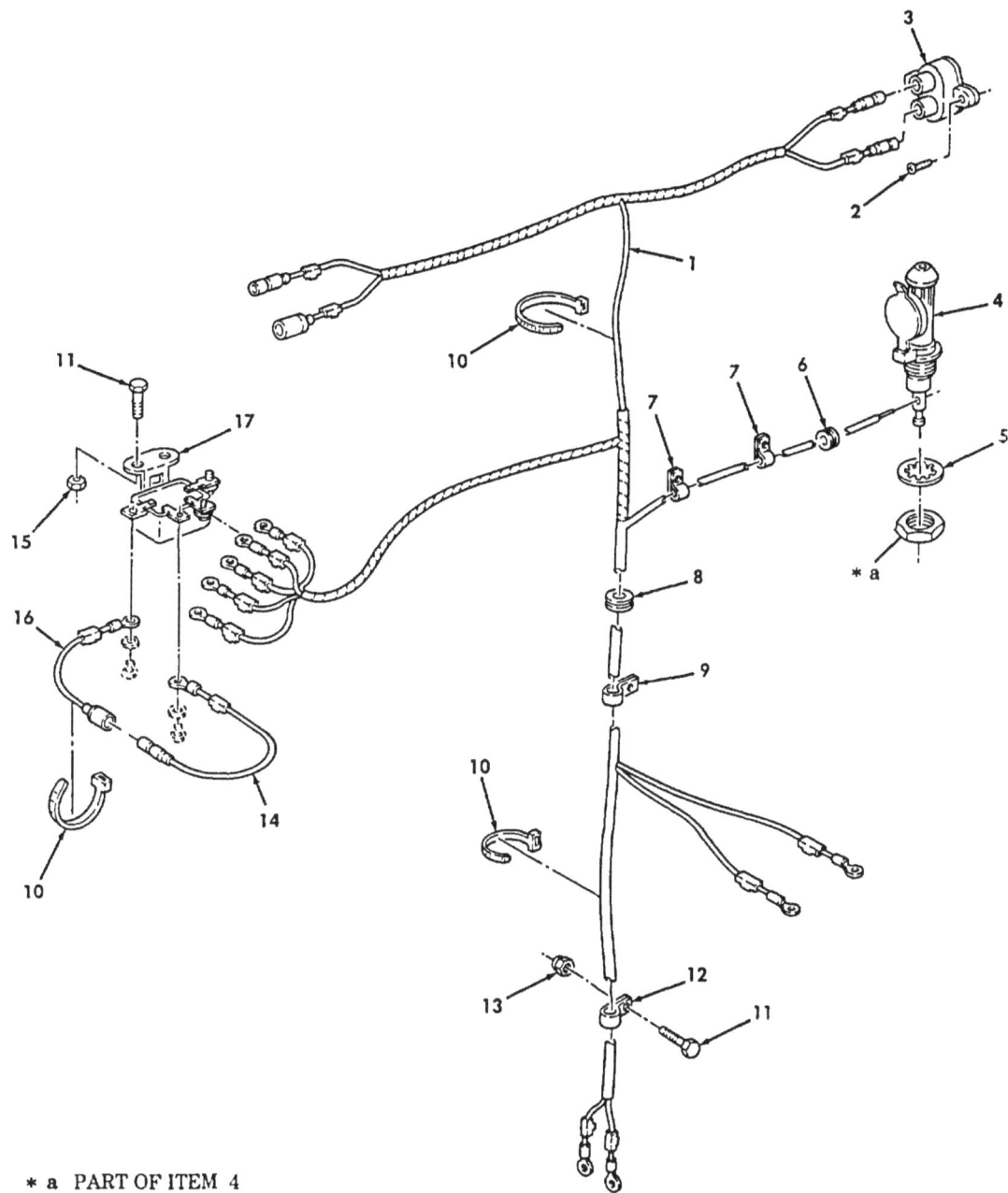

* a PART OF ITEM 4

Figure 588. Swingfire Heater Kit, Wiring Harness (M939A2).

(1) ITEM NO	(2) SMR CODE	(3) NSN	(4) CAGEC	(5) PART NUMBER	(6) DESCRIPTION AND USABLE ON CODES (UOC)	(7) QTY
					GROUP 3303 WINTERIZATION KIT	
					FIG. 588 SWINGFIRE HEATER KIT, WIRING HARNESS (M939A2)	
1	PFFZZ		19207	12356785-1	WIRING HARNESS, BRAN PART OF KIT P/N 5705626	1
2	PAFZZ	5305008550965	96906	MS24629-38	SCREW, TAPPING PART OF KIT P/N 5705626	2
3	PAFZZ	5925000264767	81349	M13516/1-1	CIRCUIT BREAKER PART OF KIT P/N 5705626	1
4	PAFZZ	5935011920627	19207	11669531-1	CONNECTOR, RECEPTACL PART OF KIT P/N 5705626	1
5	PAFZZ	5310005503714	96906	MS35333-47	WASHER, LOCK PART OF KIT P/N 5705626	1
6	PAFZZ	5325001745317	96906	MS35489-4	GROMMET, NONMETALLIC PART OF KIT P/N 5705626	1
7	PAFZZ	5340009907610	96906	MS21333-66	CLAMP, LOOP PART OF KIT P/N 5705626	2
8	PAFZZ	5325011247760	96906	MS35489-79	GROMMET, NONMETALLIC PART OF KIT P/N 5705626	1
9	PAFZZ	5340000573043	96906	MS21333-112	CLAMP, LOOP PART OF KIT P/N 5705626	1
10	PFFZZ	5975000742072	06383	SST2SC	STRAP, TIEDOWN, ELECT PART OF KIT P/N 5705626	5
11	PAFZZ	5305002678953	96906	MS90726-5	SCREW, CAP, HEXAGON H PART OF KIT P/N 5705626	2
12	PAFZZ	5340009936207	96906	MS21333-99	CLAMP, LOOP PART OF KIT P/N 5705626	2
13	PAFZZ		96906	MS51943-32	NUT, SELF-LOCKING PART OF KIT P/N 5705626	1
14	PFFZZ	5961013539187	19207	12302643-2	SEMICONDUCTOR DEVIC PART OF KIT P/N 5705626	1
15	PAFZZ	5310008775796	96906	MS21044N4	NUT, SELF-LOCKING PART OF KIT P/N 5705626	2
16	PFFZZ		19207	10938046-2	LEAD, ELECTRICAL PART OF KIT P/N 5705626	1
17	PAFZZ	2920008483292	16764	1116968	CUTOUT RELAY, ENGINE PART OF KIT P/N 5705626	1

END OF FIGURE

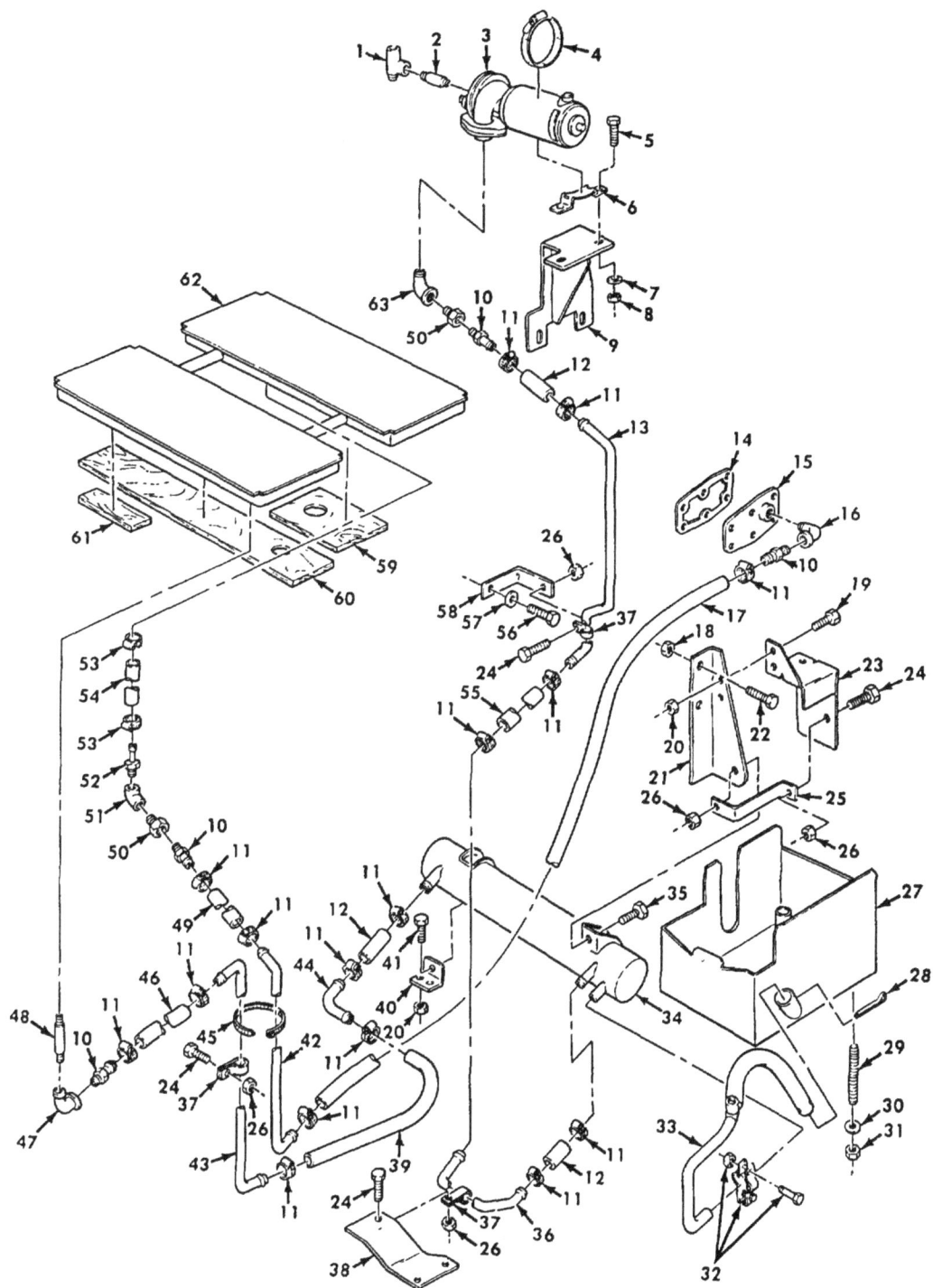

Figure 589. Swingfire Heater Mounting Kit (M939, M939A1).

(1) ITEM NO	(2) SMR CODE	(3) NSN	(4) CAGEC	(5) PART NUMBER	(6) DESCRIPTION AND USABLE ON CODES (UOC)	(7) QTY
					GROUP 3303 WINTERIZATION KITS	
					FIG. 589 SWINGFIRE HEATER MOUNTING KIT (M939, M939A1)	
1	PAFZZ	4730001388050	24617	444153	TEE, PIPE PART OF KIT P/N 12302775............................. UOC: DAA, DAB, DAC, DAD, DAE, DAF, DAG, DAH, DAJ, DAK, DAL, DAW, DAX, V12, V13, V14, V15, V16, V17, V18, V19, V20, V21, V22, V24, V25, V39	1
2	PAFZZ	4730001961493	96906	MS51953-78	NIPPLE, PIPE PART OF KIT P/N 12302775..................... UOC: DAA, DAB, DAC, DAD, DAE, DAF, DAG, DAH, DAJ, DAK, DAL, DAW, DAX, V12, V13, V14, V15, V16, V17, V18, V19, V20, V21, V22, V24, V25, V39	1
3	PFFZZ	4320009302045	19207	10946835	PUMP UNIT, CENTRIFUG PART OF KIT P/N..................... 12302775.. UOC: DAA, DAB, DAC, DAD, DAE, DAF, DAG, DAH, DAJ, DAK, DAL, DAW, DAX, V12, V13, V14, V15, V16, V17, V18, V19, V20, V21, V22, V24, V25, V39	1
4	PAFZZ	4730009086292	96906	MS35842-14	CLAMP, HOSE PART OF KIT P/N 12302775.................... UOC: DAA, DAB, DAC, DAD, DAE, DAF, DAG, DAH, DAJ, DAK, DAL, DAW, DAX, V12, V13, V14, V15, V16, V7, V178, V19, V20, V21, V22, V24, V25, V39	1
5	PAFZZ	5306000501238	96906	MS90727-32	BOLT, MACHINE PART OF KIT P/N .... 12302775.. UOC: DAA, DAB, DAC, DAD, DAE, DAF, DAG, DAH, DAJ, DAK, DAL, DAW, DAX, V12, V13, V14, V15, V16, V17, V18, V19, V20, V21, V22, V24, V25, V39	2
6	PFFZZ	5340011467667	19207	11592371	STRAP, RETAINING PART OF KIT P/N .... 12302775.. UOC: DAA, DAB, DAC, DAD, DAE, DAF, DAG, DAH, DAJ, DAK, DAL, DAW, DAX, V12, V13, V14, V15, V16, V17, V18, V19, V20, V21, V2, V2, V24, V25, V39	1
7	PAFZZ	5310000814219	96906	MS27183-12	WASHER, FLAT PART OF KIT P/N 12302775................... UOC: DAA, DAB, DAC, DAD, DAE, DAF, DAG, DAH, DAJ, DAK, DAL, DAW, DAX, V12, V13, V14, V15, V16, V17, V18, V19, V20, V21, V22, V24, V25, V39	2
8	PAFZZ	5310002416658	96906	MS51943-34	NUT, SELF-LOCKING, HE PART OF KIT P/N..................... 12302775.. UOC: DAA, DAB, DAC, DAD, DAE, DAF, DAG, DAH, DAJ, DAK, DAL, DAW, DAX, V12, V13, V14, V15, V16, V17, V18, V19, V20, V21, V22, V24, V25, V39	2
9	PFFZZ	5340012103953	19207	12302786	BRACKET, ANGLE PART OF KIT P/N .... 12302775.. UOC: DAA, DAB, DAC, DAD, DAE, DAF, DAG, DAH, DAJ, DAK, DAL, DAW, DAX, V12, V13, V14, V15, V16, V17, V18, V19, V20, V21, V22, V24, V25,	1

(1) ITEM NO	(2) SMR CODE	(3) NSN	(4) CAGEC	(5) PART NUMBER	(6) DESCRIPTION AND USABLE ON CODES (UOC)	(7) QTY
10	PFFZZ	4730011757343	19207	11656473	V39 ADAPTER, STRAIGHT, PI PART OF KIT P/N 12302775 UOC: DAA, DAB, DAC, DAD, DAE, DAF, DAG, DAH, DAJ, DAK, DAL, DAW, DAX, V12, V13, V14, V15, V16, V17, V18, V19, V20, V21, V22, V24, V25, V39	4
11	PAFZZ	4730009083193	83299	0612596-00	CLAMP, HOSE PART OF KIT P/N 12302775. UOC: DAA, DAB, DAC, DAD, DAE, DAF, DAG, DAH, DAJ, DAK, DAL, DAW, DAX, V12, V13, V14, V15, V16, V17, V18, V19, V20, V21, V22, V24, V25, V39	12
12	PFFZZ	4720011757421	19207	11656458-3	HOSE, NONMETALLIC PART OF KIT P/N 12302775 UOC: DAA, DAB, DAC, DAD, DAE, DAF, DAG, DAH, DAJ, DAK, DAL, DAW, DAX, V12, V13, V14, V15, V16, V17, V18, V19, V20, V21, V22, V24, V25, V39	3
13	PFFZZ	4710011994367	19207	12302780	TUBE, BENT, METALLIC PART OF KIT P/N 12302775 UOC: DAA, DAB, DAC, DAD, DAE, DAF, DAG, DAH, DAJ, DAK, DAL, DAW, DAX, V12, V13, V14, V15, V16, V17, V18, V19, V20, V21, V22, V24, V25, V39	1
14	PAFZZ	5330011973228	19207	12302635	GASKET PART OF KIT P/N 12302775 UOC: DAA, DAB, DAC, DAD, DAE, DAF, DAG, DAH, DAJ, DAK, DAL, DAW, DAX, V12, V13, V14, V15, V16, V17, V18, V19, V20, V21, V22, V24, V25, V39	1
15	PFFZZ	2930011890458	19207	12302789-2	WATER OUTLET, ENGINE PART OF KIT P/N 12302775 UOC: DAA, DAB, DAC, DAD, DAE, DAF, DAG, DAH, DAJ, DAK, DAL, DAW, DAX, V12, V13, V14, V15, V16, V17, V18, V19, V20, V21, V22, V24, V25, V39	1
16	PAFZZ	4730002534415	96906	MS39230-5	ELBOW, PIPE PART OF KIT P/N 12302775 UOC: DAA, DAB, DAC, DAD, DAE, DAF, DAG, DAH, DAJ, DAK, DAL, DAW, DAX, V12, V13, V14, V15, V16, V17, V18, V19, V20, V21, V22, V24, V25, V39	1
17	MFFZZ		19207	12302784-4-65	HOSE WATER LINE, MAKE FROM P/N MS52130-1A2764R, 65 INCHES LONG UOC: DAA, DAB, DAC, DAD, DAE, DAF, DAG, DAH, DAJ, DAK, DAL, DAW, DAX, V12, V13, V14, V15, V16, V17, V18, V19, V20, V21, V22, V24, V25, V39	1
18	PAFZZ	5310004883888	96906	MS51943-40	NUT, SELF-LOCKING, HE PART OF KIT P/N 12302775 UOC: DAA, DAB, DAC, DAD, DAE, DAF, DAG, DAH, DAJ, DAK, DAL, DAW, DAX, V12, V13, V14, V15, V16, V17, V18, V19, V20, V21, V22, V24, V25, V39	2
19	PAFZZ	5306000680513	60285	6893-2	BOLT, MACHINE PART OF KIT P/N 12302775	2

(1) ITEM NO	(2) SMR CODE	(3) NSN	(4) CAGEC	(5) PART NUMBER	(6) DESCRIPTION AND USABLE ON CODES (UOC)	(7) QTY
					UOC: DAA, DAB, DAC, DAD, DAE, DAF, DAG, DAH, DAJ, DAK, DAL, DAW, DAX, V12, V13, V14, V15, V16, V17, V18, V19, V20, V21, V22, V24, V25, V39	
20	PAFZZ	5310009359022	96906	MS51943-32	NUT, SELF-LOCKING PART OF KIT P/N .... 12302775....................	4
					UOC: DAA, DAB, DAC, DAD, DAE, DAF, DAG, DAH, DAJ, DAK, DAL, DAW, DAX, V12, V13, V14, V15, V16, V17, V18, V19, V20, V21, V22, V24, V25, V39	
21	PFFZZ	5340012062209	19207	12302777	BRACKET, ANGLE PART OF KIT P/N.................................. 12302775....................	1
					UOC: DAA, DAB, DAC, DAD, DAE, DAF, DAG, DAH, DAJ, DAK, DAL, DAW, DAX, V12, V13, V14, V15, V16, V17, V18, V19, V20, V21, V22, V24, V25, V39	
22	PAFZZ	5305007195238	80204	B1821BH050F200N	SCREW, CAP, HEXAGON H PART OF KIT P/N.................. 12302775....................	2
					UOC: DAA, DAB, DAC, DAD, DAE, DAF, DAG, DAH, DAJ, DAK, DAL, DAW, DAX, V12, V13, V14, V15, V16, V17, V18, V19, V20, V21, V22, V24, V25, V39	
23	PFFZZ	2540012024064	19207	12302792	SHIELD PART OF KIT P/N 12302775	1
					UOC: DAA, DAB, DAC, DAD, DAE, DAF, DAG, DAH, DAJ, DAK, DAL, DAW, DAX, V12, V13, V14, V15, V16, V17, V18, V19, V20, V21, V22, V24, V25, V39	
24	PAFZZ	5305002693234	96906	MS90727-58	SCREW, CAP, HEXAGON PART OF KIT P/N...................... 12302775....................	6
					UOC: DAA, DAB, DAC, DAD, DAE, DAF, DAG, DAH, DAJ, DAK, DAL, DAW, DAX, V12, V13, V14, V15, V16, V17, V18, V19, V20, V21, V22, V24, V25, V39	
25	PFFZZ	5340012035659	19207	12302799	SUPPORT, SHIELD PART OF KIT P/N .... 12302775....................	1
					UOC: DAA, DAB, DAC, DAD, DAE, DAF, DAG, DAH, DAJ, DAK, DAL, DAW, DAX, V12, V13, V14, V15, V16, V17, V18, V19, V20, V21, V22, V24, V25, V39	
26	PAFZZ	5310008140672	96906	MS51943-36	NUT, SELF-LOCKING, HE PART OF KIT P/N...................... 12302775....................	8
					UOC: DAA, DAB, DAC, DAD, DAE, DAF, DAG, DAH, DAJ, DAK, DAL, DAW, DAX, V12, V13, V14, V15, V16, V17, V18, V19, V20, V21, V22, V24, V25, V39	
27	KFFZZ	2990011764801	19207	11648559	SHROUD, OIL PAN OIL PAN PART OF KIT .... P/N 12302775....................	1
					UOC: DAA, DAB, DAC, DAD, DAE, DAF, DAG, DAH, DAJ, DAK, DAL, DAW, DAX, V12, V13, V14, V15, V16, V17, V18, V19, V20, V21, V22, V24, V25, V39	
28	PAFZZ	5315002981498	96906	MS24665-362	PIN, COTTER PART OF KIT P/N 12302775......................	1
					UOC: DAA, DAB, DAC, DAD, DAE, DAF, DAG, DAH, DAJ, DAK, DAL, DAW, DAX, V12, V13, V14, V15,	

(1) ITEM NO	(2) SMR CODE	(3) NSN	(4) CAGEC	(5) PART NUMBER	(6) DESCRIPTION AND USABLE ON CODES (UOC)	(7) QTY
					V16, V17, V18, V19, V20, V21, V22, V24, V25, V39	
29	PAFZZ	5307001744863	19207	11648628	STUD, CONTINUOUS THR PART OF KIT P/N .... 12302775... UOC: DAA, DAB, DAC, DAD, DAE, DAF, DAG, DAH, DAJ, DAK, DAL, DAW, DAX, V12, V13, V14, V15, V16, V17, V18, V19, V20, V21, V22, V24, V25, V39	4
30	PAFZZ	5310000877493	96906	MS27183-13	WASHER, FLAT PART OF KIT P/N 12302775................. UOC: DAA, DAB, DAC, DAD, DAE, DAF, DAG, DAH, DAJ, DAK, DAL, DAW, DAX, V12, V13, V14, V15, V16, V17, V18, V19, V20, V21, V22, V24, V25, V39	4
31	PAFZZ	5310008140673	96906	MS51943-33	NUT, SELF-LOCKING, HE PART OF KIT P/N .... 12302775... UOC: DAA, DAB, DAC, DAD, DAE, DAF, DAG, DAH, DAJ, DAK, DAL, DAW, DAX, V12, V13, V14, V15, V16, V17, V18, V19, V20, V21, V22, V24, V25, V39	4
32	PAFZZ	5340011870527	19207	11656448-1	CLAMP, LOOP PART OF KIT P/N 12302775. .... UOC: DAA, DAB, DAC, DAD, DAE, DAF, DAG, DAH, DAJ, DAK, DAL, DAW, DAX, V12, V13, V14, V15, V16, V17, V18, V19, V20, V21, V22, V24, V25, V39	1
33	PFFZZ	4710011920625	19207	12302779	TUBE, WATER SACKET PART OF KIT P/N...................... 12302775... UOC: DAA, DAB, DAC, DAD, DAE, DAF, DAG, DAH, DAJ, DAK, DAL, DAW, DAX, V12, V13, V14, V15, V16, V17, V18, V19, V20, V21, V22, V24, V25, V39	1
34	PFFZZ	2540010769286	16236	269013000000	WATER JACKET ASSEMB PART OF KIT P/N...................... 12302775... UOC: DAA, DAB, DAC, DAD, DAE, DAF, DAG, DAH, DAJ, DAK, DAL, DAW, DAX, V12, V13, V14, V15, V16, V17, V18, V19, V20, V21, V22, V24, V25, V39	1
35	PAFZZ	5305002692803	96906	MS90726-60	SCREW, CAP, HEXAGON H PART OF KIT P/N.................. 12302775... UOC: DAA, DAB, DAC, DAD, DAE, DAF, DAG, DAH, DAJ, DAK, DAL, DAW, DAX, V12, V13, V14, V15, V16, V17, V18, V19, V20, V21, V22, V24, V25, V39	2
36	PFFZZ	4710011961028	19207	12302790	TUBE, BENT, METALLIC PART OF KIT P/N...................... 12302775... UOC: DAA, DAB, DAC, DAD, DAE, DAF, DAG, DAH, DAJ, DAK, DAL, DAW, DAX, V12, V13, V14, V15, V16, V17, V18, V19, V20, V21, V22, V24, V25, V39	1
37	PAFZZ	5340011179876	19207	7397785	CLAMP, LOOP PART OF KIT P/N 12302775................. UOC: DAA, DAB, DAC, DAD, DAE, DAF, DAG, DAH, DAJ, DAK, DAL, DAW, DAX, V12, V13, V14, V15, V16, V17, V18, V19, V20, V21, V22, V24, V25, V39	4
38	PFFZZ	5340012070378	19207	12302787	BRACKET, DOUBLE ANGL PART OF KIT P/N....................	1

(1) ITEM NO	(2) SMR CODE	(3) NSN	(4) CAGEC	(5) PART NUMBER	(6) DESCRIPTION AND USABLE ON CODES (UOC)	(7) QTY
					12302775.. UOC: DAA, DAB, DAC, DAD, DAE, DAF, DAG, DAH, DAJ, DAK, DAL, DAW, DAX, V12, V13, V14, V15, V16, V17, V18, V19, V20, V21, V22, V24, V25, V39	
39	MFFZZ		19207	12302784-5-26	HOSE WATER LINE, MAKE FROM HOSE, P/N................ MS52130-1A2726R, 26 INCHES LONG............................ UOC: DAA, DAB, DAC, DAD, DAE, DAF, DAG, DAH, DAJ, DAK, DAL, DAW, DAX, V12, V13, V14, V15, V16, V17, V18, V19, V20, V21, V22, V24, V25, V39	1
40	PFFZZ	5340012059263	19207	12302785	BRACKET, ANGLE PART OF KIT P/N................................. 12302775 UOC: DAA, DAB, DAC, DAD, DAE, DAF, DAG, DAH, DAJ, DAK, DAL, DAW, DAX, V12, V13, V14, V15, V16, V17, V18, V19, V20, V21, V22, V24, V25, V39	1
41	PAFZZ	5305002693239	80204	B1821BH038F138N	SCREW, CAP, HEXAGON H PART OF KIT P/N................ 12302775.. UOC: DAA, DAB, DAC, DAD, DAE, DAF, DAG, DAH, DAJ, DAK, DAL, DAW, DAX, V12, V13, V14, V15, V16, V17, V18, V19, V20, V21, V22, V24, V25, V39	2
42	PFFZZ	4710012003255	19207	12302782	TUBE, BENT, METALLIC PART OF KIT P/N.................... 12302775 UOC: DAA, DAB, DAC, DAD, DAE, DAF, DAG, DAH, DAJ, DAK, DAL, DAW, DAX, V12, V13, V14, V15, V16, V17, V18, V19, V20, V21, V22, V24, V25, V39	1
43	PFFZZ	4710011959100	19207	12302791	TUBE, BENT, METALLIC PART OF KIT P/N.................... 12302775 UOC: DAA, DAB, DAC, DAD, DAE, DAF, DAG, DAH, DAJ, DAK, DAL, DAW, DAX, V12, V13, V14, V15, V16, V17, V18, V19, V20, V21, V22, V24, V25, V39	1
44	PFFZZ	4710011994366	19207	12302781	TUBE, BENT, METALLIC PART OF KIT P/N.................... 12302775 UOC: DAA, DAB, DAC, DAD, DAE, DAF, DAG, DAH, DAJ, DAK, DAL, DAW, DAX, V12, V13, V14, V15, V16, V17, V18, V19, V20, V21, V22, V2 4, V25, V39	1
45	PFFZZ	5975004515001	96906	MS3367-3-9	STRAP, TIEDOWN, ELECT PART OF KIT P/N................... 12302775.. UOC: DAA, DAB, DAC, DAD, DAE, DAF, DAG, DAH, DAJ, DAK, DAL, DAW, DAX, V12, V13, V14, V15, V16, V17, V18, V19, V20, V21, V22, V24, V25, V39	5
46	MFFZZ		19207	12302784-1-8	HOSE WATER LINE, MAKE FROM HOSE, P/N .... MS52130-1A278R, 8 INCHES LONG .... UOC: DAA, DAB, DAC, DAD, DAE, DAF, DAG, DAH, DAJ, DAK, DAL, DAW, DAX, V12, V13, V14, V15, V16, V17, V18, V19, V20, V21, V22, V24, V25, V39	1
47	PAFZZ	4730002491511	24617	217985	ELBOW, PIPE PART OF KIT P/N 12302775........................	1

(1) ITEM NO	(2) SMR CODE	(3) NSN	(4) CAGEC	(5) PART NUMBER	(6) DESCRIPTION AND USABLE ON CODES (UOC)	(7) QTY
					UOC: DAA, DAB, DAC, DAD, DAE, DAF, DAG, DAH, DAJ, DAK, DAL, DAW, DAX, V12, V13, V14, V15, V16, V17, V18, V19, V20, V21, V22, V24, V25, V39	
48	PAFZZ	4730001961489	96906	MS51953-55	NIPPLE, PIPE PART OF KIT P/N 12302775	1
					UOC: DAA, DAB, DAC, DAD, DAE, DAF, DAG, DAH, DAJ, DAK, DAL, DAW, DAX, V12, V13, V14, V15, V16, V17, V18, V19, V20, V21, V22, V24, V25, V39	
49	MFFZZ		19207	12302784-1-8	HOSE WATER LINE, MAKE FROM HOSE, P/N MS52130-1A2712R, 8 INCHES LONG	1
					UOC: DAA, DAB, DAC, DAD, DAE, DAF, DAG, DAH, DAJ, DAK, DAL, DAW, DAX, V12, V13, V14, V15, V16, V17, V18, V19, V20, V21, V22, V24, V25, V39	
50	PAFZZ	4730001893034	30327	24SG-12X08	REDUCER, PIPE PART OF KIT P/N 12302775	2
					UOC: DAA, DAB, DAC, DAD, DAE, DAF, DAG, DAH, DAJ, DAK, DAL, DAW, DAX, V12, V13, V14, V15, V16, V17, V18, V19, V20, V21, V22, V24, V25, V39	
51	PAFZZ	4730007222759	24617	444073	ELBOW, PIPE PART OF KIT P/N 12302775	1
					UOC: DAA, DAB, DAC, DAD, DAE, DAF, DAG, DAH, DAJ, DAK, DAL, DAW, DAX, V12, V13, V14, V15, V16, V17, V18, V19, V20, V21, V22, V24, V25, V39	
52	PAFZZ	4730009445888	96906	MS24522-7	ADAPTER, STRAIGHT, PI	1
					UOC: DAA, DAB, DAC, DAD, DAE, DAF, DAG, DAH, DAJ, DAK, DAL, DAW, DAX, V12, V13, V14, V15, V16, V17, V18, V19, V20, V21, V22, V24, V25, V39	
53	PAFZZ	4730009083194	96906	MS35842-11	CLAMP, HOSE PART OF KIT P/N 12302775. 2	
					UOC: DAA, DAB, DAC, DAD, DAE, DAF, DAG, DAH, DAJ, DAK, DAL, DAW, DAX, V12, V13, V14, V15, V16, V17, V18, V19, V20, V21, V22, V24, V25, V39	
54	MFFZZ	4720011147728	19207	8710557	HOSE, NONMETALLIC MAKE FROM HOSE, P/N MS5213014203R, LENGTH AS REQUIRED	1
					UOC: DAA, DAB, DAC, DAD, DAE, DAF, DAG, DAH, DAJ, DAK, DAL, DAW, DAX, V12, V13, V14, V15, V16, V17, V18, V19, V20, V21, V22, V24, V25, V39	
55	MFFZZ		19207	12302784-3-29	HOSE WATER LINE, MAKE FROM HOSE, P/N MS52130-1AC729R, 29 INCHES LONG	1
					UOC: DAA, DAB, DAC, DAD, DAE, DAF, DAG, DAH, DAJ, DAK, DAL, DAW, DAX, V12, V13, V14, V15, V16, V17, V18, V19, V20, V21, V22, V24, V25, V39	
56	PAFZZ	5305007320512	80204	B1821BH050C075N	SCREW, CAP, HEXAGON H PART OF KIT P/N 12302775	1
					UOC: DAA, DAB, DAC, DAD, DAE, DAF, DAG, DAH, DAJ, DAK, DAL, DAW, DAX, V12, V13, V14, V15, V16, V17, V18, V19, V20, V21, V22, V24, V25,	

(1) ITEM NO	(2) SMR CODE	(3) NSN	(4) CAGEC	(5) PART NUMBER	(6) DESCRIPTION AND USABLE ON CODES (UOC)	(7) QTY
					V39	
57	PAFZZ	5310005845272	01276	210104-8S	WASHER, LOCK PART OF KIT P/N 12302775.................. UOC: DAA, DAB, DAC, DAD, DAE, DAF, DAG, DAH, DAJ, DAK, DAL, DAW, DAX, V12, V13, V14, V15, V16, V17, V18, V19, V20, V21, V22, V24, V25, V39	1
58	PFFZZ	5340012111563	19207	12302783	BRACKET, ANGLE PART OF KIT P/N................................ 12302775.................... UOC: DAA, DAB, DAC, DAD, DAE, DAF, DAG, DAH, DAJ, DAK, DAL, DAW, DAX, V12, V13, V14, V15, V16, V17, V18, V19, V20, V21, V22, V24, V25, V39	1
59	PFFZZ	6140011256074	19207	12277133	BLOCK, SUPPORT PART OF KIT P/N .... 12302775.................... UOC: DAA, DAB, DAC, DAD, DAE, DAF, DAG, DAH, DAJ, DAK, DAL, DAW, DAX, V12, V13, V14, V15, V16, V17, V17, V119, V20, V21, V22, V24, V25, V39	1
60	PFFZZ	6140011256073	19207	12277134	BLOCK, SUPPORT PART OF KIT P/N .... 12302775.................... UOC: DAA, DAB, DAC, DAD, DAE, DAF, DAG, DAH, DAJ, DAK, DAL, DAW, DAX, V12, V13, V14, V15, V16, V17, V18, V19, V20, V21, V22, V24, V25, V39	2
61	PFFZZ	6140011256075	19207	12277132	BLOCK, SUPPORT PART OF KIT P/N .... 12302775.................... UOC: DAA, DAB, DAC, DAD, DAE, DAF, DAG, DAH, DAJ, DAK, DAL, DAW, DAX, V12, V13, V14, V15, V16, V17, V18, V19, V20, V21, V22, V24, V25, V39	4
62	PBFZZ	2540012361175	19207	12256434	PAD ASSEMBLY SWINGFIRE HEATER PART..................... OF KIT P/N 12302775 UOC: DAA, DAB, DAC, DAD, DAE, DAF, DAG, DAH, DAJ, DAK, DAL, DAW, DAX, V12, V13, V14, V15, V16, V17, V18, V19, V20, V21, V22, V24, V25, V39	1
63	PAFZZ	4730002493935	96906	MS39231-4	ELBOW, PIPE PART OF KIT P/N 12302775. .... UOC: DAA, DAB, DAC, DAD, DAE, DAF, DAG, DAH, DAJ, DAK, DAL, DAW, DAX, V12, V13, V14, V15, V16, V17, V18, V19, V20, V21, V22, V24, V25, V39	1
	PFFZZ	2540011821077	19207	12302775	MOUNTING KIT, SWINGFIRE HEATER .... UOC: DAA, DAB, DAC, DAD, DAE, DAF, DAG, DAH, DAJ, DAK, DAL, DAW, DAX, V12, V13, V14, V15, V16, V17 V18, V19, V20, V21, V22, V24, V25, V39	1

```
ADAPTER, STRAIGHT        (    4)    589-10
ADAPTER, STRAIGHT        (    1)    589-52
BOLT                     (    2)    589-19
BOLT, MACHINE            (    2)    589-5
BLOCK, SUPPORT           (    2)    589-59
BLOCK, SUPPORT           (    2)    589-60
BLOCK, SUPPORT           (    4)    589-61
BRACKET                  (    1)    587-8
```

(1) ITEM NO	(2) SMR CODE	(3) NSN	(4) CAGEC	(5) PART NUMBER	(6) DESCRIPTION AND USABLE ON CODES (UOC)			(7) QTY
					BRACKET, ANGLE	(	1)	589-9
					BRACKET, ANGLE	(	1)	589-21
					BRACKET, ANGLE	(	1)	589-40
					BRACKET, ANGLE	(	1)	589-58
					BRACKET DOUBLE ANGL	(	1)	589-38
					CIRCUIT BREAKER	(	1)	588-3
					CLAMP	(	1)	587-10
					CLAMP, HOSE	(	1)	589-4
					CLAMP, HOSE	(	12)	588-11
					CLAMP, HOSE	(	2)	589-53
					CLAMP, LOOP	(	2)	588-8
					CLAMP, LOOP	(	2)	588-14
					CLAMP, LOOP	(	1)	589-32
					CLAMP, LOOP	(	4)	589-37
					CONNECTOR, ELECTRICA	(	1)	588-11
					COVER, ASSEMBLY	(	1)	587-2
					ELBOW, PIPE	(	1)	589-16
					ELBOW, PIPE	(	1)	589-47
					ELBOW, PIPE	(	1)	589-51
					ELBOW, PIPE	(	1)	589-63
					GASKET	(	1)	589-14
					GROMMET	(	1)	588-6
					GROMMET	(	1)	588-13
					HEATER, AIR DUCT, ENG	(	1)	587-7
					HOSE	(	3)	589-12
					HOSE	(	1)	589-54
					NIPPLE, PIPE	(	1)	589-2
					NIPPLE, PIPE	(	1)	589-48
					NUT, PLAIN, HEXAGON	(	17)	587-5
					NUT, SELF-LOCKING	(	5)	87-13
					NUT, SELF-LOCKING	(	4)	587-14
					NUT, SELF-LOCKING	(	1)	588-9
					NUT, SELF-LOCKING	(	2)	589-8
					NUT, SELF-LOCKING	(	2)	589-18
					NUT, SELF-LOCKING	(	4)	589-20
					NUT, SELF-LOCKING	(	8)	589-26
					NUT, SELF-LOCKING	(	4)	589-31
					PAD ASSEMBLY	(	1)	589-62
					PIN, COTTER	(	1)	589-28
					PUMP, ASSEMBLY	(	1)	589-3
					REDUCER, PIPE	(	2)	589-50
					SCREW, CAP, HEXAGON	(	5)	587-11
					SCREW, CAP, HEXAGON	(	4)	587-16
					SCREW, CAP, HEXAGON	(	1)	588-7
					SCREW, CAP, HEXAGON	(	2)	589-19
					SCREW, CAP, HEXAGON	(	2)	589-22
					SCREW, CAP, HEXAGON	(	6)	589-24
					SCREW, CAP, HEXAGON	(	2)	589-35
					SCREW, CAP, HEXAGON	(	1)	589-41
					SCREW, CAP, HEXAGON	(	1)	589-56
					SCREW, MACHINE	(	6)	587-3
					SCREW, MACHINE	(	3)	587-9
					SCREW, TAPPING	(	2)	587-4
					SHIELD, SWINGFIRE	(	1)	589-23

(1) ITEM NO	(2) SMR CODE	(3) NSN	(4) CAGEC	(5) PART NUMBER	(6) DESCRIPTION AND USABLE ON CODES (UOC)			(7) QTY
					SHROUD, OIL PAN	(	1)	589-27
					STRAP, RETAINING	(	1)	599-6
					STRAP, TIEDOWN	(	1)	588-2
					STRAP, TIEDOWN	(	5)	589-45
					STUD	(	4)	589-29
					STUD, TURNBUTTON	(	3)	587-1
					STUD, TURNBUTTON	(	8)	587-6
					SUPPORT	(	1)	589-25
					SWITCH	(	1)	588-10
					TEE, PIPE	(	1)	589-1
					TUBE	(	1)	589-13
					TUBE	(	1)	589-33
					TUBE	(	1)	589-36
					TUBE	(	1)	589-42
					TUBE	(	1)	589-43
					TUBE	(	1)	589-44
					WASHER, FLAT	(	5)	587-12
					WASHER, FLAT	(	4)	587-15
					WASHER, FLAT	(	2)	589-7
					WASHER, FLAT	(	4)	589-30
					WASHER, LOCK	(	17)	587-4
					WASHER, LOCK	(	1)	588-5
					WASHER, LOCK	(	1)	588-12
					WASHER, LOCK	(	1)	589-57
					WATER JACKET ASSEMB	(	1)	589-34
					WATER OUTLET, ENGINE	(	1)	589-15
					WIRING HARNESS, BRAN	(	11)	588-1

END OF FIGURE

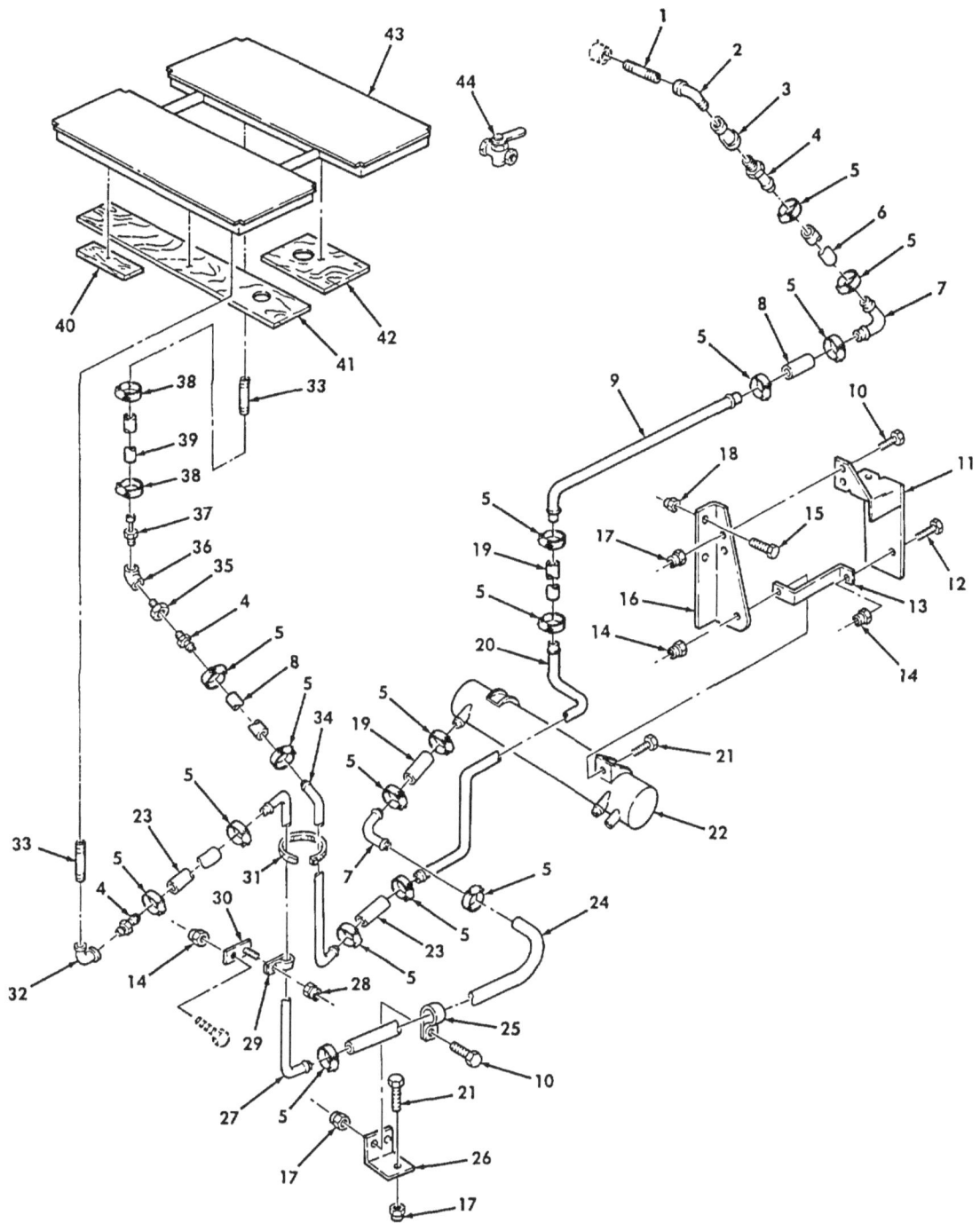

Figure 590. Swingfire Heater Mounting Kit (M939A2).

(1)	(2)	(3)	(4)	(5)	(6)	(7)
ITEM NO	SMR CODE	NSN	CAGEC	PART NUMBER	DESCRIPTION AND USABLE ON CODES (UOC)	QTY

GROUP 3303 WINTERIZATION KIT

FIG. 590 SWINGFIRE HEATER MOUNTING KIT (M939A2)

(1)	(2)	(3)	(4)	(5)	(6)	(7)
1	PAFZZ	4730001961504	24617	192075	NIPPLE, PIPE PART OF KIT P/N 5705626.	1
2	PAFZZ	4730002534414	96906	MS39230-4	ELBOW, PIPE PART OF KIT P/N 5705626	1
3	PFFZZ	4730002315650	96906	MS39232-7	REDUCER, PIPE PART OF KIT P/N 5705626	1
4	PAFZZ	4730011757343	19207	11656473	ADAPTER, STRAIGHT, PI PART OF KIT P/N 5705626	2
5	PAFZZ	4730009083193	96906	MS35842-12	CLAMP, HOSE PART OF KIT P/N 5705626	10
6	MFFZZ		19207	12302784-6	HOSE MAKE FROM HOSE, P/N MS521301A207R PART OF KIT P/N 5705626	1
7	PFFZZ	4710011994366	19207	12302781	TUBE, BENT, METALLIC PART OF KIT P/N 5705626	1
8	MFFZZ		19207	12302784-2	HOSE WATER LINE, MAKE FROM HOSE P/N MS52130-1A278R PART OF KIT P/N 5705626	1
9	PFFZZ		19207	12375599	TUBE, BENT METALLIC PART OF KIT P/N 5705626	1
10	PAFZZ	5305002678952	96906	MS90727-3	SCREW, CAP, HEXAGON H PART OF KIT P/N 5705626	2
11	PFFZZ	5340013536962	19207	12375593	COVER, ACCESS PART OF KIT P/N 5705626	1
12	PAFZZ	5305002693234	96906	MS90727-58	SCREW, CAP, HEXAGON H PART OF KIT P/N 5705626	1
13	PFFZZ	5340012035659	19207	12302799	BRACKET, DOUBLE ANGL PART OF KIT P/N 5705626	1
14	PAFZZ		96906	MS51943-36	NUT, SELF-LOCKING PART OF KIT P/N 5705626	6
15	PAFZZ	5305007195238	96906	MS90727-115	SCREW, CAP, HEXAGON H PART OF KIT P/N 5705626	2
16	PFFZZ	5340013536936	19207	12375597	BRACKET, MOUNTING PART OF KIT P/N 5705626	1
17	PAFZZ		96906	MS51943-32	NUT, SELF-LOCKING PART OF KIT P/N 5705626	2
18	PAFZZ	5310004883888	96906	MS51943-40	NUT, SELF-LOCKING, HE PART OF KIT P/N 5705626	2
19	PFFZZ	4720011757421	19207	11656458-3	HOSE, NONMETALLIC PART OF KIT P/N 5705626	1
20	PFFZZ		19207	12375600	TUBE, BENT, METALLIC PART OF KIT P/N 5705626	1
21	PAFZZ		96906	MS90726-60	SCREW, CAP, HEXAGON PART OF KIT P/N 5705626	2
22	PFFZZ	2540010769286	16236	269013000000	WATER JACKET ASSEMB PART OF KIT P/N 5705626	1
23	PFFZZ		19207	11656458-2	HOSE ASSEMBLY, NONME PART OF KIT P/N 5705626	2
24	MFFZZ		19207	12302784-3-29	HOSE MAKE FROM HOSE P/N MS521301A207R PART OF KIT P/N 5705626	1
25	PFFZZ	5340009223380	96906	MS9350-23	CLAMP, LOOP PART OF KIT P/N 5705626	1

(1) ITEM NO	(2) SMR CODE	(3) NSN	(4) CAGEC	(5) PART NUMBER	(6) DESCRIPTION AND USABLE ON CODES (UOC)	(7) QTY
26	PFFZZ	5340013536158	19207	12375596	BRACKET, MOUNTING PART OF KIT P/N 5705626	1
27	PFFZZ		19207	12302791-1	TUBE, BENT, METALLIC PART OF KIT P/N 8750177 PART OF KIT P/N 5705626	1
28	PAFZZ		96906	MS51943-36	NUT, SELF-LOCKING PART OF KIT P/N 5705626	2
29	PAFZZ	5340011179876	19207	7397785	CLAMP, LOOP PART OF KIT P/N 5705626	1
30	PFFZZ		19207	12375586	BRACKET, MOUNTING PART OF KIT P/N 5705626	1
31	PFFZZ	5975004515001	96906	MS3367-3-9	STRAP, TIEDOWN, ELECT PART OF KIT P/N 5705626	3
32	PAFZZ	4730002491511	24617	217985	ELBOW, PIPE PART OF KIT P/N 5705626	1
33	PAFZZ	4730001961489	96906	MS51953-55	NIPPLE, PIPE PART OF KIT P/N 5705626	1
34	PFFZZ		19207	12302782-1	TUBE, BENT, METALLIC PART OF KIT P/N 5705626	1
35	PAFZZ	4730001893034	24617	444019	REDUCER, PIPE PART OF KIT P/N 5705626	1
36	PAFZZ	4730007222759	24617	444073	ELBOW, PIPE PART OF KIT P/N 5705626	1
37	PAFZZ	4730009445888	96906	MS24522-7	ADAPTER, STRAIGHT, PI PART OF KIT P/N 5705626	1
38	PAFZZ	4730009083194	96906	MS35842-11	CLAMP, HOSE PART OF KIT P/N 5705626	2
39	MFFZZ		19207	8710557 X BULK	HOSE, NONMETALLIC MAKE FROM HOSE, P/N 8710557 AS REQUIRED PART OF KIT P/N 5705626	1
40	PFFZZ	6140011256075	19207	12277132	BLOCK, SUPPORT, BATTE PART OF KIT P/N 5705626	4
41	PFFZZ	6140011256073	19207	12277134	BLOCK, SUPPORT, BATTE PART OF KIT P/N 5705626	2
42	PFFZZ	6140011256074	19207	12277133	BLOCK, SUPPORT, BATTE PART OF KIT P/N 5705626	2
43	PFFZZ	2540012361175	19207	12256434	PAD ASSEMBLY PART OF KIT P/N 5705626	1
44	PAFZZ		19207	12432312	VALVE PART OF KIT P/N 5705626	1

END OF FIGURE

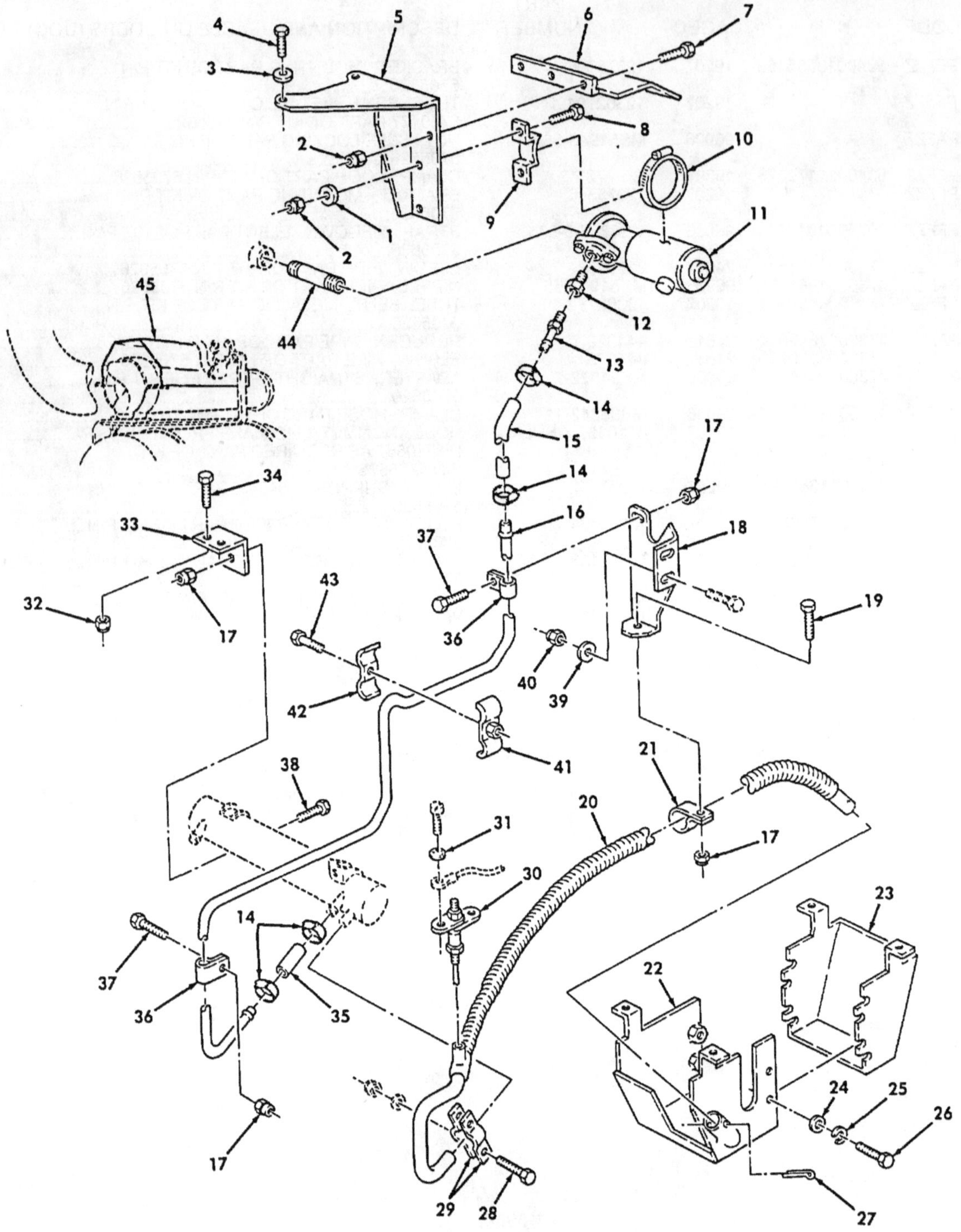

Figure 591. Swingfire Heater Mounting Kit (M939A2).

(1) ITEM NO	(2) SMR CODE	(3) NSN	(4) CAGEC	(5) PART NUMBER	(6) DESCRIPTION AND USABLE ON CODES (UOC)	(7) QTY
					GROUP 3303 WINTERIZATION KIT	
					FIG. 591 SWINGFIRE HEATER MOUNTING KIT (M939A2)	
1	PAFZZ	5310000814219	96906	MS27183-12	WASHER, FLAT PART OF KIT P/N 5705626.	2
2	PAFZZ	5310002416658	96906	MS51943-34	NUT, SELF-LOCKING, HE PART OF KIT P/N 5705626.	3
3	PAFZZ	5310011356042	73342	11500222	WASHER, LOCK PART OF KIT P/N 5705626.	2
4	PAFZZ		24617	11500928	BOLT, MACHINE PART OF KIT P/N 5705626.	2
5	PFFZZ		19207	12375589	SUPPORT ASSEMBLY, RA PART OF KIT P/N 5705626.	1
6	PFFZZ	5340013536961	19207	12375592	COVER, ACCESS PART OF KIT P/N 5705626.	1
7	PAFZZ	5306002259086	96906	MS90726-31	BOLT, MACHINE PART OF KIT P/N 5705626.	3
8	PAFZZ	5306002259087	96906	MS90726-32	BOLT, MACHINE PART OF KIT P/N 5705626.	2
9	PFFZZ	5340011467667	19207	11592371	STRAP, RETAINING PART OF KIT P/N 5705626.	1
10	PAFZZ	4730009086292	96906	MS35842-14	CLAMP, HOSE PART OF KIT P/N 5705626.	1
11	PFFZZ	4320009302045	19207	10946835	PUMP UNIT, CENTRIFUG PART OF KIT P/N 5705626.	1
12	PAFZZ		24617	444019	ADAPTER PART OF KIT P/N 5705626.	1
13	PAFZZ		19207	11656473	CONNECTOR PART OF KIT P/N 5705626.	1
14	PAFZZ		96906	MS35842-12	CLAMP PART OF KIT P/N 5705626.	4
15	MFFZZ		19207	12302784-2	HOSE WATER LINE, MAKE FROM HOSE P/N MS52130-1A278R PART OF KIT P/N 5705626.	5
16	PFFZZ		19207	12375606	TUBE, BENT, METALLIC PART OF KIT P/N 5705626.	1
17	PAFZZ	5310008140672	96906	MS51943-36	NUT, SELF-LOCKING, HE UOC:ZAA, ZAB, ZAC, ZAD, ZAE, ZAF, ZAG, ZAH, ZAJ, ZAK	17
18	PFFZZ		19207	12375595	BRACKET, MOUNTING PART OF KIT P/N 5705626.	1
19	PAFZZ	5305002693235	96906	MS90727-59	SCREW, CAP, HEXAGON H PART OF KIT P/N 5705626.	1
20	PFFZZ		19207	12375588	TUBE ASSEMBLY, METAL PART OF KIT P/N 5705626.	1
21	PAFZZ	5340011730241	19207	8707524	CLAMP, LOOP PART OF KIT P/N 5705626.	1
22	PFFZZ		19207	12375603	SHROUD, FAN, RADIATOR PART OF KIT P/N 5705626.	1
23	PFFZZ		19207	12375604	SHROUD, FAN, RADIATOR PART OF KIT P/N 5705626.	1
24	PAFZZ	5310008094058	96906	MS27183-10	WASHER, FLAT PART OF KIT P/N 5705626.	6
25	PAFZZ	5310005825965	96906	MS35338-44	WASHER, LOCK PART OF KIT P/N 5705626.	6
26	PAFZZ	5305002678952	96906	MS90727-3	SCREW, CAP, HEXAGON H PART OF KIT P/N 5705626.	6
27	PAFZZ	5315002981498	96906	MS24665-362	PIN, COTTER PART OF KIT P/N 5705626.	1
28	PAFZZ	5305002693236	96906	MS90727-60	SCREW, CAP, HEXAGON H PART OF KIT P/N 5705626.	2
29	PAFZZ	5340011870527	19207	11656448-1	CLAMP, LOOP PART OF KIT P/N 5705626.	1
30	PAFZZ	5930013182809	19207	12356766	SWITCH, THERMOSTATIC PART OF KIT P/N 5705626.	1
31	PAFZZ	5310005590070	96906	MS35333-38	WASHER, LOCK PART OF KIT P/N 5705626.	2
32	PAFZZ		96906	MS51943-32	NUT, SELF-LOCKING PART OF KIT P/N	1

(1) ITEM NO	(2) SMR CODE	(3) NSN	(4) CAGEC	(5) PART NUMBER	(6) DESCRIPTION AND USABLE ON CODES (UOC)			(7) QTY
					5705626			
33	PFFZZ		19207	12375596	BRACKET PART OF KIT P/N 5705626....................................			1
34	PAFZZ	5305000680515	96906	MS90727-8	SCREW, CAP, HEXAGON H PART OF KIT P/N..................			2
					5705626			
35	PFFZZ		19207	11656458-3	HOSE PART OF KIT P/N 5705626			1
36	PAFZZ		19207	7397785	CLAMP PART OF KIT P/N 5705626			2
37	PAFZZ		96906	MS90726-58	SCREW, CAP, HEXAGON PART OF KIT P/N.....................			2
					5705626.			
38	PAFZZ		96906	MS90726-60	SCREW, CAP, HEXAGON PART OF KIT P/N.....................			1
					5705626.			
39	PAFZZ	5310008095998	96906	MS27183-18	WASHER, FLAT PART OF KIT P/N 5705626......................			2
40	PAFZZ		96906	MS51943-40	NUT, SELF-LOCKING PART OF KIT P/N............................			2
					5705626.			
41	PFFZZ	5340013536753	19207	12375598-2	STRAP, RETAINING PART OF KIT P/N			1
					5705626			
42	PFFZZ	5340013536752	19207	12375598-1	STRAP, RETAINING PART OF KIT P/N			1
					5705626			
43	PAFZZ	5306009170900	80205	NAS1297-5-13	BOLT, SHOULDER PART OF KIT P/N			1
					5705626.			
44	PFFZZ		19207	12375591	NIPPLE, PIPE PART OF KIT P/N 5705626.			1
45	PAFZZ		19207	12432309	SHIELD, HEAT PART OF KIT P/N 5705626........................			1
	PDFZZ	2540013426810	19207	5705626	HEATER, VEHICULAR, CO			1
					ADAPTER	(1)	591-12	
					ADAPTER, STRAIGHT, PI	(2)	590-4	
					ADAPTER, STRAIGHT, PI	(1)	590-37	
					BLOCK, SUPPORT, BATTE	(4)	590-40	
					BLOCK, SUPPORT, BATTE	(2)	590-41	
					BLOCK, SUPPORT, BATTE	(2)	590-42	
					BOLT, MACHINE	(2)	591-4	
					BOLT, MACHINE	(3)	591-7	
					BOLT, MACHINE	(2)	591-8	
					BOLT, SHOULDER	(1)	591-43	
					BRACKET	(1)	591-33	
					BRACKET, DOUBLE ANGL	(1)	590-13	
					BRACKET, MOUNTING	(1)	590-16	
					BRACKET, MOUNTING	(1)	590-26	
					BRACKET, MOUNTING	(1)	590-30	
					BRACKET, MOUNTING	(1)	591-18	
					CIRCUIT BREAKER	(1)	588-3	
					CLAMP	(2)	591-36	
					CLAMP	(1)	591-14	
					CLAMP, HOSE	(1)	590-5	
					CLAMP, HOSE	(2)	590-38	
					CLAMP, HOSE	(1)	591-10	
					CLAMP, LOOP	(2)	588-7	
					CLAMP, LOOP	(1)	588-9	
					CLAMP, LOOP	(2)	588-12	
					CLAMP, LOOP	(1)	590-25	
					CLAMP, LOOP	(1)	590-29	
					CLAMP, LOOP	(1)	591-21	
					CLAMP, LOOP	(1)	591-29	
					CLAMP, RIM CLENCHING	(1)	C107-20	
					CONNECTOR	(1)	591-13	
					CONNECTOR, RECEPTACL	(1)	588-4	

(1) ITEM NO	(2) SMR CODE	(3) NSN	(4) CAGEC	(5) PART NUMBER	(6) DESCRIPTION AND USABLE ON CODES (UOC)		(7) QTY
					COVER, ACCESS	(1)	590-11
					COVER, ACCESS	(1)	591-6
					CUTOUT RELAY, ENGINE	(1)	588-17
					ELBOW, PIPE	(1)	590-2
					ELBOW, PIPE	(1)	590-32
					ELBOW, PIPE	(1)	590-36
					GROMMET, NONMETALLIC	(1)	588-6
					GROMMET, NONMETALLIC	(1)	588-8
					HOSE	(1)	590-6
					HOSE	(1)	590-8
					HOSE	(1)	590-24
					HOSE	(1)	591-15
					HOSE	(1)	591-35
					HOSE ASSEMBLY, NONME	(2)	590-23
					HOSE, NONMETALLIC	(1)	590-19
					HOSE, NONMETALLIC	(1)	590-39
					LEAD, ELECTRICAL	(1)	588-16
					NIPPLE, PIPE	(1)	590-1
					NIPPLE, PIPE	(1)	590-33
					NIPPLE, PIPE	(1)	591-44
					NUT, SELF-LOCKING, HE	(1)	588-13
					NUT, SELF-LOCKING, HE	(2)	588-15
					NUT, SELF-LOCKING, HE	(2)	590-14
					NUT, SELF-LOCKING, HE	(2)	590-17
					NUT, SELF-LOCKING, HE	(2)	590-18
					NUT, SELF-LOCKING, HE	(1)	590-28
					NUT, SELF-LOCKING, HE	(2)	591-2
					NUT, SELF-LOCKING, HE	(1)	591-32
					NUT, SELF-LOCKING, HE	(2)	591-40
					PAD ASSEMBLY	(1)	590-43
					PIN, COTTER	(1)	591-27
					PUMP UNIT, CENTRIFUG	(1)	591-11
					REDUCER, PIPE	(1)	590-3
					REDUCER, PIPE	(1)	590-35
					SCREW, CAP, HEXAGON H	(2)	588-11
					SCREW, CAP, HEXAGON H	(2)	590-10
					SCREW, CAP, HEXAGON H	(1)	590-12
					SCREW, CAP, HEXAGON H	(2)	590-15
					SCREW, CAP, HEXAGON H	(2)	590-21
					SCREW, CAP, HEXAGON H	(1)	591-19
					SCREW, CAP, HEXAGON H	(2)	591-28
					SCREW, CAP, HEXAGON H	(2)	591-34
					SCREW, CAP, HEXAGON H	(2)	591-37
					SCREW, CAP, HEXAGON H	(1)	591-38
					SCREW, TAPPING	(2)	588-2
					SEMICONDUCTOR DEVIC	(1)	588-14
					SHROUD, FAN, RADIATOR	(1)	591-22
					SHROUD, FAN, RADIATOR	(1)	591-23
					STRAP, RETAINING	(1)	591-9
					STRAP, RETAINING	(1)	591-41
					STRAP, RETAINING	(1)	591-42
					STRAP, TIEDOWN, ELECT	(5)	588-10
					STRAP, TIEDOWN, ELECT	(3)	590-31
					SUPPORT ASSEMBLY, RA	(1)	591-5

(1) ITEM NO	(2) SMR CODE	(3) NSN	(4) CAGEC	(5) PART NUMBER	(6) DESCRIPTION AND USABLE ON CODES (UOC)			(7) QTY
					SWITCH, THERMOSTATIC	(	1)	591-30
					TUBE ASSEMBLY, METAL	(	1)	591-20
					TUBE, BENT, METALLIC	(	1)	590-7
					TUBE, BENT, METALLIC	(	1)	590-9
					TUBE, BENT, METALLIC	(	1)	590-20
					TUBE, BENT, METALLIC	(	1)	590-27
					TUBE, BENT, METALLIC	(	1)	590-34
					TUBE, BENT, METALLIC	(	1)	591-16
					WASHER, FLAT	(	2)	591-1
					WASHER, FLAT	(	6)	591-24
					WASHER, FLAT	(	2)	591-39
					WASHER, LOCK	(	1)	588-5
					WASHER, LOCK	(	2)	591-3
					WASHER, LOCK	(	6)	591-25
					WASHER, LOCK	(	2)	591-31
					WATER JACKET ASSEMB	(	1)	590-22
					WIRING HARNESS, BRAN	(	1)	588-1

END OF FIGURE

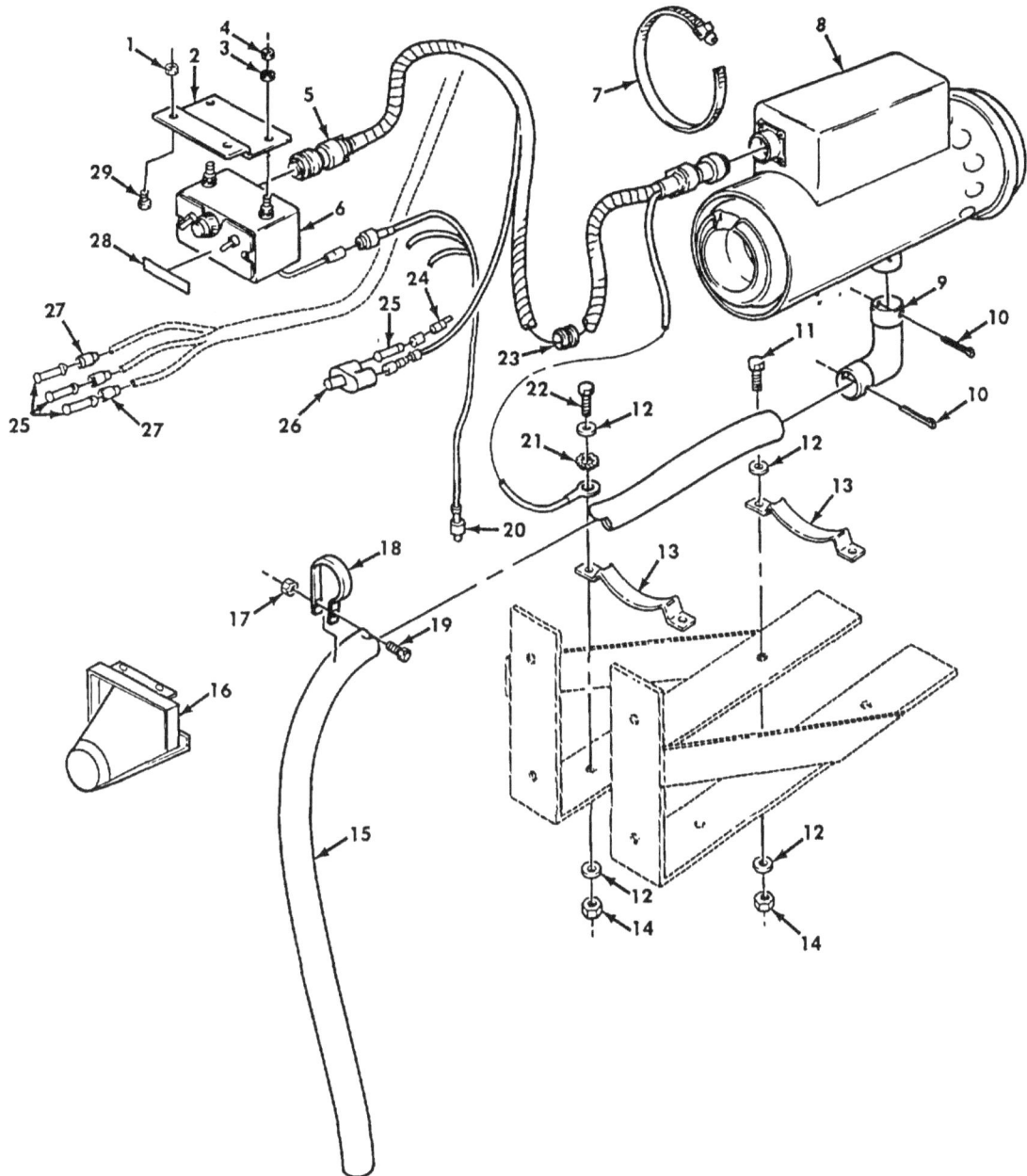

Figure 592. Fuel Burning Personnel Heater Kit, Control Box, Heater and Wiring Harness (M939, M939A1)

(1) ITEM NO	(2) SMR CODE	(3) NSN	(4) CAGEC	(5) PART NUMBER	(6) DESCRIPTION AND USABLE ON CODES (UOC)	(7) QTY
					GROUP 3303 WINTERIZATION KIT	
					FIG. 592 FUEL BURNING PERSONNEL HEATER KIT, CONTROL BOX, HEATER, AND WIRING HARNESS (M939, M939A1)	
1	PAFZZ	5310009359022	96906	MS51943-32	NUT, SELF-LOCKING, HE PART OF KIT P/N 12256443 UOC:DAA, DAB, DAC, DAD, DAE, DAF, DAG, DAH, DAJ, DAK, DAL, DAW, DAX, V12, V13, V14, V15, V16, V17, V18, V19, V20, V21, V22, V24, V25, V39	2
2	PAFZZ	5340011097950	19207	11648534	BRACKET, MOUNTING PART OF KIT P/N 12256443 UOC:DAA, DAB, DAC, DAD, DAE, DAF, DAG, DAH, DAJ, DAK, DAL, DAW, DAX, V12, V13, V14, V15, V16, V17, V18, V19, V20, V21, V22, V24, V25, V39	1
3	PAFZZ	5310002090786	96906	MS35335-33	WASHER, LOCK PART OF KIT P/N 12256443 UOC:DAA, DAB, DAC, DAD, DAE, DAF, DAG, DAH, DAJ, DAK, DAL, DAW, DAX, V12, V13, V14, V15, V16, V17, V18, V19, V20, V21, V22, V24, V25, V39	2
4	PAFZZ	5310007616882	96906	MS51967-2	NUT, PLAIN, HEXAGON PART OF KIT P/N 12256443 UOC:DAA, DAB, DAC, DAD, DAE, DAF, DAG, DAH, DAJ, DAK, DAL, DAW, DAX, V12, V13, V14, V15, V16, V17, V18, V19, V20, V21, V22, V24, V25, V39	2
5	PAFZZ	2920010958308	19207	12256430	WIRING HARNESS, BRAN PART OF KIT P/N 12256443 UOC:DAA, DAB, DAC, DAD, DAE, DAF, DAG, DAH, DAJ, DAK, DAL, DAW, DAX, V12, V13, V14, V15, V16, V17, V18, V19, V20, V21, V22, V24, V25, V39	1
6	PAOZZ	2590011256154	19207	11669705	CONTROL BOX, ELECTRI PART OF KIT P/N 12256443 UOC:DAA, DAB, DAC, DAD, DAE, DAF, DAG, DAH, DAJ, DAK, DAL, DAW, DAX, V12, V13, V134, V14, V15, V16, V17, V18, V19, V20, V21, V22, V24, V25, V39	1
7	PAFZZ	4730009086294	75160	C16054	CLAMP, HOSE PART OF KIT P/N 12256443. 2 UOC:DAA, DAB, DAC, DAD, DAE, DAF, DAG, DAH, DAJ, DAK, DAL, DAW, DAX, V12, V13, V14, V15, V16, V17, V18, V19, V20, V21, V22, V24, V25, V39	
8	PAOZZ		81349	MIL-PRF-62550/3	HEATER, VEHICULAR, CO PART OF KIT P/N 12256443 UOC:DAA, DAB, DAC, DAD, DAE, DAF, DAG, DAH, DAJ, DAK, DAL, DAW, DAX, V12, V13, V14, V15, V16, V17, V18, V19, V20, V21, V22, V24, V25, V39	1

(1) ITEM NO	(2) SMR CODE	(3) NSN	(4) CAGEC	(5) PART NUMBER	(6) DESCRIPTION AND USABLE ON CODES (UOC)	(7) QTY
9	PAFZZ	4730012146720	34623	12302645	ELBOW, TUBE PART OF KIT P/N 12256443 ...... UOC:DAA, DAB, DAC, DAD, DAE, DAF, DAG, DAH, DAJ, DAK, DAL, DAW, DAX, V12, V13, V14, V15, V16, V17, V18, V19, V20, V21, V22, V24, V25, V39	1
10	PAFZZ	5315008499854	96906	MS24665-498	PIN, COTTER PART OF KIT P/N 12256443 ...... UOC:DAA, DAB, DAC, DAD, DAE, DAF, DAG, DAH, DAJ, DAK, DAL, DAW, DAX, V12, V13, V14, V15, V16, V17, V18, V19, V20, V21, V22, V24, V25, V39	2
11	PAFZZ	5305002259091	96906	MS90726-36	SCREW, CAP, HEXAGON H PART OF KIT P/N .... 12256443 ...... UOC:DAA, DAB, DAC, DAD, DAE, DAF, DAG, DAH, DAJ, DAK, DAL, DAW, DAX, V12, V13, V14, V15, V16, V17, V18, V19, V20, V21, V22, V24, V25, V39	1
12	PAFZZ	5310000814219	96906	MS27183-12	WASHER, FLAT ...... UOC:DAA, DAB, DAC, DAD, DAE, DAF, DAG, DAH, DAJ, DAK, DAL , DAW, DAX, V12 , V13, V14 , V15, V16, V17, V18, V19, V20, V21, V22, V24, V25, V39	8
13	PAFZZ	5340014163003	19207	12256199	BRACKET, MOUNTING PART OF KIT P/N .... 12256443 ...... UOC:DAA, DAB, DAC, DAD, DAE, DAF, DAG, DAH, DAJ, DAK, DAL, DAW, DAX, V12, V13, V14, V15, V16, V17, V18, V19, V20, V21, V22, V24, V25, V39	2
14	PAFZZ	5310002416658	96906	MS51943-34	NUT, SELF-LOCKING, HE PART OF KIT P/N ...... 12256443 ...... UOC:DAA, DAB, DAC, DAD, DAE, DAF, DAG, DAH, DAJ, DAK, DAL, DAW, DAX, V12, V13, V14, V15, V16, V17, V18, V19, V20, V21, V22, V24, V25, V39	4
15	PAFZZ	2990011062296	19207	11677627-1	PIPE, EXHAUST PART OF KIT P/N ...... 12256443 ...... UOC:DAA, DAB, DAC, DAD, DAE, DAF, DAG, DAH, DAJ, DAK, DAL, DAW, DAX, V12, V13, V14, V15, V16, V17, V18, V19, V20, V21, V22, V24, V25, V39	1
16	PAFZZ	2540013104829	19207	12255940-1	NOZZLE, DEFROSTER, VE PART OF KIT P/N ...... 122564 43 ...... UOC:DAA, DAB, DAC, DAD, DAE, DAF, DAG, DAH, DAJ, DAK, DAL, DAW, DAX, V12, V13, V14, V15, V16, V17, V18, V19, V20, V21, V22, V24, V25, V39	2
17	PAFZZ	5310008140672	96906	MS51943-36	NUT, SELF-LOCKING, HE PART OF KIT P/N ...... 12256443 ...... UOC:DAA, DAB, DAC, DAD, DAE, DAF, DAG, DAH, DAJ, DAK, DAL, DAW, DAX, V12, V13, V14, V15, V16, V17, V18, V19, V20, V21, V22, V24, V25, V39	2

(1) ITEM NO	(2) SMR CODE	(3) NSN	(4) CAGEC	(5) PART NUMBER	(6) DESCRIPTION AND USABLE ON CODES (UOC)	(7) QTY
18	PAFZZ	5340011730241	19207	8707524	CLAMP, LOOP PART OF KIT P/N 12256443............... UOC:DAA, DAB, DAC, DAD, DAE, DAF, DAG, DAH, DAJ, DAK, DAL, DAW, DAX, V12, V13, V14, V15, V16, V17, V18, V19, V20, V21, V22, V24, V25, V39	2
19	PAFZZ	5305002692803	96906	MS90726-60	SCREW, CAP, HEXAGON H PART OF KIT P/N................... 12256443... UOC:DAA, DAB, DAC, DAD, DAE, DAF, DAG, DAH, DAJ, DAK, DAL, DAW, DAX, V12, V13, V14, V15, V16, V17 , 18, V19, V20, V21, V22, V24, V25, V39	2
20	PFFZZ	5995012150930	19207	12302627	WIRING HARNESS, BRAN PART OF KIT P/N................... 12256443... UOC:DAA, DAB, DAC, DAD, DAE, DAF, DAG, DAH, DAJ, DAK, DAL, DAW, DAX, V12, V13, V14, V15, V16, V17, V18 , V19 , V20, V21, V22, V24, V25, V39	1
21	PAFZZ	5310007282044	96906	MS45904-73	WASHER, LOCK PART OF KIT P/N 12256443............... UOC:DAA, DAB, DAC, DAD, DAE, DAF, DAG, DAH, DAJ, DAK, DAL, DAW, DAX, V12, V13, V14, V15, V16, V17, V18, V19, V20, V21, V22, V2 4, V25, V39	1
22	PAFZZ	5306000514077	80204	B1821BH031F113N	BOLT, MACHINE PART OF KIT P/N .... 12256443... UOC:DAA, DAB, DAC, DAD, DAE, DAF, DAG, DAH, DAJ , DAK, DAL , DAW , DAX , V12 , V13 , V14 , V15, V16, V17, V18, V19, V20, V21, V22, V24, V25, V39	4
23	PAFZZ	5325011247760	96906	MS35489-79	GROMMET, NONMETALLIC PART OF KIT P/N................... 12256443... UOC:DAA, DAB, DAC, DAD, DAE, DAF, DAG, DAH, DAJ, DAK, DAL, DAW, DAX, V12 , V13 , V14 , V15, V16, V17, V18, V19, V20, V21, V22, V24, V25, V39	1
24	PAFZZ	5935008338561	19207	8338561	SHELL, ELECTRICAL CO PART OF KIT P/N................... 12256443... UOC:DAA, DAB, DAC, DAD, DAE, DAF, DAG, DAH, DAJ, DAK, DAL, DAW, DAX, V12, V13, V14, V15, V16, V17, V18, V19 , V20, V21, V22, V24, V25, V39	1
25	PAFZZ	5935002140904	19207	7982907	DUMMY CONNECTOR, PLU PART OF KIT P/N .... 12256443... UOC:DAA, DAB, DAC, DAD, DAE, DAF, DAG, DAH, DAJ, DAK, DAL, DAW, DAX, V12, V13, V14, V15, V16, V17, V18, V19, V20, V21, V22, V24, V25, V39	4
26	PAFZA	5935009006281	96906	MS27147-1	ADAPTER, CONNECTOR PART OF KIT P/N .... 12256443 .... UOC:DAA, DAB, DAC, DAD, DAE, DAF, DAG, DAH, DAJ, DAK, DAL, DAW, DAX, V12 , V13 , V14 , V15, V16, V17, V18 , V19 20, V2, V2, V22, V24 , V25, V39	1

(1) ITEM NO	(2) SMR CODE	(3) NSN	(4) CAGEC	(5) PART NUMBER	(6) DESCRIPTION AND USABLE ON CODES (UOC)	(7) QTY
27	PAFZZ	5935005729180	19207	8338566	SHELL, ELECTRICAL CO PART OF KIT P/N 12256443 UOC:DAA, DAB, DAC, DAD, DAE, DAF, DAG, DAH, DAJ, DAK, DAL, DAW, DAX, V12, V13, V14, V15, V16, V17, V18, V19, V20, V21, V22, V24, V25, V39	3
28	PAFZZ	7690010325639	19207	10896515	DECAL PART OF KIT P/N 12256443 UOC:DAA, DAB, DAC, DAD, DAE, DAF, DAG, DAH, DAJ, DAK, DAL, DAW, DAX, V12, V13, V14, V15, V16, V17, V18, V19, V20, V21, V22, V24, V25, V39	1
29	PAFZZ	5305002678953	80204	B1821BH025F063N	SCREW, CAP, HEXAGON H PART OF KIT P/N 12256443 UOC:DAA, DAB, DAC, DAD, DAE, DAF, DAG, DAH, DAJ, DAK, DAL, DAW, DAX, V12, V13, V14, V15, V16, V17, V18, V19, V20, V21, V22, V24, V25, V39	2

END OF FIGURE

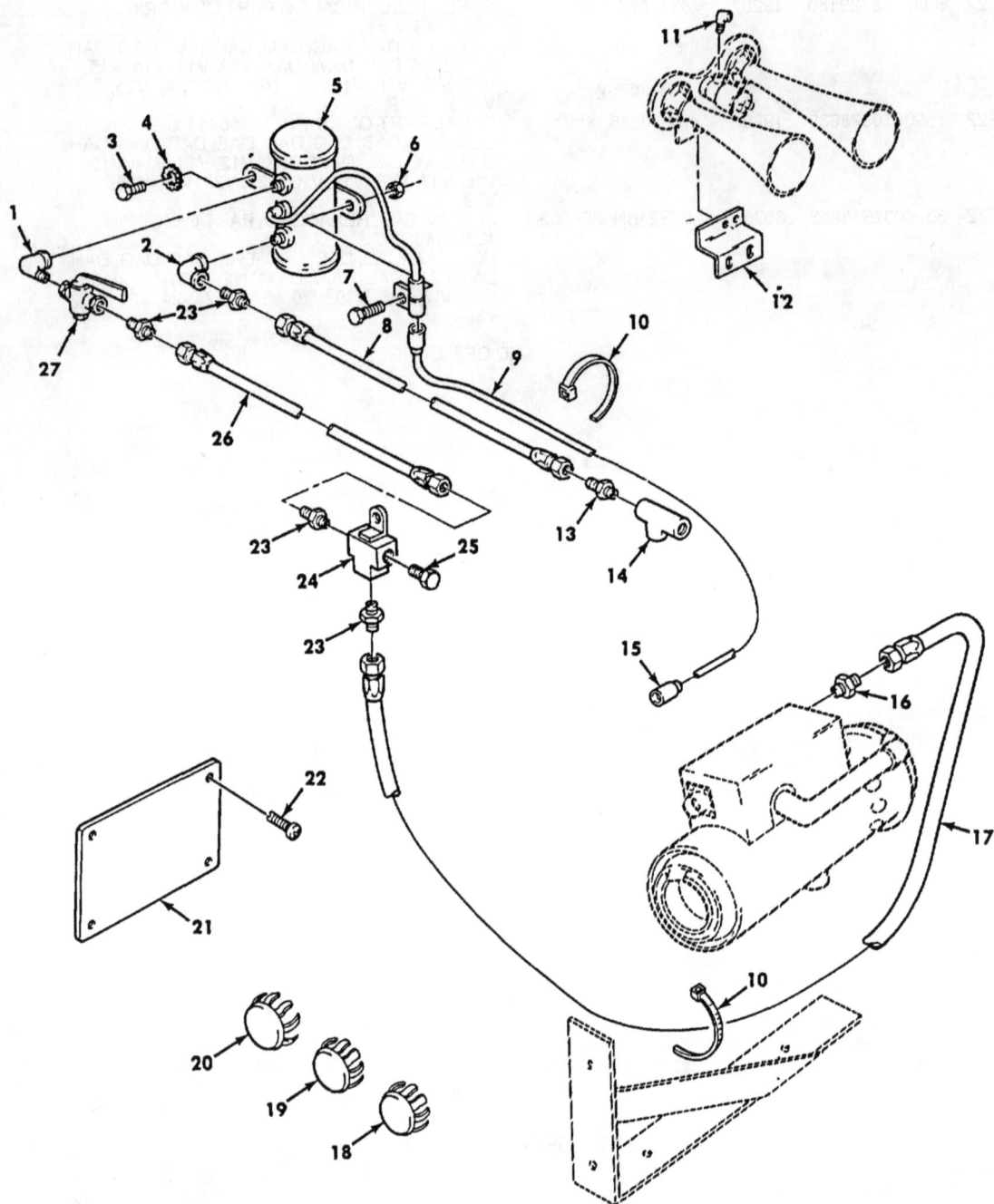

Figure 593. Fuel Burning Personnel Heater Kit (M939, M939A1).

(1) ITEM NO	(2) SMR CODE	(3) NSN	(4) CAGEC	(5) PART NUMBER	(6) DESCRIPTION AND USABLE ON CODES (UOC)	(7) QTY
					GROUP 3303 WINTERIZATION KITS	
					FIG. 593 FUEL BURNING PERSONNEL HEATER KIT(M939, M939A1)	
1	PAFZZ	4730002775533	96906	MS51952-2	ELBOW, PIPE PART OF KIT P/N 12256443 UOC:DAA, DAB, DAC, DAD, DAE, DAF, DAG, DAH, DAJ, DAK, DAL, DAW, DAX, V12, V13, V14, V15, V16, V17, V18, V19, V20, V21, V22, V24, V25, V39	1
2	PAFZZ	4730002775542	96906	MS51845-2	ELBOW, PIPE PART OF KIT P/N 12256443 UOC:DAA, DAB, DAC, DAD, DAE, DAF, DAG, DAH, DAJ, DAK, DAL, DAW, DAX, V12, V13, V14, V15, V16, V17, V18, V19, V20, V21, V22, V24, V25, V39	1
3	PAFZZ	5305002678953	80204	B1821BH025F063N	SCREW, CAP, HEXAGON H PART OF KIT P/N 12256443 UOC:DAA, DAB, DAC, DAD, DAE, DAF, DAG, DAH, DAJ, DAK, DAL, DAW, DAX, V12, V13, V14, V15, V16, V17, V18, V19, V20, V21, V22, V24, V25, V39	1
4	PAFZZ	5310008892528	96906	MS45904-68	WASHER, LOCK PART OF KIT P/N 12256443 UOC:DAA, DAB, DAC, DAD, DAE, DAF, DAG, DAH, DAJ, DAK, DAL, DAW, DAX, V12, V13, V14, V15, V16, V17, V17, V18, V19, V20, V21, V22, V24, V25, V39	2
5	PAFZZ	2910009309367	96906	MS51321-2-24N1	PUMP, FUEL, ELECTRICA UOC:DAA, DAB, DAC, DAD, DAE, DAF, DAG, DAH, DAJ, DAK, DAL, DAW, DAX, V12, V13, V14, V15, V16, V17, V18, V19, V20, V21, V22, V24, V25, V39	1
6	PAFZZ	5310009359022	96906	MS51943-32	NUT, SELF-LOCKING, HE PART OF KIT P/N 12256443 UOC:DAA, DAB, DAC, DAD, DAE, DAF, DAG, DAH, DAJ, DAK, DAL, DAW, DAX, V12, V13, V14, V15, V16, V17, V18, V19, V20, V21, V22, V24, V25, V39	2
7	PAFZZ	5305002678952	80204	B1821BH025F050N	SCREW, CAP, HEXAGON H PART OF KIT P/N 12256443 UOC:DAA, DAB, DAC, DAD, DAE, DAF, DAG, DAH, DAJ, DAK, DAL, DAW, DAX, V12, V13, V14, V15, V16, V17, V18, V19, V20, V21, V22, V24, V25, V39	1
8	XDFZZ	4720012153226	19207	12302628-2	HOSE ASSEMBLY, NONME PART OF KIT P/N 12256443 UOC:DAA, DAB, DAC, DAD, DAE, DAF, DAG, DAH, DAJ, DAK, DAL, DAW, DAX, V12, V13, V14, V15, V16, V17, V18, V19, V20, V21, V22, V24, V25, V39	1
9	PAFZZ	2920010958307	19207	12256428-2	CABLE ASSEMBLY, SPEC PART OF KIT P/N 12256443 UOC:DAA, DAB, DAC, DAD, DAE, DAF, DAG, DAH, DAJ, DAK, DAL, DAW, DAX, V12, V13, V14, V15, V16, V17, V18, V19, V20, V21, V22, V24, V25, V39	1

(1) ITEM NO	(2) SMR CODE	(3) NSN	(4) CAGEC	(5) PART NUMBER	(6) DESCRIPTION AND USABLE ON CODES (UOC)	(7) QTY
10	PAFZZ	5975008994606	96906	MS3367-2-0	STRAP, TIEDOWN, ELECT PART OF KIT P/N 12256443 UOC:DAA, DAB, DAC, DAD, DAE, DAF, DAG, DAH, DAJ, DAK, DAL, DAW, V12, V13, V14, V15, V16, V17, V18, V19, V20, V21, V22, V24, V25, V39	7
11	PAFZZ	4730012218821	24617	9409934	ELBOW, PIPE PART OF KIT P/N 12256443 UOC:DAA, DAB, DAC, DAD, DAE, DAF, DAG, DAH, DAJ, DAK, DAL, DAW, DAX, V12, V13, V14, V15, V16, V17, V18, V19, V20, V21, V22, V24, V25, V39	1
12	KFFZZ		19207	12302626	BRACKET PART OF KIT P/N 12256443 UOC:DAA, DAB, DAC, DAD, DAE, DAF, DAG, DAH, DAJ, DAK, DAL, DAW, DAX, V12, V13, V14, V15, V16, V17, V18, V19, V20, V21, V22, V24, V25, V39	1
13	PAFZZ	4730012263705	81343	8-6 010102B	ADAPTER, STRAIGHT, PI PART OF KIT P/N 12256443 UOC:DAA, DAB, DAC, DAD, DAE, DAF, DAG, DAH, DAJ, DAK, DAL, DAW, DAX, V12, V13, V14, V15, V16, V17, V18, V19, V20, V21, V22, V24, V25, V39	1
14	KFFZZ		24617	444544	TEE PART OF KIT P/N 12256443 UOC:DAA, DAB, DAC, DAD, DAE, DAF, DAG, DAH, DAJ, DAK, DAL, DAW, DAX, V12, V13, V14, V15, V16, V17, V18, V19, V20, V21, V22, V24, V25, V39	1
15	PAFZZ	5935008338561	19207	8338561	SHELL, ELECTRICAL CO PART OF KIT P/N 12256443 UOC:DAA, DAB, DAC, DAD, DAE, DAF, DAG, DAH, DAJ, DAK, DAL, DAW, DAX, V12, V13, V14, V15, V16, V17, V18, V19, V20, V21, V22, V24, V25, V39	1
16	PAFZZ	4730009003296	07295	100740	ADAPTER, STRAIGHT, PI PART OF KIT P/N 12256443 UOC:DAA, DAB, DAC, DAD, DAE, DAF, DAG, DAH, DAJ, DAK, DAL, DAW, DAX, V12, V13, V14, V15, V16, V17, V18, V19, V20, V21, V22, V24, V25, V39	1
17	PAFZZ	4720012359629	19207	11664473-1	HOSE ASSEMBLY, NONME PART OF KIT P/N 12256443. UOC:DAA, DAB, DAC, DAD, DAE, DAF, DAG, DAH, DAJ, DAK, DAL, DAW, DAX, V12, V13, V14, V15, V16, V17, V18, V19, V20, V21, V22, V24, V25, V39	1
18	KFFZZ	5340002055314	18965	1506-3	BUTTON, PLUG PART OF KIT P/N 12256443 12256443 UOC:DAA, DAB, DAC, DAD, DAE, DAF, DAG, DAH, DAJ, DAK, DAL, DAW, DAX, V12, V13, V14, V15, V16, V17, V18, V19, V20, V21, V22, V24, V25, V39	1
19	KFFZZ		24617	432527	PLUG PART OF KIT P/N 12256443 UOC:DAA, DAB, DAC, DAD, DAE, DAF, DAG, DAH, DAJ, DAK, DAL, DAW, DAX, V12, V13, V14, V15, V16, V17, V18, V19, V20, V21, V22, V24, V25, V39	1

(1) ITEM NO	(2) SMR CODE	(3) NSN	(4) CAGEC	(5) PART NUMBER	(6) DESCRIPTION AND USABLE ON CODES (UOC)	(7) QTY
20	KFFZZ		24617	432468	CAP-PLUG, PROTECTIVE PART OF KIT............................. 12256443 .... UOC:DAA, DAB, DAC, DAD, DAE, DAF, DAG, DAH, DAJ, DAK, DAL, DAW, DAX, V12, V13, V14, V15, V16, V17, V18, V19, V20, V21, V22, V24, V25, V39	1
21	KFFZZ	9905010327002	19207	10896651	PLATE, INSTRUCTION PART OF KIT P/N .... 12256443 .... UOC:DAA, DAB, DAC, DAD, DAE, DAF, DAG, DAH, DAJ, DAK, DAL, DAW, DAX, V12, V13, V14, V15, V16, V17, V18, V19, V20, V21, V22, V24, V25, V39	1
22	PAFZZ	5305007544355	21450	127851-	SCREW, TAPPING PART OF KIT P/N .... 12256443 .... UOC:DAA, DAB, DAC, DAD, DAE, DAF, DAG, DAH, DAJ, DAK, DAL, DAW, DAX, V12, V13, V14, V15, V16, , V V18, V19, V20, V21, V22, V24, V25, V39	4
23	PAOZZ	4730002660538	81343	6-4 010102B	ADAPTER, STRAIGHT, PI PART OF KIT P/N .... 12256443 .... UOC:DAA, DAB, DAC, DAD, DAE, DAF, DAG, DAH, DAJ, DAK, DAL, DAW, DAX, V12, V13, V14, V15, V16, V17, V18, V19, V20, V21, V22, V24, V25, V39	4
24	KFFZZ	4730011085103	19207	10937774-1	TEE, PIPE PART OF KIT P/N 12256443............................... UOC:DAA, DAB, DAC, DAD, DAE, DAF, DAG, DAH, DAJ, DAK, DAL, DAW, DAX, V12, V13, V14, V15, V16, V17, V18, V19, V20, V21, V22, V24, V25, V39	1
25	PAFZZ	4730000127951	21450	127951	PLUG, PIPE PART OF KIT P/N 12256443............................ UOC:DAA, DAB, DAC, DAD, DAE, DAF, DAG, DAH, DAJ, DAK, DAL, DAW, DAX, V12, V13, V14, V15, V16, V17, V18, V19, V20, V21, V22, V24, V25, V39	1
26	PBFZZ	4720012359628	19207	12302628-1	HOSE ASSEMBLY, NONME PART OF KIT P/N .... 12256443 .... UOC:DAA, DAB, DAC, DAD, DAE, DAF, DAG, DAH, DAJ, DAK, DAL, DAW, DAX, V12, V13, V14, V15, V16, V17, V18, V19, V20, V21, V22, V24, V25, V39	1
27	PAFZZ	4820002633019	96906	MS35931-2	DRAIN, PLUG PART OF KIT P/N 12256443. .... UOC:DAA, DAB, DAC, DAD, DAE, DAF, DAG, DAH, DAJ, DAK, DAL, DAW, DAX, V12, V13, V14, V15, V16, V17, V18, V19, V20, V21, V22, V24, V25, V39	1
	PBFZZ	2540010958286	19207	12256443	KIT, FUEL BURNING PERSONNEL HEATER .... UOC:DAA, DAB, DAC, DAD, DAE, DAF, DAG, DAH, DAJ, DAK, DAL, DAW, DAX, V12, V13, V14, V15, V16, V17B, 18, V19, V20, V21, V22, V24, V25, V39	1

ADAPTER, CONNECTOR	(4)	592-26
ADAPTER, STRAIGHT	(1)	593-13
ADAPTER, STRAIGHT	(1)	593-16

(1) ITEM NO	(2) SMR CODE	(3) NSN	(4) CAGEC	(5) PART NUMBER	(6) DESCRIPTION AND USABLE ON CODES (UOC)		(7) QTY
					ADAPTER, STRAIGHT	(4)	593-23
					BOLT, MACHINE	(3)	592-22
					BRACKET	(1)	593-12
					BRACKET, ANGLE	(1)	592-13
					BRACKET, CONTROL BOX	(1)	592-2
					BUTTON, PLUG	(1)	593-18
					CABLE ASSEMBLY	(1)	593-9
					CAP-PLUG	(1)	593-20
					CLAMP, HOSE	(2)	592-7
					CLAMP, LOOP	(37)	592-18
					CONTROL BOX	(1)	592-6
					DECAL	(1)	592-28
					DRAIN, PLUG	(1)	593-27
					DUMMY CONNECTOR	(4)	592-25
					ELBOW	(1)	592-9
					ELBOW, PIPE	(1)	593-1
					ELBOW, PIPE	(1)	593-2
					ELBOW, PIPE	(1)	593-11
					GROMMET	(1)	592-23
					HEATER, VEHICULAR	(1)	592-8
					HOSE ASSEMBLY	(1)	593-26
					HOSE ASSEMBLY	(1)	593-8
					HOSE ASSEMBLY	(1)	593-17
					NOZZLE AND BRACKET	(1)	592-16
					NUT, PLAIN, HEXAGON	(2)	592-4
					NUT, SELF-LOCKING	(2)	592-1
					NUT, SELF-LOCKING	(4)	592-14
					NUT, SELF-LOCKING	(2)	592-17
					NUT, SELF-LOCKING	(2)	593-6
					PIN, COTTER	(2)	592-10
					PIPE, EXHAUST	(1)	592-15
					PLATE, INSTRUCTION	(1)	593-21
					PLUG	(1)	593-19
					PLUG, PIPE	(1)	593-25
					PUMP, FUEL, ELECTRICA	(1)	593-5
					SCREW, CAP, HEXAGON	(1)	592-11
					SCREW, CAP, HEXAGON	(1)	593-3
					SCREW, CAP, HEXAGON	(2)	592-19
					SCREW, CAP, HEXAGON	(1)	593-7
					SCREW, CAP, HEXAGON	(2)	592-29
					SCREW, TAPPING	(4)	593-22
					SHELL, ELECTRICAL	(1)	592-24
					SHELL, ELECTRICAL	(3)	592-27
					SHELL, ELECTRICAL	(1)	593-15
					STRAP, TIEDOWN	(7)	593-10
					TEE	(1)	593-14
					TEE	(1)	593-24
					WASHER, FLAT	(8)	592-12
					WASHER, LOCK	(2)	592-3
					WASHER, LOCK	(1)	592-21
					WASHER, LOCK	(2)	593-4
					WIRING HARNESS	(1)	592-5
					WIRING HARNESS	(1)	592-20

END OF FIGURE

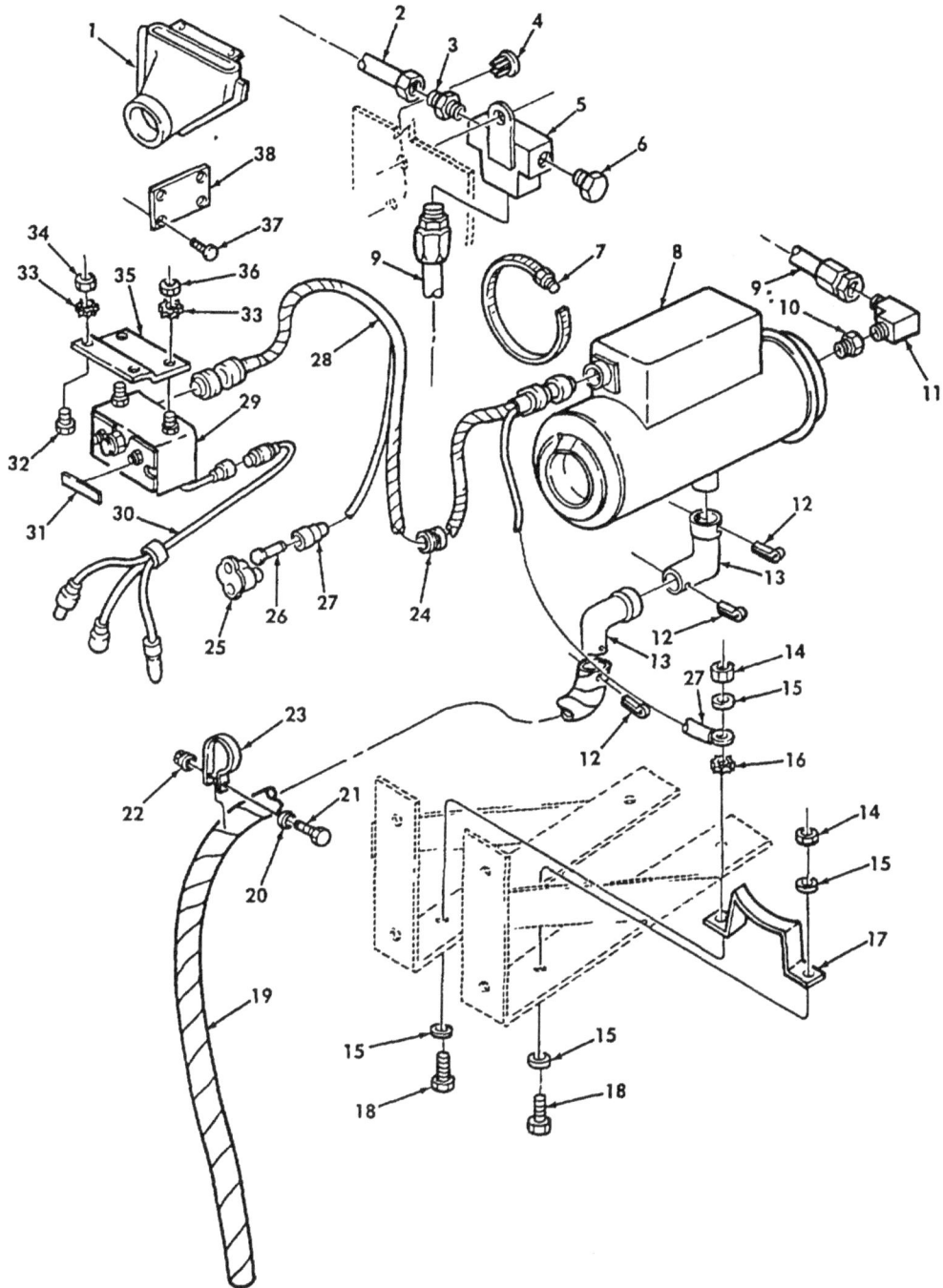

Figure 594. Fuel Burning Personnel Heater Kit, Control Box, Heater, and Wiring Harness (M939A2).

(1) ITEM NO	(2) SMR CODE	(3) NSN	(4) CAGEC	(5) PART NUMBER	(6) DESCRIPTION AND USABLE ON CODES (UOC)	(7) QTY
					GROUP 3303 WINTERIZATION KIT	
					FIG. 594 FUEL BURNING PERSONNEL HEATER KIT, CONTROL BOX, HEATER, AND WIRING HARNESS(M939A2)	
1	PAFZZ	4520013104829	19207	12255940-1	NOZZLE, SPRAY, FLUID PART OF KIT P/N 57K0243 .... UOC:ZAA, ZAB, ZAC, ZAD, ZAE, ZAF, ZAG, ZAH, ZAJ, ZAK, ZAL,	2
2	PBFZZ	4720012359628	19207	12302628-1	HOSE ASSEMBLY, NONME PART OF KIT P/N 57K0243 .... UOC:ZAA, ZAB, ZAC, ZAD, ZAE, ZAF, ZAG, ZAH, ZAJ, ZAK, ZAL	1
3	PAFZZ	4730002660538	81343	6-4 010102B	ADAPTER, STRAIGHT, PI PART OF KIT P/N .... 57K0243 UOC:ZAA, ZAB, ZAC, ZAD, ZAE, ZAF, ZAG, ZAH, ZAJ, ZAK, ZAL	1
4	PAFZZ	5340002055314	18965	1506-3	BUTTON, PLUG PART OF KIT P/N 57K0243............... UOC:ZAA, ZAB, ZAC, ZAD, ZAE, ZAF, ZAG, ZAH, ZAJ, ZAK, ZAL	1
5	PAFZZ	4730013312858	19207	10937774-3	TEE, PIPE PART OF KIT P/N 57K0243.............. UOC:ZAA, ZAB, ZAC, ZAD, ZAE, ZAF, ZAG, ZAH, ZAJ, ZAK, ZAL	1
6	PAFZZ	4730000892515	21450	444618	PLUG, PIPE PART OF KIT P/N 57K0243 .... UOC:ZAA, ZAB, ZAC, ZAD, ZAE, ZAF, ZAG, ZAH, ZAJ, ZAK, ZAL	1
7	PAFZZ	4730009086294	96906	MS35842-16	CLAMP, HOSE PART OF KIT P/N 57K0243 .... UOC:ZAA, ZAB, ZAC, ZAD, ZAE, ZAF, ZAG, ZAH, ZAJ, ZAK, ZAL	2
8	PAFZZ	2540011943323	81349	MIL-PRF-62550/3	HEATER, VEHICULAR, CO PART OF KIT P/N .... 57K0243 UOC:ZAA, ZAB, ZAC, ZAD, ZAE, ZAF, ZAG, ZAH, ZAJ, ZAK, ZAL	1
9	PAFZZ	4720012359629	19207	11664473-1	HOSE ASSEMBLY, NONME PART OF KIT P/N .... 57K0243 UOC:ZAA, ZAB, ZAC, ZAD, ZAE, ZAF, ZAG, ZAH, ZAJ, ZAK, ZAL	1
10	PAFZZ	4730004153172	46717	LA-519-9	COUPLING, PIPE PART OF KIT P/N .... 57K0243 UOC:ZAA, ZAB, ZAC, ZAD, ZAE, ZAF, ZAG, ZAH, ZAJ, ZAK, ZAL	1
11	PAFZZ	4730002546211	21450	118753	ELBOW, PIPE TO TUBE PART OF KIT P/N .... 57K0243 UOC:ZAA, ZAB, ZAC, ZAD, ZAE, ZAF, ZAG, ZAH, ZAJ, ZAK, ZAL	1
12	PAFZZ	5315008792910	96906	MS24665-427	PIN, COTTER PART OF KIT P/N 57K0243 .... UOC:ZAA, ZAB, ZAC, ZAD, ZAE, ZAF, ZAG, ZAH, ZAJ, ZAK, ZAL	3
13	PAFZZ	2990012885844	47457	20511139-1	PIPE, EXHAUST PART OF KIT P/N 57K0243 .... UOC:ZAA, ZAB, ZAC, ZAD, ZAE, ZAF, ZAG, ZAH, ZAJ, ZAK, ZAL	2

(1) ITEM NO	(2) SMR CODE	(3) NSN	(4) CAGEC	(5) PART NUMBER	(6) DESCRIPTION AND USABLE ON CODES (UOC)	(7) QTY
14	PAFZZ	5310002416658	96906	MS51943-34	NUT, SELF-LOCKING, HE PART OF KIT P/N 57K0243 UOC:ZAA, ZAB, ZAC, ZAD, ZAE, ZAF, ZAG, ZAH, ZAJ, ZAK, ZAL	4
15	PAFZZ	5310010841197	19207	8356625-2	WASHER, FLAT PART OF KIT P/N 57K0243 UOC:ZAA, ZAB, ZAC, ZAD, ZAE, ZAF, ZAG, ZAH, ZAJ, ZAK, ZAL	8
16	PAFZZ	5310007282044	96906	MS45904-73	WASHER, LOCK PART OF KIT P/N 57K0243 UOC:ZAA, ZAB, ZAC, ZAD, ZAE, ZAF, ZAG, ZAH, ZAJ, ZAK, ZAL	1
17	PAFZZ	5340012897885	47457	20511140	BRACKET, MOUNTING PART OF KIT P/N 57K0243 UOC:ZAA, ZAB, ZAC, ZAD, ZAE, ZAF, ZAG, ZAH, ZAJ, ZAK, ZAL	2
18	PAFZZ	5305002259091	96906	MS90726-36	SCREW, CAP, HEXAGON H PART OF KIT P/N 57K0243 UOC:ZAA, ZAB, ZAC, ZAD, ZAE, ZAF, ZAG, ZAH, ZAJ, ZAK, ZAL	4
19	PAFZZ	2990012878957	47457	20511145-2	PIPE, EXHAUST PART OF KIT P/N 57K0243 UOC:ZAA, ZAB, ZAC, ZAD, ZAE, ZAF, ZAG, ZAH, ZAJ, ZAK, ZAL	1
20	PAFZZ	5310010925496	24617	9422845	WASHER, FLAT PART OF KIT P/N 57K0243. UOC:ZAA, ZAB, ZAC, ZAD, ZAE, ZAF, ZAG, ZAH, ZAJ, ZAK, ZAL	1
21	PAFZZ	5305000712518	96906	MS90728-21	SCREW, CAP, HEXAGON H PART OF KIT P/N 57K0243 UOC:ZAA, ZAB, ZAC, ZAD, ZAE, ZAF, ZAG, ZAH, ZAJ, ZAK, ZAL	1
22	PAFZZ	5310000131245	21450	131245	NUT, SELF-LOCKING, HE PART OF KIT P/N 57K0243 UOC:ZAA, ZAB, ZAC, ZAD, ZAE, ZAF, ZAG, ZAH, ZAJ, ZAK, ZAL	1
23	PAFZZ	5340011278703	19207	7700263-2	CLAMP, LOOP PART OF KIT P/N 57K0243 UOC:ZAA, ZAB, ZAC, ZAD, ZAE, ZAF, ZAG, ZAH, ZAJ, ZAK, ZAL	1
24	PAFZZ	5325001749341	88044	AN931A16-22	GROMMET, NONMETALLIC PART OF KIT P/N 57K0243 UOC:ZAA, ZAB, ZAC, ZAD, ZAE, ZAF, ZAG, ZAH, ZAJ, ZAK, ZAL	1
25	PAFZZ	5935009006281	96906	MS27147-1	ADAPTER, CONNECTOR PART OF KIT P/N 57K0243 UOC:ZAA, ZAB, ZAC, ZAD, ZAE, ZAF, ZAG, ZAH, ZAJ, ZAK, ZAL	1
26	PAFZZ	5935002140904	19207	7982907	DUMMY CONNECTOR, PLU PART OF KIT P/N 57K0243 UOC:ZAA, ZAB, ZAC, ZAD, ZAE, ZAF, ZAG, ZAH, ZAJ, ZAK, ZAL	2
27	PAFZZ	5935008338561	19207	8338561	SHELL, ELECTRICAL CO PART OF KIT P/N 57K0243 UOC:ZAA, ZAB, ZAC, ZAD, ZAE, ZAF, ZAG, ZAH, ZAJ, ZAK, ZAL	1
28	PAFZZ	2920010958308	19207	12256430	WIRING HARNESS, BRAN PART OF KIT P/N	1

(1) ITEM NO	(2) SMR CODE	(3) NSN	(4) CAGEC	(5) PART NUMBER	(6) DESCRIPTION AND USABLE ON CODES (UOC)	(7) QTY
					57K0243 UOC:ZAA, ZAB, ZAC, ZAD, ZAE, ZAF, ZAG, ZAH, ZAJ, ZAK, ZAL	
29	PAFFF	2590011256154	19207	11669705	CONTROL BOX, ELECTRI PART OF KIT P/N 57K0243 UOC:ZAA, ZAB, ZAC, ZAD, ZAE, ZAF, ZAG, ZAH, ZAJ, ZAK, ZAL	1
30	PAFZZ	2590012150930	19207	12302627	WIRING HARNESS, BRAN PART OF KIT P/N 57K0243 UOC:ZAA, ZAB, ZAC, ZAD, ZAE, ZAF, ZAG, ZAH, ZAJ, ZAK, ZAL	1
31	PAFZZ	7690010325639	19207	10896515	DECAL PART OF KIT P/N 57K0243 UOC:ZAA, ZAB, ZAC, ZAD, ZAE, ZAF, ZAG, ZAH, ZAJ, ZAK, ZAL	1
32	PAFZZ	5305002678953	96906	MS90727-5	SCREW, CAP, HEXAGON H PART OF KIT P/N 57K0243 UOC:ZAA, ZAB, ZAC, ZAD, ZAE, ZAF, ZAG, ZAH, ZAJ, ZAK, ZAL	2
33	PAFZZ	5310002090786	96906	MS35335-33	WASHER, LOCK PART OF KIT P/N 57K0243. UOC:ZAA, ZAB, ZAC, ZAD, ZAE, ZAF, ZAG, ZAH, ZAJ, ZAK, ZAL	6
34	PAFZZ	5310009359022	96906	MS51943-32	NUT, SELF-LOCKING, HE PART OF KIT P/N 57K0243 UOC:ZAA, ZAB, ZAC, ZAD, ZAE, ZAF, ZAG, ZAH, ZAJ, ZAK, ZAL	2
35	PFFZZ	5340011097950	19207	11648534	BRACKET, MOUNTING PART OF KIT P/N 57K0243 UOC:ZAA, ZAB, ZAC, ZAD, ZAE, ZAF, ZAG, ZAH, ZAJ, ZAK, ZAL	1
36	PAFZZ	5310007616882	96906	MS51967-2	NUT, PLAIN, HEXAGON PART OF KIT P/N 57K0243 UOC:ZAA, ZAB, ZAC, ZAD, ZAE, ZAF, ZAG, ZAH, ZAJ, ZAK, ZAL	2
37	PAFZZ	5305000457603	96906	MS24625-42	SCREW, TAPPING PART OF KIT P/N 57K0243 UOC:ZAA, ZAB, ZAC, ZAD, ZAE, ZAF, ZAG, ZAH, ZAJ, ZAK, ZAL	4
38	PFFZZ	9905010327002	19207	10896651	PLATE, INSTRUCTION PART OF KIT P/N 57K0243	1

END OF FIGURE

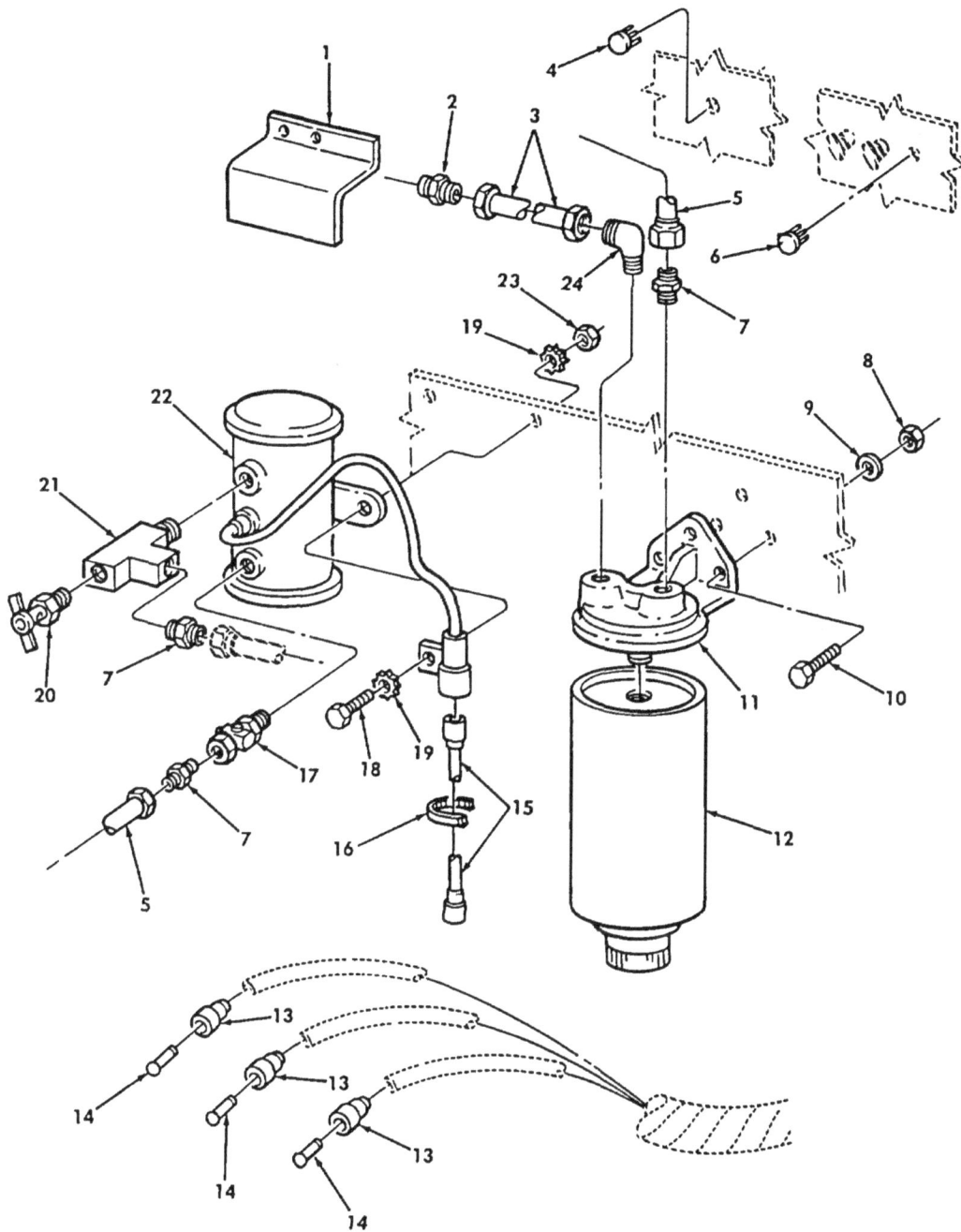

Figure 595. Fuel Burning Personnel Heater Kit, Pump, Filter, and Related Parts (M939A2).

(1) ITEM NO	(2) SMR CODE	(3) NSN	(4) CAGEC	(5) PART NUMBER	(6) DESCRIPTION AND USABLE ON CODES (UOC)	(7) QTY
					GROUP 3303 WINTERIZATION KIT	
					FIG. 595 FUEL BURNING PERSONNEL HEATER KIT, PUMP, FILTER, AND RELATED PARTS(M939A2)	
1	KFFZZ		19207	12302626	BRACKET, PART OF KIT P/N 57K0243 .... UOC:ZAA, ZAB, ZAC, ZAD, ZAE, ZAF, ZAG, ZAH, ZAJ, ZAK, ZAL	1
2	PAFZZ	4730004391722	97403	13205E3839-2	ADAPTER, STRAIGHT, PI PART OF KIT P/N .... 57K0243 .... UOC:ZAA, ZAB, ZAC, ZAD, ZAE, ZAF, ZAG, ZAH, ZAJ, ZAK, ZAL	1
3	PAFZZ	4720012874495	47457	20511138-2	HOSE ASSEMBLY, NONME PART OF KIT P/N .... 57K0243 .... UOC:ZAA, ZAB, ZAC, ZAD, ZAE, ZAF, ZAG, ZAH, ZAJ, ZAK, ZAL	1
4	KFFZZ		24617	432468	PLUG, PART OF KIT P/N 57K0243 .... UOC:ZAA, ZAB, ZAC, ZAD, ZAE, ZAF, ZAG, ZAH, ZAJ, ZAK, ZAL	1
5	PAFZZ	4720012874494	47457	20511138-1	HOSE ASSEMBLY, NONME PART OF KIT P/N .... 57K0243 .... UOC:ZAA, ZAB, ZAC, ZAD, ZAE, ZAF, ZAG, ZAH, ZAJ, ZAK, ZAL	1
6	KFFZZ		24617	432527	PLUG, PART OF KIT P/N 57K0243 .... UOC:ZAA, ZAB, ZAC, ZAD, ZAE, ZAF, ZAG, ZAH, ZAJ, ZAK, ZAL	1
7	PAFZZ	4730002660538	81343	6-4 010102B	ADAPTER, STRAIGHT, PI PART OF KIT P/N .... 57K0243 .... UOC:ZAA, ZAB, ZAC, ZAD, ZAE, ZAF, ZAG, ZAH, ZAJ, ZAK, ZAL	3
8	PAFZZ	5310002416658	96906	MS51943-34	NUT, SELF-LOCKING, HE PART OF KIT P/N .... 57K0243 .... UOC:ZAA, ZAB, ZAC, ZAD, ZAE, ZAF, ZAG, ZAH, ZAJ, ZAK, ZAL	3
9	PAFZZ	5310010841197	19207	8356625-2	WASHER, FLAT PART OF KIT P/N 57K0243 .... UOC:ZAA, ZAB, ZAC, ZAD, ZAE, ZAF, ZAG, ZAH, ZAJ, ZAK, ZAL	3
10	PAFZZ	5305002259091	96906	MS90726-36	SCREW, CAP, HEXAGON H PART OF KIT P/N .... 57K0243 .... UOC:ZAA, ZAB, ZAC, ZAD, ZAE, ZAF, ZAG, ZAH, ZAJ, ZAK, ZAL	3
11	PAFZZ	5340012882161	47457	20511143	MOUNT, FILTER, FUEL PART OF KIT P/N .... 57K0243 .... UOC:ZAA, ZAB, ZAC, ZAD, ZAE, ZAF, ZAG, ZAH, ZAJ, ZAK, ZAL	1
12	PAOZZ	2910012017719	79396	33472	FILTER ELEMENT, FLUI PART OF KIT P/N .... 57K0243 .... UOC:ZAA, ZAB, ZAC, ZAD, ZAE, ZAF, ZAG, ZAH, ZAJ, ZAK, ZAL	1
13	PAFZZ	5935005729180	19207	8338566	SHELL, ELECTRICAL CO PART OF KIT P/N .... 57K0243 .... UOC:ZAA, ZAB, ZAC, ZAD, ZAE, ZAF, ZAG, ZAH, ZAJ, ZAK, ZAL	3

(1) ITEM NO	(2) SMR CODE	(3) NSN	(4) CAGEC	(5) PART NUMBER	(6) DESCRIPTION AND USABLE ON CODES (UOC)	(7) QTY
14	PAFZZ	5935002140904	19207	7982907	DUMMY CONNECTOR, PLU PART OF KIT P/N 57K0243 .... UOC:ZAA, ZAB, ZAC, ZAD, ZAE, ZAF, ZAG, ZAH, ZAJ, ZAK, ZAL	3
15	PAFZZ	2590011531850	19207	12256427-5	LEAD, ELECTRICAL PART OF KIT P/N 57K0243 .... UOC:ZAA, ZAB, ZAC, ZAD, ZAE, ZAF, ZAG, ZAH, ZAJ, ZAK, ZAL	1
16	PAFZZ	5975001563253	96906	MS3367-2-9	STRAP, TIEDOWN, ELECT PART OF KIT P/N 57K0243 .... UOC:ZAA, ZAB, ZAC, ZAD, ZAE, ZAF, ZAG, ZAH, ZAJ, ZAK, ZAL	12
17	PAFZZ	4820002633019	96906	MS35931-2	COCK, PLUG PART OF KIT P/N 57K0243 .... UOC:ZAA, ZAB, ZAC, ZAD, ZAE, ZAF, ZAG, ZAH, ZAJ, ZAK, ZAL	1
18	PAFZZ	5305002678953	96906	MS90727-5	SCREW, CAP, HEXAGON H PART OF KIT P/N 57K0243 .... UOC:ZAA, ZAB, ZAC, ZAD, ZAE, ZAF, ZAG, ZAH, ZAJ, ZAK, ZAL	2
19	PAFZZ	5310008892528	96906	MS45904-68	WASHER, LOCK PART OF KIT P/N 57K0243. .... UOC:ZAA, ZAB, ZAC, ZAD, ZAE, ZAF, ZAG, ZAH, ZAJ, ZAK, ZAL	4
20	PAFZZ	4820008451096	96906	MS35783-2	COCK, DRAIN PART OF KIT P/N 57K0243 .... UOC:ZAA, ZAB, ZAC, ZAD, ZAE, ZAF, ZAG, ZAH, ZAJ, ZAK, ZAL	1
21	PAFZZ	4730010926442	21450	444120	TEE, PIPE PART OF KIT P/N 57K0243 .... UOC:ZAA, ZAB, ZAC, ZAD, ZAE, ZAF, ZAG, ZAH, ZAJ, ZAK, ZAL	1
22	PAFZZ	2910009309367	96906	MS51321-2	PUMP, FUEL, ELECTRICA PART OF KIT P/N 57K0243 .... UOC:ZAA, ZAB, ZAC, ZAD, ZAE, ZAF, ZAG, ZAH, ZAJ, ZAK, ZAL	1
23	PAFZZ	5310009359022	96906	MS51943-32	NUT, SELF-LOCKING, HE PART OF KIT P/N 57K0243 .... UOC:ZAA, ZAB, ZAC, ZAD, ZAE, ZAF, ZAG, ZAH, ZAJ, ZAK, ZAL	2
24	PAFZZ	4730009028991	96906	MS39162-5	ELBOW, PIPE TO TUBE PART OF KIT P/N 57K0243 .... UOC:ZAA, ZAB, ZAC, ZAD, ZAE, ZAF, ZAG, ZAH, ZAJ, ZAK, ZAL	1
	PDFZZ	2540014166784	19207	57K0243	PARTS KIT, HEATER..	1
					ADAPTER, CONNECTOR (1) 594-25	
					ADAPTER, STRAIGHT, PI (1) 594-3	
					ADAPTER, STRAIGHT, PI (1) 595-2	
					ADAPTER, STRAIGHT, PI (3) 595-7	
					BRACKET (1) 595-1	
					BRACKET, MOUNTING (2) 594-17	
					BRACKET, MOUNTING (1) 594-35	
					BUTTON, PLUG (1) 594-4	
					CLAMP, HOSE (2) 594-7	
					CLAMP, LOOP (1) 594-23	
					COCK, DRAIN (1) 595-20	
					COCK, PLUG (1) 595-17	
					CONTROL BOX, ELECTRI (1) 594-29	
					COUPLING, PIPE (1) 594-10	
					DECAL (1) 594-31	

(1) ITEM NO	(2) SMR CODE	(3) NSN	(4) CAGEC	(5) PART NUMBER	(6) DESCRIPTION AND USABLE ON CODES (UOC)	(7) QTY
					DUMMY CONNECTOR, PLU	(2) 594-26
					DUMMY CONNECTOR, PLU	(3) 595-14
					ELBOW, PIPE TO TUBE	(1) 594-11
					ELBOW, PIPE TO TUBE	(1) 595-24
					FILTER ELEMENT, FLUI	(1) 595-12
					GROMMET, NONMETALLIC	(1) 594-24
					HEATER, VEHICULAR, CO	(1) 594-8
					HOSE ASSEMBLY, NONME	(1) 594-2
					HOSE ASSEMBLY, NONME	(1) 594-9
					HOSE ASSEMBLY, NONME	(1) 595-3
					HOSE ASSEMBLY, NONME	(1) 595-5
					LEAD, ELECTRICAL	(1) 595-15
					MOUNT, FILTER, FUEL	(1) 595-11
					NOZZLE, SPRAY, FLUID	(2) 594-1
					NUT, PLAIN, HEXAGON	(2) 594-36
					NUT, SELF-LOCKING, HE	(4) 594-14
					NUT, SELF-LOCKING, HE	(1) 594-22
					NUT, SELF-LOCKING, HE	(2) 594-34
					NUT, SELF-LOCKING, HE	(3) 595-8
					NUT, SELF-LOCKING, HE	(2) 595-23
					PIN, COTTER	(3) 594-12
					PIPE, EXHAUST	(2) 594-13
					PIPE, EXHAUST	(1) 594-19
					PLATE, INSTRUCTION	(1) 594-38
					PLUG	(1) 595-4
					PLUG	(1) 595-6
					PLUG, PIPE	(1) 594-6
					PUMP, FUEL, ELECTRICA	(1) 595-22
					SCREW, CAP, HEXAGON H	(4) 594-18
					SCREW, CAP, HEXAGON H	(1) 594-21
					SCREW, CAP, HEXAGON H	(2) 594-32
					SCREW, CAP, HEXAGON H	(3) 595-10
					SCREW, CAP, HEXAGON H	(2) 595-18
					SCREW, TAPPING	(4) 594-37
					SHELL, ELECTRICAL CO	(1) 594-27
					SHELL, ELECTRICAL CO	(4) 595-13
					STRAP, TIEDOWN, ELECT	(12) 595-16
					TEE, PIPE	(1) 594-5
					TEE, PIPE	(1) 595-21
					WASHER, FLAT	(8) 594-15
					WASHER, FLAT	(1) 594-20
					WASHER, FLAT	(3) 595-9
					WASHER, LOCK	(1) 594-16
					WASHER, LOCK	(6) 594-33
					WASHER, LOCK	(4) 595-19
					WIRING HARNESS, BRAN	(1) 594-28
					WIRING HARNESS, BRAN	(1) 594-30

END OF FIGURE

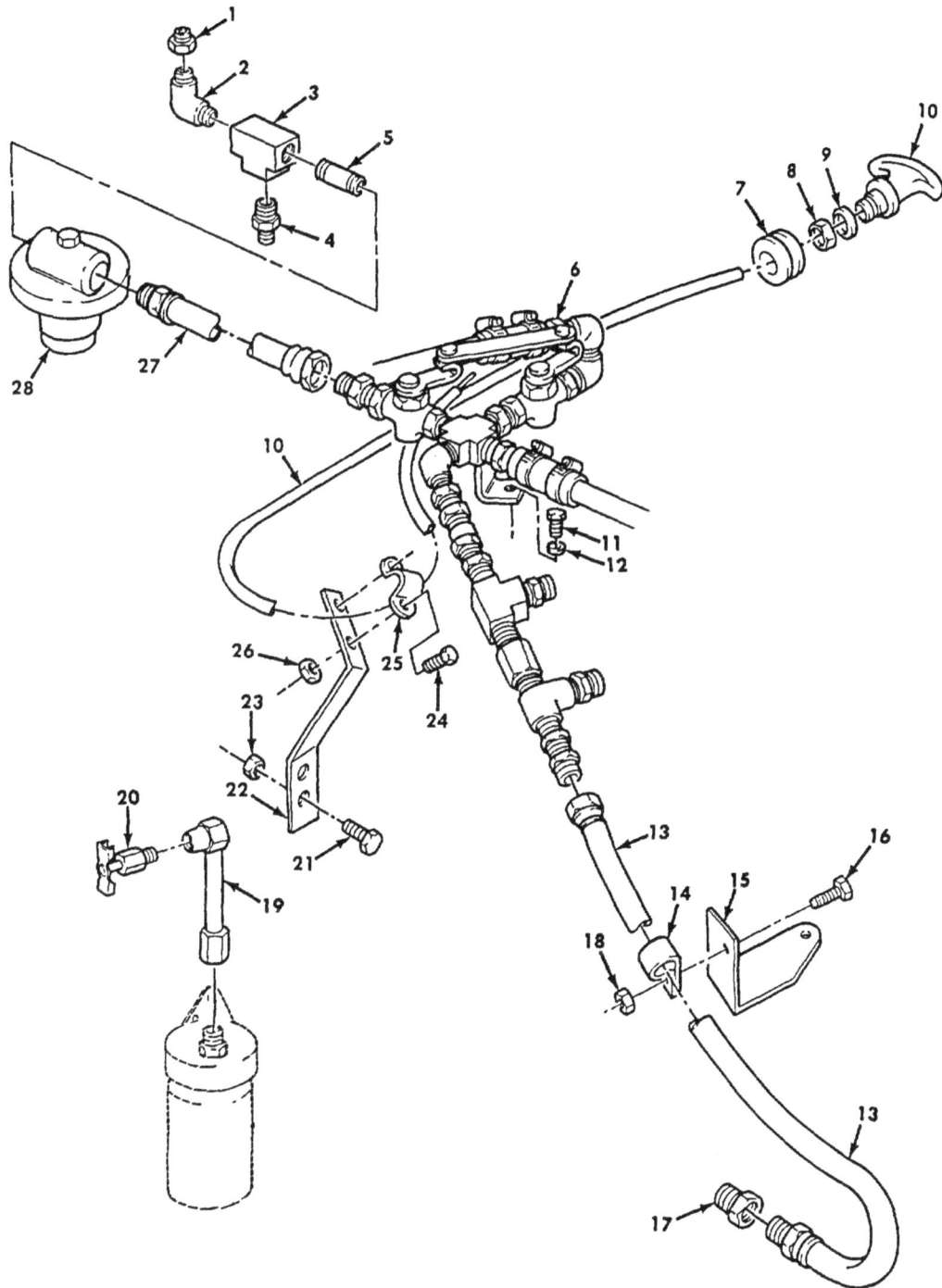

Figure 596. Deep Water Fording Kit, Valve Assembly and Related Parts (M939, M939A1)

(1) ITEM NO	(2) SMR CODE	(3) NSN	(4) CAGEC	(5) PART NUMBER	(6) DESCRIPTION AND USABLE ON CODES (UOC)	(7) QTY
					GROUP 3305 DEEP WATER FORDING KITS	
					FIG. 596 DEEP WATER FORDING KIT, VALVE ASSEMBLY AND RELATED PARTS (M939, M939A1)	
1	PAFZZ	4730008022560	21450	443987	REDUCER PIPE PART OF KIT P/N .... 12256226. UOC:DAA, DAB, DAC, DAD, DAE, DAF, DAG, DAH, DAJ, DAK, DAL, DAW, DAX, V12, V13, V14, V15, V16, V17, V18, V19, V20, V21, V22, V24, V25, V39	1
2	PAFZZ	4730002775553	96906	MS51952-2	ELBOW, PIPE REGULATOR VALVE PART OF .... KIT P/N 12256226 .... UOC:DAA, DAB, DAC, DAD, DAE, DAF, DAG, DAH, DAJ, DAK, DAL, DAW, DAX, V12, V13, V14, V15, V16, V17, V18, V19, V20, V21, V22, V24, V25, V39	1
3	PAFZZ	4730010926442	21450	444120	TEE, PIPE REGULATOR VALVE PART OF .... KIT P/N 12256226 .... UOC:DAA, DAB, DAC, DAD, DAE, DAF, DAG, DAH, DAJ, DAK, DAL, DAW, DAX, V12, V13, V14, V15, V16, V17, V18, V19, V20, V21, V22, V24, V25, V39	1
4	PAFZZ	4730000691186	81343	6-4 120102BA	ADAPTER, STRAIGHT, PI PART OF KIT P/N .... 12256226 .... UOC:DAA, DAB, DAC, DAD, DAE, DAF, DAG, DAH, DAJ, DAK, DAL, DAW, DAX, V12, V13, V14, V15, V16, V17, V18, V19, V20, V21, V22, V24, V25, V39	1
5	PAFZZ	4730001961486	96906	MS51953-33	NIPPLE, PIPE PART OF KIT P/N 12256226 .... UOC:DAA, DAB, DAC, DAD, DAE, DAF, DAG, DAH, DAJ, DAK, DAL, DAW, DAX, V12, V13, V14, V15, V16, V17, V18, V19, V20, V2 1, V22, V24, V25, V39	1
6	PAFZZ	2540012957461	19207	12277249	VALVE ASSEMBLY, FORD PART OF KIT P/N .... 12256226 .... UOC:DAA, DAB, DAC, DAD, DAE, DAF, DAG, DAH, DAJ, DAK, DAL, DAW, DAX, V12, V13, V14, V15, V16, V17, V1B, V19, V20, V21, V22, V24, V25, V39	1
7	PAFZZ	5325005432902	96906	MS35489-134	GROMMET, NONMETALLIC PART OF KIT P/N .... 12256226 .... UOC:DAA, DAB, DAC, DAD, DAE, DAF, DAG, DAH, DAJ, DAK, DAL, DAW, DAX, V12, V13, V14, V15, V16, V17, V18, V19, V20, V21, V22, V24, V25, V39	1
8	PAFZZ	5310007218000	96906	MS25082-10	NUT, PLAIN, HEXAGON PART OF KIT P/N .... 12256226 .... UOC:DAA, DAB, DAC, DAD, DAE, DAF, DAG, DAH, DAJ, DAK, DAL, DAW, DAX, V12, V13, V14, V15, V16, V17, V18, V19, V20, V21, V22, V24, V25, V39	1

(1) ITEM NO	(2) SMR CODE	(3) NSN	(4) CAGEC	(5) PART NUMBER	(6) DESCRIPTION AND USABLE ON CODES (UOC)	(7) QTY
9	PAFZZ	5310005434385	96906	MS35333-46	WASHER, LOCK PART OF KIT P/N 12256226 UOC:DAA, DAB, DAC, DAD, DAE, DAF, DAG, DAH, DAJ, DAK, DAL, DAW, DAX, V12, V13, V14, V15, V16, V17, V18, V19, V20, V21, V22, V24, V25, V39	1
10	PAFZZ	2590011003871	19207	11664701-2	CONTROL ASSEMBLY, PU PART OF KIT P/N 12256226 UOC:DAA, DAB, DAC, DAD, DAE, DAF, DAG, DAH, DAJ, DAK, DAL, DAW, DAX, V12, V13, V14, V15, V16, V17, V18, V19, V20, V21, V22, V24, V25, V39	1
11	PAFZZ	5305008213869	80204	B1821BH038C175N	SCREW, CAP, HEXAGON H PART OF KIT P/N 12256226 UOC:DAA, DAB, DAC, DAD, DAE, DAF, DAG, DAH, DAJ, DAK, DAL, DAW, DAX, V12, V13, V14, V15, V16, V17, V18, V19, V20, V21, V22, V24, V25, V39	1
12	PAFZZ	5310006379541	96906	MS35338-46	WASHER, LOCK PART OF KIT P/N 12256226 UOC:DAA, DAB, DAC, DAD, DAE, DAF, DAG, DAH, DAJ, DAK, DAL, DAW, DAX, V12, V13, V14, V15, V16, V17, V18, V19, V20, V21, V22, V24, V25, V39	1
13	KFFZZ	4720012008645	19207	11664676-1	HOSE ASSEMBLY, NONME PART OF KIT P/N 12256226 UOC:DAA, DAB, DAC, DAD, DAE, DAF, DAG, DAH, DAJ, DAK, DAL, DAW, DAX, V12, V13, V14, V15, V16, V17, V18, V19, V20, V21, V22, V24, V25, V39	1
14	PAFZZ	5340009881162	96906	MS21333-113	CLAMP, LOOP PART OF KIT P/N 12256226 UOC:DAA, DAB, DAC, DAD, DAE, DAF, DAG, DAH, DAJ, DAK, DAL, DAW, DAX, V12, V13, V14, V15, V16, V17, V18, V19, V20, V21, V22, V24, V25, V39	1
15	PAFZZ	5340011850455	19207	12302683	BRACKET, ANGLE PART OF KIT P/N 12256226 UOC:DAA, DAB, DAC, DAD, DAE, DAF, DAG, DAH, DAJ, DAK, DAL, DAW, DAX, V12, V13, V14, V15, V16, V17, V18, V19, V20, V21, V22, V24, V25, V39	1
16	PAFZZ	5306002259086	96906	MS90726-31	BOLT, MACHINE PART OF KIT P/N 12256226 UOC:DAA, DAB, DAC, DAD, DAE, DAF, DAG, DAH, DAJ, DAK, DAL, DAW, DAX, V12, V13, V14, V15, V16, V17, V18, V19, V20, V21, V22, V24, V25, V39	1
17	PAOZZ	4730002268874	30554	71-4872	REDUCER, PIPE PART OF KIT P/N 12256226 UOC:DAA, DAB, DAC, DAD, DAE, DAF, DAG, DAH, DAJ, DAK, DAL, DAW, DAX, V12, V13, V14, V15, V16, V17, V18, V19, V20, V21, V22, V24, V25, V39	1

(1) ITEM NO	(2) SMR CODE	(3) NSN	(4) CAGEC	(5) PART NUMBER	(6) DESCRIPTION AND USABLE ON CODES (UOC)	(7) QTY
18	PAFZZ	5310009824912	96906	MS21045-5	NUT, SELF-LOCKING, HE PART OF KIT P/N 12256226 UOC:DAA, DAB, DAC, DAD, DAE, DAF, DAG, DAH, DAJ, DAK, DAL, DAW, DAX, V12, V13, V14, V15, V16, V17, V18, V19, V20, V21, V22, V24, V25, V39	1
19	PAFZZ	4730011924381	19207	12302868	ELBOW, PIPE PART OF KIT P/N 12256226 UOC:DAA, DAB, DAC, DAD, DAE, DAF, DAG, DAH, DAJ, DAK, DAL, DAW, DAX, V12, V13, V14, V15, V16, V17, V18, V19, V20, V21, V22, V24, V25, V39	1
20	PAFZZ	4820007529040	96906	MS35782-4	COCK, DRAIN PART OF KIT P/N 12256226 UOC:DAA, DAB, DAC, DAD, DAE, DAF, DAG, DAH, DAJ, DAK, DAL, DAW, DAX, V12, V13, V14, V15, V16, V17, V18, V19, V20, V21, V22, V24, V25, V39	1
21	PAFZZ	5305002678953	80204	B1821BH025F063N	SCREW, CAP, HEXAGON H UOC:DAA, DAB, DAC, DAD, DAE, DAF, DAG, DAH, DAJ, DAK, DAL, DAW, DAX, V12, V13, V14, V15, V16, V17, V18, V19, V20, V21, V22, V24, V25, V39	3
22	KFFZZ		19207	12277247	BRACKET PART OF KIT P/N 12256226 UOC:DAA, DAB, DAC, DAD, DAE, DAF, DAG, DAH, DAJ, DAK, DAL, DAW, DAX, V12, V13, V14, V15, V16, V17, V18, V19, V20, V21, V22, V24, V25, V39	1
23	PAFZZ	5310009359022	96906	MS51943-32	NUT, SELF-LOCKING, HE PART OF KIT P/N 12256226 UOC:DAA, DAB, DAC, DAD, DAE, DAF, DAG, DAH, DAJ, DAK, DAL, DAW, DAX, V12, V13, V14, V15, V16, V17, V18, V19, V20, V21, V22, V24, V25, V39	2
24	PAFZZ	5305009931848	96906	MS35207-265	SCREW, MACHINE PART OF KIT P/N 12256226 UOC:DAA, DAB, DAC, DAD, DAE, DAF, DAG, DAH, DAJ, DAK, DAL, DAW, DAX, V12, V13, V14, V15, V16, V17, V18, V19, V20, V21, V22, V24, V25, V39	2
25	PAFZZ	5340011592995	19207	12277037	STRAP, RETAINING PART OF KIT P/N 12256226 UOC:DAA, DAB, DAC, DAD, DAE, DAF, DAG, DAH, DAJ, DAK, DAL, DAW, DAX, V12, V13, V14, V15, V16, V17, V18, V19, V20, V21, V22, V24, V25, V39	1
26	PAFZZ	5310008775797	96906	MS21044N3	NUT, SELF-LOCKING, HE PART OF KIT P/N 12256226 UOC:DAA, DAB, DAC, DAD, DAE, DAF, DAG, DAH, DAJ, DAK, DAL, DAW, DAX, V12, V13, V14, V15, V16, V17, V18, V19, V20, V21, V22, V24, V25, V39	2

(1) ITEM NO	(2) SMR CODE	(3) NSN	(4) CAGEC	(5) PART NUMBER	(6) DESCRIPTION AND USABLE ON CODES (UOC)	(7) QTY
27	KFFZZ		19207	11664676-5	HOSE ASSEMBLY PART OF KIT P/N .... 12256226.. UOC:DAA, DAB, DAC, DAD, DAE, DAF, DAG, DAH, DAJ, DAK, DAL, DAW, DAX, V12, V13, V14, V15, V16, V17, V18, V19, V20, V21, V22, V24, V25, V39	1
28	PAFZZ	4820011705055	19207	10922156-1	VALVE, REGULATING, FL PART OF KIT P/N .... 12256226.. UOC:DAA, DAB, DAC, DAD, DAE, DAF, DAG, DAH, DAT, DAK, DAL, DAW, DAX, V12, V13, V14, V15, V16, V17, V18, V19, V20, V21, V22, V24, V25, V39	1

END OF FIGURE

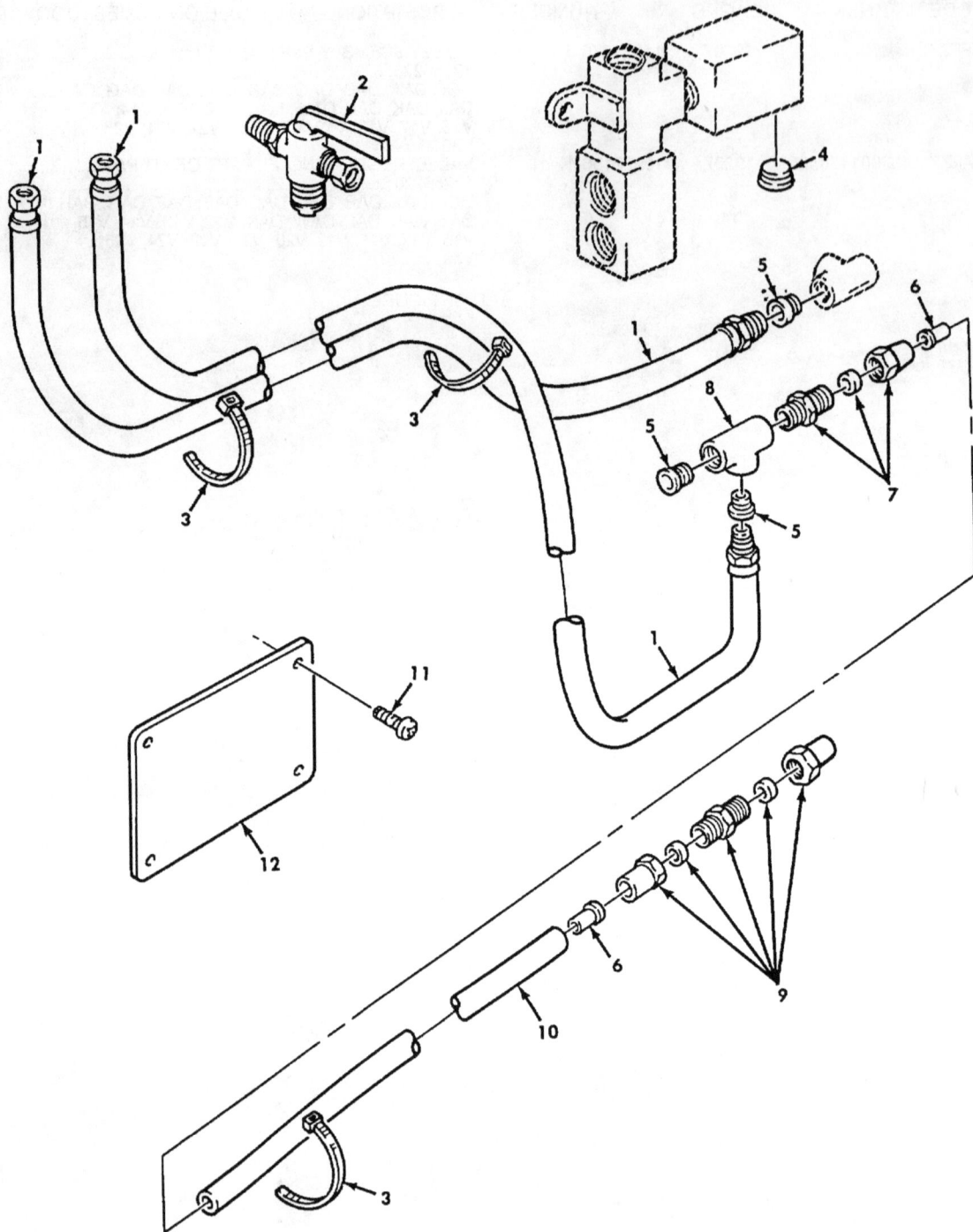

Figure 597. Deep Water Fording Kit Air Hoses and Cover Plate (M939,M939A1).

(1) ITEM NO	(2) SMR CODE	(3) NSN	(4) CAGEC	(5) PART NUMBER	(6) DESCRIPTION AND USABLE ON CODES (UOC)	(7) QTY
					GROUP 3305 DEEP WATER FORDING KITS	
					FIG. 597 DEEP WATER FORDING KIT AIR HOSES AND COVER PLATE(M939,M939A1)	
1	XFFZZ		19207	11664676-7	HOSE ASSEMBLY PART OF KIT P/N 12256226 .... UOC:DAA, DAB, DAC, DAD, DAE, DAF, DAG, DAH, DAJ, DAK, DAL, DAW, DAX, V12, V13, V14, V15, V16, 7, V18 , V19 , V20, V21, V22, V24, V25, V39	2
2	PAFZZ	4820011404298	21450	543858	COCK, DRAIN PART OF KIT P/N 12256226 .... UOC:DAA, DAB, DAC, DAD, DAE, DAF, DAG, DAH, DAJ, DAK, DAL, DAW, DAX, V12, V13, V14, V15, V16, V17, V18, V19, V20, V21, V22, V24, V25, V39	1
3	PAFZZ	5975009846582	96906	MS3367-1-0	STRAP, TIEDOWN, ELECT PART OF KIT P/N 12256226 .... UOC:DAA, DAB, DAC, DAD, DAE, DAF, DAG, DAH, DAJ, DAK, DAL, DAW, DAX, V12, V13, V14, V15, V16, V17, V18, V19, V20, V21, V22, V24, V25, V39	5
4	PAFZZ	4730009541281	81348	WW-P-471ACABCB	PLUG, PIPE PART OF KIT P/N 12256226 .... UOC:DAA, DAB, DAC, DAD, DAE, DAF, DAG, DAH, DAJ, DAK, DAL, DAW, DAX, V12, V13, V14, V15, V16, V17, V18, V19, V20, V21, V22, V24, V25, V39	1
5	PAFZZ	4730002026491	79470	3220X6X4	BUSHING, PIPE PART OF KIT P/N 12256226 .... UOC:DAA, DAB, DAC, DAD, DAE, DAF, DAG, DAH, DAJ, DAK, DAL, DAW, DAX, V12, V13, V14, V15, V16, V17, V18, V1 9, V 20, V21, V22, V24, V25, V39	2
6	PAFZZ	4730010326038	19207	CPR102321-4	INSERT, TUBE FITTING PART OF KIT P/N 12256226 .... UOC:DAA, DAB, DAC, DAD, DAE, DAF, DAG, DAH, DAJ, DAK, DAL, DAW, DAX, V12, V13, V14, V15, V16, V17, V18, V19 , V20, V21, V22, V24, V25, V39	1
7	KFFZZ	4730001423076	81343	8-6 120102BA	ADAPTER, STRAIGHT, PI PART OF KIT P/N 12256226 .... UOC:DAA, DAB, DAC, DAD, DAE, DAF, DAG, DAH, DAJ, DAK, DAL, DAW, DAX, V12, V13, V14, V15, V16, V17, V18, V19, V20, V21, V2, V22, V24, V25, V39	1
8	PAFZZ	4730002777331	21450	444122	TEE, PIPE PART OF KIT P/N 12256226 .... UOC:DAA, DAB, DAC, DAD, DAE, DAF, DAG, DAH, DAJ, DAK, DAL, DAW, DAX, V12, V13, V14, V15, V16, V17, V18, V19, V20, V21, V22, V24, V25, V39	1
9	PAFZZ	4730002783220	59206	62C-8	NIPPLE, TUBE PART OF KIT P/N 12256226 .... UOC:DAA, DAB, DAC, DAD, DAE, DAF, DAG, DAH, DAJ, DAK, DAL, DAW, DAX, V12, V13, V14, V15, V16, V17, V18, V19, V20, V21, V22, V24, V25, V39	1

(1) ITEM NO	(2) SMR CODE	(3) NSN	(4) CAGEC	(5) PART NUMBER	(6) DESCRIPTION AND USABLE ON CODES (UOC)	(7) QTY
10	KFFZZ		19207	CPR104420-3-35	TUBING, NONMETALLIC MAKE FROM HOSE, P/N CPR104420-3, 35 INCHES LONG PART OF KIT P/N 12256226 .... UOC:DAA, DAB, DAC, DAD, DAE, DAF, DAG, DAH, DAJ, DAK, DAL, DAW, DAX, V12, V13, V14, V15, V16, V17, V18, V19, V20, V21, V22, V24, V25, V39	1
11	PAFZZ	5305008550961	96906	MS24629-35	SCREW, TAPPING .... UOC:DAA, DAB, DAC, DAD, DAE, DAF, DAG, DAH, DAJ, DAK, DAL, DAW, DAX, V12, V13, V14, V15, V16, V17, V18, V19, V20, V21, V22, V24, V25, V39	
12	KFFZZ		19207	12277248	PLATE PART OF KIT P/N 12256226 .... UOC:DAA, DAB, DAC, DAD, DAE, DAF, DAG, DAH, DAJ, DAK, DAL, DAW, DAX, V12, V13, V14, V15, V16, V17, V18, V19, V20, V21, V22, V24, V25, V39	1
	PAFZZ	2540010965018	19207	12256226	KIT, DEEP WATER FORDING .... UOC:DAA, DAB, DAC, DAD, DAE, DAF, DAG, DAH, DAJ, DAK, DAL, DAW, DAX, V12, V13, V14, V15, V16, V17, V18, V19, V20, V21, V22, V24, V25, V39	1

ADAPTER, STRAIGHT, PI	(	1)	597-7
BOLT, MACHINE	(	1)	596-16
BRACKET	(	1)	596-22
BRACKET, ANGLE	(	1)	596-15
BUSHING, PIPE	(	2)	597-5
CLAMP, LOOP	(	1)	596-14
COCK, DRAIN	(	1)	596-20
COCK, DRAIN	(	1)	597-2
CONTROL ASSEMBLY, PU	(	1)	596-10
ELBOW, PIPE	(	1)	596-2
ELBOW, PIPE	(	1)	596-19
GROMMET, NONMETALLIC	(	1)	596-7
HOSE ASSEMBLY, NONME	(	1)	596-13
HOSE ASSEMBLY	(	1)	596-27
HOSE ASSEMBLY	(	2)	597-1
INSERT, TUBE FITTING	(	1)	97-6
NIPPLE, PIPE	(	1)	596-5
NIPPLE, TUBE	(	1)	597-9
NUT, PLAIN, HEXAGON	(	1)	596-8
NUT, SELF-LOCKING, HE	(	1)	596-18
NUT, SELF-LOCKING, HE	(	2)	596-26
NUT, SELF-LOCKING, HE	(	1)	596-23
PLATE	(	1)	597-12
PLUG, PIPE	(	1)	597-4
REDUCER, PIPE	(	1)	596-17
SCREW, CAP, HEXAGON H	(	1)	596-11
SCREW, MACHINE	(	2)	596-24
STRAP, RETAINING	(	1)	596-25
STRAP, TIEDOWN, ELECT	(	5)	597-3
TEE, PIPE	(	1)	596-3
TEE, PIPE	(	1)	597-8
TUBING, NONMETALLIC	(	1)	597-10

(1) ITEM NO	(2) SMR CODE	(3) NSN	(4) CAGEC	(5) PART NUMBER	(6) DESCRIPTION AND USABLE ON CODES (UOC)	(7) QTY
					VALVE ASSEMBLY,FORD	(1) 596-6
					VALVE,REGULATING,FL	(1) 596-28
					WASHER,LOCK	(1) 596-12
					WASHER,LOCK	(1) 596-9

END OF FIGURE

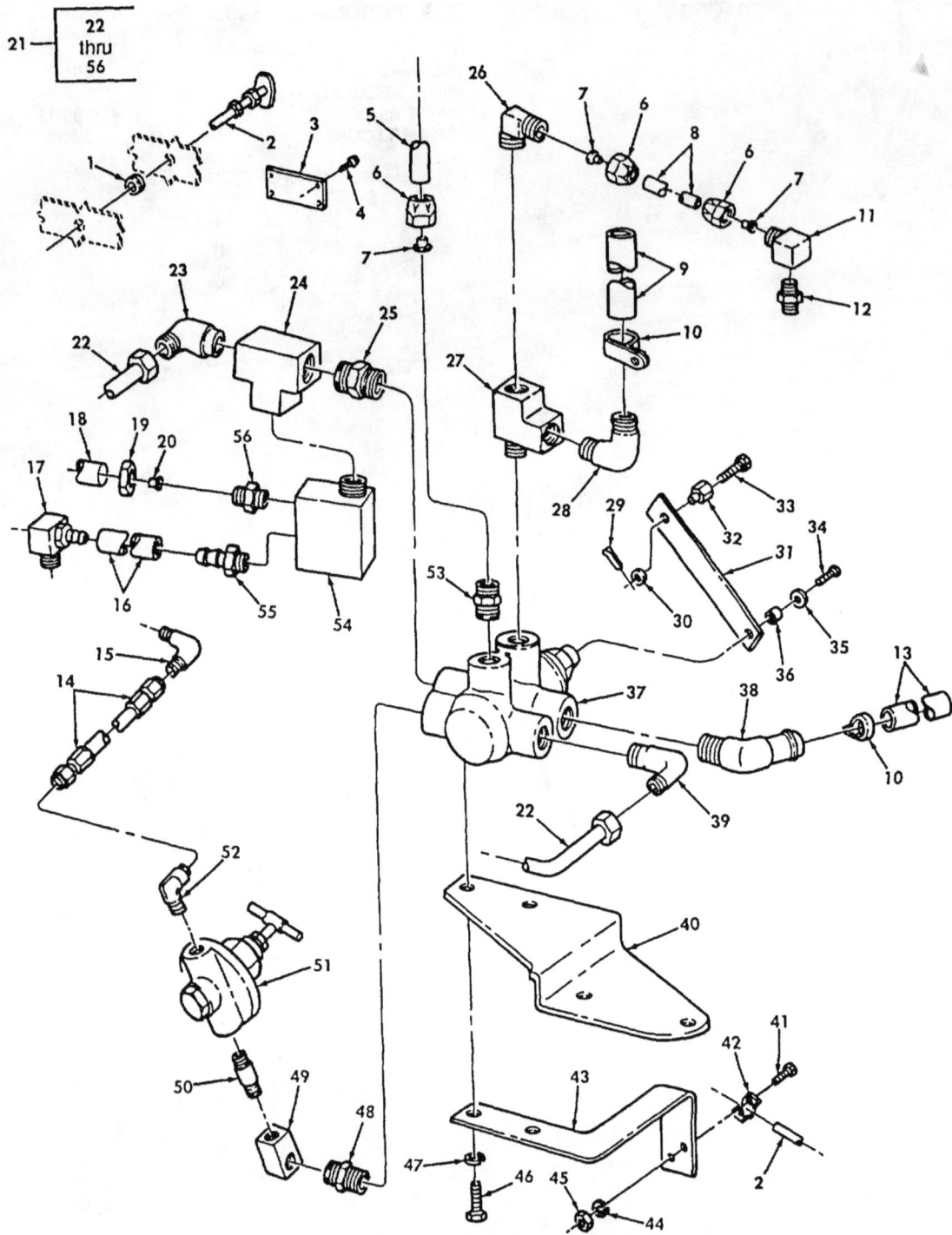

Figure 598. Deep Water Fording Kit, Regulator, Selector, and Related Parts (M939A2).

(1) ITEM NO	(2) SMR CODE	(3) NSN	(4) CAGEC	(5) PART NUMBER	(6) DESCRIPTION AND USABLE ON CODES (UOC)	(7) QTY
					GROUP 3305 DEEPWATER FORDING KIT	
					FIG. 598 DEEPWATER FORDING KIT, REGULATOR, SELECTOR, AND RELATED PARTS (M939A2)	
1	PAOZZ	5325005432902	96906	MS35489-134	GROMMET, NONMETALLIC PART OF KIT P/N 20511191 .... UOC:ZAA, ZAB, ZAC, ZAD, ZAE, ZAF, ZAG, ZAH, ZAJ, ZAK, ZAL	1
2	PAOZZ	2590012873224	60602	45752-70	CONTROL ASSEMBLY, PU PART OF KIT P/N 20511191 .... UOC:ZAA, ZAB, ZAC, ZAD, ZAE, ZAF, ZAG, ZAH, ZAJ, ZAK, ZAL	1
3	KFFZZ		19207	12277248	PLATE, INSTRUCTION, PART OF KIT P/N 20511191 .... UOC:ZAA, ZAB, ZAC, ZAD, ZAE, ZAF, ZAG, ZAH, ZAJ, ZAK, ZAL	1
4	PAOZZ	5305008550961	96906	MS24629-35	SCREW, TAPPING PART OF KIT P/N 20511191 .... UOC:ZAA, ZAB, ZAC, ZAD, ZAE, ZAF, ZAG, ZAH, ZAJ, ZAK, ZAL	4
5	MOOZZ		19207	CPR104420 X 72IN	TUBING, NYLON, MAKE FROM HOSE, P/N 06642-0000, 72 INCHES LONG PART OF KIT P/N 20511191 .... UOC:ZAA, ZAB, ZAC, ZAD, ZAE, ZAF, ZAG, ZAH, ZAJ, ZAK, ZAL	1
6	PAOZZ	4730010487874	81343	8 100110B	NUT, TUBE COUPLING PART OF KIT P/N 20511191 .... UOC:ZAA, ZAB, ZAC, ZAD, ZAE, ZAF, ZAG, ZAH, ZAJ, ZAK, ZAL	3
7	PAOZZ	4730010491559	81343	8 100115B	SLEEVE, COMPRESSION, PART OF KIT P/N 20511191 .... UOC:ZAA, ZAB, ZAC, ZAD, ZAE, ZAF, ZAG, ZAH, ZAJ, ZAK, ZAL	3
8	MOOZZ		19207	CPR104420 X 37IN	TUBING, NYLON, MAKE FROM HOSE, P/N 06642-0000, 37 INCHES LONG PART OF KIT P/N 20511191 .... UOC:ZAA, ZAB, ZAC, ZAD, ZAE, ZAF, ZAG, ZAH, ZAJ, ZAK, ZAL	1
9	MOOZZ		47457	20511200 X 16 IN	HOSE, MAKE FROM HOSE, P/N 20511200-1, 16 INCHES LONG PART OF KIT P/N 20511191 .... UOC:ZAA, ZAB, ZAC, ZAD, ZAE, ZAF, ZAG, ZAH, ZAJ, ZAK, ZAL	1
10	PAOZZ	4730010290520	55883	IPD9M7958	CLAMP, HOSE PART OF KIT P/N 20511191 UOC:ZAA, ZAB, ZAC, ZAD, ZAE, ZAF, ZAG, ZAH, ZAJ, ZAK, ZAL	2
11	PAOZZ	4730011157362	81343	8-4 100202BA	ELBOW, PIPE TO TUBE PART OF KIT P/N 20511191 .... UOC:ZAA, ZAB, ZAC, ZAD, ZAE, ZAF, ZAG, ZAH, ZAJ, ZAK, ZAL	1

(1) ITEM NO	(2) SMR CODE	(3) NSN	(4) CAGEC	(5) PART NUMBER	(6) DESCRIPTION AND USABLE ON CODES (UOC)	(7) QTY
12	PAOZZ	4730006238303	89346	444012	ADAPTER, BUSHING PART OF KIT P/N 20511191 .... UOC:ZAA, ZAB, ZAC, ZAD, ZAE, ZAF, ZAG, ZAH, ZAJ, ZAK, ZAL	1
13	MOOZZ		47457	20511200 X 32 IN	HOSE, MAKE FROM HOSE, P/N 20511200-1, 32 INCHES LONG PART OF KIT P/N 20511191 .... UOC:ZAA, ZAB, ZAC, ZAD, ZAE, ZAF, ZAG, ZAH, ZAJ, ZAK, ZAL	1
14	PAOZZ	4720012848184	98441	A3915-4	HOSE ASSEMBLY, NONME PART OF KIT P/N 20511191 .... UOC:ZAA, ZAB, ZAC, ZAD, ZAE, ZAF, ZAG, ZAH, ZAJ, ZAK, ZAL	1
15	PAOZZ	4730006473207	96906	MS51504-B4	ELBOW, PIPE TO TUBE PART OF KIT P/N 20511191	1
16	MOOZZ		47457	20511200 X 46 IN	HOSE, MAKE FROM HOSE, P/N 20511200-1, 46 INCHES LONG PART OF KIT P/N 20511191 .... UOC:ZAA, ZAB, ZAC, ZAD, ZAE, ZAF, ZAG, ZAH, ZAJ, ZAK, ZAL	1
17	PAOZZ	4730001362018	93061	129HB-6-4	ELBOW, PIPE TO HOSE PART OF KIT P/N 20511191 .... UOC:ZAA, ZAB, ZAC, ZAD, ZAE, ZAF, ZAG, ZAH, ZAJ, ZAK, ZAL	1
18	MOOZZ		19207	CPR104420-5X62IN	HOSE, NONMETALLIC MAKE FROM HOSE, P/N 247607, 62 INCHES LONG .... UOC:ZAA, ZAB, ZAC, ZAD, ZAE, ZAF, ZAG, ZAH, ZAJ, ZAK, ZAL	1
19	PAOZZ	4730012915225	81343	12 100110B	NUT, TUBE COUPLING PART OF KIT P/N 20511191 .... UOC:ZAA, ZAB, ZAC, ZAD, ZAE, ZAF, ZAG, ZAH, ZAJ, ZAK, ZAL	1
20	PAOZZ	4730012879012	81343	12 100115B	SLEEVE, COMPRESSION, PART OF KIT P/N 20511191 .... UOC:ZAA, ZAB, ZAC, ZAD, ZAE, ZAF, ZAG, ZAH, ZAJ, ZAK, ZAL	1
21	PAOZZ	4730012885882	47457	20511192	MANIFOLD ASSEMBLY, H PART OF KIT P/N 20511191 .... UOC:ZAA, ZAB, ZAC, ZAD, ZAE, ZAF, ZAG, ZAH, ZAJ, ZAK, ZAL	1
22	PAOZZ	4710012874608	47457	20511203	TUBE ASSEMBLY, METAL PART OF KIT P/N 20511191 .... UOC:ZAA, ZAB, ZAC, ZAD, ZAE, ZAF, ZAG, ZAH, ZAJ, ZAK, ZAL	1
23	PAOZZ	4730010704915	96906	MS51504A6-8	ELBOW, PIPE TO TUBE PART OF KIT P/N 20511191 .... UOC:ZAA, ZAB, ZAC, ZAD, ZAE, ZAF, ZAG, ZAH, ZAJ, ZAK, ZAL	1
24	PAOZZ	4730012870963	24617	444124	TEE, PIPE PART OF KIT P/N 20511191 .... UOC:ZAA, ZAB, ZAC, ZAD, ZAE, ZAF, ZAG, ZAH, ZAJ, ZAK, ZAL	1

(1) ITEM NO	(2) SMR CODE	(3) NSN	(4) CAGEC	(5) PART NUMBER	(6) DESCRIPTION AND USABLE ON CODES (UOC)	(7) QTY
25	PAOZZ	4730009591629	89346	443982	.NIPPLE, PIPE PART OF KIT P/N 20511191 .... UOC:ZAA, ZAB, ZAC, ZAD, ZAE, ZAF, ZAG, ZAH, ZAJ, ZAK, ZAL	1
26	PAOZZ	4730011156643	81343	8-8 100202BA	.ELBOW, PIPE TO TUBE PART OF KIT P/N 20511191 .... UOC:ZAA, ZAB, ZAC, ZAD, ZAE, ZAF, ZAG, ZAH, ZAJ, ZAK, ZAL	1
27	PAOZZ	4730005402745	24617	444152	.TEE, PIPE PART OF KIT P/N 20511191 UOC:ZAA, ZAB, ZAC, ZAD, ZAE, ZAF, ZAG, ZAH, ZAJ, ZAK, ZAL	1
28	PAOZZ	4730013090947	81343	12-8 430260BA	.ELBOW, PIPE TO HOSE PART OF KIT P/N 20511191 .... UOC:ZAA, ZAB, ZAC, ZAD, ZAE, ZAF, ZAG, ZAH, ZAJ, ZAK, ZAL	1
29	PAOZZ	5315008392326	96906	MS24665-281	.PIN, COTTER PART OF KIT P/N 20511191 .... UOC:ZAA, ZAB, ZAC, ZAD, ZAE, ZAF, ZAG, ZAH, ZAJ, ZAK, ZAL	1
30	PAOZZ	5310000814219	96906	MS27183-12	.WASHER, FLAT PART OF KIT P/N 20511191 .... UOC:ZAA, ZAB, ZAC, ZAD, ZAE, ZAF, ZAG, ZAH, ZAJ, ZAK, ZAL	1
31	PAOZZ	3040012885313	47457	20510880	.LEVER, REMOTE CONTRO PART OF KIT P/N 20511191 .... UOC:ZAA, ZAB, ZAC, ZAD, ZAE, ZAF, ZAG, ZAH, ZAJ, ZAK, ZAL	1
32	PAOZZ	5315012884584	47457	20511207	.PIN, STRAIGHT, HEADED CABLE PART OF KIT P/N 20511191 .... UOC:ZAA, ZAB, ZAC, ZAD, ZAE, ZAF, ZAG, ZAH, ZAJ, ZAK, ZAL	1
33	PAOZZ	5305012894411	24617	451695	.SCREW, MACHINE PART OF KIT P/N 20511191 .... UOC:ZAA, ZAB, ZAC, ZAD, ZAE, ZAF, ZAG, ZAH, ZAJ, ZAK, ZAL	1
34	PAOZZ	5305012373637	16764	159920	.SCREW PART OF KIT P/N 20511191 UOC:ZAA, ZAB, ZAC, ZAD, ZAE, ZAF, ZAG, ZAH, ZAJ, ZAK, ZAL	1
35	PAOZZ	5310008238804	96906	MS27183-9	.WASHER, FLAT PART OF KIT P/N 20511191 .... UOC:ZAA, ZAB, ZAC, ZAD, ZAE, ZAF, ZAG, ZAH, ZAJ, ZAK, ZAL	1
36	PAOZZ	5365012881560	47457	20510879	.SPACER, SLEEVE PART OF KIT P/N 20511191 .... UOC:ZAA, ZAB, ZAC, ZAD, ZAE, ZAF, ZAG, ZAH, ZAJ, ZAK, ZAL	1
37	PAOZZ	4820000097378	19207	11648568	.VALVE, FUEL, SIX-PORT PART OF KIT P/N 20511191 .... UOC:ZAA, ZAB, ZAC, ZAD, ZAE, ZAF, ZAG, ZAH, ZAJ, ZAK, ZAL	1
38	PAOZZ	4730013090948	81343	12-8 430360BA	.ELBOW, PIPE TO HOSE PART OF KIT P/N 20511191 .... UOC:ZAA, ZAB, ZAC, ZAD, ZAE, ZAF, ZAG, ZAH, ZAJ, ZAK, ZAL	1

(1) ITEM NO	(2) SMR CODE	(3) NSN	(4) CAGEC	(5) PART NUMBER	(6) DESCRIPTION AND USABLE ON CODES (UOC)	(7) QTY
39	PAOZZ	4730008353003	96906	MS51504A6-6	.ELBOW, PIPE TO TUBE PART OF KIT P/N 20511191 .... UOC:ZAA, ZAB, ZAC, ZAD, ZAE, ZAF, ZAG, ZAH, ZAJ, ZAK, ZAL	1
40	PAOZZ	5340012882132	47457	20510881	.BRACKET, DOUBLE ANGL PART OF KIT P/N 20511191 .... UOC:ZAA, ZAB, ZAC, ZAD, ZAE, ZAF, ZAG, ZAH, ZAJ, ZAK, ZAL	1
41	PAOZZ	5306010423586	72582	9415764	.BOLT PART OF KIT P/N 20511191 .... UOC:ZAA, ZAB, ZAC, ZAD, ZAE, ZAF, ZAG, ZAH, ZAJ, ZAK, ZAL	2
42	PAOZZ	5340009662390	96906	MS21315-3	.STRAP, RETAINING PART OF KIT P/N 20511191 .... UOC:ZAA, ZAB, ZAC, ZAD, ZAE, ZAF, ZAG, ZAH, ZAJ, ZAK, ZAL	1
43	PAOZZ	5340012895028	47457	20510878	.BRACKET, ANGLE PART OF KIT P/N 20511191 .... UOC:ZAA, ZAB, ZAC, ZAD, ZAE, ZAF, ZAG, ZAH, ZAJ, ZAK, ZAL	1
44	PAOZZ	5310000453299	96906	MS35338-42	.WASHER, LOCK PART OF KIT P/N .... 20511191 UOC:ZAA, ZAB, ZAC, ZAD, ZAE, ZAF, ZAG, ZAH, ZAJ, ZAK, ZAL	2
45	PAOZZ	5310009349757	96906	MS35649-282	.NUT, PLAIN, HEXAGON PART OF KIT P/N 20511191 .... UOC:ZAA, ZAB, ZAC, ZAD, ZAE, ZAF, ZAG, ZAH, ZAJ, ZAK, ZAL	2
46	PAOZZ	5305010102362	96906	MS18154-59	.SCREW, CAP, HEXAGON H PART OF KIT P/N 20511191 .... UOC:ZAA, ZAB, ZAC, ZAD, ZAE, ZAF, ZAG, ZAH, ZAJ, ZAK, ZAL	2
47	PAOZZ	5310002748041	90407	12084PII	.WASHER, FLAT PART OF KIT P/N 20511191 .... UOC:ZAA, ZAB, ZAC, ZAD, ZAE, ZAF, ZAG, ZAH, ZAJ, ZAK, ZAL	2
48	PAOZZ	4730002026491	24617	444028	.BUSHING, PIPE PART OF KIT P/N 20511191 .... UOC:ZAA, ZAB, ZAC, ZAD, ZAE, ZAF, ZAG, ZAH, ZAJ, ZAK, ZAL	1
49	PAOZZ	4730002775553	96906	MS51952-2	.ELBOW, PIPE PART OF KIT P/N 20511191 .... UOC:ZAA, ZAB, ZAC, ZAD, ZAE, ZAF, ZAG, ZAH, ZAJ, ZAK, ZAL	1
50	PAOZZ	4730001867798	29510	443978	.NIPPLE, PIPE PART OF KIT P/N 20511191 .... UOC:ZAA, ZAB, ZAC, ZAD, ZAE, ZAF, ZAG, ZAH, ZAJ, ZAK, ZAL	1
51	PAOZZ	4820011705055	19207	10922156-1	.VALVE, REGULATING, FL PART OF KIT P/N 20511191 .... UOC:ZAA, ZAB, ZAC, ZAD, ZAE, ZAF, ZAG, ZAH, ZAJ, ZAK, ZAL	1
52	PAOZZ	4730008778997	96906	MS51504A4-4	.ELBOW, PIPE TO TUBE PART OF KIT P/N 20511191 .... UOC:ZAA, ZAB, ZAC, ZAD, ZAE, ZAF, ZAG, ZAH, ZAJ, ZAK, ZAL	1

(1) ITEM NO	(2) SMR CODE	(3) NSN	(4) CAGEC	(5) PART NUMBER	(6) DESCRIPTION AND USABLE ON CODES (UOC)	(7) QTY
53	PAOZZ	4730010918032	79470	1468X8	.ADAPTER, STRAIGHT, PI PART OF KIT P/N 20511191 .... UOC:ZAA, ZAB, ZAC, ZAD, ZAE, ZAF, ZAG, ZAH, ZAJ, ZAK, ZAL	1
54	PAOZZ	4730011150433	19207	12255986	.CONNECTOR, MULTIPLE, PART OF KIT P/N 20511191 .... UOC:ZAA, ZAB, ZAC, ZAD, ZAE, ZAF, ZAG, ZAH, ZAJ, ZAK, ZAL	1
55	PAOZZ	4730013090949	93061	125HBL-6-8	.ADAPTER, STRAIGHT, PI PART OF KIT P/N 20511191 .... UOC:ZAA, ZAB, ZAC, ZAD, ZAE, ZAF, ZAG, ZAH, ZAJ, ZAK, ZAL	1
56	PAOZZ	4730011086410	81343	12-8 100102BA	.ADAPTER, STRAIGHT, PI PART OF KIT P/N 20511191 .... UOC:ZAA, ZAB, ZAC, ZAD, ZAE, ZAF, ZAG, ZAH, ZAJ, ZAK, ZAL	1

END OF FIGURE

VIEW A

VIEW B

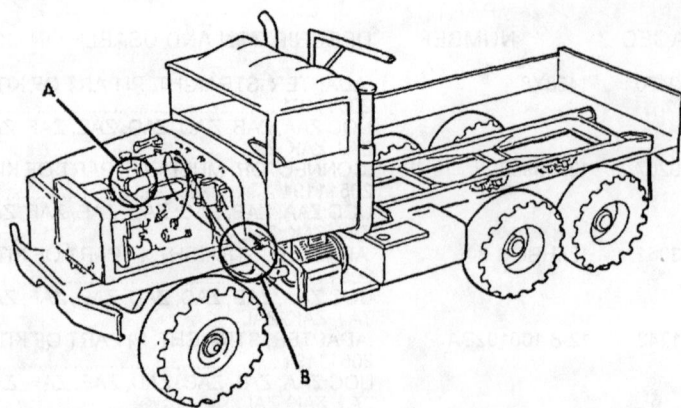

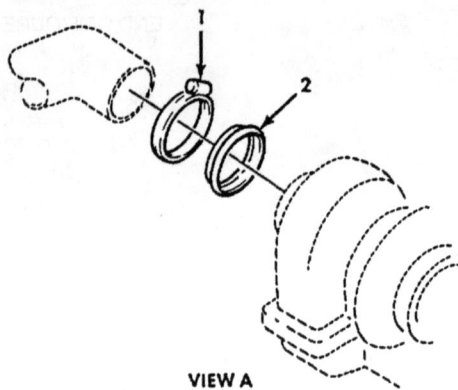

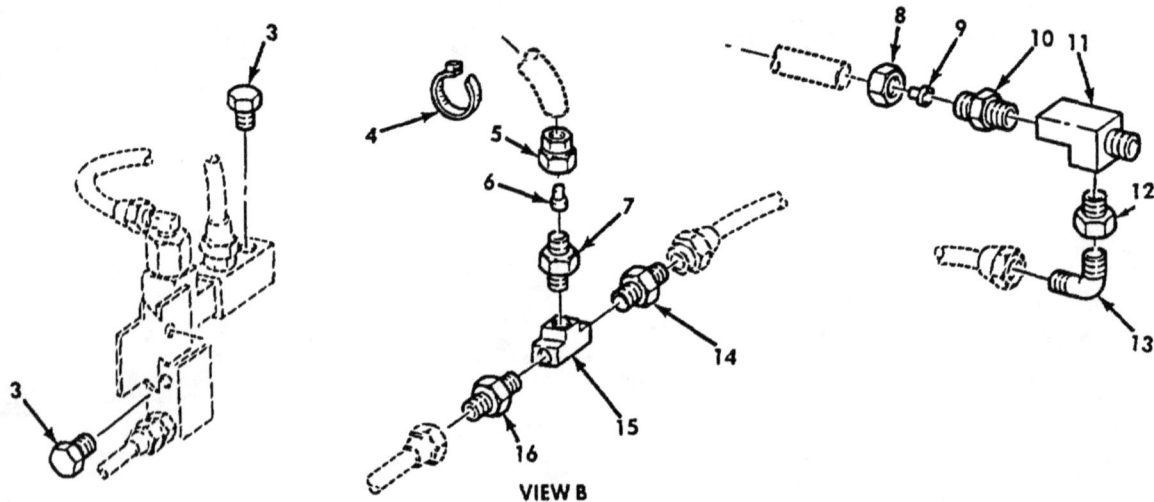

Figure 599. Deep Water Fording Kit, Connectors, and Related Parts (M939A2).

(1) ITEM NO	(2) SMR CODE	(3) NSN	(4) CAGEC	(5) PART NUMBER	(6) DESCRIPTION AND USABLE ON CODES (UOC)	(7) QTY
					GROUP 3305 DEEP WATER FORDING KIT	
					FIG. 599 DEEP WATER FORDING KIT, CONNECTORS, AND RELATED PARTS (M939A2)	
1	PAOZZ	5340012882167	63208	10446-5	COUPLING, CLAMP, GROO PART OF KIT P/N 20511191 .... UOC:ZAA, ZAB, ZAC, ZAD, ZAE, ZAF, ZAG, ZAH, ZAJ, ZAK, ZAL	1
2	PAOZZ	5330012881466	47457	20511268	GASKET PART OF KIT P/N 20511191 .... UOC:ZAA, ZAB, ZAC, ZAD, ZAE, ZAF, ZAG, ZAH, ZAJ, ZAK, ZAL UOC:ZAA, ZAB, ZAC, ZAD, ZAE, ZAF, ZAG, ZAH, ZAJ, ZAK, ZAL	1
3	PAOZZ	4730000144027	79470	3152X8	PLUG, PIPE PART OF KIT P/N 20511191 .... UOC:ZAA, ZAB, ZAC, ZAD, ZAE, ZAF, ZAG, ZAH, ZAJ, ZAK, ZAL	2
4	PAOZZ	5975001563253	96906	MS3367-2-9	STRAP, TIEDOWN, ELECT PART OF KIT P/N 20511191 .... UOC:ZAA, ZAB, ZAC, ZAD, ZAE, ZAF, ZAG, ZAH, ZAJ, ZAK, ZAL	6
5	PAOZZ	4730010487874	81343	8 100110B	NUT, TUBE COUPLING PART OF KIT P/N 20511191 .... UOC:ZAA, ZAB, ZAC, ZAD, ZAE, ZAF, ZAG, ZAH, ZAJ, ZAK, ZAL	3
6	PAOZZ	4730010491559	81343	8 100115B	SLEEVE, COMPRESSION PART OF KIT P/N 20511191 .... UOC:ZAA, ZAB, ZAC, ZAD, ZAE, ZAF, ZAG, ZAH, ZAJ, ZAK, ZAL	3
7	PAOZZ	4730010918032	79470	1468X8	ADAPTER, STRAIGHT, PI .... UOC:ZAA, ZAB, ZAC, ZAD, ZAE, ZAF, ZAG, ZAH, ZAJ, ZAK, ZAL	1
8	PAOZZ	4730012915225	81343	12 100110B	NUT, TUBE COUPLING PART OF KIT P/N 20511191 .... UOC:ZAA, ZAB, ZAC, ZAD, ZAE, ZAF, ZAH, ... ZAJ, ZAK, ZAL	1
9	PAOZZ	4730012879012	81343	12 100115B	SLEEVE, COMPRESSION PART OF KIT P/N 20511191 .... UOC:ZAA, ZAB, ZAC, ZAD, ZAE, ZAF, ZAG, ZAH, ZAJ, ZAK, ZAL	1
10	PAOZZ	4730011086410	81343	12-8 100102BA	ADAPTER, STRAIGHT, PI .... UOC:ZAA, ZAB, ZAC, ZAD, ZAE, ZAF, ZAG, ZAH, ZAJ, ZAK, ZAL	1
11	PAOZZ	4730005402745	24617	444152	TEE, PIPE .... UOC:ZAA, ZAB, ZAC, ZAD, ZAE, ZAF, ZAG, ZAH, ZAJ, ZAK, ZAL	1
12	PAOZZ	4730005806738	24617	444034	BUSHING, PIPE PART OF KIT P/N 20511191 .... UOC:ZAA, ZAB, ZAC, ZAD, ZAE, ZAF, ZAG, ZAH, ZAJ, ZAK, ZAL	1
13	PAOZZ	4730002890155	81343	6-6 120202BA	ELBOW, PIPE TO TUBE PART OF KIT P/N 20511191 .... UOC:ZAA, ZAB, ZAC, ZAD, ZAE, ZAF, ZAG, ZAH, ZAJ, ZAK, ZAL	1

(1) ITEM NO	(2) SMR CODE	(3) NSN	(4) CAGEC	(5) PART NUMBER	(6) DESCRIPTION AND USABLE ON CODES (UOC)	(7) QTY
14	PAOZZ	4730001423076	81343	8-6 120102BA	ADAPTER, STRAIGHT, PI PART OF KIT 20511191 ...... UOC:ZAA, ZAB, ZAC, ZAD, ZÆ, ZAF, ZAG, ZAH, ZAJ, ZAK, ZAL	1
15	PAOZZ	4730002777331	21450	444122	TEE, PIPE PART OF KIT P/N 20511191 ...... UOC:ZAA, ZAB, ZAC, ZAD, ZAE, ZAF, ZAG, ZAH, ZAJ, ZAK, ZAL	1
16	PAOZZ	4730002704616	81343	6-6060102B	ADAPTER, STRAIGHT, PI PART OF KIT P/N 20511191 ...... UOC:ZAA, ZAB, ZAC, ZAD, ZAE, ZAF, ZAG, ZAH, ZAJ, ZAK, ZAL	1
	PDFZZ	254001284871	847457	20511191	FORDING KIT, DEEP	1
					ADAPTER, STRAIGHT, PI (1) 598-55	
					ADAPTER, STRAIGHT, PI (1) 598-56	
					ADAPTER, STRAIGHT, PI (1) 598-53	
					ADAPTER, BUSHING (1) 598-12	
					ADAPTER, STRAIGHT, PI (1) 599-16	
					ADAPTER, STRAIGHT, PI (1) 599-14	
					BOLT (2) 598-41	
					BRACKET, ANGLE (1) 598-43	
					BRACKET, DOUBLE ANGL (1) 598-40	
					BUSHING, PIPE (1) 98-48	
					BUSHING, PIPE (1) 599-12	
					CLAMP, HOSE (2) 598-10	
					CONNECTOR, MULTIPLE, (1) 598-54	
					CONTROL ASSEMBLY, PU (1) 598-2	
					COUPLING, CLAMP, GROO (1) 599-1	
					ELBOW, PIPE TO HOSE (1) 598-17	
					ELBOW, PIPE TO TUBE (1) 598-26	
					ELBOW, PIPE TO TUBE (1) 598-11	
					ELBOW, PIPE TO HOSE (1) 598-28	
					ELBOW, PIPE TO HOSE (1) 598-38	
					ELBOW, PIPE (1) 598-49	
					ELBOW, PIPE TO TUBE (1) 599-13	
					ELBOW, PIPE TO TUBE (1) 598-15	
					ELBOW, PIPE TO TUBE (1) 598-52	
					ELBOW, PIPE TO TUBE (1) 598-39	
					ELBOW, PIPE TO TUBE (1) 598-23	
					GASKET (1) 599-2	
					GROMMET, NONMETALLIC (1) 598-1	
					HOSE ASSEMBLY, NONME (1) 598-14	
					HOSE, (1) 598-9	
					HOSE, (1) 598-13	
					HOSE, (1) 598-16	
					LEVER, REMOTE CONTRO (1) 598-31	
					MANIFOLD ASSEMBLY, H (1) 598-21	
					NIPPLE, PIPE (1) 598-50	
					NIPPLE, PIPE (1) 598-25	
					NUT, PLAIN, HEXAGON (2) 598-45	
					NUT, TUBE COUPLING (1) 599-8	
					NUT, TUBE COUPLING (3) 599-5	
					NUT, TUBE COUPLING (1) 598-19	

(1) ITEM NO	(2) SMR CODE	(3) NSN	(4) CAGEC	(5) PART NUMBER	(6) DESCRIPTION AND USABLE ON CODES (UOC)			(7) QTY
					NUT, TUBE COUPLING	(	3)	598-6
					PIN, COTTER	(	1)	598-29
					PIN, STRAIGHT, HEADED	(	1)	598-32
					PLATE, INSTRUCTION,	(	1)	598-3
					PLUG, PIPE	(	2)	599-3
					SCREW	(	1)	598-34
					SCREW, CAP, HEXAGON H	(	2)	598-46
					SCREW, MACHINE	(	1)	598-33
					SCREW, TAPPING	(	4)	598-4
					SLEEVE, COMPRESSION	(	1)	599-9
					SLEEVE, COMPRESSION	(	3)	599-6
					SLEEVE, COMPRESSION,	(	3)	598-7
					SLEEVE, COMPRESSION,	(	1)	598-20
					SPACER, SLEEVE	(	1)	598-36
					STRAP, RETAINING	(	1)	598-42
					STRAP, TIEDOWN, ELECT	(	6)	599-4
					TEE, PIPE	(	1)	599-15
					TEE, PIPE	(	1)	598-24
					TEE, PIPE	(	1)	598-27
					TUBE ASSEMBLY, METAL	(	1)	598-22
					TUBING, NYLON,	(	1)	598-8
					TUBING, NYLON,	(	1)	598-5
					VALVE, FUEL, SIX-PORT	(	1)	598-37
					VALVE, REGULATING, FL	(	1)	598-51
					WASHER, FLAT	(	2)	598-47
					WASHER, FLAT	(	1)	598-30
					WASHER, FLAT	(	1)	598-35
					WASHER, LOCK	(	2)	598-44

END OF FIGURE

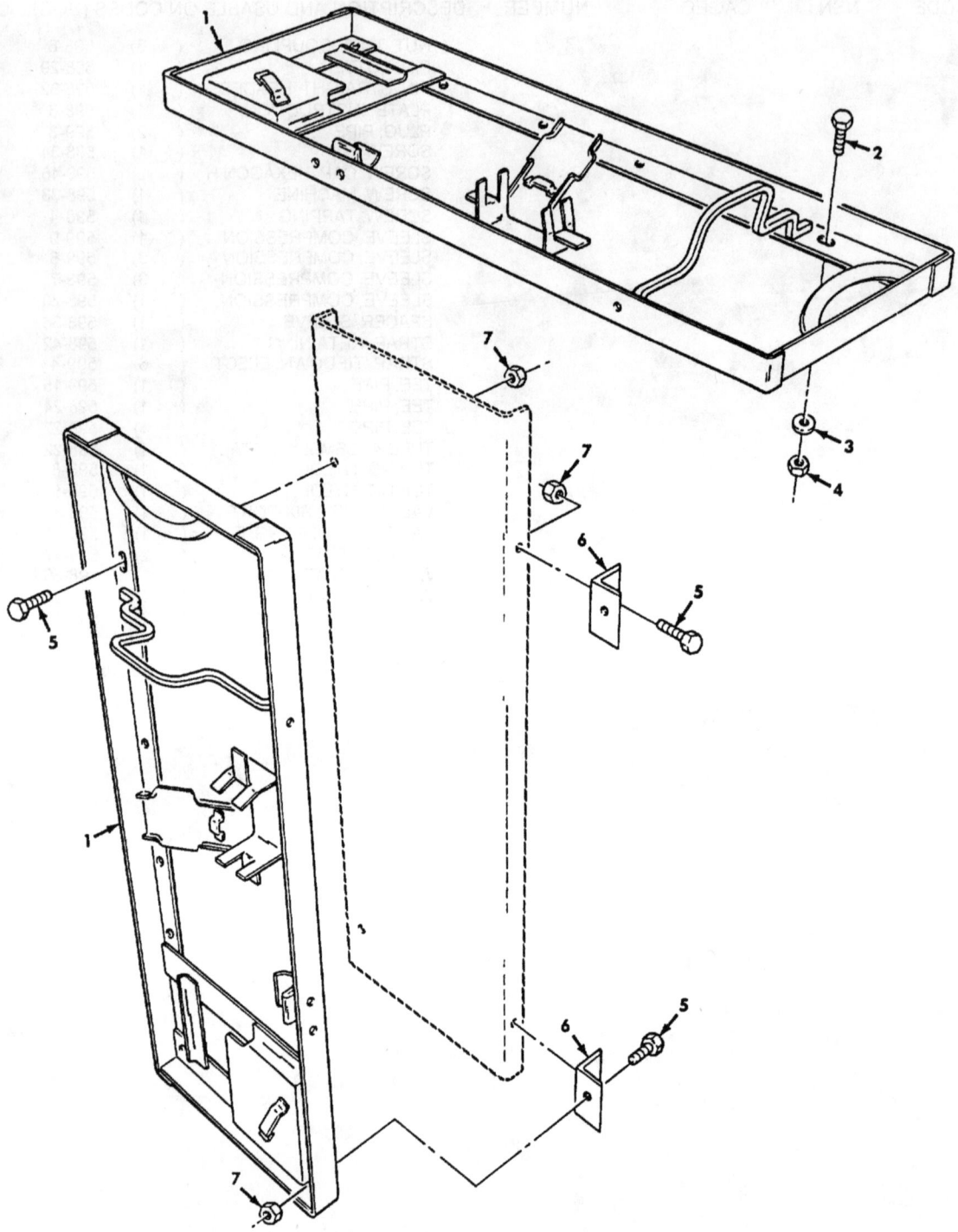

Figure 600. Pioneer Tool Bracket Kit.

(1) ITEM NO	(2) SMR CODE	(3) NSN	(4) CAGEC	(5) PART NUMBER	(6) DESCRIPTION AND USABLE ON CODES (UOC)	(7) QTY
					GROUP 3307 SPECIAL PURPOSE KITS	
					FIG. 600 PIONEER TOOL BRACKET KIT	
1	PAOZZ	4910003575494	19207	7346922	BRACKET PART OF KIT P/N 12392814 .... UOC:DAA, DAB, DAC, DAD, DAE, DAF, DAG, DAH, DAW, DAX, V12, V13, V14, V15, V16, V17, V9, V20, V21, V22, ZAA, ZAB, ZAC, ZAD, ZAE, ZAF, ZAG, ZAH	1
2	PAOZZ	5305002692803	96906	MS90726-60	SCREW, CAP, HEXAGON H PART OF KIT P/N 12302814 ...... UOC:DAA, DAB, DAC, DAD, DAW, DAX, V12, V13, V14, V15, V16, V17, ZAA, ZAB, ZAC, ZAD	4
2	PAOZZ	5305002692803	96906	MS90726-60	SCREW, CAP, HEXAGON H PART OF KIT P/N 12302814 ...... UOC:DAG, DAH, V21, V22, ZAG, ZAH	5
3	PAOZZ	5310008094061	96906	MS27183-15	WASHER, FLAT PART OF KIT P/N 12302814 .... UOC:DAA, DAB, DAC, DAD, DAW, DAX, V12, V13, V14, V15, V16, V17, ZAA, ZAB, ZAC, ZAD	4
3	PAOZZ	5310008094061	96906	MS27183-15	WASHER, FLAT PART OF KIT P/N 12302814 .... UOC:DAG, DAH, V21, V22, ZAG, ZAH	5
4	PAOZZ	5310008140672	96906	MS51943-36	NUT, SELF-LOCKING, HE PART OF KIT P/N 12302814 ...... UOC:DAA, DAB, DAC, DAD, DAW, DAX, V12, V13, V14, V15, V16, V17, ZAA, ZAB, ZAC, ZAD	4
4	PAOZZ	5310008140672	96906	MS51943-36	NUT, SELF-LOCKING, HE PART OF KIT P/N 12302814 ...... UOC:DAG, DAH, V21, V22, ZAG, ZAH	5
5	PAOZZ	5305002692803	96906	MS90726-60	SCREW, CAP, HEXAGON H PART OF KIT P/N 12302814 ...... UOC:DAE, DAF, V19, V20, ZAE, ZAF	6
6	PAOZZ	5340011867174	19207	12302816	BRACKET, ANGLE PART OF KIT P/N 12302814 ...... UOC:DAE, DAF, V19, V20, ZAE, ZAF	2
7	PAOZZ	5310008140672	96906	MS51943-36	NUT, SELF-LOCKING, HE PART OF KIT P/N 12302814 ...... UOC:DAE, DAF, V19, V20, ZAE, ZAF	6
	PDOZZ	2540011757257	19207	12302814	MOUNTING KIT, TOOL S .... UOC:DAA, DAB, DAC, DAD, DAE, DAF, DAG, DAH, DAW, DAX, V12, V13, V14, V15 6, V17, V l9, V20, V21, V22, ZAA, ZAB, ZAC, ZAD, ZAE, ZAF, ZAG, ZAH	1

BRACKET	(1)	600-1
BRACKET, ANGLE	(2)	600-6
NUT, SELF-LOCKING, HE	(4)	600-4
NUT, SELF-LOCKING, HE	(5)	600-4
NUT, SELF-LOCKING, HE	(6)	600-7
SCREW, CAP, HEXAGON H	(4)	600-2
SCREW, CAP, HEXAGON H	(5)	600-2
SCREW, CAP, HEXAGON H	(6)	600-5
WASHER, FLAT	(4)	600-3
WASHER, FLAT	(5)	600-3

END OF FIGURE

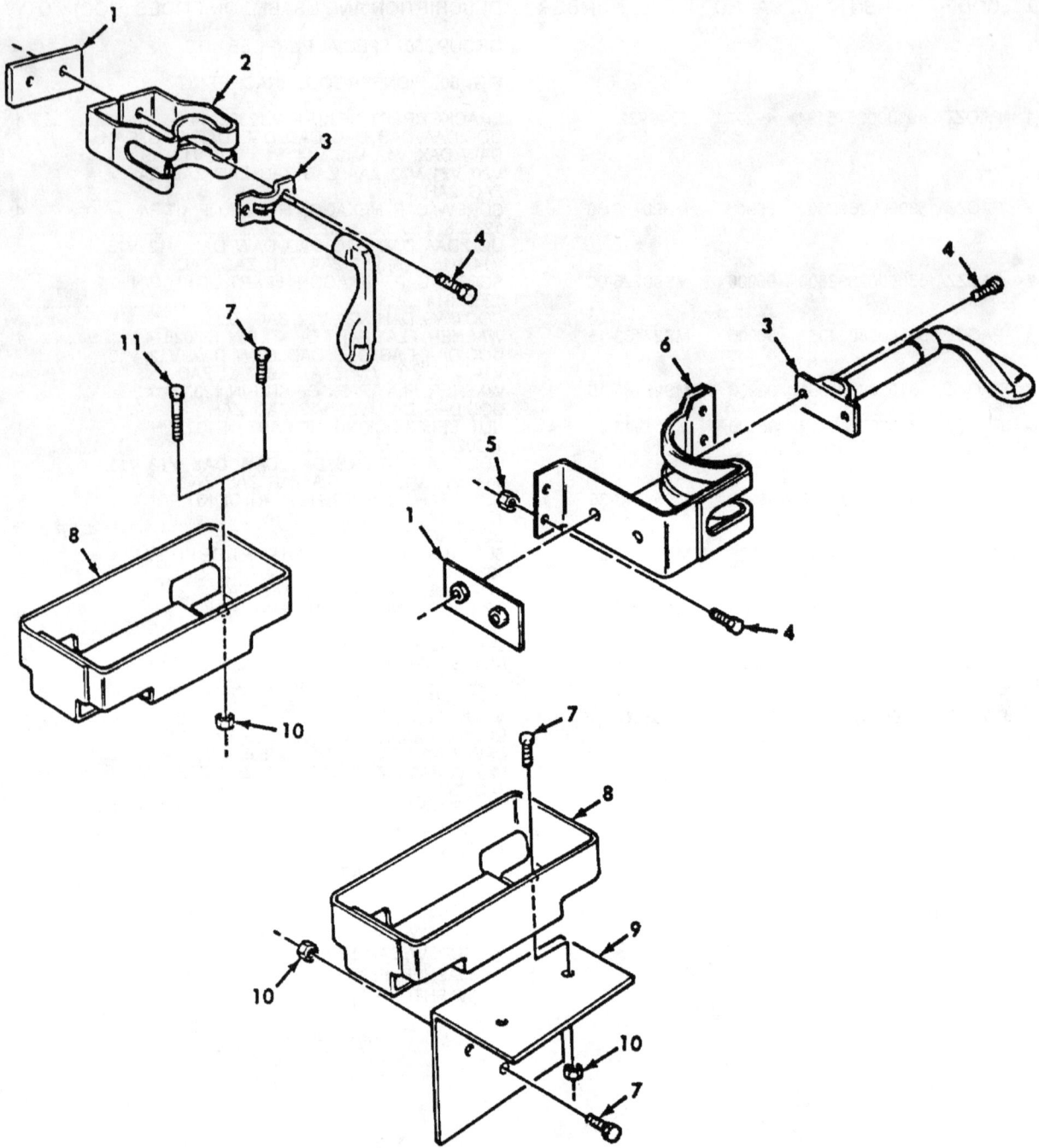

Figure 601. Rifle Mounting Kit.

(1) ITEM NO	(2) SMR CODE	(3) NSN	(4) CAGEC	(5) PART NUMBER	(6) DESCRIPTION AND USABLE ON CODES (UOC)	(7) QTY
					GROUP 3307 SPECIAL PURPOSE KITS	
					FIG. 601 RIFLE MOUNTING KIT	
1	KFOZZ		19207	12276983	REINFORCEMENT PART OF KIT P/N 12276986	3
2	KFOZZ	5340004555899	19207	11630581	BRACKET, MOUNTING, CA PART OF KIT P/N 12276986	2
3	KFOZZ		19207	10939520 KF	CATCH ASSEMBLY, RIFL PART OF KIT P/N 12276986 PART OF KIT P/N 12276986.	3
4	PAOZZ	5305002678953	80204	B1821BH025F063N	SCREW, CAP, HEXAGON H PART OF KIT P/N 12276986	8
5	PAOZZ	5310010470401	96906	MS27130-S61K	NUT, PLAIN, BLIND RIV PART OF KIT P/N 12276986	4
6	KFOZZ		19207	12276982	BRACKET, MTG, CATCH PART OF KIT P/N 12276986	1
7	PAOZZ	5306002259086	96906	MS90726-31	BOLT, MACHINE PART OF KIT P/N 12276986	7
8	KFOZZ	2590002648828	19207	11630594	SUPPORT, RIFLE MOUNT PART OF KIT P/N 12276986	3
9	KFOZZ		19207	12276984	BRACKET, GUN MTG PART OF KIT P/N 12276986	1
10	PAOZZ	5310002416658	96906	MS51943-34	NUT, SELF-LOCKING, HE PART OF KIT P/N 12276986	9
11	PAOZZ	5306002259095	96906	MS90726-40	BOLT, MACHINE PART OF KIT P/N 12276986	2
	PFFZZ	2540010965020	19207	12276986	KIT, RIFLE MOUNTING	1

BOLT, MACHINE	(7)	601-7
BOLT, MACHINE	(2)	601-11
BRACKET, GUN	(1)	601-9
BRACKET, MOUNTING	(2)	601-2
BRACKET, MOUNTING	(1)	601-6
CATCH ASSEMBLY	(3)	601-3
NUT, PLAIN	(4)	601-5
NUT, SELF-LOCKING	(9)	601-10
REINFORCEMENT	(3)	601-1
SCREW, CAP, HEXAGON	(10)	601-4
SUPPORT, RIFLE MOUNT	(3)	601-8

END OF FIGURE

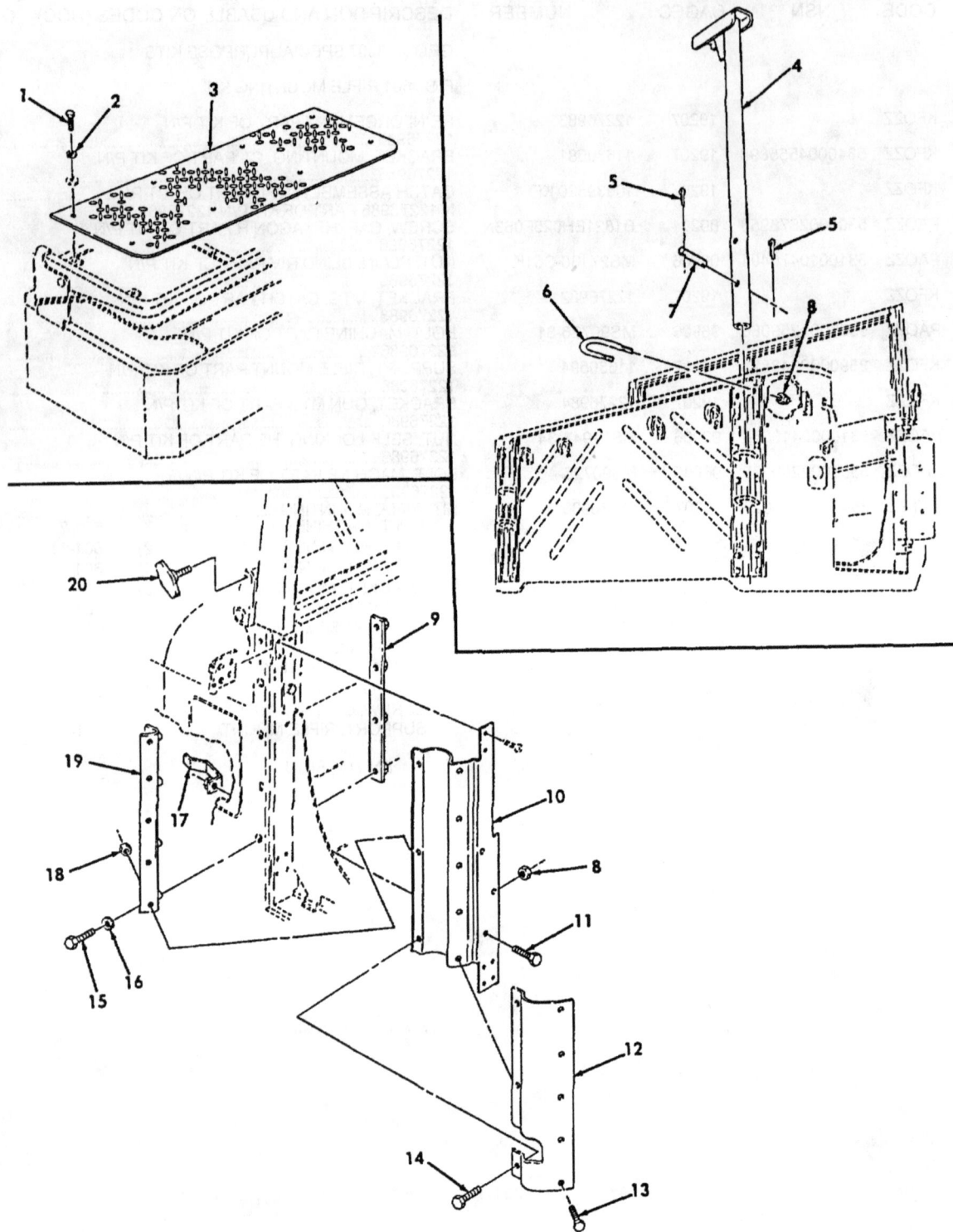

Figure 602. Lightweight Weapon Station Kit (Sheet 1 of 3).

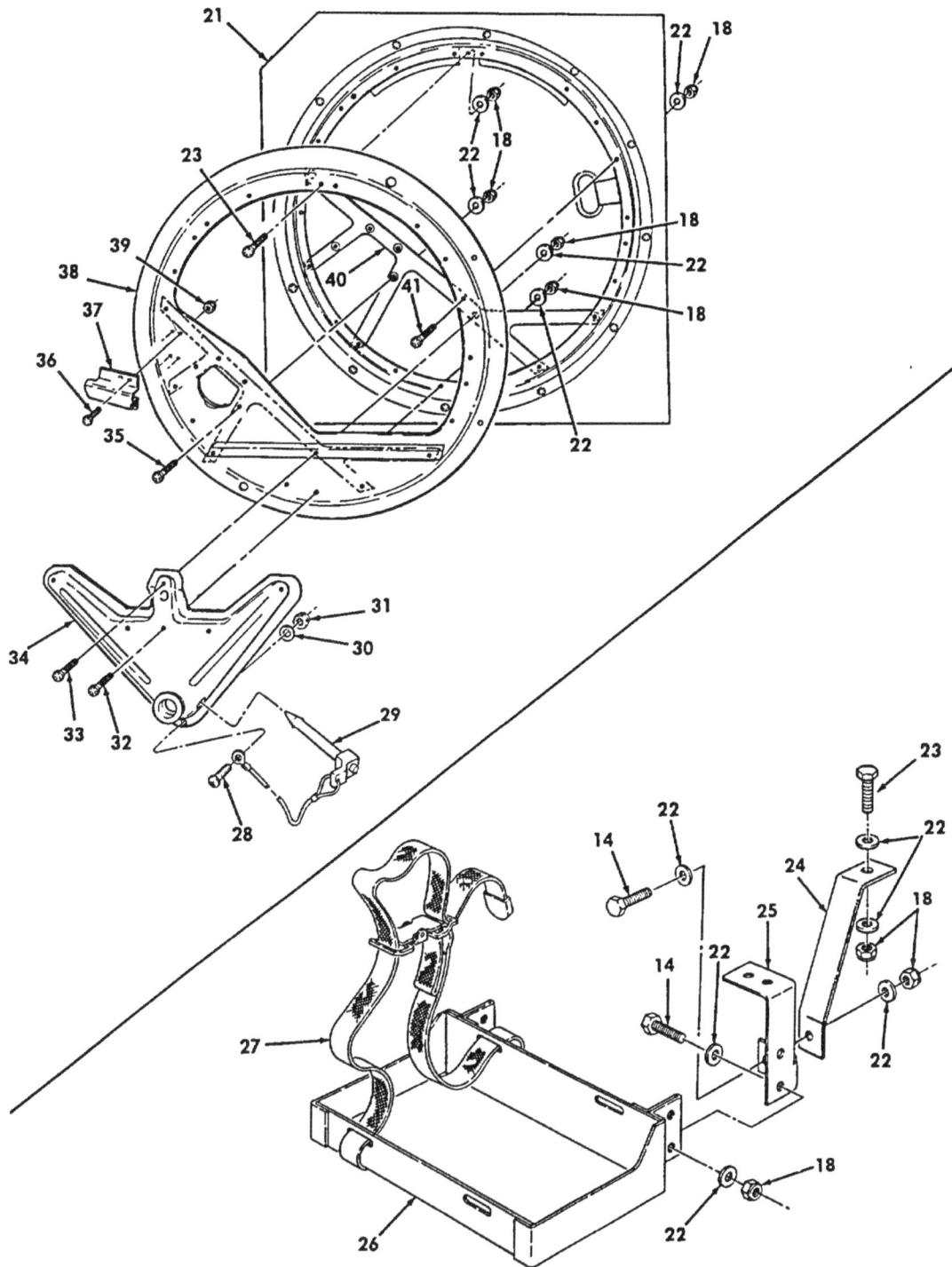

Figure 602. Lightweight Weapon Station Kit (Sheet 2 of 3).

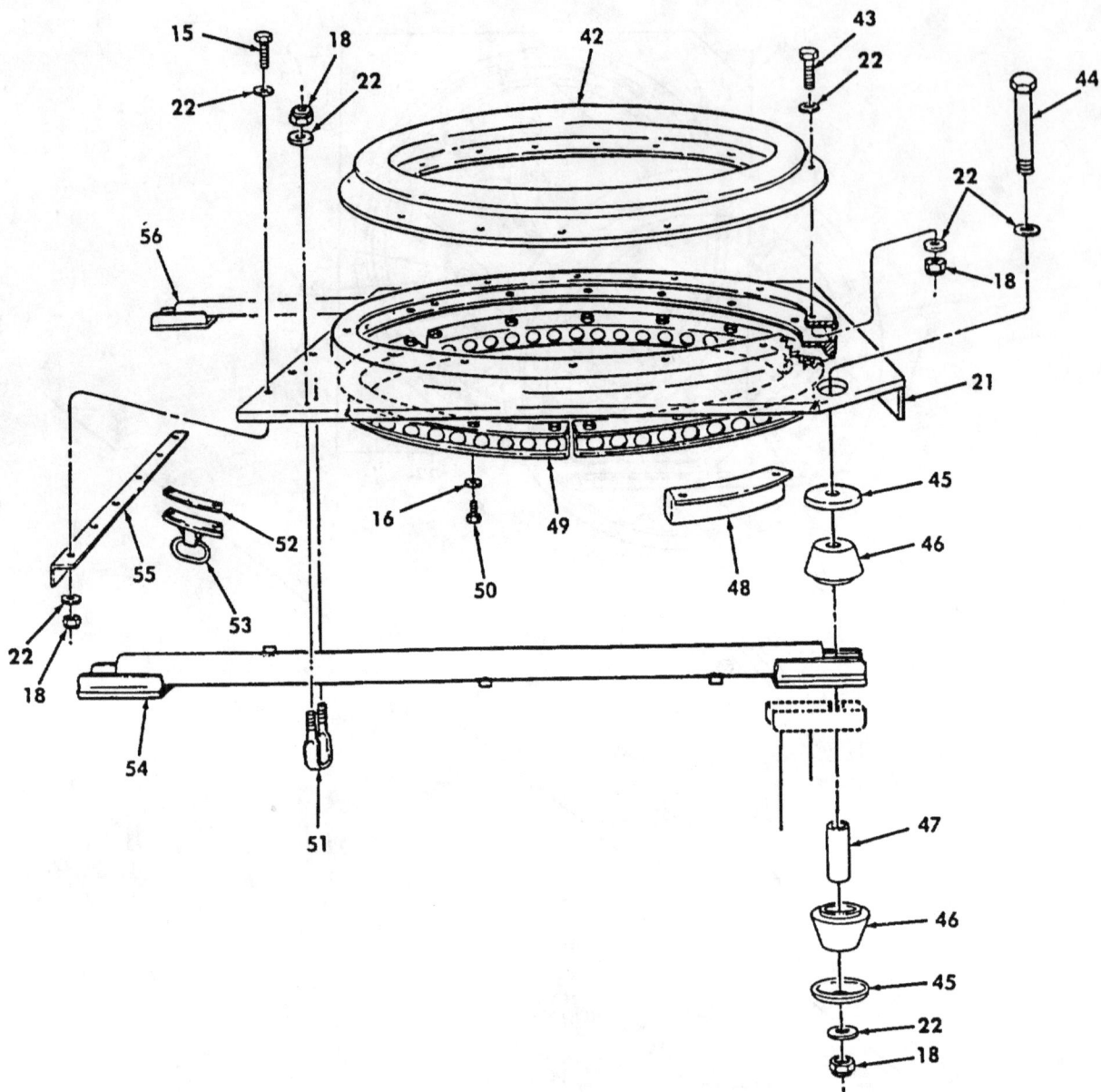

Figure 602. Lightweight Weapon Station Kit (Sheet 3 of 3).

(1) ITEM NO	(2) SMR CODE	(3) NSN	(4) CAGEC	(5) PART NUMBER	(6) DESCRIPTION AND USABLE ON CODES (UOC)	(7) QTY
					GROUP 3307 SPECIAL PURPOSE KITS	
					FIG. 602 LIGHTWEIGHT WEAPONS STATION KIT	
1	PAOZZ	5305006800509	80204	B1821BH025C125N	SCREW, CAP, HEXAGON H PART OF KIT P/N 57K0300 .... UOC:DAA, DAB, DAC, DAD, DAG, DAH, DAW, DAX, V12, V13, V14, V15, V16, V17, V21, V22, ZAA, ZAB, ZAC, ZAD, ZAG, ZAH	8
2	PAOZZ	5310005825965	96906	MS35338-14	WASHER, LOCK PART OF KIT P/N 57K0300 .... UOC:DAA, DAB, DAC, DAD, DAG, DAH, DAW, DAX, V12, V13, V14, V15, V16, V17, V21, V22, ZAA, ZAB, ZAC, ZAD, ZAG, ZAH	8
3	PAOZZ	5340044066993	19207	12300707	PANEL PART OF KIT P/N 57K0300 .... UOC:DAA, DAB, DAC, DAD, DAG, DAH, DAW, DAX, V12, V13, V14, V15, V16, V17, V21, V22, ZAA, ZAB, ZAC, ZAD, ZAG, ZAH	1
4	PAOZZ	1005014575812	19207	12450078	POST ASSEMBLY PART OF KIT P/N 57K0300 .... UOC:DAA, DAB, DAC, DAD, DAG, DAH, DAW, DAX, V12, V13, V14, V15 , V16, V17, V21, V22, ZAA, ZAB, ZAC, ZAD, ZAG, ZAH	4
5	PAOZZ	5315001879377	96906	MS24665-317	PIN, COTTER PART OF KIT P/N 57K0300 .... UOC:DAA, DAB, DAC, DAD, DAH, DAW, DAX, V12, V13, V14, V15, V16, V17, V21, V22, ZAA, ZAB, ZAC, ZAD, ZAG, ZAH	8
6	PAOZZ	5306011305994	19207	11609677	BOLT, U PART OF KIT P/N 57K0300.................. UOC:DAA, DAB, DAC, DAD, DAG, DAH, DAW, DAX, V12, V13, V14, V15 , V16, V17, V21, V22, ZAA, ZAB, ZAC, ZAD, ZAG, ZAH	4
7	PAOZZ	5315005237556	19204	5237556	PIN, STRAIGHT, HEADLE PART OF KIT P/N 57K0300 .... UOC:DAA, DAB, DAC, DAD, DAG, DAH, DAW, DAX, V12, V13, V14, V15 , V16, V17, V21, V22, ZAA, ZAB, ZAC, ZAD, ZAG, ZAH	4
8	PAOZZ	5310004883888	96906	MS51943-40	NUT, PLAIN, HEXAGON PART OF KIT P/N 57K0300 . UOC:DAA, DAB, DAC, DAD, DAG, DAH, DAW, DAX, V12, V13, V14, V15, V16, V17, V21, V22, ZAA, ZAB, ZAC, ZAD, ZAG, ZAH	18
9	PAOZZ	5340014577340	19207	12375557	PLATE, TAPPING PART OF KIT P/N .... 57K0300 . UOC:DAA, DAB, DAC, DAD, DAG, DAH, DAW, DAX, V12, V13, V14, V15, V16, V17, V21, V22, ZAA, ZAB, ZAC, ZAD, ZAG, ZAH	2
10	PAOZZ	2590014589661	19207	12375562	BRACKET, GUN MOUNT LEFT/FRONT PART OF KIT P/N 57K0300.................. UOC:DAA, DAB, DAC, DAD, DAG, DAH, DAW, DAX, V12, V13, V14, V15, V16, V17, V21, V22, ZAA, ZAB, ZAC, ZAD, ZAG, ZAH	1
10	PAOZZ	2590014589665	19207	12450083	BRACKET, GUN MOUNT RIGHT/FRONT PART OF KIT P/N 57K0300	1

(1) ITEM NO	(2) SMR CODE	(3) NSN	(4) CAGEC	(5) PART NUMBER	(6) DESCRIPTION AND USABLE ON CODES (UOC)	(7) QTY
					UOC:DAA, DAB, DAC, DAD, DAG, DAH, DAW, DAX, V12, V13, V14, V15, V16, V17, V21, V2, V22, ZAA, ZAB, ZAC, ZAD, ZAG, ZAH	
11	PAOZZ	5306011646329	96906	MS90727-57L	BOLT, SELF-LOCKING L.H. PART OF KIT P/N 57K0300	3
					UOC:DAA, DAB, DAC, DAD, DAG, DAH, DAW, DAX, V12, V13, V14, V15, V16, V17, V21, V22, ZAA, ZAB, ZAC, ZAD, ZAG, ZAH	
11	PAOZZ	5306011184888	96906	MS90727-59L	BOLT, SELF-LOCKING R.H. PART OF KIT P/N 57K0300	4
					UOC:DAA, DAB, DAC, DAD, DAG, DAH, DAW, DAX, V12, V13, V14, V15, V16, V17, V21, V22, ZAA, ZAB, ZAC, ZAD, ZAG, ZAH	
12	PAOZZ	2590014579019	19207	12375563	BRACKET, GUN MOUNT L.H. PART OF KIT P/N 57K0300	1
					UOC:DAA, DAB, DAC, DAD, DAG, DAH, DAW, DAX, V12, V13, V14, V15, V16, V17, V21, V22, ZAA, ZAB, ZAC, ZAD, ZAG, ZAH	
12	PFOZZ	2590014577394	19207	12450082	SUPPORT, POST R.H. PART OF KIT P/N 57K0300	1
					UOC:DAA, DAB, DAC, DAD, DAG, DAH, DAW, DAX, V12, V13, V14, V15, V16, V17, V21, V22, ZAA, ZAB, ZAC, ZAD, ZAG, ZAH	
13	PAOZZ	5305002693236	80204	B1821BH038F100N	SCREW, CAP, HEXAGON H PART OF KIT P/N 57K0300 .	22
					UOC:DAA, DAB, DAC, DAD, DAG, DAH, DAW, DAX, V12, V13, V14, V15, V16, V17, V21, V22, ZAA, ZAB, ZAC, ZAD, ZAG, ZAH	
14	PAOZZ	5305007195219	96906	MS90727-111	SCREW, CAP, HEXAGON H PART OF KIT P/N 57K0300	6
					UOC:DAA, DAB, DAC, DAD, DAG, DAH, DAW, DAX, V12, V13, V14, V15, V16, V17, V21, V22, ZAA, ZAB, ZAC, ZAD, ZAG, ZAH	
15	PAOZZ	5305002693239	80204	B1821BH038F138N	SCREW, CAP, HEXAGON H PART OF KIT P/N 57K0300	20
					UOC:DAA, DAB, DAC, DAD, DAG, DAH, DAW, DAX, V12, V13, V14, V15, V16, V17, V21, V22, ZAA, ZAB, ZAC, ZAD, ZAG, ZAH	
16	PAOZZ	5310006379541	96906	MS35338-46	WASHER, LOCK MOUNT PART OF KIT P/N 57K0300	26
					UOC:DAA, DAB, DAC, DAD, DAG, DAH, DAW, DAX, V12, V13, V14, V15, V16, V17, V21, V22, ZAA, ZAB, ZAC, ZAD, ZAG, ZAH	
17	PAOZZ	5310014581274	19207	12375519	NUT ASSEMBLY PART OF KIT P/N 57K0300	3
					UOC:DAA, DAB, DAC, DAD, DAG, DAH, DAW, DAX, V12, V13, V14, V15, V16, V17, V21, V22, ZAA, ZAB, ZAC, ZAD, ZAG, ZAH	
18	PAOZZ	5310008140672	96906	MS51943-36	NUT, SELF-LOCKING, HE PART OF KIT P/N 57K0300	102
					UOC:DAA, DAB, DAC, DAD, DAG, DAH, DAW, DAX, V12, V13, V14, V15, V16, V17, V21, V22, ZAA,	

(1) ITEM NO	(2) SMR CODE	(3) NSN	(4) CAGEC	(5) PART NUMBER	(6) DESCRIPTION AND USABLE ON CODES (UOC)	(7) QTY
19	PAOZZ	1005014576144	19207	12375555	ZAB, ZAC, ZAD, ZAG, ZAH BRACE ASSEMBLY L.H. PART OF KIT P/N 57K0300	1
19	PAOZZ	1005014576152	19207	12450084	UOC:DAA, DAB, DAC, DAD, DAG, DAH, DAW, DAX, V12, V13, V14, V15, V16, V17, V21, V22, ZAA, ZAB, ZAC, ZAD, ZAG, ZAH BRACE ASSEMBLY R.H. PART OF KIT P/N 57K0300	1
20	PAOZZ	5340013233081	19207	12356769	UOC:DAA, DAB, DAC, DAD, DAG, DAH, DAW, DAX, V12, V13, V14, V15, V16, V17, V21, V22, ZAA, ZAB, ZAC, ZAD, ZAG, ZAH HANDLE, MANUAL CONTR PART OF KIT P/N 57K0300	2
21	PAOZZ	1005014576176	19207	12450079	UOC:DAA, DAB, DAC, DAD, DAG, DAH, DAW, DAX, V12, V13, V14, V15, V16, V17, V21, V22, ZAA, ZAB, ZAC, ZAD, ZAG, ZAH SUPPORT ASSEMBLY PART OF KIT P/N 57K0300	124
22	PAOZZ	5310014124013	24617	2436163	UOC:DAA, DAB, DAC, DAD, DAG, DAH, DAW, DAX, V12, V13, V14, V15, V16, V17, V21, V22, ZAA, ZAB, ZAC, ZAD, ZAG, ZAH WASHER, FLAT PART OF KIT P/N 57K0300. 144	
23	PAOZZ	5305002693238	80204	B1821BH038F125N	UOC:DAA, DAB, DAC, DAD, DAG, DAH, DAW, DAX, V12, V13, V14, V15, V16, V17, V21, V22, ZAA, ZAB, ZAC, ZAD, ZAG, ZAH SCREW, CAP, HEXAGON H PART OF KIT P/N 57K0300	24
24	PAOZZ	1005014576147	19207	12450088	UOC:DAA, DAB, DAC, DAD, DAG, DAH, DAW, DAX, V12 , V13 , V14 , V15 , V16, V17 , V21, V22, ZAA, ZAB, ZAC, ZAD, ZAG, ZAH BRACE PART OF KIT P/N 57K0300 OF KIT P/N 57K0300	4
25	PAOZZ	5340014587234	19207	12450089	UOC:DAA, DAB, DAC, DAD, DAG, DAH, DAW, DAX, V12, V13, V14, V15, V16, V17, V21, V22, ZAA, ZAB, ZAC, ZAD, ZAG, ZAH SUPPORT, AMMO TRAY PART OF KIT P/N 57K0300	4
26	PAOZZ	1010014137167	19207	12251616	UOC:DAA, DAB, DAC, DAD, DAG, DAH, DAW, DAX, V12, V13, V14, V15, V16, V17, V21, V22, ZAA, ZAB, ZAC, ZAD, ZAG, ZAH TRAY, LOADING, AMMUNI PART OF KIT P/N 57K0300	2
27	PAOZZ	5340007533742	19204	8690471	UOC:DAA, DAB, DAC, DAD, DAG, DAH, DAW, DAX, V12, V13, V14, V15, V16, V17, V21, V22, ZAA, ZAB, ZAC, ZAD, ZAG, ZAH STRAP, WEBBING PART OF KIT P/N 57K0300	4
28	PAOZZ	5305009846194	96906	MS35206-246	UOC:DAA, DAB, DAC, DAD, DAG, DAH, DAW, DAX, V12, V13, V14, V15, V16, V17, V21, V22, ZAA, ZAB, ZAC, ZAD, ZAG, ZAH SCREW, MACHINE PART OF KIT P/N 57K0300	1

(1) ITEM NO	(2) SMR CODE	(3) NSN	(4) CAGEC	(5) PART NUMBER	(6) DESCRIPTION AND USABLE ON CODES (UOC)	(7) QTY
					UOC:DAA, DAB, DAC, DAD, DAG, DAH, DAW, DAX, V12, V13, V14, V15, V16, V17, V21, V22, ZAA, ZAB, ZAC, ZAD, ZAG, ZAH	
29	PAOZZ	5315014584490	19207	12340089-1	PIN ASSEMBLY PART OF KIT P/N 57K0300	1
					UOC:DAA, DAB, DAC, DAD, DAG, DAH, DAW, DAX, V12, V13, V14, V15, V16, V17, V21, V22, ZAA, ZAB, ZAC, ZAD, ZAG, ZAH	
30	PAOZZ	5310012531618	96906	MS51412-18	WASHER, FLAT PART OF KIT P/N 57K0300	1
					UOC:DAA, DAB, DAC, DAD, DAG, DAH, DAW, DAX, V12, V13, V14, V15, V16, V17, V21, V22, ZAA, ZAB, ZAC, ZAD, ZAG, ZAH	
31	PAOZZ	5310008071466	96906	MS21042-08	NUT PART OF KIT P/N 57K0300	1
					UOC:DAA, DAB, DAC, DAD, DAG, DAH, DAW, DAX, V12, V13, V14, V15, V16, V17, V21, V22, ZAA, ZAB, ZAC, ZAD, ZAG, ZAH	
32	PAOZZ	5305014077186	80204	B1821BH038F450N	SCREW, CAP, HEXAGON H PART OF KIT P/N 57K0300	2
					UOC:DAA, DAB, DAC, DAD, DAG, DAH, DAW, DAX, V12, V13, V14, V15, V16, V17, V21, V22, ZAA, ZAB, ZAC, ZAD, ZAG, ZAH	
33	PAOZZ	5305002693232	80204	B1821BH038F056N	SCREW, CAP, HEXAGON H PART OF KIT P/N 57K0300	4
					UOC:DAA, DAB, DAC, DAD, DAG, DAH, DAW, DAX, V12, V13, V14, V15, V16, V17, V21, V22, ZAA, ZAB, ZAC, ZAD, ZAG, ZAH	
34	PAOZZ	1005014576185	19207	12450086	PANEL, ARMAMENT MOUN PART OF KIT P/N 57K0300	1
					UOC:DAA, DAB, DAC, DAD, DAG, DAH, DAW, DAX, V12, V13, V14, V15, V16, V17, V21, V22, ZAA, ZAB, ZAC, ZAD, ZAG, ZAH	
35	PAOZZ	5305002693243	80204	B1821BH038F225N	SCREW, CAP, HEXAGON H PART OF KIT P/N 57K0300	6
					UOC:DAA, DAB, DAC, DAD, DAG, DAH, DAW, DAX, V12, V13, V14, V15, V16, V17, V21, V22, ZAA, ZAB, ZAC, ZAD, ZAG, ZAH	
36	PAOZZ	5306000501238	96906	MS90727-32	BOLT, MACHINE PART OF KIT P/N 57K0300	4
					UOC:DAA, DAB, DAC, DAD, DAG, DAH, DAW, DAX, V12, V13, V14, V15, V16, V17, V21, V22, ZAA, ZAB, ZAC, ZAD, ZAG, ZAH	
37	PAOZZ	5340014571778	19207	12446730	HANDLE PART OF KIT P/N 57K0300	1
					UOC:DAA, DAB, DAC, DAD, DAG, DAH, DAW, DAX, V12, V13, V14, V15, V16, V17, V21, V22, ZAA, ZAB, ZAC, ZAD, ZAG, ZAH	
38	PAOZZ	1005014575807	19207	12446728	SUPPORT ASSY, ARMAME PART OF KIT P/N 57K0300	1
					UOC:DAA, DAB, DAC, DAD, DAG, DAH, DAW, DAX, V12, V13, V14, V15, V16, V17, V21, V22, ZAA, ZAB, ZAC, ZAD, ZAG, ZAH	
39	PAOZZ	5310009353750	96906	MS51943-4	NUT, SELF-LOCKING, HE PART OF KIT P/N 57K0300	4
					UOC:DAA, DAB, DAC, DAD, DAG, DAH, DAW, DAX, V12, V13, V14, V15, V16, V17, V21, V22, ZAA,	

(1) ITEM NO	(2) SMR CODE	(3) NSN	(4) CAGEC	(5) PART NUMBER	(6) DESCRIPTION AND USABLE ON CODES (UOC)	(7) QTY
40	PAOZZ	4710011899727	19207	12340337	ZAB, ZAC, ZAD, ZAG, ZAH TUBE ASSEMBLY, WEAPO PART OF KIT P/N 57K0300	1
41	PAOZZ	5305002693241	80204	B1821BH03BF175N	UOC:DAA, DAB, DAC, DAD, DAG, DAH, DAW, DAX, V12, V13, V14, V15, V16, V17, V, V21, V22, ZAA, ZAB, ZAC, ZAD, ZAG, ZAH SCREW, CAP, HEXAGON H PART OF KIT P/N 57K0300	2
42	PAOZZ	3110011853114	19207	12340302	UOC:DAA, DAB, DAC, DAD, DAG, DAH, DAW, DAX, V12, V13 , V14 , V15, V16, V17, V21, V22, ZAA, ZAB, ZAC, ZAD, ZAG, ZAH BEARING ASSEMBLY, TU PART OF KIT P/N 57K0300	1
43	PAOZZ	5305002693240	80204	B1821BH038F150N	UOC:DAA, DAB, DAC, DAD, DAG, DAH, DAW, DAX, V12, V13 , V14 , V15, V16, V17, V21, V22, ZAA, ZAB, ZAC, ZAD, ZAG, ZAH SCREW, CAP, HEXAGON H PART OF KIT P/N 57K0300	12
44	PAOZZ	5305002693248	80204	B1821BH038F350N	UOC:DAA, DAB, DAC, DAD, DAG, DAH, DAW, DAX, V12, V13 , V14 , V15, V16, V17, V21, V22, ZAA, ZAB, ZAC, ZAD, ZAG, ZAH SCREW, CAP, HEXAGON H PART OF KIT P/N 57K0300	4
45	PAOZZ	2540011925949	19207	12339561	UOC:DAA, DAB, DAC, DAD, DAG, DAH, DAW, DAX, V12, V13 , V14 , V15, V16, V17, V21, V22, ZAA, ZAB, ZAC, ZAD, ZAG, ZAH PLATE, SNUBBER, VEHIC PART OF KIT P/N 57K0300	8
46	PAOZZ	1005014576178	19207	12450076	UOC:DAA, DAB, DAC, DAD, DAG, DAH, DAW, DAX, V12, V13 , V14 , V15, V16, V17, V21, V22, ZAA, ZAB, ZAC, ZAD, ZAG, ZAH MOUNT, ROOF PART OF KIT P/N 57K0300	1
47	PAOZZ	5365014576700	19207	12338186-67	UOC:DAA, DAB, DAC, DAD, DAG, DAH, DAW, DAX, V12, V13 , V14 , V15 , V16, V17, V21, V22, ZAA, ZAB, ZAC, ZAD, ZAG, ZAH SPACER PART OF KIT P/N 57K0300	4
48	PAOZZ	2590012047836	9207	12340313	UOC:DAA, DAB, DAC, DAD, DAG, DAH, DAW, DAX, V12, V13 , V14 , V15 , V16, V17, V21, V22, ZAA, ZAB, ZAC, ZAD, ZAG, ZAH PAD, CUSHIONING PART OD KIT P/N 57K0300	2
49	PAOZZ	2590011924525	19207	12340298	UOC:DAA, DAB, DAC, DAD, DAG, DAH, DAW, DAX, V12 , V13, V14 , V15 , V16, V17, V21, V22, ZAA, ZAB, ZAC, ZAD, ZAG, ZAH RING, TURRET LOCK PART OF KIT P/N 57K0300	3
50	PAOZZ	5305000680510	80204	B1821BH038C100N	UOC:DAA, DAB, DAC, DAD, DAG, DAH, DAW, DAX, SCREW, CAP, HEXAGON H PART OF KIT P/N 57K0300 UOC:DAA, DAB, DAC, DAD, DAG, DAH, DAW, DAX,	18

(1) ITEM NO	(2) SMR CODE	(3) NSN	(4) CAGEC	(5) PART NUMBER	(6) DESCRIPTION AND USABLE ON CODES (UOC)	(7) QTY
					V12, V13, V14, V15, V16, V17, V21, V22, ZAA, ZAB, ZAC, ZAD, ZAG, ZAH	
51	PAOZZ		19207	12450097	U-BOLT PART OF KIT P/N 57K0300 .... UOC:DAA, DAB, DAC, DAD, DAG, DAH, DAW, DAX, V12, V13, V14, V15, V16, V17, V21, V22, ZAA, ZAB, ZAC, ZAD, ZAG, ZAH	7
52	PAOZZ	5365014573364	19207	12340115-1	SPACER .050 PART OF KIT P/N 57K0300 UOC:DAA, DAB, DAC, DAD, DAG, DAH, DAW, DAX, V12, V13, V14, V15, V16, V17, V21, V22, ZAA, ZAB, ZAC, ZAD, ZAG, ZAH	3
52	PAOZZ		19207	12340115-2	SPACER .160 PART OF KIT P/N 57K0300 UOC:DAA, DAB, DAC, DAD, DAG, DAH, DAW, DAX, V12, V13, V14, V15, V16, V17, V21, V22, ZAA, ZAB, ZAC, ZAD, ZAG, ZAH	1
53	PAOZZ	5340012538933	31272	45460-10	LOCK SET, RIM PART OF KIT P/N 57K0300 UOC:DAA, DAB, DAC, DAD, DAG, DAH, DAW, DAX, V12, V13, V14, V15, V16, V17, V21, V22, ZAA, ZAB, ZAC, ZAD, ZAG, ZAH	2
54	PAOZZ	1005014576135	19207	12450077-1	CROSSMEMBER ASSY PART OF KIT P/N 57K0300 .... UOC:DAA, DAB, DAC, DAD, DAG, DAH, DAW, DAX, V12, V13, V14, V15, V16, V17, V21, V22, ZAA, ZAB, ZAC, ZAD, ZAG, ZAH	1
55	PAOZZ	1005014576150	19207	12450087	REINFORCEMENT PART OF KIT P/N .... 57K0300 UOC:DAA, DAB, DAC, DAD, DAG, DAH, DAW, DAX, V12, V13, V14, V15, V16, V17, V21, V22, ZAA, ZAB, ZAC, ZAD, ZAG, ZAH	2
56	PAOZZ	1005014576138	19207	12450077-2	CROSSMEMBER ASSY PART OF KIT P/N 57K0300 UOC:DAA, DAB, DAC, DAD, DAG, DAH, DAW, DAX, V12, V13, V14, V15, V16, V17, V21, V22, ZAA, ZAB, ZAC, ZAD, ZAG, ZAH	1
	PFOZZ	1005014323339	19207	57K0300	MODIFICATION KIT, LT WPN STA .... UOC:DAA, DAB, DAC, DAD, DAG, DAH, DAW, DAX, V12, V13, V14, V15, V16, V17, V21, V22, ZAA, ZAB, ZAC, ZAD, ZAG, ZAH	1

BEARING ASSEMBLY	(1)	602-42
BRACE	(4)	602-24
BRACE ASSEMBLY	(1)	602-19
BRACE ASSEMBLY	(1)	602-19
BRACKET, GUN MOUNT	(1)	602-10
BRACKET, GUN MOUNT	(1)	602-12
BOLT, MACHINE	(4)	602-36
BOLT, SELF-LOCKING	(3)	602-11
BOLT, SELF-LOCKING	(4)	602-11
BOLT, U	(4)	602-6
BOLT, U	(7)	602-51
CROSSMEMBER ASSY	(1)	602-54
CROSSMEMBER ASSY	(1)	602-56
HANDLE	(2)	602-37

(1) ITEM NO	(2) SMR CODE	(3) NSN	(4) CAGEC	(5) PART NUMBER	(6) DESCRIPTION AND USABLE ON CODES (UOC)			(7) QTY
					HANDLE, MANUAL	(	1)	602-20
					LOCK SET, RIM	(	2)	602-53
					MOUNT, ROOF	(	1)	602-46
					NUT	(	1)	602-31
					NUT ASSEMBLY	(	3)	602-17
					NUT, PLAIN, HEXAGON	(	18)	602-8
					NUT, SELF-LOCKING	(	102)	602-18
					NUT, SELF-LOCKING	(	4)	602-39
					PAD, CUSHIONING	(	2)	602-48
					PANEL, ARMAMENT MOUN	(	1)	602-34
					PIN ASSEMBLY	(	1)	602-29
					PIN, COTTER	(	8)	602-5
					PLATE, MOUNTING	(	1)	602-3
					PLATE, SNUBBER, VEHIC	(	8)	602-45
					PLATE, TAPPING	(	2)	602-9
					POST ASSEMBLY	(	4)	602-4
					REINFORCEMENT	(	2)	602-55
					RING, TURRET LOCK	(	3)	602-49
					PANEL	(	1)	602-22
					PIN, STRAIGHT	(	1)	602-7
					SCREW, CAP, HEXAGON, H	(	8)	602-1
					SCREW, CAP, HEXAGON, H	(	22)	602-13
					SCREW, CAP, HEXAGON, H	(	6)	602-14
					SCREW, CAP, HEXAGON, H	(	20)	602-15
					SCREW, CAP, HEXAGON, H	(	24)	602-23
					SCREW, CAP, HEXAGON, H	(	2)	602-32
					SCREW, CAP, HEXAGON, H	(	4)	602-33
					SCREW, CAP, HEXAGON, H	(	6)	602-35
					SCREW, CAP, HEXAGON, H	(	2)	602-41
					SCREW, CAP, HEXAGON, H	(	12)	602-43
					SCREW, CAP, HEXAGON, H	(	4)	602-44
					SCREW, CAP, HEXAGON, H	(	18)	602-50
					SCREW, MACHINE	(	1)	602-28
					SPACER	(	4)	602-47
					SPACER	(	3)	602-52
					SPACER	(	1)	602-52
					STRAP, WEBBING	(	4)	602-27
					SUPPORT, AMMO TRAY	(	4)	602-25
					SUPPORT ASSEMBLY	(	1)	602-21
					SUPPORT ASSY, ARMAME	(	1)	602-38
					SUPPORT, POST	(	1)	602-10
					SUPPORT, POST	(	1)	602-12
					TRAY, LOADING, AMMUNI	(	1)	602-26
					TUBE ASSEMBLY	(	1)	602-40
					WASHER, FLAT	(144)		602-22
					WASHER, FLAT	(	1)	602-30
					WASHER, LOCK	(	8)	602-2
					WASHER, LOCK	(	26	602-16

END OF FIGURE

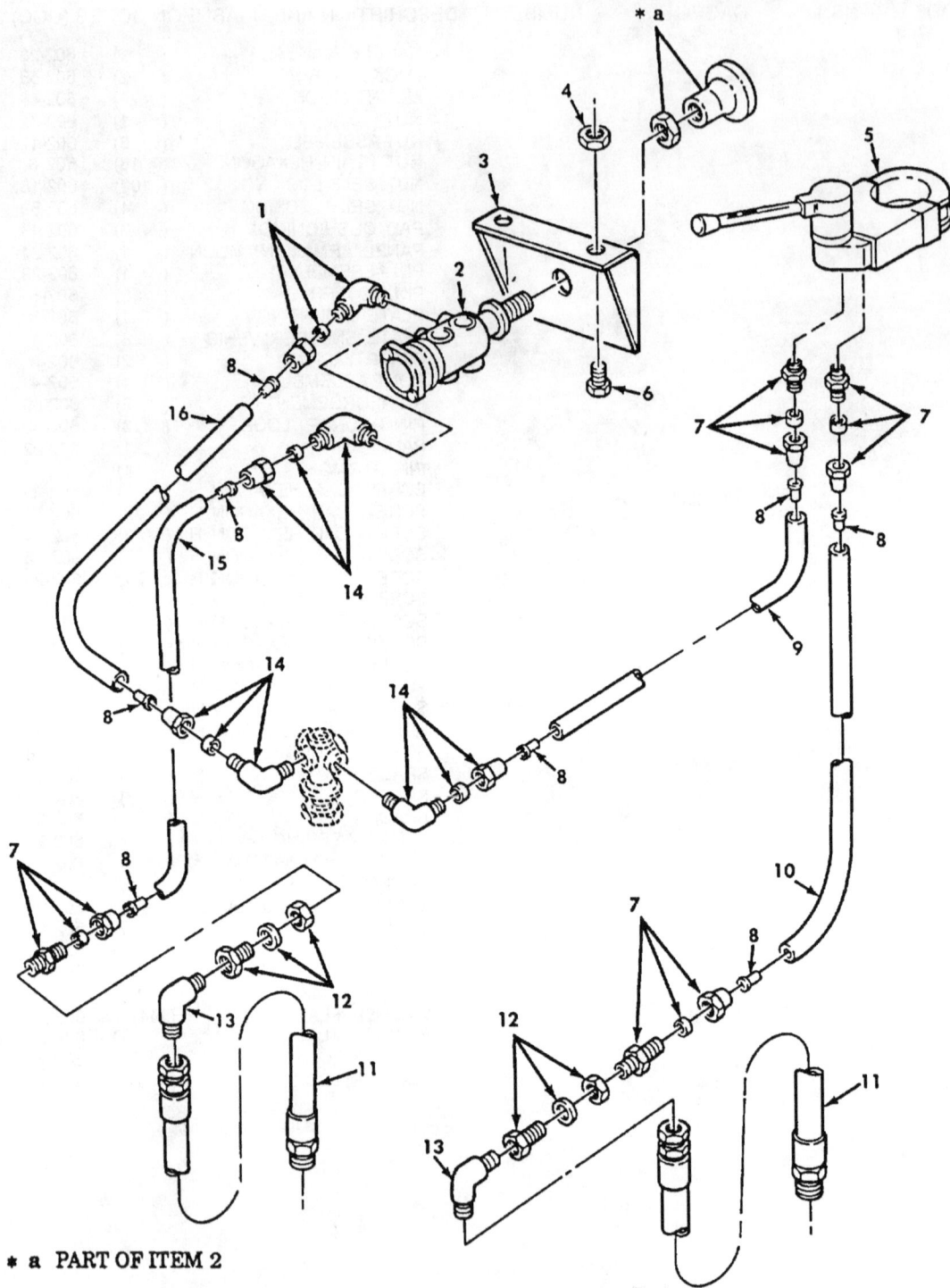

* a PART OF ITEM 2

Figure 603. Air Brake Kit, Control Valves, and Lines.

(1) ITEM NO	(2) SMR CODE	(3) NSN	(4) CAGEC	(5) PART NUMBER	(6) DESCRIPTION AND USABLE ON CODES (UOC)	(7) QTY
					GROUP 3307 SPECIAL PURPOSE KITS	
					FIG. 603 AIR BRAKE KIT, CONTROL VALVES, AND LINES	
1	PAOZZ	4730002871604	81343	6-2 120202BA	ELBOW, PIPE TO TUBE PART OF KIT P/N 12256301 UOC:DAA,DAB,DAC,DAD,DAE,DAF,DAJ,DAK, DAW,DAX,V12,V13,V14,V15,V16,V17,V19, V20,V24,V25,V39,ZAA,ZAB,ZAC,ZAD,ZAE, ZAF,ZAJ,ZAK	1
2	PAOZZ	4810011079694	19207	11669066	VALVE ASSEMBLY, WITH PART OF KIT P/N 12256301 UOC:DAA,DAB,DAC,DAD,DAE,DAF,DAJ,DAK, DAL,DAW,DAX,V12,V13,V14,VVS,V16,V17, V18,Vi9,V20,V24,V25,V39,ZAA,ZAB,ZAC, ZAD, ZAE, ZAF,ZAG, ZAH,ZAJ,ZAK,ZAL	1
3	PAOZZ	5340011079693	19207	12255848	BRACKET PART OF KIT P/N 12256301 UOC:DAA,DAB,DAC,DAD,DAE,DAF,DAJ,DAK, DAL,DAW,DAX,V12,V13,V14,V15 ,V16 ,V17, V1 B,V19,V20 ,V24 ,V25,V39,ZAA,ZAB,ZAC, ZAD,ZAE,ZAF,ZAG,ZAH,ZAJ,ZAK,ZAL	1
4	PAOZZ	5310008140672	96906	MS51943-36	NUT, SELF-LOCKING, HE PART OF KIT P/N 12256301 UOC:DAA,DAB,DAC,DAD,DAE,DAF,DAJ,DAK, DAW,DAX,V12,V13,V14,V15,V16,V17,V19, V20,V24,V25,V39,ZAA,ZAB,ZAC, ZAD,ZAE, ZAF,ZAJ,ZAK	2
5	PAOZZ	2530000629719	19207	11602159	VALVE, ROTARY, DIRECT PART OF KIT P/N 12256301 UOC:DAA,DAB,DAC,DAD,DAE,DAF,DAJ,DAK, DAW,DAX,V12,V13,V14,V15,V16,V17,V19, V20,V24,V25,V39,ZAA,ZAB,ZAC,ZAD,ZAE, ZAF,ZAJ,ZAK	1
6	PAOZZ	5305002692803	96906	MS90726-60	SCREW, CAP, REXAGON H PART OF KIT P/N 12256301 UOC:DAA,DAB,DAC,DAD,DAE,DAF,DAJ,DAK, DAW,DAX,V12,V13,V14,V15,V16,V17,V19, V20,V24,V25,V39,ZAA,ZAB,ZA C,ZAD,ZAE, ZAF,ZAJ,ZAK	2
7	PAOZZ	4730000691186	96906	MS39179-5	ADAPTER, STRAIGHT PART OF KIT P/N 12256301 UOC:DAA,DAB,DAC,DAD,DAE,DAF,DAJ,DAK, DAW,DAX,V12,V13,V14,V15 ,V16,V17,V19, V20,V24,V25,V39	4
8	PAOZZ	4730010798821	19207	CPR102321-1	INSERT, TUBE FITTING PART OF KIT P/N 12256301 UOC:DAA,DAB,DAC,DAD,DAE, DAF , DAJ,DAK, DAW,DAX,V12,V13,V14,V15,V16,V17,V19, V20,V24,V25,V39,ZAA,ZAB,ZAC,ZAD,ZAE, ZAF,ZAJ,ZAK	8

(1) ITEM NO	(2) SMR CODE	(3) NSN	(4) CAGEC	(5) PART NUMBER	(6) DESCRIPTION AND USABLE ON CODES (UOC)	(7) QTY
9	MOOZZ		19207	CPR104420-2-34	TUBING, NONMETALLIC MAKE FROM................................ TUBING, P/N CPR104420-2, 34 INCHES LONG... UOC:DAA,DAB,DAC,DAD,DAE,DAF,DAJ,DAK, DAW,DAX,V12,V13,V14,V15,V16,V17,V19, V20,V24,V25,V39,ZAA,ZAB,ZAC,ZAD,ZAE, ZAF,ZAJ,ZAK	1
10	MOOZZ		19207	CPR104420-2-22	V20,V24,V25,V39,ZAA,ZAB,ZAC,ZAD,ZAE, TUBING, NONMETALLIC MAKE FROM................................ TUBING, P/N CPR104420-2, 22 INCHES LONG... UOC:DAA,DAB,DAC,DAD,DAE,DAF,DAJ,DAR, DAW,DAX,V12,V13,V14,V15,V16,V17,V19, V20,V24,V25,V39,ZAA,ZAB,ZAC,ZAD,ZAE, ZAF,ZAJ,ZAK	1
11	PAOZZ	4720011079939	19207	12256272	HOSE ASSEMBLY, NONME PART OF KIT P/N .... 12256301 UOC:DAA,DAB,DAC,DAD,DAE,DAF,DAJ,DAK, DAW,DAX,V12, V13,V14,V15,V16,V17 ,V19, V20,V24,V25,V39,ZAA,ZAB,ZAC,ZAD,ZAE, ZAF,ZAJ,ZAK	2
12	PAOZZ	4730002786318	19207	8328782	COUPLING, PIPE PART OF KIT P/N .... 12256301.. UOC:DAA,DAB,DAC,DAD,DAE,DAF,DAJ,DAK, DAW,DAX,V12,V13,V14,V15 ,V16 , ,17,V9, V20,V24,V25,V39,ZAA,ZAB,ZAC,ZAD,ZAE, ZAF,ZAJ,ZAK	2
13	PAOZZ	4730008127999	96906	MS51504A6	ELBOW, PIPE TO TUBE PART OF KIT P/N .... 12256301 .... UOC:DAA,DAB,DAC,DAD,DAE,DAF,DAJ,DAK, DAW,DAX,V12,V13,V14,V15,V16,7,1,V9, V20,V24,V25,V39,ZAA,ZAB,ZAC,ZAD,ZAE, ZAF,ZAJ,ZAK	2
14	PAOZZ	4730000691187	81343	6-4 100202BA	ELBOW, PIPE TO TUBE PART OF KIT P/N .... 12256301 .... UOC:DAA,DAB,DAC,DAD,DAE,DAF,DAJ,DAK, DAW,DAX,V12,V13,V14,V15,V16,V1 7 ,V1 9, V20,V24,V25,V39,ZAA,ZAB,ZAC,ZAD,ZAE, ZAF,ZAJ,ZAK	3
15	MOOZZ		19207	CPR104420-2-7	TUBING, NON METALLIC MAKE FROM................................ TUBING,P/N CPR104420-2, 7 INCHES LONG .... UOC:DAA,DAB,DAC,DAD,DAE,DAF,DAJ,DAK, DAW,DAX,V12,V13,V14,V15,V16,V17,V19, V20,V24,V25,V39,ZAA,ZAB,ZAC,ZAD,ZAE, ZAF,ZAJ,ZAK	1
16	MOOZZ		19207	CPR104420-2-26	TUBING, NONMETALLIC MAKE FROM................................ TUBING,P/N CPR104420-2, 26 INCHES LONG .... UOC:DAA,DAB,DAC,DAD,DAE,DAF,DAJ,DAK, DAW,DAX,V12,V13,V14 ,V15,V16,V17,V19, V20,V24,V25,V39,ZAA,ZAB,ZAC,ZAD,ZAE, ZAF,ZAJ,ZAK	1

END OF FIGURE

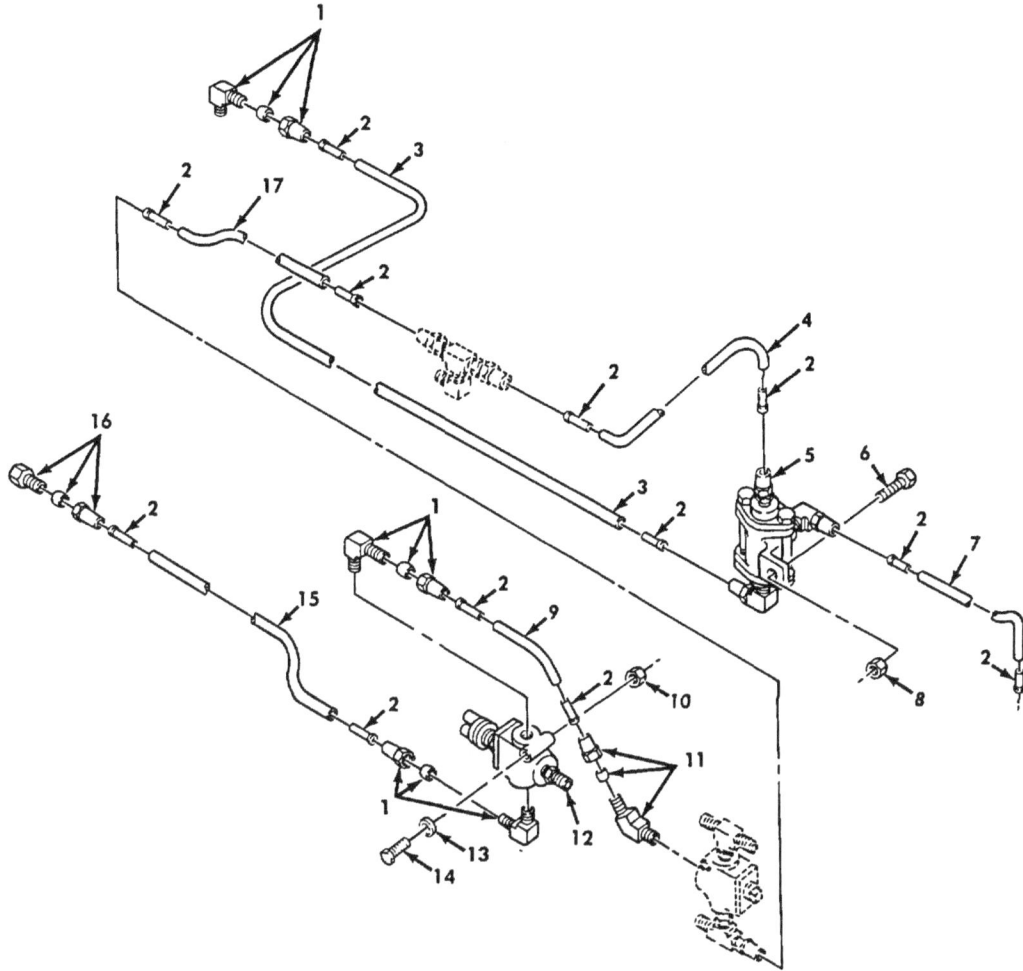

Figure 604. Air Brake Kit, Air Lines, and Double-check Values.

(1) ITEM NO	(2) SMR CODE	(3) NSN	(4) CAGEC	(5) PART NUMBER	(6) DESCRIPTION AND USABLE ON CODES (UOC)	(7) QTY
					GROUP 3307 SPECIAL PURPOSE KITS	
					FIG. 604 AIR BRAKE KIT, AIR LINES, AND DOUBLE-CHECK VALVES	
1	PAOZZ	4730001439282	02570	MS39182-8	ELBOW, PIPE TO TUBE PART OF KIT P/N .... 12256301 .... UOC:DAA,DAB,DAC,DAD,DAE,DAF,DAJ,DAK, DAW,DAX,V12,V13,V14,V15,V16,V17,V1 9, V20,V24,V25,V39,ZAA,ZAB,ZAC,ZAD,ZAE, ZAF,ZAJ,ZAK	3
2	PAOZZ	4730010798821	19207	CPR102321-1	INSERT, TUBE FITTING PART OF KIT P/N .... 12256301 .... UOC:DAA,DAB,DAC,DAD,DAE,DAF,DAJ,DAK, DAW,DAX,V12,V13,V14,V15,V16,V17,V1 9, V20,V24,V25,V39,ZAA,ZAB,ZAC,ZAD,ZAE, ZAF,ZAJ,ZAK	12
3	MFFZZ		19207	CPR104420-2-268	HOSE, NONMETALLIC MAKE FROM TUBING, ... P/N CPR104420-2,268 INCHES LONG............... UOC:DAA,DAB,DAC,DAD,DAE,DAF,DAJ,DAK, DAL,DAW,DAX,V12,V13,V14 5,V16,V17, V18,V19,V20,V21,V22,V24,V25,V39,ZAA, ZAB,ZAC,ZAD,ZAE,ZAF,ZAG,ZAH,ZAJ,ZAK, ZAL	1
4	MOOZZ		19207	CPR104420-2-90	TUBING, NONMETALLIC MAKE FROM............... TUBING P/N CPR104420-2, 90 INCHES LONG............... UOC:DAA,DAB,DAC,DAD,DAE,DAF,DAJ,DAK, DAW,DAX,V12,V13,V14,V15,V16,V17V17,V9, V20,V24,V25,V39,ZAA,ZAB,ZAC,ZAD,ZAE, ZAF,ZAJ,ZAK	1
5	PAOZZ	4820004098935	19207	5340153	VALVE, SHUTTLE PART OF KIT P/N .... 12256301 .... UOC:DAA,DAB,DAC,DAD,DAE,DAF,DAJ,DAK, DAW,DAX,V12,V13,V14,V15,,V16,V17,V19, V20,V24,V25,V39,ZAA,ZAB,ZAC,ZAD,ZAE, ZAF,ZAJ,ZAK	1
6	PAOZZ	5305002259091	96906	MS90726-34	BOLT, MACHINE PART OF KIT P/N .... 12256301 .... UOC:DAA,DAB,DAC,DAD,DAE,DAF,DAJ,DAK, DAW,DAX,V12,V13,V1,V,V15,V16,V17,V9, V20,V24,V25,V39,ZAA,ZAB,ZAC,ZAD,ZAE, ZAF,ZAJ,ZAK	1
7	MOOZZ		19207	CPR104420-2-63	TUBING, NONMETALLIC MAKE FROM................................. TUBING, P/N CPR104420-2, 63 INCHES LONG............... UOC:DAA,DAB,DAC,DAD,DAE,DAF,DAJ,DAK, DAW,DAX,V12,V13,V14,V15,V16,V17,V19, V20,V24,V25,V39,ZAA,ZAB,ZAC,ZAD,ZAE, ZAF,ZAJ,ZAK	1
8	PAOZZ	5310002416658	96906	MS51943-34	NUT, SELF-LOCKING PART OF KIT P/N............................... 12256301 .... UOC:DAA,DAB,DAC,DAD,DAE,DAF,DAJ,DAK, DAW,DAX,V12,V13,V14,V15,V16,V17,V19, V20,V24,V25,V39,ZAA,ZAB,ZAC,ZAD,ZAE, ZAF,ZAJ,ZAK	1

(1) ITEM NO	(2) SMR CODE	(3) NSN	(4) CAGEC	(5) PART NUMBER	(6) DESCRIPTION AND USABLE ON CODES (UOC)	(7) QTY
9	MOOZZ		19207	CPR104420-2-9	TUBING, NONMETALLIC MAKE FROM TUBING, P/N CPR104420-2,9 INCHES LONG UOC:DAA,DAB,DAC,DAD,DÆ,DAF,DAJ,DAK, DAW,DAX,V12,V13,V14,V15,V16,V17,V19, V20,V24,V25,V39,ZAA,ZAB,ZAC,ZAD,ZAE, ZAF,ZAJ,ZAK	1
10	PAOZZ	5310008140672	96906	MS51943-36	NUT, SELF-LOCKING PART OF KIT P/N 12256301 UOC:DAA,DAB,DAC,DAD,DAE,DAF,DAJ,DAK, DAW,DAX,V12,V1,V V14,V15,V16,V17,V19, V20,V24,V25,V39,ZAA,ZAB,ZAC,ZAD,ZAE, ZAF,ZAJ,ZAK	1
11	PAOZZ	4730002890155	81343	6-6 120202BA	ELBOW, PIPE TO TUBE PART OF KIT P/N 12256301 UOC:DAA,DAB,DAC,DAD,DAE,DAF,DAJ,DAK, DAL,DAW,DAX,V12,V13,V14,V15,V16,V17, V18,V19,V20,V24,V25,V39	1
12	PAOZZ	4820007287467	06853	278614	VALVE, SHUTTLE PART OF KIT P/N 12256301 UOC:DAA,DAB,DAC,DAD,DAE,DAF,DAJ,DAK, DAW,DAX,V12,13,V14,4,V15,V6,V17,VI9, V20,V24,V25,V39,ZAA,ZAB,ZAC,ZAD,ZAE, ZAF,ZAJ,ZAK	1
13	PAOZZ	5310008093078	96906	MS27183-11	WASHER, FLAT PART OF KIT P/N 12256301 UOC:DAA,DAB,DAC,DAD,DAE,DAF,DAJ,DAK, DAW,DAX,V12,V13,V14,V15,V16,V17,V19, V20,V24,V25,V39,ZAA,ZAB,ZAC,ZAD,ZAE, ZAF,ZAJ,ZAK	1
14	PAOZZ	5305002259093	96906	MS90726-35	BOLT, MACHINE PART OF KIT P/N 12256301 UOC:DAA,DAB,DAC,DAD,DAE,DAF,DAJ,DAK, DAW,DAX,V12,V13,V14,V15,V16,V17,V19, V20,V24,V25,V39,ZAA,ZAB,ZAC,ZAD,ZAE, ZAF,ZAJ,ZAK	1
15	MOOZZ		19207	CPR104420-2-70	TUBING, NONMETALLIC MAKE FROM TUBING, P/N CPR104420-2,70 INCHES LONG UOC:DAA,DAB,DAC,DAD,DAE,DAF,DAJ,DAK, DAW,DAX,V12,V13,V14,V15,V16,V17,V19, V20,V24,V25,V39,ZAA,ZAB,ZAC,ZAD,ZAE, ZAF,ZAJ,ZAK	1
16	PAOZZ	4730002778770	96906	MS39181-4	ADAPTER, STRAIGHT PART OF KIT P/N 12256301 UOC:DAA,DAB,DAC,DAD,DAE,DAF,DAJ,DAK, DAW,DAX,V12,V13,V14,V15,V16,V17,VI9, V20,V24,V25,V39,ZAA,ZAB,ZAC,ZAD,ZAE, ZAF,ZAJ,ZAK	1
17	MOOZZ		19207	CPR104420-2-220	TUBING, NONMETALLIC MAKE FROM TUBING, P/N CPR104420-2, 220 INCHES LONG UOC:DAA,DAB,DAC,DAD,DAE,DAF,DAJ,DAK, DAW,DAX,V12,V13,V14,V15,V16,V17,V19, V20,V24,V25,V39,ZAA,ZAB,ZAC,ZAD,ZAE, ZAF,ZAJ,ZAK	1

END OF FIGURE

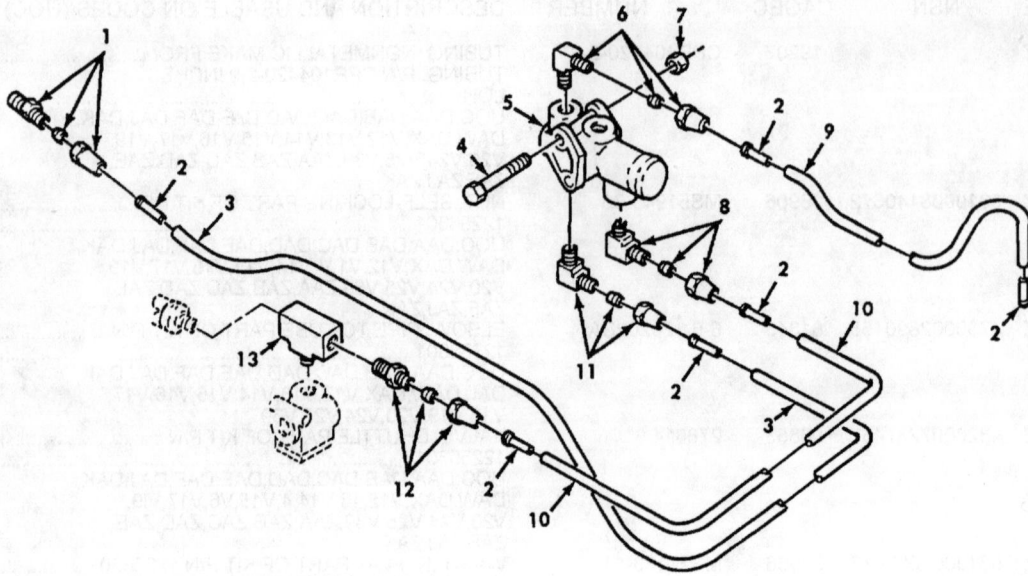

Figure 605. Air Brake Kit, Air Lines, and Tractor Protection Valve.

(1) ITEM NO	(2) SMR CODE	(3) NSN	(4) CAGEC	(5) PART NUMBER	(6) DESCRIPTION AND USABLE ON CODES (UOC)	(7) QTY
					GROUP 3307 SPECIAL PURPOSE KITS	
					FIG. 605 AIR BRAKE KIT, AIR LINES, AND TRACTOR PROTECTION VALVE	
1	PAOZZ	4730002778770	96906	MS39181-4	ADAPTER, STRAIGHT, PI PART OF KIT P/N .... 12256301 .... UOC:DAA,DAB,DAC,DAD,DAE,DAF,DAJ,DAK, DAW,DAX,V12,V13,V14,V15,V16,V17,V19, V20,V24,V25,V39,ZAA,ZAB,ZAC,ZAD,ZAE, ZAF,ZAJ,ZAK	1
2	PAOZZ	4730010798821	19207	CPR102321-1	INSERT PART OF KIT P/N 12256301 .... UOC:DAA,DAB,DAC,DAD,DAE,DAF,DAJ,DAK, DAW,DAX,V12,V13,V14,V15,V16,V17,V19, V20,V24,V25,V39,ZAA,ZAB,ZAC,ZAD,ZAE, ZAF,ZAJ,ZAK	6
3	MOOZZ		19207	CPR104420-2-130	HOSE, NONMETALLIC MAKE FROM TUBING, ... P/N CPR104420-2, 130 INCHES LONG UOC:DAA,DAB,DAC,DAD,DAE,DAF,DAJ,DAK, DAW,DAX,V12,V13,V14,V15,V16,V1V17,V9, V20,V24,V25,V39,ZAA,ZAB,ZAC,ZAD,ZAE, ZAF,ZAJ,ZAK	1
4	PAOZZ	5306002259100	96906	MS90726-45	BOLT, MACHINE PART OF KIT P/N .... 12256301 .... UOC:DAA,DAB,DAC,DAD,DAE,DAF,DAJ,DAK, DAW,DAX,V12,V13,V14,V15,V16,V17,V19, V20,V24,V25,V39,ZAA,ZAB,ZAC,ZAD,ZAE, ZAF,ZAJ,ZAK	2
5	PAOZZ	4820004363033	06853	279000	VALVE, AIR PART OF KIT P/N 12256301. .... UOC:DAA,DAB,DAC,DAD,DAE,DAF,DAJ,DAK, DAW,DAX,V12,V13,V14,V15,V16,V17,V19, V20,V24,V25,V39,ZAA,ZAB,ZAC,ZAD,ZAE, ZAF,ZAJ,ZAK	1
6	PAOZZ	4730001439282	02570	MS39182-8	ELBOW, PIPE TO TUBE PART OF KIT P/N .... 12256301 .... UOC:DAA,DAB,DAC,DAD,DAE,DAF,DAJ,DAK, DAW,DAX,V12,V13,V14,V15,V16,V17,V19, V20,V24,V25,V39,ZAA,ZAB,ZAC,ZAD,ZAE, ZAF,ZAJ,ZAK	1
7	PAOZZ	5310002416658	96906	MS51943-34	NUT, SELF-LOCKING, HE PART OF KIT P/N .... 12256301 .... UOC:DAA,DAB,DAC,DAD,DAE,DAF,DAJ,DAK, DAW,DAX,V12,V13,V14,V15,V16,V17,V19, V20,V24,V25,V39,ZAA,ZAB,ZAC,ZAD,ZAE, ZAF,ZAJ,ZAK	2
8	PAOZZ	4730002890155	81343	6-6 120202BA	ELBOW, PIPE TO TUBE PART OF KIT P/N .... 12256301 .... UOC:DAA,DAB,DAC,DAD,DAE,DAF,DAJ,DAK, DAW,DAX,V12,V13,V14,V15,V6,V1 7V1,V9, V20,V24,V25,V39,ZAA,ZAB,ZAC,ZAD,ZAE, ZAF,ZAJ,ZAK	1

(1) ITEM NO	(2) SMR CODE	(3) NSN	(4) CAGEC	(5) PART NUMBER	(6) DESCRIPTION AND USABLE ON CODES (UOC)		(7) QTY
9	MOOZZ		19207	CPR104420-2-230	HOSE, NONMETALLIC MAKE FROM TUBING.................... P/N CPR104420-2, 230 INCHES LONG UOC:DAA,DAB,DAC,DAD,DAE,DAF,DAJ,DAK, DAW,DAX,V12,V13,V14,V15,V16,V17,V19, V20,V24,V25,V39,ZAA,ZAB,ZAC,ZAD,ZAE, ZAF,ZAJ,ZAK		1
10	MOOZZ		19207	CPR104420-2-53	HOSE, NONMETALLIC MAKE FROM TUBING, ... P/N CPR104420-2, 53 INCHES LONG UOC:DAA,DAB,DAC,DAD,DAE,DAF,DAJ,DAK, DAW,DAX,V12,V13,V14,V15,V16,,V17,V9, V20,V24,V25,V39,ZAA,ZAB,ZAC,ZAD,ZAE, ZAF,ZAJ,ZAK		1
11	PAOZZ	4730000691187	81343	6-4 100202BA	ELBOW, PIPE TO TUBE PART OF KIT P/N .... 12256301 .... UOC:DAA,DAB,DAC,DAD,DAE,DAF,DAJ,DAK, DAW,DAX,V12,V13,V14,V15,V16,V17,V19, V20,V24,V25,V39,ZAA,ZAB,ZAC,ZAD,ZAE, ZAF,ZAJ,ZAK		1
12	PAOZZ	4730008371177	81343	6-8 120102BA	ADAPTER, STRAIGHT PART OF KIT P/N .... 12256301 .... UOC:DAA,DAB,DAC,DAD,DAE,DAF,DAJ,DAK, DAW,DAX,V12,V13,V14,V15,V16,V17,V19, V20,V24,V25,V39,ZAA,ZAB,ZAC,ZAD,ZAE, ZAF,ZAJ,ZAK		1
13	PAOZZ	4730012338998	19207	11648498	TEE, PIPE PART OF KIT P/N 12256301........................... UOC:DAA,DAB,DAC,DAD,DAE,DAF,DAJ,DAK, DAW,DAX,V12,V13,V14,V15,V16,V17,V19, V20,V24,V25,V39,ZAA,ZAB,ZAC,ZAD,ZAE, ZAF,ZAJ,ZAK		1
	PAOZZ	2530011147764	19207	12256301	KIT, AIR BRAKE TRAILER TOWING .... UOC:DAA,DAB,DAC,DAD,DAE,DAF,DAJ,DAKR, DAW,DAX,V12,V13,V14,V15,V16,V17,V19, V20,V24,V25,V39,ZAA,ZAB,ZAC,ZAD,ZAE, ZAF,ZAJ,ZAK		1
					ADAPTER, STRAIGHT	(4)	603-7
					ADAPTER, STRAIGHT	(1)	604-16
					ADAPTER, STRAIGHT	(1)	605-1
					ADAPTER, STRAIGHT	(I)	605-12
					BOLT, MACHINE	(1)	604-6
					BOLT, MACHINE	(1)	604-14
					BOLT, MACHINE	(2)	605-4
					BRACKET	(1)	603-3
					COUPLING, PIPE	(2)	603-12
					ELBOW	(1)	603-1
					ELBOW	(2)	603-13
					ELBOW	(3)	603-14
					ELBOW	(3)	604-1
					ELBOW	(1)	604-11
					ELBOW	(1)	605-6
					ELBOW	(1)	605-8
					ELBOW	(1)	605-11
					HOSE ASSEMBLY	(2)	603-11
					INSERT	(8)	603-8

(1) ITEM NO	(2) SMR CODE	(3) NSN	(4) CAGEC	(5) PART NUMBER	(6) DESCRIPTION AND USABLE ON CODES (UOC)		(7) QTY
					INSERT	(12)	604-2
					INSERT	(6)	605-2
					NUT, SELF-LOCKING	(2)	603-4
					NUT, SELF-LOCKING	(1)	604-8
					NUT, SELF-LOCKING	(1)	604-10
					NUT, SELF-LOCKING	(2)	605-7
					SCREW, CAP, HEXAGON	(2)	603-6
					TEE, PIPE	(1)	605-13
					VALVE, AIR	(1)	605-5
					VALVE ASSEMBLY	(1)	603-2
					VALVE, ROTARY, DIRECT	(1)	603-5
					VALVE, SHUTTLE	(1)	604-5
					VALVE, SHUTTLE	(1)	604-12
					WASHER, FLAT	(1)	604-13

END OF FIGURE

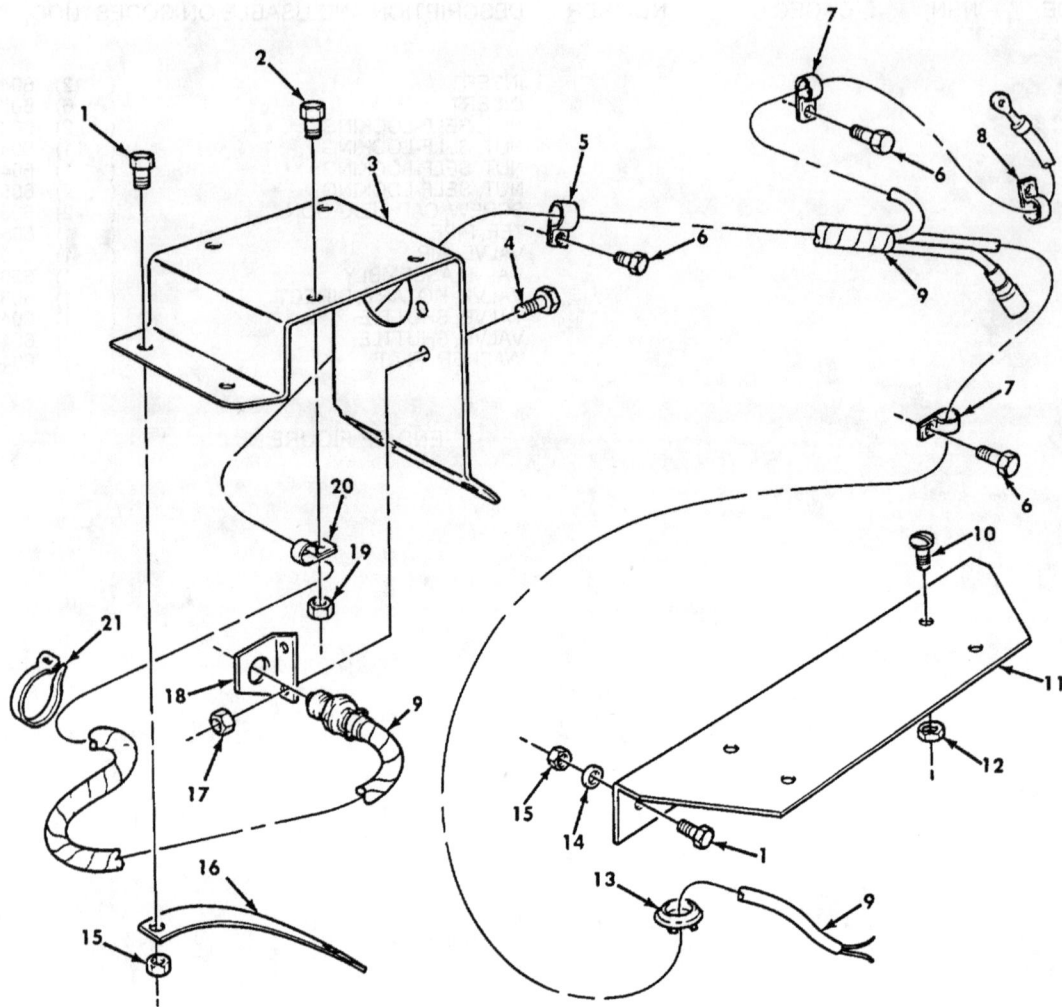

Figure 606. Chemical Agent Alarm Mounting Kit.

(1) ITEM NO	(2) SMR CODE	(3) NSN	(4) CAGEC	(5) PART NUMBER	(6) DESCRIPTION AND USABLE ON CODES (UOC)	(7) QTY
					GROUP 3307 SPECIAL PURPOSE KITS	
					FIG. 606 CHEMICAL AGENT ALARM MOUNTING KIT	
1	PAOZZ	5306000501238	96906	MS90727-32	BOLT, MACHINE PART OF KIT P/N 12256442	6
2	PAOZZ	5305000680512	80204	B1821BH025F056N	SCREW, CAP, HEXAGON H PART OF KIT P/N................... 12256442	4
3	KFOZZ	5340010965022	19207	12256391	SUPPORT, MOUNTING PART OF KIT P/N 12256442	1
4	PAOZZ	530600068051	360285	6893-2	SCREW, CAP, HEXAGON PART OF KIT P/N.................... 12256442	2
5	PAOZZ	5340007647051	96906	MS21333-69	CLAMP, LOOP PART OF KIT P/N 12256442	3
6	PAOZZ	5305008550957	96906	MS24629-46	SCREW, TAPPING PART OF KIT P/N 12256442	5
7	PAOZZ	5340000797837	96906	MS21333-67	CLAMP, LOOP PART OF KIT P/N 12256442	2
8	PAOZZ	5340000573025	96906	MS21333-108	CLAMP, LOOP PART OF KIT P/N 12256442......................	1
9	XDOZZ		19207	12257123	CABLE ASSEMBLY, SPEC PART OF KIT P/N 12256442	1
10	PAOZZ	5305009897434	96906	MS35207-263	SCREW, MACHINE PART OF KIT P/N.............................. 12256442	4
11	KFOZZ	5340010958289	19207	12257122	BRACKET, ANGLE PART OF KIT P/N.............................. 12256442	1
12	PAOZZ	5310008071467	96906	MS21042-3	NUT, SELF-LOCKING, EX PART OF KIT P/N 12256442	4
13	PAOZZ	5325007373246	70797	3104	GROMMET, NONMETALLIC PART OF KIT P/N 12256442	1
14	PAOZZ	5310008093078	96906	MS27183-11	WASHER, FLAT PART OF KIT P/N 12256442	2
15	PAOZZ	5310002416658	96906	MS51943-34	NUT, SELF-LOCKING, HE PART OF KIT P/N 12256442	6
16	KFOZZ	5340010963494	19207	12256390	STRAP, RETAINING PART OF KIT P/N 12256442	1
17	PAOZZ	5310009359022	96906	MS51943-32	NUT, SELF-LOCKING, HE PART OF KIT P/N 12256442	2
18	KFOZZ		19207	11609734	BRACKET, ANGLE PART OF KIT P/N.............................. 12256442	1
19	PAOZZ	5310008775796	96906	MS21044N4	NUT, SELF-LOCKING, HE PART OF KIT P/N...................... 12256442	4
20	PAOZZ	5340000573037	96906	MS21333-111	CLAMP, LOOP PART OF KIT P/N 12256442......................	1
21	PAOZZ	5975000742072	96906	MS3367-1-9	STRAP, TIEDOWN PART OF KIT P/N 12256442	1
	PFOZZ	6665010958285	19207	12256442	KIT, CHEMICAL AGENT ALARM MOUNTING.....................	1

```
BOLT, MACHINE                    ( 6)    606-1
BRACKET                          ( 1)    606-11
BRACKET, ANGLE                   ( 1)    606-18
CABLE ASSEMBLY                   ( 1)    606-9
CLAMP, LOOP                      ( 3)    606-5
CLAMP, LOOP                      ( 2)    606-7
CLAMP, LOOP                      ( 1)    606-8
CLAMP, LOOP                      ( 1)    606-20
GROMMET                          ( 1)    606-13
```

(1) ITEM NO	(2) SMR CODE	(3) NSN	(4) CAGEC	(5) PART NUMBER	(6) DESCRIPTION AND USABLE ON CODES (UOC)	(7) QTY
					NUT, SELF-LOCKING	(4) 606-12
					NUT, SELF-LOCKING	(6) 606-15
					NUT, SELF-LOCKING	(2) 606-17
					NUT, SELF-LOCKING	(4) 606-19
					SCREW, CAP, HEXAGON	(4) 606-2
					SCREW, CAP, HEXAGON	(2) 606-4
					SCREW, MACHINE	(4) 606-10
					SCREW, TAPPING	(5) 606-6
					STRAP, RETAINING	(1) 606-16
					STRAP, TIEDOWN	(1) 606-21
					SUPPORT	(1) 606-3
					WASHER, FLAT	(2) 606-14

SECTION II

END OF FIGURE

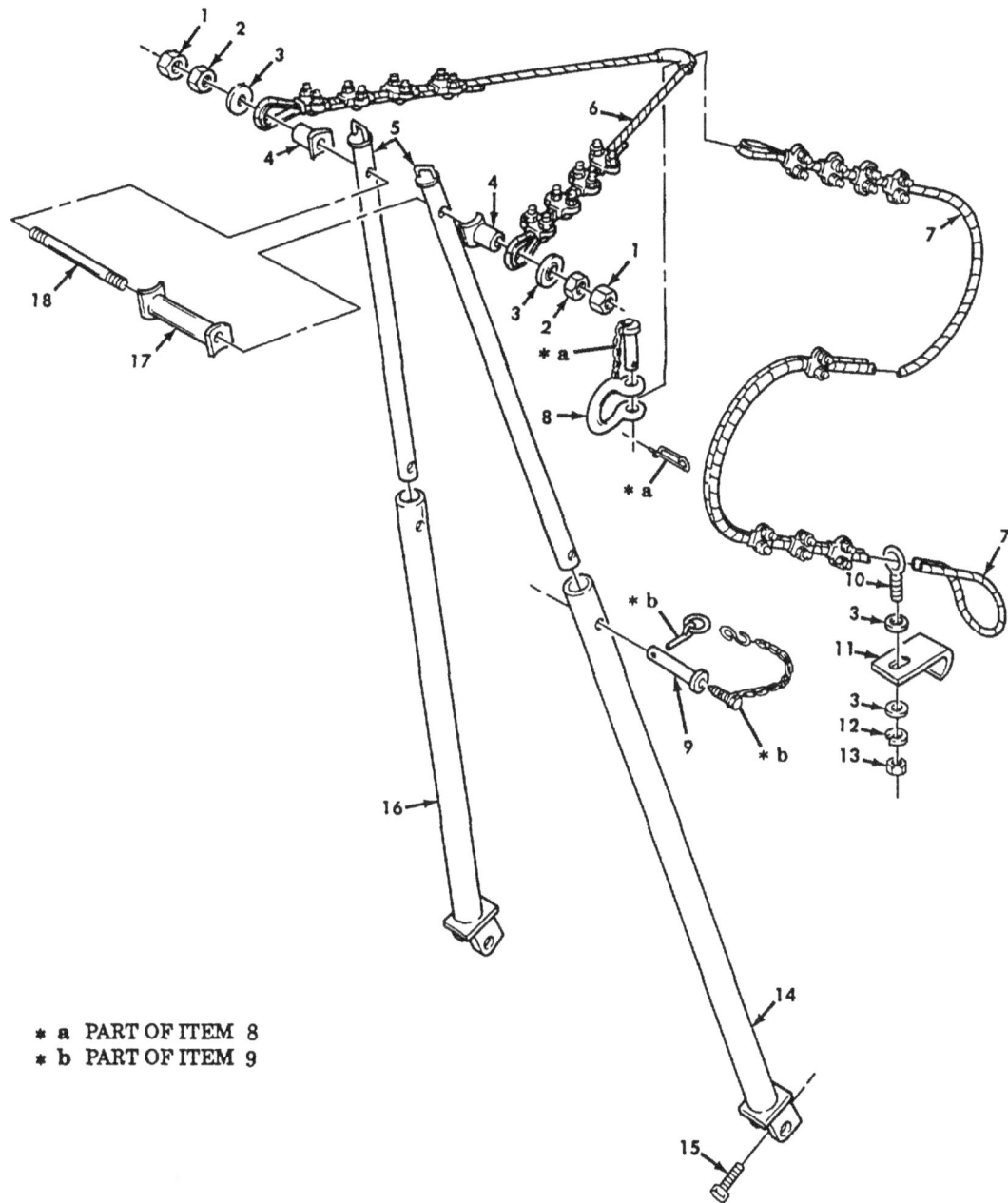

* a PART OF ITEM 8
* b PART OF ITEM 9

Figure 607. A-Frame Components Kit.

(1) ITEM NO	(2) SMR CODE	(3) NSN	(4) CAGEC	(5) PART NUMBER	(6) DESCRIPTION AND USABLE ON CODES (UOC)	(7) QTY
					GROUP 3307 SPECIAL PURPOSE KITS	
					FIG. 607 A-FRAME COMPONENTS KIT	
1	PAOZZ	5310008512677	96906	MS35691-49	NUT, PLAIN, HEXAGON PART OF KIT P/N 8390117. UOC:DAB, DAD, DAH, DAX, V12, V14, V16, V21, ZAB, ZAD, ZAH	2
2	PAOZZ	5310008913426	96906	MS35691-73	NUT, PLAIN, HEXAGON PART OF KIT P/N 8390117. UOC:DAB, DAD, DAH, DAX, V12, V14, V16, V21, ZAB, ZAD, ZAH	2
3	PAOZZ	5310009826562	96906	MS27183-28	WASHER, FLAT PART OF KIT P/N 8390117. UOC:DAB, DAD, DAX, V12, V14, V16, V21, ZAB, ZAD, ZAH	4
4	KFOZZ		19207	8338087	TUBE ASSEMBLY PART OF KIT P/N 8390117. UOC:DAB, DAD, DAH, DAX, V12, V14, V16, V21, ZAB, ZAD, ZAH	2
5	KFOZZ		19207	8337103	TUBE ASSEMBLY PART OF KIT P/N 8390117. UOC:DAB, DAD, DAH, DAX, V12, V14, V16, V21, ZAB, ZAD, ZAH	2
6	KFOZZ		19207	8668879	HARNESS ASSEMBLY PART OF KIT P/N 8390117. UOC:DAB, DAD, DAH, DAX, V12, V14, V16, V21, ZAB, ZAD, ZAH	1
7	KFOZZ		19207	8668878	CABLE ASSEMBLY PART OF KIT P/N 8390117. UOC:DAB, DAD, DAH, DAX, V12, V14, V16, V21, ZAB, ZAD, ZAH	1
8	KFOZZ	4030011730079	19207	7357965	SHACKLE ASSEMBLY COMPONENT OF A-FRAME KIT NSN2590-00-600-9035 PART OF KIT P/N 8390117. UOC:DAB, DAD, DAH, DAX, V12, V14, V16, V21, ZAB, ZAD, ZAH	1
9	PFOZZ	5325006240528	19204	7551074	PIN ASSEMBLY. UOC:DAB, DAD, DAH, DAX, V12, V14, V16, V21, ZAA, ZAB, ZAC, ZAD, ZAE, ZAF, ZAG, ZAH, ZAJ, ZAK, ZAL	2
10	PAOZZ	5306000500347	96906	MS51937-5	BOLT, EYE PART OF KIT P/N 8390117. UOC:DAB, DAD, DAH, DAX, V12, V14, V16, V21, ZAB, ZAD, ZAH	1
11	KFOZZ		19207	8689233	PLATE PART OF KIT P/N 8390117. UOC:DAB, DAD, DAH, DAX, V12, V14, V16, V21, ZAB, ZAD, ZAH	1
12	PAOZZ	5310005845272	96906	MS35338-48	WASHER, LOCK PART OF KIT P/N 8390117. UOC:DAB, DAD, DAH, DAX, V12, V14, V16, V21, ZAB, ZAD, ZAH	1
13	PAOZZ	5310010702105	96906	MS51967-14	NUT, PLAIN, HEXAGON PART OF KIT P/N 8390117. UOC:DAB, DAD, DAH, DAX, V12, V14, V16, V21, ZAB, ZAD, ZAH	1

(1) ITEM NO	(2) SMR CODE	(3) NSN	(4) CAGEC	(5) PART NUMBER	(6) DESCRIPTION AND USABLE ON CODES (UOC)	(7) QTY
14	KFOZZ		19207	8337176	LEG ASSEMBLY PART OF KIT P/N 8390117 .... UOC:DAB, DAD, DAH, DAX, V12, V14, V16, V21, ZAB, ZAD, ZAH	1
15	PAOZZ	5305011335811	21450	110093	SETSCREW PART OF KIT P/N 8390117 .... UOC:DAB, DAD, DAH, DAX, V12, V14, V16, V21, ZAB, ZAD, ZAH	2
16	KFOZZ		19207	8337175	LEG ASSEMBLY PART OF KIT P/N 8390117 .... UOC:DAB, DAD, DAH, DAX, V12, V14, V16, V21, ZAB, ZAD, ZAH	1
17	KFOZZ	2590000951484	19207	8337078	SPREADER, A-FRAME, VE PART OF KIT P/N .... 8390117 .... UOC:DAB, DAD, DAH, DAX, V12, V14, V16, V21, ZAB, ZAD, ZAH	1
18	KFOZZ	5306002181773	19207	8337079	ROD, THREADED END PART OF KIT P/N .... 8390117 .... UOC:DAB, DAD, DAH, DAX, V12, V14, V16, V21, ZAB, ZAD, ZAH	1
	PFOZZ	2590006009035	19207	8390117	KIT, A-FRAME MOUNTING .... UOC:DAB, DAD, DAH, DAX, V12, V14, V16, V21, ZAB, ZAD, ZAH	

```
BOLT, EYE                          ( 1)    607-10
CABLE ASSEMBLY                     ( 1)    607-7
HARNESS ASSEMBLY                   ( 1)    607-6
LEG ASSEMBLY                       ( 1)    607-14
LEG ASSEMBLY                       ( 1)    607-16
NUT, PLAIN, HEXAGON                ( 2)    607-1
NUT, PLAIN, HEXAGON                ( 2)    607-2
NUT, PLAIN, HEXAGON                ( 1)    607-13
PIN ASSEMBLY                       ( 2)    607-9
PLATE                              ( 1)    607-11
ROD, THREADED END                  ( 1)    607-18
SETSCREW                           ( 2)    607-15
SHACKLE ASSEMBLY                   ( 1)    607-8
SPREADER, A-FRAME                  ( 1)    607-17
TUBE ASSEMBLY                      ( 2)    607-4
TUBE ASSEMBLY                      ( 2)    607-5
WASHER, FLAT                       ( 4)    607-3
WASHER, LOCK                       ( 1)    607-12
```

END OF FIGURE

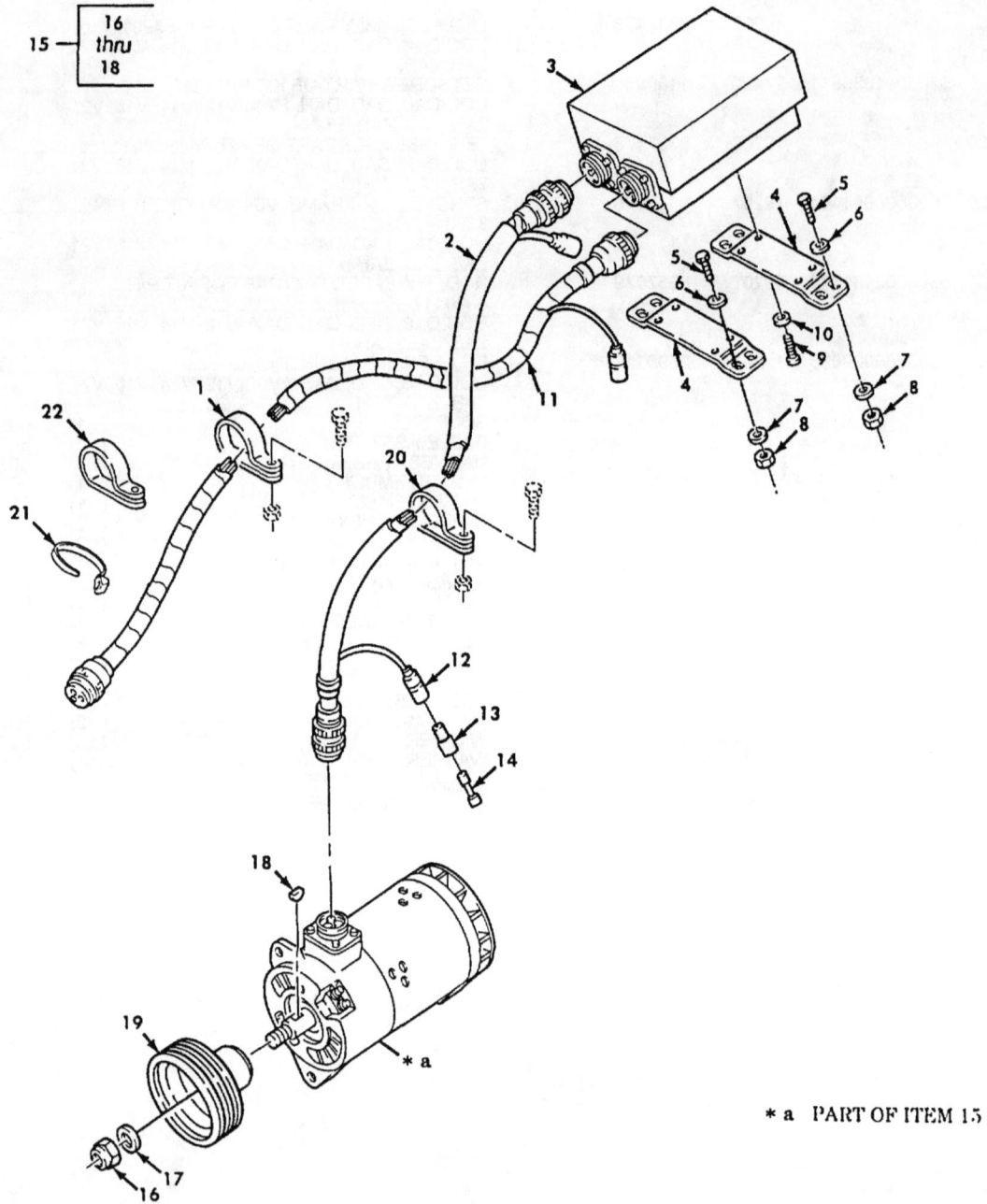

Figure 608. Alternator Kit, 60 to 100 AMP.

* a   PART OF ITEM 15

(1) ITEM NO	(2) SMR CODE	(3) NSN	(4) CAGEC	(5) PART NUMBER	(6) DESCRIPTION AND USABLE ON CODES (UOC)	(7) QTY
					GROUP 3307 SPECIAL PURPOSE KITS	
					FIG. 608 ALTERNATOR KIT, 60 TO 100 AMP	
1	PAOZZ	5340007647052	96906	MS21333-116	CLAMP, LOOP PART OF KIT P/N 12277359........................	1
2	PFOZZ	6150011312053	19207	12277356	CABLE ASSEMBLY, SPEC PART OF KIT P/N 12277359................................	1
3	PAOFF	2920009007993	19207	10947439	REGULATOR, ENGINE GE PART OF KIT P/N 12277359................................	1
4	PFOZZ	5340011949885	19207	11623670-1	PLATE, MOUNTING PART OF KIT P/N 12277359................................	2
5	PAOZZ	5305002678953	80204	B1821BH025F063N	SCREW, CAP, HEXAGON H PART OF KIT P/N 12277359................................	4
6	PAOZZ	5310005287638	19207	5287638	WASHER, FLAT PART OF KIT P/N 12277359	4
7	PAOZZ	5310008094058	96906	MS27183-10	WASHER, FLAT PART OF KIT P/N 12277359	1
8	PAOZZ	5310008775796	96906	MS21044N4	NUT, SELF-LOCKING, HE PART OF KIT P/N 12277359................................	4
9	PAOZZ	5305009896265	96906	MS35207-262	SCREW, MACHINE PART OF KIT P/N 12277359................................	4
10	PAOZZ	5310005435933	96906	MS35333-73	WASHER, LOCK PART OF KIT P/N 12277359........	4
11	PFOZZ	2920011959383	19207	12302625	WIRING HARNESS, BRAN PART OF KIT P/N 12277359................................	1
12	PAOZZ	5935008338561	19207	8338561	SHELL, ELECTRICAL CO PART OF KIT P/N 12277359................................	1
13	PAOZZ	5970008338562	19207	8338562	INSULATOR, BUSHING PART OF KIT P/N 12277359................................	1
14	PAOZZ	5940003996676	19207	8338564	TERMINAL ASSEMBLY PART OF KIT P/N 12277359................................	1
15	PAOFF	2920014198884	24617	12450173-1	GENERATOR, ENGINE AC PART OF KIT P/N 12277359................................	1
16	PAOZZ	5310009390783	96906	MS21083N12	.NUT, SELF-LOCKING, HE	1
17	PAOZZ	5310001670826	88044	AN960-1216	.WASHER, FLAT	1
18	PAOZZ	5315000124553	96906	MS35756-17	.KEY, WOODRUFF	1
19	PAOZZ	3020011312078	19207	12277358	PULLEY, GROOVE PART OF KIT P/N 12277359................................	1
20	PAOZZ	5340012323568	19207	7529133-10	CLAMP, LOOP PART OF KIT P/N 12277359........................	1
21	PAOZZ	5975001563253	96906	MS3367-2-9	STRAP, TIEDOWN, ELECT PART OF KIT P/N 12277359................................	10
22	PAOZZ	5340008333049	96906	MS21333-127	CLAMP, LOOP PART OF KIT P/N 12277359........................	1
	PAOZZ	2920011311939	19207	12277359	KIT, ALTERNATOR, 60 TO 100 AMP	1

CLAMP ASSEMBLY   (1)   608-2
CLAMP, LOOP   (1)   608-1
CLAMP, LOOP   (1)   608-20
CLAMP, LOOP   (1)   608-22
GENERATOR ASSEMBLYE   (1)   608-15
INSULATOR, BUSHING   (1)   608-13
NUT, SELF-LOCKING   (4)   608-8
PLATE, REINFORCEMENT   (2)   608-4
PULLEY, GROOVE   (1)   608-19

(1) ITEM NO	(2) SMR CODE	(3) NSN	(4) CAGEC	(5) PART NUMBER	(6) DESCRIPTION AND USABLE ON CODES (UOC)	(7) QTY
					REGULATOR (1)	608-3
					SCREW, CAP, HEXAGON (4)	608-5
					SCREW, MACHINE (4)	608-9
					SHELL, ELECTRICAL (1)	608-12
					STRAP, TIEDOWN (10)	608-21
					TERMINAL ASSEMBLY (1)	608-14
					WASHER, FLAT (4)	608-6
					WASHER, FLAT (4)	608-7
					WASHER, LOCK (4)	608-10
					WIRING HARNESS (1)	608-11

END OF FIGURE

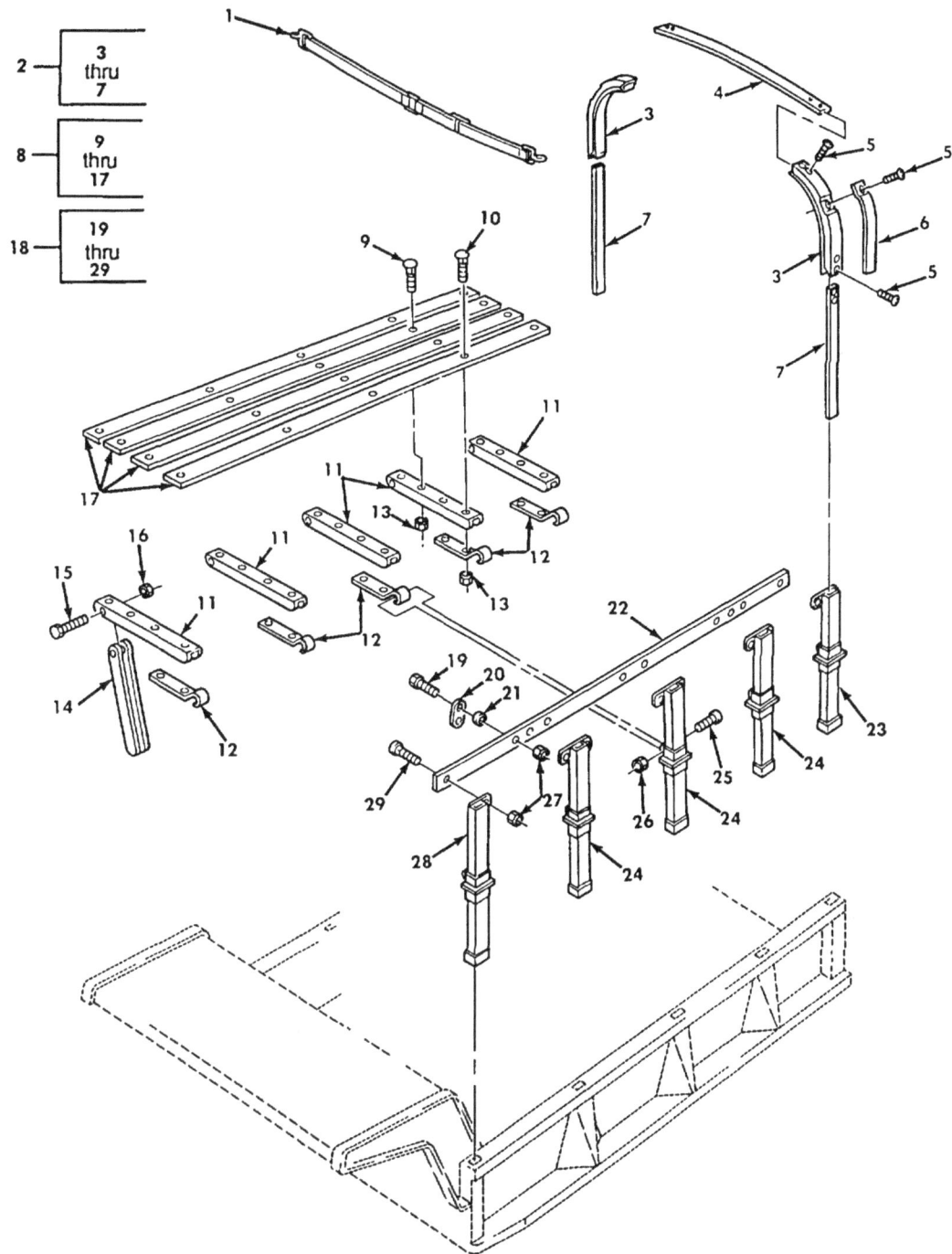

Figure 609. Dump Body Troop Seat and Tarpaulin Kit (Sheet 1 of 2).

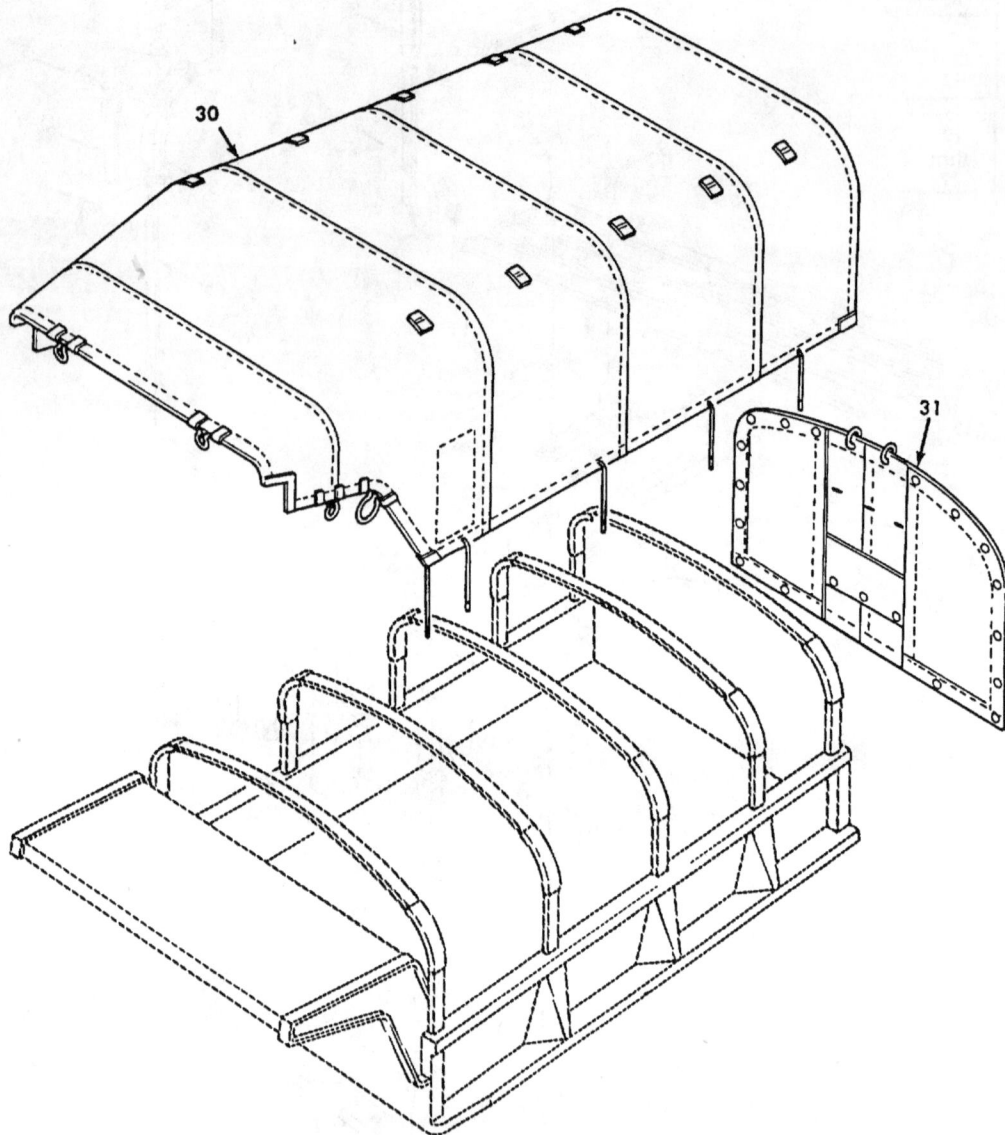

Figure 609. Dump Body Troop Seat and Tarpaulin Kit (Sheet 2 of 2).

(1) ITEM NO	(2) SMR CODE	(3) NSN	(4) CAGEC	(5) PART NUMBER	(6) DESCRIPTION AND USABLE ON CODES (UOC)	(7) QTY
					GROUP 3307 SPECIAL PURPOSE KITS	
					FIG. 609 DUMP BODY TROOP SEAT AND TARPAULIN KIT	
1	PAOZZ	5340013939372	19207	11682088-3	STRAP, WEBBING PART OF KIT P/N .... UOC:DAE, DAF, V19, V20, ZAE, ZAF 57K0171, TAN	1
1	PFOZZ	5340014580975	19207	11682088-5	STRAP, WEBBING PART OF KIT P/N .... UOC:DAE, DAF, V19, V20, ZAE, ZAF 57K0172, WHITE	1
1	PAOZZ	5340011147712	19207	11682088-1	STRAP, WEBBING PART OF KIT P/N .... 12302696, GREEN UOC:DAE, DAF, V19, V20, ZAE, ZAF	1
2	PAOZZ	2540008600519	19207	10899222	BOW, VEHICULAR TOP PART OF KIT P/N .... 12302696, GREEN UOC:DAE, DAF, V19, V20, ZAE, ZAF	5
3	PAOZZ	2540003510145	19207	7064151	.BOW ASSEMBLY GREEN .... UOC:DAE, DAF, V19, V20, ZAE, ZAF	2
4	PAOZZ	2540014579025	19207	12460216-4	.BOW, VEHICULAR TOP GREEN . ... UOC:DAE, DAF, V19, V20, ZAE, ZAF	4
5	PAOZZ	5305009845680	96906	MS35206-300	.SCREW, MACHINE .... UOC:DAE, DAF, V19, V20, ZAE, ZAF	10
6	PAOZZ		19207	12340208-6	.STRAP, WEBBING PART OF KIT P/N........................ 12302696, GREEN UOC:DAE, DAF, V19, V20, ZAE, ZAF	2
6	PAOZZ		19207	12340208-7	.STRAP, WEBBING PART OF KIT P/N........................ 57K0171, TAN UOC:DAE, DAF, V19, V20, ZAE, ZAF	2
6	PFOZZ		19207	12340208-8	.STRAP, WEBBING PART OF KIT P/N........................ 57K0172, WHITE UOC:DAE, DAF, V19, V20, ZAE, ZAF	2
7	PAOZZ	2540011739147	19207	10899223	.BOW, VEHICULAR TOP .... UOC:DAE, DAF, V19, V20, ZAE, ZAF	2
8	PAOOO	2540008600522	19207	10899217	SEAT, VEHICULAR PART OF KIT P/N....................... 12302696, GREEN L.H UOC:DAE, DAF, V19, V20, ZAE, ZAF	1
8	PAOOO	2510014227756	19207	10899217-1	SIDE RACK SET VEHIC PART OF KIT P/N .... 57K0171, TAN L.H UOC:DAE, DAF, V19, V20, ZAE, ZAF	1
8	PAOOO		19207	10899217-2	SEAT, VEHICULAR PART OF KIT P/N....................... 57K0172, WHITE L.H UOC:DAE, DAF, V19, V20, ZAE, ZAF	1
8	PAOOO	2540008600520	19207	10899218	SEAT, VEHICULAR PART OF KIT P/N....................... 12302696, GREEN R.H UOC:DAE, DAF, V19, V20, ZAE, ZAF	1
8	PAOOO	2540014231783	19207	10899218-1	SEAT, VEHICULAR PART OF KIT P/N....................... 57K0141, TAN R.H UOC:DAE, DAF, V19, V20, ZAE, ZAF	1
8	PAOOO		19207	10899218-2	SEAT, VEHICULAR PART OF KIT P/N....................... 57K0142, WHITE R.H UOC:DAE, DAF, V19, V20, ZAE, ZAF	1
9	PAOZZ	5306007536996	96906	MS35751-43	.BOLT, SQUARE NECK	12

(1) ITEM NO	(2) SMR CODE	(3) NSN	(4) CAGEC	(5) PART NUMBER	(6) DESCRIPTION AND USABLE ON CODES (UOC)	(7) QTY
					UOC:DAE, DAF, V19, V20, ZAE, ZAF	
10	PAOZZ	5306000120231	96906	MS35751-44	.BOLT, SQUARE NECK	8
					UOC:DAE, DAF, V19, V20, ZAE, ZAF	
11	PAOZZ	2540005921823	19207	7061093	.SUPPORT, SEAT, VEHICU	5
					UOC:DAE, DAF, V19, V20, ZAE, ZAF	
12	PAOZZ	5340011857768	19207	10899220	.STRAP, RETAINING	4
					UOC:DAE, DAF, V19, V20, ZAE, ZAF	
					UOC:DAE, DAF, V19, V20, ZAE, ZAF	
13	PAOZZ	5310008807744	96906	MS51967-5	.NUT, PLAIN, HEXAGON	20
					UOC:DAE, DAF, V19, V20, ZAE, ZAF	
14	PAOZZ	2540011739160	19207	10899221	.SUPPORT, SEAT, VEHICU	5
					UOC:DAE, DAF, V19, V20, ZAE, ZAF	
15	PAOZZ	5305002693244	80204	B1821BH038F250N	.SCREW, CAP, HEXAGON H	5
					UOC:DAE, DAF, V19, V20, ZAE, ZAF	
16	PAOZZ	5310009591488	96906	MS51922-21	.NUT, SELF-LOCKING	5
					UOC:DAE, DAF, V19, V20, ZAE, ZAF	
17	PAOZZ	2540012052759	19207	10899219	.SEAT, VEHICULAR	4
					UOC:DAE, DAF, V19, V20, ZAE, ZAF	
18	PFOOO	2510008600523	19207	10899201	SIDE RACK, VEHICLE B PART OF KIT P/N.......................... 12302696, GREEN L.H	1
					UOC:DAE, DAF, V19, V20, ZAE, ZAF	
18	PFOOO	2510014227752	19207	10899201-1	SIDE RACK, VEHICLE B PART OF XIT P/N.......................... 57K0171, TAN L.H	1
					UOC:DAE, DAF, V19, V20, ZAE, ZAF	
18	PFOOO		19207	10899201-2	SIDE RACK, VEHICLE B PART OF KIT P/N.......................... 57K0172, WHITE L.H	1
					UOC:DAE, DAF, V19, V20, ZAE, ZAF	
18	PFOOO	2510008600517	19207	10899202	SIDE RACK, VEHICLE B PART OF KIT P/N.......................... 12302696, GREEN R.H	1
					UOC:DAE, DAF, V19, V20, ZAE, ZAF	
18	PFOOO		19207	10899202-1	SIDE RACK, VEHICLE B PART OF KIT P/N.......................... 57K0171, TAN R.H	1
					UOC:DAE, DAF, V19, V20, ZAE, ZAF	
18	PFOOO		19207	10899202-2	SIDE RACK, VEHICLE B PART OF KIT P/N.......................... 57K0172, WHITE R.H	1
					UOC:DAE, DAF, V19, V20, ZAE, ZAF	
19	PAOZZ	5305009845680	96906	MS35206-300	.SCREW, MACHINE	2
					UOC:DAE, DAF, V19, V20, ZAE, ZAF	
20	PAOZZ	5340007372788	19207	7372788	.BRACKET, ANGLE	2
					UOC:DAE, DAF , V19, V20, ZAE, ZAF	
21	PAOZZ	5310008094058	96906	MS27183-10	.WASHER , FLAT	2
					UOC:DAE, DAF, V19, V20, ZAE, ZAF	
22	PAOZZ		19207	12450347-13	.BOARD, SIDE RACK	1
					UOC:DAE, DAF, V19, V20, ZAE, ZAF	
23	PFOZZ	2540011739172	19207	10899206	.POCKET, BOW SIDE RACK, RIGHT REAR	1
					UOC:DAE, DAF, V19, V20, ZAE, ZAF	
23	PFOZZ	2540011739173	19207	10899205	.POCKET, BOW SIDE RACK, LEFT REAR	1
					UOC:DAE, DAF, V19, V20, ZAE, ZAF	
24	PFOZZ	2540011739174	19207	10899207	.POCKET, BOW	3
					UOC:DAE, DAF, V19, V20, ZAE, ZAF	
25	PAOZZ	5305002692811	96906	MS90726-67	.SCREW, CAP, HEXAGON H	4
					UOC:DAE, DAF, V19, V20, ZAE, ZAF	
26	PAOZZ	5310009591488	96906	MS51922-21	.NUT, SELF-LOCKING	4

(1) ITEM NO	(2) SMR CODE	(3) NSN	(4) CAGEC	(5) PART NUMBER	(6) DESCRIPTION AND USABLE ON CODES (UOC)	(7) QTY
27	PAOZZ	5310008807744	96906	MS51967-5	UOC:DAE, DAF, V19, V20, ZAE, ZAF .NUT, PLAIN, HEXAGON	10
28	PFOZZ	2540011739175	19207	10899204	UOC:DAE, DAF, V19, V20, ZAE, ZAF .POCKET, TROOP SEAT SIDE RACK, RIGHT FRONT	1
28	PFOZZ	2540011739176	19207	10899203	UOC:DAE, DAF, V19, V20, ZAE, ZAF .POCKET, TROOP SEAT SIDE RACK, LEFT FRONT	1
29	PAOZZ	5306009115005	96906	MS35751-42	UOC:DAE, DAF, V19, V20, ZAE, ZAF .BOLT, SQUARE NECK	8
30	PAOFF	2540011827557	19207	12302632	UOC:DAE, DAF, V19, V20, ZAE, ZAF TARPAULIN PART OF KIT P/N 12302696, GREEN	1
30	PAOFF	2540014231966	19207	12302632-4	UOC:DAE, DAF, V19, V20, ZAE, ZAF TARPAULIN PART OF KIT P/N 57X0171, TAN	1
30	PAOFF	2540014358760	19207	12302632-5	UOC:DAE, DAF, V19, V20, ZAE, ZAF TARPAULIN PART OF KIT P/N 57K0172, WHITE	1
31	PAOFF		19207	12375769-3	UOC:DAE, DAF, V19, V20, ZAE, ZAF CURTAIN, VEHICULAR PART OF KIT P/N 12302696, GREEN	1
31	PAOFF	2540014231791	19207	12375769-1	UOC:DAE, DAF, V19, V20, ZAE, ZAF CURTAIN, VEHICULAR PART OF KIT P/N 57K0171, TAN	1
31	PFOFF	2540014231786	19207	12375769-2	UOC:DAE, DAF, V19, V20, ZAE, ZAF CURTAIN, VEHICULAR PART OF KIT P/N 57K0172, WHITE	1
	PFFFF	2540011550112	19207	12302696	UOC:DAE, DAF, V19, V20, ZAE, ZAF KIT, DUMP BODY TROOP SEAT AND TARPAULIN, GREEN.	1
					UOC:DAE, DAF, V19, V20, ZAE, ZAF	
					BOW ASSEMBLY (1) 609-2	
					CURTAIN, VEHICULAR (1) 609-31	
					SEAT, TROOP LEFT (1) 609-8	
					SEAT, TROOP RIGHT (1) 609-8	
					SIDE RACK LEFT (1) 609-18	
					SIDE RACK RIGHT (1) 609-18	
					STRAP, WEBBING (1) 609-1	
					TARPAULIN (1) 609-30	
	PFFFF	2540014577623	19207	57K0171	KIT, DUMP BODY TROOP SEAT AND TARPAULIN, TAN.	1
					UOC:DAE, DAF, V19, V20, ZAE, ZAF	
					CURTAIN, VEHICULAR (1) 609-31	
					SEAT, TROOP LEFT (1) 609-8	
					SEAT, TROOP RIGHT (1) 609-8	
					SIDE RACK LEFT (1) 609-18	
					SIDE RACK RIGHT (1) 609-18	
					STRAP, WEBBING (1) 609-1	
					TARPAULIN (1) 609-30	
	PFFFF	2540014577602	19207	57K0172	KIT, DUMP BODY TROOP SEAT AND TARPAULIN, WHITE	1
					UOC:DAE, DAF, V19, V20, ZAE, ZAF	

(1) ITEM NO	(2) SMR CODE	(3) NSN	(4) CAGEC	(5) PART NUMBER	(6) DESCRIPTION AND USABLE ON CODES (UOC)	(7) QTY
					CURTAIN, VEHICULAR ....(1)	609-31
					SEAT, TROOP LEFT ....(1)	609-8
					SEAT, TROOP RIGHT.......................................(1)	609-8
					SIDE RACK LEFT ....(1)	609-18
					SIDE RACK RIGHT ....(1)	609-18
					STRAP, WEBBING ....(1)	609-1
					TARPAULIN ....(1)	609-30

END OF FIGURE

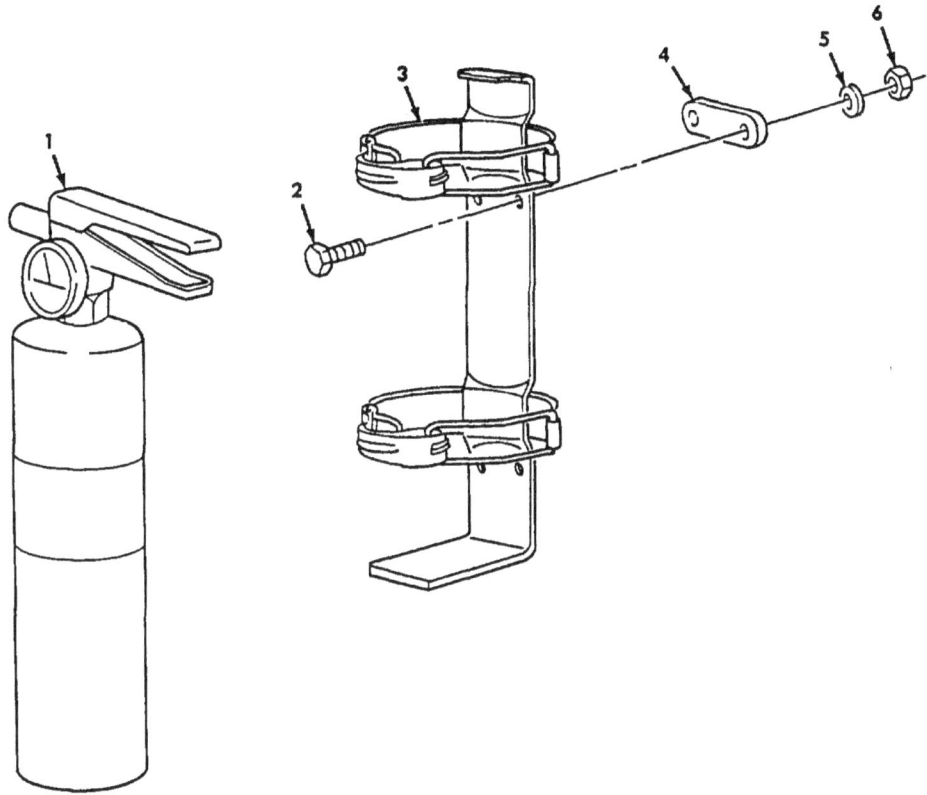

Figure 610. Fire Extinguisher Kit.

(1) ITEM NO	(2) SMR CODE	(3) NSN	(4) CAGEC	(5) PART NUMBER	(6) DESCRIPTION AND USABLE ON CODES (UOC)	(7) QTY
					GROUP 3307 SPECIAL PURPOSE KITS	
					FIG. 610 FIRE EXTINGUISHER KIT	
1	PAOZZ	4210007750127	19207	7015266	EXTINGUISHER, FIRE PART OF KIT P/N 12302876	1
2	PAOZZ	5305002678954	80204	B1821BH025F125N	SCREW, CAP, HEXAGON H PART OF KIT P/N 12302876	4
3	PFOZZ	4210011834822	19207	12255634	BRACKET, FIRE EXTING PART OF KIT P/N 12302876	1
4	PFOZZ	5965012042104	19207	12302897	PLATE, SPACER PART OF KIT P/N 12302876	2
5	PAOZZ	5310008094058	96906	MS27183-10	WASHER, FLAT PART OF KIT P/N 12302876	4
6	PAOZZ	5310009359022	96906	MS51943-32	NUT, SELF-LOCKING, HE PART OF KIT P/N 12302876	4
	PAOZZ	4210012206376	19207	12302876	PARTS KIT, FIRE EXTI	1

```
BRACKET, FIRE EXTING      ( 1)   610-3
EXTINGUISHER, FIRE        ( 1)   610-1
NUT, SELF-LOCKING, HE     ( 4)   610-6
PLATE, SPACER             ( 2)   610-4
SCREW, CAP, HEXAGON H     ( 4)   610-2
WASHER, FLAT              ( 4)   610-5
```

END OF FIGURE

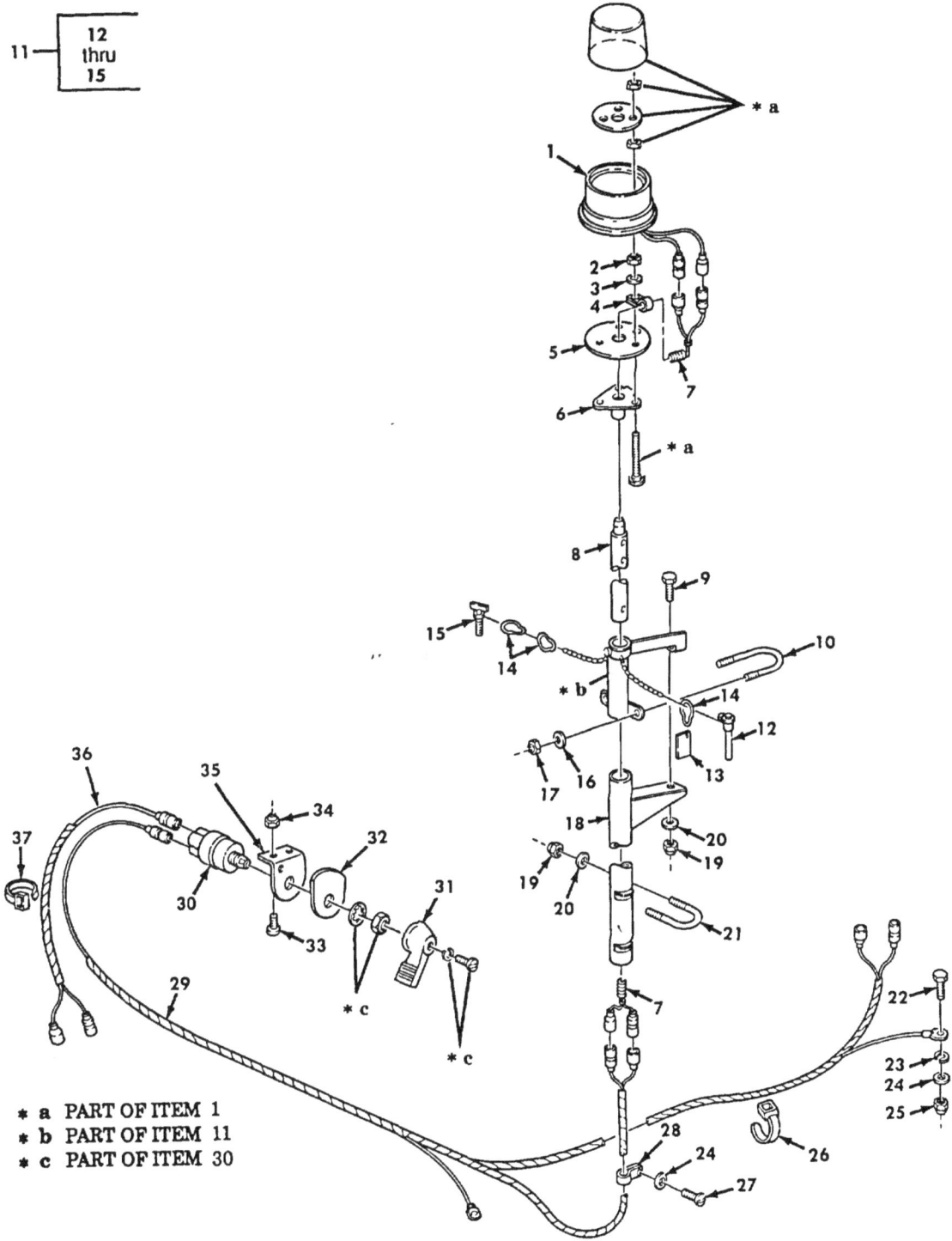

Figure 611. Convoy Warning Light Kit.

* a PART OF ITEM 1
* b PART OF ITEM 11
* c PART OF ITEM 30

(1) ITEM NO	(2) SMR CODE	(3) NSN	(4) CAGEC	(5) PART NUMBER	(6) DESCRIPTION AND USABLE ON CODES (UOC)	(7) QTY
					GROUP 3307 SPECIAL PURPOSE KITS	
					FIG. 611 CONVOY WARNING LIGHT KIT	
1	PAOZZ	6220009477570	19207	12301084	LIGHT, WARNING PART OF KIT P/N .... 12301057 .... UOC:DAA, DAB, DAC, DAD, DAG, DAH, DAL, DAW, DAX, V12, V13, V14, V15, V16, V17, V18, V21, V22, V39, ZAA, ZAB, ZAC, ZAD, ZAG, ZAH, ZAL,	2
2	PAOZZ	5310000881251	96906	MS51922-1	NUT, SELF-LOCKING, HE PART OF KIT P/N .... 12301057 .... UOC:DAA, DAB, DAC, DAD, DAG, DAH, DAL, DAW, DAX, V12, V13, V14, V15, V16, V17, V18, V21, V22, V39, ZAA, ZAB, ZAC, ZAD, ZAG, ZA, ZAL	2
3	PAOZZ	5310008094058	96906	MS27183-10	WASHER, FLAT, PART OF KIT P/N .... 12301057 .... UOC:DAA, DAB, DAC, DAD, DAG, DAH, DAL, DAW, DAX, V12, V13, V14, V15, V16, V17, V18, V21, V22, V39, ZAA, ZAB, ZAC, ZAD, ZAG, ZAH, ZAL	2
4	PAOZZ	5340000886655	96906	MS21333-101	CLAMP, LOOP PART OF KIT P/N 12301057................. UOC:DAA, DAB, DAC, DAD, DAG, DAH, DAL, DAW, DAX, V12, V13, V14, V15, V16, V17, V18, V21, V22, V39, ZAA, ZAB, ZAC, ZAD, ZAG, ZAH, ZAL	2
5	PAOZZ	5340012052504	19207	12301062	PLATE, MOUNTING PART OF KIT P/N .... 12301057 .... UOC:DAA, DAB, DAC, DAD, DAG, DAH, DAL, DAW, DAX, V12, V13, V14, V15, V16, V17, V18, V21, V22, V39, ZAA, ZAB, ZAC, ZAD, ZAE, ZAF, ZAG, ZAH, ZAL	2
6	PFOZZ	5340012178296	19207	12296642	PLATE, MOUNTING PART OF KIT P/N .... 12301057 .... UOC:DAA, DAB, DAC, DAD, DAG, DAH, DAL, DAW, DAX, V12, V13, V14, V15, V16, V17, V18, V21, V'2, V39, ZAA, ZAB, ZAC, ZAD, ZAE, ZAF, ZAG, ZAH, ZAJ, ZAK, ZAL	2
7	PBOZZ	6150012226585	19207	12296577-1	CABLE ASSEMBLY, POWE PART OF KIT P/N .... 12301057 .... UOC:DAA, DAB, DAC, DAD, DAG, DAM, DAL, DAW, DAX, V12, V13, V14, V15, V16, V17, V18 , V21, V22, V39, ZAA, ZAB, ZAC, ZAD, ZAE, ZAF, ZAG, ZAH, ZAJ, ZAK, ZAL	2
8	PFOZZ	4710012004404	19207	12301067	TUBE ASSEMBLY, METAL PART OF KIT P/N .... 12301057 .... UOC:DAA, DAB, DAC, DAD, DAG, DAH, DAL, DAW, DAX, V12, V13, V14, V15, V16, V17, V18, V21, V22, V39, ZAA, ZAB, ZAC, ZAD, ZAE, ZAF, ZAG, ZAH, ZAJ, ZAK, ZAL	2
9	PAOZZ	5305007168186	96906	MS90726-110	SCREW, CAP, HEXAGON H PART OF KIT P/N .... 12301057 .... UOC:DAA, DAB, DAC, DAD, DAG, DAH, DAL, DAW, DAX, V12, V13, V14, V15, V16, V17, V18, V21, V22, V39, ZAA, ZAB, ZAC, ZAD, ZAZ, ZAF, ZAG,	2

(1) ITEM NO	(2) SMR CODE	(3) NSN	(4) CAGEC	(5) PART NUMBER	(6) DESCRIPTION AND USABLE ON CODES (UOC)	(7) QTY
					DAX, V12, V13, V14, V15, V16, V17, V18, V21, V22, V39, ZAA, ZAB, ZAC, ZAD, ZAE, ZAF, ZAG, ZAH, ZAJ, ZAK, ZAL	
10	PAOZZ	5306012052677	19207	12301066	U-BOLT PART OF KIT P/N 12301057	2
					UOC:DAA, DAB, DAC, DAD, DAG, DAH, DAL, DAW, DAX, V12, V13, V14, V15, V16, V17, V18, V21, V22, V39, ZAA, ZAB, ZAC, ZAD, ZAE, ZAF, ZAG, ZAH, ZAJ, ZAK, ZAL	
11	KFOZZ	5340012052505	19207	12301063	BRACKET, MOUNTING PART OF KIT P/N 12301057	2
					UOC:DAA, DAB, DAC, DAD, DAG, DAH, DAL, DAW, DAX, V12, V13, V14, V15, V16, V17, V18, V21, V22, V39, ZAA, ZAB, ZAC, ZAD, ZAE, ZAF, ZAG, ZAH, ZAJ, ZAK, ZAL	
12	PAOZZ	5315002509595	96906	MS17986C619	PIN, QUICK RELEASE PART OF KIT P/N 12301057	1
					UOC:DAA, DAB, DAC, DAD, DAG, DAH, DAL, DAW, DAX, V12, V13, V14, V15, V16, V17, V18, V21, V22, V39, ZAA, ZAB, ZAC, ZAD, ZAG, ZAH, ZAL	
13	PFOZZ	9905012144053	19207	12301104	PLATE, INSTRUCTION	1
					UOC:DAA, DAB, DAC, DAD, DAG, DAH, DAL, DAW, DAX, V12, V13, V14, V15, V16, V17, V18, V21, V22, V39, ZAA, ZAB, ZAC, ZAD, ZAE, ZAF, ZAG, ZAH, ZAJ, ZAK, ZAL	
14	PAOZZ	5325012120599	19207	7331177	RING, RETAINING	3
					UOC:DAA, DAB, DAC, DAD, DAG, DAH, DAL, DAW, DAX, V12, V13, V14, V15, V16, V17, V18B, V21, V22, V39, ZAA, ZAB, ZAC, ZAD, ZAE, ZAF, ZAG, ZAH, ZAJ, ZAK, ZAL	
15	PAOZZ	5305012039064	19207	7345195-1	SCREW, WING	1
					UOC:DAA, DAB, DAC, DAD, DAG, DAH, DAL, DAW, DAX, V12, V13, V14, V15, V16, V17, V18, V21, V22, V39, ZAA, ZAB, ZAC, ZAD, ZAE, ZAF, ZAG, ZAH, ZAJ, ZAK, ZAL	
16	PAOZZ	5310000814219	96906	MS27183-12	WASHER, FLAT PART OF KIT P/N 12301057	4
					UOC:DAA, DAB, DAC, DAD, DAG, DAH, DAL, DAW, DAX, V12, V13, V14, V15, V16, V17, V18, V21, V22, V39, ZAA, ZAB, ZAC, ZAD, ZAG, ZAH, ZAL	
17	PAOZZ	5310009843807	96906	MS51922-13	NUT, SELF-LOCKING, HE PART OF KIT P/N 12301057	4
					UOC:DAA, DAB, DAC, DAD, DAG, DAH, DAL, DAW, DAX, V12, V13, V14, V15, V16, V17, V18, V21, V22, V39, ZAA, ZAB, ZAC, ZAD, ZAE, ZAF, ZAG, ZAH, ZAJ, ZAK, ZAL	
18	PFOZZ	4710012004244	19207	12301064	TUBE, METALLIC PART OF KIT P/N 12301057	2
					UOC:DAA, DAB, DAC, DAD, DAG, DAH, DAL, DAW, DAX, V12, V13, V14, V15, V16, V17, V18, V21, V22, V39, ZAA, ZAB, ZAC, ZAD, ZAE, ZAF, ZAG, ZAH, ZAJ, ZAK, ZAL	
19	PAOZZ	5310008775795	96906	MS21044-NB	NUT, SELF-LOCKING, HE PART OF KIT P/N 12301057	10
					UOC:DAA, DAB, DAC, DAD, DAG, DAB, DAL, DAW, DAX, V12, V13, V14, V15, V16, V17, V18, V21, V22, V39, ZAA, ZAB, ZAC, ZAD, ZAE, ZAF, ZAG, ZAH, ZAJ, ZAK, ZAL	

(1) ITEM NO	(2) SMR CODE	(3) NSN	(4) CAGEC	(5) PART NUMBER	(6) DESCRIPTION AND USABLE ON CODES (UOC)	(7) QTY
20	PAOZZ	5310008095998	96906	MS27183-18	WASHER, FLAT, PART OF KIT P/N 12301057 UOC:DAA, DAB, DAC, DAD, DAG, DAH, DAL, DAW, DAX, V12 V13 , V14 , V1 5, V16, V17 , V18 , V21, V22, V39, ZAA, ZAB, ZAC, ZAD, ZAE, ZAF, ZAG, ZAH, ZAJ, ZAK, ZAL	10
21	PAOZZ	5306011305994	19207	11609677	U-BOLT PART OF KIT P/N 12301057 UOC:DAA, DAB, DAC, DAD, DAG, DAH, DAL, DAW, DAX, V12, V13, V14, V15, V16, V17, V18, V21, V22, V39, ZAA, ZAB, ZAC, ZAD, ZAE, ZAF, ZAG, ZAH, ZAJ, ZAK, ZAL	4
22	PAOZZ	5305002693241	96906	MS90727-65	SCREW, CAP, HEXAGON PART OF KIT P/N 12301057 UOC:DAA, DAB, DAC, DAD, DAG, DAR, DAL, DAW, DAX, V12, V13, V14, V15, V16, V17, V18, V21, V22, V39, ZAA, ZAB, ZAC, ZAD, ZAE, ZAF, ZAG, ZAH, ZAJ, ZAK, ZAL	10
23	PAOZZ	5310005503503	96906	MS35335-36	WASHER, LOCK PART OF KIT P/N 12301057 UOC:DAA, DAB, DAC, DAD, DAG, DAH, DAL, DAW, DAX, V12, V13, V14, V15, V16, V17, V18, V21, V22, V39, ZAA, ZAB, ZAC, ZAD, ZAG, ZAH, ZAL	2
24	PAOZZ	5310000806004	96906	MS27183-14	WASHER, FLAT PART OF KIT P/N 12301057 UOC:DAA, DAB, DAC, DAD, DAG, DAH, DAL, DAW, DAX, V12, V13, V14, V15, V16, V17, V18, V21, V22, V39, ZAA, ZAB, ZAC, ZAD, ZAG, ZAH, ZAL	4
25	PAOZZ	5310009591488	96906	MS51922-21	NUT, SELF-LOCKING, HE PART OF KIT P/N 12301057 UOC:DAA, DAB, DAC, DAD, DAG, DAH, DAL, DAW, DAX, V12, V13, V14, V15, V16, V17, V18, V21, V22, V39, ZAA, ZAB, ZAC, ZAD, ZAE, ZAF, ZAG, ZAH, ZAJ, ZAK, ZAL	10
26	PAOZZ	5975004515001	96906	MS3367-3-9	STRAP, TIEDOWN, ELECT PART OF KIT P/N 12301057 UOC:DAA, DAB, DAC, DAD, DAG, DAH, DAL, DAW, DAX, V12, V13, V14, V15, V16, V17, V18, V21, V22, V39, ZAA, ZAB, ZAC, ZAD, ZAG, ZAH, ZAL	6
27	PAOZZ	5305000425648	24617	425648	SCREW, ASSEMBLED WAS PART OF KIT P/N 12301057 UOC:DAA, DAB, DAC, DAD, DAG, DAH, DAL, DAW, DAX, V12, V13, V14, V15, V16, V17, V18, V21, V22, V39, ZAA, ZAB, ZAC, ZAD, ZAE, ZAF, ZAG, ZAH, ZAJ, ZAK, ZAL	2
28	PAOZZ	5340008338476	96906	MS21333-122	CLAMP, LOOP PART OF KIT P/N 12301057 UOC:DAA, DAB, DAC, DAD, DAG, DAH, DAL, DAW, DAX, V12, V13, V14, V15, V16, V17, V18, V21, V22, V39, ZAA, ZAB, ZAC, ZAD, ZAG, ZAH, ZAL	2
29	PFOZZ	5995012002419	19207	12301085	WIRING HARNESS, BRAN PART OF KIT P/N 12301057 UOC:DAA, DAB, DAC, DAD, DAG, DAH, DAL, DAW, DAX, V12', V13 , V14, V15, V16, V17 , V18, V21, V22, V39, ZAA, ZAB, ZAC, ZAD, ZAE, ZAF, ZAG, ZAH, ZAJ, ZAK, ZAL	1

(1) ITEM NO	(2) SMR CODE	(3) NSN	(4) CAGEC	(5) PART NUMBER	(6) DESCRIPTION AND USABLE ON CODES (UOC)	(7) QTY
30	PAOZZ	5930006999438	96906	MS39060-2	SWITCH, ROTARY PART OF KIT P/N ...... 12301057 ...... UOC:DAA, DAB, DAC, DAD, DAG, DAH, DAL, DAW, DAX, V12, V13, V14, V15, V16, V17, V18, V21, V22, V39, ZAA, ZAB, ZAC, ZAD, ZAE, ZAF, ZAG,	1
31	PAOZZ	5930001305349	19207	5381088	HANDLE, SWITCH PART OF KIT P/N ...... 12301057 ...... UOC:DAA, DAB, DAC, DAD, DAG, DAH, DAL, DAW, DAX, V12, V13, V14, V15, V16, V17, V18, V21, V22, V39, ZAA, ZAB, ZAC, ZAD, ZAE, ZAF, ZAG, ZAH, ZAJ, ZAK, ZAL	1
32	PAOZZ	9905002525587	19207	7409993	PLATE, IDENTIFICATIO PART OF KIT P/N ...... 12301057 ...... UOC:DAA, DAB, DAC, DAD, DAG, DAH, DAL, DAW, DAX, V12, V13, V14, V15, V16, V17, V18, V21, V22, V39, ZAA, ZAB, ZAC, ZAD, ZAE, ZAF, ZAG, ZAH, ZAJ, ZAK, ZAL	1
33	PAOZZ	5305009906444	96906	MS35207-261	SCREW, MACHINE PART OF KIT P/N ...... 12301057 ...... UOC:DAA, DAB, DAC, DAD, DAG, DAH, DAL, DAW, DAX, V12, V13, V14, V15, V16, V17, V18, V21, V22, V39, ZAA, ZAB, ZAC, ZAD, ZAG, ZAH, ZAL	2
34	PAOZZ	5310008775797	96906	MS21044-N3	NUT, SELF-LOCKING, HE PART OF KIT P/N ...... 12301057 ...... UOC:DAA, DAB, DAC, DAD, DAG, DAH, DAL, DAW, DAX, V12, V13, V14, V15, V16, V17, V18, V21, V22, V39, ZAA, ZAB, ZAC, ZAD, ZAG, ZAH, ZAL	2
35	PFOZZ	5340001760868	19207	7059461	BRACKET, ANGLE PART OF KIT P/N ...... 12301057 ...... UOC:DAA, DAB, DAC, DAD, DAG, DAH, DAL, DAW, DAX, V12, V13, V14, V15, V16, V17, V18, V21, V22, V39, ZAA, ZAB, ZAC, ZAD, ZAE, ZAF, ZAG, ZAH, ZAJ, ZAK, ZAL	1
36	PFOZZ	5995012003203	19207	12301087	WIRING HARNESS, BRAN PART OF KIT P/N ...... 12301057 ...... UOC:DAA, DAB, DAC, DAD, DAG, DAH, DAL, DAW, DAX, V12, V13, V14, V15, V16, V17, V18, V21, V22, V39, ZABB, ZAC, ZAD, ZAE, ZAF, ZAG, ZAH, ZAJ, ZAK, ZAL	1
37	PAOZZ	5975000742072	96906	MS3367-1-9	STRAP, TIEDOWN, ELECT PART OF KIT P/N ...... 12301057 ...... UOC:DAA, DAB, DAC, DAD, DAG, DAH, DAL, DAW, DAX, V12, V13, V14, V15, V16, V17, V18, V21, V22, V39, ZAA, ZAB, ZAC, ZAD, ZAG, ZAH, ZAL	15
	PFOZZ	6220011951791	19207	12301057	KIT, CONVOY WARNING ...... UOC:DAA, DAB, DAC, DAD, DAG, DAB, DAL, DAW, DAX, V12, V13, V14, V15, V16, V17, V18, V21, V22, V39, ZAA, ZAB, ZAC, ZAD, ZAE, ZAF, ZAG, ZAH, ZAL	1

      BRACKET, ANGLE   (1)   611-35
      BRACKET ASSEMBLY   (2)   611-11
      CABLE ASSEMBLY   (2)   611-7
      CLAMP, LOOP   (2)   611-4
      CLAMP, LOOP   (2)   611-28

Description	QTY	Part Number
HANDLE, SWITCH	(1)	611-31
LIGHT, WARNING	(2)	611-1
NUT, SELF-LOCKING, HE	(2)	611-2
NUT, SELF-LOCKING, HE	(4)	611-17
NUT, SELF-LOCKING, HE	(10)	611-19
NUT, SELF-LOCKING, HE	(2)	611-25
NUT, SELF-LOCKING, HE	(2)	611-34
PIN, QUICK RELEASE	(1)	611-12
PLATE, IDENTIFICATIO	(1)	611-32
PLATE, INSTRUCTION	(1)	611-13
PLATE, MOUNTING	(2)	611-5
PLATE, MOUNTING	(2)	611-6
RING, RETAINING	(1)	611-14
SCREW, ASSEMBLED	(2)	611-27
SCREW, CAP, HEXAGON	(2)	611-9
SCREW, CAP, HEXAGON	(2)	611-22
SCREW, MACHINE	(2)	611-33
SCREW, WING	(1)	611-15
STRAP, TIEDOWN	(6)	611-26
STRAP, TIEDOWN	(15)	611-37
SWITCH, ROTARY	(1)	611-30
TUBE ASSEMBLY	(2)	611-8
TUBE ASSEMBLY	(2)	611-18
U-BOLT	(2)	611-10
U-BOLT	(4)	611-21
WASHER, FLAT	(2)	611-3
WASHER, FLAT	(4)	611-16
WASHER, FLAT	(10)	611-20
WASHER, FLAT	(4)	611-24
WASHER, LOCK	(2)	611-23
WIRING HARNESS	(1)	611-29
WIRING HARNESS	(1)	611-36

END OF FIGURE

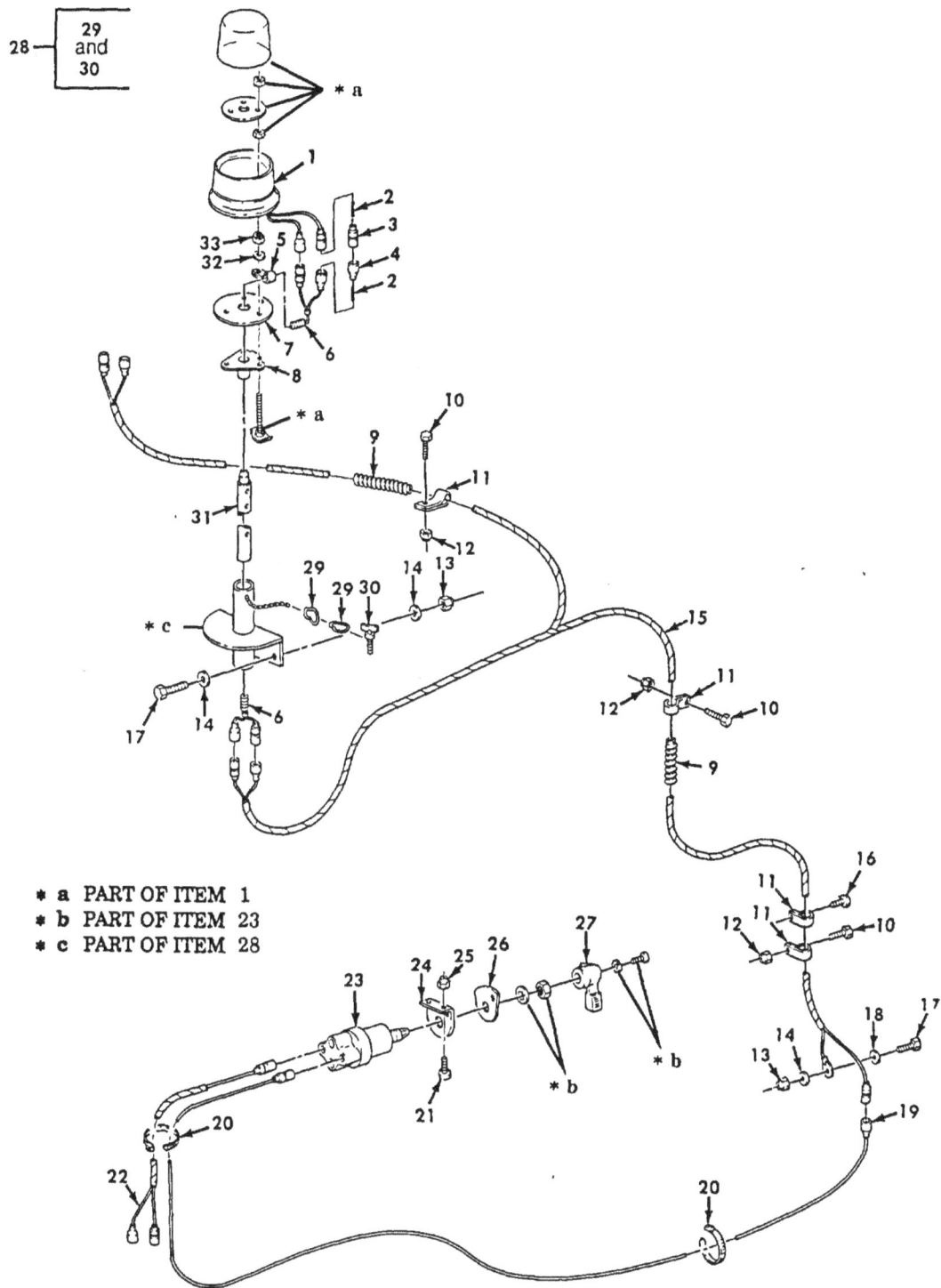

* a PART OF ITEM 1
* b PART OF ITEM 23
* c PART OF ITEM 28

Figure 612. Convoy Warning Light Kit, Dump.

(1) ITEM NO	(2) SMR CODE	(3) NSN	(4) CAGEC	(5) PART NUMBER	(6) DESCRIPTION AND USABLE ON CODES (UOC)	(7) QTY
					GROUP 3307 SPECIAL PURPOSE KITS	
					FIG. 612 CONVOY WARNING LIGHT KIT, DUMP	
1	PAOZZ	6220009477570	19207	12301084	LIGHT, WARNING PART OF KIT P/N 12301151 .... UOC:DAE, DAF, V19, V20, ZAE, ZAF	2
2	PAOZZ	5935002140904	19207	7982907	DUMMY CONNECTOR, PLU TO BE USED WHEN WARNING LIGHT IS REMOVED PART OF KIT P/N 12301151 .... UOC:DAE, DAF, V19, V20, ZAE, ZAF	4
3	PAOZZ	5935008338561	19207	8338561	SHELL, ELECTRICAL CO TO BE USED WHEN WARNING LIGHT IS REMOVED PART OF KIT P/N 12301151 .... UOC:DAE, DAF, V19, V20, ZAE, ZAF	2
4	PAOZZ	5935005729180	19207	8338566	SHELL, ELECTRICAL CO TO BE USED WHEN WARNING LIGHT IS REMOVED PART OF KIT P/N 12301151 .... UOC:DAE, DAF, V19, V20, ZAE, ZAF	2
5	PAOZZ	5340000673868	96906	MS21333-109	CLAMP, LOOP PART OF KIT P/N 12301151 .... UOC:DAE, DAF, V19, V20, ZAE, ZAF	2
6	PFOZZ	6150012237270	19207	12301167-1	CABLE ASSEMBLY, POWE PART OF KIT P/N .... 12301151 .... UOC:DAE, DAF, V19, V20, ZAE, ZAF	1
7	PAOZZ	5340012052504	19207	12301062	PLATE, MOUNTING PART OF KIT P/N .... 12301151 .... UOC:DAE, DAF, V19, V20, ZAE, ZAF	2
8	PFOZZ	5340012178296	19207	12296642	PLATE, MOUNTING PART OF KIT P/N .... 12301151 .... UOC:DAE, DAF, V19, V20, ZAE, ZAF	2
9	XBOZZ		19207	12301226	CONDUIT, NONMETALLIC MAKE FROM P/N .... L-P-390 TYPE 1 PART OF KIT P/N 12301151 .... UOC:DAE, DAF, V19, V20, ZAE, ZAF	1
10	PAOZZ	5306000680513	60285	6893-2	BOLT, MACHINE PART OF KIT P/N .... 12301151 .... UOC:DAE, DAF, V19, V20, ZAE, ZAF	30
11	PAOZZ	5340000573043	96906	MS21333-112	CLAMP, LOOP PART OF KIT P/N 12301151 .... UOC:DAE, DAF, V19, V20, ZAE, ZAF	31
12	PAOZZ	5310008775796	96906	MS21044N4	NUT, SELF-LOCKING, HE PART OF KIT P/N .... 12301151 .... UOC:DAE, DAF, V19, V20, ZAE, ZAF	30
13	PAOZZ	5310009591488	96906	MS51922-21	NUT, SELF-LOCKING PART OF KIT P/N .... 12301151 .... UOC:DAE, DAF, V19, V20, ZAE, ZAF	9
14	PAOZZ	5310000806004	96906	MS27183-14	WASHER, FLAT PART OF KIT P/N 12301151 .... UOC:DAE, DAF, V19, V20, ZAE, ZAF	17
15	PFOZZ	6150012237269	19207	12301168-1	CABLE ASSEMBLY, POWE PART OF KIT P/N .... 12301151 .... UOC:DAE, DAF, V19, V20, ZAE, ZAF	1
16	PAOZZ	5305012261945	96906	MS51850-96	SCREW, TAPPING PART OF KIT P/N .... 12301151 .... UOC:DAE, DAF, V19, V20, ZAE, ZAF	1
17	PAOZZ	5305002693239	80204	B1821BH038F138N	SCREW, CAP, HEXAGON H PART OF KIT P/N	9

(1) ITEM NO	(2) SMR CODE	(3) NSN	(4) CAGEC	(5) PART NUMBER	(6) DESCRIPTION AND USABLE ON CODES (UOC)	(7) QTY
					12301151	
					UOC:DAE, DAF, V19, V20, ZAE, ZAF	
18	PAOZZ	5310000611258	96906	MS45904-76	WASHER, LOCK PART OF KIT P/N 12301151	1
					UOC:DAE, DAF, V19, V20, ZAE, ZAF	
19	PFOZZ	6150012237251	19207	12301166-1	CABLE ASSEMBLY, POWE PART OF KIT P/N	1
					12301151	
					UOC:DAE, DAF, V19, V20, ZAE, ZAF	
20	PAOZZ	5975009846582	96906	MS3367-1-0	STRAP, TIEDOWN, ELECT PART OF KIT P/N	35
					12301151	
					UOC:DAE, DAF, V19, V20, ZAE, ZAF	
21	PAOZZ	5305009897434	96906	MS35207-263	SCREW, MACHINE PART OF KIT P/N	2
					12301151	
					UOC:DAE, DAF, V19, V20, ZAE, ZAF	
22	PAOZZ	5995012003203	19207	12301087	WIRING HARNESS, BRAN PART OF KIT P/N	1
					12301151	
					UOC:DAE, DAF, V19, V20, ZAE, ZAF	
23	PAOZZ	5930006999438	96906	MS39060-2	SWITCH, ROTARY PART OF KIT P/N	1
					12301151	
					UOC:DAE, DAF, V19, V20, ZAE, ZAF	
24	PAOZZ	5340001760868	19207	7059461	BRACKET, ANGLE PART OF KIT P/N	1
					2301151	
					UOC:DAE, DAF, V19, V20, ZAE, ZAF	
25	PAOZZ	5310008775797	96906	MS21044-N3	NUT, SELF-LOCKING, HE PART OF KIT P/N	2
					12301151	
					UOC:DAE, DAF, V19, V20, ZAE, ZAF	
26	PFOZZ	9905002525587	19207	7409993	PLATE, IDENTIFICATIO PART OF KIT P/N	1
					12301151	
					UOC:DAE, DAF, V19, V20, ZAE, ZAF	
27	PAOZZ	5930001305349	19207	5381088	HANDLE, SWITCH PART OF KIT P/N	1
					12301151	
					UOC:DAE, DAF, V19, V20, ZAE, ZAF	
28	PFOZZ	4710012267390	19207	12301165	TUBE ASSEMBLY, METAL PART OF KIT P/N	2
					12301151	
					UOC:DAE, DAF, V19, V20, ZAE, ZAF	
29	PAOZZ	5325012120599	19207	7331177	RING, RETAINING .	2
					UOC:DAE, DAF, V19, V20, ZAE, ZAF	
30	PAOZZ	5305012039064	19207	7345195-1	SCREW, WING .	1
					UOC:DAE, DAF, V19, V20, ZAE, ZAF	
31	PFOZZ	4710012267389	19207	12301067-4	TUBE ASSEMBLY, METAL PART OF KIT P/N	2
					12301151	
					UOC:DAE, DAF, V19, V20, ZAE, ZAF	
32	PAOZZ	5310008094058	96906	MS27183-10	WASHER, FLAT PART OF KIT P/N	4
					12301151	
					UOC:DAE, DAF, V19, V20, ZAE, ZAF	
33	PAOZZ	5310007616882	96906	MS51967-2	NUT, PLAIN, HEXAGON PART OF KIT P/N	4
					12301151	
					UOC:DAE, DAF, V19, V20, ZAE, ZAF	
	PFOZZ	6220012197620	19207	12301151	KIT, CONVOY WARNING	1
					UOC:DAE, DAF, V19, V20, ZAE, ZAF	
					BRACKET, ANGLE (1) 612-24	
					CABLE ASSEMBLY (2) 612-6	
					CABLE ASSEMBLY (1) 612-15	
					CABLE ASSEMBLY (1) 612-19	

SECTION II

(1) ITEM NO	(2) SMR CODE	(3) NSN	(4) CAGEC	(5) PART NUMBER	(6) DESCRIPTION AND USABLE ON CODES (UOC)	(7) QTY
					CLAMP, LOOP	(2) 612-5
					CLAMP, LOOP	(31) 612-11
					CONDUIT	(1) 612-9
					DUMMY CONNECTOR	(2) 612-2
					HANDLE, SWITCH	(1) 612-27
					LIGHT, WARNING	(2) 612-1
					NUT, PLAIN, HEXAGON	(4) 612-33
					NUT, SELF-LOCKING	(30) 612-12
					NUT, SELF-LOCKING	(9) 612-13
					NUT, SELF-LOCKING	(2) 612-25
					PLATE, IDENTIFICATIO	(1) 612-26
					PLATE, MOUNTING	(2) 612-7
					PLATE, MOUNTING	(2) 612-8
					RING, RETAINING	(2) 612-29
					SCREW, CAP, HEXAGON	(30) 612-10
					SCREW, CAP, HEXAGON	(9) 612 17
					SCREW, MACHINE	(2) 612-21
					SCREW, TAPPING	(1) 612-16
					SCREW, WING	(1) 612-30
					SHELL, ELECTRICAL	(2) 612-3
					SHELL, ELECTRICAL	(2) 612-4
					STRAP, TIEDOWN	(35) 612-20
					SWITCH, ROTARY	(1) 612-23
					TUBE ASSEMBLY	(2) 612 28
					TUBE ASSEMBLY	(2) 612-31
					WASHER, FLAT	(17) 612-14
					WASHER, FLAT	(4) 612-32
					WASHER, LOCK	(1) 612-18
					WIRING HARNESS	(1) 612 22

END OF FIGURE

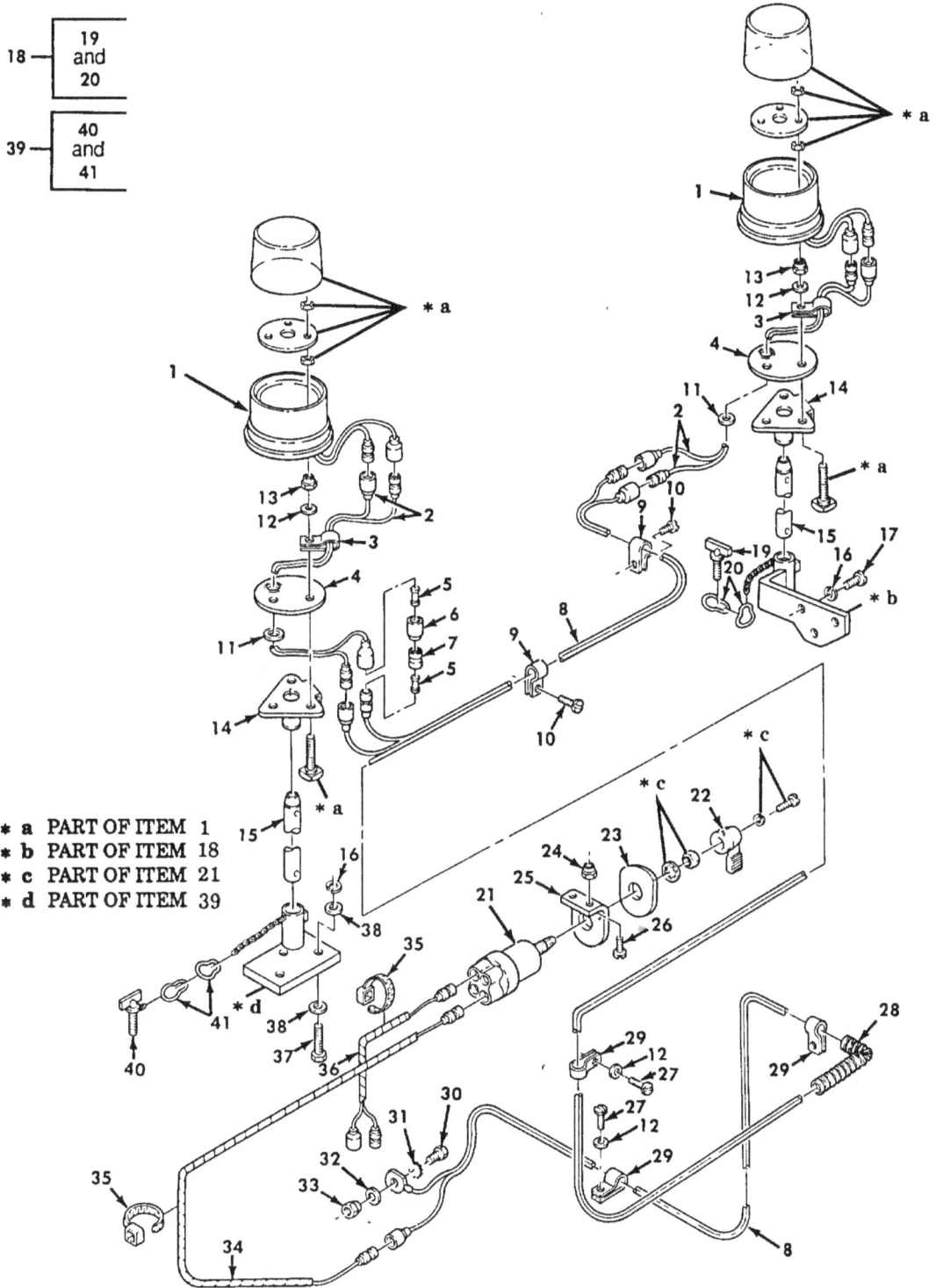

* a PART OF ITEM 1
* b PART OF ITEM 18
* c PART OF ITEM 21
* d PART OF ITEM 39

Figure 613. Convoy Warning Light Kit, Van.

(1) ITEM NO	(2) SMR CODE	(3) NSN	(4) CAGEC	(5) PART NUMBER	(6) DESCRIPTION AND USABLE ON CODES (UOC)	(7) QTY
					GROUP 3307 SPECIAL PURPOSE KITS	
					FIG. 613 CONVOY WARNING LIGHT KIT, VAN	
1	PAOZZ	6220009477570	19207	12301084	LIGHT, WARNING PART OF KIT P/N 12301193 UOC:DAJ, DAK, V24, V25, ZAJ, ZAK	2
2	PAOZZ	6150012237271	19207	12301167-2	CABLE ASSEMBLY, POWE PART OF KIT P/N 12301193 UOC:DAJ, DAK, V24, V25, ZAJ, ZAK	2
3	PAOZZ	5340000673868	96906	MS21333-109	CLAMP, LOOP PART OF KIT P/N 12301193 UOC:DAJ, DAK, V24, V25, ZAJ, ZAK	2
4	PAOZZ	5340012052504	19207	12301062	PLATE, MOUNTING PART OF KIT P/N 12301193 UOC:DAJ, DAK, V24, V25, ZAJ, ZAK	2
5	PAOZZ	5935002140904	19207	7982907	DUMMY CONNECTOR, PLU TO BE USED WHEN WARNING LIGHT IS REMOVED PART OF KIT P/N 12301193 UOC:DAJ, DAK, V24, V25, ZAJ, ZAK	4
6	PAOZZ	5935008338561	19207	8338561	SHELL, ELECTRICAL CO TO BE USED WHEN WARNING LIGHT IS REMOVED PART OF KIT P/N 12301193 UOC:DAJ, DAK, V24, V25, ZAJ, ZAK	2
7	PAOZZ	5935005729180	19207	8338566	SHELL, ELECTRICAL CO TO BE USED WHEN WARNING LIGHT IS REMOVED PART OF KIT P/N 12301193 UOC:DAJ, DAK, V24, V25, ZAJ, ZAK	2
8	PAOZZ	6150012237272	19207	12301168-2	CABLE ASSEMBLY, POWE PART OF KIT P/N 12301193 UOC:DAJ, DAK, V24, V25, ZAJ, ZAK	1
9	PAOZZ	5340000913790	96906	MS21333-72	CLAMP, LOOP PART OF KIT P/N 12301193 UOC:DAJ, DAK, V24, V25, ZAJ, ZAK	25
10	PAOZZ	5305004770124	96906	MS51861-49C	SCREW, TAPPING PART OF KIT P/N 12301193 UOC:DAJ, DAK, V24, V25, ZAJ, ZAK	25
11	PAOZZ	5325001642087	96906	MS35489-39	GROMMET, NONMETALLIC PART OF KIT P/N 12301193 UOC:DAJ, DAK, V24, V25, ZAJ, ZAK	2
12	PAOZZ	5310008094058	96906	MS27183-10	WASHER, FLAT UOC:DAJ, DAK, V24, V25, ZAJ, ZAK	13
13	PAOZZ	5310007616882	96906	MS51967-2	NUT, PLAIN, HEXAGON PART OF KIT P/N 12301193 UOC:DAJ, DAK, V24, V25, ZAJ, ZAK	2
14	PFOZZ	5340012178296	19207	12296642	PLATE, MOUNTING PART OF KIT P/N 12301193 UOC:DAJ, DAK, V24, V25, ZAJ, ZAK	2
15	PAOZZ	4730012211445	19207	12301208	COUPLING, PIPE PART OF KIT P/N 12301193 UOC:DAJ, DAK, V24, V25, ZAJ, ZAK	2
16	PAOZZ	5310008206653	80045	23MS35338-50	WASHER, LOCK PART OF KIT P/N 12301193 UOC:DAJ, DAK, V24, V25, ZAJ, ZAK	6
17	PAOZZ	5305007247222	80204	B1821BH063C200N	SCREW, CAP, HEXAGON H WARNING LIGHT BRACKET PART OF KIT P/N 12301193	3

(1) ITEM NO	(2) SMR CODE	(3) NSN	(4) CAGEC	(5) PART NUMBER	(6) DESCRIPTION AND USABLE ON CODES (UOC)	(7) QTY
18	PFOZZ	5340012233538	19207	12301194	UOC:DAJ, DAK, V24, V25, ZAJ, ZAK BRACKET, ANGLE PART OF KIT P/N 12301193	1
19	PAOZZ	5305012039064	19207	7345195-1	UOC:DAJ, DAK, V24, V25, ZAJ, ZAK SCREW, WING.	1
20	PAOZZ	5325012120599	19207	7331177	UOC:DAJ, DAK, V24, V25, ZAJ, ZAK RING, RETAINING.	2
21	PAOZZ	5930006999438	96906	MS39060-2	UOC:DAJ, DAK, V24, V25, ZAJ, ZAK SWITCH, ROTARY PART OF KIT P/N 12301193	1
22	PAOZZ	5930001305349	19207	5381088	UOC:DAJ, DAK, V24, V25, ZAJ, ZAK HANDLE, SWITCH PART OF KIT P/N 12301193	1
23	PAOZZ	9905002525587	19207	7409993	UOC:DAJ, DAK, V24, V25, ZAJ, ZAK PLATE, IDENTIFICATIO PART OF KIT P/N 12301193	1
24	PAOZZ	5310008775797	96906	MS21044-N3	UOC:DAJ, DAK, V24, V25, ZAJ, ZAK NUT, SELF-LOCKING, HE PART OF KIT P/N 12301193	2
25	PAOZZ	5340001760868	19207	7059461	UOC:DAJ, DAK, V24, V25, ZAJ, ZAK BRACKET, ANGLE PART OF KIT P/N 12301193	1
26	PAOZZ	5305009897434	96906	MS35207-263	UOC:DAJ, DAK, V24, V25, ZAJ, ZAK SCREW, MACHINE PART OF KIT P/N 12301193	2
27	PAOZZ	5305012299587	24617	9414109	UOC:DAJ, DAK, V24, V25, ZAJ, ZAK SCREW, TAPPING PART OF KIT P/N.................................. 12301193	11
28	MOOZZ		19207	12301226	UOC:DAJ, DAK, V24, V25, ZAJ, ZAK CONDUIT, NONMETALLIC MAKE FROM P/N .... L-P-390 TYPE 1 PART OF KIT P/N 12301193..........	1
29	PAOZZ	5340000573043	96906	MS21333-112	UOC:DAJ, DAK, V24, V25, ZAJ, ZAK CLAMP, LOOP PART OF KIT P/N 12301193....................	13
30	PAOZZ	5305002692803	96906	MS90726-60	UOC:DAJ, DAK, V24, V25, ZAJ, ZAK SCREW, CAP, HEXAGON H PART OF KIT P/N 12301193	1
31	PAOZZ	5310000611258	96906	MS45904-76	UOC:DAJ, DAK, V24, V25, ZAJ, ZAK WASHER, LOCK PART OF KIT P/N 12301193....................	1
32	PAOZZ	5310000806004	96906	MS27183-14	UOC:DAJ, DAK, V24, V25, ZAJ, ZAK WASHER, FLAT PART OF KIT P/N 12301193	1
33	PAOZZ	5310000881251	96906	MS51922-1	UOC:DAJ, DAK, V24, V25, ZAJ, ZAK NUT, SELF-LOCKING, HE PART OF KIT P/N.................... 12301193	1
34	PAOZZ	6150012237252	19207	12301166-2	UOC:DAJ, DAK, V24, V25, ZAJ, ZAK CABLE ASSEMBLY, POWE PART OF KIT P/N 12301193	1
35	PAOZZ	5975009846582	96906	MS3367-1-0	UOC:DAJ, DAK, V24, V25, ZAJ, ZAK STRAP, TIEDOWN, ELECT PART OF KIT P/N.................... 12301193	20
36	PAOZZ	5995012003203	19207	12301087	UOC:DAJ, DAK, V24, V25, ZAJ, ZAK WIRING HARNESS, BRAN PART OF KIT P/N	1

(1) ITEM NO	(2) SMR CODE	(3) NSN	(4) CAGEC	(5) PART NUMBER	(6) DESCRIPTION AND USABLE ON CODES (UOC)	(7) QTY
					12301193 .... UOC:DAJ, DAK, V24, V25, ZAJ, ZAK	
37	PAOZZ	5305007262559	96906	MS90727-172	SCREW, CAP, HEXAGON H PART OF KIT P/N .... 12301193 .... UOC:DAJ, DAK, V24, V25, ZAJ, ZAK	3
38	PAOZZ	5310008238803	96906	MS27183-21	WASHER, FLAT PART OF KIT P/N 12301193 .... UOC:DAJ, DAK, V24, V25, ZAJ, ZAK	6
39	PAOZZ	5340012230359	19207	12301195	BRACKET, ANGLE PART OF KIT P/N .... 12301193 .... UOC:DAJ, DAK, V24, V25, ZAJ, ZAK	1
40	PAOZZ	5305012039064	19207	7345195-1	SCREW, WING .... UOC:DAJ, DAK, V24, V25, ZAJ, ZAK	1
41	PAOZZ	5325012120599	19207	7331177	RING, RETAINING. .... UOC:DAJ, D AK, V24, V25, ZAJ, ZAK	2
	PFOZZ	6220012197621	19207	12301193	KIT, CONVOY, WARNING .... UOC:DAJ, DAK, V24, V25, ZAJ, ZAK	1

BRACKET ASSEMBLY	(1)	613-18
BRACKET ASSEMBLY	(1)	613-25
BRACKET ASSEMBLY	(1)	613-39
CABLE ASSEMBLY	(2)	613-2
CABLE ASSEMBLY	(1)	613-8
CABLE ASSEMBLY	(1)	613-34
CLAMP, LOOP	(2)	613-3
CLAMP, LOOP	(25)	613-9
CLAMP, LOOP	(13)	613-29
CONDUIT	(1)	613-28
COUPLING, PIPE	(2)	613-15
DUMMY CONNECTOR	(4)	613-5
GROMMET	(2)	613-11
HANDLE, SWITCH	(1)	613-22
LIGHT, WARNING	(2)	613-1
NUT, PLAIN, HEXAGON	(2)	613-13
NUT, SELF-LOCKING	(2)	613-24
NUT, SELF-LOCKING	(1)	613-33
PLATE, IDENTIFICATIO	(1)	613-23
PLATE, MOUNTING	(2)	613-4
PLATE, MOUNTING	(2)	613-14
RING, RETAINING	(2)	613-41
SCREW, CAP, HEXAGON	(3)	613-17
SCREW, CAP, HEXAGON	(1)	613 30
SCREW, CAP, HEXAGON	(3)	613-37
SCREW, MACHINE	(2)	613-26
SCREW, TAPPING	(25)	613-10
SCREW, TAPPING	(11)	613-27'
SCREW, WING	(1)	613-19
SCREW, WING	(1)	613-40
SHELL, ELECTRICAL	(2)	613-6
SHELL, ELECTRICAL	(2)	613-7
STRAP, TIEDOWN	(1)	613-35
SWITCH, ROTARY	(1)	613-21
WASHER, FLAT	(13)	613-12
WASHER, FLAT	(1)	613-32
WASHER, FLAT	(6)	613-38

(1) ITEM NO	(2) SMR CODE	(3) NSN	(4) CAGEC	(5) PART NUMBER	(6) DESCRIPTION AND USABLE ON CODES (UOC)	(7) QTY
					WASHER, LOCK (3) 613-16	
					WASHER, LOCK (1) 613-31	
					WIRING HARNESS (1) 613 36	
					END OF FIGURE	

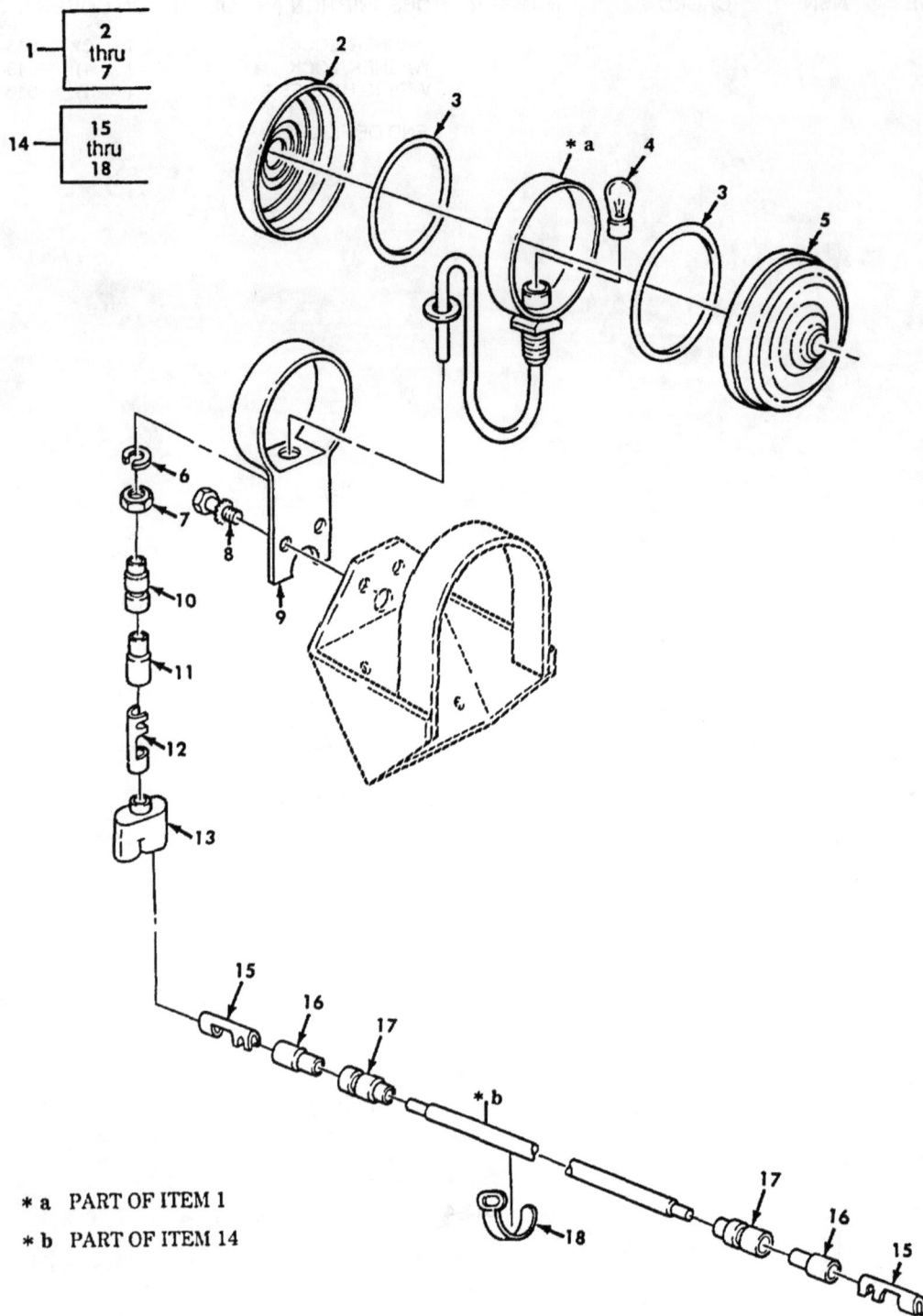

* a   PART OF ITEM 1
* b   PART OF ITEM 14

Figure 614. European Light Kit

(1) ITEM NO	(2) SMR CODE	(3) NSN	(4) CAGEC	(5) PART NUMBER	(6) DESCRIPTION AND USABLE ON CODES (UOC)	(7) QTY
					GROUP 3307 SPECIAL PURPOSE KITS	
					FIG. 614 EUROPEAN LIGHT KIT	
1	PAOZZ	6220012042597	19207	12301022	DIRECTIONAL LIGHT, V PART OF KIT P/N 12301147	2
2	PAOZZ	6220012252972	19207	12301036-2	LENS, LIGHT	1
3	PAOZZ	5330005797545	96906	MS28775-238	O-RING	2
4	PAOZZ	6240000446914	58536	A52463-2-10	LAMP, INCANDESCENT	1
5	PAOZZ	6220012131558	19207	12301036-1	LENS, LIGHT	1
6	PAOZZ	5310002617156	96906	MS35333-78	WASHER, LOCK.	1
7	PAOZZ	5310004883888	96906	MS51943-40	NUT, SELF-LOCKING, HE	1
8	PAOZZ	5306012260798	24617	423569	BOLT, ASSEMBLED WASH PART OF KIT P/N 12301147	4
9	PFOZZ	5977011959380	19207	11677047-3	HOLDER ASSEMBLY, ELE PART OF KIT P/N 12301147	2
10	PAOZZ	5935008338561	19207	8338561	SHELL, ELECTRICAL CO PART OF KIT P/N 12301147	2
11	PAOZZ	5970008338562	19207	8338562	INSULATOR, BUSHING PART OF KIT P/N 12301147	2
12	PAOZZ	5940003996676	19207	8338564	TERMINAL ASSEMBLY PART OF KIT P/N 12301147	2
13	PAOZA	5935009006281	96906	MS27147-1	ADAPTER, CONNECTOR PART OF KIT P/N 12301147	2
14	PAOZZ	5995011959405	19207	11677498-2	CABLE ASSEMBLY, SPEC PART OF KIT P/N 12301147	2
15	PAOZZ	5940003996676	19207	8338564	TERMINAL ASSEMBLY	2
16	PAOZZ	5970008338562	19207	8338562	INSULATOR, BUSHING.	2
17	PAOZZ	5935008338561	19207	8338561	SHELL, ELECTRICAL CO.	2
18	PAOZZ	9905007524649	96906	MS39020-1	BAND, MARKER	2
	PFOZZ		19207	12301147	KIT, EUROPEAN LIGHT	1

ADAPTER CONNECTOR	(2)	614-3
BOLT, ASSEMBLED	(4)	614-8
CABLE ASSEMBLY	(2)	614-14
DIRECTIONAL LIGHT	(2)	614-1
GUARD ASSEMBLY	(2)	614-9
INSULATOR, BUSHING	(2)	614-11
SHELL, ELECTRICAL	(2)	614-10
TERMINAL ASSEMBLY	(2)	614-12

END OF FIGURE

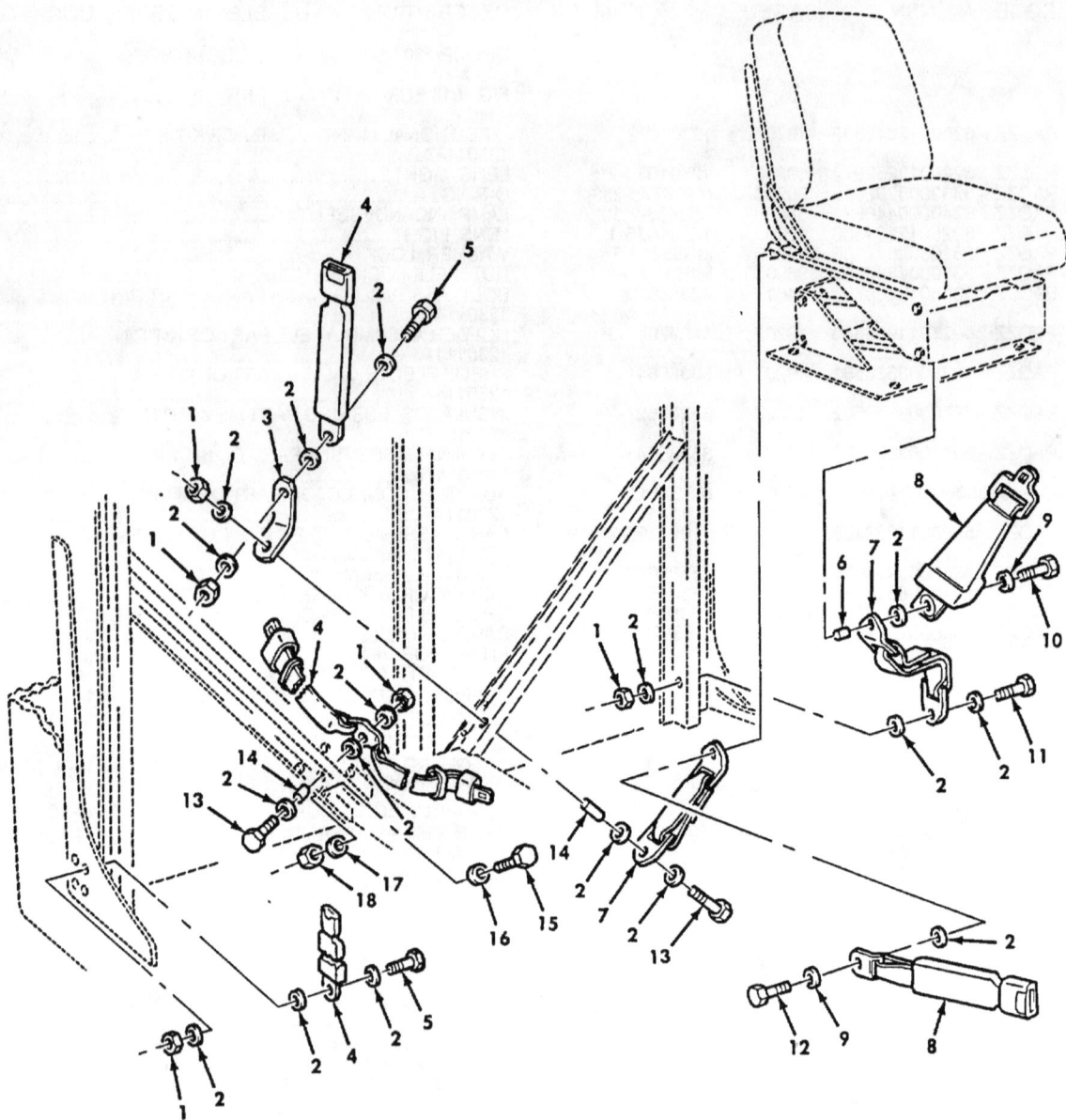

Figure 615. Seat Belt Kit.

(1) ITEM NO	(2) SMR CODE	(3) NSN	(4) CAGEC	(5) PART NUMBER	(6) DESCRIPTION AND USABLE ON CODES (UOC)	(7) QTY
					GROUP 3307 SPECIAL PURPOSE KITS	
					FIG. 615 SEAT BELT KIT	
1	PAOZZ	5310004883888	96906	MS51943-40	NUT, SELF-LOCKING, HE PART OF KIT P/N 12301340, 57K0124, 57K0125	5
2	PAOZZ	5310006560114	96906	MS15795-819	WASHER, FLAT PART OF KIT P/N 12301340 57K0124, 57K0125	17
3	PFOZZ	5340012906370	19207	12301344	BRACKET, ANGLE PART OF KIT P/N 12301340, 57K0124, 57K0125	1
4	PAOZZ	2540012867674	19207	12301346	BELT, VEHICULAR SAFE GREEN, COMPANION PART OF KIT P/N 12301340	1
4	PAOZZ		19207	12301346-1	BELT, VEHICULAR SAFE TAN, COMPANION PART OF KIT P/N 57K0124	1
4	PAOZZ		19207	12301346-2	BELT, VEHICULAR SAFE WHITE, COMPANION PART OF KIT P/N 57K0125	1
5	PAOZZ	5305007195219	96906	MS90727-111	SCREW, CAP, HEXAGON H PART OF KIT P/N 12301340, 57K0124, 57K0125	2
6	PAOZZ	5365012866180	19207	12301487	SPACER, SLEEVE PART OF KIT P/N 12301340, 57K0124, 57K0125	1
7	PAOZZ	2540012452445	19207	12301489	TETHER ASSEMBLY, GREEN, PART OF KIT 12301340	2
7	PAOZZ		19207	12301489-1	TETHER ASSEMBLY, TAN, PART OF KIT P/N 57K0124	2
7	PAOZZ		19207	12301489-2	TETHER ASSEMBLY WHITE, PART OF KIT P/N 57K0125	2
8	PAOZZ		19207	12301442-3	BELT, VEHICULAR SAFE GREEN, 7.5"-8.5" OPERATOR PART OF KIT P/N 12301340	1
8	PAOZZ	2540014188880	19207	12301442-6	BELT, VEHICULAR SAFE TAN, 7.5"-8.5" OPERATOR PART OF KIT P/N 57K0124	1
8	PAOZZ	2540014196283	19207	12301442-9	BELT, VEHICULAR SAFE WHITE, 7.5"-8.5" OPERATOR PART OF KIT P/N 57K0125	1
9	PAOZZ	5310000034094	01276	210104-8S	WASHER, LOCK PART OF KIT P/N 12301340 57K0124, 57K0125	2
10	PAOZZ	5305007098537	96906	MS90727-94	SCREW, CAP, HEXAGON H PART OF KIT P/N 12301340, 57K0124, 57K0125	1
11	PAOZZ	5305007195241	80204	B1821BH050F275N	SCREW, CAP, HEXAGON H PART OF KIT P/N 12301340, 57K0124, 57K0125.	1
12	PAOZZ	5305009146133	96906	MS18153-88	SCREW, CAP, HEXAGON H PART OF KIT P/N 12301340, 57K0124, 57K125	1
13	PAOZZ	5305007195262	96906	MS90727-121	SCREW, CAP, HEXAGON H PART OF KIT P/N 12301340, 57K0124, 57K0125	2
14	PAOZZ	5365012848152	19207	12301343	SPACER, SLEEVE PART OF KIT P/N 12301340, 57K0124, 57K0125	2
15	PAOZZ	5305000514075	96906	MS90727-33	SCREW, CAP, HEXAGON H PART OF KIT P/N 12301340, 57K0124, 57K0125	8
16	PAOZZ	5310000814219	96906	MS27183-12	WASHER, FLAT PART OF KIT P/N 12301340 57K0124, 57K0125	8
17	PAOZZ	5310011328275	24617	9418924	WASHER, FLAT PART OF KIT P/N 12301340 57K0124, 57K0125	8
18	PAOZZ	5310002416658	96906	MS51943-34	NUT, SELF-LOCKING, HE PART OF KIT P/N 12301340, 57K0124, 57K0125	8

(1) ITEM NO	(2) SMR CODE	(3) NSN	(4) CAGEC	(5) PART NUMBER	(6) DESCRIPTION AND USABLE ON CODES (UOC)			(7) QTY
	PFOZZ	2540012565331	34623	12301340	PARTS KIT, SEAT BELT GREEN			1
					BELT ASSEMBLY	(1)	615-4	
					BELT, VEHICULAR	(1)	615-8	
					BRACKET	(1)	615-3	
					NUT, SELF-LOCKING	(5)	615-1	
					NUT, SELF-LOCKING	(8)	615-18	
					SCREW, CAP, HEXAGON	(2)	615-5	
					SCREW, CAP, HEXAGON	(1)	615-10	
					SCREW, CAP, HEXAGON	(1)	615-11	
					SCREW, CAP, HEXAGON	(1	615-12	
					SCREW, CAP, HEXAGON	(2)	615-13	
					SCREW, CAP, HEXAGON	(8)	615-15	
					SLEEVE	(2)	615-14	
					SPACER	(1)	615-6	
					TETHER, BELT	(2)	615-7	
					WASHER	(8)	615-17	
					WASHER, FLAT	(17)	615-2	
					WASHER, FLAT	(8)	615-16	
					WASHER, LOCK	(2)	615-9	
	PFOZZ	2540012565331	34623	57K0124	PARTS KIT, SEAT BELT TAN			1
					BELT ASSEMBLY	(1)	615-4	
					BELT, VEHICULAR	(1)	615-8	
					BRACKET	(1)	615-3	
					NUT, SELF-LOCKING	(5)	615-1	
					NUT, SELF-LOCKING	(8)	615-18	
					SCREW, CAP, HEXAGON	(2)	615-5	
					SCREW, CAP, HEXAGON	(1)	615-10	
					SCREW, CAP, HEXAGON	(1)	615-11	
					SCREW, CAP, HEXAGON	(1)	615-12	
					SCREW, CAP, HEXAGON	(2)	615-13	
					SCREW, CAP, HEXAGON	(8)	615-15	
					SLEEVE	(2)	615-14	
					SPACER	(1)	615-6	
					TETHER, BELT	(2)	615-7	
					WASHER	(8)	615-17	
					WASHER, FLAT	(17)	615-2	
					WASHER, FLAT	(8)	615-16	
					WASHER, LOCK	(2)	615-9	
	PFOZZ	2540012565331	34623	57K0125	PARTS KIT, SEAT BELT WHITE...........................			1
					BELT ASSEMBLY	(1)	615-4	
					BELT, VEHICULAR	(1)	615-8	
					BRACKET	(1)	615-3	
					NUT, SELF-LOCKING	(5)	615-1	
					NUT, SELF-LOCKING	(8)	615-18	
					SCREW, CAP, HEXAGON	(2)	615-5	
					SCREW, CAP, HEXAGON	(1)	615-10	
					SCREW, CAP, HEXAGON	(1)	615-11	
					SCREW, CAP, HEXAGON	(1)	615-12	
					SCREW, CAP, HEXAGON	(2)	615-13	
					SCREW, CAP, HEXAGON	(8)	615-15	
					SLEEVE	(2)	615-14	
					SPACER	(1)	615-6	
					TETHER, BELT	(2)	615-7	

(1) ITEM NO	(2) SMR CODE	(3) NSN	(4) CAGEC	(5) PART NUMBER	(6) DESCRIPTION AND USABLE ON CODES (UOC)			(7) QTY
					WASHER	(8)	615-17	
					WASHER, FLAT	(17)	615-2	
					WASHER, FLAT	(8)	615-16	
					WASHER, LOCK	(2)	615-9	

END OF FIGURE

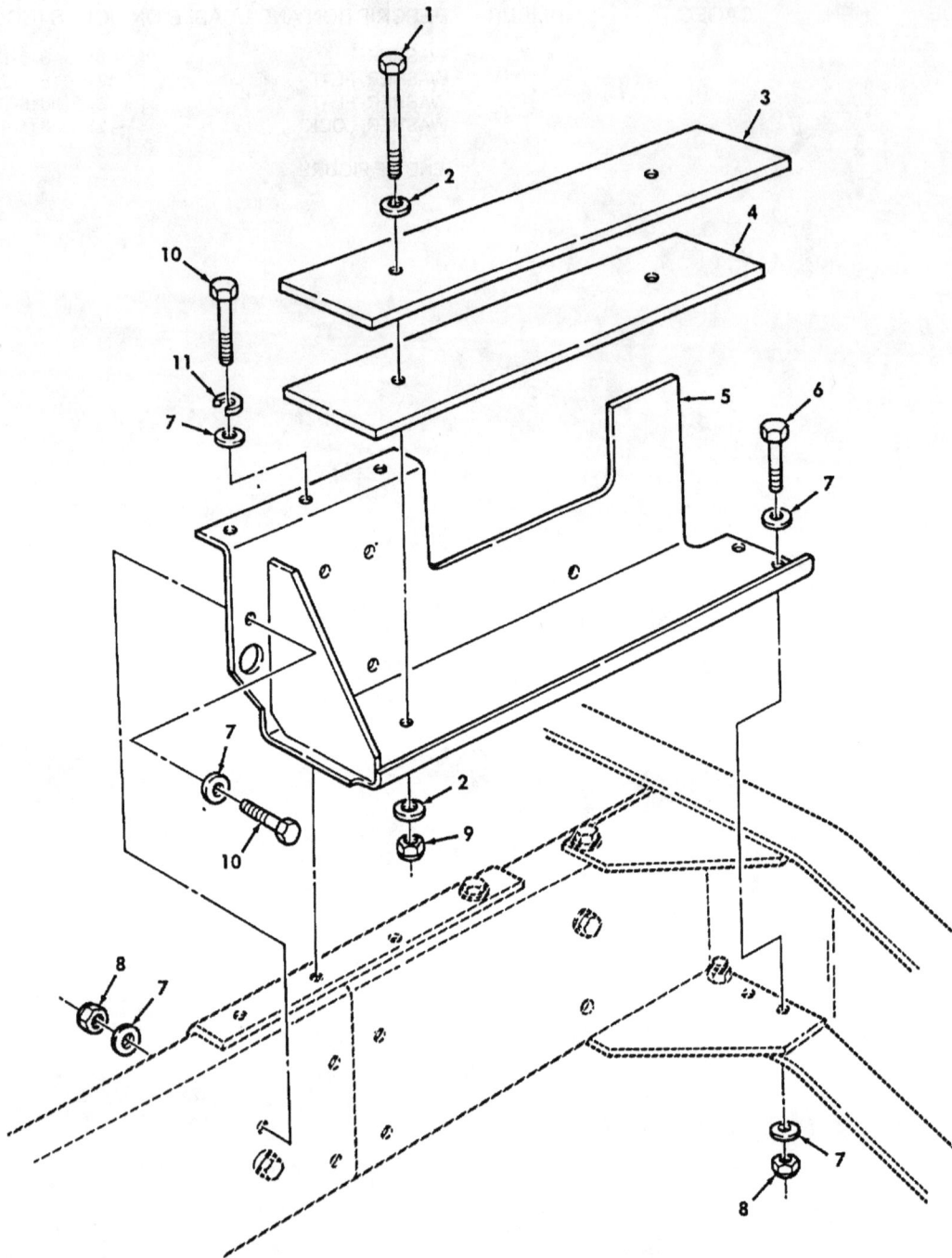

Figure 616. Wrecker Front Counterweight.

(1) ITEM NO	(2) SMR CODE	(3) NSN	(4) CAGEC	(5) PART NUMBER	(6) DESCRIPTION AND USABLE ON CODES (UOC)	(7) QTY
					GROUP 3307 SPECIAL PURPOSE KITS	
					FIG. 616 WRECKER FRONT COUNTERWEIGHT	
1	PAOZZ	5305007262567	96906	MS90727-176	SCREW, CAP, HEXAGON H .... UOC:ZAL	4
2	PAOZZ	5310011517347	24617	2436167	WASHER, FLAT .... UOC:ZAL	8
3	XDOZZ		19207	RCSK 14162-2	WEIGHT, UPPER.. UOC:ZAL	8
4	XDOZZ		19207	RCSK 14162-1	WEIGHT, LOWE R .... UOC:ZAL	4
5	XDOZZ		19207	RCSK 14159	BRACKET, MOUN TING................................... UOC:ZAL	1
5	PAOZZ	5340013531989	19207	RCSK 14160	BRACKET, MOUNTING................................... UOC:ZAL	1
6	PAOZZ	5305007195235	96906	MS90727-114	SCREW, CAP, HEXAGON H .... UOC:ZAL	15
7	PAOZZ	5310011211703	24617	2436165	WASHER, FLAT .... UOC:ZAL	41
8	PAOZZ	5310004883888	96906	MS51943-40	NUT, SELF-LOCKING, HE .... UOC:ZAL	17
9	PAOZZ	5310002416664	96906	MS51943-44	NUT, SELF-LOCKING, HE .... UOC:ZAL	4
10	PAOZZ	5305007195239	96906	MS90727-116	SCREW, CAP, HEXAGON H .... UOC:ZAL	9
11	PAOZZ	5310005845272	96906	MS35338-48	WASHER, LOCK .... UOC:ZAL	7

END OF FIGURE

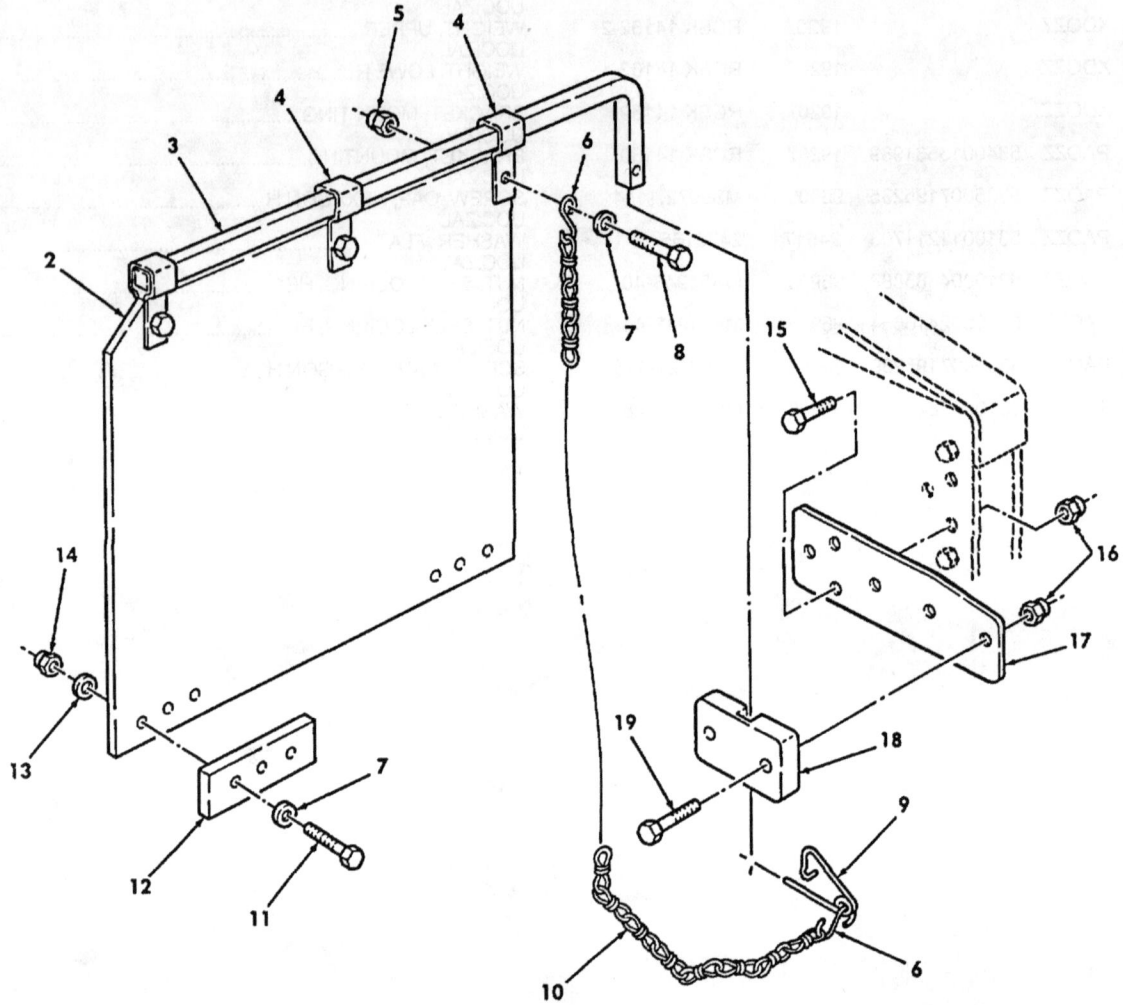

Figure 616A. Mudflap Stowage Kit (Sheet 1 of 2).

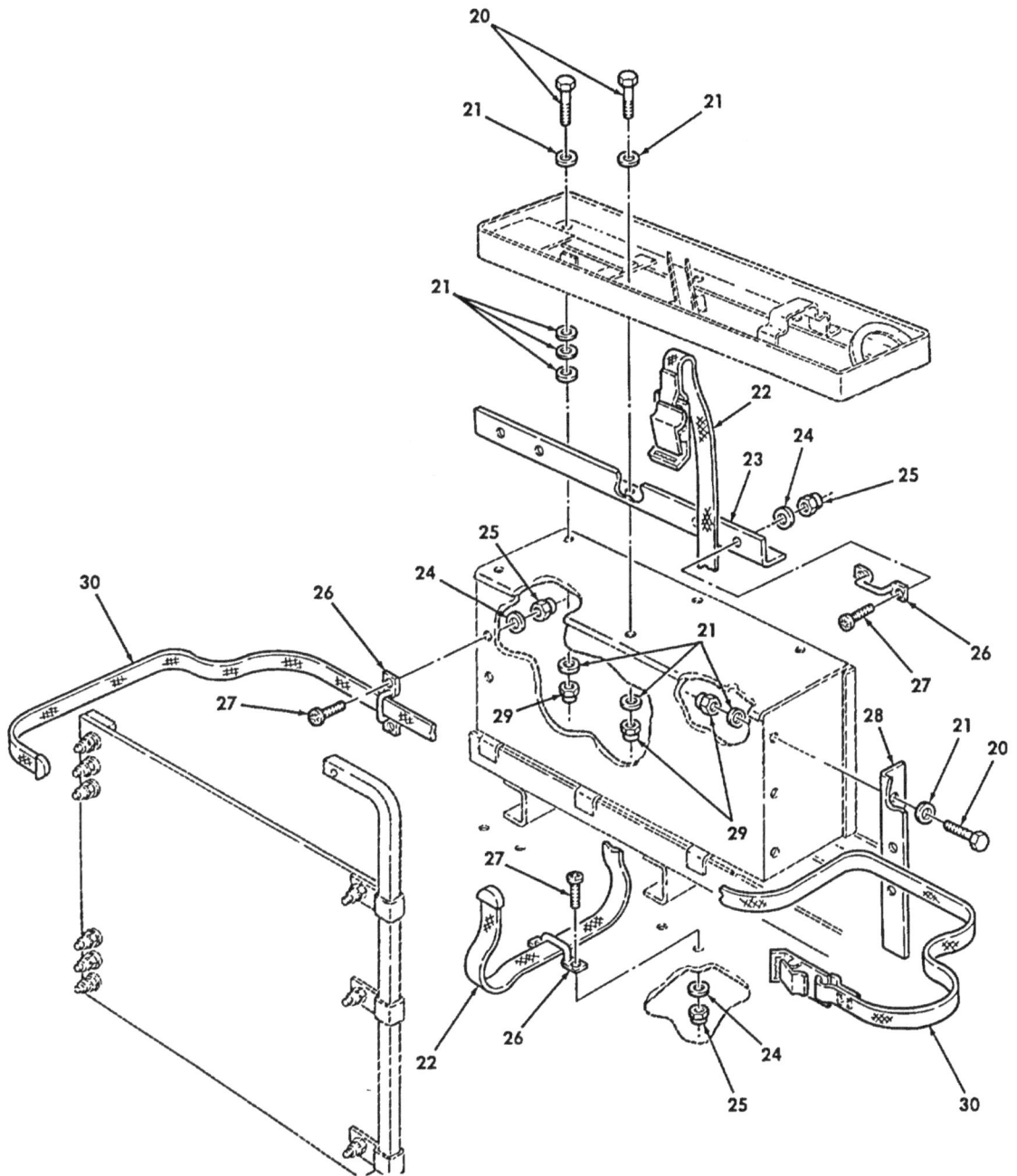

Figure 616A. Mudflap Stowage Kit (Sheet 2 of 2).

(1) ITEM NO	(2) SMR CODE	(3) NSN	(4) CAGEC	(5) PART NUMBER	(6) DESCRIPTION AND USABLE ON CODES (UOC)	(7) QTY
					GROUP 3307 SPECIAL PURPOSE KITS	
					FIG. 616A TRACTOR MUD FLAP STORAGE KIT	
1	PAOOO		19207	12300857-1	GUARD, SPLASH, VEHICU LEFT SIDE PART...... OF KIT P/N 12300848........... UOC:DAG, DAH, V21, V22, ZAG, ZAH	1
1	PAOOO		19207	12300857-2	GUARD, SPLASH, VEHICU RIGHT SIDE PART.......... OF KIT P/N 12300848 UOC:DAG, DAH, V21, V22, ZAG, ZAH	1
2	PAOZZ	2540014207925	34623	12300852	GUARD, SPLASH, VEHICU UOC:DAG, DAH, V21, V22, ZAG, ZAH	2
3	PFOZZ		19207	12300849	.ARM AND CLIP ASSEM .... UOC:DAG, DAH, V21, V22, ZAG, ZAH	2
4	PFOZZ		19207	12300851	.CLAMP, MUD FLAP .... UOC:DAG, DAH, V21, V22, ZAG, ZAH	4
5	PAOZZ	5310009359022	96906	MS51943-32	.NUT, SELF-LOCKING, HE 1/4-28...... UOC:DAG, DAH, V21, V22, ZAG, ZAH	6
6	PAOZZ	4030007809350	96906	MS87006-13	.HOOK, CHAIN, S .... UOC:DAG, DAH, V21, V22, ZAG, ZAH	4
7	PAOZZ	5310008094058	96906	MS27183-10	.WASHER, FLAT 1/4 .... UOC:DAG, DAH, V21, V22, ZAG, ZAH	14
8	PAOZZ	5305000680516	80204	B1821BH025F113N	.SCREW, CAP, HEXAGON H 1/4-28 X 1.13....... UOC:DAG, DAH, V21, V22, ZAG, ZAH	6
9	PAOZZ	5315007418971	19207	7418971	.PIN, RETAINING .... UOC:DAG, DAH, V21, V22, ZAG, ZAH	2
10	MOOZZ		74410	XB-196	.CHAIN, WELDLESS............... MAKE FROM CHAIN P/N XB-196, 17 INCHES LONG........... UOC:DAG, DAH, V21, V22, ZAG, ZAH	2
11	PAOZZ	5305002678954	80204	B1821BH025F125N	.SCREW, CAP, HEXAGON H 1/4-28 X 1.25........ UOC:DAG, DAH, V21, V22, ZAG, ZAH	12
12	PAOZZ		19207	12301456	.WEIGHT, MUD, FLAP .... UOC:DAG, DAH, V21, V22, ZAG, ZAH	4
13	PAOZZ		24617	9416904	.WASHER, FLAT 1/4 .... UOC:DAG, DAH, V21, V22, ZAG, ZAH	12
14	PAOZZ	5310000617325	96906	MS21045-4	.NUT, SELF-LOCKING, HE 1/4-28....... UOC:DAG, DAH, V21, V22, ZAG, ZAH	12
15	PAOZZ	5305007195235	80204	B1821BH050F175N	SCREW, CAP, HEXAGON H 1/2-20 X 1.75........ PART OF KIT P/N 12300848 .... UOC:DAG, DAH, V21, V22, ZAG, ZAH	6
16	PAOZZ	5310004883888	96906	MS51943-40	NUT, SELF-LOCKING, HE 1/2-20 PART OF....... KIT P/N 12300848	10
17	PFOZZ	5340012857757	19207	12300853	PLATE, MENDING PART OF KIT P/N........ 12300848......... UOC:DAG, DAH, V21, V22, ZAG, ZAH	2
18	PFOZZ	5340012859399	19207	12300854	BRACKET, MOUNTING PART OF KIT P/N .... 12300848......... UOC:DAG, DAH, V21, V22, ZAG, ZAH	2
19	PAOZZ	5305014248744	80204	B18B21BH050F225N	SCREW, CAP, HEXAGON H 1/2-20 X 2.25........ PART OF KIT P/N 12300848 .... UOC:DAG, DAH, V21, V22, ZAG, ZAH	4

(1) ITEM NO	(2) SMR CODE	(3) NSN	(4) CAGEC	(5) PART NUMBER	(6) DESCRIPTION AND USABLE ON CODES (UOC)	(7) QTY
20	PAOZZ	5305002693239	80204	B1821BH038F138N	SCREW, CAP, HEXAGON H 3/8-24 X 1.3750 PART OF KIT P/N 12300848 UOC:DAG, DAH, V21, V22, ZAG, ZAH	9
21	PAOZZ	5310008094061	96906	MS27183-15	WASHER, FLAT 7/16 PART OF KIT P/N 12300849 UOC:DAG, DAH, V21, V22, ZAG, ZAH	27
22	PZOZZ	5340005056379	21450	586689	STRAP, WEBBING 57 INCHES LONG PART OF KIT P/N 12300848 UOC:DAG, DAH, V21, V22, ZAG, ZAH	
23	PFOZZ		19207	12356744-1	BRACKET, ANGLE PART OF KIT P/N 12300848 UOC:DAG, DAH, V21, V22, ZAG, ZAH	1
24	PAOZZ	5310000145850	96906	MS27183-42	WASHER, FLAT 1/4 PART OF PART P/N 12300848 UOC:DAG, DAH, V21, V22, ZAG, ZAH	12
25	PAOZZ	5310000617326	96906	MS21045-3	NUT, SELF-LOCKING, HE 3/16-32 PART OF KIT P/N 12300848 UOC:DAG, DAH, V21, V22, ZAG, ZAH	12
26	PAOZZ		19207	10939516-2	STRAP, RETAINING PART OF KIT P/N 12300848 UOC:DAG, DAH, V21, V22, ZAG, ZAH	6
27	PAOZZ	5305009931848	96906	MS35207-265	SCREW, MACHINE 3/16-32 X .750 PART OF KIT P/N 12300848 UOC:DAG, DAH, V21, V22, ZAG, ZAH	12
28	PFOZZ		19207	12356745-1	BRACKET, ANGLE PART OF KIT P/N 12300848 UOC:DAG, DAH, V21, V22, ZAG, ZAH	1
29	PAOZZ	5310008140672	96906	MS51943-36	NUT, SELF-LOCKING, HE 3/8-24 PART OF KIT P/N 12300848 UOC:DAG, DAH, V21, V22, ZAG, ZAH	9
30	PAOZZ	5340004792947	19207	8690485	STRAP, WEBBING 76 INCHES LONG PART OF KIT P/N 12300848 UOC:DAG, DAH, V21, V22, ZAG, ZAH	1
	PFOZZ	2540012819855	19207	12300848	KIT, VE MUD FLAP STORAGE	1

BRACKET, ANGLE	(1)	616A-23
BRACKET, ANGLE	(1)	616A-28
BRACKET, MOUNTING	(2)	616A-18
NUT, SELF-LOCKING, HE	(12)	616A-14
NUT, SELF-LOCKING, HE	(10)	616A-16
NUT, SELF-LOCKING, HE	(12)	616A-25
NUT, SELF-LOCKING, HE	(9)	616A-29
PLATE, MENDING	(2)	616A-17
SCREW, CAP, HEX	(6)	616A-15
SCREW, CAP, HEX	(4)	616A-19
SCREW, CAP, HEX	(9)	616A-20
SCREW, MACHINE	(12)	616A-27
STRAP, RETAINING	(6)	616A-26
STRAP, WEBBING	(2)	616A-22
STRAP, WEBBING	(1)	616A-30
WASHER, FLAT	(27)	616A-21

END OF FIGURE

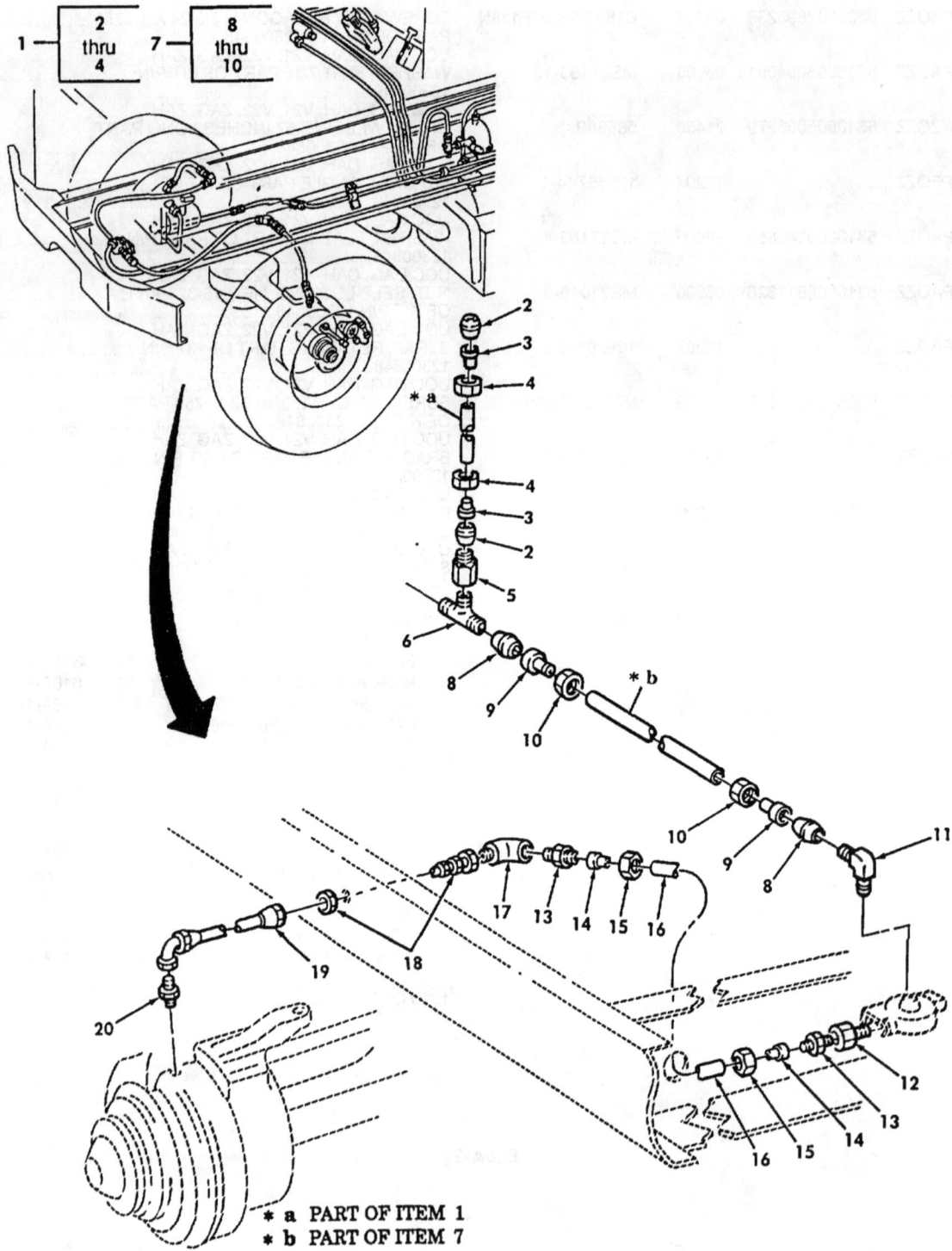

* a PART OF ITEM 1
* b PART OF ITEM 7

Figure 617. Exhaust Value Lines and Fittings, Front (M939A2).

(1) ITEM NO	(2) SMR CODE	(3) NSN	(4) CAGEC	(5) PART NUMBER	(6) DESCRIPTION AND USABLE ON CODES (UOC)	(7) QTY
					GROUP 43 CENTRAL TIRE INFLATION SYSTEM	
					4316 ASSEMBLED HOSES, FITTINGS, LINES, BREATHERS, FILTERS, AND TRAPS	
					FIG. 617 EXHAUST VALVE LINES AND FITTINGS, FRONT(M939A2)	
1	MOOZZ		19207	CPR104420-2X13	TUBE ASSEMBLY, MAKE FROM TUBING P/N .... 246115, 13 INCHES LONG........................	1
2	PAOZZ	4730002937108	16662	AC2511	SLEEVE, COMPRESSION	2
3	PAOZZ	4730010798821	19207	CPR102321-1	INSERT, TUBE FITTING	2
4	PAOZZ	4730002788825	73830	200360	NUT, TUBE COUPLING	2
5	PAOZZ	4730011340853	93061	66NTA-6-6	ADAPTER, STRAIGHT, PI	1
6	PAOZZ	4730011196895	93061	272NTA-8-6	TEE, PIPE TO TUBE	1
7	MOOZZ		19207	CPR104420-3X108	TUBE ASSEMBLY MAKE FROM TUBING P/N .... 246109, 108 INCHES LONG........................	1
8	PAOZZ	4730000542571	81343	8 120115B	SLEEVE, COMPRESSION	2
9	PAOZZ	4730010326038	19207	CPR102321-4	INSERT, TUBE FITTING	2
10	PAOZZ	4730000542572	81343	8 120111B	NUT, TUBE COUPLING	2
11	PAOZZ	4730013004104	47457	20511216	ELBOW, PIPE TO TUBE	2
12	PAOZZ	4730012842211	24617	444017	REDUCER, PIPE	2
13	PAOZZ	4730011086410	93061	68NTA-12-8	ADAPTER, STRAIGHT, PI	4
14	PAOZZ	4730001807031	06853	246091	INSERT, TUBE FITTING	4
15	PAOZZ	4730002401740	96906	MS39176-9	NUT, TUBE, COUPLING	4
16	MOOZZ		19207	CPR104420-5X42	HOSE, NONMETALLIC L.H., MAKE FROM........ TUBING, P/N CPR104420, 42 INCHES LONG	1
16	MOOZZ		19207	CPR104420-5X52	HOSE NONMETALLIC R.H., MAKE FROM........ TUBING, P/N CPR104420, 52 INCHES LONG	1
17	PAOZZ	4730013004091	47457	20511217	ELBOW, PIPE	2
18	PAOZZ	4730012808331	98441	12WGTX-WLN-S	ADAPTER, STRAIGHT, PI	2
19	PAOZZ	4720013321596	98441	A6056-5	HOSE ASSEMBLY, NONME	2
20	PAOZZ	4730000137409	93061	48IFHD-8-6	ADAPTER, STRAIGHT, PI	2

END OF FIGURE

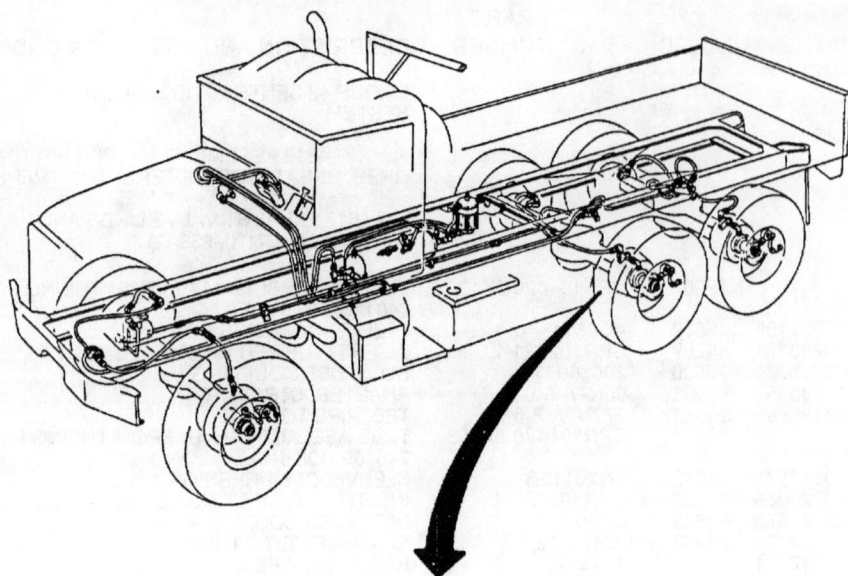

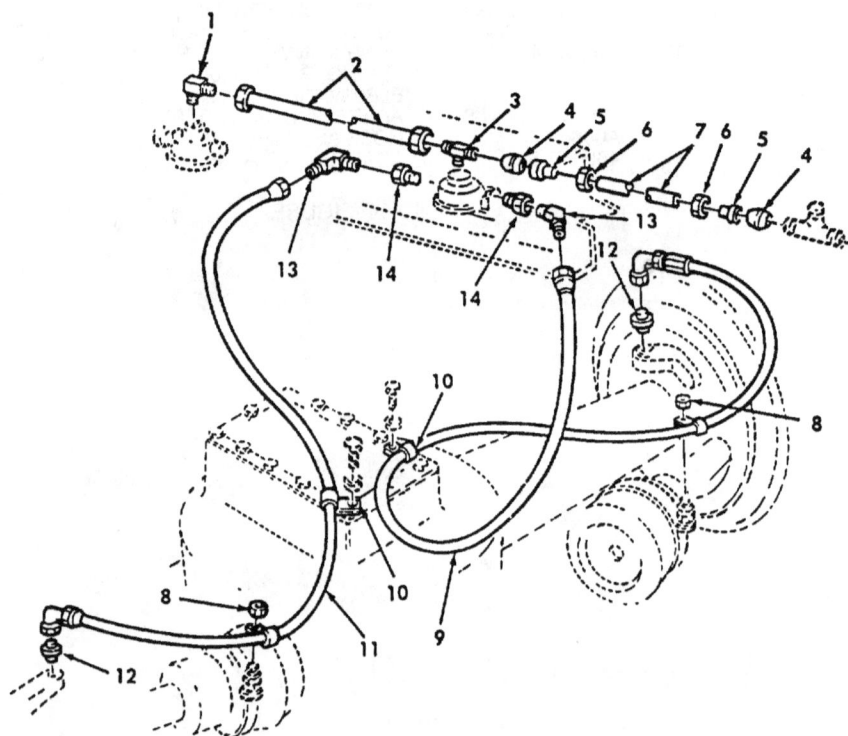

Figure 618. Exhaust Valve Lines and Fittings, Rear (M939A2).

(1) ITEM NO	(2) SMR CODE	(3) NSN	(4) CAGEC	(5) PART NUMBER	(6) DESCRIPTION AND USABLE ON CODES (UOC)	(7) QTY
					GROUP 4316 ASSEMBLED HOSES,FITTINGS, LINES,BREATHERS,FILTERS,AND TRAPS	
					FIG. 618 EXHAUST VALVE LINES AND FITTINGS,RE-AR(M939A2)	
1	PAOZZ	4730012717957	98441	30541-8-8B	ELBOW, PIPE TO TUBE	1
2	MOOZZ		19207	CPR104420-3X53	HOSE, NONMETALLIC MAKE FROM TUBING..................... P/N 06642-0000, 53 INCHES LONG....................	1
3	PAOZZ	4730013009031	93061	272NTA-8-8	TEE, PIPE TO TUBE	1
4	PAOZZ	4730000542571	81343	8 120115B	SLEEVE, COMPRESSION,	2
5	PAOZZ	4730010326038	19207	CPR102321-4	INSERT, TUBE FITTING	2
6	PAOZZ	4730000542572	81343	8 120111B	NUT, TUBE COUPLING	2
7	MOOZZ		19207	CPR104420-3X65	TUBE ASSEMBLY MAKE FROM TUBING, P/N .... 06642-0000, 65 INCHES LONG.................... UOC:ZAE, ZAF, ZAG, ZAH	1
7	MOOZZ		19207	CPR104420-3X77	TUBE ASSEMBLY MAKE FROM TUBING, P/N .... 06642-0000, 77 INCHES LONG.................... UOC:ZAA, ZAB, ZAL	1
7	MOOZZ		19207	CPR104420-3X113	TUBE ASSEMBLY MAKE FROM TUBING, P/N .... 06642-0000, 113 INCHES LONG.................... UOC:ZAC, ZAD, ZAJ, ZAK	1
8	PAOZZ	5310000874652	96906	MS51922-17	NUT, SELF-LOCKING, HE	4
9	PAOZZ	4720013105475	98441	A6057-4	HOSE ASSEMBLY, NONME	2
10	PAOZZ	5340008333049	96906	MS21333-127	CLAMP, LOOP	4
11	PAOZZ	4720013105476	98441	A6057-3	HOSE ASSEMBLY, NONME	2
12	PAOZZ	4730009277272	98441	1011-8-12B	ADAPTER, STRAIGHT, PI.....................	4
13	PAOZZ	4730012838149	81343	12-8 010302B	ELBOW, PIPE TO TUBE	4
14	PAOZZ	4730012842211	24617	444017	REDUCER, PIPE	4

END OF FIGURE

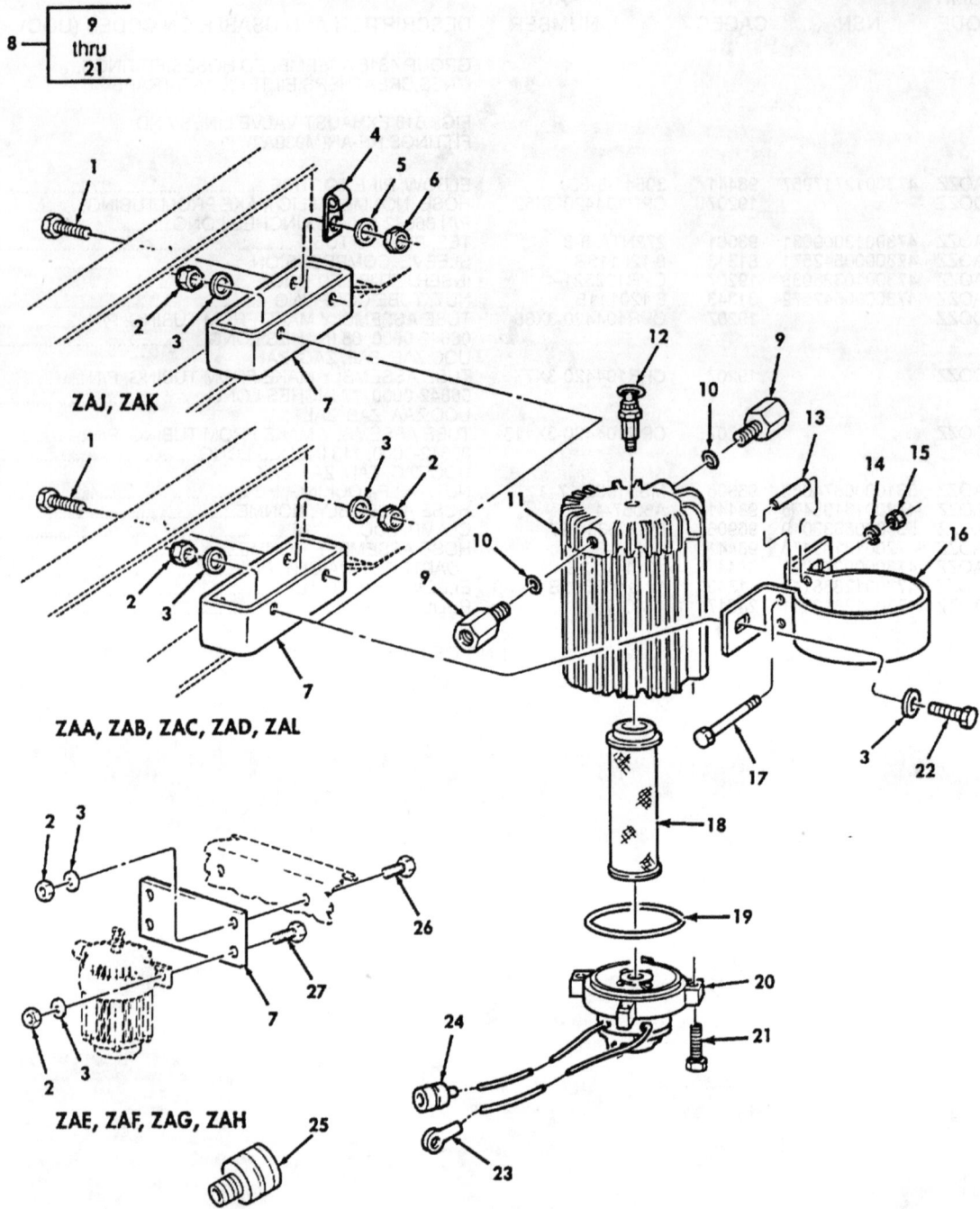

Figure 619. CTIS Air Dryer Assembly.

(1) ITEM NO	(2) SMR CODE	(3) NSN	(4) CAGEC	(5) PART NUMBER	(6) DESCRIPTION AND USABLE ON CODES (UOC)	(7) QTY
					GROUP 4316 ASSEMBLED HOSES, FITTINGS, LINES, BREATHERS, FILTERS, AND TRAPS	
					FIG. 619 CTIS AIR DRYER ASSEMBLY (M939A2)	
1	PAOZZ	5305009146131	96906	MS18153-63	SCREW, CAP, HEXAGON H .... UOC:ZAJ, ZAK	1
1	PAOZZ	5305007195238	96906	MS90727-115	SCREW, CAP, HEXAGON H .... UOC:ZAA, ZAB, ZAC, ZAD, ZAL	1
2	PAOZZ	5310004883888	96906	MS51943-40	NUT, SELF-LOCKING, HE .... UOC:ZAJ, ZAK	2
2	PAOZZ	5310004883888	96906	MS51943-40	NUT, SELF-LOCKING, HE .... UOC:ZAA, ZAB, ZAC, ZAD, ZAE, ZAF, ZAG, ZAH, ZAL	1
3	PAOZZ	5310010842362	63005	9411417	WASHER .... UOC:ZAA, ZAB, ZAC, ZAD, ZAE, ZAF, ZAG, ZAH, ZAL	1
3	PAOZZ	5310010842362	63005	9411417	WASHER .... UOC:ZAJ, ZAK	4
4	PAOZZ	5340001519651	96906	MS21333-129	CLAMP, LOOP .... UOC:ZAJ, ZAK	1
5	PAOZZ	5310002748041	90407	12084P11	WASHER, FLAT .... UOC:ZAJ, ZAK	1
6	PAOZZ	5310011505914	11862	9422298	NUT, SELF-LOCKING, HE .... UOC:ZAJ, ZAK	1
7	PAOZZ	5340012807096	47457	20511188	BRACKET, MOUNTING .... UOC:ZAA, ZAB, ZAC, ZAD, ZAL	1
7	PGOZZ	5340013303240	47457	20511221	BRACKET, MOUNTING .... UOC:ZAJ, ZAK	1
7	PAOZZ	5340013390874	47457	20511286	PLATE, MOUNTING, FLAT .... UOC:ZAE, ZAF, ZAG, ZAH	1
8	PAOFF	4730012804204	8X715	19324B	AIR DRYER AND COOLE	1
9	PAOZZ	4730012808345	8X715	18028	.ADAPTER, STRAIGHT, PI	2
10	PAOZZ	5330012806503	8X715	18048	.PACKING, PREFORMED	2
11	PAFZZ	4440012879011	8X715	19046	.HEAD, FLUID FILTER	1
12	PAFZZ	4820012803947	8X715	18152T	.VALVE, SAFETY RELIEF	1
13	PAOZZ	5365012881559	8X715	18302	.SPACER, SLEEVE	2
14	PAOZZ	5310005825965	96906	MS35338-44	.WASHER, LOCK	2
15	PAOZZ	5310002509477	96906	MS35649-2254	.NUT, PLAIN, HEXAGON	2
16	PAOZZ	5340012806995	8X715	18300	.CLAMP, LOOP	1
17	PAOZZ	5305000712510	96906	MS90728-13	.SCREW, CAP, HEXAGON H	2
18	PFOZZ	4310012808409	8X715	19130	.FILTER ELEMENT, FLUI PART OF KIT P/N 12503	1
19	PAOZZ	5330012806504	8X715	05240	.PACKING, PREFORMED PART OF KIT P/N 12503	1
20	PAOZZ	2530013671883	8X715	19174B	.VALVE, BRAKE PNEUMAT	1
21	PAOZZ	5305010314487	24617	426371	.SCREW, CAP, SOCKET HE	4
22	PAOZZ	5305007195219	96906	MS90727-111	SCREW, CAP, HEXAGON H	2
23	PAOZZ	5940001133147	96906	MS20659-127	TERMINAL, LUG	1
24	PAOZZ	5935001677775	96906	MS27144-1	CONNECTOR, PLUG, ELEC	1
25	PAOZZ	4820012803935	8X715	18164	VALVE, CHECK	1
26	PAOZZ	5305007250154	96906	MS90727-112	SCREW, CAP, HEXAGON H	1

(1) ITEM NO	(2) SMR CODE	(3) NSN	(4) CAGEC	(5) PART NUMBER	(6) DESCRIPTION AND USABLE ON CODES (UOC)	(7) QTY
27	PAOZZ	5305007195221	96906	MS90727-113	UOC:ZAE, ZAF, ZAG, ZAH SCREW, CAP, HEXAGON H .... UOC:ZAE, ZAF, ZAG, ZAH	1

END OF FIGURE

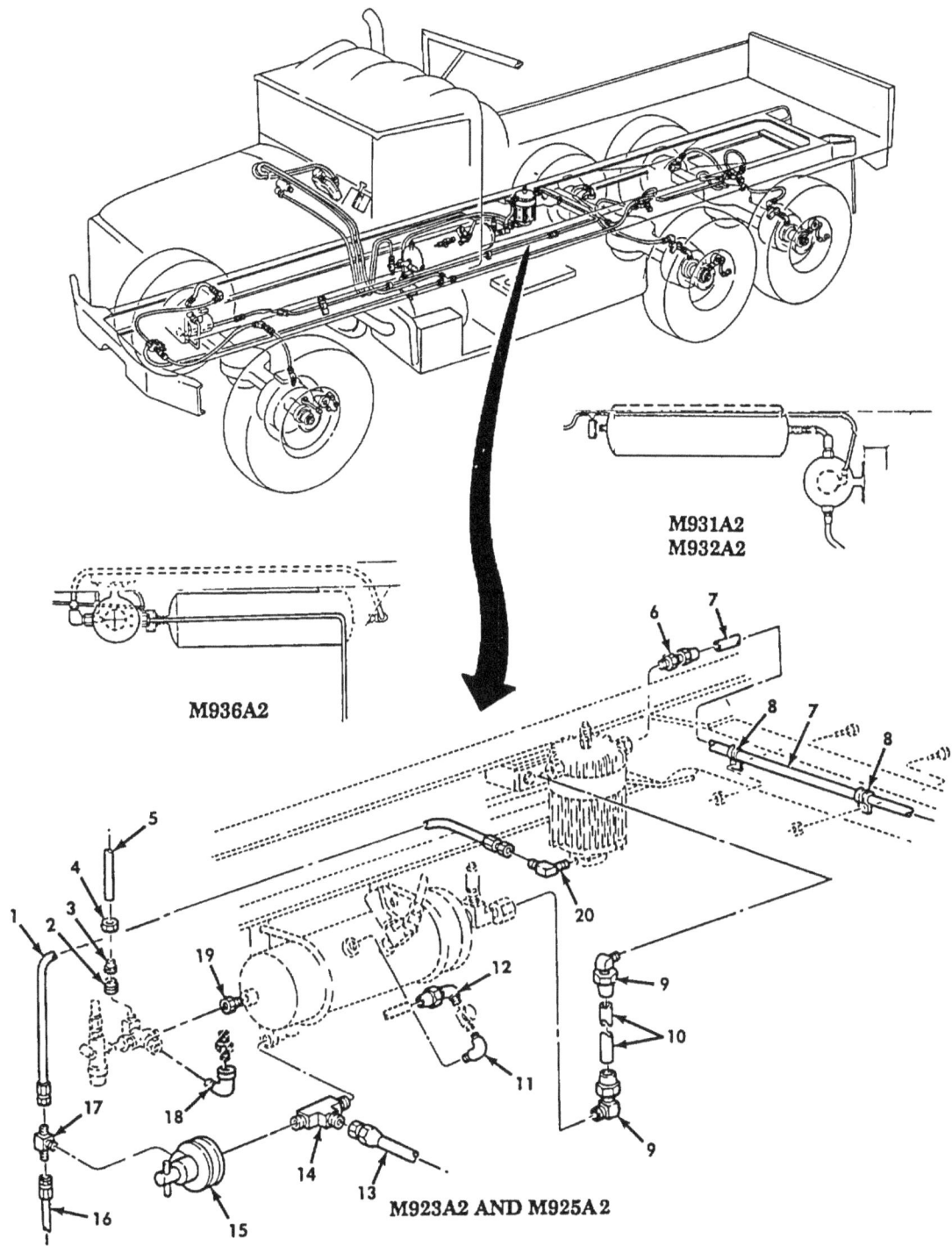

Figure 620. Air Dryer Lines and Fittings (M939A2).

(1) ITEM NO	(2) SMR CODE	(3) NSN	(4) CAGEC	(5) PART NUMBER	(6) DESCRIPTION AND USABLE ON CODES (UOC)	(7) QTY
					GROUP 4317 MANIFOLD AND/OR CONTROL VALVES	
					FIG. 620 AIR DRYER LINES AND FITTINGS (M939A2)	
1	PAOZZ	4720012842235	47457	20510558	HOSE ASSEMBLY, NONME	1
2	PAOZZ	4730002937108	16662	AC2511	SLEEVE, COMPRESSION	1
3	PAOZZ	4730010798821	19207	CPR102321-1	INSERT, TUBE FITTING	2
4	PAOZZ	4730002788825	81343	4 120111B	NUT, TUBE COUPLING	2
5	MOOZZ		47457	20510633-290	HOSE MAKE FROM HOSE, P/N 246115, 290 INCHES LONG..	1
6	PAOZZ	4730002029035	81343	12-8 120102BA	ADAPTER, STRAIGHT, PI	1
7	PAOZZ	4710012791494	4F744	20511181-1	TUBE, BENT, METALLIC .... UOC:ZAA, ZAB	1
7	PAOZZ	4710012895447	47457	20511181-2	TUBE, BENT, METALLIC .... UOC:ZAC, ZAD	1
7	PAOZZ	4710012895448	47457	20511181-3	TUBE, BENT, METALLIC .... UOC:ZAJ, ZAK	1
7	PAOZZ	4710012895449	47457	20511181-4	TUBE, BENT, METALLIC .... UOC:ZAL	1
7	PAOZZ	4710012879008	47457	20511240	TUBE ASSEMBLY, METAL .... UOC:ZAE, ZAF, ZAG, ZAH	1
8	PAOZZ	5340002827509	96906	MS21333-62	CLAMP, LOOP .... UOC:ZAG, ZAH	1
8	PAOZZ	5340002827509	96906	MS21333-62	CLAMP, LOOP .... UOC:ZAA, ZAB, ZAC, ZAD, ZAE, ZAF, ZAJ, ZAK, ZAL	2
9	PAOZZ	4730011225857	93061	269AB-12-8	ELBOW, PIPE TO TUBE	2
10	MOOZZ		19207	CPR104420-108	HOSE, NONMETALLIC MAKE FROM HOSE, P/N 247607, 108 INCHES LONG..................	1
11	PAOZZ	4730013004092	47457	20510945	ELBOW, PIPE	1
12	PAOZZ	4730012717957	98441	30541-8-8B	ELBOW, PIPE TO TUBE	1
13	PAOZZ	4720013004143	47457	20510805	HOSE ASSEMBLY, NONME	1
14	PAOZZ	4730010053262	72582	444148	TEE, PIPE	1
15	PAOZZ	2530012873131	8X715	15240-B	VALVE, BRAKE PNEUMAT	1
16	PAOZZ	4720013004144	47457	20510551	HOSE ASSEMBLY, NONME	1
17	PAOZZ	4730005221909	81343	4-4-2 070425CA	TEE, PIPE TO TUBE	1
18	PAOZZ	4730002775553	96906	MS51952-2	ELBOW, PIPE	1
19	PAOZZ	4730012999488	98441	0101-4-8	REDUCER, PIPE	1
20	PAOZZ	4730006473207	96906	MS51504-B4	ELBOW, PIPE TO TUBE	1

END OF FIGURE

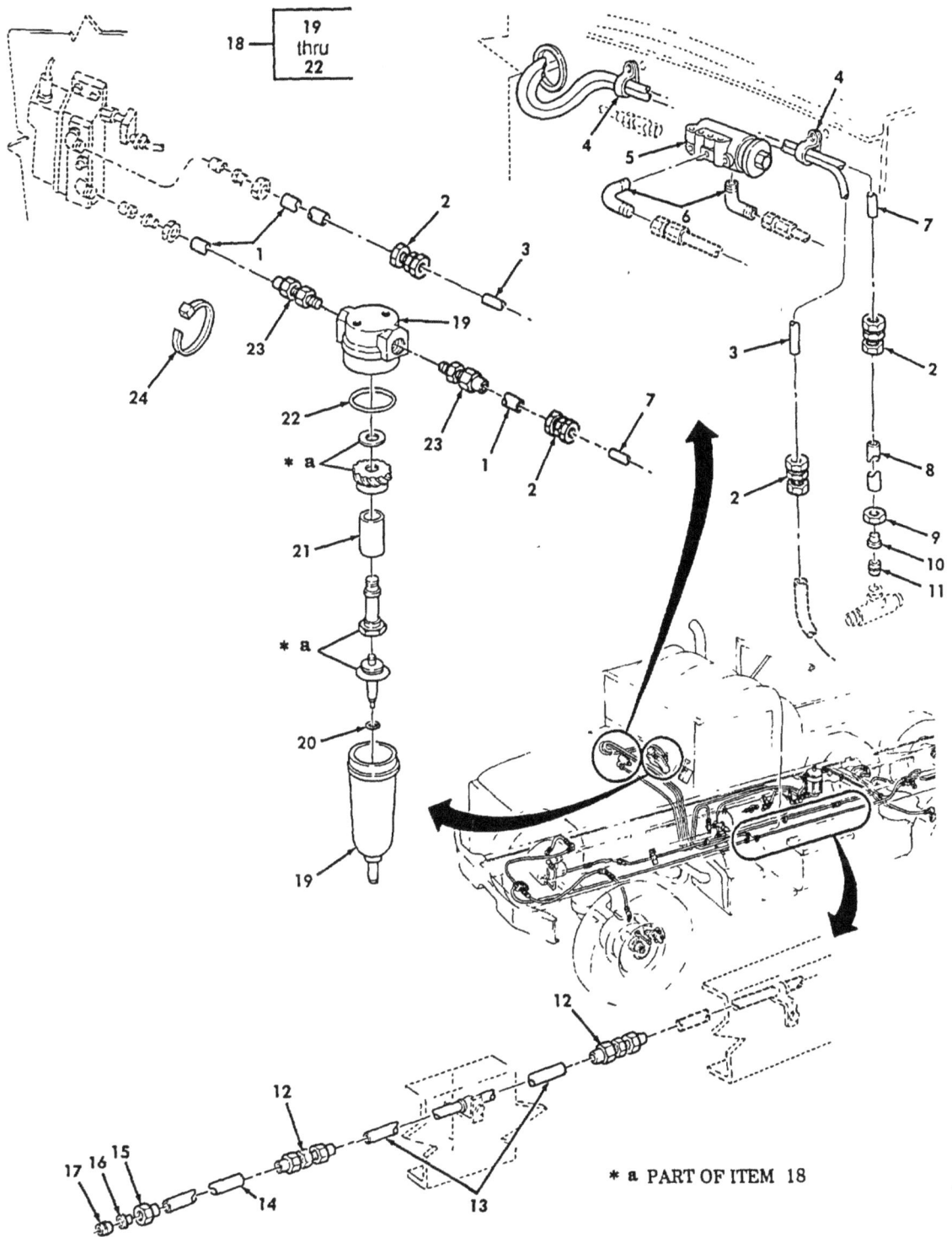

Figure 621. Air Dryer Lines and Fittings, Water Separator, and Governor (M939A2).

(1) ITEM NO	(2) SMR CODE	(3) NSN	(4) CAGEC	(5) PART NUMBER	(6) DESCRIPTION AND USABLE ON CODES (UOC)	(7) QTY
					GROUP 4317 MANIFOLD AND/OR CONTROL VALVES	
					FIG. 621 AIR DRYER LINES AND FITTINGS, WATER SEPARATOR, AND GOVERNOR (M939A2)	
1	MOOZZ		19207	CPR104420-2-22	HOSE, NONMETALLIC MAKE FROM HOSE, P/N CPR104420, 22 INCHES LONG	3
2	PAOZZ	4730011339866	93061	62NTA-6	UNION, TUBE	4
3	PAOZZ	4710012999450	47457	20511256-2	TUBE, BENT, METALLIC	1
4	PAOZZ	5340008091494	96906	MS21333-105	CLAMP, LOOP	2
5	PAOZZ	2530012874529	06853	106400	GOVERNOR ASSEMBLY, A	1
6	PAOZZ	4730009050030	98441	4EBTXB	ELBOW, T UBE	2
7	PAOZZ	4710012999451	47457	20511256-1	TUBE, BENT, METALLIC	1
8	MOOZZ		47457	20510633-176	HOSE MAKE FROM HOSE, P/N 246115, 176 INCHES LONG	1
9	PAOZZ	4730002788824	81343	4 120111B	NUT, TUBE COUPLING	2
10	PAOZZ	4730010798821	19207	CPR102321-1	INSERT, TUBE FITTING	2
11	PAOZZ	4730002937108	16662	AC2511	SLEEVE, COMP RESSIONS,	1
12	PAOZZ	4730002783214	21450	504349	NIPPLE, TUBE	2
13	PAOZZ	4710012895446	47457	20511231	TUBE, BENT, METALLIC	1
14	MOOZZ		01276	2807-12-14 1/2	HOSE MAKE FROM HOSE, P/N 2807-12, 14 1/2 INCHES LONG	1
15	PAOZZ	4730000504309	81343	12 120111B	NUT, TUBE COUPLING	1
16	PAOZZ	4730001807031	19207	CPR102321-2	INSERT, TUBE FITTING	1
17	PAOZZ	4730001818870	81341	2 120115B	SLEEVE, COMPRESSION,	1
18	PAOZZ	4730013045618	43990	F07-200-A1MA	TRAP, MOISTURE MINIATURE COMPRESSED AIR	1
19	XAOZZ		43990	F07-200-A1MA XA	.TRAP, MOISTURE	1
20	KFOZZ		43990	5938-01	.GASKET PART OF KIT P/N 3652-11	1
21	XFOZZ		43990	5726-01	.FILTER ELEMENT FLUI PART OF KIT P/N 3652-11	1
22	KFOZZ		43990	2315-24	.PACKING, PREFORMED PART OF KIT P/N 3652-11	1
23	PAOZZ	4730000691186	16662	AC2569	ADAPTER, STRAIGHT, PI PIPE TO TUBE	2
24	PAOZZ	5975001563253	96906	MS3367-2-9	STRAP, TIEDOWN, ELECT	6

END OF FIGURE

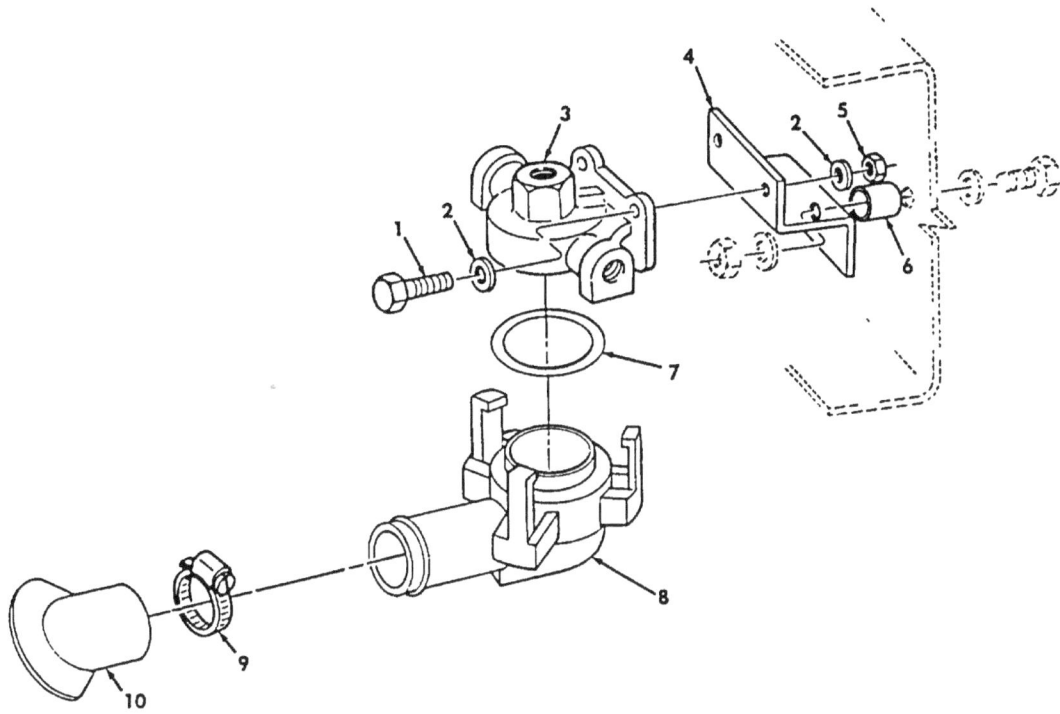

Figure 622. Relief Safety Valve Assembly .

(1) ITEM NO	(2) SMR CODE	(3) NSN	(4) CAGEC	(5) PART NUMBER	(6) DESCRIPTION AND USABLE ON CODES (UOC)	(7) QTY
					GROUP 4317 MANIFOLD AND/OR CONTROL VALVES	
					FIG. 622 RELIEF SAFETY VALVE ASSEMBLY (M939A2)	
1	PAOZZ	5306002259089	96906	MS90726-34	BOLT, MACHINE	6
2	PAOZZ	5310000814219	96906	MS27183-12	WASHER, FLAT	12
3	PAOZZ	4820012765731	52304	599802	VALVE, SAFETY RELIEF ASSEMBLY	3
4	PAOZZ	5340012884557	47457	20511248	BRACKET, DOUBLE ANGL	3
5	PAOZZ	5310009054600	96906	MS51968-6	NUT, PLAIN, HEXAGON	6
6	PAOZZ	5365013023189	4F344	20511226	SPACER, SLEEVE	2
7	PAOZZ	5330010053704	81349	M83248/1-324	PACKING, PREFORMED PROVIDED WITH ...... Q.E.V. VALVE P/N 20511265	3
8	PAOZZ	4220013074779	47457	20511265	ADAPTER, QUICK CONNE	3
9	PAOZZ	4730009083194	96906	MS35842-11	CLAMP, HOSE	3
10	PAOZZ	2530013192384	47457	20511279	BOOT, DUST AND MOIST	3

END OF FIGURE

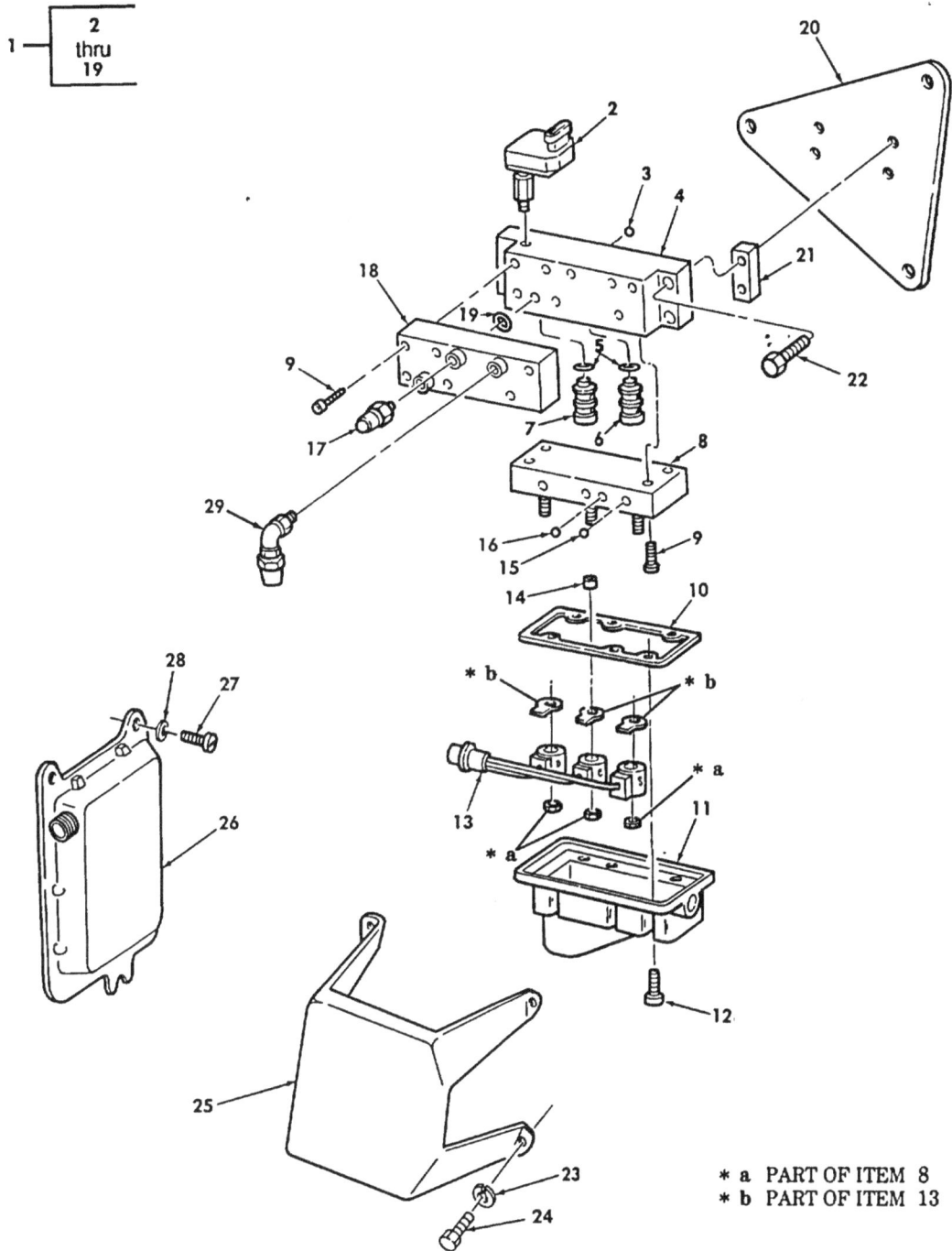

Figure 623. Electronic Control Unit and Pneumatic Controller (M939A2).

* a PART OF ITEM 8
* b PART OF ITEM 13

(1) ITEM NO	(2) SMR CODE	(3) NSN	(4) CAGEC	(5) PART NUMBER	(6) DESCRIPTION AND USABLE ON CODES (UOC)	(7) QTY
					GROUP 4317 MANIFOLD AND/OR CONTROL VALVES	
					FIG. 623 ELECTRONIC CONTROL UNIT AND PNEUMATIC CONTROLLER(M939A2)	
1	PAOFF	4820012672914	52304	599757	VALVE, REGULATING, FL CONTROLLER ASSEMBLY	1
2	PAOZZ	6695013296418	52304	599601	.TRANSDUCER, PRESSURE	1
3	XAOZZ		52304	599723 XA	.BALL	4
4	XAOZZ		52304	599706 XA	.BODY, VALVE	1
5	KFOZZ		52304	599719	.PACKING, PREFORMED, PART OF KIT P/N 599911	3
6	KFOZZ		52304	599765	.CARTRIDGE, VALVE, PART OF KIT P/N 599911	1
7	KFOZZ		52304	599615	.CARTRIDGE, VALVE, PART OF KIT P/N 5999 11	2
8	XAOZZ		52304	599707 XA	.PLATE, COVER	1
9	PAOZZ	5305012876570	52304	806738	.SCREW, CAP, SOCKET HE HEAD	10
10	KFOZZ		52304	599718	.GASKET, PART OF KIT P/N 599911	1
11	KFOZZ		52304	599729	.PROTECTOR, SOLENOID PART OF KIT P/N 599911	1
12	XAOZZ		52304	806739	.SCREW, CAP, SOCKET HE	6
13	XAOZZ		52304	599760	.WIRING HARNESS, BRAN	1
14	XAOZZ		52304	599614 XA	.SEAT	3
15	XAOZZ		52304	599721 XA	.BALL	3
16	XAOZZ		52304	599722 XA	.BALL	3
17	PAOZZ	4820012717946	52304	599766	.VALVE, SAFETY RELIEF	1
18	XAOZZ		52304	599734 XA	.PLATE, BASE	1
19	KFOZZ		52304	599720	.PACKING, PREFORMED, PART OF KIT P/N 599911	5
20	PAOZZ	5340012849656	47457	20511234	BRACE, CORNER	1
21	PAOZZ	5365013052535	47457	12256277A-2	SPACER, SLEEVE	4
22	PAOZZ	5305009836665	96906	MS16997-65	SCREW, CAP, SOCKET HE	4
23	PAOZZ	5310002090786	96906	MS35335-33	WASHER, LOCK UOC:ZAA, ZAC, ZAG, ZAJ	3
24	PAOZZ	5305002678957	96906	MS90727-13	SCREW, CAP, HEXAGON H HEAD UOC:ZAA, ZAC, ZAG, ZAJ	3
25	PAOZZ	5340012853257	47457	20511235	COVER, ACCESS UOC:ZAA, ZAC, ZAG, ZAJ	1
26	PAOZZ	6110012688739	52304	599730	ELECTRONIC CONTROL	1
27	PAOZZ	5305002253841	96906	MS90728-11	SCREW, CAP, HEXAGON H HEAD	3
28	PAOZZ	5310011023270	24617	2436161	WASHER, FLAT 1/4	3
29	PAOZZ	4730000691187	81343	6-4 100202BA	ELBOW, PIPE TO TUBE	2

END OF FIGURE

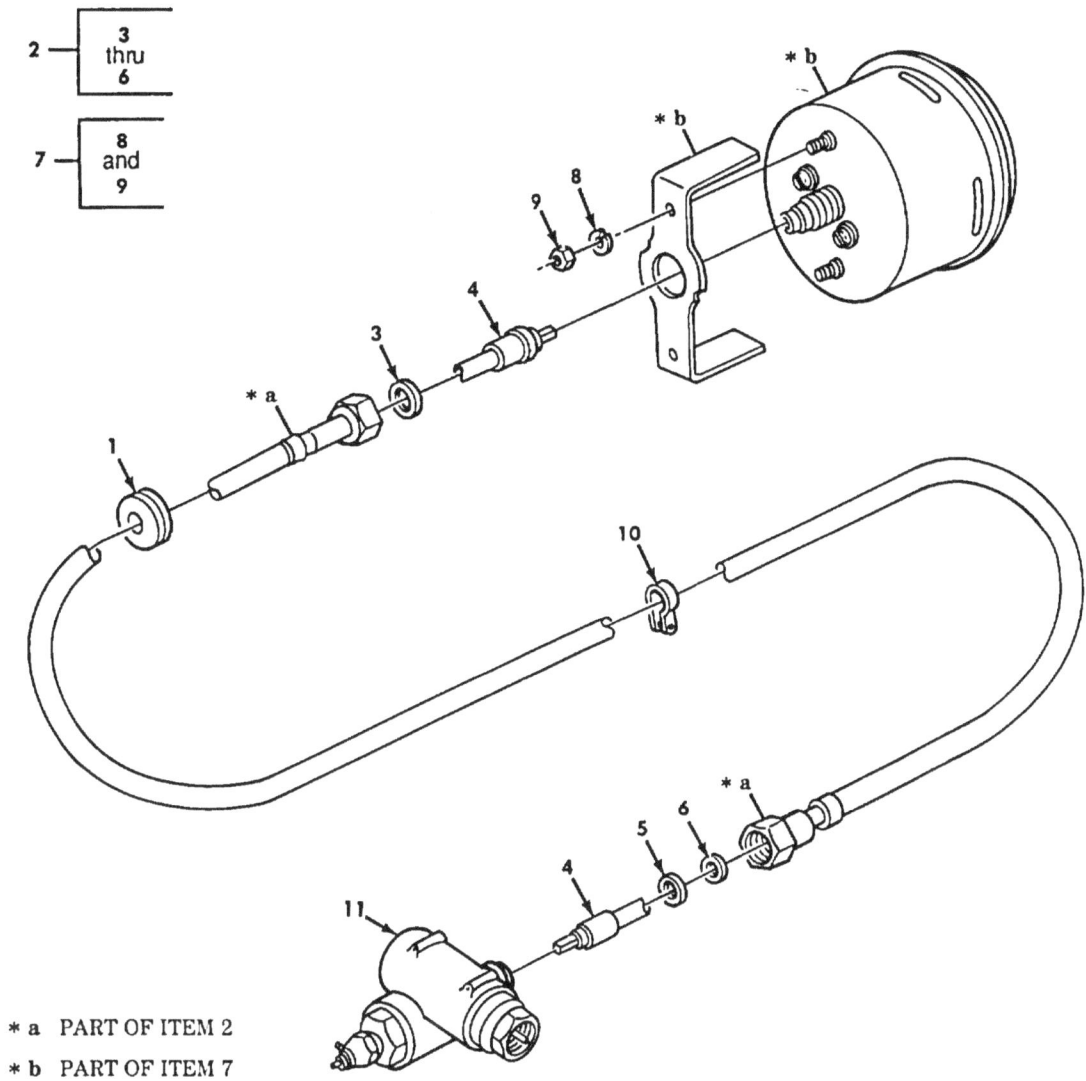

*a PART OF ITEM 2
*b PART OF ITEM 7

Figure 624. Speedometer Drive Adapter, Cable, and Gauge Assembly.

(1) ITEM NO	(2) SMR CODE	(3) NSN	(4) CAGEC	(5) PART NUMBER	(6) DESCRIPTION AND USABLE ON CODES (UOC)	(7) QTY
					GROUP 47 GAUGES 4701 SPEEDOMETER	
					FIG. 624 SPEEDOMETER DRIVE ADAPTER, CABLE, AND GAUGE ASSEMBLY	
1	PAOZZ	5325002766091	96906	MS35489-19	GROMMET, NONIETALLIC	1
2	PAOOO	6680005079992	96906	MS51071-16	SHAFT ASSEMBLY, FLEX	1
3	PAOZZ	5330007539689	19207	7539689	.GASKET	1
4	PAOZZ	3040006003860	96906	MS51072-16	.CORE, FLEXIBLE SHAFT	1
5	PAOZZ	5310005329467	19207	5329467	.WASHER, SLOTTED	1
6	PAOZZ	5310007539688	19207	7539688	.WASHER, FLAT	1
7	PAOOO	6680009333599	96906	MS39021-2	SPEEDOMETER MECHANICAL, 0-60 MPH	1
8	PAOZZ	5310000453296	96906	MS35338-43	.WASHER, LOCK	2
9	PAOZZ	5310009349751	96906	MS35650-302	.NUT, PLAIN, HEXAGON	2
10	PAOZZ	5340003379619	19207	7373381	CLAMP, LOOP UOC:V12, V13, V14, V15, V16, V17, V18, V19, V20, V21, V22, V24, V25	3
10	PAOZZ	5340003379619	19207	7313381	CLAMP, LOOP .... UOC:DAA, DAB, DAC, DAD, DAE, DAF, DAG, DAH, DAJ, DAK , DAL, DAW, DAX, V39, ZAA, ZAB, ZAC, ZAD, ZAE, ZAF, ZAG, ZAH, ZAJ, ZAK, ZAL	2
11	PAOZZ	3010012361213	57733	649-CL	DRIVE UNIT, ANGLE SPEEDOMETER SHAFT ASSEMBLY.. UOC:DAA, DAB, DAC, DAD, DAE, DAF, DAG, DAH, DAJ, DAXK, DAL, DAW, DAX, V39, ZAA, ZAB, ZAC, ZAD, ZAE, ZAF, ZAG, ZAH, ZAJ, ZAK, ZAL	1

END OF FIGURE

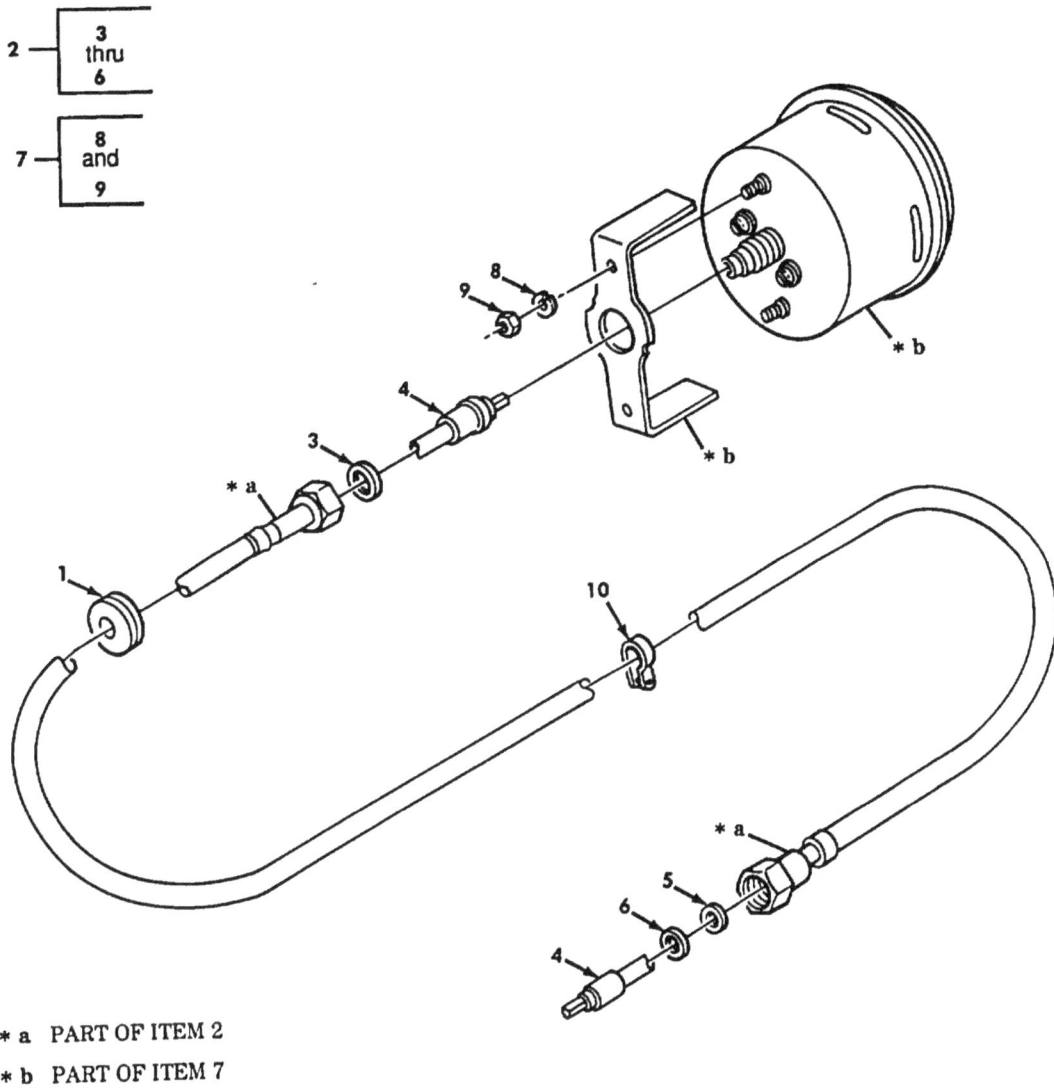

* a   PART OF ITEM 2
* b   PART OF ITEM 7

Figure 625. Tachometer Cable and Gauge Assembly.

(1) ITEM NO	(2) SMR CODE	(3) NSN	(4) CAGEC	(5) PART NUMBER	(6) DESCRIPTION AND USABLE ON CODES (UOC)	(7) QTY
					GROUP 4701 TACHOMETER	
					FIG. 625 TACHOMETER CABLE AND GAUGE ASSEMBLY	
1	PAOZZ	5325002766091	96906	MS35489-19	GROMMET, NONMETALLIC	1
2	PAOOO	6680007952641	19207	7952641	SHAFT ASSEMBLY, FLEX	1
3	PAOZZ	5330007539689	19207	7539689	.GASKET	1
4	XAOZZ		57733	418049-50IN	.SHAFT ASSEMBLY, FLEX......................	1
5	PAOZZ	5310005329467	19207	5329467	.WASHER, SLOTTED	1
6	PAOZZ	5310007539688	19207	7539688	.WASHER, FLAT	1
7	PAOOO	6680008252076	96906	MS35916-2	TACHOMETER, MECHANIC	1
8	PAOZZ	5310000453296	96906	MS35338-43	.WASHER, LOCK	2
9	PAOZZ	5310009349751	96906	MS35650-302	.NUT, PLAIN, HEXAGON	2
10	PAOZZ	5340011179876	19207	7397785	CLAMP, LOOP	1

END OF FIGURE

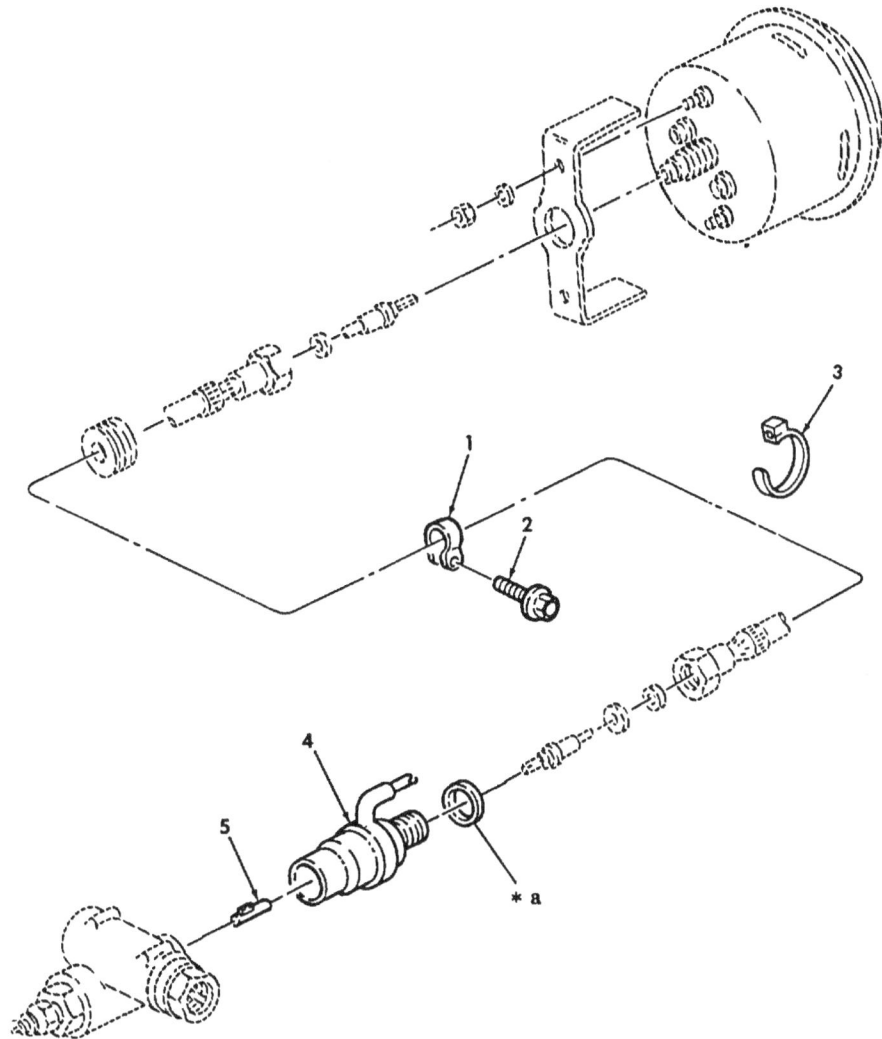

* a PART OF ITEM 4

Figure 626. Speedometer Drive Cable Clamp (M939A2).

(1) ITEM NO	(2) SMR CODE	(3) NSN	(4) CAGEC	(5) PART NUMBER	(6) DESCRIPTION AND USABLE ON CODES (UOC)	(7) QTY
					GROUP 47 GAUGES 4701 INSTRUMENTS	
					FIG. 626 SPEEDOMETER DRIVE CABLE CLAMP (M939A2)	
1	PAOZZ	5340000573043	96906	MS21333-112	CLAMP, LOOP	1
2	PAOZZ	5305000680508	96906	MS90728-6	SCREW, CAP, HEXAGON H	1
3	PAOZZ	5975005709598	96906	MS3367-7-9	STRAP, TIEDOWN, ELECT	5
4	PAOZZ	6625012892062	52304	559602	GENERATOR, SIGNAL	1
5	PAOZZ	5315013842149	78388	SA-2677	PIN, SHOULDER, HEADLE	1

END OF FIGURE

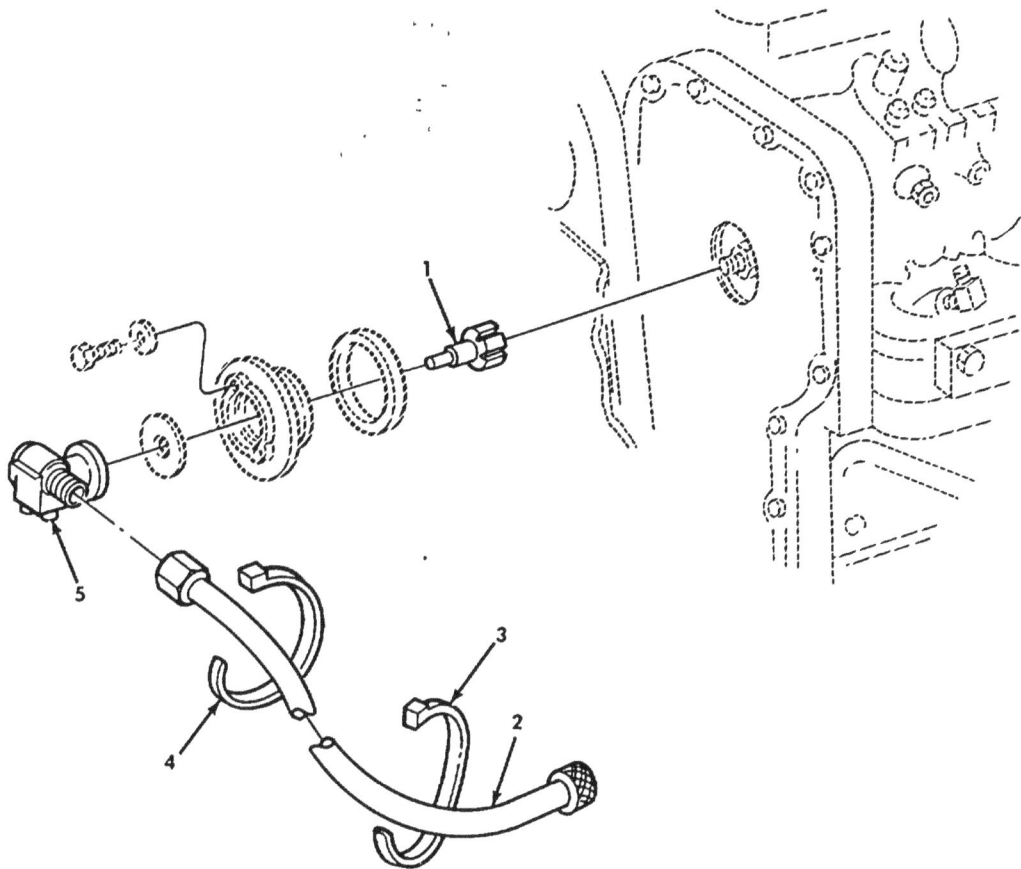

Figure 627. Tachometer Drive (M939A2).

(1) ITEM NO	(2) SMR CODE	(3) NSN	(4) CAGEC	(5) PART NUMBER	(6) DESCRIPTION AND USABLE ON CODES (UOC)	(7) QTY

GROUP 4701 INSTRUMENTS

FIG. 627 TACHOMETER DRIVE (M939A2)

(1)	(2)	(3)	(4)	(5)	(6)	(7)
1	PAFZZ	3040012715118	15434	3907757	HUB, BODY	1
2	PAFZZ	6680008820965	96906	MS51071-7	SHAFT ASSEMBLY, FLEX	1
3	PAOZZ	5975001338696	96906	MS3367-6-9	STRAP, TIEDOWN, ELECT	1
4	PAOZZ	5975012738133	96906	MS3367-3	STRAP, TIEDOWN, ELECT	1
5	PAFZZ	3040012621207	15434	3907618	ADAPTER, SPEEDOMETER	1

END OF FIGURE

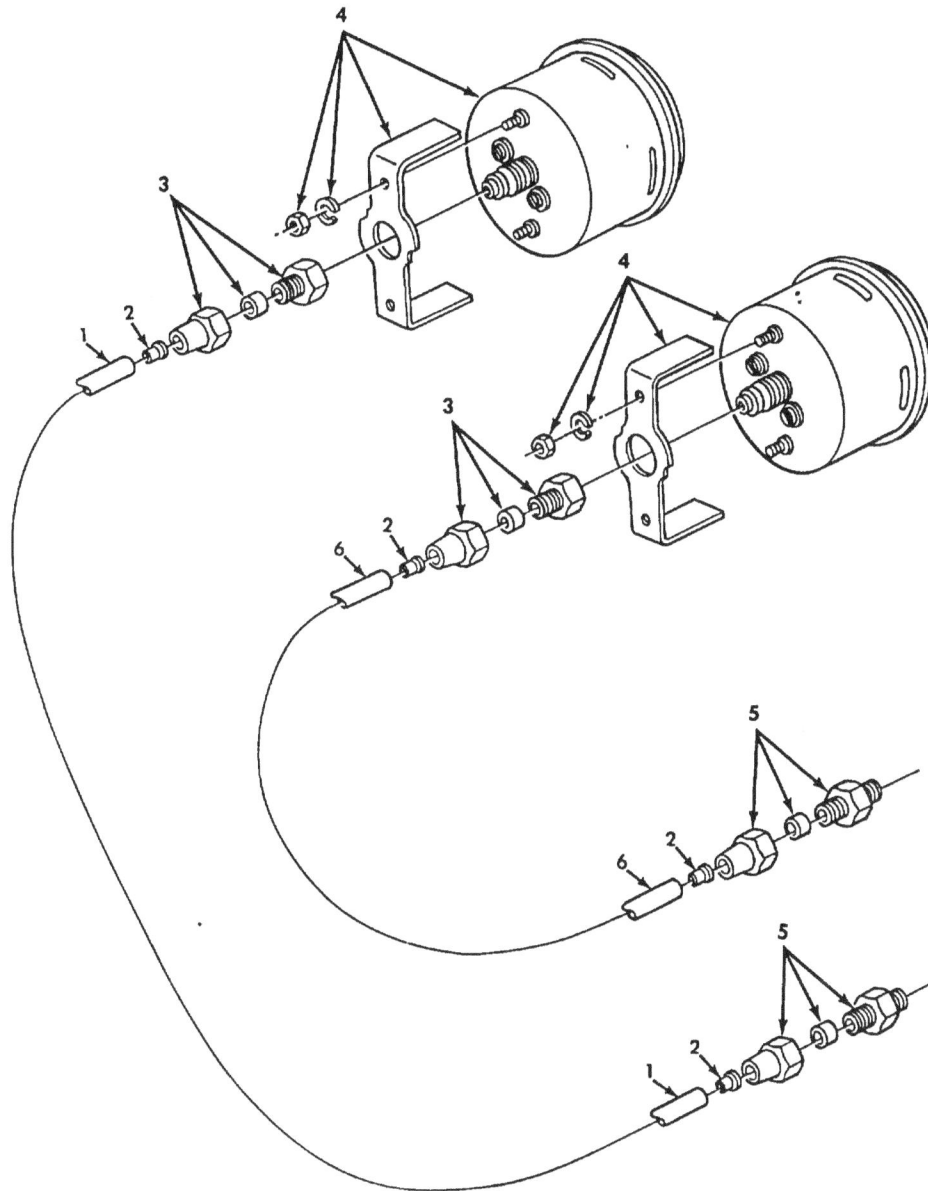

Figure 628. Primary and Secondary Air Pressure Gauges.

(1) ITEM NO	(2) SMR CODE	(3) NSN	(4) CAGEC	(5) PART NUMBER	(6) DESCRIPTION AND USABLE ON CODES (UOC)	(7) QTY
					GROUP 4702 GAUGES, MOUNTING, LINES, AND FITTINGS	
					FIG. 628 PRIMARY AND SECONDARY AIR PRESSURE GAUGES	
1	MOOZZ		19207	CPR104420-2-28	HOSE, NONMETALLIC MAKE FROM HOSE, ... P/N CPR104420-2, 28 INCHES LONG...................	1
2	PAOZZ	4730010798821	19207	CPR102321-1	INSERT, TUBE FITTING	4
3	PAOZZ	4730002000528	81343	6-2 120103BA	ADAPTER, STRAIGHT , PI.....................	2
4	PAOZZ	6685011095695	19207	12276938	GAUGE, PRESSURE , DIAL	2
5	PAOZZ	4730000691186	81343	6-4 120102BA	ADAPTER, STRAIGHT, PI.....................	2
6	MOOZZ		19207	CPR104420-2-29	HOSE, NONMETALLIC MAKE FROM HOSE, ... P/N CPR104420-2, 29 INCHES LONG..............	1

END OF FIGURE

(1) ITEM NO	(2) SMR CODE	(3) NSN	(4) CAGEC	(5) PART NUMBER	(6) DESCRIPTION AND USABLE ON CODES (UOC)			(7) QTY
	PAOZZ	2940004043057	15434	AR51480	FILTER ELEMENT,FLUI			1
					FILTER ELEMENT,FLUI	(1)	30-3	
					PACKING ASSEMBLY	(1)	30-2	
	PAFZZ	5330001336235	15434	AR51481	GASKET SET			1
					GASKET	(1)	31-8	
					GASKET	(1)	31-13	
					PACKING,PREFORMED	(2)	31-4	
	PAFZZ	5330001336236	15434	AR51482	GASKET SET			3
					GASKET	(V)	23-20	
					GASKET	(V)	23-20	
					GASKET	(V)	23-20	
					GASKET	(V)	23-20	
					GASKET	(V)	23-20	
	PAHZZ	2910011173689	15434	AR51522	PARTS KIT,FUEL INJE			1
					BALL,CHECK	(1)	46-10	
					GASKET	(1)	46-16	
					PLUG,ORIFICE	(1)	46-15	
					PLUG,ORIFICE	(1)	46-15	
					PLUG,ORIFICE	(1)	46-15	
					PLUG,ORIFICE	(1)	46-15	
					RING,RETAINING	(1)	46-13	
					STRAINER ELEMENT,SE	(1)	46-14	
	PAFZZ	2815009132074	15434	AR73350	RING SET,PISTON			1
					RING,PISTON	(1)	284-28	
					RING,PISTON	(1)	284-29	
					RING,PISTON	(1)	284-30	
	PAHZZ	3120003395642	15434	BM27253	BEARING SET,SLEEVE			1
					UOC:DAA,DAB,DAC,DAD,DAE,DAF,DAG,DAH, DAJ,DAK,DAL,DAW,DAX,V12,V13,V14 ,V15, V16,V17,V18,V19,V20,V21,V22,V24,V25, V39			
					BEARING, SLEEVE	(1)	21-3	
					BEARING, SLEEVE	(6)	21-4	
	PAFZZ	5365008295150	15434	BM56657	SHIM			1
					UOC: DAA,DAB, DAC,DAD, DAE,DAF, DAG, DAH, DAJ,DAK,DAL,DAW,DAX,V12,V13,V14,V15, V16,V17,V18,V19,V20,V21,V22,V24,V25, V39			
					SHIM	(1)	19-9	
					SHIM	(1)	19-9	
					SHIM	(1)	19-9	
	PAOZZ	2520003884197	78500	CP16NS	SPIDER,UNIVERSAL JO			2
					BEARING ASSEMBLY	(8)	219-4	
					BEARING ASSEMBLY	(4)	228-12	
					BEARING ASSEMBLY	(8)	226-7	
					CROSS ASSEMBLY	(2)	219-5	
					CROSS ASSEMBLY	(1)	228-8	
					CROSS ASSEMBLY	(2)	226-4	
					LOCKPLATE	(8)	219-3	
					LOCKPLATE	(4)	228-11	
					LOCKPLATE	(8)	226-8	
					SCREW	(16)	219-2	
					SCREW	(8)	228-10	
					SCREW	(16)	226-9	

(1) ITEM NO	(2) SMR CODE	(3) NSN	(4) CAGEC	(5) PART NUMBER	(6) DESCRIPTION AND USABLE ON CODES (UOC)			(7) QTY
					SPIDER,UNIVERSAL JO	(	2) 229-6	
					SPIDER,UNIVERSAL JO	(	8) 229-4	
					SPIDER,UNIVERSAL JO	(	8) 229-3	
					SPIDER,UNIVERSAL JO	(	16) 229-2	
	PAOZZ	2520010828619	81221	5-0280	PARTS KIT,UNIVERSAL			1
					BEARING ASSEMBLY	(	8) 227-6	
					CROSS ASSEMBLY	(	2) 227-3	
					LOCKPLATE	(	8) 227-7	
					SCREW	(	16) 227-8	
	XDOZZ		78500	CP35R-17	PARTS KIT,UNIVERSAL			1
					UOC: DAL ,V18			
					LUBRICATION FITTING	(	2) 506-9	
					NEEDLE CUP ASSY	(	8) 506-11	
					OIL SEAL	(	8) 506-12	
					RETAINING RING	(	8) 506-10	
	PAFZZ	2930011317442	50022	D-781964-A	CLAMPING STRIP KIT			1
					BOLT,MACHINE	(	8) 108-9	
					BOLT,MACHINE	(	72) 108-12	
					BRACKET,MOUNTING	(	4) 108-8	
					GASKET	(	4) 108-4	
					NUT,SELF-LOCKING,HE	(	80) 108-6	
					PLATE,MENDING	(	4) 108-7	
					PLATE,MENDING	(	4) 108-11	
					WASHER,FLAT	(	80) 108-10	
	PAFZZ	2530011375921	19954	ERS-27785	PARTS KIT,POWER STE			1
					UOC:DAA,DAB,DAC,DAD,DAE,DAF,DAG,DAH, DAJ,DAK,DAL,DAW,DAX,V12,V13,V14,V15, V16,V17,V18,V19,V20,V21,V22,V24,V25, V39			
					ELEMENT, FILTER	(	1) 311-6	
					NUT	(	2) 311-4	
					SPRING,FILTER	(	1) 311-7	
					WASHER,SEAL	(	2) 311-8	
					WASHER,SEALING	(	1) 311-5	
	PAFZZ	4330012722937	78047	ERS-28001	FILTER ELEMENT,FLUID			1
					UOC:ZAA,ZAB,ZAC,ZAD,ZAE,ZAF,ZAG,ZAH, ZAJ,ZAK,ZAL			
					FILTER CAP	(	1) 312-44	
					FILTER ELEMENT	(	1) 312-45	
					GASKET	(	1) 312-40	
					GASKET	(	1) 312-42	
					GASKET	(	2) 312-10	
					GASKET	(	2) 312-11	
					PACKING PRE.	(	1) 312-47	
	PAFZZ	4330012722937	78047	FC212028	PARTS KIT,ENGINE FA			1
					UOC:ZAA,ZAB,ZAC,ZAD,ZAE,ZAF,ZAG,ZAH, ZAJ, ZAK, ZAL			
					BEARING,ROLLER,CYLI	(	1) 122-25	
					BEARING BALL,ANNULA	(	2) 122-14	
					BUSHING, SLEEVE	(	1) 122-24	
					CAP,SPECIAL	(	1) 122-2	
					LINING,FRICTION	(	1) 122-6	
					NUT,SELF-LOCKING,HE	(	1) 122-6	
					PACKING, PREFORMED	(	1) 122-4	

(1) ITEM NO	(2) SMR CODE	(3) NSN	(4) CAGEC	(5) PART NUMBER	(6) DESCRIPTION AND USABLE ON CODES (UOC)			(7) QTY
					PACKING,PREFORMED	(	1) 122-5	
					PACKING,PREFORMED	(	1) 122-11	
					PACKING,PREFORMED	(	1) 122-13	
					RING,RETAINING	(	1) 122-23	
					SPACER,RING	(	1) 122-8	
					SPACER,RING	(	1) 122-15	
					SPACER,STEPPED	(	1) 122-16	
					WASHER,SPRING TENSI	(	1) 122-7	
	PAFZZ	4810011331459	06721	RN-32-G	PARTS KIT,PRESSURE			1
					O-RING	(	1) 282-6	
					VALVE ASSEMBLY	(	1) 282-13	
	PAHZZ	3020012141530	19207	10899232	GEAR SET,HELICAL,MA			1
					GEAR,HELICAL	(	1) 232-8	
					GEAR,HELICAL	(	1) 239-8	
					GEARSHAFT,HELICAL	(	1) 235-4	
					GEARSHAFT,HELICAL	(	1) 242-4	
	PAOFZ	2540001219081	19207	11672522	ACCESSORY KIT,VEHIC			1
					UOC:DAC,DAD,V16,V17,ZAC,ZAD			
					BOW ASSEMBLY	(	20) 532-4	
					BOW,VEHICULAR TOP	(	10) 532-2	
					COVER,FITTED,VEHICU	(	1) 532-1	
					SCREW,MACHINE	(	100) 532-3	
					STAKE,VEHICLE BODY	(	20) 532-6	
					STRAP,WEBBING	(	20) 532-5	
	PAOFZ	2540001219081	19207	11672522-1	ACCESSORY KIT,VEHIC			1
					UOC:DAC,DAD,V16,V17,ZAC,ZAD			
					BOW ASSEMBLY	(	20) 532-4	
					BOW,VEHICULAR TOP	(	10) 532-2	
					COVER,FITTED,VEHICU	(	1) 532-1	
					SCREW,MACHINE	(	100) 532-3	
					STAKE,VEHICLE BODY	(	20) 532-6	
					STRAP,WEBBING	(	20) 532-5	
	PAOFF	2540001219082	19207	11672523	ACCESSORY KIT,VEHIC			1
					UOC:DAW,DAX,V14,V15,ZAA,ZAB			
					BOW,VEHICULAR TOP	(	6) 532-2	
					CORNER,BOW,VEHICULA	(	12) 532-4	
					COVER,FITTED,VEHICU	(	1) 532-1	
					SCREW,MACHINE	(	36) 532-3	
					STAKE,VEHICLE BODY	(	12) 532-6	
					STRAP,WEBBING	(	12) 532-5	
	PAFZZ	254001096508	19207	12256226	KIT,DEEP WATER FORDING			1
					UOC:DAA,DAB,DAC,DAD,DAE,DAF,DAG,DAH, DAJ,DAK,DAL,DAW,DAX,V12,V13,V14,V15, V16,V17,V18,V19,V20,V21,V22,V24,V25, V39			
					ADAPTER,STRAIGHT,PI	(	1) 597-7	
					BOLT,MACHINE	(	1) 596-16	
					BRACKET	(	1) 596-22	
					BRACKET,ANGLE	(	1) 596-15	
					BUSHING,PIPE	(	2) 597-5	
					CLAMP,LOOP	(	1) 596-14	
					COCK,DRAIN	(	1) 596-20	
					COCK,DRAIN	(	1) 597-2	
					CONTROL ASSEMBLY,PU	(	1) 596-10	

(1) ITEM NO	(2) SMR CODE	(3) NSN	(4) CAGEC	(5) PART NUMBER	(6) DESCRIPTION AND USABLE ON CODES (UOC)			(7) QTY
					ELBOW,PIPE	(	1)	596-2
					ELBOW,PIPE	(	1)	596-19
					GROMMET,NONMETALLIC	(	1)	596-7
					HOSE ASSEMBLY,NONME	(	1)	596-13
					HOSE ASSEMBLY	(	1)	596-27
					HOSE ASSEMBLY	(	2)	597-1
					INSERT,TUBE FITTING	(	1)	597-6
					NIPPLE,PIPE	(	1)	596-5
					NIPPLE,TUBE	(	1)	597-9
					NUT,PLAIN,HEXAGON	(	1)	596-8
					NUT,SELF-LOCKING,HE	(	1)	596-18
					NUT,SELF-LOCKING,HE	(	2)	596-26
					NUT,SELF-LOCKING,HE	(	2)	596-23
					PLATE	(	1)	597-12
					PLUG,PIPE	(	1)	597-4
					REDUCER,PIPE	(	1)	596-17
					SCREW,CAP,HEXAGON H	(	1)	596-11
					SCREW,MACHINE ,	(	2)	596-24
					STRAP,RETAINING	(	1)	596-25
					STRAP,TIEDOWN,ELECT	(	5)	597-3
					TEE,PIPE	(	1)	596-3
					TEE,PIPE	(	1)	597-8
					TUBING,NONMETALLIC	(	1)	597-10
					VALVE ASSEMBLY,FORD	(	1)	596-6
					VALVE,REGULATING,FL	(	1)	596-28
					WASHER,LOCK	(	1)	596-12
					WASHER,LOCK	(	1)	596-9
	PAOZZ	2540010965023	19207	12256227	PARTS KIT,VEHICULAR			1
					BACK AND QUARTER AS	(	1)	579-7
					BOLT,HOOK	(	4)	579-12
					NUT,SELF-LOCKING, HE	(	4)	579-3
					NUT,SELF-LOCKING,HE	(	32)	579-5
					NUT,SHEET SPRING	(	2)	579-10
					ROOF ASSY(COMPLETE)	(	1)	579-4
					SCREW,TAPPING	(	2)	579-11
					SEAL,NONMETALLIC ST	(	2)	579-6
	PAOZZ	2530011147764	19207	12256301	KIT,AIR BRAKE TRAILER TOWING			1
					UOC:DAA,DAB,DAC,DAD,DAE,DAF,DAJ,DAK, DAW,DAX,V12,V13,V14,V15,V16,V1 7 ,V1 9, V20,V24,V25,V39,ZAA,ZAB,ZAC,ZAD,ZAE, ZAF,ZAJ,ZAK			
					ADAPTER,STRAIGHT	(	4)	603-7
					ADAPTER,STRAIGHT	(	1)	604-16
					ADAPTER,STRAIGHT	(	1)	605-1
					ADAPTER,STRAIGHT	(	1)	605-12
					BOLT,MACHINE	(	1)	604-6
					BOLT,MACHINE	(	1)	604-14
					BOLT,MACHINE	(	2)	605-4
					BRACKET	(	1)	603-3
					COUPLING,PIPE	(	2)	603-12
					ELBOW	(	1)	603-1
					ELBOW	(	2)	603-13
					ELBOW	(	3)	603-14
					ELBOW	(	3)	604-1

KIT-4

(1) ITEM NO	(2) SMR CODE	(3) NSN	(4) CAGEC	(5) PART NUMBER	(6) DESCRIPTION AND USABLE ON CODES (UOC)			(7) QTY
					ELBOW	(	1) 604-11	
					ELBOW	(	1) 605-6	
					ELBOW	(	1) 605-8	
					ELBOW	(	1) 605-11	
					HOSE ASSEMBLY	(	2) 603-11	
					INSERT	(	8) 603-8	
					INSERT	(	12) 604-2	
					INSERT	(	6) 605-2	
					NUT, SELF-LOCKING	(	2) 603-4	
					NUT, SELF-LOCKING	(	1) 604-8	
					NUT, SELF-LOCKING	(	1) 604-10	
					NUT, SELF-LOCKING	(	2) 605-7	
					SCREW, CAP, HEXAGON	(	2) 603-6	
					TEE, PIPE	(	1) 605-13	
					VALVE, AIR	(	1) 605-5	
					VALVE ASSEMBLY	(	1) 603-2	
					VALVE, ROTARY, DIRECT	(	1) 603-5	
					VALVE, SHUTTLE	(	1) 604-5	
					VALVE, SHUTTLE	(	1) 604-12	
					WASHER, FLAT	(	1) 604-13	
	PFOZZ	6665010958285	19207	12256442	KIT, CHEMICAL AGENT ALARM MOUNTING			1
					BOLT, MACHINE	(	6) 606-1	
					BRACKET	(	1) 606-11	
					BRACKET, ANGLE	(	1) 606-18	
					CABLE ASSEMBLY	(	1) 606-9	
					CLAMP, LOOP	(	3) 606-5	
					CLAMP, LOOP	(	2) 606-7	
					CLAMP, LOOP	(	1) 606-8	
					CLAMP, LOOP	(	1) 606-20	
					GROMMET	(	1) 606-13	
					NUT, SELF-LOCKING	(	4) 606-12	
					NUT, SELF-LOCKING	(	6) 606-15	
					NUT, SELF-LOCKING	(	2) 606-17	
					NUT, SELF-LOCKING	(	4) 606-19	
					SCREW, CAP, HEXAGON	(	4) 606-2	
					SCREW, CAP, HEXAGON	(	2) 606-4	
					SCREW, MACHINE	(	4) 606-10	
					SCREW, TAPPING	(	5) 606-6	
					STRAP, RETAINING	(	1) 606-16	
					STRAP, TIEDOWN	(	1) 606-21	
					SUPPORT	(	1) 606-3	
					WASHER, FLAT	(	2) 606-14	
	PBFZZ	2540010958286	19207	12256443	KIT, FUEL BURNING PERSONNEL HEATER .... 1 UOC: DAA, DAB, DAC, DAD, DAE, DAF, DAG, DAH, DAJ, DAK, DAL, DAW, DAX, V12, V13, V14, V15, V16, V17, V18 , V19, V20, V21, V22, V24, V25, V39			
					ADAPTER, CONNECTOR	(	4) 592-26	
					ADAPTER, STRAIGHT	(	1) 593-13	
					ADAPTER, STRAIGHT	(	1) 593-16	
					ADAPTER, STRAIGHT	(	4) 593-23	
					BOLT, MACHINE	(	3) 592-22	
					BRACKET	(	1) 593-12	
					BRACKET, ANGLE	(	1) 592-13	

(1) ITEM NO	(2) SMR CODE	(3) NSN	(4) CAGEC	(5) PART NUMBER	(6) DESCRIPTION AND USABLE ON CODES (UOC)			(7) QTY
					BRACKET, CONTROL BOX	(	1) 592-2	
					BUTTON, PLUG	(	1) 593-18	
					CABLE ASSEMBLY	(	1) 593-9	
					CAP-PLUG	(	1) 593-20	
					CLAMP, HOSE	(	2) 592-7	
					CLAMP, LOOP	(	37) 592-18	
					CONTROL BOX	(	1) 592-6	
					DECAL	(	1) 592-28	
					DRAIN, PLUG	(	1) 593-27	
					DUMMY CONNECTOR	(	4) 592-25	
					ELBOW	(	1) 592-9	
					ELBOW, PIPE	(	1) 593-1	
					ELBOW, PIPE	(	1) 593-2	
					ELBOW, PIPE	(	1) 593-11	
					GROMMET	(	1) 592-23	
					HEATER, VEHICULAR	(	1) 592-8	
					HOSE ASSEMBLY	(	1) 593-26	
					HOSE ASSEMBLY	(	1) 593-8	
					HOSE ASSEMBLY	(	1) 593-17	
					NOZZLE AND BRACKET	(	1) 592-16	
					NUT, PLAIN, HEXAGON	(	2) 592-4	
					NUT, SELF-LOCKING	(	2) 592-1	
					NUT, SELF-LOCKING	(	4) 592-14	
					NUT, SELF-LOCKING	(	2) 592-17	
					NUT, SELF-LOCKING	(	2) 593-6	
					PIN, COTTER	(	2) 592-10	
					PIPE, EXHAUST	(	1) 592-15	
					PLATE, INSTRUCTION	(	1) 593-21	
					PLUG	(	1) 593-19	
					PLUG, PIPE	(1) 593-25		
					PUMP, FUEL, ELECTRICA	(	1) 593-5	
					SCREW, CAP, HEXAGON	(	1) 592-11	
					SCREW, CAP, HEXAGON	(	1) 593-3	
					SCREW, CAP, HEXAGON	(	2) 592-19	
					SCREW, CAP, HEXAGON	(	1) 593-7	
					SCREW, CAP, HEXAGON	(	2) 592-29	
					SCREW, TAPPING	(	4) 593-22	
					SHELL, ELECTRICAL	(	1) 592-24	
					SHELL, ELECTRICAL	(	3) 592-27	
					SHELL, ELECTRICAL	(	1) 593-15	
					STRAP, TIEDOWN	(	7) 593-10	
					TEE	(	1) 593-14	
					TEE	(	1) 593-24	
					WASHER, FLAT	(	8) 592-12	
					WASHER, LOCK	(	2) 592-3	
					WASHER, LOCK	(	1) 592-21	
					WASHER, LOCK	(	2) 593-4	
					WIRING HARNESS	(	1) 592-5	
					WIRING HARNESS	(	1) 592-20	
	PAOZZ	2990010958287	19207	12256466	HEATER, COOLANT, ENGI			1

UOC: DAA, DAB, DAC, DAD, DAE, DAF, DAG, DAH,
DAJ, DAK, DAL, DAW, D AX, V12, V13, V14, V15,
V16, V17, V18, V19, V20, V21, V22, V24, V25,
V39

(1) ITEM NO	(2) SMR CODE	(3) NSN	(4) CAGEC	(5) PART NUMBER	(6) DESCRIPTION AND USABLE ON CODES (UOC)	(7) QTY
					ADAPTER, STRAIGHT, PI (3) 581-8	
					ADAPTER, STRAIGHT, PI (1) 582-23	
					ADAPTER, STRAIGHT, PI (1) 582-5	
					ADAPTER, STRAIGHT, PI (1) 582-18	
					BLOCK, SUPPORT, BATTE (4) 582-1	
					BLOCK, SUPPORT, BATTE (2) 582-2	
					BLOCK, SUPPORT, BATTE (2) 582-3	
					BOLT, MACHINE (2) 581-14	
					BRACKET (1) 581-11	
					BRACKET ASSEMBLY (1) 581-6	
					BRACKET, MOUNTING (1) 580-2	
					BUSHING, PIPE (1) 582-8	
					BUSHING, PIPE (6) 581-19	
					CABLE ASSEMBLY, SPEC (1) 580-5	
					CLAMP, HOSE (2) 581-1	
					CLAMP, HOSE (6) 582-9	
					CLAMP, HOSE (1) 581-16	
					CLAMP, HOSE { 2) 581-9	
					CLAMP, LOOP (1) 583-9	
					CLAMP, LOOP (5) 582-14	
					COCK, DRAIN (2) 582-10	
					COCK, PLUG (1) 582-6	
					COCK, PLUG (1) 581-20	
					CONTROL BOX, ELECTRI (1) 580-10	
					COUPLING, PIPE (1) 581-24	
					COVER, WATER HEADER (1) 582-11	
	PFFZZ	2540010965020	19207	12276986	KIT, RIFLE MOUNTING	1
					BOLT, MACHINE (7) 601-7	
					BOLT, MACHINE (2) 601-11	
					BRACKET, GUN (1) 601-9	
					BRACKET, MOUNTING (2) 601-2	
					BRACKET, MOUNTING (1) 601-6	
					CATCH ASSEMBLY (3) 601-3	
					NUT, PLAIN (4) 601-5	
					NUT, SELF-LOCKING (9) 601-10	
					REINFORCEMENT (3) 601-1	
					SCREW, CAP, HEXAGON (10) 601-4	
					SUPPORT, RIFLE MOUNT (3) 601-8	
	PAOZZ	2920011311939	19207	12277359	KIT, ALTERNATOR, 60 TO 100 AMP	1
					CLAMP ASSEMBLY (1) 608-2	
					CLAMP, LOOP (1) 608-1	
					CLAMP, LOOP (1) 608-20	
					CLAMP, LOOP (1) 608-22	
					GENERATOR ASSEMBLYE (1) 608-15	
					INSULATOR, BUSHING (1) 608-13	
					NUT, SELF-LOCKING (4) 608-8	
					PLATE, REINFORCEMENT (2) 608-4	
					PULLEY, GROOVE (1) 608-19	
					REGULATOR (1 608-3	
					SCREW, CAP, HEXAGON (4) 608-5	
					SCREW, MACHINE (4) 608-9	
					SHELL, ELECTRICAL (1) 608-12	
					STRAP, TIEDOWN (10) 608-21	
					TERMINAL ASSEMBLY (1) 608-14	

(1) ITEM NO	(2) SMR CODE	(3) NSN	(4) CAGEC	(5) PART NUMBER	(6) DESCRIPTION AND USABLE ON CODES (UOC)			(7) QTY
					WASHER, FLAT	(4)	608-6	
					WASHER, FLAT	(4)	608-7	
					WASHER, LOCK	(4)	608-10	
					WIRING HARNESS	(1)	608-11	
	PFOZZ	2540012819855	19207	12300848	KIT, VE MUD FLAP STORAGE			1
					BRACKET, ANGLE	(1)	616A-23	
					BRACKET, ANGLE	(1)	616A-28	
					BRACKET, MOUNTING	(2)	616A-18	
					NUT, SELF-LOCKING, HE	(12)	616A-14	
					NUT, SELF-LOCKING, HE	(10)	616A-16	
					NUT, SELF-LOCKING, HE	(12)	616A-25	
					NUT, SELF-LOCKING, HE	(9)	616A-29	
					PLATE, MENDING	(2)	616A-17	
					SCREW, CAP, HEX	(6)	616A-15	
					SCREW, CAP, HEX	(4)	616A-19	
					SCREW, CAP, HEX	(9)	616A-20	
					SCREW, MACHINE	(12)	616A-27	
					STRAP, RETAINING	(6)	616A-26	
					STRAP, WEBBING	(2)	616A-22	
					STRAP, WEBBING	(1)	616A-30	
					WASHER, FLAT	(27)	616A-21	
	PFOZZ	6220011951791	19207	12301057	KIT, CONVOY WARNING			1
					UOC: DAA, DAB, DAC, DAD, DAG, DAH, DAL, DAW, DAX, V12, V13, V14, V15, V16, V17, V18, V21, V22, V39, ZAA, ZAB, ZAC, ZAD, ZAE, ZAF, ZAG, ZAH, ZAL			
					BRACKET, ANGLE	(1)	611-35	
					BRACKET ASSEMBLY	(2)	611-11	
					CABLE ASSEMBLY	(2)	611-7	
					CLAMP, LOOP	(2)	611-4	
					CLAMP, LOOP	(2)	611-28	
					HANDLE, SWITCH	(1)	611-31	
					LIGHT, WARNING	(2)	611-1	
					NUT, SELF-LOCKING, HE	(2)	611-2	
					NUT, SELF-LOCKING, HE	(4)	611-17	
					NUT, SELF-LOCKING, HE	(10)	611-19	
					NUT, SELF-LOCKING, HE	(2)	611-25	
					NUT, SELF-LOCKING, HE	(2)	611-34	
					PIN, QUICK RELEASE	(1)	611-12	
					PLATE, IDENTIFICATIO	(1)	611-32	
					PLATE, INSTRUCTION	(1)	611-13	
					PLATE, MOUNTING	(2)	611-5	
					PLATE, MOUNTING	(2)	611-6	
					RING, RETAINING	(1)	611-14	
					SCREW, ASSEMBLED	(2)	611-27	
					SCREW, CAP, HEXAGON	(2)	611-9	
					SCREW, CAP, HEXAGON	(2)	611-22	
					SCREW, MACHINE	(2)	611-33	
					SCREW, WING	(1)	611-15	
					STRAP, TIEDOWN	(6)	611-26	
					STRAP, TIEDOWN	(15)	611-37	
					SWITCH, ROTARY	(1)	611-30	
					TUBE ASSEMBLY	(2)	611-8	
					TUBE ASSEMBLY	(2)	611-18	

(1) ITEM NO	(2) SMR CODE	(3) NSN	(4) CAGEC	(5) PART NUMBER	(6) DESCRIPTION AND USABLE ON CODES (UOC)			(7) QTY
					U-BOLT	(	2)	611-10
					U-BOLT	(	4)	611-21
					WASHER, FLAT	(	2)	611-3
					WASHER, FLAT	(	4)	611-16
					WASHER, FLAT	(	10)	611-20
					WASHER, FLAT	(	4)	611-24
					WASHER, LOCK	(	2)	611-23
					WIRING HARNESS	(	1)	611-29
					WIRING HARNESS	(	1)	611-36
	PFOZZ	6220012197620	19207	12301151	KIT, CONVOY WARNING 1			
					UOC: DAE, DAF, V19, V20, ZAE, ZAF			
					BRACKET, ANGLE	(	1)	612-24
					CABLE ASSEMBLY	(	2)	612-6
					CABLE ASSEMBLY	(	1)	612-15
					CABLE ASSEMBLY	(	1)	612-19
					CLAMP, LOOP	(	2)	612-5
					CLAMP, LOOP	(	31)	612-11
					CONDUIT	(	1)	612-9
					DUMMY CONNECTOR	(	2)	612-2
					HANDLE, SWITCH	(	1)	612-27
					LIGHT, WARNING	(	2)	612-1
					NUT, PLAIN, HEXAGON	(	4)	612-33
					NUT, SELF-LOCKING	(	30)	612-12
					NUT, SELF-LOCKING	(	9)	612-13
					NUT, SELF-LOCKING	(	2)	612-25
					PLATE, IDENTIFICATIO	(	1)	612-26
					PLATE, MOUNTING	(	2)	612-7
					PLATE, MOUNTING	(	2)	612-8
					RING, RETAINING	(	2)	612-29
					SCREW, CAP, HEXAGON	(	30)	612-10
					SCREW, CAP, HEXAGON	(	9)	612 17
					SCREW, MACHINE	(	2)	612-21
					SCREW, TAPPING	(	1)	612-16
					SCREW, WING	(	1)	612-30
					SHELL, ELECTRICAL	(	2)	612-3
					SHELL, ELECTRICAL	(	2)	612-4
					STRAP, TIEDOWN	(	35)	612-20
					SWITCH, ROTARY	(	1)	612-23
					TUBE ASSEMBLY	(	2)	612 28
					TUBE ASSEMBLY	(	2)	612-31
					WASHER, FLAT	(	17)	612-14
					WASHER, FLAT	(	4)	612-32
					WASHER, LOCK	(	1)	612-18
					WIRING HARNESS	(	1)	612 22
	PFOZZ	6220012197621	19207	12301193	KIT, CONVOY, WARNING 1			
					UOC: DAJ, DAK, V24, V25, ZAJ, ZAK			
					BRACKET ASSEMBLY	(	1)	613-18
					BRACKET ASSEMBLY	(	1)	613-25
					BRACKET ASSEMBLY	(	1)	613-39
					CABLE ASSEMBLY	(	2)	613-2
					CABLE ASSEMBLY	(	1)	613-8
					CABLE ASSEMBLY	(	1)	613-34
					CLAMP, LOOP	(	2)	613-3
					CLAMP, LOOP	(	25)	613-9

(1) ITEM NO	(2) SMR CODE	(3) NSN	(4) CAGEC	(5) PART NUMBER	(6) DESCRIPTION AND USABLE ON CODES (UOC)			(7) QTY
					CLAMP, LOOP	(| 13)	613-29	
					CONDUIT	(| 1)	613-28	
					COUPLING, PIPE	(| 2)	613-15	
					DUMMY CONNECTOR	(| 4)	613-5	
					GROMMET	(| 2)	613-11	
					HANDLE, SWITCH	(| 1)	613-22	
					LIGHT, WARNING	(| 2)	613-1	
					NUT, PLAIN, HEXAGON	(| 2)	613-13	
					NUT, SELF-LOCKING	(| 2)	613-24	
					NUT, SELF-LOCKING	(| 1)	613-33	
					PLATE, IDENTIFICATIO	(| 1)	613-23	
					PLATE, MOUNTING	(| 2)	613-4	
					PLATE, MOUNTING	(| 2)	613-14	
					RING, RETAINING	(| 2)	613-41	
					SCREW, CAP, HEXAGON	(| 3)	613-17	
					SCREW, CAP, HEXAGON	(| 1)	613 30	
					SCREW, CAP, HEXAGON	(| 3)	613-37	
					SCREW, MACHINE	(| 2)	613-26	
					SCREW, TAPPING	(| 25)	613-10	
					SCREW, TAPPING	(| 11)	613-27	
					SCREW, WING	(| 1)	613-19	
					SCREW, WING	(| 1)	613-40	
					SHELL, ELECTRICAL	(| 2)	613-6	
					SHELL, ELECTRICAL	(| 2)	613-7	
					STRAP, TIEDOWN	(| 1)	613-35	
					SWITCH, ROTARY	(| 1)	613-21	
					WASHER, FLAT	(| 13)	613-12	
					WASHER, FLAT	(| 1)	613-32	
					WASHER, FLAT	(| 6)	613-38	
					WASHER, LOCK	(| 3)	613-16	
					WASHER, LOCK	(| 1)	613-31	
					WIRING HARNESS	(| 1)	613 36	
	PFOZZ	2540012565331	34623	12301340	PARTS KIT, SEAT BELT GREEN			1
					BELT ASSEMBLY	(| 1)	615-4	
					BELT, VEHICULAR	(| 1)	615-8	
					BRACKET	(| 1)	615-3	
					NUT, SELF-LOCKING	(| 5)	615-1	
					NUT, SELF-LOCKING	(| 8)	615-18	
					SCREW, CAP, HEXAGON	(| 2)	615-5	
					SCREW, CAP, HEXAGON	(| 1)	615-10	
					SCREW, CAP, HEXAGON	(| 1)	615-11	
					SCREW, CAP, HEXAGON	(| 1)	615-12	
					SCREW, CAP, HEXAGON	(| 2)	615-13	
					SCREW, CAP, HEXAGON	(| 8)	615-15	
					SLEEVE	(| 2)	615-14	
					SPACER	(| 1)	615-6	
					TETHER, BELT	(| 2)	615-7	
					WASHER	(| 8)	615-17	
					WASHER, FLAT	(| 17)	615-2	
					WASHER, FLAT	(| 8)	615-16	
					WASHER, LOCK	(| 2)	615-9	
	PFOZZ	2540011550112	19207	12302696	KIT, DUMP BODY & TROOP SEAT....................................... & GREEN TARP .... UOC: DAE, DAF, V19, V20, ZAE, ZAF			1

(1) ITEM NO	(2) SMR CODE	(3) NSN	(4) CAGEC	(5) PART NUMBER	(6) DESCRIPTION AND USABLE ON CODES (UOC)	(7) QTY
					BOW ASSEMBLY (1) 609-2	
					CURTAIN, VEHICULAR (1 609-31	
					SEAT, TROOP LEFT (1) 609-8	
					SEAT, TROOP RIGHT (1) 609-8	
					SIDE RACK LEFT (1) 609-18	
					SIDE RACK RIGHT (1) 609-18	
					STRAP, WEBBING (1) 609-1	
					TARPAULIN (1) 609-30	
	PFFZZ	2540011821077	19207	12302775	MOUNTING KIT, SWINGFIRE HEATER	1

UOC: DAA, DAB, DAC, DAD, DAE, DAF, DAG, DAH, DAJ, DAK, DAL, DAW, DAX, V12 , V13 , V14 , V15, V16, V17, V18, V19, V20, V21, V22, V24, V25, V39

					ADAPTER, STRAIGHT (4) 589-10	
					ADAPTER, STRAIGHT (1) 589-52	
					BOLT (2) 589-19	
					BOLT, MACHINE (2) 589-5	
					BLOCK, SUPPORT (2) 589-59	
					BLOCK, SUPPORT (2) 589-60	
					BLOCK, SUPPORT (4) 589-61	
					BRACKET (1) 587-8	
					BRACKET, ANGLE (1) 589-9	
					BRACKET, ANGLE (1) 589-21	
					BRACKET, ANGLE (1) 589-40	
					BRACKET, ANGLE (1) 589-58	
					BRACKET DOUBLE ANGL (1) 589-38	
					CIRCUIT BREAKER (1) 588-3	
					CLAMP (1) 587-10	
					CLAMP, HOSE (1) 589-4	
					CLAMP, HOSE (12) 588-11	
					CLAMP, HOSE (2) 589-53	
					CLAMP, LOOP (2) 588-8	
					CLAMP, LOOP (2) 588-14	
					CLAMP, LOOP (1) 589-32	
					CLAMP, LOOP (4) 589-37	
					CONNECTOR, ELECTRICA (1) 588-11	
					COVER, ASSEMBLY (1) 587-2	
					ELBOW, PIPE (1) 589-16	
					ELBOW, PIPE (1) 589-47	
					ELBOW, PIPE (1) 589-51	
					ELBOW, PIPE (1) 589-63	
					GASKET (1) 589-14	
					GROMMET (1) 588-6	
					GROMMET (1) 588-13	
					HEATER, AIR DUCT, ENG (1) 587-7	
					HOSE (3) 589-12	
					HOSE (1) 589-54	
					NIPPLE, PIPE (1) 589-2	
					NIPPLE, PIPE (1) 589-48	
					NUT, PLAIN, HEXAGON (17) 587-5	
					NUT, SELF-LOCKING (5) 587-13	
					NUT, SELF-LOCKING (4) 587-14	
					NUT, SELF-LOCKING (1) 588-9	
					NUT, SELF-LOCKING (2) 589-8	

(1) ITEM NO	(2) SMR CODE	(3) NSN	(4) CAGEC	(5) PART NUMBER	(6) DESCRIPTION AND USABLE ON CODES (UOC)	(7) QTY
					NUT, SELF-LOCKING (2) 589-18	
					NUT, SELF-LOCKING (4) 589-20	
					NUT, SELF-LOCKING (8) 589-26	
					NUT, SELF-LOCKING (4) 589-31	
					PAD ASSEMBLY (1) 589-62	
					PIN, COTTER (1) 589-28	
					PUMP, ASSEMBLY (1) 589-3	
					REDUCER, PIPE (2) 589-50	
					SCREW, CAP, HEXAGON (5) 587-11	
					SCREW, CAP, HEXAGON (4) 587-16	
					SCREW, CAP, HEXAGON (1) 588-7	
					SCREW, CAP, HEXAGON (2) 589-19	
					SCREW, CAP, HEXAGON (2) 589-22	
					SCREW, CAP, HEXAGON (6) 589-24	
					SCREW, CAP, HEXAGON (2) 589-35	
					SCREW, CAP, HEXAGON (1) 589-41	
					SCREW, CAP, HEXAGON (1) 589-56	
					SCREW, MACHINE (6) 587-3	
					SCREW, MACHINE (3) 587-9	
					SCREW, TAPPING (2) 587-4	
					SHIELD, SWINGFIRE (1) 589-23	
					SHROUD, OIL PAN (1) 589-27	
					STRAP, RETAINING (1) 589-6	
					STRAP, TIEDOWN (1) 588-2	
					STRAP, TIEDOWN (5) 589-45	
					STUD (4) 589-29	
					STUD, TURNBUTTON (3) 587-1	
					STUD, TURNBUTTON (8) 587-6	
					SUPPORT (1) 589-25	
					SWITCH (1) 588-10	
					TEE, PIPE (1) 589-1	
					TUBE (1) 589-13	
					TUBE (1) 589-33	
					TUBE (1) 589-36	
					TUBE (1) 589-42	
					TUBE (1) 589-43	
					TUBE (1) 589-44	
					WASHER, FLAT (5) 587-12	
					WASHER, FLAT (4) 587-15	
					WASHER, FLAT (2) 589-7	
					WASHER, FLAT (4) 589-30	
					WASHER, LOCK (17) 587-4	
					WASHER, LOCK (1) 588-5	
					WASHER, LOCK (1) 588-12	
					WASHER, LOCK (1) 589-57	
					WATER JACKET ASSEMB (1) 589-34	
					WATER OUTLET, ENGINE (1) 589-15	
					WIRING HARNESS, BRAN (11) 588-1	
	PDOZZ	2540011757257	19207	12302814	MOUNTING KIT, TOOL S .... UOC: DAA, DAB, DAC, DAD, DAE, DAF, DAG, DAH, DAW, DAX, V12, V13, V14, V15 , 16, V16, V, V9, V20, V21, V22, ZAA, ZAB, ZAC, ZAD, ZAE, ZAF, ZAG, ZAH	1
					BRACKET (1) 600-1	

(1) ITEM NO	(2) SMR CODE	(3) NSN	(4) CAGEC	(5) PART NUMBER	(6) DESCRIPTION AND USABLE ON CODES (UOC)			(7) QTY
					BRACKET, ANGLE	(	2)	600-6
					NUT, SELF-LOCKING, HE	(	4)	600-4
					NUT, SELF-LOCKING, HE	(	5)	600-4
					NUT, SELF-LOCKING, HE	(	6)	600-7
					SCREW, CAP, HEXAGON H	(	4)	600-2
					SCREW, CAP, HEXAGON H	(	5)	600-2
					SCREW, CAP, HEXAGON H	(	6)	600-5
					WASHER, FLAT	(	4)	600-3
					WASHER, FLAT	(	5)	600-3
	PAOZZ	4210012206376	19207	12302876	PARTS KIT, FIRE EXTI			1
					BRACKET, FIRE EXTING	(	1)	610-3
					EXTINGUISHER, FIRE	(	1)	610-1
					NUT, SELF-LOCKING, HE	(	4)	610-6
					PLATE, SPACER	(	2)	610-4
					SCREW, CAP, HEXAGON H	(	4)	610-2
					WASHER, FLAT	(	4)	610-5
	PAFZZ	4330012846203	8X715	12503	FILTER ELEMENT, FU	(	1)	619-18
					FILTER ELEMENT, FLUI	(	1)	619-19
	KFHZZ	2530013397913	78222	1790522K	PARTS KIT, STEERING			1
					RETAINER	(	1)	307-2
					TAB, WASHER	(	1)	307-3
					WASHER, FRICTION	(	1)	307-4
	PAFZZ	5330012719544	47457	20510093-25Z	GASKET SET			1
					UOC: ZAA, ZAB, ZAC, ZAD, ZAE, ZAF, ZAG, ZAH, ZAJ, ZAK, ZAL			
					END PLATE	(	1)	312-19
					KIDNEY SEAL	(	1)	312-18
					O-RING BY-PASS	(	1)	312-16
					O-RING BOOM COVER	(	1)	312-30
					O-RING VALVE CAP	(	1)	312-26
					SEAL	(	1)	312-13
					SEAL	(	1)	312-14
	PAFZZ	2530012722910	47457	20510093-26Z	PARTS KIT, POWER STE			1
					UOC: ZAA, ZAB, ZAC, ZAD, ZAE, ZAF, ZAG, ZAH, ZAJ, ZAK, ZAL			
					CAM	(	1)	312-21
					CARRIER	(	1)	312-32
					DRIVE PIN	(	1)	312-35
					LOCATING PIN	(	1)	312-22
					PORT PLATE	(	1)	12-20
					ROLLERS	(	1)	312-31
	PDFZZ	2990012843218	47457	20511136	HEATER, COOLANT, ENGI ENGINE COOLANT			1
					HEATER			
					ADAPTER, STRAIGHT, PI	(	1)	584-9
					ADAPTER, STRAIGHT, PI	(	1)	585-37
					BELT, V	(	1)	584-16
					BLOCK, SUPPORT, BATTE	(	4)	584-29
					BLOCK, SUPPORT, BATTE	(	2)	584-30
					BLOCK, SUPPORT, BATTE	(	2)	584-31
					BOLT, MACHINE	(	1)	584-20
					BOLT, MACHINE	(	2)	585-33
					BRACKET ASSEMBLY,	(	1)	585-15
					BRACKET, MOUNTING	(	1)	584-4
					CABLE ASSEMBLY, SPEC	(	1)	584-10

(1) ITEM NO	(2) SMR CODE	(3) NSN	(4) CAGEC	(5) PART NUMBER	(6) DESCRIPTION AND USABLE ON CODES (UOC)	(7) QTY
					CABLE ASSEMBLY,	(1) 585-6
					CLAMP, HOSE	(3) 584-19
					CLAMP, HOSE	(1) 585-32
					CLAMP, HOSE	(2) 585-8
					CLAMP, HOSE	(5) 585-1
					CLAMP, LOOP	(1) 585-19
					CLAMP, LOOP	(1) 584-15
					CLAMP, LOOP	(3) 584-21
					COCK, DRAIN	(1) 585-25
					COCK, PLUG	(1) 584-8
					CONTROL BOX, ELECTRI	(1) 584-32
					COUPLING, PIPE	(1) 585-10
					COUPLING, PIPE	(2) 585-5
					DECAL	(1) 584-1
					ELBOW, PIPE TO TUBE	(1) 585-9
					ELBOW, PIPE TO HOSE	(4) 585-4
					ELBOW, PIPE	(1) 585-38
					ELBOW, PIPE	(1) 584-7
					GROMMET, NONMETALLIC	(2) 584-27
					GROMMET, NONMETALLIC	(1) 585-12
					GROMMET, NONMETALLIC	(1) 585-3
					GROMMET, NONMETALLIC	(1) 584-11
					HEATER, COOLANT, ENGI	(1) 585-13
					HOSE ASSEMBLY, NONME	(1) 584-12
					HOSE	(1) 585-26
					HOSE	(1) 585-2
					HOSE	(1) 584-22
					HOSE	(1) 584-24
					NUT, CLIP-ON	(4) 585-29
					NUT, PLAIN, HEXAGON	(2) 584-6
					NUT, SELF-LOCKING, HE	(5) 584-5
					NUT, SELF-LOCKING	(5) 585-18
					NUT, SELF-LOCKING, HE	(2) 585-35
					NUT, SELF-LOCKING	(1) 585-34
					NUT, SELF-LOCKING, HE	(2) 584-25
					PAD ASSEMBLY	(1) 584-28
					PIN, COTTER	(2) 585-20
					PIPE, EXHAUST	(1) 585-22
					PIPE, EXHAUST	(1) 585-21
					PLATE, INSTRUCTION	(1) 584-17
					PUMP UNIT, CENTRIFUG	(1) 585-24
					SCREW, CAP, HEXAGON H	(4) 585-27
					SCREW, CAP, HEXAGON H	(5) 584-2
					SCREW, CAP, HEXAGON	(2) 585-16
					SCREW, CAP, HEXAGON H	(4) 585-7
					SCREW, CAP, HEXAGON	(1) 585-36
					SCREW, CAP, HEXAGON H	(2) 584-23
					SCREW, TAPPING	(4) 584-18
					STRAP, RETAINING	(2) 585-14
					STRAP, RETAINING	(1) 585-31
					STRAP, RETAINING	(1) 585-30
					STRAP, TIEDOWN, ELECT	(14) 584-13
					WASHER, FLAT	(1) 584-14
					WASHER, FLAT	(2) 584-26

KIT-14

(1) ITEM NO	(2) SMR CODE	(3) NSN	(4) CAGEC	(5) PART NUMBER	(6) DESCRIPTION AND USABLE ON CODES (UOC)			(7) QTY
					WASHER, FLAT	(	17) 585-17	
					WASHER, LOCK	(	8) 585-28	
					WASHER, LOCK	(	4) 585-11	
					WASHER, LOCK	(	6) 584-3	
					WASHER, LOCK	(	1) 585-23	
	PAOZZ	2530012722912	47457	20511322Z	PARTS KIT, PUMP ASSE			1
					UOC: ZAA, ZAB, ZAC, ZAD, ZAE, ZAF, ZAG, ZAH, ZAJ, ZAK, ZAL			
					GASKET	(	1) 312-40	
					LID	(	1) 312-41	
	PDFZZ	2540014166784	47457	20511137	PARTS KIT, HEATER			1
					ADAPTER, CONNECTOR	(	1) 594-25	
					ADAPTER, STRAIGHT, PI	(	1) 594-3	
					ADAPTER, STRAIGHT, PI	(	1) 595-2	
					ADAPTER, STRAIGHT, PI	(	3) 595-7	
					BRACKET	(	1) 595-1	
					BRACKET, MOUNTING	(	2) 594-17	
					BRACKET, MOUNTING	(	1) 594-35	
					BUTTON, PLUG	(	1) 594-4	
					CLAMP, HOSE	(	2) 594-7	
					CLAMP, LOOP	(	1) 594-23	
					COCK, DRAIN	(	1) 595-20	
					COCK, PLUG	(	1) 595-17	
					CONTROL BOX, ELECTRI	(	1) 594-29	
					COUPLING, PIPE	(	1) 594-10	
					DECAL	(	1) 594-31	
	PDFZZ	2540012848718	47457	20511191	FORDING KIT, DEEP			1
					ADAPTER, STRAIGHT, PI	(	1) 598-55	
					ADAPTER, STRAIGHT, PI	(	1) 598-56	
					ADAPTER, STRAIGHT, PI	(	1) 598-53	
					ADAPTER, BUSHING	(	1) 598-12	
					ADAPTER, STRAIGHT, PI	(	1) 599-16	
					ADAPTER, STRAIGHT, P	(	1) 599-14	
					BOLT	(	2) 598-41	
					BRACKET, ANGLE	(	1) 598-43	
					BRACKET, DOUBLE ANGL	(	1) 598-40	
					BUSHING, PIPE	(	1) 598-48	
					BUSHING, PIPE	(	1) 599-12	
					CLAMP, HOSE	(	2) 598-10	
					CONNECTOR, MULTIPLE,	(	1) 598-54	
					CONTROL ASSEMBLY, PU	(	1) 598-2	
					COUPLING, CLAMP, GROO	(	1) 599-1	
					ELBOW, PIPE TO HOSE	(	1) 598-17	
					ELBOW, PIPE TO TUBE	(	1) 598-26	
					ELBOW, PIPE TO TUBE	(	1) 598-11	
					ELBOW, PIPE TO HOSE	(	1) 598-28	
					ELBOW, PIPE TO HOSE	(	1) 598-38	
					ELBOW, PIPE	(	1) 598-49	
					ELBOW, PIPE TO TUBE	(	1) 599-13	
					ELBOW, PIPE TO TUBE	(	1) 598-15	
					ELBOW, PIPE TO TUBE	(	1) 598-52	
					ELBOW, PIPE TO TUBE	(	1) 598-39	
					ELBOW, PIPE TO TUBE	(	1) 598-23	
					GASKET	(	1) 599-2	

(1) ITEM NO	(2) SMR CODE	(3) NSN	(4) CAGEC	(5) PART NUMBER	(6) DESCRIPTION AND USABLE ON CODES (UOC)			(7) QTY
					GROMMET, NONMETALLIC	(	1)	598-1
					HOSE ASSEMBLY, NONME	(	1)	598-14
					HOSE,	(	1)	598-9
					HOSE,	(	1)	598-13
					HOSE,	(	1)	598-16
					LEVER, REMOTE CONTRO	(	1)	598-31
					MANIFOLD ASSEMBLY, H	(	1)	598-21
					NIPPLE, PIPE	(	1)	598-50
					NIPPLE, PIPE	(	1)	598-25
					NUT, PLAIN, HEXAGON	(	2)	598-45
					NUT, TUBE COUPLING	(	1)	599-8
					NUT, TUBE COUPLING	(	3)	599-5
					NUT, TUBE COUPLING	(	1)	598-19
					NUT, TUBE COUPLING	(	3)	598-6
					PIN, COTTER	(	1)	598-29
					PIN, STRAIGHT, HEADED	(	1)	598-32
					PLATE, INSTRUCTION,	(	1)	598-3
					PLUG, PIPE	(	2)	599-3
					SCREW	(	1)	598-34
					SCREW, CAP, HEXAGON H	(	2)	598-46
					SCREW, MACHINE	(	1)	598-33
					SCREW, TAPPING	(	4)	598-4
					SLEEVE, COMPRESSION	(	1)	599-9
					SLEEVE, COMPRESSION	(	3)	599-6
					SLEEVE, COMPRESSION,	(	3)	598-7
					SLEEVE, COMPRESSION,	(	1)	598-20
					SPACER, SLEEVE	(	1)	598-36
					STRAP, RETAINING	(	1)	598-42
					STRAP, TIEDOWN, ELECT	(	6)	599-4
					TEE, PIPE	(	1)	599-15
					TEE, PIPE	(	1)	598-24
					TEE, PIPE	(	1)	598-27
					TUBE ASSEMBLY, METAL	(	1)	598-22
					TUBING, NYLON,	(	1)	598-8
					TUBING, NYLON,	(	1)	598-5
					VALVE, FUEL, SIX-PORT	(	1)	598-37
					VALVE, REGULATING, FL	(	1)	598-51
					WASHER, FLAT	(	2)	598-47
					WASHER, FLAT	(	1)	598-30
					WASHER, FLAT	(	1)	598-35
					WASHER, LOCK	(	2)	598-44
	XDOZZ	2520011341324	73342	23012502	PARTS KIT, OIL PAN			1
					GASKET	(	1)	189-7
					GASKET	(	1)	189-24
					MAGNET, OIL PAN	(	1)	189-21
					PLUG, MACHINE THREAD	(	1)	189-23
	PAHZZ	2520012116702	73342	23019201	PARTS KIT, HYDRAULIC			1
					FILTER ELEMENT, FLUI	(	1)	189-20
					GASKET	(	1)	189-25
					O-RING	(	1)	189-15
	PAHZZ	2520014287497	73342	23019596	PARTS KIT, HYDRAULIC			1
					CONRTOL VALVE	(	1)	195-1
					GOVERNOR ASSEMBLY	(	1)	198-6
					VALVE ASSEMBLY, LOW	(	1)	197-1

KIT-16

(1) ITEM NO	(2) SMR CODE	(3) NSN	(4) CAGEC	(5) PART NUMBER	(6) DESCRIPTION AND USABLE ON CODES (UOC)			(7) QTY
					VALVE, LOCKUP	(	1)	196-1
	PAHZZ	3040013374055	73342	23043043	PARTS KIT, LUBRICANT CONVERTS TRANS .... P/N 6885292 TO NEW TRANS P/N 23040127			1
					HOUSING, LIQUID PUMP	(	1)	201-4
					HOUSING, MECHANICAL	(	1)	201-9
					HOUSING, MECHANICAL	(	1)	201-11
					HUB, RANGE CLUTCH TR	(	1)	201-1
	PAOZZ	2910001522033	33457	256476	FILTER ELEMENT, FLUI			1
					ELEMENT ASSY	(	1)	93-7
					GASKET	(	1)	93-6
					O-RING	(	1)	93-4
	PAFZZ	2530011341834	06853	289352	PARTS KIT, AIR FLOW			1
					MOUNT, RESILIENT	(	1)	249-43
					NUT, SELF-LOCKING, HE	(	1)	249-39
					NUT, SLEEVE	(	1)	249-41
					O-RING	(	1)	249-19
					O-RING	(	1)	249-33
					O-RING	(	2)	249-35
					O-RING	(	1)	249-36
					O-RING	(	1)	249-34
					O-RING	(	1)	249-44
					PACKING, PREFORMED	(	1)	249-45
					SPRING, HELICAL, COMP	(	1)	249-37
					SPRING, HELICAL, COMP	(	1)	249-21
					SPRING, HELICAL, COMP	(	1)	249-32
					VALVE, INLET AND EXH	(	1)	249-24
					VALVE, INLET AND EXH	(	1)	249-5
	PAFZZ	2530011341835	06853	289353	PARTS KIT, AIR FLOW			1
					MOUNT, RESILIENT	(	1)	249-43
					O-RING	(	2)	249-35
					O-RING	(	1)	249-36
					O-RING	(	1)	249-44
					PACKING, PREFORMED	(	1)	249-45
	PAHZZ	5330006323813	15434	3010240	GASKET SET, FUEL PUMP.. UOC: DAA, DAB, DAC, DAD, DAE, DAF, DAG, DAH, DAJ, DAK, DAL, DAW, DAX, V12, V13, V14, V15, V16, V17, V18, V19, V20, V21, V22, V24, V25, V39			1
					GASKET	(	1)	49-1
					SEAL, OIL	(	2)	50-25
					PACKING, PREFORMED	(	1)	51-1
					PACKING, PREFORMED	(	1)	52-6
					PACKING, PREFORMED	(	1)	52-8
					SPACER, RING	(	1)	52-9
					GASKET	(	1)	52-24
					GASKET, FUEL PUMP	(	1)	53-8
					GASKET	(	1)	54-1
					SEAL, OIL	(	2)	55-15
					PACKING, PREFORMED	(	1)	56-1
					PACKING & RETAINER	(	1)	56-14
					GASKET	(	1)	85-13
					GASKET	(	1)	87-14
					SEAL	(	1)	95-1

(1) ITEM NO	(2) SMR CODE	(3) NSN	(4) CAGEC	(5) PART NUMBER	(6) DESCRIPTION AND USABLE ON CODES (UOC)			(7) QTY
					PACKING, PREFORMED	(	1)	129-9
					PACKING, PREFORMED	(	1)	129-10
					PACKING, PREFORMED	(	1)	129-17
	PAHZZ	5330004806133	15434	3011472	GASKET AND PREFORME			1
					UOC: DAA, DAB, DAC, DAD, DAE, DAF, DAG, DAH, DAJ, DAK, DAL, DAW, DAX, V12, V13, V14, V15, V16, V17, V18, V19 , V, V20, V21, V22, V24, V25, V39			
					GASKET	(	1)	11-1
					GASKET	(	1)	14-16
					GASKET	(	1)	19-1
					GASKET	(	3)	28-8
					GASKET	(	3)	28-9
					GASKET	(V	)	23-16
					GASKET	(V	)	23-16
					GASKET	(V	)	23-16
					GASKET	(V	)	23-16
					GASKET	(V	)	23-16
					GASKET	(1	)	16-1
					GASKET	(	1)	33-1
					GASKET	(	1)	33-39
					GASKET	(	2)	33-31
					GASKET	(	1)	37-3
					GASKET	(	1)	37-7
					GASKET	(	1)	31-13
					GASKET	(	1)	31-8
					GASKET	(	3)	41-1
					GASKET	(	1)	41-3
					GASKET	(	6)	42-5
					GASKET	(	1)	44-13
					GASKET	(	3)	46-3
					GASKET	(	1)	46-16
					GASKET	(	1)	54-1
					GASKET	(	1)	49-1
					GASKET	(	1)	114-6
					GASKET	(	6)	116-13
					GASKET	(	2)	116-9
					GASKET	(	1)	118-13
					GASKET	(	1)	118-18
					GASKET	(	1)	127-5
					GASKET	(	1)	284-46
					GASKET	(	1)	284-21
					GASKET	(	1)	284-52
					GASKET AND PREFORME	(	1)	44-16
					GASKET SET	(	1)	KITS-
					INSERT, FLEXIBLE COU	(	1)	54-2
					INSERT, FLEXIBLE COU	(	1)	49-2
					O-RING	(	6)	7-4
					O-RING	(	1)	19-8
					O-RING	(	1)	16-5
					O-RING	(	2)	31-4
					O-RING	(	6)	46-17
					O-RING	(	1)	51-11
					O-RING	(	12)	66-24

(1) ITEM NO	(2) SMR CODE	(3) NSN	(4) CAGEC	(5) PART NUMBER	(6) DESCRIPTION AND USABLE ON CODES (UOC)			(7) QTY
					O-RING	(	4) 116-2	
					O-RING	(	1) 284-7	
					O-RING	(	1) 284-16	
					O-RING	(	1) 284-8	
					PACKING ASSEMBLY	(	1) 30-2	
					PACKING WITH RETAIN	(	1) 56-14	
					PACKING, PREFORMED	(	9) 14-2	
					PACKING, PREFORMED	(	1) 284-14	
					RETAINER, PACKING	(	2) 31-3	
					SEAL	(	1) 19-13	
					SEAL	(	1) 16-4	
					SEAL, PLAIN ENCASED	(	1) 19-7	
					SEAL, THERMO	(	1) 114-4	
					SPACER, RING	(	1) 37-6	
					WASHER, KEY	(	12) 42-2	
	PAOZZ	4330012430055	43990	3652-11	PARTS KIT, FLUID PRE			1
					FILTER ELEMENT FLUI	(	1) 621-21	
					GASKET	(	1) 621-20	
					PACKING, PREFORMED	(	1) 621-22	
	PAHZZ	2815011650765	15434	3801056	RING SET, PISTON			6
					UOC: DAA, DAB, DAC, DAD, DAE, DAF, DAG, DAH, DAJ, DAK, DAL, DAW, DAX, V12, V13, V14, V15, V16, V17, V18, V19, V20, V21, V22, V24, V25, V39			
					RING, PISTON	(	1) 17-1	
					RING, PISTON	(	1) 17-2	
					RING, PISTON	(	1) 17-3	
					RING, PISTON	(	1) 17-4	
	PAHZZ	4310013049726	15434	3801607	PARTS KIT, CYLINDER			1
	PAHZZ	3120011329339	15434	3801260	BEARING SET, SLEEVE			1
					BEARING HALF, SLEEVE	(	1) 12-3	
					BEARING HALF, SLEEVE	(	1) 12-10	
					BEARING HALF, SLEEVE	(	3) 12-1	
					BEARING HALF, SLEEVE	(	3) 12-17	
					BEARING HALF, SLEEVE	(	3) 12-12	
					BEARING HALF	(	3) 12-11	
					BEARING, WASHER, THRU	(	4) 12-2	
	PAHZZ	3120011439547	15434	3801261	BEARING HALF SET, SL			1
					BEARING HALF	(	3) 12-17	
					BEARING HALF	(	3) 12-1	
					BEARING HALF	(	1) 12-3	
					BEARING HALF	(	1) 12-10	
					BEARING HALF	(	3) 12-11	
					BEARING HALF	(	3) 12-12	
					BEARING, WASHER, THRU	(	4) 12-2	
	PAHZZ	3120011448882	15434	3801262	BEARING HALF SET, SL			1
					UOC: DAA, DAB, DAC, DAD, DAE, DAF, DAG, DAH, DAJ, DAK, DAL, DAW, DAX, V12, V13, V14, V15, V16, V17, V18, V19, V20, V21, V22, V24, V25, V39			
					BEARING HALF	(	3) 12-17	
					BEARING HALF	(	3) 12-1	
					BEARING HALF	(	1) 12-3	
					BEARING HALF	(	1) 12-10	

(1) ITEM NO	(2) SMR CODE	(3) NSN	(4) CAGEC	(5) PART NUMBER	(6) DESCRIPTION AND USABLE ON CODES (UOC)			(7) QTY
					BEARING HALF	(3)	12-11	
					BEARING, WASHER, THRU	(4)	12-2	
	PAHZZ	3120011459132	15434	3801263	BEARING HALF SET, SL			1
					UOC: DAA, DAB, DAC, DAD, DAE, DAF, DAG, DAH, DAJ, DAK, DAL, DAW, DAX, V12, V13, V14, V15, V16, V17, V18, V19, V20, V21, V22, V24, V25, V39			
					BEARING HALF .030	(3)	12-1	
					BEARING HALF .030	(1)	12-3	
					BEARING HALF .030	(1)	12-10	
					BEARING HALF .030	(3)	12-11	
					BEARING HALF .030	(3)	12-12	
					BEARING HALF .030	(3)	12-17	
					BEARING, WASHER, THRU	(4)	12-2	
	PAHZZ	3120011937083	15434	3801264	BEARING SET, SLEEVE			1
					UOC: DAA, DAB, DAC, DAD, DAE, DAF, DAG, DAH, DAJ, DAK, DAL, DAW, DAX, V12, V13, V14, V15, V16, V17, V18, V19, V20, V21, V22, V24, V25, V39			
					BEARING HALF .040	(3)	12-1	
					BEARING HALF .040	(1)	12-3	
					BEARING HALF .040	(1)	12-10	
					BEARING HALF .040	(3)	12-11	
					BEARING HALF .040	(3)	12-12	
					BEARING HALF .040	(3)	12-17	
					BEARING, WASHER, THRU	(4)	12-2	
	PAHZZ	2815012781093	15434	3801535	PISTON, INTERNAL COM			1
					UOC: DAA, DAB, DAC, DAD, DAE, DAF, DAG, DAH, DAJ, DAK, DAL, DAW, DAX, V12, V13, V14, V15, V16, V17, V18, V19, V20, V21, V22, V24, V25, V39			
	PAFZZ	4310012725374	15434	3801808	REPAIR KIT, COMPRESS			1
					UOC: ZAA, ZAB, ZAC, ZAD, ZAE, ZAF, ZAG, ZAH, ZAJ, ZAK, ZAL			
					BODY, UNLOADER VALVE	(1)	286-8	
					CAP, UNLOADER	(1)	286-11	
					DISK, VALVE	(1)	286-4	
					PACKING, PREFORMED	(1)	286-7	
					PACKING, PREFORMED	(1)	286-9	
					PACKING, PREFORMED	(1	286-10	
					PACKING, PREFORMED	(1)	286-5	
					SEAT, VALVE	(1)	286-6	
					SHIM	(1)	286-2	
					SPRING, HELICAL, COMP	(1)	286-12	
					SPRING, HELICAL, COMP	(1)	286-15	
					SPRING, HELICAL, COMP	(1)	286-3	
					VALVE INTAKE COMPRE	(1)	286-14	
	PAFZZ	5330001336237	15434	3804272	GASKET			1
					UOC: DAA, DAB, DAC, DAD, DAE, DAF, DAG, DAH, DAJ, DAK, DAL, DAW, DAX, V12, V13, V14, V15, V16, V17, V18, V19, V20, V21, V22, V24, V25, V39			
					BARREL AND PLUNGER	(1)	46-6	
					CUP, INJECTOR	(1)	46-11	

KIT-20

(1) ITEM NO	(2) SMR CODE	(3) NSN	(4) CAGEC	(5) PART NUMBER	(6) DESCRIPTION AND USABLE ON CODES (UOC)		(7) QTY
					GASKET	(3) 28-8	
					GASKET	(3) 28-9	
					GASKET	(3) 41-1	
					GASKET	(1) 41-3	
					GASKET	(6) 42-5	
					GASKET	(3) 46-3	
					GASKET	(1) 46-16	
					GASKET	(1) 114-6	
					GASKET	(6) 116-13	
					O-RING	(6) 46-17	
					O-RING	(4) 116-2	
					SEAL, THERMO	(1) 114-4	
					WASHER, KEY	(12) 42-2	
	PAOZZ	2520003884197	95019	5190076	SPIDER, UNIVERSAL JO		2
					BEARING ASSEMBLY	(8) 218-4	
					BOLT, MACHINE	(16) 218-2	
					CROSS ASSEMBLY	(2) 218-6	
					LOCKING PLATE, NUT A	(8) 218-3	
	KFHZZ	5330013416583	78222	5518441	STEERING GASKET AND SEAL SET		1
					O-RING, HI-PRES SEAL	(1) 308-3	
					O-RING, COVER	(1) 308-20	
					O-RING, RELIEF PLUNG	(1) 308-11	
					O-RING, RELIEF PLUNG	(1) 308-18	
					PARTS KIT, SOLENOID	(1) 308-10	
					PIN, LOCKING	(1) 308-5	
					PLUNGER, RELIEF VALV	(1) 308-17	
					SEAL KIT	(1) 308-4	
					SEAL RING, METAL	(1) 308-6	
					SEAL, NONMETALLIC	(1) 308-9	
					SEAL, PLAIN ENCASED	(1) 308-8	
	KFHZZ	2530013400365	78222	5523281	PARTS KIT, STEERING		1
					BALL, PLUNGER RELIEF	(1) 309-11	
					BEARING, SHAFT ACTUA	(1) 308-2	
					PIN, LOCKING	(3) 308-8	
					PISTON	(1) 308-6	
					PISTON PLUG	(1) 308-1	
					SEAT, PLUNGER, RELIEF	(2) 308-13	
					SET, SCREW	(1) 308-7	
					SLIPPER, STRAINER	(1) 308-10	
					SPRING, REVERSING	(2) 308-3	
					SPRING, RELIEF, PLUNG	(1) 308-12	
					TEFLON PISTON RING	(1 308-5	
					VALVE, ADJUSTING NUT	(1) 308-2	
					VALVE POSITION PIN	(1) 308-9	
					VALVE, STEER GEAR, AC	(1) 308-4	
	PFOZZ	2540012565331	34623	57K0124	PARTS KIT, SEAT BELT TAN		1
					BELT ASSEMBLY	(1) 615-4	
					BELT, VEHICULAR	(1) 615-8	
					BRACKET	(1) 615-3	
					NUT, SELF-LOCKING	(5) 615-1	
					NUT, SELF-LOCKING	(8) 615-18	
					SCREW, CAP, HEXAGON	(2) 615-5	
					SCREW, CAP, HEXAGON	(1) 615-10	
					SCREW, CAP, HEXAGON	(1) 615-11	

(1) ITEM NO	(2) SMR CODE	(3) NSN	(4) CAGEC	(5) PART NUMBER	(6) DESCRIPTION AND USABLE ON CODES (UOC)			(7) QTY
					SCREW, CAP, HEXAGON	(1)	615-12	
					SCREW, CAP, HEXAGON	(2)	615-13	
					SCREW, CAP, HEXAGON	(8)	615-15	
					SLEEVE	(2)	615-14	
					SPACER	(1)	615-6	
					TETHER, BELT	(2)	615-7	
					WASHER	(8)	615-17	
					WASHER, FLAT	(17)	615-2	
					WASHER, FLAT	(8)	615-16	
					WASHER, LOCK	(2)	615-9	
	PFOZZ	2540012565331	34623	57K0125	PARTS KIT, SEAT BELT WHITE			1
					BELT ASSEMBLY	(1)	615-4	
					BELT, VEHICULAR	(1)	615-8	
					BRACKET	(1)	615-3	
					NUT, SELF-LOCKING	(5)	615-1	
					NUT, SELF-LOCKING	(8)	615-18	
					SCREW, CAP, HEXAGON	(2)	615-5	
					SCREW, CAP, HEXAGON	(1)	615-10	
					SCREW, CAP, HEXAGON	(1)	615-11	
					SCREW, CAP, HEXAGON	(1)	615-12	
					SCREW, CAP, HEXAGON	(2)	615-13	
					SCREW, CAP, HEXAGON	(8)	615-15	
					SLEEVE	(2)	615-14	
					SPACER	(1)	615-6	
					TETHER, BELT	(2)	615-7	
					WASHER	(8)	615-17	
					WASHER, FLAT	(17)	615-2	
					WASHER, FLAT	(8)	615-16	
					WASHER, LOCK	(2)	615-9	
	PAHZZ	2910013390423	19207	57K0144	PARTS KIT, ENGINE FU			1
					UOC: ZAA, ZAB, ZAC, ZAD, ZAE, ZAF, ZAG, ZAH, ZAJ, ZAK, ZAL			
					NUT, PLAIN, HEXAGON	(12)	59-2	
					PACKING, PREFORMED	(1)	58-11	
					SPACER, RING	(12)	59-4	
					SPRING, HELICAL, COMP	(6)	59-6	
					WASHER, FLAT	(6)	59-6	
					WASHER, LOCK	(1)	58-6	
					WASHER, LOCK	(6)	59-3	
	PFOZZ	2540014577623	19207	57K0171	KIT, DUMP BODY & TROOP SEAT .... & TAN TARP			1
					UOC: DAE, DAF, V19, V20, ZAE, ZAF			
					BOW ASSEMBLY	(1)	609-2	
					CURTAIN, VEHICULAR	(1)	609-31	
					SEAT, TROOP LEFT	(1)	609-8	
					SEAT, TROOP RIGHT	(1)	609-8	
					SIDE RACK LEFT	(1)	609-18	
					SIDE RACK RIGHT	(1)	609-18	
					STRAP, WEBBING	(1)	609-1	
					TARPAULIN	(1)	609-30	
	PFOZZ	2540014577602	19207	57K0172	KIT, DUMP BODY & TROOP SEAT .... & WHITE TARP			1
					UOC: DAE, DAF, V19, V20, ZAE, ZAF			
					BOW ASSEMBLY	(1)	609-2	

(1) ITEM NO	(2) SMR CODE	(3) NSN	(4) CAGEC	(5) PART NUMBER	(6) DESCRIPTION AND USABLE ON CODES (UOC)			(7) QTY
					CURTAIN, VEHICULAR	(	1)	609-31
					SEAT, TROOP LEFT	(	1)	609-8
					SEAT, TROOP RIGHT	(	1)	609-8
					SIDE RACK LEFT	(	1)	609-18
					SIDE RACK RIGHT	(	1)	609-18
					STRAP, WEBBING	(	1)	609-1
					TARPAULIN	(	1)	609-30
	PDFZZ	2540014166784	19207	57K0243	PARTS KIT, HEATER			1
					ADAPTER, CONNECTOR	(	1)	594-25
					ADAPTER, STRAIGHT, PI	(	1)	594-3
					ADAPTER, STRAIGHT, PI	(	1)	595-2
					ADAPTER, STRAIGHT, PI	(	3)	595-7
					BRACKET	(	1)	595-1
					BRACKET, MOUNTING	(	2)	594-17
					BRACKET, MOUNTING	(	1)	594-35
					BUTTON, PLUG	(	1)	594-4
					CLAMP, HOSE	(	2)	594-7
					CLAMP, LOOP	(	1)	594-23
					COCK, DRAIN	(	1)	595-20
					COCK, PLUG	(	1)	595-17
					CONTROL BOX, ELECTRI	(	1)	594-29
					COUPLING, PIPE	(	1)	594-10
					DECAL	(	1)	594-31
					DUMMY CONNECTOR, PLU	(	2)	594-26
					DUMMY CONNECTOR, PLU	(	3)	595-14
					ELBOW, PIPE TO TUBE	(	1)	594-11
					ELBOW, PIPE TO TUBE	(	1)	595-24
					FILTER ELEMENT, FLUI	(	1)	595-12
					GROMMET, NONMETALLIC	(	1)	594-24
					HEATER, VEHICULAR, CO	(	1)	594-8
					HOSE ASSEMBLY, NONME	(	1)	594-2
					HOSE ASSEMBLY, NONME	(	1)	594-9
					HOSE ASSEMBLY, NONME	(	1)	595-3
					HOSE ASSEMBLY, NONME	(	1)	595-5
					LEAD, ELECTRICAL	(	1)	595-15
					MOUNT, FILTER, FUEL	(	1)	595-11
					NOZZLE, SPRAY, FLUID	(	2)	594-1
					NUT, PLAIN, HEXAGON	(	2)	594-36
					NUT, SELF-LOCKING, HE	(	4)	594-14
					NUT, SELF-LOCKING, HE	(	1)	594-22
					NUT, SELF-LOCKING, HE	(	2)	594-34
					NUT, SELF-LOCKING, HE	(	3)	595-8
					NUT, SELF-LOCKING, HE	(	2)	595-23
					PIN, COTTER	(	3)	594-12
					PIPE, EXHAUST	(	2)	594-13
					PIPE, EXHAUST	(	1)	594-19
					PLATE, INSTRUCTION	(	1)	594-38
					PLUG	(	1)	595-4
					PLUG	(	1)	595-6
					PLUG, PIPE	(	1)	594-6
					PUMP, FUEL, ELECTRICA	(	1)	595-22
					SCREW, CAP, HEXAGON H	(	4)	594-18
					SCREW, CAP, HEXAGON H	(	1)	594-21
					SCREW, CAP, HEXAGON H	(	2)	594-32

(1) ITEM NO	(2) SMR CODE	(3) NSN	(4) CAGEC	(5) PART NUMBER	(6) DESCRIPTION AND USABLE ON CODES (UOC)			(7) QTY
					SCREW, CAP, HEXAGON H	(3)	595-10	
					SCREW, CAP, HEXAGON H	(2)	595-18	
					SCREW, TAPPING	(4)	594-37	
					SHELL, ELECTRICAL CO	(1)	594-27	
					SHELL, ELECTRICAL CO	(4)	595-13	
					STRAP, TIEDOWN, ELECT	(12)	595-16	
					TEE, PIPE	(1)	594-5	
					TEE, PIPE	(1)	595-21	
					WASHER, FLAT	(8)	594-15	
					WASHER, FLAT	(1)	594-20	
					WASHER, FLAT	(3)	595-9	
	PEOOO	5180013042257	19207	57K1310	TOOL KIT, DIESEL			
					INSERTER, SEAL	(1)	630-19	
					PULLER KIT, UNIVERSA	(1)	630-20	
					PUNCH, BEARING	(1)	630-18	
	PEFFF	5180013042258	19207	57K1311	TOOL KIT, INTERNAL			
					AIR GAGE ASSEMBLY	(1)	630-8	
					BARRING TOOL, GEAR	(1)	630-3	
					KIT, NOZZLE CLEANING	(1)	630-5	
					REDUCER, TUBE	(1)	630-1	
					WRENCH, BOX	(1)	630-4	
	PEHHH	5180013042164	19207	57K1312	TOOL KIT, GENERAL			
					CLAMP SET, CYLINDER	(1)	630-2	
					COMPRESSOR, VALVE SP	(1)	631-1	
					EXTENSION SHAFT	(1)	631-6	
					INSTALLING AND REMO	(1)	630-6	
					PLUNGER LIFT DEVICE	(1)	631-5	
					PULLER, MECHANICAL	(1)	630-7	
					SEPARATION TUBE	(1)	631-4	
					SIDE PLUG PULLER	(1)	631-2	
					TAPPET HOLDER	(1)	631-3	
	PFOZZ	1005014323339	19207	57K0300	MODIFICATION KIT, LT WPN STA			1
					UOC: DAA, DAB, DAC, DAD, DAG, DAH, DAW, DAX, V12, V13, V14, V15, V16, V17, V21, V22, ZAA, ZAB, ZAC, ZAD, ZAG, ZAH			
					BEARING ASSEMBLY	(1)	602-42	
					BRACE	(4)	602-24	
					BRACE ASEMBLY	(1)	602-19	
					BRACE ASSEMBLY	(1)	602-19	
					BRACKET, GUN MOUNT	(1)	602-10	
					BRACKET, GUN MOUNT	(1)	602-12	
					BOLT, MACHINE	(4)	602-36	
					BOLT, SELF-LOCKING	(3)	602-11	
					BOLT, SELF-LOCKING	(4)	602-11	
					BOLT, U	(4)	602-6	
					BOLT, U	(7)	602-51	
					CROSSMEMBER ASSY	(1)	602-54	
					CROSSMEMBER ASSY	(1)	602-56	
					HANDLE	(2)	602-37	
					HANDLE, MANUAL	(1)	602-20	
					LOCK SET, RIM	(2)	602-53	
					MOUNT, ROOF	(1)	602-46	
					NUT	(1)	602-31	
					NUT ASSEMBLY	(3)	602-17	

(1) ITEM NO	(2) SMR CODE	(3) NSN	(4) CAGEC	(5) PART NUMBER	(6) DESCRIPTION AND USABLE ON CODES (UOC)			(7) QTY
					NUT, PLAIN, HEXAGON	(18)	602-8	
					NUT, SELF-LOCKING	(102)	602-18	
					NUT, SELF-LOCKING	(4)	602-39	
					PAD, CUSHIONING	(2)	602-48	
					PANEL, ARMAMENT MOU	(1)	602-34	
					PIN ASSEMBLY	(1)	602-29	
					PIN, COTTER	(8)	602-5	
					PLATE, MOUNTING	(1)	602-3	
					PLATE, SNUBBER, VEHI	(8)	602-45	
					PLATE, TAPPING	(2)	602-9	
					POST ASSEMBLY	(4)	602-4	
					REINFORCEMENT	(2)	602-55	
					RING, TURRET LOCK	(3)	602-49	
					PANEL	(1)	602-22	
					PIN, STRAIGHT	(1)	602-7	
					SCREW, CAP, HEXAGON, H	(8)	602-1	
					SCREW, CAP, HEXAGON, H	(22)	602-13	
					SCREW, CAP, HEXAGON, H	(6)	602-14	
					SCREW, CAP, HEXAGON, H	(20	602-15	
					SCREW, CAP, HEXAGON, H	(24)	602-23	
					SCREW, CAP, HEXAGON, H	(2)	602-32	
					SCREW, CAP, HEXAGON, H	(4	602-33	
					SCREW, CAP, HEXAGON, H	(6)	602-35	
					SCREW, CAP, HEXAGON, H	(2)	602-41	
					SCREW, CAP, HEXAGON, H	(12)	602-43	
					SCREW, CAP, HEXAGON, H	(4)	602-44	
					SCREW, CAP, HEXAGON, H	(18)	602-50	
					SCREW, MACHINE	(1)	602-28	
					SPACER	(4)	602-47	
					SPACER	(3)	602-52	
					SPACER	(1)	602-52	
					STRAP, WEBBING	(4)	602-27	
					SUPPORT, AMMO TRAY	(4)	602-25	
					SUPPORT ASSEMBLY	(1)	602-21	
					SUPPORT ASSY, ARMAME	(1)	602-38	
					SUPPORT, POST	(1)	602-10	
					SUPPORT, POST	(1)	602-12	
					TRAY, LOADING, AMMUNI	(1)	602-26	
					TUBE ASSEMBLY	(1)	602-40	
					WASHER, FLAT	(144)	602-22	
					WASHER, FLAT	(1)	602-30	
					WASHER, LOCK	(8)	602-2	
					WASHER, LOCK	(26)	602-16	
	PAOZZ	2940011079689	19207	5702838	PARTS KIT, AIR FILTE			1
					FILTER ELEMENT, INTA	(1)	62-24	
					PACKING, PREFORMED	(1)	62-23	
	PAHZZ	4820010935785	19207	5704273	PARTS KIT, VALVE			1
					UOC: DAL, V18			
					O-RING	(8)	494-1	
					O-RING	(1)	494-9	
					O-RING	(1)	494-19	
					O-RING	(8)	494-2	
					O-RING	(8)	494-5	
					PACKING, PREFORMED	(16)	494-4	

(1) ITEM NO	(2) SMR CODE	(3) NSN	(4) CAGEC	(5) PART NUMBER	(6) DESCRIPTION AND USABLE ON CODES (UOC)			(7) QTY
					RETAINER, PACKING	(8)	494-3	
	PAHZZ	2590006062383	19207	5704274	PARTS KIT, WINCH			1
					UOC: DAL, V18			
					PLUG, PIPE	(1)	494-11	
					PLUNGER AND SCREEN	(1)	494-21	
					SEAT, RETAINER ASSEM	(1)	494-7	
					SPRING, SPECIAL	(1)	494-20	
					VALVE, REGULATING, FL	(1)	494-13	
	PAHZZ	2520004217229	19207	5704278	PARTS KIT, DRIVING A			1
					BEARING, WASHER, THRU(1) 234-16			
					COLLAR	(1)	234-16	
					COLLAR	(1)	234-16	
					COLLAR	(1)	234-16	
					COLLAR	(1)	241-16	
					COLLAR	(1)	241-16	
					COLLAR	(1)	241-16	
					COLLAR	(1)	241-16	
					COLLAR	(1)	241-16	
					COLLAR	(1)	241-16	
					COLLAR	(1)	241-16	
					COLLAR	(1)	241-16	
					COLLAR	(1)	241-16	
					COLLAR	(1)	241-16	
					COLLAR	(1)	241-16	
					COLLAR	(1)	241-16	
					COLLAR, BEARING	(1)	234-16	
					COLLAR, BEARING	(1)	234-16	
					COLLAR, BEARING	(1)	234-16	
					COLLAR, BEARING	(1)	234-16	
					COLLAR, BEARING	(1)	234-16	
					COLLAR, BEARING	(1)	234-16	
					COLLAR, BEARING	(1)	234-16	
					COLLAR, BEARING	(1)	234-16	
					COLLAR, BEARING	(1)	234-16	
					COLLAR, BEARING	(1)	234-16	
					COLLAR, BEARING	(1)	234-16	
					COLLAR, BEARING	(1)	241-16	
					COLLAR, SHAFT	(1)	234-16	
					COLLAR, SHAFT	(1)	234-16	
					COLLAR, SHAFT	(1)	241-16	
					COLLAR, SHAFT	(1)	241-16	
					SPACER, BEVEL PINION	(1)	234-16	
					SPACER, BEVEL PINION	(1)	241-16	
					SPACER, BEVEL PINION	(1)	241-16	
	PAFZZ	2530011259272	19207	5704510	PARTS KIT, KINGPIN			1
					UOC: V12, V13, V14, V15, V16, V17, V18, V19, V20, V21, V22, V24, V25, V39			
					BOOT, VEHICULAR COM	(2)	236-19	
					CLAMP, HOSE	(2)	236-17	
					SCREW, MACHINE	(2)	236-18	
					WIRE, NONELECTRICAL	(2)	236-21	
	XBFZZ	2910011594839	19207	5704519	PARTS KIT, ENGINE FU			1

(1) ITEM NO	(2) SMR CODE	(3) NSN	(4) CAGEC	(5) PART NUMBER	(6) DESCRIPTION AND USABLE ON CODES (UOC)			(7) QTY
					UOC: DAA, DAB, DAC, DAD, DAE, DAF, DAG, DAH, DAJ, DAK, DAL, DAW, DAX, V12, V13, V14, V15, V16, V17, V18, V19, V20, V21, V22, V24, V25, V39			
					BALL, BEARING	(1)	51-17	
					GASKET	(1)	49-1	
					GASKET	(1)	50-22	
					GASKET	(1)	85-13	
					SCREW	(2)	51-19	
					SEAL CAP	(1)	95-1	
					SEAL, SPECIAL	(1)	85-9	
					SHIM	(1)	85-16	
					SHIM	(1)	85-16	
					SHIM	(1)	85-16	
					SPACER, RING	(1)	85-16	
	PAHZZ	3020012319296	19207	5704533	GEAR SET, BEVEL, MATC			1
					GEAR, CAMSHAFT	(1)	21-2	
					GEAR, HELICAL	(1)	12-7	
					GEAR, HELICAL	(1)	44-2	
	PAOZZ	2540001081940	19207	5704495	ACCESSORY KIT, VEHIC			1
					UOC: DAC, V16, V17, ZAC			
					CURTAIN, VEHICULAR	(2)	535-9	
					PANEL, EXTENSION, COV	(1)	535-2	
					PANEL, EXTENSION COV	(1)	535-3	
					SCREW, MACHINE	(40)	535-5	
	PDFZZ	2540013426810	19207	5705626	HEATER, VEHICULAR, CO			1
					ADAPTER	(1)	591-12	
					ADAPTER, STRAIGHT, PI	(2)	590-4	
					ADAPTER, STRAIGHT, PI	(1)	590-37	
					BLOCK, SUPPORT, BATTE	(4)	590-40	
					BLOCK, SUPPORT, BATTE	(2)	590-41	
					BLOCK, SUPPORT, BATTE	(2)	590-42	
					BOLT, MACHINE	(2)	591-4	
					BOLT, MACHINE	(3)	591-7	
					BOLT, MACHINE	(2)	591-8	
					BOLT, SHOULDER	(1)	591-43	
					BRACKET	(1)	591-33	
					BRACKET, DOUBLE ANGL	(1)	590-13	
					BRACKET, MOUNTING	(1)	590-16	
					BRACKET, MOUNTING	(1)	590-26	
					BRACKET, MOUNTING	(1)	590-30	
					BRACKET, MOUNTING	(1)	591-18	
					CIRCUIT BREAKER	(1)	588-3	
					CLAMP	(2)	591-36	
					CLAMP	(1)	591-14	
					CLAMP, HOSE	(1)	590-5	
					CLAMP, HOSE	(2)	590-38	
					CLAMP, HOSE	(1)	591-10	
					CLAMP, LOOP	(2)	588-7	
					CLAMP, LOOP	(1)	588-9	
					CLAMP, LOOP	(2)	588-12	
					CLAMP, LOOP	(1)	590-25	
					CLAMP, LOOP	(1)	590-29	
					CLAMP, LOOP	(1)	591-21	

(1) ITEM NO	(2) SMR CODE	(3) NSN	(4) CAGEC	(5) PART NUMBER	(6) DESCRIPTION AND USABLE ON CODES (UOC)		(7) QTY
					CLAMP, LOOP	(1)	591-29
					CLAMP, RIM CLENCHING	(1)	586-20
					CONNECTOR	(1)	591-13
					CONNECTOR, RECEPTACL	(1)	588-4
					COVER, ACCESS	(1)	590-11
					COVER, ACCESS	(1)	591-6
					CUTOUT RELAY, ENGINE	(1)	588-17
					ELBOW, PIPE	(1)	590-2
					ELBOW, PIPE	(1)	590-32
					ELBOW, PIPE	(1)	590-36
					GROMMET, NONMETALLIC	(1)	588-6
					GROMMET, NONMETALLIC	(1)	588-8
					HOSE	(1)	590-6
					HOSE	(1)	590-8
					HOSE	(1)	590-24
					HOSE	(1)	591-15
					HOSE	(1)	591-35
					HOSE ASSEMBLY, NONME	(2)	590-23
					HOSE, NONMETALLIC	(1)	590-19
					HOSE, NONMETALLIC	(1)	590-39
					LEAD, ELECTRICAL	(1)	588-16
					NIPPLE, PIPE	(1)	590-1
					NIPPLE, PIPE	(1)	590-33
					NIPPLE, PIPE	(1)	591-44
					NUT, SELF-LOCKING, HE	(1)	588-13
					NUT, SELF-LOCKING, HE	(2)	588-15
					NUT, SELF-LOCKING, HE	(2)	590-14
					NUT, SELF-LOCKING, HE	(2)	590-17
					NUT, SELF-LOCKING, HE	(2)	590-18
					NUT, SELF-LOCKING, HE	(1)	590-28
					NUT, SELF-LOCKING, HE	(2)	591-2
					NUT, SELF-LOCKING, HE	(1)	591-32
					NUT, SELF-LOCKING, HE	(2)	591-40
					PAD ASSEMBLY	(1)	590-43
					PIN, COTTER	(1)	591-27
					PUMP UNIT, CENTRIFUG	(1)	591-11
					REDUCER, PIPE	(1)	590-3
					REDUCER, PIPE	(1)	590-35
					SCREW, CAP, HEXAGON H	(2)	588-11
					SCREW, CAP, HEXAGON H	(2)	590-10
					SCREW, CAP, HEXAGON H	(1)	590-12
					SCREW, CAP, HEXAGON H	(2)	590-15
					SCREW, CAP, HEXAGON H	(2)	590-21
					SCREW, CAP, HEXAGON H	(1)	591-19
					SCREW, CAP, HEXAGON H	(2)	591-28
					SCREW, CAP, HEXAGON H	(2)	591-34
					SCREW, CAP, HEXAGON H	(2)	591-37
					SCREW, CAP, HEXAGON H	(1)	591-38
					SCREW, TAPPING	(2)	588-2
					SEMICONDUCTOR DEVIC	(1)	588-14
					SHROUD, FAN, RADIATOR	(1)	591-22
					SHROUD, FAN, RADIATOR	(1)	591-23
					STRAP, RETAINING	(1)	591-9
					STRAP, RETAINING	(1)	591-41

(1) ITEM NO	(2) SMR CODE	(3) NSN	(4) CAGEC	(5) PART NUMBER	(6) DESCRIPTION AND USABLE ON CODES (UOC)			(7) QTY
					STRAP, RETAINING	(	1) 591-42	
					STRAP, TIEDOWN, ELECT	(	5) 588-10	
					STRAP, TIEDOWN, ELECT	(	3) 590-31	
					SUPPORT ASSEMBLY, RA	(	1) 591-5	
					SWITCH, THERMOSTATIC	(	1) 591-30	
					TUBE ASSEMBLY, METAL	(	1) 591-20	
					TUBE, BENT, METALLIC	(	1) 590-7	
					TUBE, BENT, METALLIC	(	1) 590-9	
					TUBE, BENT, METALLIC	(	1) 590-20	
					TUBE, BENT, METALLIC	(	1) 590-27	
					TUBE, BENT, METALLIC	(	1) 590-34	
					TUBE, BENT, METALLIC	(	1) 591-16	
					WASHER, FLAT	(	2) 591-1	
					WASHER, FLAT	(	6) 591-24	
					WASHER, FLAT	(	2) 591-39	
					WASHER, LOCK	(	1) 588-5	
					WASHER, LOCK	(	2) 591-3	
					WASHER, LOCK	(	6) 591-25	
					WASHER, LOCK	(	2) 591-31	
					WATER JACKET ASSEMB	(	1) 590-22	
					WIRING HARNESS, BRAN	(	1) 588-1	
	PAOZZ	2530012844287	52304	599913	PARTS KIT, RELAY VAL			1
					UOC: ZAA, ZAB, ZAC, ZAD, ZAE, ZAF, ZAG, ZAH, ZAJ, ZAK, ZAL			
					BALL	(	3) 298-4	
					DIAPHRAGM, ACTUATOR	(	1) 298-8	
					SCREW, CAP, HEXAGON H	(	3) 298-5	
					SPRING	(	1) 298-6	
	PAHZZ	2520010403541	73342	6880353	PARTS KIT, TRANSMISS			1
					GASKET	(	1) 198-8	
					PIN, GOV, ASSY	(	2) 198-7	
	PAHZZ	2520011402376	73342	6884259	KIT, BASIC-OVERHAUL			1
					BOLT, MACHINE	(	1) 189-14	
					FILTER ELEMENT, FLUI	(	1) 189-20	
					GASKET	(	1) 189-2	
					GASKET	(	1) 189-25	
					GASKET	(	1) 190-4	
					GASKET	(	1) 200-30	
					O-RING	(	1) 189-15	
					RETAINER, PACKING	(	1) 190-19	
					RING, LIP TYPE SEAL	(	1) 187-12	
					RING, LIP TYPE SEAL	(	1) 193-14	
					RING, PISTON	(	2) 192-11	
					SEAL	(	1) 200-2	
					SEAL, AIR, GAS TURBIN	(	1) 189-31	
					SEAL, AIR, GAS TURBIN	(	1) 191-6	
					SEAL, AIR, GAS TURBIN	(	1) 200-3	
					SEAL, PLAIN ENCASED	(	1) 190-16	
					SEAL, PLAIN	(	1) 193-13	
					SEAL, PLAIN	(	1) 194-8	
					SEAL, PLAIN	(	1) 191-5	
					SEAL, RING	(	2) 192-10	
					SEAL, RING	(	1) 194-7	
					WASHER, FLAT	(	1) 189-13	

(1) ITEM NO	(2) SMR CODE	(3) NSN	(4) CAGEC	(5) PART NUMBER	(6) DESCRIPTION AND USABLE ON CODES (UOC)			(7) QTY
					WASHER, SPECIAL	(12)	200-31	
	PAHZZ	4330011310279	73342	6884749	PARTS KIT, FLUID PRE			1
					O-RING	(1)	198-4	
					PARTS KIT, FLUID PRE	(1)	198-3	
	PAHZZ	2520007346959	19207	7346959	PARTS KIT, DRIVING A			1
					BEARING, WASHER, THRU	(2)	232-6	
					GEAR, BEVEL	(2)	232-7	
					GEAR, BEVEL	(4)	232-12	
					GEAR, BEVEL	(2)	239-7	
					O-RING	(1)	198-4	
					PARTS KIT, FLUID PRE	(1)	198-3	
					PARTS KIT, DRIVING A	(1)	174-9	
					SPIDER, DIFFERENTIAL	(1)	232-9	
					WASHER, THRUST	(2)	232-13	
	PAHZZ	2520010819043	73342	6885213	PARTS KIT, MANUAL SE			1
					NUT	(1)	189-33	
					SHAFT, MANUAL	(1)	189-32	
	PAHZZ	5330005131443	19207	7346807	GASKET AND SHIM SET			1
					GASKET	(1)	233-4	
					GASKET	(V)	235-5	
					GASKET	(V)	242-5	
					GASKET	(1)	241-34	
					GASKET	(1)	241-26	
	PFOZZ	2590006009035	19207	8390117	KIT, A-FRAME MOUNTING			1
					UOC: DAB, DAD, DAH, DAX, V12, V14, V16, V21, ZAB, ZAD, ZAH			
					BOLT, EYE	(1)	607-10	
					CABLE ASSEMBLY	(1)	607-7	
					HARNESS ASSEMBLY	(1)	607-6	
					LEG ASSEMBLY	(1)	607-14	
					LEG ASSEMBLY	(1)	607-16	
					NUT, PLAIN, HEXAGON	(2)	607-1	
					NUT, PLAIN, HEXAGON	(2)	607-2	
					NUT, PLAIN, HEXAGON	(1)	607-13	
					PIN ASSEMBLY	(2)	607-9	
					PLATE	(1)	607-11	
					ROD, THREADED END	(1)	607-18	
					SETSCREW	(2)	607-15	
					SHACKLE ASSEMBLY	(1)	607-8	
					SPREADER, A-FRAME	(1)	607-17	
					TUBE ASSEMBLY	(2)	607-4	
					TUBE ASSEMBLY	(2)	607-5	
					WASHER, FLAT	(4)	607-3	
					WASHER, LOCK	(1)	607-12	
	PAFZZ	4820011394888	62983	92380	PARTS KIT, LINEAR DI			1
					DUST COVER	(1)	463-20	
					DUST COVER	(2)	464-20	
					DUST COVER	(1)	511-20	
					PACKING, PREFORMED	(1)	463-24	
					PACKING, PREFORMED	(2)	464-24	
					PACKING, PREFORMED	(1)	511-24	
					RING, QUADRANT	(1)	463-25	
					RING, QUADRANT	(2)	464-25	
					RING, QUADRANT	(1)	511-25	

(1) ITEM NO	(2) SMR CODE	(3) NSN	(4) CAGEC	(5) PART NUMBER	(6) DESCRIPTION AND USABLE ON CODES (UOC)	(7) QTY
					SLEEVE	(1) 463-23
					SLEEVE	(2) 464-23
					SLEEVE	(1) 511-23

KIT-31

(1) ITEM NO	(2) SMR CODE	(3) NSN	(4) CAGEC	(5) PART NUMBER	(6) DESCRIPTION AND USABLE ON CODES (UOC)	(7) QTY
					GROUP 95 GENERAL USE STANDARDIZED PARTS	
					9501 HARDWARE SUPPLIES AND BULK MATERIAL	
					FIG. BULK	
1	PAOZZ	2590009418668	19207	7699769	CABLE ASSEMBLYXWINC	V
2	PAOZZ	6145007744579	81349	M13486/13-1	CABLE, SPECIAL PURPO	V
3	XDOZZ		16003	C43974	CHAIN, WELDLESS 12 LINKS PER FOOT, INCH DIA.080	V
4	PAOZZ	4010002904352	46156	60504	CHAIN, WELDED 11 LINKS PER FOOT, INCH DIA 1/8	V
5	PAOZZ	4010007579556	80244	42-C-16570	CHAIN, WELDED 16 LINKS PER FOOT, .037 THICKNESS	V
6	PAOZZ	4010001293221	74410	XB-196	CHAIN, WELDLESS	V
7	PAOZZ	5975012178550	80244	17-C-18035-50	CONDUIT, NONMETALLIC ID 1/4	V
8	PAOZZ	5975001771930	80244	17-C-18035-60	CONDUIT, NONMETALLIC ID 5/16	V
9	XBFZZ		80244	17-C-18035-90	CONDUIT, NONMETALLIC ID 3/8	V
10	PAOZZ	4720010144915	19207	CPR104420-2	HOSE, NONMETALLIC OD .375	V
11	PAOZZ	4720010036706	65282	06642-0000	HOSE, NONMETALLIC OD .500	V
12	PAOZZ	4720010099058	19207	CPR104420-4	HOSE, NONMETALLIC .	V
13	PAOZZ	4720010099941	19207	CPR104420-5	HOSE, NONMETALLIC OD .755, MAX, ID .566, WALL THICKNESS .092	V
14	PAOZZ	4720011147728	19207	8710557	HOSE, NONMETALLIC ID. 5/8	V
15	PAOZZ	4720002032668	34623	A12876	HOSE, NONMETALLIC	V
16	PAFZZ	4720004910102	96906	MS521301A203R	HOSE, NONMETALLIC 07, 1-1/8	V
17	PAOZZ	4720012355452	19207	CPR104420-6	HOSE, NONMETALLIC OD .125, ID .051	V
18	PAOZZ	4720006706037	01276	2565-8	HOSE, NONMETALLIC	V
19	MOOZZ		19207	12277246-1-20	HOSE, NONMETALLIC	V
20	PAOZZ	4720000805379	96906	MS521303A203R	HOSE, PREFORMED 625IDX.219-.172 WALL THICKNESS	V
21	PFFZZ	4720013714396	96906	MS500083A160360	HOSE ASSEMBLY, NONME PART OF KIT P/N 57K0228	
22	PAOZZ	4010001655607	96906	MS87008-5	LINK, CHAIN, CONNECTI 21 LINKS PER FOOT, .027 THICKNESS	V
23	PAFZZ	9390001582408	19207	10937683-2	NONMETALLIC SPECIAL MAKE FROM 10937683-2 PLISN ZARKA	V
24	PAOOO	5530002628180	81348	NN-P-530	PLYWOOD, SOFTWOOD, CO ...... 3 PLY, 1/4 INCH.	V
25	PAOZZ	4020002387734	81348	TR605	ROPE, FIBROUS MANILA, 1/4 INCH	V
26	PAOZZ	4020001328364	81348	TR571	ROPE, FIBROUS COTTON, 3/8 INCH	V
27	PAOZZ	4010009619780	19207	7699767	ROPE, WIRE	V
28	PAOZZ	4010011556142	19207	8741316	ROPE, WIRE	V
29	PAOZZ	9320004518080	19207	8380420	RUBBER STRIP	V
30	PAOZZ	9320004518080	19207	8380420	RUBBER STRIP 1/16 INCHES THICK, 1- PLY OF 16 OZ.CLOTH	V
31	PAOZZ	5330000205375	91340	10608E44S	SEAL RUBBER STRIP WIDTH .910, THICKNESS .560	V
32	PAOZZ	5330003403637	19207	11607302	SEAL, NONMETALLIC SP	V
33	PAOZZ	5330003403637	19207	11607302	SEAL, NONMETALLIC SP WIDTH .97, THICKNESS .56	V
34	XBHZZ		81349	MIL-R-3065RS510A 1BC1DF2	SEAL, NONMETALLIC SP L SHAPED, WIDTH 23/32, THICKNESS 3/32	V

(1) ITEM NO	(2) SMR CODE	(3) NSN	(4) CAGEC	(5) PART NUMBER	(6) DESCRIPTION AND USABLE ON CODES (UOC)	(7) QTY
35	PAFZZ	4710011581477	19207	7411146	TUBE, BENT, METALLIC	V
36	PAFZZ	4710004242694	95535	55229	TUBE, METALLIC DIA 500, WALL .032	V
37	PAOZZ	4710004242694	95535	55229	TUBE, METALLIC OD .125, WALL .... THICKNESS .030	V
38	PAOZZ	4710011342111	19207	8689204	TUBE, METALLIC NOM SIZE .12, DIA .... .125, THICKNESS .030 ...	V
39	PAOZZ	4720010587213	19207	CPR104420-1	TUBING, NONMETALLIC	V
40	PAOZZ	4720010587213	19207	CPR104420-1	TUBING, NONMETALLIC SPRING BRAKE .... RESERVOIR TO TEE	V
41	PAOZZ	4720008997893	19207	8675779	TUBING, NONMETALLIC ID .150, OD .250, 2500 PSI	V
42	PAOZZ	4720005229151	99227	1448	TUBING, NON METALLIC	V
43	PAOZZ	5335001411521	81348	RR-W-365ATYPE2 20X20	WIRE FABRIC	V
44	PAOZZ	6145001611609	81349	M13486/1-3	WIRE, ELECTRICAL OD .135	V
45	PAOZZ	6145007056678	81349	M13486/1-7	WIRE, ELECTRICAL 12 GUAGE	V
46	PAOZZ	6145008053354	81349	M13486/1-12	WIRE, ELECTRICAL	V
47	PAOZZ	6145007056674	81349	M13486/1-14	WIRE, ELECTRICAL DIA.865	V
48	XDOZZ	9505001913680	29510	278027R1	WIRE, NONELECTRICAL M16 GAGE	V
49	PAHZZ	5510002706031	97403	13219E0079	WOOD LAMINATE, DECKI	V

END OF FIGURE

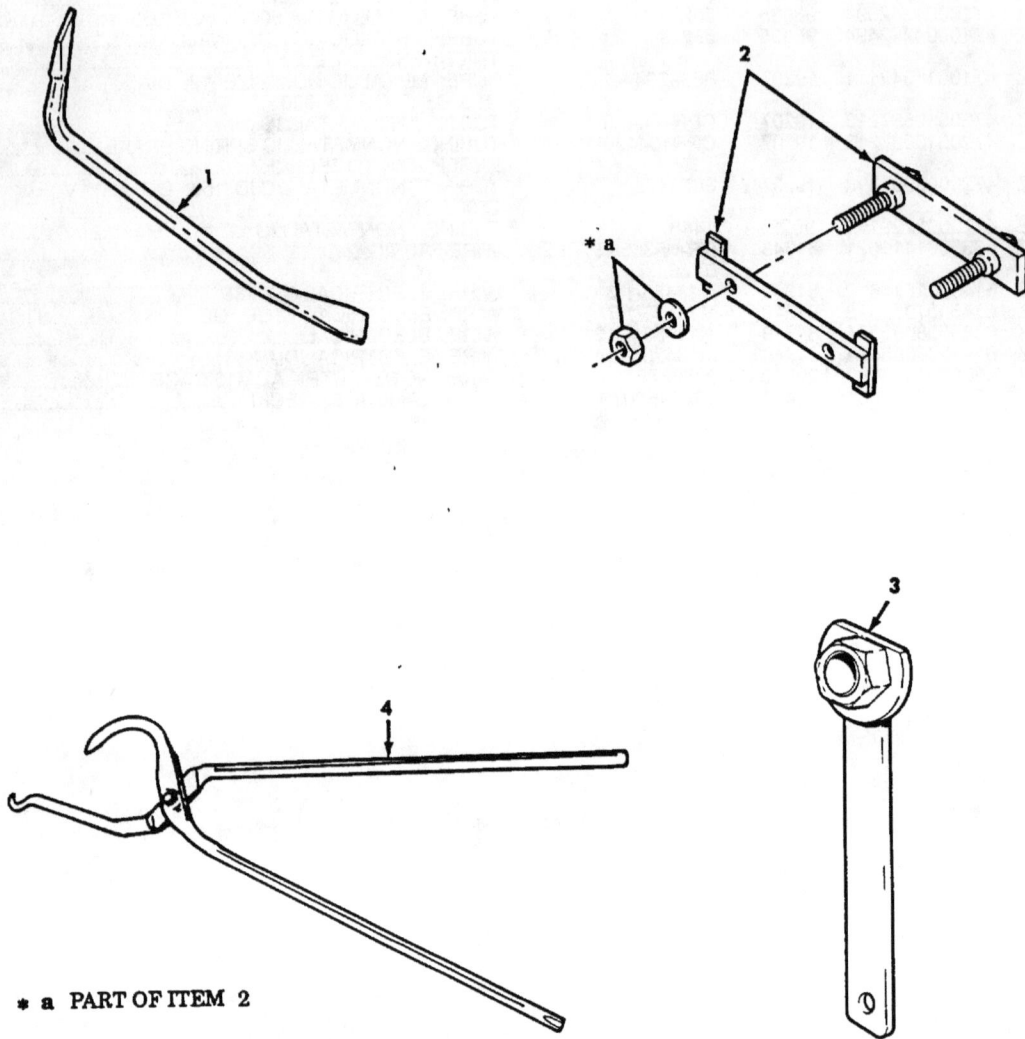

* a PART OF ITEM 2

Figure 629. Peculiar End Item Special Tools.

(1) ITEM NO	(2) SMR CODE	(3) NSN	(4) CAGEC	(5) PART NUMBER	(6) DESCRIPTION AND USABLE ON CODES (UOC)	(7) QTY
					GROUP 26 SPECIAL TOOLS AND TEST EQUIPMENT 2604 SPECIAL TOOLS	
					FIG. 629 PECULIAR END ITEM SPECIAL TOOLS	
1	PEOZZ	5120011543029	33287	J-34061	ADJUSTING TOOL, BRAK .... BOI: 1 PER AUTHORIZE NO. 1 COMMON TOOL SHOP SET BOI: 1 PER AUTHORIZE NO. 2 COMMON TOOL SHOP SET	
2	PEOZZ	4910012184490	33287	J35193	TOOL, WHEEL ASSEMBLY... BOI: 1 PER AUTHORIZE NO. 1 COMMON TOOL SHOP SET BOI: 1 PER AUTHORIZE NO. 2 COMMON TOOL SHOP SET UOC:DAA, DAB, DAC, DAD, DAE, DAF, DAG, DAH, DAJ, DAK, DAL, DAW, DAX, ZAA, ZAB, ZAC, ZAD, ZAE, ZAF, ZAG, ZAH, ZAJ, ZAK, ZAL	
3	PEOZZ	4910012001512	33287	J35198	BOLT INSERTING TOOL. (USE WITH WHEEL P/N 12301115 ONLY).. BOI: 1 PER AUTHORIZE NO. 1 COMMON TOOL SHOP SET BOI: 1 PER AUTHORIZE NO. 2 COMMON TOOL SHOP SET UOC:DAA, DAB, DAC, DAD, DA, DAF, DAG, DAH, DAJ, DAK, DAL, DAW, DAX, ZAA, ZAB, ZAC, ZAD, ZAE, ZAF, ZAG, ZAH, ZAJ, ZAK, ZAL	
4	PAOZZ	5120011522318	33287	J-33111	PLIERS, BRAKE REPAIR .... BOI: 1 PER AUTHORIZE NO. 1 COMMON TOOL SHOP SET BOI: 1 PER AUTHORIZE NO. 2 COMMON TOOL SHOP SET	

END OF FIGURE

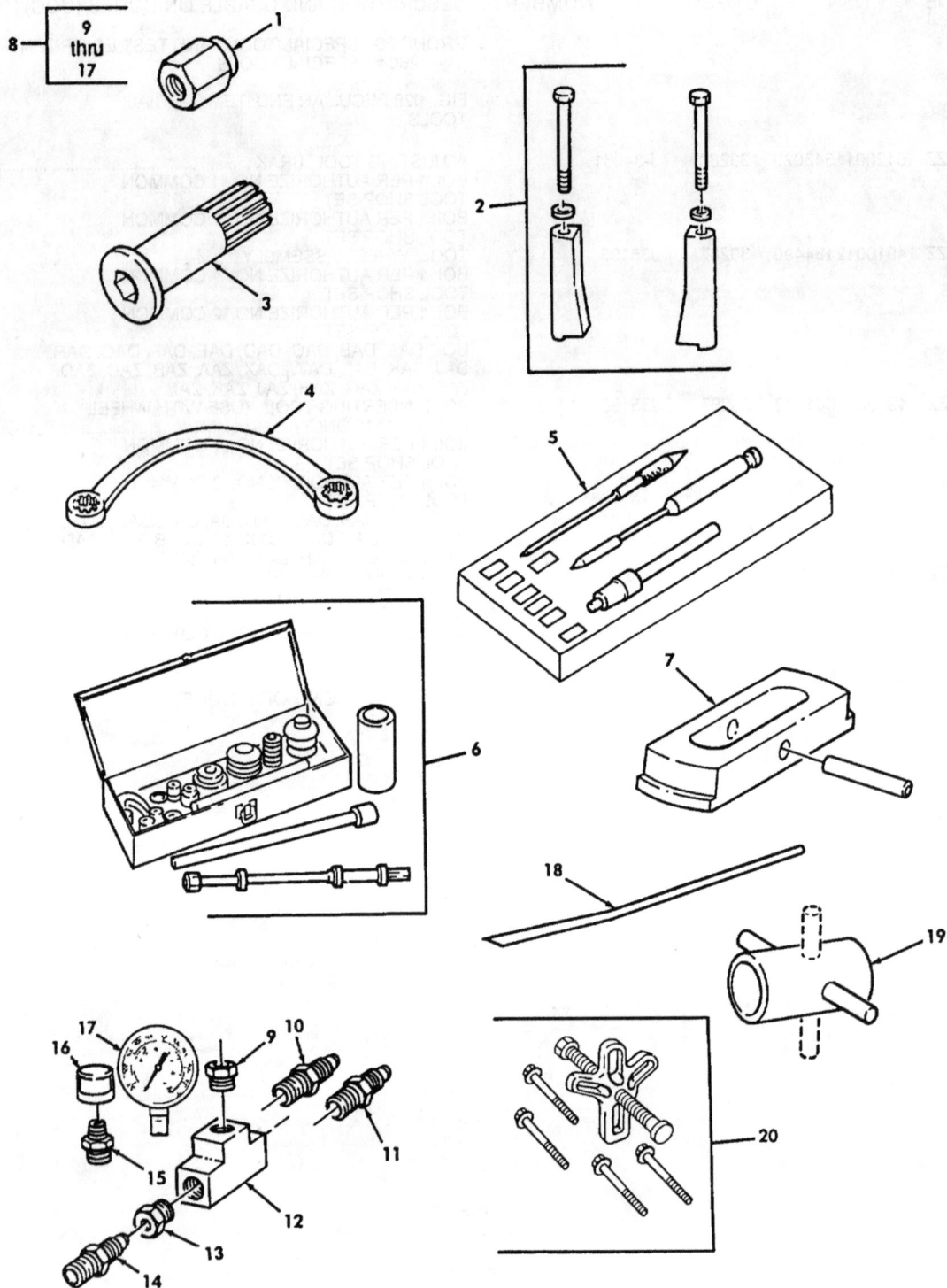

Figure 630. Special Tools.

(1) ITEM NO	(2) SMR CODE	(3) NSN	(4) CAGEC	(5) PART NUMBER	(6) DESCRIPTION AND USABLE ON CODES (UOC)	(7) QTY
					GROUP 26 TOOLS AND TEST EQUIPMENT 2604 SPECIAL TOOLS	
					FIG. 630 SPECIAL TOOLS	
1	PEFZZ	4730012849086	45225	23622	REDUCER, TUBE PART OF KIT P/N 57K1311	
2	PEHZZ	5120012627309	15434	3822503	CLAMP SET, CYLINDER PART OF KIT P/N 57K1312	
3	PEFZZ	5120012855193	15434	3377371	BARRING TOOL, GEAR PART OF KIT P/N 57K1311	
4	PEFZZ	5120011785351	55719	CXM1519	WRENCH, BOX 15 AND 19MM PART OF KIT P/N 57K1311	
5	PEFZZ	2915012852527	15434	3376947	KIT, NOZZLE CLEANING PART OF KIT P/N 57K1311.	
6	PEHZZ	5120012857829	74069 1	478	INSTALLING AND REMO PART OF KIT P/N 57K1312	
7	PEHZZ	5120012915769	15434	3822786	PULLER, MECHANICAL PART OF KIT P/N 57K1312	
8	PEOOO	5220012985730	47457	20511320	AIR GAGE ASSEMBLY TEST KIT PART OF KIT P/N 57K1311	
9	PEOZZ	4730002771896	96906	MS14315-5X	BUSHING, PIPE.	
10	PEOZZ	4730012979072	98441	0103-8-12	ADAPTER, STRAIGHT, PI..........	
11	PEOZZ	4730009277272	96906	MS39158-9	ADAPTER, STRAIGHT, PI..........	
12	PEOZZ	4730009828853	81343	8-8-8-140438B	TEE, PIPE	
13	PEOZZ	4730001422581	79470	3228X2	BUSHING, PIPE.	
14	PEOZZ	5821008636498	17875	SK20420	VALVE, TANK	
15	PEOZZ	4730002778761	81343	12-12 120102BA	ADAPTER, STRAIGHT, PI..........	
16	PEOZZ	4730002312412	78500	1898R200	CAP, PIPE	
17	PEOZZ	6685013081985	94894	1194	GAGE, PRESSURE, DIAL	
18	PEOZZ	5120012855192	47457	20511262	PUNCH, BEARING PART OF KIT P/N 57K1310	
19	PEOZZ	5120012857620	47457	20511263	INSERTER, SEAL PART OF KIT P/N 57K1310	
20	PEOZZ	5180009994053	33287	J24420-B	PULLER KIT, UNIVERSA PART OF KIT P/N 57K1310	
	PEOOO	5180013042257	19207	57K1310	TOOL KIT, DIESEL	
					INSERTER, SEAL	(1) 630-19
					PULLER KIT, UNIVERSA	(1) 630-20
					PUNCH, BEARING	(1) 630-18
	PEFFF	5180013042258	19207	57K1311	TOOL KIT, INTERNAL..........	
					AIR GAGE ASSEMBLY	(1) 630-8
					BARRING TOOL, GEAR	(1) 630-3
					KIT, NOZZLE CLEANING	(1) 630-5
					REDUCER, TUBE	(1) 630-1
					WRENCH, BOX	(1) 630-4

END OF FIGURE

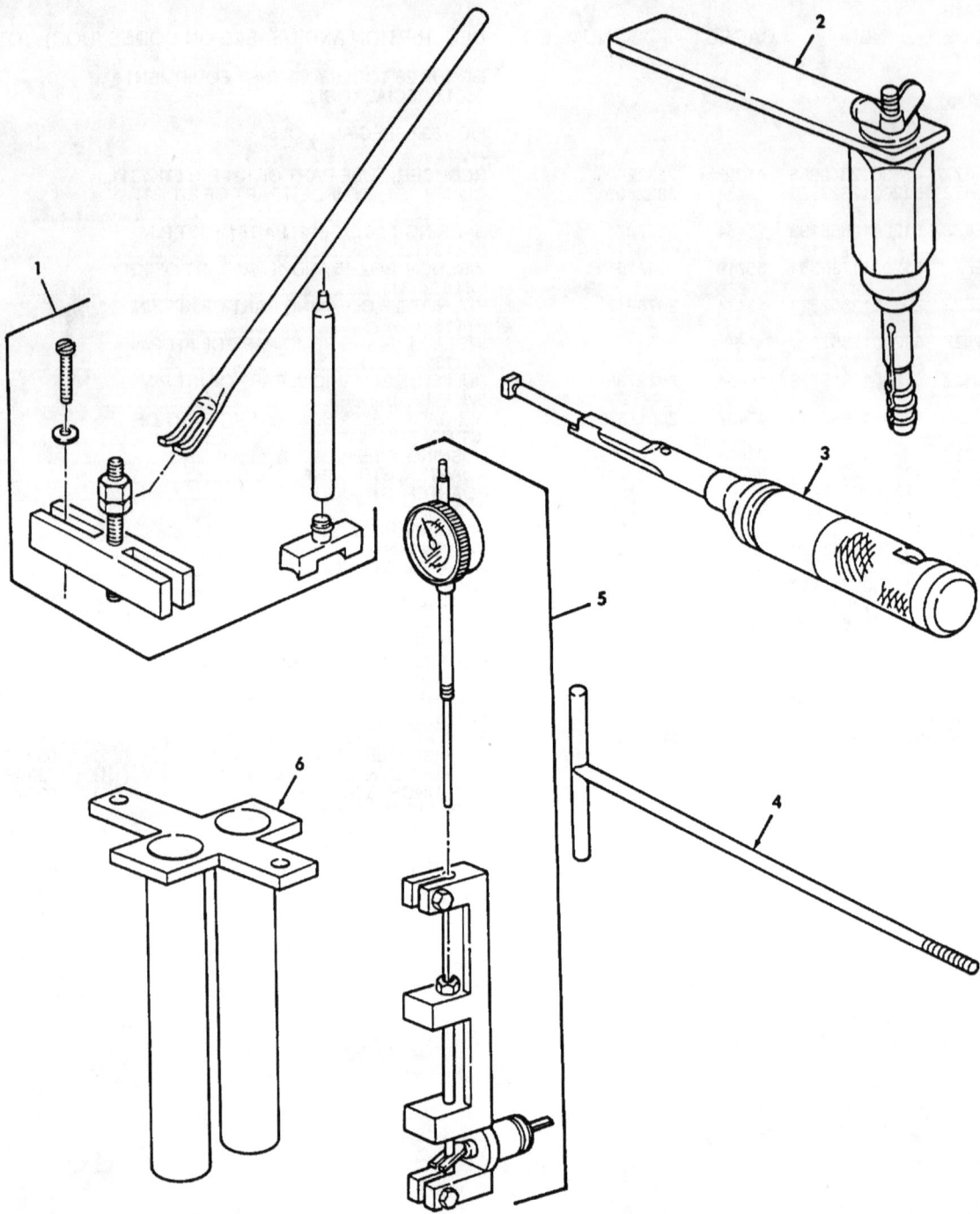

Figure 631. Special Tools.

(1) ITEM NO	(2) SMR CODE	(3) NSN	(4) CAGEC	(5) PART NUMBER	(6) DESCRIPTION AND USABLE ON CODES (UOC)	(7) QTY
					GROUP 2604 SPECIAL TOOLS	
					FIG. 631 SPECIAL TOOLS	
1	PEHZZ	5120013416000	5T151	KDEP 1505	COMPRESSOR, VALVE SP PART OF KIT P/N 57K1312	
2	PEHZZ	5120013432585	5T151	KDEP 1056	SIDE PLUG PULLER PART OF KIT P/N 57K1312	
3	PEHZZ	5120013452586	5T151	KDEP 1068	TAPPET HOLDER PART OF KIT P/N 57K1312	
4	PEHZZ	4910013368204	5T151	KDEP 1052	SEPARATION TUBE PART OF KIT P/N 57K1312	
5	PEHZZ	4910013386241	5T151	1 688 130 135	PLUNGER LIFT DEVICE PART OF KIT P/N 57K1312	
6	PEHZZ	5340013416572	5T151	9 681 233 100	EXTENSION SHAFT PART OF KIT P/N 57K1312	
	PEHHH	5180013042164	19207	57K1312	TOOL KIT, GENERAL	

```
                                         CLAMP SET, CYLINDER              ( 1)     630-2
                                         COMPRESSOR, VALVE SP            ( 1)     631-1
                                         EXTENSION SHAFT                 ( 1)     631-6
                                         INSTALLING AND REMO             ( 1)     630-6
                                         PLUNGER LIFT DEVICE             ( 1)     631-5
                                         PULLER, MECHANICAL              ( 1)     630-7
                                         SEPARATION TUBE                 ( 1)     631-4
                                         SIDE PLUG PULLER                ( 1)     631-2
                                         TAPPET HOLDER                   ( 1)     631-3
```

END OF FIGURE

631-1

CROSS-REFERENCE INDEXES

NATIONAL STOCK NUMBER INDEX

STOCK NUMBER	FIG.	ITEM	STOCK NUMBER	FIG.	ITEM
5310-00-001-4719	370	42	2930-00-004-8420	118	2
4720-00-001-7854	81	7	5330-00-005-0407	118	3
5310-00-003-4094	5	14	5305-00-005-0666	42	1
	12	15	5330-00-005-0858	16	3
	105	8	2815-00-005-7431	26	6
	106	6	5330-00-006-2529	19	13
	109	9	2530-00-006-7469	236	33
	118	20	5310-00-007-0260	197	8
	120	5	5325-00-007-2969	191	1
	177	6	5340-00-007-9442	39	8
	178	2	2815-00-008-1741	17	5
	182	29	2520-00-008-7361	189	17
	204	6	4820-00-009-7378	598	37
	235	8	5360-00-009-9270	24	3
	237	7		24	11
	242	8			
	293	5	5310-00-010-3028	316	8
	328	8	4730-00-010-3875	207	23
	351	24	5320-00-010-4131	381	15
	370	21	5306-00-010-9115	381	18
	396	7	4730-00-011-2578	60	1
	461	17	4730-00-011-3175	118	5
	467	2		129	16
	468	4		284	53
	468	4	5310-00-011-5093	487	32
	469	15	4730-00-011-6452	213	6
	470	21		266	7
	474	2		422	20
	478	17		423	18
	484	3	5315-00-012-0123	211	11
	484	30	5306-00-012-0231	382	29
	485	24		387	20
	487	10		387	40
	487	25		609	10
	489	19	5315-00-012-4553	211	18
	492	23		487	18
	493	15		108	18
	495	3	4730-00-012-7823	482	2
	501	18	4730-00-012-7951	108	13
	502	29		593	25
	503	11	5310-00-013-1245	297	6
	504	23		299	4
	505	2	5310-00-013-1498	367	39
	506	2		367	44
	509	12	5315-00-013-7214	329	2
	589	57		488	4
	615	9	5315-00-013-7228	389	10
5305-00-003-9255	420	20		487	29
4730-00-004-0732	620	19	5315-00-013-7238	379	6
5340-00-004-6854	370	6	5315-00-013-7308	487	5

CROSS-REFERENCE INDEXES

NATIONAL STOCK NUMBER INDEX

STOCK NUMBER	FIG.	ITEM	STOCK NUMBER	FIG.	ITEM
5315-00-013-7308	490	2	4730-00-018-9566	7	14
4730-00-013-7401	117	3		9	5
4730-00-013-7409	617	20		41	11
5315-00-014-1195	7	10		44	7
5315-00-014-1284	7	9		285	26
4730-00-014-2433	71	7		480	2
	74	6		482	19
	74	15	4730-00-019-0236	177	13
	75	7	6240-00-019-0877	143	3
	76	4	5305-00-019-1675	161	25
	77	10	5305-00-019-2417	533	15
	77	25	6240-00-019-3093	139	3
	79	4		140	6
	80	2	5306-00-020-0857	370	26
	80	17	5306-00-020-1058	370	28
4730-00-014-2435	71	13-	5330-00-020-5375	BULK	30
	71	16	4730-00-021-1802	456	37
	74	8		460	34
	74	12		461	34
	74	17		509	32
	75	13	5310-00-021-9760	255	5
	76	2		262	13
	77	2	5305-00-022-3843	497	2
	77	12	5310-00-022-8834	162	7
	79	2	5306-00-024-6580	189	19
	80	4	5330-00-026-2931	44	13
	80	19	4820-00-026-8473	550	1
5315-00-014-2972	472	17	7690-00-030-6615	580	11
5315-00-014-2976	507	28		584	1
4730-00-014-4027	72		5310-00-031-2673	221	4
	599	3		222	3
4730-00-014-4593	521	21	5310-00-033-6007	236	30
5310-00-014-5850	50	25	5310-00-033-6012	235	2
	55	27		242	2
	359	9	2815-00-033-9392	19	11
	528		4820-00-034-1690	463	6
	616A	24	3020-00-035-7894	530	21
5340-00-015-7560	250	11	5340-00-036-0236	81	15
	251	10	2510-00-036-0298	534	4
4730-00-017-9447	480	6	5365-00-038-9592	196	7
5305-00-018-0178	234	5		197	7
	241	5	5342-00-040-2073	358	4
5305-00-018-6475	489	54	2510-00-040-2264	390	6
5306-00-018-7527	474	18	2540-00-040-2308	347	2
	489	1	5340-00-040-2314	5	8
	492	16	5340-00-040-2321	331	9
	502	24		332	11
5305-00-018-7838	198	17	5340-00-040-2322	332	1
5315-00-018-7988	345	32	5340-00-040-2331	473	10
5320-00-018-9512	318	10	3040-00-040-2401	243	6
	381	6	5315-00-041-0916	23	9

CROSS-REFERENCE INDEXES

NATIONAL STOCK NUMBER INDEX

STOCK NUMBER	FIG.	ITEM	STOCK NUMBER	FIG.	ITEM
2530-00-041-3116	317	17	2590-00-045-4205	488	12
5340-00-041-3126	404	9	2590-00-045-4206	488	11
5315-00-042-3293	474	34	5310-00-045-5207	150	5
	489	27	5305-00-045-7603	594	37
5305-00-042-3568	139	1	2540-00-047-3926	329	1
	140	1	4730-00-047-3946	329	10
5315-00-042-4950	502	10	5306-00-050-0347	607	10
5305-00-042-5567	367	13	4730-00-050-0718	53	3
5306-00-042-5570	161	17	2640-00-050-1235	292	14
	355	6	5306-00-050-1238	97	15
	457	14		100	18
	458	24		112	9
5305-00-042-5601	168	28		140	4
5305-00-042-5603	164	32		161	13
	166	29		256	20
	167	29		263	13
5305-00-042-5648	611	27		291	2
5306-00-042-5841	543	1		351	27
5306-00-042-5859	134	77		399	11
4730-00-042-8988	481	7		533	16
5315-00-043-1789	530	16		589	5
5306-00-044-0502	111	8		606	1
	211	50	5340-00-050-1589	474	23
	513	31		489	59
	525	28	5340-00-050-1600	7	27
	528	32	5340-00-050-2622	429	8
5310-00-044-3342	6 '	17	5310-00-050-3520	329	3
4730-00-044-4655	505	11	4730-00-050-4203	218	7
6240-00-044-6914	139	4		220	9
	140	7		223	4
	614	4		224	5
5310-00-045-1031	500	17		225	3
5310-00-045-1081	473	24		329	8
5310-00-045-3296	81	4		330	8
	132	14		331	11
	132	20		332	9
	132	24		409	4
	132	27		465	7
	211	14		465	9
	505	13		512	7
	624	8		512	9
	625	8	4730-00-050-4205	351	18
5310-00-045-3299	131	2	4730-00-050-4208	221	7
	132	8		222	6
	134	5		223	9
	134	28		224	10
	134	66		226	13
	144	24		227	12
	598	44		236	24
5310-00-045-4007	586	3		303	10

CROSS-REFERENCE INDEXES

NATIONAL STOCK NUMBER INDEX

STOCK NUMBER	FIG.	ITEM	STOCK NUMBER	FIG.	ITEM
4730-00-050-4208	345	19	5305-00-052-6920	142	1
	346	11		142	19
	348	35		492	1
	390	16		544	1
	471	33		544	4
	472	3	5305-00-052-6921	410	4
	475	4	5305-00-052-6922	395	2
	476	10	5305-00-052-7492	416	12
	477	2	5305-00-052-7494	410	2
	485	11	5310-00-052-7528	418	16
	486	8	5315-00-052-8492	351	8
	486	11	4730-00-052-8502	519	10
	487	3	4730-00-053-0266	212	6
	488	9	5325-00-053-1116	163	35
	489	49		165	22
	491	18		166	21
	497	3		169	40
	500	7	5340-00-053-8994	71	25
	502	22		74	24
	507	4		75	20
	507	26		107	10
	508	11		157	15
	513	6		161	4
	525	6		164	17
4730-00-050-4309	621	15		165	20
4730-00-050-4358	269	20		167	19
4730-00-050-4652	314	8		168	17
5940-00-050-6207	144	8		243	16
	144	13		261	26
5340-00-050-9077	276	1		269	7
	287	3		271	18
5305-00-050-9221	490	6		315	5
5305-00-051-4075	615	15	4730-00-054-2027	458	11
5306-00-051-4077	204	9	4730-00-054-2571	617	8
	271	11		618	4
	290	12	4730-00-054-2572	266	34
	617	10			
	536	18		618	6
	549	30	5315-00-054-4028	472	41
	592	22	5310-00-054-4892	161	14
5306-00-051-4081	549	33	5305-00-054-6655	540	4
5306-00-051-4084	137	6	5306-00-054-8024	382	5
2610-00-051-9464	300	3		385	16
5315-00-052-0110	487	23		387	7
5305-00-052-2218	469	3	5305-00-054-9271	507	2
4730-00-052-3420	258	9	5305-00-054-9285	469	5
5305-00-052-6874	412	24	2510-00-057-1630	362	26
5305-00-052-6879	413	21	5995-00-057-1642	134	48
5305-00-052-6881	414	24	5999-00-057-2929	134	59
5305-00-052-6913	567	1		139	9
5305-00-052-6915	552	1		140	11

CROSS-REFERENCE INDEXES

NATIONAL STOCK NUMBER INDEX

STOCK NUMBER	FIG.	ITEM	STOCK NUMBER	FIG.	ITEM
5999-00-057-2929	143	6	5310-00-060-9435	348	33
	157	8	5310-00-061-1258	156	8
	166	11		161	7
	170	8		213	29
	170	18		612	18
	171	8		613	31
	172	8	5310-00-061-7325	39	13
	173	14		355	4
	174	8		356	4
	436	5		356	17
	496	6		362	29
5340-00-057-3025	606	8		405	37
5340-00-057-3037	606	20		406	30
5340-00-057-3043	155	10		536	8
				616A	14
	161	1	5310-00-061-7326	181	18
	198	13			
	263	10		257	17
	315	5		265	6
	588	9		359	19
	612	11		409	26
	613	29		457	4
	627	1		458	27
				616A	25
5340-00-057-3537	525	17	5305-00-062-4378	24	7
5330-00-057-3823	469	22	5310-00-062-4954	322	16
	470	18		346	5
	474	4		486	6
	489	36		488	1
4730-00-057-5555	33	18		492	13
	190	7	4820-00-062-9719	288	7
	473	28		603	5
2540-00-063-4730	531	6			
5310-00-057-7153	348	36	5305-00-063-5043	51	21
5305-00-058-1082	425	10		56	21
5310-00-058-1626	100	3	5315-00-063-7366	454	9
	288	6	5330-00-064-3691	489	39
	549	10	5340-00-067-3868	162	11
	551	7		173	6
	551	10		612	5
5305-00-058-9378	473	7		613	3
5305-00-058-9389	465	10	5305-00-068-0500	180	17
	512	10		549	50
5315-00-058-9931	489	37	5305-00-068-0502	370	30
5315-00-059-0029	473	23		371	12
5315-00-059-0184	234	1		408	20
	241	1	5305-00-068-0508	627	2
5315-00-059-0187	583	7	5305-00-068-0509	52	2
5315-00-059-0205	362	28		66	9
5315-00-059-0206	409	16		197	14
5315-00-059-0491	508	9	5305-00-068-0510	126	9
5305-00-059-4553	586	9	5305-00-068-0511	125	1
5305-00-059-4568	415	17		211	29
2640-00-060-3550	292	12		312	52

CROSS-REFERENCE INDEXES

NATIONAL STOCK NUMBER INDEX

STOCK NUMBER	FIG.	ITEM	STOCK NUMBER	FIG.	ITEM
5305-00-068-0511	456	28	5305-00-068-0515	169	25
	460	25		182	9
	461	22		455	2
	504	13		591	34
	505	22	5305-00-068-0516	136	6
	509	17		543	17
				616A	8
5305-00-068-0512	581	12	5305-00-068-0523	37	1
	606	2	5310-00-068-5285	183	2
5306-00-068-0513	39	9		244	17
	100	49	4010-00-068-5642	97	17
	142	15	4730-00-069-1186	182	22
	143	21		212	8
	145	6		213	22
	151	26		214	I 1
	161	20		215	9
	162	1		254	10
	163	23		255	24
	165	28		261	5
	167	33		261	28
	168	21		262	8
	169	28		264	11
	269	27		265	1
	315	12		273	2
	352	8		282	4
	353	4		288	16
	374	4		290	10
	374	9		596	4
	454	18		621	23
	513	28		626	5
	525	23	4730-00-069-1187	182	20
	543	2		199	4
	547	7		213	14
	549	3		215	1
	579	9		215	7
	586	19		254	22
	589	19		257	4
	606	4		261	14
	612	10		265	11
5306-00-068-0514	81	1		270	4
	105	1		278	15
	166	23		279	13
	255	7		279	25
	262	23		288	17
	355	2		291	10
	356	2		541	13
5305-00-068-0515	112	6		603	14
	113	2		605	11
	162	4		623	29
	167	26	3010-00-069-3047	484	22
	168	19	5305-00-069-5583	346	19

CROSS-REFERENCE INDEXES

NATIONAL STOCK NUMBER INDEX

STOCK NUMBER	FIG.	ITEM	STOCK NUMBER	FIG.	ITEM
5305-00-071-1788	106	3	3120-00-073-3163	163	26
	126	4		165	31
	286	23		168	24
	310	12		169	33
	503	14	4730-00-074-0713	505	42
	504	4	5975-00-074-2072	588	10
	505	35	5331-00-074-2692	248	12
5305-00-071-2055	107	12	2920-00-078-2390	KITS	37
5305-00-071-2056	54	4	5305-00-078-7021	362	33
	284	65	5310-00-079-1974	526	3
5305-00-071-2058	207	1	5340-00-079-7837	211	13
5305-00-071-2067	2	6		426	11
5305-00-071-2068	5	15		606	7
	105	7	5310-00-079-9573	144	31
5305-00-071-2069	5	16	5935-00-080-1020	164	2
	107	26		165	2
	176	7		166	14
	456	20		167	14
	458	17		168	2
	460	19		169	2
	461	18	5315-00-080-3503	383	5
	503	38	4720-00-080-5379	BULK	20
	504	22	5310-00-080-6004	11	4
	505	1		19	17
	509	13		34	3
5305-00-071-2070	6	14		37	15
	474	1		83	7
	489	46		100	19
	505	7		102	8
	506	5		125	19
5305-00-071-2071	235	7		153	7
	242	7		156	7
	297	11		180	15
	470	11		204	5
	602	3		211	30
5305-00-071-2073	469	24		243	4
	470	20		266	22
5305-00-071-2074	6	21		310	7
5305-00-071-2077	109	8		334	4
	297	19		336	4
5305-00-071-2083	493	8		338	4
5305-00-071-2505	66	19		339	30
5305-00-071-2509	339	15		340	16
	339	22		343	3
5305-00-071-2510	619	17		354	2
5305-00-071-2513	189	18		356	14
5305-00-071-2514	526	8		364	7
5305-00-071-2515	195	4		365	6
5305-00-071-2518	594	21		368	7
2940-00-071-2653	61	7		368	16

CROSS-REFERENCE INDEXES

NATIONAL STOCK NUMBER INDEX

STOCK NUMBER	FIG.	ITEM	STOCK NUMBER	FIG.	ITEM
5310-00-080-6004	371	7	5331-00-081-9289	56	3
	395	7	5310-00-081-9292	44	8
	396	13	5331-00-081-9299	129	10
	399	7	4730-00-081-9618	7	8
	400	7	5360-00-082-0124	85	2
	425	27		87	2
	454	11	5315-00-082-0448	84	4
	456	10		86	4
	460	12	5355-00-082-1189	129	6
	474	16	5310-00-082-1882	7	21
	486	2	5310-00-082-1888	129	14
	489	9	5340-00-084-7787	129	11
	505	27	5331-00-085-3494	249	19
	513	4	2910-00-085-7436	129	12
	513	22	2590-00-086-7459	487	21
	525	4	2590-00-086-7460	490	4
	528	8	2530-00-087-0163	250	7
	611	24		251	13
	612	14	4730-00-087-0313	480	34
	613	32	2530-00-087-0327	463	19
	629	5		464	19
2530-00-080-6572	301	5		511	19
4730-00-081-0311	456	23	5310-00-087-3946	150	4
	458	15	5310-00-087-4652	183	5
	459	3		481	4
5935-00-081-0401	158	19		618	8
5305-00-081-3728	154	23	5310-00-087-7493	62	33
5310-00-081-4219	108	10		153	22
	134	42		374	3
	205	2		375	7
	254	2		402	15
	290	14		428	19
	341	11		583	5
	382	35		589	30
	387	32	5310-00-088-1251	409	5
	397	2		544	3
	402	7		544	7
	549	31		545	3
	589	7		546	3
	591	1		594	22
	592	12		611	2
	598	30		613	33
	611	16	5340-00-088-1254	71	3
	615	16		74	29
	622	2		75	3
5315-00-081-7042	414	16		162	5
5315-00-081-7874	182	19		166	18
	454	1		168	22
5310-00-081-8500	14	3	5340-00-088-1255	100	48
5331-00-081-9289	51	4	5305-00-088-1302	362	19

CROSS-REFERENCE INDEXES

NATIONAL STOCK NUMBER INDEX

STOCK NUMBER	FIG.	ITEM	STOCK NUMBER	FIG.	ITEM
5325-00-088-6147	141	15	3110-00-100-4216	232	15
	144	19		239	15
5340-00-088-6655	611	4	3110-00-100-4220	348	21
4730-00-088-8666	213	17	3110-00-100-4223	293	16
	278	17		295	8
5305-00-088-8946	399	4		348	31
5305-00-088-9044	213	31	3110-00-100-5355	234	17
	441	8		241	17
5325-00-089-1262	188	13	3110-00-100-5825	235	13
4730-00-089-2515	594	6		242	13
4730-00-089-3406	456	18	3110-00-100-6158	471	16
	460	17	3110-00-100-6159	500	14
	461	15		507	11
	503	9	5305-00-100-6791	221	3
	504	26		222	2
	505	3	3110-00-100-9862	530	29
	509	11	3110-00-101-0836	209	14
4710-00-089-6193	481	2	2590-00-101-5594	487	27
5330-00-090-2128	252	20	5307-00-102-0962	370	37
	255	14	2510-00-106-2200	382	32
	261	23		385	39
	262	6		387	26
	291	13	9905-00-106-5744	572	2
5365-00-090-5426	160	17	9905-00-106-5745	572	3
	163	13	9905-00-106-5746	571	2
	165	15		572	7
	166	8		574	5
	167	7	9905-00-106-5750	572	8
	168	12	5330-00-106-6369	118	13
	169	12	5330-00-106-6370	118	18
5340-00-091-3790	613	9	5940-00-107-1481	440	2
5340-00-095-7146	189	30	5330-00-107-3925	189	24
4730-00-096-8756	422	4	3110-00-107-7564	118	8
	423	3	5340-00-107-7769	416	19
4720-00-096-9630	263	8	5315-00-108-1112	189	28
4720-00-096-9648	81	8	2540-00-108-1940	KITS	
5342-00-096-9662	332	1	4720-00-108-5989	505	10
			3110-00-109-1179	120	23
4730-00-097-4236	36	4	5340-00-109-8212	383	37
5305-00-097-7372	82	3	5342-00-111-3605	533	11
	145	3	9905-00-111-6662	552	14
	145	3	5940-00-113-3138	150	17
3110-00-100-0231	208	5	5940-00-113-3147	619	23
3110-00-100-0305	208	4	5940-00-113-3148	159	1
3110-00-100-0316	209	13	5940-00-113-9821	158	14
3110-00-100-0650	236	2		158	33
	296	21	5940-00-113-9825	158	10
	158	32			
	122	14	5940-00-113-9826	159	2
3110-00-100-4177	234	33	5935-00-114-0607	160	1
	241	33	4520-00-114-1055	425	1

CROSS-REFERENCE INDEXES

NATIONAL STOCK NUMBER INDEX

STOCK NUMBER	FIG.	ITEM	STOCK NUMBER	FIG.	ITEM
5940-00-114-1300	159	3	2510-00-123-0278	383	1
5940-00-114-1315	158	13	5310-00-123-2572	250	14
	158	37		251	16
9905-00-114-1334	163	10	2510-00-124-1297	385	37
	165	19	3040-00-125-2959	530	25
	167	17	3040-00-125-2961	530	15
	168	10	4730-00-125-7979	213	2
	169	10		266	3
5935-00-115-2306	158	7	2510-00-127-4768	262	2
	160	5	4010-00-129-3221	BULK	6
5935-00-115-2307	158	15	3120-00-129-9200	284	45
2540-00-115-2565	100	10	3120-00-129-9206	284	45
5940-00-115-2674	158	9	3120-00-129-9210	284	45
	158	31	5330-00-129-9389	284	46
5940-00-115-5006	151	2		286	33
	151	13	5360-00-129-9415	284	13
	151	18		286	15
	152	8	5930-00-130-5349	131	15
6620-00-115-9042	132	23		132	3
5305-00-115-9526	240	1		134	7
5930-00-116-0531	441	5		134	35
9905-00-116-5294	562	1		134	68
9905-00-116-5295	562	3		611	31
2530-00-117-9144	245	20		612	27
	246	21		613	22
2590-00-118-5551	473	29	2815-00-131-1700	33	21
3040-00-118-5554	474	12	5330-00-131-7072	284	21
	489	58		286	22
3040-00-118-8694	487	17	2815-00-132-0240	24	12
2510-00-119-3903,	383	36	5360-00-132-0245	46	8
	384	3	5365-00-132-0273	17	10
	386	11	5331-00-132-0274	129	9
	390	7	5340-00-132-3203	42	3
2510-00-119-3904	385	22	4730-00-132-4588	61	3
5340-00-119-3906	383	34		70	4
	384	9		73	1
	386	17		76	13
	390	9		77	3
2590-00-121-0707	469	2		79	13
5365-00-121-2776	530	19		80	10
5365-00-121-2780	530	27		213	4
2540-00-121-9077	445			266	5
2540-00-121-9081	KITS			266	17
2540-00-121-9082	KITS		4020-00-132-8364	BULK	25
4730-00-122-0477	70	4	5330-00-133-6235	KITS	
	73	2	5330-00-133-6236	KITS	
	76	12	5330-00-133-6237	KITS	
	77	4	5975-00-133-8696	628	3
	79	12	6150-00-134-0847	379	12
	80	11	2590-00-134-1121	490	5
			6220-01-423-0209	131	5

CROSS-REFERENCE INDEXES

NATIONAL STOCK NUMBER INDEX

STOCK NUMBER	FIG.	ITEM	STOCK NUMBER	FIG.	ITEM
4820-00-134-1122	480	18	4730-00-142-3076	259	5
	482	7		260	4
9905-00-134-3558	565	4		597	7
5310-00-134-4171	14	11		599	14
2530-00-134-4619	316	17	3110-00-142-4390	294	17
5315-00-134-4632	346	27		296	14
4720-00-134-4655	62	8	5305-00-143-3266	236	18
2940-00-134-4657	62	24	4730-00-143-3941	297	3
2510-00-134-4663	5	2		313	6
2815-00-134-4665	5	1		314	12
5930-00-134-5036	132	7	5940-00-143-4774	170	45
3020-00-134-7946	310	4	5940-00-143-4780	170	44
2940-00-134-8326	62	22	5310-00-143-6102	71	4
5330-00-135-6382	19	7		74	27
2940-00-135-6537	479	1		75	5
4730-00-136-2018	598	17		458	23
5360-00-136-4759	484	24	3040-00-143-6390	487	34
5306-00-136-9751	37	13		490	1
5305-00-138-0069	417	12	3110-00-143-7586	208	9
	492	20	5330-00-143-7737	127	3
4730-00-138-8121	480	35	5330-00-143-8371	37	7
	508	26	5330-00-143-8376	37	17
4730-00-138-8133	482	13	5331-00-143-8485	66	24
2815-00-138-8280	37	8	4730-00-143-9282	255	1
5330-00-138-8388	233	4		262	22
	240	4		264	12
5305-00-138-9848	129	13		291	6
5305-00-139-7072	348	2		604	1
5305-00-139-7074	346	18		605	6
5305-00-139-7075	348	2	5305-00-144-1514	362	31
5315-00-140-1938	211	27	5305-00-145-0602	362	24
	419	14	5330-00-145-8355	236	31
5360-00-140-2078	88	26	5310-00-147-3274	234	12
5305-00-140-4765	326	3		241	12
5305-00-140-8001	571	1	2930-00-147-5202	109	1
	572	1	5305-00-149-8610	412	14
	572	10		417	10
9905-00-141-1619	563	4		424	16
5310-00-141-1795	50	23	5310-00-149-9126	5	7
	52	4	5340-00-150-1658	429	7
	52	11	5315-00-150-4146	390	11
	55	23	3040-00-150-7145	487	31
	87	8	5306-00-151-5726	393	8
	88	21	3110-00-151-8636	530	12
5310-00-141-3062	418	16	5940-00-151-9361	430	2
4730-00-142-2581	630	13		435	10
4730-00-142-3075	214	4	5340-00-151-9651	161	15
	279	21		215	5
	538	13		269	21
4730-00-142-3076	252	8		619	4

CROSS-REFERENCE INDEXES

NATIONAL STOCK NUMBER INDEX

STOCK NUMBER	FIG.	ITEM	STOCK NUMBER	FIG.	ITEM
5340-00-151-9651	0269	21	5310-00-159-6209	66	8
5975-00-152-1127	429	11		88	24
5331-00-152-1759	479	7	3120-00-159-6992	471	18
2910-00-152-2033	KITS		2510-00-159-8822	425	17
5330-00-152-3217	408	6	2540-00-159-8823	419	2
	417	13	3110-00-160-0338	209	17
6240-00-155-7866	180	2	5305-00-161-0902	50	27
5975-00-156-3253	158	47		55	25
	161	3	6145-00-161-1609	BULK	43
	163	18	4730-00-163-0236	493	23
	165	21		505	52
	166	19		600	16
	167	18	2520-00-163-0713	191	3
	168	18	5325-00-164-2087	61	2
	169	42		613	11
	173	21	4010-00-165-5607	BULK	21
	243	18	9390-00-166-0254	492	12
	243	18	5330-00-166-4333	502	9
	584	13	5310-00-167-0680	100	44
	595	16		183	3
	599	4		460	18
	608	21	5310-00-167-0721	539	9
	621	24		540	7
3110-00-156-5453	55	11	5310-00-167-0765	409	27
3110-00-157-0531	471	27	5310-00-167-0826	608	17
5340-00-157-7938	255	19	5935-00-167-7775	158	16
	262	9		619	24
6150-00-158-0066	433	1	2540-00-169-2855	388	1
9390-00-158-2408	BULK	22	2540-00-169-2856	536	9
4030-00-158-2409	469	34	5310-00-171-1735	308	22
	470	29	3110-00-171-2489	527	2
5340-00-158-3773	90	6A	5315-00-171-2590	370	38
5340-00-158-3774	399	9	7690-00-171-2761	569	3
5340-00-158-3877	419	12	5330-00-171-2776	206	4
5340-00-158-3889	402	13	5331-00-171-3879	19	8
	404	17	5325-00-171-6387	429	9
5340-00-158-4077	405	5	5330-00-171-6600	56	14
	406	21	4730-00-172-0010	466	9
5340-00-158-4078	161	24		506	7
2640-00-158-5617	300	2	4730-00-172-0022	472	23
3110-00-158-6011	235	1		501	15
	242	1	4730-00-172-0028	316	1
5940-00-159-1292	137	8		344	11
6680-00-159-1732	34	8		466	8
4730-00-172-0031	219	6			
				221	5
				222	5
	609	6		226	5
5365-00-159-4668	414	31		228	9
5310-00-159-6209	52	3		229	7

CROSS-REFERENCE INDEXES

NATIONAL STOCK NUMBER INDEX

STOCK NUMBER	FIG.	ITEM	STOCK NUMBER	FIG.	ITEM
4730-00-172-0031	229	8	2590-00-177-9196	491	8
	486	5	3040-00-177-9266	484	21
4730-00-172-0034	303	12	5340-00-178-1036	140	16
	345	7	5330-00-178-2191	484	9
	471	31	5935-00-178-6075	158	24
	472	43	5340-00-178-6080	406	4
	487	20	5310-00-178-8631	161	12
	490	16	3040-00-179-3540	472	26
	500	26	2590-00-179-5581	487	8
4730-00-173-1867	263	7	2510-00-179-5708	363	7
	266	25	3830-00-179-6635	379	11
	267	17	2590-00-179-7053	490	3
5340-00-174-3424	310	16	2540-00-179-7094	382	17
5307-00-174-4863	583	4	5342-00-179-7123	255	20
	589	29	5305-00-180-1964	415	7
5325-00-174-5315	63	8	4730-00-180-5038	213	8
5325-00-174-5316	100	2		256	6
5325-00-174-5317	587	13		266	9
	588	6	4730-00-180-7031	271	6
5325-00-174-9341	594	24		273	13
			617	14
5306-00-175-4967	348	37		621	16
1450-00-175-9752	530	7	5310-00-180-8544	321	25
5340-00-176-0868	131	20	5340-00-181-1546	245	23
	134	33	246	17	
	134	74	5330-00-182-3489	483	2
	611	35	5342-00-182-3726	141	16
	612	24	5365-00-182-6713	463	13
	613	25		464	13
5310-00-176-6341	540	3		511	13
5306-00-182-9267	237	8			
			293	4	
5310-00-176-6690	102	4	5305-00-182-9561	495	10
	103	6	5365-00-182-9635	457	7
5310-00-177-0892	233	2		458	4
	240	2		507	8
5310-00-177-1258	250	10		507	40
	251	9	5935-00-184-6707	158	23
5975-00-177-1930	BULK	8	5310-00-184-8992	470	12
5305-00-177-5552	31	11	3040-00-184-9720	483	1
4720-00-177-6160	314	11	5325-00-184-9846	429	12
4720-00-177-6162	314	1	5325-00-185-0004	538	3
4720-00-177-6184	109	14	5365-00-185-7835	489	50
3040-00-177-7830	489	51	4730-00-186-4967	471	35
2510-00-177-7893	382	41	5310-00-186-7403	284	5
2510-00-177-7903	382	31		286	17
2540-00-177-8108	549	46	4730-00-186-7798	598	50
5995-00-177-8220	134	12	4730-00-187-0840	629	22
2510-00-177-9137	382	22	4730-00-187-4202	278	6
5365-00-177-9194	490	8	4730-00-187-4210	207	31
2590-00-177-9196	485	10		519	26

CROSS-REFERENCE INDEXES

NATIONAL STOCK NUMBER INDEX

STOCK NUMBER	FIG.	ITEM	STOCK NUMBER	FIG.	ITEM
3110-00-187-5730	527	1	4730-00-196-1504	255	10
5315-00-187-9370	354	1		261	21
5315-00-187-9414	345	30		262	4
5315-00-187-9567	350	9		266	38
5315-00-187-9591	329	15		590	1
4730-00-188-1896	521	9	4730-00-196-1506	480	4
5306-00-189-1775	543	7	4730-00-196-1534	459	4
4730-00-189-3034	589	50	4730-00-196-1964	215	4
	590	35	4730-00-196-1966	212	13
5340-00-190-6783	429	3		213	1
5355-00-191-1029	493	2		266	2
	505	18	4730-00-196-1991	255	22
5330-00-191-1161	492	14		262	17
5305-00-191-3640	341	3	4730-00-196-1993	255	22
	424	5	5305-00-196-5570	417	17-
2510-00-191-8172	348	14	2590-00-197-1739	100	7
4730-00-193-0883	482	4	9905-00-197-1741	566	16
4730-00-193-2709	264	21	9905-00-197-1742	566	4
4730-00-193-7080	36	5	9905-00-197-1743	566	12
5330-00-193-7652	19	1	9905-00-197-1744	566	8
5310-00-193-9753	528	21	9905-00-197-1746	566	6
	529	5	9905-00-197-1748	566	13
5310-00-194-1483	134	3	5310-00-197-5304	7	16
	134	30	9905-00-197-5957	557	2
	134	64	9905-00-197-5962	557	1
4730-00-194-1766	110	9	5365-00-197-9327	37	6
5360-00-194-5920	88	29		41	8
5310-00-194-9209	100	9	3110-00-198-1080	471	7
5310-00-194-9213	470	23	3110-00-198-2170	530	13
3110-00-195-0460	234	19	3110-00-198-2848	185	8
	241	19	5310-00-199-6581	282	4
5331-00-195-5757	317	20	4730-00-200-0257	273	10
4730-00-196-0888	147	2	4730-00-200-0528	626	3
4730-00-196-0930	182	14	5325-00-200-7234	500	9
	582	8	9905-00-202-3639	544	2
4730-00-196-0936	549	41		545	4
4730-00-196-1467	505	49		546	2
	581	23		547	6
4730-00-196-1468	36	11		548	2
4730-00-196-1481	61	5	5325-00-202-4005	153	11
4730-00-196-1486	596	5	4730-00-202-6491	597	5
4730-00-196-1489	589	48		598	48
	590	33	4730-00-202-6670	474	6
4730-00-196-1493	589	2		519	27
4730-00-196-1495	263	1	4730-00-202-8470	480	12
269	32	482		14	
	505	47	4730-00-202-9035	271	5
4730-00-196-1496	505-	46		273	21
4730-00-196-1504	81	11		620	6
	252	18	4730-00-202-9036	71	19

NATIONAL STOCK NUMBER INDEX

STOCK NUMBER	FIG.	ITEM	STOCK NUMBER	FIG.	ITEM
4730-00-202-9036	74	18	5310-00-209-0965	228	15
	75	14		243	23
	263	3		284	67
	263	24		286	24
	264	34		301	14
	269	13		310	13
	269	24		503	13
	271	13		504	5
	272	7		505	34
4730-00-202-9539	207	6	5340-00-209-9377	351	25
4730-00-203-0549	7	6	5360-00-211-9547	192	13
5365-00-203-1281	66	21	5305-00-213-8886	304	8
4720-00-203-2668	BULK	15	5935-00-214-0904	158	17
4730-00-203-2836	505	51		160	7
4820-00-203-3260	630	14		163	12
2610-00-204-2545	300	1		165	18
9905-00-205-2795	544	6		168	14
	545	2		169	14
	546	4		169	22
	547	1		434	6
	548	3		592	25
5340-00-205-5314	594	4		594	26
5305-00-206-4527	33	30		612	2
5307-00-206-8510	366	16		613	5
5306-00-207-4932	264	15	5305-00-215-0290	579	11
5310-00-208-1918	111	3	5307-00-215-5399	249	14
	211	47	2590-00-220-5104	488	3
	513	25	4730-00-221-2137	82	6
	525	25		261	32
	528	28	4730-00-221-2141	485	3
5310-00-209-0786	101	23	2590-00-221-4824	487	14
	141	25	5340-00-222-1653	254	16
	142	2		254	16
	180	18		261	2
	181	21	4730-00-222-1839	264	13
	184	14		264	31
	455	1	4730-00-222-1840	519	30
	584	3		521	16
	592	3	2540-00-222-8883	410	7
	594	33	2590-00-222-8906	432	9
	623	23	6150-00-222-8943	435	6
5310-00-209-0788	362	4	2510-00-222-8973	142	21
	585	23	6150-00-222-8988	434	1
5310-00-209-0965	33	34	3040-00-222-8991	473	22
	37	21	5330-00-222-8992	425	28
	49	4	2530-00-225-0680	305	4
	54	6	5305-00-225-3841	623	27
	106	2	5305-00-225-3843	371	4
	125	6		414	28
	220	15		427	3

CROSS-REFERENCE INDEXES

NATIONAL STOCK NUMBER INDEX

STOCK NUMBER	FIG.	ITEM	STOCK NUMBER	FIG.	ITEM
5305-00-225-3843	536	11	6680-00-226-4574	82	4
5310-00-225-6408	78	19		145	4
	297	14		145	4
	345	2	5306-00-226-4827	107	22
	391	8		108	12
5310-00-225-6993	419	20	5305-00-226-4831	284	57
5306-00-225-8496	109	5		285	15
	183	9		428	20
	198	9	5306-00-226-4832	382	39
	348	22		385	47
5306-00-225-8499	425	22		387	33
5305-00-225-8507	286	18	5305-00-226-7767	106	8
5306-00-225-9086	581	14		155	1
	585	33		396	8
	591	7		458	25
	596	16		474	21
	601	7		485	25
5306-00-225-9087	113	8		487	26
	591	8		489	4
5306-00-225-9088	77	21	4730-00-226-8874	35	4
	80	23		132	17
	252	2		596	17
	397	3	3110-00-227-3239	500	12
	489	41	3110-00-227-3241	472	21
5306-00-225-9089	266	41	3110-00-227-3245	471	20
	308	14		501	3
	622	1	3110-00-227-3249	472	9
5305-00-225-9091	169	30	3110-00-227-3253	500	20
	205	1	3110-00-227-3255	476	2
	592	11		477	6
	594	18	3110-00-227-3381	487	16
	604	6	3110-00-227-4667	Z09	1
5305-00-225-9092	62	28	4820-00-229-9917	312	25
	100	40	2530-00-230-3596	310	17
	536	1	3040-00-230-3598	310	19
	549	36	5365-00-230-9682	405	41
5305-00-225-9093	254	12	2530-00-231-0178	236	36
	256	19			
	308	13			
	604	14	2540-00-231-0200	406	33
5306-00-225-9095	102	1	2540-00-231-0206	406	3
	103	1	2540-00-231-0207	405	17
	601	11	5340-00-231-0210	402	3
5306-00-225-9098	341	7		417	6
5305-00-225-9099	290	5	5306-00-231-0211	393	10
5306-00-225-9100	390	21	3040-00-231-0212	409	7
	605	4	5342-00-231-0216	405	25
2590-00-226-2347	469	29	3040-00-231-0219	474	5
	472	35	3040-00-231-0221	409	17
	499	19	5340-00-231-0278	321	18

NATIONAL STOCK NUMBER INDEX

STOCK NUMBER	FIG.	ITEM	STOCK NUMBER	FIG.	ITEM
5340-00-231-0278	381	12	5310-00-236-3694	101	2
5310-00-231-0280	310	5	5315-00-236-6625	501	8
4730-00-231-2412	630	16	5315-00-236-8345	499	10
4730-00-231-3906	109	7	5315-00-236-8359	358	8
4730-00-231-5596	491	2	2540-00-237-3693	329	4
4730-00-231-5598	261	31	5340-00-237-3706	492	17
4730-00-231-5650	590	3	5315-00-237-6341	383	40
2590-00-231-7418	398	6		384	11
5340-00-231-7428	419	3		386	18
2510-00-231-7437	142	18	5330-00-237-7828	530	14
2510-00-231-7438	142	18	5315-00-238-0882	7	26
5365-00-231-7440	410	9	2910-00-238-5434	85	1
2510-00-231-7444	405	38		87	1
	406	28	5340-00-238-5435	46	22
4730-00-231-7450	478	3	4730-00-238-5594	485	7
2590-00-231-7451	487	30		491	5
2590-00-231-7452	487	24	5340-00-238-5606	416	14
5995-00-231-7454	435	1	5306-00-238-5661	310	6
2510-00-231-7465	408	17	4020-00-238-7734	BULK	24
3040-00-231-7470	409	19	5315-00-239-8032	508	20
2590-00-231-7484	409	10	5945-00-240-1684	148	2
			4730-00-240-1740	617	15
5340-00-231-7486	331	4	5340-00-240-9228	367	11
	332	4	3110-00-240-9897	530	31
2510-00-231-7488	331	5	5310-00-241-0157	213	11
2510-00-231-7489	331	1		253	11
2520-00-232-1938	530	11		256	9
2520-00-232-1944	530	17		266	12
5340-00-232-6056	414	23		272	6
5315-00-234-1848	379	15	5310-00-241-6658	77	23
	402	26		80	25
5315-00-234-1854	454	4		97	8
5315-00-234-1863	180	11		100	30
	181	19		112	8
	454	10		113	10
	513	23		134	41
5315-00-234-1864	339	2		140	14
6150-00-234-3248	440	1		169	32
3040-00-234-3250	487	28		204	14
2510-00-234-3258	414	2		205	5
	414	22		252	1
2510-00-234-3259	414	2		254	1
	414	22		256	17
2590-00-234-3262	472	31		266	39
5320-00-234-8557	419	27		290	15
4730-00-235-1483	72	5		291	1
	78	5		341	12
2510-00-235-1888	407	3		390	23
2590-00-235-4400	489	52		397	1
2530-00-235-4756	350	2		525	12
4710-00-235-4819	493	13		536	5

CROSS-REFERENCE INDEXES

NATIONAL STOCK NUMBER INDEX

STOCK NUMBER	FIG.	ITEM	STOCK NUMBER	FIG.	ITEM
5310-00-241-6658	549,	34	5330-00-243-3571	406	15
	581	17	5340-00-243-7145	34	4
	585	35	5315-00-244-1340	490	15
	589	8	4730-00-244-9848	291	9
	591	2	5310-00-246-0221	30	10
	592	14	5330-00-246-0309	11	1
	594	14	4730-00-246-9217	8	18
	601	10	5331-00-248-3847	183	8
	604	8	9505-00-248-9842	345	1
	606	15	4730-00-249-1511	589	47
	615	18		590	32
5310-00-241-6661	318	44	4730-00-249-3885	255	23
	319	45		262	11
	320	45	4730-00-249-3932	146	7
	321	46	4730-00-249-3935	271	1
	322	49		589	63
	323	48	4730-00-249-4416	493	25
	324	45	5325-00-249-6345	541	6
5310-00-241-6664	5	10	5310-00-249-6540	30	7
	207	17	6810-00-249-9354	151	23
	207	36	5310-00-250-9477	619	15
	318	22	5315-00-250-9595	611	12
	319	21	4730-00-251-6827	505	50
	320	18	5331-00-251-8839	456	1
	321	21		456	6
	322	23		460	1
	323	25		460	7
	324	19		461	1
	326	11		509	2
	326	16	5330-00-252-3274	127	5
	327	1	5325-00-252-4758	500	23
	328	2	9905-00-252-5586	552	12
	330	3	9905-00-252-5587	552	9
	331	2		611	32
	332	2		612	26
	343	13		613	23
	380	7	5330-00-252-8888	52	14
	381	10	4730-00-253-4414	505	45
	389	7		590	2
	390	20	4730-00-253-4415	589	16
	400	4	5305-00-253-5609	565	3
	616	9	5305-00-253-5618	560	2
5310-00-241-6665	460	13		563	2
5310-00-241-6921	141	12		566	3
5315-00-241-7523	190	11		567	5
5315-00-242-0818	489	34	5305-00-253-5626	329	11
5320-00-242-1580	417	2	5305-00-253-5627	569	1
5320-00-242-1582	405	40	4730-00-253-5794	81	9
	406	19	4730-00-254-6211	109	13
	416	27		582	20

CROSS-REFERENCE INDEXES

NATIONAL STOCK NUMBER INDEX

STOCK NUMBER	FIG.	ITEM	STOCK NUMBER	FIG.	ITEM
4730-00-254-6211	585	9	5305-00-267-8953	74	30
	594	11		75	1
4730-00-255-0560	503	17		102	24
	504	6		134	40
	505	33		142	8
2520-00-255-5149	207	26		181	26
	207	30		214	9
3040-00-255-5700	488	10		278	9
5325-00-256-2846	51	16		279	9
	56	16		279	17
5310-00-261-7156	614	6		352	2
5310-00-261-7340	66	11		353	3
	118	15		369	3
5310-00-262-2986	129	4		580	12
4030-00-262-3152	475	2		584	2
5320-00-262-6492	362	37		587	7
3020-00-262-7572	484	12		588	11
5530-00-262-8180	BULK	23		592	29
2610-00-262-8653	300	1		593	3
9905-00-262-9929	563	1		594	32
5310-00-263-2862	539	16		596	21
	540	13			
4820-00-263-3019	584	8		601	4
	593	27		608	5
	595	17	5305-00-267-8954	148	9
5325-00-263-6648	138	17		181	22
5325-00-263-6651	211	46		255	18
5310-00-263-9488	303	6		262	14
	303	9		610	2
				616A	11
5315-00-264-3099	462	2	5305-00-267-8955	100	38
	510	2		534	3
5320-00-264-3266	419	32	5305-00-267-8956	313	11
5310-00-264-4083	367	6	5305-00-267-8957	163	32
5340-00-264-7182	414	12		165	37
	419	9		166	26
4730-00-266-0538	108	2		167	22
	593	23		168	25
	594	3		169	34
				507	48
	131	9		623	24
	131	12	5305-00-267-8958	278	10
	132	12		405	43
5305-00-267-8951	339	27		406	25
	340	13	5305-00-269-0770	408	4
5305-00-267-8952	100	45		410	11
	101	10	5305-00-269-2803	62	1
	551	17		67	11
	590	10		68	5
	593	7		69	13
5305-00-267-8953	61	8		72	3

CROSS-REFERENCE INDEXES

NATIONAL STOCK NUMBER INDEX

STOCK NUMBER	FIG.	ITEM	STOCK NUMBER	FIG.	ITEM
5305-00-269-2803	83	3	5305-00-269-2803	613	30
	102	31		629	5
	102	36	5305-00-269-2804	6	27
	109	17		62	1
	142	10		71	26
	142	14		74	26
	143	15		75	21
	153	6		105	2
	183	10		213	34
	243	1		215	10
	253	5		220	14
	259	10		228	14
	260	10		254	17
	264	25		254	17
	264	28		257	8
	269	8		258	1
	333	19		261	25
	334	12		263	17
	337	11		266	20
	338	10		266	29
	355	16		271	17
	356	15		301	7
	364	10		302	8
	365	10		334	3
	368	1		335	15
	368	4		337	29
	374	2		338	3
	374	21		354	9
	375	6		368	6
	376	8		374	22
	377	7		397	6
	378	2		492	8
	379	1		505	29
	388	5		511	58
	454	17		513	15
	456	9	5305-00-269-2811	211	20
	457	20		243	10
	513	21		609	25
	536	16	5305-00-269-3217	177	9
	536	17	5305-00-269-3231	302	1
			5305-00-269-3232	602	33
	583	10	5305-00-269-3233	71	9
	586	10		75	8
	589	35		76	7
	590	21		77	16
	592	19		78	14
	600	2		79	7
	600	2		80	5
	600	5		154	2
	602	12	5305-00-269-3234	72	4
	603	6		74	1

NATIONAL STOCK NUMBER INDEX

STOCK NUMBER	FIG.	ITEM	STOCK NUMBER	FIG.	ITEM
5305-00-269-3234	143	11	5305-00-269-3239	165	22
	163	6		255	11
	163	24		258	7
	165	26		262	24
	365	9		267	7
	472	38		269	3
	590	12		276	7
5305-00-269-3235	71	17		310	8
	72	8		326	6
	338	25		351	2
	357	1		355	8
	365	4		355	12
	365	10		356	9
	376	16		513	34
	591	19		528	9
5305-00-269-3236	34	6		536	11
	591	28		543	9
5305-00-269-3237	107	1		589	41
5305-00-269-3238	62	6		612	17
	102	9	5305-00-269-3240	163	19
	161	5		261	29
	163	21		343	2
	185	27		368	6
	202	1		374	22
	213	34		388	4
	243	13		400	8
	254	20		401	5
	261	12		456	7
	261	12		456	8
	269	23		460	8
	273	25		461	7
	316	15		509	3
	319	34		536	15
				616A	20
	333	3	5305-00-269-3242	143	15
	334	21		243	12
	335	3		301	12
	336	3		301	13
	337	3		478	20
	338	19	5305-00-269-3243	KITS	2
	378	3	5305-00-269-3244	105	6
	388	8		106	9
	457	12		382	18
	487	33		387	22
	507	15		387	45
	513	8		460	9
	525	8		507	32
5305-00-269-3239	62	32		609	15
	72	3		629	5
	72	3	5305-00-269-3245	456	8
	93	1	5305-00-269-3246	478	12

CROSS-REFERENCE INDEXES

NATIONAL STOCK NUMBER INDEX

STOCK NUMBER	FIG.	ITEM	STOCK NUMBER	FIG.	ITEM
5305-00-269-3247	67	13	5325-00-276-6089	429	10
	68	10	5325-00-276-6091	140	15
	69	15		624	1
5305-00-269-3250	474	17		625	1
	489	10	5325-00-276-6228	432	14
5305-00-269-4528	390	14	5325-00-276-6343	582	21
5330-00-269-4953	507	17		584	27
5310-00-269-7044	243	20	3110-00-277-0476	502	5
2530-00-270-3878	255	12	4730-00-277-1896	630	9
	261	24	4730-00-277-5542	593	2
	262	7	4730-00-277-5553	273	11
4730-00-270-4580	70	6		278	16
	76	16		290	18
	79	16		426	5
	266	4		505	19
4730-00-270-4616	267	3		582	7
	268	4		596	2
	274	8		598	49
	275	8		620	18
	599	16	4730-00-277-5555	252	16
4030-00-270-5436	82	14		255	6
5510-00-270-6031	BULK	48		261	19
5305-00-270-7328	526	16		262	20
5325-00-270-8890	429	14		264	3
4820-00-272-3351	422	1		264	32
5306-00-272-3747	350	1	4730-00-277-5684	108	1
4730-00-273-6686	258	10	4730-00-277-7331	597	8
	269	9		599	15
	269	25	4730-00-277-8274	64	11
	505	39		71	18
5310-00-273-7771	293	1			
	294	18			
	295	15			
4820-00-274-3646	581	20		74	9
	582	6		76	15
5310-00-274-7721	232	1		77	8
	236	39		79	15
	239	1		80	15
5310-00-274-8041	107	5	4730-00-277-8750	256	5
	121	9		276	5
	126	8		422	13
	312	51		423	13
	584	26	4730-00-277-8761	630	15
	585	17	4730-00-277-8770	212	9
	619	5		214	11
5310-00-274-9364	125	14		257	14
4820-00-275-2224	585	25		290	2
5310-00-275-3683	345	11		604	16
5310-00-275-8264	346	15	2510-00-277-9786	348	9
5310-00-276-3012	224	11	4730-00-278-3167	519	13
5315-00-276-5385	235	3	4730-00-278-3214	621	12
	242	3	4730-00-278-3220	263	19
5340-00-276-5847	7	7		597	9

CROSS-REFERENCE INDEXES

NATIONAL STOCK NUMBER INDEX

STOCK NUMBER	FIG.	ITEM	STOCK NUMBER	FIG.	ITEM
4730-00-278-3724	629	5	5340-00-282-7521	39	12
4730-00-278-4311	495	5	4730-00-282-7609	481	9
4730-00-278-4594	81	6	5340-00-282-7793	171	14
4730-00-278-4651	276	9	5306-00-283-0407	289	12
4730-00-278-4814	472	24	5940-00-283-5281	97	10
	501	14	5325-00-285-6250	533	13
4730-00-278-4822	81	10	5310-00-285-8833	9	8
	254	24	5331-00-285-9842	204	8
	254	24		456	40
	274	5		461	31
	275	5		509	26
4730-00-278-6318	212	7	5331-00-285-9847	480	9
	270	5		505	55
	279	2	5306-00-286-1476	366	18
	279	26	5340-00-286-1868	117	11
	288	15	5310-00-286-3727	526	10
	603	12	4730-00-287-1604	148	3
4730-00-278-8824	70	3		199	1
	76	11		212	3
	77	5		213	40
	79	11		270	1
	79	11		273	6
	80	12		279	5
	148	5		279	14
	620	4		288	3
	621	9		541	11
4730-00-278-8825	617	4		603	1
4730-00-278-8888	74	13	4730-00-287-1649	132	16
	77	17	2540-00-287-2571	406	11
	80	21		416	11
4730-00-278-8889	74	10	4730-00-287-3281	454	21
	80	8	4730-00-287-3790	456	22
4730-00-278-8902	252	7	4310-00-287-8126	507	21
4730-00-278-8935	480	26	3120-00-288-1889	236	13
5306-00-281-1651	236	15	4730-00-288-7495	493	27
5310-00-281-2180	367	4	4730-00-288-8555	489	11
5315-00-281-7610	7	28	4730-00-288-9953	422	16
5315-00-281-7650	474	11		423	16
5315-00-281-7651	484	8	4730-00-289-0051	213	18
5315-00-281-7652	474	9		252	4
	489	23		253	8
5325-00-281-8643	533	8		256	10
5315-00-282-2583	499	9		266	15
5310-00-282-5661	234	2		267	11
	241	2		271	16
5325-00-282-7435	534	7	4730-00-289-0155	267	10
5325-00-282-7471	369	7		271	8
5340-00-282-7509	287	6		272	14
	620	8		274	7
	620	8		275	7

NATIONAL STOCK NUMBER INDEX

STOCK NUMBER	FIG.	ITEM	STOCK NUMBER	FIG.	ITEM
4730-00-289-0155	278	7	5310-00-298-8903	169	19
	290	7	5365-00-299-0067	409	14
	290	9	5340-00-299-0069	398	9
	599	13	6220-00-299-7425	143	4
	604	11	6220-00-299-7426	143	4
	605	8	5365-00-300-4379	341	4
4730-00-289-0382	277	7	5340-00-302-1840	402	10
	523	21	5325-00-303-4932	533	7
4730-00-289-0383	204	22	2920-00-304-3493	127	2
	276	3	5930-00-307-8856	131	28
4730-00-289-1245	480	5	2815-00-311-2521	23	5
4730-00-289-1266	482	12	4730-00-314-8366	426	9
4730-00-289-4052	212	12	2510-00-318-0959	348	11
	215	13	2510-00-318-0960	348	12
4730-00-289-4912	204	13	2520-00-318-0983	228	3
4730-00-289-5176	502	34	5360-00-321-5710	526	19
4730-00-289-5179	117	4	5310-00-321-9974	294	3
	117	9		296	23
5315-00-290-1349	339	11	4730-00-322-8339	457	10
	339	18	4730-00-322-8457	182	6
	340	2		214	12
	395	11	5310-00-325-1900	393	2
5325-00-290-1960	582	16	2815-00-327-6439	33	26
	585	12	5340-00-328-5458	427	2
4010-00-290-4352	BULK	4	5330-00-328-8656	31	8
5331-00-291-3273	293	8	3040-00-328-8862	502	7
5330-00-292-1600	502	4	3020-00-331-7672	33	17
5305-00-292-4595	185	21	2590-00-332-0095	507	36
2540-00-293-4730	532	2	5925-00-333-1584	135	3
4730-00-293-7108	148	4	5935-00-333-3088	134	13
	617	2		134	52
	620	2	5935-00-333-9414	157	14
	621	11		158	5
3110-00-293-8439	208	2	4730-00-335-1812	507	9
2520-00-294-6752	301	9	6210-00-337-7345	132	11
6240-00-295-2421	144	6	5340-00-337-9619	624	10
5970-00-296-6078	440	3		624	10
5305-00-297-4022	26	15	5315-00-338-1621	100	12
5331-00-297-9990	485	21		281	35
	503	18	5305-00-338-5162	207	24
	504	7		217	10
	505	32		526	26
5315-00-298-1481	236	9	4730-00-338-6839	37	2
	303	11	5305-00-339-1415	23	17
	391	21	3120-00-339-5642	KITS	
5315-00-298-1498	220	2	5305-00-340-3492	42	10
	316	11	5330-00-340-3637	BULK	31
	471	9		BULK	32
	589	28	5310-00-340-4953	499	24
	591	27	5325-00-343-5531	369	5
			5330-00-361-2955	16	1

CROSS-REFERENCE INDEXES

NATIONAL STOCK NUMBER INDEX

STOCK NUMBER	FIG.	ITEM	STOCK NUMBER	FIG.	ITEM
5340-00-344-2767	169	31	5310-00-374-0836	295	4
5975-00-345-8055	314	5		348	28
5325-00-349-8518	200	26	4730-00-374-2045	63	6
5305-00-350-1210	367	7	4730-00-374-4282	117	10
5310-00-350-2655	141	6	3120-00-374-4342	21	6
3110-00-350-5407	235	12	5330-00-374-4873	187	1
	242	12	2540-00-378-9049	425	9
2540-00-351-0145	532	4	3040-00-388-3126	23	18
	532	4	2520-00-388-4197	229	2
	609	3		229	3
5340-00-351-7831	469	13		229	4
	470	9		229	6
3950-00-351-7834	398	3		KITS	
2590-00-351-7835	398	2		KITS	
2590-00-351-7836	398	11	5365-00-389-0317	425	11
5306-00-351-7842	393	6	5310-00-393-6685	158	22
2510-00-351-7851	412	21		160	18
2510-00-351-7852	412	23		163	5
2590-00-351-7865	469	14		165	5
	470	10		166	9
5995-00-351-7868	431	1		167	8
2520-00-352-2168	466	3		168	5
4730-00-352-9793	213	3		169	5
5305-00-353-0969	141	23	6220-00-397-3335	144	23
5310-00-353-2427	234	11	5935-00-399-6673	131	24
	241	11	5940-00-399-6676	131	22
	293	11		134	57
	295	1		157	4
	348	26		158	36
3020-00-353-9384	33	13		170	2
5310-00-355-5453	233	11		170	10
	240	11		170	23
5310-00-356-1447	7	21		170	28
2530-00-359-1162	293	2		171	7
	295	16		182	35
4730-00-359-3872	258	6		211	33
	478	5		211	39
4730-00-359-4708	480	32		213	28
	482	6		434	2
2815-00-362-1780	23	11		496	5
4730-00-365-2690	117	1		608	14
	285	1		614	12
5365-00-369-4729	284	19		614	15
	286	2	4820-00-400-5189	33	22
2815-00-369-7846	284	33	5310-00-400-5503	161	26
	286	43	5306-00-400-5541	195	3
5325-00-371-8108	533	12	4730-00-400-6544	505	6
5310-00-374-0836	234	30	5315-00-401-4383	493	9
	241	30	2930-00-401-9531	44	18
	293	13	5315-00-402-0421	185	7

CROSS-REFERENCE INDEXES

NATIONAL STOCK NUMBER INDEX

STOCK NUMBER	FIG.	ITEM	STOCK NUMBER	FIG.	ITEM
5305-00-402-4211	363	4	5310-00-407-9566	37	10
9905-00-403-4814	572	11		112	10
	573	1		284	24
4320-00-403-5223	185	23		284	56
2590-00-404-0752	499	15		286	19
6160-00-404-2669	153	32		308	12
4730-00-404-2906	116	3		348	23
2815-00-404-2915	19	2		351	26
2805-00-404-2917	14	14		414	21
5330-00-404-2920	14	2		419	18
2815-00-404-2921	41	2		425	23
2815-00-404-2926	42	4		489	42
2815-00-404-2931	42	8		549	32
5365-00-404-2934	284	37	2510-00-408-2439	384	14
	286	29	2510-00-408-2448	384	15
5340-00-404-2940	26	10	2510-00-408-2452	384	13
5340-00-404-2944	21	1	2510-00-408-2453	384	14
5342-00-404-2946	3	1	2590-00-408-4618	365	3
5340-00-404-2947	11	2	2510-00-408-4631	83	6
2815-00-404-2954	33	2	2510-00-408-4632	83	6
2815-00-404-2956	37	18	2510-00-408-4634	365	3
2815-00-404-2957	37	16	2510-00-408-4642	343	15
2930-00-404-3050	116	7	2510-00-408-4652	343	1
2930-00-404-3053	114	2	2930-00-408-9404	118	17
2910-00-404-3054	46	1	2510-00-409-3991	385	14
2940-00-404-3057	KITS		2510-00-409-3992	536	10
9390-00-405-0215	417	27	2510-00-409-3993	366	5
2520-00-405-1842	185	16	2510-00-409-4005	383	20
3040-00-405-1931	474	8	5340-00-409-4008	319	23
	489	57	5340-00-409-4018	321	15
2510-00-405-1946	406	23		381	14
4710-00-405-1962	314	9	5340-00-409-4019	321	30
2510-00-405-1970	405	1		381	5
2510-00-405-1978	335	10	2510-00-409-4020	384	13
5365-00-405-4378	487	12	3040-00-409-4021	391	3
2590-00-405-9771	405	26	5306-00-409-4066	362	36
2520-00-406-2303	185	6		543	10
2815-00-406-8936	33	5	4730-00-409-7854	213	21
5340-00-407-2612	142	3		252	11
2510-00-407-2617	142	6		256	14
2510-00-407-5084	326	12	5340-00-409-7958	318	5
2510-00-407-5085	142	17		319	5
2510-00-407-5086	142	17		320	5
5340-00-407-5087	142	11		321	5
2510-00-407-5093	142	13		322	5
2510-00-407-5095	142	13		323	5
5340-00-407-6767	83	5		324	5
2510-00-407-6768	83	4	2510-00-409-7960	320	15
9905-00-407-7000	570	5	2510-00-409-7973	386	3
5310-00-407-9566	33	4	3040-00-409-7974	391	1

NATIONAL STOCK NUMBER INDEX

STOCK NUMBER	FIG.	ITEM	STOCK NUMBER	FIG.	ITEM
4730-00-409-8797	504	20	2590-00-418-0887	499	8
2540-00-409-8891	378	8	5320-00-418-0985	402	24
	600	1	5340-00-418-1009	347	14
5340-00-409-8936	316	12	5340-00-419-3077	169	29
7690-00-409-8937	557	5	5340-00-419-3080	330	4
9905-00-409-8948	569	2	5310-00-419-3082	485	16
5365-00-409-8987	473	16		491	14
5340-00-409-9978	36	2	5340-00-419-3084	480	29
5310-00-410-6756	55	26		481	3
5305-00-410-6957	503	12	6240-00-419-3185	136	3
5360-00-411-2511	389	14	5325-00-419-3322	493	28
5365-00-413-4371	490	10		505	20
5975-00-414-6466	413	8	3020-00-419-5854	475	11
3940-00-414-6533	475	3	5340-00-419-5859	408	16
3040-00-414-6535	475	5	5340-00-419-5860	408	10
5340-00-414-6561	484	14	5340-00-419-5866	429	5
5330-00-414-6695	406	32	5330-00-419-5872	234	9
	416	25		241	9
3040-00-415-1373	317	12	5330-00-419-5875	489	56
2520-00-415-1479	409	8	5306-00-419-5876	317	15
5330-00-415-1481	406	14	5306-00-419-5878	393	8
5330-00-415-1484	414	9	5340-00-419-5881	419	35
5330-00-415-1488	419	4	2520-00-419-9422	232	4
5342-00-415-1494	411	13		239	4
4730-00-415-3172	585	10	4730-00-419-9424	252	15
	594	10		255	3
5315-00-415-6294	411	3		261	18
	411	11		262	10
2540-00-417-2583	376	10	3040-00-419-9431	409	21
2590-00-417-2611	416	22	5310-00-419-9461	333	6
2510-00-417-2713	83	4		335	7
2540-00-417-2722	402	22		337	6
2590-00-417-2725	476	3	5340-00-419-9464	405	30
	477	7		406	37
2540-00-417-2732	492	9	5365-00-419-9465	379	8
2590-00-417-2735	405	10	5330-00-419-9468	409	25
	416	15	5330-00-419-9469	409	28
2510-00-417-2736	409	29	3110-00-419-9471	234	14
2510-00-417-2752	346	13		241	14
3040-00-417-2756	472	27	5340-00-419-9473	165	30
2540-00-417-2758	402	16	5340-00-419-9474	420	21
2590-00-417-2787	476	12	5310-00-419-9476	471	10
	477	8	5340-00-419-9484	408	5
5340-00-417-2788	408	11	5340-00-419-9487	255	15
2510-00-417-2789	408	9		262	3
5315-00-417-5223	411	16	5306-00-420-1691	12	9
5340-00-417-5800	34	9	2540-00-420-5036	532	4
2540-00-418-0603	475	6	4820-00-420-5499	252	17
2590-00-418-0668	497	7		255	9
2590-00-418-0673	497	6		261	20

CROSS-REFERENCE INDEXES

NATIONAL STOCK NUMBER INDEX

STOCK NUMBER	FIG.	ITEM	STOCK NUMBER	FIG.	ITEM
4820-00-420-5499	262	18	2510-00-425-0512	411	14
	264	14	4710-00-425-5921	36	9
	264	30	5310-00-426-3990	24	8
5310-00-420-8044	93	3	5365-00-427-2282	236	37
5331-00-420-9624	16	5	4730-00-428-5631	36	6
5325-00-420-9696	118	9	6220-00-428-5943	140	3
2590-00-421-1595	100	29	5365-00-428-6201	7	22
4730-00-421-3924	213	9	5305-00-429-1552	405	36
	253	9		406	29
	256	7	2815-00-430-2090	31	7
	266	10	2530-00-430-2392	250	12
	272	4		251	11
5315-00-421-3931	485	23	5330-00-432-2142	62	23
2590-00-421-3942	481	5	5305-00-432-4163	359	11
2510-00-421-3944	402	21		362	22
5306-00-421-3950	537	3		402	2
2510-00-421-3956	395	8		418	13
3040-00-421-3960	489	55	5305-00-432-4170	408	15
3040-00-421-3961	489	29		412	8
4730-00-421-3962	489	40		412	10
3040-00-421-3966	379	5		412	10
2910-00-421-3967	99	3		413	19
5305-00-421-3986	424	4		414	3
5340-00-421-3990	402	12		415	20
5310-00-421-3991	317	22		415	26
5340-00-421-3992	480	31		418	11
	481	12		419	6
5310-00-421-3994	471	11		420	5
5340-00-421-5083	383	18		424	13
2520-00-421-7229	KITS			425	16
5340-00-421-7235	395	4		426	12
3020-00-421-7240	411	2		429	4
5340-00-421-7254	109	11	5305-00-432-4171	410	8
	165	25		418	14
	166	22		420	15
5306-00-421-9402	414	1	5305-00-432-4172	403	11
2590-00-421-9524	28	4		404	2
5340-00-421-9696	413	18		405	24
	414	4		406	41
5365-00-421-9697	402	11		407	1
	404	16		407	19
5365-00-422-1160	487	13		413	6
5305-00-422-1161	391	2		414	11
3040-00-422-2008	391	5		415	28
5365-00-422-2017	343	7		416	30
5340-00-422-2019	391	12		421	7
5340-00-423-4080	425	6		425	19
4710-00-424-2694	BULK	35		426	3
	BULK	36		429	6
5315-00-425-0118	33	14		579	1

NATIONAL STOCK NUMBER INDEX

STOCK NUMBER	FIG.	ITEM	STOCK NUMBER	FIG.	ITEM
5305-00-432-4173	404	22	5331-00-441-0145	284	7
5305-00-432-4201	362	21		286	7
	404	11	5305-00-442-3217	182	10
	408	8	5310-00-442-6899	44	10
	409	30	6220-00-443-0589	141	13
	422	18	3040-00-443-4839	379	9
	423	4	2590-00-443-4847	480	3
	427	6	5340-00-443-6132	305	23
	429	2	5365-00-443-9901	379	10
	455	7	4730-00-444-1710	66	22
	492	18	5340-00-444-2107	393	4
	528	15	5340-00-444-2108	393	4
5305-00-432-4203	364	3	4820-00-445-0610	284	17
	402	9		286	4
	404	18	4730-00-445-4418	629	5
	405	14	5340-00-445-4555	233	17
	406	44		240	17
	407	2	5340-00-445-4557	169	26
	412	22	5340-00-445-4561	410	3
	413	26	4010-00-445-7212	402	25
	415	31	5310-00-445-7238	485	17
	416	2		491	10
	422	12	5306-00-445-7240	346	6
	423	12	3020-00-446-8616	490	13
5305-00-432-4205	405	13	5340-00-446-8732	395	13
	406	2	5315-00-446-8740	490	18
5305-00-432-4252	405	45	5306-00-446-8762	348	10
	427	9	5365-00-446-8770	343	16
	427	19	3010-00-447-9799	49	2
	548	1		54	2
5305-00-432-4253	404	10	5340-00-448-4073	262	19
	419	23	5310-00-449-2378	500	29
	419	37	5310-00-449-2381	125	11
	420	1	4820-00-449-5059	493	1
	424	2	4730-00-449-7203	629	28
5305-00-432-4254	402	23	4730-00-449-7356	493	17
5305-00-432-4390	413	14	5360-00-450-0346	197	10
5305-00-432-7956	149	4	4730-00-450-9671	493	19
5305-00-432-8027	561	2	5975-00-451-5001	110	6
	566	7		115	6
5305-00-433-3685	403	4		158	47
	418	2		161	22
5305-00-433-3711	403	14		205	10
5307-00-434-1783	230	6		589	45
	237	11		590	31
4820-00-436-3033	290	6		611	26
2590-00-436-4601	493	7		628	4
2540-00-436-8289	370	4	9320-00-451-8080	BULK	28
4730-00-439-1722	595	2		BULK	29
4730-00-439-6028	480	1	6220-00-451-8161	131	11

CROSS-REFERENCE INDEXES

NATIONAL STOCK NUMBER INDEX

STOCK NUMBER	FIG.	ITEM	STOCK NUMBER	FIG.	ITEM
2815-00-453-8994	12	5	5330-00-470-2115	409	24
5315-00-453-9349	359	15	5310-00-470-4271	5	12
5930-00-453-9367	212	5	5310-00-470-6154	118	23
5331-00-454-0364	249	10	5305-00-471-3909	463	21
5365-00-455-1382	409	9		464	21
5975-00-456-0627	438	3		511	21
2590-00-471-5343				419	11
5340-00-456-1011	383	11		419	26
	383	25	2590-00-471-5344	419	11
	383	29		419	26
	384	4	5340-00-471-8631	489	15
	384	17	5340-00-471-8635	400	5
	386	8	2540-00-472-1696	376	4
	386	12	5340-00-472-1953	100	47
	533	14	2815-00-472-2626	12	13
4730-00-456-9831	481	8	5330-00-472-2783	494	12
	482	18		494	17
4730-00-457-6295	271	9	5360-00-472-6822	530	8
	273	14	2590-00-473-6331	376	2
2815-00-457-6311	12	4		399	1
2520-00-457-6660	218	9		629	5
	224	13	9905-00-473-7260	552	1
2540-00-460-5815	370	10		553	1
2540-00-460-5826	370	3		556	2
4820-00-461-4216	478	7	5315-00-475-2574	33	12
5306-00-461-6070	88	3	4730-00-475-5168	278	4
5331-00-462-0907	139	6		585	5
5365-00-462-4504	85	16	5325-00-476-5259	306	36
	87	16	5360-00-476-9339	358	3
5935-00-462-6603	159	8	5305-00-477-0124	613	10
5331-00-463-0200	140	9	5305-00-477-0144	404	23
5310-00-463-0268	141	17		405	2
5305-00-463-0428	37	20		405	46
5305-00-463-0429	37	4		406	22
3020-00-464-4438	530	23		406	47
5330-00-465-5818	33	39		416	21
5331-00-465-6453	249	29		416	33
5340-00-466-4948	211	23		428	15
	211	52		428	21
	419	15	5325-00-477-0304	530	28
5310-00-469-4039	530	22	5305-00-477-6769	66	1
5310-00-469-4073	530	5	4930-00-477-8276	93	2
5340-00-470-1537	370	33	4720-00-477-8933	62	26
5342-00-470-1543	370	43	5310-00-478-0548	190	17
2540-00-470-1564	370	24	5320-00-478-3313	396	3
2510-00-470-2090	420	23	2530-00-478-5865	255	8
2510-00-470-2091	420	2	5340-00-478-5877	416	3
3040-00-470-2097	500	11	5340-00-478-5878	416	3
5310-00-470-2107	485	15	5307-00-478-6782	311	9
			5340-00-479-2947	616A	30
	491	15	5330-00-480-6133	44	16

CROSS-REFERENCE INDEXES

NATIONAL STOCK NUMBER INDEX

STOCK NUMBER	FIG.	ITEM	STOCK NUMBER	FIG.	ITEM
5330-00-480-6133	KITS		5310-00-488-3888	155	4
5340-00-480-7602	321	9		278	10
	324	14		304	1
				318	3
5340-00-480-7608	243	3		318	38
2540-00-481-3637	539	3		318	48
5305-00-482-1035	389	15		319	8
5340-00-482-3791	346	3		319	39
5340-00-482-4339	537	5		320	3
5360-00-482-4422	245	18		320	36
5360-00-482-4813	370	34		321	3
5360-00-482-4814	370	36		322	3
4720-00-482-8956	422	15		322	42
	423	15		323	3
5305-00-483-0554	405	31		323	40
	406	36		324	15
5307-00-483-1105	236	8		324	37
5340-00-483-1107	492	15		326	14
5340-00-483-2163	399	8		326	14
5310-00-483-2266	232	2		326	15
	239	2		327	2
5305-00-483-2339	362	7		328	1
5310-00-483-2385	311	8		330	2
5330-00-483-2408	543	19		331	6
5310-00-484-1718	50	26		332	5
	52	12		333	14
	53	7		334	9
	129	3		335	12
3020-00-484-5831	472	22		336	9
5305-00-484-6186	367	24		336	14
3040-00-484-8546	88	17		337	17
5342-00-484-8588	5	11		338	14
5306-00-485-0790	53	6		340	14
5330-00-485-0863	530	30		349	6
5330-00-485-0865	530	30		351	28
5330-00-485-0895	529	2		358	6
	530	32		365	1
5340-00-485-0945	23	19		368	11
3040-00-485-9224	33	15		368	15
5310-00-486-2505	284	60		374	23
5310-00-486-5355	506	17		381	17
5365-00-488-0799	7	2		383	33
5310-00-488-3888	5	13		384	8
	69	3		386	16
	102	12		389	4
	102	14		391	18
	102	16		456	14
	102	17		457	21
	106	11		458	8
	107	34		460	14

NATIONAL STOCK NUMBER INDEX

STOCK NUMBER	FIG.	ITEM	STOCK NUMBER	FIG.	ITEM
5310-00-488-3888	461	11	5330-00-503-5789	90	13
	468	2	3040-00-504-8913	489	32
	492	6	4310-00-504-8923	485	20
	500	1		491	13
	503	24	3110-00-504-8929	489	48
	508	21	3120-00-504-8930	486	7
	509	6		487	2
	513	11	3040-00-504-9037	489	26
	525	11	3040-00-504-9038	484	7
	589	18	2815-00-505-5116	23	4
	590	18	2815-00-505-5119	23	7
			5340-00-505-6379	616A	22
	602	6	5330-00-506-4866	50	22
	614	7	5331-00-506-4874	116	2
	615	1	5305-00-506-5722	85	7
	616	8		87	7
	616A	16			
	619	2	3040-00-506-8451	502	18
	619	2	5310-00-507-3259	85	3
5310-00-488-3889	458	18		87	3
4330-00-488-8613	457	23	5365-00-507-3260	85	16
5340-00-489-0363	389	13		87	16
2510-00-489-7104	370	25	5365-00-507-3261	85	16
7690-00-489-8322	571	5		87	16
	572	9	5365-00-507-3262	85	16
4720-00-491-0102	BULK	16		87	16
5340-00-491-0329	493	24	5365-00-507-3271	95	2
5340-00-491-0331	493	14	5365-00-507-8766	158	25
4730-00-491-9576	299	6		163	4
4730-00-492-6040	34	10		165	4
5305-00-493-3959	88	4		166	16
5310-00-493-3986	579	10		167	16
4710-00-493-8899	39	1		168	4
5340-00-494-2234	539	2		169	4
				540	18
4730-00-494-6580	214	6	6680-00-507-9992	624	2
	538	12	5330-00-508-0411	114	6
2520-00-494-6586	235	6	5325-00-508-8901	533	6
	242	6	4820-00-509-3036	505	12
5340-00-495-0236	414	30	5305-00-509-8106	129	2
5315-00-495-6497	395	9	3120-00-509-8270	490	12
5365-00-496-9706	313	10	5320-00-510-7823	558	2
5945-00-496-9708	441	4	2530-00-512-0032	303	1
5365-00-497-6718	413	22	5310-00-512-5213	489	18
5925-00-497-9661	441	9	5306-00-512-9218	380	3
3040-00-498-2386	493	3		381	4
2590-00-498-2387	493	5		400	10
2590-00-498-2388	493	4	3040-00-512-9223	507	3
3040-00-498-2389	493	6	5330-00-513-1443	KITS	
2510-00-498-2392	384	1	3040-00-513-5786	507	5
	386	9	3040-00-513-5787	507	18
5306-00-498-7209	250	8	3040-00-513-5788	507	22
	251	15	5315-00-514-2660	109	3

CROSS-REFERENCE INDEXES

NATIONAL STOCK NUMBER INDEX

STOCK NUMBER	FIG.	ITEM	STOCK NUMBER	FIG.	ITEM
5315-00-514-2660	311	3	5315-00-532-9388	7	11
5330-00-514-3289	230	3	5310-00-532-9467	624	5
	237	2	3020-00-533-3398	495	9
4030-00-514-4420	347	5	5940-00-534-0986	435	5
	402	28		436	2
5331-00-514-4804	88	12		436	9
5310-00-514-6674	156	10	5940-00-534-0991	431	4
	585	28	5940-00-534-1028	439	4
5310-00-515-7449	100	14		439	8
5310-00-515-9627	234	6	3120-00-537-0614	236	5
	241	6	5330-00-537-2382	116	9
3110-00-516-5289	50	11	4730-00-540-2745	263	21
3120-00-516-7516	207	4		264	20
				264	33
5365-00-518-6592	527	6		269	10
3110-00-519-7833	209	6		269	10
5310-00-521-4534	326	9		271	4
2540-00-521-6179	382	27		598	27
	385	8		599	11
	387	18	3040-00-541-0995	153	21
	387	42	5360-00-541-6501	502	36
5310-00-521-8595	34	1	4730-00-541-7500	182	26
	88	1		268	1
5330-00-522-1174	489	43		274	1
3020-00-522-1175	489	31		275	1
3020-00-522-1177	489	45	4730-00-541-7790	255	2
4720-00-522-1449	482	5		262	21
4730-00-522-1909	620	17	5330-00-542-1329	494	6
5975-00-522-7125	134	53	4730-00-542-5598	263	2
5330-00-522-8544	419	5		581	24
4720-00-522-9151	BULK	41	4730-00-542-5911	629	5
5330-00-523-4235	485	8	5305-00-543-2419	234	24
	491	6		241	24
5340-00-523-4305	485	18		526	15
	491	11	5325-00-543-2902	596	7
5340-00-523-5999	183	11		598	1
5315-00-523-7556	602	17	5305-00-543-4372	33	3
2590-00-525-1352	150	13		39	11
4730-00-526-0284	272	12		125	18
	273	22		189	4
5325-00-526-2663	533	5		233	1
5310-00-527-3634	101	12		399	5
3020-00-528-0511	341	10		457	2
5310-00-528-7638	608	6		471	5
4730-00-529-1487	146	8		508	2
	272	13		531	4
5325-00-530-7968	472	42	5310-00-543-4385	596	9
5940-00-531-0530	137	9	5310-00-543-5933	608	10
5310-00-531-9120	207	39	2530-00-545-1561	317	16
5340-00-532-9231	102	33	5365-00-545-3723	527	5

CROSS-REFERENCE INDEXES

NATIONAL STOCK NUMBER INDEX

STOCK NUMBER	FIG.	ITEM	STOCK NUMBER	FIG.	ITEM
2530-00-545-5406	291	3	2520-00-557-5974	188	4
2510-00-546-4759	359	14	2520-00-557-5980	188	24
			2520-00-557-5981	191	2
5330-00-548-6114	308	8	2520-00-557-6076	192	4
5940-00-549-6581	151	16	2520-00-557-6090	192	3
5940-00-549-6583	151	11	3110-00-557-6163	187	16
5330-00-549-7694	526	22	5325-00-557-6164	192	5
5310-00-550-0284	211	3	5325-00-557-6183	192	5
5310-00-550-1130	136	5	5325-00-557-6207	192	1
	145	7		192	5
	148	7	5325-00-557-6210	192	5
	151	25	2520-00-557-6211	192	19
	162	2		200	29
	163	4		200	30
	370	7	5330-00-557-6518	200	30
	370	32	2520-00-557-6549	188	8
5310-00-550-3503	611	23	5310-00-557-6568	188	1
5310-00-550-3714	587	12	5305-00-557-6612	554	2
	588	5	2520-00-557-6619	183	6
5340-00-550-8070	414	15	3110-00-557-6666	188	11
5310-00-550-8124	88	35	3110-00-557-6708	188	10
5365-00-550-8125	88	35	5310-00-559-0070	150	2
5365-00-550-8127	88	35		367	30
5305-00-551-5097	236	22		441	3
2940-00-552-3842	30	1		587	5
3110-00-554-3184	489	30		591	31
3110-00-554-3411	484	10	5305-00-559-1022	200	32
3110-00-554-3468	484	6	5325-00-559-7574	249	12
3110-00-554-3929	474	25	2540-00-562-0422	362	14
	489	35	5330-00-562-1176	85	13
4730-00-555-1764	258	4		87	14
	267	1		88	37
7690-00-555-6073	572	15	5310-00-562-6557	37	22
4730-00-555-8263	64	8		42	9
	284	48		54	5
	285	8		286	25
4730-00-555-8291	473	19	5310-00-562-6558	37	9
	502	48		284	23
5940-00-557-4344	431	8		284	55
5325-00-557-5794	191	10		286	20
	194	3	5310-00-562-6560	66	7
2520-00-557-5799	192	2	2520-00-563-6041	191	2
2520-00-557-5807	191	4	2520-00-563-8309	234	4
5325-00-557-5835	188	6		241	4
4710-00-557-5885	189	11	5325-00-566-6577	188	18
5325-00-557-5897	190	15	2510-00-567-0130	144	34
2520-00-557-5900	189	9	2990-00-567-4367	88	30
2520-00-557-5924	192	2	5310-00-568-6077	189	27
5310-00-557-5942	200	31	5310-00-568-6118	188	5
5310-00-557-5943	188	3	5305-00-569-8909	352	1
				426	16

CROSS-REFERENCE INDEXES

NATIONAL STOCK NUMBER INDEX

STOCK NUMBER	FIG.	ITEM	STOCK NUMBER	FIG.	ITEM
5306-00-570-8940	196	8	5331-00-579-7927	504	25
	197	15		505	4
5306-00-570-8942	189	14		509	22
5975-00-570-9598	167	24	5331-00-579-8108	192	18
	627	3	4730-00-580-6738	599	12
5340-00-572-8042	88	32	4730-00-580-7408	265	5
5935-00-572-9180	134	61	5330-00-582-0456	185	4
	139	11	5330-00-582-2855	502	38
	140	13		530	10
	143	22	5320-00-582-3268	404	3
	157	6		412	25
	158	18		416	23
	163	11		424	8
	165	17		428	3
	166	13		570	7
	168	15	5320-00-582-3273	558	4
	169	15	5320-00-582-3276	367	28
	170	6		405	28
	170	20		406	40
	171	10		412	3
	172	6		418	24
	173	12		424	6
	174	6	5320-00-582-3301	424	25
	434	5	5320-00-582-3302	404	19
	436	7	5320-00-582-3499	424	28
	496	8	5310-00-582-5965	100	36
	592	27		141	18
	595	13		181	5
	612	4		297	7
	613	7		352	3
5306-00-575-5419	347	13		369	2
5330-00-576-3206	494	4		371	3
5310-00-576-5752	367	38		395	3
	367	46		405	33
6220-00-577-3434	143	1		405	44
	143	1		406	34
5977-00-578-6495	150	12		408	18
5331-00-579-7544	491	9		414	29
5331-00-579-7545	614	3		416	13
5331-00-579-7916	494	2		425	4
5330-00-579-7925	456	31		507	46
	460	28		525	22
	461	26		536	7
	504	17		543	5
	505	31		547	2
	509	21		549	49
5331-00-579-7927	456	32		580	4
	460	29		591	25
	461	27		619	14
	503	10	5310-00-582-6714	213	10

NATIONAL STOCK NUMBER INDEX

STOCK NUMBER	FIG.	ITEM	STOCK NUMBER	FIG.	ITEM
5310-00-582-6714	252	13	4730-00-595-4827	237	6
	253	10	5365-00-595-4948	526	21
	255	4	5310-00-595-5839	474	14
	256	8		489	60
	261	16	5310-00-595-6153	118	28
	262	12	5310-00-595-7237	140	2
	266	11		143	9
	272	5		148	13
5320-00-584-1285	403	12		243	24
	405	39		602	9
	412	19	5310-00-596-7691	143	13
	416	28		143	13
	417	20		198	14
	424	21	5310-00-596-7693	157	10
	427	4		161	28
	428	10		579	2
	570	1	5310-00-596-7753	500	27
5310-00-584-5272	616	11	5310-00-596-7797	500	28
5310-00-584-7888	469	17	5310-00-596-9763	306	6
	470	13	5315-00-597-2723	233	9
	473	3		240	9
	488	7	5360-00-597-4075	507	12
	499	13	5360-00-597-4570	95	3
	502	2		96	2
	507	43	5315-00-597-7399	329	9
5315-00-584-9809	341	13	5340-00-598-0225	585	19
5330-00-585-3210	502	13	5365-00-598-5255	117	5
5331-00-585-8247	505	5		117	8
4820-00-588-8604	129	1		285	3
3120-00-589-3537	26	5		285	6
	26	14	5330-00-599-4517	41	6
2540-00-591-1108	385	1	9390-00-599-6405	367	15
2540-00-592-1823	382	20	3040-00-600-3860	624	4
	385	7	4310-00-603-1510	284	26
	387	17	2530-00-603-5768	292	1
	387	41	5935-00-605-9322	160	11
	609	11	2590-00-606-2383	KITS	
4730-00-595-0083	252	19	4720-00-607-7645	36	3
	255	13	5342-00-613-7784	245	6
	261	22		246	8
	262	5	5325-00-613-7796	305	14
	291	12	5305-00-614-3423	134	6
4730-00-595-2572	263	20		134	27
	264	17		134	67
	264	29	5310-00-614-3505	125	12
	266	16	5365-00-614-3903	527	4
	273	20	5935-00-614-3959	158	2
4820-00-595-2761	263	23	5320-00-616-4350	424	27
4820-00-595-3669	291	8	5315-00-616-5500	483	4
4730-00-595-4827	207	32		484	27

CROSS-REFERENCE INDEXES

NATIONAL STOCK NUMBER INDEX

STOCK NUMBER	FIG.	ITEM	STOCK NUMBER	FIG.	ITEM
5315-00-616-5500	489	33	5310-00-637-4000	316	9
5315-00-616-5501	484	28	5310-00-637-9541	19	16
5315-00-616-5514	207	41		31	12
	310	3		33	27
5315-00-616-5519	412	28		34	2
	526	5		39	10
5315-00-616-5520	391	4		62	5
	513	33		81	13
	525	32		100	4
5315-00-616-5521	473	4		120	2
5315-00-616-5526	125	10		125	2
5315-00-616-5527	21	7		141	7
5310-00-616-6354	236	40		144	21
5331-00-618-0801	494	5		153	30
4730-00-618-8497	198	2		154	3
	277	4		177	10
4730-00-620-6904	198	12		234	25
5340-00-621-2563	361	9		241	25
5330-00-621-2565	362	17		259	8
5340-00-621-2591	410	12		260	8
	420	14		264	24
	629	10		267	6
5935-00-622-2830	160	8		288	5
2510-00-622-3931	385	44		301	11
5342-00-622-3941	366	15		302	3
3040-00-622-3946	367	2		310	11
5340-00-622-7700	359	17		356	11
2540-00-622-7750	385	3		357	3
2590-00-622-7757	382	25		368	5
	387	15		380	11
4730-00-623-8303	598	12		394	3
5325-00-624-0528	607	9		395	6
5310-00-625-5756	243	8		399	6
5340-00-625-9617	362	3		405	19
5331-00-626-8281	425	12		406	9
5310-00-627-6128	105	3		416	9
	107	2		456	29
	156	14		460	26
	163	34		461	23
	165	39		469	4
	167	30		471	4
	168	29		472	39
	169	38		476	13
	585	11		477	9
2590-00-630-1567	416	6		486	3
5330-00-632-3813	KITS			495	8
5340-00-632-6239	24	4		500	4
4820-00-633-3523	264	27		504	14
	266	37		505	21
9905-00-634-5269	563	6		507	14
6210-00-635-5686	131	4			

CROSS-REFERENCE INDEXES

NATIONAL STOCK NUMBER INDEX

STOCK NUMBER	FIG.	ITEM	STOCK NUMBER	FIG.	ITEM
5310-00-637-9541	507	33	5975-00-660-5962	167	4
	509	18		168	16
	531	5		169	16
	549	11		169	23
	551	6		170	25
	551	9		170	30
	596	12		171	12
5305-00-638-0957	367	5		182	33
5305-00-638-2362	200	22		211	41
5305-00-638-8920	107	4		213	26
4730-00-640-0264	205	3	3120-00-661-6646	33	11
5330-00-641-2466	234	26		33	16
	241	26		33	25
4730-00-647-3207	276	10	3120-00-661-9026	472	30
	598	15	3120-00-662-3370	234	20
	620	20		241	20
3110-00-649-9498	306	33	5360-00-664-4374	474	15
5310-00-650-0187	284	68		489	5
2510-00-650-1015	363	1	5360-00-664-5343	32	10
4820-00-652-5548	469	7		33	8
	499	5	5340-00-664-9863	318	13
2510-00-654-4611	382	23			
5310-00-655-9544	153	31	4720-00-670-6037	BULK	18
5310-00-655-9860	158	3	2510-00-674-4487	363	1
5310-00-656-0067	97	14	5365-00-674-6831	463	3
	97	20		464	3
	144	10		511	3
	144	15	4730-00-676-7566	461	16
	151	9		509	10
	152	3	5325-00-678-8435	158	21
	158	28	2510-00-679-1420	409	22
	165	7	5342-00-679-1446	414	27
	167	12	5342-00-679-1447	425	21
	170	33	5340-00-679-1492	414	25
	170	41		419	16
	171	3	5340-00-679-1494	414	17
	180	5	5340-00-679-1495	414	17
	182	37	2510-00-679-1733	409	20
	211	7	5310-00-679-8346	223	10
	211	44	5310-00-680-5956	344	4
5310-00-656-0114	615	2		345	12
5330-00-659-3178	42	5	5315-00-682-2207	416	32
9905-00-659-7754	572	14	4820-00-684-0880	99	4
9905-00-659-7755	570	4	5340-00-685-0567	457	15
9905-00-659-7757	570	6		458	26
3120-00-659-7808	23	8	5935-00-686-2599	160	15
5975-00-660-5962	158	41	5340-00-689-7213	383	32
	163	16		384	7
	165	11		386	15
	166	4	3110-00-689-8250	293	20

CROSS-REFERENCE INDEXES

NATIONAL STOCK NUMBER INDEX

STOCK NUMBER	FIG.	ITEM	STOCK NUMBER	FIG.	ITEM
3110-00-689-8250	295	14	6210-00-699-9458	132	10
5935-00-691-5591	97	12	5365-00-700-1851	346	20
	97	19	5310-00-700-7089	234	13
	144	9		241	13
	144	14		293	12
	151	8		295	2
	152	4		348	27
	158	27	5310-00-700-7127	347	9
	165	8	2930-00-701-2091	31	5
	167	13	4730-00-701-7677	215	1
	170	32		288	8
	170	42		290	4
	171	4	5340-00-702-2848	161	10
	180	6		167	31
	182	38		582	14
	211	6		584	21
	211	45	5331-00-702-5220	494	1
5365-00-692-6119	234	16		494	9
	241	16		494	19
3040-00-692-6121	241	16	5331-00-702-5643	493	26
3040-00-692-6123	241	16	5360-00-703-6587	316	5
3120-00-692-6153	211	10	5360-00-704-4253	329	5
2590-00-693-0589	534	16	2815-00-705-2851	23	10
2510-00-693-0591	534	12	5970-00-705-6639	154	19
2510-00-693-0592	534	12	6145-00-705-6674	BULK	46
2540-00-693-0602	362	15	6145-00-705-6678	BULK	44
2540-00-693-0603	362	15	5940-00-705-6701	432	8
5340-00-693-0604	362	20	5940-00-705-6702	158	12
2510-00-693-0607	359	12		158	38
2510-00-693-0608	367	14	5940-00-705-6707	160	4
2510-00-693-0610	367	18	5940-00-705-6708	134	49
2540-00-693-0676	327	7	5940-00-705-6709	160	9
2530-00-693-1029	293	2		496	9
	295	16	5940-00-705-6711	163	15
4730-00-695-1133	35	7		165	16
5365-00-695-1247	37	5		166	10
2590-00-695-9076	169	20		167	9
5340-00-696-0264	366	8		168	13
5340-00-696-0265	366	8		169	13
5315-00-696-0789	473	18	5940-00-705-6715	158	11
5305-00-696-5285	361	4		158	39
5305-00-696-5291	282	16	5940-00-705-6730	154	14
5975-00-697-7769	160	12		154	18
5360-00-698-7100	88	31	4030-00-706-5553	499	22
5340-00-699-8463	402	1	5340-00-709-5879	376	7
5930-00-699-9438	132	2	5305-00-709-8482	125	7
	134	63	5305-00-709-8516	243	22
	611	30	5305-00-709-8523	506	14
	612	23	5305-00-709-8537	615	10
	613	21	5305-00-709-8542	509	34

NATIONAL STOCK NUMBER INDEX

STOCK NUMBER	FIG.	ITEM	STOCK NUMBER	FIG.	ITEM
4730-00-710-5571	460	6	5305-00-719-5221	324	38
2910-00-710-6054	422	5		351	21
	423	5		368	10
5340-00-711-5372	88	36		368	13
2540-00-715-7407	536	12		374	26
5340-00-716-4975	51	13		391	15
	56	11		508	6
	88	6		619	27
5305-00-716-7454	237	12	5305-00-719-5235	67	10
5305-00-716-8174	389	8		68	4
5305-00-716-8181	340	15		69	14
5305-00-716-8186	336	19		102	15
	611	9		106	17
5305-00-717-3999	348	8		120	4
4730-00-718-2621	629	27		304	3
5340-00-719-4601	66	5		318	46
5305-00-719-5184	107	28		319	22
5305-00-719-5219	67	10		320	4
	69	14		321	4
	102	10		322	4
	106	13		323	4
	107	29		324	4
	155	5		327	8
	308	23		327	8
	319	11		330	5
	326	18		331	7
	346	1		332	6
	351	23		333	15
	365	2		334	17
	383	35		336	12
	384	2		337	19
	386	10		338	16
	389	2		368	10
	458	6		374	18
	484	2		381	13
	488	2		391	26
	489	20		456	33
	492	21		460	30
	502	28		461	28
	584	23		468	3
	602	19		469	16
	615	5		503	34
	619	22		506	1
5305-00-719-5221	69	5		509	23
	110	16		616	6
				616A	15
	318	4	5305-00-719-5238	102	18
	319	4		102	19
	320	37		250	34.1
	321	39		318	16
	323	41		318	39

CROSS-REFERENCE INDEXES

NATIONAL STOCK NUMBER INDEX

STOCK NUMBER	FIG.	ITEM	STOCK NUMBER	FIG.	ITEM
5305-00-719-5238	319	18	5306-00-720-8747	227	13
	320	30		228	17
	321	16		229	12
	322	19	5320-00-721-5384	416	24
	323	21	5305-00-721-5492	81	12
	324	3		107	11
	328	7		205	9
	332	7	5306-00-721-5944	385	48
	334	7	5310-00-721-7809	125	22
	336	7	5310-00-721-8000	596	8
	337	13	4730-00-722-2759	589	51
	338	27		590	36
	376	11	5305-00-723-9383	454	5
	381	11		528	10
	456	34	5305-00-724-5830	489	53
	457	17	5305-00-724-5898	472	32
	461	29	5305-00-724-5910	402	5
	509	24		473	11
	589	22		483	6
	590	15	5305-00-724-5939	5	3
	619	1	5305-00-724-6810	469	12
5305-00-719-5239	321	23		470	8
	322	24	5340-00-724-7038	422	8
	330	12		423	8
	331	8	5305-00-724-7206	469	30
	513	9		470	5
	525	9	5305-00-724-7219	236	28
	616	10		403	15
5305-00-719-5240	320	49		469	8
	321	54		470	2
	340	9	5305-00-724-7221	127	6
	349	4		177	4
5305-00-719-5241	615	11		178	4
5305-00-719-5262	615	13		489	21
5305-00-719-5270	5	5		499	6
5305-00-719-5274	390	8	5305-00-724-7222	1	5
	503	19		206	3
5305-00-719-5275	495	2		469	25
	502	40		470	22
5305-00-719-5342	407	6		502	43
	415	10		503	23
4820-00-720-4488	108	14		613	17
5310-00-720-7627	232	11	5305-00-724-7223	236	27
	239	11	5305-00-724-7225	345	4
5306-00-720-8747	218	12	5305-00-724-7264	6	1
	220	17	5305-00-724-7265	6	10
	223	15	5305-00-724-7266	6	28
	224	2	5305-00-725-0140	318	41
	225	8		319	41
	226	2		320	43

CROSS-REFERENCE INDEXES

NATIONAL STOCK NUMBER INDEX

STOCK NUMBER	FIG.	ITEM	STOCK NUMBER	FIG.	ITEM
5305-00-725-0140	321	43	5305-00-725-4183	396	9
	322	46		468	13
	323	45		484	31
	324	40		501	17
5305-00-725-0145	318	45		503	31
	319	46	5340-00-725-5267	252	9
	320	46		549	35
	321	47	5340-00-725-5280	169	41
	322	50	5340-00-725-6033	153	4
	323	49	5340-00-726-1670	134	20
	324	46	6220-00-726-1916	143	1
5305-00-725-0154	202	5	5305-00-726-2525	305	21
	318	33		471	24
	318	35			
	319	3	5305-00-726-2543	78	21
	319	36	5305-00-726-2544	476	9
	320	33	5305-00-726-2550	318	27
	320	38		319	28
	321	35		320	26
	322	37		321	27
	322	39		322	28
	323	37		323	31
	324	34		324	23
	468	8		327	5
	487	9		328	5
	492	22		331	14
	508	23		332	13
	619	26		343	5
5305-00-725-0168	472	12		389	6
5305-00-725-0194	502	17		393	5
5305-00-725-0197	473	30		484	33
5310-00-725-1983	207	11	5305-00-726-2551	326	2
	237	13		328	4
5305-00-725-2317	310	10		330	9
5305-00-725-4138	417	7		343	11
5305-00-725-4183	67	12		390	15
	69	4		488	8
	326	10	5305-00-726-2552	303	7
	326	10		330	7
	327	6		343	10
	332	7		381	16
	333	9		400	9
	334	20	5305-00-726-2553	343	17
	335	18		400	9
	336	11	5305-00-726-2556	207	40
	337	21	5305-00-726-2559	613	37
	338	8	5305-00-726-2561	5	6
	389	5	5305-00-726-2567	616	1
	391	6	5305-00-726-3091	409	18
	391	14	4820-00-726-4719	182	15
				230	10

CROSS-REFERENCE INDEXES

NATIONAL STOCK NUMBER INDEX

STOCK NUMBER	FIG.	ITEM	STOCK NUMBER	FIG.	ITEM
4820-00-726-4719	237	15	9905-00-733-7622	555	2
	248	14	5331-00-733-9765	317	7
	278	11	5330-00-733-9766	317	2
	279	12	2520-00-734-6802	234	3
	474	7		241	3
	489	12	2530-00-734-6815	236	33
5305-00-727-2283	326	17	5306-00-734-6817	233	12
	330	11		240	12
	487	6	5365-00-734-6818	232	16
5975-00-727-5153	162	12		239	16
5310-00-727-8353	84	3	2530-00-734-6819	233	13
	86	3		240	13
5310-00-728-2044	137	4	5340-00-734-6820	233	3
	592	21		240	3
	594	16	5306-00-734-6822	232	14
5305-00-728-5475	490	17		239	14
4820-00-728-7467	254	3	3110-00-734-6877	235	9
	256	18		242	9
	290	13	5330-00-734-6878	235	5
	604	12	5365-00-734-6879	235	5
5330-00-729-4427	14	16		242	5
4030-00-729-6054	340	8	5365-00-734-6880	235	5
6220-00-729-9295	143	2	3020-00-734-6881	234	27
5305-00-732-0512	177	5		241	27
	178	1	3120-00-734-6883	233	6
	589	56		240	6
5310-00-732-0558	141	8	5330-00-734-6886	234	34
	144	20		241	34
	380	1	3040-00-734-6892	234	18
	381	1		241	18
	463	2	3110-00-734-6895	234	21
	464	2		241	21
	511	2	3040-00-734-6897	234	7
5310-00-732-0559	148	14		241	7
	301	10	5330-00-734-6899	233	16
	301	15		240	16
	402	18	5307-00-734-6900	236	34
	528	22	4320-00-734-6951	236	29
5310-00-732-0560	26	3	2510-00-734-6952	350	4
	26	7	3040-00-734-6953	350	10
	26	12	5340-00-734-6954	350	8
	318	25	5310-00-734-6957	350	7
	346	25	2520-00-734-6959	174	9
	348	7		KITS	00
	468	11	2520-00-734-6960	237	5
	469	26	2520-00-734-6970	232	3
	501	19		239	3
5340-00-732-0642	144	3	2530-00-734-6971	236	12
5365-00-732-6126	230	4	5340-00-734-6974	236	23
	237	3	2520-00-734-6975	236	20

CROSS-REFERENCE INDEXES

NATIONAL STOCK NUMBER INDEX

STOCK NUMBER	FIG.	ITEM	STOCK NUMBER	FIG.	ITEM
2530-00-734-6977	236	25	2540-00-737-3296	363	8
2530-00-734-6978	236	25	2540-00-737-3297	363	8
2530-00-734-6982	236	4	2540-00-737-3298	362	1
2520-00-734-6984	236	6	9390-00-737-3300	363	2
2520-00-734-6985	236	6	9390-00-737-3301	363	6
2520-00-734-6991	293	6	5340-00-737-3302	361	3
5340-00-734-6992	293	7	2540-00-737-3304	362	25
			9390-00-737-3317	359	18
			2510-00-737-3326	367	12
2510-00-734-6998	347	11	5340-00-737-3330	367	40
2520-00-734-7548	233	5	2540-00-737-6203	326	8
	240	5	5330-00-737-6584	177	1
2520-00-734-8101	223	12	5940-00-738-6272	151	27
2520-00-734-9606	526	17	5340-00-738-7552	249	31
5330-00-735-1272	150	11	2530-00-738-9061	292	4
5940-00-735-5520	151	7	5306-00-739-7754	153	29
	151	21	5325-00-739-7776	161	8
	152	11		334	16
5310-00-735-7460	88	7		337	18
2510-00-736-8622	366	3		338	15
5315-00-737-0134	382	42			
	383	7	2510-00-740-9002	391	25
	387	35	2510-00-740-9008	391	24
2510-00-737-2711	534	8	2510-00-740-9013	391	7
2510-00-737-2712	534	8	5315-00-740-9017	390	5
5330-00-737-2720			5315-00-740-9020	507	29
5330-00-737-2722	534	15		534	14
5340-00-737-2788	382	40	6680-00-740-9026	457	8
	385	46		458	3
	387	31		507	39
	609	20	2520-00-740-9037	507	34
2540-00-737-2790	382	19	2520-00-740-9040	507	25
	385	11	2520-00-740-9041	507	30
	387	23	3120-00-740-9042	507	27
	387	44	5315-00-740-9043	390	22
5360-00-737-2792	380	4	5315-00-740-9045	390	17
	381	7	5330-00-740-9050	507	16
	400	13	4730-00-740-9058	507	7
5360-00-737-2793	380	5	5365-00-740-9079	507	6
	381	8	2540-00-740-9150	469	28
	400	12		472	36
5325-00-737-3246	606	13		473	13
2540-00-737-3276	362	13		499	18
2540-00-737-3277	362	13			
5340-00-737-3283	362	23	2510-00-740-9212	391	20
2540-00-737-3286	362	12	3120-00-740-9299	507	31
2510-00-737-3287	362	10	5330-00-740-9312	348	17
2510-00-737-3293	361	5	5340-00-740-9335	346	4
2510-00-737-3294	361	5		347	1
2510-00-737-3295	363	3	2510-00-740-9337	318	43
				319	44
				320	44

CROSS-REFERENCE INDEXES

NATIONAL STOCK NUMBER INDEX

STOCK NUMBER	FIG.	ITEM	STOCK NUMBER	FIG.	ITEM
2510-00-740-9337	321	44	5340-00-740-9665	501	9
	322	45	3120-00-740-9666	502	12
	323	47	2520-00-740-9667	502	3
	324	44	5340-00-740-9668	501	16
2510-00-740-9340	346	16	3040-00-740-9669	502	11
5340-00-740-9341	346	16	2520-00-740-9673	502	25
2540-00-740-9343	346	26	5306-00-740-9677	473	26
2510-00-740-9344	346	7	5340-00-740-9684	473	21
5340-00-740-9361	243	11	3120-00-740-9685	473	32
5340-00-740-9366	243	14	2510-00-740-9686	534	2
3040-00-740-9372	243	19	2510-00-740-9687	534	2
5340-00-740-9391	348	24	2520-00-740-9694	472	8
2530-00-740-9445	252	21	3120-00-740-9695	472	16
5305-00-740-9507	326	3	3040-00-740-9725	472	7
2510-00-740-9508	318	50	5360-00-740-9727	469	11
	319	53		470	7
	320	51	5310-00-740-9728	472	14
	321	50	3120-00-740-9729	472	5
	322	54		475	9
	323	52		490	11
	324	49	5365-00-740-9784	398	8
2510-00-740-9534	333	4	3120-00-740-9800	473	27
	335	4		474	32
	337	4		489	13
2510-00-740-9535	333	17	5306-00-740-9803	474	30
	337	22	2520-00-740-9805	472	44
2590-00-740-9553	295	12	3120-00-740-9807	472	15
	348	19		487	19
3040-00-740-9565	474	22		488	5
	489	61	3110-00-740-9809	474	3
3120-00-740-9591	473	35	9520-00-740-9812	473	1
2510-00-740-9596	534	1	2520-00-740-9813	470	4
2540-00-740-9597	534	1	2520-00-740-9814	473	33
5330-00-740-9600	348	25	2520-00-740-9816	474	19
2510-00-740-9601	348	5		489	2
5360-00-740-9602	348	4	2530-00-740-9817	474	13
5330-00-740-9606	348	18		489	6
4710-00-740-9607	348	15	5330-00-740-9821	474	27
5306-00-740-9608	348	34		489	14
5315-00-740-9609	472	4	3020-00-740-9823	474	33
5365-00-740-9612	348	6		489	25
2510-00-740-9613	348	1	5315-00-740-9835	398	4
5310-00-740-9615	348	16	5310-00-740-9862	398	10
2510-00-740-9617	349	7	5340-00-740-9883	343	12
5365-00-740-9618	349	3	5340-00-740-9884	343	8
5315-00-740-9619	349	5	5330-00-740-9889	500	8
5310-00-740-9621	350	3	5340-00-740-9892	500	10
2540-00-740-9660	330	1	3020-00-740-9893	500	3
2510-00-740-9661	330	6	3110-00-740-9896	500	13
2530-00-740-9663	502	27	5360-00-740-9903	345	45

CROSS-REFERENCE INDEXES

NATIONAL STOCK NUMBER INDEX

STOCK NUMBER	FIG.	ITEM	STOCK NUMBER	FIG.	ITEM
2530-00-740-9925	502	31	5330-00-741-1160	471	13
5310-00-740-9928	476	4	3040-00-741-1161	472	40
	477	4	3120-00-741-1164	472	25
5330-00-740-9929	471	19	2520-00-741-1166	472	33
	501	2	3120-00-741-1170	472	19
2520-00-740-9930	501	6	2520-00-741-1410	513	7
5330-00-740-9932	502	26		525	7
5330-00-740-9933	502	45			
3020-00-740-9934	502	47	2510-00-741-2376	346	10
2520-00-740-9935	501	10	5330-00-741-2636	223	11
2520-00-740-9939	476	1	5315-00-741-2924	326	19
	477	5		327	9
3040-00-740-9945	501	13		327	9
3040-00-740-9947	476	11	2510-00-741-3458	333	2
	477	1		335	2
3020-00-740-9948	501	4		337	2
3040-00-740-9950	500	16	5325-00-741-4180	144	26
3020-00-740-9952	500	30	5315-00-741-5746	243	7
3120-00-740-9953	500	1	5310-00-741-6862	144	22
2540-00-740-9954	499	2	5306-00-741-6865	144	30
3120-00-740-9958	500	18	5310-00-741-6867	144	32
5330-00-740-9959	500	19	5340-00-741-6869	144	17
3120-00-740-9960	500	21	5306-00-741-7084	499	3
5310-00-740-9962	500	22	5330-00-741-7093	472	20
5310-00-740-9975	501	1	5330-00-741-7094	472	6
5310-00-740-9977	502	14	6250-00-741-8960	134	47
3120-00-740-9978	476	5	5315-00-741-8971	326	22
	477	3		327	11
5310-00-740-9979	502	19		339	31
2540-00-740-9980	499	1		340	17
2540-00-741-0715	362	8		383	17
				395	14
				616A	9
5342-00-741-1068	207	37	3040-00-745-7685	473	5
2530-00-741-1105	293	3	5340-00-752-1372	216	7
2510-00-741-1110	346	17	5315-00-752-1651	230	7
2510-00-741-1112	346	22		237	10
5306-00-741-1113	347	13	5305-00-752-1693	236	18
2510-00-741-1114	348	1	5305-00-752-1718	233	7
5360-00-741-1115	348	5		240	7
2510-00-741-1116	348	4	5360-00-752-1975	513	3
3020-00-741-1121	471	6		525	3
2590-00-741-112Z	469	1	9905-00-752-4649	97	11
3110-00-741-1123	471	15		97	22
3110-00-741-1124	471	17		131	25
3120-00-741-1125	471	8		134	16
2520-00-741-1142	237	5		134	62
5310-00-741-1153	471	22		143	5
5330-00-741-1154	471	2		157	5
5306-00-741-1155	471	32		158	6
3120-00-741-1156	471	12		170	5
5330-00-741-1159	471	28		170	14

CROSS-REFERENCE INDEXES

NATIONAL STOCK NUMBER INDEX

STOCK NUMBER	FIG.	ITEM	STOCK NUMBER	FIG.	ITEM
9905-00-752-4649	170	21	2910-00-753-9184	100	26
	170	26		101	3
	170	31	5340-00-753-9214	353	2
	170	43	5306-00-753-9242	266	26
	171	11	4730-00-753-9300	214	5
	172	5	2590-00-753-9545	142	9
	172	13	2510-00-753-9657	361	11
	172	19	5310-00-753-9688	624	6
	173	5	5330-00-753-9689	624	3
	173	11	2540-00-754-0419	361	10
	173	20	5325-00-754-1072	549	12
	173	26	5310-00-754-2005	346	14
	174	5		350	6
	211	5	5305-00-754-4355	580	8
	211	42		593	22
	430	4	2520-00-755-7336	301	9
	431	3	4730-00-755-7609	456	30
	433	8		460	27
	435	3		461	25
	436	4		504	16
	440	4		505	28
	440	7		509	19
	496	2	5340-00-757-5877	370	11
	614	18	5340-00-757-5901	370	14
5305-00-752-5938	144	5	5305-00-757-6567	66	14
3110-00-752-7760	500	6	4010-00-757-9556	BULK	5
4820-00-752-9040	71	20	5360-00-758-6456	244	8
	74	19	5310-00-760-7493	539	13
	75	15	5310-00-761-6882	298	13
	116	12		370	31
	124	10		383	22
	596	20		385	29
5340-00-753-3741	378	10		408	19
5340-00-753-3742	602	10		414	6
5306-00-753-6996	382	24		536	8
	385	5		580	3
	385	26		584	6
	387	14		592	4
	609	9		594	36
5310-00-753-8215	484	16		612	33
5360-00-753-8706	469	31		613	13
	472	34	5305-00-762-6041	331	12
	473	17		332	10
	499	20	5310-00-763-8905	144	33
5306-00-753-8725	473	15		303	3
2530-00-753-8726	473	36		391	27
5330-00-753-9072	82	17		417	24
	145	5	5310-00-763-8911	236	41
	145	5		345	26
2540-00-753-9114	100	13	5310-00-763-8919	236	16

CROSS-REFERENCE INDEXES

NATIONAL STOCK NUMBER INDEX

STOCK NUMBER	FIG.	ITEM	STOCK NUMBER	FIG.	ITEM
5310-00-763-8919	237	14	5935-00-773-1428	165	36
5310-00-763-8920	1	3		166	27
	206	7		167	20
	403	17		168	26
5310-00-763-8921	499	14		169	35
5310-00-763-8922	502	15	2910-00-773-2108	88	16
5340-00-764-7051	606	5	6145-00-774-4579	BULK	2
5340-00-764-7052	161	9	4210-00-775-0127	610	1
	608	1	5331-00-776-2830	204	3
4730-00-765-9102	269	31	5310-00-776-4909	329	16
3120-00-766-3327	379	7	5310-00-776-7670	185	22
	409	15	5315-00-777-3544	23	6
5340-00-766-3330	402	6		23	12
	413	23	4030-00-780-9350	326	20
5340-00-766-6336	406	27		327	13
5935-00-767-7936	160	19		339	25
5310-00-768-0318	2	2		395	10
	176	4		395	15
	370	22		419	36
				616A	6
	470	27	5315-00-781-2026	12	6
	493	16	5330-00-781-7774	189	2
5310-00-768-0319	141	19	4730-00-782-5461	214	15
	141	26	4730-00-782-7102	538	5
	148	8		539	18
				540	14
	184	13	5330-00-785-7894	31	3
	245	3	2510-00-786-4631	507	49
	314	6	2540-00-788-5637	543	22
	507	45	5945-00-789-3706	136	4
3120-00-770-2941	346	8	2540-00-789-6192	146	10
	346	24		182	25
5310-00-771-4911	329	14	2910-00-790-8736	95	4
5340-00-771-6428	578	2		96	1
5975-00-771-6634	158	20	3120-00-791-1440	23	14
	160	16	2815-00-791-1448	21	5
	164	3	2815-00-791-1453	33	9
	165	3	3120-00-792-9834	44	6
	166	15	6680-00-795-2641	625	2
	167	15	5360-00-795-6975	317	24
	168	3	5305-00-795-9308	33	37
	169	3	5305-00-795-9341	33	35
5935-00-772-0495	134	50	5305-00-795-9343	3	2
5365-00-772-2322	157	13		118	19
	158	4	5305-00-795-9345	66	10
5365-00-772-2343	134	14	5305-00-795-9352	33	36
	134	51	2540-00-797-5609	362	18
5935-00-772-2344	160	13	2530-00-797-9224	350	5
5935-00-772-2353	157	12	2510-00-797-9305	348	32
5935-00-772-3307	160	14	1015-00-798-2997	134	19
2815-00-772-9434	7	23		149	10
5935-00-773-1428	163	13		160	21

CROSS-REFERENCE INDEXES

NATIONAL STOCK NUMBER INDEX

STOCK NUMBER	FIG.	ITEM	STOCK NUMBER	FIG.	ITEM
1015-00-798-2997	163	7	5305-00-804-6318	56	19
	166	2		553	2
	167	2		555	1
	168	7		556	1
	169	9	4730-00-805-2222	313	2
	172	2	6145-00-805-3354	BULK	45
	172	10	5325-00-806-4104	317	8
	172	16	5325-00-806-4105	317	4
			5310-00-807-1466	602	31
	173	2	5310-00-807-1467	606	12
	173	8	5325-00-807-2636	85	14
	173	17		87	15
	173	23	5935-00-807-4109	141	14
	174	4		144	18
	432	2	5331-00-807-8993	317	11
	432	10	5331-00-808-0794	314	7
	436	11		503	35
5330-00-798-4635	317	6	6220-00-808-6072	136	1
5330-00-798-4637	317	10	4820-00-808-6905	182	13
5340-00-799-0843	116	10	5331-00-808-7612	480	22
5340-00-799-2218	409	3		482	9
3110-00-799-4903	197	3	5340-00-809-1492	164	23
5310-00-800-0695	236	38		165	32
	343	6		166	24
2590-00-801-2355	137	2		214	7
5330-00-801-3440	494	3		313	9
4730-00-801-8186	37	24	5340-00-809-1494	165	35
4730-00-802-2560	213	15		269	2
	596	1		621	4
4730-00-802-2818	629	5	5340-00-809-1500	145	11
	629	24		156	17
2910-00-803-2631	85	4		584	15
4730-00-803-5765	182	1	4720-00-809-2430	549	1
4730-00-803-5776	277	2	5331-00-809-2667	52	6
4730-00-803-6266	584	9	5310-00-809-3078	50	20
5325-00-803-7299	489	44		55	20
5310-00-804-1209	539	11		109	6
5325-00-804-2025	249	30		370	12
	516	23		411	7
5306-00-804-2468	7	20		579	8
5325-00-804-2784	17	8		604	13
5325-00-804-4997	200	8		606	14
5331-00-804-5694	480	14	5310-00-809-3079	176	5
	493	18		383	6
	493	20		419	21
	493	22		602	2
	505	54	5331-00-809-3276	52	8
5331-00-804-5695	297	2	5310-00-809-4058	100	46
				299	7
	313	5		145	12
	508	13		161	19
5305-00-804-6318	51	19		181	25
				255	16

CROSS-REFERENCE INDEXES

NATIONAL STOCK NUMBER INDEX

STOCK NUMBER	FIG.	ITEM	STOCK NUMBER	FIG.	ITEM
5310-00-809-4058	543	13	2510-00-809-8046	407	12
	549	6		415	4
	616A	7			
5310-00-809-4061	125	3	5310-00-809-8533	211	12
	310	14		279	1
	354	8		297	13
	376	9		389	12
	377	6		508	12
	511	56	5310-00-809-8536	349	2
	600	3		471	29
	600	3	5310-00-809-8540	304	5
	600	3	5310-00-809-8541	499	25
	616A	21			
5310-00-809-4085	125	5	5310-00-809-8546	364	4
	540	10	4730-00-809-9427	254	5
5310-00-809-5997	125	15		257	2
5310-00-809-5998	106	10		269	30
	144	29		505	43
	198	19		508	5
	318	36	5315-00-810-3701	187	7
	319	37		333	21
	320	34		334	25
	321	36		335	19
	322	40		336	20
	323	38		337	31
	324	35		338	29
	333	10	5970-00-811-0640	154	15
	334	11	5315-00-811-1241	528	1
	335	17	5360-00-811-1609	312	27
	336	18	5315-00-812-3427	525	18
	337	14	4730-00-812-7999	212	16
	338	9		278	14
	339	3		288	14
	389	3		603	13
	391	16	5320-00-813-4144	417	21
	396	6		424	7
	437	20	4730-00-813-7811	279	24
	439	20	4730-00-813-9611	253	12
	458	9		254	4
	468	5		278	8
	468	5	5310-00-814-0672	6	23
	495	4		83	1
	502	39		102	6
	503	25		102	11
	508	22		102	30
	591	39		105	14
5365-00-809-6292	345	40		106	19
4010-00-809-6294	390	2		142	20
2540-00-809-7792	416	31		153	23
2540-00-809-7793	405	15		157	16
2540-00-809-7796	406	45		161	6
5306-00-809-7824	416	8		163	1

CROSS-REFERENCE INDEXES

NATIONAL STOCK NUMBER INDEX

STOCK NUMBER	FIG.	ITEM	STOCK NUMBER	FIG.	ITEM
5310-00-814-0672	163	20	5310-00-814-0672	351	5
	163	22		354	6
	165	24		357	4
	168	20		359	1
	202	6		364	8
	211	21		365	7
	213	36		368	17
	215	12		374	7
	218	13		375	8
	220	16		376	15
	222	10		377	5
	223	1		378	5
	224	1		379	3
	225	1		388	3
	226	1		388	7
	227	1		390	19
	228	16		397	8
	229	13		399	12
	243	2		400	2
	253	7		401	3
	254	19		454	16
	254	19		456	12
	257	7		457	13
	258	3		457	22
	261	6		460	11
	261	27		461	9
	263	15		478	11
	264	6		480	28
	264	23		481	11
	266	31		492	4
	267	14		509	5
	269	5		513	5
	271	19		513	19
	273	24		513	24
	301	1		525	5
	302	4		525	30
	316	16		536	13
	318	19		536	20
	319	32		537	8
	320	29		581	6
	324	31		583	8
	333	7		584	25
	334	6		586	12
	335	8		589	26
	336	6		590	14
	337	7		592	17
	337	26		600	4
	338	6		600	4
	338	23		600	4
	343	14		600	7

NATIONAL STOCK NUMBER INDEX

STOCK NUMBER	FIG.	ITEM	STOCK NUMBER	FIG.	ITEM
5310-00-814-0672	603	4	5310-00-820-6653	502	44
	604	10		503	22
				616A	29
	629	5		552	45
5310-00-814-0673	108	6		613	16
	263	11	3020-00-820-7914	44	2
	583	6	3020-00-820-7915	33	33
	589	31	2540-00-821-2277	329	6
5315-00-814-3530	100	33	5340-00-821-2364	34	5
	101	12	5305-00-821-3869	596	11
6685-00-814-5271	146	4	4730-00-822-5609	503	36
	146	6	5310-00-822-8525	369	6
	147	1	5325-00-823-5999	586	6
2815-00-815-0355	118	11	5325-00-823-6002	96	5
5315-00-815-1405	454	8	5310-00-823-8803	297	12
	528	4		358	2
	KITS	11		383	27
5315-00-816-1794	341	18		384	12
	414	13		386	19
5306-00-816-2441	380	10		391	10
5331-00-816-3546	480	10		503	33
	493	12		613	38
	503	1	5310-00-823-8804	112	3
4730-00-817-6578	97	4		113	3
	549	45		315	13
	550	2		413	24
	581	19		598	35
5310-00-820-6653	1	2	6680-00-825-2076	625	7
	5	4	5315-00-826-3251	101	16
	5	9	5330-00-826-5202	484	5
	14	4	5330-00-826-5203	484	15
	177	3	4720-00-826-5606	480	33
	178	5	4720-00-826-5607	480	30
	206	6	4720-00-826-5610	480	25
	236	26		482	3
	303	4	3120-00-826-5630	485	9
	345	5		491	7
	402	4	5340-00-827-2453	422	10
	403	16		423	10
	417	8	4730-00-827-5852	504	2
	469	9		504	11
	470	3	5365-00-827-6452	341	8
	471	23	5340-00-827-8314	100	50
	473	12		101	7
	474	29	2815-00-828-7013	33	6
	476	8	6685-00-828-7126	52	1
	483	5	5365-00-829-5150	KITS	
	484	32	2815-00-829-5227	42	6
	487	7	5310-00-829-5238	12	8
	489	22	4820-00-829-5600	129	7
	499	7	2910-00-829-5616	52	5

CROSS-REFERENCE INDEXES

NATIONAL STOCK NUMBER INDEX

STOCK NUMBER	FIG.	ITEM	STOCK NUMBER	FIG.	ITEM
5340-00-829-5617	52	10	5935-00-833-8561	496	3
5310-00-829-9981	62	11		592	24
3110-00-830-8802	185	15		593	15
5360-00-832-0178	370	23		594	27
3030-00-832-4312	310	9		608	12
5325-00-832-5650	141	21		612	3
5305-00-832-5743	141	20		613	6
5310-00-832-6852	134	4		614	10
	134	29		614	17
	134	65	5970-00-833-8562	131	23
2540-00-832-7027	330	10		134	18
2530-00-832-7123	236	19		134	56
4820-00-832-8077	212	4		149	9
	279	20		157	3
5310-00-832-9719	349	1		158	35
5340-00-833-0342	163	18		160	20
5342-00-833-1236	62	22		163	8
5340-00-833-3049	608	22		166	3
	618	10		167	3
5340-00-833-7966	33	7		168	8
4730-00-833-8230	458	30		169	8
5340-00-833-8476	611	28		170	3
5935-00-833-8561	134	17		170	11
	134	31		170	24
	134	55		170	29
	149	8		171	6
	157	2		172	3
	158	34		172	11
	160	6		172	17
	163	9		173	3
	166	6		173	9
	167	10		173	18
	168	9		173	24
	169	7		174	3
	170	4		182	34
	170	12		211	34
	171	5		211	40
	172	4		213	27
	172	12		432	3
	172	18		432	11
	173	4		434	3
	173	10		436	12
	173	19		496	4
	173	25		608	13
	174	2		614	11
	211	35		614	16
	432	4	5310-00-833-8567	134	60
	432	12		139	10
	434	4		140	12
	436	13		143	23

NATIONAL STOCK NUMBER INDEX

STOCK NUMBER	FIG.	ITEM	STOCK NUMBER	FIG.	ITEM
5310-00-833-8567	157	7	5315-00-839-5821	501	7
	166	12	5315-00-839-5822	383	26
	170	7		384	10
	170	19		386	20
	171	9		419	17
	172	7	5315-00-841-4443	370	39
	173	13	9905-00-841-4445	431	7
	174	7		434	7
	436	6		435	8
	496	7		436	15
4730-00-833-9315	508	24			
4730-00-834-6187	507	13		439	2
5310-00-834-7606	351	20		439	7
5310-00-834-8732	414	19	5315-00-842-3044	100	20
	469	6		243	9
	500	24		371	6
5310-00-834-8734	383	3		382	43
	537	4		383	8
5310-00-834-8736	339	14		387	36
	339	21		402	20
5310-00-835-2140	236	10		469	19
	303	15		470	16
4730-00-835-3003	273	7		602	16
	598	39	3110-00-842-6572	209	2
5330-00-835-7712	494	10	5935-00-843-4561	438	5
	494	18	5340-00-843-7825	57	6
5325-00-836-2131	249	11	4820-00-845-1096	31	10
4730-00-837-1177	213	33		204	12
	271	22		595	20
	605	12	5315-00-846-0126	475	7
4730-00-837-7073	629	23		490	7
5310-00-838-1490	530	3	5325-00-846-1637	88	33
5310-00-838-1702	473	25	5935-00-846-3884	163	14
5340-00-839-0098	405	18		165	14
	406	8		166	7
	416	4		167	6
5310-00-839-2066	313	3		168	11
5315-00-839-2325	100	17		169	11
	182	16	3040-00-847-3169	530	4
	411	17	2920-00-848-3292	588	17
	528	18	4820-00-848-4361	316	4
	534	11		317	25
5315-00-839-2326	525	21	4820-00-849-1220	107	36
	598	29		265	7
5315-00-839-5820	341	2	5315-00-849-5582	501	5
	359	13	5310-00-849-6874	507	23
5315-00-839-5821	411	1	5315-00-849-9854	485	1
	415	8		583	1
	419	22		592	10
	425	20	5310-00-851-2677	499	4

CROSS-REFERENCE INDEXES

NATIONAL STOCK NUMBER INDEX

STOCK NUMBER	FIG.	ITEM	STOCK NUMBER	FIG.	ITEM
5310-00-851-2677	607	1	5330-00-861-8592	28	8
2815-00-851-7637	26	11	5330-00-864-5422	114	4
4730-00-852-5654	620	11	4730-00-865-9251	271	2
5330-00-852-7347	284	36		630	10
	286	51	5340-00-865-9496	182	18
5310-00-853-9335	494	16		454	2
2530-00-854-4457	282	1	5315-00-866-2673	405	16
5305-00-855-0957	170	15		406	46
	171	13	5315-00-866-5015	24	10
	211	15	5330-00-866-6236	473	8
	362	9	5310-00-867-1465	463	1
	379	13		464	1
	539	1		511	1
	540	19			
	606	6	5935-00-868-2606	432	5
5305-00-855-0958	82	16	5306-00-869-6549	347	3
5305-00-855-0960	366	2	5305-00-873-6946	473	20
	366	7	5331-00-873-7214	479	3
	578	3	5365-00-876-6862	317	5
5305-00-855-0961	367	17	3120-00-877-2213	19	3
	441	2	5310-00-877-5795	67	3
	552	3		68	12
	597	11		69	16
	598	4		202	7
5305-00-855-0965	135	1		318	26
	413	16		319	27
	587	4		320	23
	588	2		321	26
5305-00-855-0968	573	2		322	27
	574	1		323	30
5305-00-855-0971	559	3		324	26
	593	4		345	17
5305-00-855-0973	362	5		365	1
4820-00-857-2737	256	21		368	11
4710-00-857-2782	487	22		374	27
5305-00-857-3367	257	15		376	14
2540-00-857-6332	388	2		390	10
9520-00-857-6344	536	14		391	9
2910-00-858-3522	85	12		391	13
	87	13		396	11
2530-00-859-7335	276	6		468	12
2540-00-860-0516	609	31		602	23
2510-00-860-0517	609	18		611	19
2540-00-860-0519	609	2	5310-00-877-5796	100	35
2510-00-860-0523	609	18		163	31
2540-00-860-2355	531	2		165	33
2540-00-860-2356	531	3		166	28
2540-00-860-2357	531	1		167	21
2540-00-860-2358	531	8		168	27
2520-00-860-7340	473	6		169	36
5310-00-861-2316	317	3		184	8

CROSS-REFERENCE INDEXES

NATIONAL STOCK NUMBER INDEX

STOCK NUMBER	FIG.	ITEM	STOCK NUMBER	FIG.	ITEM
5310-00-877-5796	214	8	5310-00-880-7746	100	21
	266	33		101	13
	352	10		351	22
	353	11		540	6
	425	7	6680-00-882-0965	628	2
	505	30	3120-00-882-7960	19	4
	534	5	5331-00-883-2799	249	33
	579	5	5305-00-885-7252	530	2
	588	15	5310-00-885-7734	424	26
	606	19	5315-00-886-1432	284	44
	608	8	5310-00-887-8325	42	2
	612	12	5310-00-889-2527	211	31
	629	12	5310-00-889-2528	143	19
5310-00-877-5797	131	26		587	5
	134	32		593	4
	134	73		595	19
	153	8	5320-00-889-2632	383	10
	180	13		383	24
	183	1		384	5
	249	39		384	18
	355	5		386	7
	374	16		386	13
	542	9		534	9
	596	26	5305-00-889-3000	586	1
	611	34	5305-00-889-3002	100	27
	612	25		101	4
	613	24		144	25
4730-00-877-8997	598	52		150	1
6220-00-878-7301	144	1	5305-00-889-3116	415	12
5315-00-879-2910	585	20	5342-00-889-5209	463	22
	594	12		464	22
5305-00-879-7941	534	13		511	22
			2930-00-890-2440	118	1
5310-00-880-7744	370	29	5310-00-891-1709	472	37
	382	1	5310-00-891-1711	346	21
	382	28		348	3
	382	33	5310-00-891-1733	211	53
	385	23		305	22
	385	32	5310-00-891-1751	100	16
	387	1		153	20
	387	19	5310-00-891-3426	607	2
	387	43	5310-00-891-3428	252	12
	414	20		261	15
	419	19	2540-00-891-7830	534	4
	469	27	5940-00-892-3151	149	11
	473	14		211	36
	499	17		213	24
	609	13		436	10
	609	27	5310-00-893-3381	207	2
5310-00-880-7745	301	15		208	11

CROSS-REFERENCE INDEXES

NATIONAL STOCK NUMBER INDEX

STOCK NUMBER	FIG.	ITEM	STOCK NUMBER	FIG.	ITEM
5310-00-893-3381	217	11	5310-00-901-1339	155	4
	526	27		155	6
9905-00-893-3570	151	6	5305-00-901-2099	571	3
	151	15	5305-00-901-2135	407	11
	151	20	9905-00-901-2942	569	5
	152	10	5340-00-901-8132	455	9
	154	16	3110-00-902-3757	475	10
	154	20	4730-00-902-8991	595	24
	431	2	5305-00-902-9338	183	4
	431	6	4310-00-903-7174	284	31
	433	3		286	46
	433	9	5305-00-903-7794	233	14
	437	3		240	14
			5340-00-904-0933	161	16
	438	4	5315-00-904-1633	339	7
5360-00-895-3216	284	18	5315-00-904-3408	508	15
	286	3		559	18
5330-00-895-3424	474	20	5315-00-904-3412	508	19
	489	3		559	22
5930-00-898-0500	131	14	3120-00-904-9595	50	10
	134	26		55	9
2510-00-898-5415	383	28	4730-00-905-0030	621	6
5305-00-899-2049	305	3	5325-00-905-1492	367	10
3110-00-899-4353	463	15	5331-00-905-2679	284	16
	464	15		286	5
	511	15	5310-00-905-4600	622	5
5975-00-899-4606	456	42	5970-00-906-0159	150	6
	460	38	4730-00-906-0982	218	8
	461	36	4730-00-908-3193	589	11
	580	7		590	5
	582	13	4730-00-908-3194	39	2
	593	10		39	5
4720-00-899-7893	BULK	40		581	9
5305-00-899-8054	88	15		582	9
5305-00-900-0576	502	33		584	19
5305-00-900-1118	473	2		589	53
2590-00-900-1640	148	1		590	38
5305-00-900-3243	572	13		622	9
4730-00-900-3296	582	5	4730-00-908-6292	136	2
	593	16		581	16
5935-00-900-6281	172	14		585	32
	173	15		589	4
	173	27		591	10
	211	37	4730-00-908-6294	62	17
	592	26		63	2
	594	25		581	1
	614	13		585	8
2920-00-900-7993	608	3		592	7
5310-00-901-0279	161	18	3130-00-908-8589	220	6
	162	9	2920-00-909-2483	125	9

NATIONAL STOCK NUMBER INDEX

CROSS-REFERENCE INDEXES

NATIONAL STOCK NUMBER INDEX

STOCK NUMBER	FIG.	ITEM	STOCK NUMBER	FIG.	ITEM
5999-00-926-3144	152	2	2510-00-933-9577	318	12
	158	29		321	13
	165	6	3040-00-933-9579	486	10
	167	11	3020-00-933-9585	484	1
	170	34	4710-00-933-9587	481	6
	170	40	5360-00-934-0089	494	20
	171	2	5310-00-934-9747	586	4
	180	4	5310-00-934-9751	132	15
	182	36		132	21
	211	8		132	25
	211	43		132	28
5310-00-926-5885	347	8		624	9
4730-00-927-7272	618	12		625	9
	630	11	5310-00-934-9754	163	3
	630	12	5310-00-934-9755	134	11
2910-00-928-3505	66	2	5310-00-934-9757	157	11
3020-00-929-0713	118	12		161	23
4730-00-930-0982	270	6		412	15
5340-00-930-1754	170	16		427	17
	171	15		428	2
4320-00-930-2045	581	15		598	45
	585	24	5310-00-934-9758	143	14
	589	3		143	14
	591	11		299	1
2510-00-930-2714	383	9	5310-00-935-3569	106	16
5340-00-930-2716	537	2		107	25
5340-00-930-2717	537	7		297	14
2540-00-930-3139	532	2		302	5
4730-00-930-5392	480	15		318	20
	505	53		319	9
4730-00-930-6354	274	4		320	19
	275	4		321	22
5310-00-930-7013	118	21		322	21
2510-00-930-7778	385	27		323	26
3040-00-930-7864	230	2		324	20
5320-00-930-7865	410	17		331	3
2910-00-930-9367	593	5		332	3
	595	22		380	9
5342-00-931-4527	249	43	9905-00-935-7777	166	17
5360-00-932-7452	33	23		430	5
5340-00-933-3009	24	2	5310-00-935-8984	155	2
6680-00-933-3599	624	7	5310-00-935-9021	107	8
6680-00-933-3600	132	26		205	12
9905-00-933-4632	566	10	5310-00-935-9022	61	10
9905-00-933-4635	566	15		81	14
2530-00-933-4941	350	2		100	42
2540-00-933-6262	543	15		101	9
2540-00-933-6263	543	16		105	17
2540-00-933-6267	543	4		112	2
2540-00-933-8645	532	1		113	6

NATIONAL STOCK NUMBER INDEX

STOCK NUMBER	FIG.	ITEM	STOCK NUMBER	FIG.	ITEM
5310-00-935-9022	142	7	5310-00-935-9022	593	6
	142	12		594	34
	142	16		596	23
	143	18		606	17
	145	13		610	6
				616A	5
	162	8	6625-00-936-2139	132	13
	163	25	3120-00-937-1164	484	23
	165	29	5305-00-939-0576	472	13
	166	25	5310-00-939-0783	366	14
	167	35		608	16
	168	23	5305-00-939-9204	469	18
	169	27		470	14
	182	12		499	16
	255	17		502	6
	255	21	4730-00-940-0947	585	4
	262	1	5305-00-940-8069	107	32
	262	16	5305-00-940-9517	346	12
	269	1	5331-00-941-3762	284	8
	276	8		286	10
	278	18	3110-00-941-3830	185	12
	279	10	2590-00-941-8668	BULK	1
	279	15	5360-00-941-8684	100	31
	289	11		101	21
	313	8	5305-00-941-9460	235	10
	315	14		242	10
	326	7	5305-00-942-2196	19	14
	339	29		259	7
	340	10		260	7
	353	9	5310-00-943-2141	348	13
	361	1	4730-00-944-5888	581	8
	374	6		582	18
	374	17		589	52
	454	23		590	37
	513	16	5365-00-946-2231	539	10
	525	15		540	8
	543	6	5305-00-947-3437	26	4
	543	14		26	8
	545	6		26	13
	547	5	5305-00-947-4360	106	14
	549	5	6220-00-947-7570	611	1
	551	18		612	1
	561	5		613	1
	580	1	5305-00-948-0803	331	13
	581	22		332	12
	584	5	5331-00-948-6482	463	4
	586	17		464	4
	587	9		511	4
	588	13	4030-00-948-7315	326	23
	589	20		327	10
	592	1		383	16

CROSS-REFERENCE INDEXES

NATIONAL STOCK NUMBER INDEX

STOCK NUMBER	FIG.	ITEM	STOCK NUMBER	FIG.	ITEM
4030-00-948-7315	390	4	5320-00-956-7355	419	7
3110-00-948-9796	469	10	5305-00-957-1497	353	7
	470	6		361	7
5310-00-949-6280	244	18	5305-00-957-1538	281	5
2940-00-950-8410	457	24	5315-00-957-2399	402	14
	508	28	5315-00-957-9386	339	23
3110-00-950-9700	236	3		340	6
	296	10	5305-00-958-5258	410	16
3120-00-951-1850	485	12	5305-00-958-5267	474	31
5340-00-951-3536	52	7		489	16
5330-00-951-3538	129	17	5310-00-959-1488	62	2
5310-00-951-7209	380	6		67	9
	380	6		68	3
	381	9		69	6
	400	11		71	10
5315-00-951-7542	354	3		72	8
5360-00-953-5205	463	16		74	4
	464	16		75	10
	511	16		76	8
4730-00-954-1281	7	29		80	20
	31	6		93	10
	41	10		109	18
	214	17		142	4
	230	9		143	8
	473	9		164	28
	597	4		248	1
	629	21		337	9
5305-00-954-4295	370	9		368	8
5340-00-954-6014	71	8		374	20
	72	4	4730-00-959-1629	598	25
	74	3	5305-00-959-2703	426	15
	75	9	5305-00-959-2723	470	26
	76	6	5305-00-959-2739	211	4
	77	15	5360-00-960-9326	463	26
	78	3		464	26
	79	6		511	26
	80	6	4820-00-960-9329	464	6
	164	27		511	6
	165	27	5330-00-961-3596	295	7
	184	9	5330-00-961-6314	33	31
	254	18	5330-00-961-9470	95	1
	254	18	4010-00-961-9780	BULK	26
	257	6	2815-00-962-5623	24	14
	257	9	5305-00-964-0565	33	28
	261	11	5305-00-964-0589	460	31
	261	11	5340-00-964-5267	157	17
	287	2		429	13
3805-00-955-5320	284	11	5365-00-965-0870	52	9
	286	14	5340-00-966-2390	598	42
5320-00-956-7355	417	4	6240-00-966-3831	141	22

CROSS-REFERENCE INDEXES

NATIONAL STOCK NUMBER INDEX

STOCK NUMBER	FIG.	ITEM	STOCK NUMBER	FIG.	ITEM
5340-00-968-4060	376	1	5310-00-984-3806	347	10
	399	2		382	34
4730-00-968-6129	230	5		385	9
	484	20		385	41
	489	28		387	28
5340-00-969-6407	528	2	5310-00-984-3807	100	25
4730-00-969-6941	426	8		101	20
5331-00-970-3461	51	11		370	17
5310-00-971-7989	51	14		425	14
	56	13		533	17
	66	16		611	17
	88	8	5305-00-984-5676	382	26
	88	25		387	16
2590-00-972-2632	491	12	5305-00-984-5677	609	5
2590-00-972-2634	485	19	5305-00-984-5680	532	3
5330-00-972-2635	485	6		532	3
	491	4		532	3
3040-00-972-2638	485	4		535	3
				609	5
3040-00-972-2639	491	1		609	19
3040-00-972-2640	484	29	5305-00-984-6189	131	3
3020-00-972-2641	484	19		132	9
2540-00-972-2642	531	7		150	10
4730-00-974-7313	456	5	5305-00-984-6191	412	18
	461	5		427	16
	509	27		428	4
	605	5	5305-00-984-6192	157	9
2590-00-974-9670	488	6		367	29
5365-00-974-9851	41	5	5305-00-984-6193	161	27
4730-00-976-0981	629	5	5305-00-984-6208	81	3
2530-00-981-8736	278	5		279	18
5331-00-982-4259	479	9		552	13
5310-00-982-4908	310	18	5305-00-984-6210	505	14
	322	36	5305-00-984-6212	125	23
	376	13		143	7
5310-00-982-4912	62	29		143	7
	596	18		549	15
5310-00-982-4935	138	16		551	15
5310-00-982-5009	380	7	5305-00-984-6226	150	14
	393	3	5975-00-984-6582	70	2
5310-00-982-5012	380	9		73	4
	490	9		76	14
5310-00-982-5014	304	4		77	19
	497	4		79	14
5310-00-982-6562	607	3		80	9
3030-00-983-2873	125	8		97	6
5305-00-983-6665	623	22		158	47
5305-00-983-8082	495	11		166	20
5331-00-984-3756	26	17		167	23
5310-00-984-3806	107	20		169	24
	346	2		182	7

CROSS-REFERENCE INDEXES

NATIONAL STOCK NUMBER INDEX

STOCK NUMBER	FIG.	ITEM	STOCK NUMBER	FIG.	ITEM
5975-00-984-6582	587	2	5330-00-990-5804	134	21
	597	3	5305-00-990-6444	131	19
	606	21		134	34
	611	37		134	72
	612	20		611	33
	613	35	5305-00-990-7168	502	42
5305-00-984-7342	100	8	5340-00-990-7610	587	14
5340-00-984-8540	513	26		588	7
5310-00-984-8818	490	14	5995-00-991-6707	144	7
5340-00-985-0823	402	17	5935-00-991-6708	144	35
5975-00-985-6630	204	20	5930-00-991-6713	144	12
5340-00-988-1162	596	14	5305-00-993-1848	163	5
5305-00-988-1170	383	21		183	15
	385	30		418	20
5305-00-988-1723	181	6		596	24
				616A	27
5305-00-988-1724	137	1	5305-00-993-1851	359	8
	405	34	5305-00-993-2461	425	8
	406	35		547	3
	544	5		549	7
	545	1	5305-00-993-2463	314	4
5305-00-988-1725	370	8	5305-00-993-2738	374	13
	546	1		544	8
5305-00-988-1726	370	13		545	5
5305-00-988-9106	629	11		547	4
5340-00-989-1771	71	11		551	4
	72	9		561	4
	74	2	6620-00-993-5546	35	5
	75	11		146	9
	76	5	5340-00-993-6207	162	10
	77	14		587	8
	79	5		588	12
	80	7	5310-00-994-1006	456	13
	253	4		461	10
	263	14		509	25
5305-00-989-6265	608	9	4730-00-995-1559	278	13
5305-00-989-7434	374	11	5305-00-995-3442	539	14
	540	12			
	606	10	5305-00-995-3444	367	41
	612	21		367	47
	613	26	5305-00-995-3569	362	2
5305-00-989-7435	61	9	5320-00-995-8907	369	10
	100	6	2990-00-997-1532	581	2
	131	27		585	13
	132	30	5310-00-998-0608	389	11
	153	5	2530-00-998-4711	472	18
	542	7'	6220-00-998-6142	141	24
	586	14	5330-00-999-3752	190	16
5305-00-990-0695	182	28	5330-00-999-3760	185	2
	478	16	5180-00-999-4053	630	20
5315-00-990-2889	454	12	5315-00-999-4238	398	5
	513	12		398	7

NATIONAL STOCK NUMBER INDEX

STOCK NUMBER	FIG.	ITEM	STOCK NUMBER	FIG.	ITEM
5340-00-999-6467	390	18	9905-01-013-8723	166	5
3110-00-999-6469	507	42		167	5
9905-00-999-7369	559	2		168	6
	559	2		169	6
9905-00-999-7370	559	1		182	32
	559	1		213	25
5342-00-999-8591	243	15		432	6
4720-01-003-6706	BULK	11		432	13
5315-01-004-4835	197	18		433	7
5315-01-004-4836	197	9		434	8
5360-01-004-4863	197	20		435	4
4730-01-005-0623	272	1		435	9
4730-01-005-3262	620	14		436	3
5935-01-005-3579	433	6		436	16
5331-01-005-3704	622	7	4720-01-014-4915	BULK	10
	541	14			
5305-01-006-2052	409	31	3120-01-016-4883	286	31
2520-01-006-7116	187	19	5331-01-019-2448	485	22
	193	4	5305-01-020-0709	463	14
2520-01-006-7118	193	3		464	14
2520-01-006-7120	192	12		511	14
5365-01-006-9622	200	19	5360-01-020-7063	289	14
4730-01-006-9629	197	21	5940-01-021-1874	158	8
2520-01-007-0345	200	9		158	30
2520-01-007-0346	200	11	2510-01-022-2580	407	17
4820-01-007-0350	197	22		415	27
4730-01-007-0802	186	7	5330-01-023-0269	207	9
4820-01-007-0962	196	6			22
3110-01-007-2609	186	6			34
3020-01-008-2769	188	7		234	10
3020-01-008-2770	188	14		234	23
5340-01-008-6448	144	28		241	10
5315-01-008-7084	211	17		241	23
4720-01-009-9058	BULK	12	5305-01-023-2428	410	10
4720-01-009-9941	BULK	13	2510-01-024-3618	407	16
5305-01-010-2362	598	46		415	30
5365-01-010-9687	187	9	2510-01-024-3619	407	18
5365-01-010-9688	187	3		415	29
5365-01-010-9689	187	12			
	193	14	5305-01-024-4775	6	13
5331-01-010-9693	189	15	4730-01-026-0929	314	10
5315-01-010-9777	188	9		315	2
3110-01-010-9779	196	2	2510-01-027-0203	343	9
2520-01-011-1067	200	25	4010-01-027-0356	469	35
2520-01-011-1068	200	13		470	30
4820-01-011-1069	200	15	5306-01-027-4634	458	12
9905-01-013-4599	571	4	4730-01-027-6590	313	4
9905-01-013-8723	151	4	4730-01-027-8943	182	4
	152	5	4730-01-028-0342	629	29
	163	6	5306-01-028-2443	425	26
	165	13	5305-01-028-4831	441	6

NATIONAL STOCK NUMBER INDEX

STOCK NUMBER	FIG.	ITEM	STOCK NUMBER	FIG.	ITEM
5310-01-028-4848	539	12	9905-01-046-4676	574	3
4730-01-028-8147	204	18	9905-01-046-4677	572	6
5305-01-028-8869	284	64	2540-01-046-9402	415	21
5320-01-029-7722	134	24	5310-01-047-0401	601	5
4730-01-030-4950	19	18	9340-01-047-4100	363	5
	248	17	2815-01-048-6702	23	13
	279	19	4730-01-048-7874	598	6
	474	26	4730-01-049-1559	598	7
5305-01-031-4487	619	21	5330-01-049-7374	149	3
2920-01-031-9027	128	1	5330-01-051-1053	55	22
7690-01-032-5639	592	28	5330-01-051-4243	96	4
	594	31	2910-01-051-4292	96	3
4730-01-032-6038	213	19	2910-01-051-9444	198	6
	252	5	5306-01-052-2402	200	21
	253	2	5340-01-052-9022	414	26
	259	2	5340-01-052-9023	414	14
	260	2	5340-01-052-9024	414	5
	264	16	5310-01-053-1936	337	27
	266	13		338	24
	267	12	5310-01-054-2568	236	11
	269	14	9905-01-054-3827	572	5
	271	14	9905-01-054-3828	574	4
	272	8	9905-01-054-4002	572	4
	273	19		574	2
	597	6	5306-01-054-4485	286	41
	617	9	5320-01-055-4452	153	3
	618	5	5365-01-055-8769	236	11
5320-01-032-6530	376	5	5307-01-055-8843	528	23
	377	2	3110-01-056-0031	190	14
	383	30	5340-01-056-0037	185	26
9905-01-032-7002	580	9	5305-01-057-4265	195	2
	584	17	5310-01-057-5518	494	14
	594	38	5315-01-057-8371	549	14
4010-01-035-0159	499	23		551	12
5940-01-035-4212	151	3	5310-01-058-3353	385	13
	152	7	5315-01-058-3487	186	9
5340-01-038-7759	377	4	5315-01-058-4551	7	13
4010-01-039-4831	134	23	4720-01-058-7213	BULK	38
5935-01-040-0463	431	9		BULK	39
2520-01-040-3541	KITS		5315-01-058-7268	234	32
5306-01-042-3586	598	41		241	32
5360-01-042-9532	473	37		293	15
2510-01-042-9692	415	1		295	6
9905-01-043-5322	558	3		348	29
5330-01-044-2096	30	6	4720-01-058-9489	482	15
5935-01-044-8382	154	12	4720-01-058-9490	480	27
5970-01-044-8391	154	4	4710-01-058-9494	480	17
5306-01-046-0553	458	13	5340-01-059-0114	154	6
5330-01-046-3300	489	8	5330-01-059-4286	154	5
	526	13	5330-01-060-0992	534	14

NATIONAL STOCK NUMBER INDEX

STOCK NUMBER	FIG.	ITEM	STOCK NUMBER	FIG.	ITEM
5330-01-060-9061	284	14	9905-01-069-7222	437	2
5306-01-062-3148	245	8		438	2
	246	6			
5310-01-062-3384	245	21		439	3
	246	20		439	6
5342-01-062-4715	494	8		440	6
5310-01-063-2299	245	7	5315-01-070-2168	534	10
	246	7	4730-01-070-4915	598	23
2530-01-064-2630	250	7	4730-01-070-6667	66	18
	251	12	3020-01-070-9003	50	13
5310-01-064-3422	387	27		55	13
2520-01-064-8847	191	2	3040-01-070-9004	50	17
2520-01-064-8849	191	9		55	16
	194	4	5330-01-071-5727	312	54
2520-01-065-0077	187	20	4730-01-071-5740	97	3
	193	5	5305-01-072-4270	9	9
2520-01-065-0078	192	14	5331-01-072-4436	46	17
3020-01-065-0183	187	10	5305-01-072-8816	41	9
2520-01-065-0841	187	15	5305-01-072-8826	51	6
	193	7		56	5
3020-01-065-0871	188	2	5330-01-072-8828	50	4
2530-01-065-1828	245	14		55	4
3110-01-065-2469	188	21	5330-01-072-8829	50	15
2520-01-065-2530	188	12		55	15
3110-01-065-5842	193	15	5330-01-072-8830	85	9
	200	28		87	11
5340-01-065-7287	427	8		88	20
4730-01-066-1282	480	11	5331-01-072-8983	51	5
	503	3		56	4
5310-01-066-2942	46	21	5331-01-072-8984	53	5
4730-01-066-3071	582	24	4820-01-073-0080	35	6
5340-01-066-6086	106	15	5360-01-074-8305	197	5
	107	31	4330-01-074-9642	198	3
5365-01-066-7554	177	2	5306-01-075-8519	37	11
5305-01-066-8646	419	34	4730-01-076-2735	494	11
4730-01-066-9484	278	1	7510-01-076-4238	568	2
	603	7	2910-01-076-8632	46	7
5330-01-067-1740	224	12	2540-01-076-9286	589	34
4730-01-067-3932	503	2		590	22
	503	6	5935-01-077-2622	160	10
4730-01-067-4711	549	48	2520-01-077-4009	120	18
5330-01-067-8567	289	5		122	17
2840-01-068-1713	200	3	3110-01-077-7134	241	16
5320-01-068-2340	406	31	5365-01-077-8564	234	15
5930-01-069-2776	97	2	3020-01-078-0627	200	6
5320-01-069-6364	411	15	4730-01-078-2731	66	12
5320-01-069-6365	153	18	4730-01-078-2732	192	8
9905-01-069-7222	169	21	4730-01-078-4703	51	12
	430	3		56	15
	433	2	2520-01-078-5564	192	16
	437	2			

CROSS-REFERENCE INDEXES

NATIONAL STOCK NUMBER INDEX

STOCK NUMBER	FIG.	ITEM	STOCK NUMBER	FIG.	ITEM
7510-01-078-5855	568	1	4730-01-079-8821	271	20
2520-01-078-6123	192	16		271	23
4710-01-078-8748	189	16		272	10
4730-01-078-9859	53	4		273	3
2590-01-079-1506	183	7		274	2
2520-01-079-1615	200	23		275	2
3040-01-079-1799	26	19		279	3
5360-01-079-3096	187	14		279	22
	193	8		288	2
5360-01-079-3097	191	8		291	4
	194	5		538	11
2815-01-079-3290	286	48		603	8
4310-01-079-3319	284	34		604	2
	286	42		605	2
4310-01-079-3383	284	43		617	3
3010-01-079-3461	44	12		620	3
5342-01-079-4678	46	12		621	10
3120-01-079-5208	26	9		626	2
4310-01-079-5245	286	49	3120-01-079-9882	345	41
5315-01-079-6506	46	9	2520-01-080-0448	185	30
5331-01-079-6513	120	21	5365-01-080-0482	185	9
5330-01-079-6514	31	13	5365-01-080-0483	185	9
5325-01-079-6526	185	20	5365-01-080-0484	185	9
5360-01-079-6702	200	14	5365-01-080-0485	185	9
5360-01-079-6703	200	24	5365-01-080-0486	185	9
5360-01-079-6704	200	10	5307-01-080-0492	345	42
5310-01-079-6708	46	4	2990-01-080-0533	284	1
5306-01-079-7027	17	12	4730-01-080-0930	345	9
5305-01-079-7028	46	5	3010-01-080-1529	50	19
5310-01-079-7036	120	27		55	19
3110-01-079-7049	185	29	2910-01-080-3149	51	3
5310-01-079-7059	345	18		56	2
5315-01-079-8059	345	39	5331-01-080-3254	198	4
5340-01-079-8097	66	20	5306-01-080-4757	345	31
3110-01-079-8190	51	17	5330-01-080-5021	9	7
	56	17	2510-01-080-6424	345	35
4730-01-079-8821	182	21	5310-01-080-9007	345	3
	199	2	5360-01-080-9008	345	13
	212	2	5325-01-081-0662	50	12
	213	13		55	12
	214	2	5315-01-081-4275	345	38
	215	2	2520-01-081-9043	KITS	
	254	8	2510-01-081-9226	396	14
	257	11	2510-01-081-9227	533	1
	261	4	5315-01-081-9991	351	9
	264	9	5330-01-082-1818	222	8
	265	2		227	9
	267	9	5330-01-082-1906	16	4
	268	2	5340-01-082-2510	457	19
	270	2	2990-01-082-2511	102	25

CROSS-REFERENCE INDEXES

NATIONAL STOCK NUMBER INDEX

STOCK NUMBER	FIG.	ITEM	STOCK NUMBER	FIG.	ITEM
5340-01-082-2512	109	16	2510-01-082-7458	374	8
2930-01-082-2513	112	5	2510-01-082-7460	368	2
2930-01-082-2514	112	1	2510-01-082-7508	384	16
2510-01-082-2515	468	7	2540-01-082-7510	371	13
5340-01-082-2516	396	4	2520-01-082-8619	KITS	
5340-01-082-2517	328	9	3120-01-082-9008	471	21
2590-01-082-2520	468	15	2990-01-082-9009	102	23
2590-01-082-2521	468	15	6150-01-082-9040	175	2
5340-01-082-2522	543	12	6150-01-082-9042	175	1
5340-01-082-2523	543	11	2510-01-083-1105	536	19
5306-01-082-2524	513	2	2510-01-083-1106	536	6
2590-01-082-2526	469	32	5340-01-083-1107	396	5
2510-01-082-2642	382	30			
2590-01-082-2644	470	1	2540-01-083-1109	549	4
2510-01-082-2645	386	6	4730-01-083-1110	541	3
2540-01-082-3592	387	11			
5340-01-082-3593	549	40	2540-01-083-1113	551	20
5340-01-082-3594	549	38	2590-01-083-1115	351	29
5342-01-082-3595	536	3	2540-01-083-1116	551	3
5340-01-082-3596	366	6	5340-01-083-1117	102	13
4030-01-082-3597	382	15	5340-01-083-1120	359	20
3020-01-082-3598	471	3	5340-01-083-1121	359	20
3040-01-082-3599	471	26	2930-01-083-1122	109	15
2510-01-082-3603	373	5	2990-01-083-1123	102	34
2510-01-082-3604	373	8	2510-01-083-1125	354	4
2590-01-082-3606	497	5	2510-01-083-1126	354	4
2590-01-082-3607	495	1	2510-01-083-1140	373	2
3950-01-082-3608	497	1	2930-01-083-1141	352	4
5340-01-082-3609	394	5	2990-01-083-1142	102	32
5340-01-082-3610	394	4	5340-01-083-1143	351	15
5340-01-082-3611	492	11	2510-01-083-1145	359	4
5342-01-082-3612	492	10	2510-01-083-1146	373	3
2510-01-082-3619	364	11	5340-01-083-1147	369	1
2510-01-082-3620	364	11	2510-01-083-1149	351	1
2510-01-082-3621	373	7	6150-01-083-1152	167	1
2510-01-082-3622	373	9	6150-01-083-1153	175	1
2510-01-082-3623	364	12	6150-01-083-1154	151	17
2540-01-082-3624	371	5	2590-01-083-1155	525	13
2510-01-082-3625	364	1	2510-01-083-1156	387	24
2510-01-082-3629	364	1	2510-01-083-1158	383	23
2510-01-082-3630	543	18	4010-01-083-1159	383	15
2930-01-082-3631	112	4	6150-01-083-1161	174	1
2510-01-082-3788	387	24	2920-01-083-1187	164	1
6150-01-082-3818	175	6	2920-01-083-1188	169	1
2510-01-082-3824	353	1	4820-01-083-2127	197	6
2590-01-082-3828	165	1	5340-01-083-3015	353	10
5330-01-082-6985	14	17	2930-01-083-3016	352	5
5340-01-082-7448	543	8	2930-01-083-3017	352	6
2510-01-082-7455	373	6	2520-01-083-4404	221	8
2540-01-082-7457	387	37	5340-01-083-5402	396	1

NATIONAL STOCK NUMBER INDEX

STOCK NUMBER	FIG.	ITEM	STOCK NUMBER	FIG.	ITEM
5340-01-083-5403	396	2	3120-01-085-3338	489	47
2510-01-083-5404	468	6	2590-01-085-3584	507	37
5340-01-083-5406	543	23	2540-01-085-3588	382	17
3990-01-083-5407	472	10	2540-01-085-3589	342	1
2920-01-083-5408	125	4	2530-01-085-3592	471	25
2510-01-083-5442	353	5	5340-01-085-3593	351	13
5340-01-083-5443	177	7	5340-01-085-3594	351	12
6150-01-083-5479	175	3	3130-01-085-3595	472	29
5306-01-083-5536	141	10	2590-01-085-3596	398	1
2990-01-083-5715	102	5	2990-01-085-3786	102	35
2590-01-083-5728	158	1	2530-01-085-3787	498	1
2590-01-083-5729	166	1	2590-01-085-3806	507	1
5315-01-083-6351	197	23	2990-01-085-3833	102	5
5315-01-083-6352	197	2	2590-01-085-3834	164	1
5340-01-083-6420	249	25	4730-01-085-4156	284	54
2530-01-083-8102	249	24	4820-01-085-4762	494	21
5315-01-083-9387	383	19	2990-01-085-5349	102	20
4730-01-083-9925	197	17	2990-01-085-5350	102	20
2510-01-084-0446	373	10	2590-01-085-5352	457	5
5310-01-084-1197	107	21	2510-01-085-5353	468	6
	594	15	4730-01-085-7328	116	5
5340-01-084-1232	469	23	5340-01-085-8136	543	20
	470	19	2930-01-085-8137	352	5
5310-01-084-1768	189	29	2930-01-085-8138	352	6
5310-01-084-2362	6	19	5305-01-085-8197	11	3
	107	24	3040-01-086-1449	51	10
	190	12		56	9
	205	8	5360-01-086-3480	284	10
	286	34		286	12
	619	3	5330-01-086-3523	41	1
	619	3	3020-01-086-4158	50	18
5305-01-084-5370	197	25		55	18
5306-01-084-5389	208	10	2815-01-086-4508	17	9
2530-01-084-6975	251	7	5340-01-086-6193	9	3
4310-01-084-7148	284	9	2510-01-086-6802	364	12
	286	11	5305-01-086-7036	114	1
2540-01-084-9630	375	1	2910-01-086-7715	50	2
5340-01-084-9631	457	1		55	2
	508	1	3020-01-086-8780	50	7
2590-01-084-9632	455	3		55	7
	528	7	6625-01-086-9580	132	19
2510-01-084-9633	362	27	5315-01-087-0534	50	16
2590-01-084-9634	168	1		55	17
2540-01-084-9644	371	1	5340-01-087-0681	9	2
2815-01-085-2569	26	1	5340-01-087-0682	9	4
3040-01-085-2616	129	8	3120-01-087-2539	50	6
2815-01-085-2618	24	5		55	6
3010-01-085-2732	44	9	3120-01-087-3004	17	13
	284	41	2540-01-087-4741	551	19
3120-01-085-3338	471	1	3010-01-088-5727	50	14

CROSS-REFERENCE INDEXES

NATIONAL STOCK NUMBER INDEX

STOCK NUMBER	FIG.	ITEM	STOCK NUMBER	FIG.	ITEM
3010-01-088-5727	55	14	5310-01-088-9298	301	6
5310-01-088-5851	306	1	2530-01-088-9357	236	36
2510-01-088-5914	318	34	3040-01-088-9401	211	19
	319	35	4720-01-088-9650	109	12
	320	32		110	4
	321	34	4720-01-088-9651	456	24
	322	38	4720-01-088-9680	551	1
	323	36	4720-01-088-9681	551	16
	324	33	4720-01-089-0766	456	4
5305-01-088-6019	31	1		460	4
	33	41		461	4
	116	8	4720-01-089-1108	104	11
	284	40	5340-01-089-1355	337	10
	286	35	4720-01-089-2016	456	36
	310	15		460	33
2540-01-088-6036	388	2		461	33
2520-01-088-8172	120	20		509	31
	122	12	4710-01-089-2059	480	13
5340-01-088-9152	333	12		495	6
	334	14	4710-01-089-2060	456	26
	337	15		460	23
	338	12		461	20
5340-01-088-9153	318	9		509	15
	319	12	4720-01-089-2061	105	12
	320	9	6350-01-089-2987	145	8
	321	10	5340-01-089-2988	207	38
	323	19	5340-01-089-2989	207	16
	324	9	5342-01-089-2990	211	51
2510-01-088-9156	318	30	3040-01-089-2991	211	22
	319	31	5330-01-089-2992	211	48
	320	27	6210-01-089-3037	547	8
	321	33	5330-01-089-3046	211	49
	322	33	2530-01-089-3047	301	2
	323	35	5340-01-089-3057	376	3
	324	29	4140-01-089-3058	120	3
2510-01-088-9157	318	29	5340-01-089-3067	335	9
	319	30	5340-01-089-3068	319	33
	320	24		320	28
	321	32		322	35
	322	31		324	30
	323	34	5305-01-089-3070	346	1
	324	28	4730-01-089-3071	204	2
3040-01-088-9161	318	40	4730-01-089-3072	204	21
	319	40	5330-01-089-3073	316	13
	320	41	4720-01-089-3074	204	23
	321	40	5340-01-089-3076	549	17
	322	43	3040-01-089-3081	302	6
	323	44	5340-01-089-3126	333	1
	324	39		334	1
2540-01-088-9162	335	6		335	1

NATIONAL STOCK NUMBER INDEX

STOCK NUMBER	FIG.	ITEM	STOCK NUMBER	FIG.	ITEM
5340-01-089-3126	336	1	5340-01-090-6407	388	6
	337	1	4730-01-090-6468	71	23
	338	1		74	22
5365-01-089-3573	189	23		75	18
4730-01-089-3829	456	21	4730-01-090-6474	456	38
	460	20		460	35
4730-01-089-4370	541	1		461	35
4730-01-089-4596	104	1		509	33
2540-01-089-5017	397	5	2520-01-090-6673	218	11
2540-01-089-5018	397	9		220	13
3020-01-089-7333	486	12	3020-01-090-6695	125	13
4720-01-089-9049	104	8	4720-01-090-7617	62	35
5340-01-089-9128	318	18	4720-01-090-7618	71	27
	319	19		74	25
	320	16		75	22
	321	20	5340-01-090-7619	62	34
	322	20	4710-01-090-7620	62	30
	323	23	4730-01-090-7621	62	21
	324	17	5340-01-090-7622	62	18
5340-01-089-9130	318	17	5305-01-090-7625	100	39
	319	17	5305-01-090-7626	134	44
	321	17		351	11
	322	18		352	11
	323	22		359	7
	324	16		364	2
3040-01-089-9326	528	12	5340-01-090-7627	62	3
4710-01-089-9375	456	25	5340-01-090-7628	62	7
4720-01-089-9887	460	16	5340-01-090-7629	62	19
4730-01-090-0258	541	5	5340-01-090-7631	506	3
5365-01-090-2074	513	29	4320-01-090-7632	506	4
	525	27	2520-01-090-7633	180	1
5305-01-090-3012	420	22	2530-01-090-7636	302	2
5340-01-090-4479	419	8	5340-01-090-7637	304	2
2940-01-090-4480	62	27	5340-01-090-7638	304	6
3040-01-090-4482	100	24	2510-01-090-7639	318	49
5340-01-090-4484	183	16		319	49
4710-01-090-4486	62	16		321	49
4710-01-090-4487	62	14		322	52
5306-01-090-4544	333	20		323	51
	335	16	5340-01-090-7640	374	25
	337	30	2510-01-090-7641	368	9
4710-01-090-4584	62	14		374	24
4730-01-090-4919	456	17	5340-01-090-7643	134	75
	457	3	5340-01-090-7644	134	45
	461	14	5340-01-090-7645	69	7
	509	9	5340-01-090-7649	67	8
4730-01-090-4924	456	27		68	11
	460	24		69	12
	461	21	5340-01-090-7650	62	15
	509	16	5340-01-090-7657	67	1

CROSS-REFERENCE INDEXES

NATIONAL STOCK NUMBER INDEX

STOCK NUMBER	FIG.	ITEM	STOCK NUMBER	FIG.	ITEM
5340-01-090-7657	69	1	5305-01-091-2498	118	22
5342-01-090-7658	71	2	4720-01-091-5169	71	5
	74	31		74	28
	75	2		75	4
3010-01-090-7747	506	6	2540-01-091-5449	397	4
5365-01-090-7769	301	4	2540-01-091-5450	536	4
5306-01-090-8620	383	4	5340-01-091-7627	364	6
5340-01-090-9331	180	7	2510-01-091-7628	339	9
	183	14	2530-01-091-7814	250	5
5340-01-090-9332	100	41	4730-01-091-8032	253	1
5360-01-090-9333	100	34		598	53
5342-01-090-9334	183	18	4730-01-091-9212	70	7
5360-01-090-9335	183	17		73	7
	184	3	4730-01-091-9370	314	2
3040-01-090-9341	100	23		315	8
3040-01-090-9342	100	28	2540-01-092-1264	533	3
5306-01-090-9344	100	32	5310-01-092-5495	107	18
2910-01-090-9345	55	1		110	18
4010-01-090-9352	390	3		126	3
4730-01-091-0266	146	5	5310-01-092-5496	584	14
	204	1		594	20
5340-01-091-1608	211	28	4730-01-092-6442	213	16
5340-01-091-1609	368	14		595	21
5340-01-091-1610	374	19		596	3
2530-01-091-1611	316	14	2540-01-092-9323	364	9
2540-01-091-1612	368	12	2540-01-092-9324	364	9
2540-01-091-1613	355	11	4310-01-092-9815	44	1
2540-01-091-1614	355	13	4310-01-092-9816	44	4
3940-01-091-1617	339	10	5360-01-093-0644	339	24
	339	17		340	7
	340	1	6160-01-093-4256	153	19
4710-01-091-1618	339	8	2520-01-093-4274	228	3
7125-01-091-1619	337	12	2540-01-093-4305	355	1
2590-01-091-1620	335	10	4820-01-093-5785	KITS	
	337	8	6160-01-093-5836	153	15
5340-01-091-1621	335	14	4310-01-094-0791	310	1
2590-01-091-1622	335	11		312	53
9520-01-091-1624	333	13	3950-01-094-1381	339	1
	334	15	2520-01-094-4751	185	11
2540-01-091-1625	335	13	2510-01-094-6714	343	4
5365-01-091-1630	183	13	4730-01-095-2034	71	22
5340-01-091-1633	324	51		72	
5340-01-091-1634	351	19		74	21
2590-01-091-1635	333	16		75	17
	334	18	2530-01-095-3561	251	8
	337	20	5360-01-095-3661	84	2
	338	17	4730-01-095-5833	272	3
2520-01-091-1659	230	1	6665-01-095-8285	606	
2520-01-091-1660	230	1	2990-01-095-8287	583	
2510-01-091-1687	355	15		593	1

CROSS-REFERENCE INDEXES

NATIONAL STOCK NUMBER INDEX

STOCK NUMBER	FIG.	ITEM	STOCK NUMBER	FIG.	ITEM
2920-01-095-8307	593	9	3040-01-101-0086	489	24
2920-01-095-8308	592	5	5340-01-101-0090	484	18
	594	28	3040-01-101-0096	491	17
6150-01-095-8309	173	1	5305-01-101-3312	582	15
6150-01-095-8310	173	16	3010-01-101-6712	469	21
2540-01-096-5018	592			470	15
	597				
5910-01-096-5021	150	15	5340-01-101-6713	471	34
2540-01-096-5023	579		6150-01-101-6741	175	4
6150-01-096-5053	580	5	2510-01-101-8358	404	1
	584	10	2510-01-101-8359	404	1
2815-01-096-9198	26	2	2540-01-101-8453	415	19
			5310-01-102-2711	292	7
			5310-01-102-2711	297	17
5340-01-097-8094	536	2	5310-01-102-2715	586	11
			5310-01-102-3270	623	28
2910-01-097-8591	93	5	5935-01-102-7124	160	2
5935-01-097-9974	154	1	5365-01-103-6006	476	6
2640-01-098-2029	300	5	3040-01-103-6009	525	10
2910-01-098-5093	88	22	5340-01-103-6010	368	3
2520-01-098-5108	185	17	5310-01-103-6438	502	30
6685-01-098-5110	132	22	5340-01-103-7587	180	19
4320-01-098-5115	51	18	5342-01-103-7589	102	26
	56	18	5342-01-103-7839	369	9
2520-01-098-5117	185	10	2510-01-103-8688	351	6
2520-01-098-5124	202	2	5340-01-103-8772	455	6
5330-01-098-6668	236	32		528	13
2910-01-098-6741	88	2	2540-01-103-9128	248	8
3020-01-098-6742	200	4	5306-01-104-1048	530	6
5930-01-098-6743	134	1	3020-01-104-3803	200	7
5330-01-098-9210	KITS		5310-01-104-3804	127	4
5310-01-099-0397	207	28	4730-01-104-3805	351	17
	244	9	5340-01-104-3828	456	11
5330-01-099-0554	149	1		461	8
5310-01-099-2550	381	19		509	4
	400	3	5340-01-104-3832	533	10
4730-01-100-0109	266	42	3040-01-104-3851	513	10
5310-01-100-2067	141	5	4730-01-104-4314	71	24
2590-01-100-3871	596	10		74	23
5310-01-100-5199	97	17		75	19
1680-01-100-5608	30	5	5340-01-104-4326	456	15
5342-01-101-0005	105	10		460	15
	106	4		461	12
	107	6		509	7
5340-01-101-0007	489	38	5340-01-104-4327	351	3
2540-01-101-0010	542	1	5340-01-104-4328	351	4
5342-01-101-0075	106	12	5330-01-104-4329	352	7
	107	33	5330-01-104-4330	352	7
3120-01-101-0076	484	17	5340-01-104-7400	366	4
5310-01-101-0077	484	11	5330-01-104-7702	153	2
4820-01-101-0080	549	42	5342-01-104-7843	351	14
	582	10	5365-01-104-7846	109	10
2530-01-101-0084	303	13	5306-01-104-8389	366	12
3040-01-101-0085	491	3	5315-01-104-8942	525	2

NATIONAL STOCK NUMBER INDEX

STOCK NUMBER	FIG.	ITEM	STOCK NUMBER	FIG.	ITEM
2530-01-104-8943	513	30	4820-01-106-1760	56	10
	525	26	4320-01-106-2061	462	1
5340-01-104-8944	106	7		510	1
	107	35	4810-01-106-2062	496	1
5340-01-104-8946	148	11	9535-01-106-2065	417	22
2510-01-104-8966	408	1	5330-01-106-2067	405	27
5340-01-104-9005	454	13	5340-01-106-2068	143	20
	513	13	2990-01-106-2296	592	15
5340-01-104-9012	525	24	5306-01-106-3850	127	1
5340-01-104-9013	528	31	4720-01-106-3982	509	20
5305-01-104-9018	364	13	4720-01-106-3986	582	4
	366	17		584	12
5305-01-104-9019	369	8	5315-01-106-4036	485	2
5340-01-104-9075	366	13	5325-01-106-4125	180	16
5340-01-104-9076	366	15		528	33
4710-01-104-9099	177	11	5365-01-106-4282	472	2
3040-01-104-9151	180	10	5365-01-106-4286	472	1
3040-01-104-9152	454	7	5342-01-106-5488	102	2
3040-01-104-9154	513	1		103	2
3040-01-104-9155	525	1	5342-01-106-6015	202	8
4820-01-104-9159	456	35	5365-01-106-6060	137	3
	460	32	5330-01-106-6735	375	4
4820-01-104-9313	248	11	5340-01-106-6751	549	16
4820-01-104-9992	289	4		551	14
5340-01-105-0992	106	5	2540-01-106-7121	541	7
5340-01-105-0993	105	15	4720-01-106-8289	457	11
5315-01-105-3318	407	7	5340-01-106-8346	204	4
	415	9		205	6
4720-01-105-3564	508	25	4730-01-107-2027	478	8
4720-01-105-4067	456	16		484	13
	461	13	5310-01-107-3570	110	13
	509	8	5307-01-107-3675	396	15
4720-01-105-4068	456	24	5355-01-107-4178	454	6
4720-01-105-4069	456	24		528	11
2910-01-105-6457	46	7	5331-01-107-4950	629	18
2520-01-105-6465	530	1	5340-01-107-5219	397	10
4810-01-105-6966	504	27	5340-01-107-5220	397	7
5340-01-105-9143	257	16	5340-01-107-5221	454	15
4730-01-105-9466	111	22		513	20
5315-01-105-9475	248	10	5340-01-107-6971	134	46
4730-01-106-0202	7	25	5340-01-107-9688	62	10
	8	10	2940-01-107-9689	KITS	
	14	15	4730-01-107-9690	254	21
	15	10		261	7
	32	3		261	30
	33	19	5340-01-107-9691	252	3
	37	12	4730-01-107-9692	213	37
2520-01-106-0826	189	6		273	1
4710-01-106-0914	456	25	5342-01-107-9693	288	4
	460	22		603	3

CROSS-REFERENCE INDEXES

NATIONAL STOCK NUMBER INDEX

STOCK NUMBER	FIG.	ITEM	STOCK NUMBER	FIG.	ITEM
4810-01-107-9694	289	1	5325-01-108-7375	369	4
	603	2	5330-01-108-7567	551	21
4820-01-107-9695	281	1	5340-01-108-9109	328	6
2590-01-107-9917	100	1	2540-01-108-9114	370	2
	101	8	4720-01-108-9118	104	9
3040-01-107-9928	100	22	2590-01-108-9121	359	10
	101	18	2510-01-108-9122	366	1
	525	29	5342-01-108-9123	353	8
5340-01-107-9929	252	14		361	2
	261	17	2540-01-108-9124	356	1
2520-01-107-9932	301	8	6210-01-108-9125	131	1
4720-01-107-9939	212	15	2540-01-108-9129	377	1
	214	13	9905-01-108-9187	565	1
	288	13	5365-01-108-9258	143	17
	603	11	5340-01-108-9268	143	12
5365-01-107-9967	374	14	5305-01-109-1292	'88	9
	376	6	5315-01-109-1443	371	9
	377	3	5365-01-109-2472	397	12
5360-01-108-0828	371	8	5365-01-109-3353	454	24
5340-01-108-0996	441	10		513	14
9905-01-108-1032	560	1	5342-01-109-4013	88	28
9905-01-108-1034	552	10	6685-01-109-5695	628	4
5306-01-108-3185	382	6	9320-01-109-5696	367	42
	387	4	9340-01-109-5934	367	45
5306-01-108-3186	382	12	5310-01-109-6056	526	11
	387	9	5310-01-109-6753	358	5
5365-01-108-4811	105	13	5310-01-109-6754	358	10
	106	18	5310-01-109-6755	328	10
5365-01-108-4814	359	3	5310-01-109-6756	328	11
5365-01-108-4815	528	30	5315-01-109-6846	469	20
4730-01-108-5103	254	9		470	17
4810-01-108-5107	289	17	5340-01-109-7553	397	11
6220-01-108-5108	131	7	5340-01-109-7950	580	2
4330-01-108-5109	61	6		584	4
9905-01-108-5164	552	8		592	2
5355-01-108-5184	289	2		594	35
5310-01-108-5236	131	18	2590-01-109-7990	551	5
	132	4	2590-01-109-7991	551	8
	134	8	2590-01-109-7992	549	9
	134	36	4730-01-109-8001	72	1
	134	71		78	1
4720-01-108-5346	104	4		110	2
4720-01-108-5347	104	5		538	4
4730-01-108-6410	598	56		539	17
				540	15
	599	10	5340-01-109-8013	552	11
	617	13	5305-01-109-9307	51	15
5340-01-108-7263	353	6		56	12
	361	6	5306-01-109-9384	525	20
5342-01-108-7265	328	3	5330-01-109-9410	102	22
5315-01-108-7340	347	7	5365-01-109-9490	513	32

CROSS-REFERENCE INDEXES

NATIONAL STOCK NUMBER INDEX

STOCK NUMBER	FIG.	ITEM	STOCK NUMBER	FIG.	ITEM
5365-01-109-9490	525	31	5315-01-112-4507	248	13
4730-01-110-0342	64	1	2530-01-112-6435	248	2
5342-01-110-1505	88	18	5340-01-112-6436	264	7
5330-01-110-2462	364	5	2940-01-112-6438	508	27
2940-01-110-2489	204	17	5360-01-112-6546	248	20
2540-01-110-4057	401	7	5340-01-112-6554	264	5
4730-01-110-4059	204	7	5306-01-112-6560	180	8
2510-01-110-4060	367	32		181	24
				265	8
2510-01-110-4061	382	36	4730-01-112-6561	77	22
5365-01-110-8163	528	6		80	24
5365-01-110-8183	359	3		213	35
3020-01-110-8251	191	11		215	11
4330-01-110-9054	93	8		273	23
5340-01-110-9205	468	14	4010-01-112-6562	454	14
5340-01-111-1351	513	17	4720-01-112-6577	267	2
5365-01-111-1520	525	14	4320-01-112-8365	504	24
2530-01-111-2260	264	2	2520-01-112-8366	218	1
2530-01-111-2261	264	8	5340-01-112-8909	263	5
	266	27	5305-01-112-9021	88	19
2815-01-111-2262	1	1		116	11
2520-01-111-2280	465	1	5305-01-112-9110	50	24
	512	1		55	24
2590-01-111-5391	149	2		85	8
2640-01-111-5467	292	13		87	12
5330-01-111-9291	189	7		88	27
5340-01-111-9878	513	35	5305-01-112-9698	14	12
5365-01-112-1524	10	5	5305-01-113-1179	49	5
2520-01-112-2156	223	1	5360-01-113-9615	197	19
2520-01-112-2157	220	1	5305-01-114-0895	12	16
2530-01-112-2161	310	2	2510-01-114-3689	367	32
4810-01-112-2162	509	20	2520-01-114-3691	190	18
5340-01-112-2164	258	2	2510-01-114-3693	367	21
2590-01-112-2167	243	21	5342-01-114-3694	367	23
4730-01-112-2168	266	40	6150-01-114-3697	150	9
	271	10	5970-01-114-3753	358	9
5340-01-112-2169	67	4	5935-01-114-5354	160	3
	68	6	2815-01-114-7397	14	7
	69	8	2815-01-114-7398	23	3
2510-01-112-2170	62	13	2815-01-114-7399	33	40
5340-01-112-2292	62	31	2920-01-114-7538	180	3
4720-01-112-2325	257	1	4820-01-114-7539	283	1
4720-01-112-2326	273	9	4730-01-114-7541	182	27
4720-01-112-2328	269	4	2510-01-114-7542	266	30
4720-01-112-2329	254	6	4820-01-114-7543	264	19
4720-01-112-2332	267	8	2510-01-114-7545	339	16
4720-01-112-2333	258	8	5340-01-114-7546	339	28
5340-01-112-4280	20	17		340	11
5310-01-112-4307	49	3	5340-01-114-7550	478	18
	284	66	2590-01-114-7551	478	9
5305-01-112-4312	286	30	9905-01-114-7589	557	6

CROSS-REFERENCE INDEXES

NATIONAL STOCK NUMBER INDEX

STOCK NUMBER	FIG.	ITEM	STOCK NUMBER	FIG.	ITEM
9905-01-114-7598	552	6	5340-01-117-0606	68	1
9905-01-114-7603	565	2	5330-01-117-0608	67	5
9905-01-114-7606	558	1		68	7
9905-01-114-7607	563	7		69	10
4010-01-114-7614	354	7	5330-01-117-0609	67	6
5935-01-114-7615	433	5		68	8
7690-01-114-7620	564	1		69	9
7690-01-114-7621	567	7	5310-01-117-0610	345	23
5340-01-114-7655	351	7	4730-01-117-1614	256	15
5340-01-114-7656	362	34		259	1
5340-01-114-7657	362	35		260	1
2520-01-114-7690	226	2		266	43
4720-01-114-7703	503	37		269	28
	504	28	2520-01-117-3010	176	1
5340-01-114-7712	537	1	2520-01-117-3013	206	5
	609	1			
4720-01-114-7728	BULK	14	2520-01-117-3014	237	1
2590-01-114-7741	487	11	2520-01-117-3015	237	1
4730-01-114-7742	504	3	2540-01-117-3025	370	18
	505	36	2910-01-117-3689	KITS	
4730-01-114-7743	503	39	5340-01-117-3793	370	15
	505	58	4730-01-117-3837	105	11
4710-01-114-7759	104	3	2540-01-117-4882	370	16
4710-01-114-7761	177	12	3110-01-117-4893	192	7
5325-01-114-7763	454	22	5306-01-117-4900	345	28
2530-01-114-7764	604	7	2520-01-117-4933	237	1
	605	1	8145-01-117-4978	176	3
	605	7		576	1
4730-01-115-0433	269	12	2930-01-117-5238	31	2
	269	26	2520-01-117-6592	237	1
	271	12	2590-01-117-6596	523	1
	598	54	2590-01-117-6597	522	3
5925-01-115-0557	441	7	5365-01-117-6655	137	5
9905-01-115-0569	552	5	6105-01-117-7944	519	21
9905-01-115-0570	552	6	4820-01-117-7949	81	2
5330-01-115-0604	467	5	5340-01-117-9876	151	24
5340-01-115-0615	351	10		162	3
3020-01-115-0619	486	9		253	15
	487	4		454	20
4730-01-115-0646	493	21		589	37
4720-01-115-2274	204	24		590	29
2520-01-115-2285	120	7		626	10
4730-01-115-6643	598	26	3030-01-118-1318	120	32
4730-01-115-7362	598	11	5306-01-118-2300	30	9
4730-01-116-1658	629	5	5306-01-118-4889	198	18
4730-01-116-5969	456	2	5360-01-118-5596	30	8
	460	2	2520-01-118-5971	526	1
	461	2	4730-01-118-5972	236	17
	509	1	9905-01-118-6092	567	6
4730-01-116-7886	266	36	5305-01-118-8826	28	2
4730-01-117-0602	104	7	5340-01-119-2642	345	27

CROSS-REFERENCE INDEXES

NATIONAL STOCK NUMBER INDEX

STOCK NUMBER	FIG.	ITEM	STOCK NUMBER	FIG.	ITEM
5340-01-119-2643	345	44	2590-01-120-8451	502	16
5365-01-119-2644	345	10	5340-01-120-8452	505	17
2530-01-119-2645	345	6	5330-01-120-8454	579	6
5340-01-119-2646	345	43	6150-01-120-8483	438	6
5315-01-119-2733	344	13	4730-01-120-8495	505	8
5340-01-119-2735	345	25	2590-01-120-8506	487	1
5360-01-119-2736	345	8	5310-01-120-8507	105	4
5360-01-119-2737	345	29	5340-01-120-8510	339	26
5365-01-119-2739	345	20		340	12
2510-01-119-4094	322	13	2590-01-120-8512	499	11
2590-01-119-4103	479	8	4720-01-120-8516	480	7
5306-01-119-4271	284	59	4720-01-120-8517	482	1
5365-01-119-4954	85	15	4720-01-120-8518	480	16
	87	17	4720-01-120-8519	504	19
5315-01-119-5239	502	46	4720-01-120-8521	505	38
9510-01-119-5679	533	2	4720-01-120-8522	503	5
5306-01-119-5681	225	6	4710-01-120-8538	276	2
5340-01-119-5682	223	7	4710-01-120-8539	504	10
	225	7	4710-01-120-8540	503	7
9905-01-119-5788	552	4	4710-01-120-8541	504	1
5330-01-119-5801	473	34	4710-01-120-8542	504	19
5306-01-119-5834	579	12	4710-01-120-8543	505	24
4720-01-119-5843	270	8	4710-01-120-8544	505	37
4720-01-119-5844	270	7	4730-01-120-8547	493	11
4730-01-119-6895	617	6	9520-01-120-8551	502	8
5305-01-119-8621	44	14	9390-01-120-9864	367	16
2530-01-119-8710	304	7	5310-01-121-1703	120	6
3040-01-119-8711	502	37		616	7
5306-01-119-8870	50	21	2510-01-121-2541	404	13
	55	21	5305-01-121-2696	413	25
5305-01-119-8889	500	25	5365-01-121-3068	19	10
3040-01-120-2164	345	22	5940-01-121-4280	150	8
2990-01-120-2883	16	2	5970-01-121-6587	150	7
5340-01-120-2891	322	9	4710-01-121-7600	104	10
3040-01-120-3054	345	24	5340-01-121-7601	143	20
2520-01-120-3673	526	18	5910-01-121-7603	211	38
5330-01-120-3733	249	27	4730-01-121-7604	504	12
5935-01-120-3744	163	2	2590-01-121-7605	503	16
5360-01-120-4610	149	12	2590-01-121-7606	503	8
5330-01-120-8090	189	25	5315-01-121-7688	505	16
5305-01-120-8438	120	11	5315-01-121-7689	505	15
2530-01-120-8441	145	2	4730-01-121-7718	522	19
5365-01-120-8442	339	4	3040-01-121-7741	211	26
5340-01-120-8443	333	8	3040-01-121-7742	211	54
5342-01-120-8444	359	2	3040-01-121-7744	211	16
			4720-01-121-7749	104	2
2590-01-120-8447	403	13	5307-01-121-9883	207	18
2510-01-120-8448	403	1	5330-01-121-9886	367	15
2510-01-120-8449	403	3	4730-01-121-9889	504	15
3120-01-120-8450	502	20		505	23

CROSS-REFERENCE INDEXES

NATIONAL STOCK NUMBER INDEX

STOCK NUMBER	FIG.	ITEM	STOCK NUMBER	FIG.	ITEM
5331-01-122-0492	523	10	3020-01-122-5906	186	13
	523	33	2930-01-122-5982	120	12
5360-01-122-0634	517	15	5310-01-122-5992	517	7
5330-01-122-0864	523	4	5310-01-122-5993	522	15
	523	27	2530-01-122-6016	245	19
5315-01-122-2014	516	16	5310-01-122-6109	516	1
5342-01-122-2015	519	3		518	4
5315-01-122-2031	524	3		519	24
5315-01-122-2033	518	8	5310-01-122-6110	521	23
5310-01-122-2060	549	13	5310-01-122-6111	519	17
	551	11	5310-01-122-6112	519	34
2590-01-122-2419	157	1	5310-01-122-6130	517	17
2510-01-122-2822	516	4	5310-01-122-6148	515	3
4720-01-122-3656	153	13		516	2
2910-01-122-4015	51	7		518	3
	56	6		519	25
5342-01-122-4489	515	5	5310-01-122-6149	521	2
5315-01-122-4490	517	4	5310-01-122-6150	516	31
5342-01-122-4491	516	25		519	16
5310-01-122-4492	519	40	5310-01-122-6151	519	35
5310-01-122-4595	514	3	5310-01-122-6152	516	9
5340-01-122-4596	516	8	5310-01-122-6153	519	39
5315-01-122-4602	517	5	4720-01-122-6166	505	59
	518	11	5310-01-122-7676	516	32
5330-01-122-4609	523	7		517	10
	523	30	5310-01-122-7677	516	18
5331-01-122-4612	522	17	5340-01-122-8002	14	18
5331-01-122-4613	523	14	5315-01-122-8482	517	13
	523	37	5305-01-122-8516	570	9
5340-01-122-4632	518	2	2520-01-122-9928	188	20
5305-01-122-4634	521	3	2530-01-122-9933	249	17
5305-01-122-4635	521	6	5310-01-123-0905	523	12
5325-01-122-4661	522	5		523	35
	523	3	5340-01-123-1333	213	30
	523	26	4730-01-123-1516	503	15
5305-01-122-4692	524	4	5365-01-123-1638	120	22
5305-01-122-4693	519	4		120	24
5305-01-122-4702	516	22	2520-01-123-1648	186	5
5305-01-122-4703	518	6	2520-01-123-2649	204	15
2520-01-122-5169	186	8			
3040-01-122-5201	189	26	2530-01-123-3105	248	16
5680-01-122-5214	367	22	2990-01-123-3454	33	10
5330-01-122-5392	522	18	5331-01-123-4536	249	36
5305-01-122-5474	516	17	4730-01-123-4546	457	9
5330-01-122-5636	522	9		458	2
5330-01-122-5637	523	15		507	38
	523	38	5999-01-123-4557	134	22
5305-01-122-5638	519	36	4710-01-123-4576	214	10
5305-01-122-5639	519	23	2940-01-123-4875	33	20
4730-01-122-5857	620	9	5360-01-123-5483	189	10

NATIONAL STOCK NUMBER INDEX

STOCK NUMBER	FIG.	ITEM	STOCK NUMBER	FIG.	ITEM
2520-01-123-5556	527	16	4710-01-125-3608	214	16
5330-01-123-6409	248	3	9520-01-125-4082	376	12
5315-01-123-6812	508	16	2530-01-125-4280	246	5
2540-01-123-6823	539	2	4720-01-125-4467	521	5
5331-01-123-7038	249	9	5340-01-125-4688	339	13
5325-01-123-7063	523	9		339	20
	523	32		340	4
5365-01-123-7082	516	5	4720-01-125-5521	519	33
5342-01-123-8045	521	7	4720-01-125-5865	519	19
5340-01-123-8324	521	8		521	12
2520-01-123-8704	120	26	6140-01-125-6073	153	28
	122	9		582	2
4730-01-123-8824	273	15		584	31
3040-01-123-9681	211	1		589	60
2815-01-124-0232	17	11		590	41
	44	15	6140-01-125-6074	582	3
4730-01-124-0293	26	18		584	30
5306-01-124-1225	248	15		589	59
5360-01-124-1402	249	32		590	42
5365-01-124-2831	190	3	6140-01-125-6075	153	26
5340-01-124-3189	316	2		582	1
5360-01-124-3373	120	17		584	29
3020-01-124-3421	188	19		589	61
4930-01-124-3523	30	11		590	40
4730-01-124-3762	51	20	2530-01-125-6076	251	5
	56	20	5340-01-125-6078	391	19
2590-01-124-5052	508	14	5360-01-125-6118	508	17
5340-01-124-5054	508	18	2540-01-125-6154	580	10
9515-01-124-5055	406	39		584	32
4810-01-124-5056	97	9		592	6
5331-01-124-5720	249	44		594	29
5305-01-124-5779	189	5	5330-01-125-6280	249	8
3020-01-124-6079	120	8	3040-01-125-6491	522	6
5330-01-124-6405	249	45	5315-01-125-9084	354	5
5310-01-124-6463	44	19	2530-01-125-9272	KITS	
2520-01-124-6469	189	20	3040-01-125-9678	518	13
2530-01-124-6530	246	5	4730-01-125-9679	523	18
5360-01-124-6811	249	7	4730-01-125-9680	523	19
5325-01-124-7760	580	6	4730-01-125-9681	521	19
	584	11		523	16
	587	6	4820-01-125-9698	523	22
	588	8	4310-01-125-9714	523	13
	592	23		523	36
6210-01-124-9301	131	10	4720-01-125-9759	521	13
3020-01-124-9417	120	15	4720-01-125-9767	519	42
5305-01-125-0929	249	18	4720-01-125-9768	519	8
5305-01-125-1181	134	25	2590-01-125-9770	522	8
4720-01-125-1365	521	18	3040-01-125-9828	522	16
5360-01-125-1670	249	37	2520-01-125-9837	524	1
5360-01-125-1671	249	21	4730-01-125-9965	519	1

CROSS-REFERENCE INDEXES

NATIONAL STOCK NUMBER INDEX

STOCK NUMBER	FIG.	ITEM	STOCK NUMBER	FIG.	ITEM
4730-01-125-9965	521	17	3020-01-126-8444	207	42
4730-01-125-9969	519	31	2530-01-126-8445	306	7
	521	15	5365-01-126-8689	120	25
4730-01-125-9975	516	12	5305-01-126-8811	33	29
	522	14	5310-01-126-8834	249	41
	524	2	2530-01-126-9209	303	14
2590-01-126-0093	523	6	2530-01-126-9303	311	2
	523	29	2520-01-126-9363	207	25
5342-01-126-0176	374	12		244	11
4730-01-126-0218	516	3	2520-01-126-9364	207	21
	517	12	2815-01-126-9367	33	32
	518	7	5310-01-126-9404	107	16
4730-01-126-0265	519	32		121	10
	521	11	5315-01-126-9412	516	29
4730-01-126-0266	521	4	5315-01-126-9484	522	2
5430-01-126-0268	519	29	5315-01-126-9485	522	1
2520-01-126-0443	524	6	4720-01-126-9555	217	8
5330-01-126-0565	527	8	4710-01-126-9565	526	25
5310-01-126-0566	250	1	5360-01-127-0858	517	8
	251	1	5360-01-127-0953	289	8
	296	8	2815-01-127-1060	24	12
	526	9	2520-01-127-1676	208	3
5315-01-126-0601	23	15	2530-01-127-1677	249	1
5310-01-126-1045	44	20	2540-01-127-1812	248	2
3120-01-126-1097	186	11	2520-01-127-2336	209	4
5305-01-126-1128	50	28	2530-01-127-2337	249	2
	85	11	2520-01-127-2623	209	11
	87	9	2520-01-127-2624	207	5
	88	23	2520-01-127-2625	208	6
5331-01-126-1233	249	34	2520-01-127-2626	526	7
3110-01-126-1287	120	14	2520-01-127-3595	207	29
2520-01-126-1493	526	20	2530-01-127-3596	245	4
5306-01-126-1618	527	10	2815-01-127-3597	24	12
4730-01-126-1723	519	45	2815-01-127-3598	24	12
5305-01-126-2619	190	13	5330-01-127-3803	289	19
5365-01-126-3334	50	5	2520-01-127-3969	198	10
	55	5	2530-01-127-3971	249	6
5330-01-126-3469	473	31	2520-01-127-5005	197	16
4730-01-126-3541	521	14	2530-01-127-5006	305	16
4730-01-126-3845	519	22	2520-01-127-5766	209	16
2520-01-126-3847	516	28	2520-01-127-6254	14	10
3830-01-126-3886	523	24	4730-01-127-6697	311	11
5365-01-126-5192	120	9	4730-01-127-6900	188	16
4820-01-126-5379	190	10	2520-01-127-6901	210	1
5325-01-126-7264	526	23	5340-01-127-6920	391	17
2815-01-126-7404	14	13	5340-01-127-6921	508	7
5330-01-126-7512	412	20	4820-01-127-6922	504	18
2530-01-126-7869	250	5	2520-01-127-6950	219	1
5342-01-126-7910	177	8	2520-01-127-6951	221	1
	178	6	2520-01-127-6952	226	2

CROSS-REFERENCE INDEXES

NATIONAL STOCK NUMBER INDEX

STOCK NUMBER	FIG.	ITEM	STOCK NUMBER	FIG.	ITEM
2520-01-127-6953	229	3	6220-01-129-5740	131	8
2520-01-127-6954	229	3	2510-01-129-6074	387	11
5305-01-127-7087	560	4	4720-01-129-6082	503	20
5330-01-127-7134	289	15	5365-01-129-6590	207	12
4730-01-127-7346	217	7	5365-01-129-6777	217	6
	526	28	5305-01-129-6842	306	4
5310-01-127-7599	549	18	5325-01-129-6849	207	43
	551	13	5305-01-129-6901	28	3
2520-01-127-7790	526	14	2590-01-129-7523	478	1
5331-01-127-8550	289	16	3120-01-129-7659	284	42
5340-01-127-8703	549	2	5315-01-129-9190	501	11
	594	23	2910-01-130-1535	97	23
4720-01-128-0179	508	8	2530-01-130-2339	284	38
5340-01-128-0191	153	17	5305-01-130-3267	207	15
	586	15	4710-01-130-3437	105	16
4720-01-128-0331	36	1	5935-01-130-3536	134	15
3950-01-128-1549	339	5	5310-01-130-4274	6	7
4730-01-128-1554	504	9	4730-01-130-5158	526	24
3120-01-128-1565	306	32	2520-01-130-5770	187	2
5310-01-128-1592	236	11	2520-01-130-5772	193	11
5999-01-128-2755	155	3	2520-01-130-5821	197	12
	156	3	5315-01-130-5963	602	24
5331-01-128-3954	629	17	5306-01-130-5994	602	25
5305-01-128-4095	499	26		611	21
5365-01-128-5061	207	19	5305-01-130-6130	16	6
5365-01-128-5062	207	35		118	16
5365-01-128-5063	207	35	4810-01-130-7930	493	10
5365-01-128-5064	207	35	3040-01-130-7931	100	5
3120-01-128-5222	516	13	5340-01-130-7932	102	28
5340-01-128-5302	454	19	5340-01-130-7933	105	9
	513	27	2930-01-130-7934	105	5
5410-01-128-5529	407	8	2590-01-130-7935	132	29
	415	11	2510-01-130-7936	323	10
2530-01-128-5552	280	1	9520-01-130-7937	323	14
2910-01-128-9537	97	7	5340-01-130-7938	318	28
5340-01-128-9558	97	16	5340-01-130-7939	339	6
5310-01-129-0193	306	37	2540-01-130-7940	356	13
2940-01-129-0261	62	9	5340-01-130-7941	356	10
2510-01-129-0278	508	10	2510-01-130-7942	367	8
5306-01-129-0327	244	14	2510-01-130-7943	367	20
5330-01-129-0384	311	10	2510-01-130-7944	367	20
5365-01-129-0399	361	8	2510-01-130-7945	367	33
5340-01-129-0474	143	10	2540-01-130-7946	370	40
5305-01-129-4214	284	58	2540-01-130-7947	371	11
5310-01-129-4227	311	4	2540-01-130-7948	371	11
3040-01-129-4302	44	5	5340-01-130-7949	374	5
5310-01-129-4373	207	27	2510-01-130-7950	382	11
	244	10		387	2
5305-01-129-4384	49	7	2510-01-130-7951	382	7
5315-01-129-4621	289	3		387	8

CROSS-REFERENCE INDEXES

NATIONAL STOCK NUMBER INDEX

STOCK NUMBER	FIG.	ITEM	STOCK NUMBER	FIG.	ITEM
2510-01-130-7952	382	2	5365-01-131-3396	412	29
6150-01-130-8042	172	1	4730-01-131-4884	115	18
	173	7	5330-01-131-5416	502	32
6160-01-130-8045	153	1	4730-01-131-5944	519	14
4720-01-130-8086	277	5	4730-01-131-5945	519	41
6150-01-130-8102	175	7		521	10
2930-01-131-0107	106	1	4730-01-131-6099	519	44
5340-01-131-0108	134	43	4820-01-131-6123	249	16
5340-01-131-0109	323	17	2530-01-131-6172	284	3
5340-01-131-0110	323	13	2930-01-131-7442	KITS	
2540-01-131-0111	382	10	5340-01-131-7443	108	7
	387	3	5340-01-131-7444	108	11
2510-01-131-0112	387	25	2530-01-131-7445	305	5
2510-01-131-0113	387	25	3040-01-131-7446	306	40
2510-01-131-0114	385	35	5340-01-131-7447	337	23
2510-01-131-0115	385	34	5340-01-131-7452	365	8
9520-01-131-0117	385	24	2590-01-131-7453	494	25
2510-01-131-0118	385	17	7690-01-131-7499	557	4
2510-01-131-0119	385	21	4710-01-131-7529	148	10
2510-01-131-0120	385	43	2530-01-131-7741	245	4
2510-01-131-0121	385	38	2530-01-131-7855	316	3
2510-01-131-0122	382	4	2540-01-131-9639	248	9
2590-01-131-0123	393	7	4820-01-132-0582	463	6
5340-01-131-0124	393	12	4820-01-132-0606	249	42
5340-01-131-0125	396	10	5306-01-132-0834	526	2
2590-01-131-0126	398	12	5365-01-132-1984	118	10
2590-01-131-0128	470	24	5365-01-132-2015	295	17
2590-01-131-0129	470	24	5930-01-132-3247	97	18
5340-01-131-0130	470	25	5315-01-132-3569	390	12
5340-01-131-0131	470	25	2590-01-132-4654	317	14
6150-01-131-0141	175	5	5330-01-132-4734	249	35
6150-01-131-0147	172	9	4730-01-132-4852	519	9
6150-01-131-0148	172	15	4730-01-132-4858	504	8
6150-01-131-0150	173	22	4720-01-132-4859	460	21
4330-01-131-0279	KITS		4720-01-132-4868	153	25
2920-01-131-1939	608		5510-01-132-4879	393	13
2590-01-131-1940	480	21	3120-01-132-5579	248	7
	482	10	2520-01-132-6847	527	3
6150-01-131-2053	608	2	4730-01-132-6848	207	44
7690-01-131-2061	569	4	4730-01-132-6849	197	13
5360-01-131-2063	480	20	5510-01-132-7138	393	11
	482	11	5510-01-132-7139	393	11
5330-01-131-2066	176	6	5306-01-132-8271	244	3
3020-01-131-2078	608	19	3040-01-132-8272	244	15
5310-01-131-2085	480	23			
2520-01-131-2694	188	15	5307-01-132-8273	295	9
4710-01-131-2729	190	8			
2520-01-131-2831	526	4	5306-01-132-8274	295	9
4720-01-131-2857	519	18	5310-01-132-8275	112	7
5330-01-131-2967	41	3		113	9

NATIONAL STOCK NUMBER INDEX

STOCK NUMBER	FIG.	ITEM	STOCK NUMBER	FIG.	ITEM
5310-01-132-8275	528	27	2520-01-134-0899	216	5
	615	17	2530-01-134-0900	207	8
5330-01-132-8346	216	1	2530-01-134-0901	245	2
5305-01-132-8387	316	7	2530-01-134-0902	282	3
	316	10	2530-01-134-0903	282	14
5305-01-132-8390	528	26	3020-01-134-1015	209	8
	529	10	2520-01-134-1089	465	2
4730-01-132-8700	519	20		512	2
5510-01-132-8746	393	9	3040-01-134-1184	295	11
3020-01-132-8860	530	20	3040-01-134-1318	216	4
2530-01-132-8943	317	26	2530-01-134-1332	244	13
3120-01-132-9339	KITS		2530-01-134-1333	245	2
5365-01-133-0069	209	18	2530-01-134-1336	250	13
5305-01-133-0163	530	24	3020-01-134-1830	190	2
5330-01-133-0205	528	24	2530-01-134-1834	KITS	
	529	9	2530-01-134-1835	KITS	
	568	24	4820-01-134-1836	249	4
4810-01-133-1459	KITS		2805-01-134-1837	249	23
4730-01-133-1461	197	4	2530-01-134-1838	249	5
4710-01-133-1538	284	50	2520-01-134-1839	527	13
5305-01-133-2060	52	13	3830-01-134-1946	514	1
2930-01-133-2143	108	3	5306-01-134-1966	361	12
2540-01-133-2150	365	5	3040-01-134-2023	317	18
5310-01-133-3614	209	9	4710-01-134-2111	BULK	37
5365-01-133-3749	207	19	2930-01-134-2192	118	6
5365-01-133-3750	207	19	2930-01-134-2238	116	1
5310-01-133-4481	529	7	2520-01-134-3418	306	13
5325-01-133-4679	186	4	2520-01-134-3455	465	3
3040-01-133-4821	523	5		512	3
	523	28	4820-01-134-3457	519	11
5305-01-133-5811	607	15	2520-01-134-3471	465	8
5310-01-133-5847	82	10		466	7
5330-01-133-5858	297	8		512	8
4730-01-133-6209	71	21	4720-01-134-3475	526	29
	74	20	3040-01-134-3480	118	7
	75	16	3040-01-134-3605	217	3
4710-01-133-6947	34	7	2520-01-134-3665	207	33
2930-01-133-6948	120	16	2530-01-134-3667	282	5
5305-01-133-7193	248	5	2530-01-134-3668	305	1
5305-01-133-7201	264	4	2530-01-134-3669	306	28
5310-01-133-7216	248	4	2530-01-134-3670	317	1
5330-01-133-7262	295	3	2520-01-134-3706	465	5
2520-01-133-7876	228	6		512	5
3020-01-133-9037	526	6	6150-01-134-3777	438	1
5310-01-133-9186	306	3	4820-01-134-3788	282	2
4730-01-133-9866	621	2	4720-01-134-3847	277	1
5310-01-134-0209	236	11	5342-01-134-3860	118	24
5365-01-134-0522	306	9	5970-01-134-5093	153	10
4730-01-134-0853	617	5	5365-01-134-5534	120	31
2520-01-134-0898	185	1	5305-01-134-5659	50'	29

CROSS-REFERENCE INDEXES

NATIONAL STOCK NUMBER INDEX

STOCK NUMBER	FIG.	ITEM	STOCK NUMBER	FIG.	ITEM
5310-01-134-5791	236	11	5305-01-135-4754	514	2
5340-01-134-6530	345	46	9510-01-135-4762	367	27
2520-01-134-6534	200	12	5310-01-135-4797	538	8
5340-01-134-6535	476	7	5310-01-135-4798	506	18
8145-01-134-6538	577	1	5331-01-135-4807	306	10
5306-01-134-6540	358	7	5330-01-135-4808	306	16
2530-01-134-6570	293	18	5330-01-135-4809	306	18
	295	10			
4820-01-134-6619	306	27	5310-01-135-4835	541	10
2530-01-134-6626	244	2	5305-01-135-4839	515	2
4730-01-134-6988	217	9		516	7
	526	30		519	6
4730-01-134-6991	256	1	5305-01-135-5344	14	19
5340-01-134-7635	375	3	4820-01-135-5372	521	1
2520-01-134-7657	216	6	5305-01-135-5446	51	9
5365-01-134-8660	198	5		56	8
5320-01-134-8671	562	2	5305-01-135-5447	282	11
3040-01-134-8909	526	12	5310-01-135-6042	591	3
2590-01-134-9834	317	21	5342-01-135-6350	306	25
5310-01-135-0049	295	5	4720-01-135-6694	284	49
2520-01-135-0085	210	2	5310-01-135-6699	220	3
5330-01-135-0682	305	17	5310-01-135-6700	220	4
5365-01-135-2077	282	10	5340-01-135-6743	517	3
5365-01-135-2829	207	13	5310-01-135-6752	228	4
5365-01-135-2830	207	13	5310-01-135-6754	305	18
5365-01-135-2831	207	13	5310-01-135-6755	305	9
4730-01-135-3009	523	20	5310-01-135-6758	306	20
5330-01-135-3376	219	7	5306-01-135-7202	528	20
	226	10		529	4
	229	11	9905-01-135-7474	566	1
5310-01-135-3464	541	9	9905-01-135-7476	563	3
5305-01-135-3536	207	10	5307-01-135-9290	295	9
5310-01-135-3554	296	7	5307-01-135-9291	295	9
5330-01-135-4064	305	7	5315-01-135-9402	522	20
5360-01-135-4065	306	29	5310-01-135-9506	228	3
5330-01-135-4066	306	15	2530-01-136-1098	149	5
5330-01-135-4067	306	17	2590-01-136-1438	374	10
5330-01-135-4068	306	30	5305-01-136-1618	306	19
5331-01-135-4069	306	31	5305-01-136-1619	306	39
5340-01-135-4292	249	38	5315-01-136-1659	228	2
6150-01-135-4401	432	1	5340-01-136-1660	306	21
2920-01-135-4402	440	5	5340-01-136-1665	282	12
2920-01-135-4403	440	5	5305-01-136-1686	306	41
6150-01-135-4404	437	1	5305-01-136-1688	522	4
2590-01-135-4405	437	4		523	2
6150-01-135-4478	430	1		523	25
6150-01-135-4479	436	1	2815-01-136-1986	23	2
6150-01-135-4480	436	8	2815-01-136-1987	23	1
6150-01-135-4481	439	5	2510-01-136-4442	359	5
6150-01-135-4482	433	4	4720-01-136-4454	204	10

NATIONAL STOCK NUMBER INDEX

STOCK NUMBER	FIG.	ITEM	STOCK NUMBER	FIG.	ITEM
5320-01-136-4495	318	14	3040-01-139-1611	293	21
5315-01-136-4542	507	24	2540-01-139-2947	535	9
5306-01-136-5331	602	21	2520-01-139-3124	207	20
5315-01-136-7307	522	21	4730-01-139-4231	277	6
5945-01-136-7640	213	23	2520-01-139-4233	207	14
5330-01-136-8569	53	8	4820-01-139-4346	306	26
2520-01-136-8717	216	2	4820-01-139-4888	KITS	
2520-01-136-8718	216	8	5305-01-139-6484	519	15
2520-01-136-8719	216	12	5315-01-139-6568	502	23
2815-01-136-8720	217	5	5310-01-139-9856	494	15
2590-01-136-8721	243	17	2530-01-140-0107	244	4
4720-01-137-1453	204	19	5340-01-140-1156	517	14
5330-01-137-4487	128	2	2520-01-140-2376	KITS	
5330-01-137-4799	217	2	9520-01-140-2378	323	11
	527	15	9515-01-140-2379	394	1
2520-01-137-4843	207	7	4820-01-140-4298	597	2
2530-01-137-5921	KITS		9535-01-140-6470	333	18
3040-01-137-6264	485	13	9905-01-140-8219	561	3
4320-01-137-6293	467	3	4810-01-140-8221	198	15
2520-01-137-6294	484	4	2940-01-140-8227	192	17
5340-01-137-6302	478	13	5945-01-140-8242	198	16
5305-01-137-6706	37	14	3110-01-140-8880	530	18
5360-01-137-6707	84	5	5340-01-141-0832	367	40
5315-01-137-6858	516	14	6685-01-141-0907	114	5
5310-01-137-7062	579	3	3020-01-141-1554	209	3
5330-01-137-7089	494	24	2910-01-141-4337	50	8
5330-01-137-7090	1	4	3040-01-141-5032	518	10
5325-01-137-8828	494	23	2590-01-141-6305	494	7
2815-01-137-9707	42	7	5930-01-141-8414	211	2
3110-01-137-9725	306	14	5310-01-141-8464	209	12
2530-01-138-0922	306	24	5310-01-141-8465	306	35
			2840-01-141-9503	189	31
			5330-01-141-9579	185	24
2540-01-138-0925	549	8	2510-01-142-1294	367	43
2530-01-138-2015	244	12	2590-01-142-1310	513	18
2530-01-138-2016	244	1	2815-01-142-1732	28	7
5305-01-138-5115	306	5	2520-01-142-1979	217	1
3040-01-138-5391	524	5	5305-01-142-2526	305	19
5310-01-138-5516	500	2	5305-01-142-2792	217	4
5305-01-138-6295	306	2		527	14
5365-01-138-7102	282	7	5305-01-142-2793	216	3
5305-01-138-7624	516	30	5305-01-142-2794	216	9
3040-01-138-8578	244	16	9905-01-142-3114	563	3
5340-01-139-1002	172	21	3040-01-142-3991	523	8
4730-01-139-1585	456	3		523	31
	458	21	5307-01-142-4783	293	17
	460	3	5307-01-142-4784	293	17
	461	3	2530-01-142-7939	284	6
	503	40		286	8
	509	30	3020-01-142-8090	208	7

CROSS-REFERENCE INDEXES

NATIONAL STOCK NUMBER INDEX

STOCK NUMBER	FIG.	ITEM	STOCK NUMBER	FIG.	ITEM
4730-01-142-8524	285	27	5342-01-145-1549	86	6
3110-01-142-9490	527	7	5320-01-145-3183	102	27
5310-01-143-0542	192	15	5320-01-145-3184	319	16
2920-01-143-1263	189	21		320	39
2510-01-143-1265	389	1		321	14
7690-01-143-1270	557	3		322	14
2590-01-143-1455	516	15		323	9
5331-01-143-2780	305	15		324	13
4820-01-143-4172	521	20	5320-01-145-3185	319	14
5330-01-143-4185	305	20		320	11
5330-01-143-4186	306	38		321	12
2530-01-143-4203	306	22		322	10
5315-01-143-4220	209	19		323	12
9905-01-143-4526	572	12		324	11
6150-01-143-4527	431	5	5320-01-145-3186	320	14
2990-01-143-5489	84	1		321	29
	86	1		321	31
5330-01-143-6013	306	34		323	33
5342-01-143-6045	8	16		324	25
5342-01-143-6046	85	10	5320-01-145-3187	322	11
	87	10	5320-01-145-3188	319	13
5342-01-143-6048	24	1		320	10
5331-01-143-6322	522	12		321	11
5330-01-143-6486	317	13		322	12
5305-01-143-6534	393	1		323	18
5340-01-143-8312	375	5		324	10
3110-01-143-9242	208	8	5320-01-145-3189	323	15
3120-01-143-9249	209	15	5320-01-145-3191	371	10
6150-01-143-9543	439	1	3130-01-145-3943	516	6
3120-01-143-9547	KITS		5310-01-145-3991	209	7
2520-01-144-1528	206	1	5320-01-145-4621	318	6
5305-01-144-1625	402	8		319	7
5340-01-144-3233	502	41		320	6
5970-01-144-4841	153	9		321	6
5325-01-144-4871	149	6		322	6
4330-01-144-5557	494	22		323	6
5305-01-144-6233	284	4		324	6
	286	16	5330-01-145-5381	116	13
5315-01-144-8675	180	12	2520-01-145-6820	530	9
	181	20	5325-01-145-6921	187	22
5340-01-144-8676	542	8		193	6
3120-01-144-8882	KITS		5310-01-145-6923	189	13
2940-01-145-0350	62	25	5305-01-145-8359	284	39
6150-01-145-0361	131	21	5305-01-145-8381	19	15
4720-01-145-0371	253	13	5305-01-145-8385	516	11
5305-01-145-0777	46	19	3120-01-145-9132	KITS	
5310-01-145-1114	46	2	2940-01-145-9398	32	13
5305-01-145-1196	591	4	2910-01-146-0048	46	18
5365-01-145-1310	323	16	2815-01-146-1024	28	1
5342-01-145-1549	84	6	2520-01-146-1034	191	7

CROSS-REFERENCE INDEXES

NATIONAL STOCK NUMBER INDEX

STOCK NUMBER	FIG.	ITEM	STOCK NUMBER	FIG.	ITEM
4820-01-146-1048	129	15	5365-01-147-9802	7	2
4730-01-146-1075	189	1	5340-01-147-9822	549	37
	190	9	5320-01-148-3706	367	36
2815-01-146-1092	28	6	4730-01-148-7397	71	14
3130-01-146-1150	7	19		77	18
3130-01-146-1228	7	12	5365-01-148-8353	7	2
2910-01-146-1999	50	1	4820-01-148-9278	187	4
2520-01-146-3039	200	1	2520-01-148-9279	187	18
2520-01-146-3447	207	3	3110-01-149-0876	200	20
4730-01-146-4113	182	3	3040-01-149-1111	216	11
2815-01-146-4164	66	23	3020-01-149-1239	186	12
2815-01-146-4182	9	1	4820-01-149-1317	277	3
3020-01-146-4198	209	5	9905-01-149-1343	566	2
3130-01-146-4504	7	17	2520-01-149-1785	187	6
2520-01-146-5254	189	8	2520-01-149-3438	187	13
2590-01-146-5257	487	15		193	10
2590-01-146-5258	492	5	2520-01-149-3439	194	6
4310-01-146-5921	284	20	2520-01-149-3808	187	13
	286	21		193	10
3040-01-146-5952	517	6	3020-01-149-3836	187	21
5330-01-146-6053	187	11	5310-01-149-4407	6	18
	193	13		107	23
6695-01-146-7132	145	10	5330-01-149-4415	305	11
3110-01-146-7144	185	28	4710-01-149-5078	82	5
3120-01-146-7196	286	44	4710-01-149-5079	82	7
5340-01-146-7667	581	13	5935-01-149-5165	158	46
	585	31		159	4
	588	6	5365-01-149-5416	367	26
	591	9	6695-01-149-5830	132	1
2530-01-146-8941	301	3	5330-01-149-7229	306	8
5320-01-146-9582	244	5	2520-01-150-0899	190	5
5330-01-147-0748	37	3	2520-01-150-0900	187	13
2590-01-147-1517	408	12		193	10
4730-01-147-2223	7	18	2520-01-150-0901	193	1
	8	5	2520-01-150-3690	194	6
	284	47	5310-01-150-4003	107	19
6680-01-147-2421	478	2	2910-01-150-4925	46	20
5365-01-147-2495	7	2	3040-01-150-4926	51	8
5365-01-147-2496	7	2		56	7
5365-01-147-2497	7	2	5310-01-150-5914	619	6
5305-01-147-4033	19	5	5365-01-150-6257	7	15
5330-01-147-4071	33	1	5320-01-150-7745	319	43
5360-01-147-4787	216	10		320	42
2510-01-147-4992	367	35		321	29
3120-01-147-5275	44	3		321	42
5995-01-147-5423	136	7		322	48
3110-01-147-6681	186	10	5999-01-150-8808	158	43
5340-01-147-6759	162	6		159	5
	263	12	5330-01-150-9691	474	24
4730-01-147-7954	505	56		489	7

CROSS-REFERENCE INDEXES

NATIONAL STOCK NUMBER INDEX

STOCK NUMBER	FIG.	ITEM	STOCK NUMBER	FIG.	ITEM
5360-01-150-9693	187	5	5340-01-158-7095	378	7
5310-01-151-0113	89	21	5342-01-158-7096	378	7
2520-01-151-2628	194	6	2540-01-158-7169	401	1
3040-01-151-3568	186	2	2540-01-158-7170	401	1
5340-01-151-6145	102	21	2540-01-158-7171	401	6
4730-01-151-6316	115	15	2815-01-159-1789	24	6
5310-01-151-7347	616	2		24	9
5120-01-152-2318	629	4	5340-01-159-2995	596	25
2520-01-152-2384	308	17	2815-01-159-9538	7	5
5330-01-152-5943	367	19	4720-01-160-0733	291	7
2510-01-152-8812	367	1	4820-01-160-0759	505	48
3020-01-152-8895	188	23	5365-01-160-1832	51	1
5340-01-153-0890	391	23	3120-01-160-1891	88	11
2590-01-153-1850	595	15	5340-01-160-2299	102	29
4720-01-153-8240	272	2	4730-01-160-3579	37	23
5120-01-154-3029	629	1	5340-01-160-6919	549	39
5935-01-154-6233	158	42	5342-01-160-7381	86	5
	159	7	5331-01-160-7458	31	4
6220-01-155-2357	144	2	5330-01-160-7460	54	1
3120-01-155-4442	17	13	2540-01-160-7915	401	2
4730-01-155-5449	204	11	4820-01-160-9597	289	13
3040-01-155-5795	249	40	2530-01-160-9652	249	20
4010-01-155-6142	BULK	27	5315-01-160-9818	*283	3
5315-01-155-7302	200	27	5330-01-161-0289	88	14
3120-01-155-8707	17	13	4730-01-161-5115	7	24
5340-01-155-9468	519	38		19	6
5340-01-155-9469	519	37		116	6
5340-01-156-0421	326	5	5305-01-161-6000	120	10
6140-01-156-6187	153	12	5310-01-161-6131	305	8
5315-01-156-6314	158	44	2510-01-161-7631	415	2
	159	6	5410-01-161-7703	415	13
5325-01-156-9497	538	2	3020-01-161-7710	409	13
2510-01-157-1306	321	8	5930-01-161-9580	145	9
3120-01-157-3316	17	13	2510-01-162-0561	417	18
5340-01-157-3752	378	9	2510-01-162-0562	417	19
4820-01-157-4138	505	44	2510-01-162-0564	405	9
2590-01-157-6240	134	54	5310-01-163-2472	539	6
2940-01-157-6309	32	12	4330-01-163-2733	479	4
5305-01-157-6794	420	12	5340-01-163-4777	536	10
5340-01-157-9898	31	9	9520-01-163-4902	409	32
5360-01-157-9921	185	13	2510-01-163-4903	409	2
2510-01-158-0773	344	5			
4710-01-158-1477	BULK	34	3110-01-163-4932	306	23
6620-01-158-3125	114	3	2540-01-163-5170	536	9
5340-01-158-3126	344	12	4730-01-163-7823	482	17
5330-01-158-6289	403	7	2510-01-163-9752	427	7
	421	3	2590-01-163-9755	427	14
2540-01-158-7092	401	2	5340-01-163-9912	427	5
2540-01-158-7093	401	7	5310-01-164-1642	351	16
5340-01-158-7094	378	6	5305-01-164-2310	539	5

CROSS-REFERENCE INDEXES

NATIONAL STOCK NUMBER INDEX

STOCK NUMBER	FIG.	ITEM	STOCK NUMBER	FIG.	ITEM
4820-01-164-7002	53	1	5340-01-170-4940	321	37
5342-01-164-7597	318	39		322	41
	319	42		323	39
	320	40		324	36
	321	41	5340-01-170-5007	182	31
	322	47	4820-01-170-5055	596	28
	323	43		598	51
5342-01-164-7598	318	52	2520-01-171-2360	225	2
	319	52	5340-01-171-8267	353	12
	320	52	9905-01-172-2393	552	14
	321	53		552	15
	322	55	9905-01-172-2394	566	14
	323	54	2510-01-173-0091	416	1
5340-01-165-0469	414	10	2510-01-173-0092	416	1
2815-01-165-0765	KITS		5340-01-173-0133	88	5
5310-01-165-2184	27	5	5340-01-173-0241	583	9
5330-01-165-2314	128	4		591	21
5325-01-165-2352	529	12		592	18
5306-01-165-3272	492	7	5315-01-173-0397	182	23
5305-01-165-3892	19	12	9905-01-173-2183	570	8
5340-01-165-4506	412	16	5306-01-173-3524	505	25
5340-01-165-4546	543	3	4720-01-173-4609	505	57
4730-01-165-4647	426	7	2540-01-173-9147	609	7
2540-01-165-4677	543	21	2540-01-173-9160	609	14
5910-01-165-5255	182	30	2540-01-173-9172	609	23
5305-01-165-7541	529	11	2540-01-173-9173	609	23
4730-01-165-9491	14	1	2540-01-173-9174	609	24
	32	2	2540-01-173-9175	609	28
	57	2	2540-01-173-9176	609	28
4720-01-165-9531	505	40	7690-01-173-9197	570	10
2510-01-166-2016	405	23	2590-01-173-9200	420	16
5330-01-166-3662	479	5			
3020-01-166-5647	53	2	5320-01-174-4726	318	31
5970-01-174-9449	151	5			
				151	14
				151	19
				152	9
5306-01-166-6895	180	14	2590-01-175-7230	536	10
5310-01-167-1964	58	23	2540-01-175-7257	600	
5340-01-167-5535	182	11	4730-01-175-7343	589	10
2520-01-168-1632	192	6		590	4
5340-01-168-3102	528	29	4720-01-175-7421	589	12
4330-01-168-6891	457	6		590	19
	458	5	5340-01-176-5923	505	26
	507	41	5305-01-176-8018	284	61
5340-01-168-7904	356	16	3990-01-176-9359	408	2
5340-01-168-9267	412	2	5340-01-176-9443	586	20
5342-01-169-5686	366	13	5680-01-177-1525	407	14
4520-01-169-8680	425	5	5680-01-177-1526	407	15
2510-01-169-9850	534	6	2590-01-177-4455	420	3
5945-01-170-0553	148	6	2590-01-177-4456	420	3
5340-01-170-4940	318	37	2510-01-177-4457	420	10
	319	38	2510-01-177-4458	419	1
	320	35	5670-01-177-4460	404	15

CROSS-REFERENCE INDEXES

NATIONAL STOCK NUMBER INDEX

STOCK NUMBER	FIG.	ITEM	STOCK NUMBER	FIG.	ITEM
2520-01-177-4462	411	8	5330-01-181-0630	284	52
3040-01-177-4463	411	10		285	13
2510-01-177-4466	416	29	5330-01-181-0631	284	63
2510-01-177-4467	416	29	5340-01-181-0840	378	12
3040-01-177-4468	411	5	5306-01-181-5018	278	13
5315-01-177-7507	44	11		358	11
5340-01-177-7580	284	62	4730-01-181-5777	584	7
2910-01-177-8816	88	10	2590-01-181-6059	417	9
2590-01-178-0759	406	43	5342-01-181-9445	402	19
5340-01-178-3734	404	12	5340-01-181-9449	422	17
5120-01-178-5351	630	4	3040-01-181-9509	88	13
4520-01-178-6680	418	19	2540-01-182-1077	587	
2590-01-178-7043	407	4	2540-01-182-1309	388	1
	415	24	2590-01-182-1847	413	2
2510-01-178-7044	412	5	5365-01-182-5468	92	2
2510-01-178-7046	405	29	2540-01-182-7557	609	30
2510-01-178-7086	404	4	2530-01-183-0651	284	2
3040-01-178-7087	404	4	5330-01-183-0985	507	19
3110-01-179-1161	185	19	2920-01-183-2693	430	1
5340-01-179-1476	379	2	4210-01-183-4822	421	1
5305-01-179-2380	7	20		610	3
5360-01-179-3105	344	9	5306-01-183-5953	419	24
2510-01-179-4083	407	5	5306-01-183-5954	419	13
	415	25	2540-01-183-6796	418	32
2510-01-179-4084	406	6	5306-01-183-6970	358	1
2510-01-179-4112	419	1	5680-01-183-9754	415	23
5340-01-179-4303	378	4	5340-01-184-3464	379	4
5360-01-179-4599	344	10	2590-01-184-3938	479	2
2590-01-179-4911	405	8	2510-01-184-4765	404	6
2590-01-179-5802	417	11	5310-01-184-5784	110	10
3020-01-179-7515	190	1	5340-01-185-0455	596	15
2815-01-179-7516	55	8	4330-01-185-1226	479	6
2510-01-180-1004	408	14	2590-01-185-1278	413	3
2510-01-180-1005	412	7	5342-01-185-3561	6	3
2590-01-180-1006	404	21	5310-01-185-4675	415	18
2590-01-180-1092	417	9	5680-01-185-4944	415	22
5325-01-180-2448	305	6	9905-01-185-5762	492	2
4520-01-180-3577	418	9	9905-01-185-5778	570	2
2540-01-180-3579	418	33	9905-01-185-5779	570	3
5340-01-180-4819	379	17	9905-01-185-5781	561	1
5961-01-180-5634	149	7	9905-01-185-5782	567	4
	211	32	9905-01-185-5783	566	5
2510-01-180-8503	418	10	9905-01-185-5784	566	9
2590-01-180-8571	424	15	9905-01-185-5785	567	2
4720-01-181-0098	583	2	9905-01-185-5786	567	3
2590-01-181-0309	412	13	9905-01-185-5787	552	2
			5310-01-185-7191	90	15
2590-01-181-0310	417	25	5340-01-185-7768	609	12
2590-01-181-0311	413	1	2510-01-186-0867	412	30
	413	12	5320-01-186-1277	411	9

CROSS-REFERENCE INDEXES

NATIONAL STOCK NUMBER INDEX

STOCK NUMBER	FIG.	ITEM	STOCK NUMBER	FIG.	ITEM
5340-01-186-3505	404	12	5670-01-187-3638	406	1
2590-01-186-3720	413	4	5670-01-187-3639	406	1
2590-01-186-3721	413	7	4810-01-187-4925	129	5
2590-01-186-3722	403	2	2590-01-188-0316	413	11
	413	9	2590-01-188-0317	413	9
2590-01-186-3723	413	1	2590-01-188-0318	416	16
	413	10	5315-01-188-0761	8	14
	413	12	5365-01-188-0954	20	9
2590-01-186-3724	413	5	5310-01-188-0997	38	10
2590-01-186-3725	420	18	5365-01-188-1054	47	5
5365-01-186-3773	412	31	5365-01-188-1055	47	5
4520-01-186-5897	421	8	5365-01-188-1056	47	5
2590-01-186-6096	413	5	5340-01-188-5086	259	9
5305-01-186-7137	411	12		260	9
5305-01-186-7140	503	29	5340-01-188-5087	161	2
5340-01-186-7174	600	6		269	18
5365-01-186-7247	412	28	5340-01-188-5088	264	7
9905-01-186-7948	570	8	4720-01-188-5139	258	5
2590-01-186-9584	412	7	4720-01-188-5140	273	8
2590-01-186-9585	403	5	2930-01-189-0458	589	15
2590-01-187-0363	427	12	5930-01-189-0494	132	18
2510-01-187-0364	427	18		146	1
2510-01-187-0366	428	9	2910-01-189-0901	85	1
2510-01-187-0367	428	9		87	1
2510-01-187-0368	428	5	3040-01-189-1760	20	8
2510-01-187-0369	428	18	5310-01-189-5407	47	5
5340-01-187-0370	428	14	5310-01-189-5408	47	5
2510-01-187-0371	428	16	5310-01-189-5409	47	5
2510-01-187-0372	424	14	5310-01-189-5410	47	5
2510-01-187-0373	424	24	5310-01-189-5411	47	5
2510-01-187-0374	424	19	5310-01-189-5412	47	5
2510-01-187-0375	424	12	5310-01-189-5413	47	5
5340-01-187-0376	424	3	5331-01-189-6351	141	9
2510-01-187-0377	424	20	2510-01-189-6401	133	3
2510-01-187-0378	424	18		134	39
2510-01-187-0379	424	1	5340-01-189-6402	426	4
2510-01-187-0380	426	6	5340-01-189-6403	426	14
2510-01-187-0381	426	2	3040-01-189-6406	100	37
2510-01-187-0382	425	18	2590-01-189-6415	417	26
5340-01-187-0527	589	32	2590-01-189-6416	417	26
	591	29	2510-01-189-6417	418	1
5305-01-187-0531	91	19	5310-01-189-7499	47	5
2510-01-187-0566	427	1	5365-01-189-7804	47	5
2510-01-187-0567	428	1	5365-01-189-9026	47	5
2510-01-187-0568	428	1	5365-01-189-9027	47	5
5365-01-187-3589	412	32	5365-01-189-9028	47	5
5365-01-187-3590	412	9	5365-01-189-9029	47	5
5365-01-187-3591	412	11	5365-01-189-9030	47	5
2590-01-187-3614	417	15	5340-01-190-0373	420	17
3040-01-187-3620	367	3	5330-01-190-1905	64	5

CROSS-REFERENCE INDEXES

NATIONAL STOCK NUMBER INDEX

STOCK NUMBER	FIG.	ITEM	STOCK NUMBER	FIG.	ITEM
5365-01-190-2131	47	5	4730-01-192-9590	460	37
2530-01-190-8376	313	1		461	32
2540-01-190-8484	549	19		509	29
2520-01-190-8485	225	2	2990-01-192-9724	582	11
4710-01-190-8490	313	7	5305-01-193-6839	38	2
5330-01-190-9555	60	7		130	3
5340-01-191-0596	413	27	5310-01-193-6884	110	17
5365-01-191-0774	47	5	3120-01-193-7083	KITS	
5365-01-191-0775	47	5	4710-01-193-7944	64	7
5365-01-191-0776	47	5	2590-01-194-2149	413	7
5365-01-191-2411	47	5	2540-01-194-3323	592	8
5365-01-191-2413	47	5		594	8
5310-01-191-2494	47	5	5320-01-194-5034	411	4
5360-01-191-2949	86	2	5340-01-194-5298	403	9
5360-01-191-2950	87	18		421	4
5365-01-191-3532	47	5	5340-01-194-6474	403	8
5365-01-191-3533	47	5		421	6
5365-01-191-3534	47	5	2590-01-194-6994	420	9
5365-01-191-3535	47	5	5340-01-194-7036	525	19
5365-01-191-3575	418	12	5340-01-194-8936	8	6
5310-01-191-6333	47	5		10	6
4730-01-191-6433	539	15	3040-01-194-9884	471	14
5310-01-191-7512	47	5	5340-01-194-9885	608	4
5330-01-191-8047	45	3	4730-01-195-0095	254	23
2910-01-191-8470	46	6		257	12
4710-01-192-0625	589	33	4730-01-195-0825	48	12
5340-01-192-0626	586	8	4730-01-195-1884	259	4
5935-01-192-0627	587	11		260	5
	588	4		267	4
9390-01-192-1610	67	2	5330-01-195-5268	47	2
	69	2		48	13
5305-01-192-2036	15	18	3020-01-195-6990	44	17
	40	4	4730-01-195-7339	480	8
	179	3	4720-01-195-7604	154	21
5330-01-192-2037	15	16	4710-01-195-7644	198	1
4720-01-192-3504	461	19	4710-01-195-7645	198	11
	509	14	4710-01-195-9100	589	43
4730-01-192-4381	596	19	5977-01-195-9380	614	9
4730-01-192-4729	263	22	2920-01-195-9383	608	11
	264	22	5995-01-195-9405	614	14
	269	11	4720-01-195-9518	271	3
	269	29	4720-01-196-1166	505	41
5305-01-192-5677	15	4	5340-01-196-3680	48	3
5310-01-192-5760	211	25	5307-01-196-4246	94	1
5330-01-192-5788	412	33	7690-01-196-6355	110	20
2590-01-192-6089	413	10		557	7
3040-01-192-8356	424	17	5306-01-196-6632	250	15
	425	13		251	17
4730-01-192-9578	504	29	2930-01-196-7521	118	27
4730-01-192-9590	456	41	5340-01-196-8113	428	12

CROSS-REFERENCE INDEXES

NATIONAL STOCK NUMBER INDEX

STOCK NUMBER	FIG.	ITEM	STOCK NUMBER	FIG.	ITEM
5306-01-197-1492	156	12	4730-01-201-0717	117	6
5306-01-197-1513	110	15	2540-01-201-0968	475	8
5305-01-197-1663	27	6	5331-01-201-3623	124	14
4310-01-197-1882	286	50	2910-01-201-7719	94	2
5340-01-197-2183	125	16		595	12
5365-01-197-3179	41	7	2510-01-202-0965	428	17
5330-01-197-3228	582	12	4730-01-202-3351	213	7
	589	14		266	8
5305-01-197-3449	23	21		285	24
2510-01-197-4200	404	7		287	7
5365-01-197-4545	193	9	2540-01-202-4064	589	23
2590-01-197-7239	420	8	4730-01-202-4101	425	2
5340-01-197-7597	215	6	2990-01-202-4128	586	7
5340-01-197-8214	269	6	2590-01-202-5769	413	17
	315	16	5310-01-202-6775	414	18
5340-01-197-8215	456	11	2815-01-202-9715	12	14
	460	10	5305-01-203-2609	563	5
3040-01-197-8555	405	21	5330-01-203-3612	65	14
	406	5	5340-01-203-5659	589	25
3040-01-197-8556	405	22		590	13
	406	7	4010-01-203-5687	411	6
2510-01-197-8562	418	31	5306-01-203-6299	14	6
5930-01-197-8661	131	16	5305-01-203-9064	611	15
	132	6		612	30
	134	10		613	19
	134	38		613	40
	134	69	5340-01-203-9261	66	3
			9390-01-204-1161	102	37
5340-01-197-9300	253	6	5310-01-204-2002	539	8
5340-01-197-9350	269	16	5365-01-204-2104	610	4
	269	16	6220-01-204-2597	614	1
2590-01-198-0964	337	7	2930-01-204-4475	116	4
5340-01-198-2415	266	21	2510-01-204-7704	404	13
3040-01-198-2652	516	19	5342-01-204-8720	515	1
3040-01-198-2653	517	1	5340-01-205-2503	492	3
4710-01-199-4366	589	44	5340-01-205-2504	611	5
	590	7		612	7
9905-01-199-6809	566	4		613	4
	566	11	5306-01-205-2677	611	10
5310-01-200-1318	3	3	2540-01-205-2759	609	17
5995-01-200-2419	611	29	5310-01-205-2830	237	13
5995-01-200-3203	611	36		237	14
	612	22	5305-01-205-3407	14	5
	613	36	5340-01-205-3548	418	23
5325-01-200-4035	586	24	5340-01-205-3549	418	25
4710-01-200-4244	611	18	5340-01-205-5957	507	35
4710-01-200-4404	611	8	5340-01-205-9057	517	11
3030-01-200-6004	118	26	5340-01-205-9262	145	1
5310-01-200-9879	250	2	5340-01-205-9263	589	40
	251	2	3040-01-205-9264	416	7
	527	9	5340-01-205-9479	515	4

CROSS-REFERENCE INDEXES

NATIONAL STOCK NUMBER INDEX

STOCK NUMBER	FIG.	ITEM	STOCK NUMBER	FIG.	ITEM
5340-01-206-2209	589	21	9520-01-210-2159	338	5
5315-01-206-2239	347	4	9520-01-210-2160	334	5
2530-01-206-3911	306	12		336	5
5315-01-206-5207	518	12		338	5
4730-01-206-6162	186	3	5340-01-210-2162	334	19
	264	1		338	18
	266	28	2510-01-210-2163	334	22
	457	16	5340-01-210-2164	334	23
	458	29		338	26
5340-01-206-8586	585	14	2510-01-210-2166	336	13
2520-01-206-9575	189	22	2510-01-210-2167	336	16
5340-01-207-0378	589	38	2590-01-210-2168	339	8
5305-01-207-7243	57	23	2590-01-210-2174	507	1
5305-01-207-7447	35	9	5340-01-210-3953	589	9
	119	4	5305-01-210-4595	120	1
9390-01-207-8125	68	2	2990-01-210-4650	102	20
5365-01-208-3878	419	28	2510-01-210-4651	336	13
5340-01-208-5371	418	27	2510-01-210-4652	338	7
4730-01-208-5859	77	20	2510-01-210-4653	338	11
	79	17	5340-01-210-4654	341	1
	80	22	3040-01-210-4655	341	5
3040-01-208-5897	180	9	2510-01-210-4656	341	6
5340-01-208-8080	420	11	5340-01-210-4657	341	16
5310-01-209-0508	10	7	5340-01-210-4658	365	8
	38	13	5340-01-210-4659	394	4
5305-01-209-7068	118	29	9905-01-210-4733	552	6
5305-01-209-7112	426	13	9905-01-210-4734	552	6
5306-01-209-7114	414	8	4010-01-210-6196	341	9
5330-01-209-7354	427	11	2510-01-210-6197	389	1
5340-01-209-7447	425	24	4710-01-210-6198	423	2
5325-01-209-7625	30	4	2590-01-210-6199	458	1
5310-01-210-0199	292	7	4720-01-210-6200	458	22
			4720-01-210-6201	458	20
5340-01-210-0201	478	9	2510-01-210-6228	386	6
5340-01-210-0202	478	13	2510-01-210-6229	383	23
2590-01-210-0203	507	37	2510-01-210-6230	384	16
9905-01-210-0217	552	6	2540-01-210-6231	327	7
9905-01-210-0219	570	3	5340-01-210-6238	458	10
9905-01-210-0220	570	3	2815-01-210-6947	26	16
9905-01-210-0229	552	5	5340-01-210-7510	454	15
9905-01-210-0230	552	6	2510-01-210-8799	338	20
9905-01-210-0231	552	6	4820-01-210-8821	292	11
5306-01-210-0264	334	24	2950-01-211-0163	65	23
	336	17	5340-01-211-1563	589	58
	338	28	5340-01-211-1622	517	16
5305-01-210-1608	118	14	2815-01-211-5270	32	4
5331-01-210-2155	292	9	2510-01-211-6611	334	13
5340-01-210-2158	102	7	2510-01-211-6612	340	5
9520-01-210-2159	334	5	2510-01-211-6613	424	20
	336	5	4710-01-211-6614	458	14

CROSS-REFERENCE INDEXES

NATIONAL STOCK NUMBER INDEX

STOCK NUMBER	FIG.	ITEM	STOCK NUMBER	FIG.	ITEM
2520-01-211-6702	KITS		4910-01-218-4490	629	2
2530-01-211-8405	292	16	2910-01-218-5155	46	1
5325-01-212-0599	611	14	2910-01-218-5158	54	3
	612	29	2990-01-218-6935	85	5
	613	20	4710-01-218-6949	66	17
	613	41	4710-01-218-6950	66	6
5340-01-212-2464	154	22	4710-01-218-6951	66	13
5305-01-212-5210	37	19	4710-01-218-6952	66	13
2510-01-212-7620	468	1	5340-01-218-7159	321	15
5340-01-212-8476	336	15	5365-01-218-9917	341	17
2530-01-213-1040	418	15	5340-01-219-0163	341	15
5330-01-213-1258	65	25	2910-01-219-2086	46	6
6220-01-213-1558	614	5	5330-01-219-2375	200	2
5305-01-213-9852	455	10	5330-01-219-2555	186	1
2610-01-214-1344	300	6		190	4
3020-01-214-1530	KITS		5325-01-219-2580	187	17
9905-01-214-4053	611	13		193	2
4730-01-214-6720	592	9	5315-01-219-3665	341	22
5995-01-215-0930	592	20	5305-01-219-5386	341	14
	594	30	3040-01-219-5697	56	1
2815-01-215-1705	28	5			
4730-01-215-3218	508	3	5365-01-219-9172	405	12
4720-01-215-4295	508	4		406	42
5360-01-215-5766	84	5	4810-01-220-1185	196	3
3010-01-215-6598	503	26	4910-01-220-1512	629	3
3020-01-215-6599	503	28	5331-01-220-2389	7	4
2910-01-215-6721	49	6	5330-01-220-5498	72	
3020-01-215-8827	503	30	4210-01-220-6376	610	
3040-01-215-8853	503	21	4730-01-221-1445	613	15
3020-01-216-2332	503	27	4720-01-221-1448	481	1
3040-01-216-5337	341	19	3040-01-221-2092	196	4
3120-01-216-6699	409	12	4730-01-221-3565	581	7
2510-01-216-6834	334	2		582	17
	336	2	4820-01-221-5415	196	1
	338	2	5365-01-221-8749	66	4
3040-01-216-6982	341	20	4730-01-221-8821	593	11
3950-01-216-9319	341	21	2520-01-221-8893	410	15
5330-01-217-0734	414	7	4030-01-222-6037	326	1
5340-01-217-0818	417	16	6150-01-222-6585	611	7
5365-01-217-1995	10	8	5360-01-222-6879	85	17
	38	14	5360-01-222-6880	88	34
5315-01-217-2269	409	11	4730-01-222-7575	629	30
			5340-01-223-0359	613	39
5340-01-217-4069	321	30	5340-01-223-3538	613	18
5340-01-217-8296	611	6	6150-01-223-7251	612	19
	612	8	6150-01-223-7252	613	34
	613	14	6150-01-223-7269	612	15
5975-01-217-8550	BULK	7	6150-01-223-7270	612	6
5340-01-217-9141	458	28	6150-01-223-7271	613	2
2990-01-218-2100	102	20	6150-01-223-7272	613	8

CROSS-REFERENCE INDEXES

NATIONAL STOCK NUMBER INDEX

STOCK NUMBER	FIG.	ITEM	STOCK NUMBER	FIG.	ITEM
5340-01-223-8037	321	18	3040-01-233-7768	416	5
5340-01-223-9799	525	16	4730-01-233-8998	290	16
4730-01-224-4152	256	22		605	13
5310-01-224-9142	211	24	5310-01-234-1411	48	2
5320-01-224-9157	375	2	5305-01-234-3714	57	4
5360-01-224-9218	196	5		60	4
2510-01-225-1000	417	3	5305-01-234-3755	20	16
2510-01-225-1001	409	1		60	5
2590-01-225-1071	428	6		115	21
5305-01-225-2106	362	6	5305-01-234-3756	48	1
6220-01-225-2972	614	2	4730-01-235-3007	598	24
2540-01-225-5863	370	5	5315-01-235-4688	22	7
2510-01-225-5864	424	11	3020-01-235-5055	286	39
4730-01-225-9003	104	12	5430-01-235-5442	108	5
5306-01-226-0798	614	8	4720-01-235-5452	BULK	17
5305-01-226-1945	612	16	4730-01-235-9617	503	41
2520-01-226-3399	194	1	4720-01-235-9627	481	10
4730-01-226-3705	593	13	4720-01-235-9628	593	26
2590-01-226-4587	409	23		594	2
5306-01-226-5917	467	1	4720-01-235-9629	593	17
4710-01-226-7389	612	31		594	9
4710-01-226-7390	612	28	2540-01-236-1175	584	28
5340-01-226-9161	408	7		589	62
2510-01-226-9407	384	1		590	43
	386	9	3010-01-236-1213	624	11
2510-01-226-9408	383	28	5305-01-236-6157	123	5
9905-01-226-9437	552	2	5306-01-237-1166	20	1
5305-01-227-6249	33	38	5310-01-237-2615	246	10
5340-01-227-6479	379	16	2510-01-237-2945	417	23
5360-01-228-0747	87	4	5305-01-237-3637	598	34
9905-01-229-3443	553	3	5305-01-237-4915	20	14
5310-01-229-8029	293	1			
	295	15			
5340-01-229-8365	266	19		64	6
5305-01-229-9587	613	27		124	2
2910-01-230-1919	84	6	5310-01-237-5224	92	15
5342-01-230-6702	492	19	4730-01-237-6950	15	8
5340-01-231-5357	267	15	5306-01-237-7531	123	4
5340-01-231-5516	374	15	3040-01-238-0863	518	1
5325-01-231-5963	161	21	4730-01-238-6443	503	4
5325-01-231-7594	292	15	5330-01-238-8316	58	11
5342-01-231-9151	202	4	5305-01-238-8438	58	12
5365-01-231-9280	425	25	3110-01-239-1253	188	22
5340-01-231-9290	418	3	5340-01-239-7078	27	10
5340-01-231-9291	143	16	5342-01-239-7140	48	15
3020-01-231-9296	KITS		5305-01-239-7202	124	12
5342-01-231-9313	425	15	5340-01-239-8606	8	3
5330-01-232-1487	108	4	5340-01-239-8607	35	3
5340-01-232-3568	608	20	5365-01-239-9381	406	12
4730-01-232-7159	505	9	5360-01-240-1626	85	6
5340-01-233-4202	334	8		87	6
	336	8	4730-01-240-6112	422	14

CROSS-REFERENCE INDEXES

NATIONAL STOCK NUMBER INDEX

STOCK NUMBER	FIG.	ITEM	STOCK NUMBER	FIG.	ITEM
4730-01-240-6112	423	14	5325-01-265-7251	312	33
5325-01-241-3888	7	22	3120-01-266-1530	13	3
5340-01-241-6939	87	5	5340-01-266-3023	45	4
5325-01-242-7083	153	14	5330-01-266-3294	15	20
	422	19	5305-01-266-8568	38	9
	423	17	4820-01-267-2914	623	1
	429	1	5305-01-268-5558	4	1
4730-01-244-8434	111	5	2910-01-268-8736	60	6
	549	29	2815-01-268-8737	124	11
	598	10	6110-01-268-8739	623	26
2540-01-245-2445	615	7	2530-01-268-8740	286	1
5305-01-245-3192	48	8	2930-01-268-8751	119	1
	124	13	2930-01-268-8752	122	1
5305-01-245-3193	119	5	2815-01-268-8753	35	8
	126	1	2910-01-268-8757	57	11
5305-01-245-3817	40	5	5305-01-269-6274	115	32
			5310-01-270-5463	293	1
				295	15
5315-01-246-4339	248	6	5330-01-270-8144	45	5
5340-01-246-6172	66	15	5310-01-270-8229	57	17
4730-01-247-3140	581	21	5310-01-270-8244	64	14
5305-01-248-3222	92	19	5310-01-270-8245	65	5
3040-01-248-3995	208	1	5310-01-270-8246	18	3
5365-01-249-9707	629	5	5310-01-270-8251	38	4
5315-01-270-8268				246	12
4820-01-251-1699	278	12	5315-01-270-8269	246	12
5330-01-254-6377	489	17	5315-01-270-8284	8	7
3040-01-255-4406	185	3	5315-01-270-8285	13	6
5305-01-256-3046	122	20	5365-01-270-8290	32	8
2540-01-256-5331	615		5310-01-270-8341	57	9
5330-01-257-0750	367	31	5310-01-270-8342	120	28
2815-01-257-0853	41	4		122	6
	41	4	5310-01-270-8343	57	18
4710-01-257-2193	39	4	5325-01-270-8360	27	9
4730-01-257-3328	39	7	5325-01-270-8361	65	16
5310-01-257-7590	107	37	5325-01-270-8362	122	26
4520-01-257-8938	418	30	5325-01-270-8363	122	23
5310-01-259-0296	113	7	5365-01-270-8376	286	38
5315-01-259-2517	370	41	5310-01-270-8386	27	8
5310-01-260-4937	372	4	5310-01-270-8387	286	40
5340-01-260-7895	372	5	5310-01-270-8388	32	9
5340-01-260-9942	629	13	5310-01-270-8389	45	2
3040-01-262-1207	628	5	5310-01-270-8390	38	8
5310-01-270-8391				38	3
	286	27	5310-01-270-8392	107	15
5120-01-262-7309	630	2	5310-01-270-8405	57	16
5305-01-263-2708	20	10	5310-01-270-8406	312	6
5305-01-263-2733	45	1	5310-01-270-8417	47	6
5330-01-263-6179	20	11	5310-01-270-8422	27	7
4820-01-263-6410	93	9	5310-01-270-8423	38	7
5305-01-264-5886	89	19	5310-01-270-8425	294	2
	90	5		296	24

I-98

CROSS-REFERENCE INDEXES

NATIONAL STOCK NUMBER INDEX

STOCK NUMBER	FIG.	ITEM	STOCK NUMBER	FIG.	ITEM
5365-01-270-8459	122	16	4710-01-271-3837	48	19
5365-01-270-8472	122	8	4710-01-271-3838	48	7
5365-01-270-8473	122	22	4710-01-271-3839	48	11
5365-01-270-8481	128	3	4710-01-271-3840	48	10
5365-01-270-8482	38	16	4710-01-271-3841	57	24
5310-01-271-1796	122	7	4710-01-271-3843	48	9
5365-01-271-1837	294	16	5330-01-271-4308	15	14
	296	15	5330-01-271-4311	10	9
5365-01-271-1852	22	4	5305-01-271-4326	27	1
5365-01-271-1854	122	15	2815-01-271-5074	27	12
7690-01-271-1926	124	9	5342-01-271-5088	126	2
5342-01-271-2340	57	13	5342-01-271-5089	126	7
2950-01-271-2341	65	4	2815-01-271-5096	13	4
2950-01-271-2342	65	10	2815-01-271-5098	27	3
5342-01-271-2343	65	29	2930-01-271-5102	32	6
5342-01-271-2346	285	14	4310-01-271-5103	286	47
5342-01-271-2347	286	36	3040-01-271-5118	628	1
5342-01-271-2348	286	28	2815-01-271-5119	18	1
5342-01-271-2349	47	8	2815-01-271-5120	27	16
5340-01-271-2415	8	15	3020-01-271-5126	122	21
5340-01-271-2416	8	2	5330-01-271-5151	236	1
5340-01-271-2417	8	4	5305-01-271-5168	312	5
5340-01-271-2418	8	8	5342-01-271-5704	27	2
5340-01-271-2419	8	17	5340-01-271-5705	48	14
	10	4	5310-01-271-5706	57	14
5340-01-271-2420	10	3	5342-01-271-5709	312	24
5340-01-271-2441	124	1	5342-01-271-5711	312	3
5310-01-271-2467	294	4	5360-01-271-5767	122	18
	296	22	5330-01-271-5791	32	7
			5306-01-271-5842	293	17
5340-01-271-2470	40	6		294	6
5340-01-271-2471	65	6		296	20
			5306-01-272-5843	293	17
5340-01-271-2475	130	2		294	6
5340-01-271-2485	25	1		296	20
5340-01-271-2496	15	19	5305-01-271-5848	8	13
5340-01-271-2497	15	15	5305-01-271-5850	32	1
5340-01-271-2504	65	17	5305-01-271-5851	15	13
5340-01-271-2505	65	28		57	21
2835-01-271-2510	20	12	5305-01-271-5852	10	10
2835-01-271-2511	65	15	5342-01-271-5862	48	21
3040-01-271-3597	122	27	5310-01-271-5872	296	4
3040-01-271-3649	312	34	3110-01-271-6353	122	25
4730-01-271-3739	40	7	5306-01-271-6362	18	2
3030-01-271-3754	123	1	5330-01-271-6404	32	5
2815-01-271-3763	27	11	5305-01-271-6446	312	2
4310-01-271-3807	286	32	5305-01-271-6448	43	1
4310-01-271-3808	286	52	5305-01-271-6449	285	11
3020-01-271-3812	13	5	5305-01-271-6450	294	10
3040-01-271-3820	57	12	4820-01-271-6918	312	28
3040-01-271-3821	57	10	4720-01-271-6945	64	4
3020-01-271-3832	126	5	4720-01-271-6946	64	9

CROSS-REFERENCE INDEXES

NATIONAL STOCK NUMBER INDEX

STOCK NUMBER	FIG.	ITEM	STOCK NUMBER	FIG.	ITEM
4720-01-271-6950	40	8	5360-01-271-8282	25	2
4720-01-271-6951	64	3	5331-01-271-8289	286	9
2520-01-271-7007	238	3	5330-01-271-8306	285	22
2530-01-271-7074	245	1	5330-01-271-8307	29	5
	247	1	5305-01-271-8332	138	14
2530-01-271-7075	246	18	5305-01-271-8337	181	10
	247	3	3120-01-271-8353	294	14
2815-01-271-7076	13	2		296	19
5342-01-271-7081	48	16	5305-01-271-8375	65	2
2990-01-271-7086	15	21	5305-01-271-8379	65	18
3020-01-271-7114	15	2	5305-01-271-8381	63	16
2520-01-271-7152	122	19	5305-01-271-8383	222	11
3040-01-271-7165	15	3	5330-01-271-9347	246	14
2815-01-271-7171	25	7	5331-01-271-9371	120	30
				122	4
			5331-01-271-9372	120	29
	25	11		122	5
4730-01-271-7181	8	11	5331-01-271-9373	122	11
4730-01-271-7187	247	2	5331-01-271-9374	120	19
4710-01-271-7198	48	18		122	13
4710-01-271-7199	60	3	5330-01-271-9375	15	17
4720-01-271-7202	111	10	5331-01-271-9376	312	9
2940-01-271-7203	32	14	5330-01-271-9407	312	4
2590-01-271-7861	194	2	5330-01-271-9408	312	10
4730-01-271-7874	124	3	5330-01-271-9409	312	11
4730-01-271-7903	40	1	5330-01-271-9410	294	7
	124	4		296	9
4710-01-271-7921	27	13	5307-01-271-9511	57	19
4710-01-271-7922	40	3	2815-01-271-9792	18	9
4710-01-271-7938	64	2	2815-01-271-9794	25	6
4710-01-271-7939	296	3	2815-01-271-9802	18	6
4710-01-271-7940	124	6	2530-01-271-9809	294	1
4710-01-271-7941	64	10	2990-01-271-9816	123	3
4710-01-271-7942	285	10	5342-01-271-9818	123	2
4710-01-271-7943	38	17	2815-01-271-9822	KITS	20
4820-01-271-7946	623	17	2530-01-271-9824	236	33
4730-01-271-7955	40	10	2910-01-271-9826	47	3
4730-01-271-7956	285	4	2815-01-271-9838	KITS	17
4730-01-271-7957	618	1	4720-01-272-0504	205	4
	620	12	2530-01-272-0542	236	4
4730-01-271-7969	285	7	2815-01-272-0547	43	3
4730-01-271-7977	115	13	4710-01-272-0572	312	7
4730-01-271-7978	294	12	4730-01-272-0582	294	11
4720-01-271-7995	285	25	5342-01-272-1038	122	2
4720-01-271-7996	205	11	5330-01-272-1108	20	15
4720-01-271-7997	205	7	5331-01-272-1120	40	9
4720-01-271-8000	285	12	5331-01-272-1121	57	15
2530-01-271-8023	294	8	5331-01-272-1122	57	20
	296	12	5331-01-272-1123	119	2
4710-01-271-8031	285	2	5330-01-272-1124	15	11
4710-01-271-8032	285	5	5331-01-272-1125	65	9
4710-01-271-8033	115	14	5330-01-272-1138	124	8
			5330-01-272-1142	115	8

CROSS-REFERENCE INDEXES

NATIONAL STOCK NUMBER INDEX

STOCK NUMBER	FIG.	ITEM	STOCK NUMBER	FIG.	ITEM
5330-01-272-1143	38	1	3120-01-273-4653	13	7
5330-01-272-1144	10	9	3120-01-273-4654	13	7
5330-01-272-1145	10	9	3120-01-273-4655	13	7
5330-01-272-1146	64	13	3120-01-273-4656	13	7
5330-01-272-1147	294	15	5310-01-274-0041	586	18
	296	17	3120-01-274-3377	13	7
5330-01-272-1148	294	13	3120-01-274-3378	18	4
5330-01-272-1246	29	6	5305-01-274-4404	20	13
5330-01-272-1250	65	11	5305-01-274-4405	312	23
5330-01-272-1282	8	19	5305-01-274-4406	128	7
5305-01-272-1334	48	20	5305-01-274-4407	10	11
5340-01-272-1471	63	11	2530-01-274-4457	312	1
6620-01-272-1716	115	12	5305-01-274-5655	13	1
6680-01-272-1867	35	1	5342-01-275-0384	20	6
4730-01-272-2827	57	25	5340-01-275-0487	40	11
4720-01-272-2841	285	9	5360-01-275-0545	246	23
4710-01-272-2882	124	7	5340-01-275-3403	48	5
2530-01-272-2911	312	36			
2510-01-272-2923	325	4	3120-01-275-7664	18	4
2530-01-272-2926	236	33	3120-01-275-7665	18	4
3120-01-272-3269	65	22	5305-01-276-0859	556	4
3120-01-272-3270	65	20	4720-01-276-1252	182	5
3120-01-272-3271	286	37	4720-01-276-1253	278	9
3120-01-272-3272	18	4	5306-01-276-1601	121	2
3120-01-272-3273	18	4	2510-01-276-5729	360	2
3120-01-272-3276	57	1	4820-01-276-5731	622	3
5305-01-272-3305	65	27	4720-01-276-5923	458	19
2930-01-272-3925	KITS		5325-01-276-8488	58	28
2815-01-272-3954	29	4		89	8
2815-01-272-3980	25	13	5305-01-277-0423	100	43
5305-01-272-4809	27	15	5305-01-277-0461	263	6
5305-01-272-4810	115	1	3120-01-277-1036	122	24
5305-01-272-4811	47	7	4820-01-277-1486	279	6
	115	9		279	7
5305-01-272-4812	38	15	5340-01-277-3023	279	11
			3020-01-277-4635	200	6
2815-01-272-5538	25	5	3020-01-277-4636	200	7
2815-01-272-5539	25	12	3020-01-277-4637	200	7
2815-01-272-6679	25	9	3020-01-277-4638	200	7
2815-01-272-6714	38	12	5307-01-277-5053	236	35
2930-01-272-6716	115	7	2590-01-277-9100	518	1
2815-01-272-6719	15	9	5315-01-277-9765	209	10
5342-01-272-6724	4	2	5365-01-277-9878	192	2
5342-01-272-6725	4	3	2815-01-278-1093	KITS	
2520-01-272-7767	466	1	5340-01-278-2190	472	11
6680-01-272-9204	147	3	4820-01-278-7385	284	15
6220-01-273-0177	138	15	4710-01-279-1490	63	3
2815-01-273-0571	15	12	4710-01-279-1493	63	17
5305-01-273-1594	29	2	4710-01-279-1494	620	7
2590-01-273-3321	29	1	4730-01-279-1510	111	15
4730-01-278-1001	297	3			

CROSS-REFERENCE INDEXES

NATIONAL STOCK NUMBER INDEX

STOCK NUMBER	FIG.	ITEM	STOCK NUMBER	FIG.	ITEM
4730-01-279-1519	299	9	5340-01-280-6996	110	19
4720-01-279-3034	297	4	5340-01-280-7096	619	7
4720-01-279-3038	315	3	5340-01-280-7097	184	4
2510-01-279-3089	6	4	5340-01-280-7098	63	14
4710-01-279-3158	287	5	5340-01-280-7099	297	5
4710-01-279-3159	297	9	5340-01-280-7100	107	13
4710-01-279-3162	287	1	5340-01-280-7101	107	7
4710-01-279-3163	72	9	5340-01-280-7102	299	8
4710-01-279-3164	72	6	4730-01-280-8331	617	18
4710-01-279-3165	315	7	4730-01-280-8345	619	9
4720-01-279-3168	111	14	2815-01-280-8961	25	8
4720-01-279-3169	115	5	2530-01-280-8989	297	1
4720-01-279-3170	111	2	3020-01-280-9024	57	22
4720-01-279-3171	63	1	5305-01-280-9367	110	21
2540-01-279-4593	357	2	5365-01-280-9432	63	12
2805-01-279-4630	63	9			
3040-01-279-4658	101	19	4730-01-281-0812	8	9
3040-01-279-4659	101	24		20	4
9905-01-279-4690	65	3	4730-01-281-0840	57	7
9905-01-279-4691	556	3	4720-01-281-0994	459	2
5945-01-279-4802	130	4	2815-01-281-1125	8	20
5905-01-280-3388	98	2	4710-01-281-1148	78	2
9390-01-280-3408	110	22	2590-01-281-1271	318	23
5365-01-280-3668	63	13		468	9
5365-01-280-3690	297	15	2540-01-281-1272	468	10
5340-01-280-3716	110	11	5306-01-281-2333	296	5
4820-01-280-3935	619	25	5306-01-281-3387	15	1
4820-01-280-3947	619	12	5340-01-281-5150	110	14
2520-01-280-4129	221	2	2815-01-281-5206	25	10
3040-01-280-4153	101	14	5306-01-281-6560	101	22
2510-01-280-4155	133	1	5340-01-281-7205	107	30
3040-01-280-4156	245	10	5307-01-281-7219	236	35
4730-01-280-4204	619	8	5340-01-281-7792	25	3
2510-01-280-4244	6	15	5331-01-281-8997	29	3
2510-01-280-4245	6	25	5330-01-281-9013	43	4
2510-01-280-4246	6	22	3130-01-281-9164	8	12
2990-01-280-4284	103	3	2590-01-281-9716	35	2
			2540-01-281-9855	616A	KIT
3040-01-280-4382	297	18	5340-01-282-0452	181	8
5325-01-280-5592	18	7	5305-01-282-1528	298	5
5365-01-280-5643	192	16	5305-01-282-1529	298	2
4730-01-280-6402	78	4	5340-01-282-2229	455	8
5331-01-280-6503	619	10	2510-01-282-2526	6	20
3120-01-280-6566	13	11	4710-01-282-2585	78	17
5342-01-280-6639	48	17	4710-01-282-2586	78	16
5340-01-280-6869	111	12	5310-01-282-2807	15	5
5340-01-280-6874	184	11	5310-01-282-3391	128	6
5340-01-280-6875	111	4	5340-01-282-3589	101	13
5365-01-280-6924	128	5	3040-01-282-4336	455	4
5340-01-280-6947	360	3	5330-01-282-5653	25	4
5340-01-280-6995	619	16	2530-01-282-8619	294	5

NATIONAL STOCK NUMBER INDEX

STOCK NUMBER	FIG.	ITEM	STOCK NUMBER	FIG.	ITEM
5306-01-283-4199	15	6	5340-01-285-0575	181	17
4730-01-283-8148	110	5	5330-01-285-1601	181	11
4730-01-283-8149	618	13	5306-01-285-1703	184	12
4730-01-283-8150	121	4	2915-01-285-2527	630	5
	287	4	4710-01-285-3007	459	1
5305-01-283-8462	121	8	4720-01-285-3008	111	17
2530-01-283-9694	296	13	5340-01-285-3239	107	27
6240-01-284-1925	141	3	5340-01-285-3257	623	25
4730-01-284-2211	617	12	2835-01-285-3269	98	1
	618	14	2530-01-285-3563	296	18
4720-01-284-2235	620	1	2510-01-285-4592	133	3
2815-01-284-2284	222	12	2590-01-285-4600	455	5
2530-01-284-2310	294	5	5330-01-285-4827	28	9
2530-01-284-2311	296	11	2930-01-285-5027	121	6
4460-01-284-2344	298	11	5120-01-285-5192	630	18
6220-01-284-2709	139	5	5120-01-285-5193	630	3
2990-01-284-3218	585		5315-01-285-5562	184	16
	598	47	5306-01-285-5976	370	44
5340-01-284-3789	181	13	5360-01-285-6010	101	15
2530-01-284-4399	294	9	4820-01-285-6159	297	16
2590-01-284-4547	184	10	2910-01-285-6253	111	18
2530-01-284-4566	307	1	2590-01-285-6261	179	2
5445-01-284-6173	181	12	4320-01-285-6262	222	9
3040-01-284-6230	101	23	2520-01-285-6282	466	2
3040-01-284-6232	184	15	2520-01-285-6283	224	1
2530-01-284-7446	296	11	2520-01-285-6295	529	1
5365-01-284-8138	184	7	4720-01-285-6312	63	4
			5980-01-285-6688	131	6
5365-01-284-8152	615	14	5120-01-285-7620	630	19
			5340-01-285-7757	616A	17
4720-01-284-8184	121	5	5340-01-285-7762	78	20
	598	14	5120-01-285-7829	630	6
			5340-01-285-9399	616A	18
2520-01-284-8240	455	5	9390-01-285-9623	362	16
2540-01-284-8718	599	5		405	35
	599	6	2910-01-285-9850	82	1
	599	8	4820-01-285-9851	461	30
	599	9	9515-01-285-9855	410	1
4710-01-284-9029	111	9	9515-01-285-9856	410	1
4710-01-284-9032	115	4	2530-01-286-0108	466	5
4730-01-284-9071	110	7	2530-01-286-3257	245	14
4730-01-284-9086	630	1		246	1
5342-01-284-9246	181	1	2510-01-286-3328	6	12
5340-01-284-9247	181	7	2510-01-286-3329	6	5
5315-01-284-9583	101	5	4730-01-286-4611	582	23
5340-01-284-9652	178	3		585	37
5340-01-284-9653	299	3	5310-01-286-5452	121	3
5340-01-284-9654	181	14		126	6
5340-01-284-9655	181	9		184	5
5340-01-284-9656	623	20	2520-01-286-5650	181	2
5306-01-284-9663	110	12	5310-01-286-6075	6	9
5315-01-284-9812	101	1		63	15
	184	2	5365-01-286-6180	615	6
			2540-01-286-7673	370	20

NATIONAL STOCK NUMBER INDEX

STOCK NUMBER	FIG.	ITEM	STOCK NUMBER	FIG.	ITEM
2540-01-286-7674	615	4	3040-01-288-5313	598	31
2930-01-286-8357	121	1	5310-01-288-5690	65	7
5340-01-287-0751	38	5	2990-01-288-5844	585	22
5307-01-287-0854	64	12	2530-01-288-5877	292	3
5307-01-287-0855	285	23	5330-01-288-6304	308	4
5305-01-287-1585	101	24	6625-01-289-2062	138	11
				627	4
6680-01-287-2153	179	1	5330-01-289-3135	38	6
3030-01-287-3155	584	16	2530-01-289-3962	292	1
2930-01-287-3180	113	4	5305-01-289-4411	598	33
2990-01-287-3189	585	21	5365-01-289-4434	125	20
2590-01-287-3224	598	2	5340-01-289-5028	598	43
4820-01-287-3963	298	1	5340-01-289-5030	250	4
5930-01-287-3965	138	13		251	4
5995-01-287-3988	158	26			
5935-01-287-4286	138	5	4710-01-289-5446	621	13
4720-01-287-4494	595	5	4710-01-289-5447	620	7
4720-01-287-4495	595	3	4710-01-289-5448	620	7
2815-01-287-4502	18	5	4710-01-289-5449	620	7
2530-01-287-4529	621	5	5365-01-289-7852	125	21
2930-01-287-4545	113	5	2530-01-289-8359	250	9
4710-01-287-4608	598	22	5306-01-289-9197	184	6
5340-01-287-5699	285	19		584	20
5310-01-287-5742	285	20	5995-01-290-1293	138	1
5365-01-287-5762	285	16	5995-01-290-1294	138	1
5305-01-287-6570	623	9	5340-01-290-1738	146	3
5340-01-287-6660	107	9	4720-01-290-2510	254	6
5330-01-287-8656	115	11	2540-01-290-3017	629	5
5310-01-287-8812	111	13	5930-01-290-5677	279	15
2990-01-287-8957	594	19	5320-01-290-6360	245	15
4710-01-287-9008	620	7		246	3
4440-01-287-9011	619	11	5340-01-290-6370	615	3
4730-01-287-9012	598	20	5310-01-290-7456	585	29
4730-01-287-9084	286	26	5340-01-290-8882	251	3
4730-01-287-9100	57	26	5340-01-290-8883	250	3
2910-01-287-9119	82	1	5340-01-290-8884	250	3
2910-01-287-9120	82	1	5306-01-291-0136	6	8
	82	1	5305-01-291-0156	6	29
5340-01-288-0607	6	24	2590-01-291-4598	110	3
5306-01-288-1011	285	17	5305-01-291-5138	107	14
5305-01-288-1012	285	18	4730-01-291-5225	598	19
5305-01-288-1417	585	27	5120-01-291-5769	630	7
5330-01-288-1466	599	2	2930-01-291-5867	115	10
			5330-01-291-6537	20	7
			7690-01-291-8971	565	5
5365-01-288-1559	619	13	2540-01-291-9037	370	19
5365-01-288-1560	598	36	5340-01-291-9205	347	12
5340-01-288-2132	598	40	5340-01-291-9212	320	21
5342-01-288-2161	595	11		321	24
5342-01-288-2167	599	1	2520-01-291-9992	231	1
5340-01-288-4557	622	4	2520-01-291-9993	238	1
5315-01-288-4584	598	32	2520-01-291-9994	238	1
5340-01-288-5156	585	30	5935-01-292-2336	158	40
			5305-01-292-3182	65	26

CROSS-REFERENCE INDEXES

NATIONAL STOCK NUMBER INDEX

STOCK NUMBER	FIG.	ITEM	STOCK NUMBER	FIG.	ITEM
4710-01-292-3701	57	5	5365-01-300-7159	59	25
2910-01-292-5663	47	1	5365-01-300-7160	59	25
2590-01-292-5707	373	2	5365-01-300-7161	59	25
5310-01-292-7251	107	17	5365-01-300-7162	59	25
5310-01-292-9504	6	6	5365-01-300-7163	59	25
5330-01-292-9573	245	5	5365-01-300-7164	59	25
	246	11	5365-01-300-7165	59	25
5331-01-292-9575	296	1	5365-01-300-7166	59	25
2540-01-295-7461	596	6	5365-01-300-7167	59	25
6680-01-296-2758	177	14	5365-01-300-7168	59	25
5305-01-296-2849	584	18	5365-01-300-7169	59	25
5340-01-297-1187	103	4	5365-01-300-7170	59	25
6220-01-297-3217	139	6	5365-01-300-7171	59	25
5342-01-297-4454	251	3	5365-01-300-7172	59	25
5340-01-298-4861	72	11	5365-01-300-7173	59	25
5365-01-298-4877	82	11	5365-01-300-7174	59	25
5220-01-298-5730	630	8	5365-01-300-7175	59	25
5330-01-299-6616	82	12	5365-01-300-7176	59	25
4710-01-299-9450	621	3	5365-01-300-7177	59	25
4710-01-299-9451	621	7	5365-01-300-7178	59	25
4820-01-300-2759	24	12	5365-01-300-7179	59	25
6220-01-300-3643	141	1	5365-01-300-7180	59	25
4820-01-300-4051	58	1	5365-01-300-7181	59	25
4730-01-300-4091	617	17	5365-01-300-7182	59	25
4730-01-300-4104	617	11	5365-01-300-7183	59	25
4730-01-300-4112	111	23	5365-01-300-7184	59	25
4720-01-300-4143	620	13	5365-01-300-7185	59	25
4720-01-300-4144	620	16	5365-01-300-7186	59	25
4720-01-300-4146	111	25	5365-01-300-7187	59	25
4820-01-300-4257	59	1	5365-01-300-7188	59	25
2910-01-300-4271	59	19	5340-01-300-7203	58	18
3040-01-300-5309	92	14	5340-01-300-7295	58	17
2530-01-300-5918	246	13	4730-01-300-8819	58	10
2930-01-300-5943	115	25	4730-01-300-9024	115	17
2930-01-300-5944	115	29	4730-01-300-9030	115	16
3110-01-300-6796	58	5	4730-01-300-9031	618	3
3110-01-300-6804	58	21	5310-01-301-0455	91	12
5365-01-300-6885	58	3	5305-01-301-0533	58	16
5360-01-300-6888	59	17	5365-01-301-0554	59	25
5360-01-300-6889	59	27	5330-01-301-1761	115	26
5330-01-300-6898	58	25	5330-01-301-1763	58	27
5331-01-300-6907	58	14	5310-01-301-1798	58	7
5365-01-300-7021	59	23	5310-01-301-1802	59	5
5310-01-300-7037	58	6	5331-01-301-1825	92	13
3120-01-300-7135	58	13	5330-01-301-1828	115	23
3110-01-300-7136	58	22	5330-01-301-1829	115	30
5365-01-300-7149	59	21	5310-01-301-1875	92	16
5340-01-300-7153	59	18	5310-01-301-1885	90	29
5340-01-300-7154	59	15	5310-01-301-3885	90	2
5365-01-300-7158	59	25	5315-01-301-3892	58	4

CROSS-REFERENCE INDEXES

NATIONAL STOCK NUMBER INDEX

STOCK NUMBER	FIG.	ITEM	STOCK NUMBER	FIG.	ITEM
5365-01-301-4006	59	25	4730-01-304-5618	621	18
5365-01-301-4007	59	25	5365-01-304-9529	318	51
	320	50		321	51
4730-01-301-4286	60	9	5365-01-304-9530	59	25
4710-01-301-4358	115	19	4310-01-304-9726	KITS	
5306-01-301-4881	92	18	2520-01-305-0457	315	1
5305-01-301-5034	92	1	4720-01-305-2440	63	10
5305-01-301-5112	92	17	5365-01-305-2535	623	21
5331-01-301-5992	59	20	2930-01-305-3303	115	22
5330-01-301-5995	58	20	5961-01-305-8848	139	5
5310-01-301-7807	59	28		140	8
5310-01-301-7811	89	10	7690-01-305-9103	552	7
	89	20		552	15
	90	6	4220-01-307-4779	622	8
5307-01-301-7815	59	6	5330-01-308-0175	296	16
5305-01-301-7817	58	15			
	59	12	6685-01-308-1985	630	17
	92	9	5307-01-308-5081	529	8
5305-01-301-7818	58	24			
5310-01-301-7863	6	16	5935-01-308-7866	158	45
5331-01-301-7867	59	24	4730-01-309-0947	598	28
5365-01-301-7871	6	26	4730-01-309-0948	598	38
5365-01-301-7872	6	26	4730-01-309-0949	598	55
	6	26	4820-01-309-3799	82	8
5315-01-301-7912	58	8	2530-01-309-3803	303	5
5305-01-301-9756	115	20	2530-01-309-6203	303	8
5305-01-301-9757	115	24	5340-01-309-7782	458	16
2910-01-301-9936	59	14	5310-01-309-8575	6	2
6620-01-302-0045	115	27	2590-01-310-1166	82	2
5330-01-302-0780	115	28	2540-01-310-4829	594	1
5365-01-302-3189	622	6	2540-01-310-4854	540	5
5330-01-302-5738	58	19	4720-01-310-5475	618	9
5365-01-302-5848	59	25	4720-01-310-5476	618	11
5365-01-302-6555	115	31	3020-01-310-5493	486	1
4720-01-302-6687	115	2	5340-01-310-7070	418	7
5310-01-302-9472	297	12	5340-01-310-7071	418	6
5365-01-302-9485	6	26	5340-01-311-0225	414	25
5330-01-302-9948	529	3	2590-01-311-7213	336	10
5365-01-302-9953	59	25	5340-01-312-1136	183	12
2530-01-303-0801	292	6	5340-01-312-7652	48	4
5365-01-303-0937	59	25	2510-01-313-0011	373	1
5365-01-303-0938	59	25	5340-01-314-2423	315	9
2910-01-303-1195	51	2	4730-01-314-5825	82	13
5365-01-303-1612	59	4	5306-01-314-6742	292	10
5331-01-303-1635	59	29	5331-01-314-7598	292	9
5310-01-303-5531	285	21	4730-01-315-5596	121	7
5995-01-303-6428	587	1	5340-01-315-8921	326	4
5365-01-304-0652	6	26	5220-01-317-1436	63	5
4720-01-304-0688	182	2	9905-01-317-2715	554	1
4720-01-304-1439	299	5	5330-01-317-3213	20	5
5365-01-304-1802	59	25			

NATIONAL STOCK NUMBER INDEX

STOCK NUMBER	FIG.	ITEM	STOCK NUMBER	FIG.	ITEM
5340-01-317-8144	107	3	5340-01-331-9625	32	11
2520-01-317-8288	185	5	4720-01-332-1596	617	19
5930-01-318-2809	587	10	2815-01-332-5462	65	12
	591	30	5310-01-332-7265	184	1
5355-01-318-2811	279	8	4730-01-333-0133	64	11
5640-01-318-2812	372	1	5305-01-333-5382	8	13
2510-01-318-2814	373	1	5340-01-333-5564	315	15
5640-01-318-2815	372	3	5315-01-333-6603	327	9
5640-01-318-2816	372	6	5340-01-333-8147	243	5
2510-01-318-2817	373	11	5310-01-335-4861	60	8
2510-01-318-2818	373	4	5310-01-335-6455	KITS	1
4730-01-318-2821	296	2	5315-01-335-9940	91	18
5340-01-319-2349	296	6	5315-01-335-9941	59	7
6680-01-319-2354	146	2	5360-01-335-9947	59	9
5640-01-319-2376	372	2	5360-01-335-9948	90	19
2530-01-319-2384	622	10	5360-01-335-9949	91	5
5342-01-319-8630	312	37	5360-01-335-9950	91	4
4320-01-320-4744	483	3	5360-01-335-9951	91	3
5330-01-321-2053	20	3	5305-01-335-9965	89	17
	538	6	5305-01-335-9966	90	10
5340-01-321-6196	KITS		5305-01-335-9967	90	14
3040-01-321-6365	20	2	5305-01-335-9968	90	11
			5305-01-335-9969	91	11
5340-01-322-2906	181	23	5340-01-335-9996	89	11
2520-01-323-6342	227	2	5305-01-336-0006	89	9
2520-01-323-6343	227	2	5305-01-336-0007	89	25
4730-01-324-5071	267	5	5305-01-336-0008	92	6
			5305-01-336-0015	90	16
5999-01-328-0524	156	13	5340-01-336-0016	91	16
4820-01-329-3245	212	14	5340-01-336-0020	91	7
	266	1	5340-01-336-0021	91	2
6695-01-329-6418	623	2	5305-01-336-0026	90	12
2815-01-330-3036	10	1	5340-01-336-0046	92	5
5340-01-330-3240	619	7	2810-01-336-2199	90	20
4120-01-330-6543	426	1			
2815-01-330-8069	8	1	5340-01-336-2292	90	4
5310-01-330-8313	43	2	5340-01-336-2588	92	4
4730-01-331-2670	8	11	5330-01-336-3150	92	3
4730-01-331-2858	594	5	5340-01-336-3889	59	11
4730-01-331-2913	38	11	5315-01-336-5185	90	23
4730-01-331-6630	104	6	5365-01-336-5187	58	26
	104	6	5365-01-336-5212	91	26
	111	1	5365-01-336-5213	91	26
	115	3	5365-01-336-5214	91	26
3130-01-331-7182	65	19	5365-01-336-5215	91	26
5342-01-331-7985	6	11	5365-01-336-5216	91	26
4720-01-331-8717	64	4	5365-01-336-5217	91	26
5331-01-331-9293	64	15	5365-01-336-5218	91	26
5305-01-331-9479	10	10	5365-01-336-5219	91	26
5305-01-331-9480	10	11	5365-01-336-5220	91	9
5340-01-331-9487	130	5	5365-01-336-5221	91	6

CROSS-REFERENCE INDEXES

NATIONAL STOCK NUMBER INDEX

STOCK NUMBER	FIG.	ITEM	STOCK NUMBER	FIG.	ITEM
5365-01-336-5224	90	21	3040-01-338-2546	92	20
5365-01-336-5910	91	26	4820-01-338-4642	89	5
5365-01-336-5911	91	26	5330-01-338-4829	49	1
			4910-01-338-6241	631	5
5365-01-336-5912	91	8	5330-01-339-0141	90	1
5365-01-336-5913	92	7	4710-01-339-0584	48	6
5310-01-336-6683	90	9	5360-01-339-0684	89	7
5310-01-336-6684	91	22	5330-01-339-0708	201	3
5310-01-336-6687	91	20	5340-01-339-0839	59	10
5315-01-336-6689	92	21	5330-01-339-0859	308	9
5340-01-336-6704	90	25	5340-01-339-0868	90	26
5331-01-336-6717	92	25	5340-01-339-0874	619	7
5325-01-336-6738	92	26	2520-01-339-1648	222	1
5310-01-336-6747	90	28	5365-01-339-3835	91	26
5310-01-336-6748	59	8	5365-01-339-3836	91	26
5340-01-336-6761	92	11	5365-01-339-3837	91	26
5365-01-336-6782	91	26	5365-01-339-3838	91	26
5340-01-336-6783	91	26	5365-01-339-3839	92	23
5340-01-336-6784	92	24			
5310-01-336-7368	91	13	3040-01-339-8572	90	17
3040-01-336-8168	90	27	2540-01-339-8594	542	4
4910-01-336-8204	631	4	2910-01-339-8598	59	26
3040-01-336-8217	193	10	2530-01-339-8673	307	5
3040-01-336-8293	91	14	2815-01-340-0377	201	2
5310-01-336-8721	59	2	3020-01-340-0402	201	7
5310-01-336-8722	91	1	3020-01-340-0403	201	7
5310-01-336-8865	59	3	3020-01-340-0404	201	8
5365-01-336-8895	89	1	3020-01-340-0407	201	8
5365-01-336-8908	90	7	3020-01-340-0408	201	8
5365-01-336-8909	91	26	5310-01-340-8469	57	3
5365-01-336-8910	91	26		60	2
5365-01-336-8911	91	26	5999-01-341-2972	308	10
5365-01-336-8916	90	26	8125-01-341-3838	542	3
5360-01-336-9366	92	8	5120-01-341-6000	631	1
5305-01-336-9380	90	3	3120-01-341-6519	201	5
5331-01-336-9559	89	14	5340-01-341-6572	631	6
4320-01-337-0608	201	4	5340-01-341-6655	90	26
2910-01-337-2983	91	27	5310-01-341-8953	89	3
2910-01-337-2984	91	23	5305-01-342-5171	89	18
5340-01-337-3760	90	8	2540-01-342-6810	590	17
2910-01-337-4142	92	10		590	28
3040-01-337-4155	193	11		591	
3040-01-337-4161	201	9		591	12
3040-01-337-4167	193	10		591	13
3040-01-337-4168	193	10		591	14
3040-01-337-4368	91	17		591	15
	92	22		591	17
3040-01-337-4379	201	1		591	26
3010-01-337-8937	91	24		591	32
2910-01-338-2335	59	16		591	33
3040-01-338-2545	90	22		591	35

NATIONAL STOCK NUMBER INDEX

STOCK NUMBER	FIG.	ITEM	STOCK NUMBER	FIG.	ITEM
2540-01-342-6810	591	36	4720-01-371-4396	613	20
	591	37		629	5
	591	38	2920-01-371-6064	KIT	38
	591	40	4730-01-371-7700	629	5
5120-01-343-2585	631	2			
5120-01-343-2586	631	3	6220-01-372-3883	139	2
3020-01-347-2855	201	8	2930-01-372-9135	591	22
4730-01-348-9510	111	11	4720-01-373-5652	40	2
4730-01-349-2436	58	30		124	5
5330-01-350-6119	2	4	2610-01-373-7294	300	1
5310-01-351-7542	89	12	5310-01-374-0508	184	8
3120-01-351-7779	91	21	6680-01-374-2083	629	5
5315-01-352-8225	91	15	2815-01-374-7539	2	1
			5310-01-375-3107	293	1
				294	18
				295	15
5340-01-353-6158	590	26	6685-01-376-2290	629	5
			5330-01-377-9775	237	4
				293	10
5340-01-353-6752	591	42		629	8
5340-01-353-6753	591	41	5330-01-379-4345	102	3
5340-01-353-6936	590	16		103	7
5340-01-353-6961	591	6	2815-01-379-4920	22	1
5340-01-353-6962	590	11	4910-01-380-9029	629	5
9905-01-353-8846	555	3	5340-01-381-3862	629	5
5961-01-353-9187	588	14	5315-01-384-2149	138	12
				626	5
5320-01-354-2548	586	5	2590-01-384-6244	586	2
4730-01-354-2645	591	44	5315-01-385-2731	528	17
2815-01-354-2702	24	13	5355-01-386-8877	538	9
2930-01-354-5433	591	23	2815-01-388-8596	22	2
4710-01-354-8064	590	9	4730-01-391-8301	297	10
4710-01-354-8065	590	20	2510-01-394-6119	345	34
4710-01-354-8066	591	16	5310-01-407-3451	156	11
4710-01-354-8067	590	34	6220-01-411-3584	141	2
4710-01-354-8068	590	27	3020-01-414-8008	22	6
			5315-01-416-1809	507	47
4720-01-354-9029	590	23	2520-01-416-4538	629	5
4720-01-354-9585	591	20	5340-01-416-6503	629	6
5340-01-355-8746	590	30	4730-01-416-6783	629	16
5340-01-356-0011	591	18	2540-01-416-6784	594	7
5305-01-357-1656	245	22		594	13
	246	19		595	7
5330-01-359-2316	111	19		595	8
6220-01-359-2870	139	7		595	9
6150-01-360-1517	588	1		595	10
5330-01-361-5600	103	5		595	14
6685-01-361-7552	111	20		595	18
6150-01-361-8128	588	16		595	23
2510-01-363-8981	373	1			
2530-01-367-1883	619	20	2540-01-416-9644	629	9
4710-01-369-4814	285	10	5320-01-417-1019	629	14
2520-01-369-5335	301	8	5340-01-417-2432	629	8
5306-01-370-6947	62	12	6680-01-417-3281	629	15
2930-01-371-1389	591	5	6695-01-417-4453	133	2
4720-01-371-4395	629	5	2540-01-417-6379	533	4
	629	19	9905-01-417-8434	552	5
			9905-01-417-8435	552	5

NATIONAL STOCK NUMBER INDEX

STOCK NUMBER	FIG.	ITEM	STOCK NUMBER	FIG.	ITEM
5340-01-418-1936	133	1	4720-01-442-2533	72	14
4730-01-418-4254	629	25	9905-01-442-3562	552	5
9905-01-418-6621	552	5	4730-01-442-6120	273	17
9905-01-418-6622	552	5	4730-01-443-2888	63	19
9905-01-418-6628	552	5	2590-01-443-8097	101	6
9905-01-418-6633	552	5	6220-01-443-8805	140	9
9905-01-418-7651	552	5	6220-01-443-8813	140	5
9905-01-418-7652	552	5	5340-01-444-1016	101	17
9905-01-418-8324	552	5	6110-01-444-2546	137	7
9905-01-418-8326	552	5	5340-01-444-5359	97	21
9905-01-418-8327	552	5	5340-01-444-5366	97	20
2540-01-418-8880	615	8			
9905-01-418-9772	552	5	5340-01-444-6658	108	8
9905-01-418-9773	552	5	3040-01-444-7878	101	26
9905-01-418-9774	552	5	5330-01-444-8350	293	19
9905-01-418-9776	552	5		295	13
9905-01-418-9777	552	5		348	20
2540-01-419-6283	615	8			
2920-01-419-8884	608	15	3040-01-445-1651	100	23
9905-01-420-0672	552	5	5315-01-446-9007	184	2
			2990-01-446-5359	65	1
9905-01-420-0674	552	5	5306-01-447-2565	108	9
9905-01-420-1451	552	5	2510-01-447-4754	390	13
9905-01-420-1764	552	5	2990-01-448-1098	591	45
2530-01-420-4221	280	1	5340-01-457-1778	602	37
9905-01-420-6755	552	5	5365-01-457-3364	602	52
2540-01-420-7925	616A	2	1005-01-457-5807	602	38
2540-01-423-1786	609	31	1005-01-457-5812	602	4
			1005-01-457-6135	602	54
2540-01-423-1791	609	31	1005-01-457-6138	602	56
2540-01-423-1964	532		1005-01-457-6144	602	19
2540-01-423-1966	509	30	1005-01-457-6147	602	24
2590-01-423-1967	507	49	1005-01-457-6150	602	55
2540-01-423-1968	445		1005-01-457-6152	602	19
2815-01-424-4736	22	3	1005-01-457-6176	602	21
5305-01-424-8744	616A	19	1005-01-457-6178	602	46
2540-01-424-9440	532	1	1005-01-457-6185	602	34
2530-01-425-9520	250	5	1005-01-457-6700	602	47
2520-01-428-7497	KITS		5340-01-457-7340	602	9
5925-01-430-2318	135	2	2510-01-457-9011	382	13
	445	11		387	10
	587	3	2590-01-457-9019	602	12
	588	3	2540-01-457-9025	609	4
6140-01-431-1172	151	22	2510-01-457-9780	387	29
2540-01-434-8725	532	1	2510-01-457-9788	382	9
2540-01-435-0568	532	1		387	6
5306-01-435-3269	101	25	2510-01-457-9827	382	8
5365-01-435-4806	484	25		387	5
2540-01-435-4924	533	4			
2540-01-435-4928	532	1			
2540-01-435-4931	533	4			
2540-01-435-4936	532	1			
2540-01-435-8760	509	30			
5310-01-437-8728	101	11			
9905-01-438-7063	552	5			
4720-01-441-2922	72	23			
9905-01-442-2472	552	5			

CROSS-REFERENCE INDEXES

NATIONAL STOCK NUMBER INDEX

STOCK NUMBER	FIG.	ITEM	STOCK NUMBER	FIG.	ITEM
5340-01-458-0975	609	1			
5310-01-458-1274	602	17			
5315-01-458-4490	602	29			
2510-01-458-4512	382	16			
5340-01-458-7234	602	25			
2590-01-458-9661	602	10			
2590-01-458-9665	602	10			
2510-01-458-9668	387	30			
	78	4			
2510-01-458-9670	385	18			
2510-01-459-0259	382	3			
5910-01-459-0289	126	10			
2510-01-459-0564	385	33			
2540-01-459-0567	387	13			
5305-12-142-0188	90	30			
5315-12-156-4502	90	18			
5330-12-156-4523	58	2			
5305-12-168-9310	22	5			
3120-12-198-1758	90	24			
5365-12-305-0652	58	29			

CROSS-REFERENCE INDEXES

PART NUMBER INDEX

CAGEC	PART NUMBER	STOCK NUMBER	FIG.	ITEM
96906	A-A-52418-28-Y	6220-00-947-7570	611	1
			612	1
			613	1
58536	A-A-52427	5310-01-270-5463	293	1
			295	15
58536	A-A-52483	9905-00-999-7369	559	2
			559	2
58536	A-A-55487-8	5315-01-446-9007	184	2
78500	A-1205-D-2162	5330-01-308-0175	296	16
78500	A-1205-N-2120	5330-01-272-1147	294	15
			296	17
78500	A-1205-P-1758	5330-01-132-8346	216	1
78500	A-1205-U-1633	5330-01-126-0565	527	8
			234	10
			234	23
			241	10
			241	23
78500	A-1898-T-1164	4730-01-130-5158	526	24
78500	A-2206-D-56	4730-01-132-6848	207	44
78500	A-2206-V-1010	4730-01-271-7978	294	12
78500	A-2206-Y-1013	4730-01-318-2821	296	2
78500	A-2296-C-81	4720-01-126-9555	217	8
78500	A-2296-E-83	4720-01-134-3475	526	29
78500	A-2297-V-5326		245	12
			246	15
78500	A-2747-H-112	2530-00-117-9144	245	20
			246	21
78500	A-3107-M-39	2520-01-135-0085	210	2
78500	A-3211-D-2994	2530-01-125-4280	246	5
78500	A-3211-H-2868	2530-01-134-0901	245	2
78500	A-3211-K-2871	2530-01-124-6530	246	5
78500	A-3211-L-2872	2530-01-134-1333	245	2
78500	A-3213X1558	2530-01-272-0542	236	4
78500	A-3260-A-79	2520-01-127-3595	207	29
78500	A-3261-S-253	3040-01-149-1111	216	11
78500	A-3266-Y-623	2520-01-134-3665	207	33
78500	A-3270-R-1032	2530-01-271-9809	294	1
78500	A-3280-E-6817		251	6
78500	A-3280-U-7535	2520-01-126-9363	207	25
			244	11
78500	A-3280-V-8186	3040-01-280-4156	245	10
78500	A-3280-Z-8190		245	9
			246	16
78500	A-3289-G-189	2520-01-139-4233	207	14
78500	A-3297-K-63	2520-01-127-1676	208	3
78500	A-3297-T-72	2520-01-127-2623	209	11
78500	A-3299-C-6139	5340-01-319-2349	296	6
78500	A-3303-E-5	2520-01-127-2626	526	7
78500	A-3303-J-10	2815-01-136-8720	217	5
78500	A-333-U-801	3040-01-134-1184	295	11
78500	A-333-X-3534	2530-01-284-4399	294	9

CROSS-REFERENCE INDEXES

PART NUMBER INDEX

CAGEC	PART NUMBER	STOCK NUMBER	FIG.	ITEM
78500	A-3722-V-360		245	14
78500	A-3787-N-14	2530-01-134-1332	244	13
78500	A-3892-D-4554	2520-01-127-5766	209	16
78500	A-3892-E-4555	3020-01-134-1015	209	8
58536	AA55571/01-001	5925-01-430-2318	135	2
			445	11
			587	3
			588	3
58536	AA55571/01-004	5925-00-333-1584	135	3
56501	AB53	5940-01-021-1874	158	30
83879	ACV-938	5975-00-152-1127	429	11
16662	AC2511	4730-00-293-7108	617	2
			620	2
			621	11
16662	AC2569	4730-00-069-1186	621	23
16662	AD17544	2530-00-430-2392	250	12
			251	11
90030	AD64H	5320-00-956-7355	417	4
			419	7
88044	AN365-1024A	5310-00-208-1918	111	3
			211	47
			513	25
			525	25
			528	28
81352	AN415-4	5315-01-057-8371	549	14
			551	12
88044	AN415-6	5315-01-081-9991	351	9
88044	AN565F428H24	5305-00-063-5043	51	21
			56	21
88044	AN742-26	5340-00-598-0225	585	19
88044	AN914-4	4730-00-231-5598	261	31
88044	AN914-8	4730-00-231-5596	491	2
88044	AN915-3	4730-00-231-3906	109	7
88044	AN931A16-22	5325-00-174-9341	594	24
01496	AN960-C416L	5310-00-515-7449	100	14
88044	AN960-1216	5310-00-167-0826	608	17
88044	AN960-416	5310-00-141-1795	50	23
			52	4
			52	11
			55	23
			87	8
			88	21
88044	AN970-3	5310-00-167-0765	409	27
15434	AR-2308	2815-00-005-7431	26	6
15434	AR-40230	5342-01-160-7381	86	5
15434	AR-51264	2930-00-404-3053	114	2
15434	AR10922	4310-01-079-3383	284	43
15434	AR41010	2910-01-080-3149	51	3
			56	2
15434	AR4284	2930-00-890-2440	118	1
15434	AR45247	2815-00-404-2956	37	18

CROSS-REFERENCE INDEXES

PART NUMBER INDEX

CAGEC	PART NUMBER	STOCK NUMBER	FIG.	ITEM
15434	AR45724	3040-01-129-4302	44	5
15434	AR51480	2940-00-404-3057	KITS	
15434	AR51481	5330-00-133-6235	KITS	
15434	AR51482	5330-00-133-6236	KITS	
15434	AR51522	2910-01-117-3689	KITS	
15434	AR73350	2815-00-913-2074	KITS	
15434	AS-1603506MS	4720-01-128-0331	36	1
81346	ASTM A641	9505-00-248-9842	345	1
15434	AS1602906MS	4720-00-607-7645	36	3
81343	AS3551-12	5331-00-776-2830	204	3
01265	AX-1597	2815-00-962-5623	24	14
78500	A1-3211-D-2994		246	9
78500	A1-3211-K-2871		246	9
78500	A1-3261-D-290	2520-01-134-0899	216	5
78500	A1-333-Z-3536	2530-01-283-9694	296	13
70403	A11	4730-00-555-8263	64	8
			284	48
			285	8
78500	A1244J556	4320-00-734-6951	236	29
34623	A12876	4720-00-203-2668	BULK	15
78500	A1805H60	5330-00-145-8355	236	31
78500	A2-2297-N-3212		245	14
78500	A2-3111-D-2604	2530-01-272-2926	236	33
78500	A2-3111-E-2605	2530-01-271-9824	236	33
78500	A2-333-X-3534	2530-01-282-8619	294	5
78500	A2-333-Z-3536	2530-01-284-2311	296	11
78500	A2-3780-J-62	2530-01-289-8359	250	9
78500	A2-3800E473	2520-00-734-7548	233	5
			240	5
78500	A2244-U-21	3040-01-134-8909	526	12
78500	A2257M1079	5340-00-015-7560	250	11
			251	10
98441	A2697 X 15 1/2IN		72	13
			78	2
78500	A3-333-X-3534	2530-01-284-2310	294	5
78500	A3-333-Z-3536	2530-01-284-7446	296	11
78500	A3-3722-D-420	2530-01-140-0107	244	4
78500	A3102Y3431	2530-00-512-0032	303	1
78500	A3107-B-28	2520-01-134-1839	527	13
78500	A3144U437	2530-01-309-6203	303	8
78500	A3266-A-625	2520-01-126-9364	207	21
78500	A3266-Q-745	3110-01-142-9490	527	7
78500	A3266-W-621	2520-01-146-3447	207	3
78500	A3268-R-1084	2530-01-134-0900	207	8
78500	A3280G1957	2520-00-734-6802	234	3
			241	3
78500	A3290Y233	3040-01-248-3995	208	1
78500	A333-T-2334	3040-01-139-1611	293	21
99199	A335	4730-00-254-6211	109	13
99199	A335	4730-00-254-6211	582	20
			585	9

PART NUMBER INDEX

CAGEC	PART NUMBER	STOCK NUMBER	FIG.	ITEM
			594	11
41947	A337	4730-00-902-8991	595	24
70960	A3722-D-420		244	6
78500	A3736-K-375	2530-01-134-6626	244	2
78500	A3800E473	2520-00-734-6970	232	3
			239	3
98441	A3915-3	4720-01-271-7997	205	7
98441	A3915-4	4720-01-284-8184	121	5
			598	14
98441	A3916-2	4720-01-271-7996	205	11
98441	A3917-1	4720-01-272-0504	205	4
98441	A3994-2	4720-01-279-3034	297	4
98441	A3995-5	4720-01-304-1439	299	5
78500	A4-3280-F-5232		250	6
			251	14
78500	A4-3866-J-556	3040-00-734-6897	234	7
			241	7
78500	A45-1779-V-230	2530-01-064-2630	250	7
			251	12
41947	A4739	4730-00-927-7272	618	12
			630	12
78500	A5-3280-F-5232		250	6
58536	A52431-1	2530-00-859-7335	276	6
58536	A52432-2	2540-00-788-5637	543	22
58536	A52463-1-08	6240-00-019-0877	143	3
58536	A52463-1-09	6240-00-019-3093	139	3
			140	6
58536	A52463-2-10	6240-00-044-6914	139	4
			140	7
			614	4
58536	A52484-1	4730-00-595-0083	252	19
			255	13
			261	22
			262	5
			291	12
79470	A555		423	1
72464	A5711	5310-01-133-5847	82	10
78500	A59-1779-V-230		250	13
70494	A6S	4730-00-374-2045	63	6
98441	A6036-2	4720-01-281-0994	459	2
98441	A6056-5	4720-01-332-1596	617	19
98441	A6057-3	4720-01-310-5476	618	11
98441	A6057-4	4720-01-310-5475	618	9
78500	A7-3222S2021	2530-01-065-1828	245	14
78500	A71-1779-V-230	2530-01-134-1336	250	13
96152	A82-1	5315-00-839-2325	100	17
78500	A9-3280-F-5232		250	6
81205	BACW10P53S	5310-00-725-1983	207	11
			237	13
13475	BD809990	2520-00-163-0713	191	3'
B1821	BH038F225N		100	15

CROSS-REFERENCE INDEXES

PART NUMBER INDEX

CAGEC	PART NUMBER	STOCK NUMBER	FIG.	ITEM
15434	BM-37634	2815-00-505-5116	23	4
15434	BM-76340	6685-00-828-7126	52	1
15434	BM-77410	4310-00-603-1510	284	26
15434	BM-95162	2815-00-851-7637	26	11
15434	BM-98685	4310-01-079-3319	286	42
15434	BM27253	3120-00-339-5642	KITS	
15434	BM30245	5360-00-194-5920	88	29
15434	BM56657	5365-00-829-5150	KITS	
15434	BM57837	2930-01-117-5238	31	2
15434	BM66076	2990-01-123-3454	33	10
15434	BM94080		33	24
15434	BM94081	2940-01-123-4875	33	20
15434	BM94082	2815-00-404-2954	33	2
15434	BM95161	2815-01-096-9198	26	2
61465	BP2415	5340-00-456-1011	383	11
			383	25
			383	29
			384	17
			386	12
13940	BS 78505	5325-00-371-8108	533	12
80204	B1821BH038F138N	5305-00-269-3239	165	22
80204	B1821BH038F056N	5305-00-269-3232	602	33
80204	B1821BH038F250N	5305-00-269-3244	385	12
80204	B1821BH025C088N	5305-00-071-2505	66	19
80204	B1821BH025C100N	5305-00-225-3843	371	4
			414	28
			427	3
			536	11
80204	B1821BH025C125N	5305-00-068-0509	52	2
			66	9
			197	14
80204	B1821BH025C138N	5305-00-225-3841	623	27
80204	B1821BH025C150N	5305-00-071-2509	339	15
			339	22
80204	B1821BH025C175N	5305-00-071-2510	619	17
80204	B1821BH025C250N	5305-00-071-2513	189	18
80204	B1821BH025C275N	5305-00-071-2514	526	8
80204	B1821BH025C300N	5305-00-071-2515	195	4
80204	B1821BH025C375N	5305-00-071-2518	594	21
80204	B1821BH025F044N	5305-00-267-8951	339	27
			340	13
80204	B1821BH025F050N	5305-00-267-8952	100	45
			101	10
			551	17
			590	10
			593	7
80204	B1821BH025F056N	5305-00-068-0512	581	12
			606	2
80204	B1821BH025F063N	5305-00-267-8953	61	8
			71	1
			181	26

PART NUMBER INDEX

CAGEC	PART NUMBER	STOCK NUMBER	FIG.	ITEM
80204	B1921BH025F063N	5305-00-267-8953	74	30
			75	1
			102	24
			134	40
			142	8
			181	21
			214	9
			278	9
			279	9
			279	17
			352	2
			353	3
			369	3
			580	12
			584	2
			587	7
			588	11
			592	29
			593	3
			594	32
			596	21
			601	4
			608	5
80204	B1821BH025F088N	5306-00-068-0514	81	1
			105	1
			166	23
			255	7
			262	23
			355	2
			356	2
80204	B1821BH025F100N	5305-00-068-0515	112	6
			113	2
			162	4
			167	26
			168	19
			169	25
			182	9
			455	2
			591	34
80204	B1821BH025113N	5305-00-068-0516	136	6
			543	17
			616A	8
80204	B1821BH025F125N	5305-00-267-8954	148	9
			181	22
			255	18
			262	14
			610	2
			616A	11
80204	B121BNH025F138N	5305-00-267-8955	100	38
			534	3
80204	B1821BH025F175N	5305-00-267-8957	163	32

PART NUMBER INDEX

CAGEC	PART NUMBER	STOCK NUMBER	FIG.	ITEM
			165	37
			166	26
80204	B1821BH025F175N	5305-00-267-8957	167	22
			168	25
			169	34
			507	48
			623	24
80204	B1821BH025F200N	5305-00-267-8958	278	10
			405	43
			406	25
80204	B1821BH031C100N	5306-00-226-4827	107	22
			108	12
80204	B1821BH031C150N	5305-00-226-4831	284	57
			285	15
			428	20
80204	B1821BH031C175N	5306-00-226-4832	382	39
			385	47
			387	33
80204	B1821BH031F113N	5306-00-051-4077	204	9
			271	11
			290	12
			536	18
			549	30
			592	22
80204	B1821BH031F175N	5306-00-051-4081	549	33
80204	B1821BH038C063N	5305-00-721-5492	81	12
			107	11
			205	9
80204	B1821BH038C075D	5305-00-115-9526	240	1
80204	B1821BH038C075N	5305-00-543-4372	33	3
			39	11
			125	18
			189	4
			233	1
			399	5
			457	2
			471	5
			508	2
			531	4
80204	B1821BH038C100D	5305-00-942-2196	19	14
			259	7
			260	7
80204	B1821BH038C100N	5305-00-068-0510	126	9
80204	B1821BH038C113N	5305-00-543-2419	234	24
			241	24
			526	15
80204	B1821BH038C125N	5305-00-068-0511	125	1
			211	29
			312	52
			456	28
			460	25

CROSS-REFERENCE INDEXES

PART NUMBER INDEX

CAGEC	PART NUMBER	STOCK NUMBER	FIG.	ITEM
			461	22
			504	13
80204	B1821BH038C125N	5305-00-068-0511	505	22
			509	17
80204	B1821BH038C150N	5305-00-725-2317	310	10
80204	B1821BH038C175N	5305-00-821-3869	596	11
80204	B1821BH038C225N	5305-00-638-8920	107	4
80204	B1821BH038F050N	5305-00-269-3231	302	1
80204	B1821BH038F063N	5305-00-269-3233	71	9
			75	8
			76	7
			77	16
			78	14
			79	7
			80	5
			154	2
80204	B1821BH038F075N	5305-00-269-3234	72	4
			74	1
			143	11
			163	6
			163	24
			165	26
			365	9
			405	20
			472	38
			590	12
80204	B1821BH038F088N	5305-00-269-3235	71	17
			72	8
			338	25
			357	1
			365	4
			365	10
			376	16
			591	19
80204	B1821BH038F100N	5305-00-269-3236	34	6
			591	28
80204	B1B21BH038F125N	5305-00-269-3238	62	6
			102	9
			161	5
			163	21
			185	27
			202	1
			213	34
			243	13
			254	20
			261	12
			261	12
			269	23
			273	25
			316	15
			319	34

PART NUMBER INDEX

CAGEC	PART NUMBER	STOCK NUMBER	FIG.	ITEM
			333	3
			334	21
			335	3
80204	B1821BH038F125N	5305-00-269-3238	336	3
			337	3
			338	19
			378	3
			388	8
			457	12
			487	33
			507	15
			513	8
			525	8
80204	B1821BH038F138N	5305-00-269-3239	62	32
			72	3
			72	3
			93	1
			255	11
			258	7
			262	24
			267	7
			269	3
			276	7
			310	8
			326	6
			351	2
			355	8
			355	12
			356	9
			513	34
			528	9
			536	11
			543	9
			589	41
			612	17
			616A	20
80204	B1821BH038F150N	5305-00-269-3240	163	19
			261	29
			343	2
			368	6
			374	22
			388	4
			400	8
			401	5
			456	7
			456	8
			460	8
			461	7
			509	3
			536	15
80204	B1821BH038F200N	5305-00-269-3242	143	15

PART NUMBER INDEX

CAGEC	PART NUMBER	STOCK NUMBER	FIG.	ITEM
			243	12
			301	12
			301	13
			478	20
80204	B1821BH038F225N	5305-00-269-3243	KITS	2
80204	B1821BH038F250N	5305-00-269-3244	105	6
			106	9
			382	18
			387	22
			387	45
			460	9
			507	32
			609	15
			629	5
80204	B1821BH038F275N	5305-00-269-3245	456	8
80204	B1821BH044C125N	5305-00-071-1788	106	3
			126	4
			286	23
			310	12
			503	14
			504	4
			505	35
80204	B1821BHO44C150N	5305-00-071-2055	107	12
80204	B1821BH044C175N	5305-00-071-2056	54	4
			284	65
80204	B1821BH044C225N	5305-00-071-2058	207	1
80204	B1921BH044F125N	5305-00-709-8523	506	14
80204	B1821BH044F200N	5305-00-709-8542	509	34
80204	B1821BH050C075L	5306-01-118-4889	198	18
80204	B1821BH050C075N	5305-00-732-0512	177	5
			178	1
			589	56
80204	B1821BH050C125N	5305-00-071-2067	2	6
80204	B1821BH050C138N	5305-00-071-2068	5	15
			105	7
80204	B1821BH050C150N	5305-00-071-2069	5	16
			107	26
			176	7
			456	20
			458	17
			460	19
			461	18
			503	38
			504	22
			505	1
			509	13
80204	B1821BH050C175N	5305-00-071-2070	6	14
			474	1
			489	46
			505	7
			506	5

PART NUMBER INDEX

CAGEC	PART NUMBER	STOCK NUMBER	FIG.	ITEM
80204	B1821BH050C200N	5305-00-071-2071	235	7
			242	7
			297	11
			470	11
80204	B1821BH050C200N	5305-00-071-2071	602	3
80204	B1821BH050C250N	5305-00-071-2073	469	24
			470	20
80204	B1821BH050C275N	5305-00-071-2074	6	21
80204	B1821BH050C350N	5305-00-071-2077	109	8
			297	19
80204	B1821BH050C500N	5305-00-071-2083	493	8
80205	B1821BH050F088N	5305-00-990-0695	182	28
			478	16
80204	B1821BH050F100N	5305-00-719-5184	107	28
80204	B1821BH050F138N	5305-00-725-0154	202	5
			322	37
			619	26
80204	B1821BH050F150N	5305-00-719-5221	69	5
			110	16
			318	4
			319	4
			320	37
			321	39
			323	41
			324	38
			351	21
			368	10
			368	13
			374	26
			391	15
			508	6
			619	27
80204	B1821BH050F175N	5305-00-719-5235	67	10
			68	4
			69	14
			102	15
			106	17
			120	4
			304	3
			318	46
			319	22
			320	4
			321	4
			322	4
			323	4
			324	4
			327	8
			327	8
			330	5
			331	7
			332	6

PART NUMBER INDEX

CAGEC	PART NUMBER	STOCK NUMBER	FIG.	ITEM
			333	15
			334	17
			336	12
			337	19
80204	B1821BH050F175N	5305-00-719-5235	338	16
			368	10
			374	18
			381	13
			391	26
			456	33
			460	30
			461	28
			468	3
			469	16
			503	34
			506	1
			509	23
			616	6
			616A	15
80204	B1821BH050F200N	5305-00-719-5238	102	18
			102	19
			250	34.1
			318	16
			318	39
			319	18
			320	30
			321	16
			322	19
			323	21
			324	3
			328	7
			332	7
			334	7
			336	7
			337	13
			338	27
			376	11
			381	11
			456	34
			457	17
			461	29
			509	24
			589	22
			590	15
			619	1
80204	B1821BH050F225N	5305-01-424-8744	616A	19
80204	B1821BH050F275N	5305-00-719-5241	615	11
80204	B1821BH050F350N	5305-00-719-5262	615	13
80204	B1821BH050F475N	5305-00-410-6957	503	12
80204	B1821BH056C175N	5305-00-716-7454	237	12
80204	B1821BH056F200N	5305-00-725-0145	318	45

CROSS-REFERENCE INDEXES

PART NUMBER INDEX

CAGEC	PART NUMBER	STOCK NUMBER	FIG.	ITEM
			319	46
			320	46
			321	47
			322	50
			323	49
			324	46
80204	B1821BH063C125N	5305-00-724-7219	236	28
			403	15
			469	8
			470	2
80204	B1821BH063C175N	5305-00-724-7221	127	6
			177	4
			178	4
			489	21
			499	6
80204	B1821BH063C200N	5305-00-724-7222	1	5
			206	3
			469	25
			470	22
			502	43
			503	23
			613	17
80204	B1821BH063C225N	5305-00-724-7223	236	27
80204	B1821BH063C275N	5305-00-724-7225	345	4
80204	B1821BH063C450N	5305-00-724-7264	6	1
80204	B1821BH063C475N	5305-00-724-7265	6	10
80204	B1821BH063C500N	5305-00-724-7266	6	28
80204	B1821BH063F138N	5305-00-726-2544	476	9
80204	B1821BH063F150N	5305-00-727-2283	326	17
			330	11
			487	6
80204	B1821BH063F175N	5305-00-726-2550	318	27
			319	28
			320	26
			321	27
			322	28
			323	31
			324	23
			327	5
			328	5
			331	14
			332	13
			343	5
			389	6
			393	5
			484	33
80204	B1821BH063F200N	5305-00-726-2551	326	2
			328	4
			330	9
			343	11
			390	15

PART NUMBER INDEX

CAGEC	PART NUMBER	STOCK NUMBER	FIG.	ITEM
			488	8
80204	B1821BH063F225N	5305-00-726-2552	303	7
			330	7
			343	10
			381	16
			400	9
80205	B1821BH063F250N	5305-00-726-2553	343	17
			400	9
80204	B1821BH063F450N	5305-00-726-2561	5	6
80204	B1821BH075C150N	5305-00-900-1118	473	2
80204	B1821BH075C225N	5305-00-900-0576	502	33
80204	B1821BH075C250N	5305-00-922-7994	344	3
			499	12
			502	1
80204	B1821BH075C450N	5305-00-947-4360	106	14
80204	B1821BH075F200N	5305-00-916-2345	318	21
			319	20
			320	17
			321	19
			322	22
			323	24
			324	18
			507	44
80204	B1821BH075F250N	5305-00-762-6041	331	12
			332	10
80204	B1821BH075F275N	5305-00-926-1826	331	10
			332	8
			380	8
80204	B1821BH075F450N	5305-00-940-8069	107	32
80205	B1821BH088F250N	5305-00-022-3843	497	2
80204	B1821BH088F400N	5305-00-213-8886	304	8
80204	B18234B06016N	5306-01-435-3269	101	25
80204	B18244B06	5310-01-437-8728	101	11
72286	B410-1L	2510-00-674-4487	363	1
93742	B52-1111-2	5340-00-532-9231	102	33
82484	C-2029	5355-01-386-8877	538	9
73740	C-334101-1531	2530-00-041-3116	317	17
77640	C-363833-A-3-165 6		317	23
77640	C-364007		317	19
77640	C-364007-A-1-153 1	2530-00-545-1561	317	16
77640	C-364007-A-2	3040-01-134-2023	317	18
77640	C-365509		317	9
77640	C-366005	2590-01-134-9834	317	21
21102	CA200011	2930-01-285-5027	121	6
21102	CA200046	4820-01-149-1317	277	3
81348	CMDX2-3PT573036	5340-00-809-1492	313	9
62983	CM11-ND2-R22-BL-21-066	4820-01-104-9159	460	32
62983	CM11-N02-R12-DL5	4820-01-285-9851	461	30

CROSS-REFERENCE INDEXES

PART NUMBER INDEX

CAGEC	PART NUMBER	STOCK NUMBER	FIG.	ITEM
	-21-233			
78500	CPL6240	2520-01-134-3455	465	3
			512	3
78500	CPL6522-P2	2520-00-755-7336	301	9
19207	CPR102321-1	4730-01-079-8821	182	21
			199	2
19207	CPR102321-1	4730-01-079-8821	212	2
			213	13
			214	2
			215	2
			254	8
			257	11
			261	4
			264	9
			265	2
			267	9
			268	2
			270	2
			271	20
			271	23
			272	10
			273	3
			274	2
			275	2
			279	3
			279	22
			288	2
			291	4
			538	11
			603	8
			604	2
			605	2
			617	3
			620	3
			621	10
			628	2
19207	CPR102321-2	4730-00-180-7031	271	6
			273	13
			621	16
19207	CPR102321-4	4730-01-032-6038	213	19
			252	5
			253	2
			259	2
			260	2
			264	16
			266	13
			267	12
			269	14
			271	14
			272	8
			273	19

PART NUMBER INDEX

CAGEC	PART NUMBER	STOCK NUMBER	FIG.	ITEM
			597	6
			617	9
			618	5
19207	CPR104420.X 37IN		598	8
19207	CPR104420 X 42IN		617	16
19207	CPR104420 X 52IN		617	16
19207	CPR104420 X 72IN		598	18
19207	CPR104420-B-3		618	2
19207	CPR104420-1	4720-01-058-7213	BULK	38
			BULK	39
19207	CPR104420-1-105		76	9
19207	CPR104420-1-108		70	5
19207	CPR104420-1-122		73	6
19207	CPR104420-1-123		77	6
19207	CPR104420-1-128		80	14
19207	CPR104420-1-142		80	13
19207	CPR104420-1-144		70	1
19207	CPR104420-1-150		79	9
19207	CPR104420-1-160		73	5
19207	CPR104420-1-36		76	17
			79	18
19207	CPR104420-1-37		76	10
19207	CPR104420-1-6		213	5
			266	6
19207	CPR104420-1-60		79	10
19207	CPR104420-1-92		77	7
19207	CPR104420-2	4720-01-014-4915	BULK	10
			617	1
19207	CPR104420-2-100		272	11
19207	CPR104420-2-101		261	9
19207	CPR104420-2-11		265	3
19207	CPR104420-2-110		265	10
19207	CPR104420-2-113		254	13
19207	CPR104420-2-115		265	10
19207	CPR104420-2-13		254	14
19207	CPR104420-2-130		290	1
			605	3
19207	CPR104420-2-14		268	3
			273	4
			279	23
			290	11
19207	CPR104420-2-146		215	3
19207	CPR104420-2-148		261	9
19207	CPR104420-2-15		213	32
19207	CPR104420-2-159		215	3
19207	CPR104420-2-16		265	4
			268	5
19207	CPR104420-2-169		261	9
19207	CPR104420-2-17		213	38
			274	3
			275	3

PART NUMBER INDEX
STOCK NUMBER

CAGEC	PART NUMBER	FIG.	ITEM
19207	CPR104420-2-172	215	3
19207	CPR104420-2-174	215	3
19207	CPR104420-2-178	261	13
		267	18
19207	CPR104420-2-18	212	1
		268	3
		279	4
19207	CPR104420-2-19	268	6
		291	5
19207	CPR104420-2-191	261	13
		267	18
19207	CPR104420-2-20	268	5
		268	6
19207	CPR104420-2-203	261	13
19207	CPR104420-2-206	267	18
19207	CPR104420-2-21	275	6
		288	12
19207	CPR104420-2-22	182	24
		288	11
		538	10
		603	10
19207	CPR104420-2-220	604	17
19207	CPR104420-2-229	261	13
		267	18
19207	CPR104420-2-23	274	6
		291	11
19207	CPR104420-2-230	605	9
19207	CPR104420-2-24	199	3
		213	39
		215	8
19207	CPR104420-2-26	212	11
		265	9
		288	1
		603	16
19207	CPR104420-2-268	604	3
19207	CPR104420-2-28	254	14
		628	1
19207	CPR104420-2-29	270	3
		628	6
19207	CPR104420-2-30	214	3
19207	CPR104420-2-32	538	10
19207	CPR104420-2-33	273	5
19207	CPR104420-2-34	288	10
		603	9
19207	CPR104420-2-35	257	5
19207	CPR104420-2-37	257	5
19207	CPR104420-2-38	261	1
19207	CPR104420-2-39	264	10
19207	CPR104420-2-41	254	14
19207	CPR104420-2-42	541	13
19207	CPR104420-2-43	254	11

PART NUMBER INDEX
STOCK NUMBER

CAGEC	PART NUMBER		FIG.	ITEM
19207	CPR104420-2-45		212	10
			271	21
19207	CPR104420-2-49		264	10
			290	8
19207	CPR104420-2-50		257	10
19207	CPR104420-2-53		605	10
19207	CPR104420-2-54		254	13
19207	CPR104420-2-55		213	12
19207	CPR104420-2-56		257	13
19207	CPR104420-2-6		256	4
19207	CPR104420-2-60		182	8
19207	CPR104420-2-63		597	7
19207	CPR104420-2-64		214	14
19207	CPR104420-2-66		257	13
			261	8
19207	CPR104420-2-7		603	15
19207	CPR104420-2-70		254	13
			290	17
			604	15
19207	CPR104420-2-73		254	14
19207	CPR104420-2-77		254	13
			261	9
19207	CPR104420-2-78		257	10
19207	CPR104420-2-80		254	7
			254	14
19207	CPR104420-2-83		254	13
19207	CPR104420-2-89		264	10
19207	CPR104420-2-9		279	27
			288	18
			604	9
19207	CPR104420-2-90		604	4
19207	CPR104420-2-94		254	14
19207	CPR104420-2X22IN		621	1
19207	CPR104420-2X42IN	4720-01-014-4915	541	12
19207	CPR104420-3		256	13
			618	7
19207	CPR104420-3-107		269	19
19207	CPR104420-3-120		273	18
19207	CPR104420-3-124		273	18
19207	CPR104420-3-126		269	19
19207	CPR104420-3-130		269	22
19207	CPR104420-3-136		273	18
19207	CPR104420-3-143		269	15
19207	CPR104420-3-145		269	19
19207	CPR104420-3-166		269	15
19207	CPR104420-3-180		273	18
19207	CPR104420-3-187		269	19
19207	CPR104420-3-19		260	6
19207	CPR104420-3-192		269	15
19207	CPR104420-3-20		266	24
19207	CPR104420-3-21		259	3

PART NUMBER INDEX

CAGEC	PART NUMBER	STOCK NUMBER	FIG.	ITEM
19207	CPR104420-3-22		271	15
19207	CPR104420-3-23		252	6
			259	6
			260	3
19207	CPR104420-3-25		252	6
			266	14
19207	CPR104420-3-35		597	10
19207	CPR104420-3-37		253	3
19207	CPR104420-3-38		253	14
19207	CPR104420-3-44		267	13
19207	CPR104420-3-45		252	10
			264	26
19207	CPR104420-3-46		252	10
19207	CPR104420-3-48		213	20
19207	CPR104420-3-54		264	26
19207	CPR104420-3-64		264	18
19207	CPR104420-3-68		264	18
19207	CPR104420-3-69		256	13
19207	CPR104420-3-7		266	18
19207	CPR104420-3-74		256	13
19207	CPR104420-3-80		269	22
			272	9
19207	CPR104420-3-83		256	12
19207	CPR104420-3-9		256	16
19207	CPR104420-3-92		269	22
19207	CPR104420-3-95		269	19
19207	CPR104420-4	4720-01-009-9058	BULK	12
19207	CPR104420-5	4720-01-009-9941	BULK	13
19207	CPR104420-5-140		273	16
19207	CPR104420-5-144		273	16
19207	CPR104420-5-15		271	7
19207	CPR104420-5-151		273	16
19207	CPR104420-5-200		273	16
19207	CPR104420-5-37		273	12
19207	CPR104420-5X62IN		598	18
19207	CPR104420-6	4720-01-235-5452	BULK	17
19207	CPR104420-6-20		97	1
19207	CPR104420-6-30		97	5
19207	CPR104420X 108IN		620	10
19207	CPR104420X42 IN LH &52 IN RH		617	16
78500	CP16NS	2520-00-388-4197	229	2
			229	3
			229	4
			229	6
			KITS	
78500	CP35R-17		KITS	
78500	CP85WB62	2520-01-280-4129	221	2
78500	CP85WB62 KF		222	4
11815	CR9163-6-6	5320-00-582-3499	424	28
70960	CS-H5-24-15		219	2

CROSS-REFERENCE INDEXES

PART NUMBER INDEX

CAGEC	PART NUMBER	STOCK NUMBER	FIG.	ITEM
			226	9
			228	10
70960	CS-H6-24-49		227	8
78500	CS-H8-20-16	5305-00-100-6791	221	3
			222	2
59342	CT-2900-2	5330-01-359-2316	111	19
7Z588	CT9444	4730-01-348-9510	111	11
55719	CXM1519	5120-01-178-5351	630	4
78500	C02-40-4	5315-01-136-1659	228	2
75160	C16054	4730-00-908-6294	62	17
75160	C16054	4730-00-908-6294	63	2
			581	1
			585	8
			592	7
78500	C3-3201-W-7589		238	2
81349	C3030	5325-00-184-9846	429	12
79470	C3159X2	4730-00-081-9618	7	8
79470	C3309X8	4730-00-542-5598	263	2
			581	24
77640	C36549-1-0900	2530-00-134-4619	316	17
16003	C43974		BULK	3
16003	C43974 X 15 LG		82	15
16003	C43974 X 8		326	21
16003	C43974 X-8		402	27
66295	C44PS		551	2
79470	C5165X4	4730-00-969-6941	426	8
79470	C5205X20	4730-00-202-8470	482	14
50022	D-781964-A	2930-01-131-7442	KITS	
78500	DCL6N-3	2520-01-134-3706	465	5
			512	5
21969	DIN933 MI10X1.5X2 0-8.8	5305-01-145-1196	591	4
99953	DT-1417	5340-01-314-2423	315	9
78500	E 5-3276L-12	2530-01-425-9520	250	5
78500	E-3276-L-12	2530-01-091-7814	250	5
01212	E-450121VG	5330-01-131-5416	502	32
19954	ER 28105		312	41
19954	ER 42504		312	31
19954	ER 42505		312	21
19954	ER 42550		312	32
19954	ER 82150		312	47
19954	ER 82413		312	19
19954	ER 8309		312	40
19954	ER 8412		312	42
19954	ER 85349		312	14
19954	ER 92739-069		312	35
19954	ER 92814		312	45
19954	ER 99562		312	20
19954	ER 99635		312	22
19954	ER 99658		312	13
19954	ER 99690		312	44

CROSS-REFERENCE INDEXES

PART NUMBER INDEX

CAGEC	PART NUMBER	STOCK NUMBER	FIG.	ITEM
19954	ER-15490-1		311	12
19954	ER-17319		312	29
19954	ER-22663	4730-01-127-6697	311	11
19954	ER-27512	2530-01-126-9303	311	2
19954	ER-27804		311	1
19954	ER-27856-150	4820-01-271-6918	312	28
19954	ER-72269		311	7
19954	ER-82141	5330-01-129-0384	311	10
19954	ER-93978		311	5
19954	ER-97043	5307-00-478-6782	311	9
19954	ER-99401	5310-01-129-4227	311	4
19954	ER-99420		311	6
19954	ER15996-1	2530-01-112-2161	310	2
81118	ER7614	5360-00-811-1609	312	27
19954	ER93765	5310-00-483-2385	311	8
19954	ERS-27785	2530-01-137-5921	KITS	
19954	ERS-28103		312	38
			312	39
			312	43
			312	46
			312	48
			312	49
			312	50
97902	ESM-3302-S-26021	5945-01-136-7640	213	23
13445	EX2235	6250-00-741-8960	134	47
78500	E1-3276-L-12	2530-01-126-7869	250	5
33334	E1-8B	4730-00-235-1483	72	5
			78	5
60703	F-1343-1ZF	5310-01-135-4797	538	8
17284	FB0037	9390-01-204-1161	102	37
21102	FC030002	5365-01-126-5192	120	9
21102	FC030003	5365-01-270-8472	122	8
21102	FC030004	5310-01-271-1796	122	7
21102	FC030005	5310-01-270-8342	120	28
			122	6
21102	FC030007	5360-01-271-5767	122	18
21102	FC030009	2520-01-077-4009	120	18
			122	17
21102	FC030010	2930-01-122-5982	120	12
21102	FC030011	5305-01-120-8438	120	11
21102	FC030020	5331-01-271-9372	120	29
			122	5
21102	FC030021	5331-01-271-9371	120	3
			122	4
21102	FC030024	3110-00-100-2368	122	14
21102	FC030025	2520-01-088-8172	122	12
21102	FC030027	5331-01-271-9373	122	11
21102	FC030028	5331-01-271-9374	122	13
21102	FC030029	5365-01-126-8689	120	25
21102	FC030035	2930-01-133-6948	120	16
21102	FC030043	5365-01-271-1854	122	15

CROSS-REFERENCE INDEXES

PART NUMBER INDEX

CAGEC	PART NUMBER	STOCK NUMBER	FIG.	ITEM
21102	FC030047	5305-01-256-3046	122	20
21102	FC030060	5365-01-270-8459	122	16
21102	FC030061	2520-01-271-7152	122	19
21102	FC030064	5325-01-270-8363	122	23
21102	FC030065	3120-01-277-1036	122	24
21102	FC030066	3110-01-271-6353	122	25
21102	FC030067	5325-01-270-8362	122	26
21102	FC030098	5365-01-270-8473	122	22
21102	FC200002	5342-01-272-1038	122	2
21102	FC200002		122	3
21102	FC200006	3020-01-124-9417	120	15
21102	FC200007	3020-01-124-6079	120	8
91929	FC200008	2520-01-123-8704	120	26
			122	9
			122	10
21102	FC200161	3020-01-271-5126	122	21
21102	FC200162	3040-01-271-3597	122	27
21102	FC212000	2520-01-115-2285	120	7
21102	FC212028	2930-01-272-3925	KITS	
21102	FC212143	5365-01-134-5534	120	31
21102	FD0077	5331-01-079-6513	120	21
21102	FD019	5360-01-124-3373	120	17
21102	FD0202	5365-01-123-1638	120	22
			120	24
21102	FD145A	5331-01-271-9374	120	19
21102	FD613	5305-01-161-6000	120	10
21102	FD70A	5310-01-079-7036	120	27
01276	FF9311-36	4730-00-909-8627	111	24
			153	24
01276	FG2722HHH0270	4720-01-441-2922	72	23
			78	
60380	FG42642	3110-01-065-5842	193	15
			200	28
98349	FP1411	2930-00-004-8420	11	2
1Z155	FRS-4-100	4730-01-315-5596	121	7
43990	F07-200-A1MA	4730-01-304-5618	621	18
			621	19
44185	F100753	4730-00-202-6670	474	6
60703	F1343-1ZT	5310-01-135-3464	541	9
64203	G-18711	5305-01-135-4754	514	2
64203	G-20151	5305-01-122-8516	570	9
64203	G-25102	5305-01-122-4692	524	4
64203	G-31051	5305-01-122-4693	519	4
64203	G-39200	5310-01-122-6149	521	2
64203	G-39203	5310-01-122-6151	519	35
64203	G-39250	5310-01-122-6148	515	3
			516	2
			518	3
			519	25
64203	G-39300	5310-01-122-6150	516	31
			519	16

CROSS-REFERENCE INDEXES

PART NUMBER INDEX

CAGEC	PART NUMBER	STOCK NUMBER	FIG.	ITEM
64203	G-39350	5310-01-122-6153	519	39
64203	G-39475	5310-01-122-7676	516	32
			517	10
64203	G-39500	5310-01-122-6152	516	9
64203	G-40-2	3830-01-134-1946	514	1
64203	G-40254	5315-01-122-2031	524	3
64203	G-43028	5325-00-804-2025	516	23
64203	G-46609		519	12
64203	G-46659	4730-00-278-3167	519	13
64203	G-46662	4730-00-202-6670	519	27
64203	G-6055	5315-01-206-5207	518	12
64203	G-6107	5315-01-122-4602	517	5
			518	11
64203	G-61513	5331-01-122-4612	522	17
64203	G-6161	5315-01-137-6858	516	14
64203	G-63517	5331-01-122-0492	523	10
			523	33
64203	G-63533	5331-01-143-6322	522	12
64203	G-6365	5315-01-135-9402	522	20
64203	G-63707	5331-01-122-4613	523	14
			523	37
64203	G-7009	4730-01-125-9975	516	12
			522	14
			524	2
64203	G-7030	4730-01-126-0218	516	3
			517	12
			518	7
64203	G-74004	5310-01-122-7677	516	18
64203	G-8061		521	22
64203	G-8154	4730-01-126-1723	519	45
64203	G-8200	4730-01-131-5945	519	41
			521	10
64203	G-8204	4730-01-125-9965	519	1
			521	17
64203	G-8211	4730-01-125-9969	519	31
			521	15
64203	G-9075	5310-01-122-6110	521	23
64203	G-9083	5310-01-122-6112	519	34
64203	G-9125	5310-01-122-6109	516	1
			518	4
			519	24
64203	G-9177	5310-01-122-6111	519	17
64203	G-9178	5310-01-122-5992	517	7
64203	G-9185	5310-01-122-6130	517	17
64204	G-9225	5310-01-122-4492	519	40
64203	G-9333	5310-01-122-4595	514	3
64203	G-9377	5340-01-122-4596	516	8
73740	G-9415992	5310-00-421-3991	317	22
60703	GK-9		538	7
81348	GP3STYLXTYRBCLA/T/11.00-R20/H/TB	2610-01-373-7294	300	1

PART NUMBER INDEX

CAGEC	PART NUMBER	STOCK NUMBER	FIG.	ITEM
24617	G1251	4730-00-018-9566	9	5
			285	26
77640	G179810	5305-01-136-1618	306	19
64731	G40-5	3110-00-554-3468	484	6
77640	G9410358		306	11
77640	G9429710	5305-01-142-2526	305	19
79146	HD-169-6X4	4730-00-069-1187	215	7
79146	HD-169-6X4	4730-00-069-1187	278	15
			279	13
			279	25
77640	HFB523001-A1	2520-01-134-3418	306	13
77640	HFB523001-A11	2530-01-206-3911	306	12
07367	HFB642010-A1		305	12
77640	HFB644100-A3-397	2530-01-134-3668	305	1
77640	HFB645001-A1	2530-01-131-7445	305	5
77640	HFB646013-A1	3040-01-131-7446	306	40
07367	HFB647000-A1	2530-01-138-0922	306	24
77640	HFB649000	5330-01-149-4415	305	11
60038	HM807010	3110-00-227-4667	209	1
60038	4807049	3110-00-842-6572	209	2
79146	HO-159-4	4730-00-132-4588	61	3
79146	H0-159-4	4730-00-132-4588	213	4
			266	17
55061	H813-1	5945-01-170-0553	148	6
89749	IF316	5315-00-816-1794	341	18
			414	13
96139	IRS-WP-1254AP	6220-01-273-0177	138	15
33287	J-33111	5120-01-152-2318	629	4
33287	J-34061	5120-01-154-3029	629	1
33287	J-35198	4910-01-200-1512	629	3
76005	J-9813-1	2815-00-134-4665	5	1
60380	JF45495	3110-00-557-6708	188	10
60380	JF45496	3110-00-557-6666	188	11
02032	JHP85-430	6620-00-993-5546	35	5
78500	JM-207049	3110-00-519-7833	209	6
60038	JM207010	3110-00-160-0338	209	17
33287	J24420-C	5180-00-999-4053	630	20
33287	J35193	4910-01-218-4490	629	2
13038	J950112-6	2520-00-734-6984	236	6
5T151	KDEP 1052	4910-01-336-8204	631	4
5T151	KDEP 1056	5120-01-343-2585	631	2
5T151	KDEP 1068	5120-01-343-2586	631	3
53867	KDEP1505	5120-01-341-6000	631	1
31033	L-28-VC-126	2530-01-126-9209	303	14
46717	LA-519-9	4730-00-415-3172	585	10
			594	10
78500	LM102910	3110-00-171-2489	527	2
78500	LM102949	3110-00-187-5730	527	1
61112	LP-5146-10	4730-01-071-5740	97	3
70960	LPHC37-1		219	3
			226	8

CROSS-REFERENCE INDEXES

PART NUMBER INDEX

CAGEC	PART NUMBER	STOCK NUMBER	FIG.	ITEM
			228	11
70960	LPHC39-3		227	7
97902	LSC-3270-S26022-1	4820-01-329-3245	212	14
			266	1
70960	L16SYS20-42		506	8
77640	L28SV5000G15	2530-01-309-3803	303	5
10988	L32618	5342-01-062-4715	494	8
70960	L6NLS20-22		465	6
			466	6
70960	L6NLS20-24		512	6
78500	L6NYR14-19	2520-01-134-3471	465	8
			466	7
			512	8
78500	L6NYR20-4	2520-01-134-1089	465	2
			512	2
70960	L6NYR20-87	2520-01-285-6282	466	2
46717	L6451-101	5315-00-187-9591	329	15
78500	M-1240-RDAX-14-644	2520-01-291-9992	231	1
78500	M-1240-RDAX-29-644	2520-01-291-9993	238	1
78500	M-1240-RDAX-30-644	2520-01-291-9994	238	1
34623	MA207-21139	4730-00-338-6839	37	2
34623	MA207-22-742	2520-01-088-8172	120	20
63702	MH-40-MP	4120-01-330-6543	426	1
81349	MIL-PRF-62550/3	2540-01-194-3323	592	8
			594	8
81349	MIL-R-3065RS510A 1BC1DF2		BULK	33
81349	MIL-S-13623/1-1	5930-00-898-0500	131	14
			134	26
81348	MIL-T-12459/CLCL /SA/1100-20/F/CC	2610-00-262-8653	300	1
78500	MPS 2813	2520-01-118-5971	526	1
78500	MPS516	3020-00-734-6881	234	27
			241	27
96906	MS122032	5310-00-159-6209	52	3
			66	8
			88	24
96906	MS14307-2	4730-00-277-5684	108	1
96906	MS14314-2X	4730-00-011-2578	60	1
96906	MS14315-1XA	4730-00-196-0930	582	8
96906	MS14315-4	4730-00-817-6578	97	4
			549	45
			581	19
96906	MS14315-5X	4730-00-277-1896	630	9
96906	MS14315-6X	4730-01-133-6209	71	21
			74	20
			75	16

CROSS-REFERENCE INDEXES

PART NUMBER INDEX

CAGEC	PART NUMBER	STOCK NUMBER	FIG.	ITEM
96906	MS15001-1	4730-00-050-4203	218	7
			220	9
			223	4
			224	5
			225	3
			329	8
			330	8
			331	11
			332	9
			409	4
96906	MS15001-1	4730-00-050-4203	465	7
			465	9
			512	7
			512	9
96906	MS15001-3	4730-00-050-4205	351	18
96906	MS15002-1	4730-00-172-0010	466	9
			506	7
96906	MS15003-1	4730-00-050-4208	221	7
			222	6
			223	9
			224	10
			226	13
			227	12
			236	24
			303	10
			345	19
			346	11
			348	35
			390	16
			471	33
			472	3
			475	4
			476	10
			477	2
			485	11
			486	8
			486	11
			487	3
			488	9
			489	49
			491	18
			497	3
			500	7
			502	22
			507	4
			507	26
			508	11
			513	6
			525	6
96906	MS15003-2	4730-00-172-0022	472	23
			501	15

CROSS-REFERENCE INDEXES

PART NUMBER INDEX

CAGEC	PART NUMBER	STOCK NUMBER	FIG.	ITEM
96906	MS15003-4	4730-00-172-0028	316	1
			344	11
			466	8
96906	MS15003-5	4730-00-172-0031	219	6
			221	5
			222	5
			226	5
			228	9
			229	7
			229	8
96906	MS15003-5	4730-00-172-0031	486	5
96906	MS15003-6	4730-00-172-0034	303	12
			345	7
			471	31
			472	43
			497	20
			490	16
			500	26
96906	MS15573-3	6240-00-155-7866	180	2
96906	MS15795-812	5310-00-625-5756	243	8
96906	MS15795-819	5310-00-656-0114	615	2
96906	MS15795-820	5310-00-614-3505	125	12
96906	MS15795-908	5310-00-045-5207	150	5
96906	MS16207-26	5306-01-370-6947	62	12
96906	MS16536-243	5320-01-290-6360	245	15
			246	3
96906	MS16562-223	5315-00-826-3251	101	16
96906	MS16562-225	5315-00-841-4443	370	39
96906	MS16562-35	5315-00-814-3530	100	33
			101	12
96906	MS16562-36	5315-00-810-3701	187	7
			333	21
			334	25
			335	19
			336	20
			337	31
			338	29
96906	MS16624-1087	5325-00-804-2025	249	30
96906	MS16624-1100	5325-00-530-7968	472	42
96906	MS16624-1150	5325-00-803-7299	489	44
96906	MS16624-1200	5325-00-252-4758	500	23
96906	MS16624-1275	5325-00-200-7234	500	9
96906	MS16625-1077	5325-00-804-4997	200	8
96906	MS16625-1100	5325-00-807-2636	85	14
			87	15
96906	MS16625-1200	5325-00-804-2784	17	8
96906	MS16627-1093	5325-00-846-1637	88	33
96906	MS16627-1137	5325-00-559-7574	249	12
96906	MS16632-1050	5325-00-256-2846	51	16
			56	16
96906	MS16998-112	5305-00-983-8082	495	11

CROSS-REFERENCE INDEXES

PART NUMBER INDEX

CAGEC	PART NUMBER	STOCK NUMBER	FIG.	ITEM
96906	MS16998-48	5305-00-983-6665	623	22
96906	MS17131-29	3110-00-157-0531	471	27
96906	MS17131-54	3110-00-198-1080	471	7
96906	MS17830-06C	5310-00-176-6341	540	3
96906	MS17981-1	5945-00-240-1684	148	2
96906	MS17986C619	5315-00-250-9595	611	12
96906	MS17989-C624	5315-01-125-9084	354	5
96906	MS18006-4572	6240-00-295-2421	144	6
96906	MS18153-63	5305-00-914-6131	100	15
			118	25
			318	1
			320	31
			322	34
			324	32
			333	5
			335	5
			337	5
			619	1
96906	MS18153-88	5305-00-914-6133	125	17
			456	39
			460	36
			461	6
			615	12
96906	MS18154-59	5305-01-010-2362	598	46
96906	MS19059-2422	3110-00-100-6158	471	16
96906	MS19059-2424	3110-00-100-6159	500	14
			507	11
96906	MS19061-10007	3110-01-117-4893	192	7
96906	MS19061-20013	3110-00-948-9796	469	10
			470	6
96906	MS19081-137	3110-00-100-4220	348	21
96906	MS19081-181	3110-00-100-4223	293	16
			295	8
			348	31
96906	MS19081-182	3110-00-689-8250	293	20
			295	14
96906	MS20002-10	5310-00-149-9126	5	7
96906	MS20066-187	5315-00-264-3099	462	2
			510	2
96906	MS20067-221	5315-00-242-0818	489	34
96906	MS20067-270	5315-00-042-3293	474	34
			489	27
96906	MS20067-305	5315-00-042-4950	502	10
96906	MS20067-493	5315-01-119-5239	502	46
96906	MS20068-271	5315-00-781-2026	12	6
96906	MS20073-06-12	5306-00-720-8747	218	12
			220	17
			223	15
			224	2
			225	8
			226	2

PART NUMBER INDEX

CAGEC	PART NUMBER	STOCK NUMBER	FIG.	ITEM
			227	13
			228	17
			229	12
96906	MS20220-A1	3020-00-528-0511	341	10
96906	MS20230-GB4	5325-00-526-2663	533	5
96906	MS20230-WB4	5325-00-508-8901	533	6
96906	MS20392-10C57	5315-00-904-3412	508	19
			559	22
96906	MS20392-10C67	5315-00-904-3408	508	15
			559	18
96906	MS20392-10C91	5315-01-008-7084	211	17
96906	MS20392-2C17	5315-00-811-1241	528	1
96906	MS20392-3C17	5315-00-063-7366	454	9
96906	MS20392-3C23	5315-00-081-7874	182	19
			454	1
96906	MS20392-4C25	5315-00-812-3427	525	18
96906	MS20392-4C97	5315-00-052-8492	351	8
96906	MS20392-5C31	5315-00-990-2889	454	12
			513	12
96906	MS20392-5C35	5315-00-957-2399	402	14
96906	MS20392-5C67	5315-00-951-7542	354	3
96906	MS20392-5C69	5315-00-957-9386	339	23
			340	6
96906	MS20392-7C-117	5315-00-904-1633	339	7
96906	MS20392-7C37	5315-00-081-7042	414	16
96906	MS20470-A6-4	5320-00-242-1582	405	40
96906	MS20470A5-8	5320-00-234-8557	419	27
96906	MS20470A6-4	5320-00-242-1582	406	19
			416	27
96906	MS20470A6-6	5320-00-242-1580	417	2
96906	MS20470A6-9	5320-00-264-3266	419	32
96906	MS20600-AD6W7	5320-00-813-4144	417	21
96906	MS20600-MP8NW/4	5320-01-068-2340	406	31
96906	MS20600AD5W2	5320-00-582-3302	404	19
96906	MS20600AD6-W7	5320-00-813-4144	424	7
96906	MS20600AD6W4	5320-00-582-3276	367	28
			405	28
			406	40
			418	24
			424	6
96906	MS20600AD8W7	5320-00-721-5384	416	24
96906	MS20604R5W2	5320-01-354-2548	586	5
96906	MS20659-102	5940-00-113-3138	150	17
96906	MS20659-104	5940-00-107-1481	440	2
96906	MS20659-105	5940-00-114-1300	159	3
96906	MS20659-108	5940-00-115-2674	158	9
			158	31
96906	MS20659-127	5940-00-113-3147	619	23
96906	MS20659-141	5940-00-113-9825	158	10
			158	32
96906	MS20659-142	5940-00-114-1315	158	13

PART NUMBER INDEX

CAGEC	PART NUMBER	STOCK NUMBER	FIG.	ITEM
			158	37
96906	MS20659-164	5940-00-113-3148	159	1
96906	MS20659-166	5940-00-113-9821	158	14
			158	33
96906	MS20822-16K	4730-00-282-7609	481	9
96906	MS20823-6B	4730-00-930-0982	270	6
96906	MS20825-16	4730-00-278-8935	480	26
96906	MS20913-2S	4730-00-221-2137	82	6
			261	32
96906	MS20913-8S	4730-00-221-2141	485	3
96906	MS20995F91-12		485	14
			491	16
96906	MS21042-08	5310-00-807-1466	602	31
96906	MS21042-3	5310-00-807-1467	606	12
96906	MS21044-N3	5310-00-877-5797	131	26
			134	32
			134	73
			153	8
			180	13
			183	1
			374	16
			542	9
			611	34
			612	25
			613	24
96906	MS21044-N4	5310-00-877-5796	184	8
			214	8
			629	12
96906	MS21044-N8	5310-00-877-5795	67	3
			68	12
			69	16
			202	7
			318	26
			319	27
			320	23
			321	26
			322	27
			323	30
			324	26
			345	17
			365	1
			368	11
			374	27
			376	14
			390	10
			391	9
			391	13
			396	11
			468	12
			602	23
			611	19

CROSS-REFERENCE INDEXES

PART NUMBER INDEX

CAGEC	PART NUMBER	STOCK NUMBER	FIG.	ITEM
96906	MS21044N3	5310-00-877-5797	249	39
			355	5
			596	26
96906	MS21044N4	5310-00-877-5796	100	35
			163	31
			165	33
			166	28
			167	21
			168	27
			169	36
			266	33
96906	MS21044N4	5310-00-877-5796	352	10
			353	11
			425	7
			505	30
			534	5
			579	5
			588	15
			606	19
			608	8
			612	12
96906	MS21045-C3	5310-00-263-2862	539	16
96906	MS21045-10	5310-00-982-5009	380	7
			393	3
96906	MS21045-12	5310-00-982-5012	380	9
			490	9
96906	MS21045-14	5310-00-982-5014	304	4
			497	4
96906	MS21045-18	5310-00-057-7153	348	36
96906	MS21045-3	5310-00-061-7326	181	18
			198	13
			257	17
			265	6
			359	19
			409	26
			457	4
			458	27
			616A	25
96906	MS21045-4	5310-00-061-7325	39	13
			355	4
			356	4
			356	17
			362	29
			405	37
			406	30
			536	8
			616A	14
96906	MS21045-5	5310-00-982-4912	62	29
			596	18
96906	MS21045-6	5310-00-982-4908	310	18
			322	36

CROSS-REFERENCE INDEXES

PART NUMBER INDEX

CAGEC	PART NUMBER	STOCK NUMBER	FIG.	ITEM
			376	13
96906	MS21045-7	5310-00-274-9364	125	14
96906	MS21045-8	5310-00-062-4954	322	16
			346	5
			486	6
			488	1
			492	13
96906	MS21083C12	5310-00-923-4219	471	30
96906	MS21083N12	5310-00-939-0783	366	14
			608	16
96906	MS21245-L10	5310-00-449-2381	125	11
96906	MS21245-L8	5310-00-449-2378	500	29
96906	MS21315-3	5340-00-966-2390	598	42
96906	MS21318-13	5305-00-253-5609	565	3
96906	MS21318-27	5305-00-253-5618	563	3
			566	3
			567	5
96906	MS21318-47	5305-00-253-5626	329	11
96906	MS21318-48	5305-00-253-5627	569	1
96906	MS21333-100	5340-00-809-1492	164	23
			166	24
96906	MS21333-101	5340-00-088-6655	611	4
96906	MS21333-102	5340-00-984-8540	513	26
96906	MS21333-104	5340-00-088-1254	71	3
			74	29
			75	3
			162	5
			166	18
			168	22
96906	MS21333-105	5340-00-809-1494	165	35
			269	2
			621	4
96906	MS21333-107	5340-00-809-1500	145	11
			156	17
			584	15
96906	MS21333-108	5340-00-057-3025	606	8
96906	MS21333-109	5340-00-067-3868	162	11
			173	6
			612	5
			613	3
96906	MS21333-111	5340-00-057-3037	606	20
96906	MS21333-112	5340-00-057-3043	155	10
			161	1
			263	10
			315	5
			588	9
			612	11
			613	29
			626	1
96906	MS21333-113	5340-00-988-1162	596	14
96906	MS21333-115	5340-00-725-5267	252	9

CROSS-REFERENCE INDEXES

PART NUMBER INDEX

CAGEC	PART NUMBER	STOCK NUMBER	FIG.	ITEM
			549	35
96906	MS21333-116	5340-00-764-7052	161	9
			608	1
96906	MS21333-119	5340-00-050-9077	276	1
			287	3
96906	MS21333-120	5340-00-964-5267	157	17
			429	13
96906	MS21333-121	5340-00-954-6014	71	8
			72	4
			74	3
			75	9
			76	6
			77	15
			78	3
96906	MS21333-121	5340-00-954-6014	79	6
			80	6
			164	27
			165	27
			184	9
			254	18
			254	18
			257	6
			257	9
			261	11
			261	11
			287	2
96906	MS21333-122	5340-00-833-8476	611	28
96906	MS21333-123	5340-00-989-1771	71	11
			72	9
			74	2
			75	11
			76	5
			77	14
			79	5
			80	7
			253	4
			263	14
96906	MS21333-125	5340-00-725-5280	169	41
96906	MS21333-126	5340-00-053-8994	71	25
			74	24
			75	20
			107	10
			157	15
			161	4
			164	17
			165	20
			167	19
			168	17
			243	16
			261	26
			269	7

CROSS-REFERENCE INDEXES

PART NUMBER INDEX

CAGEC	PART NUMBER	STOCK NUMBER	FIG.	ITEM
			271	18
			315	5
96906	MS21333-127	5340-00-833-3049	608	22
			618	10
96906	MS21333-128	5340-00-702-2848	161	10
			167	31
			582	14
			584	21
96906	MS21333-129	5340-00-151-9651	161	15
			215	5
			269	21
			619	4
			0269	21
96906	MS21333-33	5340-00-827-8314	100	50
96906	MS21333-33	5340-00-827-8314	101	7
96906	MS21333-40	5340-00-282-7521	39	12
96906	MS21333-62	5340-00-282-7509	287	6
			620	8
			620	8
96906	MS21333-66	5340-00-990-7610	587	14
			588	7
96906	MS21333-67	5340-00-079-7837	211	13
			426	11
			606	7
96906	MS21333-68	5340-00-843-7825	57	6
96906	MS21333-69	5340-00-764-7051	606	5
96906	MS21333-72	5340-00-091-3790	613	9
96906	MS21333-76	5340-00-724-7038	422	8
			423	8
96906	MS21333-77	5340-00-922-6300	111	7
96906	MS21333-80	5340-00-904-0933	161	16
96906	MS21333-96	5340-00-088-1255	100	48
96906	MS21333-99	5340-00-993-6207	162	10
			587	8
			588	12
96906	MS21334-2	5340-00-930-1754	170	16
			171	15
96906	MS21334-26	5340-00-901-8132	455	9
96906	MS21334-3	5340-00-494-2234	539	2
			540	18
96906	MS21334-36	5340-00-050-2622	429	8
96906	MS21334-42	5340-01-105-9143	257	16
96906	MS24394-6	4730-00-805-2222	313	2
96906	MS24519-2	4730-01-191-6433	539	15
96906	MS24519-7	4730-00-940-0947	585	4
96906	MS24522-2	4730-00-782-7102	538	5
			539	18
			540	14
96906	MS24522-23	4730-01-286-4611	582	23
			585	37
96906	MS24522-7	4730-00-944-5888	581	8

PART NUMBER INDEX

CAGEC	PART NUMBER	STOCK NUMBER	FIG.	ITEM
			582	18
			589	52
			590	37
96906	MS24532-2REVG	6625-01-086-9580	132	19
96906	MS24537	6685-00-814-5271	146	6
96906	MS24537-1	6685-00-814-5271	146	4
			147	1
96906	MS24539	6620-00-993-5546	146	9
96906	MS24540-2	6620-00-115-9042	132	23
96906	HS24543-2	6625-00-936-2139	132	13
96906	MS24544-2	6680-00-933-3600	132	26
96906	MS24617-31	5305-00-879-7941	534	13
96906	MS24621-42	5305-00-068-0523	37	1
96906	MS24625-42	5305-00-045-7603	594	37
96906	MS24625-47	5305-01-296-2849	584	18
96906	MS24627-50	5305-00-052-6874	412	24
96906	MS24627-55	5305-00-052-6879	413	21
96906	MS24627-62	5305-01-157-6794	420	12
96906	MS24627-63	5305-00-052-6881	414	24
96906	MS24629-10	5305-00-855-0968	573	2
			574	1
96906	MS24629-21	5305-00-855-0971	559	3
			593	4
96906	MS24629-22	5305-00-995-3569	362	2
96906	MS24629-24	5305-00-855-0973	362	5
96906	MS24629-28	5305-00-957-1538	281	5
96906	M524629-33	5305-00-052-6913	567	1
96906	MS24629-35	5305-00-855-0961	367	17
			441	2
			552	3
			597	11
			598	4
96906	MS24629-36	5305-00-855-0960	366	2
			366	7
			578	3
96906	MS24629-38	5305-00-855-0965	135	1
			413	16
			587	4
			588	2
96906	MS24629-44	5305-00-052-6915	552	1
96906	MS24629-45	5305-00-855-0958	82	16
96906	MS24629-46	5305-00-855-0957	170	15
			171	13
			211	15
			362	9
			379	13
			539	1
			540	19
			606	6
96906	MS24629-56	5305-00-052-6920	142	1
			142	19

PART NUMBER INDEX

CAGEC	PART NUMBER	STOCK NUMBER	FIG.	ITEM
			492	1
			544	1
			544	4
96906	MS24629-57	5305-00-052-6921	410	4
96906	MS24629-58	5305-00-052-6922	395	2
96906	MS24629-61	5305-00-052-7492	416	12
96906	MS24629-63	5305-00-052-7494	410	2
96906	MS24629-66	5305-00-442-3217	182	10
96906	MS24662-234	5320-00-930-7865	410	17
96906	MS24665-132	5315-00-839-2325	182	16
			411	17
			528	18
96906	MS24665-134	5315-00-839-5820	341	2
			359	13
96906	MS24665-151	5315-00-815-1405	454	8
			528	4
			KITS	11
96906	MS24665-153	5315-00-234-1854	454	4
96906	MS24665-172	5315-00-187-9370	354	1
96906	MS24665-214	5315-00-080-3503	383	5
96906	MS24665-281	5315-00-839-2326	525	21
			598	29
96906	MS24665-283	5315-00-842-3044	100	20
			243	9
			371	6
			382	43
			383	8
			387	36
			402	20
			469	19
			470	16
			602	16
96906	MS24665-300	5315-00-234-1863	180	11
			181	19
			454	10
			513	23
96906	MS24665-302	5315-00-234-1864	339	2
96906	MS24665-351	5315-00-839-5821	411	1
			415	8
			419	22
			425	20
			501	7
96906	MS24665-353	5315-00-839-5822	383	26
			384	10
			386	20
			419	17
96906	MS24665-355	5315-00-012-0123	211	11
96906	MS24665-357	5315-00-298-1481	236	9
			303	11
			391	21
96906	MS24665-359	5315-00-013-7214	329	2

CROSS-REFERENCE INDEXES

PART NUMBER INDEX

CAGEC	PART NUMBER	STOCK NUMBER	FIG.	ITEM
			488	4
96906	MS24665-361	5315-00-059-0184	234	1
			241	1
96906	MS24665-362	5315-00-298-1498	316	11
			471	9
			589	28
			591	27
96906	MS24665-363	5315-00-059-0187	583	7
96906	MS24665-370	5315-00-236-8359	358	8
96906	MS24665-372	5315-00-059-0491	508	9
96906	MS24665-423	5315-00-013-7228	389	10
			487	29
96906	MS24665-425	5315-00-013-7238	379	6
96906	MS24665-427	5315-00-879-2910	585	20
			594	12
96906	MS24665-49	5315-01-136-4542	507	24
96906	M524665-490	5315-00-059-0205	362	28
96906	MS24665-491	5315-00-059-0206	409	16
96906	MS24665-493	5315-00-018-7988	345	32
96906	MS24665-498	5315-00-849-9854	485	1
			583	1
			592	10
96906	MS24665-5	5315-00-236-8345	499	10
96906	MS24665-500	5315-00-187-9567	350	9
96906	MS24665-502	5315-00-849-5582	501	5
96906	MS24665-513	5315-00-239-8032	508	20
96906	MS24665-627	5315-00-013-7308	487	5
			490	2
96906	M524665-628	5315-00-846-0126	475	7
			490	7
96906	MS24665-629	5315-00-234-1848	379	15
			402	26
96906	MS24665-631	5315-00-597-7399	329	9
96906	M524665-655	5315-00-187-9414	345	30
96906	MS24667-41	5305-00-050-9221	490	6
96906	MS24667-85	5305-00-990-7168	502	42
96906	MS24679-63	5310-00-013-1498	367	39
			367	44
96906	MS24679-66	5310-01-057-5518	494	14
96906	MS25036-108	5940-00-143-4780	170	44
96906	MS25036-109	5940-00-283-5281	97	10
96906	MS25036-114	5940-00-113-9826	159	2
96906	MS25036-120	5940-00-557-4344	431	8
96906	MS25036-133	5940-00-115-5006	151	2
			151	13
			151	18
			152	8
96906	MS25036-153	5940-00-143-4774	170	45
96906	MS25082-10	5310-00-721-8000	596	8
96906	MS25082-12	5310-00-087-3946	150	4
96906	MS25231-1829	6240-00-266-9940	131	9

CROSS-REFERENCE INDEXES

PART NUMBER INDEX

CAGEC	PART NUMBER	STOCK NUMBER	FIG.	ITEM
			131	12
			132	12
96906	MS25231-1873	6240-00-419-3185	136	3
96906	MS27039-0825	5305-00-143-3266	236	18
96906	MS27040-6	5310-00-982-4935	138	16
96906	MS27130-S61K	5310-01-047-0401	601	5
96906	MS27142-2	5935-00-462-6603	159	8
96906	MS27142-3	5935-00-115-2306	158	7
			160	5
16528	MS27143-1	5935-00-114-0607	160	1
96906	MS27144-1	5935-00-167-7775	158	16
			619	24
96906	MS27144-2	5935-00-115-2307	158	15
96906	MS27144-3	5935-00-184-6707	158	23
96906	MS27145-1	5935-00-767-7936	160	19
96906	MS27147-1	5935-00-900-6281	172	14
96906	MS27147-1	5935-00-900-6281	173	15
			173	27
			211	37
			592	26
			594	25
			614	13
96906	MS27148-3	5999-00-926-3144	97	13
			97	21
			144	11
			144	16
			151	10
			152	2
			158	29
			165	6
			167	11
			170	34
			170	40
			171	2
			180	4
			182	36
			211	8
			211	43
96906	MS27183-10	5310-00-809-4058	100	46
			145	12
			161	19
			181	25
			255	16
			543	13
			549	6
			616A	7
96906	MS27183-11	5310-00-809-3078	50	20
			109	6
			370	12
			411	7
			579	8

CROSS-REFERENCE INDEXES

PART NUMBER INDEX

CAGEC	PART NUMBER	STOCK NUMBER	FIG.	ITEM
			604	13
			606	14
96906	MS27183-12	5310-00-081-4219	108	10
			134	42
			205	2
			254	2
			290	14
			341	11
			382	35
			387	32
			397	2
			402	7
			549	31
			589	7
			591	1
			592	12
			598	30
			611	16
96906	MS27183-12	5310-00-081-4219	615	16
			622	2
96906	MS27183-13	5310-00-087-7493	62	33
			153	22
			374	3
			375	7
			402	15
			428	19
			583	5
			589	30
96906	MS27183-14	5310-00-080-6004	11	4
			19	17
			34	3
			37	15
			83	7
			100	19
			102	8
			125	19
			153	7
			156	7
			180	15
			204	5
			211	30
			243	4
			266	22
			310	7
			334	4
			336	4
			338	4
			339	30
			340	16
			343	3
			354	2

PART NUMBER INDEX

CAGEC	PART NUMBER	STOCK NUMBER	FIG.	ITEM
			356	14
			364	7
			365	6
			368	7
			368	16
			371	7
			395	7
			396	13
			399	7
			400	7
			425	27
			454	11
			456	10
			460	12
			474	16
			486	2
			489	9
			505	27
96906	MS27183-14	5310-00-080-6004	513	4
			513	22
			525	4
			528	8
			611	24
			612	14
			613	32
			629	5
96906	MS27183-15	5310-00-809-4061	125	3
			310	14
			354	8
			376	9
			377	6
			511	56
			600	3
			600	3
			600	3
			616A	21
96906	MS27183-16	5310-00-809-4085	125	5
			540	10
96906	MS27183-17	5310-00-809-5997	125	15
96906	MS27183-18	5310-00-809-5998	106	10
			144	29
			198	19
			318	36
			319	37
			320	34
			321	36
			322	40
			323	38
			324	35
			333	10
			334	11

PART NUMBER INDEX
STOCK NUMBER

CAGEC	PART NUMBER	STOCK NUMBER	FIG.	ITEM
			335	17
			336	18
			337	14
			338	9
			339	3
			389	3
			391	16
			396	6
			437	20
			439	20
			458	9
			468	5
			468	5
			495	4
			502	39
			503	25
			508	22
			591	39
96906	MS27183-19	5310-00-809-3079	176	5
96906	MS27183-19	5310-00-809-3079	383	6
			419	21
			602	2
96906	MS27183-20	5310-00-068-5285	183	2
			244	17
96906	MS27183-21	5310-00-823-8803	297	12
			358	2
			383	27
			384	12
			386	19
			391	10
			503	33
			613	38
96906	MS27183-22	5310-00-951-7209	380	6
			380	6
			381	9
			400	11
96906	MS27183-23	5310-00-809-8533	211	12
			279	1
			297	13
			389	12
			508	12
96906	MS27183-24	5310-00-809-8536	349	2
			471	29
96906	MS27183-25	5310-00-809-8540	304	5
96906	MS27183-27	5310-00-809-8541	499	25
96906	MS27183-28	5310-00-982-6562	607	3
96906	MS27183-32	5310-00-984-8818	490	14
96906	MS27183-42	5310-00-014-5850	50	25
			55	27
			359	9
			528	3

CROSS-REFERENCE INDEXES

PART NUMBER INDEX

CAGEC	PART NUMBER	STOCK NUMBER	FIG.	ITEM
			616A	24
96906	MS27183-8	5310-00-809-8546	364	4
96906	MS27183-9	5310-00-823-8804	112	3
			113	3
			315	13
			413	24
			598	35
96906	MS27980-21B	5325-00-905-1492	367	10
96906	MS27980-6B	5325-00-285-6250	533	13
96906	MS28774-114	5330-00-576-3206	494	4
96906	MS28774-115	5330-00-801-3440	494	3
96906	MS28774-117	5330-00-835-7712	494	10
			494	18
96906	MS28775-032	5330-01-049-7374	149	3
96906	MS28775-111	5331-00-579-8108	192	18
96906	MS28775-113	5330-00-582-2855	502	38
			530	10
96906	MS28775-114	5331-00-618-0801	494	5
96906	MS28775-115	5331-00-579-7916	494	2
96906	MS28775-117	5331-00-702-5220	494	1
96906	MS28775-117	5331-00-702-5220	494	9
			494	19
96906	MS28775-120	5330-00-542-1329	494	6
96906	MS28775-128	5331-00-702-5643	493	26
96906	MS28775-131	5331-00-808-7612	480	22
			482	9
96906	MS28775-206	5330-01-133-5858	297	8
96906	MS28775-219	5330-00-579-7925	456	31
			460	28
			461	26
			504	17
			505	31
			509	21
96906	MS28775-222	5331-00-297-9990	485	21
			503	18
			504	7
			505	32
96906	MS28775-225	5331-00-579-7927	456	32
			460	29
			461	27
			503	10
			504	25
			505	4
			509	22
96906	MS28775-228	5331-00-807-8993	317	11
96906	MS28775-232	5331-00-585-8247	505	5
96906	MS28775-238	5331-00-579-7545	614	3
96906	MS28775-243	5331-00-579-7544	491	9
96906	MS28775-249	5331-01-019-2448	485	22
96906	MS28778-10	5331-00-285-9842	204	8
			456	40

CROSS-REFERENCE INDEXES

PART NUMBER INDEX

CAGEC	PART NUMBER	STOCK NUMBER	FIG.	ITEM
			461	31
			509	26
96906	MS28778-12	5331-00-251-8839	456	1
			456	6
			460	1
			460	7
			461	1
			509	2
96906	MS28778-14	5330-00-472-2783	494	12
			494	17
96906	MS28778-16	5331-00-804-5694	480	14
			493	18
			493	20
			493	22
			505	54
96906	MS28778-20	5331-00-816-3546	480	10
			493	12
			503	1
96906	MS28778-32	5331-00-285-9847	480	9
			505	55
96906	MS28778-6	5331-00-804-5695	297	2
			299	7
			313	5
			508	13
96906	MS28778-8	5331-00-808-0794	314	7
			503	35
96906	MS29513-115	5331-00-248-3847	183	8
96906	MS29513-229	5331-00-291-3273	293	8
96906	MS29523-1	5315-00-514-2660	109	3
			311	3
96906	MS3108R20-4S	5935-00-843-4561	438	5
96906	MS3367-1-0	5975-00-984-6582	70	2
			73	4
			76	14
			77	19
			79	14
			80	9
			97	6
			158	47
			166	20
			167	23
			169	24
			182	7
			587	2
			597	3
			606	21
			611	37
			612	20
			613	35
96906	MS3367-2-0	5975-00-899-4606	456	42
			460	38

CROSS-REFERENCE INDEXES

PART NUMBER INDEX

CAGEC	PART NUMBER	STOCK NUMBER	FIG.	ITEM
			461	36
			580	7
			582	13
			593	10
96906	MS3367-2-9	5975-00-156-3253	158	47
			161	3
			163	18
			165	21
			166	19
			167	18
			168	18
			169	42
			173	21
			243	18
			243	18
			584	13
			595	16
			599	4
			608	21
			621	24
96906	MS3367-3	5975-01-273-8133	627	4
96906	MS3367-3-0	5975-00-985-6630	204	20
96906	MS3367-3-9	5975-00-451-5001	110	6
			115	6
			158	47
			161	22
			205	10
			589	45
			590	31
			611	26
96906	MS3367-4-9	5975-00-727-5153	162	12
96906	MS3367-6-9	5975-00-133-8696	627	3
96906	MS3367-7-9	5975-00-570-9598	167	24
			626	3
96906	MS3456W18-1S	5935-00-622-2830	160	8
96906	MS3456W18-8S	5935-01-077-2622	160	10
96906	MS3456W24-22S	5935-01-040-0463	431	9
96906	MS35140-4	5340-00-190-6783	429	3
96906	MS35140-6	5340-00-282-7793	171	14
96906	MS35150-5	5340-00-827-2453	422	10
			423	10
96906	MS35190-238	5305-00-059-4553	586	9
96906	MS35190-253	5305-00-059-4568	415	17
96906	MS35190-287	5305-00-954-4295	370	9
96906	MS35190-317	5305-00-958-5258	410	16
96906	MS35190-319	5305-00-959-2723	470	26
96906	MS35190-343	5305-00-958-5267	474	31
			489	16
96906	MS35191-237	5305-00-959-2739	211	4
96906	MS35191-274	5305-00-984-7342	100	8
96906	MS35191-292	5305-00-988-9106	629	11

CROSS-REFERENCE INDEXES

PART NUMBER INDEX

CAGEC	PART NUMBER	STOCK NUMBER	FIG.	ITEM
96906	MS35191-293	5305-00-957-1497	353	7
			361	7
96906	MS35191-322	5305-00-959-2703	426	15
96906	MS35206-213	5305-00-889-3116	415	12
96906	MS35206-230	5305-00-889-3000	586	1
96906	MS35206-240	5305-00-984-6226	150	14
96906	MS35206-241	5305-00-984-6189	131	3
			132	9
			150	10
96906	MS35206-242	5305-00-889-3002	100	27
			101	4
			144	25
			150	1
96906	MS35206-243	5305-00-984-6191	412	18
			427	16
			428	4
96906	MS35206-244	5305-00-984-6192	157	9
			367	29
96906	MS35206-245	5305-00-984-6193	161	27
96906	MS35206-261	5305-00-984-6208	81	3
96906	MS35206-261	5305-00-984-6208	279	18
			552	13
96906	MS35206-263	5305-00-984-6210	505	14
96906	MS35206-265	5305-00-984-6212	125	23
			143	7
			143	7
			549	15
			551	15
96906	MS35206-279	5305-00-988-1723	181	6
96906	MS35206-280	5305-00-988-1724	137	1
			405	34
			406	35
			544	5
			545	1
96906	MS35206-281	5305-00-988-1725	370	8
			546	1
96906	MS35206-282	5305-00-988-1726	370	13
96906	MS35206-284	5305-00-988-1170	383	21
			385	30
96906	MS35206-296	5305-00-984-5676	382	26
			387	16
96906	MS35206-300	5305-00-984-5680	532	3
			532	3
			532	3
			535	3
			609	5
			609	19
96906	MS35207-242	154	11	
96906	MS35207-260	5305-00-088-9044	213	31
			441	8
96906	MS35207-261	5305-00-990-6444	131	19

CROSS-REFERENCE INDEXES

PART NUMBER INDEX

CAGEC	PART NUMBER	STOCK NUMBER	FIG.	ITEM
			134	34
			134	72
			611	33
96906	MS35207-262	5305-00-989-6265	608	9
96906	MS35207-263	5305-00-989-7434	374	11
			606	10
			612	21
			613	26
96906	MS35207-264	5305-00-989-7435	61	9
			100	6
			131	27
			132	30
			153	5
			542	7
			586	14
96906	MS35207-265	5305-00-993-1848	163	5
			183	15
			418	20
			596	24
			616A	27
96906	MS35207-266	5305-00-995-3444	367	41
96906	MS35207-266	5305-00-995-3444	367	47
96906	MS35207-267	5305-00-993-1851	359	8
96906	MS35207-268	5305-00-995-3442	539	14
96906	MS35207-278	5305-00-088-8946	399	4
96906	MS35207-279	5305-00-993-2463	314	4
96906	MS35207-280	5305-00-993-2738	374	13
			544	8
			545	5
			547	4
			551	4
			561	4
96906	MS35207-281	5305-00-993-2461	425	8
			547	3
			549	7
96906	MS35237-176	5305-00-018-6475	489	54
96906	MS35265-107	5305-00-052-2218	469	3
96906	MS35265-43	5305-00-614-3423	134	6
			134	27
			134	67
96906	MS35265-94	5305-00-551-5097	236	22
96906	MS35333-108	5310-00-022-8834	162	7
96906	MS35333-35		154	10
96906	MS35333-38	5310-00-559-0070	150	2
			367	30
			441	3
			587	5
			591	31
96906	MS35333-39	5310-00-576-5752	367	38
			367	46
96906	MS35333-40	5310-00-550-1130	136	5

CROSS-REFERENCE INDEXES

PART NUMBER INDEX

CAGEC	PART NUMBER	STOCK NUMBER	FIG.	ITEM
			145	7
			148	7
			151	25
			162	2
			163	4
			370	7
			370	32
96906	MS35333-41	5310-00-167-0721	539	9
			540	7
96906	MS35333-42	5310-00-595-7237	140	2
			143	9
			148	13
			243	24
			602	9
96906	MS35333-44	5310-00-194-1483	134	3
			134	30
			134	64
96906	MS35333-46	5310-00-543-4385	596	9
96906	MS35333-47	5310-00-550-3714	587	12
			588	5
96906	MS35333-49	5310-00-582-6714	213	10
96906	MS35333-49	5310-00-582-6714	252	13
			253	10
			255	4
			256	8
			261	16
			262	12
			266	11
			272	5
96906	MS35333-73	5310-00-543-5933	608	10
96906	MS35333-75	5310-00-178-8631	161	12
96906	MS35333-78	5310-00-261-7156	614	6
96906	MS35335-30	5310-00-209-0788	362	4
			585	23
96906	MS35335-31	5310-00-596-7693	157	10
			161	28
			579	2
96906	MS35335-32	5310-00-596-7691	143	13
			143	13
			198	14
96906	MS35335-33	5310-00-209-0786	101	23
			141	25
			142	2
			180	18
			181	21
			184	14
			455	1
			584	3
			592	3
			594	33
			623	23

CROSS-REFERENCE INDEXES

PART NUMBER INDEX

CAGEC	PART NUMBER	STOCK NUMBER	FIG.	ITEM
96906	MS35335-34	5310-00-514-6674	156	10
			585	28
96906	MS35335-35	5310-00-627-6128	105	3
			107	2
			156	14
			163	34
			165	39
			167	30
			168	29
			169	38
			585	11
96906	MS35335-36	5310-00-550-3503	611	23
96906	MS35335-38	5310-00-616-6354	236	40
96906	MS35335-39	5310-00-800-0695	236	38
			343	6
96906	MS35335-40	5310-00-275-3683	345	11
96906	MS35335-61	5310-00-527-3634	101	12
96906	MS35336-21	5310-00-194-9209	100	9
96906	MS35336-39	5310-00-194-9213	470	23
96906	MS35336-9	5310-00-550-0284	211	3
96906	M535338-41	5310-00-045-4007	586	3
96906	MS35338-42	5310-00-045-3299	131	2
			132	8
			134	5
			134	28
			134	66
			144	24
			598	44
96906	MS35338-43	5310-00-045-3296	81	4
			132	14
			132	20
			132	24
			132	27
			211	14
			505	13
			624	8
			625	8
96906	MS35338-44	5310-00-582-5965	100	36
			141	18
			181	5
			297	7
			352	3
			369	2
			371	3
			395	3
			405	33
			405	44
			406	34
			408	18
			414	29
			416	13

CROSS-REFERENCE INDEXES

PART NUMBER INDEX

CAGEC	PART NUMBER	STOCK NUMBER	FIG.	ITEM
			425	4
			507	46
			525	22
			536	7
			543	5
			547	2
			549	49
			580	4
			591	25
			619	14
96906	MS35338-45	5310-00-407-9566	33	4
			112	10
			284	56
			286	19
			308	12
			414	21
			419	18
			425	23
			549	32
96906	MS35338-46	5310-00-637-9541	19	16
			31	12
96906	MS35338-46	5310-00-637-9541	33	27
			34	2
			39	10
			62	5
			81	13
			100	4
			120	2
			125	2
			141	7
			144	21
			153	30
			154	3
			177	10
			234	25
			241	25
			259	8
			260	8
			264	24
			267	6
			288	5
			301	11
			302	3
			310	11
			356	11
			357	3
			368	5
			380	11
			394	3
			395	6
			399	6

PART NUMBER INDEX
STOCK NUMBER

CAGEC	PART NUMBER	STOCK NUMBER	FIG.	ITEM
			405	19
			406	9
			416	9
			456	29
			460	26
			461	23
			469	4
			471	4
			472	39
			476	13
			477	9
			486	3
			495	8
			500	4
			504	14
			505	21
			507	14
			507	33
			509	18
			531	5
			549	11
96906	MS35338-46	5310-00-637-9541	551	6
			551	9
			596	12
96906	MS35338-47	5310-00-209-0965	33	34
			37	21
			49	4
			54	6
			106	2
			125	6
			220	15
			228	15
			243	23
			284	67
			286	24
			301	14
			310	13
			503	13
			504	5
			505	34
96906	MS35338-48	5310-00-584-5272	616	11
96906	MS35338-49	5310-00-167-0680	100	44
			183	3
			460	18
96906	MS35338-50	5310-00-820-6653	1	2
			14	4
			178	5
			206	6
			402	4
			403	16
			417	8

PART NUMBER INDEX

CAGEC	PART NUMBER	STOCK NUMBER	FIG.	ITEM
			489	22
			502	44
			503	22
96906	MS35338-51	5310-00-584-7888	469	17
			470	13
			473	3
			488	7
			499	13
			502	2
			507	43
96906	MS35338-52	5310-00-754-2005	346	14
			350	6
96906	MS35338-55	5310-00-060-9435	348	33
96906	MS35338-65	5310-00-011-5093	487	32
96906	MS35338-8	5310-00-261-7340	66	11
96906	MS35340-43	5310-00-721-7809	125	22
96906	MS35340-48	5310-00-834-7606	351	20
96906	MS35340-53	5310-00-926-5885	347	8
96906	MS35356-33	5306-01-447-2565	108	9
96906	MS35387-1	9905-00-205-2795	544	6
			545	2
96906	MS35387-1	9905-00-205-2795	546	4
			547	1
			548	3
96906	MS35387-2	9905-00-202-3639	544	2
			545	4
			546	2
			547	6
			548	2
96906	MS35421-1	6220-00-299-7425	143	4
96906	MS35421-2	6220-00-299-7426	143	4
96906	MS35422-1	6220-00-729-9295	143	2
96906	MS35423-1	6220-00-577-3434	143	1
			143	1
96906	MS35423-2	6220-00-726-1916	143	1
96906	MS35426-27	5310-01-100-5199	97	17
96906	MS35436-10	5940-00-534-1028	439	4
			439	8
96906	MS35436-4	5940-00-534-0986	435	5
			436	2
			436	9
96906	MS35436-44	5940-00-151-9361	430	2
			435	10
96906	MS35436-6	5940-00-534-0991	431	4
96906	MS35436-9	5940-00-892-3151	149	11
			211	36
			213	24
			436	10
96906	MS35458-76	5305-00-182-9561	495	10
96906	MS35489-110	5325-00-202-4005	153	11
96906	MS35489-134	5325-00-543-2902	596	7

PART NUMBER INDEX

CAGEC	PART NUMBER	STOCK NUMBER	FIG.	ITEM
			598	1
96906	MS35489-135	5325-00-263-6648	138	17
96906	MS35489-138	5325-00-754-1072	549	12
96906	MS35489-16	5325-00-276-6089	429	10
96906	1S35489-17	5325-01-242-7083	153	14
			422	19
			423	17
			429	1
96906	MS35489-19	5325-00-276-6091	140	15
			624	1
			625	1
96906	MS35489-2	5325-00-174-5316	100	2
96906	MS35489-22	5325-00-270-8890	429	14
96906	MS35489-23	5325-00-276-6343	584	27
96906	MS35489-27	5325-00-290-1960	582	16
			585	12
96906	MS35489-39	5325-00-164-2087	613	11
96906	MS35489-4	5325-00-174-5317	587	13
			588	6
96906	MS35489-40	5325-00-185-0004	538	3
96906	MS35489-51	5325-00-171-6387	429	9
96906	MS35489-7	5325-00-174-5315	63	8
96906	MS35489-79	5325-01-124-7760	580	6
			584	11
			587	6
			588	8
			592	23
96906	MS35489-9	5325-00-276-6228	432	14
96906	MS35493-1	5305-00-900-3243	572	13
96906	MS35493-51	5305-00-901-2099	571	3
96906	MS35493-52	5305-00-901-2135	407	11
96906	MS35493-56	5305-00-180-1964	415	7
96906	MS35648-3	5340-00-050-1589	474	23
			489	59
96906	MS35648-8	5340-00-050-1600	7	27
96906	MS35649-202	5310-00-934-9758	143	14
			143	14
			299	1
96906	MS35649-2254	5310-00-250-9477	619	15
96906	MS35649-2312	5310-00-829-9981	62	11
96906	MS35649-262	5310-00-934-9747	586	4
96906	MS35649-282	5310-00-934-9757	157	11
			161	23
			412	15
			427	17
			428	2
			598	45
96906	MS35650-302	5310-00-934-9751	132	15
			132	21
			132	25
			132	28

CROSS-REFERENCE INDEXES

PART NUMBER INDEX

CAGEC	PART NUMBER	STOCK NUMBER	FIG.	ITEM
			624	9
			625	9
96906	MS35650-3254	5310-00-400-5503	161	26
96906	MS35650-3312	5310-00-054-4892	161	14
96906	MS35650-3382	5310-00-058-1626	100	3
			288	6
			549	10
			551	7
			551	10
96906	MS35650-362	5310-00-934-9755	134	11
96906	MS35650-382	5310-00-934-9754	163	3
96906	MS35671-42	5315-00-014-2972	472	17
96906	MS35672-46	5315-01-132-3569	390	12
96906	MS35672-47	5315-00-999-4238	398	5
			398	7
96906	MS35677-46	5315-00-682-2207	416	32
96906	MS35677-48	5315-00-866-2673	405	16
			406	46
96906	MS35690-504	5310-01-064-3422	387	27
96906	MS35690-604	5310-00-655-9544	153	31
96906	MS35690-824	5310-00-010-3028	316	8
96906	MS35691-13	5310-00-853-9335	494	16
96906	MS35691-2	5310-00-834-8736	339	14
			339	21
96906	MS35691-22	5310-00-891-1751	100	16
			153	20
96906	MS35691-25	5310-00-891-1711	346	21
			348	3
96906	MS35691-33	5310-00-834-8732	414	19
			469	6
			500	24
96906	MS35691-37	5310-00-834-8734	383	3
			537	4
96906	MS35691-38	5310-00-891-1733	211	53
			305	22
96906	MS35691-45	5310-00-839-2066	313	3
96906	MS35691-49	5310-00-851-2677	499	4
			607	1
96906	MS35691-5	5310-00-971-7989	51	14
			56	13
			66	16
			88	8
			88	25
96906	MS35691-57	5310-00-838-1702	473	25
96906	MS35691-73	5310-00-891-3426	607	2
96906	MS35691-77	5310-00-891-3428	252	12
			261	15
96906	MS35691-9	5310-00-891-1709	472	37
96906	MS35692-61	5310-00-998-0608	389	11
96906	MS35692-69	5310-00-835-2140	236	10
			303	15

PART NUMBER INDEX

CAGEC	PART NUMBER	STOCK NUMBER	FIG.	ITEM
96906	MS35692-93	5310-00-849-6874	507	23
96906	MS35743-16	5320-01-069-6365	153	18
96906	MS35743-36	5320-01-032-6530	376	5
			377	22
			383	30
96906	MS35743-38	5320-00-889-2632	383	10
			383	24
			384	5
			384	18
			386	7
			386	13
			534	9'
96906	MS35743-39	5320-00-478-3313	396	3
96906	MS35743-55	5320-01-145-3191	371	10
96906	MS35743-57	5320-01-069-6364	411	15
96906	MS35743-60	5320-01-186-1277	411	9
96906	MS35743-73	5320-01-145-3183	102	27
96906	MS35743-79	5320-01-136-4495	318	14
96906	MS35751-41	5306-00-721-5944	385	48
96906	MS35751-42	5306-00-911-5005	382	14
			382	38
96906	MS35751-42	5306-00-911-5005	385	2
			385	15
			385	31
			387	34
			609	29
96906	MS35751-43	5306-00-753-6996	382	24
			385	5
			385	26
			387	14
			609	9
96906	MS35751-44	5306-00-012-0231	382	29
			387	20
			387	40
			609	10
96906	MS35751-47	5306-00-054-8024	382	5
			385	16
			387	7
96906	MS35751-71	5306-00-816-2441	380	10
96906	MS35751-72	5306-00-010-9115	381	18
96906	MS35756-1	5315-00-616-5519	472	28
			526	5
96906	MS35756-13	5315-00-616-5521	473	4
96906	MS35756-14	5315-00-616-5520	391	4
			513	33
			525	32
96906	MS35756-17	5315-00-012-4553	211	18
			487	18
			608	18
96906	MS35756-18	5315-00-616-5527	21	7
96906	MS35756-20	5315-00-616-5501	484	28

CROSS-REFERENCE INDEXES

PART NUMBER INDEX

CAGEC	PART NUMBER	STOCK NUMBER	FIG.	ITEM
96906	MS35756-21	5315-00-616-5500	483	4
			484	27
			489	33
96906	MS35756-38	5315-00-043-1789	530	16
96906	MS35756-6	5315-00-616-5514	207	41
			310	3
96906	MS35756-8	5315-00-616-5526	125	10
96906	MS35764-1297	5306-01-052-2402	200	21
96906	MS35769-21	5330-00-514-3289	230	3
			237	2
96906	MS35782-1	4820-00-684-0880	99	4
96906	MS35782-2	4820-00-720-4488	108	14
96906	MS35782-4	4820-00-752-9040	71	20
			74	19
			75	15
			116	12
			124	10
			596	20
96906	MS35782-5	4820-00-849-1220	107	36
			265	7
96906	MS35783-1	4820-00-275-2224	585	25
96906	MS35783-2	4820-00-845-1096	31	10
			204	12
			595	20
96906	MS35810-1	5315-00-417-5223	411	16
96906	MS35810-11	5315-01-070-2168	534	10
96906	MS35810-6	5315-00-140-1938	211	27
			419	14
96906	MS35812-1	5340-00-969-6407	528	2
96906	MS35812-2	5340-00-865-9496	182	18
			454	2
96906	MS35812-3	5340-00-057-3537	525	17
96906	MS35812-4	5340-00-985-0823	402	17
96906	MS35812-6	5340-00-550-8070	414	15
96906	MS35842-11	4730-00-908-3194	39	2
			39	5
			581	9
			582	9
			584	19
			589	53
			590	38
			622	9
96906	MS35842-14	4730-00-908-6292	136	2
			581	16
			585	32
			589	4
			591	10
96906	MS35914-148		154	8
96906	MS35916-2	6680-00-825-2076	625	7
96906	MS35931-2	4820-00-263-3019	584	8
			593	27

CROSS-REFERENCE INDEXES

PART NUMBER INDEX

CAGEC	PART NUMBER	STOCK NUMBER	FIG.	ITEM
			595	17
96906	MS39020-1	9905-00-752-4649	614	18
96906	MS39021-2	6680-00-933-3599	624	7
96906	MS39060-2	5930-00-699-9438	132	2
			134	63
			611	30
			612	23
			613	21
96906	M539176-9	4730-00-240-1740	617	15
96906	M539179-2	4730-00-270-4580	70	6
			76	16
			79	16
			266	4
02570	MS39182-8	4730-00-143-9282	262	22
			264	12
			291	6
			604	1
			605	6
96906	MS39185-1	4730-00-701-7677	215	1
			288	8
			290	4
906	MS39189-2	4730-00-278-8902	252	7
906	MS39191-3	730-00-930-6354	274	4
			275	4
96906	MS39230-4	4730-00-253-4414	505	45
			590	2
96906	MS39230-5	4730-00-253-4415	589	16
96906	MS39230-7	4730-00-012-7823	482	2
96906	MS39231-1	4730-00-249-3932	146	7
96906	MS39231-4	4730-00-249-3935	271	1
			589	63
96906	MS39231-9	4730-00-017-9447	480	6
96906	MS39232-7	4730-00-231-5650	590	3
96906	MS45904-68	5310-00-889-2528	143	19
			587	5
			593	4
			595	19
96906	MS45904-72	5310-00-889-2527	211	31
96906	MS45904-73	5310-00-728-2044	137	4
			592	21
			594	16
96906	MS45904-74	5310-00-901-0279	161	18
			162	9
96906	MS45904-76	5310-00-061-1258	156	8
			161	7
			213	29
			612	18
			613	31
96906	MS45904-84	5310-00-935-8984	155	2
96906	MS45904-87	5310-00-901-1339	155	4

CROSS-REFERENCE INDEXES

PART NUMBER INDEX

CAGEC	PART NUMBER	STOCK NUMBER	FIG.	ITEM
			155	6
02951	MS49005-8	4730-00-289-5176	502	34
96906	MS49005-9	4730-00-288-8555	489	11
96906	MS49006-10	4730-00-968-6129	230	5
			484	20
			489	28
96906	MS500040-6	6680-00-226-4574	82	4
			145	4
			145	4
96906	MS500083A041200	4720-01-371-4395	629	5
			629	19
96906	MS5000B3A160360	4720-01-371-4396	613	20
			629	5
96906	MS51000-131-2	5330-01-254-6377	489	17
96906	MS51031-130	5305-01-186-7140	503	29
96906	MS510504A8-8	4730-00-273-6686	269	9
96906	MS51065RP48-2	3030-00-983-2873	125	8
96906	MS51066-48-2	3030-00-832-4312	310	9
96906	MS51069RC49-2	3030-01-118-1318	120	32
96906	MS51071-16	6680-00-507-9992	624	2
96906	MS51071-7	6680-00-882-0965	627	2
96906	MS51072-16	3040-00-600-3860	624	4
96906	MS51095-374	5305-00-964-0565	33	28
96906	MS51095-386	5305-00-941-9460	235	10
			242	10
96906	MS51095-410	5305-00-903-7794	233	14
			240	14
96906	MS51095-416	5305-00-964-0589	460	31
96906	MS51096-359	5305-00-912-5113	62	4
			83	2
			142	5
			148	12
			161	11
			164	26
			202	3
			267	16
			315	4
			337	28
			356	12
			365	4
			537	6
96906	MS51106-421	5305-00-940-9517	346	12
96906	MS51113-1	5930-00-307-8856	131	28
96906	MS51301-1	2590-00-900-1640	148	1
96906	MS51321-1-24N1	2910-00-710-6054	422	5
			423	5
96906	MS51321-2-24N1	2910-00-930-9367	593	5
			595	22
96906	MS51412-6	5310-01-259-0296	113	7
96906	MS51412-7	5310-01-257-7590	107	37
96906	MS51500-A12-8	4730-00-865-9251	271	2

PART NUMBER INDEX

CAGEC	PART NUMBER	STOCK NUMBER	FIG.	ITEM
96906	MS51500-A4	4730-00-289-0383	204	22
			276	3
96906	MS51500-AS-4	4730-00-620-6904	198	12
96906	MS51500-A6-6	4730-00-322-8457	182	6
96906	MS51500-AS-8	4730-00-809-9427	257	2
96906	MS51500-B8	4730-00-813-9611	253	12
			278	8
96906	MS51500A10	4730-00-278-4311	495	5
96906	MS51500A12-8	4730-00-865-9251	630	10
96906	MS51500A16-12	4730-00-021-1802	456	37
			460	34
			461	34
			509	32
96906	MS51500A20	4730-00-202-8470	480	12
96906	MS51500A24	4730-00-676-7566	461	16
			509	10
96906	MS51500A6	4730-00-995-1559	278	13
96906	MS51500A6-2	4730-00-803-5776	277	2
96906	MS51500A6-6	4730-00-322-8457	214	12
96906	MS51500A6-8	4730-00-289-0382	277	7
			523	21
96906	MS51500A8	4730-00-813-9611	254	4
96906	MS51500AS-8	4730-00-809-9427	254	5
			269	30
			505	43
			508	5
96906	MS51504-A10	4730-00-833-8230	458	30
96906	MS51504-A6	4730-00-812-7999	278	14
			288	14
96906	MS51504-A6-2	4730-00-803-5765	182	1
96906	MS51504-B4	4730-00-647-3207	598	15
			620	20
96906	MS51504A10S	4730-00-322-8339	457	10
96906	MS51504A12	4730-00-289-4912	204	13
96906	MS51504A16	4730-00-833-9315	508	24
96906	MS51504A4	4730-00-647-3207	276	10
96906	MS51504A4-4	4730-00-877-8997	598	52
96906	MS51504A6	4730-00-812-7999	212	16
			603	13
96906	MS51504A6-6	4730-00-835-3003	273	7
			598	39
96906	MS51504A6-8	4730-01-070-4915	598	23
96906	MS51504A8	4730-00-555-1764	258	4
			267	1
96906	MS51506-A20	4730-01-163-7823	482	17
96906	MS51506-B24	4730-00-081-0311	459	3
96906	MS51506A24	4730-00-081-0311	456	23
			458	15
96906	MS51508-B12	4730-01-028-8147	204	18
96906	MS51508A6-6S	4730-01-005-0623	272	1
96906	MS51508A8-8	4730-00-052-3420	258	9

CROSS-REFERENCE INDEXES

PART NUMBER INDEX

CAGEC	PART NUMBER	STOCK NUMBER	FIG.	ITEM
96906	MS51508A8Z	4730-01-324-5071	267	5
96906	MS51514A6	4730-01-146-4113	182	3
96906	MS51518B10	5365-01-249-9707	629	5
96906	MS51521-B10	4730-01-139-1585	458	21
96906	MS51521-B24	4730-00-409-8797	504	20
96906	MS51521A10	4730-01-139-1585	456	3
			460	3
			461	3
			503	40
			509	30
96906	MS51521A16	4730-00-163-0236	505	52
			600	16
96906	MS51521A20	4730-01-238-6443	503	4
96906	MS51522-A16	4730-01-192-9578	504	29
96906	MS51522-A5	4730-01-139-4231	277	6
96906	MS51522A32	4730-01-232-7159	505	9
96906	MS51523-B16	4730-00-827-5852	504	2
96906	MS51523-B20	4730-01-067-3932	503	6
96906	MS51523A16	4730-00-827-5852	504	11
96906	MS51523A20	4730-01-067-3932	503	2
96906	MS51523A8	4730-00-074-0713	505	42
96906	MS51525-A6	4730-00-491-9576	299	6
96906	MS51525-A12	4730-00-710-5571	460	6
96906	MS51525A10-12S	4730-01-116-5969	456	2
			460	2
			461	2
			509	1
96906	MS51525A16	4730-00-930-5392	480	15
			505	53
96906	MS51525A20	4730-01-066-1282	480	11
			503	3
96906	MS51525A6	4730-00-491-9576	299	6
96906	MS51526A10	4730-01-192-9590	456	41
			460	37
			461	32
			509	29
96906	MS51527-A10	4730-00-974-7313	461	5
96906	MS51527-A6Z	4730-01-278-1001	297	3
96906	MS51527A10	4730-00-974-7313	456	5
			509	27
			605	5
96906	MS51527A32	4730-01-147-7954	505	56
96906	MS51527A6	4730-00-143-3941	313	6
			314	12
96906	MS51527A8	4730-00-822-5609	503	36
96906	MS51528A32	4730-01-195-7339	480	8
96906	MS51531-B4	4730-00-314-8366	426	9
96906	MS51532B10	4730-00-542-5911	629	5
96906	MS51815-3	4730-00-096-8756	422	4
			423	3
96906	MS51817-3	4730-00-278-4651	276	9

PART NUMBER INDEX

CAGEC	PART NUMBER	STOCK NUMBER	FIG.	ITEM
96906	MS51838-147	5315-00-054-4028	472	41
96906	MS51843-6P	4730-00-492-6040	34	10
96906	MS51845-2	4730-00-277-5542	593	2
96906	MS51845-4	4730-00-249-3885	255	23
			262	11
96906	MS51846-13	4730-00-196-1966	212	13
			213	1
			266	2
96906	MS51846-24	4730-00-193-2709	264	21
96906	MS51846-58	4730-00-222-1839	264	13
			264	31
96906	MS51846-64	4730-00-196-1991	255	22
			262	17
96906	MS51846-67	4730-00-196-1993	255	22
96906	MS51846-77	4730-00-222-1840	519	30
			521	16
96906	MS51848-14	5310-00-171-1735	308	22
96906	MS51849-65	5305-01-006-2052	409	31
96906	MS51850-96	5305-01-226-1945	612	16
96906	MS51851-106	5305-01-090-3012	420	22
96906	MS51851-108	5305-00-003-9255	420	20
96906	MS51851-112	5305-01-144-1625	402	8
96906	MS51851-126	5305-01-101-3312	582	15
96906	MS51851-128	5305-01-219-5386	341	14
96906	MS51851-129	5305-01-209-7112	426	13
96906	MS51851-85	5305-00-191-3640	341	3
			424	5
96906	MS51861-12	5305-00-140-8001	571	1
			572	1
			572	10
96906	MS51861-15	5305-00-432-4173	404	22
96906	MS51861-22	5305-00-432-8027	561	2
			566	7
96906	MS51861-24	5305-00-432-4163	359	11
			362	22
			402	2
			418	13
96906	MS51861-34	5305-00-058-1082	425	10
96906	MS51861-35	5305-00-432-4170	408	15
			412	8
			412	10
			412	10
			413	19
			414	3
			415	20
			415	26
			418	11
			419	6
			420	5
			424	13
			425	16

CROSS-REFERENCE INDEXES

PART NUMBER INDEX

CAGEC	PART NUMBER	STOCK NUMBER	FIG.	ITEM
			426	12
			429	4
96906	MS51B61-35C	5305-00-433-3711	403	14
96906	MS51861-36	5305-00-432-4171	410	8
			418	14
			420	15
96906	MS51861-37	5305-00-432-4172	403	11
			404	2
			405	24
			406	41
			407	1
			407	19
			413	6
			414	11
			415	28
			416	30
			421	7
			425	19
			426	3
			429	6
			579	1
96906	MS51861-37C	5305-00-433-3685	403	4
96906	MS51861-37C	5305-00-433-3685	418	2
96906	MS51861-40	5305-00-432-7956	149	4
96906	MS51861-44	5305-00-138-0069	417	12
			492	20
96906	MS51861-45	5305-00-432-4201	362	21
			404	11
			408	8
			409	30
			422	18
			423	4
			427	6
			429	2
			455	7
			492	18
			528	15
96906	MS51861-47	5305-00-432-4203	364	3
			402	9
			404	18
			405	14
			406	44
			407	2
			412	22
			413	26
			415	31
			416	2
			422	12
			423	12
96906	MS51861-49	5305-00-432-4205	405	13
			406	2

PART NUMBER INDEX

CAGEC	PART NUMBER	STOCK NUMBER	FIG.	ITEM
96906	MS51961-49C	5305-00-477-0124	613	10
96906	MS51861-65		351	30
			425	3
			528	5
			549	47
96906	MS51861-66	5305-00-432-4252	405	45
			427	9
			427	19
			548	1
96906	MS51861-67	5305-00-432-4253	404	10
			419	23
			419	37
			420	1
			424	2
96906	MS51861-68	5305-00-477-0144	404	23
			405	2
			405	46
			406	22
			406	47
			416	21
			416	33
			428	15
96906	MS51861-68	5305-00-477-0144	428	21
96906	MS51861-69	5305-00-432-4254	402	23
96906	MS51861-72	5305-01-121-2696	413	25
96906	MS51862-12	5305-01-028-4831	441	6
96906	MS51862-16	5305-00-432-4390	413	14
96906	MS51862-23	5305-00-402-4211	363	4
96906	MS51862-26	5305-00-483-0554	405	31
			406	36
96906	MS51862-26C	5305-01-023-2428	410	10
96906	M351862-33	5305-01-066-8646	419	34
96906	MS51862-36	5305-00-149-8610	412	14
			417	10
			424	16
96906	MS51862-56	5305-01-186-7137	411	12
			420	13
96906	MS51862-56C	5305-00-269-0770	408	4
			410	11
96906	MS51862-61	5305-00-196-5570	417	17
96906	MS51873-32B	4730-01-027-8943	182	4
96906	MS51874-10	4730-00-019-0236	177	13
96906	MS51884-9	4730-00-187-4210	207	31
			519	26
96906	MS51887-7C	4730-01-100-0109	266	42
96906	NS51922-1	5310-00-088-1251	409	5
			544	3
			544	7
			545	3
			546	3
			594	22

CROSS-REFERENCE INDEXES

PART NUMBER INDEX

CAGEC	PART NUMBER	STOCK NUMBER	FIG.	ITEM
			611	2
			613	33
96906	MS51922-13	5310-00-984-3807	100	25
			101	20
			370	17
			425	14
			533	17
			611	17
96906	MS51922-17	5310-00-087-4652	183	5
			481	4
			618	8
96906	MS51922-21	5310-00-959-1488	62	2
			67	9
			68	3
			69	6
			71	10
			72	8
			74	4
			75	10
			76	8
			80	20
			93	10
96906	MS51922-21	5310-00-959-1488	109	18
			142	4
			143	8
			164	28
			248	1
			337	9
			368	8
			374	20
96906	MS51922-33	5310-00-225-6993	419	20
96906	MS51922-53	5310-00-225-6408	78	19
			297	14
			345	2
			391	8
96906	MS51922-6	5310-00-143-6102	71	4
			74	27
			75	5
			458	23
96906	MS51922-61	5310-00-832-9719	349	1
96906	MS51922-9	5310-00-984-3806	107	20
			346	2
			347	10
			382	34
			385	9
			385	41
			387	28
96906	MS51923-361	5315-01-129-4621	289	3
96906	MS51928-1	5340-01-107-6971	134	46
96906	MS51932-103	5315-00-236-6625	501	8
96906	MS51932-135	5315-01-129-9190	501	11

PART NUMBER INDEX

CAGEC	PART NUMBER	STOCK NUMBER	FIG.	ITEM
96906	MS51932-154	5315-00-237-6341	383	40
			384	11
			386	18
96906	MS51937-5	5306-00-050-0347	607	10
96906	MS51943-12	5310-01-137-7062	579	3
96906	MS51943-2	5310-01-374-0508	184	8
96906	MS51943-32	5310-00-935-9022	61	10
			81	14
			100	42
			101	9
			105	17
			112	2
			113	6
			142	7
			142	12
			142	16
			143	18
			145	13
			162	8
			163	25
			165	29
			166	25
96906	MS51943-32	5310-00-935-9022	167	35
			168	23
			169	27
			182	12
			255	17
			255	21
			262	1
			262	16
			269	1
			276	8
			278	18
			279	10
			279	15
			289	11
			313	8
			315	14
			326	7
			339	29
			340	10
			353	9
			361	1
			374	6
			374	17
			454	23
			513	16
			525	15
			543	6
			543	14
			545	6

PART NUMBER INDEX

CAGEC	PART NUMBER	STOCK NUMBER	FIG.	ITEM
			547	5
			549	5
			551	18
			561	5
			580	1
			581	22
			584	5
			586	17
			587	9
			588	13
			589	20
			592	1
			593	6
			594	34
			596	23
			606	17
			610	6
			616A	5
96906	MS51943-33	5310-00-814-0673	108	6
			263	11
			583	6
			589	31
96906	MS51943-34	5310-00-241-6658	77	23
96906	MS51943-34	5310-00-241-6658	80	25
			97	8
			100	30
			112	8
			113	10
			134	41
			140	14
			169	32
			204	14
			205	5
			252	1
			254	1
			256	17
			266	39
			290	15
			291	1
			341	12
			390	23
			397	1
			525	12
			536	5
			549	34
			581	17
			585	35
			589	8
			591	2
			592	14
			594	14

PART NUMBER INDEX

CAGEC	PART NUMBER	STOCK NUMBER	FIG.	ITEM
			601	10
			604	8
			606	15
			615	18
96906	MS51943-35	5310-00-935-9021	107	8
			205	12
96906	MS51943-36	5310-00-814-0672	6	23
			83	1
			102	6
			102	11
			102	30
			105	14
			106	19
			142	20
			153	23
			157	16
			161	6
			163	1
			163	20
			163	22
			165	24
			168	20
			202	6
96906	MS51943-36	5310-00-814-0672	211	21
			213	36
			215	12
			218	13
			220	16
			222	10
			223	1
			224	1
			225	1
			226	1
			227	1
			228	16
			229	13
			243	2
			253	7
			254	19
			254	19
			257	7
			258	3
			261	6
			261	27
			263	15
			264	6
			264	23
			266	31
			267	14
			269	5
			271	19

CAGEC	PART NUMBER	STOCK NUMBER	FIG.	ITEM
			273	24
			301	1
			302	4
			316	16
			318	19
			319	32
			320	29
			324	31
			333	7
			334	6
			335	8
			336	6
			337	7
			337	26
			339	6
			338	23
			343	14
			351	5
			354	6
			357	4
			359	1
			364	8
			365	7
96906	MS51943-36	5310-00-814-0672	368	17
			374	7
			375	8
			376	15
			377	5
			378	5
			379	3
			388	3
			388	7
			390	19
			397	8
			399	12
			400	2
			401	3
			454	16
			456	12
			457	13
			457	22
			460	11
			461	9
			478	11
			480	28
			481	11
			492	4
			509	5
			513	5
			513	19
			513	24

PART NUMBER INDEX

CAGEC	PART NUMBER	STOCK NUMBER	FIG.	ITEM
			525	5
			525	30
			536	13
			536	20
			537	8
			581	6
			583	8
			584	25
			586	12
			589	26
			590	14
			592	17
			600	4
			600	4
			600	4
			600	7
			603	4
			604	10
			616A	29
			629	5
96906	MS51943-38	5310-00-994-1006	456	13
			461	10
			509	25
96906	MS51943-39	5310-00-488-3889	458	18
96906	MS51943-40	5310-00-488-3888	5	13
			69	3
			102	12
			102	14
			102	16
			102	17
			106	11
			107	34
			155	4
			278	10
			304	1
			318	3
			318	38
			318	48
			319	8
			319	39
			320	3
			320	36
			321	3
			322	3
			322	42
			323	3
			323	40
			324	15
			324	37
			326	14
			326	14

CROSS-REFERENCE INDEXES

PART NUMBER INDEX

CAGEC	PART NUMBER	STOCK NUMBER	FIG.	ITEM
			326	15
			327	2
			328	1
			330	2
			331	6
			332	5
			333	14
			334	9
			335	12
			336	9
			336	14
			337	17
			338	14
			340	14
			349	6
			351	28
			358	6
			365	1
			368	11
			368	15
			374	23
			381	17
			383	33
			384	8
96906	MS51943-40	5310-00-488-3888	386	16
			389	4
			391	18
			456	14
			457	21
			458	8
			460	14
			461	11
			468	2
			492	6
			500	1
			503	24
			508	21
			509	6
			513	11
			525	11
			589	18
			590	18
			602	6
			614	7
			615	1
			616	8
			616A	16
			619	2
			619	2
96906	MS51943-42	5310-00-241-6661	318	44
			319	45

CROSS-REFERENCE INDEXES

PART NUMBER INDEX

CAGEC	PART NUMBER	STOCK NUMBER	FIG.	ITEM
			320	45
			321	46
			322	49
			323	48
			324	45
96906	MS51943-44	5310-00-241-6664	5	10
			207	17
			207	36
			318	22
			319	21
			320	18
			321	21
			322	23
			323	25
			324	19
			326	11
			326	16
			327	1
			328	2
			330	3
			331	2
			332	2
			343	13
			380	7
			381	10
96906	MS51943-44	5310-00-241-6664	389	7
			390	20
			400	4
			616	9
96906	MS51943-46	5310-00-935-3569	106	16
			107	25
			297	14
			302	5
			318	20
			319	9
			320	19
			321	22
			322	21
			323	26
			324	20
			331	3
			332	3
			380	9
96906	MS51943-48	5310-00-241-6665	460	13
96906	MS51943-50	5310-00-340-4953	499	24
96906	MS51952-1	4730-00-053-0266	212	6
96906	MS51952-2	4730-00-277-5553	278	16
			290	18
			582	7
			596	2
			598	49

PART NUMBER INDEX

CAGEC	PART NUMBER	STOCK NUMBER	FIG.	ITEM
			620	18
96906	MS51952-4	4730-00-277-5555	252	16
			255	6
			261	19
			262	20
			264	3
			264	32
96906	MS51953-124	4730-00-196-1506	480	4
96906	MS51953-132	4730-00-138-8121	480	35
96906	MS51953-154	4730-00-288-7495	493	27
96906	MS51953-175	4730-00-196-1534	459	4
96906	MS51953-181	4730-00-054-2027	458	11
96906	MS51953-33	4730-00-196-1486	596	5
96906	MS51953-55	4730-00-196-1489	589	48
			590	33
96906	MS51953-78	4730-00-196-1493	589	2
96906	MS51953-80	4730-00-196-1495	263	1
			269	32
			505	47
96906	MS51953-81	4730-00-196-1496	505	46
96906	MS51953-9	4730-00-196-1481	61	5
96906	MS51953-97	4730-00-196-1468	36	11
96906	MS51955-36	5305-00-054-9271	507	2
96906	MS51955-71	5305-00-054-9285	469	5
96906	MS51957-31	5305-00-054-6655	540	4
96906	MS51961-15	3110-00-227-3245	501	3
96906	MS51961-22	3110-00-227-3249	472	9
96906	MS51961-30	3110-00-227-3253	500	20
96906	MS51961-32	3110-00-227-3255	476	2
			477	6
96906	MS51961-40	3110-00-752-7760	500	6
96906	MS51961-6	3110-00-227-3239	500	12
96906	MS51961-9	3110-00-227-3241	472	21
96906	MS51963-101	5305-00-724-6810	469	12
			470	8
96906	MS51963-137	5305-00-725-0168	472	12
96906	MS51963-155	5305-00-724-7206	469	30
			470	5
96906	MS51963-170	5305-00-725-0194	502	17
96906	MS51963-172	5305-00-725-0197	473	30
96906	MS51963-34	5305-00-719-5342	407	6
			415	10
96906	MS51963-67	5305-00-723-9383	454	5
			528	10
96906	MS51963-83	5305-00-724-5898	472	32
96906	MS51965-43	5305-00-726-3091	409	18
96906	MS51965-90	5305-00-724-5830	489	53
96906	MS51967-14	5310-00-768-0318	2	2
			176	4
96906	MS51967-18	5310-00-763-8919	236	16
			237	14

CROSS-REFERENCE INDEXES

PART NUMBER INDEX

CAGEC	PART NUMBER	STOCK NUMBER	FIG.	ITEM
96906	MS51967-2	5310-00-761-6882	298	13
			370	31
			383	22
			385	29
			408	19
			414	6
			536	8
			580	3
			584	6
			592	4
			594	36
			612	33
			613	13
96906	MS51967-20	5310-00-763-8920	1	3
			206	7
			403	17
96906	MS51967-23	5310-00-763-8921	499	14
96906	MS51967-24	5310-00-763-8922	502	15
96906	MS51967-5	5310-00-880-7744	370	29
			382	1
			382	28
			382	33
			385	23
			385	32
			387	1
96906	MS51967-5	5310-00-880-7744	387	19
			387	43
			414	20
			419	19
			469	27
			473	14
			499	17
			609	13
			609	27
96906	MS51967-8	5310-00-732-0558	141	8
			144	20
			380	1
			381	1
96906	MS51968-11	5310-00-880-7745	301	15
96906	MS51968-14	5310-00-732-0560	26	3
			26	7
			26	12
			318	25
			346	25
			348	7
			468	11
			469	26
			501	19
96906	MS51968-15	5310-00-943-2141	348	13
96906	MS51968-17	5310-00-763-8911	236	41
			345	26

CROSS-REFERENCE INDEXES

PART NUMBER INDEX

CAGEC	PART NUMBER	STOCK NUMBER	FIG.	ITEM
96906	MS51968-2	5310-00-768-0319	141	19
			141	26
			148	8
			184	13
			245	3
			314	6
			507	45
96906	MS51968-20	5310-00-763-8905	144	33
			303	3
			391	27
			417	24
96906	MS51968-5	5310-00-880-7746	100	21
			101	13
			351	22
			540	6
96906	MS51968-6	5310-00-905-4600	622	5
96906	MS51968-8	5310-00-732-0559	148	14
			301	10
			301	15
			402	18
			528	22
96906	MS51975-28	5305-00-914-4171	409	6
96906	MS51977-50	5305-00-058-9378	473	7
96906	MS51977-74	5305-00-058-9389	465	10
			512	10
96906	MS51977-84	5305-00-939-0576	472	13
96906	MS51983-8	5310-01-229-8029	293	1
			295	15
96906	MS521301A203R	4720-00-491-0102	BULK	16
96906	MS521303A203R	4720-00-080-5379	BULK	20
96906	MS52150-37HE	5340-01-272-1471	63	11
96906	MS53000-1	2590-00-801-2355	137	2
96906	MS53007-1	9905-00-999-7370	559	1
			559	1
96906	MS53011-1	2920-00-304-3493	127	2
96906	MS53011-2	2920-01-031-9027	128	1
96906	MS53046-3	6160-00-404-2669	153	32
96906	MS53053-1	2540-00-409-8891	378	8
			600	1
96906	MS53063-3	2940-00-071-2653	61	7
96906	MS75004-1	5940-00-549-6581	151	16
96906	MS75004-2	5940-00-549-6583	151	11
96906	MS75021-2	5935-00-846-3884	163	14
			165	14
			166	7
			167	6
			168	11
			169	11
96906	MS77074-3	5940-00-159-1292	137	8
96906	MS77074-7	5940-00-531-0530	137	9
96906	MS87006-13	4030-00-780-9350	326	20

PART NUMBER INDEX

CAGEC	PART NUMBER	STOCK NUMBER	FIG.	ITEM
			327	13
			339	25
			395	10
			395	15
			419	36
			616A	6
96906	MS87006-3	4030-00-270-5436	82	14
96906	MS87006-33	4030-00-948-7315	326	23
			327	10
			383	16
			390	4
96906	MS87006-53	4030-00-916-2141	339	12
			339	19
			340	3
96906	MS87006-63	4030-00-729-6054	340	8
96906	MS87008-5	4010-00-165-5607	BULK	21
96906	MS9024-18	5340-00-685-0567	457	15
			458	26
96906	MS90707-1050	5325-00-419-3322	493	28
			505	20
96906	MS90723-38	5310-01-127-7599	549	18
			551	13
96906	MS90725-103	5305-00-717-3999	348	8
96906	MS90725-162	5305-00-724-5910	402	5
			473	11
			483	6
96906	MS90725-173	5305-00-724-5939	5	3
96906	MS90725-187	5305-00-939-9204	469	18
			470	14
			499	16
			502	6
96906	MS90725-3	5305-00-068-0500	180	17
			549	50
96906	MS90725-31	5306-00-225-8496	109	5
			183	9
			198	9
			348	22
96906	MS90725-34	5306-00-225-8499	425	22
96906	MS90725-36	5306-01-075-8519	37	11
96906	MS90725-43	5305-00-225-8507	286	18
96906	MS90725-6	5305-00-068-0502	370	30
			371	12
			408	20
96906	MS90725-67	5305-00-269-3217	177	9
96906	MS90725-99	5305-00-069-5583	346	19
96906	MS90726-109	5305-00-226-7767	106	8
			155	1
			396	8
			458	25
			474	21
			485	25

CROSS-REFERENCE INDEXES

PART NUMBER INDEX

CAGEC	PART NUMBER	STOCK NUMBER	FIG.	ITEM
			487	26
			489	4
96906	MS90726-110	5305-00-716-8186	336	19
			611	9
96906	MS90726-113	5305-00-725-4183	67	12
			69	4
			326	10
			326	10
			327	6
			332	7
			333	9
			334	20
			335	18
			336	11
			337	21
			338	8
			389	5
			391	6
			391	14
			396	9
			468	13
			484	31
			501	17
			503	31
96906	MS90726-118	5305-00-716-8181	340	15
96906	MS90726-124	5305-00-716-8174	389	8
96906	MS90726-134	5305-00-902-9338	183	4
96906	MS90726-170	5305-00-725-4138	417	7
96906	MS90726-p37	5305-01-128-4095	499	26
96906	MS90726-31	5306-00-225-9086	581	14
			585	33
			591	7
			596	16
			601	7
96906	MS90726-32	5306-00-225-9087	113	8
			591	8
96906	MS90726-33	5306-00-225-9088	77	21
			80	23
			252	2
			397	3
			489	41
96906	MS90726-34	5306-00-225-9089	266	41
			308	14
			622	1
96906	MS90726-36	5305-00-225-9091	169	30
			205	1
			592	11
			594	18
			604	6
96906	MS90726-37	5305-00-225-9092	62	28
			100	40

PART NUMBER INDEX

CAGEC	PART NUMBER	STOCK NUMBER	FIG.	ITEM
			536	1
			549	36
96906	MS90726-38	5305-00-225-9093	254	12
			256	19
			308	13
			604	14
96906	MS90726-40	5306-00-225-9095	102	1
			103	1
			601	11
96906	MS90726-43	5306-00-225-9098	341	7
96906	MS90726-44	5305-00-225-9099	290	5
96906	MS90726-45	5306-00-225-9100	390	21
			605	4
96906	MS90726-60	5305-00-269-2803	62	1
			67	11
			68	5
			69	13
			72	3
			72	13
			83	3
			102	31
			102	36
			109	17
			142	10
			142	14
			143	15
			153	6
96906	MS90726-60	5305-00-269-2803	183	10
			243	1
			253	5
			259	10
			260	10
			264	25
			264	28
			269	8
			333	19
			334	12
			337	11
			338	10
			355	16
			356	15
			364	10
			365	10
			368	1
			368	4
			374	2
			374	21
			375	6
			376	8
			377	7
			378	2

PART NUMBER INDEX

CAGEC	PART NUMBER	STOCK NUMBER	FIG.	ITEM
			379	1
			388	5
			454	17
			456	9
			457	20
			513	21
			536	16
			536	17
			583	10
			586	10
			589	35
			590	21
			592	19
			600	2
			600	2
			600	5
			602	12
			603	6
			613	30
			629	5
96906	MS90726-61	5305-00-269-2804	6	27
			62	1
			71	26
			74	26
			75	21
			105	2
			213	34
96906	MS90726-61	5305-00-269-2804	215	10
			220	14
			228	14
			254	17
			254	17
			257	8
			258	1
			261	25
			263	17
			266	20
			266	29
			271	17
			301	7
			302	8
			334	3
			335	15
			337	29
			338	3
			354	9
			368	6
			374	22
			397	6
			492	8
			505	29

PART NUMBER INDEX

CAGEC	PART NUMBER	STOCK NUMBER	FIG.	ITEM
			511	58
			513	15
96906	MS90726-67	5305-00-269-2811	211	20
			243	10
			609	25
96906	MS90727-111	5305-00-719-5219	67	10
			69	14
			102	10
			106	13
			107	29
			155	5
			308	23
			319	11
			326	18
			346	1
			351	23
			365	2
			383	35
			384	2
			386	10
			389	2
			458	6
			484	2
			488	2
			489	20
			492	21
			502	28
96906	MS90727-111	5305-00-719-5219	584	23
			602	19
			615	5
			619	22
96906	MS90727-112	5305-00-725-0154	318	33
			318	35
			319	3
			319	36
			320	33
			320	38
			321	35
			322	39
			323	37
			324	34
			468	8
			487	9
			492	22
			508	23
96906	MS90727-116	5305-00-719-5239	321	23
			322	24
			330	12
			331	8
			513	9
			525	9

CROSS-REFERENCE INDEXES

PART NUMBER INDEX

CAGEC	PART NUMBER	STOCK NUMBER	FIG.	ITEM
			616	10
96906	MS90727-12	5305-00-267-8956	313	11
96906	MS90727-123	5305-00-719-5270	5	5
96906	MS90727-125	5305-00-719-5274	390	8
			503	19
96906	MS90727-128	5305-00-719-5275	495	2
			502	40
96906	MS90727-130	5305-01-277-0461	263	6
96906	MS90727-132	5305-01-277-0423	100	43
96906	MS90727-139	5305-00-725-0140	318	41
			319	41
			320	43
			321	43
			322	46
			323	45
			324	40
96906	MS90727-158	5305-00-726-2525	305	21
			471	24
96906	MS90727-160	5305-00-726-2543	78	21
96906	MS90727-169	5305-00-726-2556	207	40
96906	MS90727-172	5305-00-726-2559	613	37
96906	MS90727-176	5305-00-726-2567	616	1
96906	MS90727-191	5305-00-948-0803	331	13
			332	12
96906	MS90727-200	5305-00-728-5475	490	17
96906	MS90727-32	5306-00-050-1238	97	15
			100	18
96906	MS90727-32	5306-00-050-1238	112	9
			140	4
			161	13
			256	20
			263	13
			291	2
			351	27
			399	11
			533	16
			589	5
			606	1
96906	MS90727-33	5305-00-051-4075	615	15
96906	MS90727-42	5306-00-051-4084	137	6
96906	MS90727-61	5305-00-269-3237	107	1
96906	MS90727-70	5305-00-269-3246	478	12
96906	MS90727-71	5305-00-269-3247	67	13
			68	10
			69	15
96906	MS90727-74	5305-00-269-3250	474	17
			489	10
96906	MS90727-75	5305-00-269-4528	390	14
96906	MS90727-86	5305-00-709-8516	243	22
96906	MS90727-94	5305-00-709-8537	615	10
96906	MS90727-96	5305-00-709-8482	125	7

PART NUMBER INDEX

CAGEC	PART NUMBER	STOCK NUMBER	FIG.	ITEM
96906	MS90728-6	5305-00-068-0858	626	2
96906	MS9316-08	5305-00-081-3728	154	23
96906	MS9316-13	5305-00-857-3367	257	15
96906	MS9320-14	5310-00-184-8992	470	12
96906	MS9350-23	5340-00-922-3380	590	25
96906	MS9380-09		502	35
96906	MS9845-34	5315-01-219-3665	341	22
81349	M12133/1-6P	5310-01-p53-1936	337	27
			338	24
81349	M13486/1-12	6145-00-805-3354	BULK	45
81349	M13486/1-14	6145-00-705-6674	BULK	46
81349	M13486/1-3	6145-00-161-1609	BULK	43
81349	M13486/1-7	6145-00-705-6678	BULK	44
81349	M13486/13-1	6145-00-774-4579	BULK	2
61465	M2020817	2520-00-740-9037	507	34
81349	M23053/5 X 3 IN		152	6
81349	M24066/2-321	5340-00-725-6033	153	4
81349	M24240-4-20916	5306-01-226-5917	467	1
81349	M24243/1-B304	5320-01-055-4452	153	3
81349	M24243/1-B402	5320-00-510-7823	558	2
81349	M25988/1-246	5331-01-189-6351	141	9
61465	M305815	2520-00-740-9667	502	3
61465	M305820K	2590-00-404-0752	499	15
61465	M305849	2520-00-740-9939	476	1
			477	5
01212	M39807	5330-00-005-0858	16	3
81349	M43436/1-1	9905-00-752-4649	97	11
			97	22
81349	M43436/1-1	9905-00-752-4649	131	25
			134	16
			134	62
			143	5
			157	5
			158	6
			170	5
			170	14
			170	21
			170	26
			170	31
			170	43
			171	11
			172	5
			172	13
			172	19
			173	5
			173	11
			173	20
			173	26
			174	5
			211	5
			211	42

CAGEC	PART NUMBER	STOCK NUMBER	FIG.	ITEM
			430	4
			431	3
			433	8
			435	3
			436	4
			440	4
			440	7
			496	2
81349	M43436/1-2	9905-00-841-4445	431	7
			434	7
			435	8
			436	15
			439	2
			439	7
81349	M43436/1-3	9905-00-893-3570	151	6
			151	15
			151	20
			152	10
			154	16
			154	20
			431	2
			431	6
			433	3
			433	9
			437	3
			438	4
81349	M43436/2-1	9905-01-069-7222	169	21
			430	3
			433	2
			437	2
			438	2
			439	3
			439	6
			440	6
81349	M43436/3-1	9905-01-013-8723	151	4
			152	5
			163	6
			165	13
			166	5
			167	5
			168	6
			169	6
			182	32
			213	25
			432	6
			432	13
			433	7
			434	8
			435	4
			435	9
			436	3

CROSS-REFERENCE INDEXES

PART NUMBER INDEX

CAGEC	PART NUMBER	STOCK NUMBER	FIG.	ITEM
			436	16
81349	M43436/4-2	9905-00-935-7777	166	17
			430	5
D8286	M5X15DIN933-STL8 .8 CADMIUM PL	5305-01-248-3222	92	19
2P971	M504130	5340-01-280-6947	360	3
81349	M52525/16-16	4730-00-755-7609	456	30
			460	27
			461	25
			504	16
			505	28
			509	19
81349	M52525/16-20	4730-00-255-0560	503	17
			504	6
			505	33
81349	M52525/16-24	4730-00-089-3406	456	18
			460	17
			461	15
			503	9
			504	26
			505	3
			509	11
81349	M52525/16-40	4730-00-400-6544	505	6
D8286	M6X22DIN933-8-8A 2P	5305-12-142-0188	90	30
81349	M83248/1-324	5331-01-005-3704	622	7
81349	M83461/1-012	5330-01-046-3300	489	8
			526	13
81349	M83461/1-020	5331-01-107-4950	629	18
81349	M83461/1-219	5331-01-128-3954	629	17
81349	M83461/1-427	5330-01-183-0985	507	19
60703	M874-44EC	2540-00-481-3637	539	3
			540	1
40342	N-11728-BY		282	6
40342	N-13484-C		282	15
06721	N-13556-E	2530-01-134-3667	282	5
06721	N-13807-C	2530-01-134-0902	282	3
06721	N-13808	2530-01-134-0903	282	14
78500	N-14-C	5310-01-237-2615	246	10
40342	N-14370	5365-01-135-2077	282	10
40342	N-20856-D	2530-00-854-4457	282	1
06721	N-30252	4820-00-633-3523	264	27
			266	37
78500	N-35-P	5310-00-177-1258	250	10
78500	N-35P	5310-00-177-1258	251	9
80205	NAS1021-N17	5310-00-325-1900	393	2
80205	NAS1201B10A6A	4010-00-068-5642	97	17
80205	NAS1297-5-10	5306-01-028-2443	425	26
80205	NAS1297-5-13	5306-00-917-0900	591	43
80205	NAS1329A3-130	5310-00-141-3062	418	16
80205	NAS1329A3-80	5310-00-052-7528	418	16

CROSS-REFERENCE INDEXES

PART NUMBER INDEX

CAGEC	PART NUMBER	STOCK NUMBER	FIG.	ITEM
80205	NAS1399B4-4	5320-00-995-8907	369	10
80205	NAS3104-12-12	5306-01-046-0553	458	13
80205	NAS3104-14-14	5306-01-027-4634	458	12
80205	NAS3104-9-10	5306-01-173-3524	505	25
80205	NAS427W34	5315-00-584-9809	341	13
80205	NAS43HT4-24	5365-00-300-4379	341	4
80205	NAS43HT5-32	5365-00-827-6452	341	8
80205	NAS561P6-12	5315-01-058-3487	186	9
78500	NL-25-1-C	5310-01-126-0566	250	1
			251	1
			296	8
			526	9
81348	NN-P-530	5530-00-262-8180	BULK	23
78500	NU-HX7-20-1	5310-01-135-4798	506	18
78500	NU-PT20-18	5310-01-135-9506	228	3
79136	N5002-500MD	5325-00-914-5837	120	13
03038	N72006	5331-00-873-7214	479	3
05779	N72259	5331-00-152-1759	479	7
05779	N72260	5330-01-166-3662	479	5
62983	OFM101-25	4330-00-488-8613	457	23
78500	P-26-C	5365-01-129-6777	217	6
70040	PF297	2940-00-950-8410	457	24
			508	28
73370	PH3519	2940-01-110-2489	204	17
66131	PS-566	5360-01-090-9335	183	17
			184	3
70960	PS40-16-34	4320-01-285-6262	222	9
43334	QH20312N01	3110-00-277-0476	502	5
61112	QS4577	5930-01-069-2776	97	2
1HT81	Q33562	2990-01-280-4284	103	3
1HT81	Q45405	2805-01-279-4630	63	9
1HT81	Q46545	4710-01-279-1490	63	3
19207	RCSK 14159		616	5
19207	RCSK 14160		616	5
19207	RCSK 14162-1		616	4
19207	RCSK 14162-2		616	3
06721	RN-32-G	4810-01-133-1459	KITS	
81348	RR-W-365ATYPE2 2 0X20		BULK	42
70960	RR-1R22-2		506	10
78500	RV-876	5320-01-146-9582	244	5
53551	RV200-6-1	5320-00-616-4350	424	27
53551	RV200-6-2	5320-00-584-1285	403	12
			405	39
			412	19
			416	28
			417	20
			424	21
			427	4
			428	10
			570	1

CAGEC	PART NUMBER	STOCK NUMBER	FIG.	ITEM
53551	RV200-6-3	5320-00-582-3268	404	3
			412	25
			416	23
			424	8
			428	3
			570	7
53551	RV200-6-4	5320-00-582-3276	412	3
53551	RV200-6-5	5320-00-582-3301	424	25
53551	RV250-4-1	5320-00-582-3273	558	4
15434	S-1002-A	4730-00-365-2690	285	1
15434	S-1003-A	5365-00-598-5255	285	3
			285	6
15434	S-1005-A	4730-00-374-4282	117	10
15434	S-1014-1	4730-00-277-8274	64	11
15434	S-1097	4730-01-131-4884	115	18
15434	S-112-A	5305-01-114-0895	12	16
15434	S-1315	5305-00-757-6567	66	14
72210	S-13227	4820-00-848-4361	316	4
72210	S-13353D		316	6
72210	3-13354D	2530-01-131-7855	316	3
78500	S-146-C	5306-01-062-3148	245	8
			246	6
72210	S-15079H	5340-01-124-3189	316	2
15434	S-16255	2815-00-815-0355	118	11
15434	S-176	5305-00-795-9343	3	2
15434	S-176	5305-00-795-9343	118	19
15434	S-190-C	5305-00-795-9345	66	10
15434	S-199-A	5305-01-126-8811	33	29
15434	S-205	5310-01-165-2184	27	5
15434	S-223	5310-00-521-8595	88	1
82484	S-2280	5365-00-946-2231	540	8
15434	S-2286	5305-00-804-6318	51	19
			56	19
			553	2
			555	1
			556	1
78500	S-256-1-C	5306-01-126-1618	527	10
78500	S-265-1-C	5305-01-130-3267	207	15
78500	3-2710-1-C	5305-00-338-5162	207	24
			217	10
			526	26
78500	S-2712-1	5306-01-084-5389	208	10
15434	S-285	5310-00-470-6154	118	23
70960	S-2914-C-1	5305-01-135-3536	207	10
15434	S-315	5315-00-886-1432	284	44
82484	S-3723		540	11
82484	S-3725		540	9
19691	S-4055C-15T	2540-01-201-0968	475	8
77210	S-4281	5310-00-637-4000	316	9
15434	S-604	5310-00-261-7340	118	15
15434	S-608	5310-01-200-1318	3	3

CROSS-REFERENCE INDEXES

PART NUMBER INDEX

CAGEC	PART NUMBER	STOCK NUMBER	FIG.	ITEM
15434	S-622	5310-00-562-6557	54	5
			286	25
15434	S-626	5310-00-562-6558	286	20
15434	S-631	5310-00-562-6560	66	7
78500	S-853	5305-01-142-2792	217	4
			527	14
15434	S-910-B	4730-01-160-3579	37	23
15434	S-965-E	5365-00-404-2934	286	29
15434	S-987	4730-00-193-7080	36	5
78388	SA-2677	5315-01-384-2149	138	12
			626	5
70960	SE-RUR14-2		506	12
78500	SE-RUR25-2	5330-01-135-3376	219	7
			226	10
			229	11
78500	SE-RUR33-4	5330-01-082-1818	227	9
70960	SERUR14-16	2530-01-286-0108	466	5
78500	SERUR33-4	5330-01-082-1818	222	8
78500	SERUR40-3	2520-01-083-4404	221	8
65884	SIA-2017		138	10
17875	SK20420	4820-00-203-3260	630	14
78500	SN-921-1-C	5307-01-277-5053	236	35
78500	SN-925-1C	5307-01-281-7219	236	35
78500	SN1020-1	5307-00-434-1783	230	6
			237	11
70411	SP2346CM	4820-00-808-6905	182	13
70411	SP2483CM,REV C	4820-01-117-7949	81	2
06383	SST2SC	5975-00-074-2072	588	10
15434	S1002A	4730-00-365-2690	117	1
15434	S1003A	5365-00-598-5255	117	5
			117	8
15434	S1004-1	4730-01-151-6316	115	15
15434	S1004A	4730-00-289-5179	117	4
			117	9
15434	S1090	4730-01-201-0717	117	6
15434	S126	5305-00-177-5552	31	11
15434	S130A	5305-00-206-4527	33	30
15434	S147B	5306-00-136-9751	37	13
72210	S14955T	5305-01-132-8387	316	7
			316	10
15434	S152	5305-00-340-3492	42	10
15434	S155	5305-01-028-8869	284	64
15434	S159B	5305-00-493-3959	88	4
15434	S16052	3110-00-516-5289	50	11
15434	S16053	3110-00-156-5453	55	11
15434	S189-C	5305-00-509-8106	129	2
15434	S217	5310-00-650-0187	284	68
15434	S223	5310-00-521-8595	34	1
27737	S2414	3110-00-902-3757	475	10
78500	S255Z	5305-01-357-1656	245	22
			246	19

PART NUMBER INDEX

CAGEC	PART NUMBER	STOCK NUMBER	FIG.	ITEM
15434	S606	5310-00-410-6756	55	26
15434	S622	5310-00-562-6557	37	22
			42	9
15434	S626	5310-00-562-6558	37	9
			284	23
			284	55
15434	S719	5340-00-276-5847	7	7
15434	S911B	4730-00-018-9566	7	14
			44	7
15434	S915A	4730-00-801-8186	37	24
15434	S965E	5365-00-404-2934	284	37
59875	TD97203	5325-00-263-6651	211	46
56501	TG15	5940-00-705-6715	158	11
56501	TG2	5940-00-705-6702	158	38
17875	TRVC-8	2640-01-098-2029	300	5
81348	TR571	4020-00-132-8364	BULK	25
81348	TR605	4020-00-238-7734	BULK	24
91265	TS33-016 70 DURO BUNA N	5330-00-951-3538	129	17
81348	TY IV/CL1/TR VC3	2640-00-050-1235	292	14
81348	TYV/CL2/TR C1/ST ANDARD LENGTH	2640-01-111-5467	292	13
78500	UCB-104	2520-01-133-7876	228	6
77640	UC28254	2530-01-146-8941	301	3
92878	UH68D	4520-00-114-1055	425	1
43334	U1218TAM	3110-00-158-6011	235	1
43334	U121BTAM	3110-00-158-6011	242	1
17875	VS-1072	4820-01-210-8821	292	11
70960	WA-LH8-3	5310-00-031-2673	221	4
			222	3
70960	WA-LM7-2	5310-00-486-5355	506	17
78500	WA-14-C	5310-01-063-2299	245	7
			246	7
78500	WAR-21-3	5310-01-135-6752	228	4
82679	WA132	5310-00-079-9573	144	31
81348	WC596/13-3	5935-01-005-3579	433	6
79370	WPC-4946	2520-01-006-7116	187	19
			193	4
81348	WW-P-471AASBCC	4730-00-187-4202	278	6
81348	WW-P-471AASBUD	4730-00-834-6187	507	13
81348	WW-P-471ACABCA	4730-00-018-9566	41	11
			480	2
			482	19
81348	WW-P-471ACABCB	4730-00-954-1281	214	17
			230	9
			473	9
			597	4
			629	21
81348	WW-P-471ACABCC	4730-01-224-4152	256	22
81349	WW-P-471ACABCD	4730-01-206-6162	186	3
			264	1

CROSS-REFERENCE INDEXES

PART NUMBER INDEX

CAGEC	PART NUMBER	STOCK NUMBER	FIG.	ITEM
			266	28
			457	16
			458	29
81348	WW-P-471ACABCE	4730-00-359-3872	258	6
			478	5
81348	WW-P-471ACBBUE	4730-00-010-3875	207	23
81348	WW-P-471ACBBUH	4730-00-203-0549	7	6
81348	WW-P-471BDQBCDB	4730-01-192-4729	263	22
			264	22
			269	11
			269	29
81349	WW-P-471BDQBUFD	4730-00-203-2836	505	51
74410	XB-196	4010-00-129-3221	BULK	6
			616A	10
13940	XB78323-05001	5325-00-823-5999	586	6
78500	X76-3276-L-12	2530-01-125-6076	251	5
81348	ZZ-I-550E/GP2/14 .00-20/TR179A/OC	2610-00-051-9464	300	3
81348	ZZ-T-381M/GRP3/1 4.00-20/F/TBCC	2610-00-204-2545	300	1
81348	ZZ-V-25/TYPEIV/ CLASS1/TR-VC-2	2640-00-060-3550	292	12
53867	0 403 436 109	2910-01-268-8757	57	11
64678	000 824 08 71	5305-01-264-5886	89	19
			90	5
64678	000125 008413	5310-01-237-5224	92	15
93568	00265-0014	5330-00-246-0309	11	1
78222	0031832K		308	15
D8046	007603014106	5330-12-156-4523	58	2
19954	008913-015		312	16
19954	008913-041		312	30
19954	008913-211		312	26
30780	0101-4-8	4730-00-004-0732	620	19
78222	0130491K		309	5
78222	0170121		309	13
77640	020196	5306-00-419-5876	317	15
12195	02021050	2610-01-214-1344	300	6
77640	020251	5305-01-136-1619	306	39
77640	020252	5305-01-138-5115	306	5
77640	021200	5305-00-899-2049	305	3
77640	021318	5305-01-138-6295	306	2
77640	021333	5305-01-129-6842	306	4
77640	021336	5305-01-136-1686	306	41
77640	025121	5310-01-088-5851	306	1
07367	025122	5310-01-133-9186	306	3
07367	025124	5310-01-129-0193	306	37
14153	02764	5310-01-377-9775	237	4
			293	10
77640	028271	5310-00-861-2316	317	3
77640	028272	5365-00-876-6862	317	5
77640	028426	5310-01-135-6758	306	20

CROSS-REFERENCE INDEXES

PART NUMBER INDEX

CAGEC	PART NUMBER	STOCK NUMBER	FIG.	ITEM
77640	028430	3120-01-128-1565	306	32
77640	028433	5310-01-135-6754	305	18
77640	028434	5310-01-135-6755	305	9
77640	028435	5310-01-161-6131	305	8
77640	028445	5310-01-141-8465	306	35
77640	032156	5977-00-578-6495	150	12
77640	032200-17	5331-00-733-9765	317	7
77640	032205	5330-00-733-9766	317	2
77640	032229	5331-01-135-4807	306	10
77640	032249	5330-00-798-4635	317	6
77640	032272	5330-00-798-4637	317	10
77640	032361	5331-00-195-5757	317	20
77640	032460	5330-01-143-6486	317	13
77640	032536	5330-01-135-4066	306	15
77640	032552	5330-01-135-4808	306	16
77640	032570	5330-01-135-4067	306	17
77640	032571	5330-01-135-4809	306	18
77640	032577-A1	5330-01-143-6013	306	34
77640	032579	5330-01-143-4186	306	38
77640	032586	5331-01-143-2780	305	15
77640	032590	5331-01-135-4069	306	31
77640	032591	5330-01-143-4185	305	20
77640	032615	5330-01-135-4068	306	30
77640	032616	5330-01-149-7229	306	8
77640	032634-A1	5330-01-135-0682	305	17
77640	032791-A1	5330-01-135-4064	305	7
IHT81	033545	4710-01-279-1493	63	17
77640	036141	5340-00-443-6132	305	23
78222	0400511		309	4
77640	040124	4820-01-134-6619	306	27
77640	040125	2530-01-134-3669	306	28
77640	040127	3110-01-137-9725	306	14
8X715	05240 KIT		619	19
32828	05830-001	3020-00-533-3398	495	9
78222	0591304		309	7
30780	0603-10-4	4730-01-116-1658	629	5
83299	0612596-00	4730-00-908-3193	589	11
			590	5
77640	062005	2530-00-225-0680	305	4
65282	06642-0000	4720-01-003-6706	BULK	11
77640	067026	3110-00-649-9498	306	33
77640	071016		305	10
77640	071018		305	13
78222	0751921		309	12
78222	0752021		309	3
14371	08-203713	5305-01-127-7087	560	4
78222	0856121		308	5
78222	0856311K		309	9
78222	0856401		309	8
78222	0857112		309	6
78222	0924734K		308	21

CROSS-REFERENCE INDEXES

PART NUMBER INDEX

CAGEC	PART NUMBER	STOCK NUMBER	FIG.	ITEM
5T151	1 190 200 000	5365-01-336-8908	90	7
5T151	1 200 102 624	5365-01-336-5220	91	9
53867	1 410 100 002	5310-01-301-7807	59	28
53867	1 410 107 005	5330-01-238-8316	58	11
53867	1 410 149 001	5310-01-301-1802	59	5
5T151	1 410 151 002	5310-01-336-8865	59	3
53867	1 410 200 019	5365-01-303-1612	59	4
5T151	1 410 210 014	5331-01-336-9559	89	14
53867	1 410 210 030	5330-01-302-5738	58	19
53867	1 410 210 041	5331-01-303-1635	59	29
53867	1 410 210 501	5331-01-301-7867	59	24
53867	1 410 283 005	5330-01-300-6898	58	25
53867	1 410 290 005	5365-01-300-7149	59	21
53867	1 410 422 031	2910-01-300-4271	59	19
53867	1 410 505 012	5365-01-300-6885	58	3
53867	1 410 505 015	5340-01-300-7153	59	18
53867	1 410 505 023	5365-01-300-7021	59	23
53867	1 410 520 007	5340-01-300-7154	59	15
53867	1 410 900 015	3110-01-300-6796	58	5
53867	1 410 910 005	3110-01-300-6804	58	21
5T151	1 411 030 000	5365-01-336-5224	90	21
53867	1 411 030 134	5365-01-300-7158	59	25
53867	1 411 030 135	5365-01-300-7159	59	25
53867	1 411 030 136	5365-01-300-7160	59	25
53867	1 411 030 137	5365-01-301-0554	59	25
53867	1 411 030 138	5365-01-300-7161	59	25
53867	1 411 030 139	5365-01-300-7162	59	25
53867	1 411 030 140	5365-01-300-7163	59	25
53867	1 411 030 141	5365-01-300-7164	59	25
53867	1 411 030 142	5365-01-300-7165	59	25
53867	1 411 030 143	5365-01-300-7166	59	25
53867	1 411 030 144	5365-01-300-7167	59	25
53867	1 411 030 145	5365-01-300-7168	59	25
53867	1 411 030 146	5365-01-300-7169	59	25
58367	1 411 030 147	5365-01-303-0937	59	25
53867	1 411 030 148	5365-01-304-9530	59	25
53867	1 411 030 149	5365-01-302-9953	59	25
53867	1 411 030 150	5365-01-303-0938	59	25
53867	1 411 030 151	5365-01-304-1802	59	25
53867	1 411 030 152	5365-01-300-7170	59	25
53867	1 411 030 153	5365-01-300-7171	59	25
53867	1 411 030 154	5365-01-300-7172	59	25
53867	1 411 030 155	5365-01-300-7173	59	25
53867	1 411 030 156	5365-01-300-7174	59	25
53867	1 411 030 157	5365-01-300-7175	59	25
53867	1 411 030 158	5365-01-300-7176	59	25
53867	1 411 030 159	5365-01-300-7177	59	25
53867	1 411 030 160	5365-01-300-7178	59	25
53867	1 411 030 161	5365-01-300-7179	59	25
53867	1 411 030 162	5365-01-300-7180	59	25
53867	1 411 030 163	5365-01-300-7181	59	25

CROSS-REFERENCE INDEXES

PART NUMBER INDEX

CAGEC	PART NUMBER	STOCK NUMBER	FIG.	ITEM
53867	1 411 030 164	5365-01-300-7182	59	25
53867	1 411 030 165	5365-01-302-5848	59	25
53867	1 411 030 166	5365-01-300-7183	59	25
53867	1 411 030 167	5365-01-300-7184	59	25'
53867	1 411 030 168	5365-01-300-7185	59	25
53867	1 411,030 169	5365-01-300-7186	59	25
53867	1 411 030 170	5365-01-300-7187	59	25
53867	1 411 030 171	5365-01-301-4006	59	25
53867	1 411 030 172	5365-01-300-7188	59	25
53867	1 411 030 173	5365-01-301-4007	59	25
5T151	1 411 032 004		59	13
53867	1 413 105 008	5315-01-335-9941	59	7
53867	1 413 106 002	5315-01-301-3892	58	4
53867	1 413 300 014	5310-01-301-1798	58	7
5T151	1 413 300 023	5310-01-336-8721	59	2
53867	1 413 356 040	4820-01-300-4257	59	1
53867	1 413 457 001	4730-01-300-8819	58	10
53867	1 413 500 006	5307-01-301-7815	59	6
5T151	1 414 601 004		59	22
53867	1 414 613 002	5360-01-300-6889	59	27
53867	1 414 618 030	5360-01-300-6888	59	17
53867	1 415 511 047	3110-01-300-7136	58	22
53867	1 415 616 005	5340-01-300-7203	58	18
53867	1 415 800 014	3120-01-300-7135	58	13
5T151	1 416 016 013	5340-01-336-3889	59	11
5T151	1 416 116 342		58	9
53867	1 417 413 047	4820-01-300-4051	58	1)
5T151	1 418 415 082	2910-01-338-2335	59	16
5T151	1 418 512 225	2910-01-339-8598	59	26
5T151	1 420 026 004		91	25
5T151	1 420 100 602	5365-01-339-3839	92	23
5T151	1 420 101 023	5365-01-336-5221	91	6
5T151	1 420 113 004	5330-01-336-3150	92	3
5T151	1 420 210 022		89	22
5T151	1 420 301 002	3120-12-198-1758	90	24
53867	1 420 500 025	3010-01-337-8937	91	24
5T151	1 420 505 036	5340-01-335-9996	89	11
5T151	1 420 505 057	5340-01-336-6784	92	24
5T151	1 420 505 062	5340-01-339-0839	59	10
5T151	1 420 555 000	5340-01-336-6704	90	25
5T151	1 420 560 004	5340-01-336-2588	92	4
5T151	1 421 015 082		92	27
5T151	1 421 036 010	5365-01-336-8916	90	26
5T151	1 421 036 012	5340-01-339-0868	90	26
5T151	1 421 036 013	5340-01-341-6655	90	26
5T151	1 421 331 009	5310-01-336-6687	91	20
5T151	1 421 389 008	5340-01-336-6761	92	11
53867	1 421 933 132	3040-01-336-8293	91	14
5T151	1 422 002 216	3040-01-338-2546	92	20
5T151	1 422 033 068	5340-01-336-0016	91	16
53867	1 422 120 023	3040-01-338-2545	90	22

PART NUMBER INDEX

CAGEC	PART NUMBER	STOCK NUMBER	FIG.	ITEM
53867	1 422 130 011	3040-01-337-4368	91	17
			92	22
53867	1 423 103 016	3040-01-336-8168	90	27
53867	1 423 124 108	5306-01-301-4881	92	18
53867	1 423 300 045	5310-01-301-0455	91	12
5T151	1 423 314 003	5310-01-351-7542	89	12
5T151	1 423 412 025	5305-01-336-0026	90	12
53867	1 423 414 021	5305-01-301-5034	92	1
5T151	1 423 414 031	5305-01-335-9968	90	11
5T151	1 423 415 003	5305-01-335-9967	90	14
5T151	1 423 421 007		92	12
53867	1 423 450 056	5305-01-301-5112	92	17
5T151	1 423 450 900		91	28
5T151	1 423 452 000	5305-01-336-0008	92	6
5T151	1 423 463 007	5365-01-336-5187	58	26
5T151	1 423 521 014	5315-01-335-9940	91	18
5T151	1 424 610 053	5360-01-335-9947	59	9
5T151	1 424 618 035	5360-01-335-9951	91	3
5T151	1 424 619 179	5360-01-339-0684	89	7
53867	1 425 100 323	2910-01-337-4142	92	10
53867	1 425 703 013	3040-01-300-5309	92	14
5T151	1 427 133 313	5365-01-336-8895	89	1
53867	1 428 199 009	2910-01-337-2984	91	23
5T151	1 688 130 135	4910-01-338-6241	631	5
5T151	1 810 210 147	5331-01-336-6717	92	25
5T151	1 900 023 002	5315-01-336-6689	92	21
53867	1 900 023 011	5315-01-301-7912	58	8
95019	1-P-455		529	6
30780	1-4X1-BFGS	4730-00-226-8874	132	17
53867	1-410-137-021	5330-01-301-5995	58	20
5T151	1-422-010-015	5340-01-336-0046	92	5
53867	1-423-462-099	5365-01-182-5468	92	2
30780	1/4CR-B	4730-01-181-5777	584	7
23382	1AB2568	5315-00-532-9388	7	11
11083	1F7960	5310-00-595-6153	118	28
79500	1JH4350REVCPC	5331-00-514-4804	88	12
81343	10 070110C	4730-00-976-0981	629	5
64203	10-11641A	5325-01-122-4661	522	5
			523	3
			523	26
30780	10-12FBTXS	4730-00-445-4418	629	5
64203	10-12295A	5325-01-123-7063	523	9
			523	32
64203	10-17117-19	4730-01-126-0265	519	32
			521	11
64203	10-17118B-18		519	7
64203	10-17118B-20	4730-01-131-6099	519	44
64203	10-27728	5305-01-136-1688	522	4
			523	2
			523	25
64203	10-30326A	5330-01-122-5392	522	18

CROSS-REFERENCE INDEXES

PART NUMBER INDEX

CAGEC	PART NUMBER	STOCK NUMBER	FIG.	ITEM
64203	10-31091B	5330-01-122-0864	523	4
			523	27
64203	10-31103A	5330-01-122-5637	523	15
			523	38
64203	10-40133-A	5342-01-122-4489	515	5
64203	10-40178A-28	3040-01-138-5391	524	5
64203	10-40214A	5315-01-122-2014	516	16
64203	10-40397A	3130-01-145-3943	516	6
64203	10-40436-A	3040-01-198-2652	516	19
64203	10-40446-A		516	24
64203	10-40447-A	5315-01-122-8482	517	13
64203	10-40450-A	5315-01-126-9412	516	29
77820	10-40457-12S	5975-00-522-7125	134	53
64203	10-40723-A	5340-01-155-9468	519	38
64203	10-40724-A	3040-01-146-5952	517	6
64203	10-40726-A	5315-01-126-9485	522	1
64203	10-40730-A	3040-01-141-5032	518	10
64203	10-40771A		519	2
64203	10-40772A		519	5
64203	10-40897-A		522	7
64203	10-40953-A	5310-01-122-5993	522	15
64203	10-41045A	2590-01-126-0093	523	6
			523	29
64203	10-41058A	4730-00-052-8502	519	10
64203	10-41247	5315-01-126-9484	522	2
64203	10-41311A	5330-01-122-4609	523	7
			523	30
64203	10-41317B	3040-01-142-3991	523	8
64203	10-41317B	3040-01-142-3991	523	31
64203	10-41318B	4310-01-125-9714	523	13
			523	36
64203	10-41320A	5330-01-122-5636	522	9
64203	10-41321A	2590-01-125-9770	522	8
64203	10-41360C	3040-01-125-9828	522	16
64203	10-41361C		522	10
64203	10-41535	9905-01-173-2183	570	8
64203	10-41540-C	4730-01-121-7718	522	19
64203	10-41638	9905-01-186-7948	570	8
64203	10-42090-B	5342-01-122-4491	516	25
64203	10-42163-A		516	26
64203	10-42253	5305-01-135-4839	515	2
			516	7
			519	6
64203	10-42256-A	5305-01-122-4634	521	3
64203	10-42258-A	3120-01-128-5222	516	13
64203	10-42259-A	5305-01-145-8385	516	11
64203	10-42264-A	5360-01-127-0858	517	8
64203	10-42265-A	5315-01-122-2033	518	8
64203	10-42266-A	5340-01-122-4632	518	2
64203	10-42276-A	5305-01-138-7624	516	30
64203	10-42284A	5305-01-139-6484	519	15

CROSS-REFERENCE INDEXES

PART NUMBER INDEX

CAGEC	PART NUMBER	STOCK NUMBER	FIG.	ITEM
64203	10-42286A	2520-01-125-9837	524	1
64203	10-42288A	2520-01-126-0443	524	6
64203	10-42289A	4730-01-126-3845	519	22
64203	10-42296-A		517	9
64203	10-42297-A	5360-01-122-0634	517	15
64203	10-42322A	5310-01-123-0905	523	12
			523	35
64203	10-42351	5340-01-123-8324	521	8
64203	10-42417A	5305-01-122-5639	519	23
64203	10-42426	4730-01-126-0266	521	4
64203	10-42427-D	2590-01-143-1455	516	15
64203	10-42451-A	5305-01-122-4703	518	6
64203	10-42460-A	5340-01-205-9057	517	11
64203	10-42466-A		519	28
64203	10-42469-A	5305-01-122-5474	516	17
64203	10-42473-A	5315-01-122-4490	517	4
64203	10-42481	5305-01-122-4635	521	6
64203	10-42505A	4730-01-131-5944	519	14
64203	10-42507A	4730-01-132-4852	519	9
64203	10-42508A	4820-01-134-3457	519	11
64203	10-42509A	5305-01-122-5638	519	36
64203	10-42512	5342-01-123-8045	521	7
64203	10-42513	4730-01-125-9681	521	19
64203	10-42513-A	4730-01-125-9681	523	16
64203	10-42516-A	4730-01-135-3009	523	20
64203	10-42517-A	4730-01-125-9680	523	19
64203	10-42518A	4730-01-125-9679	523	18
64203	10-42521	7690-01-173-9197	570	10
64203	10-42599A	4730-01-132-8700	519	20
77820	10-42622-235	5935-00-614-3959	158	2
64203	10-42651A	5365-01-123-7082	516	5
64203	10-42685-A	5315-01-136-7307	522	21
64203	10-42723-A	5305-01-122-4702	516	22
64203	10-9115A		522	11
78500	10X-1250	5306-01-132-8271	244	3
78500	10X-507-C	5306-01-129-0327	244	14
15434	100099	5331-00-809-2667	52	6
24491	10015	4730-00-042-8988	481	7
15434	100478	5331-00-081-9289	51	4
			56	3
15434	100670		21	3
07295	100740	4730-00-900-3296	582	5
			593	16
15434	100764	5330-00-506-4866	50	22
06721	100880	4820-01-134-3788	282	2
85105	101301104	5310-01-260-4937	372	4
40342	101374		282	13
18876	10161958	2520-00-232-1938	530	11
18876	10164117	4730-00-251-6827	505	50
60602	10166	5315-01-284-9812	101	1
			184	2

PART NUMBER INDEX

CAGEC	PART NUMBER	STOCK NUMBER	FIG.	ITEM
15434	101662	2815-00-327-6439	33	26
14205	10217-3736	2520-01-305-0457	315	1
06853	102352	2530-01-112-6435	248	2
79470	103-60047	4730-01-233-8998	290	16
6N171	1039234	5305-01-280-9367	110	21
06853	103998	4820-01-251-1699	278	12
21450	104130 XD		318	11
63208	10446-5	5342-01-288-2167	599	1
06853	104689	4820-01-114-7539	283	1
89346	104842R92	2510-00-741-1110	346	17
56161	10505055	5310-01-167-1964	58	23
15434	105199	2815-00-829-5227	42	6
15434	105375	9905-00-733-7622	555	2
15434	105574	5305-00-463-0429	37	4
15434	105953	5306-00-804-2468	7	20
91340	10608E44S	5330-00-020-5375	BULK	30
91340	10608E44S BULK		404	20
			408	13
			410	6
			412	6
			412	12
			412	17
			412	27
			418	22
			418	26
91340	10608E44S-4		403	10
06853	106400	2530-01-287-4529	621	5
15434	106549	5305-01-210-1608	118	14
21450	106912	5320-01-194-5034	411	4
52304	10709	2530-00-693-1029	293	2
			295	16
60602	10718	3040-01-284-6230	101	23
15434	107738	2815-00-505-5119	23	7
15434	108330	5310-00-486-2505	284	60
15434	108392	5365-01-197-3179	41	7
24617	108686	4730-00-246-9217	8	18
19207	10871206	5340-01-120-2891	322	9
19207	10871207	5340-00-231-0278	321	18
			381	12
19207	10871208	5340-00-409-4019	321	30
			381	5
19207	10871210	5340-00-409-4018	321	15
			381	14
19207	10871251	5306-00-869-6549	347	3
19207	10871253	5306-00-175-4967	348	37
19207	10871273		346	9
19207	10871282	2510-01-131-0112	387	25
19207	10871283	2510-01-131-0113	387	25
19207	10871284	2510-01-083-1156	387	24
19207	10871285	2510-01-082-3788	387	24
19207	10871287	2540-01-082-7457	387	37

PART NUMBER INDEX

CAGEC	PART NUMBER	STOCK NUMBER	FIG.	ITEM
19207	10871289	2510-00-408-2439	384	14
19207	10871290	2510-00-408-2453	384	14
19207	10871293	2540-01-082-3592	387	11
19207	10871294	2510-01-129-6074	387	11
19207	10871295	2510-00-409-4020	384	13
19207	10871296	2510-00-408-2452	384	13
19207	10871297	2510-00-408-2448	384	15
19207	10871302-1	2510-01-138-1157	532	6
			535	6
19207	10871356-1	2510-01-082-7508	384	16
19207	10871356-2	2510-01-210-6230	384	16
19207	10871456	2540-00-417-2583	376	10
19207	10871458-1	9520-01-125-4082	376	12
19207	10871462	2540-00-472-1696	376	4
19204	10872111	5340-01-157-3752	378	9
19207	10872159	5310-00-180-8544	321	25
19207	10872414	9520-01-163-4902	409	32
19207	10875107-7	5330-01-119-5801	473	34
19207	10875529	6810-00-249-9354	151	23
19207	10876131	5330-00-522-1174	489	43
19207	10876132	5330-00-826-5203	484	15
19207	10876133	5330-00-826-5202	484	5
19207	10876134	3020-00-522-1175	489	31
19207	10876135	3020-00-522-1177	489	45
19207	10876137-1	4720-01-235-9627	481	10
19207	10876139	4720-00-826-5607	480	30
19207	10876142	4720-00-826-5606	480	33
19207	10876144	4720-00-826-5610	480	25
			482	3
19207	10876145	4720-00-522-1449	482	5
19207	10876146-1	4720-01-221-1448	481	1
19207	10876153	4310-00-504-8923	485	20
			491	13
19207	10876157	3110-00-504-8929	489	48
19207	10876158	3120-00-504-8930	486	7
			487	2
19207	10876159	3120-00-826-5630	485	9
			491	7
19207	10876163	3040-00-504-9038	484	7
19207	10876164	3040-00-504-9037	489	26
19207	10876166	5310-00-753-8215	484	16
19207	10876173	2590-01-120-8506	487	1
19207	10876174	2590-00-235-4400	489	52
19207	10876182	2590-00-086-7460	490	4
19207	10876186		484	26
19207	10876190	2590-00-086-7459	487	21
19207	10876197	2590-00-974-9670	488	6
19207	10876198	3020-00-933-9585	484	1
19207	10876202	2540-00-417-2732	492	9
19207	10876205	2590-00-179-7053	490	3
19207	10876213	3040-00-972-2638	485	4

CROSS-REFERENCE INDEXES

PART NUMBER INDEX

CAGEC	PART NUMBER	STOCK NUMBER	FIG.	ITEM
19207	10876226	5340-00-414-6561	484	14
19207	10876227	2590-00-179-5581	487	8
19207	10876231	2590-00-221-4824	487	14
19207	10876232	3040-00-177-9266	484	21
19207	10876240	3950-00-351-7834	398	3
19207	10876243	2590-01-131-0126	398	12
19207	10876244	2590-00-231-7418	398	6
19207	10876246	3040-00-421-3961	489	29
19207	10876247	5340-01-101-0007	489	38
19207	10876249	2540-00-418-0603	475	6
19207	10876250	3940-00-414-6533	475	3
19207	10876252	4730-00-421-3962	489	40
19207	10876253	5340-00-471-8631	489	15
19207	10876254	2590-01-085-3596	398	1
19207	10876261-1	5340-01-205-2503	492	3
19207	10876266	5340-01-082-3611	492	11
19207	10876279	3020-01-310-5493	486	1
19207	10876292	3040-00-933-9579	486	10
19207	10876307	3040-01-101-0085	491	3
19207	10876309	3040-01-101-0096	491	17
19207	10876310	2590-00-177-9196	485	10
			491	8
19207	10876311		485	5
19207	10876312	3040-01-137-6264	485	13
19207	10876324	5510-01-132-4879	393	13
19207	10876325	5510-01-132-7138	393	11
19207	10876326	5510-01-132-7139	393	11
19207	10876329	2590-01-131-0123	393	7
19207	10876331	5510-01-132-8746	393	9
19207	10876332	5306-00-151-5726	393	8
19207	10876333	5306-00-419-5878	393	8
19207	10876355	3040-01-101-0086	489	24
19207	10876356	3040-00-177-7830	489	51
19207	10876357	3040-00-421-3960	489	55
19207	10876359	2590-01-146-5257	487	15
19207	10876360	3020-00-972-2641	484	19
19207	10876361	3040-00-972-2640	484	29
19207	10876364	3040-00-504-8913	489	32
19207	10876370	3040-00-118-8694	487	17
19207	10876382	2520-01-137-6294	484	4
19207	10876383	3020-00-446-8616	490	13
19207	10876391-1	5342-01-082-3612	492	10
19207	10876395	2590-00-351-7835	398	2
19207	10876396	5340-00-444-2107	393	4
19207	10876397	5340-00-444-2108	393	4
19207	10876401	2540-00-972-2642	531	7
19207	10876402	2540-00-063-4730	531	6
19207	10876412	2590-01-114-7741	487	11
19207	10876416	2590-00-351-7836	398	11
19207	10876418	5340-00-299-0069	398	9
19207	10876423	3020-00-419-5854	475	11

CROSS-REFERENCE INDEXES

PART NUMBER INDEX

CAGEC	PART NUMBER	STOCK NUMBER	FIG.	ITEM
19207	10876428	4710-00-933-9587	481	6
19207	10876430	4710-00-089-6193	481	2
19207	10876433	2540-00-860-2355	531	2
19207	10876437	4710-00-857-2782	487	22
19207	10876453	2590-00-972-2632	491	12
19207	10876454	5310-00-445-7238	485	17
			491	10
19207	10876461	4730-00-238-5594	485	7
			491	5
19207	10876502	2510-00-421-3956	395	8
19207	10876503	5340-00-446-8732	395	13
19207	10876506	5310-00-419-3082	485	16
			491	14
19207	10876527		170	17
19207	10876543		170	17
19207	10876552	5340-00-237-3706	492	17
19207	10876554	2590-00-417-2725	476	3
			477	7
19207	10876563	2590-00-045-4206	488	11
19207	10876565	2540-00-860-2357	531	1
19207	10876566	2540-00-860-2356	531	3
19207	10876570	2590-00-972-2634	485	19
19207	10876579	2510-01-081-9226	396	14
19207	10876599	2540-01-085-3589	342	1
19207	10876599-1		342	1
19207	10882221	5340-00-495-0236	414	30
19207	10882484	2540-00-809-7796	406	45
19207	10883101	2590-00-408-4618	365	3
19207	10883102	2510-00-408-4634	365	3
19207	10883106-1	5340-00-407-5087	142	11
19207	10883109		223	13
19207	10883113	2510-00-407-5084	326	12
19207	10883114	5342-01-108-7265	328	3
19207	10883115	5340-01-108-9109	328	6
19207	10883116-1	5340-01-082-2517	328	9
19207	10883118	5305-00-140-4765	326	3
19207	10883119	5305-00-740-9507	326	3
19207	10883130	5342-00-999-8591	243	15
19207	10883157	5305-00-139-7074	346	18
19207	10883158	5305-01-089-3070	346	1
19207	10883167		224	15
19207	10883168		224	15
19207	10883199	5305-00-139-7075	348	2
19207	10883200	5305-00-139-7072	348	2
19207	10883203	5306-00-741-1113	347	13
19207	10883319	5365-00-496-9706	313	10
19207	10883328	5342-00-179-7123	255	20
19207	10883329	5340-00-448-4073	262	19
19207	10883331	5340-00-157-7938	255	19
			262	9
19207	10883335	2510-00-408-4642	343	15

CROSS-REFERENCE INDEXES

PART NUMBER INDEX

CAGEC	PART NUMBER	STOCK NUMBER	FIG.	ITEM
19207	10883351	9905-00-933-4632	566	10
19207	10883360	5340-00-419-3080	330	4
19207	10883437	2510-00-127-4768	262	2
19207	10883438	5340-00-419-9487	255	15
			262	3
19207	10883445	3020-01-089-7333	486	12
19207	10883450	2510-00-222-8973	142	21
19207	10891476	5365-00-159-4668	414	31
19207	10896277	2590-00-421-1595	100	29
19207	10896285-1	5999-01-328-0524	156	13
19207	10896477		581	4
		5340-01-206-8586	585	14
19207	10896514	7690-00-030-6615	580	11
			584	1
19207	10896515	7690-01-032-5639	592	28
			594	31
19207	10896651	9905-01-032-7002	580	9
			584	17
19207	10896651		593	21
		9905-01-032-7002	594	38
19207	10896726	5306-00-182-9267	237	8
			293	4
19207	10896774	5340-00-419-5881	419	35
19207	10896779-1	2510-01-161-7631	415	2
19207	10896780		415	3
19207	10896781		415	6
19207	10896789	5315-01-105-3318	407	7
			415	9
19207	10896799-1	2510-00-809-8046	407	12
			415	4
19207	10896812	5410-01-161-7703	415	13
19207	10896813	5410-01-128-5529	407	8
			415	11
19207	10896815	2510-01-179-4083	407	5
			415	25
19207	10896816-1		407	9
19207	10896841	2540-01-101-8453	415	19
19207	10896845	2540-01-046-9402	415	21
19207	10896856		415	16
19207	10896867		415	15
19207	10896868	5310-01-185-4675	415	18
19207	10896870-1	5680-01-183-9754	415	23
19207	10896870-2	5680-01-185-4944	415	22
19207	10896881		407	13
19207	10896883		407	10
19207	10896885-1	5680-01-177-1526	407	15
19207	10896885-2	5680-01-177-1525	407	14
19207	10896925-2	5340-01-158-7095	378	7
19207	10896925-3	5342-01-158-7096	378	7
19207	10897028-1	2520-01-221-8893	410	15
19207	10899201	2510-00-860-0523	609	18

PART NUMBER INDEX

CAGEC	PART NUMBER	STOCK NUMBER	FIG.	ITEM
19207	10899202	2510-00-860-0517	609	18
19207	10899203	2540-01-173-9176	609	28
19207	10899204	2540-01-173-9175	609	23
19207	10899205	2540-01-173-9173	609	23
19207	10899206	2540-01-173-9172	609	23
19207	10899207	2540-01-173-9174	609	24
19207	10899208		609	2
19207	10899217		609	8
19207	10S99218		609	8
19207	10899219	2540-01-205-2759	609	17
19207	10899220	5340-01-185-7768	609	12
19207	10899221	2540-01-173-9160	609	14
19207	10899222	2540-00-860-0519	609	2
19207	10899223	2540-01-173-9147	609	7
19207	10899232	3020-01-214-1530	KITS	
19207	10899255	5340-00-407-6767	83	5
19207	10899256	2510-00-407-6768	83	4
19207	10899258	2510-00-408-4631	83	6
19207	10899259	2510-00-408-4632	83	6
19207	10899366	5315-00-052-0110	487	23
19207	10899372	2510-00-417-2713	83	4
19207	10899416	5340-00-158-3774	399	9
19207	10899434	5340-00-158-4078	161	24
19207	10899976	5315-00-446-8740	490	18
19207	10899978	2590-00-231-7451	487	30
19207	10899979	3040-00-234-3250	487	'28
19207	10899980	3040-00-150-7145	487	31
19207	10899983	2590-00-231-7452	487	24
19207	10899986	2590-00-134-1121	490	5
19207	10899989	5315-00-244-1340	490	15
19207	10899991	5315-00-421-3931	485	23
19207	10899994	3120-00-937-1164	484	23
19207	10899995	5330-00-182-3489	483	2
19207	10899996	3010-00-069-3047	484	22
96151	109-1094-106	4320-01-137-6293	467	3
19207	10900003	9905-00-197-1746	566	6
19207	10900004	5365-00-405-4378	487	12
19207	10900005	5365-00-422-1160	487	13
19207	10900013	5315-01-106-4036	485	2
19207	10900017	2590-00-220-5104	488	3
19207	10900018	3040-00-255-5700	488	10
19207	10900021	2590-00-045-4205	488	12
19207	10900021	9905-00-197-1744	566	8
19207	10900027	9905-00-197-1742	566	4
19207	10900051-2	9905-01-185-5762	492	2
19207	10900055	5365-01-103-6006	476	6
19207	10900056	3020-01-115-0619	486	9
			487	4
19207	10900080	4820-00-134-1122	480	18
			482	7
19207	10900081		480	24

CROSS-REFERENCE INDEXES

PART NUMBER INDEX

CAGEC	PART NUMBER	STOCK NUMBER	FIG.	ITEM
19207	10900082		480	19
19207	10900083	3040-00-414-6535	475	5
19207	10900086	5340-00-421-7235	395	4
19207	10900089	3040-00-184-9720	483	1
19207	10900090	5330-00-064-3691	489	39
19207	10900091	2590-01-131-1940	480	21
			482	10
19207	10900101	5365-00-185-7835	489	50
19207	10900102	5310-00-512-5213	489	18
19207	10900106	4730-00-231-7450	478	3
19207	10900107	5360-01-131-2063	480	20
			482	11
19207	10900108	4730-00-439-6028	480	1
19207	10900112	2590-00-443-4847	480	3
19207	10900113	5310-01-131-2085	480	23
			482	8
19207	10900129	5340-01-082-2516	396	4
19207	10900130	5306-00-231-0211	393	10
19207	10900133	5340-01-131-0125	396	10
19207	10900134	5306-00-351-7842	393	6
19207	10900137	5340-01-131-0124	393	12
19207	10900151	5310-00-470-2107	485	15
			491	15
19207	10900171-1	5307-01-107-3675	396	15
19207	10900172	5355-00-191-1029	493	2
			505	18
19207	10900175	5340-01-083-1107	396	5
19207	10900181	9905-00-933-4635	566	15
19207	10900182	9905-00-197-1743	566	12
19207	10900183	9905-00-197-1741	566	16
19207	10900184	9905-01-172-2394	566	14
19207	10900185	4820-00-461-4216	478	7
19207	10900186	9905-00-197-1748	566	13
19207	10900189	5360-00-136-4759	484	24
19207	10900202	3120-01-101-0076	484	17
19207	10900203	3040-00-143-6390	487	34
			490	1
19207	10900204	5365-01-435-4806	484	25
19207	10900212	5340-01-101-0090	484	18
19207	10900232	5310-01-101-0077	484	11
19207	10900237		475	1
19207	10900247	5365-00-177-9194	490	8
19207	10900249	2540-00-860-2358	531	8
19207	10900266	4730-00-456-9831	481	8
			482	18
19207	10900278	2590-00-417-2787	476	12
			477	8
19207	10900282	5340-01-134-6535	476	7
19207	10900300	5330-00-523-4235	485	8
			491	6
19207	10900304	5340-00-523-4305	485	18

CROSS-REFERENCE INDEXES

PART NUMBER INDEX

CAGEC	PART NUMBER	STOCK NUMBER	FIG.	ITEM
			491	11
19207	10900396	5330-00-419-5872	234	9
			241	9
19207	10900408	5310-00-355-5453	233	11
			240	11
19207	10900409	5310-00-483-2266	232	2
			239	2
19207	10900472	9390-00-166-0254	492	12
19207	10905840	5975-00-345-8055	314	5
19207	10906322	2590-00-421-3942	481	5
19207	10906350	2510-00-179-5708	363	7
19207	10910268	5330-00-191-1161	492	14
19207	10910298	5340-00-419-3084	480	29
			481	3
19207	10910299	5340-00-421-3992	480	31
			481	12
19207	10910303	2590-00-418-0668	497	7
19207	10910304	2590-00-418-0673	497	6
19207	10910347	2590-00-101-5594	487	27
19207	10910956-1	5342-01-230-6702	492	19
19207	10911036-1	2590-00-630-1567	416	6
19207	10911036-2	2540-00-918-4184	416	6
19207	10911102	3120-00-951-1850	485	12
19207	10913149	4730-00-087-0313	480	34
19207	10913152		482	16
19207	10915159	9390-00-405-0215	417	27
19207	10915207	5315-00-495-6497	395	9
19207	10919646	4520-01-169-8680	425	5
19207	10919647	5365-00-389-0317	425	11
19207	10919648	2540-00-378-9049	425	9
19207	10919652	5340-00-423-4080	425	6
19207	10921633	2940-00-135-6537	479	1
19207	10921722-2-52		415	14
19207	10921898-1	2530-01-136-1098	149	5
19207	10922156-1	4820-01-170-5055	596	28
			598	51
19207	10923475	5331-00-626-8281	425	12
19207	10923515	4820-01-101-0080	549	42
			582	10
19207	10924753	4820-00-832-8077	212	4
			279	20
19207	10924916		399	3
19207	10929868	2920-00-909-2483	125	9
			126	11
15434	109319	2815-00-406-8936	33	5
19207	10931962	5340-00-036-0236	81	15
19207	10932085	5330-00-171-2776	206	4
19207	10932088		206	2
19207	10932088-1	8145-01-134-6538	577	1
19207	10932091	7690-00-171-2761	569	3
19207	10937462	2510-00-930-2714	383	9

CROSS-REFERENCE INDEXES

PART NUMBER INDEX

CAGEC	PART NUMBER	STOCK NUMBER	FIG.	ITEM
19207	10937495	2510-01-131-0120	385	43
19207	10937496	2510-01-131-0121	385	38
19207	10937519	5995-00-351-7868	431	1
19207	10937530	9905-00-407-7000	570	5
19207	10937532	6150-00-222-8943	435	6
19207	10937533	5995-00-231-7454	435	1
19207	10937537	5925-00-497-9661	441	9
19207	10937541	5930-00-116-0531	441	5
19207	10937542	5945-00-496-9708	441	4
19207	10937550	5340-01-108-0996	441	10
19207	10937560	5342-01-181-9445	402	19
19207	10937568	2510-00-421-3944	402	21
19207	10937579-1	5320-00-418-0985	402	24
19207	10937607	2540-00-417-2758	402	16
19207	10937619	2510-01-178-7046	405	29
19207	10937627-1	5330-01-106-2067	405	27
19207	10937627-2	9515-01-124-5055	406	39
19207	10937632	4010-00-445-7212	402	25
19207	10937640-204		412	4
19207	10937640-206		419	33
19207	10937677	5340-01-090-4479	419	8
19207	10937682-81		415	5
19207	10937683-2	9390-00-158-2408	BULK	22
19207	10937683-2X35		404	5
19207	10937683-3-35		419	29
19207	10937687	5340-00-328-5458	427	2
19207	10937691-35		419	30
19207	10937693-2	2510-01-184-4765	404	6
19207	10937693-3	2510-01-197-4200	404	7
19207	10937693-4	5365-01-208-3878	419	28
19207	10937727-206		419	31
19207	10937760	9905-00-111-6662	552	14
19207	10937774	4730-01-107-9690	254	21
			261	7
			261	30
19207	10937774-1	4730-01-108-5103	254	9
			593	24
19207	10937774-3	4730-01-331-2858	594	5
19207	10937879	2540-00-420-5036	532	4
19207	10937880	2510-00-124-1297	385	37
19207	10937881	5340-00-930-2716	537	2
19207	10937882	5340-00-930-2717	537	7
19207	10937885	2510-00-123-0278	383	1
19207	10937889	5315-01-083-9387	383	19
19207	10937897	5306-00-421-3950	537	3
19207	10937932	2510-00-119-3904	385	22
19207	10937933	5306-01-090-8620	383	4
19207	10937936		383	2
19207	10938046-1	6150-01-101-6741	175	4
19207	10938046-2	6150-01-361-8128	588	16
19207	10938295	5360-00-941-8684	100	31

CROSS-REFERENCE INDEXES

PART NUMBER INDEX

CAGEC	PART NUMBER	STOCK NUMBER	FIG.	ITEM
			101	21
19207	10938304-2	5310-00-263-9488	303	6
			303	9
19207	10938449	5340-00-483-2163	399	8
19207	10938454	5330-00-419-5875	489	56
19207	10939516-2		616A	26
19207	10939520 KF		601	3
81601	109417	3040-00-541-0995	153	21
19207	10942521	5940-00-738-6272	151	27
19207	10944429-1	5340-00-417-2788	408	11
19207	10944429-2	2510-00-417-2789	408	9
19207	10946835	4320-00-930-2045	581	15
			585	24
			589	3
19207	10946835	4320-00-930-2045	591	11
19207	10947439	2920-00-900-7993	608	3
19207	10949040-4	5340-01-151-6145	102	21
15434	109557	5310-00-186-7403	284	5
			286	17
15434	10951608	3110-00-109-1179	120	23
15434	109686	5360-01-222-6880	88	34
15434	109917	5305-01-109-1292	88	9
50153	11M011	2530-01-095-3561	251	8
98076	11M012	2530-01-084-6975	251	7
15434	110058	5340-00-716-4975	51	13
			56	11
			88	6
21450	110093	5305-01-133-5811	607	15
15434	110453	5330-00-143-8371	37	7
78500	1105B	2530-01-300-5918	246	13
15434	110827	5330-00-785-7894	31	3
15434	110848	2930-00-701-2091	31	5
79470	1110X4	4730-00-011-6452	422	20
			423	18
79470	1110X4	4730-00-011-6452	213	6
24617	111274	5320-01-150-7745	319	43
			320	42
			321	29
			321	42
			322	48
16764	1116968	2920-00-848-3292	588	17
17773	11176106-5	5340-00-150-1658	429	7
15434	112076	5365-01-160-1832	51	1
15434	112302	5325-00-420-9696	118	9
15434	112408	5310-00-735-7460	88	7
06853	112442	2530-01-083-8102	249	24
15434	112700	2815-01-136-1987	23	1
89346	112877	4730-00-196-0930	182	14
21450	113038	5305-01-119-8889	500	25
78222	1132375K		308	16
27618	11345P11	4730-00-289-1266	482	12

CROSS-REFERENCE INDEXES

PART NUMBER INDEX

CAGEC	PART NUMBER	STOCK NUMBER	FIG.	ITEM
15434	114421	5340-00-409-9978	36	2
15434	114638	5310-00-887-8325	42	2
15434	114850	5365-00-203-1281	66	21
72582	11500661	5310-01-290-7456	585	29
24617	11500713	5306-01-289-9197	184	6
			584	20
24617	11500869	5306-01-276-1601	121	2
73342	11500878	5305-01-291-5138	107	14
73342	11500899	5305-01-291-0156	6	29
24617	11501033		189	33
24617	11501092	5305-01-271-8381	63	16
73342	11502456	5306-01-291-0136	6	8
24617	11502788	5306-01-197-1513	110	15
24617	11503669	5306-01-284-9663	110	12
72582	11505185	5305-01-288-1417	585	27
24617	11508687	5306-01-197-1492	156	12
11331	115112		*283	4
24617	11511514	5310-01-286-5452	121	3
			126	6
			184	5
24617	11511515	5310-01-270-8392	107	15
24617	11511516	5310-01-286-6075	6	9
			63	15
21450	115224	4730-00-196-0936	549	41
15434	115519	3110-00-107-7564	118	8
19207	11592371	5340-01-146-7667	581	13
			585	31
			588	6
			591	9
19207	11592462	2540-00-809-7792	416	31
19207	11592566	5330-00-414-6695	406	32
			416	25
19207	11592573-1	2510-01-179-4084	406	6
19207	11592573-2	2510-01-166-2016	405	23
19207	11592574	5340-01-168-9267	412	2
19207	11593204	5340-00-483-1107	492	15
19207	11593217	5340-00-158-3773	90	6A
19207	11593218	5340-00-480-7602	321	9
			324	14
19207	11593258	5340-00-471-8635	400	5
19207	11593266		412	26
19207	11593267		412	1
19207	11593275-2		401	6
19207	11593277	2540-01-110-4057	401	7
19207	11593371	2530-00-478-5865	255	8
19207	11593372-1	4010-01-083-1159	383	15
19207	11593374-1	2510-01-083-1158	383	23
19207	11593374-2	2510-01-210-6229	383	23
19207	11599001		134	76
19207	11601643	6150-00-134-0847	379	12
19207	11601651	5310-00-079-1974	526	3

CROSS-REFERENCE INDEXES

PART NUMBER INDEX

CAGEC	PART NUMBER	STOCK NUMBER	FIG.	ITEM
19207	11601698	2990-00-997-1532	581	2
			585	13
19207	11602023	5342-00-833-1236	62	22
19207	11602155	4820-00-857-2737	256	21
19207	11602159	4820-00-062-9719	288	7
			603	5
19207	11602160	2540-00-789-6192	146	10
			182	25
19207	11604520-5	5330-00-432-2142	62	23
19207	11604545	2940-00-134-4657	62	24
19207	11604607	9905-00-197-5962	557	1
19207	11604608	9905-00-197-5957	557	2
19207	11607262	5330-00-243-3571	406	15
19207	11607263	5330-00-415-1481	406	14
19207	11607265-1	5340-00-478-5877	416	3
19207	11607265-2	5340-00-478-5878	416	3
19207	11607265-3	2540-00-231-0206	406	3
19207	11607265-4	2540-00-231-0207	405	17
19207	11607266	5306-00-809-7824	416	8
19207	11607267-2-203		410	14
19207	11607269	2540-00-809-7793	405	15
19207	11607302	5330-00-340-3637	BULK	31
			BULK	32
19207	11607302-24		428	13
19207	11607302-78		405	4
			406	24
			416	20
			428	7
19207	11607334	2510-01-178-7044	412	5
19207	11607335	5365-00-497-6718	413	22
19207	11607374	2540-00-417-2722	402	22
19207	11607383-3	2510-01-173-0092	416	1
19207	11607383-4	2510-01-173-0091	416	1
19207	11607385	2510-00-405-1970	405	1
19207	11607385-1		405	1
19207	11607388		405	7
19207	11607392	2510-01-162-0564	405	9
19207	11607393-1		406	16
19207	11607394	2590-01-178-0759	406	43
19207	11607395	2590-01-179-4911	405	8
19207	11607401	3990-01-176-9359	408	2
19207	11607402	5340-00-178-6080	406	4
19207	11607430	5340-00-232-6056	414	23
19207	11608763	2540-01-088-6036	388	2
19207	11608772	5340-00-480-7608	243	3
19207	11608778	2540-00-169-2855	388	1
19207	11608788	9905-00-116-5295	562	3
19207	11608789	9905-00-116-5294	562	1
19207	11608806-1		389	16
19207	11608806-2		389	16
19207	11608809		389	9

PART NUMBER INDEX

CAGEC	PART NUMBER	STOCK NUMBER	FIG.	ITEM
19207	11608818	2510-00-409-7960	320	15
19207	11608826		324	8
19207	11608831	9520-01-140-2378	323	11
19207	11608871	5340-01-181-9449	422	17
19207	11608875	2590-00-222-8906	432	9
19207	11608882	5365-00-422-2017	343	7
19207	11608925	2540-00-933-6267	543	4
19207	11608931	5306-00-409-4066	362	36
			543	10
19207	11608933	2540-00-933-6263	543	16
19207	11608936	5306-00-189-1775	543	7
19207	11608938	2540-00-933-6262	543	15
19207	11608950-16	4730-01-067-4711	549	48
19207	11608950-2	4730-01-091-9370	314	2
			315	8
19207	11609215	5330-00-269-4953	507	17
19207	11609301	5930-00-453-9367	212	5
19207	11609348-11	5342-01-106-5488	102	2
			103	2
19207	11609358-2	5305-00-097-7372	82	3
			145	3
			145	3
19207	11609666	5340-00-109-8212	383	37
19207	11609677	5306-01-130-5994	602	25
			611	21
19207	11609727-2	5310-00-176-6690	102	4
			103	6
19207	11609734		606	18
19207	11611570	2510-00-898-5415	383	28
19207	11611570-1	2510-01-226-9408	383	28
19207	11611587		412	1
19207	11611588		412	26
19207	11611597		385	28
19207	11611598	2510-00-930-7778	385	27
19207	11611602-1	9520-01-131-0117	385	24
19207	11611602-2	2510-01-131-0118	385	17
19207	11611609-1	2510-01-131-0114	385	35.
19207	11611609-2	2510-01-131-0115	385	34
19207	11611612	2510-00-409-3991	385	14
19207	11611613	5340-00-482-4339	537	5
19207	11611614	5340-00-421-5083	383	18
19207	11611615	2510-00-409-4005	383	20
19207	11611618	2510-01-131-0119	385	21
19207	11611632	2540-00-169-2856	536	9
19207	11611633	2510-00-409-3992	536	10
19207	11611644		319	6
		2510-01-157-1306	321	8
			323	8
19207	11611648	5340-00-409-7958	318	5
			319	5
			320	5

PART NUMBER INDEX

CAGEC	PART NUMBER	STOCK NUMBER	FIG.	ITEM
			321	5
			322	5
			323	5
			324	5
19207	11611648-1	5340-01-089-9128	318	18
			319	19
19207	11611648-1	5340-01-089-9128	320	16
			321	20
			322	20
			323	23
			324	17
19207	11611656	5365-00-413-4371	490	10
19207	11611789		321	7
			323	7
			324	7
19207	11611804-1	5340-01-131-0109	323	17
19207	11611804-2	5340-01-131-0110	323	13
19207	11611805-1	9520-01-130-7937	323	14
19207	11611806	2510-01-130-7936	323	10
19207	11613631	5945-00-789-3706	136	4
19207	11613632-3	6220-00-808-6072	136	1
19207	11614131	5930-00-134-5036	132	7
19207	11621056-3	5320-01-134-8671	562	2
19207	11621116	4810-01-130-7930	493	10
19207	11621117	4820-00-449-5059	493	1
19207	11621118	5315-00-401-4383	493	9
19207	11621118-1	5315-01-121-7689	505	15
19207	11621118-2	5315-01-121-7688	505	16
19207	11621120	4710-00-235-4819	493	13
19207	11621121-1	3040-00-498-2386	493	3
19207	11621121-2	3040-00-498-2389	493	6
19207	11621122-2	2590-00-436-4601	493	7
19207	11621123	5340-00-491-0329	493	24
19207	11621124	5340-00-491-0331	493	14
19207	11621624-2	5340-00-240-9228	367	11
19207	11621890-1	4730-01-115-0646	493	21
19207	11621891	4730-00-249-4416	493	25
19207	11621892	4730-01-120-8547	493	11
19207	11621894-1	4730-00-450-9671	493	19
19207	11621894-2	4730-00-449-7356	493	17
19207	11623670-1	5340-01-194-9885	608	4
19207	11630581		601	2
19207	11630585-1	9905-01-043-5322	558	3
19207	11630594		601	8
19207	11637822-1	5315-00-415-6294	411	3
			411	11
15434	116391	3120-00-792-9834	44	6
19207	11639519-1	5331-00-463-0200	140	9
19207	11639519-2	5331-00-462-0907	139	6
19207	11640313	5330-01-126-3469	473	31
19207	11640341	5365-00-809-6292	345	40

PART NUMBER INDEX

CAGEC	PART NUMBER	STOCK NUMBER	FIG.	ITEM
19207	11640362-1	5310-01-117-0610	345	23
19207	11640377-9	4520-01-178-6680	418	19
19207	11640398-1	2590-00-498-2388	493	4
19207	11640398-2	2590-00-498-2387	493	5
19207	11640433	5310-01-139-9856	494	15
19207	11640442-3		370	1
19207	11640446	2590-01-131-7453	494	25
19207	11640447	5330-01-137-7089	494	24
19207	11640504-1	5340-00-757-5877	370	11
19207	11640504-2	5340-00-757-5901	370	14
19207	11640506	5315-00-171-2590	370	38
19207	11640507	5340-00-004-6854	370	6
19207	11640508	5360-00-832-0178	370	23
19207	11640510	5307-00-102-0962	370	37
19207	11640512	2540-01-130-7946	370	40
19207	11640513	2540-00-436-8289	370	4
19207	11640515-1	5360-00-482-4813	370	34
19207	11640515-2	5360-00-482-4814	370	36
19207	11640516	5342-00-470-1543	370	43
19207	11640520	2540-00-470-1564	370	24
19207	11640521-1	2540-01-225-5863	370	5
19207	11640522	2540-00-460-5815	370	10
19207	11640523-3	2540-01-291-9037	370	19
19207	11640524-1	2540-01-286-7673	370	20
19207	11640525-1		370	27
19207	11640526	2540-00-460-5826	370	3
19207	11641043	5315-01-259-2517	370	41
19207	11648420		320	7
19207	11648456	5340-01-107-5220	397	7
19207	11648457	5342-01-082-3595	536	3
19207	11648458-1	5340-01-107-5219	397	10
19207	11648458-2	5340-01-109-7553	397	11
19207	11648460	2540-01-091-5450	536	4
19207	11648461	2510-01-083-1105	536	19
19207	11648462-1	5365-01-109-2472	397	12
19207	11648463	5340-01-097-8094	536	2
19207	11648464	2510-01-083-1106	536	6
19207	11648465-1	2540-01-089-5017	397	5
19207	11648465-2	2540-01-091-5449	397	4
19207	11648466	2540-01-089-5018	397	9
19207	11648493	4730-01-027-6590	313	4
19207	11648494	7690-00-409-8937	557	5
19207	11648495	4710-00-493-8899	39	1
19207	11648497		39	3
			39	6
19207	11648498	4730-01-233-8998	605	13
19207	11648500	4710-00-405-1962	314	9
19207	11648534	5340-01-109-7950	580	2
			584	4
			592	2
			594	35

CROSS-REFERENCE INDEXES

PART NUMBER INDEX

CAGEC	PART NUMBER	STOCK NUMBER	FIG.	ITEM
19207	11648559		583	3
			589	27
19207	11648560-1	4720-01-088-9680	551	1
19207	11648560-2	4720-01-088-9681	551	16
19207	11648568	4820-00-009-7378	598	37
19207	11648582	2940-01-129-0261	62	9
19207	11648617	2940-00-134-8326	62	22
19207	11648628	5307-00-174-4863	583	4
			589	29
19207	11648643	5342-00-484-8588	5	11
19207	11648644	4720-00-096-9630	263	8
19207	11648657	5330-01-137-7090	1	4
19207	11648729	5340-00-007-9442	39	8
19207	11648730	5340-01-112-8909	263	5
19207	11648745	2930-00-147-5202	109	1
19207	11648745-1		109	2
19207	11656448-1	5340-01-187-0527	589	32
			591	29
19207	11656458-2	4720-01-354-9029	590	23
19207	11656458-3	4720-01-175-7421	589	12
			590	19
19207	11656473	4730-01-175-7343	589	10
			590	4
19207	11658679	5340-00-407-2612	142	3
19207	11658682-1	2510-00-407-5085	142	17
19207	11658682-2	2510-00-407-5086	142	17
19207	11658683-1	2510-00-407-5093	142	13
19207	11658683-2	2510-00-407-5095	142	13
19207	11658685	2510-00-407-2617	142	6
19207	11658686-1	2510-00-231-7437	142	18
19207	11658686-2	2510-00-231-7438	142	18
19207	11658687	6220-00-428-5943	140	3
19207	11658696	5365-00-446-8770	343	16
19207	11658697	2510-00-408-4652	343	1
19207	11662389-2	2640-00-158-5617	300	2
19207	11662487	5306-00-020-1058	370	28
19207	11662489	5340-00-470-1537	370	33
19207	11662536	2590-01-141-6305	494	7
			494	13
19207	11662537		494	11
19207	11662758	4730-01-076-2735	494	21
19207	11662760	4820-01-085-4762	345	28
19207	11662860	5306-01-117-4900	72	1
19207	11662913	4730-01-109-8001	78	1
			110	2
			538	4
			539	17
			540	15
19207	11663036	5310-00-001-4719	370	42
19207	11663037	5340-01-107-9688	62	10
19207	11663070	5305-01-104-9018	364	13

CROSS-REFERENCE INDEXES

PART NUMBER INDEX

CAGEC	PART NUMBER	STOCK NUMBER	FIG.	ITEM
			366	17
19207	11663070-4	5306-01-285-5976	370	44
19207	11663288	5935-01-114-7615	433	5
19207	11663341	5306-00-020-0857	370	26
19207	11663369	5975-00-456-0627	438	3
19207	11663385-1	2540-01-108-9114	370	2
19207	11664234-3	2940-01-090-4480	62	27
19207	11664241	5310-00-231-0280	310	5
19207	11664243	5306-00-238-5661	310	6
19207	11664244	3020-00-134-7946	310	4
19207	11664271-1		318	47
			319	47
			320	47
			321	48
			322	51
			323	50
			324	47
19207	11664286	5330-01-089-3073	316	13
19207	11664287	5340-00-409-8936	316	12
19207	11664388	2590-00-197-1739	100	7
19207	11664388-3	2590-01-107-9917	100	1
			101	8
19207	11664388-4	2590-01-443-8097	101	6
19207	11664431	5330-00-143-7737	127	3
19207	11664432-1	5310-01-104-3804	127	4
19207	11664437	3040-01-107-9928	100	22
			101	18
			525	29
19207	11664472-1	4720-01-088-9650	109	12
			110	4
19207	11664473	4720-00-177-6184	109	14
19207	11664473-1	4720-01-235-9629	593	17
			594	9
19207	11664479-1	5306-01-106-3850	127	1
19207	11664480	5330-00-252-3274	127	5
19207	11664491	2940-01-145-0350	62	25
19207	11664493	2510-00-134-4663	5	2
19207	11664534	5315-00-134-4632	346	27
19207	11664542-1	2520-01-107-9932	301	8
19207	11664542-2	2520-01-369-5335	301	8
19207	11664545	4720-00-134-4655	62	8
19207	11664549	3040-00-230-3598	310	19
19207	11664558	4720-00-177-6162	314	1
19207	11664560	4720-00-177-6160	314	11
19207	11664569	2510-00-417-2752	346	13
19207	11664570	5306-00-445-7240	346	6
19207	11664571	5310-00-470-4271	5	12
19207	11664574		575	1
19207	11664597-1	2530-01-088-9357	236	36
19207	11664598	2530-00-231-0178	236	36
19207	11664599	5307-00-483-1105	236	e

CROSS-REFERENCE INDEXES

PART NUMBER INDEX

CAGEC	PART NUMBER	STOCK NUMBER	FIG.	ITEM
19207	11664613	5340-00-472-1953	100	47
19207	11664614	4720-00-001-7854	81	7
19207	11664615	4720-00-096-9648	81	8
19207	11664675	2530-00-230-3596	310	17
19207	11664676-1		596	13
19207	11664676-5		596	27
19207	11664676-7		597	1
19207	11664701-2	2590-01-100-3871	596	10
19207	11664706	2910-00-421-3967	99	3
19207	11664720	4730-01-026-0929	314	10
			315	2
19207	11665738		380	2
19207	11665759	9535-01-106-2065	417	22
19207	11665760	2510-01-162-0562	417	19
19207	11665761	2510-01-162-0561	417	18
19207	11665767	5340-01-217-0818	417	16
19207	11668054	4730-01-090-6468	71	23
			74	22
			75	18
19207	11668054-1	4730-01-104-4314	71	24
			74	23
			75	19
19207	11668950	2990-01-202-4128	586	7
19207	11668979	5310-01-100-2067	141	5
19207	11668993	2530-01-119-8710	304	7
19207	11669021-1	4730-01-104-3805	351	17
19207	11669053	5340-01-117-3793	370	15
19207	11669066	4810-01-107-9694	289	1
			603	2
19207	11669076	4810-01-108-5107	289	17
19207	11669079-1	4730-01-089-4596	104	1
19207	11669079-2	4730-01-117-3837	105	11
19207	11669079-4	4730-01-225-9003	104	12
19207	11669079-5	4730-01-331-6630	104	6
			104	6
			111	1
			115	3
19207	11669081	4730-01-114-7541	182	27
19207	11669082	4730-01-112-6561	77	22
			80	24
			213	35
			215	11
			273	23
19207	11669094	2530-01-120-8441	145	2
19207	11669104	4820-01-114-7543	264	19
19207	11669105	4820-01-107-9695	281	1
19207	11669108	2990-01-085-3786	102	35
19207	11669109	5342-01-101-0005	105	10
			106	4
			107	6
19207	11669112	2520-01-117-3013	206	5

PART NUMBER INDEX

CAGEC	PART NUMBER	STOCK NUMBER	FIG.	ITEM
19207	11669131	2530-01-101-0084	303	13
19207	11669142	6350-01-089-2987	145	8
19207	11669144	2520-01-114-7690	226	2
19207	11669164	2520-01-098-5124	202	2
19207	11669165	2930-01-133-2143	108	3
19207	11669167-1	2540-01-117-3025	370	18
19207	11669167-2	2540-01-117-4882	370	16
19207	11669173-1	6680-01-296-2758	177	14
19207	11669174	5342-01-126-7910	177	8
			178	6
19207	11669185	4140-01-089-3058	120	3
19207	11669206	4720-01-058-9489	482	15
19207	11669207	4720-01-058-9490	480	27
19207	11669214	5930-01-098-6743	134	1
19207	11669313	2520-01-105-6465	530	1
19207	11669319-1	5970-01-114-3753	358	9
19207	11669322	3020-01-090-6695	125	13
19207	11669323	2920-01-083-5408	125	4
19207	11669324-1	5340-01-197-2183	125	16
19207	11669329	4810-01-105-6966	504	27
19207	11669329-2	5340-01-416-6503	629	6
19207	11669330	4820-01-127-6922	504	18
19207	11669330-1	5340-01-417-2432	629	8
19207	11669333	3010-01-090-7747	506	6
19207	11669334	5340-01-090-7631	506	3
19207	11669335	4320-01-090-7632	506	4
19207	11669342-1	2590-01-083-1155	525	13
19207	11669345	5315-01-144-8675	180	12
			181	20
19207	11669346	3040-01-104-9151	180	10
19207	11669350	4320-01-112-8365	504	24
19207	11669352	4810-01-106-2062	496	1
19207	11669355	6685-01-098-5110	132	22
19207	11669413-1	2510-01-094-6714	343	4
19207	11669424	4820-01-073-0080	35	6
19207	11669426-1	2520-01-112-2156	223	1
19207	11669428	2520-01-114-3691	190	18
19207	11669456	2940-01-112-6438	508	27
19207	11669457	4820-01-104-9159	456	35
19207	11669457-1	4810-01-112-2162	509	20
19207	11669461	2520-01-111-2280	465	1
			512	1
19207	11669463		528	16
19207	11669464	4010-01-112-6562	454	14
19207	11669464-1	2590-01-142-1310	513	18
19207	11669531-1	5935-01-192-0627	587	11
			588	4
19207	11669618-1		539	7
19207	11669624	2540-01-123-6823	539	2
19207	11669679	5930-01-141-8414	211	2
19207	11669705	2540-01-125-6154	580	10

CROSS-REFERENCE INDEXES

PART NUMBER INDEX

CAGEC	PART NUMBER	STOCK NUMBER	FIG.	ITEM
			584	32
			592	6
			594	29
19207	11669772	5930-01-132-3247	97	18
19207	11669790	4730-01-165-4647	426	7
19207	11669805	5940-01-021-1874	158	8
19207	11669826	4810-01-140-8221	198	15
19207	11669856	3020-01-216-2332	503	27
19207	11669857	3020-01-215-6599	503	28
19207	11669858	3020-01-215-8827	503	30
19207	11669859	3010-01-215-6598	503	26
19207	11672521	2540-00-121-9077	445	
19207	11672522	2540-00-121-9081	KITS	
19207	11672523	2540-00-121-9082	KITS	
19207	11672543-1	2510-01-101-8359	404	1
19207	11672543-2	2510-01-101-8358	404	1
19207	11674728	5935-01-097-9974	154	1
19207	11674729	5330-01-059-4286	154	5
19207	11674730	5970-01-044-8391	154	4
19207	11675004	5340-01-059-0114	154	6
19207	11675004-1		154	7
19207	11677047-3	5977-01-195-9380	614	9
19207	11677306	2590-01-111-5391	149	2
19207	11677309	4710-01-058-9494	480	17
19207	11677310	5315-01-130-5963	602	24
19207	11677498-2	5995-01-195-9405	614	14
19207	11677580	5365-01-191-3575	418	12
19207	11677584	9905-01-054-3827	572	5
19207	11677607	5325-00-164-2087	61	2
19207	11677627-1	2990-01-106-2296	592	15
19207	11677663	9905-01-054-4002	572	4
			574	2
19207	11677668	9905-01-046-4676	574	3
19207	11677675	9905-01-046-4677	572	6
19207	11677676	9905-01-054-3828	574	4
19207	11677677	5340-01-184-3464	379	4
19207	11677678	2540-01-158-7171	401	6
19207	11677679	2540-01-158-7093	401	7
19207	11677680	2510-01-120-8449	403	3
19207	11677681	2540-01-158-7169	401	1
19207	11677686	6150-00-158-0066	433	1
19207	11677688	5340-01-163-4777	536	10
19207	11677690	2540-01-163-5170	536	9
19207	11677693	5340-01-179-4303	378	4
19207	11677693-1	5340-01-179-1476	379	2
19207	11677696	6150-01-134-3777	438	1
19207	11677701	5340-01-181-0840	378	12
19207	11677708	6150-01-135-4482	433	4
19207	11677709	2590-01-135-4405	437	4
19207	11677712	2510-01-120-8448	403	1
19207	11677713	2590-01-175-7230	536	10

PART NUMBER INDEX

CAGEC	PART NUMBER	STOCK NUMBER	FIG.	ITEM
19207	11677714	2530-01-213-1040	418	15
19207	11677717	6150-01-135-4481	439	5
19207	11677718	2540-01-158-7170	401	1
19207	11677719	2920-01-135-4403	440	5
19207	11677720-1	5340-01-180-4819	379	17
19207	11677720-2	5340-01-227-6479	379	16
19207	11677723	6150-01-135-4480	436	8
19207	11677724	2540-01-158-7092	401	2
19207	11677726	2540-01-160-7915	401	2
19207	11677728	2920-01-135-4402	440	5
19207	11677729	6150-01-135-4401	432	1
19207	11677730	6150-01-143-9543	439	1
19207	11677731	6150-01-135-4479	436	1
19207	11677736	2510-01-180-8503	418	10
19207	11677737	4520-01-180-3577	418	9
19207	11677738	2920-01-183-2693	430	1
19207	11677740	6150-01-135-4478	430	1
19207	11677741	2540-01-180-3579	418	33
19207	11677744		441	1
19207	11677745-5	9515-01-285-9855	410	1
19207	11677745-6	9515-01-285-9856	410	1
19207	11677747	2540-01-183-6796	418	32
19207	11677749	5925-01-115-0557	441	7
19207	11677754-9	2510-01-104-8966	408	1
19207	11677758	2510-01-197-8562	418	31
19207	11677828-1		400	6
19207	11677828-2		400	6
19207	11677829-1		400	14
19207	11677829-2		400	14
19207	11681630	7690-00-555-6073	572	15
19207	11681649	4730-01-202-4101	425	2
19207	11682088-1	5340-01-114-7712	537	1
			609	1
19207	11682088-3	5340-01-393-9372	509	30
			609	1
19207	11682088-5	5340-01-458-0975	509	30
			609	1
19207	11682319	5340-01-089-9130	318	17
			321	17
			322	18
			323	22
			324	16
19207	11682323-1		381	3
19207	11682323-2		381	3
19207	11682324-1		381	2
19207	11682324-2		381	2
19207	11682345	5935-01-044-8382	154	12
40670	11682888	4730-00-244-9848	291	9
06853	117055		*283	3
24617	117212	5310-00-568-6077	189	27

CROSS-REFERENCE INDEXES

PART NUMBER INDEX

CAGEC	PART NUMBER	STOCK NUMBER	FIG.	ITEM
37239	1180	4820-00-274-3646	581	20
15434	118226	5305-00-161-0902	50	27
			55	25
15434	118377	3120-00-791-1440	23	14
15434	118378	3120-00-659-7808	23	8
15434	118939	5315-00-777-3544	23	6
			23	12
94894	1194	6685-01-308-1985	630	17
15434	119810	4310-00-903-7174	284	31
			286	46
15434	119859	5325-00-922-9101	284	27
			286	45
78500	1199-A-3927	5306-01-281-2333	296	5
78500	1199-M-1105	3120-00-516-7516	207	4
78500	1199-N-1106	4730-00-202-9539	207	6
78500	1199-N-3082	9905-01-353-8846	555	3
78500	1199G111	5310-00-273-7771	293	1
			294	18
			295	15
78500	1199J114C	2530-00-359-1162	293	2
			295	16
81343	12 100110B	4730-01-291-5225	598	19
81343	12 100115B	4730-01-287-9012	598	20
81343	12 120111B	4730-00-050-4309	621	15
81343	12-12 070202BA	4730-00-640-0264	205	3
81343	12-12 120102BA	4730-00-277-8761	630	15
81343	12-12 120202BA	4730-01-123-8824	273	15
81343	12-12 140137C	4730-00-278-3724	629	5
81343	12-12-12010424C		629	5
81343	12-8 100102BA	4730-01-108-6410	598	56
			599	10
81343	12-8 120102BA	4730-00-202-9035	271	5
			273	21
			620	6
81343	12-8 430260BA	4730-01-309-0947	598	28
81343	12-8 430360BA	4730-01-309-0948	598	38
30780	12WGTX-WLN-S	4730-01-280-8331	617	18
10001	12Z329PC93	4730-00-555-8291	473	19
			502	48
10001	12Z48PC611	5315-00-839-2325	534	11
30327	120-B-04X02	4730-00-529-1487	146	8
21450	120322	4730-00-817-6578	550	2
78500	1205-B-2004	5330-01-292-9573	245	5
			246	11
78500	1205-C-2005		245	13
		5330-01-271-9347	246	14
78500	1205-Y-1663	5330-01-137-4799	217	2
			527	15
15434	120819		23	16
27618	12084P11	5310-00-274-8041	107	5
			121	9

PART NUMBER INDEX

CAGEC	PART NUMBER	STOCK NUMBER	FIG.	ITEM
			126	8
			312	51
			584	26
			585	17
			619	5
72582	121208	4730-00-196-1467	505	49
			581	23
15434	121933	3020-00-820-7914	44	2
78500	1224-D-551	5310-01-128-1592	236	11
19207	12250498	5340-01-176-9443	586	20
19207	12251616		602	14
19207	12253104-4	4010-01-027-0356	469	35
			470	30
19207	12253104-5	4010-01-035-0159	499	23
19207	12253105-13		470	28
19207	12253105-15		469	33
19207	12253105-19		499	21
19207	12255630	5305-01-143-6534	393	1
19207	12255634	4210-01-183-4822	421	1
			610	3
19207	12255644-1	2590-01-136-8721	243	17
19207	12255648	5340-01-082-2510	457	19
19207	12255651-2	4730-01-090-4919	456	17
			457	3
			461	14
			509	9
19207	12255660	5360-01-120-4610	149	12
19207	12255677	5330-01-099-0554	149	1
19207	12255711	9905-01-140-8219	561	3
19207	12255731-1	5340-01-090-7643	134	75
19207	12255731-2	5340-01-090-7644	134	45
19207	12255736	5340-01-083-5443	177	7
19207	12255752	2530-01-089-3047	301	2
19207	12255753	2530-01-090-7636	302	2
19207	1225575B	5340-01-090-7637	304	2
19207	12255760	5340-01-090-7638	304	6
19207	12255761	3040-01-089-3081	302	6
19207	12255764	5365-01-090-7769	301	4
19207	12255776	2510-01-082-3824	353	1
19207	12255797	2930-01-083-1141	352	4
19207	12255807-1	2540-01-130-7947	371	11
19207	12255807-2	2540-01-130-7948	371	11
19207	12255817 NON-ASB ESTOS	5330-01-379-4345	102	3
			103	7
19207	12255820	5340-01-085-3593	351	13
19207	12255829-1	2540-01-092-9323	364	9
19207	12255829-2	2540-01-092-9324	364	9
19207	12255837	5340-01-104-8944	106	7
			107	35
19207	12255840	9905-01-108-1034	552	10

CROSS-REFERENCE INDEXES

PART NUMBER INDEX

CAGEC	PART NUMBER	STOCK NUMBER	FIG.	ITEM
19207	12255844	5340-01-105-0992	106	5
19207	12255847-1		355	9
19207	12255848	5342-01-107-9693	288	4
			603	3
19207	12255849	2540-01-093-4305	355	1
19207	12255850		355	10
19207	12255852	2930-01-131-0107	106	1
19207	12255855	5340-01-066-6086	106	15
			107	31
19207	12255857	2590-01-136-1438	374	10
19207	12255858	2510-01-082-7460	368	2
19207	12255862	5340-01-090-7622	62	18
19207	12255868	2990-01-083-1123	102	34
19207	12255871	5340-01-090-7640	374	25
19207	12255874		374	1
19207	12255880	6160-01-130-8045	153	1
19207	12255881	6160-01-093-5836	153	15
19207	12255882		153	16
19207	12255886		359	6
19207	12255890	2510-01-083-1149	351	1
19207	12255895		355	7
19207	12255897	2510-01-083-5442	353	5
19207	12255901	5340-01-104-4328	351	4
19207	12255901-1	5340-01-104-4327	351	3
19207	12255902	5340-01-083-3015	353	10
19207	12255909	5342-01-126-0176	374	12
19207	12255911	2510-01-090-7641	368	9
			374	24
19207	12255913	5330-01-110-2462	364	5
19207	12255915	5315-01-109-1443	371	9
19207	12255916	5340-01-110-9205	468	14
19207	12255917-1	2590-01-082-2520	468	15
19207	12255917-2	2590-01-082-2521	468	15
19207	12255918-1	2510-01-082-3629	364	1
19207	12255918-2	2510-01-082-3625	364	1
19207	12255919-1	2510-01-085-5353	468	6
19207	12255919-2	2510-01-083-5404	468	6
19207	12255920	2990-01-083-5715	102	5
19207	12255921	5340-01-131-0108	134	43
19207	12255922	2540-01-082-3624	371	5
19207	12255924		549	21
19207	12255925	5360-01-108-0828	371	8
19207	12255926	2510-01-082-2515	468	7
19207	12255935	2540-01-138-0925	549	8
19207	12255937	2540-01-092-1264	533	3
19207	12255938	2510-01-081-9227	533	1
19207	12255939-2	9510-01-119-5679	533	2
19207	12255940	2540-01-083-1116	551	3
19207	12255940-1		592	16
		2540-01-310-4829	594	1
19207	12255941	2540-01-083-1109	549	4

CROSS-REFERENCE INDEXES

PART NUMBER INDEX

CAGEC	PART NUMBER	STOCK NUMBER	FIG.	ITEM
19207	12255944	5340-01-103-7587	180	19
19207	12255947		533	9
19207	12255949	2540-01-083-1113	551	20
19207	12255959	4710-01-130-3437	105	16
19207	12255960	2540-01-084-9644	371	1
19207	12255961	2540-01-082-7510	371	13
19207	12255962		263	9
19207	12255963	4720-01-108-9118	104	9
19207	12255964	2990-01-082-9009	102	23
19207	12255965-1	3040-01-090-4482	100	24
19207	12255967-1		263	25
19207	12255967-3		263	25
19207	12255968-1		263	18
19207	12255968-2		263	18
19207	12255968-3		263	18
19207	12255969-1	5340-01-112-6554	264	5
19207	12255970	5340-01-107-9691	252	3
19207	12255972	3040-01-090-9341	100	23
19207	12255972-2	3040-01-445-1651	100	23
19207	12255974	5340-01-112-6436	264	7
19207	12255975	5360-01-112-6546	248	20
19207	12255978	4720-01-089-9049	104	8
47457	12255978A	4720-01-279-3168	111	14
19207	12255981	5340-01-090-7650	62	15
19207	12255982	2530-01-111-2260	264	2
19207	12255983	2530-01-111-2261	264	8
			266	27
19207	12255984	5306-01-112-6560	180	8
			181	24
			265	8
19207	12255985	4730-01-107-9692	213	37
			273	1
19207	12255986	4730-01-115-0433	269	12
			269	26
			271	12
			598	54
19207	12255988		318	32
			319	29
			320	25
			321	28
			322	29
			323	32
			324	24
19207	12255989-1	2540-01-190-8484	549	19
19207	12255991	2510-01-083-1145	359	4
19207	12255991-2	2510-01-276-5729	360	2
19207	12255994	2510-01-082-3630	543	18
19207	12255995	5340-01-082-7448	543	8
19207	12255996	5340-01-085-8136	543	20
19207	12255997	5340-01-082-2523	543	11
19207	12255998	5340-01-083-5406	543	23

CROSS-REFERENCE INDEXES

PART NUMBER INDEX

CAGEC	PART NUMBER	STOCK NUMBER	FIG.	ITEM
19207	12255999	5340-01-082-2522	543	12
19207	12256002-1	2920-01-083-1187	164	1
19207	12256002-2	2590-01-085-3834	164	' 1
19207	12256002-3	2590-01-084-9634	168	1
19207	12256002-4	2920-01-083-1188	169	1
19207	12256003	2590-01-083-1115	351	29
19207	12256005	5330-01-108-7567	551	21
19207	12256006	2590-01-083-5728	158	1
19207	12256007	6150-01-083-1161	174	1
19207	12256008	6150-01-083-1154	151	17
19207	12256009	5340-01-089-3076	549	17
19207	12256012-1	5340-01-083-1121	359	20
19207	12256012-2	5340-01-083-1120	359	20
19207	12256013-1		163	1
19207	12256013-2		163	1
19207	12256018	4710-01-104-9099	177	11
19207	12256018-1	4710-01-114-7761	177	12
19207	12256019	2930-01-083-1122	109	15
19207	12256023	5340-01-090-9331	180	7
47547	12256023	5340-01-285-0575	181	12
19207	12250623	5340-01-090-9331	183	14
19207	12256024	5365-01-091-1630	183	13
19207	12256025	5340-01-090-4484	183	16
19207	12256026-1	5340-01-312-1136	183	12
19207	12256028	2510-01-088-9157	318	29
			319	30
			320	24
			321	32
			322	31
			323	34
			324	28
19207	12256030-1	5340-01-091-1609	368	14
19207	12256030-2	5340-01-091-1610	374	19
19207	12256031-2	5340-01-090-7619	62	34
19207	12256032	5340-01-091-7627	364	6
19207	12256033	2510-01-088-5914	318	34
			319	35
			320	32
			321	34
			322	38
			323	36
			324	33
19207	12256036	2510-01-088-9156	318	30
			319	31
			320	27
			321	33
			322	33
			323	35
			324	29
19207	12256040	5340-01-090-7627	62	3
19207	12256041	5330-01-104-7702	153	2

PART NUMBER INDEX

CAGEC	PART NUMBER	STOCK NUMBER	FIG.	ITEM
19207	12256042	5340-01-112-2292	62	31
19207	12256043	2530-01-091-1611	316	14
19207	12256044	2540-01-091-1612	368	12
19207	12256045	4710-01-090-7620	62	30
19207	12256046-1	2510-01-086-6802	364	12
19207	12256046-2	2510-01-082-3623	364	12
19207	12256051	5995-01-147-5423	136	7
19207	12256063-1	5365-01-106-6060	137	3
19207	12256063-3	5365-01-117-6655	137	5
19207	12256065	4710-01-131-7529	148	10
19207	12256068	4720-01-090-7617	62	35
19207	12256070	5342-01-103-7839	369	9
19207	12256075	2990-01-083-1142	102	32
19207	12256080	4730-01-090-0258	541	5
19207	12256082-1	5330-01-109-9410	102	22
19207	12256083	5340-01-089-2988	207	38
19207	12256084	2540-01-087-4741	551	19
19207	12256088	9905-01-115-0570	552	6
19207	12256092	5340-01-104-8946	148	11
19207	12256094	5342-01-090-9334	183	18
19207	12256095	5340-01-085-3594	351	12
19207	12256096	4710-01-090-4486	62	16
19207	12256097	6150-01-083-1152	167	1
19207	12256098	6150-01-082-9040	175	2
19207	12256099	6150-01-083-5479	175	3
19207	12256101		263	25
19207	12256102	2590-01-082-3828	165	1
19207	12256104		318	42
19207	12256106	5330-01-120-8454	579	6
19207	12256108		318	8
19207	12256136	2540-01-091-1613	355	11
19207	12256137	2540-01-091-1614	355	13
19207	12256140		579	4
19207	12256145		579	7
19207	12256148	9905-01-108-1032	560	1
19207	12256150-1		318	2
			319	2
			322	2
19207	12256150-2		320	2
			321	2
			323	2
			324	2
19207	12256156-1	2520-01-117-6592	237	1
19207	12256156-2	2520-01-117-3014	237	1
19207	12256156-3	2520-01-117-4933	237	1
19207	12256156-4	2520-01-117-3015	237	1
19207	12256158		71	15
19207	12256159		71	12
			74	16
19207	12256160		71	6
19207	12256165	5342-01-106-6015	202	8

CROSS-REFERENCE INDEXES

PART NUMBER INDEX

CAGEC	PART NUMBER	STOCK NUMBER	FIG.	ITEM
19207	12256166	2510-01-082-2645	386	6
19207	12256166-1	2510-01-210-6228	386	6
19207	12256169	5340-01-090-6407	388	6
19207	12256176-1	5340-01-090-7645	69	7
19207	12256176-2	5340-01-090-7649	67	8
			68	11
			69	12
19207	12256177		75	12
47457	12256177A	4710-01-282-2586	78	16
19207	12256178		77	24
19207	12256179		77	9
19207	12256180		75	6
47457	12256180A	4710-01-282-2585	78	17
19207	12256181		77	1
19207	12256182		77	11
19207	12256183		76	3
19207	12256184		76	1
19207	12256185	4720-01-091-5169	71	5
			74	28
			75	4
19207	12256186	4720-01-090-7618	71	27
			74	25
			75	22
19207	12256187	5342-01-090-7658	71	2
			74	31
			75	2
19207	12256188	5340-01-089-2989	207	16
19207	12256199 KIT		592	13
19207	12256205	5365-01-108-4811	105	13
			106	18
19207	12256205-1	5310-01-109-6755	328	10
19207	12256205-2	5310-01-109-6756	328	11
19207	12256206-1		318	1
19207	12256206-2		318	1
19207	12256206-3		318	1
19207	12256206-4		318	1
19207	12256213	2510-01-082-7458	374	8
19207	12256214	5306-01-119-5834	579	12
19207	12256226	2540-01-096-5018	592	
			597	
19207	12256227	2540-01-096-5023	579	
19207	12256233	3940-01-091-1617	339	10
			339	17
			340	1
19207	12256234	5340-01-114-7546	339	28
			340	11
19207	12256239	3950-01-094-1381	339	1
19207	12256239-1	5340-01-130-7939	339	6
19207	12256239-2	3950-01-128-1549	339	5
19207	12256239-3	5365-01-120-8442	339	4
19207	12256241-1	2510-01-091-7628	339	9

PART NUMBER INDEX

CAGEC	PART NUMBER	STOCK NUMBER	FIG.	ITEM
19207	12256241-2	2510-01-114-7545	339	16
19207	12256244	4710-01-091-1618	339	8
19207	12256245	5360-01-093-0644	339	24
			340	7
19207	12256248	7125-01-091-1619	337	12
19207	12256249	2590-01-091-1620	335	10
			337	8
19207	12256251-1	5340-01-125-4688	339	13
			339	20
			340	4
19207	12256252	5340-01-091-1621	335	14
19207	12256253	2590-01-091-1622	335	11
19207	12256255-2	9520-01-091-1624	333	13
			334	15
19207	12256255-3	9535-01-140-6470	333	18
19207	12256256	2540-01-091-1625	335	13
19207	12256261	4710-01-123-4576	214	10
19207	12256262	4710-01-125-3608	214	16
19207	12256263	4720-01-112-2325	257	1
19207	12256264	4710-01-120-8538	276	2
19207	12256265	5340-01-107-9929	252	14
			261	17
19207	12256266	5340-01-147-6759	162	6
			263	12
19207	12256269-1	4720-01-112-2328	269	4
19207	12256269-3	4720-01-112-2329	254	6
19207	12256270-1	4720-01-145-0371	253	13
19207	12256271-1	4720-01-112-2333	258	8
19207	12256271-2	4720-01-112-2332	267	8
19207	12256271-5	4720-01-112-6577	267	2
19207	12256271-6	4720-01-188-5139	258	5
19207	12256272	4720-01-107-9939	212	15
			214	13
			288	13
			603	11
19207	12256272-1	4720-01-119-5843	270	8
19207	12256272-2	4720-01-119-5844	270	7
19207	12256277	5365-01-110-8163	528	6
47457	12256277A-2	5365-01-305-2535	623	21
19207	12256279	2510-01-082-7455	373	6
19207	12256280	2930-01-082-2514	112	1
19207	12256280-1	2930-01-082-3631	112	4
19207	12256280-2	2930-01-082-2513	112	5
19207	12256281-3	4720-01-130-8086	277	5
19207	12256283-1	2930-01-085-8137	352	5
19207	12256283-2	2930-01-083-3016	352	5
19207	12256283-3	2930-01-085-8138	352	6
19207	12256283-4	2930-01-083-3017	352	6
19207	12256283-5	5330-01-104-4329	352	7
19207	12256283-6	5330-01-104-4330	352	7
19207	12256284	5340-01-082-2512	109	16

CROSS-REFERENCE INDEXES

PART NUMBER INDEX

CAGEC	PART NUMBER	STOCK NUMBER	FIG.	ITEM
19207	12256285	5340-01-128-5302	454	19
			513	27
19207	12256286	5365-01-109-3353	454	24
			513	14
19207	12256288	5340-01-112-2164	258	2
19207	12256292-1	2990-01-082-2511	102	25
19207	12256293	2510-01-084-0446	373	10
19207	12256294-1	2510-01-082-3622	373	9
19207	12256294-2	2510-01-082-3621	373	7
19207	12256295-1	2510-01-082-3604	373	8
19207	12256295-2	2510-01-082-3603	373	5
19207	12256296	2510-01-083-1140	373	2
19207	12256297		372	3
19207	12256298	2990-01-085-3833	102	5
19207	12256299		322	44
19207	12256301	2530-01-114-7764	604	7
			605	1
			605	7
19207	12256303	2510-01-083-1146	373	3
19207	12256307	5340-01-090-7628	62	7
19207	12256310	4730-01-090-7621	62	21
19207	12256312	5340-01-090-7629	62	19
19207	12256314	4710-01-090-4487	62	14
19207	12256315	4710-01-090-4584	62	14
19207	12256316-1	2510-01-082-3620	364	11
19207	12256316-2	2510-01-082-3619	364	11
19207	12256338	9905-01-114-7589	557	6
19207	12256340		323	46
19207	12256341		320	13
19207	12256344		319	48
19207	12256345		321	45
19207	12256346		324	41
19207	12256350	5365-01-111-1520	525	14
19207	12256352	3040-01-090-9342	100	28
19207	12256353	5306-01-090-9344	100	32
19207	12256356-1	3040-01-104-9154	513	1
19207	12256356-2	3040-01-104-9155	525	1
19207	12256357	5340-01-090-9332	100	41
19207	12256358		319	10
19207	12256359		320	48
19207	12256360		321	52
19207	12256361		322	53
19207	12256362		323	53
19207	12256366	5305-01-090-7625	100	39
19207	12256370	5340-01-130-7949	374	5
19207	12256372	5340-01-104-9012	525	24
19207	12256376	5340-01-111-9878	513	35
19207	12256377	5365-01-090-2074	513	29
			525	27
19207	12256378	2530-01-104-8943	513	30
			525	26

PART NUMBER INDEX

CROSS-REFERENCE INDEXES

PART NUMBER INDEX

CAGEC	PART NUMBER	STOCK NUMBER	FIG.	ITEM
19207	12256467		318	7
			322	7
19207	12256468	4720-01-089-1108	104	11
19207	12256469-1		324	43
19207	12256469-2	5340-01-091-1633	324	51
19207	12256489	4720-01-089-3074	204	23
19207	12256491	5365-01-104-7846	109	10
19207	12256518		578	1
19207	12256531	4730-01-091-0266	146	5
			204	1
19207	12256540	2520-01-091-1659	230	1
19207	12256540-1	2520-01-091-1660	230	1
19207	12256547-1	5342-01-089-2990	211	51
19207	12256547-2	3040-01-089-2991	211	22
19207	12256548-1	3040-01-121-7742	211	54
19207	12256548-2	3040-01-121-7741	211	26
19207	12256551	3040-01-088-9401	211	19
19207	12256552	5330-01-089-2992	211	48
19207	12256555	5330-01-089-3046	211	49
19207	12256559	4720-01-120-8518	480	16
19207	12256561	4720-01-120-8517	482	1
19207	12256562	5340-01-114-7550	478	18
19207	12256563	2590-01-114-7551	478	9
19207	12256563-2	5340-01-210-0201	478	9
19207	12256568	4710-01-089-2059	480	13
			495	6
19207	12256570-1	2510-01-313-0011	373	1
19207	12256570-2		373	4
19207	12256574	5365-01-145-1310	323	16
19207	12256577	5340-01-091-1634	351	19
19207	12256578-4	5325-01-114-7763	454	22
19207	12256578-5	5325-01-106-4125	180	16
			528	33
19207	12256579	2510-01-103-8688	351	6
19207	12256581	5365-01-108-4815	528	30
19207	12256582	5340-01-104-9013	528	31
19207	12256589	5340-01-114-7655	351	7
19207	12256590	5340-01-115-0615	351	10
19207	12256591-1		400	1
19207	12256591-3		400	1
19207	12256592	5340-01-137-6302	478	13
19207	12256592-2	5340-01-210-0202	478	13
19207	12256594	4730-01-112-2168	266	40
			271	10
19207	12256595		79	1
19207	12256596		79	3
19207	12256597		80	1
19207	12256598		80	16
19207	12256599		80	18
19207	12256600		80	3
19207	12256602-1	4720-01-120-8516	480	7

PART NUMBER INDEX

CAGEC	PART NUMBER	STOCK NUMBER	FIG.	ITEM
19207	12256603	2590-01-129-7523	478	1
19207	12256604		478	4
19207	12256611	4720-01-089-2061	105	12
19207	12256615		392	1
19207	12256615-1		392	1
19207	12256616		395	1
19207	12256619	5340-01-083-5402	396	1
19207	12256620	5340-01-083-5403	396	2
19207	12256622	3950-01-082-3608	497	1
19207	12256623	2590-01-082-3607	495	1
19207	12256629-1	2510-01-083-1126	354	4
19207	12256629-2	2510-01-083-1125	354	4
19207	12256632-1	2990-01-085-5349	102	20
19207	12256632-2	2990-01-085-5350	102	20
19207	12256637	4710-01-120-8543	505	24
19207	12256638	4710-01-120-8544	505	37
19207	12256639	4710-01-120-8539	504	10
19207	12256640	4730-01-128-1554	504	9
19207	12256642	4710-01-120-8541	504	1
19207	12256644	4710-01-120-8542	504	19
19207	12256647	4710-01-120-8540	503	7
19207	12256650	5340-01-083-1117	102	13
19207	12256657	5315-01-104-8942	525	2
19207	12256658-2	4720-01-120-8521	505	38
19207	12256658-3	4720-01-173-4609	505	57
19207	12256659	4720-01-122-6166	505	59
19207	12256660	4720-01-120-8522	503	5
19207	12256661-1	4730-01-121-9889	504	15
			505	23
19207	12256661-2	4730-01-114-7742	504	3
			505	36
19207	12256661-3	4730-01-114-7743	503	39
			505	58
19207	12256665	4730-01-132-4858	504	8
19207	12256665-1	4730-01-121-7604	504	12
19207	12256668	5340-01-089-3067	335	9
19207	12256669	2590-01-091-1635	333	16
			334	18
			337	20
			338	17
19207	12256670	5340-01-089-3068	319	33
			320	28
			322	35
			324	30
19207	12256674-3	5306-01-181-5018	278	13
			358	11
19207	12256680		395	12
19207	12256683	5306-01-109-9384	525	20
19207	12256689	3040-01-103-6009	525	10
19207	12256694	9515-01-140-2379	394	1
19207	12256696	2590-01-082-3606	497	5

CROSS-REFERENCE INDEXES

PART NUMBER INDEX

CAGEC	PART NUMBER	STOCK NUMBER	FIG.	ITEM
19207	12256699-1		154	17
19207	12256699-2		154	13
19207	12256705	5340-01-083-1143	351	15
19207	12256707	5342-01-104-7843	351	14
19207	12256720	5340-01-082-3610	394	4
19207	12256720-2	5340-01-210-4659	394	4
19207	12256727		320	8
19207	12256736	5342-01-120-8444	359	2
19207	12256737	2510-01-136-4442	359	5
19207	12256738-1	5365-01-110-8183	359	3
19207	12256738-2	5365-01-108-4814	359	3
19207	12256740	5340-01-120-8452	505	17
19207	12256796		503	32
19207	12256800	5340-01-160-2299	102	29
19207	12256801	5340-01-130-7932	102	28
19207	12256802	5342-01-103-7589	102	26
19207	12256808-1	2590-01-121-7606	503	8.
19207	12256808-2	2590-01-121-7605	503	16
19207	12256810	4720-01-120-8519	504	19
19207	12256811	4720-01-114-7703	503	37
			504	28
19207	12256812	5340-01-082-3609	394	5
19207	12256962	5310-01-109-6754	358	10
19207	12256962-1	5310-01-109-6753	358	5
19207	12256997	5365-01-066-7554	177	2
19207	12257052		324	48
19207	12257053		324	1
19207	12257055		74	11
19207	12257056	9905-01-115-0569	552	5
19207	12257057		74	7
19207	12257058		74	5
19207	12257059		74	14
19207	12257060	9905-01-114-7598	552	6
19207	12257062	5330-01-106-6735	375	4
19207	12257065	2540-01-084-9630	375	1
19207	12257069	5970-01-144-4841	153	9
19207	12257070	5970-01-134-5093	153	10
19207	12257108	4710-01-089-2060	456	26
			460	23
			461	20
			509	15
19207	12257109-2	4720-01-105-4067	456	16
			461	13
			509	8
19207	12257109-3	4720-01-089-9887	460	16
19207	12257110-1	4720-01-089-0766	456	4
			460	4
			461	4
19207	12257110-2	4720-01-106-3982	509	20
19207	12257112	4710-01-106-0914	456	25
			460	22

PART NUMBER INDEX

CAGEC	PART NUMBER	STOCK NUMBER	FIG.	ITEM
19207	12257113	4730-01-090-6474	456	38
			460	35
			461	35
			509	33
19207	12257114	4730-01-120-8495	505	8
19207	12257119	4730-01-089-3829	456	21
			460	20
19207	12257120	4730-01-090-4924	456	27
			460	24
			461	21
19207	12257120	4730-01-090-4924	509	16
19207	12257122		606	11
19207	12257123		606	9
19207	12257125	5340-01-084-9631	457	1
			508	1
19207	12257126	2590-01-085-5352	457	5
19207	12257127	5340-01-104-4326	456	15
			460	15
			461	12
			509	7
19207	12257128	5340-01-104-3828	456	11
			461	8
			509	4
7Z588	12258931-3	6680-01-319-2354	146	2
19207	12258932-7	6695-01-146-7132	145	10
19207	12258939-1	5315-01-156-6314	158	44
			159	6
19207	12258939-2	5999-01-150-8808	158	43
			159	5
19207	12258940-2	5935-01-149-5165	158	46
19207	12258940-4	5935-01-154-6233	158	42
			159	7
19207	12258941	5935-01-102-7124	160	2
19207	12267583	5365-00-038-9592	196	7
			197	7
19207	12267602	5315-01-004-4836	197	9
19207	12269868		154	9
78500	1227-C-939	5310-01-109-6056	526	11
78500	1227-D-1356	5310-01-271-2467	294	4
			296	22
78500	1227-E-1357	5310-01-270-8425	294	2
			296	24
14262	1227-K-1051	5310-01-099-0397	207	28
			244	9
78500	1227-X-1350	5310-01-271-5872	296	4
19207	12276902-1		457	18
19207	12276902-2		457	18
19207	12276904	5340-01-084-1232	469	23
			470	19
19207	12276906	5340-01-107-5221	454	15
			513	20

CROSS-REFERENCE INDEXES

PART NUMBER INDEX

CAGEC	PART NUMBER	STOCK NUMBER	FIG.	ITEM
19207	12276907	2590-01-084-9632	455	3
			528	7
19207	12276908	5340-01-103-8772	455	6
			528	13
19207	12276913	3040-01-089-9326	528	12
19207	12276914	3040-01-104-9152	454	7
19207	12276915	3010-01-101-6712	469	21
			470	15
19207	12276916	5340-01-104-9005	454	13
			513	13
19207	12276917	4720-01-105-3564	508	25
19207	12276920	4710-01-089-9375	456	25
19207	12276921-1	5355-01-107-4178	454	6
			528	11
19207	12276922	9905-01-114-7603	565	2
19207	12276924	9905-01-108-9187	565	1
19207	12276929	2510-01-114-7542	266	30
19207	12276930	5340-01-083-1147	369	1
19207	12276933	5340-01-103-6010	368	3
19207	12276938	6685-01-109-5695	628	4
19207	12276939	5340-01-091-1608	211	28
19207	12276944-1	4720-01-112-2326	273	9
19207	12276944-4	4720-01-153-8240	272	2
19207	12276944-5	4720-01-188-5140	273	8
19207	12276945	5340-01-105-0993	105	15
19207	12276950	4730-01-116-7886	266	36
19207	12276953-1	6150-01-082-9042	175	1
19207	12276953-2	6150-01-083-1153	175	1
19207	12276954	3040-01-104-3851	513	10
19207	12276955	5340-01-111-1351	513	17
19207	12276958	2510-01-091-1687	355	15
19207	12276959	5340-01-089-3126	333	1
			334	1
			335	1
			336	1
			337	1
			338	1
19207	12276960-1	5306-01-090-4544	333	20
			335	16
			337	30
19207	12276960-6	5306-01-210-0264	334	24
			336	17
			338	28
19207	12276967	5340-01-089-3057	376	3
19207	12276970-1	4720-01-088-9651	456	24
19207	12276970-2	4720-01-105-4068	456	24
19207	12276970-3	4720-01-105-4069	456	24
19207	12276970-4	4720-01-132-4859	460	21
19207	12276970-5	4720-01-276-5923	458	19
19207	12276971	7690-01-114-7620	564	1
19207	12276972-1	4720-01-106-8289	457	11

CROSS-REFERENCE INDEXES

PART NUMBER INDEX

CAGEC	PART NUMBER	STOCK NUMBER	FIG.	ITEM
19207	12276973	4720-01-089-2016	456	36
			460	33
			461	33
			509	31
19207	12276976	5340-01-109-8013	552	11
19207	12276977	5340-01-130-7938	318	28
19207	12276978	5315-01-109-6846	469	20
			470	17
19207	12276982		601	6
19207	12276983		601	1
19207	12276984		601	9
19207	12276998-1	5340-01-117-0606	68	1
19207	12276998-2	5340-01-090-7657	67	1
			69	1
19207	12277002	5340-01-108-9268	143	12
19207	12277002-1	5340-01-231-9291	143	16
19207	12277004-1	6150-01-131-0147	172	9
19207	12277004-2	6150-01-131-0148	172	15
19207	12277004-4	6150-01-131-0150	173	22
19207	12277004-5	6150-01-095-8309	173	1
19207	12277004-6	6150-01-095-8310	173	16
19207	12277005-2	6150-01-130-8042	172	1
			173	7
19207	12277026-1	5330-01-117-0608	67	5
			68	7
			69	10
19207	12277026-2	5330-01-117-0609	67	6
			68	8
			69	9
19207	12277027-1	9390-01-207-8125	68	2
19207	12277027-2	9390-01-192-1610	67	2
			69	2
19207	12277030	5340-01-112-2169	67	4
			68	6
			69	8
19207	12277032	2590-01-108-9121	359	10
19207	12277033	5310-01-120-8507	105	4
19207	12277034	5340-01-130-7933	105	9
19207	12277035	2930-01-130-7934	105	5
19207	12277037	5340-01-159-2995	596	25
19207	12277040	2590-01-122-2419	157	1
19207	12277042-1	5340-01-121-7601	143	20
19207	12277042-2	5340-01-106-2068	143	20
19207	12277043	5325-01-108-7375	369	4
19207	12277044	4720-01-129-6082	503	20
19207	12277045	4730-01-123-1516	503	15
19207	12277047	5365-01-108-9258	143	17
19207	12277050		151	1
19207	12277051		151	1
19207	12277052	5935-01-120-3744	163	2
19207	12277055	5910-01-096-5021	150	15

CROSS-REFERENCE INDEXES

PART NUMBER INDEX

CAGEC	PART NUMBER	STOCK NUMBER	FIG.	ITEM
19207	12277055-1		150	16
19207	12277058	2510-01-108-9122	366	1
19207	12277059		367	9
19207	12277060-1		367	37
19207	12277060-2		367	37
19207	12277061	2510-01-130-7942	367	8
19207	12277062-1	2510-01-110-4060	367	32
19207	12277062-2	2510-01-114-3689	367	32
19207	12277066	9320-01-109-5696	367	42
19207	12277069-1	2510-01-130-7943	367	20
19207	12277069-2	2510-01-130-7944	367	20
19207	12277072	9340-01-109-5934	367	45
19207	12277073		367	25
19207	12277074	5342-01-114-3694	367	23
19207	12277075	2510-01-114-3693	367	21
19207	12277085-1		539	21
19207	12277085-2		539	19
19207	12277085-3		538	1
19207	12277085-4		538	1
19207	12277085-5		540	15
19207	12277088	9905-01-185-5787	552	2
19207	12277090	9905-01-210-4733	552	6
19207	12277091	9905-01-210-4734	552	6
19207	12277094	3040-01-121-7744	211	16
19207	12277111	2510-01-112-2170	62	13
19207	12277124	5680-01-122-5214	367	22
19207	12277125	9905-01-114-7606	558	1
19207	12277126	5340-01-108-7263	353	6
			361	6
19207	12277127	5342-01-108-9123	353	8
			361	2
19207	12277128	5330-01-115-0604	467	5
19207	12277131	9905-01-114-7607	563	7
19207	12277132	6140-01-125-6075	153	26
			582	1
			584	29
			589	61
			590	40
19207	12277133	6140-01-125-6074	153	27
			582	3
			584	30
			589	59
			590	42
19207	12277134	6140-01-125-6073	153	28
			582	2
			584	31
			589	60
			590	41
19207	12277135	6140-01-156-6187	153	12
19207	12277138	4720-01-122-3656	153	13
19207	12277147	9905-01-119-5788	552	4

CROSS-REFERENCE INDEXES

PART NUMBER INDEX

CAGEC	PART NUMBER	STOCK NUMBER	FIG.	ITEM
19207	12277149	3040-01-130-7931	100	5
19207	12277150	4330-01-108-5109	61	6
19207	12277152		131	13
19207	12277153-1	6210-01-108-9125	131	1
19207	12277166	6220-01-129-5740	131	8
19207	12277167	6220-01-108-5108	131	7
19207	12277168	4730-01-117-0602	104	7
19207	12277170	4710-01-114-7759	104	3
19207	12277171	4710-01-121-7600	104	10
19207	12277172-1	4720-01-121-7749	104	2
19207	12277172-3	4720-01-108-5346	104	4
19207	12277173	4720-01-108-5347	104	5
19207	12277175	2590-01-130-7935	132	29
19207	12277175-1	5340-01-418-1936	133	1
47457	12277175A	2510-01-280-4155	133	1
19207	12277176-1	6695-01-149-5830	132	1
19207	12277176-3	6695-01-417-4453	133	2
47457	12277176A		133	2
19207	12277177	2510-01-189-6401	133	3
			134	39
47457	12277177A	2510-01-285-4592	133	3
19207	12277179	4720-01-115-2274	204	24
19207	12277182	9510-01-135-4762	367	27
19207	12277189-1	2590-01-109-7990	551	5
19207	12277189-2	2590-01-109-7991	551	8
19207	12277189-3	2590-01-109-7992	549	9
19207	12277192	5340-01-129-0474	143	10
19207	12277227	4720-01-134-3847	277	1
19207	12277229-2	5999-01-128-2755	155	3
			156	3
19207	12277230	2590-01-157-6240	134	54
19207	12277231	6150-01-145-0361	131	21
19207	12277233		356	8
19207	12277234	2540-01-130-7940	356	13
19207	12277235		356	5
19207	12277236		356	7
19207	12277237	5340-01-130-7941	356	10
19207	12277238		356	6
19207	12277239	2540-01-108-9124	356	1
19207	12277240	5910-01-121-7603	211	38
19207	12277246 X 4 IN		63	7
19207	12277246-1-20		BULK	19
19207	12277246-1X 37IN		541	14
19207	12277246-2X 20IN		541	2
19207	12277246-3X 33IN		541	4
19207	12277247		596	22
19207	12277248		597	12
			598	3
19207	12277249	2540-01-295-7461	596	6
19207	12277326-3		170	1
19207	12277327-1		170	27

CROSS-REFERENCE INDEXES

PART NUMBER INDEX

CAGEC	PART NUMBER	STOCK NUMBER	FIG.	ITEM
19207	12277328		170	22
19207	12277329		170	39
19207	12277330	5306-01-165-3272	492	7
19207	12277332		171	1
19207	12277333	2590-01-146-5258	492	5
19207	12277334 XB		170	9
19207	12277338	5340-01-212-2464	154	22
19207	12277340	4720-01-132-4868	153	25
19207	12277340-1	4720-01-195-7604	154	21
19207	12277352-1	5342-01-164-7597	318	39
19207	12277352-1	5342-01-164-7597	319	42
			320	40
			321	41
			322	47
			323	43
19207	12277352-2	5342-01-164-7598	318	52
			319	52
			320	52
			321	53
			322	55
			323	54
19207	12277353	5306-01-134-6540	358	7
19207	12277356	6150-01-131-2053	608	2
19207	12277358	3020-01-131-2078	608	19
19207	12277359	2920-01-131-1939	608	
19207	12277361	5340-01-123-1333	213	30
19207	12277362		211	9
19207	12277363	3040-01-123-9681	211	1
19207	12277367	5640-01-318-2812	372	1
19207	12277368-1	5640-01-318-2816	372	6
19207	12277368-2	5640-01-318-2815	372	3
19207	12277369	5640-01-319-2376	372	2
19207	12277371	5340-01-144-8676	542	8
19207	12277374	5325-01-156-9497	538	2
19207	12277375	5340-01-260-7895	372	5
19207	12277376	9905-01-172-2393	552	14
			552	15
19207	12277378	2910-01-128-9537	97	7
19207	12277379	2910-01-130-1535	97	23
19207	12277380	4810-01-124-5056	97	9
19207	12277381	5340-01-128-9558	97	16
19207	12277382	8145-01-117-4978	176	3
			576	1
19207	12277387	2530-01-190-8376	313	1
19207	12277388	4710-01-190-8490	313	7
19207	12277391	5340-01-168-3102	528	29
47457	12277391A	5340-01-282-2229	455	8
19207	12277392-1	5315-01-385-2731	528	17
19207	12277393	5340-01-131-7452	365	8
19207	12277394	2540-01-133-2150	365	5
19207	12277395-1	5340-01-082-3594	549	38

PART NUMBER INDEX

CAGEC	PART NUMBER	STOCK NUMBER	FIG.	ITEM
19207	12277395-2	5340-01-082-3593	549	40
19207	12277396	5340-01-160-6919	549	39
19207	12277397	5340-01-147-9822	549	37
19207	12277398-1	4720-01-136-4454	204	10
19207	12277398-2	4720-01-137-1453	204	19
19207	12287561-1	5306-01-083-5536	141	10
19207	12288013	5330-00-781-7774	189	2
78500	1229-A-3095	5365-01-271-1837	294	16
			296	15
78500	1229-E-1669-C	5310-01-200-9879	250	2
78500	1229-E-1669-C	5310-01-200-9879	251	2
			527	9
78500	1229-H-1022		232	6
			239	6
78500	1229-H-2816	5325-01-126-7264	526	23
78500	1229-N-2796	5310-01-133-3614	209	9
78500	1229-Q-1031		232	13
			239	13
78500	1229-R-2800	3120-01-143-9249	209	15
78500	1229-S-513-C	5310-01-062-3384	245	21
			246	20
78500	1229-T-1450	5310-01-129-4373	207	27
			244	10
78500	1229-U-1009	5310-00-321-9974	294	3
			296	23
78500	1229-U-2803	5310-01-145-3991	209	7
78500	1229-U-2829	5325-01-129-6849	207	43
78500	1229-Y-1507	5310-01-205-2830	237	13
			237	14
78500	1229E1331	5310-00-286-3727	526	10
19207	12296577-1	6150-01-222-6585	611	7
19207	12296642	5340-01-217-8296	611	6
			612	8
			613	14
19207	12300575	4730-01-118-5972	236	17
19207	12300579	2510-01-158-0773	344	5
19207	12300580	5340-01-119-2646	345	43
19207	12300582	5340-01-158-3126	344	12
19207	12300583	5340-01-119-2735	345	25
19207	12300584	3040-01-120-3054	345	24
19207	12300585	5315-01-119-2733	344	13
19207	12300586		344	2
19207	12300587		344	8
19207	12300588	3120-01-079-9882	345	41
19207	12300590	5360-01-080-9008	345	13
19207	12300591	5315-01-079-8059	345	39
19207	12300592	2530-01-119-2645	345	6
19207	12300593	5340-01-134-6530	345	46
19207	12300594	5340-01-119-2643	345	44
19207	12300595	5306-01-080-4757	345	31
19207	12300596	3040-01-120-2164	345	22

CROSS-REFERENCE INDEXES

PART NUMBER INDEX

CAGEC	PART NUMBER	STOCK NUMBER	FIG.	ITEM
19207	12300600		345	21
19207	12300601	5315-01-081-4275	345	38
19207	12300603	5365-01-119-2739	345	20
19207	12300604	5307-01-080-0492	345	42
19207	12300605		345	15
			345	37
19207	12300606		344	1
19207	12300608	5310-01-079-7059	345	18
19207	12300609	5310-01-080-9007	345	3
19207	12300610	5365-01-119-2644	345	10
19207	12300611	5340-01-119-2642	345	27
19207	12300612	4730-01-080-0930	345	9
19207	12300613		345	16
			345	36
19207	12300614	5360-01-119-2737	345	29
19207	12300615-1	2510-01-080-6424	345	35
19207	12300615-2		345	14
19207	12300617		345	33
19207	12300626	5360-01-119-2736	345	8
19207	12300634	5340-01-127-6920	391	17
19207	12300635	5340-01-124-5054	508	18
19207	12300638	5315-01-123-6812	508	16
19207	12300639	5340-01-127-6921	508	7
19207	12300640	5360-01-125-6118	508	17
19207	12300641	5340-01-125-6078	391	19
19207	12300643	2510-01-129-0278	508	10
19207	12300644	4720-01-128-0179	508	8
19207	12300659	9905-01-135-7474	566	1
19207	12300665-1	2590-01-131-0128	470	24
19207	12300665-2	2590-01-131-0129	470	24
19207	12300666-1	5340-01-131-0130	470	25
19207	12300666-2	5340-01-131-0131	470	25
19207	12300679	2590-01-124-5052	508	14
19207	12300691-2	9905-01-142-3114	563	3
19207	12300691-3	9905-01-135-7476	563	3
19207	12300707		602	22
19207	12300722	5340-01-208-5371	418	27
19207	12300723	5340-01-205-3548	418	23
19207	12300724	5340-01-205-3549	418	25
19207	12300727-2		408	1
19207	12300731	5365-01-239-9381	406	12
19207	12300732	5365-01-219-9172	405	12
			406	42
19207	12300738		344	7
19207	12300739		344	6
19207	12300740	9905-01-143-4526	572	12
19207	12300743	6150-01-143-4527	431	5
19207	12300744	5365-01-231-9280	425	25
19207	12300745	3040-01-192-8356	424	17
			425	13
19207	12300747	5340-01-209-7447	425	24

I-247

PART NUMBER INDEX

CAGEC	PART NUMBER	STOCK NUMBER	FIG.	ITEM
19207	12300751	5340-01-189-6403	426	14
19207	12300756	2510-01-187-0379	424	1
19207	12300828	5340-01-165-4546	543	3
19207	12300829	2540-01-165-4677	543	21
19207	12300849		616A	3
19207	12300851		616A	4
19207	12300852	2540-01-420-7925	616A	2
19207	12300853	5340-01-285-6657	616A	17
19207	12300854	5340-01-285-9399	616A	18
19207	12300857-1		616A	1
19207	12300857-2		616A	1
19207	12300865		418	29
19207	12300866		418	21
19207	12300867	2510-01-237-2945	417	23
19207	12300868		417	1
19207	12300869		418	5
19207	12300870		418	4
19207	12300871	4520-01-257-8938	418	30
19207	12300872		418	18
19207	12300873	5340-01-310-7071	418	6
19207	12300875		418	28
19207	12300881	2510-01-225-1000	417	3
19207	12300882		417	5
19207	12300895	2590-01-187-3614	417	15
19207	12300897		417	14
19207	12300914		406	17
19207	12300920		406	38
19207	12300923-1	5670-01-187-3639	406	1
19207	12300923-2	5670-01-187-3638	406	1
19207	12300934-1	2510-01-177-4466	416	29
19207	12300934-2	2510-01-177-4467	416	29
19207	12300936-1		404	15
19207	12300936-2	5670-01-177-4460	404	15
19207	12300937	3040-01-177-4463	411	10
19207	12300938	3040-01-177-4468	411	5
19207	12300939	2520-01-177-4462	411	8
19207	12300947	5340-01-190-0373	420	17
19207	12300960	2510-01-177-4457	420	10
19207	12300961-1		420	4
19207	12300961-2		420	4
19207	12300963-1		420	19
19207	12300963-2		420	19
19207	12300967-1		404	14
19207	12300967-2		404	14
19207	12300968-1		416	17
19207	12300968-2		416	17
19207	12300971-1		419	10
19207	12300971-2		419	10
19207	12300973-1	2590-01-177-4456	420	3
19207	12300973-2	2590-01-177-4455	420	3
19207	12300975-1	2510-01-179-4112	419	1

PART NUMBER INDEX

CAGEC	PART NUMBER	STOCK NUMBER	FIG.	ITEM
19207	12300975-2	2510-01-177-4458	419	1
19207	12300989	5360-01-179-3105	344	9
19207	12300990	5360-01-179-4599	344	10
19207	12301004	5306-01-209-7114	414	8
19207	12301007	5330-01-192-5788	412	33
19207	12301022	6220-01-204-2597	614	1
19207	12301036-1	6220-01-213-1558	614	5
19207	12301036-2	6220-01-225-2972	614	2
19207	12301040	4030-01-222-6037	326	1
19207	12301062	5340-01-205-2504	611	5
			612	7
			613	4
19207	12301063		611	11
19207	12301064	4710-01-200-4244	611	18
19207	12301066	5306-01-205-2677	611	10
19207	12301067	4710-01-200-4404	611	8
19207	12301067-4	4710-01-226-7389	612	31
19207	12301081	5340-01-444-6658	108	8
19207	12301084	6220-00-947-7570	611	1
			612	1
			613	1
19207	12301085	5995-01-200-2419	611	29
19207	12301087	5995-01-200-3203	611	36
			612	22
			613	36
19207	12301092	5340-01-210-2158	102	7
19207	12301094	9520-01-210-2159	334	5
			336	5
			338	5
19207	12301095	9520-01-210-2160	334	5
			336	5
			338	5
19207	12301096	5340-01-212-8476	336	15
19207	12301098	2510-01-210-2167	336	16
19207	12301100	2590-01-210-2168	339	8
19207	12301101	2510-01-216-6834	334	2
			336	2
			338	2
19207	12301103	2510-01-211-6612	340	5
19207	12301104	9905-01-214-4053	611	13
19207	12301119	2530-01-211-8405	292	16
19207	12301128-1		292	5
19207	12301152	4730-01-235-9617	503	41
19207	12301165	4710-01-226-7390	612	28
19207	12301166-1	6150-01-223-7251	612	19
19207	12301166-2	6150-01-223-7252	613	34
19207	12301167-1	6150-01-223-7270	612	6
19207	12301167-2	6150-01-223-7271	613	2
19207	12301168-1	6150-01-223-7269	612	15
19207	12301168-2	6150-01-223-7272	613	8
19207	12301194	5340-01-223-3538	613	18

CROSS-REFERENCE INDEXES

PART NUMBER INDEX

CAGEC	PART NUMBER	STOCK NUMBER	FIG.	ITEM
19207	12301195	5340-01-223-0359	613	39
19207	12301208	4730-01-221-1445	613	15
19207	12301226		612	9
			613	28
19207	12301264	5340-01-290-1738	146	3
19207	12301265	5340-01-291-9212	320	21
			321	24
19207	12301297-4		82	9
19207	12301297-5	2910-01-287-9119	82	1
19207	12301297-6	2910-01-285-9850	82	1
19207	12301298-1		82	9
19207	12301298-3	2910-01-287-9120	82	1
			82	1
34623	12301340	2540-01-256-5331	615	
19207	12301343	5365-01-284-8152	615	14
19207	12301344	5340-01-290-6370	615	3
19207	12301346	2540-01-286-7674	615	4
19207	12301346-1		615	4
19207	12301346-2		615	4
19207	12301442-1	2540-01-250-9842	615	8
19207	12301444	7690-01-291-8971	565	5
19207	12301456		616A	12
19207	12301487	5365-01-286-6180	615	6
19207	12301489	2540-01-245-2445	615	7
19207	12301489-1		615	7
19207	12301489-2		615	7
19207	12302605	4710-01-149-5078	82	5
19207	12302606	4710-01-149-5079	82	7
19207	12302607	5330-01-152-5943	367	19
19207	12302611	7690-01-131-2061	569	4
19207	12302613	5330-01-131-2066	176	6
24617	12302615		263	4
19207	12302618	5430-01-235-5442	108	5
19207	12302621	5330-01-232-1487	108	4
19207	12302623	5340-01-131-7444	108	11
19207	12302624	5340-01-131-7443	108	7
19207	12302625	2920-01-195-9383	608	11
19207	12302626		593	12
			595	1
19207	12302627	5995-01-215-0930	592	20
			594	30
19207	12302628-1	4720-01-235-9628	593	26
			594	2
19207	12302628-2		593	8
19207	12302629	5342-01-231-9151	202	4
19207	12302630	5340-01-168-7904	356	16
19207	12302632	2540-01-182-7557	609	30
19207	12302632-4	2540-01-423-1966	509	30
			609	30
19207	12302632-5	2540-01-435-8760	509	30
			609	30

CROSS-REFERENCE INDEXES

PART NUMBER INDEX

CAGEC	PART NUMBER	STOCK NUMBER	FIG.	ITEM
19207	12302634	9905-01-149-1343	566	2
19207	12302635	5330-01-197-3228	582	12
			589	14
19207	12302637		581	11
		5340-01-288-5156	585	30
19207	12302639	2990-01-192-9724	582	11
19207	12302640	7690-01-131-7499	557	4
19207	12302641	7690-01-143-1270	557	3
19207	12302643	5961-01-180-5634	149	7
			211	32
19207	12302643-2	5961-01-353-9187	588	14
34623	12302645	4730-01-214-6720	592	9
19207	12302646	5340-01-167-5535	182	11
19207	12302647	5315-01-173-0397	182	23
19207	12302648	5910-01-165-5255	182	30
19207	12302654	5340-01-131-7447	337	23
19207	12302654-1	2510-01-210-8799	338	20
19207	12302657	2590-01-112-2167	243	21
19207	12302660	4710-01-195-7644	198	1
19207	12302661	4710-01-195-7645	198	11
19207	12302662	5945-01-140-8242	198	16
19207	12302663	4720-01-165-9531	505	40
19207	12302664	4720-01-196-1166	505	41
19207	12302665	5340-01-176-5923	505	26
19207	12302666	4820-01-160-0759	505	48
19207	12302667	4820-01-157-4138	505	44
19207	12302669	5310-01-164-1642	351	16
19207	12302671	3120-01-280-6050	370	35
19207	12302672	5340-01-170-4940	318	37
			319	38
			320	35
			321	37
			322	41
			323	39
			324	36
19207	12302674-1	2510-01-090-7639	318	49
			319	49
			321	49
			322	52
			323	51
19207	12302674-2	2510-01-212-7620	468	1
19207	12302675	4730-01-195-1884	259	4
			260	5
			267	4
19207	12302678	5340-01-156-0421	326	5
19207	12302679-1	3040-01-189-6406	100	37
19207	12302683	5340-01-185-0455	596	15
19207	12302684	5930-01-161-9580	145	9
19207	12302690-6		261	3
19207	12302690-6.		254	15
19207	12302691	5340-01-188-5086	259	9

PART NUMBER INDEX

CAGEC	PART NUMBER	STOCK NUMBER	FIG.	ITEM
			260	9
19207	12302693	5340-01-197-8214	269	6
			315	16
19207	12302693-1	5340-01-333-5564	315	15
19207	12302694	5340-01-197-9300	253	6
19207	12302700	5306-01-183-6970	358	1
19207	12302702	5340-01-171-8267	353	12
19207	12302705-1	2510-01-187-0372	424	14
19207	12302705-2	2510-01-187-0378	424	18
19207	12302706-1	2510-01-187-0373	424	24
19207	12302706-2	2510-01-187-0374	424	19
19207	12302720		424	23
19207	12302720-1		424	23
19207	12302722		424	9
19207	12302730	2510-01-187-0377	424	20
19207	12302730-1	2510-01-211-6613	424	20
19207	12302731	2510-01-225-5864	424	11
19207	12302732	5340-01-187-0376	424	3
19207	12302738		427	13
19207	12302739		427	15
19207	12302740	2510-01-187-0566	427	1
19207	12302742	2590-01-187-0363	427	12
19207	12302744	5330-01-209-7354	427	11
19207	12302747	5342-01-231-9313	425	15
19207	12302748	2510-01-187-0375	424	12
19207	12302750	2510-01-187-0364	427	18
19207	12302753	5340-01-210-7510	454	15
19207	12302754	5340-01-197-8215	456	11
			460	10
19207	12302755	5340-01-188-5088	264	7
19207	12302756	5340-01-198-2415	266	21
19207	12302759	4730-01-195-0095	254	23
			257	12
19207	12302763	2510-01-187-0380	426	6
19207	12302764	5340-01-189-6402	426	4
19207	12302765	2510-01-187-0381	426	2
19207	12302766	2510-01-187-0382	425	18
19207	12302768	5340-01-188-5087	161	2
			269	18
19207	12302774		424	10
19207	12302775	2540-01-182-1077	587	
19207	12302777	5340-01-206-2209	589	21
19207	12302779	4710-01-192-0625	589	33
19207	12302780		589	13
19207	12302781	4710-01-199-4366	589	44
			590	7
19207	12302782		589	42
19207	12302782-1	4710-01-354-8067	590	34
19207	12302783	5340-01-211-1563	589	58
19207	12302784-1-8		589	46
			589	49

CROSS-REFERENCE INDEXES

PART NUMBER INDEX

CAGEC	PART NUMBER	STOCK NUMBER	FIG.	ITEM
19207	12302784-2		590	8
19207	12302784-3-29		589	55
			590	24
19207	12302784-4-65		589	17
19207	12302784-5-26		589	39
19207	12302784-6		590	6
19207	12302785	5340-01-205-9263	589	40
19207	12302786	5340-01-210-3953	589	9
19207	12302787	5340-01-207-0378	589	38
19207	12302789-2	2930-01-189-0458	589	15
19207	12302790		589	36
19207	12302791	4710-01-195-9100	589	43
19207	12302791-1	4710-01-354-8068	590	27
19207	12302792	2540-01-202-4064	589	23
19207	12302799	5340-01-203-5659	589	25
			590	13
19207	12302805	2510-01-187-0368	428	5
19207	12302809	2510-01-187-0369	428	18
19207	12302810	2510-01-202-0965	428	17
19207	12302811-1	2510-01-187-0366	428	9
19207	12302811-2	2510-01-187-0367	428	9
19207	12302812	2590-01-384-6244	586	2
19207	12302814	2540-01-175-7257	600	
19207	12302816	5340-01-186-7174	600	6
19207	12302817-1		428	8
19207	12302817-2		428	8
19207	12302819	5340-01-196-8113	428	12
19207	12302822	5340-01-192-0626	586	8
19207	12302825		586	13
19207	12302838	5340-01-223-9799	525	16
19207	12302840	3040-01-215-8853	503	21
19207	12302842	5340-01-187-0370	428	14
19207	12302844	2510-01-187-0371	428	16
19207	12302846	2590-01-225-1071	428	6
19207	12302848-1	2510-01-187-0567	428	1
19207	12302848-2	2510-01-187-0568	428	1
19207	12302850	9905-01-185-5781	561	1
19207	12302851	9905-01-185-5783	566	5
19207	12302852	9905-01-185-5784	566	9
19207	12302853	2510-01-189-6417	418	1
19207	12302854	5340-01-231-9290	418	3
19207	12302855-1	2590-01-189-6415	417	26
19207	12302855-2	2590-01-189-6416	417	26
19207	12302856	9905-01-185-5778	570	2
19207	12302857	9905-01-185-5779	570	3
19207	12302859	5340-01-205-9262	145	1
19207	12302860	9905-01-185-5785	567	2
19207	12302861	9905-01-185-5786	567	3
19207	12302862	5340-01-194-7036	525	19
19207	12302864		278	2
19207	12302865		278	3

CROSS-REFERENCE INDEXES

PART NUMBER INDEX

CAGEC	PART NUMBER	STOCK NUMBER	FIG.	ITEM
19207	12302866	4720-01-304-0688	182	2
19207	12302867	4720-01-276-1252	182	5
19207	12302868	4730-01-192-4381	596	19
19207	12302869	7690-01-196-6355	110	20
			557	7
19207	12302870	5340-01-229-8365	266	19
19207	12302871	4720-01-192-3504	461	19
			509	14
19207	12302873	4720-01-195-9518	271	3
19207	12302874	5310-01-192-5760	211	25
19207	12302875	5310-01-224-9142	211	24
19207	12302876	4210-01-220-6376	610	
19207	12302879	5365-01-204-2104	610	4
19207	12302881	9905-01-199-6809	566	4
			566	11
19207	12302883	6160-01-093-4256	153	19
19207	12302893	4720-01-290-2510	254	6
19207	12302894	4720-01-276-1253	278	9
19207	12302896	5930-01-290-5677	279	16
19207	12302898	5340-01-277-3023	279	11
19207	12302899	4730-01-215-3218	508	3
19207	12302900	4720-01-215-4295	508	4
19207	12302904-1	5355-01-318-2811	279	8
97902	12302904-2	4820-01-277-1486	279	6
			279	7
19207	12302918	2510-01-210-4652	338	7
19207	12302919	5340-01-210-2162	334	19
			338	18
19207	12302920	2510-01-210-4653	338	11
19207	12302935	9905-01-210-0229	552	5
19207	12302944	9905-01-210-0230	552	6
19207	12302946	9905-01-210-0217	552	6
19207	12302947	9905-01-210-0231	552	6
19207	12302949-1	2990-01-210-4650	102	20
19207	12302949-2	2990-01-218-2100	102	20
19207	12302956	5340-01-210-4658	365	8
19207	12302957	5340-01-217-9141	458	28
19207	12302962	9905-01-210-0219	570	3
19207	12302963	9905-01-210-0220	570	3
19207	12302965	5340-01-210-2164	334	23
			338	26
19207	12302966	2510-01-210-2163	334	22
19207	12302970	2540-01-210-6231	327	7
19207	12302980		389	9
19207	12302988	5340-01-309-7782	458	16
19207	12302992	2590-01-210-6199	458	1
19207	12302993	4710-01-211-6614	458	14
19207	12302994	5340-01-210-6238	458	10
19207	12303002		395	1
19207	12303014	4720-01-210-6200	458	22
19207	12303015	4720-01-210-6201	458	20

CROSS-REFERENCE INDEXES

PART NUMBER INDEX

CAGEC	PART NUMBER	STOCK NUMBER	FIG.	ITEM
19207	12303016	5325-01-231-7594	292	15
19207	12303018	5340-01-218-7159	321	15
19207	12303020	5340-01-217-4069	321	30
19207	12303021	5340-01-223-8037	321	18
19207	12303028		334	10
19207	12303032	2510-01-211-6611	334	13
19207	12303034	2510-01-210-6197	389	1
19207	12303038	2510-01-210-4656	341	6
19207	12303039	5340-01-219-0163	341	15
19207	12303040	3040-01-216-5337	341	19
19207	12303043	5365-01-218-9917	341	17
19207	12303045	5340-01-210-4657	341	16
19207	12303046	4710-01-210-6198	423	2
19207	12303050	3040-01-216-6982	341	20
19207	12303051	5340-01-210-4654	341	1
19207	12303055	4010-01-210-6196	341	9
19207	12303056	3040-01-210-4655	341	5
19207	12303057	3950-01-216-9319	341	21
19207	12303061	9905-01-226-9437	552	2
19207	12303064	5340-01-233-4202	334	8
			336	8
19207	12338186-67	5365-01-457-6700	602	47
19207	12340089-1	5315-01-458-4490	602	4
19207	12340115-2	5365-01-457-3364	602	52
19207	12340208-6		532	5
			535	5
			609	6
19207	12340208-7		532	5
			535	5
			609	6
19207	12340208-8		532	5
			535	5
			609	6
19207	12363523	4730-01-391-8301	297	10
15434	123558	5315-00-866-5015	24	10
19207	12356716	7690-01-305-9103	552	7
19207	12356716	7690-01-305-9103	552	15
19207	12356744-1		616A	23
19207	12356745-1		616A	28
19207	12356765	5340-01-291-9205	347	12
19207	12356766	5930-01-318-2809	587	10
			591	30
19207	12356772	5365-01-298-4877	82	11
19207	12356774	2590-01-310-1166	82	2
19207	12356775	5330-01-299-6616	82	12
19207	12356776	4730-01-314-5825	82	13
19207	12356785	5995-01-303-6428	587	1
19207	12356785-1	6150-01-360-1517	588	1
19207	12356821	2530-01-289-3962	292	1
19207	12356823	2530-01-288-5877	292	3
19207	12356832-1	2510-01-318-2814	373	1

CROSS-REFERENCE INDEXES

PART NUMBER INDEX

CAGEC	PART NUMBER	STOCK NUMBER	FIG.	ITEM
19207	12356832-2	2510-01-319-2817	373	11
19207	12356832-3	2510-01-318-2819	373	4
19207	12356832-4	2510-01-363-8981	373	1
19207	12356842	2590-01-281-1271	318	23
			468	9
19207	12356845	2540-01-281-1272	468	10
19207	12356896		245	16
			246	2
19207	12356897		245	17
			246	4
19207	12356901	5365-01-304-9529	318	51
			320	50
			321	51
60703	12356924	5930-01-321-4866	538	6
19207	12356925	2540-01-310-4854	540	5
19207	12357116	5365-01-289-4434	125	20
19207	12357126	5365-01-289-7852	125	21
19207	12360840-1	6240-01-284-1925	141	3
19207	12360850-1	6220-01-284-2709	139	5
19207	12360865	5961-01-305-8848	139	5
			140	8
19207	12360870-2	6220-01-297-3217	139	6
19207	12360890-1	5980-01-285-6688	131	6
19207	12360910-1	6220-01-300-3643	141	1
19207	12360911		141	4
19207	12360912	6220-01-411-3584	141	2
19207	12363294-2	5340-01-444-1016	101	17
19207	12363340	5995-01-287-3988	158	26
19207	12363411		325	5
19207	12363412	2510-01-272-2923	325	4
19207	12363413		325	3
19207	12363414		325	2
19207	12363415		325	3
19207	12363416		325	3
19207	12363417		325	3
19207	12363418		325	3
19207	12363420		325	2
19207	12363421		325	2
19207	12363422		325	2
19207	12363423		325	2
19207	12363425		325	3
19207	12363426		325	2
19207	12363433		360	1
19207	12363487-1		325	1
19207	12363487-2		325	1
19207	12363488-1		325	1
19207	12363488-2		325	1
19207	12363489		325	1
19207	12363490		325	1
19207	12363491-1		325	1
19207	12363491-2		325	1

CROSS-REFERENCE INDEXES

PART NUMBER INDEX

CAGEC	PART NUMBER	STOCK NUMBER	FIG.	ITEM
19207	12363492-1		325	1
19207	12363492-2		325	1
19207	12363593	4730-01-443-2888	63	19
19207	12363603	2530-01-303-0801	292	6
19207	12363604	5306-01-314-6742	292	10
19207	12363606	5331-01-314-7598	292	9
19207	12363607		292	8
19207	12363633-1	3040-01-444-7878	101	26
19207	12363654	5330-01-350-6119	2	4
19207	12368265	9390-01-285-9623	362	16
			405	35
19207	12368334	5340-01-381-3862	629	5
19207	12368335	2540-01-290-3017	629	5
19207	12368440	5340-01-315-8921	326	4
19207	12375352	5315-01-333-6603	327	9
19207	12375362	5340-01-333-8147	243	5
19207	12375377	2510-01-210-4651	336	13
19207	12375378	2590-01-311-7213	336	10
19207	12375379	2510-01-210-2166	336	13
19207	12375388	4320-01-320-4744	483	3
19207	12375453	5930-01-189-0494	132	18
			146	1
19207	12375495		134	2
19207	12375519	5310-01-458-1274	602	17
19207	12375544	5340-01-444-5366	97	20
19207	12375545	5340-01-444-5359	97	21
19207	12375555	1005-01-457-6144	602	19
19207	12375557	5340-01-457-7340	602	9
19207	12375562	2590-01-458-9661	602	10
19207	12375563	2590-01-457-9019	602	12
19207	12375586	5340-01-355-8746	590	30
19207	12375588	4720-01-354-9585	591	20
19207	12375589	2930-01-371-1389	591	5
19207	12375591	4730-01-354-2645	591	44
19207	12375592	5340-01-353-6961	591	6
19207	12375593	5340-01-353-6962	590	11
19207	12375595	5340-01-356-0011	591	18
19207	12375596	5340-01-353-6158	590	26
19207	12375597	5340-01-353-6936	590	16
19207	12375598-1	5340-01-353-6752	591	42
19207	12375598-2	5340-01-353-6753	591	41
19207	12375599	4710-01-354-8064	590	9
19207	12375600	4710-01-354-8065	590	20
19207	12375688	5315-00-740-9024	507	47
19207	12375603	2930-01-372-9135	591	22
19207	12375604	2930-01-354-5433	591	23
19207	12375606	4710-01-354-8066	591	16
19207	12375625		542	5
19207	12375626		542	6
19207	12375627	8125-01-341-3838	542	3
19207	12375628	2540-01-339-8594	542	4

CROSS-REFERENCE INDEXES

PART NUMBER INDEX

CAGEC	PART NUMBER	STOCK NUMBER	FIG.	ITEM
19207	12375629		542	2
19207	12375658	6685-01-376-2290	629	5
			629	8
19207	12375659	6680-01-374-2083	629	5
19207	12375660-1	2540-01-416-9644	629	9
19207	12375661		629	5
19207	12375718	9905-01-418-9773	552	5
19207	12375719	9905-01-418-9774	552	5
19207	12375720	9905-01-420-6755	552	5
19207	12375722	9905-01-418-9777	552	5
19207	12375724	9905-01-418-6628	552	5
19207	12375725	9905-01-420-0672	552	5
19207	12375726	9905-01-418-7652	552	5
19207	12375727	9905-01-418-9776	552	5
19207	12375728	9905-01-418-8327	552	5
19207	12375729	9905-01-420-1451	552	5
19207	12375730	9905-01-420-1764	552	5
19207	12375731	9905-01-418-9772	552	5
19207	12375732	9905-01-418-8324	552	5
19207	12375733	9905-01-417-8434	552	5
19207	12375734	9905-01-438-7063	552	5
19207	12375735	9905-01-417-8435	552	5
19207	12375736	9905-01-442-2472	552	5
19207	12375737	9905-01-442-3562	552	5
19207	12375738	9905-01-418-6622	552	5
19207	12375739	9905-01-418-8326	552	5
19207	12375740	9905-01-418-6633	552	5
19207	12375741	9905-01-420-0674	552	5
19207	12375742	9905-01-418-6621	552	5
19207	12375743	9905-01-418-7651	552	5
19207	12375769-1	2540-01-423-1791	609	31
19207	12375769-2	2540-01-423-1786	609	31
19207	12375769-3		609	31
19207	12375801	5330-01-444-8350	293	19
			295	13
			348	20
19207	12375829	6680-01-417-3281	629	15
19207	12375830	4730-01-416-6783	629	16
19207	12375837	6220-01-372-3883	139	2
19207	12375841	6220-01-359-2870	139	7
19207	12432245	5910-01-459-0289	126	10
19207	12432286	4710-01-279-3159	297	9
19207	12432309	2990-01-448-1098	591	45
19207	12432312		590	44
19207	12432371	4720-01-442-2533	72	14
19207	12432437-1	6220-01-443-8813	140	5
19207	12432440	6220-01-443-8805	140	9
19207	12432457	2510-01-143-1265	389	1
78500	1244-D-1590	5365-01-129-6590	207	12
78500	1244-G-553	5365-01-055-8769	236	11
78500	1244-Y-1585	5365-01-133-0069	209	18

CROSS-REFERENCE INDEXES

PART NUMBER INDEX

CAGEC	PART NUMBER	STOCK NUMBER	FIG.	ITEM
78500	1244B548 XD		236	11
19207	12446183		2	3
19207	12446728	1005-01-457-5807	602	38
19207	12446730	5340-01-457-1778	602	37
19207	12450076	1005-01-457-6178	602	46
19207	12450077-1	1005-01-457-6135	602	54
19207	12450077-2	1005-01-457-6138	602	56
19207	12450078	1005-01-457-5812	602	4
19207	12450083	2590-01-458-9665	602	10
19207	12450087	1005-01-457-6150	602	55
19207	12450079	1005-01-457-6176	602	21
19207	12450082	2590-01-457-7394	602	12
19207	12450084	1005-01-457-6152	602	19
19207	12450086	1005-01-457-6185	602	34
19207	12450088	1005-01-457-6147	602	24
19207	12450089	6340-01-458-7234	602	25
24617	12450173-1	2920-01-419-8884	608	15
19207	12450217		532	7
			535	7
19207	12450238-1	2540-01-417-6379	533	4
19207	12450238-2	2540-01-435-4924	533	4
19207	12450238-3	2540-01-435-4931	533	4
19207	12450242-1	2540-01-434-8725	532	1
19207	12450242-2	2540-01-435-4928	532	1
19207	12450242-3	2540-01-435-0568	532	1
19207	12450243-1	2540-00-933-8645	532	1
19207	12450243-2	2540-01-435-4936	532	1
19207	12450243-3	2540-01-424-9440	532	1
19207	12450242-6		535	1
19207	12450242-7		535	1
19207	12450242-8		535	1
19207	12450333	6110-01-444-2546	137	7
19207	12450345-6		385	4
19207	12450345-2	2540-01-459-0567	387	13
19207	12450345-4		382	23
19207	12450345-5	2510-00-654-4611	382	23
19207	12450345-6	2540-00-622-7750	385	3
19207	12450345-8	2510-01-177-4434	385	45
19207	12450345-11	2510-00-622-3931	385	44
19207	12450346-1	2510-01-459-0259	382	3
19207	12450346-2	2510-01-457-9827	382	8
			387	5
19207	12450346-3	2510-01-457-9788	382	9
			387	6
19207	12450346-5		385	19
19207	12450346-6		385	36
19207	12450346-7	2510-01-457-9011	387	13
			382	10
19207	12450346-10		387	38
19207	12450346-11		385	20
19207	12450346-14		387	12

PART NUMBER INDEX

CAGEC	PART NUMBER	STOCK NUMBER	FIG.	ITEM
19207	12450347-4		382	22
19207	12450347-5	2510-00-177-9137	387	22
19207	12450347-9	2510-01-458-9668	387	30
19207	12450347-10		382	41
19207	12450347-11		385	25
19207	12450347-12	2510-01-458-9670	385	18
19207	12450347-14	2510-00-177-7893	382	41
19207	12450347-15	2510-01-457-9780	387	29
19207	12450347-16		382	37
19207	12450347-17		382	37
19207	12450354	2510-01-459-0564	385	33
19207	12460216-1	2540-01-435-8208	532	2
			535	2
19207	12460216-4	2540-01-457-9025	609	4
78500	1246-Q-615	5315-01-277-9765	209	10
19207	12470105	5330-01-023-0269	207	9
			207	22
			207	34
			241	10
			241	23
93061	125HBL-6-8	4730-01-309-0949	598	55
78500	1250C3	5365-00-595-4948	526	21
78500	1250Z182	4730-00-595-4827	237	6
21450	125915	4730-00-193-0883	482	4
15434	126304	2815-00-828-7013	33	6
15434	127316	5310-00-081-8500	14	3
15434	127459	5340-00-174-3424	310	16
15434	127554	5342-01-143-6048	24	1
15434	127558	2815-00-791-1453	33	9
21450	127851	5305-00-754-4355	580	8
			593	22
15434	127863	2815-00-430-2090	31	7
15434	127935	4820-01-300-2759	24	12
15434	127936	5331-00-941-3762	284	8
			286	10
15434	127940	4820-00-445-0610	284	17
			286	4
72582	127950	4730-00-287-3281	454	21
21450	127951	4730-00-012-7951	108	13
			593	25
21450	127960	4730-00-287-1649	132	16
15434	128080	5360-00-895-3216	284	18
			286	3
15434	128085	5331-00-905-2679	284	16
			286	5
15434	128086	5331-00-441-0145	284	7
			286	7
15434	128807	5325-00-823-6002	96	5
93061	129HB-6-4	4730-00-136-2018	598	17
15434	129768	5310-00-082-1888	129	14
15434	129826	4820-00-829-5600	129	7

CROSS-REFERENCE INDEXES

PART NUMBER INDEX

CAGEC	PART NUMBER	STOCK NUMBER	FIG.	ITEM
15434	129838	5355-00-082-1189	129	6
15434	129839	5340-00-084-7787	129	11
15434	129888	5331-00-081-9299	129	10
15434	130226	5330-00-106-6370	118	18
15434	130227	2930-00-408-9404	118	17
15434	130240	5330-00-106-6369	118	13
15434	130394	4730-00-404-2906	116	3
79470	131-60250	4820-00-026-8473	550	1
15434	131026	5331-00-143-8485	66	24
21450	131245	5310-00-013-1245	297	6
			299	4
18876	13142983-2/48	5310-01-151-0113	89	21
15434	131622	2930-01-196-7521	118	27
15434	132019	5340-00-799-0843	116	10
97403	13205E3839-2	4730-00-439-1722	595	2
97403	13207E6882-1	9905-00-114-1334	163	10
			165	19
			167	17
			168	10
			169	10
97403	13219E0079	5510-00-270-6031	BULK	48
97403	13219E0079 XX		385	4
97403	13222E0109	5305-01-210-4595	120	1
15434	133538	5305-01-133-2060	52	13
15434	133848	3040-00-485-9224	33	15
15434	134074	2910-00-085-7436	129	12
15434	134276	5330-00-193-7652	19	1
15434	134285	5330-00-465-5818	33	39
15434	134596	5340-00-833-7966	33	7
19554	13521G2	2590-01-178-7043	407	4
			415	24
21450	135290	5320-01-148-3706	367	36
15434	135308	2815-01-146-4164	66	23
15434	136403	9905-00-473-7260	552	1
			553	1
			556	2
15434	136521	4730-01-257-3328	39	7
15434	137075	5331-00-420-9624	16	5
79470	1372X6X6X6	4730-01-095-5833	272	3
72582	137401	4730-00-013-7401	117	3
21450	137415	4730-00-278-8888	74	13
			77	17
			80	21
21450	137417	4730-00-278-8889	74	10
			80	8
21450	137423	4730-00-277-8274	71	18
			74	9
			76	15
			77	8
			79	15
			80	15

CROSS-REFERENCE INDEXES

PART NUMBER INDEX

CAGEC	PART NUMBER	STOCK NUMBER	FIG.	ITEM
15434	138042	5305-01-205-3407	14	5
15434	138608	5365-00-974-9851	41	5
15434	138609	5330-00-599-4517	41	6
15434	138795	5360-01-215-5766	84	5
15434	138862	2910-00-238-5434	85	1
			87	1
15434	138988	2815-00-033-9392	19	11
15434	138999	5360-01-137-6707	84	5
34623	139395	5340-00-243-7145	34	4
15434	139988	5331-00-809-3276	52	8
11757	14-P-36	5365-00-121-2776	530	19
75755	14M7303	5310-01-335-6455	KITS	1
15434	140218	5310-00-082-1882	7	21
15434	140330	3120-00-589-3537	26	5
			26	14
15434	140410	5306-00-420-1691	12	9
15434	140411	5310-00-829-5238	12	8
24617	14042	4730-00-196-0888	147	2
11862	14079550	5330-00-107-3925	189	24
53867	1410210503	5331-01-301-5992	59	20
24617	141231	5315-00-241-7523	190	11
53867	1413335001	5340-01-300-7295	58	17
53867	1413453025	5305-01-238-8438	58	12
D8015	1413462898	5365-12-305-0652	58	29
15434	141634	2910-01-189-0901	85	1
			87	1
53867	1418-710-019	2910-01-301-9936	59	14
15434	142234	5330-00-659-3178	42	5
21450	142341	5305-01-203-2609	563	5
24617	142390	5320-01-145-3187	322	11
24617	142391	5320-01-145-3185	319	14
			320	11
			321	12
			322	10
			323	12
			324	11
24617	142392	5320-01-145-3184	319	16
			320	39
			321	14
			322	14
			323	9
			324	13
24617	142393	5320-01-145-3189	323	15
21450	142433	4730-00-014-2433	71	7
			74	6
			74	15
			75	7
			76	4
			77	10
			77	25
			79	4

CROSS-REFERENCE INDEXES

PART NUMBER INDEX

CAGEC	PART NUMBER	STOCK NUMBER	FIG.	ITEM
			80	2
			80	17
12204	142435	4730-00-014-2435	71	13
			71	16
			74	8
			74	12
			74	17
12204	142435	4730-00-014-2435	75	13
			76	2
			77	2
			77	12
			79	2
			80	4
			80	19
15434	142689		44	2
15434	142804		12	7
24617	142851	4730-00-196-1964	215	4
21450	142976	5315-00-014-2976	507	28
15434	143066	5340-01-271-2415	8	15
15434	143848	5360-01-095-3661	84	2
15434	143854	5360-01-191-2949	86	2
21450	144077	4730-00-541-7790	255	2
			262	21
21450	144089	4730-00-765-9102	269	31
15434	144178	5315-00-082-0448	84	4
			86	4
15434	144179	5310-00-727-8353	84	3
			86	3
15434	144195	5360-00-082-0124	85	2
			87	2
24617	144387	4730-01-442-6120	273	17
15434	144714	4820-00-909-4174	286	6
99227	1448	4720-00-522-9151	BULK	41
15434	144948	3805-00-955-5320	284	11
			286	14
79470	145	4820-01-263-6410	93	9
15434	145028	4820-00-909-4175	284	12
			286	13
62983	1454	5310-00-732-0558	463	2
			464	2
			511	2
15434	145504	5330-01-051-4243	96	4
15434	145530	5331-01-201-3623	124	14
15434	146160	5310-00-809-3078	55	20
15434	146483	2910-00-790-8736	95	4
			96	1
79470	1468X8	4730-01-091-8032	598	53
62983	146835	5305-01-020-0709	463	14
			464	14
			511	14
15434	147100	2910-00-928-3505	66	2

CROSS-REFERENCE INDEXES

PART NUMBER INDEX

CAGEC	PART NUMBER	STOCK NUMBER	FIG.	ITEM
15434	147389	5305-00-062-4378	24	7
15434	147588	5342-01-134-3860	118	24
24617	147601	5305-01-133-7201	264	4
15434	147610		284	35
		3120-01-016-4883	286	31
74069	1478	5120-01-285-7829	630	6
78500	15-X-1557	5306-01-132-0834	526	2
79780	15X-725	5306-00-498-7209	251	15
79780	15X725	5306-00-498-7209	250	8
18965	1506-3		593	18
		5340-00-205-5314	594	4
15434	151623	5330-01-082-1906	16	4
15434	151911	5330-00-961-6314	33	31
15434	151917	2815-01-126-9367	33	32
77060	15300027	5935-01-308-7866	158	45
15434	153336	2910-00-829-5616	52	5
15434	153338	5340-00-829-5617	52	10
15434	153516	4930-01-124-3523	30	11
33457	153518	5330-01-044-2096	30	6
15434	153519-S	5360-01-118-5596	30	8
15434	153520	5310-00-249-6540	30	7
15434	153521	1680-01-100-5608	30	5
15434	153526	2815-00-131-1700	33	21
15434	153528		30	2
15434	153964	4310-01-146-5921	284	20
			286	21
15434	153966	2530-01-142-7939	284	6
			286	8
62983	154008		463	17
			464	17
			511	17
15434	154018	5330-00-852-7347	284	36
			286	51
15434	154088	5330-00-961-9470	95	1
62983	154129	5331-00-948-6482	463	4
			464	4
			511	4
15434	154916	5330-01-071-5727	312	54
15434	154966	3020-00-929-0713	118	12
56161	15557258	5330-01-220-5498	72	
15434	156075	5340-01-271-2419	8	17
			10	4
24617	15666151	4730-00-457-6295	271	9
			273	14
15434	157088	5365-00-507-3271	95	2
15434	157551	5330-00-143-8376	37	17
15434	157870		21	4
15434	158139		30	3
15434	158145	5340-00-404-2947	11	2
93061	159F-4-4	4730-01-283-8150	121	4
			287	4

PART NUMBER INDEX

CAGEC	PART NUMBER	STOCK NUMBER	FIG.	ITEM
24617	159358	5305-01-271-8332	138	14
16764	159920	5305-01-237-3637	598	34
81343	16 070109C	4730-01-418-4254	629	25
81343	16-12 070102C	4730-01-371-7700	629	5
81343	16-16 070102CA	4730-00-718-2621	629	27
70960	16NF3		226	11
			228	13
70960	16NF3		229	10
70960	16NLS32-2		226	6
			229	5
70960	16NYS28-21		228	5
70960	16N-1A		219	5
			226	4
			228	8
70960	16N4-3A		219	4
			226	7
			228	12
83058	160403	6210-00-699-9458	132	10
15434	160514	5365-00-965-0870	52	9
18028	161	5360-01-285-6010	101	15
24446	161A820P1	5310-00-596-9763	306	6
45152	1618210	5935-01-292-2336	158	40
15434	162426	3010-00-447-9799	49	2
			54	2
21450	162684	5305-00-215-0290	579	11
15434	163944	3120-00-904-9595	50	10
			55	9
70960	1641		506	9
15434	164164	5315-00-425-0118	33	14
15434	165430		284	32
62983	1656	3110-00-899-4353	463	15
			464	15
			511	15
15434	166009	5360-00-132-0245	46	8
19207	1665775	5340-01-310-7070	418	7
15434	166777	5305-01-091-2498	118	22
15434	167157		46	10
29510	167961R1	4820-00-848-4361	317	25
15434	168306	5305-00-947-3437	26	4
			26	8
			26	13
15434	168319	5305-00-297-4022	26	15
15434	169657	5305-00-795-9308	33	37
80244	17-C-18035-50	5975-01-217-8550	BULK	7
80244	17-C-18035-50-119		170	13
80244	17-C-18035-50-50		432	7
80244	17-C-18035-60		435	7
		5975-00-177-1930	BULK	8
80244	17-C-18035-60-38		422	3

PART NUMBER INDEX

CAGEC	PART NUMBER	STOCK NUMBER	FIG.	ITEM
80244	17-C-18035-60-52		422	7
80244	17-C-18035-60-68		423	7
80244	17-C-18035-60-90		423	9
80244	17-C-18035-60-92		422	9
80244	17-C-18035-90		436	14
			BULK	9
80244	17-I-1728-651	5970-00-811-0640	154	15
94222	17-10015-13	5310-00-885-7734	424	26
70960	17NF1		227	10
70960	17NLS40-24		227	5
70960	17NLS40-43		227	5
70960	17N1-19A		227	3
70960	17N4-15A		227	6
15434	170226	5342-00-404-2946	3	1
15434	170296	5340-00-933-3009	24	2
78500	1707-C-3	5340-00-181-1546	245	23
78500	1707C3	5340-00-181-1546	246	17
21450	171104	5305-00-429-1552	405	36
			406	29
30076	17176	4730-00-852-5654	620	11
78500	1718-D-134	5342-00-613-7784	245	6
			246	8
15434	172034	5340-00-632-6239	24	4
15434	172648	5330-00-404-2920	14	2
78500	1727-N-40	5310-00-123-2572	250	14
			251	16
15434	173086		46	16
15434	174299		46	13
80064	1755683	5935-00-605-9322	160	11
15434	175831	5340-00-485-0945	23	19
15434	176027	5330-00-129-9389	284	46
			286	33
80201	17657/55-542465	5330-01-150-9691	474	24
			489	7
15434	177734	5305-01-072-4270	9	9
78500	1779-Q-433	2530-00-087-0163	250	7
			251	13
78500	1779-Z-260	5310-00-949-6280	244	18
78222	1790422		307	2
73342	179397	5305-00-559-1022	200	32
15434	179688	2815-00-404-2957	37	16
72582	179827	5305-01-024-4775	6	13
15434	180175	5305-01-112-9698	14	12
24617	180178	5305-00-018-0178	234	5
			241	5
15434	180371	5340-01-079-8097	66	20
15434	180372	5340-00-719-4601	66	5
8X715	18048	5331-01-280-6503	619	10
15434	180626	5310-01-066-2942	46	21
15434	180810		284	30
2815-01-079-3290			286	48

CROSS-REFERENCE INDEXES

PART NUMBER INDEX

CAGEC	PART NUMBER	STOCK NUMBER	FIG.	ITEM
15434	181213	4730-00-444-1710	66	22
15434	181466	5310-00-484-1718	50	26
			52	12
			53	7
			129	3
78222	1820651K	5999-01-341-2972	308	10
78500	1828-S-149	3110-00-293-8439	208	2
78500	1829S149	3110-00-350-5407	235	12
			242	12
8X715	18300	5340-01-280-6995	619	16
8X715	18302	5365-01-288-1559	619	13
15434	183429	5365-00-369-4729	284	19
			286	2
78500	1844-J-634	2520-01-132-6847	527	3
78500	1846-M-351	5315-01-143-4220	209	19
78500	1846-X-258	2520-01-120-3673	526	18
78500	1850-Z-78	5340-00-752-1372	216	7
15434	185012	3040-00-484-8546	88	17
15434	185138	5342-01-079-4678	46	12
15434	185139	2910-01-105-6457	46	7
15434	185574	5365-01-121-3068	19	10
15434	185804	5305-00-463-0428	37	20
62983	186580	5365-00-674-6831	463	3
			464	3
			511	3
15434	186780	5330-00-864-5422	114	4
62983	187000		463	24
			464	24
			511	24
78222	1870213		308	19
15434	187126	2910-01-098-6741	88	2
15434	187127	5306-00-461-6070	88	3
21450	187343	4730-00-288-9953	423	16
15434	187350		284	29
		4310-01-197-1882	286	50
15434	187420	5365-00-132-0273	17	10
21450	187527	5306-00-018-7527	474	18
			489	1
			492	16
			502	24
15434	187556	5305-00-138-9848	129	13
24617	187838	5305-00-018-7838	198	17
24617	187995	5305-00-696-5285	361	4
15434	188040	3120-00-129-9200	284	45
15434	188042	3120-00-129-9206	284	45
15434	188044	3120-00-129-9210	284	45
15434	188318	5330-01-301-1761	115	26
24617	189510K	5320-01-174-4726	318	31
19207	189512	5320-00-018-9512	318	10
			381	6
78500	1898-H-34	2520-01-126-1493	526	20

PART NUMBER INDEX

CAGEC	PART NUMBER	STOCK NUMBER	FIG.	ITEM
78500	1898R200	4730-00-231-2412	630	16
15434	189800	5365-00-462-4504	85	16
			87	16
D8015	1900023005	5315-12-156-4502	90	18
15434	190334	5360-00-129-9415	284	13
			286	15
15434	190397	2930-00-401-9531	44	18
8X715	19046	4440-01-287-9011	619	11
15434	190876	5331-00-132-0274	129	9
8X479	191037	4310-01-084-7148	284	9
			286	11
8X715	19130-KF		619	18
15434	191517	5310-00-442-6899	44	10
21450	191561	4730-00-278-4594	81	6
24617	191675	5305-00-019-1675	161	25
8X715	19174B	2530-01-367-1883	619	20
15434	191916	5340-00-238-5435	46	22
15434	191970		17	7
24617	192075	4730-00-196-1504	81	11
			255	10
			261	21
			262	4
			266	38
			590	1
21450	192417	5305-00-019-2417	533	15
72582	192481	5310-01-058-3353	385	13
15434	193136	5310-01-124-6463	44	19
8X715	19324B	4730-01-280-4204	619	8
15434	193736		46	3
24617	193780	5320-01-224-9157	375	2
15434	194037	5340-00-404-2940	26	10
15434	194921	2815-00-404-2926	42	4
15434	194923	2815-00-404-2931	42	8
60703	1951-7-ED	5305-01-164-2310	539	5
15434	195755	5305-01-109-9307	51	15
			56	12
19954	19622		312	18
15434	196282	4730-01-085-4156	284	54
15434	196844	5365-01-132-1984	118	10
15434	196845		118	4
15434	199349	3010-01-079-3461	44	12
15434	199410	3040-01-134-3480	118	7
53867	2 410 113 004	5331-01-300-6907	58	14
5T151	2 420 101 027	5365-01-336-5912	91	8
5T151	2 420 101 056	5365-01-339-3835	91	26
5T151	2 420 101 057	5365-01-339-3836	91	26
5T151	2 420 101 058	5340-01-336-6783	91	26
5T151	2 420 101 059	5365-01-336-5212	91	26
5T151	2 420 101 060	5365-01-336-6782	91	26
5T151	2 420 101 061	5365-01-336-5213	91	26
5T151	2 420 101 062	5365-01-336-5910	91	26

CROSS-REFERENCE INDEXES

PART NUMBER INDEX

CAGEC	PART NUMBER	STOCK NUMBER	FIG.	ITEM
5T151	2 420 101 063	5365-01-336-5911	91	26
5T151	2 420 101 064	5365-01-336-5214	91	26
5T151	2 420 101 065	5365-01-339-3837	91	26
5T151	2 420 101 066	5365-01-336-5215	91	26
5T151	2 420 101 067	5365-01-336-5216	91	26
5T151	2 420 101 068	5365-01-336-5217	91	26
5T151	2 420 101 069	5365-01-336-5218	91	26
5T151	2 420 101 070	5365-01-339-3838	91	26
5T151	2 420 101 071	5365-01-336-8909	91	26
5T151	2 420 101 072	5365-01-336-8910	91	26
5T151	2 420 101 073	5365-01-336-8911	91	26
5T151	2 420 101 074	5365-01-336-5219	91	26
5T151	2 420 200 009	5365-01-336-5913	92	7
5T151	2 420 328 034	5340-01-336-0020	91	7
5T151	2 420 360 004		89	2
5T151	2 420 503 013	4820-01-338-4642	89	5
5T151	2 420 520 005	5340-01-336-0021	91	2
5T151	2 420 551 005		89	4
ST151	2 421 015 066	5330-01-339-0141	90	1
ST151	2 421 321 009	5310-01-336-6747	90	28
ST151	2 421 332 022	5340-01-336-2292	90	4
ST151	2 421 335 025		89	23
ST151	2 422 060 021	2810-01-336-2199	90	20
5T151	2 423 050 056		89	6
5T151	2 423 061 006	3040-01-339-8572	90	17
5T151	2 423 121 021	5315-01-352-8225	91	15
5T151	2 423 202 001	5315-01-336-5185	90	23
5T151	2 423 300 008	5310-01-336-6683	90	9
5T151	2 423 315 004		89	16
ST151	2 423 345 006	5310-01-336-8722	91	1
ST151	2 423 345 015	5310-01-336-6684	91	22
ST151	2 423 400 031	5305-01-336-9380	90	3
ST151	2 423 400 034		89	13
5T151	2 423 410 026	5305-01-335-9966	90	10
5T151	2 423 412 003	5305-01-335-9969	91	11
53867	2 423 450 002	5305-01-336-0007	89	25
5T151	2 423 450 005	5305-01-336-0006	89	9
5T151	2 423 457 003	5325-01-336-6738	92	26
5T151	2 424 611 024	5360-01-335-9948	90	19
5T151	2 424 619 172	5360-01-335-9949	91	5
5T151	2 424 651 025	5360-01-336-9366	92	8
5T151	2 424 680 003		89	24
			91	10
5T151	2 425 650 759	5340-01-337-3760	90	8
5T151	2 425 703 005	3120-01-351-7779	91	21
53867	2 426 449 018	2910-01-337-2983	91	27
5T151	2 910 022 197	5305-01-342-5171	89	18
5T151	2 910 172 197	5305-01-335-9965	89	17
53867	2 912 732 191	5305-01-301-0533	58	16
5T151	2 912 732 203	5305-01-336-0015	90	16
53867	2 912 742 196	5305-01-301-7818	58	24

PART NUMBER INDEX

CAGEC	PART NUMBER	STOCK NUMBER	FIG.	ITEM
53867	2 914 552 158	5305-01-301-7817	58	15
			59	12
			92	9
53867	2 915 012 007		89	15
		5310-01-301-3885	90	2
5T151	2 916 020 010	5310-01-336-6748	59	8
5T151	2 916 069 007	5310-01-336-7368	91	13
5T151	2 916 069 083	5310-01-341-8953	89	3
53867	2 916 690 005	5310-01-301-1885	90	29
53867	2 916 699 083	5310-01-301-7811	89	10
			89	20
			90	6
53867	2 916 699 085	5310-01-301-1875	92	16
53867	2 916 699 092	5310-01-300-7037	58	6
53867	2 916 710 619	5330-01-301-1763	58	27
11757	2-P-283	3020-00-464-4438	530	23
40342	2-X-39	5310-00-199-6581	282	4
53867	2-915-011-007	5310-01-185-7191	90	15
53867	2-916-710-605	5330-00-503-5789	90	13
45152	2DR636	2815-01-272-3980	25	13
45152	2DR671	2815-01-271-9822	KITS	20
45152	2DR672	2815-01-271-9802	18	6
11083	2D1683	4730-00-278-4814	472	24
			501	14
70960	2WCS24-75		506	16
81343	20-16 140140C	4730-00-449-7203	629	28
81343	20-20 140239C	4730-01-028-0342	629	29
64203	20-23214-D		516	21
64203	20-24761-21	4720-01-125-4467	521	5
64203	20-24762-13	4720-01-125-9759	521	13
64203	20-24762-14	4720-01-131-2857	519	18
64203	20-24765B-15	4720-01-125-9768	519	8
64203	20-24765B-20	4720-01-125-9767	519	42
64203	20-24768B-25	4720-01-125-5521	519	33
64203	20-33064-C	2520-01-126-3847	516	28
64203	20-33069C	5430-01-126-0268	519	29
64203	20-33138	4820-01-143-4172	521	20
64203	20-33139-11	4720-01-125-5865	519	19
			521	12
64203	20-33139B-10		519	43
64203	20-33169	4820-01-135-5372	521	1
64203	20-33170-A	4820-01-125-9698	523	22
64203	20-33191C	6105-01-117-7944	519	21
64203	20-33281-D		516	10
64203	20-33433B	5342-01-204-8720	515	1
64203	20-33479-B		516	20
64203	20-33506-B		516	27
64203	20-33607-A	5340-01-205-9479	515	4
64203	20-33623-A		518	9
64203	20-33660-A	5340-01-140-1156	517	14
64203	20-33662	4730-01-126-3541	521	14

CROSS-REFERENCE INDEXES

PART NUMBER INDEX

CAGEC	PART NUMBER	STOCK NUMBER	FIG.	ITEM
64203	20-33726A	3040-01-125-9678	518	13
64203	20-33736-A	5340-01-135-6743	517	3
64203	20-33738-A	5340-01-211-1622	517	16
64203	20-33739-A	3040-01-198-2653	517	1
64203	20-33762-B	2510-01-122-2822	516	4
64203	20-33820B	5342-01-122-2015	519	3
64203	20-33821B	5340-01-155-9469	519	37
64203	20-33822	2590-01-117-6596	523	1
64203	20-33823-B		518	5
64203	20-34197-B	2590-01-277-9100	518	1
64203	20-34198-B	3040-01-238-0863	518	1
64203	20-34206-11	4720-01-125-1365	521	18
64203	20-34219-A		523	23
64203	20-34220-A		523	17
64203	20-34366B		523	11
			523	34
64203	20-34409B	3040-01-125-6491	522	6
64203	20-34410D	2590-01-117-6597	522	3
64203	20-34412B		522	13
64203	20-34431B	3040-01-133-4821	523	5
			523	28
64203	20-34455	3830-01-126-3886	523	24
78500	20X-1336-Z	5306-01-132-8274	295	9
78500	20X-1337Z	5307-01-132-8273	295	9
78500	20X-1526	5307-01-135-9291	295	9
78500	20X-1527	5307-01-135-9290	295	9
78500	20X-1528	5307-01-142-4784	293	17
78500	20X-1529	5307-01-142-4783	293	17
78500	20X-1815-Z	5306-01-271-5842	293	17
			294	6
			296	20
78500	20X-1816-Z	5306-01-271-5843	293	17
			294	6
			296	20
01276	2000-8-12B	4730-00-927-7272	630	11
15434	200064	5340-00-417-5800	34	9
73830	200360	4730-00-278-8825	617	4
15434	200517	4720-01-135-6694	284	49
40342	200743-A		282	8
15434	200819	4820-00-400-5189	33	22
72447	200825-3	2520-00-318-0983	228	3
15434	200861	5310-00-134-4171	14	11
15434	200908	5305-00-005-0666	42	1
15434	200919	5340-00-132-3203	42	3
06853	200981	5315-01-105-9475	248	10
15434	200998	5330-01-051-1053	55	22
15434	201007	2910-01-051-4292	96	3
61465	2012993-4	5330-00-972-2635	485	6
			491	4
15434	201737	6685-01-141-0907	114	5
15434	201806	4820-01-106-1760	56	10

PART NUMBER INDEX

CAGEC	PART NUMBER	STOCK NUMBER	FIG.	ITEM
79470	202X5X4	4730-00-288-9953	422	16
01276	2021-2-53	4730-00-618-8497	277	4
15434	202334	2930-00-404-3050	116	7
15434	202356	5340-01-177-7580	284	62
15434	202603	4710-00-425-5921	36	9
15434	202890	2815-00-404-2921	41	2
15434	202891	2815-00-472-2626	12	13
15434	202897	5340-00-951-3536	52	7
15434	202903	5315-01-058-4551	7	13
06853	202998	5307-00-215-5399	249	14
79470	2030X4A	4730-00-132-4588	70	4
			73	1
			76	13
			77	3
			79	13
			80	10
			266	5
15434	203131	5310-00-426-3990	24	8
15434	203346	5365-01-119-4954	85	15
			87	17
15434	203426	5315-01-079-6506	46	9
15434	203502	2815-01-114-7399	33	40
15434	203762	5360-00-140-2078	88	26
15434	203849	4730-01-078-9859	53	4
15434	203933	5340-01-239-7078	27	10
78222	2040081		309	11
15434	204531	2815-00-138-8280	37	8
72447	204581-3	2520-01-171-2360	225	2
15434	204587	4710-01-133-6947	34	7
61465	2046068	3040-00-972-2639	491	1
15434	204898		117	2
15434	204904	4710-01-133-1538	284	50
15434	204966	5340-00-821-2364	34	5
15434	204995	4730-01-271-7956	285	4
47457	20510006	2510-01-280-4245	6	25
47457	20510087	4720-01-305-2440	63	10
47457	20510093	2530-01-274-4457	312	1
47457	20510093-10Z	5330-01-271-9408	312	10
47457	20510093-11Z	5330-01-271-9409	312	11
47457	20510093-12Z		312	12
47457	20510093-13Z	5305-01-274-4405	312	23
47457	20510093-14Z		312	15
47457	20510093-15Z		312	17
47457	20510093-16Z	5342-01-271-5709	312	24
47457	20510093-17Z	4820-00-229-9917	312	25
47457	20510093-2Z	5305-01-271-6446	312	2
47457	20510093-20Z	5325-01-265-7251	312	33
47457	20510093-21Z	3040-01-271-3649	312	34
47457	20510093-22Z	2530-01-272-2911	312	36
47457	20510093-23Z	5342-01-319-8630	312	37
47457	20510093-3Z	5342-01-271-5711	312	3

CROSS-REFERENCE INDEXES

PART NUMBER INDEX

CAGEC	PART NUMBER	STOCK NUMBER	FIG.	ITEM
47457	20510093-4Z	5330-01-271-9407	312	4
47457	20510093-5Z	5305-01-271-5168	312	5
47457	20510093-6Z	5310-01-270-8406	312	6
47457	20510093-7Z	4710-01-272-0572	312	7
47457	20510093-8Z		312	8
47457	20510093-9Z	5331-01-271-9376	312	9
2W567	20510104	2540-01-279-4593	357	2
47457	20510131	2510-01-282-2526	6	20
47457	20510132	2510-01-286-3328	6	12
47457	20510133	2510-01-280-4246	6	22
47457	20510134	2510-01-280-4244	6	15
47457	20510266-1	2930-01-287-3180	113	4
47457	20510266-2		113	1
		2930-01-287-4545	113	5
47457	20510270	5340-01-317-8144	107	3
47457	20510277	5340-01-285-3239	107	27
47457	20510296	2510-01-279-3089	6	4
47457	20510302	5340-01-284-9652	178	3
3T063	20510303	5310-01-282-3391	128	6
33477	20510304	5365-01-280-6924	128	5
3T063	20510312	5306-01-281-3387	15	1
04827	20510329	4720-01-279-3170	111	2
47457	20510333	5310-01-292-7251	107	17
47457	20510341	5340-01-281-7205	107	30
3T063	20510370	5310-01-292-9504	6	6
4F744	20510396	2590-01-285-6261	179	2
3T063	20510400	5310-01-309-8575	6	2
7A964	20510403	5340-01-322-2906	181	23
34805	20510404	4710-01-279-3162	287	1
16251	20510409	4710-01-284-9029	111	9
47457	20510429	5340-01-280-7100	107	13
47457	20510460	5340-01-298-4861	72	11
47457	20510473	4720-01-302-6687	115	2
47457	20510476	5340-01-280-7097	184	4
47457	20510477	4710-01-281-1148	78	2
47457	20510480-1	5340-01-280-7101	107	7
04827	20510481	4720-01-279-3169	115	5
47457	20510492-1	5365-01-280-3668	63	13
47457	20510492-2	5365-01-280-9432	63	12
3T063	20510498	5306-01-283-4199	15	6
47457	20510518	4710-01-284-9032	115	4
33477	20510540	3040-01-279-4658	101	19
47457	20510551	4720-01-300-4144	620	16
47457	20510558	4720-01-284-2235	620	1
47457	20510601 X 108IN		617	7
47457	20510601-5X170IN		618	7
47457	20510633 X 176IN		621	8
47457	20510633 X 290IN		620	5
16567	20510669-1	5340-01-280-7102	299	8
16567	20510669-2	5340-01-284-9653	299	3
4F744	20510674	5365-01-280-3690	297	15

PART NUMBER INDEX

CAGEC	PART NUMBER	STOCK NUMBER	FIG.	ITEM
47457	20510675	2530-01-280-8989	297	1
47457	20510684	5340-01-280-7099	297	5
47457	20510736	4730-01-279-1519	299	9
47457	20510805	4720-01-300-4143	620	13
47457	20510807	5340-01-280-7098	63	14
24825	20510859	2590-01-292-5707	373	2
47457	20510863	4710-01-285-3007	459	1
47457	20510878	5340-01-289-5028	598	43
47457	20510879	5365-01-288-1560	598	36
47457	20510880	3040-01-288-5313	598	31
47457	20510881	5340-01-288-2132	598	40
47457	20510893	4720-01-279-3038	315	3
47457	20510894 X 44 IN		315	11
7T637	20510895	4710-01-279-3165	315	7
3T063	20510912	5310-01-282-2807	15	5
08277	20510914	3040-01-280-4153	101	14
47457	20510922	2910-01-285-6253	111	18
47457	20510922-5	6685-01-361-7552	111	20
5U403	20510923	2590-01-291-4598	110	3
4F744	20510924	4730-01-279-1510	111	15
47457	20510929	5340-01-280-6996	110	19
47457	20510933	5340-01-285-7762	78	20
04827	20510979	4720-01-285-3008	111	17
47457	20510980	5340-01-280-6869	111	12
08627	20510993	5340-01-280-6875	111	4
47457	20510996	5310-01-287-8812	111	13
16567	20511102	3040-01-279-4659	101	24
47457	20511127	5340-01-280-6874	184	11
47457	20511136	2990-01-284-3218	585	
			598	47
47457	20511138-1	4720-01-287-4494	595	5
47457	20511138-2	4720-01-287-4495	595	3
47457	20511139-1	2990-01-288-5844	585	22
47457	20511143	5342-01-288-2161	595	11
47457	20511145-1	2990-01-287-3189	585	21
47457	20511145-2	2990-01-287-8957	594	19
47457	20511155	3040-01-280-4382	297	18
47457	20511162	5340-01-281-5150	110	14
47457	20511166	5340-01-280-3716	110	11
16567	20511172	5340-01-287-6660	107	9
4F744	20511181-1	4710-01-279-1494	620	7
47457	20511181-2	4710-01-289-5447	620	7
47457	20511181-3	4710-01-289-5448	620	7
47457	20511181-4	4710-01-289-5449	620	7
57643	20511185	9390-01-280-3408	110	22
05MG0	20511188	5340-01-280-7096	619	7
47457	20511190	5305-01-287-1585	101	24
47457	20511191	2540-01-284-8718	599	5
			599	6
			599	8
			599	9

CROSS-REFERENCE INDEXES

PART NUMBER INDEX

CAGEC	PART NUMBER	STOCK NUMBER	FIG.	ITEM
47457	20511192		598	21
47457	20511200 X 16 IN		598	13
47457	20511200 X 32 IN		598	13
47457	20511200 X 46 IN		598	16
4F744	20511201	4710-01-279-3158	287	5
47457	20511203	4710-01-287-4608	598	22
47457	20511207	5315-01-288-4584	598	32
47457	20511216	4730-01-300-4104	617	11
47457	20511217	4730-01-300-4091	617	17
47457	20511221	5340-01-330-3240	619	7
47457	20511224-1	5365-01-301-7871	6	26
47457	20511224-2	5365-01-301-7872	6	26
			6	26
47457	20511224-3	5365-01302-9485	6	26
47457	20511225-1	5365-01-304-0652	6	26
4F344	20511226	5365-01-302-3189	622	6
47457	20511231	4710-01-289-5446	621	13
47457	20511234	5340-01-284-9656	623	20
47457	20511235	5340-01-285-3257	623	25
47457	20511240	4710-01-287-9008	620	4
47457	20511248	5340-01-288-4557	622	4
47457	20511256-1	4710-01-299-9451	621	7
47457	20511256-2	4710-01-299-9450	621	3
47457	20511258 X 22IN		110	8
47457	205112580 X 25IN		110	1
05MG0	20511262	5120-01-285-5192	630	18
47457	20511263	5120-01-285-7620	630	19
47457	20511265	4220-01-307-4779	622	8
47457	20511268	5330-01-288-1466	599	2
47457	20511272	5310-01-302-9472	297	12
4F744	20511273	4710-01-279-3163	72	9
47457	20511274	4710-01-279-3164	72	6
47457	20511279	2530-01-319-2384	622	10
47457	20511284 X BULK		585	6
47457	20511285	4730-01-300-4112	111	23
47457	20511286	5340-01-339-0874	619	7
47457	20511320	5220-01-298-5730	630	8
47457	20511338	6680-01-272-9204	147	3
OUBK6	20511420	5330-01-361-5600	103	5
15434	2063260	5305-01-113-1179	49	5
61465	2069245	5305-00-270-7328	526	16
15434	2069245	2590-00-421-9524	28	4
15434	208069	5330-00-006-2529	19	13
15434	208346	5305-01-179-2380	7	20
15434	208460	2815-00-453-8994	12	5
15434	209726	4730-01-078-2731	66	12
11757	21-P-131	3110-00-240-9897	530	31
15434	210036	5340-01-112-4280	20	17
95019	210084-3X	3130-00-908-8589	220	6
01276	210104-8S	5310-00-003-4094	5	14
			12	15

PART NUMBER INDEX

CAGEC	PART NUMBER	STOCK NUMBER	FIG.	ITEM
01276	210104-8S	5310-00-003-4094	105	8
			106	6
			109	9
			118	20
			120	5
			177	6
			178	2
			182	29
			204	6
			235	8
			237	7
			242	8
			293	5
			328	8
			351	24
			370	21
			396	7
			461	17
			467	2
			468	4
			468	4
			469	15
			470	21
			474	2
			478	17
			484	3
			484	30
			485	24
			487	10
			487	25
			489	19
			492	23
			493	15
			495	3
			501	18
			502	29
			503	11
			504	23
			505	2
			506	2
			509	12
			589	57
			615	9
06853	210492	5315-01-246-4339	248	6
15434	210879	2815-01-137-9707	42	7
15434	210884	5365-01-150-6257	7	15
15434	211255	5330-00-135-6382	19	7
15434	211305	5342-01-110-1505	88	18
15434	211315	5330-01-060-9061	284	14
15434	211662	3120-01-129-7659	284	42
15434	211999	5360-00-009-9270	24	3

CROSS-REFERENCE INDEXES

PART NUMBER INDEX

CAGEC	PART NUMBER	STOCK NUMBER	FIG.	ITEM
			24	11
06853	212193	4730-00-050-4358	269	20
06853	212227	2530-00-545-5406	291	3
15434	212601	3040-01-070-9004	50	17
			55	16
15434	212602	3020-01-086-4158	50	18
			55	18
15434	212603	5330-01-072-8828	50	4
			55	4
15434	212604	5325-01-081-0662	50	12
			55	12
15434	212605	3020-01-070-9003	50	13
			55	13
15434	212609	3120-01-087-2539	50	6
			55	6
15434	212610	3020-01-086-8780	50	7
			55	7
15434	212613	3010-01-080-1529	50	19
			55	19
15434	212639	3010-01-088-5727	50	14
			55	14
15434	212668	5315-01-087-0534	50	16
			55	17
15434	212954	5305-01-079-7028	46	5
15434	213394	5340-01-087-0682	9	4
15434	213395	5340-01-087-0681	9	2
15434	213559	2815-01-048-6702	23	13
06853	213630	5330-00-090-2128	252	20
			255	14
			261	23
			262	6
			291	13
15434	213768	5331-01-072-8983	51	5
			56	4
15434	213769	3110-01-079-8190	51	17
			56	17
24617	2146061	5340-00-456-1011	386	8
15434	214730		17	3
15434	214950	3120-01-087-3004	17	13
15434	214951	3120-01-155-4442	17	13
15434	214952	3120-01-157-3316	17	13
15434	214953	3120-01-155-8707	17	13
15434	215090		7	3
7Y635	21606	2520-00-557-6549	188	8
15434	216165	2990-01-120-2883	16	2
77640	216191-X1	3110-01-163-4932	306	23
06853	216310	4730-00-595-2572	263	20
			264	17
			264	29
			266	16
			273	20

CROSS-REFERENCE INDEXES

PART NUMBER INDEX

CAGEC	PART NUMBER	STOCK NUMBER	FIG.	ITEM
15434	216524	5340-01-086-6193	9	3
15434	216908		50	3
			55	3
15434	216983		17	2
15434	217381		14	8
15434	217385	2520-01-127-6254	14	10
24617	2173982	5340-00-456-1011	384	4
			533	14
15434	217632	4730-01-085-7328	116	5
06853	217690	4730-00-526-0284	272	12
			273	22
24617	217985	4730-00-249-1511	589	47
			590	32
15434	218025		17	1
15434	218153	3120-01-079-5208	26	9
78222	2182413K		308	7
44674	218406	4730-00-014-4593	521	21
21450	218709	4730-00-138-8133	482	13
15434	218732		17	4
15434	218736	4730-01-124-0293	26	18
15434	218793		284	22
74080	219132	2930-01-286-8357	121	1
15434	219153	5306-01-079-7027	17	12
24617	219682	4730-00-138-8121	508	26
21450	219760	5310-00-021-9760	262	13
21450	219831	4730-00-287-3790	456	22
11757	22-P-24-1	5330-00-485-0863	530	30
11757	22-P-24-2	5330-00-485-0865	530	30
10001	22-W-1642-100-40		236	21
72962	22NA797-82	5310-00-596-7753	500	27
80244	22W1642125X12 IN		233	15
			235	11
			240	15
			242	11
80244	22W1642125X36 IN		232	10
			239	10
81860	22003-14	5342-01-185-3561	6	3
78500	2203-L-7786	5365-01-128-5061	207	19
78500	2203-M-7787	5365-01-133-3749	207	19
78500	2203-N-7788	5365-01-133-3750	207	19
78500	2203-P-7790	5365-01-128-5062	207	35
78500	2203-Q-7791	5365-01-128-5063	207	35
78500	2203-R-7792	5365-01-128-5064	207	35
78500	2203-V-7848	5365-01-135-2829	207	13
78500	2203-W-7849	5365-01-135-2830	207	13
78500	2203-X-7850	5365-01-135-2831	207	13
78500	2205-Q-43	2520-01-127-7790	526	14
78500	2206-F-58	4730-01-134-6988	217	9
			526	30
78500	2206-J-88	4730-01-127-7346	217	7
			526	28

CROSS-REFERENCE INDEXES

PART NUMBER INDEX

CAGEC	PART NUMBER	STOCK NUMBER	FIG.	ITEM
78500	2206-W-1011	5305-01-271-6450	294	10
78500	2207-K-11	3040-01-134-3605	217	3
78500	2208-S-1033	5330-01-272-1148	294	13
78500	2208-U-697	5330-00-549-7694	526	22
78500	2221-B-2	5310-01-141-8464	209	12
62983	222640	2530-00-087-0327	463	19
			464	19
			511	19
78500	2233-W-101		232	12
			239	12
62983	223388	5360-00-953-5205	463	16
			464	16
			511	16
62983	223489		463	9
			464	9
			511	9
62983	223493		463	10
			464	10
			511	10
60602	2242-54	2590-01-284-4547	184	10
78500	2244-D-30	2520-01-134-7657	216	6
78500	2244-Y-25	2520-01-136-8719	216	12
78500	2244-Z-26	3040-01-134-1318	216	4
78500	2245-H-1022	3120-01-271-8353	294	14
			296	19
78500	2255-H-86	5340-01-289-5030	250	4
			251	4
78500	2258-P-640	5360-01-147-4787	216	10
78500	2258-S-1033	5360-01-275-0545	246	23
62983	226161		463	11
			464	11
			511	11
78222	2262021		307	4
78222	2262121		307	3
81860	22696-5	5342-01-101-0075	106	12
			107	33
68505	229	2920-00-078-2390	KITS	37
78500	2296-N-1002	4710-01-271-7939	296	3
97286	2297-C-627	2520-00-255-5149	207	26
			207	30
78500	2297-J-3832	2520-01-142-1979	217	1
78500	2297-K-3885	2530-01-127-3596	245	4
78500	2297-L-3886	2530-01-131-7741	245	4
78500	2297-N-5630	4730-01-272-0582	294	11
78500	2297-S-3815	2520-01-127-2624	207	5
78500	2297-T-3816	3020-01-126-8444	207	42
78500	2297-Y-4341	5306-01-196-6632	250	15
			251	17
78500	2297-Y-5329	5315-01-270-8268	246	12
78500	2297-Z-5330	5315-01-270-8269	246	12
80045	23MS35338-50	5310-00-820-6653	5	4

PART NUMBER INDEX

CAGEC	PART NUMBER	STOCK NUMBER	FIG.	ITEM
80045	23MS35338-50	5310-00-820-6653	5	9
			177	3
			236	26
			303	4
			345	5
			469	9
			470	3
			471	23
			473	12
			474	29
			476	8
			483	5
			484	32
			487	7
			499	7
			552	45
			613	16
73342	23010610	2840-01-141-9503	189	31
73342	23010620	4320-00-403-5223	185	23
73342	23010654	2940-01-140-8227	192	17
73342	23011471		192	11
73342	23011665	2520-01-146-1034	191	7
73342	23011670		198	8
73342	23011821	2520-01-146-5254	189	8
73342	23011825	2520-01-130-5772	193	11
73342	23011924	4730-01-078-2732	192	8
73342	23012036	4730-01-146-1075	189	1
			190	9
95019	230123-8	5310-01-135-6700	220	4
73342	23012444		195	1
73342	23012446	4820-01-148-9278	187	4
73342	23012502		KITS	
63005	23012925	5360-01-157-9921	185	13
73342	23013398	5306-00-570-8942	189	14
73342	23013437		200	17
73342	23013668	2520-01-206-9575	189	22
73342	23013746	2520-01-149-1785	187	6
73342	23013747	5360-01-150-9693	187	5
73342	23013789	2520-01-150-0901	193	1
73342	23014094	5310-01-145-6923	189	13
73342	23014097	4820-01-007-0350	197	22
73342	23014221	5330-00-557-6518	200	30
73342	23014632	2520-00-557-6211	192	19
			200	29
73342	23015398	3020-01-277-4637	200	7
73342	23015399	3020-01-277-4638	200	7
73342	23015400	3020-01-104-3803	200	7
73342	23015799	3110-00-557-6163	187	16
73342	23015870	2520-01-146-3039	200	1
73342	23015872	2520-01-134-6534	200	12
73342	23015874		200	16

PART NUMBER INDEX

CAGEC	PART NUMBER	STOCK NUMBER	FIG.	ITEM
73342	23015876		200	18
72582	23015880	5330-01-146-6053	187	11
			193	13
95019	230160-1	5310-01-135-6699	220	3
73342	23016018	2520-01-150-0899	190	5
73342	23016019		190	6
73342	23016106	3020-01-078-0627	200	6
73342	23016107	3020-01-277-4635	200	6
73342	23016347	2840-01-068-1713	200	3
73342	23016366	5360-01-224-9218	196	5
73342	23016413	4820-01-221-5415	196	1
73342	23016606	2520-00-557-5807	191	4
73342	23016608	2520-00-557-6076	192	4
73342	23016610	2520-01-065-0077	187	20
73342	23016643	5330-01-219-2375	200	2
73342	23016723	3040-01-337-4379	201	1
62860	23016745	5310-01-135-6042	591	3
73342	23016763	3040-01-337-4161	201	9
			201	11
73342	23016862	2520-01-123-1648	186	5
73342	23016957	2520-00-406-2303	185	6
73342	23016995	2520-01-134-0898	185	1
73342	23017112	3020-01-340-0403	201	7
73342	23017113	3020-01-340-0402	201	7
73342	23017114	4320-01-337-0608	201	4
73342	23017115	2815-01-340-0377	201	2
			201	6
73342	23017172		187	8
73342	23017257	3040-01-337-4155	193	11
73342	23017260	3040-01-336-8217	193	10
73342	23017261	3040-01-337-4168	193	10
73342	23017262	3040-01-337-4167	193	10
73342	23017311	3020-01-340-0407	201	8
73342	23017312	3020-01-347-2855	201	8
73342	23017314	3020-01-340-0404	201	8
73342	23017361		192	9
73342	23017446	2520-01-098-5108	185	17
73342	23017610 KIT		189	32
73342	23017696	2520-01-130-5770	187	2
73342	23017699	2520-01-131-2694	188	15
73342	23017700		188	17
73342	23018593	2520-01-168-1632	192	6
73342	23018867	5325-01-219-2580	187	17
			193	2
73342	23019201	2520-01-211-6702	KITS	
73342	23019596	2520-01-428-7497	KITS	
06853	230250	5310-01-133-7216	248	4
73342	23040890	5310-00-478-0548	190	17
73342	23042126		201	10
73342	23042210	2590-01-271-7861	194	2
73342	23045053	3040-01-151-3568	186	2

CROSS-REFERENCE INDEXES

PART NUMBER INDEX

CAGEC	PART NUMBER	STOCK NUMBER	FIG.	ITEM
73342	23045054	2520-01-226-3399	194	1
73342	23045085	4710-00-557-5885	189	11
73342	23045099	5330-01-219-2555	186	1
			190	4
73342	23045281	3020-01-008-2769	188	7
73342	23045372	2520-01-122-5169	186	8
73342	23045496	2520-01-151-2628	194	6
73342	23045497	2520-01-150-3690	194	6
73342	23045498	2520-01-149-3439	194	6
73342	23045556	5330-01-339-0708	201	3
61038	23046	5305-00-253-5618	560	2
73342	23046152	3120-01-341-6519	201	5
73342	23046718	5365-01-277-9878	192	2
73342	23046720	2520-00-557-5981	191	2
73342	23046721	2520-01-064-8847	191	2
73342	23046722	2520-00-563-6041	191	2
73342	23046745	2520-00-557-5799	192	2
73342	23046746	2520-00-557-5924	192	2
73342	23046747	5365-01-280-5643	192	16
73342	23046748	2520-01-078-5564	192	16
73342	23048028	3110-01-147-6681	186	10
73342	23048030	3110-01-179-1161	185	19
73342	23048511	2520-01-317-8288	185	5
06032	2310-0143-001	5310-01-121-1703	120	6
43990	2315-24		621	22
62983	232797	4820-00-960-9329	464	6
			511	6
62963	232798	4820-00-034-1690	463	6
62983	233019	4820-01-132-0582	463	6
06853	233955	5331-00-074-2692	248	12
78222	2350961	5330-00-548-6114	308	8
78222	2351081	5330-01-339-0859	308	9
26151	236-0317	4720-00-477-8933	62	26
78222	2360451 KF		308	11
78222	2360651		308	20
78222	2360831		308	3
45225	23622	4730-01-284-9086	630	1
78222	2369001		307	6
78222	2370291		308	6
78222	2370392K		309	10
78222	2370461K	5330-01-288-6304	308	4
62983	237736		463	12
			464	12
			511	12
06853	239029	5331-00-454-0364	249	10
06853	239136	5331-00-883-2799	249	33
71843	2392A	5340-00-679-1492	419	16
06853	239219	5331-00-465-6453	249	29
06853	239330	5360-01-127-0953	289	8
06853	239643	5331-00-085-3494	249	19
81343	24 070109C	4730-01-222-7575	629	30

CROSS-REFERENCE INDEXES

PART NUMBER INDEX

CAGEC	PART NUMBER	STOCK NUMBER	FIG.	ITEM
30327	24SG-12X08	4730-00-189-3034	589	50
06853	240273	4820-01-104-9992	289	4
06853	240345	5340-00-738-7552	249	31
06853	240445	5315-01-112-4507	248	13
78222	2411521		308	1
78222	2411531K		309	2
06853	241559	5342-00-931-4527	249	43
53867	2420210034	5331-01-301-1825	92	13
24617	2436161	5310-01-102-3270	623	28
24617	2436165	5310-01-121-1703	616	7
24617	2436167	5310-01-151-7347	616	2
06853	243633		289	18
06853	243635	5360-01-020-7063	289	14
06853	243637	4820-01-160-9597	289	13
06853	243890	5305-01-133-7193	248	5
06853	244282	5355-01-108-5184	289	2
06853	244435	5340-01-083-6420	249	25
06853	244680	4820-01-104-9313	248	11
06853	244682	2540-01-103-9128	248	8
06853	245118	3120-01-132-5579	248	7
95879	2460-44	4730-00-194-1766	110	9
06853	246091	4730-00-180-7031	617	14
62983	246632	5360-00-960-9326	463	26
			464	26
			511	26
06853	247213		249	3
06853	247216	2530-01-127-3971	249	6
06853	247217	5360-01-124-6811	249	7
06853	247221	5331-01-124-5720	249	44
06853	247222		249	22
06853	247224	5360-01-124-1402	249	32
06853	247226	5331-01-126-1233	249	34
06853	247231	5360-01-125-1671	249	21
06853	247233	5330-01-132-4734	249	35
06853	247234	5330-01-124-6405	249	45
06853	247235	5331-01-123-4536	249	36
06853	248114		249	15
70040	25011151		204	16
70040	25011159	2520-01-123-2649	204	15
15434	251081	5310-00-420-8044	93	3
15434	251152	5360-00-932-7452	33	23
15434	251390		93	6
79470	252X5	4730-01-240-6112	422	14
			423	14
33457	252916	2940-00-552-3842	30	1
81860	25295-1	5342-01-331-7985	6	11
15434	256172	4930-00-477-8276	93	2
15434	256416		93	7
33457	256424	4330-01-110-9054	93	8
33457	256476	2910-00-152-2033	KITS	
01276	2565-8	4720-00-670-6037	BULK	18

CROSS-REFERENCE INDEXES

PART NUMBER INDEX

CAGEC	PART NUMBER	STOCK NUMBER	FIG.	ITEM
01276	2565-8 BULK		314	3
78222	2574418		308	2
15434	257548	2910-01-097-8591	93	5
78222	2582563	2530-01-339-8673	307	5
62983	259871	5360-00-918-1920	463	5
			464	5
			511	5
62983	26VQ14A-1C-20	4320-01-106-2061	462	1
			510	1
78500	26X-230	5305-01-142-2794	216	9
78500	26X-235	5305-01-142-2793	216	3
80064	2601167	5935-01-114-5354	160	3
98168	261-3	4820-01-309-3799	82	8
01212	2622PN	2815-00-008-1741	17	5
10001	265850PC88	5310-00-193-9753	528	21
			529	5
06848	2666715-060	5310-01-135-3554	296	7
93061	269AB-12-8	4730-01-122-5857	620	9
16236	269013000000	2540-01-076-9286	589	34
			590	22
81348	27-W-918-20		478	14
81348	27-W-918-31		478	15
81348	27-W-921-19		478	19
81348	27-W-921-40		478	10
71843	2708-6A	5340-00-466-4948	211	23
			211	52
			419	15
78500	2710-U-151	3040-01-132-8272	244	15
24617	271500	5310-01-107-3570	110	13
62983	271722		463	18
			464	18
			511	18
04164	272M0075P004	4730-00-580-7408	265	5
93061	272NTA-8-6	4730-01-119-6895	617	6
93061	272NTA-8-8	4730-01-300-9031	618	3
24617	272977	4730-01-030-4950	19	18
			248	17
			279	19
			474	26
73342	273340	5306-00-400-5541	195	3
78500	2740-D-1122		244	7
73342	274613	5325-00-349-8518	200	26
70485	2758	5325-00-249-6345	541	6
78500	2758-A-53	5360-00-758-6456	244	8
78500	2758-W-127	5360-00-482-4422	245	18
29510	278027R1		BULK	47
06853	278599	2530-00-981-8736	278	5
06853	278614	4820-00-728-7467	254	3
			256	18
			290	13
			604	12

PART NUMBER INDEX

CAGEC	PART NUMBER	STOCK NUMBER	FIG.	ITEM
06853	279000	4820-00-436-3033	290	6
11757	28-P-52	5330-00-237-7828	530	14
78500	2803-L-220	5365-00-614-3903	527	4
78500	2803-M-221	5365-00-545-3723	527	5
01276	2807-12X14 1/21N		621	14
62983	282027	5305-00-471-3909	463	21
			464	21
			511	21
62983	284154	5342-00-889-5209	463	22
			464	22
			511	22
62983	284155		463	23
			464	23
			511	23
62983	284156	5310-00-867-1465	463	1
			464	1
			511	1
78500	2849-N-92	2520-00-734-9606	526	17
06853	285172	4820-00-420-5499	252	17
			255	9
			261	20
			262	18
			264	14
			264	30
06853	285696	4820-00-595-3669	291	8
78500	2858-T-20	5360-00-321-5710	526	19
62983	286669		463	7
			464	7
			511	7
06853	288251	4820-00-857-2737	280	1
06853	289022		249	13
06853	289024	2540-01-127-1812	248	2
06853	289352	2530-01-134-1834	KITS	
06853	289353	2530-01-134-1835	KITS	
06853	289849	2530-01-128-5552	280	1
06853	290184	2530-01-160-9652	249	20
06853	290185	5360-01-125-1670	249	37
06853	290186	5305-01-125-0929	249	18
06853	290187	4820-01-132-0606	249	42
06853	290188	5310-01-126-8834	249	41
06853	290189	5340-01-135-4292	249	38
06853	290202	2530-01-122-9933	249	17
53867	2911061196	5305-01-187-0531	91	19
06853	291111	5315-01-160-9818	*283	3
53867	2916710613	5325-01-276-8488	58	28
			89	8
06853	291882	5330-01-123-6409	248	3
06853	291883	2530-01-123-3105	248	16
78222	292SAF61	2530-01-284-4566	307	1
06853	292532	2540-01-131-9639	248	9
06853	292894	5330-01-120-3733	249	27

CROSS-REFERENCE INDEXES

PART NUMBER INDEX

CAGEC	PART NUMBER	STOCK NUMBER	FIG.	ITEM
06853	292896	2530-01-134-1838	249	5
06853	292897	5325-00-836-2131	249	11
06853	292898	5330-01-125-6280	249	8
06853	292899	5331-01-123-7038	249	9
06853	293337	5306-01-124-1225	248	15
06853	293514		289	7
06853	294515	3040-01-155-5795	249	40
73342	29501160	5330-01-120-8090	189	25
73342	29502387	3020-01-179-7515	190	1
73342	29503795	3020-01-134-1830	190	2
73342	29503808	3040-01-255-4406	185	3
73342	29516333	2520-01-006-7118	193	3
73342	29527485	3020-01-340-0408	201	8
73342	29527492	3020-01-277-4636	200	7
73342	29527509	3020-01-098-6742	200	4
11757	3-P-202	3040-00-125-2961	530	15
06721	3-X-285	5305-01-135-5447	282	11
1E045	3-10	5340-01-260-9942	629	13
60528	3-82-6	5365-00-700-1851	346	20
24614	3A7313	5310-01-375-3107	293	1
			294	18
			295	15
70960	3NDCA-4		506	13
70960	3NF20		506	15
88032	3008149X1	5310-00-680-5956	344	4
			345	12
15434	3000171	4310-01-092-9816	44	4
15434	3000173	5306-01-119-4271	284	59
15434	3000174	3010-01-085-2732	44	9
			284	41
15434	3000266	3040-01-085-2616	129	8
15434	3000446	4320-01-098-5115	51	18
			56	18
15434	3000464	2910-01-076-8632	46	7
15434	3000465	5310-01-079-6708	46	4
15434	3000940	5360-01-222-6879	85	17
15434	3001155	5360-01-191-2950	87	18
15434	3002110	2910-01-086-7715	50	2
			55	2
15434	3903156	5330-01-072-8830	85	9
			87	11
			88	20
15434	3004258	5340-01-271-2416	8	2
15434	3004293	4730-01-078-4703	51	12
			56	15
15434	3004724	5365-01-126-3334	50	5
			55	5
15434	3005133	4310-01-092-9815	44	1
15434	3005152	2530-01-130-2339	284	38
15434	3005543	5342-01-109-4013	88	28
15434	3006182	5305-01-118-8826	28	2

CROSS-REFERENCE INDEXES

PART NUMBER INDEX

CAGEC	PART NUMBER	STOCK NUMBER	FIG.	ITEM
15434	3006183	2815-01-142-1732	28	7
15434	3006343	3040-01-086-1449	51	10
			56	9
15434	3006344	5305-01-135-5446	51	9
			56	8
15434	3006350	3040-01-150-4926	51	8
			56	7
15434	3006358	2815-01-146-1024	28	1
15434	3006430	2910-01-122-4015	51	7
			56	6
15434	3006456	2815-01-085-2618	24	5
15434	3007024	4730-01-070-6667	66	18
15434	3007025	5340-01-246-6172	66	15
15434	3007148	2815-01-126-7404	14	13
15434	3007279	2815-01-114-7397	14	7
15434	3007759	5331-01-072-4436	46	17
15434	3007835	2815-01-146-4182	9	1
15434	3008017	5330-01-079-6514	31	13
15434	3008047	3130-01-146-1150	7	19
15434	3008048	3130-01-146-4504	7	17
15434	3008049	3130-01-146-1228	7	12
15434	3008069	5305-01-212-5210	37	19
15434	3008464	5365-01-112-1524	10	5
15434	3008465	4730-01-165-9491	14	1
			32	2
			57	2
15434	3008466	4730-00-954-1281	7	29
			31	6
			41	10
15434	3008468	4730-01-147-2223	7	18
			8	5
			284	47
15434	3008469	4730-01-106-0202	7	25
			8	10
			14	15
			15	10
			32	3
			33	19
			37	12
15434	3008470	5342-01-143-6045	8	16
15434	3008591	5330-01-086-3523	41	1
15434	3008595	2815-01-146-1092	28	6
15434	3008706		46	14
15434	3009213	5310-00-356-1447	7	21
15434	3009846	4710-01-218-6950	66	6
15434	3010240	5330-00-632-3813	KITS	
15434	3010589	5305-01-129-6901	28	3
15434	3010590	5305-01-119-8621	44	14
15434	3010592	5305-01-176-8018	284	61
15434	3010593	5305-01-197-3449	23	21
15434	3010594	5305-01-130-6100	16	6

PART NUMBER INDEX

CAGEC	PART NUMBER	STOCK NUMBER	FIG.	ITEM
			118	16
15434	3010595	5305-01-085-8197	11	3
15434	3010596	5305-01-088-6019	31	1
			33	41
			116	8
			284	40
			286	35
15434	3010596	5305-01-088-6019	310	15
15434	3010597	5305-01-086-7036	114	1
15434	3010810	5342-01-145-1549	84	6
			86	6
15434	3010937		93	4
15434	3010941	3120-01-160-1891	88	11
15434	3010942	3040-01-181-9509	88	13
15434	3010945	5330-01-161-0289	88	14
15434	3011272		23	16
15434	3011273		23	16
15434	3011315	5306-01-203-6299	14	6
15434	3011342	5305-01-135-5344	14	19
15434	3011472	5330-00-480-6133	44	16
			KITS	
15434	3011711	5305-01-147-4033	19	5
15434	3011713	5305-01-145-8381	19	15
15434	3011714	5305-01-165-3892	19	12
15434	3011715	5305-01-072-8816	41	9
15434	3011934	2910-01-146-0048	46	18
15434	3012297	2815-01-215-1705	28	5
15434	3012472	5305-Oi-112-4312	286	30
15434	3012473	5305-01-137-6706	37	14
15434	3012479	5305-00-795-9352	33	36
15434	3012480	5305-01-227-6249	33	38
15434	3012497	2910-01-098-5093	88	22
15434	3012526	5310-01-126-1045	44	20
15434	3012537		46	11
15434	3012972	5330-01-131-2967	41	3
15434	3013001	2930-01-134-2238	116	1
15434	3013607	6620-01-158-3125	114	3
15434	3013786	4730-01-161-5115	7	24
			19	6
			116	6
15434	3013904	5305-01-112-9021	88	19
			116	11
15434	3013930	2815-01-086-4508	17	9
15434	3014103	5310-00-081-9292	44	8
15434	3014622	2815-00-132-0240	24	12
15434	3014623	2815-01-127-1060	24	12
15434	3014624	2815-01-127-3597	24	12
15434	3014625	2815-01-127-3598	24	12
15434	3014979	2805-00-404-2917	14	14
15434	3015282	5305-01-129-4384	49	7
15434	3015469	5310-01-145-1114	46	2

PART NUMBER INDEX

CAGEC	PART NUMBER	STOCK NUMBER	FIG.	ITEM
15434	3016021	2910-01-177-8816	88	10
15434	3017051	5305-01-112-9110	50	24
			55	24
			85	8
			87	12
			88	27
15434	3017052	5305-01-126-1128	50	28
15434	3017052	5305-01-126-1128	85	11
			87	9
			88	23
15434	3017065	3020-00-331-7672	33	17
15434	3017544	5315-01-126-0601	23	15
15434	3017750	5330-00-861-8592	28	8
15434	3017946	2815-01-202-9715	12	14
15434	3018049	2815-00-705-2851	23	10
15434	3018051	2815-01-114-7398	23	3
15434	3018153	3120-01-146-7196	286	44
15434	3018491	2530-01-183-0651	284	2
15434	3018527	2990-01-080-0533	284	1
15434	3018682	5305-01-134-5659	50	29
15434	3018888	4730-01-300-9024	115	17
15434	3019061	3030-01-200-6004	118	26
15434	3019077		14	9
15434	3019116	5331-01-160-7458	31	4
15434	3019174		12	17
15434	3019175		12	17
15434	3019176		12	17
15434	3019177		12	17
15434	3019178		12	17
15434	3019180		12	12
15434	3019181		12	12
15434	3019182		12	7
15434	3019183		12	12
15434	3019184		12	12
15434	3019186		12	1
15434	3019187		12	1
15434	3019188		12	1
15434	3019189		12	1
15434	3019190		12	1
15434	3019192		12	11
15434	3019193		12	11
15434	3019194		12	11
15434	3019195		12	11
15434	3019196		12	11
15434	3019198		12	3
15434	3019199		12	3
15434	3019200		12	3
15434	3019201		12	3
15434	3019202		12	3
15434	3019204		12	10
15434	3019205		12	10

CROSS-REFERENCE INDEXES

PART NUMBER INDEX

CAGEC	PART NUMBER	STOCK NUMBER	FIG.	ITEM
15434	3019206		12	10
15434	3019207		12	10
15434	3019208		12	10
15434	3019218		12	2
15434	3019301	5365-01-221-8749	66	4
15434	3019573	5305-01-145-8359	284	39
15434	3019955	5365-01-147-9802	7	2
15434	3019956	5365-00-488-0799	7	2
15434	3019957	5365-01-147-2496	7	2
15434	3019958	5365-01-147-2497	7	2
15434	3019959	5365-01-148-8353	7	2
15434	3019960	5365-01-147-2495	7	2
78500	302-Z-676	3040-01-138-8578	244	16
15434	3020479	2930-01-134-2192	118	6
15434	3020523	5340-01-173-0133	88	5
15434	3021470	5305-01-144-6233	284	4
			286	16
15434	3021735	5330-01-082-6985	14	17
60602	30223-3	5315-01-284-9583	101	5
15434	3022589	5306-01-119-8870	50	21
			55	21
15434	3022590	5305-01-129-4214	284	58
15434	3023101	5360-01-086-3480	284	10
			286	12
15434	3023198	4730-01-300-9030	115	16
15434	3023451		55	10
15434	3024416	2815-00-404-2915	19	2
15434	3024709	5330-01-145-5381	116	13
15434	3024989	2990-01-143-5489	84	1
			86	1
15434	3025198	3020-00-820-7915	33	33
15434	3025458	4730-00-011-3175	129	16
15434	3025459	5342-01-143-6046	85	10
			87	10
15434	3025460	4730-01-124-3762	51	20
			56	20
15434	3025516		17	6
15434	3026396	4730-01-110-0342	64	1
15434	3026556	3120-01-147-5275	44	3
15434	3026624	4730-00-050-0718	53	3
15434	3027646	5340-01-239-8606	8	3
15434	3027685	5305-01-263-2733	45	1
15434	3028279	5305-01-145-0777	46	19
15434	3028302		50	9
15434	30283770	2815-01-179-7516	55	8
15434	3029852	3120-00-877-2213	19	3
15434	3030038	2815-01-159-1789	24	6
			24	9
15434	3030269	2910-01-146-1999	50	1
15434	3030276	2910-01-090-9345	55	1
15434	3030464	5340-01-157-9896	31	9

CROSS-REFERENCE INDEXES

PART NUMBER INDEX

CAGEC	PART NUMBER	STOCK NUMBER	FIG.	ITEM
15434	3030506	3020-01-195-6990	44	17
15434	3030866	2930-01-204-4475	116	4
15434	3030970	4820-01-146-1048	129	15
15434	3031007	5330-01-165-2314	128	4
15434	3031137	2910-01-150-4925	46	20
15434	3031619		117	7
15434	3032014	5360-01-240-1626	85	6
15434	3032014	5360-01-240-1626	87	6
15434	3032682	2815-01-136-1986	23	2
15434	3032693	5340-01-271-2420	10	3
15434	3032861	5330-01-147-0748	37	3
15434	3032874	5331-01-220-2389	7	4
15434	3033098	5340-01-203-9261	66	3
15434	3033740	4820-01-164-7002	53	1
15434	3034217	3020-01-166-5647	53	2
15434	3034438	5315-01-177-7507	44	11
15434	3034578	2940-01-145-9398	32	13
15434	3035053	5330-01-160-7460	54	1
15434	3035194		21	2
15434	3035344	4810-01-187-4925	129	5
15434	3035612	4710-01-218-6951	66	13
15434	3035614	4710-01-218-6949	66	17
15434	3035616	4710-01-218-6952	66	13
15434	3035806	5310-01-303-5531	285	21
15434	3035961	2815-01-085-2569	26	1
15434	3036285	2815-01-210-6947	26	16
15434	3036472	2990-01-218-6935	85	5
15434	3036474	5340-01-241-6939	87	5
15434	3037045	5325-01-241-3888	7	22
15434	3037236	5331-01-331-9293	64	15
15434	3037625	4720-01-285-6312	63	4
15434	3038215	2910-00-803-2631	85	4
15434	3038216	5360-01-228-0747	87	4
15434	3038668	3040-01-219-5697	56	1
15434	3040760	2910-01-230-1919	84	6
15434	3042320		17	6
15434	3042763		7	1
15434	3043254	2910-01-303-1195	51	2
15434	3043947	4820-01-278-7385	284	15
15434	3043995	5331-01-271-8289	286	9
15434	3044873	4710-01-257-2193	39	4
15434	3044876	2815-01-257-0853	41	4
			41	4
15434	3044992		46	15
15434	3044993		46	15
15434	3044994		46	15
15434	3044995		46	15
15434	3045173	5330-01-072-8829	50	15
			55	15
15434	3045424-9254	2910-01-218-5158	54	3
15434	3045552	9905-01-229-3443	553	3

PART NUMBER INDEX

CAGEC	PART NUMBER	STOCK NUMBER	FIG.	ITEM
15434	3045670	4310-01-271-5103	286	47
15434	3046201	5331-01-072-8984	53	5
15434	3046281	2910-01-218-5155	46	1
61465	304677	5360-00-740-9727	469	11
			470	7
15434	3047159	5330-00-131-7072	284	21
			286	22
			286	27
15434	3050367	5342-01-271-2347	286	36
15434	3050924	3120-01-272-3271	286	37
15434	3050926	5342-01-271-2348	286	28
15434	3050927	5365-01-270-8376	286	38
15434	3053093	5310-01-270-8387	286	40
98441	30541-8-8B	4730-01-271-7957	618	1
			620	12
15434	3054250	2910-00-404-3054	46	1
15434	3054532	2910-01-191-8470	46	6
15434	3054533	2910-01-219-2086	46	6
61465	305478	2520-00-740-9805	472	44
15434	3054841	5330-01-285-4827	28	9
15434	3060711-4144	2910-01-215-6721	49	6
15434	3060912	5330-01-272-1142	115	8
15434	3065125	3040-00-388-3126	23	18
15434	3067613	5330-01-147-4071	33	1
15434	3067616	5330-00-361-2955	16	1
15434	3069017	5330-01-136-8569	53	8
15434	3069101	5330-00-026-2931	44	13
15434	3069103	5330-01-338-4829	49	1
15434	3071085	5330-00-005-0407	118	3
62983	307198	5365-00-182-6713	463	13
			464	13
			511	13
15434	3076040	5305-01-072-8826	51	6
			56	5
15434	3076189	5330-01-080-5021	9	7
62983	307951		463	20
			464	20
			511	20
15434	3081346	5305-01-271-6448	43	1
95019	308594-2	2520-01-285-6295	529	1
11757	31-P-27	5310-00-469-4039	530	22
1GF04	31R82-2-4B	4730-01-284-9071	110	7
56232	310-8	5340-00-809-1492	165	32
			214	7
70797	3104	5325-00-737-3246	606	13
78500	3107-E-31	2520-01-127-6901	210	1
78500	3107-V-22	2520-01-123-5556	527	16
41625	310777-000-0060.0	2590-01-079-1506	183	7
79470	3152X8	4730-00-014-4027	72	
			599	3

CROSS-REFERENCE INDEXES

PART NUMBER INDEX

CAGEC	PART NUMBER	STOCK NUMBER	FIG.	ITEM
10001	319119PC12	5320-00-010-4131	381	15
78500	3196-H-8	4710-01-126-9565	526	25
15434	320-1850	5330-01-181-0631	284	63
15434	3201386	5330-01-181-0630	284	52
			285	13
78500	3202J2610	2520-00-734-6960	237	5
78500	3211-H-2868		245	11
78500	3211-L-2872		245	11
43334	3212BAXR1A	3110-01-056-0031	190	14
78500	3219-C-4345	2530-01-134-6570	293	18
			295	10
78500	3219-E-4815	2530-01-271-8023	294	8
			296	12
78500	3219-H-4064	2530-01-138-2015	244	12
79470	3220X6X4	4730-00-202-6491	597	5
79470	3228X2	4730-00-142-2581	630	13
05840	323W231-J	3110-00-100-5355	234	17
			241	17
78500	3236-K-2013	2530-01-122-6016	245	19
85757	326C-78002	4720-01-160-0733	291	7
78500	3264-A-1067	2530-01-271-7075	246	18
			247	3
78500	3264-B-1068	2530-01-271-7074	245	1
			247	1
78500	3264-V-100	2530-01-138-2016	244	1
78500	3264-W-101	2520-01-137-4843	207	7
78500	3266-F-864	2520-01-139-3124	207	20
78500	3268-A-1067	2520-01-131-2831	526	4
78500	3268-L-1338	2520-01-271-7007	238	3
78500	3280-A-1977	5365-01-132-2015	295	17
78500	3280-B-1978	2530-00-741-1105	293	3
78500	3280-V-7536	2520-01-127-2336	209	4
78500	3280-W-8551	4730-01-271-7187	247	2
11757	328024X	3110-01-140-8880	530	18
78500	3282-K-63		527	12
11757	328273X	2520-00-232-1944	530	17
78500	3286-N-1054	5330-01-271-5151	236	1
78500	3286-P-1056	2530-01-285-3563	296	18
78500	3296-C-107	2520-01-136-8717	216	2
78500	3296-V-74	2520-01-136-8718	216	8
78500	3297-C-55		527	11
78500	3297-L-64	2520-01-127-2625	208	6
78500	3299-F-5362	5340-01-290-8882	251	3
78500	3299-G-5363	5342-01-297-4454	251	3
78500	3299-M-5369	5340-01-290-8884	250	3
78500	3299-N-5370	5340-01-290-8883	250	3
10001	33G1724	5325-00-276-6343	582	21
15434	3301954	5306-01-118-2300	30	9
33457	3301956	5325-01-209-7625	30	4
15434	33069826	5330-01-137-4487	128	2
15434	3313281	2940-01-157-6309	32	12

CROSS-REFERENCE INDEXES

PART NUMBER INDEX

CAGEC	PART NUMBER	STOCK NUMBER	FIG.	ITEM
78222	3331081K		309	1
79396	33472	2910-01-201-7719	94	2
			595	12
15434	3353977	5330-01-272-1108	20	15
15434	3376947	2915-01-285-2527	630	5
11757	34P17	1450-00-175-9752	530	7
98441	34982-14-6	4730-01-349-2436	58	30
95019	35-P-41	5330-01-133-0205	528	24
95019	35-P-41	5330-01-133-0205	529	9
			568	24
11757	35-P-8	5330-00-485-0895	529	2
			530	32
45152	35H113	2520-01-152-2384	308	17
70960	35R4-2A		506	11
15434	3501102	5340-01-271-2505	65	28
15434	3501188	5342-01-271-2343	65	29
15434	3502066	5340-01-271-2504	65	17
15434	3502449	5330-01-272-1250	65	11
15434	3503100	3120-01-272-3270	65	20
15434	3503347	5305-01-292-3182	65	26
15434	3503562	5310-01-288-5690	65	7
15434	3503662		65	24
15434	3503668	2950-01-211-0163	65	23
15434	3518980		65	21
15434	3519163	5305-01-271-8375	65	2
15434	3519302	2835-01-271-2511	65	15
15434	3519905	9905-01-279-4690	65	3
15434	3522801	2950-01-271-2342	65	10
15434	3522827	5305-01-271-8379	65	18
15434	3522879		65	13
15434	3523958	5331-01-272-1125	65	9
15434	3525358		65	8
15434	3525359	2950-01-271-2341	65	4
15434	3525739	3120-01-272-3269	65	22
15434	3527122	2815-01-332-5462	65	12
15434	3529372	5310-01-270-8245	65	5
15434	3530592	3130-01-331-7182	65	19
08627	3545	4730-01-280-6402	78	4
15434	3558006	2530-01-268-8740	286	1
7U263	3558516	4730-01-287-9084	286	26
15434	3558653	4310-01-079-3319	284	34
15434	3558655	2815-00-369-7846	284	33
			286	43
15434	3558749	4310-01-271-3807	286	32
15434	3558762	4310-01-271-3808	286	52
08627	3564	5340-01-288-0607	6	24
60602	35841-42	3040-01-282-4336	455	4
77640	365509-S-1	2530-01-134-3670	317	1
11757	37-P-20	5360-00-472-6822	530	8
99696	375N300-001-21	5305-00-719-5240	320	49
			321	54

PART NUMBER INDEX

CAGEC	PART NUMBER	STOCK NUMBER	FIG.	ITEM
			340	9
			349	4
15434	3756754	5330-01-203-3612	65	14
15434	3758848	5330-01-213-1258	65	25
15434	3759917	5305-01-272-3305	65	27
15434	3762259	5325-01-270-8361	65	16
78500	3780-Q-381	5331-01-292-9575	296	1
11757	378003	5310-00-838-1490	530	3
11757	378004	5310-00-469-4073	530	5
95019	378041-4	5306-01-135-7202	528	20
			529	4
11757	378391	5325-00-477-0304	530	28
11757	378430-10	5306-01-104-1048	530	6
11757	378452-3	5305-01-133-0163	530	24
95019	378766	5305-01-165-7541	529	11
8N900	378767	5325-01-165-2352	529	12
95019	379423-15	5307-01-055-8843	528	23
8N900	379423-18	5307-01-308-5081	529	8
15434	3801030	2815-00-791-1448	21	5
15434	3801056	2815-01-165-0765	KITS	
15434	3801260	3120-01-132-9339	KITS	
15434	3801261	3120-01-143-9547	KITS	
15434	3801262	3120-01-144-8882	KITS	
15434	3801263	3120-01-145-9132	KITS	
15434	3801264	3120-01-193-7083	KITS	
15434	3801314	2815-01-159-9538	7	5
15434	3801408	2815-00-457-6311	12	4
15434	3801433	3040-01-079-1799	26	19
15434	3801535	2815-01-278-1093	KITS	
15434	3801607	4310-01-304-9726	KITS	
15434	3801792	2530-01-131-6172	284	3
15434	3802081	2930-01-268-8751	119	1
			119	3
15434	3802085	2815-01-271-9794	25	6
15434	3802091	2910-01-292-5663	47	1
15434	3802110	2815-01-271-9792	18	9
15434	3802210	3120-01-273-4653	13	7
15434	3802211	3120-01-273-4654	13	7
15434	3802212	3120-01-273-4655	13	7
15434	3802213	3120-01-273-4656	13	7
15434	3802214	3120-01-274-3377	13	7
15434	3802257	2990-01-446-5359	65	1
15434	3802275	2815-01-272-6679	25	9
15434	3802278	2815-01-268-8753	35	8
15434	3802370	2815-01-271-9838	KITS	17
15434	3802467	2815-01-330-3036	10	1
15434	3802549	2815-01-330-8069	8	1
15434	3803431	5340-01-271-2471	65	6
15434	3803512	2815-01-354-2702	24	13
15434	3803676	2910-01-141-4337	50	8
15434	3804272	5330-00-133-6237	KITS	

PART NUMBER INDEX

CAGEC	PART NUMBER	STOCK NUMBER	FIG.	ITEM
7U263	3818823	5307-01-287-0854	64	12
15434	3818824	5310-01-270-8244	64	14
98441	38221-64-01RN	4720-01-279-3171	63	1
15434	3822503	5120-01-262-7309	630	2
15434	3822786	5120-01-291-5769	630	7
15434	3824591	5120-01-285-5193	630	3
73342	3829139	5306-00-024-6580	189	19
78500	3866-K-557	5340-00-734-6820	233	3
			240	3
59150	3875-425	5220-01-317-1436	63	5
78500	3880M533	3040-00-734-6892	241	18
78500	3892-A-4473	3020-01-142-8090	208	7
78500	3892-B-4474	3020-01-146-4198	209	5
78500	3892-J-4430	3020-01-133-9037	526	6
78500	3892-Y-4471	3020-01-141-1554	209	3
15434	3900216	5310-01-198-0997	38	10
15434	3900257	5315-01-188-0761	8	14
15434	3900620	5305-12-168-9310	22	5
15434	3900621	5305-01-245-3193	119	5
			126	1
15434	3900627	5305-01-193-6839	38	2
			130	3
15434	3900628	5305-01-272-4812	38	15
15434	3900629	5305-01-237-4915	20	14
			64	6
			124	2
15434	3900630	5305-01-234-3755	20	16
			60	5
			115	21
15434	3900632	5305-01-239-7202	124	12
15434	3900633	5306-01-237-1166	20	1
15434	3900677	5305-01-207-7447	35	9
			119	4
15434	3900706	5305-01-197-1663	27	6
15434	3900955	5340-01-239-8607	35	3
15434	3900965	5340-01-194-8936	8	6
			10	6
15434	3901177	2815-01-271-7171	25	7
			25	11
15434	3901249	5305-01-245-3817	40	5
15434	3901380	5306-01-271-6362	18	2
15434	3901381	5310-01-270-8246	18	3
15434	3901383	2815-01-271-5119	18	1
15434	3901395	5305-01-192-5677	15	4
15434	3901430	3120-01-274-3378	18	4
15434	3901431	3120-01-272-3272	18	4
15434	3901432	3120-01-272-3273	18	4
15434	3901433	3120-01-275-7664	18	4
15434	3901434	3120-01-275-7665	18	4
15434	3901445	5305-01-271-5850	32	1
15434	3901590		13	10

CROSS-REFERENCE INDEXES

PART NUMBER INDEX

CAGEC	PART NUMBER	STOCK NUMBER	FIG.	ITEM
15434	3901597	2815-01-287-4502	18	5
15434	3901617	5305-01-271-4326	27	1
15434	3901685	5365-01-271-1852	22	4
15434	3901693	5342-01-271-5704	27	2
15434	3901717	2815-01-271-5098	27	3
			27	4
15434	3901757	5306-01-237-7531	123	4
15434	3901764	5325-01-270-8360	27	9
15434	3901798	5310-01-270-8388	32	9
15434	3901846	5315-01-270-8284	8	7
15434	3901969	5340-01-271-2418	8	8
15434	3901996	5325-01-280-5592	18	7
15434	3902089	5331-01-272-1123	119	2
15434	3902112	5305-01-272-4811	47	7
			115	9
15434	3902114	5305-01-269-6274	115	32
15434	3902116	5305-01-274-4404	20	13
15434	3902253	2815-01-281-5206	25	10
15434	3902254	2815-01-280-8961	25	8
15434	3902332	5315-01-235-4688	22	7
15434	3902338	2815-01-211-5270	32	4
15434	3902425	5310-01-209-0508	10	7
			38	13
15434	3902466	5330-01-272-1246	29	6
15434	3902468	2590-01-273-3321	29	1
15434	3902595	3020-01-235-5055	286	39
15434	3902662	5310-01-270-8343	57	18
15434	3903035	5305-01-207-7243	57	23
15434	3903095	5305-01-236-6157	123	5
7U263	3903112	5306-01-288-1011	285	17
15434	3903118	5305-01-301-9756	115	20
15434	3903200	5305-01-268-5558	4	1
15434	3903380	5330-01-195-5268	47	2
			48	13
7U263	3903464	5305-01-288-1012	285	18
15434	3903475	5330-01-191-8047	45	3
15434	3903609	5305-01-234-3756	48	1
15434	3903652	5340-01-297-1187	103	4
15434	3903723	5310-01-234-1411	48	2
15434	3903744	4710-01-193-7944	64	7
15434	3903745	4720-01-271-6946	64	9
15434	3903845	5307-01-196-4246	94	1
15434	3903924	3040-01-189-1760	20	8
15434	3904181	4730-01-237-6950	15	8
15434	3904361	3040-01-271-7165	15	3
15434	3904386	5365-01-217-1995	10	8
			38	14
15434	3904483	5315-01-270-8285	13	6
15434	3904519	5340-01-196-3680	48	3
60602	39048	5365-01-284-8138	184	7
15434	3904849	5365-01-188-0954	20	9

CROSS-REFERENCE INDEXES

PART NUMBER INDEX

CAGEC	PART NUMBER	STOCK NUMBER	FIG.	ITEM
15434	3905156	5365-01-188-1054	47	5
15434	3905157	5365-01-190-2131	47	5
15434	3905158	5365-01-188-1055	47	5
15434	3905159	5365-01-191-3532	47	5
15434	3905160	5365-01-191-2413	47	5
15434	3905161	5365-01-191-3533	47	5
15434	3905162	5365-01-191-0774	47	5
15434	3905163	5365-01-189-7804	47	5
15434	3905164	5365-01-191-3534	47	5
15434	3905165	5310-01-189-5407	47	5
15434	3905166	5310-01-191-2494	47	5
15434	3905167	5365-01-189-9026	47	5
15434	3905168	5365-01-191-0775	47	5
15434	3905169	5310-01-191-7512	47	5
15434	3905170	5310-01-191-6333	47	5
15434	3905171	5365-01-189-9027	47	5
15434	3905172	5365-01-189-9028	47	5
15434	3905173	5365-01-189-9029	47	5
15434	3905174	5365-01-189-9030	47	5
15434	3905175	5365-01-188-1056	47	5
15434	3905176	5310-01-189-5408	47	5
15434	3905177	5365-01-191-0776	47	5
15434	3905178	5365-01-191-3535	47	5
15434	3905180	5365-01-191-2411	47	5
15434	3905181	5310-01-189-5409	47	5
15434	3905182	5310-01-189-5410	47	5
15434	3905183	5310-01-189-7499	47	5
15434	3905184	5310-01-189-5411	47	5
15434	3905185	5310-01-189-5412	47	5
15434	3905186	5310-01-189-5413	47	5
15434	3905194	2815-01-271-5120	27	16
15434	3905307	4730-01-195-0825	48	12
15434	3905401	5340-01-271-2417	8	4
15434	3905449	5330-01-271-8307	29	5
15434	3905779	2590-01-281-9716	35	2
15434	3905860	5305-01-234-3714	57	4
			60	4
15434	3905870	2815-01-271-5074	27	12
15434	3905928	4730-01-271-7181	8	11
15434	3906081	3120-01-266-1530	13	3
15434	3906100	5310-01-270-8422	27	7
15434	3906216	5310-01-270-8251	38	4
15434	3906299	5365-01-270-8290	32	8
15434	3906412	5360-01-271-8282	25	2
15434	3906436	4710-01-272-2882	124	7
15434	3906439	5340-01-271-2441	124	1
15434	3906440	5342-01-272-6724	4	2
15434	3906619	4730-01-281-0812	8	9
			20	4
15434	3906655	5305-01-271-5848	8	13
15434	3906659	5310-01-270-8417	47	6

CROSS-REFERENCE INDEXES

PART NUMBER INDEX

CAGEC	PART NUMBER	STOCK NUMBER	FIG.	ITEM
15434	3906715	5305-01-272-4810	115	1
15434	3906720	2815-01-272-0547	43	3
15434	3906733	5305-01-274-5655	13	1
15434	3906747	4720-01-271-7995	285	25
15434	3907167	4730-01-271-7969	285	7
15434	3907206	5310-01-270-8386	27	8
15434	3907233	5305-01-274-4407	10	11
15434	3907234	5305-01-271-5852	10	10
15434	3907242	6620-01-272-1716	115	12
15434	3907535	5340-01-271-2497	15	15
15434	3907546	5340-01-271-5705	48	14
15434	3907618	3040-01-262-1207	627	5
15434	3907734	2815-01-272-3954	29	4
15434	3907740	4730-01-271-3739	40	7
15434	3907741	5340-01-275-0487	40	11
15434	3907757	3040-01-271-5118	627	1
15434	3907792	2815-01-281-1125	8	20
15434	3907860	5305-01-266-8568	38	9
15434	3907861	5342-01-272-6725	4	3
15434	3907978	5307-01-271-9511	57	19
15434	3907998	5305-01-263-2708	20	10
15434	3908095	5340-01-271-2496	15	19
15434	3908096	5330-01-266-3294	15	20
15434	3908110	3120-01-272-3276	57	1
15434	3908316	5310-01-270-8390	38	8
15434	3908321	5305-01-274-4406	128	7
15434	3908328	5365-01-270-8481	128	3
15434	3908402	4710-01-271-7940	124	6
15434	3908513		47	4
15434	3908612	5305-01-276-0859	556	4
15434	3908738	5365-01-270-8482	38	16
15434	3908750		18	8
15434	3908763	4730-01-271-7977	115	13
15434	3908830	2815-01-272-5539	25	12
11862	3909063	5310-01-143-0542	192	15
15434	3909397	5331-01-272-1120	40	9
15434	3909410	5330-01-192-2037	15	16
15434	3909416		10	2
15434	3909545	4720-01-271-6945	64	4
15434	3909552	4730-01-272-2827	57	25
15434	3909556	4710-01-271-3841	57	24
7U263	3909557	4730-01-287-9100	57	26
15434	3909582	5305-01-271-6449	285	11
15434	3909669	4710-01-271-7922	40	3
15434	3909886	2910-01-271-9826	47	3
15434	3909897	5342-01-271-9818	123	2
15434	3909898	5342-01-271-5088	126	2
15434	3909899	5342-01-271-5089	126	7
15434	3910037	5305-01-272-1334	48	20
15434	3910248	2815-01-273-0571	15	12
15434	3910260	5330-01-272-1124	15	11

CROSS-REFERENCE INDEXES

PART NUMBER INDEX

CAGEC	PART NUMBER	STOCK NUMBER	FIG.	ITEM
15434	3910266	5310-01-270-8423	38	7
15434	3910279	5342-01-271-2349	47	8
15434	3910503	5331-01-272-1121	57	15
15434	3910540		15	7
15434	3910685	5340-01-275-3403	48	5
15434	3910687	5342-01-271-7081	48	16
15434	3910749	4710-01-271-3837	48	19
15434	3910750	4710-01-271-7198	48	18
15434	3910751	4710-01-271-3838	48	7
15434	3910752	4710-01-271-3843	48	9
15434	3910753	4710-01-271-3839	48	11
15434	3910754	4710-01-271-3840	48	10
15434	3910778	7690-01-271-1926	124	9
15434	3910824	5331-01-281-8997	29	3
15434	3910911	4710-01-271-7921	27	13
			27	14
15434	3910959	5310-01-270-8341	57	9
15434	3910960	5310-01-270-8391	38	3
15434	3910981	5305-01-272-4809	27	15
15434	3911258	2990-01-271-7086	15	21
15434	3911260	3020-01-271-7114	15	2
15434	3911456	4710-01-271-8032	285	5
15434	3911493	2930-01-291-5867	115	10
15434	3911533	5342-01-271-2346	285	14
15434	3911537	5330-01-271-9375	15	17
15434	3911604	2815-01-272-6719	15	9
15434	3911617	4710-01-271-7943	38	17
15434	3911630	6680-01-272-1867	35	1
15434	3911638	4730-01-331-2913	38	11
15434	3911934	4710-01-271-7942	285	10
15434	3911935	4720-01-272-2841	285	9
15434	3911936	4720-01-271-8000	285	12
15434	3911937	4710-01-271-8031	285	2
15434	3911941	5330-01-272-1146	64	13
15434	3911942	5330-01-281-9013	43	4
15434	3912004	3030-01-271-3754	123	1
15434	3912072	5305-01-192-2036	15	18
			40	4
			179	3
15434	3912484	3040-01-271-3821	57	10
15434	3912487	5340-01-271-2475	130	2
15434	3912627	4710-01-369-4814	285	10
15434	3912800	2930-01-272-6716	115	7
15434	3912888	5342-01-271-2340	57	13
15434	3912889	5310-01-271-5706	57	14
15434	3912996	5310-01-270-8229	57	17
15434	3912897	5310-01-270-8405	57	16
15434	3912925	4710-01-292-3701	57	5
15434	3912976	5340-01-271-2485	25	1
15434	3913024	2930-01-300-5943	115	25
15434	3913025	5330-01-302-0780	115	28

CROSS-REFERENCE INDEXES

PART NUMBER INDEX

CAGEC	PART NUMBER	STOCK NUMBER	FIG.	ITEM
15434	3913027	5330-01-301-1829	115	30
15434	3913028	6620-01-302-0045	115	27
15434	3913029	5365-01-302-6555	115	31
15434	3913030	2930-01-305-3303	115	22
15434	3913032	5330-01-301-1828	115	23
15434	3913033	4710-01-301-4358	115	19
15434	3913034	5305-01-301-9757	115	24
15434	3913326	3040-01-271-3820	57	12
7U263	3913370	5307-01-287-0855	285	23
7U263	3913371	5310-01-287-5742	285	20
15434	3913638	5305-01-271-5851	15	13
			57	21
15434	3913994	5330-01-291-6537	20	7
15434	3914011	2815-01-272-6714	38	12
7U263	3914017	5330-01-289-3135	38	6
15434	3914036	4710-01-271-7199	60	3
15434	3914037	4730-01-301-4286	60	9
15434	3914301	5330-01-271-4308	15	14
15434	3914302	5330-01-272-1143	38	1
15434	3914308	5330-01-271-6404	32	5
7U263	3914310	5330-01-287-8656	115	11
15434	3914311	5330-01-272-1138	124	8
15434	3914338	5342-01-239-7140	48	15
15434	3914339	5340-01-312-7652	48	4
15434	3914341	5342-01-280-6639	48	17
15434	3914388	5330-01-190-1905	64	5
15434	3914456	2815-01-271-7076	13	2
15434	3914501	2815-01-268-8737	124	11
15434	3914708	5310-01-330-8313	43	2
7U263	3914722	5340-01-287-5699	285	19
7U263	3914723	5365-01-287-5762	285	16
15434	3914943	4710-01-271-7938	64	2
15434	3915074	2835-01-271-2510	20	12
15434	3915416	2815-01-271-3763	27	11
15434	3915707	5340-01-281-7792	25	3
15434	3915772	5330-01-263-6179	20	11
15434	3915800	5330-01-270-8144	45	5
15434	3916042	5330-01-271-8306	285	22
15434	3916165	4720-01-271-6951	64	3
15434	3916193	3040-01-321-6365	20	2
15434	3916284	5331-01-272-1122	57	20
15434	3916369	5305-01-333-5382	8	13
15434	3916585	5305-01-273-1594	29	2
15434	3916830		13	9
15434	3916840		13	8
15434	3916857	4730-01-333-0133	64	11
15434	3917313	3130-01-281-9164	8	12
15434	3917328	3020-01-414-8008	22	6
15434	3917394	4730-01-271-7874	124	3
15434	3917417	4710-01-271-7941	64	10
15434	3917728	5305-01-331-9480	10	11

CROSS-REFERENCE INDEXES

PART NUMBER INDEX

CAGEC	PART NUMBER	STOCK NUMBER	FIG.	ITEM
15434	3917729	5305-01-331-9479	10	10
15434	3917737	5330-01-272-1282	8	19
15434	3917748	5342-01-271-5862	48	21
15434	3917780	5330-01-321-2053	20	3
15434	3918109	5305-01-245-3192	48	8
			124	13
15434	3918163	4730-01-271-7903	40	1
			124	4
15434	3918174	5330-01-271-5791	32	7
15434	3918175	2930-01-271-5102	32	6
15434	3918191	5310-01-340-8469	57	3
			60	2
15434	3918192		57	8
		5310-01-335-4861	60	8
15434	3918215	5340-01-266-3023	45	4
15434	3918290	2940-01-271-7203	32	14
15434	3918532	5340-01-331-9625	32	11
15434	3918611	4720-01-373-5652	40	2
			124	5
15434	3918616	4720-01-271-6950	40	8
15434	3918776	3020-01-271-3812	13	5
15434	3918986	2815-01-271-5096	13	4
15434	3919003	4730-01-331-2670	8	11
15434	3919038	5330-01-282-5653	25	4
15434	3919390	3020-01-271-3832	126	5
15434	3919683	5342-01-275-0384	20	6
15434	392049700	4730-01-271-7955	40	10
15434	3920703	9905-01-279-4691	556	3
15434	3920762	4720-01-271-7202	111	10
15434	3921530	4720-01-331-8717	64	4
15434	3921850	5330-01-271-4311	10	9
15434	3921852	5330-01-272-1144	10	9
15434	3921853	5330-01-272-1145	10	9
15434	3921980	5945-01-279-4802	130	4
73342	3921988	5365-01-089-3573	189	23
15434	3922117	4710-01-339-0584	48	6
15434	3922901	2990-01-271-9816	123	3
15434	3923054	5330-01-190-9555	60	7
15434	3924471	2815-01-388-8596	22	2
15434	3924725	4730-01-281-0840	57	7
15434	3924912	5340-01-331-9487	130	5
15434	3925031	2815-01-379-4920	22	1
15434	3925626	3120-01-280-6566	13	11
15434	3926094	2930-01-300-5944	115	29
15434	3926545	4710-01-271-8033	115	14
15434	3927155	2815-01-424-4736	22	3
7U263	3927611	5340-01-287-0751	38	5
15434	3929005	2815-01-272-5538	25	5
15434	3929253	5330-01-317-3213	20	5
15434	39317900	3020-01-280-9024	57	22
15434	3932226	2910-01-268-8736	60	6

CROSS-REFERENCE INDEXES

PART NUMBER INDEX

CAGEC	PART NUMBER	STOCK NUMBER	FIG.	ITEM
78500	394A	3110-00-100-0305	208	4
78500	39520	3110-00-143-7586	208	9
78500	39578	3110-01-143-9242	208	8
81343	4 120111B	4730-00-278-8824	70	3
			76	11
			77	5
			79	11
			79	11
			80	12
			148	5
			620	4
			621	9
81343	4 120115B	4730-00-122-0477	70	4
			73	2
			76	12
			77	4
			79	12
			80	11
11757	4-P-45	5365-00-121-2780	530	27
81343	4-010110B	4730-00-011-6452	266	7
72447	4-14-19	2520-00-457-6660	218	9
			224	13
81343	4-2 070102C	4730-00-187-0840	629	22
81343	4-2 100302BA	4730-01-134-6991	256	1
81343	4-2 120102BA	4730-00-277-8750	256	5
			276	5
			422	13
			423	13
81343	4-2 120103BA	4730-00-352-9793	213	3
81343	4-2 120202BA	4730-00-921-3240	70	3
			99	2
81343	4-2 120203BA	4730-00-912-9114	61	4
81343	4-4 070102SA	4730-00-837-7073	629	23
81343	4-4 070103CA	4730-00-802-2818	629	5
			629	24
81343	4-4 100102BA	4730-01-091-9212	70	7
			73	7
81343	4-4-2 070425CA	4730-00-522-1909	620	17
56442	4B4H27	2910-00-753-9184	100	26
			101	3
11083	4B4280	5310-00-930-7013	118	21
30780	4EBTXB	4730-00-905-0030	621	6
77640	400122-X1	2530-01-143-4203	306	22
77640	401212	5360-00-795-6975	317	24
77640	401233	5325-00-806-4105	317	4
77640	401264	5325-00-806-4104	317	8
77640	401309	5325-00-613-7796	305	14
77640	401314	5325-00-476-5259	306	36
77640	401375	5360-01-135-4065	306	29
77640	401379	5342-01-135-6350	306	25
77640	401445	5325-01-180-2448	305	6

CROSS-REFERENCE INDEXES

PART NUMBER INDEX

CAGEC	PART NUMBER	STOCK NUMBER	FIG.	ITEM
56152	401569		237	14
77640	402230	3040-00-415-1373	317	12
1DD64	402235-A1	4320-00-922-4933	150	3
77640	402341	2590-01-132-4654	317	14
77640	402368-A1	2530-01-127-5006	305	16
07367	402377	2530-01-126-8445	306	7
77640	402434	5340-01-136-1660	306	21
77640	403490	2530-01-132-8943	317	26
98441	40483-4-6S25	4730-01-283-8148	110	5
77640	415437-AI	5365-01-134-0522	306	9
77640	415442	4820-01-139-4346	306	26
57733	418049-50IN		625	4
80244	42-C-14490		383	39
80244	42-C-16560-9		109	4
80244	42-C-16570	4010-00-757-9556	BULK	5
80244	42-C-16570-6		379	14
80244	42C15120-205-6		347	6
0U276	42124	6680-01-287-2153	179	1
21450	423533	5305-00-484-6186	367	24
21450	423568	5305-00-042-3568	139	1
			140	1
24617	423569	5306-01-226-0798	614	8
98441	4244-24	4720-01-300-4146	111	25
21450	425346	5306-00-421-9402	414	1
21450	425544	5305-01-125-1181	134	25
24617	425567	5305-00-042-5567	367	13
21450	425570	5306-00-042-5570	161	17
			355	6
			457	14
			458	24
24617	425601	5305-00-042-5601	168	28
24617	425603	5305-00-042-5603	164	32
			166	29
			167	29
21450	425647	5306-01-134-1966	361	12
24617	425648	5305-00-042-5648	611	27
24617	425734	5306-01-136-5331	602	21
21450	425736	5306-01-166-6895	180	14
24617	425841	5306-00-042-5841	543	1
24617	425859	5306-00-042-5859	134	77
24617	425923	5320-01-145-3186	320	14
			321	29
			321	31
			323	33
			324	25
24617	425924	5320-01-145-4621	318	6
			319	7
			320	6
			321	6
			322	6
			323	6

PART NUMBER INDEX

CAGEC	PART NUMBER	STOCK NUMBER	FIG.	ITEM
			324	6
24617	426371	5305-01-031-4487	619	21
21450	426687	5325-00-281-8643	533	8
24617	432468		593	20
24617	432468		595	4
24617	432527		593	19
			595	6
60602	4333-60	3040-01-208-5897	180	9
15434	44035		23	20
21450	440502	5306-00-044-0502	111	8
			211	50
		5305-01-213-9852	455	10
		5306-00-044-0502	513	31
			525	28
			528	32
24617	441294	3110-00-100-0231	208	5
24617	442340	4730-01-148-7397	71	14
			77	18
29510	443978	4730-00-186-7798	598	50
89346	443982	4730-00-959-1629	598	25
21450	443987	4730-00-802-2560	213	15
			596	1
24617	444004	4730-00-475-5168	278	4
			585	5
24617	444012	4730-00-529-1487	272	13
		4730-00-623-8303	598	12
24617	444014	4730-00-200-0257	273	10
24617	444017	4730-01-284-2211	617	12
			618	14
24617	444019	4730-00-189-3034	590	35
24617	444026	4730-00-180-5038	213	8
			256	6
			266	9
24617	444028	4730-00-202-6491	598	48
24617	444034	4730-00-580-6738	599	12
24617	444040	4730-00-277-5553	273	11
			505	19
72582	444042	4730-00-278-4822	254	24
			254	24
21450	444072	4730-00-253-5794	81	9
24617	444073	4730-00-722-2759	589	51
			590	36
21450	444120	4730-01-092-6442	213	16
			595	21
			596	3
21450	444122	4730-00-277-7331	597	8
			599	15
24617	444124	4730-01-235-3007	598	24
19207	444134	4730-00-088-8666	213	17
			278	17
29510	444136	4730-00-125-7979	213	2

CROSS-REFERENCE INDEXES

PART NUMBER INDEX

CAGEC	PART NUMBER	STOCK NUMBER	FIG.	ITEM
			266	3
24617	444152	4730-00-540-2745	263	21
			264	20
			264	33
24617	444152	4730-00-540-2745	269	10
			269	10
			271	4
			598	27
			599	11
19207	444166	4730-00-277-5553	426	5
24617	444544		593	14
21450	444618	4730-00-089-2515	594	6
19207	444655	4730-00-044-4655	505	11
21450	444673	4730-01-107-2027	478	8
			484	13
29930	444697	4730-00-057-5555	33	18
			190	7
			473	28
61849	444958C1	2520-01-065-2530	188	12
61849	444965C1	2520-00-557-5980	188	24
61849	444969C1	3020-01-065-0871	188	2
73342	445090	5340-00-095-7146	189	30
15434	44678	2910-00-858-3522	85	12
			87	13
21450	451031	5310-00-045-1031	500	17
21450	451081	5310-00-045-1081	473	24
60602	45114-1	2520-01-284-8240	455	5
60602	45114-60	2590-01-285-4600	455	5
24617	451695	5305-01-289-4411	598	33
24617	451956	5320-01-145-3188	319	13
			320	10
			321	11
			322	12
			323	18
			324	10
78500	454	3110-00-100-0316	209	13
24617	454086	4730-01-066-3071	582	24
11862	454147	4730-01-247-3140	581	21
63005	454815	5305-01-057-4265	195	2
73342	454817	5306-00-570-8940	196	8
			197	15
39428	4549K572	4730-00-196-1504	252	18
78500	4555	3110-00-101-0836	209	14
21450	455172	5305-00-696-5291	282	16
21450	455176	5305-01-225-2106	362	6
73342	457118	3110-00-830-8802	185	15
60602	45737-40	2520-01-286-5650	181	2
60602	45752-70	2590-01-287-3224	598	2
01212	4591SCR	5330-01-271-9410	294	7
			296	9
27996	48B2071	3040-00-512-9223	507	3

PART NUMBER INDEX

CAGEC	PART NUMBER	STOCK NUMBER	FIG.	ITEM
93061	48IFHD-8-6	4730-00-013-7409	617	20
76599	4820SS	4730-01-244-8434	111	5
			549	29
			598	10
11757	5-A-062	5330-01-302-9948	529	3
11757	5-P-320	3020-00-035-7894	530	21
11757	5-P-569	3020-01-132-8860	530	20
78500	5-X-633	5330-01-133-7262	295	3
81221	5-0280	2520-01-082-8619	KITS	
81221	5-0280 KF		225	5
95019	5-170X	2520-00-294-6752	301	9
95019	5-2-629	2520-01-090-6673	218	11
			220	13
95019	5-3-2751X		218	5
95019	5-4-1721		220	5
95019	5-60-249		220	7
95019	5-60-354		218	10
95019	5-92X	2520-00-352-2168	466	3
11083	5M6214	4730-00-924-7886	30	12
			530	26
78500	5X625	5330-00-740-9312	348	17
38597	50-4-18-17-7	5310-00-515-9627	234	6
			241	6
11757	500118-3	4730-00-924-7886	569	26
21450	500163	5330-00-178-2191	484	9
95019	500168-2	4730-00-906-0982	218	8
21450	500207	5330-00-585-3210	502	13
95019	500398-12	5305-01-132-8390	528	26
			529	10
95019	500398-30		528	25
11757	500409-6	5305-00-885-7252	530	2
11757	501146-3	5310-01-133-4481	529	7
60602	50161-2	5330-01-285-1601	181	11
21450	504349	4730-00-278-3214	621	12
60602	-50451	5315-01-285-5562	184	16
60602	50452-2	3040-01-284-6232	184	15
21450	506207	5940-00-050-6207	144	8
			144	13
60602	50662	5306-01-285-1703	184	12
11757	51-P-22	3040-00-847-3169	530	4
19207	5139123	5310-00-700-7089	234	13
			241	13
			293	12
			295	2
			348	27
19207	5167785	5310-01-088-9298	301	6
30327	51807	4820-00-588-8604	129	1
19207	5186592	5365-00-518-6592	527	6
95019	5190076	2520-00-388-4197	KITS	
19207	5196397	4820-00-726-4719	182	15
			230	10

CROSS-REFERENCE INDEXES

PART NUMBER INDEX

CAGEC	PART NUMBER	STOCK NUMBER	FIG.	ITEM
			237	15
			248	14
			278	11
			279	12
			297	16
57733	5196397	4820-00-726-4719	474	7
			489	12
19207	5214479		220	11
			224	8
19207	5214483		220	10
			224	9
19207	5214534	5310-00-521-4534	326	9
19207	5225875	5306-00-281-1651	236	15
19204	5237556	5315-00-523-7556	602	17
60602	52442	5445-01-284-6173	181	12
60602	52443	5340-01-282-0452	181	8
60602	52456	5340-01-284-9654	181	14
60602	52478	5340-01-284-3789	181	13
60602	52479	5340-01-284-9655	181	9
19207	5277992	5360-00-664-4374	474	15
			489	5
19207	5287638	5310-00-528-7638	608	6
19207	5294507	5310-00-350-2655	141	6
19207	5298551	5310-00-275-8264	346	15
79136	5305-18	5310-01-122-2060	549	13
19207	5310615	5310-00-463-0268	141	17
19207	5329467	5310-00-532-9467	624	5
			625	5
19207	5331179	5310-00-241-0157	213	11
			253	11
			256	9
			266	12
			272	6
19207	5381051	5930-01-197-8661	131	16
			132	6
			134	10
			134	38
			134	69
19207	5381087	5310-01-108-5236	131	18
			132	4
			134	8
			134	36
			134	71
19207	5381088	5930-00-130-5349	131	15
			132	3
			134	7
			134	35
			134	68
			611	31
			612	27
			613	22

PART NUMBER INDEX

CAGEC	PART NUMBER	STOCK NUMBER	FIG.	ITEM
19207	5381233	5310-00-832-6852	134	4
			134	29
			134	65
17576	538174	5310-00-285-8833	9	8
19207	5416501	5360-00-541-6501	502	36
21450	543852	4820-00-274-3646	582	6
21450	543858	4820-01-140-4298	597	2
21450	549176		586	6
21450	549182	5342-00-111-3605	533	11
21450	549222	5325-00-303-4932	533	7
11757	550221	3110-00-198-2170	530	13
11757	550397	3110-00-151-8636	530	12
95019	550532	3110-00-100-9862	530	29
95535	55229	4710-00-424-2694	BULK	35
			BULK	36
60602	55303-2342	2520-01-090-7633	180	1
15434	554316	5306-01-054-4485	286	41
89346	55602H	3120-00-770-2941	346	8
			346	24
34623	5590560	5310-01-126-9404	107	16
52304	559602	6625-01-289-2062	626	4
90005	569020-02	4310-01-094-0791	310	1
			312	53
19207	57K0108	5340-01-321-6196	KITS	
34623	57K0138	2590-01-423-1967	507	49
19207	57K0139	4910-01-380-9029	629	5
34623	57K0164	2540-01-423-1968	445	
34623	57K0165	2540-01-423-1964	532	
19207	57K0208	2920-01-371-6064	KIT	38
19207	57K0228	2520-01-416-4538	629	5
19207	57K0243	2540-01-416-6784	594	7
			594	13
			595	7
			595	8
			595	9
			595	10
			595	14
			595	18
			595	23
19207	57K1310		630	
19207	57K1311		630	
19207	57K3194	2815-01-374-7539	2	1
19207	57K3237	2530-01-420-4221	280	1
19207	5702838	2940-01-107-9689	KITS	
19207	5704273	4820-01-093-5785	KITS	
19207	5704274	2590-00-606-2383	KITS	
19207	5704278	2520-00-421-7229	KITS	
19207	5704495	2540-00-108-1940	KITS	
19207	5704507	2815-01-111-2262	1	1
19207	5704510	2530-01-125-9272	KITS	
19207	5704512	2520-01-117-3010	176	1

PART NUMBER INDEX

CAGEC	PART NUMBER	STOCK NUMBER	FIG.	ITEM
19207	5704517	2520-01-144-1528	206	1
19207	5704519		KITS	
19207	5704533	3020-01-231-9296	KITS	
19207	5705626	2540-01-342-6810	590	17
			590	28
			591	
			591	12
19207	5705626	2540-01-342-6810	591	13
			591	14
			591	15
			591	17
			591	26
			591	32
			591	33
			591	35
			591	36
			591	37
			591	38
			591	40
19207	5705696	2530-01-286-3257	245	14
			246	1
43990	5726-01		621	21
19204	572929	5999-00-057-2929	134	59
			139	9
			140	11
			143	6
			157	8
			166	11
			170	8
			170	18
			171	8
			172	8
			173	14
			174	8
			436	5
			496	6
31007	58253D	5310-00-021-9760	255	5
21450	582818	5330-01-217-0734	414	7
21450	582826	5330-00-522-8544	419	5
19207	583244	5325-00-282-7435	534	7
21450	586689	5340-00-505-6379	616A	22
16567	587-12	2510-01-286-3329	6	5
21450	587646	5325-01-200-4035	586	24
21450	589931	5315-00-058-9931	489	37
21450	589965	5315-01-139-6568	502	23
21450	590029	5315-00-059-0029	473	23
72540	5900370	2510-01-394-6119	345	34
60602	59188	5342-01-284-9246	181	1
60602	59189	5340-01-284-9247	181	7
78500	592A	3110-00-142-4390	294	17
			296	14

CROSS-REFERENCE INDEXES

PART NUMBER INDEX

CAGEC	PART NUMBER	STOCK NUMBER	FIG.	ITEM
21450	592786	4820-00-272-3351	422	1
21450	593416	4030-00-514-4420	347	5
			402	28
43990	5938-01		621	20
60038	594A	3110-00-950-9700	236	3
			296	10
60038	598	3110-00-100-0650	236	2
			296	21
52304	599601	6695-01-329-6418	623	2
52304	599602	6625-01-289-2062	138	11
52304	599603	5930-01-287-3965	138	13
52304	599614		623	14
52304	599615		623	7
52304	599618		298	3
52304	599622		298	7
52304	599624		298	4
52304	599669	5995-01-290-1293	138	1
			138	2
52304	599706		623	4
52304	599707		623	8
52304	599718		623	10
52304	599719		623	5
52304	599720		623	19
52304	599721		623	15
52304	599722		623	16
52304	599723		623	3
52304	599728		298	12
52304	599730	6110-01-268-8739	623	26
52304	599731		298	9
52304	599734		623	18
52304	599735	4820-01-287-3963	298	1
52304	599747		138	3
52304	599751		138	8
52304	599752		138	4
52304	599753	5935-01-287-4286	138	5
52304	599754		138	6
52304	599756		298	10
52304	599757	4820-01-267-2914	623	1
52304	599760		623	13
52304	599763	5995-01-290-1294	138	1
52304	599765		623	6
52304	599766	4820-01-271-7946	623	17
52304	599791	4460-01-284-2344	298	11
52304	599802	4820-01-276-5731	622	3
52304	599805		298	8
52304	599810	5305-01-282-1528	298	5
52304	599811	5305-01-282-1529	298	2
52304	599812KF		298	6
52304	599971		138	7
52304	599980		138	3
52304	599982		138	9

PART NUMBER INDEX

CAGEC	PART NUMBER	STOCK NUMBER	FIG.	ITEM
81343	6 120115B	4730-00-293-7108	148	4
81343	6-2 120102BA	4730-00-142-3075	214	4
			279	21
			538	13
81343	6-2 120103BA	4730-00-200-0528	628	3
81343	6-2 120202BA	4730-00-287-1604	148	3
			199	1
			212	3
81343	6-2 120202BA	4730-00-287-1604	213	40
			270	1
			273	6
			279	5
			279	14
			288	3
			541	11
			603	1
81343	6-2 120203BA	4730-00-289-4052	212	12
			215	13
81343	6-4 010102B	4730-00-266-0538	108	2
			593	23
			594	3
81343	6-4 100202BA	4730-00-069-1187	182	20
			199	4
			213	14
			215	1
			254	22
			257	4
			261	14
			265	11
			270	4
			288	17
			291	10
			541	13
			603	14
			605	11
			623	29
81343	6-4 120102BA	4730-00-069-1186	182	22
			212	8
			213	22
			214	1
			215	9
			254	10
			255	24
			261	5
			261	28
			262	8
			264	11
			265	1
			273	2
			282	4
			288	16

CROSS-REFERENCE INDEXES

PART NUMBER INDEX

CAGEC	PART NUMBER	STOCK NUMBER	FIG.	ITEM
			290	10
			596	4
			628	5
81343	6-4 120302BA	4730-01-066-9484	278	1
			603	7
81343	6-4-6 120424BA	4730-00-813-7811	279	24
81221	6-5-228X KF	225	4	
81343	6-6 120103BA	4730-00-277-8770	212	9
81343	6-6 120103BA	4730-00-277-8770	214	11
			257	14
			290	2
			604	16
81343	6-6 120202BA	4730-00-289-0155	267	10
			271	8
			272	14
			274	7
			275	7
			278	7
			290	7
			290	9
			599	13
			604	11
			605	8
81343	6-6 120302BA(LONG NUT)	4730-00-541-7500	182	26
			268	1
			274	1
			275	1
81343	6-6-2 120425BA	4730-00-782-5461	214	15
81343	6-6-4 120425BA	4730-00-494-6580	214	6
			538	12
81343	6-6060102B	4730-00-270-4616	267	3
			268	4
			274	8
			275	8
			599	16
81343	6-8 120102BA	4730-00-837-1177	213	33
			271	22
			605	12
04055	6TLFP	6140-01-431-1172	151	22
77820	60-37398-12	5340-00-726-1670	134	20
83259	600-001 1-4	5330-00-171-6600	56	14
40342	601068		282	9
52676	6017-2RS	3110-01-126-1287	120	14
15434	60408	5315-00-238-0882	7	26
46156	60504	4010-00-290-4352	BULK	4
15434	60575	5365-00-428-6201	7	22
09990	612668	5360-00-934-0089	494	20
6B719	615116	2510-00-489-7104	370	25
15434	61554	4720-00-918-9634	284	51
59206	62C-8	4730-00-278-3220	597	9

CROSS-REFERENCE INDEXES

PART NUMBER INDEX

CAGEC	PART NUMBER	STOCK NUMBER	FIG.	ITEM
93061	62NTA-6	4730-01-133-9866	621	2
15434	62392	5310-01-270-8389	45	2
11757	63-P-16	2520-01-145-6820	530	9
09990	635439	4330-01-144-5557	494	22
89346	63942HA	3120-00-692-6153	211	10
57328	65003-S	5310-01-099-2550	381	19
			400	3
15434	650330		284	28
15434	650330	4310-01-079-5245	286	49
19207	6566675	2590-00-473-6331	376	2
			399	1
			629	5
93061	66NTA-6-6	4730-01-134-0853	617	5
15434	66292	5310-00-197-5304	7	16
73342	6700213	3110-00-799-4903	197	3
5Y952	670168	5340-01-282-3589	101	13
5Y952	670170	5306-01-281-6560	101	22
5Y952	670263	5305-01-283-8462	121	8
15434	67270	5331-00-171-3879	19	8
73342	6750020	5325-01-079-6526	185	20
73342	6757563	5330-00-923-1409	190	19
73342	6758740	5330-00-582-0456	185	4
73342	6762127	5331-01-010-9693	189	15
73342	6762187	5365-01-006-9622	200	19
15434	67684	5310-00-262-2986	129	4
73342	6769319	5325-00-089-1262	188	13
73342	6769636	5310-00-776-7670	185	22
73342	6770492	5330-00-999-3760	185	2
73342	6771070	3110-00-916-4286	185	25
73342	6771366	5330-00-911-9411	185	18
73342	6772552	5315-00-402-0421	185	7
73342	6773311	5330-00-999-3752	190	16
73342	6774322	2520-00-919-7240	189	3
73342	6775703	2920-01-143-1263	189	21
73342	6778156	5360-00-450-0346	197	10
15434	67944	4730-00-097-4236	36	4
15434	67946	5365-00-197-9327	37	6
			41	8
11757	68 P 2	9905-00-134-3558	565	4
93061	68HB-10-8	4730-01-105-9466	111	22
93061	68NTA-12-8	4730-01-108-6410	617	13
			618	13
15434	68061-A	5331-00-970-3461	51	11
62983	680701		463	8
			464	8
			511	8
62983	680702		463	25
			464	25
			511	25
15434	68139	4730-01-142-8524	285	27
15434	68152	5340-00-286-1868	117	11

CROSS-REFERENCE INDEXES

PART NUMBER INDEX

CAGEC	PART NUMBER	STOCK NUMBER	FIG.	ITEM
15434	68192A		19	9
15434	68192B		19	9
15434	68192C		19	9
15434	68193	5340-00-404-2944	21	1
15434	68210	5330-00-328-8656	31	8
15434	68226-1	3120-00-882-7960	19	4
15434	68274	5360-00-664-5343	32	10
			33	8
73342	6830187	2520-00-405-1842	185	16
73342	6831656	5360-00-211-9547	192	13
73342	6831774	5315-00-108-1112	189	28
73342	6833896	4820-01-007-0962	196	6
73342	6833949	5310-00-007-0260	197	8
73342	6833981	5365-01-010-9689	187	12
			193	14
73342	6833993	5325-00-557-5794	191	10
			194	3
73342	6833999	5365-01-010-9688	187	3
73342	6834230	2520-01-006-7120	192	12
73342	6834339	2520-01-064-8849	191	9
			194	4
73342	6834354	2520-01-065-0078	192	14
73342	6834369	2520-01-065-0841	187	15
			193	7
73342	6834374	2520-01-065-0077	193	5
73342	6834389	5310-00-568-6118	188	5
73342	6834410	2520-01-011-1068	200	13
73342	6834412	2520-01-011-1067	200	25
73342	6834413	2520-01-007-0345	200	9
73342	6834414	2520-01-007-0346	200	11
73342	6834512	5325-00-557-5835	188	6
73342	6834556	5365-01-124-2831	190	3
73342	6834567	5325-00-557-5897	190	15
73342	6834583	5325-00-566-6577	188	18
73342	6834624	2520-00-557-5900	189	9
73342	6834908	5310-00-557-5942	200	31
73342	6834940	5315-01-010-9777	188	9
73342	6835386	5310-00-557-5943	188	3
73342	6835561	3020-01-008-2770	188	14
73342	6835568	3020-01-110-8251	191	11
73342	6835720	2520-00-557-6090	192	3
73342	6835921	4730-01-006-9629	197	21
73342	6836202	4820-01-011-1069	200	15
73342	6836265	5325-00-557-6164	192	5
73342	6836266	5325-00-557-6183	192	5
73342	6836267	5325-00-557-6207	192	1
			192	5
73342	6836268	5325-00-557-6210	192	5
73342	6836277	5360-01-079-6704	200	10
73342	6836773	5360-01-079-3096	187	14
			193	8

PART NUMBER INDEX

CAGEC	PART NUMBER	STOCK NUMBER	FIG.	ITEM
73342	6836928	5360-01-123-5483	189	10
73342	6837167	2520-01-094-4751	185	11
73342	6837976	3110-00-941-3830	185	12
73342	6838278	3040-01-122-5201	189	26
73342	6838364	5365-01-010-9687	187	9
73342	6838494	9905-01-317-2715	554	1
73342	6838750	4730-01-083-9925	197	17
73342	6839122	5315-01-004-4835	197	18
73342	6839163	5330-00-374-4873	187	1
73342	6839214	5360-01-074-8305	197	5
73342	6839271	5360-01-004-4863	197	20
73342	6839364	5310-00-557-6568	188	1
73342	6839761	5310-01-084-1768	189	29
73342	6839975	2520-01-098-5117	185	10
73342	6839976	5365-01-080-0482	185	9
73342	6839977	5365-01-080-0483	185	9
73342	6839978	5365-01-080-0484	185	9
73342	6839979	5365-01-080-0485	185	9
73342	6839980	5365-01-080-0486	185	9
73342	6839985	2520-01-080-0448	185	30
15434	68445	5315-00-281-7610	7	28
15434	68513	5315-00-041-0916	23	9
15434	68585	5315-00-014-1195	7	10
15434	68586	3120-00-661-6646	33	11
			33	16
			33	25
15434	68588	3020-00-353-9384	33	13
73342	6880008	2520-01-122-9928	188	20
73342	6880024	3120-01-126-1097	186	11
73342	6880152	4730-01-132-6849	197	13
73342	6880251	5360-01-079-3097	191	8
			194	5
73342	6880353	2520-01-040-3541	KITS	
73342	6880389	5330-01-141-9579	185	24
73342	6880899	5340-01-056-0037	185	26
73342	6880967		189	12
73342	6881007	4810-01-220-1185	196	3
73342	6881009		197	11
73342	6881072	5360-01-079-6702	200	14
73342	6881138	4820-01-083-2127	197	6
73342	6881227	2520-01-106-0826	189	6
73342	6881380	2520-01-079-1615	200	23
73342	6881381	5360-01-079-6703	200	24
73342	6881386	2520-01-127-5005	197	16
73342	6881387		197	24
73342	6881580	3020-01-122-5906	186	13
73342	6881581	5325-01-133-4679	186	4
73342	6881645		200	5
73342	6882321	2520-01-078-6123	192	16
73342	6882565	2520-00-557-5974	188	4
73342	6882586	5305-01-124-5779	189	5

CROSS-REFERENCE INDEXES

PART NUMBER INDEX

CAGEC	PART NUMBER	STOCK NUMBER	FIG.	ITEM
73342	6882687	4330-01-074-9642	198	3
73342	6882689	5331-01-080-3254	198	4
73342	6882811	4820-01-126-5379	190	10
73342	6883020	3020-01-149-3836	187	21
73342	6883031		191	5
			194	8
73342	6883033		191	6
73342	6883035		192	10
			194	7
73342	6883044	2520-01-124-6469	189	20
73342	6883046	4710-01-078-8748	189	16
73342	6883090	2520-01-148-9279	187	18
73342	6883577	3040-01-221-2092	196	4
73342	6883579		197	1
73342	6883581	2520-01-130-5821	197	12
73342	6883707	4730-01-127-6900	188	16
73342	6883901	3020-01-152-8895	188	23
73342	6883974	4710-01-131-2729	190	8
73342	6883999	3020-01-149-1239	186	12
73342	6884259	2520-01-140-2376	KITS	
73342	6884275	5325-00-007-2969	191	1
73342	6884653	5365-01-197-4545	193	9
73342	6884749	4330-01-131-0279	KITS	
73342	6884872	5330-01-111-9291	189	7
73342	6885146	3020-01-065-0183	187	10
73342	6885151	3020-01-124-3421	188	19
73342	6885153	2520-01-149-3808	187	13
			193	10
73342	6885154	2520-01-150-0900	187	13
			193	10
73342	6885155	2520-01-149-3438	187	13
			193	10
73342	6885156	5325-01-145-6921	187	22
			193	6
73342	6885166	5360-01-113-9615	197	19
73342	6885188	4730-01-133-1461	197	4
73342	6885213	2520-01-081-9043	KITS	
73342	6885571	2910-01-051-9444	198	6
60285	6893-2	5306-00-068-0513	39	9
			100	49
			142	15
			143	21
			145	6
			151	26
			161	20
			162	1
			163	23
			165	28
			167	33
			168	21
			169	28

CROSS-REFERENCE INDEXES

PART NUMBER INDEX

CAGEC	PART NUMBER	STOCK NUMBER	FIG.	ITEM
			269	27
			315	12
			352	8
			353	4
			374	4
			374	9
			454	18
			513	28
			525	23
60285	6893-2	5306-00-068-0513	543	2
			547	7
			549	3
			579	9
			586	19
			589	19
			606	4
			612	10
30327	69FL3-4	4730-00-428-5631	36	6
9F512	691-10014	5330-00-252-8888	52	14
15434	69324	5310-01-112-4307	49	3
			284	66
15434	69519	5315-00-475-2574	33	12
15434	69736	5305-00-339-1415	23	17
15434	69952	5305-01-209-7068	118	29
15434	69960	5305-00-795-9341	33	35
15434	69962	5365-00-695-1247	37	5
40342	7-X-106	5365-01-138-7102	282	7
11083	7B5049	5320-00-262-6492	362	37
11083	7N9738	5935-01-149-5165	159	4
95097	70000-362	5315-00-298-1498	220	2
21450	700287	3110-00-554-3929	474	25
			489	35
19207	7003615	4820-00-509-3036	505	12
21450	700536	3110-00-554-3184	489	30
19207	7005602	2510-01-169-9850	534	6
19207	7005603		534	6
19207	7005638	2510-00-036-0298	534	4
19207	7005798	2540-00-891-7830	534	4
19207	7007127	5310-00-700-7127	347	9
15434	70089-1	5330-00-537-2382	116	9
19207	7014965	2530-00-270-3878	255	12
			261	24
			262	7
19207	7015266	4210-00-775-0127	610	1
19207	7017002	5306-00-446-8762	348	10
19207	7017087-1		67	7
			68	9
			69	11
19207	7017190	3040-00-745-7685	473	5
19207	7017195	2520-00-860-7340	473	6
19207	7017450		413	13

CROSS-REFERENCE INDEXES

PART NUMBER INDEX

CAGEC	PART NUMBER	STOCK NUMBER	FIG.	ITEM
19207	7018109	5975-00-414-6466	413	8
06853	7022-23	2530-01-127-2337	249	2
06853	7022-24	4820-01-134-1836	249	4
06853	7022-25	4820-01-131-6123	249	16
06853	7022-26	2805-01-134-1837	249	23
06853	7022-27	2530-01-127-1677	249	1
15434	70295	4730-00-011-3175	118	5
			284	53
21450	703189	3110-00-100-4177	234	33
			241	33
19207	7035447	5340-00-264-7182	419	9
19207	7036587	5360-00-703-6587	316	5
15434	70441	5330-00-508-0411	114	6
19207	7044253	5360-00-704-4253	329	5
19207	7045151	5340-00-445-4561	410	3
19207	7047096	2510-00-235-1888	407	3
19207	7047097	2510-00-679-1420	409	22
19207	7047098	2510-00-679-1733	409	20
19207	7047876	5340-01-134-7635	375	3
19207	7053776	7690-00-489-8322	571	5
			572	9
15434	70550	2815-01-124-0232	17	11
			44	15
19207	7056639	5970-00-705-6639	154	19
19207	7056640	5970-01-174-9449	151	5
			151	14
			151	19
			152	9
19207	7056701	5940-00-705-6701	432	8
10001	7056702	5940-00-705-6702	158	12
19207	7056707	5940-00-705-6707	160	4
19207	7056708	5940-00-705-6708	134	49
19207	7056709	5940-00-705-6709	160	9
			496	9
19207	7056711	5940-00-705-6711	163	15
			165	16
			166	10
			167	9
			168	13
			169	13
19207	7056715	5940-00-705-6715	158	39
19207	7056730	5940-00-705-6730	154	14
			154	18
19207	7059241	5340-00-419-5866	429	5
19207	7059461	5340-00-176-0868	131	20
			134	33
			134	74
			611	35
			612	24
			613	25
19207	7059462	9905-00-252-5586	552	12

CROSS-REFERENCE INDEXES

PART NUMBER INDEX

CAGEC	PART NUMBER	STOCK NUMBER	FIG.	ITEM
19207	7060039	5940-01-121-4280	150	8
19207	7060040	6150-01-114-3697	150	9
19207	7060041	5970-01-121-6587	150	7
19207	7060081-2	5330-01-127-3803	289	19
19207	7060081-3	5330-01-127-7134	289	15
19207	7061093	2540-00-592-1823	382	20
			385	7
19207	7061093	2540-00-592-1823	387	17
			387	41
			609	11
19207	7061094	2540-00-521-6179	382	27
			385	8
			387	18
			387	42
19207	7061871	5310-00-050-3520	329	3
15434	70622	5340-01-271-2470	40	6
15434	70624	5331-00-506-4874	116	2
19207	7064151	2540-00-351-0145	532	4
			532	4
			535	4
			609	3
19207	7064165	2510-00-119-3903	383	36
			384	3
			386	11
			390	7
19207	7064597	3120-00-509-8270	490	12
15434	70653	2815-00-772-9434	7	23
19207	7065553	4030-00-706-5553	499	22
15434	70657	5340-01-122-8002	14	18
19207	7066008	4820-00-595-2761	263	23
19207	7066092		224	14
19207	7066093		224	6
19207	7066097		218	6
			220	8
			224	4
19207	7066101		218	4
			220	12
			224	7
34623	7066104	5330-01-067-1740	224	12
15434	70700	5360-00-597-4570	95	3
			96	2
15434	70705	5330-00-562-1176	85	13
			87	14
			88	37
19207	7071098	5305-00-752-1693	236	18
15434	70715	5310-00-507-3259	85	3
			87	3
15434	70716	5305-00-506-5722	85	7
			87	7
15434	70717	5365-00-507-3260	85	16
			87	16

PART NUMBER INDEX

CAGEC	PART NUMBER	STOCK NUMBER	FIG.	ITEM
15434	70717-A	5365-00-507-3261	85	16
			87	16
15434	70717-B	5365-00-507-3262	85	16
			87	16
19207	7071882	4030-00-262-3152	475	2
19207	7074517	4030-00-158-2409	469	34
			470	29
15434	70772	5305-00-477-6769	66	1
21450	707728	3110-00-227-3245	471	20
15434	70778	5360-00-698-7100	88	31
15434	70790	5306-00-485-0790	53	6
15434	70798	5340-00-572-8042	88	32
15434	70811	5310-00-550-8124	88	35
15434	70811-A	5365-00-550-8125	88	35
15434	70811-B	5365-00-550-8127	88	35
15434	70813	5305-00-899-8054	88	15
15434	70820	2990-00-567-4367	88	30
15434	70834	2910-00-773-2108	88	16
15434	70836	5340-00-711-5372	88	36
19207	7084716	2510-00-470-2090	420	23
19207	7084724	5340-01-208-8080	420	11
19207	7084725	5340-00-799-2218	409	3
19207	7084726	2510-00-470-2091	420	2
19207	7084729	5306-01-183-5954	419	13
19207	7084730	5340-00-419-9474	420	21
19207	7084732		420	6
19207	7084738	2590-00-471-5343	419	11
			419	26
19207	7084744		410	5
			410	13
19207	7084768	2510-01-121-2541	404	13
19207	7084769	2510-01-204-7704	404	13
19207	7084770	5340-01-178-3734	404	12
19207	7084771	3040-01-178-7087	404	4
19207	7084772	5365-00-421-9697	402	11
			404	16
19207	7084792	2510-01-024-3618	407	16
			415	30
19207	7084793	2510-01-024-3619	407	18
			415	29
19207	7084794	2510-01-022-2580	407	17
			415	27
19207	7084799	5340-00-238-5606	416	14
19207	7084828	2540-00-231-0200	405	35
			406	33
19207	7084840	2590-00-405-9771	405	26
19207	7084860	5342-00-231-0216	405	25
19207	7084861	5340-00-419-9464	405	30
			406	37
19207	7084876	5365-00-231-7440	410	9
19207	7084881	2510-00-234-3258	414	2

PART NUMBER INDEX

CAGEC	PART NUMBER	STOCK NUMBER	FIG.	ITEM
			414	22
19207	7084882	2510-00-234-3259	414	2
			414	22
19207	7084886-139		413	20
19207	7084887-140		413	15
19207	7084889		413	28
19207	7084890		413	28
19207	7084891-139		413	20
19207	7084892	5340-01-191-0596	413	27
19207	7084893	5340-01-052-9022	414	26
19207	7084894	5340-01-052-9023	414	14
19207	7084895	5340-01-052-9024	414	5
19207	7084896	5340-00-421-9696	413	18
			414	4
19207	7084939	2510-01-163-9752	427	7
19207	7084942	5340-01-163-9912	427	5
19207	7084949	2510-00-159-8822	425	17
19207	7084952	2590-01-180-8571	424	15
19207	7084955	2590-01-181-0310	417	25
19207	7084956	2590-01-180-1092	417	9
19207	7084957	2590-01-181-6059	417	9
19207	7084958	2590-01-179-5802	417	11
19207	7084960	2590-01-181-0309	412	13
19207	7084961	2510-00-425-0512	411	14
19207	7084969	3020-00-421-7240	411	2
19207	7084984	5340-01-186-3505	404	12
19207	7084985	2510-01-178-7086	404	4
19207	7084987	5340-00-158-3877	419	12
19207	7084988	5340-00-041-3126	404	9
19207	7084989	5340-00-107-7769	416	19
19207	7084990	5340-00-158-4077	405	5
			406	21
19207	7084996	5340-01-311-0225	414	25
19207	7084997	5340-00-679-1494	414	17
19207	7085367	5310-00-493-3986	579	10
30554	71-4872	4730-00-226-8874	35	4
			596	17
19207	7106057		602	11
19207	7106058		602	13
78500	712148	3110-00-195-0460	234	19
			241	19
21450	712627	3110-00-100-5825	235	13
			242	13
21450	712699	3110-00-100-4216	232	15
			239	15
21450	713806	3110-00-227-3381	487	16
21450	714042	3110-00-554-3411	484	10
19207	7212871		144	4
08752	725197	5325-01-137-8828	494	23
19207	7264749	2540-00-287-2571	406	11
			416	11

CROSS-REFERENCE INDEXES

PART NUMBER INDEX

CAGEC	PART NUMBER	STOCK NUMBER	FIG.	ITEM
19207	7320642	5340-00-732-0642	144	3
19207	7326126	5365-00-732-6126	230	4
			237	3
19207	7331177	5325-01-212-0599	611	14
			612	29
			613	20
			613	41
19207	7335051		223	6
19207	7335052		223	5
19207	7335053	5340-01-119-5682	223	7
			225	7
19207	7335054		223	8
		5306-01-119-5681	225	6
19207	7336058	5305-00-421-3986	424	4
19207	7336402-1	4730-01-202-3351	213	7
			266	8
			285	24
			287	7
19207	7339964	5315-01-108-7340	347	7
19207	7339966-1	5315-01-206-2239	347	4
19207	7339982	4730-00-421-3924	213	9
			253	9
			256	7
			266	10
			272	4
19207	7345195-1	5305-01-203-9064	611	15
			612	30
			613	19
			613	40
19207	7346807	5330-00-513-1443	KITS	
19207	7346812	5365-00-427-2282	236	37
19207	7346813	5310-00-274-7721	232	1
			236	39
			239	1
19207	7346815	2530-00-734-6815	236	33
19207	7346816-1	2530-00-006-7469	236	33
19207	7346817	5306-00-734-6817	233	12
			240	12
19207	7346818	5365-00-734-6818	232	16
			239	16
19207	7346819	2530-00-734-6819	233	13
			240	13
19207	7346822	5306-00-734-6822	232	14
			239	14
19207	7346824		232	8
			239	8
19207	7346825		235	4
			242	4
19207	7346877	3110-00-734-6877	235	9
			242	9
19207	7346878	5330-00-734-6878	235	5

CROSS-REFERENCE INDEXES

PART NUMBER INDEX

CAGEC	PART NUMBER	STOCK NUMBER	FIG.	ITEM
			242	5
19207	7346879	5365-00-734-6879	235	5
			242	5
19207	7346880	5365-00-734-6880	235	5
			242	5
19207	7346882	5310-00-033-6012	235	2
			242	2
19207	7346883	3120-00-734-6883	233	6
19207	7346883	3120-00-734-6883	240	6
19207	7346886	5330-00-734-6886	234	34
			241	34
19207	7346888	5365-01-077-8564	234	15
			241	15
19207	7346889		234	15
			241	15
19207	7346890		234	15
			241	15
19207	7346892	3040-00-734-6892	234	18
19207	7346893	5310-00-282-5661	234	2
			241	2
19207	7346894	3120-00-662-3370	234	20
			241	20
19207	7346895	3110-00-734-6895	234	21
			241	21
19207	7346896	5330-00-641-2466	234	26
			241	26
19207	7346899	5330-00-734-6899	233	16
			240	16
19207	7346900	5307-00-734-6900	236	34
19207	7346953	3040-00-734-6953	350	10
19207	7346954	5340-00-734-6954	350	8
19207	7346956	5306-00-272-3747	350	1
19207	7346959	2520-00-734-6959	174	9
			KITS	00
19207	7346962	2510-00-191-8172	348	14
19207	7346971	2530-00-734-6971	236	12
19207	7346972	3120-00-288-1889	236	13
19207	7346974	5340-00-734-6974	236	23
19207	7346975	2520-00-734-6975	236	20
19207	7346977	2530-00-734-6977	236	25
19207	7346978	2530-00-734-6978	236	25
19207	7346982	2530-00-734-6982	236	4
19207	7346983	3120-00-537-0614	236	5
19207	7346986	5310-00-033-6007	236	30
19207	7346991	2520-00-734-6991	293	6
19207	7346992	5340-00-734-6992	293	7
15434	7348-2	2815-00-362-1780	23	11
19207	7348101	2520-00-734-8101	223	12
19207	7348215		223	14
19207	7351272	5330-00-735-1272	150	11
19207	7351285	5340-00-409-4008	319	23

CROSS-REFERENCE INDEXES

PART NUMBER INDEX

CAGEC	PART NUMBER	STOCK NUMBER	FIG.	ITEM
19207	7351289	5310-00-281-2180	367	4
19207	7355520	5940-00-735-5520	151	7
			151	21
			152	11
19207	7357965		607	8
19207	7358098	5315-00-290-1349	339	11
			339	18
			340	2
19207	7358098	5315-00-290-1349	395	11
19207	7358621	6210-00-337-7345	132	11
19204	7358622-1	6220-00-451-8161	131	11
19204	7358672-4	6220-01-423-0209	131	5
19207	7359274	5340-01-104-3832	533	10
19207	7363002	5995-00-177-8220	134	12
19207	7363004-1	6150-01-082-3818	175	6
19207	7363005-1	6150-01-130-8102	175	7
19207	7364214	4730-00-278-4822	81	10
			274	5
			275	5
19207	7368622	2510-00-736-8622	366	3
19207	7368623	2510-00-409-3993	366	5
19207	7368624	5340-01-082-3596	366	6
19207	7368625	5340-01-104-7400	366	4
19207	7370134	5315-00-737-0134	382	42
			383	7
			387	35
19207	7370149	2590-00-622-7757	382	25
			387	15
19207	7370383	2540-00-591-1108	385	1
19207	7370384	2510-00-106-2200	382	32
			385	39
			387	26
19207	7371886	5310-00-264-4083	367	6
19207	7372083	5305-00-569-8909	352	1
			426	16
19207	7372083-1	5305-01-090-7626	134	44
			351	11
			352	11
			359	7
			364	2
19207	7372705	5330-00-483-2408	543	19
19207	7372711	2510-00-737-2711	534	8
19207	7372712	2510-00-737-2712	534	8
19207	7372716	2590-00-693-0589	534	16
19207	7372718	2510-00-693-0591	534	12
19207	7372719	2510-00-693-0592	534	12
19207	7372720	5330-00-737-2720	534	14
19207	7372721	5330-01-060-0992	534	14
19207	7372722	5330-00-737-2722	534	15
19207	7372781		532	6
			532	6

CROSS-REFERENCE INDEXES

PART NUMBER INDEX

CAGEC	PART NUMBER	STOCK NUMBER	FIG.	ITEM
19207	7372788	5340-00-737-2788	382	40
			385	46
			387	31
			609	20
19207	7372790	2540-00-737-2790	382	19
			385	11
			387	23
19207	7372790	2540-00-737-2790	387	44
19207	7372792	5360-00-737-2792	380	4
			381,	7
			400	13
19207	7372793	5360-00-737-2793	380	5
			381	8
			400	12
19207	7373206	5340-00-664-9863	318	13
19207	7373218	5340-00-523-5999	183	11
19207	7373244	5310-00-269-7044	243	20
19207	7373276	2540-00-737-3276	362	13
19207	7373277	2540-00-737-3277	362	13
19207	7373279	5305-01-104-9019	369	8
56442	7373283	5340-00-737-3283	362	23
19207	7373286	2540-00-737-3286	362	12
19207	7373287	2510-00-737-3287	362	10
19207	7373289	2540-00-693-0602	362	15
19207	7373290	2540-00-693-0603	362	15
19207	7373291-2	5330-01-126-7512	412	20
19207	7373291-3	5330-00-152-3217	408	6
			417	13
19207	7373293	2510-00-737-3293	361	5
5U403	7373294	2510-00-737-3294	361	5
19207	7373295	2510-00-737-3295	363	3
19207	7373296	2540-00-737-3296	363	8
19207	7373297	2540-00-737-3297	363	8
16662	7373298	2540-00-737-3298	362	1
19207	7373299	5340-00-693-0604	362	20
19207	7373300	9390-00-737-3300	363	2
19207	7373301	9390-00-737-3301	363	6
19207	7373302	5340-00-737-3302	361	3
19207	7373316	2510-00-693-0607	359	12
19207	7373317	9390-00-737-3317	359	18
19207	7373318	5306-01-104-8389	366	12
19207	7373319	5307-00-206-8510	366	16
19207	7373321	5340-00-696-0264	366	8
19207	7373322	5340-00-696-0265	366	8
19207	7373324	2510-00-693-0608	367	14
19207	7373325-1	5330-01-257-0750	367	31
19207	7373326	2510-00-737-3326	367	12
19207	7373327-1	2510-01-152-8812	367	1
19207	7373330	5340-00-737-3330	367	40
19207	7373330-1	5340-01-141-0832	367	40
19207	7373332	5330-01-121-9886	367	15

CROSS-REFERENCE INDEXES

PART NUMBER INDEX

CAGEC	PART NUMBER	STOCK NUMBER	FIG.	ITEM
19207	7373333	9390-00-599-6405	367	15
19207	7373334	9390-01-120-9864	367	16
19207	7373335		366	10
19207	7373336	2510-00-693-0610	367	18
19207	7373338		366	9
19207	7373381	5340-00-337-9619	624	10
			624	10
19207	7374401	5331-00-984-3756	26	17
19207	7374809	5305-00-350-1210	367	7
19207	7376203	2540-00-737-6203	326	8
19207	7376584	5330-00-737-6584	177	1
19207	7385905	6220-01-155-2357	144	2
19207	7388359	5935-00-178-6075	158	24
19207	7388366	5365-00-507-8766	158	25
			163	4
			165	4
			166	16
			167	16
			168	4
			169	4
19207	7388820	2530-00-603-5768	292	1
19207	7389061	2530-00-738-9061	292	4
19207	7389493		292	2
19207	7392454		131	17
			132	5
			134	9
			134	37
			134	70
19207	7397673		362	32
19207	7397674		362	32
19207	7397723	5315-00-453-9349	359	15
19207	7397725		359	16
19207	7397726	5340-00-622-7700	359	17
19207	7397729		366	11
84324	7397744-1	5325-01-231-5963	161	21
19207	7397754	5306-00-739-7754	153	29
19207	7397758	5340-00-178-1036	140	16
19207	7397775	3120-00-073-3163	163	26
			165	31
			168	24
			169	33
19207	7397776	5325-00-739-7776	161	8
			334	16
			337	18
			338	15
19207	7397785	5340-01-117-9876	151	24
			162	3
			253	15
			454	20
			589	37
			590	29

PART NUMBER INDEX

CAGEC	PART NUMBER	STOCK NUMBER	FIG.	ITEM
19207	7397788	5340-01-120-8510	339	26
			340	12
19207	7397790	5325-00-053-1116	163	35
			165	22
			166	21
			169	40
19207	7397798	2540-00-115-2565	100	10
19207	7397843	5365-01-107-9967	374	14
			376	6
			377	3
19207	7397852	5330-00-621-2565	362	17
19207	7397853	5340-00-621-2591	410	12
			420	14
			629	10
19207	7409002	2510-00-740-9002	391	25
19207	7409008	2510-00-740-9008	391	24
19207	7409009	5340-01-153-0890	391	23
19207	7409010	5340-00-422-2019	391	12
19207	7409011	3040-00-422-2008	391	5
19207	7409012	3040-00-409-4021	391	3
19207	7409013	2510-00-740-9013	391	7
19207	7409017	5315-00-740-9017	390	5
19207	7409020	5315-00-740-9020	507	29
19207	7409023	2510-00-786-4631	507	49
19207	7409025	4330-01-168-6891	457	6
			458	5
			507	41
19207	7409026	6680-00-740-9026	457	8
			458	3
			507	39
19207	7409027	4730-01-123-4546	457	9
			458	2
			507	38
19207	7409029	5360-00-411-2511	389	14
19207	7409030	5340-00-489-0363	389	13
19207	7409038	5340-01-205-5957	507	35
19207	7409040	2520-00-740-9040	507	25
19207	7409041	2520-00-740-9041	507	30
19207	7409042	3120-00-740-9042	507	27
19207	7409043	5315-00-740-9043	390	22
19207	7409045	5315-00-740-9045	390	17
19207	7409050	5330-00-740-9050	507	16
19207	7409054	5360-00-597-4075	507	12
19207	7409058	4730-00-740-9058	507	7
19207	7409079	5365-00-740-9079	507	6
19207	7409080	5365-00-182-9635	457	7
			458	4
			507	8
			507	40
19207	7409150	2540-00-740-9150	469	28
			472	36

PART NUMBER INDEX

CAGEC	PART NUMBER	STOCK NUMBER	FIG.	ITEM
			473	13
			499	18
19207	7409212	2510-00-740-9212	391	20
19207	7409213	2510-00-040-2264	390	6
19207	7409299	3120-00-740-9299	507	31
19207	7409335	5340-00-740-9335	346	4
			347	1
19207	7409337	2510-00-740-9337	318	43
			319	44
			320	44
			321	44
			322	45
			323	47
			324	44
19207	7409340	2510-00-740-9340	346	16
19207	7409341	5340-00-740-9341	346	16
19207	7409343	2540-00-740-9343	346	26
19207	7409344	2510-00-740-9344	346	7
19207	7409361	5340-00-740-9361	243	11
19207	7409366	5340-00-740-9366	243	14
19207	7409372	3040-00-740-9372	243	19
19207	7409445	2530-00-740-9445	252	21
19207	7409508	2510-00-740-9508	318	50
			319	53
			320	51
			321	50
			322	54
			323	52
			324	49
19207	7409524		326	13
19207	7409534	2510-00-740-9534	333	4
			335	4
			337	4
19207	7409534-2	2540-01-088-9162	335	6
19207	7409535	2510-00-740-9535	333	17
			337	22
19207	7409553	2590-00-740-9553	295	12
			348	19
19207	7409565	3040-00-740-9565	474	22
			489	61
19207	7409591	3120-00-740-9591	473	35
19207	7409596	2510-00-740-9596	534	1
19207	7409597	2540-00-740-9597	534	1
19207	7409601	2510-00-740-9601	348	5
19207	7409602	5360-00-740-9602	348	4
19207	7409606	5330-00-740-9606	348	18
19207	7409607	4710-00-740-9607	348	15
19207	7409609	5315-00-740-9609	472	4
19207	7409612	5365-00-740-9612	348	6
19207	7409613	2510-00-740-9613	348	1
19207	7409617	2510-00-740-9617	349	7

PART NUMBER INDEX

CAGEC	PART NUMBER	STOCK NUMBER	FIG.	ITEM
19207	7409618	5365-00-740-9618	349	3
19207	7409619	5315-00-740-9619	349	5
19207	7409623-1	3010-01-236-1213	624	11
19207	7409660	2540-00-740-9660	330	1
19207	7409661	2510-00-740-9661	330	6
19207	7409663	2530-00-740-9663	502	27
19207	7409665	5340-00-740-9665	501	9
19207	7409666	3120-00-740-9666	502	12
19207	7409668	5340-00-740-9668	501	16
19207	7409669	3040-00-740-9669	502	11
19207	7409671	9520-01-120-8551	502	8
19207	7409673	2520-00-740-9673	502	25
19207	7409677	5306-00-740-9677	473	26
19207	7409684	5340-00-740-9684	473	21
19207	7409685	3120-00-740-9685	473	32
19207	7409686	2510-00-740-9686	534	2
19207	7409687	2510-00-740-9687	534	2
19207	7409694	2520-00-740-9694	472	8
19207	7409695	3120-00-740-9695	472	16
19207	7409725	3040-00-740-9725	472	7
19207	7409728	5310-00-740-9728	472	14
19207	7409729	3120-00-740-9729	472	5
			475	9
			490	11
19207	7409784	5365-00-740-9784	398	8
19207	7409800	3120-00-740-9800	473	27
			474	32
			489	13
19207	7409803	5306-00-740-9803	474	30
19207	7409804-1	5340-01-278-2190	472	11
19207	7409807	3120-00-740-9807	472	15
			487	19
			488	5
19207	7409809	3110-00-740-9809	474	3
19207	7409812	9520-00-740-9812	473	1
19207	7409813	2520-00-740-9813	470	4
19207	7409814	2520-00-740-9814	473	33
19207	7409816	2520-00-740-9816	474	19
			489	2
19207	7409817	2530-00-740-9817	474	13
			489	6
19207	7409821	5330-00-740-9821	474	27
			489	14
19207	7409822	5330-00-057-3823	469	22
			470	18
			474	4
			489	36
19207	7409823	3020-00-740-9823	474	33
			489	25
19207	7409835	5315-00-740-9835	398	4
19207	7409862	5310-00-740-9862	398	10

CROSS-REFERENCE INDEXES

PART NUMBER INDEX

CAGEC	PART NUMBER	STOCK NUMBER	FIG.	ITEM
19207	7409883	5340-00-740-9883	343	12
19207	7409884	5340-00-740-9884	343	8
19207	7409889	5330-00-740-9889	500	8
19207	7409892	5340-00-740-9892	500	10
19207	7409893	3020-00-740-9893	500	3
19207	7409896	3110-00-740-9896	500	13
19207	7409903	5360-00-740-9903	345	45
19207	7409925	2530-00-740-9925	502	31
19207	7409928	5310-00-740-9928	476	4
			477	4
19207	7409929	5330-00-740-9929	471	19
			501	2
19207	7409930	2520-00-740-9930	501	6
19207	7409931	5330-00-166-4333	502	9
19207	7409932	5330-00-740-9932	502	26
19207	7409933	5330-00-740-9933	502	45
19207	7409934	3020-00-740-9934	502	47
19207	7409935	2520-00-740-9935	501	10
19207	7409940	5330-00-292-1600	502	4
19207	7409945	3040-00-740-9945	501	13
19207	7409947	3040-00-740-9947	476	11
			477	1
19207	7409948	3020-00-740-9948	501	4
19207	7409950	3040-00-740-9950	500	16
19207	7409952	3020-00-740-9952	500	30
19207	7409953	3120-00-740-9953	500	1
19207	7409954	2540-00-740-9954	499	2
19207	7409955	5310-00-596-7797	500	28
19207	7409958	3120-00-740-9958	500	18
19207	7409959	5330-00-740-9959	500	19
19207	7409960	3120-00-740-9960	500	21
19207	7409962	5310-00-740-9962	500	22
19207	7409975	5310-00-740-9975	501	1
19207	7409977	5310-00-740-9977	502	14
19207	7409978	3120-00-740-9978	476	5
			477	3
19207	7409979	5310-00-740-9979	502	19
19207	7409980	2540-00-740-9980	499	1
19207	7409990-1	9905-01-108-5164	552	8
19207	7409993	9905-00-252-5587	552	9
			611	32
			612	26
			613	23
19207	7410218	5310-00-407-9566	37	10
			284	24
			348	23
			351	26
			489	42
19207	7410715	2540-00-741-0715	362	8
19207	7411068	5342-00-741-1068	207	37
19207	7411111		346	23

PART NUMBER INDEX

CAGEC	PART NUMBER	STOCK NUMBER	FIG.	ITEM
19207	7411112	2510-00-741-1112	346	22
19207	7411114	2510-00-741-1114	348	1
19207	7411115	5360-00-741-1115	348	5
19207	7411116	2510-00-741-1116	348	4
19207	7411121	3020-00-741-1121	471	6
19207	7411122	2590-00-741-1122	469	1
19207	7411123	3110-00-741-1123	471	15
19207	7411124	3110-00-741-1124	471	17
19207	7411125	3120-00-741-1125	471	8
19207	7411142	2520-00-741-1142	237	5
19207	7411146	4710-01-158-1477	BULK	34
19207	7411152	3120-00-159-6992	471	18
19207	7411153	5310-00-741-1153	471	22
19207	7411154	5330-00-741-1154	471	2
19207	7411155	5306-00-741-1155	471	32
19207	7411156	3120-00-741-1156	471	12
19207	7411159	5330-00-741-1159	471	28
19207	7411160	5330-00-741-1160	471	13
19207'	7411161	3040-00-741-1161	472	40
19207	7411162	5365-01-106-4286	472	1
19207	7411163	3120-00-661-9026	472	30
19207	7411164	3120-00-741-1164	472	25
19207	7411166	2520-00-741-1166	472	33
19207	7411170	3120-00-741-1170	472	19
19207	7411410	2520-00-741-1410	513	7
			525	7
19207	7412376	2510-00-741-2376	346	10
19207	7412382-1	2590-01-082-2644	470	1
19207	7412636	5330-00-741-2636	223	11
19207	7412924	5315-00-741-2924	326	19
			327	9
			327	9
19207	7413375	5365-01-149-5416	367	26
19207	7413447	5330-00-961-3596	295	7
19207	7413456	2510-01-119-4094	322	13
19207	7413458	2510-00-741-3458	333	2
			335	2
			337	2
19207	7413565	5340-01-128-0191	153	17
			586	15
19207	7413565-1	5340-01-143-8312	375	5
19207	7414180	5325-00-741-4180	144	26
19207	7414388	2510-00-567-0130	144	34
19207	7414917	5340-00-419-9473	165	30
19207	7415746	5315-00-741-5746	243	7
19207	7415752	5340-00-222-1653	254	16
			254	16
			261	2
19207	7416861	5365-00-923-4253	144	27
19207	7416862	5310-00-741-6862	144	22
19207	7416865	5306-00-741-6865	144	30

PART NUMBER INDEX

CAGEC	PART NUMBER	STOCK NUMBER	FIG.	ITEM
19207	7416866	5340-01-008-6448	144	28
19207	7416867	5310-00-741-6867	144	32
19207	7416869	5340-00-741-6869	144	17
19207	7417084	5306-00-741-7084	499	3
19207	7417093	5330-00-741-7093	472	20
19207	7417094	5330-00-741-7094	472	6
19207	7418774	5330-00-866-6236	473	8
19207	7418959	5995-00-057-1642	134	48
19207	7418962	5310-00-531-9120	207	39
19207	7418971	5315-00-741-8971	326	22
			327	11
			339	31
			340	17
			383	17
			395	14
			616A	9
75543	747R	5340-00-264-7182	414	12
19207	7521436	5342-00-040-2073	358	4
19207	7521651	5315-00-752-1651	230	7
			237	10
19207	7521718	5305-00-752-1718	233	7
			240	7
19207	7521743	4730-00-595-4827	207	32
19207	7521975	5360-00-752-1975	513	3
			525	3
19207	7525938	5305-00-752-5938	144	5
19207	7527645	5975-00-697-7769	160	12
19207	7529133-10	5340-01-232-3568	608	20
19207	75-9133-5	5340-01-106-8346	204	4
			205	6
19207	7529133-7	5340-01-197-7597	215	6
19207	7529133-8	5340-01-197-9350	269	16
			269	16
19207	7529133-9	5340-01-231-5357	267	15
19207	7529293	5360-00-476-9339	358	3
19207	7529296	5340-00-689-7213	383	32
			384	7
			386	15
19207	7529300	5365-01-129-0399	361	8
19207	7529301	5340-01-114-7656	362	34
19207	7529302	5340-01-114-7657	362	35
19207	7529305	2510-00-650-1015	363	1
19207	7529306	5340-00-625-9617	362	3
19207	7529307	9340-01-047-4100	363	5
19207	7529309	2540-00-562-0422	362	14
19207	7529310	5340-00-621-2563	361	9
19207	7529312	2510-00-546-4759	359	14
19207	7529317	5340-01-104-9076	366	15
19207	7529318	5342-00-622-3941	366	15
19207	7529319	5340-01-104-9075	366	13
19207	7529320	5342-01-169-5686	366	13

PART NUMBER INDEX

CAGEC	PART NUMBER	STOCK NUMBER	FIG.	ITEM
19207	7529322	5305-00-638-0957	367	5
19207	7529324-1	3040-01-187-3620	367	3
19207	7529327	3040-00-622-3946	367	2
19207	7529332	2510-01-142-1294	367	43
19207	7534653-4		421	5
19207	7534654	5340-01-065-7287	427	8
19207	7534663	5306-01-183-5953	419	24
19207	7534717	5315-00-597-2723	233	9
			240	9
19207	7534781	5315-00-276-5385	235	3
			242	3
19207	7535048	2520-00-563-8309	234	4
			241	4
19207	7535079	5330-00-138-8388	233	4
			240	4
19207	7535564	2510-00-405-1946	406	23
19207	7535572	5340-01-165-0469	414	10
19207	7535583	5330-00-415-1484	414	9
19207	7535589	6150-00-222-8988	434	1
19207	7535590	5330-01-158-6289	403	7
			421	3
19207	7535591		403	6
			421	2
19207	7535592	5340-01-194-5298	403	9
			421	4
19207	7535593	5340-01-194-6474	403	8
			421	6
19207	7535597	9905-00-106-5750	572	8
19207	7535600	9905-00-659-7757	570	6
19207	7535601	9905-00-106-5746	571	2
			572	7
			574	5
19207	7535603	9905-00-106-5744	572	2
19207	7535608	9905-00-403-4814	572	11
			573	1
19207	7535612	5342-00-679-1447	425	21
19207	7535615	9905-00-659-7755	570	4
19207	7535620	5340-00-766-6336	406	27
19207	7535620-1	5365-00-230-9682	405	41
19207	7535627		405	3
19207	7535631	5315-01-217-2269	409	11
19207	7535639	6150-00-234-3248	440	1
19207	7535641	9905-00-106-5745	572	3
19207	7535642	9905-01-013-4599	571	4
19207	7535643	5340-00-766-3330	402	6
			413	23
19207	7535644	9905-00-659-7754	572	14
19207	7535647	4520-01-186-5897	421	8
19207	7538146	5305-00-353-0969	141	23
19207	7538688	3020-00-262-7572	484	12
19207	7538706	5360-00-753-8706	469	31

PART NUMBER INDEX

CAGEC	PART NUMBER	STOCK NUMBER	FIG.	ITEM
			472	34
			473	17
			499	20
19207	7538712	5360-01-042-9532	473	37
19207	7538725	5306-00-753-8725	473	15
19207	7538726	2530-00-753-8726	473	36
19207	7538733	5305-00-873-6946	473	20
19207	7539072	5330-00-753-9072	82	17
			145	5
			145	5
19207	7539107-2	4720-00-482-8956	422	15
			423	15
19207	7539108	5340-01-139-1002	172	21
19207	7539114	2540-00-753-9114	100	13
19207	7539131	5315-00-338-1621	100	12
			281	35
19207	7539185	5340-00-699-8463	402	1
19207	7539214	5340-00-753-9214	353	2
19207	7539221	5340-00-709-5879	376	7
19207	7539222-1	5340-01-231-5516	374	15
19207	7539242	5306-00-753-9242	266	26
19207	7539300	4730-00-753-9300	214	5
19207	7539545	2590-00-753-9545	142	9
19207	7539657	2510-00-753-9657	361	11
19207	7539688	5310-00-753-9688	624	6
			625	6
19207	7539689	5330-00-753-9689	624	3
			625	3
19207	7540419	2540-00-754-0419	361	10
19200	7550233-1	5340-00-753-3741	378	10
19200	7550233-2		378	11
19204	7551074	5325-00-624-0528	607	9
19204	7551080	9905-01-118-6092	567	6
19207	7551081	7690-01-114-7621	567	7
60703	76115J	5310-00-804-1209	539	11
19207	7699767	4010-00-961-9780	BULK	26
19207	7699769	2590-00-941-8668	BULK	1
19207	7700246	5340-01-106-6751	549	16
			551	14
19207	7700263-2	5340-01-127-8703	549	2
			594	23
82484	77121-1	5310-00-760-7493	539	13
60703	77121-3JD	5310-01-028-4848	539	12
19207	7714880	2540-00-237-3693	329	4
19207	7714911	5310-00-771-4911	329	14
19207	7716428	5340-00-771-6428	578	2
19207	7716634	5975-00-771-6634	158	20
			160	16
			164	3
			165	3
			166	15

CROSS-REFERENCE INDEXES

PART NUMBER INDEX

CAGEC	PART NUMBER	STOCK NUMBER	FIG.	ITEM
			167	15
			168	3
			169	3
19207	7720497	5935-01-130-3536	134	15
19207	7722322	5365-00-772-2322	157	13
			158	4
19207	7722333	5365-00-090-5426	160	17
			163	13
			165	15
			166	8
			167	7
			168	12
			169	12
19207	7722343	5365-00-772-2343	134	14
			134	51
19207	7722344	5935-00-772-2344	160	13
19207	7722353	5935-00-772-2353	157	12
19207	7723306	5935-00-333-3088	134	13
			134	52
19204	7723307	5935-00-772-3307	160	14
19207	7723308	5935-00-333-9414	157	14
			158	5
19207	7723309	5310-00-393-6685	158	22
			160	18
			163	5
			165	5
			166	9
			167	8
			168	5
			169	5
19207	7728780	5940-01-035-4212	151	3
			152	7
19207	7731428	5935-00-773-1428	163	13
			165	36
			166	27
			167	20
			168	26
			169	35
19207	7748089	2510-01-147-4992	367	35
19207	7748743	5310-00-177-0892	233	2
			240	2
19207	7748744	5310-00-893-3381	207	2
			208	11
			217	11
			526	27
19207	7748911	5340-00-839-0098	405	18
			406	8
			416	4
19207	7762738		151	12
19207	7764909	5310-00-776-4909	329	16
56529	78011-170-1	5310-01-103-6438	502	30

CROSS-REFERENCE INDEXES

PART NUMBER INDEX

CAGEC	PART NUMBER	STOCK NUMBER	FIG.	ITEM
77915	79210	2530-00-080-6572	301	5
19207	7951057	2540-00-177-8108	549	46
19207	7951084-3		583	11
19207	7951713	7510-01-076-4238	568	2
19207	7951714	7510-01-078-5855	568	1
19207	7951891	5310-01-122-2060	551	11
19207	7952641	6680-00-795-2641	625	2
19207	7954454	3130-00-919-2915	473	38
19207	7954470	3990-01-083-5407	472	10
19207	7954471	5340-00-351-7831	469	13
			470	9
19207	7954473	3040-00-921-0478	474	10
19207	7954475	3040-00-118-5554	474	12
			489	58
19207	7954476	5340-00-923-4233	474	28
19207	7954477	2590-00-118-5551	473	29
19207	7954478	3040-00-231-0219	474	5
19207	7954482	3040-00-222-8991	473	22
19207	7954483	2590-00-121-0707	469	2
19207	7954484	3040-00-405-1931	474	8
			489	57
19207	7954486	2590-00-351-7865	469	14
			470	10
19207	7954534	3120-01-120-8450	502	20
19207	7954539-1	5310-01-138-5516	500	2
19207	7954562	3040-00-328-8862	502	7
19207	7954563	3040-00-506-8451	502	18
19207	7954574	2590-00-418-0887	499	8
19207	7954584		501	12
19207	7954585		502	21
19207	7954586	2590-01-120-8451	502	16
19207	7954587	5340-01-144-3233	502	41
19207	7954588	3040-01-119-8711	502	37
19207	7954590	2590-01-120-8512	499	11
19207	7954727	4730-00-289-1245	480	5
19207	7971111	6210-00-635-5686	131	1
19207	7971324	5305-00-422-1161	391	2
19207	7971653	5306-00-512-9218	380	3
			381	4
			400	10
			586	22
19207	7971783		507	36
19207	7971949	2590-00-332-0095	569	2
19207	7973325	9905-00-409-8948	569	5
19207	7973326	9905-00-901-2942	474	20
19207	7973339	5330-00-895-3424	489	3
19207	7974745-1	2530-01-085-3787	498	1
19207	7975607	2540-00-737-3304	362	25
19207	7975609	2540-00-797-5609	362	18
19207	7979183	5310-00-740-9621	350	3
19207	7979224	2530-00-797-9224	350	5

CROSS-REFERENCE INDEXES

PART NUMBER INDEX

CAGEC	PART NUMBER	STOCK NUMBER	FIG.	ITEM
19207	7979263	5310-00-353-2427	234	11
			241	11
			293	11
			295	1
			348	26
19207	7979265	5310-00-740-9615	348	16
19207	7979272	2510-00-734-6952	350	4
19207	7979274	5330-00-740-9600	348	25
19207	7979305	2510-00-797-9305	348	32
19207	7979306	5340-00-740-9391	348	24
19207	7979308	5310-00-374-0836	234	30
			241	30
			293	13
			295	4
			348	28
19207	7979309		234	31
			241	31
			293	14
		5310-01-135-0049	295	5
			348	30
19207	7979310	5315-01-058-7268	234	32
			241	32
			293	15
			295	6
			348	29
19207	7979327	5310-00-734-6957	350	7
19207	7979329	5306-00-740-9608	348	34
19207	7979363	2510-00-734-6998	347	11
19207	7982399	5970-00-906-0159	150	6
19207	7982401	5935-00-399-6673	131	24
18876	7982403	5935-00-868-2606	432	5
19207	7982907	5935-00-214-0904	158	17
			160	7
			163	12
			165	18
			168	14
			169	14
			169	22
			434	6
			592	25
			594	26
			612	2
			613	5
19207	7982997	1015-00-798-2997	134	19
			149	10
			160	21
			163	7
			166	2
			167	2
			168	7
			169	9

PART NUMBER INDEX

CAGEC	PART NUMBER	STOCK NUMBER	FIG.	ITEM
			172	2
			172	10
			172	16
19207	7982997	1015-00-798-2997	173	2
			173	8
			173	17
			173	23
			174	4
			432	2
			432	10
			436	11
19207	7986268	4720-01-181-0098	583	2
19207	7992548	5340-00-482-3791	346	3
19207	7994440	5310-00-276-3012	224	11
19207	7994954	2530-01-085-3592	471	25
19207	7994965	3040-00-417-2756	472	27
19207	7994966	9905-00-634-5269	563	6
19207	7994976-1	3040-01-194-9884	471	14
81343	8 100110B	4730-01-048-7874	598	6
81343	8 100115B	4730-01-049-1559	598	7
81343	8 120111B	4730-00-054-2572	266	34
			617	10
			618	6
81343	8 120115B	4730-00-054-2571	617	8
			618	4
06721	8-X-19	5340-01-136-1665	282	12
81343	8-2 070102CA	4730-00-618-8497	198	2
81343	8-4 100202BA	4730-01-115-7362	598	11
81343	8-4 120202BA	4730-00-409-7854	213	21
			252	11
			256	14
81343	8-6 010102B	4730-01-226-3705	593	13
81343	8-6 100102BA	4730-01-091-8032	253	1
81343	8-6 120102BA	4730-00-142-3076	252	8
			259	5
			260	4
			597	7
			599	14
81343	8-6 120103BA	4730-00-173-1867	263	7
			266	25
			267	17
81343	8-6 120202BA	4730-00-289-0051	213	18
			252	4
			253	8
			256	10
			266	15
			267	11
			271	16
81343	8-6 120302BA	4730-01-117-1614	256	15
			259	1
			260	1

PART NUMBER INDEX

CAGEC	PART NUMBER	STOCK NUMBER	FIG.	ITEM
			266	43
			269	28
81343	8-8 070202CA(CAD)	4730-00-273-6686	258	10
81343	8-8 070202CA(CAD)	4730-00-273-6686	269	25
			505	39
81343	8-8 100202BA	4730-01-115-6643	598	26
81343	8-8 120101BA	4730-00-278-3220	263	19
81343	8-8 120102BA	4730-00-202-9036	71	19
			74	18
			75	14
			263	3
			263	24
			264	34
			269	13
			269	24
			271	13
			272	7
81343	8-8 130239C	4730-01-221-3565	581	7
81343	8-10 010220CA		72	16
□				
81343	8-10 010220CA		78	22
			582	17
□				
70434	8BCO	4730-00-050-4652	314	8
11083	8H9620	5310-00-236-3694	101	2
47457	8PK1730	3030-01-287-3155	584	16
78388	SA-2677	5315-01-384-2149	626	5
89619	8QD1038 KF		308	18
70411	8X29		81	5
52304	806738	5305-01-287-6570	623	9
52304	806739		623	12
02413	811E1139	2520-00-734-6985	236	6
49185	816701	4730-00-143-9282	255	1
15434	8265	5310-00-246-0221	30	10
82484	82845-JD	5310-01-163-2472	539	6
6Y402	829-5220	5905-01-280-3388	98	2
6Y402	8295204	2835-01-285-3269	98	1
19207	8327011-4	4720-00-108-5989	505	10
19207	8327027	2540-00-832-7027	330	10
19207	8327139	2510-00-277-9786	348	9
19207	8327209	2510-00-318-0959	348	11
19207	8327210	2510-00-318-0960	348	12
19207	8327443	5315-00-281-7650	474	11
19207	8327444	5315-00-281-7652	474	9
			489	23
19207	8327473	5306-00-207-4932	264	15
19207	8327979	3040-00-513-5786	507	5
19207	8327980	3040-00-513-5787	507	18
19207	8327982	4730-00-335-1812	507	9

CROSS-REFERENCE INDEXES

PART NUMBER INDEX

CAGEC	PART NUMBER	STOCK NUMBER	FIG.	ITEM
19207	8327983	3040-00-513-5788	507	22
19207	8327985	4310-00-287-8126	507	21
19207	8327986		507	20
19207	8328341	5315-00-281-7651	484	8
19207	8328777	4730-01-208-5859	77	20
			79	17
			80	22
19207	8328782	4730-00-278-6318	212	7
			270	5
19207	8328782	4730-00-278-6318	279	2
			279	26
			288	15
			603	12
19207	8330265	2540-00-040-2308	347	2
19207	8330333	5340-00-040-2314	5	8
19207	8330338	5340-00-344-2767	169	31
19207	8330341	5340-00-445-4557	169	26
19207	8330342	5340-00-833-0342	163	18
19207	8330343	5340-00-419-3077	169	29
19207	8330414	5310-00-595-5839	474	14
			489	60
19207	8330478	5315-00-282-2583	499	9
19207	8330737	2510-00-231-7488	331	5
19207	8330741	2510-00-231-7489	331	1
19207	8330742	5340-00-040-2321	331	9
			332	11
19207	8330743	5340-00-040-2322	332	1
19207	8330743-1	5342-00-096-9662	332	1
19207	8331198	2510-00-933-9577	318	12
			319	15
			321	13
			323	20
			324	12
19207	8331199		320	12
19207	8331200	5340-00-231-7486	331	4
			332	4
19207	8331202	5340-00-040-2331	473	10
19207	8331791-1	5307-01-121-9883	207	18
19207	8331864	5340-00-209-9377	351	25
19207	8332248	2520-01-093-4274	228	3
19207	8337078		607	17
19207	8337079		607	18
19207	8337103		607	5
19207	8337175		607	16
19207	8337176		607	14
19207	8338087		607	4
19207	8338561	5935-00-833-8561	134	17
			134	31
			134	55
			149	8
			157	2

PART NUMBER INDEX

CAGEC	PART NUMBER	STOCK NUMBER	FIG.	ITEM
			158	34
			160	6
			163	9
			166	6
			167	10
			168	9
			169	7
			170	4
			170	12
19207	8338561	5935-00-833-8561	171	5
			172	4
			172	12
			172	18
			173	4
			173	10
			173	19
			173	25
			174	2
			211	35
			432	4
			432	12
			434	4
			436	13
			496	3
			592	24
			593	15
			594	27
			608	12
			612	3
			613	6
			614	10
			614	17
19207	8338562	5970-00-833-8562	131	23
			134	18
			134	56
			149	9
			157	3
			158	35
			160	20
			163	8
			166	3
			167	3
			168	8
			169	8
			170	3
			170	11
			170	24
			170	29
			171	6
			172	3
			172	11

PART NUMBER INDEX

CAGEC	PART NUMBER	STOCK NUMBER	FIG.	ITEM
			172	17
			173	3
			173	9
			173	18
			173	24
			174	3
			182	34
			211	34
			211	40
19207	8338562	5970-00-833-8562	213	27
			432	3
			432	11
			434	3
			436	12
			496	4
			608	13
			614	11
			614	16
19207	8338564	5940-00-399-6676	131	22
			134	57
			157	4
			158	36
			170	2
			170	10
			170	23
			170	28
			171	7
			182	35
			211	33
			211	39
			213	28
			434	2
			496	5
			608	14
			614	12
			614	15
19207	8338566	5935-00-572-9180	134	61
			139	11
			140	13
			143	22
			157	6
			158	18
			163	11
			165	17
			166	13
			168	15
			169	15
			170	6
			170	20
			171	10
			172	6

PART NUMBER INDEX

CAGEC	PART NUMBER	STOCK NUMBER	FIG.	ITEM
			173	12
			174	6
			434	5
			436	7
			496	8
			592	27
			595	13
			612	4
			613	7
19207	8338567	5310-00-833-8567	134	60
			139	10
			140	12
			143	23
			157	7
			166	12
			170	7
			170	19
			171	9
			172	7
			173	13
			174	7
			436	6
			496	7
19207	8338569	2590-00-695-9076	169	20
19207	8338570	5310-00-298-8903	169	19
19207	8342292-3	4010-01-203-5687	411	6
19207	8343681	5365-00-409-8987	473	16
19207	8344234	2590-01-082-2526	469	32
19207	8344235	9905-00-262-9929	563	1
19207	8344240	2590-00-226-2347	469	29
			472	35
			499	19
19207	8344241	5310-00-419-9476	471	10
19207	8344242	3120-01-085-3338	471	1
			489	47
19207	8344243	3120-01-082-9008	471	21
19207	8344244	4730-00-186-4967	471	35
19207	8344245	3040-01-082-3599	471	26
19207	8344246	5340-01-101-6713	471	34
19207	8344247	5365-01-106-4282	472	2
19207	8344248	3040-00-179-3540	472	26
19207	8344250	3020-00-484-5831	472	22
19207	8344251	2590-00-234-3262	472	31
19207	8344505	3020-01-082-3598	471	3
19207	8344506	3130-01-085-3595	472	29
19207	8345033	5340-01-158-7094	378	6
19207	8352678	5340-01-170-5007	182	31
19207	8352679	5340-00-421-7254	109	11
			165	25
			166	22
19207	8356625-2	5310-01-084-1197	107	21

CROSS-REFERENCE INDEXES

PART NUMBER INDEX

CAGEC	PART NUMBER	STOCK NUMBER	FIG.	ITEM
			594	15
			595	9
19207	8365663	3040-00-040-2401	243	6
19207	8366166-1	4730-01-095-2034	71	22
			72	
			74	21
			75	17
19207	8376311	4730-00-419-9424	252	15
			255	3
			261	18
			262	10
19207	8376499-1	6210-01-124-9301	131	10
19207	8376629-1	5340-01-038-7759	377	4
19207	8376986	5340-00-302-1840	402	10
19207	8378730	4730-00-695-1133	35	7
19207	8380401	2540-00-222-8883	410	7
19207	8380402		410	13
19207	8380412	2590-01-120-8447	403	13
19207	8380419	2540-00-159-8823	419	2
19207	8380420	9320-00-451-8080	BULK	28
			BULK	29
19207	8380420-206		419	25
19207	8380420-27		428	11
19207	8380420-51		427	10
19207	8380420-77		405	6
			406	20
			416	18
19207	8380420-78		404	8
19207	8380421	2590-01-180-1006	404	21
19207	8380422	5340-01-165-4506	412	16
19207	8380423	2510-01-180-1004	408	14
19207	8380424	2590-01-147-1517	408	12
19207	8380425	2510-01-180-1005	412	7
19207	8380425-1	2590-01-186-9584	412	7
19207	8380431	5330-00-415-1488	419	4
19207	8380437	2590-01-181-0311	413	1
			413	12
19207	8380438	2590-01-182-1847	413	2
19207	8380439	2590-01-185-1278	413	3
19207	8380440	2590-01-186-3720	413	4
19207	8380441	2590-01-186-3724	413	5
19207	8380442	2590-01-186-6096	413	5
19207	8380443	2540-00-924-1296	405	11
			406	18
19207	8380444	2590-00-417-2735	405	10
			416	15
19207	8380445	2590-00-924-1425	406	13
19207	8380446	2590-01-188-0316	413	11
19207	8380447	2590-01-186-3725	420	18
19207	8380448	2590-01-188-0318	416	16
19207	8380453-1	2590-01-173-9200	420	16

CROSS-REFERENCE INDEXES

PART NUMBER INDEX

CAGEC	PART NUMBER	STOCK NUMBER	FIG.	ITEM
19207	8380454	2590-01-202-5769	413	17
19207	8380462	2590-01-186-9585	403	5
19207	8380463	2590-01-194-6994	420	9
19207	8380464	2590-01-198-0964	337	7
19207	8380465	2590-01-197-7239	420	8
19207	8380467	5340-00-679-1495	414	17
19207	8380468	5340-00-679-1492	414	25
19207	8380469	5365-01-187-3591	412	11
19207	8380470	5365-01-187-3590	412	9
19207	8380471	5365-01-187-3589	412	32
19207	8380472	5365-01-186-3773	412	31
19207	8380473	2510-01-186-0867	412	30
19207	8380474	5365-01-131-3396	412	29
19207	8380475	5365-01-186-7247	412	28
19207	8380476	2590-01-186-3723	413	1
			413	10
			413	12
19207	8380477	2590-01-192-6089	413	10
19207	8380479	2590-01-194-2149	413	7
19207	8380480	2590-01-186-3721	413	7
19207	8380481	2590-01-188-0317	413	9
19207	8380482	2590-01-186-3722	403	2
			413	9
19207	8380484	2590-00-417-2611	416	22
19207	8380498	2590-00-471-5344	419	11
			419	26
19207	8380499	5340-00-231-0210	402	3
			417	6
19204	8419961	5970-00-296-6078	440	3
94222	85-15-140-16	5325-00-343-5531	369	5
94222	85-34-101-20	5310-00-822-8525	369	6
94222	85-35-295-15	5325-00-282-7471	369	7
70960	85WBLS40-34		222	7
70960	85WBL548-19		221	6
70960	85WBYSM48-68		221	9
55783	851-204993	6680-00-159-1732	34	8
11862	8622361	5305-00-557-6612	554	2
73342	8622757		193	12
73342	8623232		198	7
11862	8623262	2520-01-127-3969	198	10
73342	8623484	4730-01-007-0802	186	7
73342	8625431	2520-00-008-7361	189	17
73342	8627650	2520-00-557-6619	183	6
60703	86324-1J	5365-00-946-2231	539	10
19207	8668878		607	7
19207	8668879		607	6
19207	8675779	4720-00-899-7893	BULK	40
19207	8675779-28		99	1
19207	8675779-32		61	1
19207	8684206-68		426	10
19207	8686994		367	34

CAGEC	PART NUMBER	STOCK NUMBER	FIG.	ITEM
19207	8689204	4710-01-134-2111	BULK	37
19207	8689206-26		539	20
			540	17
19207	8689206-34		276	4
19207	8689206-40		422	2
19207	8689206-54		422	6
19207	8689206-70		423	6
19207	8689206-92		422	11
			423	11
			607	11
19207	8689233		616A	30
19207	8690485	5340-00-479-2947	376	1
19207	8690527	5340-00-968-4060	399	2
19207	8690884	2530-00-998-4711	472	18
19207	8690892	5310-00-421-3994	471	11
19207	8690893	3040-00-470-2097	500	11
19207	8690894		500	15
19207	8690899	9905-00-913-6879	563	7
19207	8690900	9905-00-141-1619	563	4
19207	8693035-1	5330-01-067-8567	289	5
19207	8693035-2	5331-01-127-8550	289	16
19207	8698433	5340-00-231-7428	419	3
19207	8698434	5342-00-679-1446	414	27
19207	8701223	5320-01-029-7722	134	24
19207	8701226	5330-00-990-5804	134	21
19207	8701233	4010-01-039-4831	134	23
19207	8701287	5999-01-123-4557	134	22
19207	8701325	5310-00-655-9860	158	3
19207	8707524	5340-01-173-0241	583	9
			591	21
			592	18
19207	8710557		589	54
		4720-01-114-7728	BULK	14
19207	8710557 X BULK		590	39
19207	8710557 X 107 IN		584	24
19207	8710557 X 22 IN		585	26
19207	8710557 X 25IN		111	21
19207	8710557 X 3 IN		111	16
19207	8710557 X 44 IN		585	2
19207	8710557 X 44IN		111	6
19207	8710557 X 72 IN		584	22
19207	8710557-23		581	10
19207	8710557-28		549	44
19207	8710557-32		549	43
19207	8710557-46		582	19
19207	8710557-84		582	22
19207	8710626	2540-00-921-0481	329	13
19207	8710627	2540-00-821-2277	329	6
19207	8710628	4730-00-047-3946	329	10
19207	8710629		329	7
19207	8710630	2540-00-047-3926	329	1

PART NUMBER INDEX

CAGEC	PART NUMBER	STOCK NUMBER	FIG.	ITEM
19207	8712058	2910-00-911-5635	61	11
			63	18
19204	8712289-5	5310-00-044-3342	6	17
19207	8720328	2510-01-130-7945	367	33
19207	8720887	4010-00-809-6294	390	2
19207	8720887-1	4010-01-090-9352	390	3
19207	8720944		507	10
19207	8722186-20	3040-01-205-9264	416	7
19207	8722186-21	3040-01-197-8555	405	21
19207	8722186-21	3040-01-197-8555	406	5
19207	8722186-22	3040-01-197-8556	405	22
			406	7
19207	8722186-23	3040-01-233-7768	416	5
19207	8723828	5315-00-696-0789	473	18
19207	8724198	5935-00-772-0495	134	50
19207	8724255	5935-00-080-1020	164	2
			165	2
			166	14
			167	14
			168	2
			169	2
19207	8724258	5935-00-686-2599	160	15
19207	8724260	5935-00-081-0401	158	19
19207	8724264	5325-00-678-8435	158	21
19207	8724494	5975-00-660-5962	158	41
			163	16
			165	11
			166	4
			167	4
			168	16
			169	16
			169	23
			170	25
			170	30
			171	12
			182	33
			211	41
			213	26
19207	8724495	5935-00-691-5591	97	12
			97	19
			144	9
			144	14
			151	8
			152	4
			158	27
			165	8
			167	13
			170	32
			170	42
			171	4
			180	6

CROSS-REFERENCE INDEXES

PART NUMBER INDEX
STOCK NUMBER

CAGEC	PART NUMBER	STOCK NUMBER	FIG.	ITEM
			182	38
			211	6
			211	45
19207	8724497	5310-00-656-0067	97	14
			97	20
			144	10
			144	15
			151	9
19207	8724497	5310-00-656-0067	152	3
			158	28
			165	7
			167	12
			170	33
			170	41
			171	3
			180	5
			182	37
			211	7
			211	44
19207	8728126	2530-00-933-4941	350	2
19207	8728128	2530-00-235-4756	350	2
19207	8728234	2590-00-525-1352	150	13
19207	8729078	2510-01-042-9692	415	1
19207	8734961	3830-00-179-6635	379	11
19207	8734990	2510-00-231-7465	408	17
19207	8735018-1	2510-01-225-1001	409	1
19207	8735022	2520-00-415-1479	409	8
19207	8735023	3040-00-231-7470	409	19
19207	8735024	5365-00-455-1382	409	9
19207	8735025	3020-01-161-7710	409	13
19207	8735026	3120-01-216-6699	409	12
19207	8735027	2510-01-163-4903	409	2
19207	8735028	5365-00-299-0067	409	14
19207	8735029	3040-00-231-0221	409	17
19207	8735030	3040-00-419-9431	409	21
19207	8735033	2510-00-417-2736	409	29
19207	8735034	5330-00-419-9468	409	25
19207	8735035	5330-00-470-2115	409	24
19207	8735036	5330-00-419-9469	409	28
19207	8735037	5340-00-419-5859	408	16
19207	8735038	5340-00-419-5860	408	10
19207	8735038-1	5340-01-226-9161	408	7
19207	8735050	5365-00-419-9465	379	8
19207	8735056	3040-00-231-0212	409	7
19207	8735057	3120-00-766-3327	379	7
			409	15
19207	8735058	5365-00-443-9901	379	10
19207	8735059	3040-00-421-3966	379	5
19207	8735060	3040-00-443-4839	379	9
19207	8735074	2590-00-231-7484	409	10
19207	8735084	5342-00-415-1494	411	13

CROSS-REFERENCE INDEXES

PART NUMBER INDEX

CAGEC	PART NUMBER	STOCK NUMBER	FIG.	ITEM
19207	8735124		424	22
19207	8735125		424	22
19207	8735151		418	17
19207	8735226		417	28
19207	8735318	5330-00-222-8992	425	28
19207	8735338	2510-00-351-7851	412	21
19207	8735349	2510-00-351-7852	412	23
19207	8735423-1	2590-01-226-4587	409	23
19207	8735425	5340-00-158-3889	402	13
			404	17
19207	8735435	5340-00-421-3990	402	12
19207	8735437	5340-00-419-9484	408	5
19207	8737999	4730-01-083-1110	541	3
19207	8738000	4730-01-089-4370	541	1
19207	8739551	6220-00-878-7301	144	1
19207	8739552	5995-00-991-6707	144	7
19207	8739556	5935-00-991-6708	144	35
19207	8739559	6220-00-397-3335	144	23
19207	8739573	5930-00-991-6713	144	12
19207	8741316	4010-01-155-6142	BULK	27
19207	8741435	5310-00-241-6921	141	12
19207	8741437	5305-00-832-5743	141	20
19207	8741441	5342-00-182-3726	141	16
19207	8741442	5325-00-088-6147	141	15
			144	19
19207	8741446	5325-00-832-5650	141	21
19207	8741447	6220-00-998-6142	141	24
19207	8741461	6220-00-443-0589	141	13
19207	8741491	6240-00-966-3831	141	22
19207	8741492	5935-00-807-4109	141	14
			144	18
19207	8741539	5306-00-575-5419	347	13
19207	8741768	4820-00-652-5548	469	7
			499	5
19207	8743908	5305-00-483-2339	362	7
19207	8743909	5305-00-088-1302	362	19
19207	8743910	5305-00-145-0602	362	24
19207	8743911	5305-00-078-7021	362	33
19207	8743917	5305-00-144-1514	362	31
19207	8754124	5325-01-144-4871	149	6
82484	87549-1	5310-01-135-4835	541	10
19207	8757842-1	2510-01-084-9633	362	27
19207	8757843	2510-00-057-1630	362	26
19207	8758038	2510-00-498-2392	384	1
			386	9
19207	8758038-1	2510-01-226-9407	384	1
			386	9
19207	8758065	2510-01-130-7952	382	2
19207	8758067	2510-01-131-0122	382	4
19207	8758068	4030-01-082-3597	382	15
19207	8758070		382	16

PART NUMBER INDEX

CAGEC	PART NUMBER	STOCK NUMBER	FIG.	ITEM
19207	8758083	2510-01-130-7950	382	11
			387	2
19207	8758086	2510-01-130-7951	382	7
			387	8
19207	8758090	2540-01-131-0111	382	10
			387	3
19207	8758095	5306-01-108-3186	382	12
			387	9
19207	8758096	5306-01-108-3185	382	6
			387	4
19207	8758106	5340-00-119-3906	383	34
			384	9
			386	17
			390	9
19207	8758125	2540-00-179-7094	382	17
19207	8758129	2510-00-177-7903	382	31
19207	8758132	2510-01-110-4061	382	36
19207	8758139	2540-01-085-3588	382	17
19207	8758143	2510-01-082-2642	382	30
19207	8758156		536	20
19207	8758157	2540-00-715-7407	536	12
19207	8758158	9520-00-857-6344	536	14
19207	8758159	2540-01-182-1309	388	1
19207	8758160	2540-00-857-6332	388	2
19207	8758177	2510-00-409-7973	386	3
19207	8758194	5310-00-419-9461	333	6
			335	7
			337	6
19207	8758195	2510-00-405-1978	335	10
19207	8758195-1	5340-01-120-8443	333	8
19207	8758198	6210-01-089-3037	547	8
19207	8758202	3110-00-419-9471	234	14
			241	14
19207	8758203		232	7
			239	7
19207	8758204		234	28
			241	28
19207	8758205		234	29
			241	29
19207	8758206 KF		232	9
19207	8758206-KF		239	9
19207	8758207		234	22
			241	22
19207	8758208	2520-00-494-6586	235	6
19207	8758208	2520-00-494-6586	242	6
19207	8758209	5365-00-692-6119	234	16
			241	16
19207	8758211		234	16
			241	16
19207	8758213		234	16
			241	16

PART NUMBER INDEX

CAGEC	PART NUMBER	STOCK NUMBER	FIG.	ITEM
19207	8758215		234	16
			241	16
19207	8758217	3040-00-692-6121	234	16
			241	16
19207	8758219		234	16
			241	16
19207	8758221	3110-01-077-7134	234	16
			241	16
19207	8758223		234	16
			241	16
19207	8758225	3040-00-692-6123	234	16
			241	16
19207	8758227		234	16
			241	16
19207	8758229		234	16
			241	16
19207	8758231		234	16
			241	16
19207	8758233		234	16
			241	16
19207	8758235		234	16
			241	16
19207	8758237		234	16
			241	16
19207	8758239		234	16
			241	16
19207	8758241		234	16
			241	16
19207	8758243		234	16
			241	16
19207	8758245		234	16
			241	16
19207	8758248	5340-00-445-4555	233	17
			240	17
19207	8758249		233	10
			240	10
19207	8758251		234	8
			241	8
19207	8758252		232	5
			239	5
19207	8758253		236	14
19207	8758258	5310-00-147-3274	234	12
			241	12
19207	8758260		233	8
19207	8758260		240	8
19207	8758268	5330-01-098-6668	236	32
19207	8758275	5340-00-418-1009	347	14
19207	8758278		293	9
19207	8758280	5310-01-054-2568	236	11
19207	8758282	5310-01-134-0209	236	11
19207	8758283	5310-01-134-5791	236	11

CROSS-REFERENCE INDEXES

PART NUMBER INDEX

CAGEC	PART NUMBER	STOCK NUMBER	FIG.	ITEM
19207	8758285	2520-00-419-9422	232	4
			239	4
19207	8758322		237	9
19207	8758325		303	2
19207	8758343		230	8
19207	8758344		237	9
19207	8758345	3040-00-930-7864	230	2
19207	8758365	2510-01-027-0203	343	9
19207	8758374	3110-00-999-6469	507	42
19207	8758403	3040-01-088-9161	318	40
			319	40
			320	41
			321	40
			322	43
			323	44
			324	39
19207	8758409	5340-01-088-9153	318	9
			319	12
			320	9
			321	10
			323	19
			324	9
19207	8758436-1	2540-01-108-9129	377	1
19207	8758442	5340-01-088-9152	333	12
			334	14
			337	15
			338	12
19207	8758443	5340-01-089-1355	337	10
19207	8758598	3040-00-409-7974	391	1
34623	8758628-1	2510-01-447-4754	390	13
19207	8758705	5340-00-999-6467	390	18
19207	8758706	5315-00-150-4146	390	11
19207	8758719	5305-00-482-1035	389	15
19207	8759398	2590-01-163-9755	427	14
19207	8759465	2510-00-231-7444	405	38
			406	28
19207	8761279	5310-00-720-7627	232	11
			239	11
19207	8764802		223	3
60703	87900-112	2540-01-101-0010	542	1
60703	88944-17		541	8
72582	8924145	4730-00-803-6266	584	9
72582	8924178	4730-01-005-3262	620	14
41885	89481		292	8
60703	89515-16	2540-01-106-7121	541	7
5T151	9 428 270 053	5360-01-335-9950	91	4
5T151	9 681 233 100	5340-01-341-6572	631	6
11757	9P35	3040-00-125-2959	530	25
05779	900719	2590-01-184-3938	479	2
05779	900877	2590-01-119-4103	479	8
05779	901422	4330-01-163-2733	479	4

CROSS-REFERENCE INDEXES

PART NUMBER INDEX

CAGEC	PART NUMBER	STOCK NUMBER	FIG.	ITEM
32537	902466-113	4730-00-188-1896	521	9
78500	903-04-48-785	2520-01-272-7767	466	1
70960	903-55		466	4
70960	9035544862		465	4
			512	4
72447	908930-3716	2520-01-112-2157	220	1
95019	912257-0416	2520-01-112-8366	218	1
82484	91522-J	5310-01-204-2002	539	8
72447	916210-4823	2520-01-323-6342	227	2
62983	920021	5330-01-098-9210	KITS	
61822	920095-15		218	2
1CW61	920095-4	5310-00-679-8346	223	10
13445	92113-06	2920-01-114-7538	180	3
15434	9226	5315-00-014-1284	7	9
62983	923080	4820-01-139-4888	KITS	
15434	9235-1	3120-00-374-4342	21	6
15434	9260-1	2815-00-311-2521	23	5
15434	9266		23	16
15434	9266A		23	16
05779	927110	4330-01-185-1226	479	6
64829	9303557	6680-01-147-2421	478	2
15434	9333-1	5330-00-729-4427	14	16
79470	9405-16-16	4730-00-359-4708	480	32
			482	6
24617	9406129	5305-01-271-8337	181	10
73342	9409037	5305-00-292-4595	185	21
24617	9409106	5306-00-283-0407	289	12
24617	9409115	5306-00-286-1476	366	18
24617	9409225	5305-00-638-2362	200	22
24617	9409934	4730-01-221-8821	593	11
24617	9410285	4730-00-163-0236	493	23
24617	9410360	5365-01-134-8660	198	5
63005	9411417	5310-01-084-2362	6	19
			107	24
			190	12
			205	8
			286	34
			619	3
			619	3
24617	9413509	5310-00-768-0318	370	22
			470	27
			493	16
24617	9414109	5305-01-229-9587	613	27
24617	9415757	5305-01-271-8383	222	11
72582	9415764	5306-01-042-3586	598	41
19207	12257242	5310-01-102-2711	292	7
			297	17
24617	9416095	5310-01-102-2715	586	11
24617	9416904		616A	13
73342	9417722	3110-00-916-9273	185	14
24617	9417794	5310-01-202-6775	414	18

CROSS-REFERENCE INDEXES

PART NUMBER INDEX

CAGEC	PART NUMBER	STOCK NUMBER	FIG.	ITEM
73342	9418483	3110-01-007-2609	186	6
24617	9418910	5315-01-155-7302	200	27
24617	9418924	5310-01-132-8275	112	7
			113	9
			528	27
			615	17
73342	9420965	5315-01-083-6352	197	2
24617	9421393	5310-01-274-0041	586	18
24617	9421861	5310-01-332-7265	184	1
24617	9421887	5320-01-417-1019	629	14
24617	9422277	5310-01-126-9404	121	10
11862	9422279	5310-01-193-6884	110	17
11862	9422298	5310-01-150-5914	619	6
24617	9422299	5310-01-150-4003	107	19
24617	9422301	5310-01-149-4407	6	18
			107	23
11862	9422302	5310-01-184-5784	110	10
24617	9422305	5310-01-130-4274	6	7
24617	9422772	5310-01-407-3451	156	11
24617	9422845	5310-01-092-5496	584	14
			594	20
			594	15
24617	9422848	5310-01-092-5495	107	18
			110	18
			126	3
24617	9422849	5310-01-301-7863	6	16
73342	9422961	5305-01-126-2619	190	13
73342	9423346	5315-01-083-6351	197	23
63005	9424899	5305-01-084-5370	197	25
73342	9428493	3110-01-010-9779	196	2
24617	9434057	3110-01-065-2469	188	21
73342	9436116	3110-01-239-1253	188	22
73342	9436972	3110-01-079-7049	185	29
73342	9436973	3110-01-146-7144	185	28
24617	9438017	3110-01-149-0876	200	20
73342	954528	3110-00-198-2848	185	8
70960	963-92-48-799	2520-01-285-6283	224	1
74480	9711085-002	2815-01-284-2284	222	12
34623	976683	4720-00-809-2430	549	1
95019	98-741		218	3
97907	981072	5331-00-982-4259	479	9
78500	983-04-44757	2520-01-127-6953	229	3
78500	983-04-48-798	2520-01-339-1648	222	1
70960	983-55-44478		226	12
70960	983-55-44757		226	12
70960	983-55-44760		219	9
70960	983-65-44-761		228	7
78500	9830444478	2520-01-127-6954	229	3
78500	9830444760	2520-01-127-6950	219	1
78500	9836444761	2520-01-127-6951	221	1
78500	984-04-44758	2520-01-190-8485	225	2

CROSS-REFERENCE INDEXES

PART NUMBER INDEX

CAGEC	PART NUMBER	STOCK NUMBER	FIG.	ITEM
78500	984-04-44759	2520-01-127-6952	226	2
78500	984-04-48-325	2520-01-323-6343	227	2
70960	984-55-44758		227	11
70960	984-55-44759		227	11
92679	98883R91	2530-00-832-7123	236	19
31875	990127	2930-01-268-8752	122	1
89346	99311R1	2540-00-693-0676	327	7
8X715	995250	4730-01-280-8345	619	9
8X715	995253	4820-01-280-3947	619	12
31875	995257	4820-01-280-3935	619	25